ELECTRIC MACHINE THEORY FOR POWER ENGINEERS

ELECTRIC MACHINE THEORY FOR POWER ENGINEERS

Van E. Mablekos
Emeritus, University of Washington

HARPER & ROW, PUBLISHERS, New York
Cambridge, Hagerstown, Philadelphia, San Francisco
London, Mexico City, São Paulo, Sydney

1817

Sponsoring Editor: Charlie Dresser
Project Editors: Eleanor Castellano/Céline Keating
Designer: T. R. Funderburk
Production Manager: Marion A. Palen
Compositor: Science Typographers, Inc.
Printer and Binder: Halliday Lithograph Corporation
Art Studio: Vantage Art Inc.

ELECTRIC MACHINE THEORY FOR POWER ENGINEERS

Library of Congress Cataloging in Publication Data

Mablekos, Van E Date-
Electric machine theory for power engineers.

Includes index.
1. Electric machinery. 2. Electric power systems.
I. Title.
TK2211.M32 621.31′042 80-12219
ISBN 0-06 044149-6

To my father

CONTENTS

PREFACE

This text is especially designed to be used as a one-year introductory course for electrical engineering students whose major field is power systems engineering. The text is unique in that it serves as an electric machines text with a natural extension into power systems engineering and that the material is presented from the systems point of view. Therefore, this text is of value to the electrical engineering student interested in power systems and also to the nonelectrical engineering student interested in excitation systems, synchronous devices, and the basics of power systems engineering.

This text assumes no special mathematical background in power systems engineering nor prior knowledge of electric machine theory or power systems analysis. It can be used as a first level electric machinery text. However, it contains enough material to be used in a two-semester (or three-quarter) sequence. The subject of electric machines is treated thoroughly, but the text does not stop there. Rather, it leads into the area of power systems as a natural extension of electric machine theory. In this way the text distinguishes itself from other electric machines textbooks because it prepares the reader not only for advanced electric machine theory but also for the area of power systems analysis.

The text is carefully designed and considerable effort has been made so that no essential principles are omitted; rather, a deliberate attempt has been made to optimize the presentation of fundamental concepts. The many numerical examples demonstrate the theory and principles set forth in the text, while the abundance of figures and diagrams further serves to illustrate the important concepts. If a given curriculum does not allow sufficient time to cover the entire text, proper screening of the material in the text enables one to design an excellent one-semester or one-quarter course in electric machine theory. The remaining material can then be

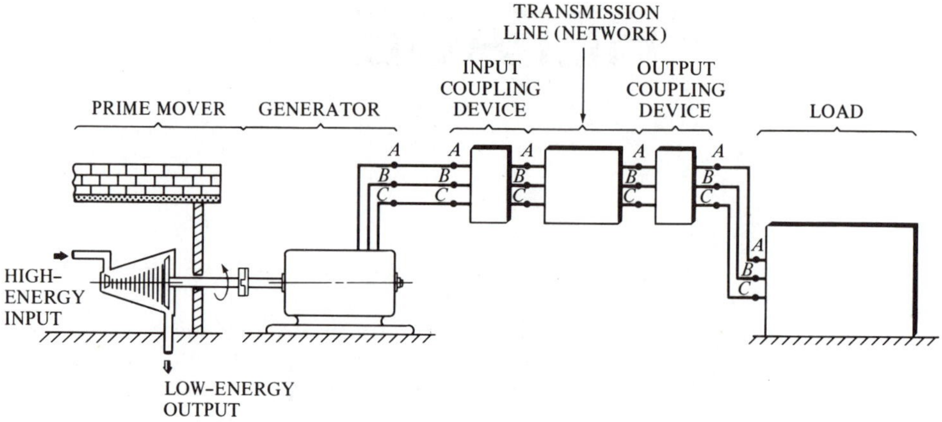

Plate 1 Model Elementary Power System.

used as a comprehensive reference source by the student (or the practicing engineer) interested in power systems engineering.

For the purposes of this discussion the elementary *power system* exhibited in Plate 1 is considered. It is comprised of six basic components: (1) the prime mover, (2) the generator, (3) the load, (4) the input coupling device, (5) the transmission line, and (6) the output coupling device.

The *prime mover*, Plate 1, is a system that converts any of the readily available forms of nonelectrical energy into mechanical energy. Such forms of energy, at least at the time of the writing of this text, are (1) thermal energy extracted from the burning of fossil fuels (coal, natural gas, or oil), (2) geothermal energy, (3) controlled nuclear reaction thermal energy, and (4) potential energy extracted from the controlled flow of stored water in large dams (hydroenergy).

No matter what kind of energy enters the prime mover, energy always leaves the prime mover and reaches the input of the electrical generator, Plate 1, in the form of mechanical energy. Specifically, it reaches the input of this generator in the form of rotational motion, measured in units of torque. In the meter-kilogram-second (MKS) system of measurement, torque is measured in newton meters. The classical name of the electrical generator is *mechanoelectrical* (the word usually encountered, however, is *electromechanical*) *energy conversion device*, or simply *generator*. Depending on the structure of the generator and its principle of operation, such a device can generate either (1) direct current (termed *dc generator**) or (2) alternating current (termed *ac generator*).

For the explicit purpose of this text the *load* (Plate 1) is assumed to utilize electrical energy to do mechanical work. That is, the load converts electrical energy into useful mechanical energy. Thus, the classical name *electromechanical energy conversion device* or simply *motor* is used for such a device. As in the case of the generator, depending on the structure of the device and its principle of operation, the device can utilize either (1) dc current (hence *dc motor*) or (2) ac current (hence *ac motor*).

Although the uses of dc generators and motors are somewhat limited, the *opposite* is true with ac generators and motors. Practically all the electrical energy

*The generator exhibited in Plate 1 is designed specifically for ac energy, as are the remaining components connected to the generator.

used in the United States today is generated by ac generators known as *synchronous generators* or *alternators*. Similarly, the ac motor is the work horse of our industry and is used extensively in our daily activity. The operating potential level of these devices varies and is determined by the design specifications of each device. As a rule, however, ac energy is generated in the form of three-phase voltages at a potential level of 25 kV, and ac motors operate at the potential levels of 208 or 120 V.

Motors, or electromechanical energy conversion devices, are encountered in homes, offices, workshops, and industrial processes, all of which are found in or near urban areas. But urban areas are usually located far away from generation sites, especially in the cases of hydroplants, coal-fired plants, and nuclear plants.

Since, as a rule, the electrical energy generated at 25 kV is utilized at large distances from the generation sites, it has to be transmitted to the load site *economically* and *reliably*. Research has shown that electrical energy can be transmitted over large distances in the ac form,* utilizing specifically designed *transmission lines* (Plate 1), if the voltage level is raised to 765 kV (for transmission purposes). At these potential levels transmission allows tremendous savings by (1) reducing the amount of copper used and (2) reducing the power losses in the transmission lines. The copper cost reduction is a result of capitalizing on the *skin effect phenomenon* in energy transmission through conductors. Such transmission allows the manufacturers to (1) utilize steel strands at the center of transmission line conductors and (2) use copper strands *only* on the surface of these conductors. Reliability of energy transmission is insured by the practice of good engineering design of transmission lines and terminal equipment by all concerned.

For the effective ac energy transmission, the elevation of the potential from the level of generation of 25 kV to the level of transmission of 765 kV is achieved with the aid of the *input coupling device* (Plate 1). Since this device increases the potential level of the generated voltage, it is defined as a *step-up transformer*. Similarly, the reduction of the potential from the level of transmission of 765 kV to the level of utilization of 240 or 120 V is achieved with the aid of the *output coupling device* (Plate 1). Since this device decreases the potential level of the transmitted voltage, it is defined as a *step-down transformer*.

It is logical from the engineering point of view to assume that the input and output coupling devices incorporate *switching systems*. Therefore, when the generated energy is of the dc type (it is at much lower levels in potential), the input and output coupling devices perform *only* direct coupling (switching) functions and no transformation of potential levels.†

The previously defined devices which constitute the elementary power system of Plate 1 (except the prime mover), as well as additional subjects associated with these devices, are described here in the following sequence.

In Chapter 1 the fundamental magnetic circuit, an essential concept in transformer, dc- and ac-electromechanical energy converter analysis, is developed with the aid of Biot-Savart's law and Ampere's law. Also set forth are the fundamentals of transformer operations. The describing equations of the *real* and *ideal* transformers are developed and their corresponding equivalent circuits drawn. In

*dc energy transmission is still in the research stages at the writing of this text.

† Under such circumstances, the entire model power system shown in Plate 1 is comprised of components designed specifically for dc energy; and only two conductors are present throughout the system.

addition, methods for measuring the transformer parameters are presented and procedures for simplifying the equivalent circuit of the real transformer are given. The chapter continues with a discussion of the applications of the ideal and real transformer. Finally, the balanced three-phase transformer, the autotransformer, and the Scott transformer are introduced.

Chapter 2 introduces the fundamental principles of motoring and generation and thus Kron's primitive machine (KPM) and the commonly encountered classical dc devices that employ these principles: the electromechanical energy converter (EMEC) or motor as well as the mechanoelectrical energy converter (MEEC) or generator. The basic physical structure as well as the mathematical equations describing the dynamic behavior of these devices are developed. These equations are appropriately simplified and the mathematical equations describing the steady-state behavior of these same devices are derived. Once the steady-state mathematical equations of a given device are derived, the circuit analog for the device at hand is drawn. These ideas lead to the development of (1) the equivalent circuit, (2) the describing equations, and (3) the terminal characteristics, that is, the torque vs speed or the voltage vs current, of the commonly encountered dc energy converters. Also presented is an account of the applications of dc devices. For the benefit of the student who studies automatic control theory and is interested in modeling the separately excited dc motor, the *transfer function* and the *state variable* modeling techniques are set forth. Finally, methods for measuring the parameters of the equivalent circuits for the separately excited dc device operating in both its modes are presented.

Chapter 3 is devoted to the development of the theories concerned with the generation of a sinusoidal field in the rotor structure of balanced three-phase generators as well as the voltages induced in the windings of the balanced three-phase systems comprised of ideal Y- or Δ-connected generators interconnected to balanced, three-phase, Y- or Δ-connected loads via balanced three-phase transmission lines that are of long, medium, or short length. Therefore, the relationships of (1) line-to-line and line-to-neutral voltages, (2) line and phase currents, and (3) phase and total complex power of balanced three-phase systems are derived. In addition, the chapter presents the three- and two-wattmeter methods of measuring the total real power, the one-wattmeter method of measuring the total reactive power, and the two-wattmeter method for measuring the power-factor angle of a balanced three-phase system. Finally, the student is introduced to the properties that unbalanced three-phase Y- or Δ-connected loads exhibit when connected to balanced three- phase Y- or Δ-connected ideal generators.

Chapter 4 develops the theory of the rotating field in the stator and rotor structures of the smooth airgap balanced, three-phase induction device. Consequently, the torque developed by the rotor of the induction device is derived. Also derived is the per phase equivalent circuit of the balanced three-phase induction device, essential in steady state analysis. This equivalent circuit is used to derive the equations that describe the performance of the balanced three-phase induction device as well as its terminal characteristics, that is, its torque or power vs slip (speed) characteristics. The available methods of speed control are discussed and the procedure for the parameter measurement of the per phase equivalent circuit of the balanced three-phase induction device is presented. In addition, the two-phase induction device is studied in detail. For the benefit of the student who studies automatic control theory and is interested in modeling the two-phase induction motor in addition to the steady state analysis, the mechanical output to the electrical input transfer function is derived. Furthermore, the steady-state behavior of the single-phase induction motor is studied comprehensively. The chapter concludes with a thorough account of applications of the three types of induction devices.

Chapter 5 is devoted to the derivation of the per phase equivalent circuit that describes the smooth-rotor synchronous device operating under steady-state conditions in both the generator and motor modes, and with loads or forcing functions with unity, lagging, and leading power factors. Also derived are the loading voltages vs current characteristics of the synchronous generator for loads with various power factors, and the real and reactive power vs power angle characteristics of the synchronous device. Finally, the two- and single-phase synchronous devices are discussed in detail. The chapter then concludes with a thorough account of applications of the three types of synchronous devices.

Chapter 6 presents a thorough account of the behavior of the normally excited, underexcited, and overexcited synchronous devices connected to an infinite bus. Consequently, the most popular excitation systems of the balanced three-phase synchronous devices are presented. The parallel operation of synchronous generators connected to an infinite bus is set forth and the effects of underexcitation and speed variation to the synchronization of these generators are discussed. The effects of variable prime mover output and dc excitation of synchronous generators connected in parallel are also presented. Next, the swing equation of the three-phase synchronous device is derived, and the stability analysis of the synchronous device connected to an infinite bus is carried out by use of either linear (differential equations) or nonlinear (equal area criterion) techniques. The chapter continues with the fairly complex analysis of the *salient-rotor* synchronous device via the d- and q-axis variable components. Finally, balanced three-phase faults on synchronous generators (as well as balanced three-phase systems) are considered and the *subtransient*, *transient*, and *synchronous* reactances are derived. Also derived are the per phase equivalent circuits associated with these reactances and used for post-fault studies.

Chapter 7 sets forth the detailed theory pertaining to the measurement of the circuit parameters R_S and X_S of the smooth-rotor synchronous device, and the subtransient X_d'' and X_q'', transient X_d' and X_q' and steady-state X_d and X_q reactances as well as R_S of the salient-rotor synchronous device. The chapter closes with the presentation of the method used to measure the d-axis short-circuit subtransient and transient time constants, τ_d'' and τ_d', respectively, as well as the armature time constant τ_a.

Appendix A provides the fundamental definitions of complex numbers and complex quantities. It sets forth the principles governing the mathematical operations of complex numbers, and presents the concepts of impedance, admittance, complex power, imaginary power, and average and effective values of the essential quantities used in the field of power systems engineering.

Appendix B describes the principle of operation of the electrodynamometer movement from which the four basic instruments—ammeter, voltmeter, wattmeter, and varmeter—are derived. Also, it presents the moving-iron ammeter and voltmeter, and the principle of operation of the D'Arsonval movement from which the ohmmeter is derived.

Appendix C deals with the correspondence of the time response and the root location in the complex plane, of the characteristic equation of 1st, 2nd, and nth order systems and thus with stability considerations. The appendix concludes with

Routh's stability criterion and its applications to determine the stability, marginal stability, or instability of a system.

Appendix D comprises a set of definitions and Laplace Transform pairs pertinent to the formulation of the stability theory set forth in Appendix C.

Appendix E, which provides the reader with the basic theory of operation of the thyristor and its most basic circuit, and a comprehensive list of references on solid-state power electronics, closes the text.

Van E. Mablekos

Editor's Note

The late Van E. Mablekos completed writing this book shortly before his death on January 22, 1979. However, much work still needed to be done in preparing the manuscript for publication: illustrations needed to be finalized; galley proofs and page proofs needed to be read. Professor Demosthenes P. Gelopulos of Arizona State University and Eugenio Villaseca, a doctoral student at the same institution, finished the necessary work conscientiously and admirably. On behalf of the late author and the publisher, I wish to acknowledge their contributions.

1 TRANSFORMERS

1.1 INTRODUCTION

The transformer may be defined as a device with two (or more) nonmoving (i.e., stationary) electric circuits that are conductively disjointed but magnetically coupled. That is, the windings are separate physically but are linked by a common time-varying magnetic field.

One of the transformer windings is connected to a source of ac electrical energy and is defined as the **input winding**, or as the **primary**. The second winding of the transformer delivers energy to the load and is defined as **output winding** or as the **secondary**. The voltage levels at the primary and secondary are **usually different**.

Although the transformer involves the interchange of electrical energy between two or more electrical systems, some of its principles of operation form the foundation for the study of electromechanical energy conversion. Even though electromechanical energy conversion deals with the **interchange of energy between an electrical system and a mechanical system**, the coupling device between these two systems (as between the two circuits in the transformers) is the magnetic field, and the behavior of the field in each case is fundamentally the same. Therefore, many of the pertinent equations and conclusions of transformer theory are equally applicable to electromechanical energy conversion theory.

Beyond serving as a basis to electromechanical energy conversion theory, transformer theory is important in its own right because of the many useful functions the transformer performs in prominent areas of electrical engineering. In power systems engineering the transformer makes it possible to convert from generation voltage levels of 25 kV to transmission voltage levels of 765 kV, to subtransmission voltage levels of 11–138 kV, to distribution voltage levels of 240/120 kV. Also, the Scott transformer makes it possible to convert two- (three-)

phase balanced voltages to three- (two-) phase balanced voltages. In communication systems ranging in frequency from **audio** to **radio** to **video**, the transformer is used as an (1) input transformer, (2) inter-stage transformer, (3) output transformer, (4) impedance matching device, (5) insulation device between a number of electric circuits, and (6) isolation device of direct current, while maintaining ac continuity between a number of electric circuits.

This chapter, therefore, is dedicated to transformer theory and relevant information. The topics discussed are arranged in the following sequence.

First, the fundamentals of transformer operation are set forth. Then the Biot-Savart law and Ampere's law, in conjunction with some of the fundamental concepts of electromagnetic fields, are used to establish the fundamental magnetic circuit, a very important concept in understanding transformer and electromechanical energy converter operation. The describing equations of the real and ideal transformers are then developed and their equivalent circuits are drawn. Next, methods for measuring the transformer parameters are discussed, and procedures for simplifying the equivalent circuit of the real transformer are given. Finally, the three-phase transformer, the autotransformer, and the Scott transformer are introduced.

1.2 PRELUDE: MAGNETIC CIRCUITS

The magnetic field produced by a current-carrying conductor at a given point P a distance R meters away from the conductor, Fig. 1.1, is given by the Biot-Savart law as

$$d\mathbf{H} = \frac{I\,d\mathbf{L}\times\mathbf{a}_R}{4\pi R^2} \tag{1.1}$$

Utilizing vector calculus techniques the solution of Eq. (1.1) for N current-carrying conductors in the same direction yields

$$\mathbf{H} = H_\phi \mathbf{a}_\phi = \frac{NI}{2\pi r} \text{ ampere-turns/meter} \tag{1.2}$$

Eq. (1.2) suggests that the electromagnetic field is **solenoidal**, that is, the magnetic field lines close upon themselves, Fig. 1.2. The field direction is given by the right-hand rule. When grasping the conductor with the thumb pointing in the direction of the current (into the page in Fig. 1.2), the fingers give the direction of the field. The tail of an arrowhead is used to indicate the direction of the current as seen. If the current were coming out of the page, the cross inside the small circle would be replaced by a dot.

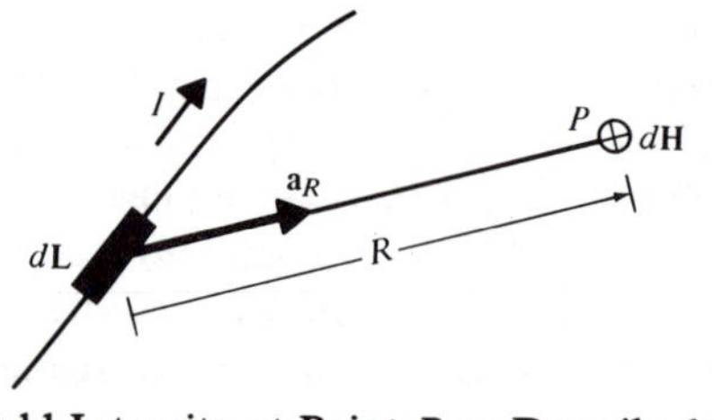

Fig. 1.1 Magnetic Field Intensity at Point P as Described by the Biot-Savart Law

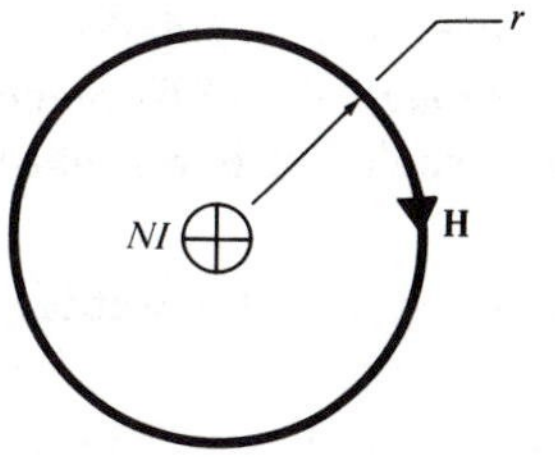

Fig. 1.2 Magnetic Field Produced by a Current-Carrying Conductor

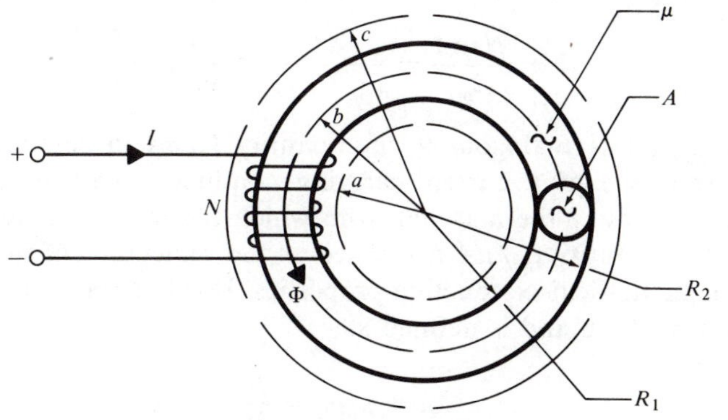

Fig. 1.3 Excited N-turn Toroid

The magnetic field intensity of Fig. 1.2 is related to its source NI by Ampere's law, $\oint \mathbf{H} \cdot d\mathbf{L} = \Sigma I = NI$. When the variables of the integrand are expressed in the cylindrical coordinate system and the integration is performed, Ampere's law yields*

$$\mathbf{H} \equiv H_\phi = \frac{NI}{2\pi r} \tag{1.3}$$

Eq. (1.3) shows a very important mathematical relationship that provides the foundation of magnetic circuit analysis. (Throughout this text the symbol $\equiv$ is used to mean "defined as.") This assertion is demonstrated by comparing it to the results obtained when Ampere's law is applied to Fig. 1.3, which exhibits an energized toroid. It should be noted that this toroid divides the space in which it lies in three distinct regions defined by three ranges of r, namely, $0<r<R_1$; $R_1<r<R_2$ and $R_2<r<\infty$.

When Ampere's law is applied to Fig. 1.3 and the integration is performed over the three closed-line integrals designated by the closed paths a, b, and c for which $\Sigma I=0$, NI, and $+NI-NI=0$, respectively, it yields

1. $$0 < r < R_1; \quad \oint \mathbf{H} \cdot d\mathbf{L} = 0 \quad \therefore H = 0 \tag{1.4}$$
2. $$R_1 < r < R_2; \quad \oint \mathbf{H} \cdot d\mathbf{L} = NI \rightarrow H 2\pi r_m = NI \quad \therefore H = \frac{NI}{2\pi r_m} \equiv \frac{NI}{\ell_m} \tag{1.5}$$
3. $$R_2 < r < \infty; \quad \oint \mathbf{H} \cdot d\mathbf{L} = +NI - NI = 0 \quad \therefore H = 0 \tag{1.6}$$

*Since the magnetic field components are always the H_ϕ and B_ϕ components, and these are always confined to the magnetic path provided by the structure of the device at hand, the subscript ϕ will not be used and the variables H and B will simply designate scalers unless otherwise stated.

where $r_m = (R_1 + R_2)/2$ and $\ell_m = 2\pi r_m$ are defined as the *mean radius* and the *mean path*, respectively. *Eq.* (1.4) *through Eq.* (1.6) *suggest that the field is confined within the toroid.* Since the field is circular, it is normal to the cross-sectional area of the toroid A. Thus,

$$\oint \mathbf{B} \cdot d\mathbf{S} = \Phi = BA \text{ webers} = BA \text{ Wb} \tag{1.7}$$

Since

$$B = \mu H \tag{1.8}$$

substitution of Eq. (1.5) and Eq. (1.7) into Eq. (1.8) and rearranging yields

$$\Phi = \frac{NI\mu A}{\ell_m} = \frac{NI}{(\ell_m/\mu A)} \equiv \frac{NI}{\mathcal{R}} \equiv \frac{\mathcal{F}}{\mathcal{R}} \tag{1.9}$$

The quantity $\ell_m/\mu A$ is analogous to the quantity $L/\sigma A$ in circuit analysis, which defines the resistance of a current-carrying conductor, with σ representing the conducting properties of the material from which the wire is made. Thus utilizing this analogy, the quantity $\ell_m/\mu A$ is defined as the reluctance of the toroid, with μ defining its magnetic-field-conducting properties. It is known as the *permeability* of the conducting medium and is defined as

$$\mu = \mu_r \mu_0 = \frac{B}{H} \tag{1.10}$$

where $\mu_0 = 4\pi \times 10^{-7}$ H/m *is the permeability of free space* (and air) and $1^- < \mu_r < \infty$ *is the relative permeability of the material used.* The quantity μ_r takes four distinct ranges of numerical values, and thus it separates all materials into four distinct groups: (a) *diamagnetic* ($\mu_r = 1.0^-$), (b) *nonmagnetic* ($\mu_r = 1.0$), (c) *paramagnetic* ($\mu_r = 1.0^+$), and (d) *ferromagnetic* ($1.0^+ < \mu_r < \infty$).

The symbol for *reluctance* is $\mathcal{R}$ and its dimensions are ampere-turns/weber. Thus,

$$\mathcal{R} = \frac{\ell_m}{\mu A} \tag{1.11}$$

The quantity $\mathcal{F}$ is recognized as the generating source of the magnetic field Φ. Thus $\mathcal{F}$ is analogous to the generating source (the emf or E) of the current I in an electric circuit comprised of a series connection of a voltage source (emf) and a resistor (R). Such an electric circuit is described by Ohm's law, that is,

$$I = \frac{E}{R} \tag{1.12}$$

Comparison of Eq. (1.9) to Eq. (1.12) leads to the following analogies:

$$\Phi \supset I \tag{1.13}$$

$$\mathcal{F} = NI \supset E \tag{1.14}$$

$$\mathcal{R} \supset R \tag{1.15}$$

Thus $NI = \mathcal{F}$ is defined as the *mmf* of the toroid. This analogy enables drawing a circuit analog, defined as the magnetic circuit, of the toroid as shown in Fig. 1.4, where **the plus sign signifies the exit of the field** with respect to the N-turns coil.

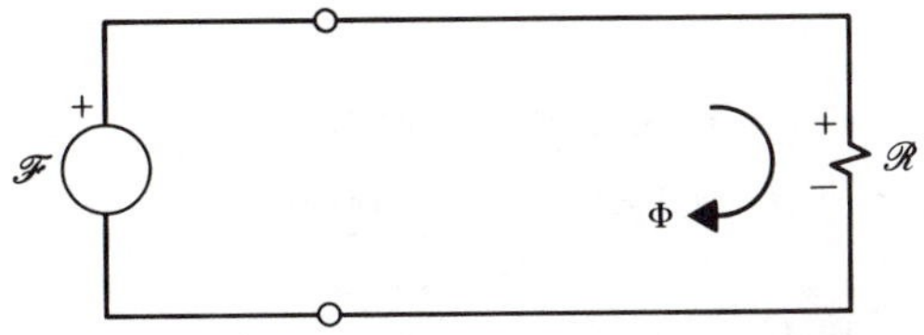

Fig. 1.4 The Circuit Analog of the Toroid of Fig. 1.3

The analogies of Eq. (1.13) to Eq. (1.15) suggest that **the flux in a magnetic circuit behaves as the current in an electric circuit, the reluctance of the various paths in a magnetic circuit can be treated as the resistance in an electric circuit, and the mmf's in the magnetic circuit can be treated as the emf's in an electric circuit**. Thus the loop-current method, the node-voltage method, the current-dividing equation, and so on, of circuit analysis can be used for magnetic circuit analysis.

Generally, there are two types of magnetic circuit problems that are encountered in the field of electromechanical energy conversion: (1) the magnetic circuit is completely defined, that is, the mmf's, the permeabilities of all paths, and the physical dimensions are clearly defined, but the magnetic field is to be found; (2) the geometry of the magnetic circuit is completely defined but the manufacturer in addition provides the graphical relationship of the flux density (B) versus the field intensity (H), that is, the permeability curve of the magnetic material comprising the magnetic circuit.

Airgaps are essential in many of the magnetic circuits used in practice. Every electromechanical energy converter is comprised of two parts: (1) the stator and (2) the rotor embedded in the airgap of the stator. Thus the two following examples encompass airgaps in the magnetic circuits to be studied.

Example 1.1 (type 1)

Given: the configuration shown in Fig. 1.1.1. For the given magnetic circuit, $\ell = 5\times10^{-2}$ m, $N_1 = N_2 = 1000$ turns, $I_1 = I_2 = 5$ A, $\ell_g = 5\times10^{-3}$ m. Calculate the airgap field quantities, Φ_g, B_g, and H_g.

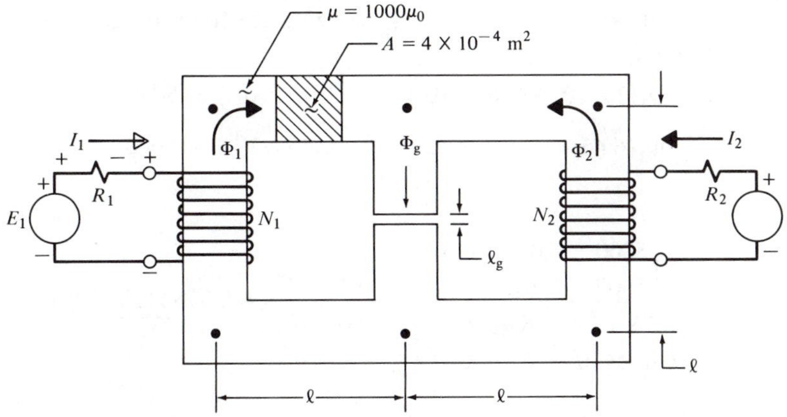

Fig. 1.1.1 A Two-Window Magnetic Circuit With Two Forcing Functions in the Outer Legs and a Small Airgap in the Middle Leg

Solution

$$\mathcal{F}_1 = N_1 I_1 = 1000 \times 5 = \underline{5000} \text{ At}$$
$$\mathcal{F}_2 = N_2 I_2 = 1000 \times 5 = \underline{5000} \text{ At}$$
$$\mathcal{R}_1 = \frac{\ell}{\mu A} = \frac{5\times 10^{-2}}{10^3 \times 4\pi \times 10^{-7} \times 4 \times 10^{-4}} = \frac{5\times 10^{-2}}{16\pi \times 10^{-8}}$$
$$= \underline{1\times 10^5} \text{ At/Wb}$$
$$\mathcal{R}_{LEG} = 3\mathcal{R}_1 = \underline{3\times 10^5} \text{ At/Wb}$$
$$\mathcal{R}_g = \frac{\ell_g}{\mu A_g}$$

i. $$\ell_g = 5 \times 10^{-3}\text{m}$$

ii. $$A_g = \left[(2+0.5^*)\times 10^{-2}\right] \times \left[(2+0.5)\times 10^{-2}\right]$$
$$= \underline{6.25\times 10^{-4} m^2}$$

Note that in order to account for the bulging effect of the field in the airgap, each linear dimension of the cross-sectional area of the magnetic material is increased by the length of the gap, ℓ_g. This increased area gives the *effective area* of the airgap used in the reluctance calculation for the airgap. You should be aware that various authors use various corrective schemes to adjust the airgap area. When the opening of the airgap is extremely high, you should consult design manuals in order to obtain the correct airgap adjustment empirical formula.

$$\mathcal{R}_g = \frac{5\times 10^{-3}}{4\pi \times 10^{-7} \times 6.25 \times 10^{-4}} = \underline{6.37 \times 10^6} \text{ At/Wb}$$
$$\mathcal{R}_{NON\ GAP} = \frac{\ell - \ell_g}{\mu A} = \frac{4.5\times 10^{-2}}{16\pi \times 10^{-8}} = \underline{0.09\times 10^6} \text{ At/Wb}$$
$$\mathcal{R}_{CENTER\ LEG} = \mathcal{R}_g + \mathcal{R}_{NON\ GAP} = 6.37 \times 10^6 + 0.09 \times 10^6$$
$$= \underline{6.46 \times 10^{+6}} \text{ At/Wb}$$

Drawing an analog electric circuit utilizing the quantities $\mathcal{F}_1$, $\mathcal{F}_2$, $\mathcal{R}_{LEG}$, and $\mathcal{R}_{CENTER\ LEG}$ and solving for $\Phi_g = \Phi_1 + \Phi_2$, B_g and H_g utilizing the loop method for fluxes, we obtain (see Fig. 1.1.2)

$$\Phi_1(3\times 10^5 + 6.46 \times 10^6) + \Phi_2(6.46\times 10^6) = 5000$$
$$\Phi_1(6.46\times 10^6) + \Phi_2(3\times 10^5 + 6.46 \times 10^6) = 5000$$

Now

$$\Phi_1 = \frac{\begin{vmatrix} 5000 & 6.46\times 10^6 \\ 5000 & 6.76\times 10^6 \end{vmatrix}}{\begin{vmatrix} 6.76\times 10^6 & 6.46\times 10^6 \\ 6.46\times 10^6 & 6.76\times 10^6 \end{vmatrix}} = \frac{1.5\times 10^9}{4.0\times 10^{12}} = \underline{0.378\times 10^{-3}} \text{ Wb}$$

*0.5×10^{-2} designates the opening of the airgap ℓ_g by which each linear dimension of the iron core is increased to yield the effective cross-sectional area A_{EF} at the center of the airgap; that is, $A_{EF} = (b+\ell_g) \times (h+\ell_g)$, where b = base and h = height of the cross-sectional area of the iron core.

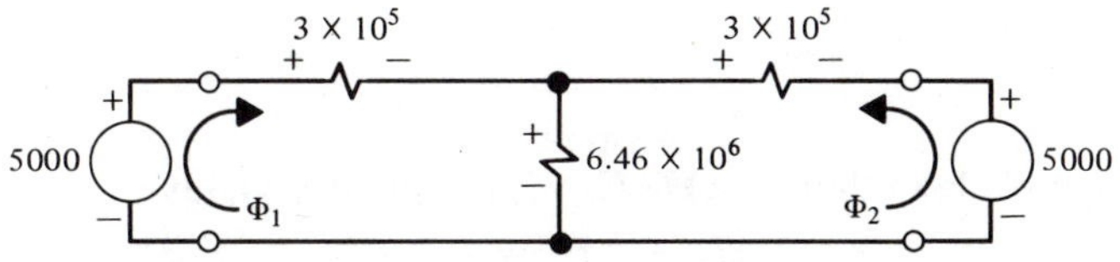

Fig. 1.1.2 Electric Circuit Analog of Fig. 1.1.1

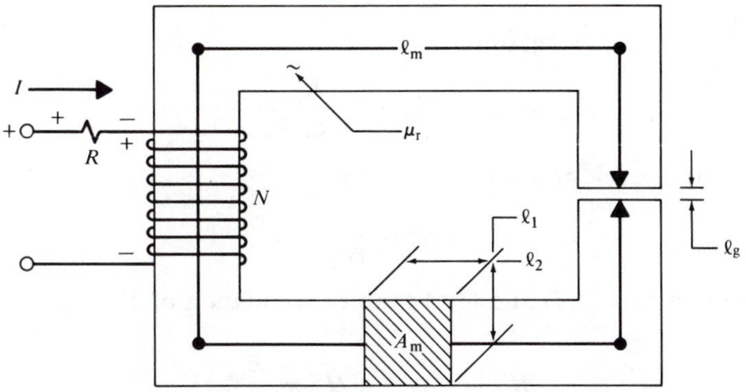

Fig. 1.2.1 One-Window Magnetic Circuit with One Forcing Function and a Small Airgap

and

$$\Phi_2 = \frac{\begin{vmatrix} 6.76\times 10^6 & 5000 \\ 6.46\times 10^6 & 5000 \end{vmatrix}}{4.0\times 10^{12}} = \frac{1.5\times 10^9}{4.0\times 10^{12}} = \underline{0.378\times 10^{-3}\ \text{Wb}}$$

Therefore

$$\Phi_g = \Phi_1 + \Phi_2 = \underline{\underline{0.76\times 10^{-3}}}\ \text{Wb}$$

Now

$$B_g = \frac{\Phi_g}{A_g} = \frac{0.78\times 10^{-3}}{6.25\times 10^{-4}} = \underline{\underline{1.22}}\ \text{Wb/m}^2$$

The unit 1 Wb/m^2 is also known as 1 Tesla (1 T) and

$$H_g = \frac{B_g}{\mu_0} = \frac{1.22}{4\pi\times 10^{-7}} = \underline{\underline{0.97\times 10^6}}\ \text{At/m}$$

Example 1.2 (type 2)*

Given: the configuration shown in Fig. 1.2.1. For the given magnetic circuit, $I, N, \ell_m, \ell_g, \mu_r, A_m, A_g = (\ell_1 + \ell_g)\times(\ell_2 + \ell_g)$, and μ_0 are given. The magnetic field Φ_m, B_m, H_m, Φ_g, B_g, and H_g can be calculated if one of the following procedures is followed.

*Perhaps before you proceed you might want to refer to p. 5, second paragraph.

Solution

Procedure No. 1.

Use of Ampere's law, that is, $\oint \mathbf{H} \cdot d\mathbf{L} = \Sigma I$. This law yields

$$\ell_m H_m + \ell_g H_g = NI = \mathcal{F}$$

Assuming that

$$\Phi_m \equiv \Phi_g \equiv \Phi$$

(this is true, why?) and calculating

$$\mathcal{R}_T = \mathcal{R}_m + \mathcal{R}_g = \frac{\ell_m}{\mu_0 \mu_R A_m} + \frac{\ell_g}{\mu_0 A_g}$$

enables calculation of Φ utilizing Eq. (1.9) as

$$\Phi = \frac{\mathcal{F}}{\mathcal{R}_T}$$

Now utilization of Eqs. (1.7) and (1.8) enables calculation of B_m, H_m, B_g, and H_g as follows:

$$B_m = \frac{\Phi}{A_m} \qquad H_m = \frac{B_m}{\mu}$$

$$B_g = \frac{\Phi}{A_g} \qquad H_g = \frac{B_g}{\mu_0}$$

Procedure No. 2.

B_m versus H_m is provided either as data or as a graph. Utilizing Ampere's law yields

$$H_m \ell_m + H_g \ell_g = NI$$

Substitution in this equation for

$$H_g = \frac{B_g}{\mu_0}$$

yields

$$H_m \ell_m + B_g \left(\frac{\ell_g}{\mu_0} \right) = NI$$

Since $\Phi_m \equiv \Phi_g$,

$$B_m A_m = B_g A_g$$

Thus

$$B_g = \frac{B_m A_m}{A_g}$$

and

$$H_m \ell_m + B_m \left(\frac{A_m \ell_g}{A_g \mu_0} \right) = NI$$

Solving this equation for $B_m = \psi\{H_m\}$ yields

$$B_m = \left(\frac{A_g \mu_0}{A_m \ell_g} \right) NI - \left(\frac{A_g \ell_m \mu_0}{A_m \ell_g} \right) H_m$$

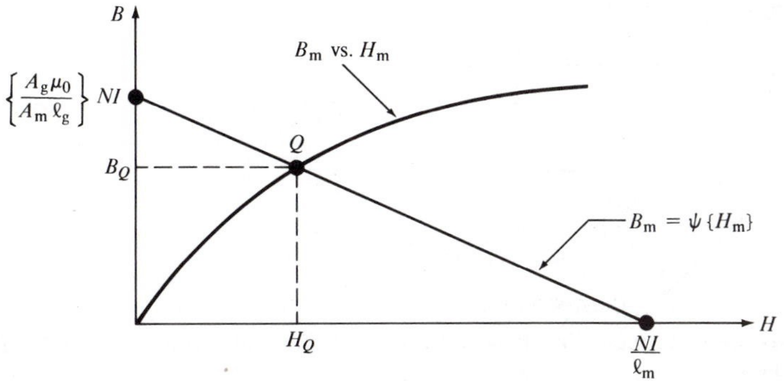

Fig. 1.2.2 Graphical Solution of the Magnetic Circuit Exhibited in Fig. 1.2.1

In order to plot this equation as a load line on the nonlinear B versus H characteristic for the iron, assume that

1.
$$H_m = 0$$
$$B_m = \left(\frac{A_g \mu_0}{A_m \ell_g}\right) NI$$

and

2.
$$H_m = \frac{NI}{\ell_m} \quad B_m = 0$$

Thus plotting $B_m = \psi\{H_m\}$ on the same set of axes as B_m versus H_m for the iron yields the graph shown in Fig. 1.2.2. Therefore, the desired flux density and field intensity are given by the point of intersection, that is, B_Q and H_Q, of the curve $B_m = \psi\{H_m\}$ or the load line and the B_m versus H_m characteristic of the iron.

Example 1.3

Given:* the configuration shown in Fig. 1.3.1. The dimensions of the magnetic circuit given in Fig. 1.3.1 are tabulated in the adjacent table. Determine the current that must flow in the exciting winding in order to produce the given flux Φ if the exciting winding has 200 turns and produces a flux of $\Phi = 3 \times 10^{-4}$ Wb.

For the solution of this problem, assume that the flux is confined in the path provided by the magnetic circuit (i.e., there is no flux leakage), and use the magnetization curves provided in Fig. 1.3.2.

Solution

Since all the flux is confined in the magnetic circuit, Φ_{IRON} is identical to Φ_{STEEL}. In addition, since by definition the cross-sectional areas of the cast iron A_{IRON} and the cast steel A_{STEEL} are identical, Eq. (1.7) enables us to utilize the input data and calculate the flux densities of the cast iron B_{IRON} and the steel B_{STEEL} as

$$B \equiv B_{IRON} = B_{STEEL} = \frac{\Phi}{A} = \frac{3 \times 10^{-4}}{0.002} = \underline{0.15 \text{ Wb/m}^2} \qquad (1)$$

If we were to plot the results of Eq. (1) in Fig. 1.3.2, we would observe that the $B = 0.15$ W/m^2 curve intersects the B versus H curves for the cast iron and the

*Note that the dimensional unit of ℓ is the meter and that of A is the meter squared.

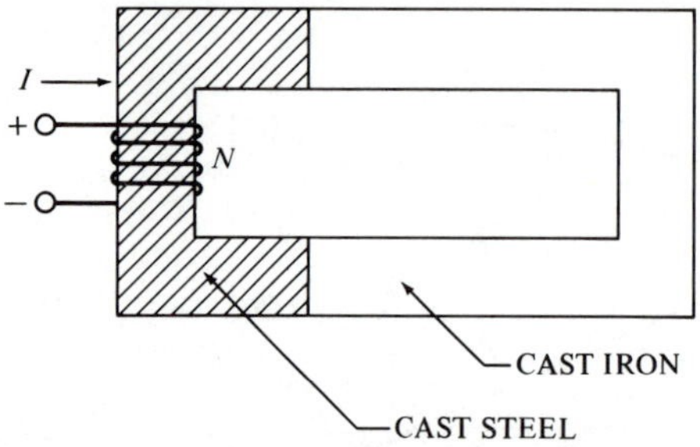

	ℓ	A
CAST STEEL	0.25	0.002
CAST IRON	0.50	0.002

Fig. 1.3.1 Magnetic Circuit Comprised from Two Ferromagnetic Materials and Pertinent Data

steel at points that yield the following magnetic field intensities:

$$H_{\text{IRON}} = \underline{345} \text{ A-t/m} \tag{2}$$

and

$$H_{\text{STEEL}} = \underline{122} \text{ A-t/m} \tag{3}$$

The utilization of Ampere's law, $\oint \mathbf{H} \cdot d\mathbf{L} = H_1\ell_1 + H_2\ell_2 + \cdots + H_N\ell_N = \Sigma I = NI$, in conjunction with input data, that is, $\ell_{\text{IRON}} = 0.50$ m and $\ell_{\text{STEEL}} = 0.25$ m, and Eqs. (2) and (3) enables us to write the following:

$$\sum I = NI = H_{\text{IRON}}\ell_{\text{IRON}} + H_{\text{STEEL}}\ell_{\text{STEEL}} = 345 \times 0.50 + 122 \times 0.25$$
$$= 172.50 + 30.50 = \underline{203.0} \text{ At} \tag{4}$$

The solution of Eq. (4) for the current I, with the aid of input data, $N = 200$ turns, yields

$$I = \frac{203}{200} = \underline{\underline{1.01 \text{ A}}} \tag{5}$$

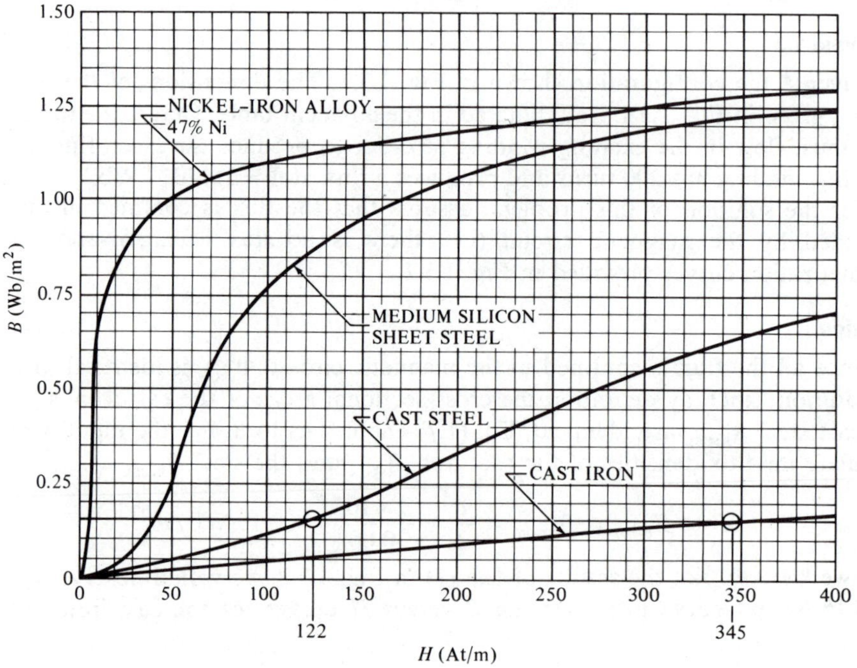

Fig. 1.3.2 Magnetization Curves for Various Ferromagnetic Materials

Please note that this approach is not the only one that can be used to solve this type of problem. It is, however, the simplest approach if the magnetization curves are available.

Example 1.4

A very popular and useful application of magnetic circuit principles is the so-called weight lifting with a magnetic crane.

For the given magnetic circuit, Fig. 1.4.1, with N, I, ℓ_m, ℓ_g, A, and ℓ_b completely defined, calculate the weight that can be lifted by the crane for the given parameters.

Solution

The magnetic energy stored in an incremental volume of space, where a given field $\phi(B,H)$ exists, is given by

$$dW_{MAG} = \tfrac{1}{2}(\mathbf{B}\cdot\mathbf{H})\,dv \tag{1}$$

Also, the definition of work yields

$$dW = \mathbf{F}\cdot d\mathbf{L} \tag{2}$$

For one pole of the magnetic circuit of Fig. 1.4.1, the volume is given as

$$v = A\ell_g \tag{3}$$

and the incremental change of the volume becomes

$$dv = A\,d\ell_g \tag{4}$$

Since **B** and **H** coincide, Eq. (1) yields

$$dW_{MAG} = \tfrac{1}{2}BH\,dv = \tfrac{1}{2}BHA\,d\ell_g = \frac{1}{2}\left(\frac{B^2}{\mu_0}\right)A\,d\ell_g \tag{5}$$

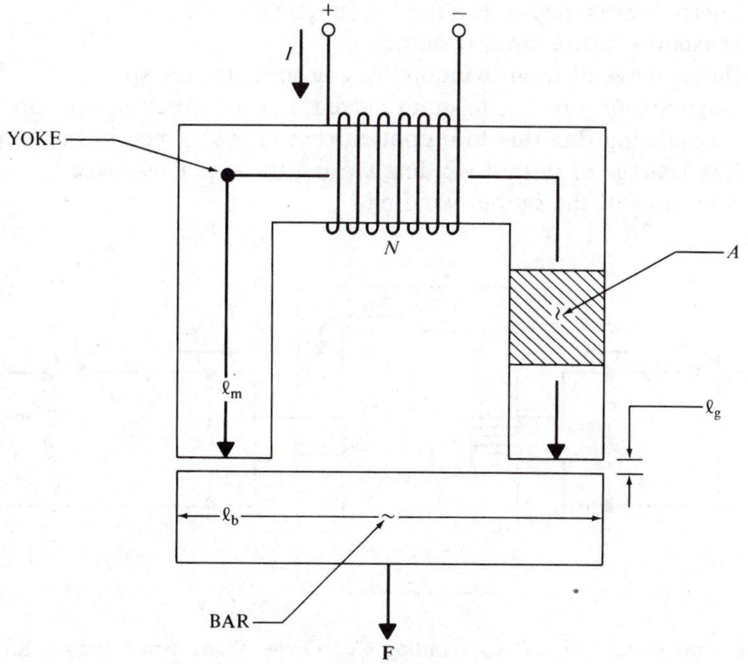

Fig. 1.4.1 Magnetic Circuit Comprised of Two Distinct Pieces of the Same Magnetic Material and Two Airgaps

The system tends to reach its minimum reluctance. Thus a pull is exerted upon the bar [you may think of the magnetic field, i.e., the flux lines, as elastic (rubber) bands pulling on the bar] and an energy dW equal to the magnetic energy dW_{MAG} stored in the magnetic field is expended. Therefore, for $|d\mathbf{L}| = d\ell_g$ and $d\mathbf{L}$ and $\mathbf{F}$ coinciding,

$$dW_{\text{MAG}} = \frac{1}{2}\left(\frac{B^2}{\mu_0}\right)A\,d\ell_g \equiv F\,d\ell_g = dW \tag{6}$$

Thus the pulling force per pole on the bar of Fig. 1.4.1 is

$$F = \frac{1}{2}\left(\frac{B^2}{\mu_0}\right)A \qquad \text{N/pole} \tag{7}$$

The total pulling force on the bar is

$$F_{\text{TOTAL}} = 2\left(\frac{1}{2}\right)\left(\frac{B^2}{\mu_0}\right)A = \left(\frac{B^2}{\mu_0}\right)A \qquad \text{N} \tag{8}$$

1.3 TRANSFORMERS

Structurally a transformer is comprised of (a) a highly permeable metallic core, (b) a winding with N_1-turns to which the forcing function is connected, known as the input winding, and (c) a winding with N_2-turns to which the load is connected, known as the output winding. The core usually has a rectangular cross section. The diagram of a transformer is shown in Fig. 1.5. The parameters and variables on the transformer in the figure are identified as follows:

r_S = internal resistance of the forcing function
r_1 = resistance of the input winding
$\Phi_{\ell 1}$ = flux leakage of input winding closing through free space
Φ_{m1} = magnetizing flux due to input current $i_1(t)$ confined in the core
Φ_{m2} = magnetizing flux due to output current $i_2(t)$ confined in the core
$\Phi_{\ell 2}$ = flux leakage of output winding closing through free space
r_2 = resistance of the output winding

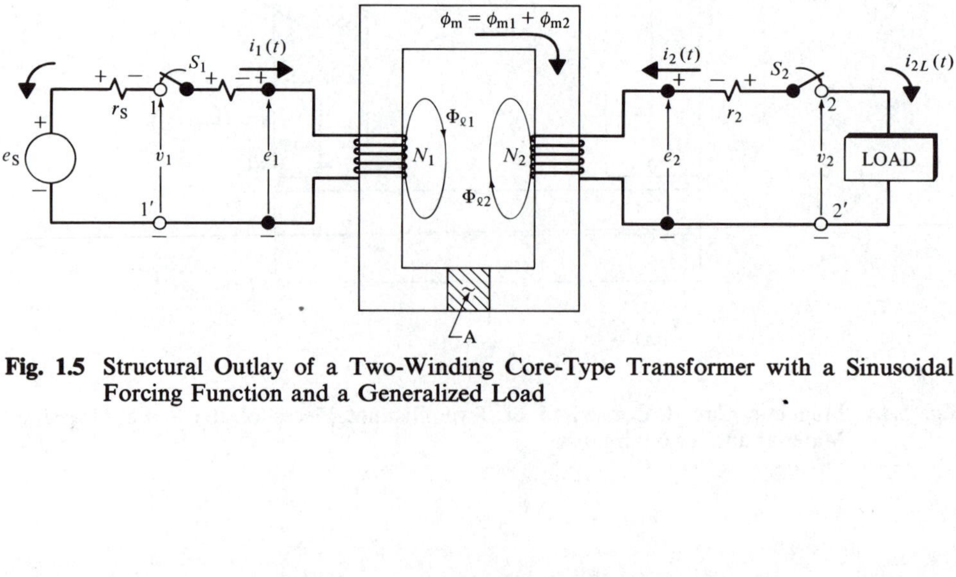

Fig. 1.5 Structural Outlay of a Two-Winding Core-Type Transformer with a Sinusoidal Forcing Function and a Generalized Load

e_1 = potential rise across the ideal coil N_1
e_2 = potential rise across the ideal coil N_2
e_S = forcing function
v_1 = potential drop across the input winding
v_2 = potential drop across the output winding
LOAD = electrical load
S_1 = switch at the input
S_2 = switch at the output

Note that each winding is *broken up* into a *resistor* and an *ideal coil* for their mathematical treatment.

The right-hand rule establishes the direction of the flux Φ_{m1} for the given instantaneous direction of the current $i_1(t)$. If the current $i_1(t)$ is varying, then Φ_{m1} is varying. The type of variation is not important in the present discussion. Now for a varying Φ_{m1} and S_2 closed, the following phenomenon takes place: For the sense of the output winding wrapping and say Φ_{m1} increasing, a current $i_{2L}(t)$ will be generated (induced). This current is a consequence of Lenz's law and will be called a *Lenzian current* at the output.

Lenz's law is stated as follows: *If a field of a certain direction is impressed on a wire loop, then the current induced in the wire loop will have a direction such that its field opposes the change in the impressed field.* This current exists in terminal 2 at the output. Thus the plus sign is at terminal 2.

A similarly induced *Lenzian current* $i_{1L}(t)$ is present at the input terminal 1. However, since the *impressed current* $i_1(t)$ is the current of importance *at the input* and it opposes $i_{1L}(t)$, a similar type of current is considered at the output that opposes the Lenzian current $i_{2L}(t)$. Such a current is $i_2(t)$. The current $i_2(t)$ is looked upon as an *impressed current at the output. Thus both* $i_1(t)$ *and* $i_2(t)$ *are impressed currents.* Utilizing Kirchhoff's voltage law to write the describing equations of the transformer between terminals 1 and 1′ and between 2 and 2′ yields

$$v_1 = r_1 i_1 + e_1 \tag{1.16}$$
$$v_2 = r_2 i_2 + e_2 \tag{1.17}$$

According to Faraday's law,

$$e_1 = \frac{d\lambda_1}{dt} \tag{1.18}$$

$$e_2 = \frac{d\lambda_2}{dt} \tag{1.19}$$

Where the flux linkage λ is defined as the product of the flux Φ* linking a coil of N-turns times N,

$$\lambda \equiv N\Phi \tag{1.20}$$

the flux linkages of the input and output coil can be written as

$$\lambda_1 = N_1\Phi_1 = N_1(\Phi_{m1}+\Phi_{m2}+\Phi_{\ell 1}) \tag{1.21}$$
$$\lambda_2 = N_2\Phi_2 = N_2(\Phi_{m1}+\Phi_{m2}+\Phi_{\ell 2}) \tag{1.22}$$

*Whenever this letter appears, it should be interpreted as $\Phi \equiv \Phi(t)$. The capital letter Φ is used, for the sake of consistency as well as appearance, to designate a time-varying flux, instead of the lowercase ϕ, as is usually the case. The letter Φ is mostly associated with the vector quantities **B**, **H**, and **A** (or **S**) and is directly related to **B** through $\Phi = \int_S \mathbf{B}\cdot d\mathbf{S}$.

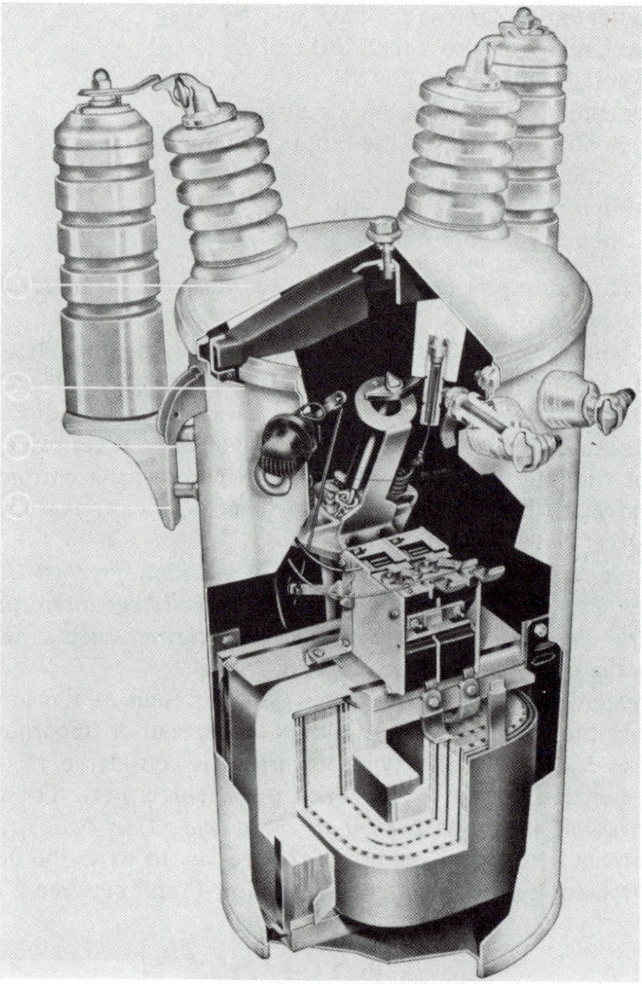

Plate 1.1 Transformer. Cutaway View of a 60-Hz, 5–500-kVA, 2400–34400-V, Single-phase, Liquid-immersed, Pole-mounted Distribution Transformer: (1) Cover, (2) Tank, (3) Lifting Lugs, and (4) Arrester Mounting. (Courtesy of Westinghouse Electric Co., Pittsburgh, Pa.)

Thus Eq. (1.18) and Eq. (1.19) can be written as

$$e_1 = \frac{d}{dt}(N_1\Phi_{\ell 1}) + N_1\frac{d}{dt}(\Phi_m) \tag{1.23}$$

and

$$e_2 = \frac{d}{dt}(N_2\Phi_{\ell 2}) + N_2\frac{d}{dt}(\Phi_m) \tag{1.24}$$

where

$$\Phi_m = \Phi_{m1} + \Phi_{m2} \tag{1.25}$$

Also, it is known that

$$\lambda = Li \tag{1.26}$$

Eqs. (1.20) and (1.26) enable us to write

$$\lambda = N\Phi = Li \tag{1.27}$$

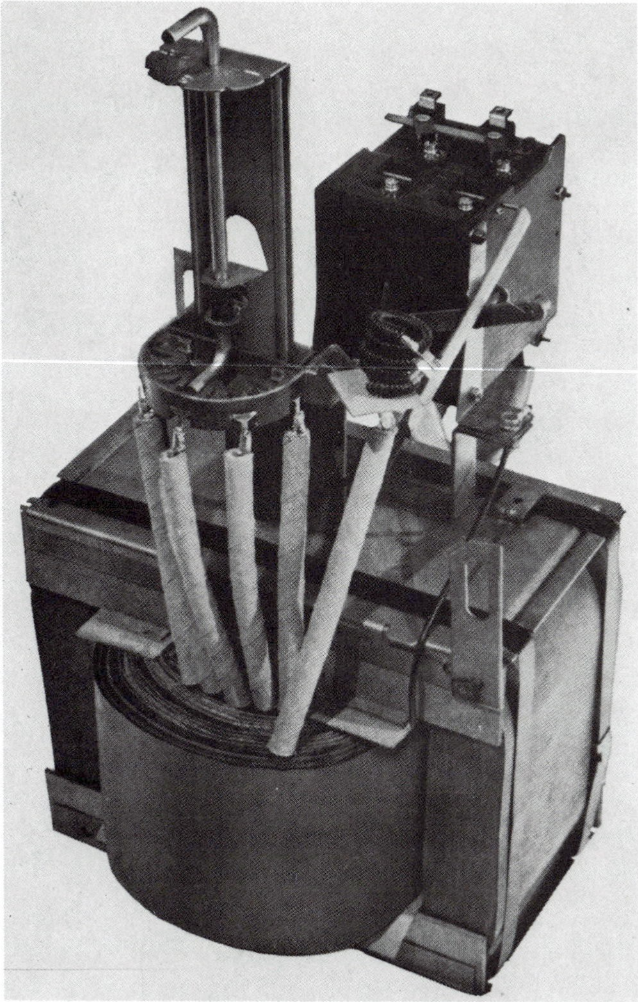

Plate 1.2 Core and Coil Assembly of a 60-Hz, 5–500 kVA, 2400–34400-V, Single-phase, Liquid-immersed, Pole-mounted Distribution Transformer. (Courtesy of Westinghouse Electric Co., Pittsburgh, Pa.)

Thus utilization of the corresponding leakage flux Φ_ℓ and the leakage inductance L_ℓ in conjunction with Eq. (1.27) yields

$$N_1\Phi_{\ell 1} = L_{\ell 1} i_1 \tag{1.28}$$

and

$$N_2\Phi_{\ell 2} = L_{\ell 2} i_2 \tag{1.29}$$

Substitution of Eq. (1.28) and Eq. (1.29) in Eqs. (1.23) and (1.24), respectively, enables us to write Eq. (1.16) and Eq. (1.17) as

$$v_1 = r_1 i_1 + L_{\ell 1}\frac{d}{dt}(i_1) + N_1\frac{d}{dt}(\Phi_m) \tag{1.30}$$

and

$$v_2 = r_2 i_2 + L_{\ell 2}\frac{d}{dt}(i_2) + N_2\frac{d}{dt}(\Phi_m) \tag{1.31}$$

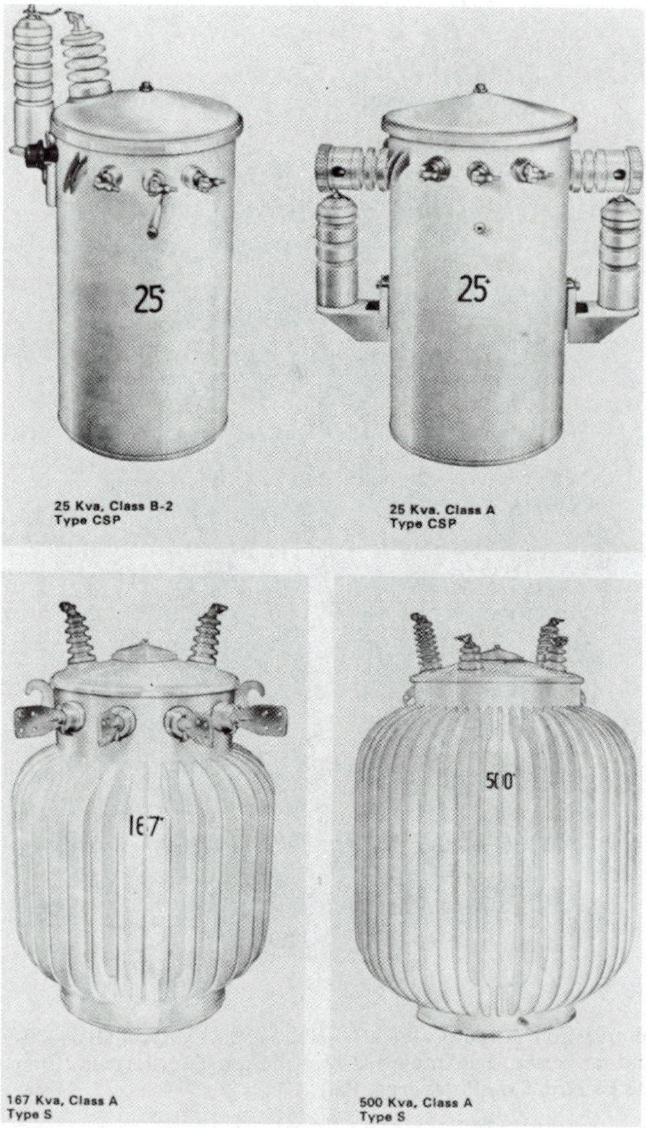

Plate 1.3 Various 60-Hz, 5–500-kVA, 2400–34400-V, Single-phase, Liquid-immersed, Pole-mounted Distribution Transformers. (Courtesy of Westinghouse Electric Co., Pittsburgh, Pa.)

Utilizing Eq. (1.9) enables us to write Eq. (1.25) as

$$\Phi_m = \frac{N_1 i_1}{\mathcal{R}} + \frac{N_2 i_2}{\mathcal{R}} = \frac{N_1}{\mathcal{R}}\left[i_1 + \left(\frac{N_2}{N_1}\right)i_2\right] \tag{1.32}$$

Also, from Eq. (1.27) and Eq. (1.9),

$$\lambda_{m1} = N_1 \Phi_{m1} = L_{m1} i_1 = N_1\left(\frac{N_1 i_1}{\mathcal{R}}\right) \tag{1.33}$$

Thus,

$$\frac{N_1}{\mathcal{R}} = \frac{L_{m1}}{N_1} \tag{1.34}$$

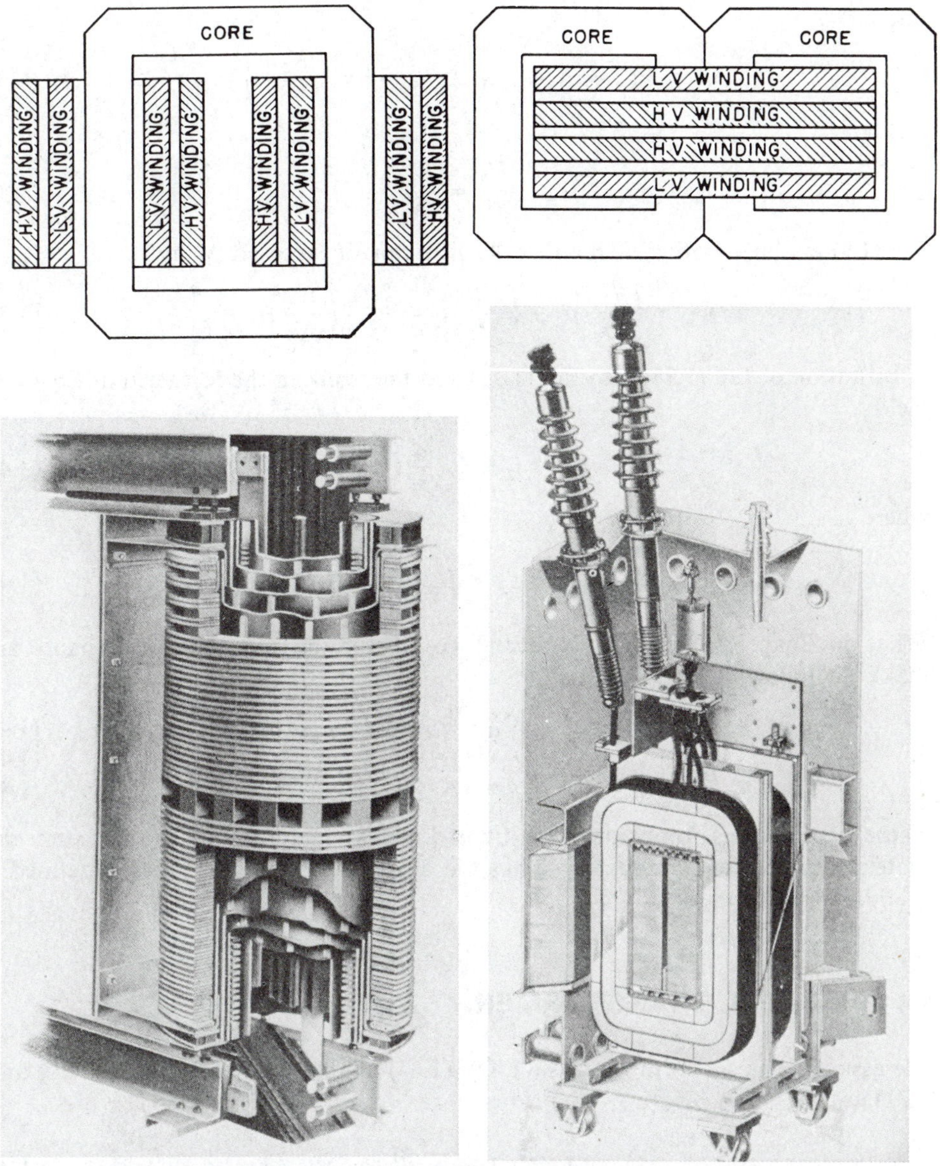

Plate 1.4 Construction of Single-phase Transformers: (a) Core-Form and (b) Shell-Form. (Reproduced from *Electric Transmission and Distribution Book*, by Central Station Engineers, Westinghouse Electric Co., Pittsburgh, Pa.)

Substitution of Eq. (1.34) into Eq. (1.32) and the result in the last term of Eq. (1.30) yields

$$v_1 = r_1 i_1 + L_{\ell 1}\frac{d}{dt}(i_1) + e_{m1} \tag{1.35}$$

where

$$e_{m1} \equiv N_1 \frac{d}{dt}(\Phi_m) = L_{m1}\frac{d}{dt}\left[i_1 + \left(\frac{N_2}{N_1}\right)i_2\right] \tag{1.36}$$

Similarly,

$$\lambda_{m2} = N_2\Phi_{m2} = L_{m2}i_2 = N_2\left(\frac{N_2 i_2}{\mathcal{R}}\right) \tag{1.37}$$

yields

$$\frac{N_2}{\mathcal{R}} = \frac{L_{m2}}{N_2} \tag{1.38}$$

Eq. (1.9) enables writing of the flux Φ_m in terms of factored N_2 as

$$\Phi_m = \left(\frac{N_1}{\mathcal{R}}\right)i_1 + \left(\frac{N_2}{\mathcal{R}}\right)i_2 = \frac{N_2}{\mathcal{R}}\left[\left(\frac{N_1}{N_2}\right)i_1 + i_2\right] \tag{1.39}$$

Substitution of Eq. (1.38) into Eq. (1.39) and the result in the last term of Eq. (1.31) yield

$$v_2 = r_2 i_2 + L_{\ell 2}\frac{d}{dt}(i_2 + e_{m2}) \tag{1.40}$$

where

$$e_{m2} \equiv N_2\frac{d}{dt}(\Phi_m) = L_{m2}\frac{d}{dt}\left[\left(\frac{N_1}{N_2}\right)i_1 + i_2\right] \tag{1.41}$$

What do Eqs. (1.36) and (1.41) mean? To answer this question let us pause and make the following assumptions:

1. $$\Phi_{\ell 1} = \Phi_{\ell 2} = 0 \tag{1.42}$$
2. $$r_1 = r_2 = 0 \tag{1.43}$$
3. $$\mu = \infty \tag{1.44}$$

These assumptions imply that the transformer is made up of highly permeable material and perfect conductors. Thus the device is an ideal one and is defined as the ideal transformer.

1.4 THE IDEAL TRANSFORMER

The assumptions set forth by Eqs. (1.42)–(1.44) enable us to write Eqs. (1.30) and (1.31) as

$$v_1 \equiv e_{m1} = N_1\frac{d}{dt}(\Phi_m) \tag{1.45}$$

$$v_2 \equiv e_{m2} = N_2\frac{d}{dt}(\Phi_m) \tag{1.46}$$

Dividing Eq. (1.46) into Eq. (1.45) yields

$$\frac{v_1}{v_2} = \frac{N_1}{N_2} \equiv \frac{e_{m1}}{e_{m2}} \tag{1.47}$$

In accordance with the ideas of Section 1.1, the magnetic circuit of the ideal transformer is as shown in Fig. 1.6. The describing equation of the magnetic circuit (remembering that $\mathcal{R} = \ell_m/\mu A$ and $\mu = \infty$) is

$$N_1 i_1 + N_2 i_2 = N_1 i_\alpha\big|_{i_\alpha \equiv i_1} + N_2 i_\beta\big|_{i_\beta \equiv i_2} = 0 \tag{1.48}$$

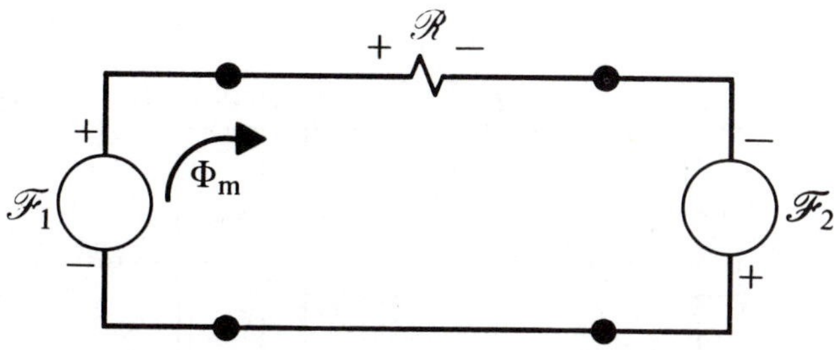

Fig. 1.6 The Magnetic Circuit of the Ideal Transformer

Eq. 1.48 yields

$$\frac{i_1}{i_2} = -\frac{N_2}{N_1} \equiv \frac{i_\alpha}{i_\beta} \tag{1.49}$$

The circuit representation of the mathematical analog defined by Eqs. (1.47) and (1.49) is as given in Fig. 1.7. It is known as the equivalent circuit of the ideal transformer for sinusoidal excitation. The two *dots indicate the ends of the windings that are on a positive potential rise at the same time.*

If the forcing function of the transformer is sinusoidal in nature, $e_S(t) = \sqrt{2}\,|\mathbf{E}_S|\cos(\omega_E t + \phi°)$, then all voltages and currents throughout the transformer are sinusoidal in nature. Thus the phasors $\mathbf{V}_1$, $\mathbf{V}_2$, $\mathbf{I}_1$, and $\mathbf{I}_2$ will have more meaning in Eqs. (1.47) and (1.49) instead of the sinusoidally varying voltages and currents. Thus Eqs. (1.47) and (1.49) can be written as

$$\frac{\mathbf{V}_1}{\mathbf{V}_2} = \frac{N_1}{N_2} \tag{1.50}$$

and

$$\frac{\mathbf{I}_1}{\mathbf{I}_2} = -\frac{N_2}{N_1} \tag{1.51}$$

Solving Eq. (1.50) for $\mathbf{V}_1$ and Eq. (1.51) for $\mathbf{I}_1$ [remembering that the current through the load is the Lenzian current $i_{L2}(t)$, which is defined as the loading current $\mathbf{I}_L \equiv \mathbf{I}_{L2} = -\mathbf{I}_2$] and taking the ratio $\mathbf{V}_1/\mathbf{I}_1$ yields

$$\frac{\mathbf{V}_1}{\mathbf{I}_1} = \left(\frac{N_1}{N_2}\right)^2\left(\frac{\mathbf{V}_2}{\mathbf{I}_L}\right) \tag{1.52}$$

Eq. (1.52) relates the *input impedance* of the transformer, $\mathbf{Z}_1 = \mathbf{V}_1/\mathbf{I}_1$ ohms, to the *loading impedance* of the transformer, $\mathbf{Z}_L = \mathbf{V}_2/\mathbf{I}_L$. These ideas enable us to write

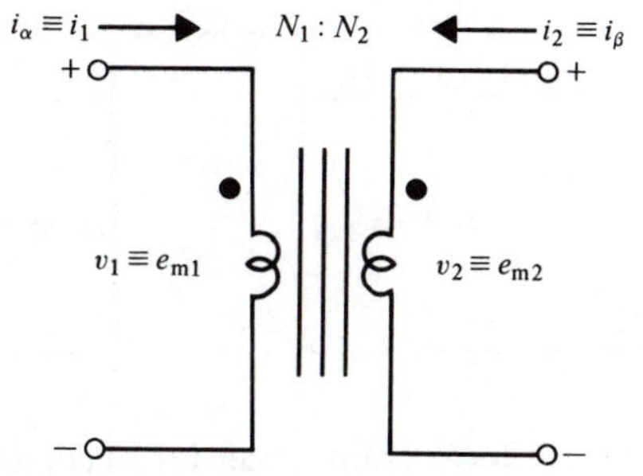

Fig. 1.7 The Equivalent Circuit of the Ideal Transformer

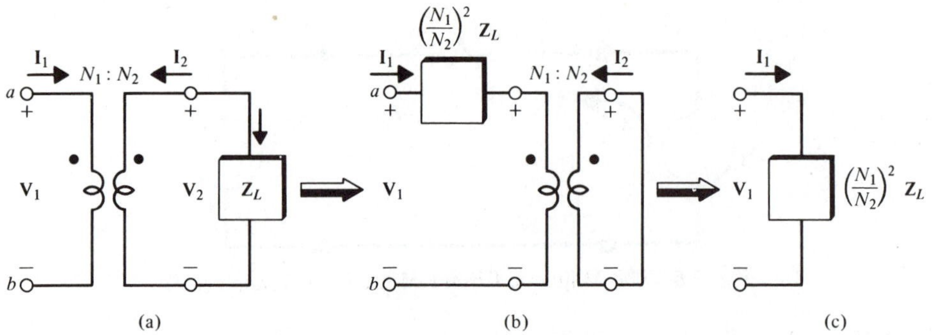

Fig. 1.8 Loading of the Ideal Transformer. (All networks are equivalent at the terminals *ab*)

Eq. (1.52) as

$$\mathbf{Z}_1 = \left(\frac{N_1}{N_2}\right)^2 \mathbf{Z}_L \tag{1.53}$$

The circuit representation of these ideas is exhibited in Fig. 1.8.

Solving Eq. (1.50) for $\mathbf{V}_1$, conjugating Eq. (1.51), solving for $\mathbf{I}_1^*$, and taking the product $\mathbf{V}_1\mathbf{I}_1^*$ yields

$$\mathbf{V}_1\mathbf{I}_1^* = -\mathbf{V}_2\mathbf{I}_2^* \tag{1.54}$$

Eq. (1.54) relates the *input complex power* to the transformer, that is, $\mathbf{W}_1 = \mathbf{V}_1\mathbf{I}_1^*$, to the *output complex power* of the transformer, $\mathbf{W}_2 = -\mathbf{V}_2\mathbf{I}_2^* = \mathbf{V}_2\mathbf{I}_L^*$. Note the significance of the *sign* at the output power.

1.5 USES OF THE IDEAL TRANSFORMER

The ideal transformer, however fictitious it may be, is a very useful device in communication and power system analysis. It is used to represent the impedance matching effect and the step-up and step-down of voltage and current levels in a real transformer. Note that an ideal transformer does not block dc. Equations (1.50), (1.51), and (1.53) are self explanatory for the step-up or step-down of the various quantities. However, the concept of impedance matching is explained further here.

Assume that an impedance of 5 Ω must be connected to a source defined by $E_{TH} = 100\ e^{j0°}$ V and $R_{TH} = 20$ Ω, with the expectation that maximum power is transferred to the load, Fig. 1.9. The maximum power transfer theorem states that

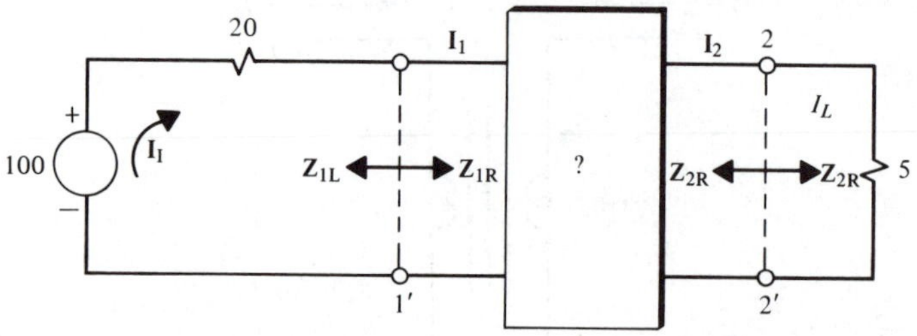

Fig. 1.9 Utilization of the Ideal Transformer for Maximum Power Transfer

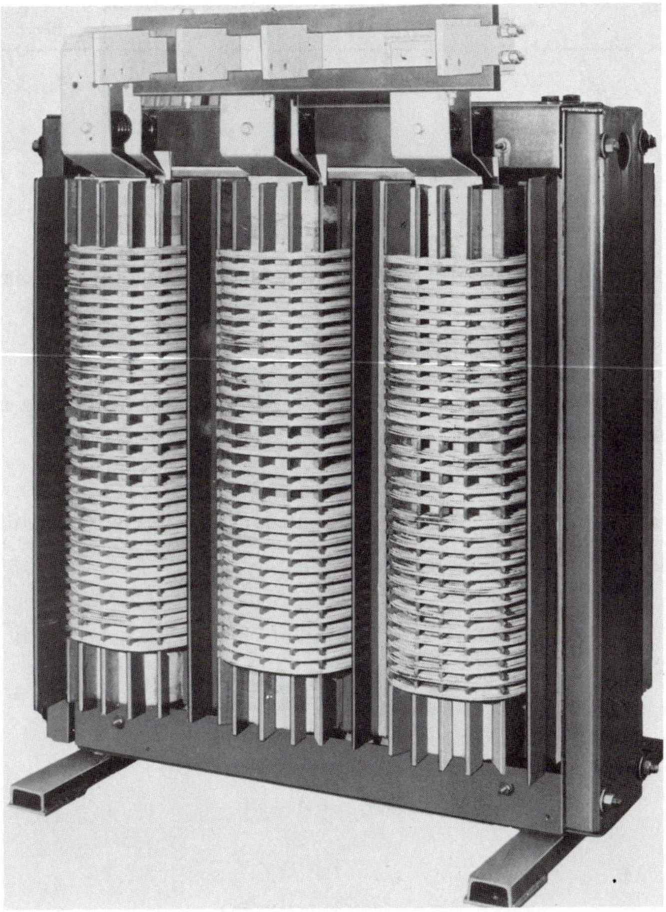

Plate 1.5 Core and Coil Assembly of Ventilated Dry-type Transformers. (Courtesy of Westinghouse Electric Co., Pittsburgh, Pa.)

the impedance connected at the terminals of Thevenin's equivalent should be equal to the conjugate of the Thevenin impedance, that is, $\mathbf{Z}_L = \mathbf{Z}_{TH}^*$, *for maximum power transfer*. Utilization of the maximum power transfer theorem suggests that the impedances looking left and right at the reference line drawn at the terminals 1–1′ should be complex conjugates. Using Eq. (1.53) at 1–1′ yields

$$20 = \mathbf{Z}_1 = \left(\frac{N_1}{N_2}\right)^2 5 \tag{1.55}$$

Impedance matching could also take place at the terminals 2–2′. Thus

$$5 = \mathbf{Z}_L = \left(\frac{N_2}{N_1}\right)^2 20 \tag{1.56}$$

Although this is a very simple-minded example, nevertheless it points out that the given source can be connected to the given load *through an ideal transformer* with a turns ratio of $N_1/N_2 = 2$. Any combination of turns that gives a ratio of 2 is valid, such as $N_1 = 200$ and $N_2 = 100$, and so on.

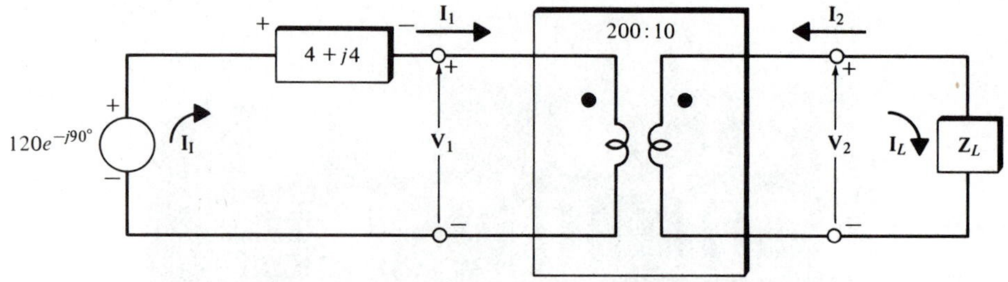

Fig. 1.5.1 Use of the Ideal Transformer for Matching the Loading Impedance Z_L to the Source Impedance $Z_S = 4 + j4$ ohms.

Example 1.5

Given: the configuration shown in Fig. 1.5.1. Find the following quantities for the conditions of maximum power transfer: $\mathbf{Z}_L, \mathbf{I}_1, \mathbf{I}_2, \mathbf{V}_1, \mathbf{V}_2, \mathbf{W}_1, \mathbf{W}_2$.

Solution

The utilization of Eq. (1.53) and pertinent data enable us to calculate $\mathbf{Z}_1$ as

$$\mathbf{Z}_1 = \left(\tfrac{200}{10}\right)^2 \mathbf{Z}_L = 400\mathbf{Z}_L$$

Since the maximum power transfer theorem states that $\mathbf{Z}_0 = \mathbf{Z}_{TH}^*$, $\mathbf{Z}_1$ can be calculated by use of pertinent data as

$$\mathbf{Z}_1 = \mathbf{Z}_{TH}^* = 4 - j4$$

Therefore,

$$400\mathbf{Z}_L = 4 - j4$$

and

$$\underline{\underline{\mathbf{Z}_L = 10^{-2} - j10^{-2}\ \Omega}}$$

See Fig. 1.5.2.

$$\mathbf{I}_1 = \frac{120e^{-j90°}}{(4+j4)+(4-j4)} = \frac{120e^{-j90°}}{8} = \underline{\underline{15e^{-j90°}}}\ \text{A}$$

$$\mathbf{I}_2 = -\left(\frac{N_1}{N_2}\right)\mathbf{I}_1 = -\left(\tfrac{200}{10}\right)15e^{-j90°} = \underline{\underline{300e^{+j90°}}}\ \text{A}$$

$$\mathbf{I}_L = -\mathbf{I}_2 = \underline{\underline{300e^{j270°}}}\ \text{A}$$

$$\mathbf{V}_2 = \mathbf{Z}_L\mathbf{I}_L = (10^{-2} - j10^{-2})(300e^{j270°}) = \underline{\underline{3\sqrt{2}\ e^{j225°}}}\ \text{V}$$

$$\mathbf{V}_1 = \left(\frac{N_1}{N_2}\right)\mathbf{V}_2 = \left(\tfrac{200}{10}\right) \times 3\sqrt{2}\ e^{j225°} = \underline{\underline{60\sqrt{2}\ e^{j225°}}}\ \text{V}$$

$$\mathbf{W}_1 = \mathbf{V}_1\mathbf{I}_1^* = (60\sqrt{2}\ e^{j225°})(15e^{j90°}) = \underline{\underline{900\sqrt{2}\ e^{j315°}}}\ \text{VA}$$

$$\mathbf{W}_L = \mathbf{V}_2\mathbf{I}_L^* = (3\sqrt{2}\ e^{j225°})(300e^{-j270°}) = \underline{\underline{900\sqrt{2}\ e^{-j45°}}}\ \text{VA}$$

$$\mathbf{W}_2 = -\mathbf{V}_2\mathbf{I}_2^* = -(3\sqrt{2}\ e^{j225°})(300e^{-j90°}) = \underline{\underline{900\sqrt{2}\ e^{j315°}}}\ \text{VA}$$

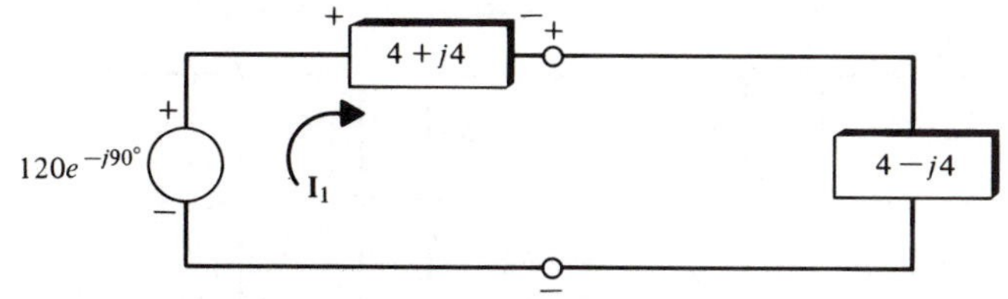

Fig. 1.5.2 The Equivalent Circuit of Fig. 1.5.1

1.6 THE REAL TRANSFORMER

Eqs. (1.36) and (1.41) can be written as

$$e_{m1} = L_{m1}\frac{d}{dt}\left[i_1 + \left(\frac{N_2}{N_1}\right)i_2\right] \equiv L_{m1}\frac{d}{dt}[i_1 - i_a] \equiv L_{m1}\frac{d}{dt}(i_{m1})\Big|_{i_{m1}\equiv(i_1 - i_a)} \quad (1.57)$$

and

$$e_{m2} = L_{m2}\frac{d}{dt}\left[\left(\frac{N_1}{N_2}\right)i_1 + i_2\right] \equiv L_{m1}\frac{d}{dt}[-i_\beta + i_2] \equiv L_{m2}\frac{d}{dt}(i_{m2})\Big|_{i_{m2}\equiv(-i_\beta + i_2)} \quad (1.58)$$

respectively. It is important to notice the subtle difference between the currents i_1 and i_α, and i_2 and i_β*. Therefore, when the constraints set forth by Eqs. (1.42) and (1.43) are **not valid**, the circuit analog of the transformer can be drawn with the aid of Eqs. (1.30), (1.57), and (1.40), as exhibited in Fig. 1.10. This equivalent circuit describes the transformer more realistically than does Fig. 1.7. Thus from the circuit theory point of view, L_{m1}, connected as in Fig. 1.10, or L_{m2}, connected across ε_2, **represents the effect of the field $\mathbf{B}_m$ within the core structure of the transformer,** Fig. 1.5. Consequently, the inductances L_{m1} and L_{m2} are defined as **magnetizing inductances**; the currents i_{m1} and i_{m2} are defined as **magnetizing currents.**

As the core structure's field $\mathbf{B}_m$ increases with time, as exhibited in Fig. 1.11, circulating currents, that is, $i_E(t)$, which are known as eddy currents, are induced in the core structure. These currents produce **heating** of the structure due to the $[i_E(t)]^2R_C$ losses. To minimize these losses, the core structure is made from **laminated sheets** rather than as a solid structure. The insulation between each sheet reduces the effective area, $A_{EF} = K_S A$, of the core. Thus a multiplying factor K_S, known as the *stacking factor*, is used to effect the reduction of the core area A when calculations utilizing the quantity area are performed. The numerical value of the stacking factor K_S changes from manufacturer to manufacturer. However, its usual numerical value is $K_S = 0.95$.

As the forcing function changes in strength and direction, the field changes in strength and direction. If we were to plot $B_m = f(H_m)$ as B versus H ($H\ell = Ni$), the phenomena of Fig. 1.12 would be observed. The dimensional unit of the incremental area under the curve, $H\,dB$, is the joule per cubic meter, that is,

$$W_I = \int_0^{B_S} H\,dB \equiv A_{01B_S} \quad (1.59)$$

*As a rule $i_1 \neq i_\alpha$ and $i_2 = i_\beta$ or $i_2 \neq i_\beta$ and $i_1 = i_\alpha$. For example, see Figs. 1.10 and 1.13.

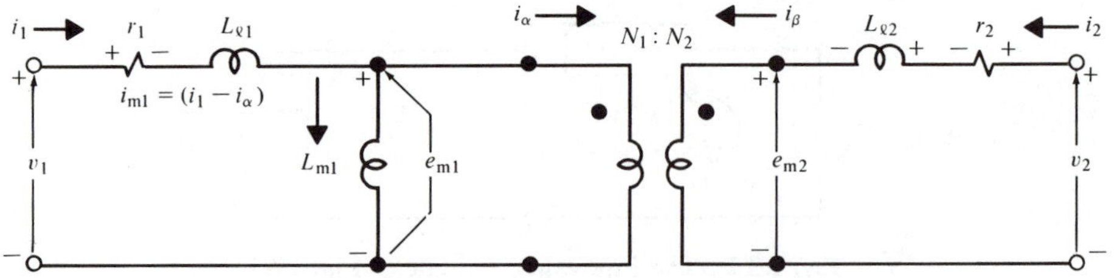

Fig. 1.10 First Step in the Development of the Equivalent Circuit of the Real Transformer

represents the **energy input** into the system. This takes place during the increase of the forcing function. When the forcing function begins to decrease a new area is traced,

$$W_O = \int_{B_S}^{B_R} H\,dB \equiv A_{1B_SB_R} \tag{1.60}$$

representing the **energy given up** by the system. Thus the energy represented by the area A_{01B_R} **is lost as heat** or **is not recoverable. Over an entire cycle of the forcing function, the energy represented by the area bounded by the loop 1B$_R$(−H$_C$)2(−B$_R$)H$_C$1, known as the magnetization curve or hysteresis loop, is not recoverable. The word "hysteresis" is derived from the Greek verb "ὑστερέω," meaning that *B* always lags *H*** (**B_R**≡*residual field* or *retentivity* and H_C≡*coercive force* or *coercivity*). The amount of this energy decreases as the area of the hysteresis loop decreases, that is, as the area gets thinner and thinner. Ideally, this area could be a flattened "S" with the center going through the BH-origin. A material having such a magnetization curve is known as a **soft magnetic material** and is the ideal material for the structure of transformers and rotating machinery.

The nonrecoverable energy is expended as heat due to friction of the magnetic domains. A *magnetic domain* is defined as a region of an iron crystal comprised of 10^6 atoms whose spins and orbital fields are aligned similarly. Any iron crystal forms a cubic structure that has 3(+) and 3(−) axes of each magnetization. The magnetic domains behave as small magnets and normally are oriented randomly along the axis of easy magnetization of the crystal. This random orientation of the domains yields a zero net field. However, when the crystal is subjected to an externally applied magnetic field, the domain whose field is not parallel to the applied field becomes unstable. The field of the domain rotates (in steps, i.e., erratically, yielding what is known as *Barkhausen's steps*), tending to become parallel to the applied field. This yields a *linearly* increased field, $B=f(H)$, up to a certain value of H, say H_P. The region defined by H_P, $0<H<H_P$, is known as the *easy magnetization region*. Beyond H_P, H must be increased very much for a small motion of the domain fields towards parallelism, that is, B increases slightly. In this region, $H_P<H<\infty$, $B=f(H)$ is *nonlinear*, and the region is known as the *hard magnetization region*.

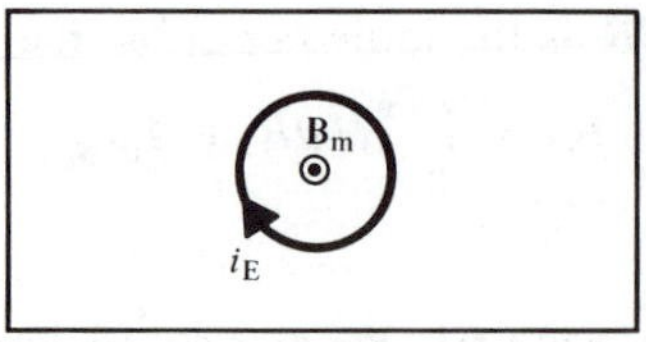

Fig. 1.11 Eddy Current Generation. (**B$_m$** Is Assumed to be Coming Out of the Page and Increasing)

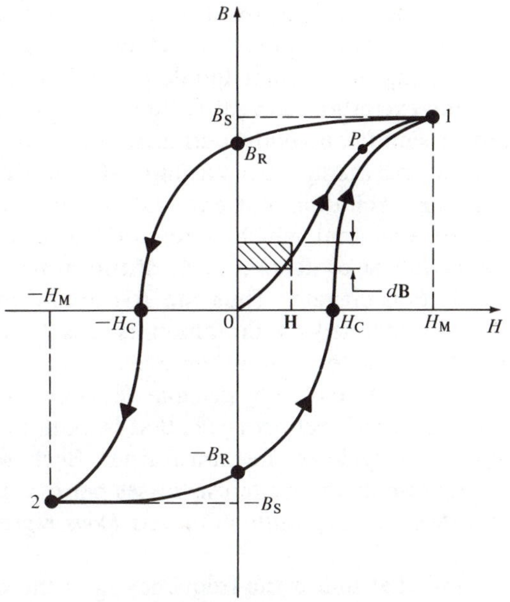

Fig. 1.12 The Magnetization Curve or Hysteresis Loop of a Magnetic Material

Finally, for very high values of H, all domains have aligned themselves with the externally applied field, which exhibits a net increase. This phenomenon is known as **magnetization.** This means that if the external field were to be removed, the sample would exhibit a magnetic field. Also, in this state (as compared to the state of the sample before it was subjected to the external field) the sample exhibits a change in its physical dimensions. This phenomenon is known as **magnetostriction.** If the direction of the applied field changes, the domains will tend to follow the field so that field parallelism is maintained. The motion of the domains against their surroundings results in *expenditure of energy* that also appears in the form of *heat*. This heat energy can be calculated by $P_H = K_H f_E \int_V B^{1.6} dv$ watts. Here f_E is the frequency of the externally applied field in hertz, K_H is a material constant and is known as the *Steinmetz constant*, B is the maximum flux density, V is the volume of the finite material, and P_H is the *average* **power loss due to hysteresis.** The energy lost due to eddy currents can be calculated by $P_E = K_E t^2 f_E^2 B^2 V$ watts, where K_E is a material constant, t is the thickness of the laminations, and P_E is the *average* **power loss due to eddy currents.**

Clearly then, the total power loss in the core of the transformer at hand is the sum of the hysteresis and eddy current losses. This sum, defined as the **core loss, iron loss,** or **excitation loss,** is designated as P_c. Graphically this power is represented by the magnetization curve or hysteresis loop of Fig. 1.12. However, because the physical significance of Fig. 1.12 is often *misunderstood or misinterpreted*, it will be further explained here.

Experience has shown that the size of this loop changes for a defined excitation current variation, that is, $|i| \leqslant i_{MAX}$, as the frequency of this current changes. When, for example, the frequency f_E of the excitation current is zero, or very close to it, the eddy currents in the core of the transformer are zero. Therefore, Fig. 1.12 represents only the *hysteresis losses per cycle* of the transformer at hand and the loop is known as its **dc** or **static hysteresis loop.** However, when the frequency f_E of the excitation current is high, the eddy currents in the core are nonzero and affect the dc or static hysteresis loop in a positive (increasing) manner. To understand this effect, suppose that the current required to set up a given flux $\Phi_{m1} = B_{m1}/A$ is $i_1 = \pm H_{m1}\ell_m/N_1$, depending on whether the dc or static hysteresis loop is ascending or descending. If the excitation current is time-varying and of sufficiently high frequency f_E to induce significant eddy currents, the eddy currents will tend to prevent this flux from increasing (decreasing). If the flux at hand is to be maintained **constant**, the excitation current will increase (decrease) by an appropriate amount Δi beyond that which is required to establish the dc or static hysteresis loop. This **increment** or **decrement** Δi of the current is provided from (to) the source of the excitation current. Thus the BH-axes dimensions of the dc or static hysteresis loop are changed by the amounts $\pm\Delta B$ and $\pm\Delta H$, correspondingly. The changed hysteresis loop now has a *larger area* than the dc or static hysteresis loop. It is defined as the *ac* or *dynamic hysteresis loop* of the transformer at hand and represents its *core losses per cycle*, that is, both the hysteresis losses and the eddy current losses per cycle of the transformer. Because, however, the dc or static hysteresis loop represents the hysteresis losses per cycle, it is obvious that *the difference between the dynamic and static hysteresis loops represents the eddy current losses.*

It should be clear now that unless the frequency f_E of the excitation current of a given magnetic core is known, we are unable to state with certainty (simply by examining its physical appearance) whether a given magnetization curve or hysteresis loop is static (dc) or dynamic (ac) in nature.

The core losses $P_c = P_H + P_E$ of the transformer at hand are represented as i^2R losses in a resistive element connected in parallel with the inductor L_{m1} in Fig. 1.10 or the inductor L_{m2} connected across ε_2. Note that the resistance of such an element, designated as r_c ($g_c = 1/r_c$), and the current through it, designated as i_c, are defined as *core-loss* parameters. Usually the magnetizing inductance is connected across the input of the ideal transformer, Fig. 1.10; therefore, the subscript 1 in L_{m1} is *omitted.*

Thus when we take into account not only the **magnetization** of the *iron core* and the **core losses** of the core structure of the transformer but also the **wire capacitance** (designated by C_W), we can draw the **equivalent circuit of the transformer operating in steady state**, as given in Fig. 1.13. Note that in Fig. 1.13 the sum of the steady-state magnetizing current I_m through the susceptance $-jb_m = 1/j\omega_E L_m$ (which designates the magnetizing effects of the field $\mathbf{B}_m$ within the core structure of the transformer), and the core-loss current I_c through the conductance $g_c = 1/r_c$

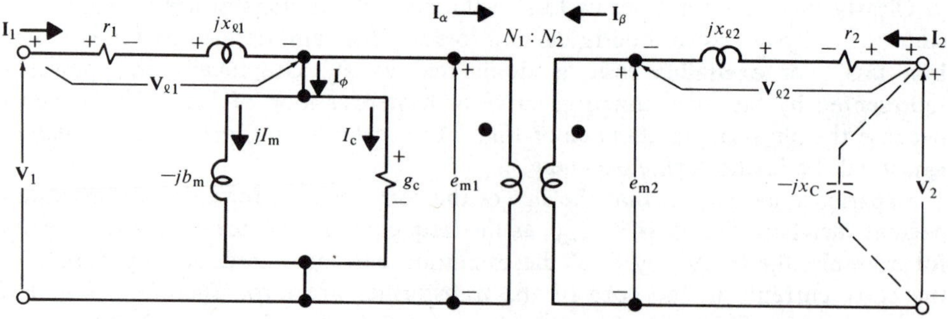

Fig. 1.13 Second Step in the Development of the Equivalent Circuit of the Real Transformer

(which designates the core losses in the core structure of the transformer), constitutes the **exciting current $\mathbf{I}_\phi$**, that is, $\mathbf{I}_\phi = I_c + jI_m$.

The exciting current is determined by the magnetic properties of the transformer's core structure. This current must adjust itself so that it produces the mmf required to create the flux demanded by Eqs. (1.23) through (1.25). It can be graphically reconstructed using Fig. 1.5 with the following constraints imposed: (1) the switch S_2 is set in the open position, (2) $(r_S + r_1)$ is negligible, (3) $e_S(t) = \sqrt{2}\,|\mathbf{E}_S|\sin\omega_E t = N_1 d/dt\Phi_m \rightarrow \Phi_m = -\sqrt{2}\,|\Phi_{RMS}|\cos\omega_E t$, (4) $\Phi_m = B_m A$, and (5) $i_\phi = H_m \ell_m / N_1$ if we were to reflect Φ_m versus t off the B_m versus H_m magnetization curve, that is, the hysteresis loop of the core material after rescaling the BH-axes to depict the Φi-axes, respectively. The detailed graphical construction of i_ϕ versus t, for a given Φ_m versus t, is exhibited in Fig. 1.14. Note that since the hysteresis loop, Fig. 1.14(b), is **nonlinear** and is **double valued**, the waveform of the extracted exciting current i_ϕ versus t, Fig. 1.14(c), is **nonsinusoidal**, even though the flux Φ_m versus t, Fig. 1.14(a), is **sinusoidal**. Fourier series analysis of this exciting current yields only components whose frequencies are the fundamental ω_E, as well as odd harmonics of it, the principal harmonic being the third. For example, for typical power transformers the third harmonic is about 40 percent of the exciting current, which in turn is about 5 percent of the full-load current. Consequently, the effects of the harmonics are ignored in the usual transformer analysis, except in problems concerned directly with the effect of harmonics. Thus the exciting current i_ϕ is represented by an **equivalent sine wave**, which has the fundemental frequency of Fig. 1.14(c) and an RMS value which produces the same average power as the graphically reconstructed waveform of Fig. 1.14(c).

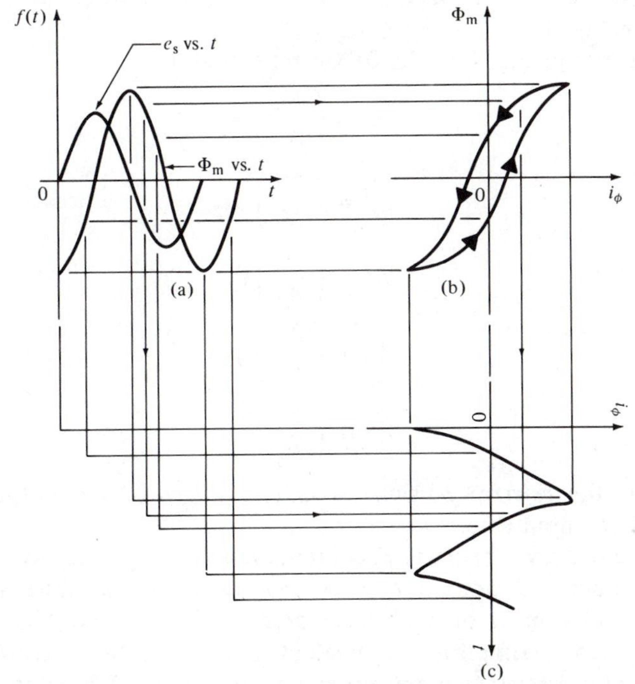

Fig. 1.14 Excitation Phenomena of the Transformer of Fig. 1.5 with the Output Open-circuited: (a) Waveforms of Impressed Voltage and Response Flux, (b) Hysteresis Loop of the Transformer Core, and (c) Waveform of the Exciting Current

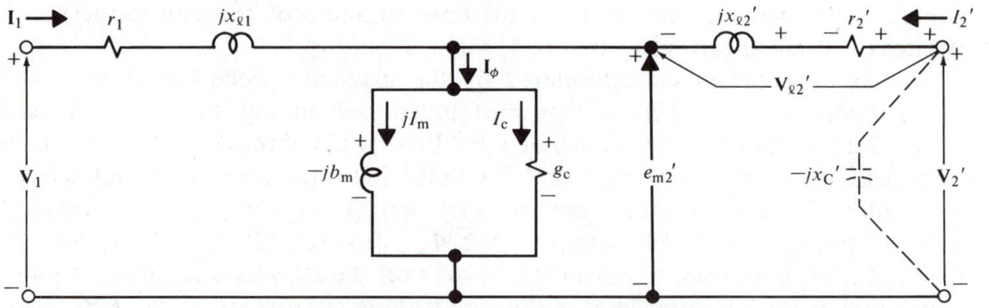

Fig. 1.15 The Referred-to-the-Primary Equivalent Circuit of the Transformer

It is of interest to note that the leakage flux is considered to be entirely through air. Since air does not saturate, the associated **leakage reactance is linear.** This is not the case for the excitation flux that passes through iron, which may saturate. However, the analysis is carried out in the linear region of λ_m versus i_m. Thus the model of the transformer is linear. Note that *reactances are used as circuit parameters* in Fig. 1.13 rather than reluctances because reactances lend themselves to circuit analysis.

Utilizing the properties of the ideal transformer enables us to draw the equivalent circuit of the transformer of Fig. 1.13 with *all variables and parameters of the output referred to the primary*, as in Fig. 1.15. Similarly, the variables and parameters of the input could be referred to the output side.

Application of the voltage transformation equation [Eq. (1.50)] for $\mathbf{V}_1 \equiv \mathbf{E}_1$ and $\mathbf{V}_2 \equiv \mathbf{E}_2$ enables us to obtain $\mathbf{E}_2' = \mathbf{V}_2' - \mathbf{V}_{\ell 2}'$ in Fig. 1.15 from $\mathbf{E}_2 = \mathbf{V}_2 - \mathbf{V}_{\ell 2}$ in Fig. 1.13. Similarly, we may obtain $\mathbf{E}_1' = \mathbf{V}_1' - \mathbf{V}_{\ell 1}'$ (not drawn in Fig. 1.15) from $\mathbf{E}_1 = \mathbf{V}_1 - \mathbf{V}_{\ell 1}$ in Fig. 1.13.

Please note that in Fig. 1.15 the following equations

$$r_2' = \left(\frac{N_1}{N_2}\right)^2 r_2 \tag{1.61}$$

$$x_{l2}' = \left(\frac{N_1}{N_2}\right)^2 x_{l2} \tag{1.62}$$

$$\mathbf{V}_2' = \left(\frac{N_1}{N_2}\right)\mathbf{V}_2 \tag{1.63}$$

$$\mathbf{I}_2' = \left(\frac{N_2}{N_1}\right)\mathbf{I}_2 \tag{1.64}$$

$$x_C' = \left(\frac{N_1}{N_2}\right)^2 x_C \tag{1.65}$$

are defined as the *referred parameters and variables* of the output side of the transformer to its input side.

It is possible to move the parallel combination, $\mathbf{Y}_\phi = g_c - jb_m$, to the input or the output and *introduce* 2 *percent to* 6 *percent error* in quantities that would be influenced by such a move of $\mathbf{Y}_\phi$. This error is within expected tolerances for many calculations. Thus a simplified equivalent circuit of the transformer with an inherent error of 2 percent to 6 percent can be drawn as exhibited in Fig. 1.16.

The parameters $|r_1 + r_2'| \ll |j\,x_{\ell 1} + j\,x_{\ell 2}'|$, thus $r_1 + r_2'$, can be neglected. Since the value of I_ϕ is only a few percent of the nominal value of I_1, Y_ϕ may be omitted. For the frequencies of interest, $-jx_c$ is large enough to be ignored. Thus the

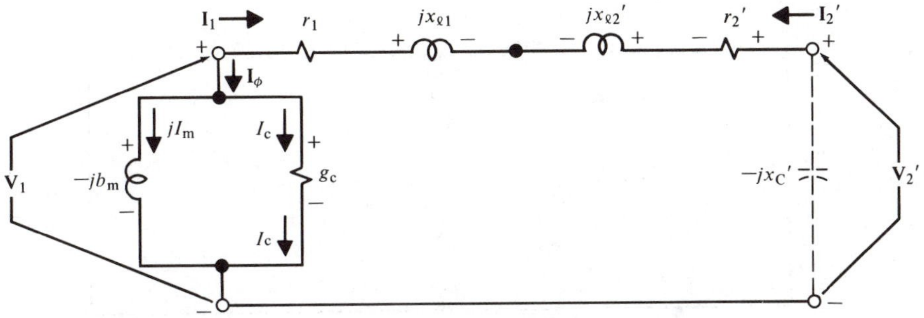

Fig. 1.16 The Simplified Equivalent Circuit of the Real Transformer with an Inherent Error of 2%–6%

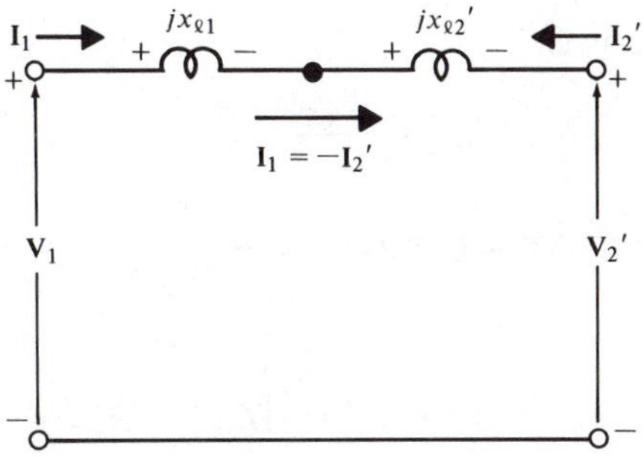

Fig. 1.17 The Oversimplified Equivalent Circuit of the Real Transformer

equivalent circuit of the transformer can be drawn in an oversimplified manner, as in Fig. 1.17.

1.7 PARAMETER MEASUREMENT OF THE REAL TRANSFORMER

The parameters of a transformer, r_1, $x_{\ell 1}$, r_2, $x_{\ell 2}$, r_c, and x_m, can be determined by measuring the *input voltage*, *current*, and *power* to the primary, first with the *secondary short-circuited*, and then with *the secondary open-circuited.*

To strengthen the knowledge of the student on instrumentation, Appendix B has been added, which deals with fundamentals of operation of the following instruments: (1) ammeter, (2) voltmeter, (3) wattmeter, (4) varmeter, and (5) ohmmeter. It is strongly recommended that students familiarize themselves with Appendix B before proceeding with the *short- and open-circuit tests of transformers.*

Short-Circuit Test

With the secondary short-circuited, a primary voltage of only *2 percent to 12 percent of the rated primary value* need be impressed to obtain full-load primary or secondary current. **The high-voltage side is usually taken as the primary in this test.** This very low input voltage produces a very low magnetizing flux, which suggests that the exciting current is negligible, that is, $\mathbf{I}_\phi \doteq 0$. Thus $\mathbf{Z}_\phi = \mathbf{E}_\phi / \mathbf{I}_\phi = \infty$, and the equivalent circuit of the transformer becomes as shown in Fig. 1.18.

Thus if $\mathbf{E}_{SC}$, $\mathbf{I}_{SC}$, and $\mathbf{P}_{SC}$ are the impressed primary **voltage, current,** and **power,** the short-circuit impedance $\mathbf{Z}_{SC}$ and its resistance R_{SC} and reactance X_{SC} components referred to the primary are calculated utilizing Ohm's law and the

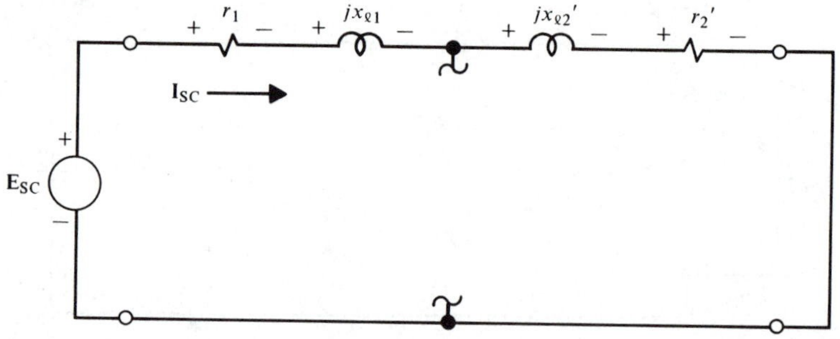

Fig. 1.18 The Equivalent Circuit of the Transformer Under Short-Circuit Conditions of Operation

Pythagorean theorem as follows:

$$|\mathbf{Z}_{SC}| = \frac{|\mathbf{E}_{SC}|}{|\mathbf{I}_{SC}|} \tag{1.66}$$

$$r_T \equiv R_{SC} \doteq 2r_1 \doteq 2r_2' = r_1 + r_2' = \frac{P_{SC}}{|\mathbf{I}_{SC}|^2}; r_1 \doteq r_2' \tag{1.67}$$

and

$$x_{\ell T} \equiv X_{SC} \doteq 2x_{\ell 1} \doteq 2x_{\ell 2}' = x_{\ell 1} + x_{\ell 2}' = (|\mathbf{Z}_{SC}|^2 - R_{SC}{}^2)^{1/2}; x_{\ell 1} \doteq x_{\ell 2}' \tag{1.68}$$

Utilizing Eqs. (1.61) and (1.62) enables calculation of r_2 and $x_{\ell 2}$ as follows:

$$r_2 = \left(\frac{N_2}{N_1}\right)^2 r_2' \tag{1.69}$$

and

$$x_{\ell 2} = \left(\frac{N_2}{N_1}\right)^2 x_{\ell 2}' \tag{1.70}$$

Note: If there is no other information available, the assumptions of Eqs. (1.67) and (1.68) are reasonable, that is, $r_1 \doteq r_2'$ and $x_{\ell 1} \doteq x_{\ell 2}'$. However, if other information is available, appropriate adjustments should be made to Eqs. (1.66) to (1.68). If the rated current $\mathbf{I}_{RTD}$ of each winding is known, an impressed voltage $\mathbf{E}_S$ of appropriate magnitude to yield the rated current in each winding, with the other winding open-circuited, yields the necessary data, $|\mathbf{I}_{RTD}|$ and $|\mathbf{E}_S|$, for the calculation of the value of the resistance of each winding.

Open-Circuit Test

With the secondary open-circuited and rated voltage impressed on the primary, an exciting current, $\mathbf{I}_1 = \mathbf{I}_\phi$, of only *2 percent to 6 percent of full-load current* is obtained. **The low-voltage side is usually taken as the primary in this test.** This small exciting current creates a negligible voltage drop across the primary impedance, resulting in negligible primary copper losses. Thus the impressed voltage appears across $\mathbf{Y}_\phi = g_c - jb_m$ and the input power is very nearly equal to the core losses P_c. The equivalent circuit of the transformer becomes as shown in Fig. 1.19.

Thus, if E_{OC}', I_{OC}, and P_{OC} are the impressed primary **voltage, current**, and **power**, the exciting admittance $\mathbf{Y}_\phi$ and its conductance and susceptance are

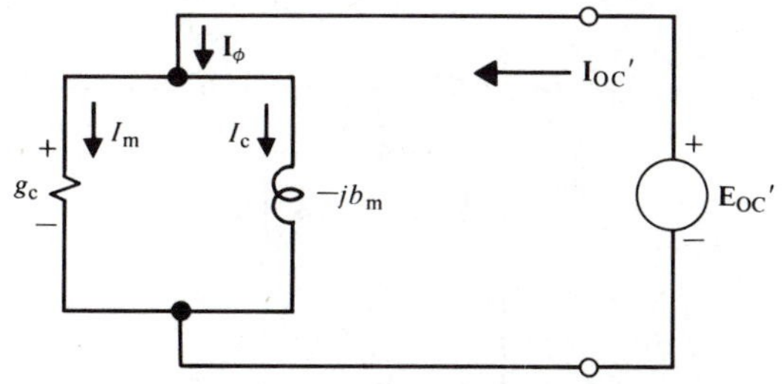

Fig. 1.19 The Equivalent Circuit of the Transformer Under Open-Circuit Conditions of Operation

calculated utilizing Ohm's law and the Pythagorean theorem as follows:

$$|\mathbf{Y}_{OC}| \equiv |\mathbf{Y}_\phi| = \frac{|\mathbf{I}_{OC}'|}{|\mathbf{E}_{OC}'|} \tag{1.71}$$

$$g_{OC} \equiv g_c \equiv \frac{1}{r_c} = \frac{P_{OC}}{|\mathbf{E}_{OC}'|^2} \tag{1.72}$$

and

$$b_{OC} \equiv b_m \equiv \frac{1}{x_m} = \left\{|\mathbf{Y}_{OC}|^2 - g_{OC}^2\right\}^{1/2} \tag{1.73}$$

The values provided by Eqs. (1.71) through (1.73) are referred to the side which is used as the primary in this test. However, when the much simplified equivalent circuit of the transformer, Fig. 1.17, is used, the open-circuit test is used only to obtain core loss for efficiency computations and to check the magnitude of the exciting current. (Note: The voltage at the terminals of the open-circuited secondary is measured as a *check of the turns ratio*.)

There are two other quantities that provide information about the performance of a transformer. They are (1) the *efficiency* and (2) the *voltage regulation* of the transformer. The definition of efficiency is the classical definition of efficiency ($\eta = P_{OUT}/P_{IN}$) and warrants no further discussion. However, regulation is a new quantity and should be clearly understood.

Voltage regulation is defined as the measure of the ability of the transformer to limit deviations of the output voltage within small limits of the rated value for a given input voltage and output power factor while the output current is varied from zero to full load. **The better the voltage regulation is, the lower its numerical value.** Mathematically these ideas are given as follows:

$$\begin{aligned}\text{voltage regulation} &\equiv \left.\frac{|\mathbf{V}_2'|_{NL} - |\mathbf{V}_2'|_{FL}}{|\mathbf{V}_2'|_{FL}}\right|_{\mathbf{E}_I = k;\ \cos\theta_Z = k_1;\ \text{Fig. 1.20}} \\ &= \left.\frac{|\mathbf{V}_2'|_{NL} - |\mathbf{V}_2'|_{FL}}{|\mathbf{V}_2'|_{FL}}\right|_{\mathbf{V}_1 = k;\ \cos\theta_Z = k_1;\ \text{Fig. 1.16}} \\ &= \left.\frac{|\mathbf{V}_2|_{NL} - |\mathbf{V}_2|_{FL}}{|\mathbf{V}_2|_{FL}}\right|_{\mathbf{V}_1 = k;\ \cos\theta_Z = k_1;\ \text{Fig. 1.13}}\end{aligned} \tag{1.74}$$

The primed quantities are referred quantities, with $\mathbf{V}_2'$ being referred to the primary side of the ideal transformer.

To understand the meaning of $|\mathbf{V}_2'|_{NL}$ and $|\mathbf{V}_2'|_{FL}$, refer to Figs. 1.20 and 1.21. Note that the impressed voltage ($\mathbf{E}_I$) in these two circuits is maintained constant.

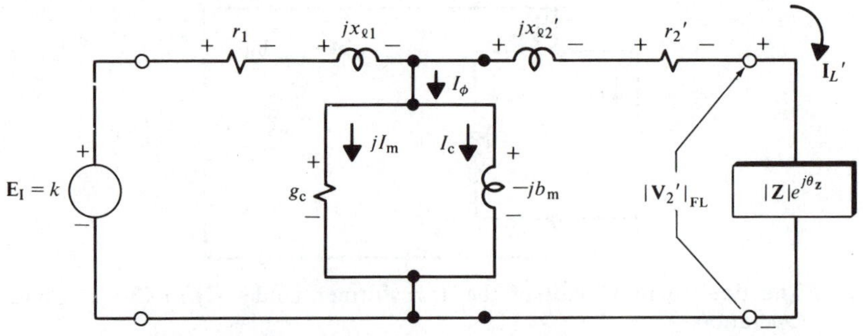

Fig. 1.20 Equivalent Circuit of a Transformer Fully Loaded

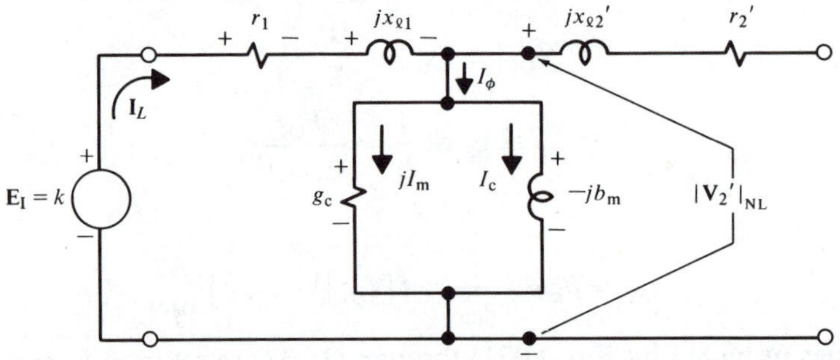

Fig. 1.21 Equivalent Circuit of a Transformer Unloaded

Example 1.6

Given: a 50-kVA, 2400:240-V transformer feeding a load with a PF=0.8 lagging.* The load draws rated transformer secondary current.

Calculate: (1) the parameters of the equivalent circuit of the transformer, (2) the efficiency, and (3) the voltage regulation of the transformer. The energized sides of the transformer are instrumented. The readings are as follows:

1. The short-circuit measurements on the high-voltage side: $|\mathbf{E}_{SC}| = 48$ V, $|\mathbf{I}_{SC}| = 20.8$ A, $P_{SC} = 617$ W.
2. The open-circuit measurements on the low-voltage side: $|E_{OC}| = 240$ V, $|I_{OC}| = 5.41$ A, $P_{OC} = 186$ W.

Solution

1. Short-circuit parameter evaluations:

$$|\mathbf{Z}_{SC}| = \frac{|\mathbf{E}_{SC}|}{|\mathbf{I}_{SC}|} = \frac{48}{20.8} = \underline{2.31\ \Omega}$$

$$r_T = \frac{P_{SC}}{|\mathbf{I}_{SC}|^2} = \frac{617}{20.8^2} = \underline{1.43\ \Omega}$$

$$x_{\ell T} = \left(|\mathbf{Z}_{SC}|^2 - r_T^2\right)^{1/2} = (2.31^2 - 1.43^2)^{1/2} = \underline{1.81\ \Omega}$$

*If you have not studied Appendix A, please refer to pages 604–609 for the definitions of *power factor* and the term *lagging*.

Thus

$$r_1 = 1.43/2 = \underline{\underline{0.71\ \Omega}}$$

$$r_2' = 1.43/2 = \underline{0.710\ \Omega}$$

$$r_2 = \left(\tfrac{240}{2400}\right)^2 0.71 = \underline{\underline{0.0071\ \Omega}}$$

$$x_{\ell 1} = 1.81/2 = \underline{\underline{0.91\ \Omega}}$$

$$x_{\ell 2}' = 1.81/2 = \underline{0.91\ \Omega}$$

$$x_{\ell 2} = \left(\tfrac{240}{2400}\right)^2 0.91 = \underline{\underline{0.0091\ \Omega}}$$

2. Open-circuit parameter evaluation:

$$|\mathbf{Y}_\phi| = \frac{|\mathbf{I}_{OC}'|}{|\mathbf{E}_{OC}'|} = \frac{5.41}{2400} = \underline{2.25 \times 10^{-3}} \text{ mho}$$

$$g_c = \frac{P_{OC}}{|\mathbf{E}_{OC}'|^2} = \frac{186}{2400^2} = \underline{\underline{0.323 \times 10^{-4}}} \text{ mho}$$

$$b_m = \left(|\mathbf{Y}_\phi|^2 - g_c^{\,2}\right)^{1/2} = \underline{\underline{2.25 \times 10^{-3} \text{ mho}}}$$

3. The equivalent circuit of the transformer is shown in Fig. 1.6.1.

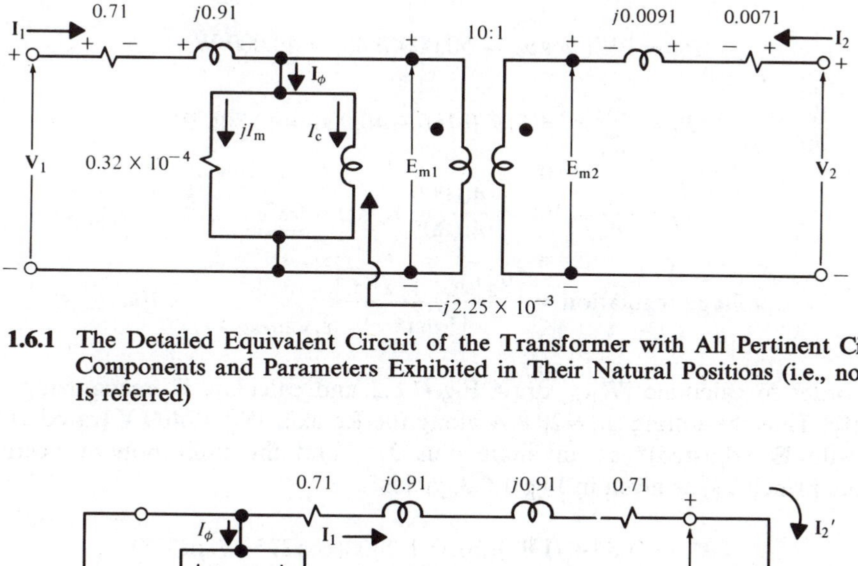

Fig. 1.6.1 The Detailed Equivalent Circuit of the Transformer with All Pertinent Circuit Components and Parameters Exhibited in Their Natural Positions (i.e., nothing Is referred)

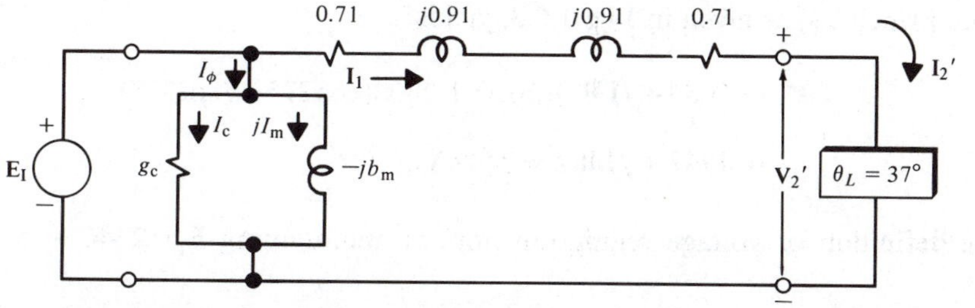

Fig. 1.6.2 The Simplified Equivalent Circuit of Fig. 1.6.1 with the Circuit Parameters and Variables of the Secondary Side and the Load Referred to the Primary Side of the Transformer

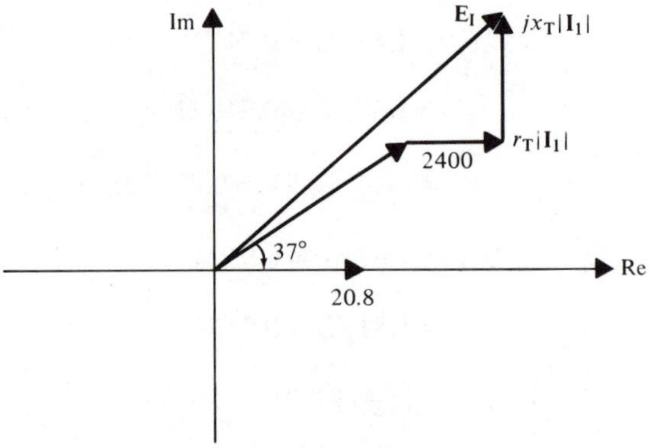

Fig. 1.6.3 Phasor Diagram Exhibiting the Variables of Fig. 1.6.2

4. The approximate calculations utilizing the circuit parameters referred to the primary side are as shown next. Since at full-load operation $|\mathbf{W}_O| = 50$ kVA $\equiv$ 50,000 VA,

$$|\mathbf{I}_2'^*| = \frac{|\mathbf{W}_O|}{|\mathbf{V}_2'|} = \frac{50{,}000}{2{,}400} = \underline{20.8} \text{ A}$$

$$P_T = r_T|\mathbf{I}_2'^*|^2 = 1.43(20.8)^2 = \underline{619} \text{ W}$$

$$P_c = g_c|\mathbf{E}_c|^2 = 0.323 \times 10^{-4}(2{,}400)^2 = \underline{186} \text{ W}$$

$$P_O = |\mathbf{W}_O| \cos\theta_L = 50{,}000(0.80) = \underline{40{,}000} \text{ W}$$

$$P_I = \sum_{j=1}^{4} P_j = 619 + 186 + 40{,}000 = \underline{40{,}805} \text{ W}$$

$$\eta = \frac{P_O}{P_I} \times 100 = \frac{40{,}000}{40{,}805} \times 100 = \underline{\underline{98\%}}$$

$$\text{voltage regulation} = \frac{|\mathbf{V}_2'|_{NL} - |\mathbf{V}_2'|_{FL}}{|\mathbf{V}_2'|_{FL}}\Bigg|_{\mathbf{E}_I = k;\, \cos\theta_Z = k_1} \times 100$$

In order to calculate $|\mathbf{V}_2'|_{NL}$ draw Fig. 1.6.2 and calculate $\mathbf{E}_I$ recognizing that $|\mathbf{I}_1| = |\mathbf{I}_2'|$. Thus the setting $|\mathbf{I}_1| = 20.8$ A along the Re-axis, $|\mathbf{V}_2'| = 2400$ V (rated at full load with $\mathbf{E}_I$ adjusted)* at an angle plus 37°, and the utilization of vectorial diagram principles, as given in Fig. 1.6.3, yields:

$$\mathbf{E}_I = (1.43 + j1.81)(20.8) + 2400(\cos 37° + j\sin 37°)$$

$$= 1947 + j1482 = \underline{2447} \text{ V}e^{j\,37.28°}$$

As the definition of voltage regulation implies, maintaining $\mathbf{E}_I = 2446$ V constant

*One can look upon the potential drop $|\mathbf{V}_2'| = 2400$ V at an angle of 37° as $\mathbf{V}_2' = 2400e^{j37°} \equiv R_L'|\mathbf{I}_2'| + jX_L'|\mathbf{I}_2'|$.

and removing the load yields an output no-load voltage of 2446 V, that is, $|\mathbf{V}_2'|_{NL}=|\mathbf{E}_I|=2446$ V. Thus the voltage at the secondary from full-load to no-load, with the input remaining constant, changes from 2400 V to 2446 V. Therefore,

$$\text{voltage regulation} = \frac{2447-2400}{2400}\times 100 = \underline{\underline{1.96\%}}$$

Example 1.7

When the load impedance is explicitly defined, the approach is not much different; in fact, circuit analysis techniques can be used more readily.

Solution

The referred loading impedance $\mathbf{Z}_L'=5+j5\ \Omega$ connected to the referred output of the transformer's equivalent circuit, Fig. 1.6.1, for example, yields the newly drawn equivalent circuit of the transformer, Fig. 1.7.1. The pertinent analysis proceeds as follows. Note that in this case the input voltage $\mathbf{V}_1$ is given and the output voltage $\mathbf{V}_2'$ is not known.

$$\mathbf{Z}_I = 0.71 + j0.91 + \mathbf{Z}_P$$

where

$$\mathbf{Z}_P = \frac{\mathbf{Z}_\phi(\mathbf{Z}_2'+\mathbf{Z}_L')}{\mathbf{Z}_\phi+\mathbf{Z}_2'+\mathbf{Z}_L'}$$

$$\mathbf{Z}_2' = \underline{0.71+j0.91}\ \Omega$$

$$\mathbf{Z}_\phi = \frac{1}{0.323\times 10^{-4}-j2.25\times 10^{-3}} = \underline{444.4e^{j\,89.18^\circ}}\ \Omega$$

and

$$\mathbf{Z}_L' = \underline{5+j5}\ \Omega$$

Thus,

$$\mathbf{Z}_P = \underline{\underline{8.11e^{j\,46.7^\circ}}}\ \Omega$$

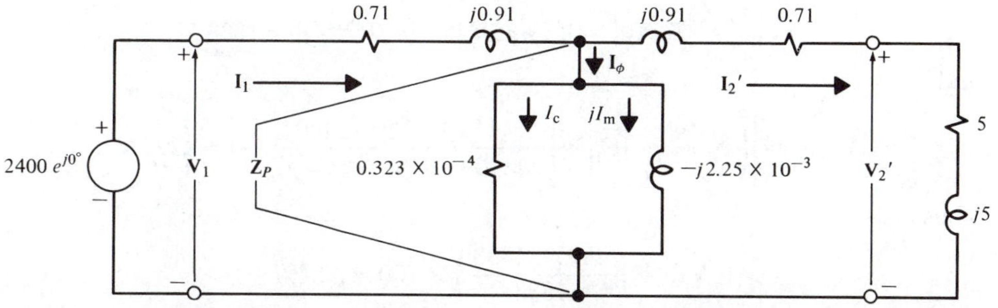

Fig. 1.7.1 The Equivalent Circuit of Fig. 1.6.1 with the Loading Impedance Explicitly Defined and, Along with All the Circuit Parameters and Variables of the Secondary Side, Referred to the Primary Side of the Transformer

and

$$\mathbf{Z}_{\mathrm{I}} = \underline{\underline{9.26e^{j47.37°}}}\ \Omega$$

$$\mathbf{I}_1 = \frac{2400e^{j0°}}{\mathbf{Z}_{\mathrm{I}}} = \frac{2400e^{j0°}}{9.26e^{j47.37°}} = \underline{\underline{259.2e^{-j47.37°}}}\ \Omega$$

$$\mathbf{I}_2{}' = \left(\frac{\mathbf{Z}_\phi}{\mathbf{Z}_\phi + \mathbf{Z}_2{}' + \mathbf{Z}_L{}'}\right)\mathbf{I}_1 = \left(\frac{444.4e^{j89.18°}}{450.43e^{j88.46°}}\right)(259.2e^{-j47.37°})$$

$$= \underline{\underline{257.7e^{-j46.65°}}}\ \mathrm{A}$$

$$\mathbf{I}_\phi = \left(\frac{\mathbf{Z}_2{}' + \mathbf{Z}_L{}'}{\mathbf{Z}_\phi + \mathbf{Z}_2{}' + \mathbf{Z}_L{}'}\right)\mathbf{I}_1 = \left(\frac{8.218e^{j45.99°}}{450.43e^{j88.46°}}\right)(259.2e^{-j47.37°})$$

$$= \underline{\underline{4.73e^{-j89.84°}}}\ \mathrm{A}$$

$$\mathbf{I}_c = \left(\frac{g_c}{g_c - jb_m}\right)\mathbf{I}_\phi = \left(\frac{0.323\times 10^{-4}}{(0.323 - j22.5)\times 10^{-4}}\right)(4.73e^{-j89.84°})$$

$$= \underline{\underline{0.068e^{-j0.67°}}}\ \mathrm{A}$$

$$P_1 = r_1|\mathbf{I}_1|^2 = (0.71)(259.2)^2 = \underline{\underline{47.7}}\ \mathrm{kW}$$

$$P_2 = r_2{}'|\mathbf{I}_2{}'|^2 = (0.71)(255.7)^2 = \underline{\underline{46.4}}\ \mathrm{kW}$$

$$P_c = r_c|\mathbf{I}_c|^2 = \left(\frac{1}{0.323\times 10^{-4}}\right)(0.068)^2 = \underline{\underline{143}}\ \mathrm{W}$$

$$P_{\mathrm{O}} = r_L{}'|\mathbf{I}_2{}'|^2 = (5)(255.7)^2 = \underline{\underline{326.9}}\ \mathrm{kW}$$

$$P_{\mathrm{I}} = \mathrm{Re}(\mathbf{E}_{\mathrm{I}}\mathbf{I}_1{}^*) = \mathrm{Re}\left[(2400)(259.2e^{j47.37°})\right] = \underline{\underline{421.3}}\ \mathrm{kW}$$

$$\eta = \frac{P_{\mathrm{O}}}{P_{\mathrm{I}}}\times 100 = \underline{\underline{77.6\%}}$$

$$\mathrm{VR} = \left.\frac{|\mathbf{V}_2{}'|_{\mathrm{NL}} - |\mathbf{V}_2{}'|_{\mathrm{FL}}}{|\mathbf{V}_2{}'|_{\mathrm{FL}}}\right|_{\mathbf{E}_{\mathrm{I}} = k;\ \cos\theta_{\mathbf{Z}} = k_1} \times 100$$

where

$$|\mathbf{V}_2{}'|_{\mathrm{FL}} = |\mathbf{Z}_L{}'\mathbf{I}_2{}'| = |(5 + j5)(255.7e^{-j46.65°})| = \underline{\underline{1808}}\ \mathrm{V}$$

and

$$|\mathbf{V}_2{}'|_{\mathrm{NL}} = \left|\left(\frac{\mathbf{Z}_\phi}{\mathbf{Z}_\phi + \mathbf{Z}_1}\right)\mathbf{E}_{\mathrm{I}}\right| = \left|\frac{444.4e^{j89.18°}(2400)}{444.4e^{j89.18°} + 0.71 + j0.91}\right| = \underline{\underline{2395}}\ \mathrm{V}$$

Thus

$$\mathrm{VR} = \left(\frac{2395 - 1833}{1833}\right)\times 100 = \underline{\underline{32.4\%}}$$

Using the cricuit in Figure 1.7.1, if the input voltage $\mathbf{V}_1$ were not given, then the output voltage $\mathbf{V}_2$ would be assumed to be 240 V, (or $\mathbf{V}_2{}' = 2400$ V). The input

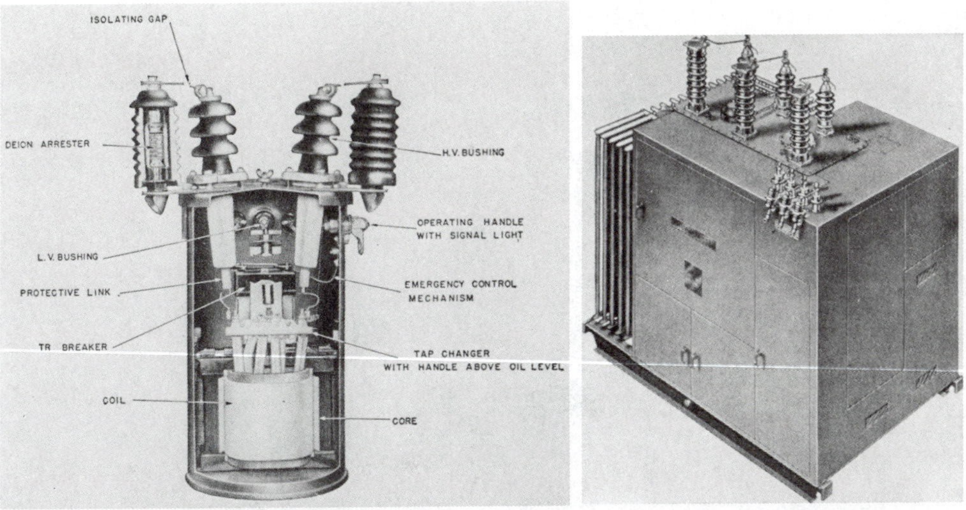

Plate 1.6 CSP (Completely Self-protected) Transformers: (a) Sectional View of a Transformer Used for Operation on Grounded-Neutral Circuits and (b) Fully Assembled, 300-kVA, 33–4.16-kV Power Transformer. (Reproduced from *Electric Transmission and Distribution Reference Book*, by Central Station Engineers, Westinghouse Electric Co., Pittsburgh, Pa.)

voltage $\mathbf{V}_1$ would be determined as follows. For the given load:

$$\mathbf{I}_2' = \frac{\mathbf{V}_2'}{\mathbf{Z}_2'} = \frac{2400e^{j0^\circ}}{5+j5} = \underline{\underline{339e^{-j45^\circ}}} \text{ A}$$

$$\mathbf{V}_\phi = \mathbf{I}_2'(r_2'+jx_{\ell 2}'+\mathbf{Z}_L') = (339e^{-j45^\circ})(0.71+j0.91+5+j5)$$

$$= \underline{\underline{2786e^{j1^\circ}}} \text{ V}$$

$$\mathbf{I}_\phi = \mathbf{Y}_\phi \mathbf{V}_\phi = (0.323\times 10^{-4} - j2.25\times 10^{-3})(2786e^{j1^\circ})$$

$$= \underline{\underline{6.27e^{-j88.18^\circ}}} \text{ A}$$

$$\mathbf{I}_1 = \mathbf{I}_\phi + \mathbf{I}_2' = 6.27e^{-j88.18^\circ} + 339e^{-j45^\circ} = \underline{\underline{344e^{-j45.7^\circ}}} \text{ A}$$

$$\mathbf{V}_1 = \mathbf{I}_1\mathbf{Z}_\mathrm{I} = (344e^{-j45.7^\circ})(9.26e^{j47.37^\circ}) = \underline{\underline{3182e^{j1.66^\circ}}} \text{ V}$$

Under no-load conditions $\mathbf{I}_2' = 0$. Then for no-load:

$$\mathbf{V}_2' = \frac{\mathbf{Z}_\phi \mathbf{V}_1}{r_1 + x_{\ell 1} + \mathbf{Z}_\phi} = \frac{(444.4e^{j89.18^\circ})(3182e^{j1.66^\circ})}{(0.71+j0.91)+444.4e^{j89.18^\circ}} = \underline{\underline{3175e^{j1.75^\circ}}} \text{ V}$$

$$|\mathbf{V}_2'|_{\mathrm{NL}} = \underline{\underline{317.5}} \text{ V}$$

$$\text{voltage regulation} = \left.\frac{|\mathbf{V}_2'|_{\mathrm{NL}} - |\mathbf{V}_2'|_{\mathrm{FL}}}{|\mathbf{V}_2'|_{\mathrm{FL}}}\right|_{\mathbf{V}_1 = k;\, \cos\theta_\mathbf{Z} = k_1} \times 100\%$$

$$= \left(\frac{317.5-240}{240}\right) \times 100\% = \underline{\underline{32.3\%}}$$

Plate 1.7 Three-Phase Primary Substation Transformer Without Load-Tap-changing Equipment for Steady-Load Applications. (Courtesy of General Electric Co., Schenectady, N.Y.)

Plate 1.8 Three-Phase Power Transformer. (Courtesy of General Electric Co., Schenectady, N.Y.)

Notice that the numerical value of voltage regulation (32.3%) is close to that value calculated when $\mathbf{V}_1$ was given (32.4%). The calculation of the various power losses is exactly the same as in the example above and will not be repeated here. It should be noted that the power rating of a transformer is given in kilovoltamperes. This means that the magnitude of the rated output current is independent of the power factor of the load.

1.8 USES AND FREQUENCY RESPONSE OF TRANSFORMERS

There are two types of transformers: (1) *power transformers* operating at a fixed frequency (usually) and (2) *variable-frequency transformers* operating in the audio- or video-frequency ranges and thus we have (a) audio- and (b) video-frequency transformers. The variable-frequency transformers may have a tiny sintered iron core.

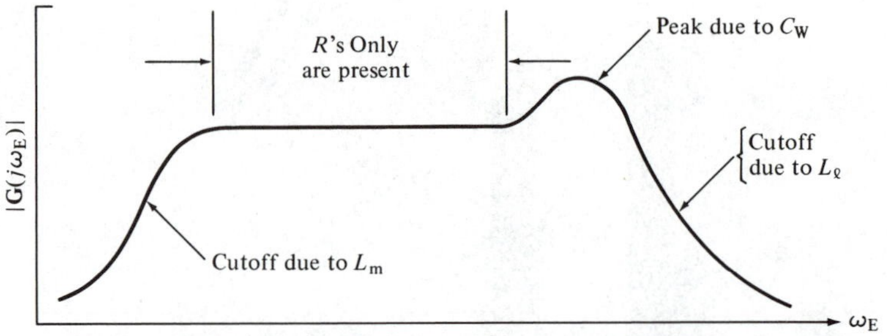

Fig. 1.22 Frequency Response of a Typical Real Transformer

Power transformers are used as *step-down* or *step-up* devices of (1) *voltage levels*, (2) *current levels*, and (3) *impedance levels*.

Variable-frequency transformers are used as (1) *gain or attenuation devices*, (2) *impedance-matching devices*, and (3) *dc-blocking devices*.

The performance of a transformer can be observed by studying its frequency response. The frequency response of a transformer is a plot of the magnitude of the voltage ratio, that is, $|\mathbf{G}(j\omega_E)| = |\mathbf{E}_2'|/|\mathbf{E}_I|$, as a function of radian frequency ω_E, that is, as the frequency of the forcing function ω_E changes for a fixed load. The magnitude of the forcing function is maintained constant during the process of taking data for the $|\mathbf{G}(j\omega_E)|$ versus ω_E plot. *For power systems the information provided by* $|\mathbf{G}(j\omega_E)|$ *versus* ω_E *is of no importance because power systems operate at one frequency, for example, 60* Hz *or 50* Hz. However, in communication systems the frequency of the source may vary over a wide range. Fig. 1.22 provides information about which parameters of the transformer affect the $|\mathbf{G}(j\omega_E)|$ versus ω_E plot.

1.9 THREE-PHASE TRANSFORMERS

The preceding transformer analysis is complete for a single-phase transformer that operates either over a **range** of frequencies, as in communication systems, or at a **single** frequency, as in single-phase power systems. In both of these applications, the transformed power level is usually in the order of kilowatts or less.

Electrical energy, however, is generated in the **form of the three-phase voltages** at a voltage level in the neighborhood of 25 kV and is transmitted in the *same form* but at a voltage level usually in the neighborhood of 765 kV. Consequently, the transformed power level is raised to the order of megawatts or more. Therefore, **three-phase transformers, capable of handling the given three-phase voltages at these high levels of potential and power, are essential in power systems engineering.**

In addition to the high voltages that are deliberately impressed on the three-phase transformers, overvoltages, imposed on the transmission lines of power systems by atmospheric disturbances, end up impressed on these transformers. Therefore, they should be designed to withstand insulation breakdown and thus *avoid customer service interruption*. As an added quality to their performance, the three-phase power transformers should be designed to yield very low *core* and *copper losses*. This quality is highly desirable because the *large blocks of power* handled by the power transformers over *long periods of time* yield *large amounts of energy loss*, and therefore *loss in revenues*. Thus *the performance indices of the three-phase power transformers are* **reliable service** and economic operation.

Plate 1.9 Typical Grounding Transformer, Used in Conjunction with Δ-connected Ungrounded Systems. Its Objective Is to Secure Enough Ground Current for Relaying. (Reproduced from *Electric Transmission and Distribution Reference Book*, by Central Station Engineers, Westinghouse Electric Co., Pittsburgh, Pa.)

Structurally, there are two types of transformers: (1) the **bank arrangement,*** consisting of three separate single-phase units properly wound, each transforming one-third of the throughpower, Fig. 1.23, and (2) the *core* or *three-phase arrangement*, properly wound on a two-window (three-legged) single core, Fig. 1.24.

Three **single-phase** transformers may be connected to form a **three-phase bank** in any of the four ways shown in Fig. 1.25, known as the (1) Y-Δ-connection, (2) Δ-Y-connection, (3) Δ-Δ-connection, and (4) Y-Y-connection, respectively.

In Figs. 1.25(a) through 1.25(d), the windings at the **left** are defined as the **primaries** and those at the **right** are the **secondaries**. Every primary winding is mated in one transformer with the secondary winding drawn parallel to it. The windings of every primary and secondary are drawn equally displaced from each other; that is, their physical displacement from each other is 120°. *Each winding of every primary or secondary has the same number of turns.* The primary of each transformer is excited by *voltages of* **equal magnitudes but displaced from each other by 120°**, that is,

$$e_S(t) = \sqrt{2}\,|\mathbf{E}_S|\sin(\omega_E t - \phi_i);\ \phi_i = 0°,\ 120°,\ 240° \tag{1.75}$$

*Also known as shell-type or shell-form three-phase transformer.

Plate 1.10 1300-MVA Generator Step-up Transformer with a Forced-Oil, Forced-Air Cooling System. (Courtesy of Westinghouse Electric Co., Pittsburgh, Pa.)

Such voltages are termed *balanced voltages. Usually currents are also balanced*; that is, *they are of equal magnitude but displaced from each other by 120°*. There is the usual *phase angle* between the input voltages and currents, which depends on the load connected to the output of the transformer.

The secondary voltages and currents are related to the primary voltages and currents by an appropriate turns ratio. Thus, *the response voltages and currents are*

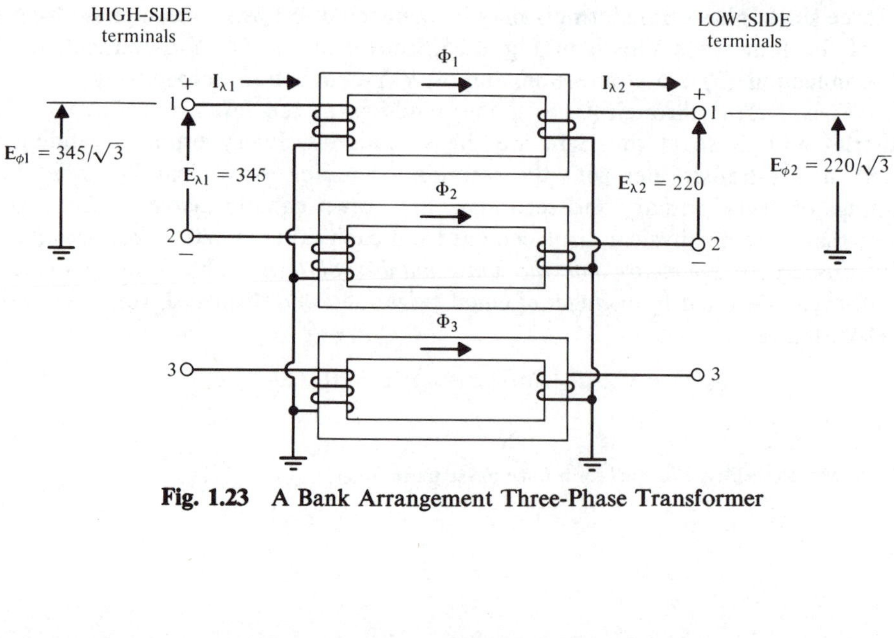

Fig. 1.23 A Bank Arrangement Three-Phase Transformer

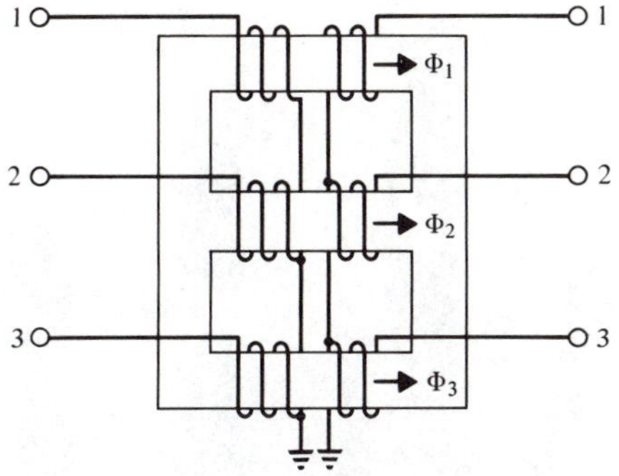

Fig. 1.24 A Core Arrangement Three-Phase Transformer

balanced; that is, **they are of equal magnitude and are displaced from each other by 120°**. As above, there is the usual *phase angle* between the output voltages and currents, which depends on the load connected to the secondary. Such setups are known as *three-phase transformer banks with balanced primaries and secondaries*. The relationships between line and phase currents and line-to-line and phase voltages are as follows: (1) Δ-connection, $\mathbf{I}_\lambda = \sqrt{3}\,\mathbf{I}_\phi$ and $\mathbf{V}_\lambda = \mathbf{V}_\phi$, (2) Y-connection, $\mathbf{I}_\lambda = \mathbf{I}_\phi$ and $\mathbf{V}_\lambda = \sqrt{3}\,\mathbf{V}_\phi$. For further information about three-phase balanced systems, please see Chapter 3. Note that for fixed line-to-line voltages and total kilovoltamperes, *the kilovoltampere rating of each transformer is one-third of the kilovoltampere rating of the bank, regardless of the connection used*, while the voltage and current ratings of the individual transformers depend on the connections.

The uses of the various connections of Fig. 1.25 are as follows.

1. The Y-Δ-connection is commonly used in **stepping down** from a high to a medium or low voltage (e.g., 138 kV to 13.8 kV). One reason for using this connection is that the neutral of the Y-connected high side of the transformer is provided for *grounding*, a desirable procedure in most cases.

Plate 1.11 Installation View of a 25,000-kVA, 115–12-kV, 60-Hz, Three-Phase, OA/FA (Oil-immersed Self-cooled/Forced-Air-cooled) Transformer. (Reproduced from *Electric Transmission and Distribution Reference Book*, by Central Station Engineers, Westinghouse Electric Co., Pittsburgh, Pa.)

Plate 1.12 Power Transformer Equipment with Forced-Oil, Forced-Air Cooling Package. (Courtesy of General Electric Co., Schenectady, N.Y.)

2. The Δ-Y-connection is commonly used for *stepping up* from low to high voltages. Usually it is used to connect a generator to a transmission network, with the generator connected to the Δ-side of the transformer and the transmission network connected to its Y-side. This connection is preferable because the neutral of the high voltage side, the Y-connection, can be **grounded**. The grounding could be (a) effective—the neutral could be directly or solidly connected to ground, (b) resistive—a resistor could be inserted between neutral and ground, and (c) reactive —an inductor could be inserted between neutral and ground. Note: *Solid or effective grounding* is the most prevalent grounding procedure in the U.S. on transmission systems that have voltages higher than the generated voltages, particularly 115 kV and above. This type of grounding leads to **great savings** in system cost by having the transformer insulation graded from the line to the neutral. It also leads to **additional cost reduction** by eliminating the grounding resistor cost and the space required, as such resistors are huge and can occupy perhaps thousands of cubic feet of space. *Resistive or inductive grounding is used to limit the flow of large amounts of current to ground during a fault in the system.* Sometimes the coil is *tuned to parallel resonance*, with the distributed capacitances from the line to ground. This causes the current flow to ground to become nearly zero and thus *the fault to be cleared.* This inductor is known as a **ground-fault neutralizer** or **Petersen coil.** The Δ-connection assures balanced line-to-neutral voltages on the Y-side and provides

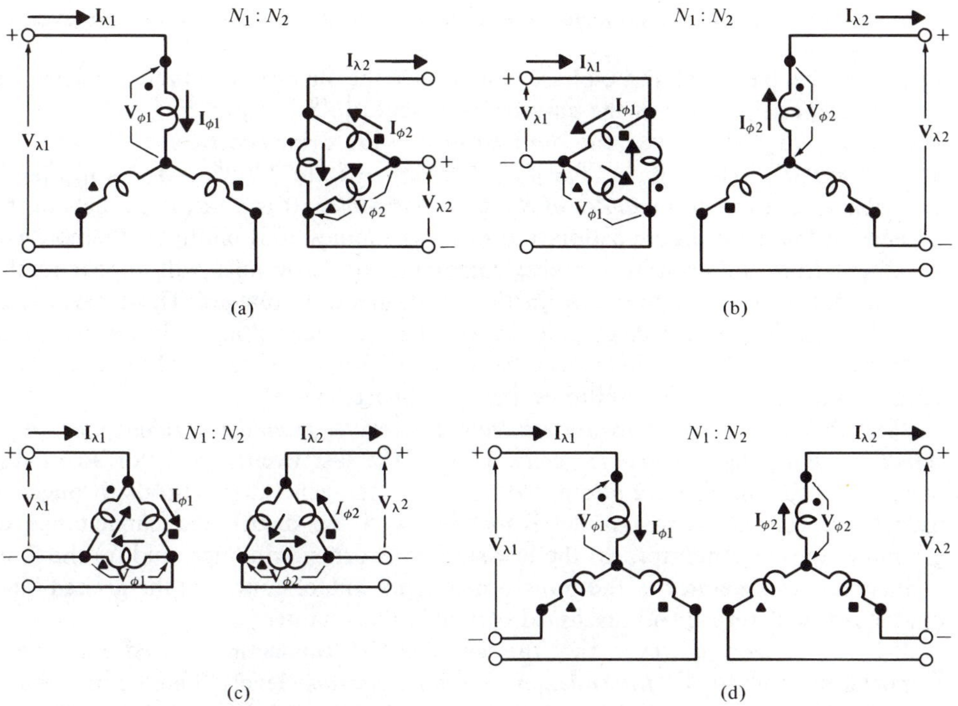

Fig. 1.25 Common Three-Phase Transformer Connections

a path for the *circulation of the third harmonics and their multiples* without the use of the neutral wire on the Y-side. This procedure is desirable in most cases. (You might want to refer to Chapter 3 for some of these definitions, such as "line-to-neutral.")

The Δ-Y-connection is also found in the distribution networks whereby the Δ-side is connected to the high-voltage transmission network and *the Y-side is connected to provide low-voltage to users*, for example, 208/120 V. In this connection the neutral of the Y-side is solidly grounded, and single loads are connected **line-to-neutral** for 120-V operation, while three-phase equipment is connected **line-to-line** for 208-V operation.

3. The Δ-Δ-connection has the advantages that *one transformer may be removed* for repair or maintenance, while the remaining two continue to function as a three-phase bank, with the rating reduced to 58 percent of that of the original bank. For obvious reasons the disturbed Δ-Δ-connection is known as an open-delta, or V-, connection.

4. The Y-Y-connection is seldom used* because of the difficulties with exciting current i_ϕ phenomena. When this transformer connection is used to connect a generator to a transmission network, for instance, *a fourth wire connected to the neutrals of* **both sides** of the Y-Y-connected transformer, and therefore to the neutral of the generator, must be used. (In many cases the *neutral is grounded*, and therefore the fourth wire is eliminated.) This connection necessitates that the windings of the generator be Y-connected. The neutral connection between the primary of the transformer and the generator assures balanced **line-to-neutral**

*The opposite is true, however, with the modified Y-Y-connected transformer.

voltages and *provides a path for the third-harmonic components* of the exciting currents.

Under the balanced three-phase conditions the neutral conductor carries no *fundamental* components of the current because $|\mathbf{I}_A{}^0|e^{j0°}+|\mathbf{I}_B{}^0|e^{-j120°}+|\mathbf{I}_C{}^0|e^{-j240°}=\mathbf{I}_N{}^0\equiv 0$. (Why?) However, the *third harmonics* are displaced from each other by $3\times 120°=360°$. Thus $|\mathbf{I}_A{}^3|e^{j0°}+|\mathbf{I}_B{}^3|e^{-j3\times 120°}+|\mathbf{I}_C{}^3|e^{-j3\times 240°}=\mathbf{I}_N{}^3=3|\mathbf{I}_A{}^3|e^{j0°}$. This is also true for *all multiples of the third harmonics*. It is clear, then, that in the absence of the neutral connection, the third harmonics, and multiples thereof, will be absent from the exciting current; corresponding harmonics will appear in the flux waveforms ϕ_m and therefore in the line-to-neutral voltages. Third harmonics and their multiples are negligible in the line-to-line voltages, however; these voltages are the phasor differences between the line-to-neutral voltages, and the third harmonics and their multiples, being in phase, cancel.

Note that, as a rule, *generator neutrals are usually grounded through fairly high resistance or through inductance coils*. During the last twenty years or so in this country, some utilities have attempted to ground the generator neutrals through an inductance coil tuned to parallel resonance with the distributed capacitance to ground of (a) the generator, (b) the low-side transformer windings, and (c) the lines connecting the generator to the transformer. This coil, of course, is the **ground-fault neutralizer** or **Petersen coil** discussed earlier in this chapter.

It is emphasized, however, that the **modified Y-Y-connection** is used extensively in power systems at the *transmission* or *subtransmission* levels. This connection is comprised of the classical Y-Y-connected transformer with an added **Δ-tertiary**. *A Δ-tertiary is comprised of three windings connected in a δ-configuration, with no external connections, and with each of its legs wound in parallel with the corresponding leg of the Y-connected transformer. The function of the δ-tertiary is to provide a low-impedance path for the circulation of the third-harmonic currents.**

Analysis of systems that contain three-phase balanced transformers can be carried out on a per-phase Y-connection (ie., line-to-neutral) basis. Thus for Y-Δ- or Δ-Y-connections, all parameters must be referred to the Y-connected side, and for Δ-Δ-connections, all parameters must be replaced by equivalent Y-connected parameters. When dealing with equivalent Δ-Y-impedance transformations, the following equation is useful:

$$\mathbf{Z}_\Delta = 3\mathbf{Z}_\mathrm{Y} \tag{1.76}$$

Eq. (1.76) can be verified if impedance measurements are taken at the terminals of networks (a) and (b), Fig. 1.26, *two at a time*, that is, terminals 1–2, 2–3, and 3–1. For Δ-Y-equivalence the measurements of both networks are equated. Thus an impedance measurement at terminals 1–2 of both networks yields

$$\frac{\mathbf{Z}_{12}\{\mathbf{Z}_{13}+\mathbf{Z}_{23}\}}{\mathbf{Z}_{12}+\mathbf{Z}_{13}+\mathbf{Z}_{23}}\Big|\Delta=(\mathbf{Z}_1+\mathbf{Z}_2)\Big|_\mathrm{Y} \tag{1.77}$$

The process of measuring should be repeated for terminals 2–3 and 3–1. *Add two sets of measurements and subtract the third set*. Repeat until all possible combinations are considered. This process yields the **Z**-parameters of the Y-network as functions of the **Z**-parameters of the Δ-network. In a balanced system the parameters of the Δ- and Y-networks are equal. When this equality is imposed on the

*For some of this information I am indebted to the Central Station Engineers of the Westinghouse Corporation, **Electrical Transmission and Distribution Reference Book**, 4th ed. (Pittsburgh, Pa, 1964), pp. 643–665, and **Electrical Utility Engineering Reference Book—Distribution Systems**, *vol.* 3 (*Pittsburgh, Pa*, 1965), *pp.* 201–246.

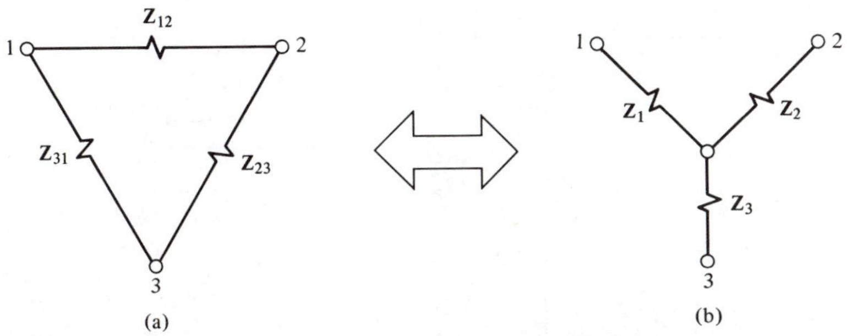

Fig. 1.26 Δ-Network to Y-Network Equivalence is Obtained if the Impedance Measurements at Terminals 1–2, 2–3, and 3–1 of Both Networks Are the Same

resulting equations, Eq. (1.76) results. Note that *the core or three-phase arrangement has all six windings on a common core and contained within a common tank*.

Clearly, *the bank arrangement has the advantages* that, in case of breakdown, only one-third of the equipment needs to be replaced, while the core or three-phase arrangement saves transformer core and consequently core losses, weighs less, requires less floor space, and has a somewhat higher efficiency than the three single-phase transformers of equal rating. Therefore, the one-core transformer has the *economic advantage* over a bank of three single-phase units. Finally, it should be noted that three-phase transformers are *vital* to the convenient transport of high-quality energy on which our quality of life depends so much.

An example involving three-phase transformers will be postponed until Chapter 3 where three-phase balanced systems analysis is undertaken.

1.10 AUTOTRANSFORMERS

The ordinary transformer, whether it is single-phase or three-phase, is characterized by the *complete electrical isolation* (not magnetic) between the primary and secondary windings. This property is of particular interest and advantage in the case of distribution transformers, where large transformation ratios are involved and safety plays a dominant role. However, when the transformation ratio is close to unity and isolation between coils is less critical, a transformer with a single tapped winding can be used with distinct advantages.

Such a transformer is known as an **autotransformer**, Fig. 1.27. The three currents in an ideal autotransformer are related by the following equations:

$$\delta \mathbf{I}_H = N_L \mathbf{I}_C \tag{1.78}$$

and

$$\mathbf{I}_C + \mathbf{I}_H = \mathbf{I}_L \tag{1.79}$$

Substitution of $\delta = N_H - N_L$ in Eq. (1.78) and solving for $\mathbf{I}_C/\mathbf{I}_H$ yields

$$\frac{\mathbf{I}_C}{\mathbf{I}_H} = \frac{N_H - N_L}{N_L} = \frac{N_H}{N_L} - 1 = \nu - 1 \equiv \alpha \tag{1.80}$$

where

$$\nu \equiv \frac{N_H}{N_L} \tag{1.81}$$

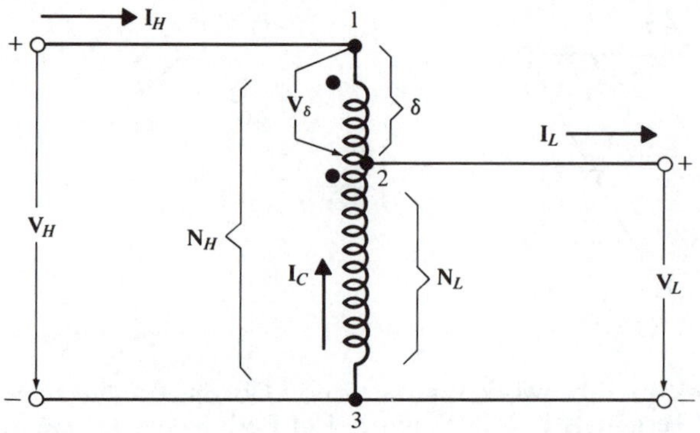

Fig. 1.27 The Autotransformer: $N_H > N_L$, thus $\delta = N_H - N_L$

Solving Eq. (1.79) for $\mathbf{I}_C = \mathbf{I}_L - \mathbf{I}_H$ and substituting into Eq. (1.80) yields the relationship between the low- and high-side currents, that is,

$$\frac{\mathbf{I}_C}{\mathbf{I}_H} \equiv \frac{\mathbf{I}_L - \mathbf{I}_H}{\mathbf{I}_H} = \nu - 1 \tag{1.82}$$

or

$$\mathbf{I}_L - \mathbf{I}_H = \nu \mathbf{I}_H - \mathbf{I}_H \tag{1.83}$$

Thus

$$\frac{\mathbf{I}_L}{\mathbf{I}_H} = \nu \tag{1.84}$$

Finally, substitution of $\mathbf{I}_H = \mathbf{I}_L / \nu$ into Eq. (1.80) yields

$$\frac{\mathbf{I}_C}{\mathbf{I}_L} = \frac{\nu - 1}{\nu} \tag{1.85}$$

Taking the voltage ratio of the windings yields

$$\frac{\mathbf{V}_\delta}{\mathbf{V}_L} = \frac{\mathbf{V}_H - \mathbf{V}_L}{\mathbf{V}_L} = \frac{\mathbf{V}_H}{\mathbf{V}_L} - 1 = \nu - 1 \equiv \alpha \tag{1.86}$$

where

$$\nu \equiv \frac{\mathbf{V}_H}{\mathbf{V}_L} \tag{1.87}$$

Note that Eqs. (1.80) and (1.86) indicate that the ratio of the voltages and currents in the coils 1–2 and 2–3 are the same as if the turns $\delta = N_{12}$ and $N_2 = N_{23}$ formed the primary and secondary windings of an ordinary transformer having a transformation ratio α, $\alpha = \nu - 1$.

Thus *the autotransformer can be treated as an equivalent ordinary transformer with a turns ratio of* α. The established theory of ordinary transformers is immediately applicable.

Due to the physical connection between its input and output, *the autotransformer has two kinds of voltampere power associated with it*. Since transformer action in the autotransformer takes place in windings 1–2 and 2–3, Fig. 1.27, it follows that the number of voltamperes in winding 1–2 is equal to the number of voltamperes in winding 2–3 and specifically represents the *transformed apparent power*. Thus the transformed power for negligible winding resistance and leakage reactance can be written as

$$\mathbf{V}_\delta \mathbf{I}_H^* = \mathbf{V}_L \mathbf{I}_C^* \equiv \text{transformed voltamperes (TVA)} \tag{1.88}$$

However, the input and output to the autotransformer are given in terms of input and output voltages and currents as

$$\mathbf{S}_I = \mathbf{V}_H\mathbf{I}_H^* \tag{1.89}$$

$$\mathbf{S}_O = \mathbf{V}_L\mathbf{I}_L^* \tag{1.90}$$

The difference between Eq. (1.89) and the left-hand side of Eq. (1.88), or Eq. (1.90) and the right-hand side of Eq. (1.88), yields the untransformed or *conducted voltamperes*,

$$\mathbf{V}_H\mathbf{I}_H^* - \mathbf{V}_\delta\mathbf{I}_H^* = (\mathbf{V}_H - \mathbf{V}_\delta)\mathbf{I}_H^* = \mathbf{V}_L\mathbf{I}_H^* \equiv \text{conducted voltamperes (CVA)} \tag{1.91}$$

or

$$\mathbf{V}_L\mathbf{I}_L^* - \mathbf{V}_L\mathbf{I}_C^* = \mathbf{V}_L(\mathbf{I}_L^* - \mathbf{I}_C^*) = \mathbf{V}_L\mathbf{I}_H^* \equiv \text{conducted voltamperes (CVA)} \tag{1.92}$$

The ratio of the output voltamperes of the autotransformer to the transformed voltamperes is given, with the aid of Eq. (1.85), as follows:

$$\frac{\mathbf{S}_O}{\text{TVA}} = \frac{\mathbf{V}_L\mathbf{I}_L^*}{\mathbf{V}_L\mathbf{I}_C^*} = \frac{\mathbf{I}_L^*}{\mathbf{I}_C^*} = \frac{\nu}{\nu - 1} \tag{1.93}$$

The ratio $\nu/(\nu-1)$ is much larger than unity, since for many autotransformer applications $\nu \doteq 1$. This emphasizes the fact that *most of the power in autotransformers is delivered to the load through direct conduction and only a small portion is delivered through transformation.*

The *advantages* that result from the use of autotransformers are primarily related to *economics* due to savings in copper: (1) the primary and secondary of the autotransformer share one winding; (2) the current of the common winding is always the difference between the currents of the primary and secondary and thus is a low current—accordingly, *the cross-sectional area of the common winding is much smaller than the primary or secondary winding of an ordinary transformer of the same current density*; (3) the efficiency of the autotransformer for the same output is much higher; (4) the voltage regulation of the autotransformer is superior to the reduced resistance drop and lower leakage reactance drop; and (5) the autotransformer is smaller in size for the same output.

Example 1.8

Connect a 2400:240-V, 50-kVA ordinary transformer as a step-up autotransformer and calculate the (1) high-side current, (2) autotransformer kilovoltampere rating, (3) common current, (4) low-side current, (5) conducted kilovoltamperes, and (6) equivalent ordinary transformer.

Solution

See Fig. 1.8.1.

1. $|240|\,|\mathbf{I}_H^*| = 50{,}000$,

$$|\mathbf{I}_H^*| = \frac{50{,}000}{240} = \underline{\underline{208}}\text{ A}$$

2.

$$|\mathbf{V}_H| = 2{,}400 + 240 = \underline{2{,}640}\text{ V}$$

$$\text{kVA}_{AT} = |\mathbf{V}_H|\,|\mathbf{I}_H^*| = (2.640)(208) = \underline{\underline{549}}\text{ kVA}$$

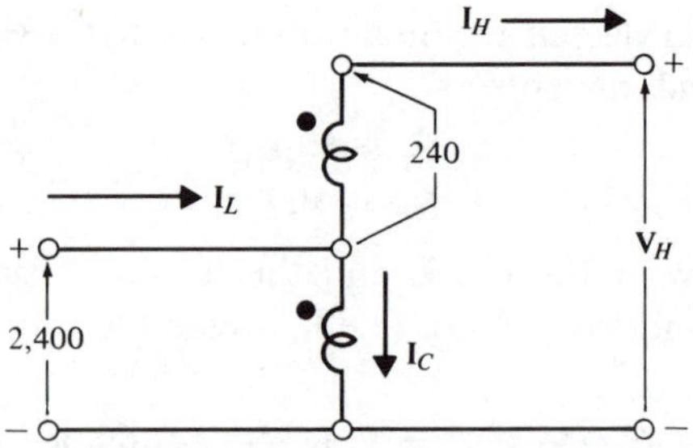

Fig. 1.8.1 The Equivalent Circuit of the Autotransformer

3. $|2.400||\mathbf{I_C}^*| = 50{,}000$,

$$|\mathbf{I_C}^*| = \frac{50{,}000}{2{,}400} = \underline{\underline{20.8}}\text{ A}$$

4. $|\mathbf{I_H}| + |\mathbf{I_R}| = |\mathbf{I_L}|$

$$|\mathbf{I_L}| = 208 + 20.8 = \underline{\underline{228.8}}\text{ A}$$

Also,

5.
$$\text{kVA}_{\text{AT}} = |\mathbf{V_L}||\mathbf{I_L}^*| = (2.400)(228.8) = \underline{\underline{549}}\text{ kVA}$$

$$\text{CVA} = |\mathbf{V_L}||\mathbf{I_H}^*| = (2.400)(208) = \underline{\underline{500}}\text{ kVA}$$

6. The equivalent ordinary transformer is

$$\alpha = \nu - 1 = \frac{|\mathbf{I_C}|}{|\mathbf{I_H}|} = \frac{20.8}{208} = \underline{\underline{0.1}}$$

See Fig. 1.8.2. Then

$$\nu = \alpha + 1 = 0.1 + 1 = \underline{1.1}$$

or

$$\nu = \frac{|\mathbf{I_L}|}{|\mathbf{I_H}|} = \frac{228.8}{208} = \underline{1.1}$$

and

$$\alpha = \frac{|\mathbf{I_L}| - |\mathbf{I_H}|}{|\mathbf{I_H}|} = \frac{228.8 - 208}{208} = \underline{\underline{0.1}}$$

or

$$\alpha = \frac{|\mathbf{V_\delta}|}{|\mathbf{V_L}|} = \frac{|\mathbf{V_H}| - |\mathbf{V_L}|}{|\mathbf{V_L}|} = \frac{240}{2{,}400} = \frac{2{,}640 - 2{,}400}{2{,}400} = \underline{\underline{0.1}}$$

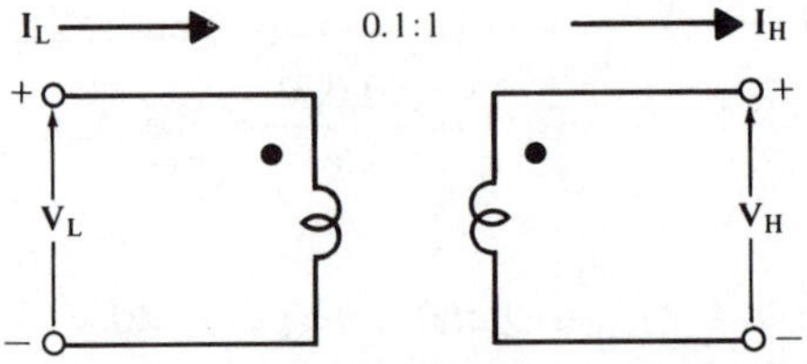

Fig. 1.8.2 Ordinary Transformer Equivalent Circuit for Fig. 1.8.1.

1.11 THE SCOTT TRANSFORMER

The Scott transformer, Fig. 1.28(a), named after its inventor C. F. Scott, is wound in such a manner as **to transform a set of balanced three-phase voltages*** (the three voltages are of equal magnitude but are displaced 120 electrical degrees from each other) **to a set of balanced two-phase voltages** (the two voltages are of equal magnitude but are displaced 90 electrical degrees from each other) and **vice versa.**

The voltages $\mathbf{E}_{\mathcal{A}\mathcal{O}}$ and $\mathbf{E}_{AO}$, and $\mathbf{E}_{\mathcal{B}\mathcal{O}}$ and $\mathbf{E}_{BC}$, Fig. 1.28(a), are related to the appropriate turn ratios of the windings, in accordance with Eq. (1.50), as

$$\frac{\mathbf{E}_{\mathcal{A}\mathcal{O}}}{\mathbf{E}_{AO}} = \frac{N_{\mathcal{A}\mathcal{O}}}{N_{AO}} \tag{1.94}$$

and

$$\frac{\mathbf{E}_{\mathcal{B}\mathcal{O}}}{\mathbf{E}_{BC}} = \frac{N_{\mathcal{B}\mathcal{O}}}{N_{BC}} \tag{1.95}$$

Solution of these two equations for the voltages $\mathbf{E}_{\mathcal{A}\mathcal{O}}$ and $\mathbf{E}_{\mathcal{B}\mathcal{O}}$ yields

$$\mathbf{E}_{\mathcal{A}\mathcal{O}} = \frac{N_{\mathcal{A}\mathcal{O}}}{N_{AO}} \times \mathbf{E}_{AO} \tag{1.96}$$

and

$$\mathbf{E}_{\mathcal{B}\mathcal{O}} = \frac{N_{\mathcal{B}\mathcal{O}}}{N_{BC}} \times \mathbf{E}_{BC} \tag{1.97}$$

If the voltages $\mathbf{E}_{\mathcal{A}\mathcal{O}}$ and $\mathbf{E}_{\mathcal{B}\mathcal{O}}$ are to constitute a set of balanced voltages, then

$$|\mathbf{E}_{\mathcal{A}\mathcal{O}}| = |\mathbf{E}_{\mathcal{B}\mathcal{O}}| \tag{1.98}$$

and the ratio of the magnitudes of the voltages $|\mathbf{E}_{\mathcal{A}\mathcal{O}}|/|\mathbf{E}_{\mathcal{B}\mathcal{O}}|$, Eq. (1.96) and Eq. (1.97), yields

$$1 = \frac{N_{\mathcal{A}\mathcal{O}} N_{BC}}{N_{\mathcal{B}\mathcal{O}} N_{AO}} \times \frac{|\mathbf{E}_{AO}|}{|\mathbf{E}_{BC}|}\Bigg|_{|\mathbf{E}_{\mathcal{A}\mathcal{O}}| = |\mathbf{E}_{\mathcal{B}\mathcal{O}}|} \tag{1.99}$$

The voltage $|\mathbf{E}_{AO}|$ of Eq. (1.99) can be written,† from inspection of Fig. 1.28(b), as

$$|\mathbf{E}_{AO}| = |\mathbf{E}_{AB}| \cos 30° = \frac{\sqrt{3}}{2} |\mathbf{E}_{AB}| \tag{1.100}$$

However, since the voltages $\mathbf{E}_{AB}$, $\mathbf{E}_{BC}$, and $\mathbf{E}_{CA}$ constitute a balanced set of voltages, their magnitudes are identical, that is, $|\mathbf{E}_{AB}| \equiv |\mathbf{E}_{BC}|$. Therefore, Eq. (1.99) can be written as

$$1 = \frac{N_{\mathcal{A}\mathcal{O}} N_{BC}}{N_{\mathcal{B}\mathcal{O}} N_{AO}} \times \frac{\sqrt{3}/2}{1}\Bigg|_{\substack{|\mathbf{E}_{\mathcal{A}\mathcal{O}}| = |\mathbf{E}_{\mathcal{B}\mathcal{O}}| \\ \text{and} \\ |\mathbf{E}_{AB}| = |\mathbf{E}_{BC}|}} \tag{1.101}$$

If we were to impose the additional constraint

$$N_{\mathcal{A}\mathcal{O}} = N_{\mathcal{B}\mathcal{O}} \tag{1.102}$$

*Refer to Chapter 3, and in particular to pp. 167–170, for additional information about and the mathematics of this topic.

†Perhaps you should read the paragraph that follows Eq. (1.104).

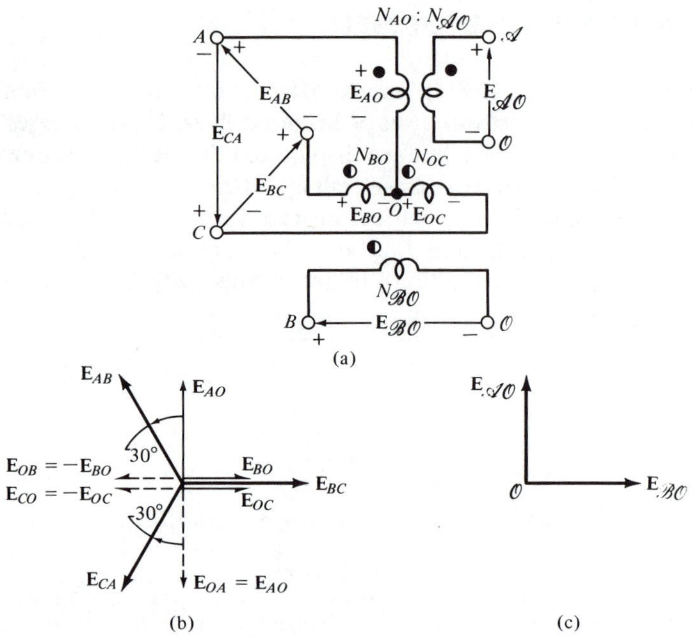

Fig. 1.28 The Scott Transformer: (a) Winding Geometry, (b) Voltage Relationship Among the Voltages of the Three-Phase Winding, and (c) Voltage Relationship Among the Voltages of the Two-Phase Winding

on Eq. (1.101), then we could write this equation as

$$1 = \frac{\sqrt{3}}{2} \times \frac{N_{BC}}{N_{AO}}\Bigg|_{\substack{|\mathbf{E}_{\mathcal{AO}}| = |\mathbf{E}_{\mathcal{BO}}| \\ |\mathbf{E}_{AB}| = |\mathbf{E}_{BC}| \\ \text{and} \\ N_{\mathcal{AO}} = N_{\mathcal{BO}}}} \tag{1.103}$$

That is, for the sets of the voltages at the input and output of the Scott transformer to be balanced—in addition to the constraint given by Eq. (1.102)—the following relationship must exist between the number of turns of the windings AO and BC:

$$N_{AO} = \frac{\sqrt{3}}{2} N_{BC} = \frac{\sqrt{3}}{2}(N_{BO} + N_{OC})\Big|_{N_{BO} = N_{OC}} \tag{1.104}$$

Eq. (1.104) justifies the drawing of the voltage phasors $\mathbf{E}_{AO}$, $\mathbf{E}_{BO}$, and $\mathbf{E}_{OC}$ of Fig. 1.28(a) as exhibited in Fig. 1.28(b). Now the terminal-to-terminal voltages $\mathbf{E}_{AB}$, $\mathbf{E}_{BC}$, and $\mathbf{E}_{CA}$, Fig. 1.28(d), can be written by direct application of Kirchhoff's voltage law to be designated paths as

$$\mathbf{E}_{AB} - \mathbf{E}_{AO} + \mathbf{E}_{BO} = 0 \tag{1.105}$$

$$\mathbf{E}_{BC} - \mathbf{E}_{BO} - \mathbf{E}_{OC} = 0 \tag{1.106}$$

$$\mathbf{E}_{CA} + \mathbf{E}_{OC} + \mathbf{E}_{AO} = 0 \tag{1.107}$$

Solution of these three equations for the terminal-to-terminal voltage phasors $\mathbf{E}_{AB}$, $\mathbf{E}_{BC}$, and $\mathbf{E}_{CA}$ yields

$$\mathbf{E}_{AB} = \mathbf{E}_{AO} + (-\mathbf{E}_{BO}) = \mathbf{E}_{AO} + \mathbf{E}_{OB} \tag{1.108}$$

$$\mathbf{E}_{BC} = \mathbf{E}_{BO} + \mathbf{E}_{OC} \tag{1.109}$$

$$\mathbf{E}_{CA} = -\mathbf{E}_{AO} - \mathbf{E}_{OC} = \mathbf{E}_{OA} + \mathbf{E}_{CO} \tag{1.110}$$

The vectorial sum of the voltage phasors of the right-hand side of Eqs. (1.108) through (1.110) yields the desired terminal-to-terminal voltages of the three-phase winding of Fig. 1.28(a). These voltages are drawn as *heavy vectors* in Fig. 1.28(b). Note that the negative phasor voltages of Eqs. (1.108) through (1.110) are drawn as *dashed vectors*.

With the voltages $\mathbf{E}_{AB}$, $\mathbf{E}_{BC}$, and $\mathbf{E}_{CA}$ at hand, by inspection of Fig. 1.28(a), we can deduce that the voltages $\mathbf{E}_{\mathcal{A}\mathcal{O}}$ and $\mathbf{E}_{\mathcal{B}\mathcal{O}}$ across the windings $\mathcal{A}\mathcal{O}$ and $\mathcal{B}\mathcal{O}$ are explicitly known. They are, in fact, related to the voltages $\mathbf{E}_{AO}$ and $\mathbf{E}_{BC}$ of the three-phase winding in accordance with Eqs. (1.96) and (1.97) and have identical polarity with these two voltages. The voltage phasor diagram of the two-phase voltages $\mathbf{E}_{\mathcal{A}\mathcal{O}}$ and $\mathbf{E}_{\mathcal{B}\mathcal{O}}$ is exhibited in Fig. 1.28(c).

If we were interested in the **conversion of a set of balanced two-phase voltages into a set of balanced three-phase voltages**, the process of identifying the appropriate phasors and determining their magnitudes necessitates solution of Eqs. (1.96) and (1.97) for the voltages $\mathbf{E}_{AO}$ and $\mathbf{E}_{BC}$ in terms of the known voltages $\mathbf{E}_{\mathcal{A}\mathcal{O}}$ and $\mathbf{E}_{\mathcal{B}\mathcal{O}}$. Now the three-phase voltages are explicitly known and the writing of their mathematical equations is much simpler. However, as in the case of the three-phase–to–two-phase voltage conversion, the process requires the comprehension of the material covered in Chapter 3, in particular, pages 167–170.

Usually this requirement does not occur in practice, as the balanced three-phase voltages are available. However, this assertion cannot be made for balanced two-phase voltages.

Example 1.9

If a set of balanced three-phase voltages, given as $e_{AB}(t)=\sqrt{2}\,(120)\cos(\omega_{\mathrm{E}}t-240°)$, $e_{CA}(t)=\sqrt{2}\,(120)\cos(\omega_{\mathrm{E}}t-120°)$, and $e_{BC}(t)=\sqrt{2}\,(120)\cos\omega_{\mathrm{E}}t$, is impressed in the three-phase winding of a Scott transformer which has $N_{AO}=140$ turns, calculate the balanced two-phase voltage output $e_{\mathcal{A}\mathcal{O}}(t)$ and $e_{\mathcal{B}\mathcal{O}}(t)$ of this transformer if $N_{\mathcal{A}\mathcal{O}}=N_{\mathcal{B}\mathcal{O}}=200$ turns.

Solution

Utilization of Fig. 1.28(b) enables us to calculate the voltage $|\mathbf{E}_{AO}|$ as

$$|\mathbf{E}_{AO}| = \mathbf{E}_{AB}\cos 30° = 120\times 0.866 = \underline{\underline{103.92\text{ V}}} \tag{1}$$

Similarly, utilization of Eq. (1.104) enables us to calculate the number of turns $N_{BO}+N_{CO}$ as

$$N_{BO}+N_{OC} = N_{AO}\left(\frac{2}{\sqrt{3}}\right) \tag{2}$$

Now, utilization of Eqs. (1) and (2), and Eqs. (1.96) and (1.97), and pertinent data, enables us to calculate the magnitudes of the voltage phasors $|\mathbf{E}_{\mathcal{A}\mathcal{O}}|$ and $|\mathbf{E}_{\mathcal{B}\mathcal{O}}|$ as

$$|\mathbf{E}_{\mathcal{A}\mathcal{O}}| = \frac{N_{\mathcal{A}\mathcal{O}}}{N_{AO}}\times\mathbf{E}_{AO} = \tfrac{200}{140}\times 103.92 = \underline{\underline{148.45\text{ V}}} \tag{3}$$

and

$$|\mathbf{E}_{\mathcal{B}\mathcal{O}}| = \frac{N_{\mathcal{B}\mathcal{O}}}{N_{BO}+N_{CO}}\times\mathbf{E}_{BC} = \frac{200}{(2/\sqrt{3}\,)\times 140}\times 120 = \underline{\underline{148.45\text{ V}}} \tag{4}$$

Again, through Fig. 1.28(b), the voltage phasors $\mathbf{E}_{AO}$ and $\mathbf{E}_{BC}$ can be written as

$$\mathbf{E}_{AO} = |\mathbf{E}_{AO}|e^{j\theta_{AO}} = \underline{\underline{148.45e^{j90^\circ}}} \text{ V} \tag{5}$$

$$\mathbf{E}_{BC} = |\mathbf{E}_{BC}|e^{j\theta_{BC}} = \underline{\underline{120e^{j\,0^\circ}}} \text{ V} \tag{6}$$

Since the voltage $\mathbf{E}_{\mathscr{A}\mathscr{O}}$ is parallel to the voltage $\mathbf{E}_{AO}$ and the voltage $\mathbf{E}_{\mathscr{B}\mathscr{O}}$ is parallel to the voltage $\mathbf{E}_{BC}$, Figs. 1.28(b) and 1.28(c), the voltage phasors $\mathbf{E}_{\mathscr{A}\mathscr{O}}$ and $\mathbf{E}_{\mathscr{B}\mathscr{O}}$ can be written as

$$\mathbf{E}_{\mathscr{A}\mathscr{O}} = \underline{\underline{148.45e^{j90^\circ}}} \text{ V} \tag{7}$$

and

$$\mathbf{E}_{\mathscr{B}\mathscr{O}} = \underline{\underline{148.45e^{j0^\circ}}} \text{ V} \tag{8}$$

Finally, Eqs. (7) and (8) and the theory set forth in Appendix A enable us to write the time-domain voltage $e_{\mathscr{A}\mathscr{O}}(t)$ and $e_{\mathscr{B}\mathscr{O}}(t)$ as

$$e_{\mathscr{A}\mathscr{O}}(t) = \sqrt{2}\,(148.45)\cos(\omega_{\mathrm{E}}t + 90^\circ) = \underline{\underline{-\sqrt{2}\,(148.45)\sin\omega_{\mathrm{E}}t}} \text{ V} \tag{9}$$

and

$$e_{\mathscr{B}\mathscr{O}}(t) = \underline{\underline{\sqrt{2}\,(148.45)\cos\omega_{\mathrm{E}}t}} \text{ V} \tag{10}$$

1.12 SUMMARY

This chapter describes the fundamentals of magnetic circuit theory required for transformer analysis and rotating machine theory.

The generalized theory of transformers is set forth, and through it the equivalent circuits of the ideal and real transformers are derived. For the real transformer the effects of flux leakages, copper losses, hysteresis losses, eddy current losses, and winding capacitances are taken into account for a realistic circuit representation of the transformer. Methods for measuring the transformer equivalent circuit parameters, that is, the short-circuit and open-circuit tests, are set forth. Also, simplified equivalent circuits of the transformer are derived. In addition, the function of the transformer in communications and single-phase power systems is discussed. Finally, an introduction is given to the three-phase transformer, the autotransformer, and the Scott transformer.

Before leaving this chapter the student is urged to master this material because the rest of the chapters, particularly the chapter on induction devices, relies heavily upon the material discussed here.

PROBLEMS

1.1 For the magnetic circuit of Example 1.1, $\ell = 6\times10^{-2}$ m, $N_1 = N_2 = 1200$ turns, $I_1 = I_2 = 6$ A and $\ell_g = 6\times10^{-3}$ m. Calculate the airgap field quantities, Φ_g, B_g, and H_g.

1.2 For the magnetic circuit of Example 1.2, $I = 100$ A, $N = 1000$ turns, $\ell_m = 1$ m, $\ell_1 = \ell_2 = 0.05$ m, and $\ell_g = 0.01$ m. Calculate the magnetic field quantities Φ_m, B_m, H_m, Φ_g, B_g, and H_g following each of these procedures:
   a. Use of Ampere's law, $\oint \mathbf{H}\cdot d\mathbf{L} = \Sigma\, I$.
   b. Use of the medium silicon sheet-steel curve of Fig. 1.3.2.

1.3 Use the magnetic circuit of Example 1.3 and determine the current that must flow in the exciting winding in order to produce the given flux Φ, if the exciting winding has 400 turns and produces a flux of $\Phi=2\times10^{-4}$ Wb. Note: For the solution of this problem, assume that the flux is confined in the path provided by the magnetic circuit, (i.e., there is no flux leakage), and use the magnetization curves provided in Example 1.3.

1.4 A very popular and useful application of magnetic circuit principles is the so-called lifting of a magnetic crane. For the magnetic circuit of Example 1.4, with the parameters $N=1000$ turns, $I=100$ A, $\ell_m=4$ m, $\ell_g=0.01$ m, $A=4\times 10^{-2}$ m^2, $\ell_b=1$ m, and $\mu_r=10^6$ completely defined, calculate the force that can be lifted by the crane.

1.5 For the circuit of Example 1.5 and $\mathbf{E}_S=240e^{-j50°}$, $\mathbf{Z}_S=6+j6$, and $N_1:N_2=300:15$, calculate the quantities $\mathbf{Z}_L$, $\mathbf{I}_1$, $\mathbf{I}_2$, $\mathbf{V}_1$, $\mathbf{V}_2$, $\mathbf{S}_1$, and $\mathbf{S}_2$ for the conditions of maximum power transfer.

1.6 Given: a 100-kVA, 4800:480-V transformer feeding a load with a PF=0.85 LGG. The load draws rated transformer secondary current. Calculate: (1) the parameters of the equivalent circuit of the transformer, (2) the efficiency, and (3) the regulation of the transformer. The energized sides of the transformer are instrumented and the readings are as follows:
   a. The short-circuit measurements on the high-voltage side are $|\mathbf{E}_{SC}|=96$ V, $|\mathbf{I}_{SC}|=41.6$ A, and $P_{SC}=1234$ W.
   b. The open-circuit measurements on the low-voltage side are $|\mathbf{E}_{OC}|=480$ V, $|\mathbf{I}_{OC}|=10.82$ A, and $P_{OC}=372$ W.

1.7 Refer to Fig. 1.7.1 of Example 1.7. For a referred $\mathbf{Z}_L'=10+j10$-Ω load, use circuit analysis techniques and calculate $\mathbf{I}_1$, $\mathbf{I}_2'$, $\mathbf{I}_\phi$, $\mathbf{I}_c$, P_1, P_2, P_c, P_O, P_I, η, $|\mathbf{V}_2'|_{NL}$, $|\mathbf{V}_2'|_{FL}$, and VR.

1.8 Connect a 4800:480-V, 80-kVA ordinary transformer as a step-up autotransformer and calculate the (1) high-side current, (2) autotransformer kilovoltampere rating, (3) common current, (4) low-side current, (5) conducted kilovoltamperes, and (6) equivalent ordinary transformer.

1.9 If a set of balanced three-phase voltages, given as $e_{AB}(t)=\sqrt{2}\,(240)\cos(\omega_E t-240°)$, $e_{CA}(t)=\sqrt{2}\,(240)\cos(\omega_E t-120°)$, and $e_{BC}(t)=\sqrt{2}\,(240)\cos\omega_E t$, is impressed in the three-phase winding of the Scott transformer, Fig. 1.27, which has $N_{AO}=200$ turns, calculate the number of turns $N_{\mathcal{AO}}$ and $N_{\mathcal{BO}}$ if the balanced two-phase voltages $e_{\mathcal{AO}}(t)$ and $e_{\mathcal{BO}}(t)$ at the output are to be $e_{\mathcal{AO}}(t)=-\sqrt{2}\,(100)\sin\omega_E t$ and $e_{\mathcal{BO}}(t)=\sqrt{2}\,(100)\cos\omega_E t$.

REFERENCES

1. Boast, W. B., *Principles of Electric and Magnetic Circuits*. Harper & Row, New York, 1950.
2. ———, *Principles of Electric and Magnetic Circuits*, 2nd ed. Harper & Row, New York, 1956.
3. Del Toro, V., *Electromechanical Devices for Energy Conversion and Control Systems*. Prentice-Hall, Englewood Cliffs, N.J., 1968.
4. Elgerd, O. I., *Electric Energy Systems Theory*: *An Introduction*. McGraw-Hill, New York, 1971.
5. Fitzgerald, A. E., and Kingsley, C., Jr., *Electric Machinery*, 2nd ed. McGraw-Hill, New York, 1961.
6. Hayt, W. H., Jr., *Engineering Electromagnetics*. McGraw-Hill, New York, 1958.
7. Majmudar, H., *Electromechanical Energy Converters*. Allyn & Bacon, Boston, 1965.
8. Matsch, L. W., *Electromagnetic and Electromechanical Machines*. Intext, New York, 1972.

2 dc ENERGY CONVERTERS

2.1 INTRODUCTION

A dc device is defined as the electromechanical energy converter that structurally is comprised of the stationary member, **the stator**, with protrusions known as saliencies in its inner surface and a rotating member, **the rotor**, with a smooth outer surface. The stator has concentrated windings and the rotor distributed windings, and both are connected to external circuitry designed always to send dc current to the stator winding and either to send dc current to, or to receive it from, the rotor winding.

Depending on the way in which the dc device is excited, it functions in either the motor or the generator mode of operation. The corresponding mode is characterized by the development of the electromagnetic torque produced by the interaction of the magnetic fields only when present in both the stator and the rotor of the device. In the *motor mode of operation*, this torque acts in the direction of rotation and balances the opposing torque that results mainly from the motor's mechanical load. In the *generator mode of operation*, the developed electromagnetic torque opposes rotation, and mechanical torque must be applied from the prime mover in order to sustain the rotation. As the developed electromagnetic torque increases, so does the mechanical torque applied from the prime mover. Therefore, this developed electromagnetic torque is the means through which the electric power output from the generator is related to its mechanical power input.

The dc device is manufactured to yield a wide variety of dynamic and steady state **torque versus speed** characteristics. These characteristics can be modified with the addition of external **feedback** and **thyristor**-controlled* circuitry to provide

*See Appendix E for pertinent information on this subject.

Plate 2.1 Cutaway View of a 300/600-r/min, 600-V dc Device. (Courtesy of General Electric Co., Schenectady, N.Y.)

adjustable speed drives, known as **dc drives**, with unique capabilities. These capabilities make the dc drive highly desirable in specific fields of engineering, such as steel mills, machine tools, and transportation systems.

Thus the objectives of this chapter are sixfold: (1) to set forth the principles of operation of the most common types of the dc *electromechanical* and *mechanoelectrical* energy converters, dc motors and generators; (2) to describe the principle of commutation as it applies to dc motors and generators; (3) to develop mathematical models that describe the dynamic and steady state performance of dc energy converters; (4) to obtain and plot terminal characteristics under steady state for both modes (i.e., motor and generator) of operation of this class of devices; (5) to introduce dynamic analysis via the s-domain and t-domain techniques (for the s-domain analysis the concept of transfer function is utilized; however, for the t-domain analysis the concepts of the state-variable method—with time-varying and time-invariant coefficients—and simulation diagram techniques are utilized); and (6) to provide experimental procedures for the measurement of the equivalent circuit parameters for both the motor and the generator models.

2.2 FUNDAMENTAL CONCEPTS OF ELECTROMECHANICAL ENERGY CONVERTERS

The operation of electromechanical energy converters is based upon *two fundamental principles*: (1) the developed force on a current-carrying conductor moving within a magnetic field and (2) the potential generated (induced) on a conductor moving within a magnetic field. Both principles are based on the expression*

$$\mathbf{F} = q\mathbf{u}\times\mathbf{B} \tag{2.1}$$

Eq. (2.1) is an experimental observation that gives the relationship among the developed force **F** (newtons) on a positive charge q (coulombs) moving within a magnetic field **B** (webers/meter^2) with a velocity **u** (meters/second). The geometric relationship among the vectors **F**, **u**, and **B** is given in Fig. 2.1.

The charge q of Eq. (2.1) can represent an incremental charge dQ. If a conductor CD is in the magnetic field **B**, Fig. 2.2, and dQ is moving with velocity **u** an incremental distance $d\mathbf{L}$ along the conductor CD, then Eq. (2.1) can be written as

$$d\mathbf{F} = dQ\,\mathbf{u}\times\mathbf{B} \tag{2.2}$$

Since $\mathbf{u}\equiv d\mathbf{L}/dt$, Eq. (2.2) can be written as

$$d\mathbf{F} = dQ\frac{d\mathbf{L}}{dt}\times\mathbf{B} \tag{2.3}$$

or

$$d\mathbf{F} = \frac{dQ}{dt}\,d\mathbf{L}\times\mathbf{B} \tag{2.4}$$

Since $dQ/dt\equiv i$, Eq. (2.4) can be written as

$$d\mathbf{F} = i\,d\mathbf{L}\times\mathbf{B} \equiv -\mathbf{B}\times d\mathbf{L} \tag{2.5}$$

Because the force $d\mathbf{F}$ is a consequence of the interaction of the electric current i (amperes) through the conductor CD and the magnetic field **B**, the force is termed an *electromagnetic* force. Henceforth it is abbreviated as $\mathbf{F}_{\text{em}}$ where em is the abbreviation of electromagnetic. Thus the total electromagnetic force on the current-carrying conductor CD residing in a field **B** is found by integration of Eq. (2.5). Therefore,

$$\mathbf{F}_{\text{em}} = -\int_C^D i\mathbf{B}\times d\mathbf{L} \tag{2.6}$$

The vectors **B** and $d\mathbf{L}$ should be expressed in the same coordinate system before integration is initiated. Thus for **B** and $d\mathbf{L}$ expressed in the Cartesian coordinate system, the cross product $\mathbf{B}\times d\mathbf{L}$ is defined as†

$$\mathbf{B}\times d\mathbf{L} \equiv \begin{vmatrix} \mathbf{a}_x & \mathbf{a}_y & \mathbf{a}_z \\ B_x & B_y & B_z \\ dx & dy & dz \end{vmatrix}$$
$$\equiv |\mathbf{B}|\quad|d\mathbf{L}|\sin\theta\,\mathbf{a}_N \tag{2.7}$$

where θ is the angle between the vectors **B** and $d\mathbf{L}$, and $\mathbf{a}_N$ is a unit vector normal to the plane determined by vectors **B** and $d\mathbf{L}$.

Example 2.1

Given: the configuration shown in Fig. 2.1.1. Assume that (1) $\mathbf{B}=|\mathbf{B}|(-\mathbf{a}_x)$ webers/meter^2, (2) $d\mathbf{L}=dz\,\mathbf{a}_z$ meters, (3) $D\rightarrow z_1=-L/2$ meters, and (4) $C\rightarrow z_2=$

*Lowercase letters imply time dependence. For example, the quantity $u(t)$ is written simply as u for convenience.

†$\mathbf{a}_x$, $\mathbf{a}_y$, and $\mathbf{a}_z$ designate the unit vectors on a right-handed Cartesian coordinate system, and $\mathbf{a}_N$ designates a unit vector, normal to the plane determined by **B** and $d\mathbf{L}$.

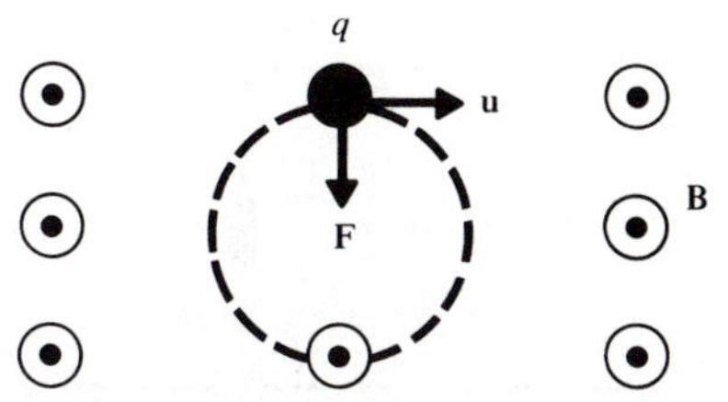

Fig. 2.1 The Effects of the Field **B** on a Positive Charge q Moving with Velocity **u**

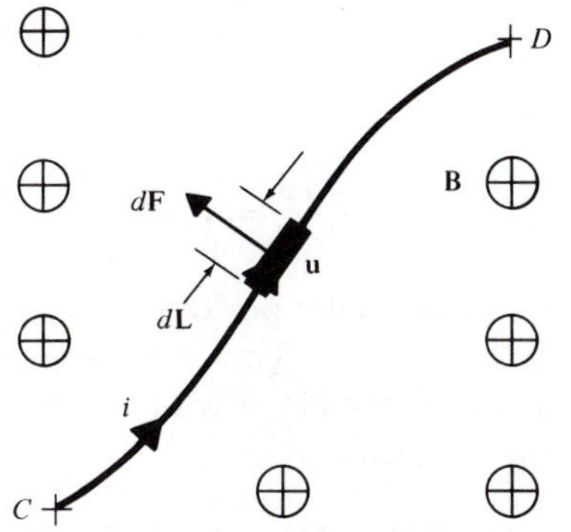

Fig. 2.2 Force on a Current-Carrying Conductor Imbedded in a Magnetic Field **B**

$L/2$ meters. Calculate the electromagnetic force developed on the current-carrying conductor of length L meters residing within the field **B** webers/meter^2.

Solution

$$\mathbf{F}_{em} = -\int_D^C i\mathbf{B} \times d\mathbf{L} = -i\int_{-L/2}^{L/2} |\mathbf{B}|(-\mathbf{a}_x) \times dz\,\mathbf{a}_z = -i|\mathbf{B}|\mathbf{a}_y \int_{-L/2}^{L/2} dz$$
$$= \underline{\underline{i|\mathbf{B}|L(-\mathbf{a}_y)}} \text{ newtons}$$

Now consider the system shown in Fig. 2.3. The charge q of Eq. (2.1) represents an incremental charge dQ within a conductor L meters long. The conductor slides without friction along two infinitely long bars No. 1 and No. 2 with a velocity **u**. The entire system is imbedded in a magnetic field **B**. Eq. (2.1) can now be written as

$$d\mathbf{F} = dQ\,\mathbf{u} \times \mathbf{B} \tag{2.8}$$

Dividing both sides of Eq. (2.8) by dQ yields a **potential-producing field intensity** $\boldsymbol{\varepsilon}$

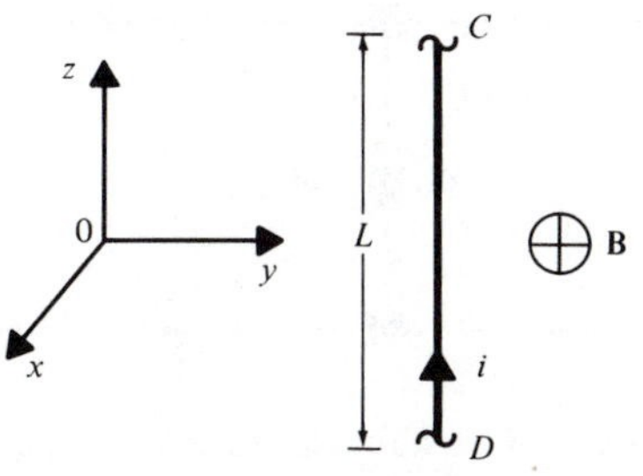

Fig. 2.1.1 Current-Carrying Conductor Imbedded in a Constant Magnetic Field **B**

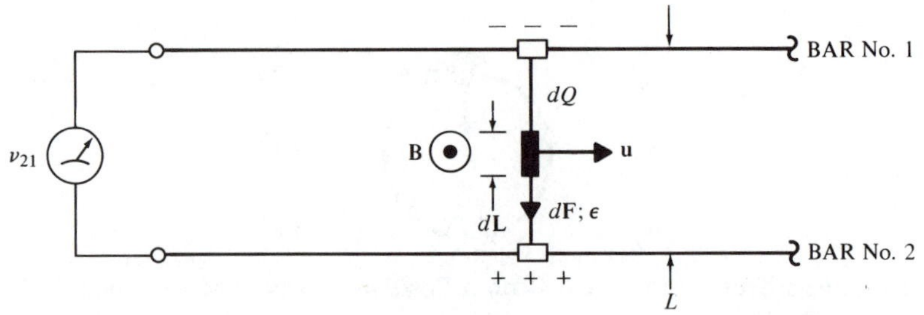

Fig. 2.3 Potential Generated on a Conductor Moving with Constant Speed **u** Within a Field **B**

(newtons/coulomb),

$$\boldsymbol{\varepsilon} \equiv \frac{d\mathbf{F}}{dQ} = \mathbf{u} \times \mathbf{B} \tag{2.9}$$

This potential-producing field causes positive charges to move toward bar No. 2 and negative charges toward bar No. 1. Thus measurement of the *motional potential difference* between bar No. 2 and bar No. 1 with an ideal voltmeter, that is, a meter with infinite internal resistance yields

$$v_{21} \equiv \oint \boldsymbol{\varepsilon} \cdot d\mathbf{L} = \oint (\mathbf{u} \times \mathbf{B}) \cdot d\mathbf{L} \tag{2.10}$$

The potential difference v_{21} provided by Eq. (2.10) is known as a **motional** or **speed** potential because it depends on the velocity **u** with which the moving conductor cuts the field **B**. In general, v_{21} depends on the path of integration, and it is a multivalued function.

Here again the vectors **B**, **u**, $\boldsymbol{\varepsilon}$, and $d\mathbf{L}$ should be expressed in the same coordinate system before integration is initiated. Thus for $\boldsymbol{\varepsilon}$ and $d\mathbf{L}$ expressed in the Cartesian coordinate system, the dot product $\boldsymbol{\varepsilon} \cdot d\mathbf{L}$ is defined as

$$\boldsymbol{\varepsilon} \cdot d\mathbf{L} \equiv \varepsilon_x \, dx + \varepsilon_y \, dy + \varepsilon_z \, dz \equiv |\boldsymbol{\varepsilon}| \quad |d\mathbf{L}| \cos\theta \tag{2.11}$$

where θ is the angle between the vectors $\boldsymbol{\varepsilon}$ and $d\mathbf{L}$

Example 2.2

Given: the configuration shown in Fig. 2.2.1. Assume that (1) friction = 0 N, (2) $R_{\text{BAR}} = 0\ \Omega$, (3) $\mathbf{u} = U\mathbf{a}_y$ meters/second, and (4) $R_{\text{CONDUCTOR}} = 0\ \Omega$. Calculate the motional potential v_{21} between bars No. 2 and No. 1.

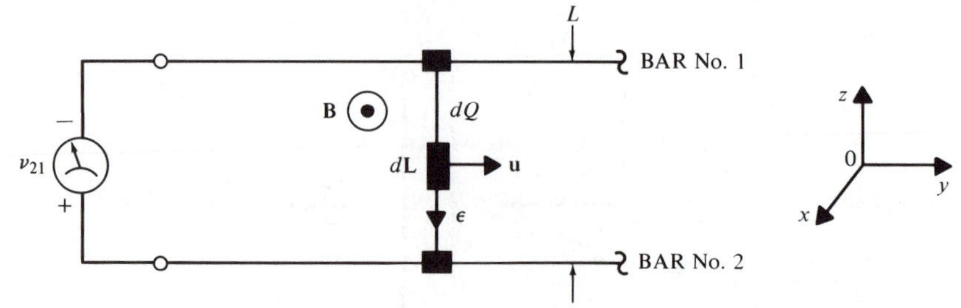

Fig. 2.2.1 Conductor Moving at Constant Speed **u** Within a Constant Magnetic Field **B**

Solution

$$v_{21} = \int_0^L (\mathbf{u}\times\mathbf{B})\cdot d\mathbf{L} = \int_{-L/2}^{L/2} (U\mathbf{a}_y \times |\mathbf{B}|\mathbf{a}_x)\cdot dz(-\mathbf{a}_z)$$
$$= \int_{-L/2}^{L/2} U|\mathbf{B}|(-\mathbf{a}_z)\cdot dz(-\mathbf{a}_z) = \underline{\underline{U|\mathbf{B}|L \textit{ volts}}}$$

Eqs. (2.10) and (2.6) are the key equations in the study of electromechanical energy conversion. Note that in both equations the magnetic field **B** is present. This field can be provided either by a *permanent magnet* (a name that was given to an ore that was discovered by the Greeks near Μαγνησία, (Magnesia) one of their Asian cities) or by an *electromagnet*. If in addition to the magnetic field represented as *field energy* (w_F), *mechanical energy* (w_M) is provided to the system in the form of linear or rotational velocity, Eq. (2.10), *electrical energy* (w_E) is generated by the system in the form of a potential difference, that is, $v_{21} = dw_E/dq$. A device based upon Eq. (2.10) for its operation is termed a **mechanoelectrical energy converter** (MEEC), that is, a **generator of electrical energy**. Its mode of operation is known as the **generator mode of operation**, and its black box representation is given in Fig. 2.4.

On the other hand, if in addition to the *field energy* (w_F), *electrical energy* (w_E) is provided to the system in the form of current, Eq. (2.6), *mechanical energy* (w_{em}) is produced by the system in the form of torque, that is, $d\tau_{em} \equiv R\mathbf{a}_R \times d\mathbf{F}_{em}$. R is the radial distance of the current-carrying conductor, L meters long, from an axis about which the conductor is allowed to rotate. A device based upon Eq. (2.6) for its operation is termed an **electromechanical energy converter** (EMEC), that is, a **motor**. Its mode of operation is known as the **motor mode of operation**, and its black box representation is given in Fig. 2.5.

The cross section of a generalized device with equivalent winding representation and its describing equation in matrix form is exhibited in Table 2.1. The definitions of the entries to the elements of Table 2.1 are as follows: (1) e_j, $j = \text{ds, dr, qr, qs}$ designate externally impressed voltages; (2) L_{jk};$jk \equiv \text{ds, dr, qr, qs}$ designate self-inductances of corresponding coils; (3) R_{jk},$jk \equiv \text{ds, dr, qr, qs}$ designate resistances of corresponding coils; (4) M_{jk},$j \neq k \equiv \text{ds, dr, qr, qs}$ designate mutual inductances between corresponding coils defined by the appropriate subscripts; (5) the product $M_{jk}\omega_R \equiv \mathfrak{L}_{jk}$, $jk = \text{drqr, drqs, qrds, qrdr}$, that is, $\mathfrak{L}_{drqr} \equiv -M_{drqr}\omega_R$, $\mathfrak{L}_{drqs} \equiv -M_{drqs}\omega_R$, $\mathfrak{L}_{qrds} \equiv M_{qrds}\omega_R$, and $\mathfrak{L}_{qrdr} \equiv M_{qrdr}\omega_R$, designate speed-dependent inductances (henrys/second), which are linear functions of the angular velocity of the rotor, ω_R; and (6) i^k, $k = \text{ds, dr, qr, qs}$ designate the response currents. This device is known as *Kron's Primitive Machine* and is usually abbreviated as KPM. It was

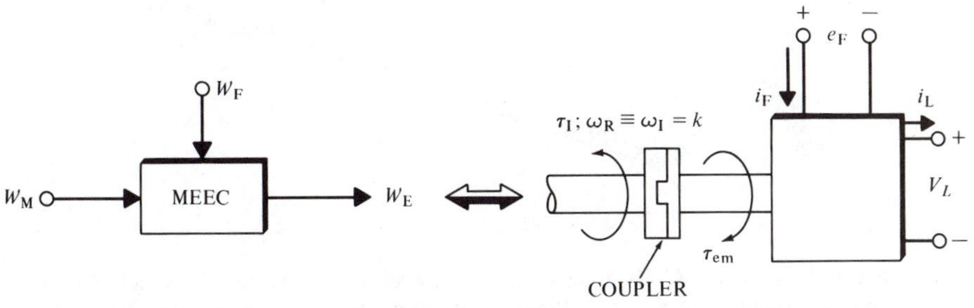

Fig. 2.4 Black Box Representation, Generator Mode of Operation

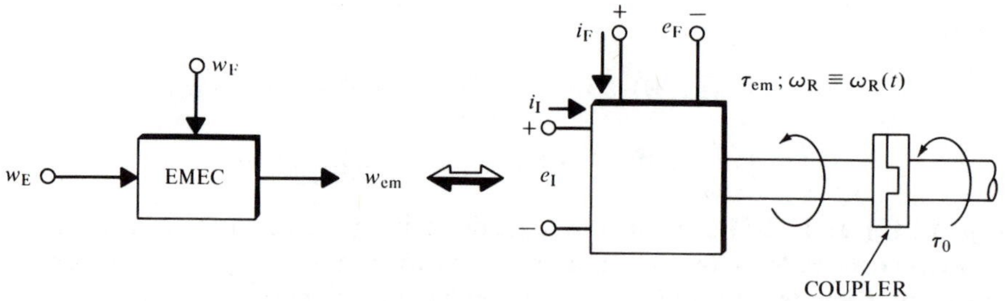

Fig. 2.5 Black Box Representation, Motor Mode of Operation

originally studied by a General Electric researcher named Gabriel Kron. Kron's primitive machine, originally invented in 1935, was more recently designed and tested by the Massachusetts Institute of Technology. In turn, it was revised and manufactured by the Westinghouse Electric Corporation and others for engineering schools to be used as a teaching tool. For a thorough understanding of the function of the device of Table 2.1, when the figure of Table 2.1 is compared to Fig. 2.5, the following analogies should be made: (1) $\{e_{ds}, e_{qs}\} \rightarrow w_F$, (2) $\{e_{dr}, e_{qr}\} \rightarrow w_E$, and (3) $\omega_R \rightarrow w_{em}$.

This device has the following properties (see Table 2.1, Figs. 2.6, 2.11, 2.12, and 2.32 for terminology): (1) it has a *salient* stator (stationary member of the device with protrusions along the *d*- and *q*-axes) with a concentrated main winding on the *d*-axis (*direct axis*) and a concentrated winding on the *q*-axis (*quadrature axis*); (2) it has a *smooth* rotor (rotating member of the device) with a distributed winding; (3) it has two sets of *brushes* always in contact with the rotor windings—one set is located along the *d*-axis and the second set is located $\Delta\theta_R$ radians off the *q*-axis in the positive sense of rotation; (4) *it is linear electrically and magnetically*. (Note that each winding of the stator and rotor has a magnetic field associated with it.) Thus

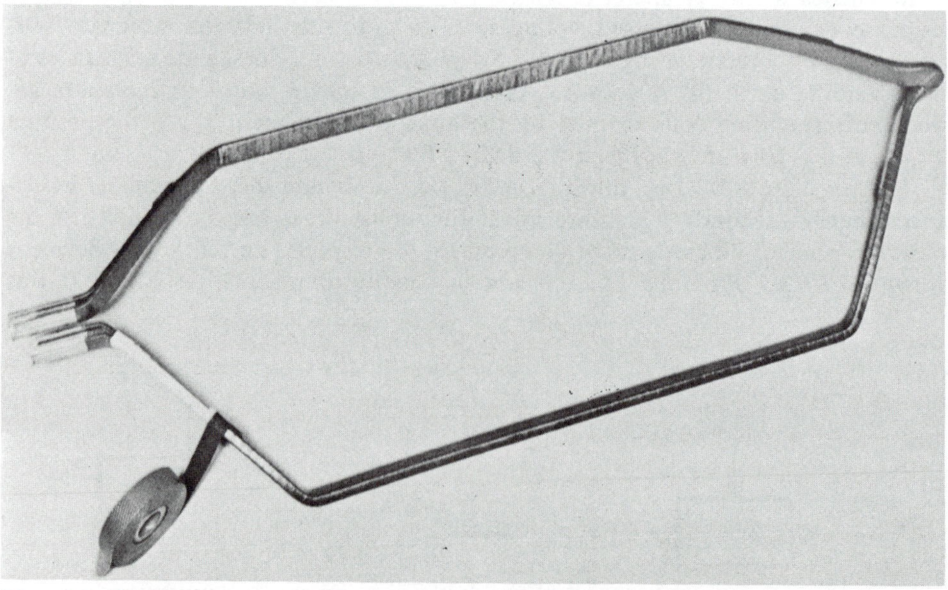

Plate 2.2 Fully Laminated Coils Comprised of Conductors, Each of Which Consists of Several Strands of Copper Wire in Parallel. Both Strands and Conductors Are Insulated with Polyamide Tape to Reduce Eddy Current Losses and Improve Commutation. (Courtesy of Westinghouse Electric Co., Pittsburgh, Pa.)

Table 2.1

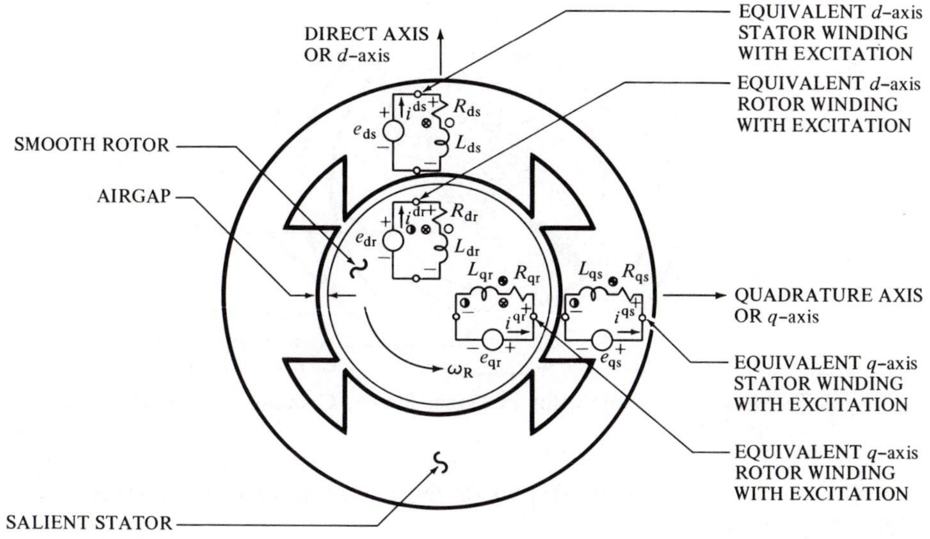

$$
\begin{bmatrix} e_{ds} \\ e_{dr} \\ e_{qr} \\ e_{qs} \end{bmatrix} = \begin{bmatrix} R_{ds} + L_{ds}\frac{d}{dt} & M_{dsdr}\frac{d}{dt} & 0 & 0 \\ M_{drds}\frac{d}{dt} & R_{dr} + L_{dr}\frac{d}{dt} & -M_{drqr}\omega_R & -M_{drqs}\omega_R \\ M_{qrds}\omega_R & M_{qrdr}\omega_R & R_{qr} + L_{qr}\frac{d}{dt} & M_{qrqs}\frac{d}{dt} \\ 0 & 0 & M_{qsqr}\frac{d}{dt} & R_{qs} + L_{qs}\frac{d}{dt} \end{bmatrix} \begin{bmatrix} i^{ds} \\ i^{dr} \\ i^{qr} \\ i^{qs} \end{bmatrix}
$$

the device can be analyzed as (1) a static network through utilization of tensor theory* of circuit analysis and (2) a single pole-pair device (*poles* are protrusions on the stator structure where the main winding is wound; when the field $\mathbf{B}_S$ *exits* the protrusion, it is known as the *north pole*, and when the field $\mathbf{B}_S$ *enters* the protrusion, it is known as the *south pole*), because the magnetic fields are added together to form a single resultant field for both the stator and the rotor.

Structurally, the windings of the stator are *wound on the poles* of the *d*- and *q*-axes so that *concentrated fields* are produced and directed along the *positive sense* of the *d*- and *q*-axes. These windings are represented by the ds- and qs-windings in Table 2.1. The windings of the rotor are *positioned along slots* parallel to the axis of rotation, Fig. 2.6. The continuous contact of the two sets of brushes on different rotor wires as the rotor rotates produces *stationary fields* along the *positive sense* of the *d*- and *q*-axes represented by the dr- and qr-windings, Table 2.1.

The describing equation of KPM, Table 2.1, can be written, using *matrix* notation, in compact form as

$$[\mathbf{E}] = [\mathbf{R}][\mathbf{I}] + [\mathfrak{L}]\frac{d}{dt}[\mathbf{I}] + \omega_R[\mathbf{M}][\mathbf{I}] \tag{2.12}$$

***Tensor theory, a sophisticated extension of matrix theory, will not be used in this text.**

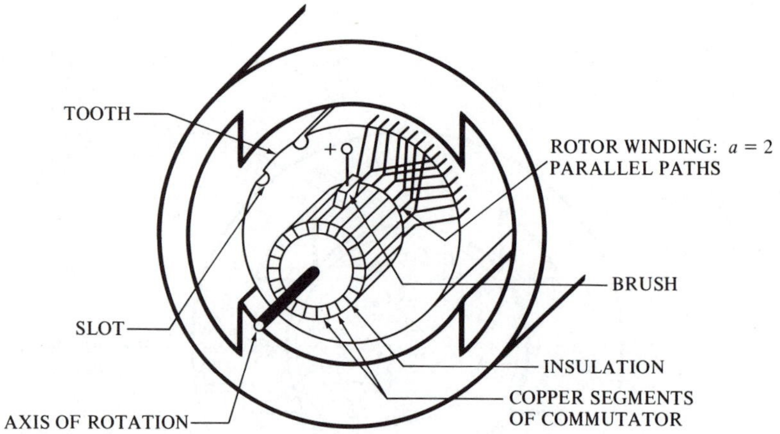

Fig. 2.6 A Perspective View of Rotor in Place: Stator, Commutator, Brushes, and Rotor Structure Details

Multiplication of Eq. (2.12) by $[\mathbf{I}^{\mathrm{T}}]$ yields the power input to the system*, that is, the KPM, given as

$$p_{\text{INPUT}} = [\mathbf{I}^{\mathrm{T}}][\mathbf{E}] = [\mathbf{I}^{\mathrm{T}}][\mathbf{R}][\mathbf{I}] + [\mathbf{I}^{\mathrm{T}}][\mathfrak{L}]\frac{d}{dt}[\mathbf{I}] + \omega_{\mathrm{R}}[\mathbf{I}^{\mathrm{T}}][\mathbf{M}][\mathbf{I}] \quad (2.13)$$

A careful study of Eq. (2.13) reveals the following:

$$p_{\text{INPUT}} = [\mathbf{I}^{\mathrm{T}}][\mathbf{E}] \quad (2.14)$$

$$p_{\text{DISSIPATED}} = [\mathbf{I}^{\mathrm{T}}][\mathbf{R}][\mathbf{I}] \quad (2.15)$$

$$p_{\text{STORED}} = [\mathbf{I}^{\mathrm{T}}][\mathfrak{L}]\frac{d}{dt}[\mathbf{I}] \quad (2.16)$$

$$p_{\text{ELECTROMAGNETIC}} = \omega_{\mathrm{R}}[\mathbf{I}^{\mathrm{T}}][\mathbf{M}][\mathbf{I}] \equiv p_{\mathrm{em}} \quad (2.17)$$

Thus the power *input* to KPM is equal to the sum of (1) the power *dissipated* in the resistances of the coils, (2) the power *stored* in its coils, that is, the time rate of change of the energy stored in the coils, and (3) the electromagnetic power, p_{em}, *developed* by the interaction of the rotor and stator coil fields. The last term, p_{em}, is of great importance. It is defined as $p_{\mathrm{em}} \equiv \tau_{\mathrm{em}}\omega_{\mathrm{R}}$. Comparison of the definition of p_{em} with Eq. (2.17) enables writing the electromagnetic torque, τ_{em} (whose dimensional unit is the newton meter) as

$$\tau_{\mathrm{em}} = [\mathbf{I}^{\mathrm{T}}][\mathbf{M}][\mathbf{I}] \quad (2.18)$$

The quantity τ_{em} takes the following signs: (1) **positive for motor action,** meaning that the KPM *develops* and *delivers* torque, and (2) **negative for generator action,** meaning that the KPM *absorbs* (*needs*) that much power from the *prime mover.*†

*Note that $[\mathbf{I}^{\mathrm{T}}]$ is the transpose of the matrix $[\mathbf{I}]$. The brackets are used here in conjunction with bold face letter notation *to stress* the fact that Eq. (2.12) results from the *extraction* of the appropriate terms from Table 2.1.

† A prime mover is usually a steam or gas *turbine rotating at constant* speed n_{R}, measured in revolutions per minute, abbreviated as RPM, used to drive a generator at the same constant speed. The symbol that designates the speed of the prime mover is ω_{R} and the MKS dimensional unit is the radian per second (rad/s).

Table 2.2

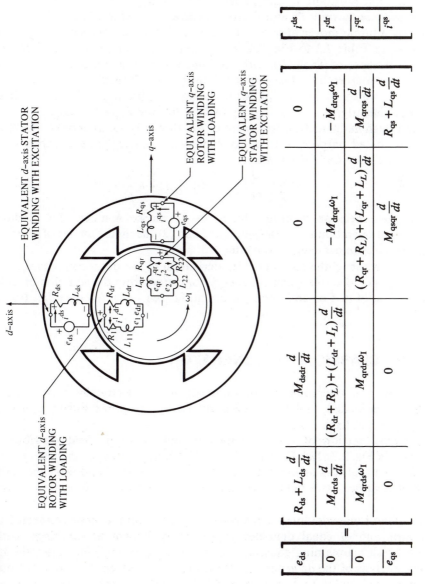

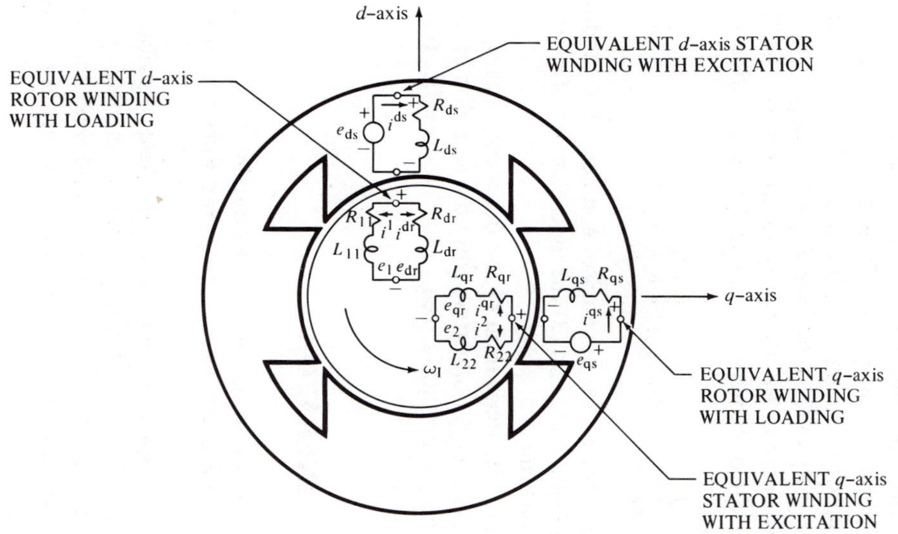

$$\begin{bmatrix} e_{ds} \\ \hline 0 \\ \hline 0 \\ \hline e_{qs} \end{bmatrix} = \begin{bmatrix} R_{ds}+L_{ds}\frac{d}{dt} & M_{dsdr}\frac{d}{dt} & 0 & 0 \\ \hline M_{drds}\frac{d}{dt} & (R_{dr}+R_L)+(L_{dr}+I_L)\frac{d}{dt} & -M_{drqr}\omega_I & -M_{drqs}\omega_I \\ \hline M_{qrds}\omega_I & M_{qrdr}\omega_I & (R_{qr}+R_L)+(L_{qr}+L_L)\frac{d}{dt} & M_{qrqs}\frac{d}{dt} \\ \hline 0 & 0 & M_{qsqr}\frac{d}{dt} & R_{qs}+L_{qs}\frac{d}{dt} \end{bmatrix} \begin{bmatrix} i^{ds} \\ \hline i^{dr} \\ \hline i^{qr} \\ \hline i^{qs} \end{bmatrix}$$

To understand the distinct differences between the motor and the generator modes of operation, please refer to Table 2.1. When all four windings are excited, that is, when the sources e_{ds} and e_{qs} produce w_F and the sources e_{dr} and e_{qr} provide w_E, the device produces τ_{em}, and thus a rotational motion ω_R, viewed as w_{em}.

For the device to function as a generator, the following conditions may be met: (1) the stator windings are excited, that is, the sources e_{ds} and e_{qs} produce w_F; (2) the rotor sources e_{dr} and e_{qr} are replaced by loading coils, that is, series *RL* loads whose response currents i^{dr} and i^{qr} are viewed as w_E, Table 2.2; and (3) the rotor is driven by a prime mover at a constant speed $\omega_R \equiv \omega_I \equiv k$, viewed as w_M. Compare the figure in Table 2.2 to Fig. 2.4. The definitions of the entries of Table 2.2 are identical to those of Table 2.1 with the exception of the entries e_{dr} and e_{qr}. These two voltage sources have been replaced by series *RL* loading coils (see figure in Table 2.2), which cause the following changes. Entry R_{dr} of Table 2.1 becomes $R_{dr}+R_L$. Similarly, $L_{dr}\rightarrow L_{dr}+L_L$, $R_{qr}\rightarrow R_{qr}+R_L$, and $L_{qr}\rightarrow L_{qr}+L_L$.

A thorough mathematical treatment of this device is not undertaken here, as this is not the stated purpose of this text. However, the describing equation of Table 2.1, which can be derived from

$$e_j = \sum_k \left[\frac{d}{dt}(M_{jk} i^k) + R_{jk} i^k \right] + \frac{d}{dt} L_k i^k \tag{2.19}$$

is used here because it brings forth, in a concise manner, the fundamental principles of operation of (1) the generator and (2) the motor modes of operation of an electromechanical energy converter.

Use of proper mathematics enables derivation of the describing equations of all known types of electromechanical energy converters from the equation of Table 2.1 or extensions of it. For additional information, see the references [8, 10, 11, 12, and 13].

2.3 THE CLASSICAL dc GENERATOR

In this mode of operation the stator windings are electrically excited, the rotor windings are loaded, and the rotor is driven by the prime mover at a constant speed $\omega_R \equiv \omega_i$ radians per second.

The cross section of the dc generator and its stator field distribution are exhibited in Fig. 2.7. *This field distribution is achieved by proper shaping of the poles.* The $\mathbf{B}_S$ field is always assumed to *exit* the north (*N*) poles and *enter* the south (*S*) poles of a device. The field direction from stator to rotor is always considered *positive*.

Isolation of one wire loop of the rotor, Fig. 2.8* (with its ends connected to two concentric and of equal diameter copper rings, known as **slip rings**, with one carbon brush maintaining constant contact with each ring), facilitates the mathematical analysis of the dc generator. If the axis about which the wire loop rotates is placed perpendicular to a magnetic field $\mathbf{B}_S = B_0\, \mathbf{a}_y$, the analysis proceeds as follows. Use of Eq. (2.10) enables us to write the **motional potential difference across the wire loop** as

$$v_m = \oint (\mathbf{u} \times \mathbf{B}_S) \cdot d\mathbf{L} \tag{2.20}$$

*Note that for convenience Fig. 2.8 is not drawn in a one-to-one correspondence with Fig. 2.7.

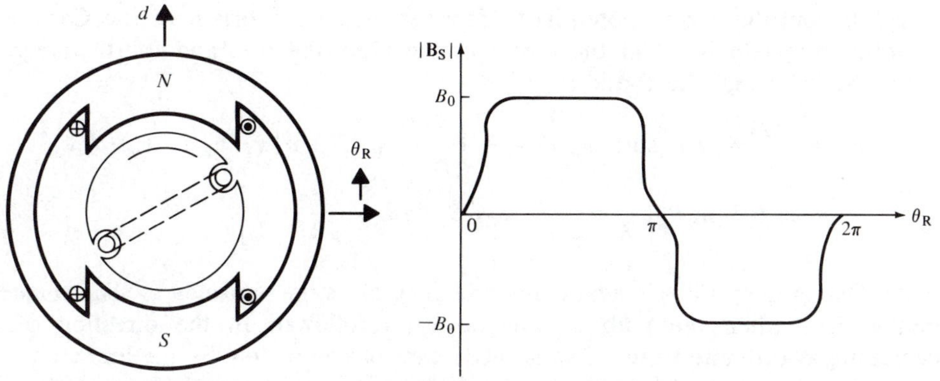

Fig. 2.7 The Cross Section of the dc Energy Converter and Its Stator Field $|\mathbf{B}_S|$ Distribution

For a clockwise rotation of the loop, Fig. 2.8, and $\omega_R = \omega_I = k$,

$$\mathbf{u} = \omega_R R \mathbf{a}_u \tag{2.21}$$

Substitution of Eq. (2.21) into Eq. (2.20) yields

$$v_m = \oint (\omega_R R \mathbf{a}_u \times B_0 \mathbf{a}_y) \cdot d\mathbf{L} = \oint (\omega_R R B_0 \sin\theta_R \mathbf{a}_L) \cdot d\mathbf{L}\mathbf{a}_L \tag{2.22}$$

Referring to Fig. 2.8 and taking the closed-line integral 123456 yields

$$v_m = \oint \boldsymbol{\varepsilon} \cdot d\mathbf{L} = \int_4^5 + \int_5^6 + \int_1^2 + \int_2^3 + \int_3^4 \tag{2.23}$$

Since the resulting potential-producing field intensity $\boldsymbol{\varepsilon}$ and the path of integration are normal to each other over the paths 12, 34, and 56, the second, third, and fifth

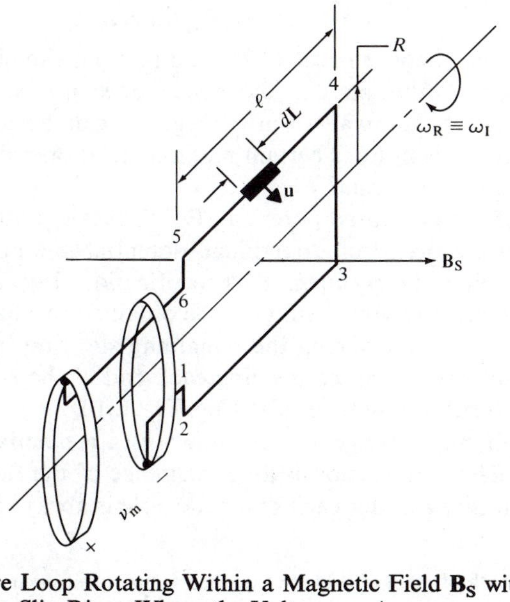

Fig. 2.8 Isolated Wire Loop Rotating Within a Magnetic Field $\mathbf{B}_S$ with Ends Permanently Connected to Slip Rings Where the Voltage v_m Appears

integrals contribute zero potential.* However, when the origin of the Cartesian coordinate system is set at the center of the loop, the first and fourth integrals make the following contribution:

$$v_m = \int_{-\ell/2}^{\ell/2} \omega_R R B_0 \sin\theta_R \mathbf{a}_y \cdot dy\, \mathbf{a}_y + \int_{\ell/2}^{-\ell/2} \omega_R R B_0 \sin(\pi - \theta_R)(-\mathbf{a}_y)\cdot dy\, \mathbf{a}_y$$
$$= \omega_R R B_0 \sin\theta_R\left[\frac{\ell}{2}+\frac{\ell}{2}\right] - \omega_R R B_0 \sin\theta_R\left[-\frac{\ell}{2}-\frac{\ell}{2}\right] \tag{2.24}$$

Note: One of the pitfalls in evaluating line integrals is the tendency to use *too many* minus signs when the path of integration is followed in the direction of a decreasing coordinate value. This is taken care of completely by the limits on the integral, and no misguided attempt should be made to change the sign of $d\mathbf{L}$ [6].

Thus the motional potential difference (*speed potential*) that appears across the slip rings due to a one-turn loop is

$$v_m = 2R\ell\omega_R B_0 \sin\theta_R \tag{2.25}$$

If the one-turn loop of Fig. 2.8 is replaced by a *coil with N turns*, and if θ_R is expressed in terms of ω_R (i.e., $\theta_R = \omega_R t$), Eq. (2.25) can be written as

$$v_C = N2R\ell\omega_R B_0 \sin\omega_R t \tag{2.26}$$

The speed potential, Eq. (2.26), is sinusoidal in nature. It should be noted that if the field were to enter the poles normally and if the field were distributed sinusoidally in the airgap of the device, Eq. (2.26) would be written (since $\mathbf{u}\times\mathbf{B} \equiv |\mathbf{u}|\,|\mathbf{B}|\,\mathbf{a}_L$ and $\mathbf{u} = \omega_R R \mathbf{a}_u$) as

$$v_C = N2R\ell\omega_R|\mathbf{B}_S| \tag{2.27}$$

For the sinusoidally distributed field, that is,

$$\mathbf{B}_S = B_M \sin\theta_R \mathbf{a}_B \tag{2.28}$$

Eq. (2.27) could be written as

$$v_C = N2R\ell\omega_R B_M \sin\omega_R t \tag{2.29}$$

Note that the difference between Eqs. (2.26) and (2.29) is the quantity B, that is, B_0 and B_M, respectively. Although this difference does not seem significant (and indeed it is not, because the waveform of Fig. 2.7 can be approximated with a sinusoidal waveform), it simplifies certain mathematical operations, as will be seen shortly, and thus its use is attractive.

The waveform of the generated potential, Eq. (2.29), is plotted in Fig. 2.9. Since it is desired to obtain a dc, or unidirectional, potential, the potential waveform of Fig. 2.9 should somehow be *rectified*. This rectification, Fig. 2.9, can be achieved electronically. However, the same result can be obtained mechanically by removing one of the slip rings, Fig. 2.8, slicing the remaining slip ring in half, insulating the two halves from each other, and connecting each end of the N-turn coil (wire loop if there is only one turn) to a semicircular ring, Fig. 2.10.

The resulting half-ring arrangement is known as a **commutator**, or a mechanical switch or rectifier. The rectification is a consequence of the fact that the potential-producing field intensity $\boldsymbol{\varepsilon}$ under each magnetic pole is always in the same direction

*In practice the axial depth of the $\mathbf{B}_S$ field is shorter than ℓ, that is, segments 12, 34, and 56 are not in the $\mathbf{B}_S$ field environment.

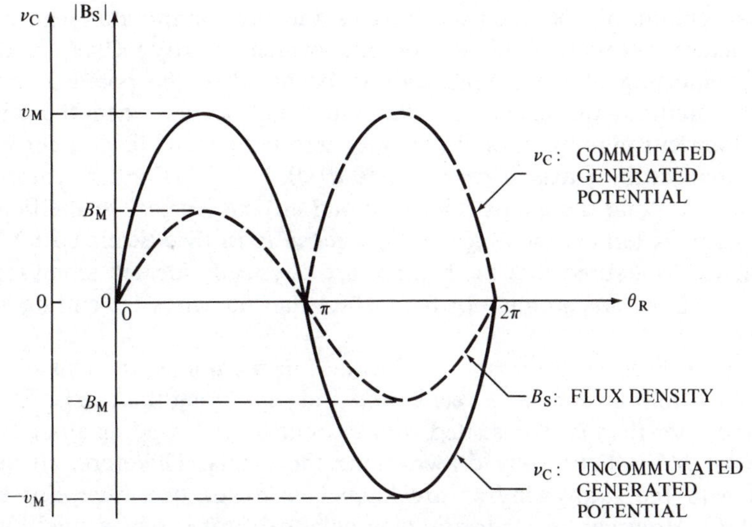

Fig. 2.9 Stator Field and Generated Potential of a Classical dc Generator

regardless of the side of the loop that is cutting the field of the given pole. Thus, as a coil side changes position from pole to pole, the current through it changes direction by 180°. Further discussion is deferred until the structure of the commutator is completely developed.

The motional potential difference v_C is made to appear on a pair of brushes made of hard-wearing carbon material, which are in constant contact with the commutator. However, as the commutator rotates the brushes short the two half-rings for a small instant of time if the spacing between the two half-rings is less than the width of the brushes. This shorting causes arcing, which might lead to erosion of the commutator and metallic bridging of the insulating gap between segments, thus causing short circuits between several commutator segments. This

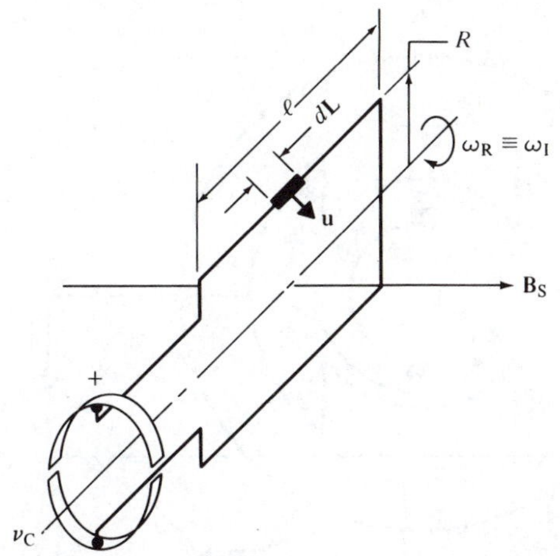

Fig. 2.10 Isolated N-turn Coil Rotating Within a Magnetic Field with Ends Permanently Connected to a Commutator, That Is, a Mechanical Rectifier; Compare to Fig. 2.8

dimensional relationship between the brushes and the commutator segment opening is a structural necessity. (This will become evident shortly.) Thus it is desirable to force the shorting of the commutator to occur when the potential difference between the commutator halves is minimum. This occurs when the wires that define the loop cut minimum, or if possible, zero field. Zero field is cut when the two wires intersect the q-axis at $q = \pm R$ and $d = 0$, Fig. 2.7. For zero rotor current, there is no field along the q-axis, that is, along an axis normal to the field of the stator. This axis is termed the *magnetically neutral axis*. (See Section 2.8.) Thus by proper design it is assured that the brushes are physically located along the d-axis, Fig. 2.11, and short the commutator the instant that the wires are cutting *minimum* or zero field.

Although the brushes are physically located on the d-axis, the axis of the field due to all the rotor currents under each pole is always along the q-axis. Thus in the literature the rotor field is represented with a concentrated winding along the q-axis and the brushes are fictitiously drawn along the q-axis. Therefore, all quantities associated with the rotor winding are known as q-axis quantities, for example, $R_q, L_q, i_q(t), E_q$. However, in schematic diagrams the rotor is represented by a circle with an M inside representing motor, or a G representing generator (Figs. 2.29 and 2.44) with the brushes on the q-axis. While in the actual devices the brushes are **always mounted** on the d-axis, in schematic diagrams they are drawn so that they **appear to be** mounted on the q-axis.

The d-axis position of the brushes also insures a maximum average potential appearing across the brushes. This average potential is calculated from Eq. (2.29), recognizing that the period of a sinusoidal waveform is $T = 2\pi$, as follows:

$$v_{AVG} = \frac{1}{T/2}\int_0^{T/2} v_C\,dt = \frac{1}{\pi}\int_0^{\pi} 2R\ell\omega_R N B_M \sin\theta_R\,d\theta_R = \frac{2}{\pi}(2R\ell)NB_M\omega_R \tag{2.30}$$

The potential v_{AVG} of Eq. (2.30) can be looked upon as the strength of a voltage

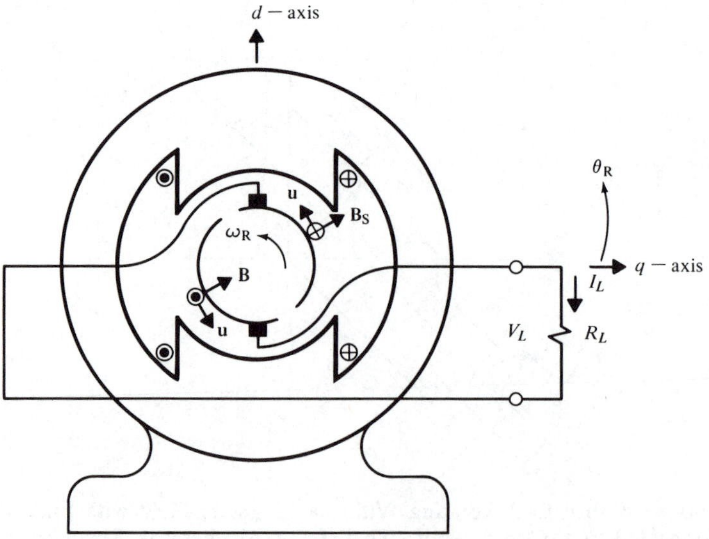

Fig. 2.11 Commutator, Wire Loop, and Brush Pair Arrangement for a dc Generator

source representing one coil. To make the induced voltage as large as possible and as constant as possible, the commutator segments are divided into many small segments. Also, an equal (to the segment number) number of coils, with the same number of turns each, is connected in series. The ends of each coil are connected to **consecutive commutator segments**, Figs. 2.12(a), 2.12(b), and 2.12(c), and each coil spans one *pole pitch*,* that is, π radians. Usually two *layers* of coils are allowed in the rotor slots. The coils are placed in such a manner that one side occupies the lower position of the slot and the other side the upper position of the slot. This winding layout is known as *lap winding*. (There are other types of windings in use.) Note that the interconnecting wires in Fig. 2.12 between commutator segments and coil sides are all of equal length in a real device. They are drawn unequal here for simplification purposes only. The entire number of N-turn coils is divided into two parallel paths, Fig. 2.12(a), by the two brushes, which are located along the d-axis with the stated purpose to insure (1) maximum generated voltage and (2) minimum sparking across consecutive commutator segments.

As the rotor turns counterclockwise, the brushes,† Fig. 2.12(a), shift position from complete contact with one commutator segment to complete contact with the commutator segment that follows, and so on.

When the brush P, Fig. 2.12(b), is in complete contact with a commutator segment, say No. 1, the brush receives a total current $I_B = a(I_B/a)$ from (1) the coil that connects commutator segment 0 to 1 and (2) the coil that connects commutator segment 2 to 1. These currents are designated as $I_{0\text{ to }1}$ and $I_{2\text{ to }1}$ and represent the current I_C that flows through each coil. Its maximum value is equal to the brush current divided by the number of parallel paths as seen by each brush. However, when the brush is in full contact with commutator segment No. 2, Fig. 2.12(c), the same brush receives currents from (1) the coil that connects commutator segment 1 to 2 and (2) the coil that connects commutator segment 3 to 2. These currents are designated as $I_{1\text{ to }2}$ and $I_{3\text{ to }2}$. Note that the direction of the current of the coil connecting commutator segments 1 and 2 to the brush P changes its direction from $I_{2\text{ to }1}$ to $I_{1\text{ to }2}$ as the brush P moves from commutator segment No. 1 to commutator segment No. 2. This reversal takes place in τ_{BW} (BW≡brush width) seconds, Fig. 2.12(d). If the commutation depends **only** on the resistance R_B of the brushes, the current in the coil undergoing commutation changes linearly during the commutation period τ_{BW}, Fig. 2.12(d), curve 1. Note that the coil current I_C is exactly zero when the brush is in a position such that the currents $I_{2\text{ to }1}$ and $I_{1\text{ to }2}$ are equal in magnitude but opposite in direction. This happens when the consecutive commutator segments No. 1 and No. 2 are shorted by the brush P. The controlling parameters are (1) the generated emf in the coil undergoing commutation, (2) the coil parameters R_C and L_C, and (3) the brush resistance $R_B = \rho_B L_B / A_B$. Now the brush current I_B is comprised exclusively of currents $I_{0\text{ to }1}$ and $I_{3\text{ to }2}$. Since each coil undergoing commutation has a self inductance L_C, it opposes the attempted change in the direction of the coil current I_C. This delay, Fig. 2.12(d), curve 2, is known as *undercommutation*. Undercommutation can also be a result of the speed voltage generated in the coil undergoing commutation due to the mmf of the rotor winding, or it can be a result of the combination of both

*One pole pitch is the angular displacement between two opposite poles.

† The brushes in the physical device rest on the outside of the commutator segment; however, they appear in the inside of the commutator in Fig. 2.12(a) so that the coil connections can be shown clearly.

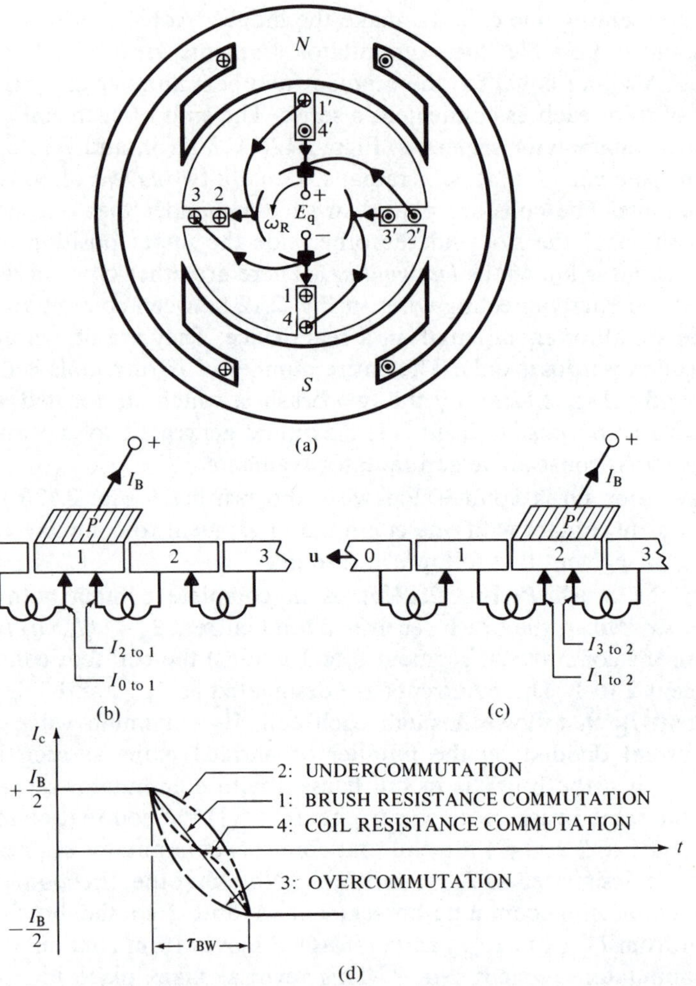

Fig. 2.12 Structural Details of a dc Energy Converter: (a) Stator Poles and Winding, Rotor Winding, Commutator Outlay, and Brushes; (b) and (c) Commutator Segment-Rotor Coil-Brush Relationships; and (d) Current vs. Time Relationship of Rotor Coil Undergoing Commutation

these effects, that is, speed and self-inductance induced voltages. Since compensation is possible (Section 2.8) for the self-inductance induced voltage and speed voltages in the coil undergoing commutation, the total induced voltage in that coil can be made to equal twice the brush voltage drop, $2R_BL_C$, when the coil resistance R_C is neglected. This overcompensation causes what is known as *overcommutation*. Thus the change in current direction may be faster than in any of the previous cases, Fig. 2.12(d), curve 3. This is the *most desirable* type of commutation, as the current reverses direction before the commutator has left the brush. Thus **sparking can be virtually eliminated by overcommutating.** When the resistance R_C of the coil undergoing commutation is not neglected, its effects on commutation are more severe than one would think. These effects are described by Fig. 2.12(d), curve 4.

Returning to Fig. 2.8 and Eq. (2.24), it can be seen that two motional potential differences are developed: one in segment 2–3, the other in segment 4–5. The total potential difference v_m developed is equal to the series connection of these *two* voltages. Each segment is referred to as a loop or coil side. Similarly, if there is more than one turn, the loop is considered as a *coil*, designated by Z. Each coil is composed of two *coil sides* designated by $\mathfrak{Z}$, and for an N-turn coil, each coil side consists of N conductors. It may also be convenient to *think of a coil side of an N-turn coil as an N-conductor bundle*. In Fig. 2.12 there are a total of *four coils*, that is, $Z=4$, on the rotor. Since there are four coils, the total number of *coil sides* is eight, that is, $\mathfrak{Z}=8$.

Referring to Fig. 2.12(a), for $\mathfrak{Z}=8$ N-turn coil sides uniformly distributed in the rotor slots, the simplified equivalent source representation of the coils at a moment when the brushes are in full contact with two commutator segments π radians apart is exhibited in Fig. 2.13. Note that Fig. 2.13(a) represents the rotor in motion.

It is important that Figs. 2.12(a) and 2.13 be understood thoroughly. Note that current converges on the plus brush from two different directions—from coil side 2′ and coil side 4′. Further investigation reveals that coil side 2′ is associated with a path that consists of two coils: 2′–2 and 1′–1. These coils are represented by small series-connected voltage sources of strength v_{Ci}, $\underline{i = 1,2,3,4}$, in Fig. 2.13(a). Similarly, coil side 4′ is associated with a path that consists of two coils: 4′–4 and 3′–3. These coils are also denoted by small series-connected voltage sources of strength v_{Ci} in Fig. 2.13(a). It should be understood from Fig. 2.12(a) [i.e., Fig. 2.13(a)] that coils No. 1 and No. 2 are in series, as are coils No. 3 and No. 4. (Note that coil No. 1 consists of coil sides 1 and 1′, which are connected in series "behind the page.") The brushes "see" the parallel combination of the two series connections. In the orientation shown the voltage developed across coils No. 1 and No. 4 is the

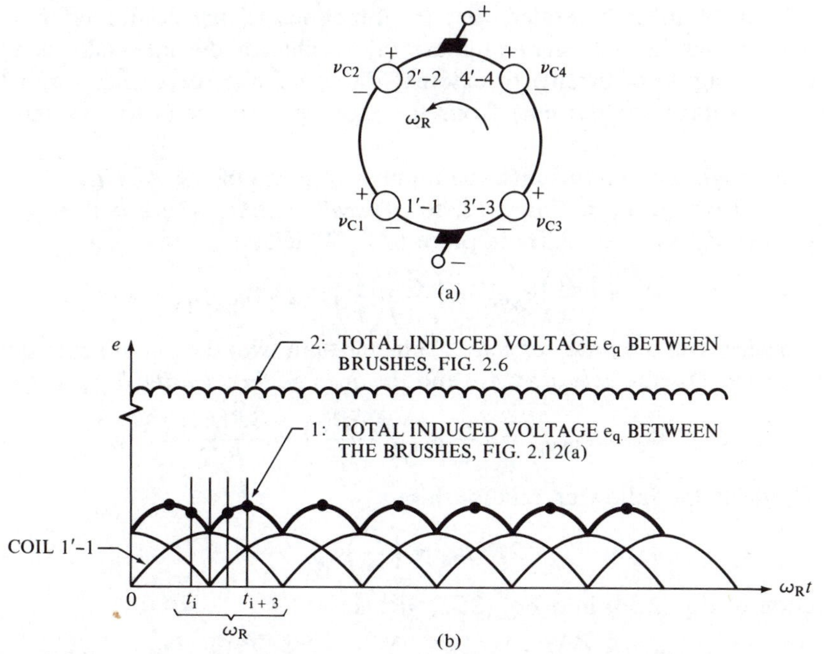

Fig. 2.13 Generation of dc Voltage: (a) Equivalent Source Representation of the Rotor Coils and (b) Evolution of the Total Induced Voltage as a Function of Time

maximum value and the voltage developed across coils No. 2 and No. 3 is the minimum value. After the rotor rotates 90° counterclockwise, the brushes now see the following coil configuration. Coils No. 1 and No. 3 are in series, as are coils No. 2 and No. 4. The brushes see the two series connections in parallel. In this new orientation coils No. 2 and No. 3 are developing maximum voltage, and coils No. 1 and No. 4 are generating miminum voltage.

A partial time history of these coil voltages for various positions of the rotor relative to the field distribution, starting with coil 1′–1, is exhibited in Fig. 2.13(b). Note that each coil voltage over any full cycle appears as a *rectified* wave because of the action of the commutator. The total induced emf E_q appearing between the brushes at any t_i is the sum of the instantaneous coil voltages. Summation at $t_i, \ldots, t_{i+3}$, and at every 90° thereafter, is shown, Fig. 2.13(b), curve 1, to establish the shape of the total emf E_q, between the brushes, Fig. 2.12(a).

Inspection of Fig. 2.13(b) makes it obvious that an almost constant and high voltage can appear between the brushes if each commutator segment is divided into two parts indefinitely, the two parts are maintained properly insulated, and the number of coils is increased correspondingly. The limit of the number of commutator segments to be used in each case is strictly a design problem. Fig. 2.6 gives a perspective view of such a commutator and brush details. Fig. 2.13(b), curve 2, exhibits a total induced voltage that can be expected from a commutator such as the one exhibited in Fig. 2.6.

Mathematically, the total potential that appears across the brushes is

$$E_q = \left(\frac{Z}{2}\right)v_{AVG} = \left(\frac{Z}{2}\right)\left(\frac{2}{\pi}\right)N(2R\ell)\omega_R B_M \tag{2.31}$$

In practical machines the ripple, Fig. 2.13(b), curve 2, is hardly detectable, even with sensitive instruments, and in the limit $e_q \rightarrow E_q$.

In addition, it must be stated that the functions of the commutator and its associated brushes in a dc generator are (1) to change the internally generated alternating voltage (or current) to external direct voltage (or current) and (2) to transfer the voltage (or current) from the moving rotor coils to the stationary brushes.

To obtain high-density currents, the number of paths of the Z *N-turn coils* may be increased from 2, as in Fig. 2.12, to *a parallel paths*, where a is *even*. Thus substitution of Z/a in Eq. (2.31) in place of $Z/2$ yields

$$E_q = \left(\frac{Z}{a}\right)v_{AVG} = \left(\frac{Z}{a}\right)\left(\frac{2}{\pi}\right)N(2R\ell)\omega_R B_M \tag{2.32}$$

For a device with a *number of poles P* greater than two, the relationship between the *flux per pole* Φ_P, the *pole area* A_P, and the *average flux density* B_{AVG} is given by

$$\Phi_P = B_{AVG} = \left(\frac{2}{\pi}B_M\right)\left(\frac{2\pi B\ell}{P}\right) = \frac{2R\ell B_M}{P/2} \tag{2.33}$$

Eq. (2.33) yields the following relationship:

$$2R\ell B_M = \left(\frac{P}{2}\right)\Phi_P \tag{2.34}$$

Substitution of Eq. (2.34) into Eq. (2.32) yields

$$E_q = \left(\frac{Z}{a}\right)\left(\frac{2}{\pi}\right)N\left(\frac{P}{2}\Phi_P\right)\omega_R = \left(\frac{ZNP}{a\pi}\right)\Phi_P\omega_R \tag{2.35}$$

Depending on our interest in this equation, it may appear in either of the following two forms:

$$E_q = \mathcal{K}\Phi_p\omega_R = K_V\omega_R \tag{2.36}$$

In Eq. (2.36)* $\mathcal{K}$ is the *design constant*, and K_V is the *voltage constant* of interest mainly to the *control engineer*. Note that

$$\mathcal{K} = \frac{ZNP}{a\pi} = \frac{\mathcal{Z}NP}{2a\pi} \tag{2.37}$$

and

$$K_V = \mathcal{K}\Phi_P \tag{2.38}$$

Where

$Z \equiv$ number of coils (i.e., $Z = \mathcal{Z}/2$; where $\mathcal{Z}$ = number of coil sides uniformly distributed in the rotor surface
$N \equiv$ number of conductors per coil side (bundle)
$P \equiv$ number of poles in the stator
$a \equiv$ number of parallel paths that the brushes see
$\pi \equiv$ radial displacement of brushes for maximum voltage generation
$\Phi_P \equiv$ flux per stator pole
$\mathcal{K} \equiv$ design constant
$\mathcal{K}_F \equiv$ field constant
$\mathcal{M} \equiv$ design coefficient
$K_V \equiv$ voltage constant
$K_E \equiv$ excitation constant
$\omega_R \equiv$ radian velocity of the rotor
$E_q \equiv$ brush voltage or generated voltage or generated emf

Eq. (2.36) gives the mathematical relationship between the generated voltage E_q, the design constant $\mathcal{K}$, the flux per pole Φ_P, and the angular speed of the rotor ω_R. It should be remembered that the flux per pole depends on the reluctance of the magnetic circuit, which consists of linear parts (the airgap) and nonlinear parts (the

*It is known, Eq. (1.9), that $\Phi_P = N_F i_F(t)/\mathcal{R}$, where $\mathcal{R}$ is the reluctance of the path of the flux Φ_P. Thus Eq. (2.36) can be written as

$$E_q = \mathcal{K}\left(\frac{N_F i_F(t)}{\mathcal{R}}\right)\omega_R \equiv \mathcal{K}\mathcal{K}_F\omega_R i_F(t) \equiv K_V\omega_R \tag{2.36a}$$

where $\mathcal{K}_F = N_F/\mathcal{R}$ is defined as the *field constant*. However, defining $\mathcal{K}\mathcal{K}_F\omega_R$ the *excitation constant* as K_E, that is,

$$\mathcal{K}\mathcal{K}_F\omega_R \equiv K_E \tag{2.36b}$$

enables writing $E_q = \psi[i_F(t)]$, that is,

$$E_q = K_E i_F(t) \equiv K_E i_F \tag{2.36c}$$

Note that Eq. (2.36c) describes the curves of Fig. 2.14(c), that is, the linearized curves of Fig. 2.14(b). Thus Eq. (2.36c) is a very powerful equation in generator analysis.

If $\mathcal{K}\mathcal{K}_F$ in Eq. (2.36a) is defined as the *design coefficient* $\mathcal{M}$, that is,

$$\mathcal{K}\mathcal{K}_F \equiv \mathcal{M} \tag{2.36d}$$

then the equation for the generated emf (voltage) E_q can take any one of the following forms (depending on one's interest):

$$E_q = \mathcal{K}\Phi_P\omega_R = K_V\omega_R = \mathcal{M}\omega_R i_F(t) = K_E i_F \tag{2.36e}$$

frame, poles, and the core of the rotor). The effects of these linearities or nonlinearities on the generated flux per pole, Φ_P, and the generated potential, E_q, are exhibited in Figs. 2.14(a) through 2.14(d). Since $\Phi_P = N_F i_F/\mathcal{R}$, the Φ_P versus i_F characteristic, Fig. 2.14(a), is *nonlinear* and *double-valued* due to hysteresis. Unless the poles are excited for the first time, there is always a residual magnetic field in the structure represented by point Φ_R in the characteristic, Fig. 2.14(a). For a mathematical analysis the *mean* of the two curves should be used.

The hysteresis that creates the double-valued effect of Fig. 2.14(a) has similar effects in the E_q versus i_F characteristics. Known as the no-load magnetization curves, Figs. 2.14(b) and 2.14(c) are taken with the rotor circuit open, with the field circuit excited with a voltage $e_F \rightarrow E_F$, which produces a current $i_F \rightarrow I_F$, and with the rotor velocity $\omega_{Rj} \equiv \omega_{Ij} (j = 1, 2, \ldots)$ as the running parameter. Only the means of the double-valued curves are drawn here for the various ω_R's. The effects of the residual magnetization are indicated by the point E_R in Fig. 2.14(b). However, if the structure is assumed to be linear, the E_q versus i_F characteristic becomes that of Fig. 2.14(c). Finally, if the generated emf is looked upon as a function of ω_R with Φ_P the running parameter, the results are as exhibited in Fig. 2.14(d).

Note that the device under investigation, Figs. 2.11 and 2.12, has two distinct windings. The stator winding is excited from a distinct external source of dc-type electrical energy $e_F(t)$, and the rotor winding across the brushes (in reality this winding is made up of different coils connected across the brushes from instant to

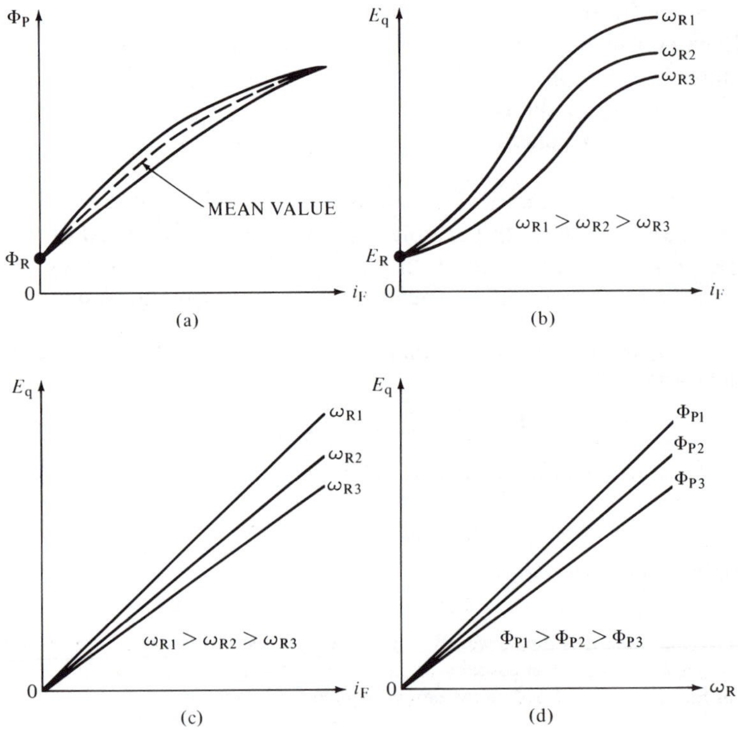

Fig. 2.14 (a) Pole Flux vs. Field Current, (b) and (c) Generated Voltage vs. Field Current, and (d) Generated Voltage vs. Mechanical Speed

instant as the rotor rotates) is loaded with a distinct external load R_L, Fig. 2.11. Such a device is defined as a **separately excited dc generator.**

By accounting for the inductance and resistance of the stator and rotor windings, that is, L_F and R_F, and L_q and R_q, the generated emf, $E_q(\omega_I)$, the stator winding excitation emf, $e_F \rightarrow E_F$, and the load, the equivalent circuit for the *generator mode of operation* of the separately excited dc energy converter can be drawn as in Fig. 2.15. **Note the direction of the current $i_q(t)$ and the polarity of the voltage source E_q.** It should be noted that the subscript *F*, for example, R_F and L_F, designates field quantities, while the subscript q, for example, R_q and L_q, designates rotor quantities. (q stands for *quadrature* axis and represents the fictitious brush position in the magnetically neutral axis.) Many times a subscript a will be encountered in the literature instead of q. The subscript a designates *armature*, that is, the winding in which the potential E_q is induced.

The dynamic equations that describe the equivalent circuit, Fig. 2.15, and thus the **generator mode of operation** of the separately excited dc electromechanical energy converter are

$$e_F(t) = R_F i_F(t) + L_F + L_F \frac{d}{dt}[i_F(t)] \tag{2.39}$$

$$E_q = R_q i_q(t) + L_q \frac{d}{dt}[i_q(t)] + v_L(t) \tag{2.40}$$

where $e_F(t)$ represents the dc forcing function of the field winding externally impressed and E_q is defined in Eq. (2.36e) as $E_q = K_E i_F(t)$. An electromagnetic torque τ_{em} is developed by the interaction of the fields of the stator (field) and rotor (armature) windings. However, since the current $i_F(t)$, which is *impressed* on the field winding, and the current $i_q(t)$, which is the *response* current in the armature winding, are opposite to each other in direction, electromagnetic torque τ_{em} developed by the generator is *negative* in sign when compared to the torque developed when the device functions as a motor. The torque τ_{em} developed by the

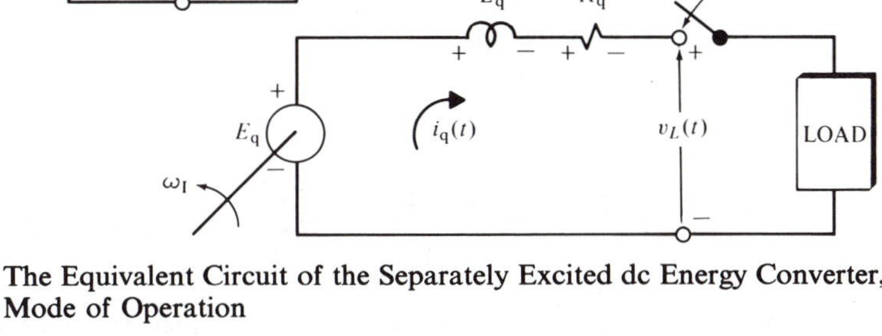

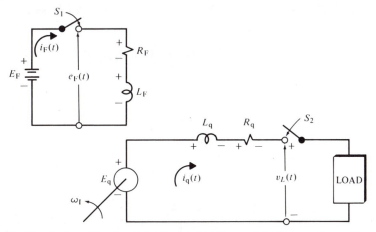

Fig. 2.15 The Equivalent Circuit of the Separately Excited dc Energy Converter, Generator Mode of Operation

generator, opposing the externally applied torque by the prime mover τ_I, is defined as*

$$\tau_{em} = -|\mathfrak{M} i_F(t) i_q(t)| \tag{2.41}$$

It should be emphasized that Eq. (2.41) is just as essential as Eqs. (2.39) and (2.40) in describing the separately excited generator. It provides the minimum prime mover requirements (τ_I) for preset voltage or current output requirements by a given generator.

In addition, it should be stated that Eqs. (2.39), (2.40), and (2.41) describe the dynamic performance of the generalized, separately excited dc generator. Note that Eqs. (2.39) and (2.40) are coupled by $E_q = K_E i_F(t)$. These equations, in addition to the means they provide for analytical techniques of analysis, can be used for (1) computer simulation by utilizing transfer function, block diagram, or simulation diagram techniques and (2) providing, upon proper simplification (Section 2.9), the describing equations for steady state studies of the separately excited generator.

Finally, it should be emphasized that the above-mentioned equations are the key equations for the analysis of the entire class of dc generators to be presented in this text. Every other dc type of generator to be studied here is properly derived from the separately excited dc generator.

Example 2.3

Assume that the switch S_1 in Fig. 2.15 has been closed for some time (while the switch S_2 is open) and the field circuit has reached steady state, that is, $L_F d[i_F(t)]/dt \to 0$. At $t=0$ s, close the switch S_2 and calculate (1) the response current $i_q(t)$ and (2) the developed electromagnetic torque τ_{em}. What can you say about the turning effect of the torque? Data: $E_F=100$, $R_F=20$, $L_q=1$, $R_q=0.5$, $R_L=9.5$, all MKS units, and for a rated speed of 1800 rpm, the no-load, linearized, magnetization curve, that is, E_q versus i_F, yields $K_E=100$ V/A.

Solution

The performance of the generator for $L_F\, d[i_F(t)]/dt \to 0$ is given, by utilizing Eqs. (2.39), (2.40), (2.36c), and (2.41), and letting $e_F \to E_F$ and $i_F \to I_F$, as

$$E_F = R_F I_F \tag{1}$$

$$E_q = L_q \frac{d}{dt}[i_q(t)] + (R_q + R_L) i_q(t) \tag{2}$$

$$E_q = K_E i_F \tag{3}$$

and

$$\tau_{em} = -\mathfrak{M} i_F i_q(t) \tag{4}$$

Since $E_q = K_E i_F$ and $K_E = 100$. Thus for

$$I_F = \frac{E_F}{R_F} \tag{5}$$

$$E_q = K_E I_F = K_E \times \left(\frac{E_F}{R_F}\right) \tag{6}$$

*For the derivation of this equation, please see Eqs. (2.12) through (2.18) and the paragraph following Eq. (2.18), Eqs. (2.51) through (2.59), and Eq. (2.69), and the paragraph following Eq. (2.69).

Eq. (6) enables us to write Eq. (2) as a function of the E_q versus i_F readings and the parameters of the field circuit, that is, Eq. (1). Thus

$$L_q \frac{d}{dt}[i_q(t)] + (R_q + R_L)i_q(t) = K_E \frac{E_F}{R_F} \equiv E_q \tag{7}$$

or

$$\frac{d}{dt}[i_q(t)] + \left(\frac{R_q + R_L}{L_q}\right)i_q(t) \equiv \frac{E_q}{L_q} \tag{8}$$

The solution of this first-order linear differential equation in $i_q(t)$ with constant coefficients and forcing function (this type of equation is known as a *nonhomogeneous* linear first-order differential equation with constant coefficients) is

$$i_q(t) = i_q(t)_t + i_q(t)_{ss} \tag{9}$$

where (1) $i_q(t)_t$, known as the transient response of the system, and (2) $i_q(t)_{ss}$, known as the steady state response of the system, depends only on the final value of forcing function of the system. Thus $i_q(t)_{ss}$ is similar in nature to the forcing function present in the system at hand. The response current $i_q(t)$ can be found utilizing the following procedure.

1. Assume the solution of $i_q(t)_t$ to be of the type

$$i_q(t)_t = Ke^{st} \tag{10}$$

Substitute Eq. (10) and its first derivative into Eq. (8) and set the right-hand side of Eq. (8) equal to zero, that is, $E_q/L_q = 0$. This operation allows calculation of one of the arbitrary constants of Eq. (10), namely, the constant s, which is known as *complex frequency*, in terms of the parameters of the system. In general, s is defined as

$$s \equiv \sigma \pm j\omega_d \tag{11}$$

where σ (in seconds^{-1}) is defined as the *damping coefficient* of the system and ω_d (in radians/second) is defined as the *damped frequency* of the system. The quantity σ may be *negative*, *zero*, or *positive*, with positive values highly undesirable because positive σ's define unstable systems. The imaginary part of s, if present, always appears as a conjugate pair, $\pm j\omega_d$, for equations of order two or more. The time derivative of Eq. (10) is

$$\frac{d}{dt}[i_q(t)_t] = sKe^{st} \tag{12}$$

Substitution of Eqs. (10) and (12) into Eq. (8), that is, allowing $i_q(t) \rightarrow i_q(t)_t$, and setting the right-hand side of Eq. (8) equal to zero yields

$$sKe^{st} + \left(\frac{R_q + R_L}{L_q}\right)Ke^{st} = 0 \tag{13}$$

Factoring Ke^{st} in Eq. (13) yields

$$\left[s + \left(\frac{R_q + R_L}{L_q}\right)\right]Ke^{st} = 0 \tag{14}$$

since Ke^{st} is the assumed solution for $i_q(t)_t$ and it cannot be zero. Thus only the bracketed expression of Eq. (14) can be set equal to zero,

$$s + \left(\frac{R_q + R_L}{L_q}\right) = 0 \tag{15}$$

Solution of Eq. (15) for s yields

$$s = -\left(\frac{R_q + R_L}{L_q}\right) \tag{16}$$

Eq. (16) enables us to write Eq. (10) in terms of the physical parameters of the generator. Thus

$$i_q(t)_t = Ke^{-[(R_q+R_L)/L_q]t} \tag{17}$$

2. Assuming the solution, designated as $\mathcal{I}$, of $i_q(t)_{ss}$ to be similar in nature to the forcing function of the system, that is, a constant in this case, yields

$$i_q(t)_{ss} = \mathcal{I} \tag{18}$$

Substitution of Eq. (18) and its first derivative into Eq. (8) enables solving for the assumed solution in terms of the physical parameters of the system and the forcing function. The time derivative of Eq. (18) is

$$\frac{d}{dt}\left[i_q(t)_{ss}\right] = 0 \tag{19}$$

Substitution of Eqs. (18) and (19) in Eq. (8), that is, allowing $i_q(t) \rightarrow i_q(t)_{ss}$, yields

$$0 + \left(\frac{R_q + R_L}{L_q}\right)\mathcal{I} = \frac{E_q}{L_q} \tag{20}$$

Solution of Eq. (20) for $\mathcal{I}$ yields

$$\mathcal{I} = \frac{E_q}{R_q + R_L} \tag{21}$$

Substitution of Eqs. (17) and (21) into Eq. (9) yields

$$i_q(t) = i_q(t)_t + i_q(t)_{ss} = Ke^{-[(R_q+R_L)/L_q]t} + \frac{E_q}{R_q + R_L} \tag{22}$$

3. The constant K of Eq. (22) can be found by utilizing the initial conditions of the current $i_q(t)$. Thus a closed form of solution of Eq. (8) can be determined. If the current $i_q(t)$ is known at $t=0^-$, then the current $i_q(t)$ is also known at $t=0^+$. This is so because the current through an inductor cannot change instantaneously in the *absence of impulses* in voltage. Thus if the current $i_q(t)$ is known at $t=0^-$, for example, $i_q(0^-)=0$, then the current at $t=0^+$ is also known. It is found by setting $i_q(0^+) \equiv i_q(0^-)=0$. It is said that the current at $t=0$ is equal to the current at $t=0^-$ or 0^+, that is, $i_q(0) \equiv i_q(0^-) = i_q(0^+)$. This identity is a consequence of the *principle of conservation of flux linkages* in an inductor (i.e., flux linkages cannot be changed instantaneously in a given system). Utilizing the known initial conditions and Eq. (22) enables calculation of the only constant, Eqs. (10) and (18), that remains unknown, namely K. Thus for time $t=0$, Eq. (22) becomes

$$i_q(t)\big|_{t=0} = 0 = Ke^{-[(R_q+R_L)/L_q](0)} + \frac{E_q}{R_q + R_L} \tag{23}$$

Solution of Eq. (23) for K yields

$$K = -\frac{E_q}{R_q + R_L} \tag{24}$$

Substitution of Eq. (24) into Eq. (23) and appropriate factoring yield

$$i_q(t) = \left(\frac{E_q}{R_q + R_L}\right)\{1 - e^{-[(R_q+R_L)/L_q]t}\} \tag{25}$$

where R_q and L_q are known as design parameters of the generator, R_L is the given load, and E_q is the generated voltage that can also be calculated from design parameters and characteristics information. Eq. (25) gives the rotor current $i_q(t)$ for the conditions of operation specified in this example. It can be plotted as shown in Fig. 2.3.1.

The quantity τ_E is defined as the *electric time constant* (to distinguish it from the *electromechanical time constant* $\tau_{\mathfrak{M}}$ to be encountered in Example 2.4) of the generator. The time constant, τ_E, is defined as the time required to make the exponent of Eq. (25) equal to unity, that is,

$$\left(\frac{R_q + R_L}{L_q}\right)\tau_E \equiv 1 \tag{26}$$

or the time on the time axis determined by the perpendicular drawn from the point of intersection between the final value of the i_q versus t response and its slope at $t=0$. The value of the response i_q versus t at $t=\tau_E$ is 63 percent of its final value, that is, $i_q(\infty) \equiv E_q/R_q + R_L$. This final value of the i_q versus t response should be reached within a time interval of five time constants if the system is to be stable. If the final value of the response is not reached within five time constants, the behavior of the system is questionable. For example, if the response of the system grows with time, the system is **unstable**. However, if the response reaches its final value and stays there (deviations of ± 0.5 percent of this value are considered normal), the system is **stable. Stable systems are desirable. Unstable systems are of no use and may be dangerous.**

Since the developed electromagnetic torque, as given by Eq. (2.41), is $\tau_{em} = -\mathfrak{M}\, i_F i_q(t)$, substitution for $i_q(t)$ from Eq. (25) yields

$$\tau_{em} = -\left(\frac{\mathfrak{M} i_F E_q}{R_q + R_L}\right)\{1 - e^{-[(R_q+R_L)/L_q]t}\} \tag{27}$$

substitution of the numerical values of the design data and the given parameters

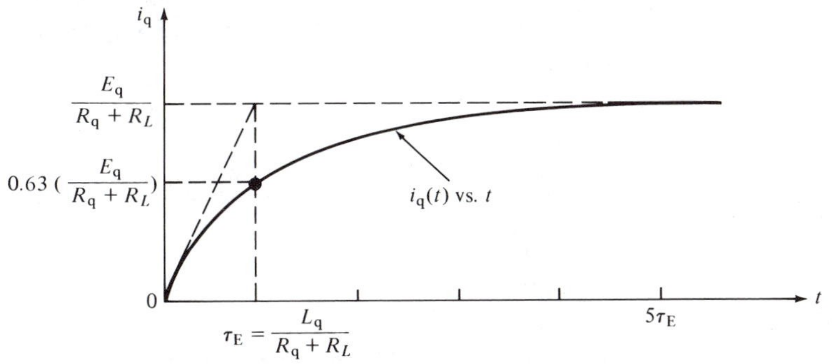

Fig. 2.3.1 Rotor Current vs. Time Plot for the Separately Excited Generator with the Field Current in Steady State at All Times

into Eqs. (6) and (25) yield the rotor current as

$$i_q(t) = \underline{\underline{50(1-e^{-10t})\,A}} \tag{28}$$

From Eqs. (2.36d) and (2.36e) and data,

$$E_q = K_E i_F = \mathfrak{M}\,\omega_R i_F = 100 i_F \tag{29}$$

Eq. (29) yields

$$\mathfrak{M}\,\omega_R = \underline{100} \tag{30}$$

Since

$$\omega_R = 2\pi f_R \tag{31}$$

premultiplication of Eq. (31) by 60 s/min yields

$$60\omega_R = 2\pi 60 f_R \equiv 2\pi n_R \tag{32}$$

Here n_R is defined as the speed of the rotor in revolutions per minute (abbreviated as rpm). However, since revolutions per minute are not standard MKS units of speed measurement, for calculation purposes ω_R measured in radians per second should always be used. Solution of Eq. (32) for ω_R yields

$$\omega_R = \frac{2\pi n_R}{60} \tag{33}$$

Note that Eq. (33) relates ω_R in radians per second, an MKS unit, to n_R in revolutions per minute, a non-MKS unit.

Substitution for ω_R into Eq. (30) and solution for $\mathfrak{M}$ yields

$$\mathfrak{M} = \frac{100}{\omega_R} = \frac{100}{(2\pi \times 1800)/60} = \underline{\frac{10}{6\text{p}}}\ \text{MKS units} \tag{34}$$

From Eq. (1)

$$i_F = \frac{E_F}{R_F} = \frac{100}{20} = \underline{5}\ \text{A} \tag{35}$$

Proper substitution into Eqs. (6) and (27) yields

$$\tau_{em} = \underline{\underline{-132.5(1-e^{-10t})}}\ \text{N}\cdot\text{m} \tag{36}$$

Eq. (36) can be plotted as shown in Fig. 2.3.2. Thus the turning effect of the torque is to oppose rotation. The torque reaches its final value, $\tau_{em}(\infty) = -132.5$, in five time constants, that is, $t_F = 5\tau_E$.

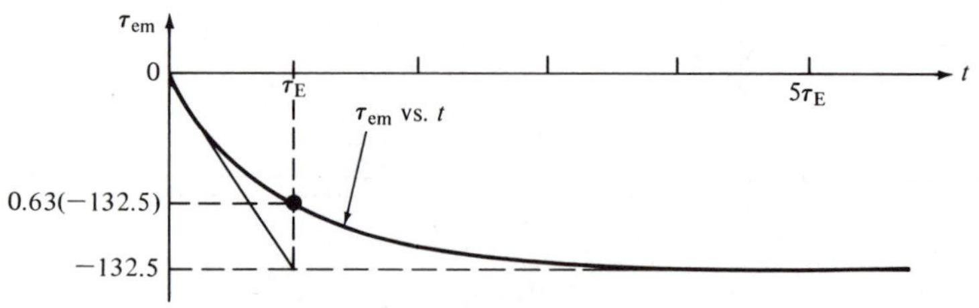

Fig. 2.3.2 Torque vs. Time Plot for the Separately Excited Generator with the Field Current in Steady State at All Times

2.4 THE CLASSICAL dc Motor

In this mode of operation both the stator and rotor windings are electrically excited.

The cross section of the dc generator, Fig. 2.7, is valid for the explanation of the principles of operation of the dc motor if the following conditions are imposed: (1) the device is not driven mechanically and (2) the rotor coils are excited from external electrical dc energy sources through the *d*-axis brushes and the commutator system, Fig. 2.11. The functions of the commutator and its associated brushes in the dc motor are (1) to change the externally applied direct voltage (or current) to an internally utilized alternating voltage (or current) as the conductors alternately move under opposite poles (thus a rotation in the same direction is produced) and (2) to transfer the externally applied voltage (or current) from the stationary brushes to the moving rotor coils. The principle of operation becomes clearer if we utilize Fig. 2.16, that is, observe the rotor winding, the injected dc current to it $i_q \mathbf{dL}$,* the radial dimension of the rotor **R**, the radial stator field $\mathbf{B}_\theta$, and the developed tangential force $d\mathbf{F} = i_q d\mathbf{L} \times \mathbf{B}$, Eq. (2.5).

Torque is defined as the *turning effect* of a given force about a given axis of rotation and is a vector quantity parallel to the axis of rotation. Thus the torque on the given coil side of Fig. 2.16 is

$$d\tau_w = \mathbf{R} \times d\mathbf{F} \tag{2.42}$$

or

$$d\tau_w = \mathbf{R} \times (i_q d\mathbf{L} \times \mathbf{B}) \tag{2.43}$$

The torque on the one-turn rotor coil (or loop) is $d\tau_c = 2d\tau_w$ (why?), that is,

$$d\tau_c = 2\mathbf{R} \times (i_q d\mathbf{L} \times \mathbf{B}) \tag{2.44}$$

Examination of Fig. 2.16 yields (1) $i_q d\mathbf{L} \perp \mathbf{R}$ and (2) $\mathbf{B} \| \mathbf{R}$. These two relationships enable us to write the magnitude of Eq. (2.44), keeping in mind that the vector torque is oriented **out of the page**, as

$$d\tau_c = 2RBi_q dL \tag{2.45}$$

Integration of Eq. (2.45) yields the total torque on the one-turn rotor coil. Thus

$$\int_0^{\tau_C} d\tau_c = \int_0^{\ell} 2RBi_q dL \tag{2.46}$$

yields

$$\tau_C = (2R\ell)Bi_q \tag{2.47}$$

If Z coils are distributed on the circumference of the rotor, the number of coils located within a small sector of the rotor circumference defined by $d\theta$ is† $(Z/2\pi)d\theta$. The torque developed on the $(Z/2\pi)d\theta$ coils due to a sinusoidally distributed field is as given in Eq. (2.48). (Remember that the field of a dc energy converter has a square waveform, Fig. 2.7. However, decomposition of the square

*Lowercase letters for currents i_q and i_F are used throughout this chapter in order to emphasize the generality of the equations used.

†θ is strictly a measure of the angular displacement along the rotor circumference, used *only* to locate coil positions.

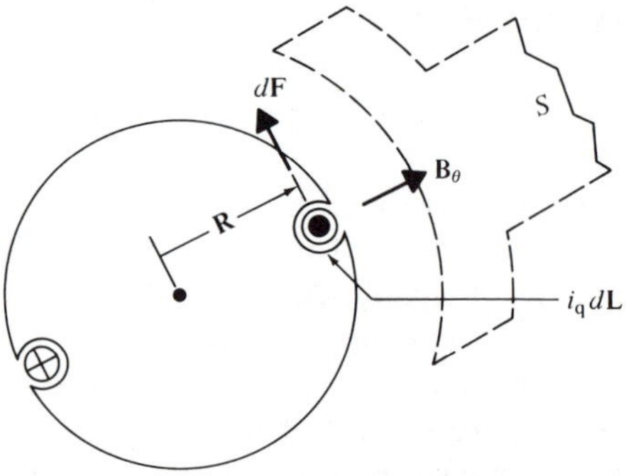

Fig. 2.16 The Rotor Cross Section, Motor Mode of Operation

waveform to Fourier components yields the fundamental waveform, which is sinusoidal, and odd space harmonics. It can be shown experimentally that 95 percent of the torque developed by a dc motor is due to the fundamental component of the rectangular field. This is an acceptable approximation for the worst case. Thus the trade-off of the square waveform for the sinusoidal waveform for the added advantages in the mathematical analysis is highly desirable.)

$$d\tau_\theta = (2R\ell)B_M i_q (Z/2\pi)\sin\theta\, d\theta \tag{2.48}$$

The electromagnetic torque developed on all the one-turn rotor coils, occupying π radians of the rotor circumference, is obtained by integrating Eq. (2.48). Thus

$$\int_0^{\tau_{em}} d\tau_\theta = \int_0^{\pi} (2R\ell)B_M i_q (Z/2\pi)\sin\theta\, d\theta \tag{2.49}$$

However, when each coil has N-turns instead of one turn, the total electromagnetic torque on the N-turn coils is obtained by multiplying Eq. (2.49) by N. Thus

$$\begin{aligned}\tau_{em} &= -N(2R\ell)(Z/2\pi)B_M i_q \cos\theta\Big|_0^{\pi} = -N(2R\ell)(Z/2\pi)B_M i_q(-1-1)\\ &= 2N(2R\ell)(Z/2\pi)B_M i_q\end{aligned} \tag{2.50}$$

Looking upon the Z coils as constituting two parallel paths, each path containing $(Z/2)$ coils, enables transformation of the 2 parallel path distribution to an a-parallel path distribution (a being an even number). Thus every $(Z/2)$ in Eq. (2.49) becomes (Z/a)e. Also, utilizing Eq. (2.34) enables us to write Eq. (2.50) for a P-pole device as

$$\tau_{em} = 2N\left(\frac{P}{2}\right)(\Phi_P)\left(\frac{Z}{a\pi}\right)i_q = \left[\frac{ZNP}{a\pi}\right]\Phi_P i_q \tag{2.51}$$

Recognizing the quantity within the brackets as the design constant $\mathcal{K}$ of Eq. (2.37) enables us to write Eq. (2.51) as

$$\tau_{em} = \mathcal{K}\Phi_P i_q \tag{2.52}$$

Remembering, Eq. (1.9), that

$$\Phi_P = \left(\frac{N_F}{\mathcal{R}_F}\right)i_F \equiv \mathcal{K}_F i_F \tag{2.53}$$

enables the writing of Eq. (2.51) as

$$\tau_{em} = \mathcal{K}\Phi_P i_q = \mathcal{K}\mathcal{K}_F i_F i_q \tag{2.54}$$

Defining $\mathcal{K}\mathcal{K}_F$ of Eq. (2.54) as a *torque coefficient* $\mathfrak{M}$, that is,

$$\mathfrak{M} \equiv \mathcal{K}\mathcal{K}_F \tag{2.55}$$

enables the writing of Eq. (2.54) as

$$\tau_{em} = \mathfrak{M} i_F i_q \tag{2.56}$$

Control engineers are primarily interested in the torque developed by the dc motor and its dependence on the field current. Thus they absorb $\mathfrak{M} i_q$ into a new constant K_T, which they define as a *torque constant*,

$$K_T \equiv \mathfrak{M}\, i_q \tag{2.57}$$

This enables writing Eq. (2.56) as

$$\tau_{em} = K_T i_F \tag{2.58}$$

Thus, by holding the current i_q constant, the torque developed is nearly linearly dependent on the field current.

In summary, it should be stated that, depending on one's interest, the electromagnetic torque equation can be written as

$$\tau_{em} = \mathcal{K}\Phi_p i_q = \mathcal{K}\mathcal{K}_F i_F i_q = \mathfrak{M}\, i_F i_q = K_T i_F \tag{2.59}$$

Note that particular attention should be paid to the various constants of Eq. (2.59), their meaning, and the physical significance in their use.

It is clear that the effect of the torque on the rotor is that of rotation in the positive direction; that is, for an *out-of-the-page* electromagnetic torque, Fig. 2.16, the device rotates *counterclockwise*, with a speed $\omega_R(t)$ radians per second. **This speed is variable and one must solve for it.**

This rotation causes a speed-dependent voltage, Eq. (2.36), to appear across the brushes located along the *d*-axis, used to inject the current i_q into the rotor coils. This voltage is generated by the developed speed $\omega_R(t)$ instead of an externally supplied speed, ω_R, as in Eq. (2.36). For this reason, this voltage is termed the *back emf*. It is related to the developed speed $\omega_R(t)$ by Eq. (2.36), which is rewritten here, after E_q is defined, as $E_q \rightarrow E_b = \psi[\omega_R(t)] \equiv$ back emf. Thus

$$E_b = \mathcal{K}\Phi_p \omega_R(t) \tag{2.60}$$

Using Eqs. (2.53) and (2.55) enables us to write Eq. (2.60) as

$$E_b = \mathcal{K}\mathcal{K}_F i_F \omega_R(t) = \mathfrak{M}\, i_F \omega_R(t) \tag{2.61}$$

Defining $\mathfrak{M} i_F$ as the voltage constant K_V [note also that $K_V = \mathcal{K}\mathcal{K}_F i_F \mathcal{K}\Phi_p$, Eq. (2.38)], that is,

$$K_V \equiv \mathfrak{M}\, i_F \tag{2.62}$$

enables writing Eq. (2.61) as

$$E_b = \mathcal{K}\mathcal{K}_F i_F \omega_R(t) = \mathfrak{M}\, i_F \omega_R(t) = K_V \omega_R(t) \tag{2.63}$$

Summarizing, depending on one's interest, the back emf equation can take any one of the following forms $\{E_b = f(\omega_R)\}$ is used henceforth rather than $E_b = \psi[\omega_R(t)]$ for consistency with the footnote on p. 75.

$$E_b = \mathcal{K}\Phi_p \omega_R = \mathcal{K}\mathcal{K}_F i_F \omega_R = \mathfrak{M}\, i_F \omega_R = K_V \omega_R \tag{2.64}$$

Again, particular attention should be paid to the various constants of Eq. (2.64), their meaning, and the physical significance in their use. Also, a comparison of Eqs. (2.59) and (2.64) should be made and the significance of i_q and ω_R should be noticed.

By accounting for the resistance and inductance of the rotor and stator windings, the back emf, and the external sources of electrical energy used for excitation of the stator and rotor windings, $e_F(t)$ and $e_S(t)$, the equivalent circuit for the **motor mode of operation** of the separately excited dc energy converter can be drawn as in Fig. 2.17. **Note the direction of the current i_q(t) and the polarity of the voltage source E_b.**

The dynamic equations that describe the equivalent circuit of Fig. 2.17 and thus the **motor mode of operation** of the separately excited dc electromechanical energy converter are

$$e_F(t) = R_F i_F(t) + L_F \frac{d}{dt}[i_F(t)] \quad (2.65)$$

$$e_S(t) = R_S i_q(t) + R_q i_q(t) + L_q \frac{d}{dt}[i_q(t)] + E_b \quad (2.66)$$

$$E_b = K_V \omega_R(t) \quad (2.67)$$

and

$$\tau_{em} = \mathfrak{M}\, i_F i_q = \mathrm{K_T i_F(t)} \quad (2.68)$$

where $e_F(t)$ and $e_S(t)$ are dc voltage sources.

The mechanical portion of Fig. 2.17 can be described by using *D'Alembert's principle* stated mathematically as follows:

$$\tau_{em} = J\frac{d^2}{dt^2}[\theta_R(t)] + B\frac{d}{dt}[\theta_R(t)] + K\theta_R(t) + \tau_L \quad (2.69)$$

where (1) J is the moment of inertia in kilogram meters^2; (2) B is the rotational damping coefficient in Newton meters per radian per second; (3) K is the rotational stiffness (analogous to a spring constant) in newton meters per radian; (4) τ_L is the loading torque in newton meters; (5) τ_{em} is the developed electromag-

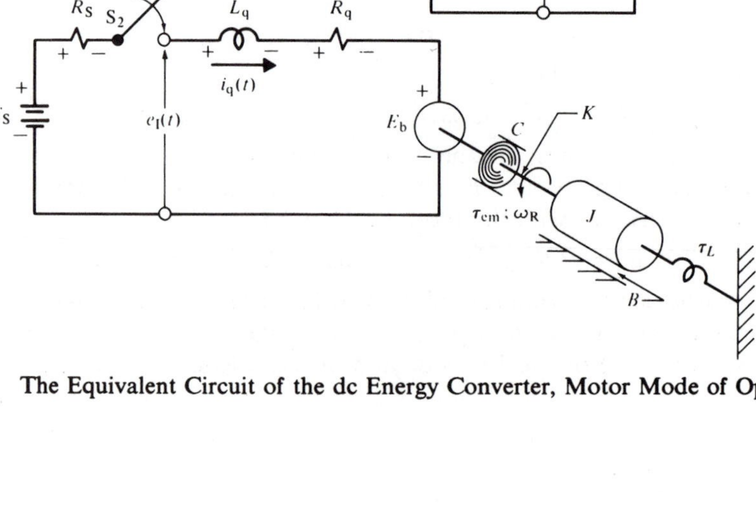

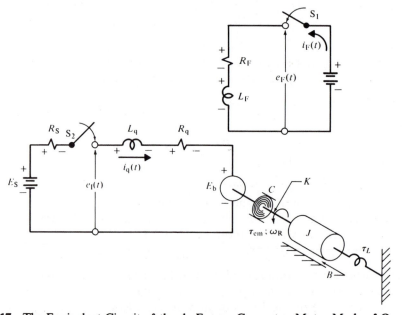

Fig. 2.17 The Equivalent Circuit of the dc Energy Converter, Motor Mode of Operation

netic torque, Eq. (2.59), in newton meters, which takes the following signs: (a) **positive for motor action** and (b) **negative for generator action**; and (6) $\theta_R(t)$ is the angular displacement of the shaft in radians.

It must be noted that the term **developed electromagnetic torque** means the torque produced by the device and given by Eq. (2.59). It is this torque that must overcome all the opposing torques, Eq. (2.69), due to the moment of inertia, the rotational damping, the rotational stiffness, and the loading torque, if the rotor is to rotate.

Finally, Eqs. (2.65), (2.66), (2.67), (2.68), and (2.69) describe the dynamic performance of the generalized separately excited dc motor. Note that Eqs. (2.66) and (2.69) are coupled by $E_b = K_V\omega_R(t)$. These equations, in addition to the means they provide for analytical techniques of analysis, can be used for (1) computer simulation by utilizing transfer function, block diagram, or simulation diagram techniques; (2) providing, upon proper simplification (Section 2.6), the describing equations for steady state studies of the separately excited dc motor.

Note that these equations are the equations for the analysis of the entire class of dc motors to be presented in this text. Every other motor to be studied is properly derived from the separately excited dc motor.

Example 2.4

Assume that in Fig. 2.17 the switch S_2 is in the open position and the switch S_1 has been closed for some time, and the field circuit has reached steady state, that is, $I_F\, d[i_F(t)]/dt \to 0$. In addition, assume that $\tau_L = K = B \equiv 0$; that is, (1) the loading torque, (2) the rotational stiffness, and (3) the rotational damping are zero. Now as the switch S_2 closes the mechanical coupler, C, goes to its ON position, that is, the motor is loaded with the moment of inertia J in kilogram meters^2.

For $L_q \cong 0$, that is, $L_q\, d[i_q(t)]/dt \ll R_q i_q(t)$, calculate the speed of the motor as a function of time.

Solution

The electrical performance of the motor is described by Eqs. (2.65), (2.66), and (2.67), which can be modified to describe the performance of the motor under the present conditions of operation. Thus

$$E_F = R_F I_F \tag{1}$$

$$E_I \equiv E_S - R_S i_q(t) = R_q i_q(t) + E_b \tag{2}$$

$$E_b = K_V \omega_R(t) \tag{3}$$

With the mechanical coupler on, Eq. (2.69) can be written [with the aid of Eq. (2.68)] as

$$\tau_{em} = J\frac{d^2}{dt^2}[\theta_R(t)] \equiv J\frac{d}{dt}[\omega_R(t)] = \mathfrak{M}\, I_F i_q(t) \tag{4}$$

Solution of Eq. (4) for $i_q(t)$ yields

$$i_q(t) = \left(\frac{J}{\mathfrak{M}\, I_F}\right)\frac{d}{dt}[\omega_R(t)] \tag{5}$$

Substitution for $i_q(t)$ from Eq. (5) into Eq. (2) and utilization of Eq. (2.63) yields

$$\left(\frac{R_q J}{\mathfrak{M}\, I_F}\right)\frac{d}{dt}[\omega_R(t)] + \mathfrak{M}\, I_F \omega_R(t) = E_I \tag{6}$$

or after simplification,

$$\frac{d}{dt}[\omega_R(t)] + \left[\frac{(\mathcal{M}\, I_F)^2}{R_q J}\right]\omega_R(t) = \left(\frac{\mathcal{M}\, I_F}{R_q J}\right)E_I \tag{7}$$

The solution of this first-order linear differential equation in $\omega_R(t)$ with constant coefficients and forcing function is obtained by following the procedure outlined in Example 2.3. Thus

$$\omega_R(t) = \omega_R(t)_t + \omega_R(t)_{ss} \tag{8}$$

where the assumed solutions for $\omega_R(t)_t$ and $\omega_R(t)_{ss}$ are

$$\omega_R(t)_t = Ke^{st} \tag{9}$$

and

$$\omega_R(t)_{ss} = \Omega \tag{10}$$

The constant K in Eq. (9) should not be confused with the rotational stiffness constant.

Substitution of Eq. (9) and its first derivative into Eq. (7), that is, allowing $\omega_R(t) \rightarrow \omega_R(t)_t$, setting the right-hand side of Eq. (7) equal to zero, and factoring Ke^{st} yields

$$\left[s + \frac{(\mathcal{M}\, I_F)^2}{R_q J}\right]Ke^{st} = 0 \tag{11}$$

Therefore,

$$s = -\left[\frac{(\mathcal{M}\, I_F)^2}{R_q J}\right] \tag{12}$$

Eq. (12) enables us to write Eq. (9) in terms of the physical parameters of the motor as

$$\omega_R(t)_n = Ke^{[(\mathcal{M}\, I_F)^2/R_q J]t} \tag{13}$$

Now substitution of Eq. (10) and its first derivative in Eq. (7), that is, allowing $\omega_R(t) \rightarrow \omega_R(t)_{ss}$, yields

$$0 + \left[\frac{(\mathcal{M}\, I_F)^2}{R_q J}\right]\Omega = \left(\frac{\mathcal{M}\, I_F}{R_q J}\right)E_I \tag{14}$$

Solution of Eq. (14) for Ω yields

$$\Omega = \frac{E_I}{\mathcal{M}\, I_F} \tag{15}$$

Substitution of Eqs. (13) and (15) into Eq. (8) yields

$$\omega_R(t) = Ke^{-[(\mathcal{M}\, I_F)^2/R_q J]t} + \frac{E_I}{\mathcal{M}\, I_F} \tag{16}$$

For the assumed initial conditions, $\omega_R(t) = \omega_R(0) \equiv \omega_R(0^-) \equiv \omega_R(0^+) = 0$, K becomes

$$K = -\frac{E_I}{\mathcal{M}\, I_F} \tag{17}$$

Thus the angular velocity of the motor [for the conditions of operation specified in

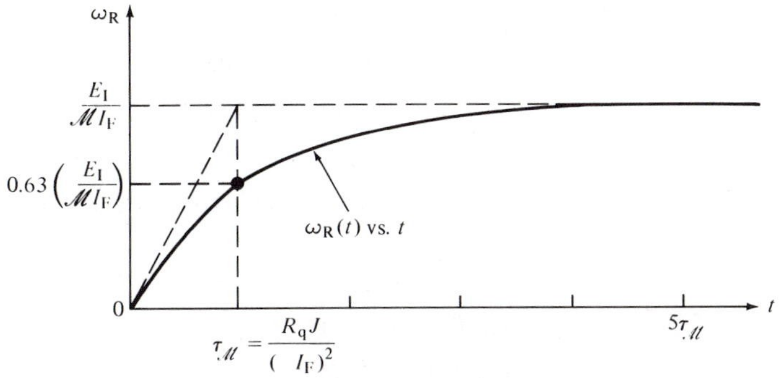

Fig. 2.4.1 Speed vs. Time Plot for the Separately Excited Motor with the Field Current in Steady State at All Times

the given example and $\omega_R(0)=0$] becomes

$$\omega_R(t) = \frac{E_I}{\mathfrak{M}\, I_F}\{1-e^{-[(\mathfrak{M}\, I_F)^{2/R_qJ}]t}\}\ \text{rd/s} \tag{18}$$

where $\mathfrak{M}$, R_q, and J are system parameters, E_I is the forcing function of the motor, and I_F is the steady state field current of the motor that can be calculated from Eq. (1).

Eq. (18) can be plotted as shown in Fig. 2.4.1, where $\tau_{\mathfrak{M}}$ is defined as the *electromechanical time constant* of the motor, and the final value of the response $\omega_R(t)$ versus t is $\omega_R(\infty)\equiv E_I/(\mathfrak{M}\, I_F)$.

Again, time constant $\tau_{\mathfrak{M}}$ is defined as the time required to make the exponent of Eq. (16) [or Eq. (18)] equal to unity, that is,

$$\left[\frac{(\mathfrak{M}\, I_F)^2}{R_qJ}\right]\tau_{\mathfrak{M}} = 1 \tag{19}$$

or the time on the time axis determined by the perpendicular drawn from the point of intersection between the final value of the $\omega_R(t)$ versus t response and its slope at $t=0$. The value of the response $\omega_R(t)$ versus t at $t=\tau_{\mathfrak{M}}$ is 63 percent of its final value, $\omega_R(\infty)\equiv E_I/\mathfrak{M}\, I_F$. This final value of the $\omega_R(t)$ versus t response should be reached within a time interval of five time constants if the system is to be *stable*. If the final value of the response is not reached within five time constants, the behavior of the system is questionable. For example, if the response of the system departs from its final value with time, the system is *unstable*. However, if the response reaches and stays within a margin of ± 0.5 percent of the final value, the system is **stable**. Unstable systems are intolerable.

2.5 POWER BALANCE EQUATIONS OF THE CLASSICAL dc MOTOR

The work done on a body in rotational motion is given by

$$dw_M = \tau_{em}\, d\theta_R \tag{2.70}$$

Dividing Eq. (2.70) by dt yields

$$p_{em} \equiv \frac{dw_M}{dt} = \tau_{em}\frac{d\theta_R}{dt} \tag{2.71}$$

Utilizing Eq. (2.56) yields

$$p_{em} = \tau_{em}\omega_R = (\mathfrak{M}\, i_F i_q)\omega_R \tag{2.72}$$

Substituting from Eq. (2.61) for E_b into Eq. (2.72) yields

$$p_{em} = \tau_{em}\omega_R = E_b i_q \tag{2.73}$$

Utilizing Kirchhoff's voltage law for the motor equivalent circuit, Fig. 2.17, yields

$$e_I(t) = L_q \frac{d}{dt}\left[i_q(t)\right] + R_q i_q(t) + E_b = v_L + R_q i_q(t) + E_b \tag{2.74}$$

Solution of Eq. (2.74) for E_b, and dropping the "(t)", yield

$$E_b = e_I - R_q i_q - v_L \tag{2.75}$$

Multiplying Eq. (2.75) by i_q yields

$$E_b i_q = (e_I - R_q i_q - v_L) i_q = e_I i_q - R_q i_q^2 - v_L i_q \tag{2.76}$$

Comparison of Eq. (2.76) with Eq. (2.73) yields

$$p_{em} = \tau_{em}\omega_R = E_b i_q = e_I i_q - R_q i_q^2 - v_L i_q \equiv p_I - p_R - p_L \tag{2.77}$$

where (1) p_I is the *input power* to the rotor winding; (2) p_R is the *power dissipated* in the rotor resistance; (3) p_L is the *power stored* in the rotor inductance, and (4) p_{em} is the *developed* electromagnetic power. p_{em} provides the *useful* power to the shaft, that is, p_{SH}, and the following losses: (1) *bearing friction* p_{BF}; (2) *wind resistance* p_{WR}, and (3) *core losses* in the rotor iron p_{CL}.

The wind resistance power, p_{WR}, sometimes referred to as windage loss, is the power that is dissipated for the most part *to rotate a fan* attached to the shaft of the motor. The fan draws air axially along the device for cooling purposes.

These ideas enable us to write the power balance equation as follows:

$$p_{SH} = p_I - p_R - p_L - p_{BF} - p_{WR} - p_{CL} \equiv p_O \tag{2.78}$$

Usually, the bearing friction, wind resistance, and core losses are identified under the common name of power losses, p_{LOSSES}. Thus the quantity, power losses,

$$p_{LOSSES} = p_{BF} + p_{WR} + p_{CL} \tag{2.79}$$

is usually given by the manufacturer as one value. The MKS dimensional unit of power is the watt, defined as one joule/second. However, quite often the shaft power (many times the names "useful" or "output power" are used instead of shaft, abbreviated as p_U or p_O, correspondingly), p_{SH} of a motor is measured by a non-MKS dimensional unit known as horsepower (hp). If this is the case, you should convert the horsepower (hp) to MKS units, watts, before performing any calculations.

Finally, the performance of a motor is measured by an *index of performance* known as the percentage *efficiency*, which is defined as $\eta = (p_U/p_I) \times 100 \equiv (p_O/p_I) \times 100 \equiv (p_{SH}/p_I) \times 100$.

Example 2.5

The dc motor of Fig. 2.17 is connected in series, to which the commutating or interpole winding (see Fig. 2.32) is also connected in series. The rating* of the

*The "rated" or nameplate values of an energy converter are such that when rated electrical input is applied to the device and rated output is delivered at the rated speed, rated current will flow into (motor) or out (generator) of the device.

motor is 40 hp, 600 V. When tested at 500 r/min, it exhibits the following results: $I_q=80$, $R_q=0.3$, $R_F=0.2$, $R_C=0.1$, $p_{CL}=1000$, $p_{BF}=50$, $p_{WR}=150$, all in MKS units. Calculate: (1) the electromagnetic power p_{em}, (2) the electromagnetic torque τ_{em}, (3) the shaft power p_{SH}, (4) the shaft torque τ_{SH}, (5) the efficiency η, and (6) the shaft power p_{SH} in horsepower.

Solution

1. Use $p_{em}=\tau_{em}\omega_R=p_I-p_{\text{TOTAL LOSSES}}$. Since

$$p_I = E_I I_q$$
$$p_{\text{TOTAL LOSSES}} = p_{\text{RTOTAL}} + p_{BL}$$

then*

$$p_I = (600)(80) = \underline{48{,}000}\text{ W}$$
$$p_{R[\text{TOTAL}]} = R_q I_q^2 + R_F I_q^2 + R_C I_q^2$$
$$= (0.3)(80)^2 + (0.2)(80)^2 + (0.1)(80)^2$$
$$= 1{,}920 + 1{,}280 + 640 = \underline{3{,}840}\text{ W}$$
$$p_{\text{BRUSH LOSSES}} = 2I_q = (2)(80) = \underline{160}\text{ W}$$
$$p_{\text{TOTAL LOSSES}} = 3{,}840 + 160 = \underline{4{,}000}\text{ W}$$

Thus

$$p_{em} = 48{,}000 - 4{,}000 = \underline{\underline{44{,}000}}\text{ W}$$

2. Use $p_{em}=\omega_R\tau_{em}$. Since $2\pi f_R=\omega_R$, then

$$60 f_R 2\pi = 60\omega_R \equiv 2\pi n_R$$

or

$$\omega_R = \frac{2\pi n_R}{60} = \frac{2\pi \times 500}{60} = \underline{52.4}\text{ rad/s}$$

Therefore,

$$\tau_{em} = \frac{p_{em}}{\omega_R} = \frac{44{,}000}{52.4} = \underline{\underline{840}}\text{ N}\cdot\text{m}$$

3. Use $p_{SH}=p_{em}-p_{\text{LOSSES}}$. Since

$$p_{\text{LOSSES}} = p_{BF} + p_{WR} + p_{CL}$$
$$= 50 + 150 + 1{,}000 = \underline{1{,}200}\text{ W}$$

Therefore,

$$p_{SH} = 44{,}000 - 1{,}200 = \underline{\underline{42{,}800}}\text{ W}$$

4. Use $p_{SH}=\omega_R\tau_{SH}$. Therefore,

$$\tau_{SH} = \frac{42{,}800}{52.4} = \underline{\underline{817}}\text{ N}\cdot\text{m}$$

5. Use $\eta=(p_{SH}/p_I)\times 100$. Therefore,†

$$\eta = \frac{42{,}800}{48{,}000} \times 100 = \underline{\underline{89.2\%}}$$

*The IEEE recommends the formula $p_{BL}\equiv 2I_q$ for the calculation of *brush losses*, where 2 represents the voltage drop across both brushes, that is, one volt per brush.

†The conversion factor between horsepower and the MKS unit of watts is 1 hp$\equiv$746 W.

2.6 TORQUE SPEED CHARACTERISTICS OF VARIOUS dc MOTOR TYPES UNDER STEADY STATE OPERATION

Depending on the excitation of the field and rotor windings, the dc motor assumes corresponding names such as given in the following paragraphs.

Separately Excited Motor

As the name implies, the field and rotor windings of Fig. 2.17 are excited from the separate dc sources, $e_F(t)$ and $e_I(t)$, or from rectified ac sources. For steady state operation, where $L_F d[i_F(t)]/dt = L_q d[i_q(t)]/dt \equiv 0$ and $e_F(t) \rightarrow E_F$, $e_I(t) \rightarrow E_I$, then $i_F(t) \rightarrow I_F$, and $i_q(t) \rightarrow I_q$. With the mechanical loading included, the equivalent circuit of Fig. 2.17 is drawn as in Fig. 2.18.

The describing steady-state equation of the rotor circuit becomes

$$E_I = R_q I_q + E_b \tag{2.80}$$

and the developed electromagnetic torque given by Eq. (2.56) can be written as

$$\tau_{em} = \mathcal{M}\, I_F I_q \tag{2.81}$$

Solution of Eq. (2.80) for I_q yields

$$I_q = \frac{E_I - E_b}{R_q} \tag{2.82}$$

Substitution of Eq. (2.82) into Eq. (2.81) yields

$$\tau_{em} = \mathcal{M}\, I_F \left(\frac{E_I - E_b}{R_q} \right) \tag{2.83}$$

Since $E_b = \mathcal{M}\, I_F \omega_R$ [Eq. (2.61)], Eq. (2.83) can be written as

$$\tau_{em} = \mathcal{M}\, I_F \left(\frac{E_I}{R_q} - \frac{\mathcal{M}\, I_F \omega_R}{R_q} \right) \tag{2.84}$$

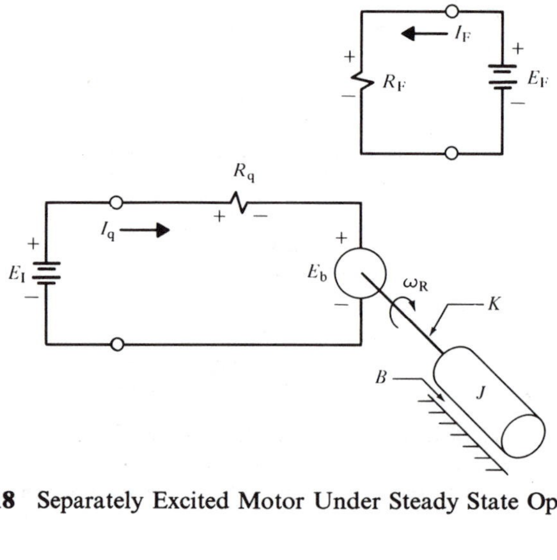

Fig. 2.18 Separately Excited Motor Under Steady State Operation

or regrouping

$$\tau_{em} = \left(\frac{\mathfrak{M}\, I_F E_I}{R_q}\right) - \left[\frac{(\mathfrak{M}\, I_F)^2}{R_q}\right]\omega_R \qquad (2.85)$$

Thus the developed electromagnetic torque, Eq. (2.85), $\tau_{em} = \psi(\omega_R, E_I, I_F, R_q, \mathfrak{M})$, can be examined with ω_R treated as the variable and with the remaining parameters (either none or just one at a time) treated as running parameters, with the exception of $\mathfrak{M}$. The four conditions are as follows:

1. Torque versus speed characteristic with the angular velocity ω_R treated as the variable and E_I, I_F, and R_q all treated as constants. Eq. (2.85) can be written as follows:

$$\tau_{em} = \tau_0 - \mu\omega_R \qquad (2.86)$$

where $\tau_0 \equiv \mathfrak{M}\, I_F E_I / R_q$ is defined as the starting torque and $\mu = \mathfrak{M}^2 I_F^2 / R_q$ is the negative slope of the τ_{em} versus ω_R characteristic. Eq. (2.86) is in the form of the well-known algebraic equation with a y-intercept, $y = b + mx$, and is plotted as in Fig. 2.19.

In practice it is found that the starting torque τ_0 is very high. Thus the slope of the τ_{em} versus ω_R characteristic is very steep, almost infinite. In fact, for a loading torque, τ_L, it is found that $\omega_0 - \omega_{RL} = 5\%(\omega_0)$. Thus the device is known as a *nearly constant speed* motor. The constancy of the speed of a motor is measured by its percentage **speed regulation** (SR), which is defined as

$$SR \equiv \frac{\omega_0 - \omega_{RL}}{\omega_{RL}} \times 100 = \frac{\omega_{\text{NO LOAD}} - \omega_{\text{FULL LOAD}}}{\omega_{\text{FULL LOAD}}} \times 100 \qquad (2.87)$$

It is obvious from this equation that speed regulation is defined as the ability of the device to maintain constant speed as the loading conditions change from *no-load* to *full-load* and R_q, E_I, and I_F are held constant. Note that **the closer the SR is to zero, the more constant the speed is.**

2. Torque versus speed characteristic with the rotor winding voltage E_I treated as the running parameter and I_F and R_q treated as constants. Under these conditions Eq. (2.85) can be written as

$$\tau_{em} = \nu_0 E_I - \mu\omega_R \qquad (2.88)$$

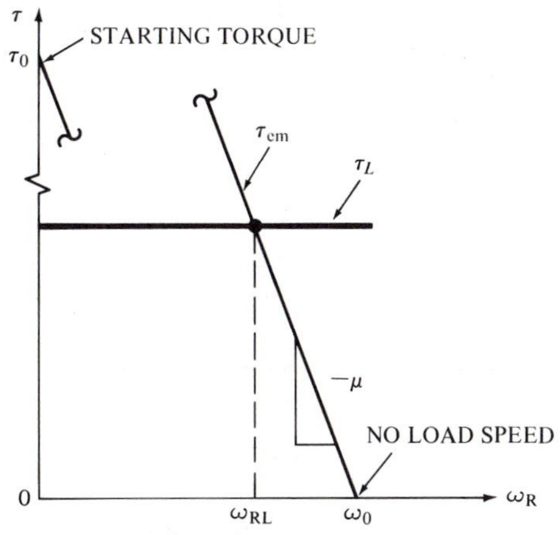

Fig. 2.19 Torque vs. Speed Characteristic of a Separately Excited Motor and Load

where $\nu_0 \equiv \mathfrak{M} I_F / R_q$ and μ retains the definition given in paragraph 1. Since E_I is variable, Eq. (2.88) can be plotted as shown in Fig. 2.20.

The nature of this characteristic is similar to that of Fig. 2.19. However, by increasing E_I we can raise the starting torque and the no-load speed and thus the operating speed for a given loading torque τ_L. The opposite is also true.

The variable E_I can be obtained from an ac source through silicon-controlled rectifiers (SCR) or using ac motors to drive dc generators with a variable external field resistance, R_{EXT}.

3. Torque versus speed characteristics with the field current I_F treated as the running parameter and E_I and R_q treated as constants. Under these conditions Eq. (2.85) can be written as

$$\tau_{em} = \xi_0 I_F - (\mu_0 I_F^2)\omega_R \tag{2.89}$$

where $\xi_0 \equiv \mathfrak{M} E_I / R_q$ and $\mu_0 \equiv \mathfrak{M}^2 / R_q$. Each change in I_F will change both the τ_{em}-axis intercept, that is, $\tau_0 = \xi_0 I_F$, and the slope of the characteristic, that is, $\mu \equiv \mu_0 I_F^2$. Thus for $I_{F1} < I_{F2}$, the torque versus speed characteristics can be plotted as shown in Fig. 2.21. The effect that a change in I_F has on the starting torque, no-load speed, and operating speed for a given load torque τ_L should be obvious from Fig. 2.21.

The field current I_F can be changed by either varying E_F or using a variable external resistance R_{EXT} in series with R_F.

4. Torque versus speed characteristics with R_q treated as the running parameter (how?) and E_I and I_F treated as constants. Under these conditions Eq. (2.85) can be written as

$$\tau_{em} = \frac{\xi_1}{R_q} - \left[\frac{\mu_1}{R_q}\right]\omega_R \tag{2.90}$$

where $\xi_1 \equiv \mathfrak{M} E_I I_F$ and $\mu_1 (\mathfrak{M} I_F)^2$. As R_q varies, both the τ_{em}-axis intercept, $\tau_0 = \xi_1 / R_q$, and the slope of the characteristic, $\mu = \mu_1 / R_q$, will change. Thus for $R_{q1} < R_{q2}$ the torque versus speed characteristic can be plotted as shown in Fig. 2.22. The effects that a change in R_q has on the starting torque, no-load speed, and operating speed for a given loading torque τ_L are evident in Fig. 2.22. For example, as $R_q \to 0$, $\tau_0 \to \infty$, $\mu \to \infty$, and the device under these conditions of operation tends to become an *absolutely constant speed* motor with $\omega_{RL} = \omega_0$. Also, for $R_q \to 0$ under steady state operation, the current I_q is completely independent of the speed. It is determined only by the loading torque. When $K = B \equiv 0$ in Eq. (2.69) and

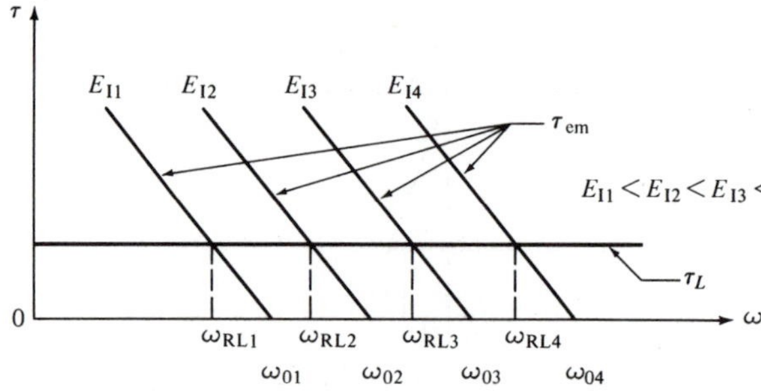

Fig. 2.20 Torque vs. Speed Characteristics of a Separately Excited Motor and Load; E_I Is the Running Parameter

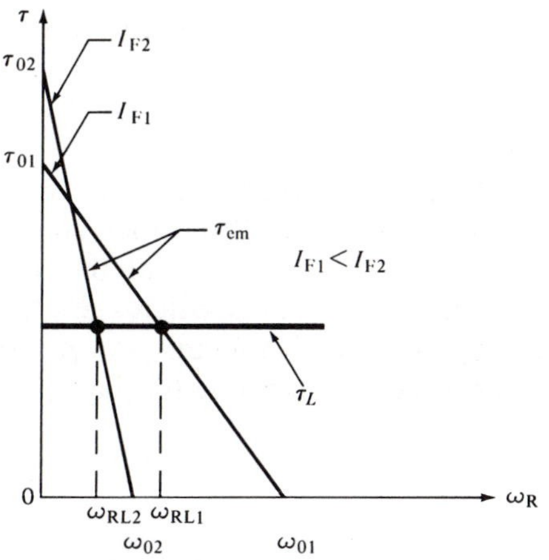

Fig. 2.21 Torque vs. Speed Characteristics of a Separately Excited Motor and Load; I_F Is the Running Parameter

$d^2\theta_R(t)/dt^2 = d\omega_R(t)/dt = 0,$

$$\tau_{em} = J\frac{d^2}{dt^2}[\theta_R(t)] \pm \tau_L \rightarrow \pm \tau_L \tag{2.91}$$

Since

$$\tau_{em} = \mathfrak{M}\, I_F I_q \tag{2.92}$$

$$I_q = \frac{\pm \tau_L}{\mathfrak{M}\, I_F} \tag{2.93}$$

In summary, it is stated that because of its excellent torque versus speed characteristics, the separately excited motor has a wide use in (1) control systems (refer to Section 2.10 for additional information), where speed control is achieved

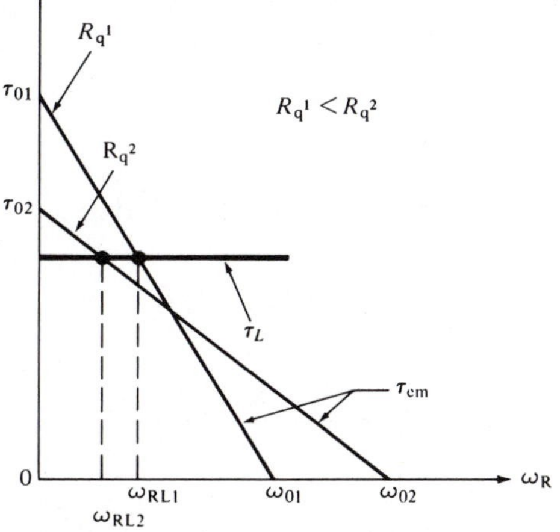

Fig. 2.22 Torque vs. Speed Characteristics of a Separately Excited Motor and Load; R_q Is the Running Parameter

by either maintaining $i_q(t)$ constant and varying $i_F(t)$ or by maintaining $i_F(t)$ constant and varying $i_q(t)$, and (2) constant speed applications since $\omega_0 - \omega_{RL} \simeq 0$.

Example 2.6

The rotor resistance of a separately excited motor, Fig. 2.18, operating at steady state is 0.20 Ω, and its flux per pole is 1.62 Wb. When connected to a 125-V source, the rotor draws 50 A. For the design constants of the motor, $P=2$ poles, $Z=40$ coils, one turn each, and $a=2$ parallel paths, calculate: (1) the generated back emf E_b, (2) the speed acquired by the rotor in radians per second and in revolutions per minute, and (3) the developed electromagnetic torque τ_{em}.

Solution

1. Use $E_b = E_I - R_q I_q$. Therefore,

$$E_b = 125 - (0.2)(50) = \underline{\underline{115}}\ \text{V}$$

2. Use $E_b = \mathcal{K}\Phi_P \omega_R$. Since

$$\mathcal{K} = \frac{NZP}{\pi a} = \frac{1\times 40\times 2}{\pi \times 2} = \frac{40}{\pi}\ \text{MKS units}$$

$$\omega_R = \frac{E_b}{\mathcal{K}\Phi_P} = \frac{115}{(40/\pi)\times 1.62} = \underline{\underline{5.58}}\ \text{rad/s}$$

$$n_R = \left(\frac{60}{2\pi}\right)\omega_R = \frac{60\times 5.58}{6.28} = \underline{\underline{53.2}}\ \text{r/min}$$

3. Use $\tau_{em} = \mathcal{K}\Phi_P I_q$. Therefore,

$$\tau_{em} = \left(\frac{40}{\pi}\right)\times(1.62)\times(50) = \underline{\underline{1031.32\ \text{N}\cdot\text{m}}}$$

Direction: The turning effect of τ_{em} is in the direction of the rotation that it causes.

Example 2.7

For Example 2.6, (1) obtain a τ_{em} versus ω_R relationship; (2) calculate the no load speed, ω_0, and the standstill torque, τ_0; (3) sketch the τ_{em} versus ω_R characteristic; and (4) calculate the speed regulation of the motor, $\omega_{RATED} = 0.95\omega_0$.

Solution

1. Since

$$E_I = \mathcal{K}\mathcal{K}_F I_F \omega_R + R_q I_q$$

and

$$\tau_{em} = \mathcal{K}\mathcal{K}_F I_F I_q$$

then

$$I_q = \frac{E_I - \mathcal{K}\mathcal{K}_F I_F \omega_R}{R_q}$$

Therefore,

$$\tau_{em} = \mathcal{K}\mathcal{K}_F I_F \left(\frac{E_I - \mathcal{K}\mathcal{K}_{<} I_F \omega_R}{R_q}\right)$$

or

$$\tau_{em} = \left(\frac{\mathcal{K}\mathcal{K}_F I_F}{R_q}\right)E_I - \left[\frac{(\mathcal{K}\mathcal{K}_F I_F)^2}{R_q}\right]\omega_R$$

2. Since $E_b = \mathcal{K}\Phi_p\omega_R \equiv \mathcal{K}\mathcal{K}_F I_F \omega_R$, then

$$\mathcal{K}\mathcal{K}_F I_F = \mathcal{K}\Phi_p = \frac{40}{\pi} \times 1.62 = \underline{20.6} \text{ MKS units}$$

$$\tau_{em} \equiv 0;\ \omega_0 \equiv \omega_R|_{\tau_{em}\equiv 0} = \frac{E_I}{\mathcal{K}\mathcal{K}_F I_F} = \frac{125}{20.7} = \underline{\underline{6.05}} \text{ rad/s}$$

$$\omega_R \equiv 0;\ \tau_0 \equiv \tau_{em}|_{\omega_R\equiv 0} = \left(\frac{\mathcal{K}\mathcal{K}_F I_F}{R_q}\right)E_I = \frac{20.6}{0.2} \times 125 = \underline{\underline{12{,}900}} \text{ N}\cdot\text{m}$$

3. Plotting τ_{em} versus ω_R yields the graph shown in Fig. 2.7.1.
4.

$$\text{SR} = \frac{\omega_0 - 0.95\omega_0}{0.95\omega_0} \times 100 = \underline{\underline{5.26\%}}$$

Series-connected dc Motor

As the name suggests, the field and rotor windings of Fig. 2.17 are connected in series and are excited by a common dc source $e_I(t)$. For steady state operation, where $L_F\, d[i_F(t)]/dt = L_q\, d[i_q(t)]/dt \equiv 0$ and $e_I(t) \rightarrow E_I$, then $i_F(t) \rightarrow I_F$, $i_q(t) \rightarrow I_q$, and $I_F \equiv I_q \equiv I_I$. With the mechanical loading included, the equivalent circuit of Fig. 2.17 is drawn as in Fig. 2.23.

The describing steady state equation of Fig. 2.23 is

$$E_I = (R_F + R_q)I_I + E_b \tag{2.94}$$

Using $E_b = \mathcal{M}\,\omega_R i_F$ enables writing Eq. (2.94) as

$$E_I = (R_q + R_F)I_I + \mathcal{M}\,\omega_R I_I \tag{2.95}$$

or

$$E_I = [(R_q + R_F) + \mathcal{M}\,\omega_R]I_I \tag{2.96}$$

Thus

$$I_I = \frac{E_I}{(R_q + R_F) + \mathcal{M}\,\omega_R} \tag{2.97}$$

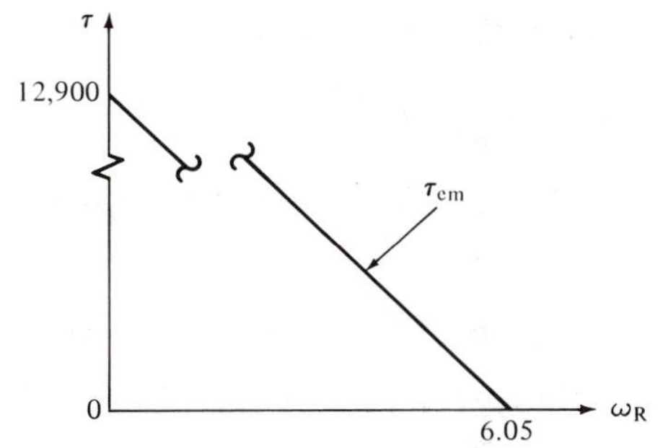

Fig. 2.7.1 Torque vs. Speed Characteristic for the Separately Excited Motor Under Steady State Operation

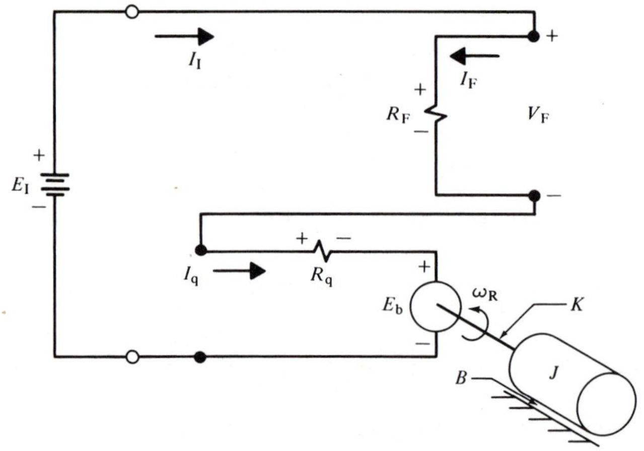

Fig. 2.23 Series-connected Motor Under Steady State Conditions of Operation

Since it is known that $\tau_{em} = \mathfrak{M}\, I_F I_q$ and $I_q = I_F \equiv I_I$, the developed electromagnetic torque of this device becomes

$$\tau_{em} = \frac{\mathfrak{M}\, E_I^2}{\left[(R_q + R_F) + \mathfrak{M}\,\omega_R\right]^2} \tag{2.98}$$

It can be verified that the torque versus speed characteristic of the series motor, that is, Eq. (2.98), can be plotted as shown in Fig. 2.24, with $\tau_0 = \mathfrak{M}\, E_I^2/(R_q + R_F)^2$ when $\omega_R = 0$. The effect of a change in E_I on the torque versus speed characteristic of the series motor should be evident from Eq. (2.98).

Example 2.8

A 250-V series motor, Fig. 2.23, has a rotor resistance of 0.1 Ω and a field resistance of 0.05 Ω. When the rotor current is 85 A, the speed of the motor is 600 r/min. Calculate: (1) the generated back emf E_b, (2) the developed electromagnetic torque τ_{em} at this speed, (3) the τ_{em} versus ω_R relationship and calculate τ_{em} for $\omega_R = 0$, 3.38, 20π and ∞ rad/s, and (4) the speed when the rotor current is 100 A.

Solution

1. Use $E_I = E_b + (R_q + R_F)I_I$. Therefore,

$$E_b = E_I - (R_q + R_F)I_I = 250 - (0.15)(85) = \underline{\underline{237.3}}\text{ V}$$

2. Use $\tau_{em} = \mathcal{K}\mathcal{K}_F I_F I_q$. Since $E_b = \mathcal{K}\mathcal{K}_F I_F \omega_R$ and

$$\omega_R = \frac{2\pi n_R}{60} = \frac{2\pi \times 600}{60} = \underline{20\pi}\text{ rad/s}$$

$$\mathcal{K}\mathcal{K}_F I_F = \frac{237.3}{20\pi} = \underline{3.78}\text{ MKS units}$$

Therefore,

$$\tau_{em} = 3.78 \times 85 = \underline{\underline{321.3}}\text{ N}\cdot\text{m}$$

3. Since $E_I = (R_q + R_F)I_I + \mathfrak{M}\,\omega_R I_I$ and $\mathcal{K}\mathcal{K}_F I_F = \mathfrak{M}\, I_F = 3.78$, then,

$$\mathfrak{M} = \frac{3.78}{I_F} = \frac{3.78}{85} = \underline{0.0445}\text{ MKS units}$$

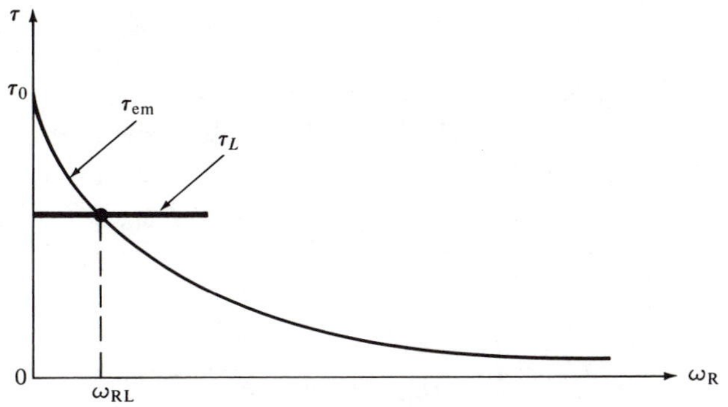

Fig. 2.24 Torque vs. Speed Characteristic of a Series Motor and Load

and

$$I_I = \frac{E_I}{0.15+0.0445\omega_R}$$

Thus,

$$\tau_{em} = \mathfrak{M} I_I^2 = \frac{0.0445(250)^2}{(0.15+0.0445\omega_R)^2} \text{ N}\cdot\text{m}$$

$$\tau_{em}\big|_{\omega_R=0} = \frac{4.45\times10^{-2}\times6.25\times10^4}{2.25\times10^{-2}} = \frac{27.8\times10^2}{2.25\times10^{-2}} = \underline{\underline{12.36\times10^4}} \text{ N}\cdot\text{m}$$

$$\tau_{em}\big|_{\omega_R=3.38} = \frac{27.8\times10^2}{9\times10^{-2}} = \underline{\underline{3.08\times10^4}} \text{ N}\cdot\text{m}$$

$$\tau_{em}\big|_{\omega_R=20\pi} = \frac{27.8\times10^2}{8.68} = \underline{\underline{3.20\times10^4}}\text{N}\cdot\text{m}$$

$$\tau_{em}\big|_{\omega_R=\infty} = \underline{\underline{0}} \text{ N}\cdot\text{m}$$

4. $$E_{b[100]} = E_I-(R_q+R_F)I_I = 250-100(0.15) = \underline{235.0}\ V.$$

Now since

$$E_{b[100]} = \mathcal{K}\mathcal{K}_F I_F \omega_{R[100]} = 235.0 = \mathcal{K}\mathcal{K}_F 100 n_{R[100]}$$

and

$$E_{b[85]} = \mathcal{K}\mathcal{K}_F I_F \omega_{R[85]} = 237.3 = \mathcal{K}\mathcal{K}_F 85 n_{R[85]}$$

taking the ratio $E_{b[100]}/E_{b[85]}$ yields

$$\frac{E_{b[100]}}{E_{b[85]}} = \frac{235.0}{237.5} = \frac{\mathcal{K}\mathcal{K}_F 100 n_{R[100]}}{\mathcal{K}\mathcal{K}_F 85 n_{R[85]}}$$

Thus

$$n_{R[100]} = \frac{235.0}{237.3}\times\frac{85}{100}\times n_{R[85]} = 0.99\times0.85\times600 = \underline{\underline{505}} \text{ rpm}$$

Shunt-connected dc Motor

Here the field and rotor windings of Fig. 2.17 are connected in parallel and are excited by a common source $E_I(t)$. For steady state operation, where $I_F\, d[i_F(t)]/dt = L_q\, d[i_q(t)]/dt \equiv 0$ and $e_I(t)\rightarrow E_I$, then $i_F(t)\rightarrow I_F$, $i_q(t)\rightarrow I_q$, and $V_F = V_q \equiv E_I$. With the mechanical load included, the equivalent circuit of Fig. 2.17 is drawn as in Fig. 2.25.

From Eqs. (2.59), (2.82), and (2.61), it is known that (1) $\tau_{em} = \mathfrak{M} I_F I_q$, (2) $I_q = (E_I - E_b)/R_q$, and (3) $E_b = \mathfrak{M} I_F \omega_R$. Thus the developed electromagnetic torque equation can be written as

$$\tau_{em} = \mathfrak{M} I_F I_q = \mathfrak{M} I_F \left(\frac{E_I - \mathfrak{M} I_F \omega_R}{R_q} \right) \tag{2.99}$$

or

$$\tau_{em} = \left(\frac{\mathfrak{M} E_I I_F}{R_q} \right) - \left(\frac{\mathfrak{M}^2 I_F^{\,2}}{R_q} \right) \omega_R \tag{2.100}$$

Defining $\tau_1 \equiv (\mathfrak{M} E_I I_F / R_q)$ and $\mu_1 \equiv (\mathfrak{M}^2 I_F^{\,2} / R_q)$ enables writing Eq. (2.100) as

$$\tau_{em} = \tau_1 - \mu_1 \omega_R \tag{2.101}$$

which can be plotted as shown in Fig. 2.26.

For $\tau_{em} = 0$ the no-load speed ω_0 becomes

$$\omega_0 = \frac{E_I}{\mathfrak{M} I_F} \tag{2.102}$$

The speed of operation ω_{RL} is defined by the loading torque τ_L. The effect of a change in E_I on the torque versus speed characteristic of the shunt motor should be evident from Eq. (2.100).

Example 2.9

A 120-V, 40-A shunt motor, Fig. 2.25, has a rotor resistance of 0.2 Ω and a field resistance of 60 Ω. The full load speed is 1800 rpm. Calculate: (1) the back emf E_b, (2) the torque developed by the motor and its full load speed, (3) the τ_{em} versus ω_R characteristic and plot it, (4) the speed at half load, (5) the rotor speed at an overload (what does that mean?) of 125%, and (6) the speed when the motor is connected to a 66-A source with R_F decreased to 50 Ω to allow for an increase of flux by 12% (which is used to produce the extra torque).

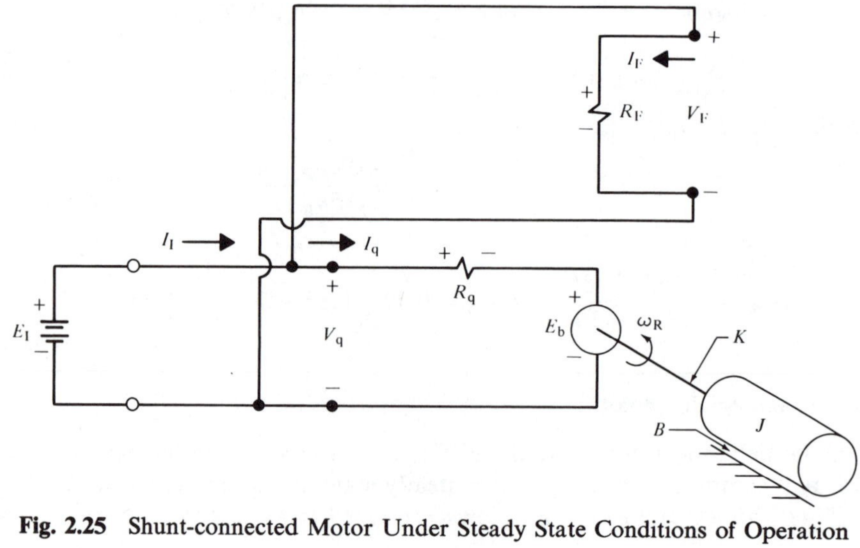

Fig. 2.25 Shunt-connected Motor Under Steady State Conditions of Operation

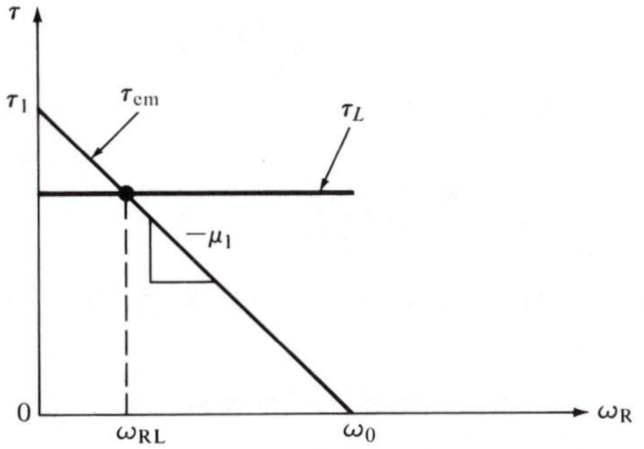

Fig. 2.26 Torque vs. Speed Characteristic of the Shunt Motor and Load

Solution

1. Use $I_I = I_q + I_F$. Since $I_I = \underline{40}$ A and

$$I_F = \frac{V_F}{R_F} = \frac{120}{60} = \underline{2}\text{ A}$$

then

$$I_q = I_I - I_F = 40 - 2 = \underline{38}\text{ A}$$

Use

$$E_I = E_b + R_q I_q$$

Therefore,

$$E_b = E_I - R_q I_q = 120 - (0.2)(38) = 120 - 7.6 = \underline{\underline{112.4}}\text{ V}$$

2. Use $\tau_{em} = \mathcal{K}\mathcal{K}_F I_F I_q$. Since $E_b = \mathcal{K}\mathcal{K}_F I_F \omega_R$ and

$$\omega_R = \frac{2\pi n_R}{60} = \frac{2\pi \times 1800}{60} = \underline{60\pi}\text{ rad/s}$$

then

$$\mathcal{K}\mathcal{K}_F I_F \doteq \frac{E_b}{\omega_R} = \frac{112.40}{60\pi} = \underline{0.60}\text{ MKS units}$$

Therefore

$$\tau_{em} = 0.60 \times 38 = \underline{\underline{22.70}}\text{ N}\cdot\text{m}$$

3. Use $\tau_{em} = \mathcal{M} I_F I_q$. Since $E_I = E_b + R_q I_q$, then

$$I_q = \frac{E_I - E_b}{R_q} = \frac{E_I - \mathcal{M} I_F \omega_R}{R_q}$$

Therefore,

$$\tau_{em} = \mathcal{M} I_F \left(\frac{E_I - \mathcal{M} I_F \omega_R}{R_q} \right) = \left(\frac{\mathcal{M} I_F}{R_q} \right) E_I - \left[\frac{(\mathcal{M} I_F)^2}{R_q} \right] \omega_R$$

Since $\mathcal{M} I_F \equiv \mathcal{K}\mathcal{K}_F I_F = 0.60$ MKS units,

$$\tau_{em} = \left(\frac{0.60}{0.2} \right) E_I - \left[\frac{(0.60)^2}{0.2} \right] \omega_R$$

or

$$\tau_{em} = \underline{\underline{3E_I - 1.8\omega_R}}$$

and the graph is shown in Fig. 2.9.1.

4. Half-load speed:

$$I_{qHL} = \frac{I_q}{2} = \frac{38}{2} = \underline{19}\text{ A}$$

$$E_{bHL} = E_I - R_q I_{HL} = 120 - (0.2)(19) = \underline{116.2}\text{ V}$$

Now

$$E_{bFL} = K_V \omega_{RFL}$$

and

$$E_{bHL} = K_V \omega_{RHL}$$

Take the ratio of E_{bFL}/E_{bHL}. Thus

$$\frac{E_{bFL}}{E_{bHL}} = \frac{K_V \omega_{RFL}}{K_V \omega_{RHL}} = \frac{n_{RFL}}{n_{RHL}}$$

Therefore,

$$n_{RHL} = \left(\frac{E_{bHL}}{E_{bFL}}\right) n_{RFL} = \frac{116.2}{112.4} \times 1800 = \underline{\underline{1860}}\text{ rpm}$$

5. 125% overload speed:

$$I_{q[125]} = I_q + \frac{I_q}{4} = 38 + \frac{38}{4} = \underline{47.5}\text{ A}$$

$$E_{b[125]} = E_I - R_q I_{q[125]} = 120 - (0.2)(47.5) = \underline{110.5}\text{ V}$$

From part 4,

$$n_{R[125]} = \left(\frac{E_{b[125]}}{E_{bFL}}\right) n_{RFL} = \frac{110.7}{112.4} \times 1800 = \underline{\underline{1773}}\text{ rpm}$$

6. Use $I_q = I_I - I_F$. Since $I_I = 66$ A and

$$I_F = \frac{V_F}{R_F} = \frac{120}{50} = \underline{2.4}\text{ A}$$

then

$$I_q = I_I - I_F = 66 - 2.4 = \underline{63.6}\text{ A}$$

$$E_{b[66]} = E_I - R_q I_q = 120 - (0.2)(63.6) = \underline{107.3}\text{ V}$$

$$E_{bFL} = K_V \omega_{RFL} \simeq K_V n_{RFL} = 112.4$$

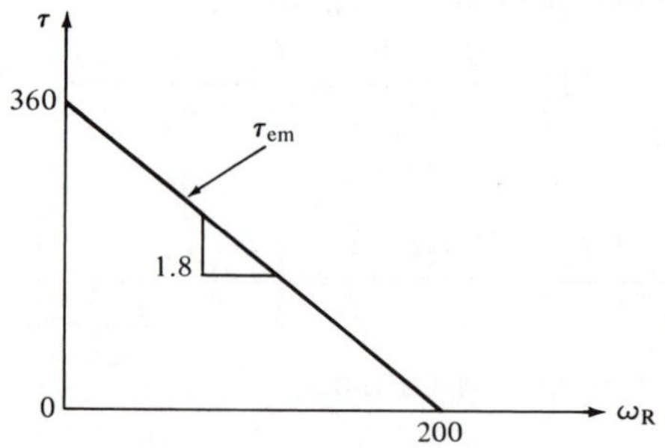

Fig. 2.9.1 Torque vs. Speed Characteristic for the Shunt Motor Under Steady State Operation

Therefore,

$$K_V = \frac{112.4}{1800}$$

$$E_{b[66]} = K_V \omega_{R[66]} \simeq K_V n_{R[66]}$$

Therefore

$$n_{R[66]} = \frac{E_{b[66]}}{K_V} = \frac{107.3}{112.4/1800} = \underline{\underline{1718}} \text{ rpm}$$

2.7 STABLE VERSUS UNSTABLE MODES OF OPERATION OF dc MOTORS

For normal operation (i.e., the device operates in the linear region of the B versus H curve of its magnetic circuit) of shunt motors, the speed regulation may vary from 2 percent to 8 percent. However, when the device operates in or near the knee of the magnetization curve of the magnetic circuit, the **armature reaction** (see Section 2.8) may reduce considerably the flux per pole, Φ_P. Since $E_b = \mathcal{K}\Phi_P\omega_R$ and $\tau_{em} = \mathcal{K}\Phi_P I_q$ [Eqs. (2.60) and (2.59)], an increase or reduction of Φ_P yields an increase or decrease of τ_{em} and E_b. These changes may lead to a more **stable** or **unstable** operation of the shunt motor.

The concept of stable or unstable operation of a motor can be understood from the following discussion.

It should be known by now that *the function of a motor is to supply mechanical power to be used to drive a mechanically operated load.* In such a task it is desired to drive the mechanical device at a given speed or, perhaps, at different speeds over different intervals of time. Such requirements lead to the desire to be able to *control* the speed of a motor.

The problem of **speed control** is to devise a scheme whereby the point of intersection of the two torque versus speed curves, that is, τ_{em} versus ω_R and τ_L versus ω_R, is made to lie anywhere within the first quadrant of the torque versus speed plane.

Fig. 2.27 is similar to Fig. 2.26, with a superimposed load versus speed characteristic, τ_L versus ω_R. Eq. (2.100) suggests that the τ_{em} versus ω_R characteristic may be shifted left or right by varying I_F, or E_I, or both, in such a manner that the resulting torque versus speed characteristic of the motor (τ_{em} versus ω_R) can be made to intersect with the torque versus speed characteristic of the load (τ_L versus ω_R) at any desired point Q.

The current I_F can be changed by varying an external resistance R_{EXT} purposely connected in series with R_F. However, changing R_{EXT} to values greater than zero, thus decreasing I_F, results in shifting the τ_{em} versus ω_R characteristic **only to the right** of the point Q. The resulting new speed ω_{RLH} is limited by the structural limitations of the device. Otherwise, the device might destroy itself due to excessive centrifugal forces created by high speeds.

To make the τ_{em} versus ω_R characteristic move to the left, E_I is supplied from a device capable of producing a variable voltage E_I (either automatically or manually). Thus smaller values of E_I, yielding point Q in Fig. 2.27, will effect a shift of the τ_{em} versus ω_R characteristic to the left. This shift of the τ_{em} versus ω_R characteristic provides speeds lower than ω_{RL}. In contrast, higher values of E_I,

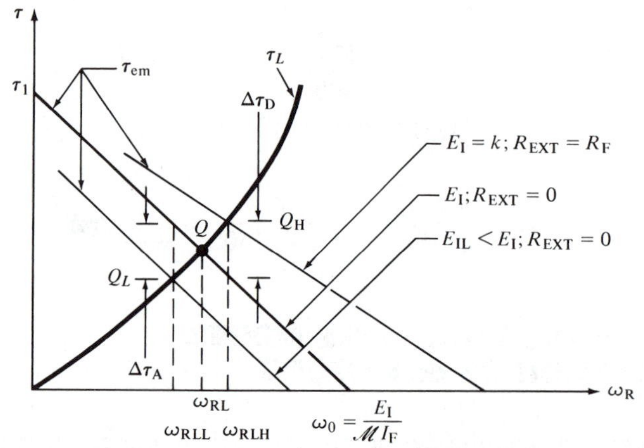

Fig. 2.27 Torque vs. Speed Characteristics of a Shunt Motor and Load, Stable Operation

yielding a point Q in Fig. 2.27, will provide speeds higher than ω_{RL}. Thus, changing E_I can shift the τ_{em} versus ω_R characteristic **either to the left or right** of the operating point Q, yielding new points of intersection, Q_L or Q_H, Fig. 2.27, and corresponding speeds, ω_{RLL} or ω_{RLH}. In contrast, I_F shifts the τ_{em} versus ω_R characteristic only to the ritht of the operating point if the field supply voltage is held constant, because resistance may be added to the field winding, but not removed from the winding itself.

Note that (1) the speed range that can be attained with field current (I_F) control *extends from 1.5:1 to 6:1* and (2) the speed range that can be attained with rotor (armature) voltage (E_I) control *extends from 10:1 to 30:1*. These ranges depend on the base speed ω_{RL}. *Base speed is defined as the maximum speed obtained with maximum rotor voltage and field current*.

Now assume that a disturbance in the system forces the motor to increase its speed from ω_{RL} to ω_{RLH}, Fig. 2.27. What would happen? Would the motor return to its original operating speed ω_{RL}? At the new speed ω_{RLH}, $\tau_L > \tau_{em}$. Thus $\Delta\tau_D = \tau_{em} - \tau_L < 0$. In order to develop more torque to meet the demand of the load, the motor must **decelerate**. The end result of this process is that the motor will return to the operating point Q. Similarly, a disturbance that forces the motor to decrease its speed to ω_{RLL} yields $\tau_{em} > \tau_L$. This excessive accelerating torque, $\Delta\tau_A = \tau_{em} - \tau_L > 0$, tends to bring the motor to the operation point Q. Thus the equilibrium point Q in Fig. 2.27 is defined as a *stable equilibrium point* since the device *always returns* to this point of operation regardless of the nature of the disturbance.

However, if the slope of τ_{em} versus ω_R in Fig. 2.27 changes sign, the situation becomes entirely different. Consider Fig. 2.28. If a disturbance in the system forces the motor to increase its speed to ω_{RLH}, what would happen? Would the device return to its original operating speed ω_{RL}? At the new speed ω_{RLH}, $\tau_{em} > \tau_L$. Thus $\Delta\tau_A = \tau_{em} - \tau_L > 0$. Since the motor is producing more torque than the load requires, the motor accelerates and ω_R must increase. The increase in speed results in an increasing increment of torque. The cumulative result is that the motor tends to run away; a destructive consequence. Similarly, a disturbance that forces the motor to decrease its speed to ω_{RLL} yields $\tau_L > \tau_{em}$, thus $\Delta\tau_D = \tau_L - \tau_{em} < 0$. In this case the motor stops; an undesirable consequence. Thus the equilibrium point Q in Fig.

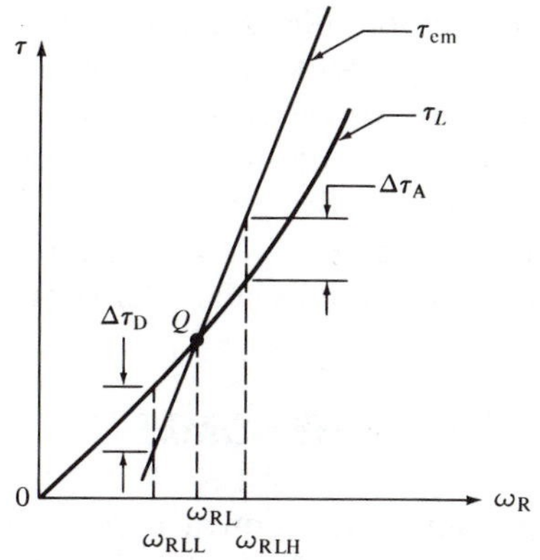

Fig. 2.28 Torque vs. Speed Characteristic of a Shunt Motor and Load, Unstable Operation

2.28 is defined as an **unstable equilibrium point** since the device **never returns** to this point of operation once the device is disturbed.

It must be noted that the nature of the equilibrium, that is, stable or unstable, is determined by the slopes of τ_{em} versus ω_R and τ_L versus ω_R in the neighborhood of the operating point Q, Figs. 2.27 and 2.28.

To avoid instability a distinctly separate winding $R_S L_S$ is added in series with the $R_q L_q$-winding and is wound in a manner so that its mmf is either *additive* or *subtractive* to the mmf of the field winding $R_F L_F$.

This new winding gives rise to two new motor types: (1) *cumulative compound*, that is, when the fields of the windings $R_F L_F$ and $R_S L_S$ are *additive*, and (2) *differential compound*, that is, when the fields of the windings $R_F L_F$ and $R_S L_S$ are *subtractive*. The schematic diagram of the cumulative and differential compound motor is given in Fig. 2.29. Note the distinct differences between the schematic

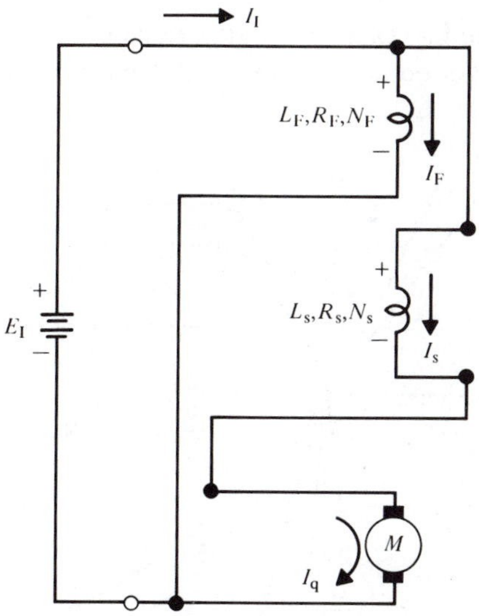

Fig. 2.29 Schematic Diagram for Compound-type Motor; $R_S L_S$ and $R_F L_F$ Windings Are on the Same Stator Axis, but $N_F \gg N_S$: (1) Cumulative Compound, Fields Are Aiding; (2) Differential Compound, Fields Are Opposing

diagram of the compounded motor, Fig. 2.29, and the equivalent circuits of the series and shunt motors, Figs. 2.23 and 2.25. The mathematical analysis of these two devices *is not undertaken* here. However, the torque versus speed characteristics of these two motor types along with the torque versus speed characteristics of the motors that are fully discussed in Section 2.6, are exhibited in Fig. 2.30 for *comparative* purposes. The scales of the axes are in percentage of the rated values to simplify comparison among the various characteristics.

2.8 ARMATURE REACTION PHENOMENON

The term armature reaction that is mentioned in Section 2.7 is a very important concept and warrants special attention. To understand its significance, refer to Fig. 2.31. Fig. 2.31(a) exhibits the dc motor structure, its windings, and its various magnetic fields, B_S, B_R, and B_P. Note that *the field direction from stator to rotor is assumed to be positive*. Fig. 2.31(b) shows the waveforms of the same magnetic fields. The resultant field of the rotor coils B_R is along the q-axis (when $B_S=0$), and the resultant field of the stator B_S is along the d-axis (when $B_R=0$). Fig. 2.31(a) also shows the resultant field B_P (heavy lines) of the system. This field is a result of the interaction of the stator and rotor fields when both fields coexist in the system, that is, $B_R \neq 0$ and $B_S \neq 0$ at the same time. The two fields *add* when their directions are the same and *subtract* when their directions oppose. The flux distribution of the stator, which was originally uniform in the stator (with zero rotor current), is now nonuniform. It is *dense* where the fields are additive and is *less dense* where the fields are subtractive. The net result is distortion of the field B_P, Fig. 2.31(a), and skewing of the flux lines between the poles. This skewing of the flux lines causes the *magnetically neutral axis* (abbreviated as MNA), that is, the axis normal to the stator field when the rotor field is zero, to shift $\Delta\theta_R$ in the direction of rotation. Compare the skewing of B_P in Fig. 2.31(a) to the deformation of the waveform of B_P in Fig. 2.31(b).

For a linear system the increase of the flux density at the points where the fields are additive is offset by an equal amount of decrease in flux density at the points

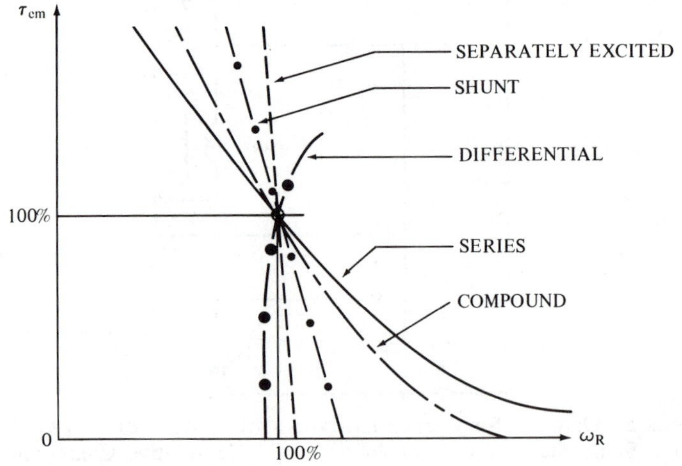

Fig. 2.30 Torque vs. Speed Characteristics of Various Types of dc Motors, a Comparative Exposé

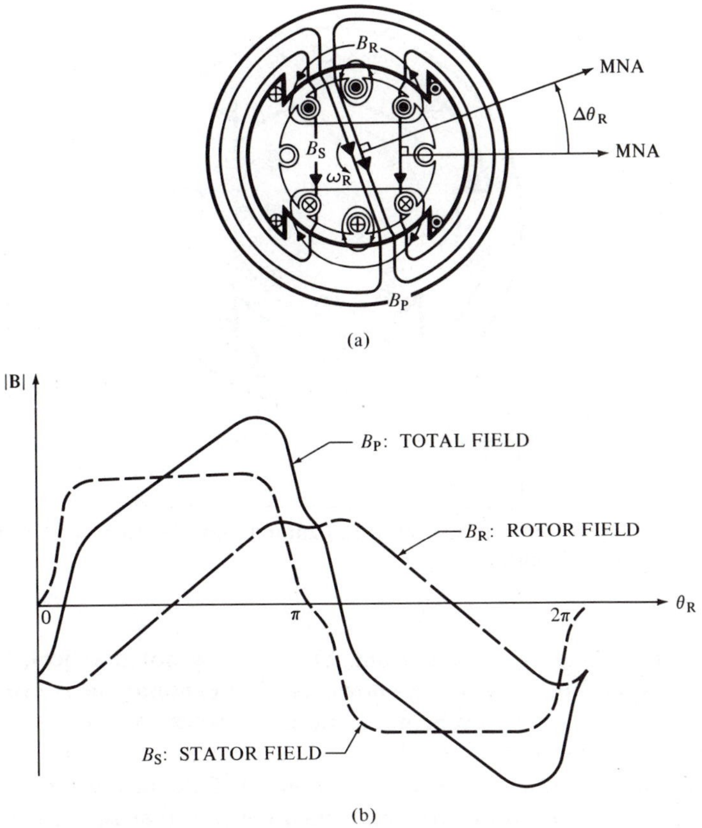

Fig. 2.31 Armature Reaction Effects (that is, effects of rotor field on stator field). Note (a) the Shifting of the Magnetically Neutral Axis (MNA) and (b) the Deformation and Shift of the Total Waveform

where the fields are subtractive. Thus the total flux between the poles remains more or less the same. However, the magnetic circuits of most practical devices are made of ferromagnetic material, with operating flux densities near the knee of the B versus H characteristic of the material. Thus for an operating point at the knee of the B versus H curve or close to it, it should be obvious that the incremental rise in B_S due to the additive B_R is smaller than the incremental decrease of B_S due to the subtractive B_R. It should be clear, therefore, that this effect is totally due to saturation, and the result is a reduction in the pole flux Φ_P.

These two effects of the rotor current on the field distribution, that is, (1) the distortion of the flux distribution and, consequently, the shifting of the magnetically neutral axis and (2) the decrease in the total flux, constitute the so-called **armature reaction** effect.

Armature reaction has two distinct effects on the performance of dc devices, as described in the following paragraphs.

Cross-magnetizing Effect

This affects the commutation. Since the rotor current changes magnitude, the magnetically neutral axis shifts position. Thus the brushes, which are placed perpendicular to the coils on this axis to avoid sparking, should constantly shift

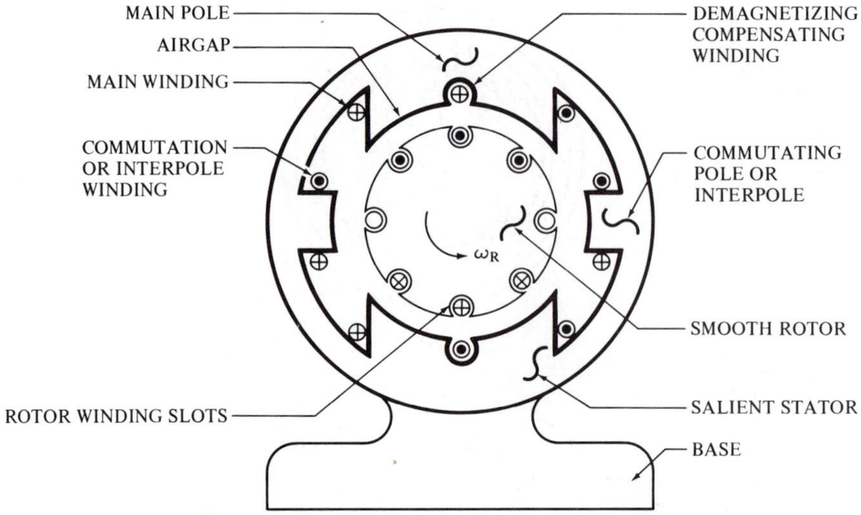

Fig. 2.32 Detailed Outlay of a dc Device with Commutation Winding and Demagnetizing Compensating Winding

position with the changing rotor current. Since this is not practical, the brushes remain fixed in space and a winding known as a **commutation or interpole winding** is placed on two poles (halfway between the main poles), known as **interpoles** or **commutating poles**. The polarity of this winding is such that *it counteracts* (1) the emf generated in the rotor coils (due to a nonzero field in the **interpolar region**) undergoing commutation and (2) the reactance voltage, that is, the emf generated due to self-induction in the rotor coils undergoing commutation.

Demagnetizing Effect

This affects the generated emf. Since $E_q = \mathcal{K}\, \Phi_P \omega_R$, any change in Φ_P affects the generated emf E_q. Of course, since Φ_P is dependent on the rotor current, the generated emf also depends on the rotor current. To avoid variations of the generated emf E_q, a compensating winding $R_D L_D$, known as a **demagnetizing compensating winding**, is located in the main pole faces and is connected in series with the rotor circuit in such a way that the magnetic field it sets up counteracts the demagnetizing effect of the armature reaction.

The detailed outlay of a dc device with the main, interpolar or commutation, and compensating windings present is exhibited in Fig. 2.32.

2.9 VOLTAGE VERSUS CURRENT CHARACTERISTICS OF VARIOUS dc GENERATOR TYPES UNDER STEADY STATE OPERATION

As with the case of motor excitation, depending on the connection of the field and rotor windings to the excitation or load, the dc generator assumes corresponding names, such as those given in the following paragraphs.

Separately Excited dc Generator

As the name implies, the field and rotor circuits are not physically connected. For simplicity, the equivalent circuit of Fig. 2.15 is redrawn here, Fig. 2.33, for steady state operation, that is, $L_F\ d[i_F(t)/dt = L_q\ d[i_q(t)/dt \equiv 0$, $i_F(t) \to I_F$, and $i_q(t) \to I_q$.

The describing equation of the rotor circuit for steady state operation becomes

$$E_q = R_q I_q + V_L \tag{2.103}$$

Solution of Eq. (2.103) for the terminal voltage V_L yields

$$V_L = E_q - R_q I_q \tag{2.104}$$

Eq. (2.104) is identical to the describing equation of a simple battery, that is, $v = E_0 - R_0 i$, and provides the same information about this device. Eq. (2.104) is plotted in Fig. 2.34, and its circuit analog is drawn in Fig. 2.35. Note that in Fig. 2.34 R_q designates the slope of the voltage versus current characteristic of the device, E_q designates the v-axis intercept, and V_L and $I_L \equiv I_q$ designate the terminal voltage and current of the device. However, the use of Eqs. (2.36) and (2.53), that is, $E_q = \mathcal{K}\Phi_P\omega_R$ and $\Phi_P = (N_F/\mathcal{R})I_F = \mathcal{K}_F I_F$, also enables us to write the generated voltage E_q, Eq. (2.36), as

$$E_q = \mathcal{K}\mathcal{K}_F I_F \omega_R \equiv \mathcal{M} I_F \omega_R \tag{2.105}$$

where $\mathcal{M}$ is constant if saturation is neglected, the field current I_F is obtained from the excitation source under steady state operation, that is,

$$E_F = R_F I_F \tag{2.106}$$

and the velocity ω_R is provided by the prime mover. This velocity is always assumed to be maintained constant, that is, $\omega_R \equiv \omega_I = k$, or very close to being constant.

The characteristic of Fig. 2.34, curve 1, is known as the **external characteristic** of the separately excited generator, plotted for (1) constant field current I_F; (2) constant speed ω_R, and (3) armature reaction ignored. Such a device is defined as a *real* voltage source. Should armature reaction be taken into account, the external characteristic will be slightly modified (reduced), Fig. 2.34, curve 2, because of the reduction of the net flux due to saturation (see Section 2.8).

Ideally, it would be desirable to have a device for which V_L remains independent of I_q. Such a device should have $R_q = 0$ and would be looked upon as an *ideal* voltage source. The external characteristic of such an ideal voltage source is exhibited in Fig. 2.34, curve 3.

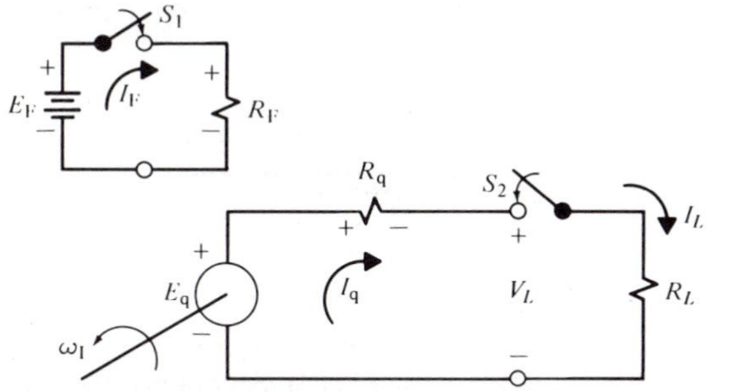

Fig. 2.33 Separately Excited Generator Under Steady State Operation

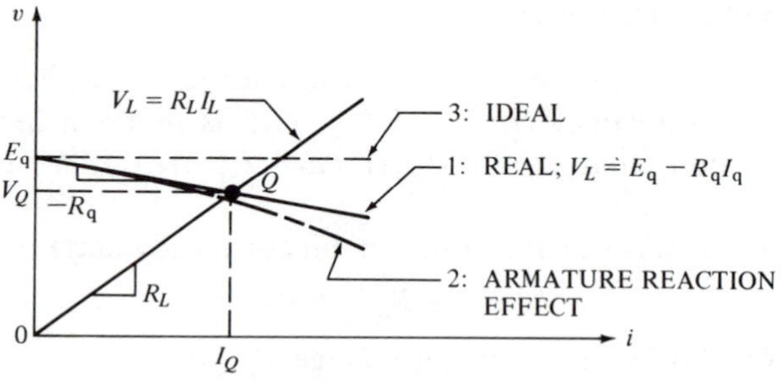

Fig. 2.34 Voltage vs. Current Characteristics for a Separately Excited Generator and Load Under Steady State Operation

The nature of this device can be better understood if the power balance is examined. The power input to the field,

$$p_F = E_F I_F \tag{2.107}$$

is dissipated in the field resistor R_F. However, the mechanical power supplied by the prime mover is

$$p_I = \omega_R \tau_I \equiv \omega_I \tau_I \tag{2.108}$$

A portion of this mechanical power p_I is lost because of frictional losses, p_{BF}, and core losses, p_{CL}. The remainder of p_I is available to the rotor circuit to be converted to electromagnetic power, that is,

$$p_{em} = \omega_R \tau_{em} \equiv \omega_I \tau_{em} \tag{2.109}$$

Thus

$$p_{em} = p_I - (\text{frictional losses} + \text{core losses}) \tag{2.110}$$

or

$$\omega_I \tau_{em} = \omega_I \tau_I - (\text{total losses}) \tag{2.111}$$

The electromagnetic power, p_{em}, can be calculated from Eq. (2.104) by multiplying the equation by $I_L \equiv I_q$ and utilizing Eq. (2.36). Thus

$$V_L I_L = E_q I_q - R_q I_q^2 = \mathcal{K} \Phi_P \omega_R I_q - R_q I_q^2 \equiv p_O \tag{2.112}$$

Eq. (2.52) enables writing Eq. (2.112) as

$$V_L I_L = \tau_{em} \omega_R - R_q I_q^2 \equiv p_O \tag{2.113}$$

or

$$p_{em} = \tau_{em} \omega_R = V_L I_L + R_q I_q^2 = p_O + R_q I_q^2 \tag{2.114}$$

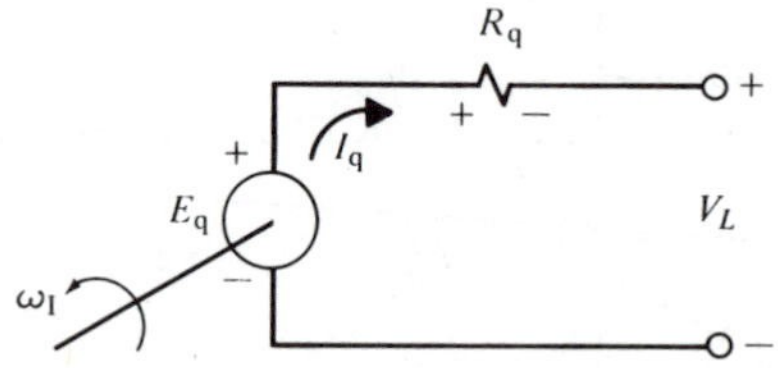

Fig. 2.35 Voltage Source Equivalent Circuit of a Separately Excited Generator, Steady State Operation

Thus the electromagnetic power converted from mechanical power provides for the I^2R-losses in the rotor and the rest avails itself as an electric power output. Therefore, the separately excited generator can be looked upon as an *electric power amplifier* with a power amplification, $G \equiv p_O/p_F$, given as

$$G = \frac{V_L I_L}{E_F I_F} \gg 1 \tag{2.115}$$

It should be obvious that since the field power $E_F I_F$ is dissipated in the field resistance R_F, the output power $V_L I_L$ is coming directly from the mechanical power supplied to the shaft by the prime mover. On account of this arrangement, the dc generator is often referred to as the *rotating amplifier*.

Since the objective of the separately excited generator is to supply voltage to a load, it is essential to have a measure of the constancy of this voltage. The quantity **percentage voltage regulation** (VR) is used as the index of measurement of the constancy of the terminal voltage. It is defined as

$$\mathrm{VR} \equiv \frac{E_q - V_L}{V_L} \times 100 \equiv \frac{E_{\text{NO LOAD}} - V_{\text{FULL LOAD}}}{V_{\text{FULL LOAD}}} \times 100 \tag{2.116}$$

Eq. (2.116) gives a measure of the ability of the separately excited generator to maintain constant terminal voltage as the loading conditions change from no-load to full-load.

Example 2.10

A two-pole separately excited generator, Fig. 2.33, has a rotor containing 40 one-turn coils connected in two parallel paths. The flux per pole is 3.24 Wb and the steady state speed n_R of the prime mover is 30 rpm. The resistance of each coil side is 0.01 Ω and its current-carrying capacity is 10 A. Calculate: (1) the generated voltage E_q, (2) the full-load current delivered to an external load I_q, (3) the rotor resistance R_q, (4) the terminal voltage of the generator, V_L, (5) the average voltage generated per coil, v_{AVG}, (6) the loading resistance for the given conditions, R_L, (7) the developed electromagnetic torque, τ_{em}, magnitude and direction, (8) the voltage regulation VR of the generator, and (9) the power gain G of the generator.

Solution

1. Use $E_q = \mathcal{K}\Phi_p\omega_R$.

$$\mathcal{K} = \frac{NPZ}{\pi a} = \frac{1\times2\times40}{\pi\times2} = \underline{\frac{40}{\pi}} \text{ MKS units}$$

$$\omega_R = \frac{2\pi n_R}{60} = \frac{2\pi\times30}{60} = \underline{\pi} \text{ rad/s}$$

Therefore,

$$E_q = \frac{40}{\pi}\times3.24\times\pi = \underline{\underline{129.60 \text{ V}}}$$

2. Use

$$I_q = I_L = (2 \text{ paths})(10 \text{ A/path}) = \underline{\underline{20}} \text{ A}$$

3. Use

$$R_q = \frac{1}{G_q} = \frac{1}{G_1 + G_2}$$

where

$$R_1 = \frac{1}{G_1} \equiv \text{resistance per path}$$

$$= \left(\frac{40}{2}\right) \times \frac{\text{coil sides}}{\text{path}} \times 0.01\left(\frac{\text{ohm}}{\text{coil side}}\right) = \underline{0.20}\ \frac{\Omega}{\text{path}}$$

Therefore,

$$G_1 = \frac{1}{0.2} = \underline{5}\ \text{mho}$$

Since $R_1 = R_2$ and $G_1 = G_2$, that is, the number of coil sides per path is the same,

$$R_q = \frac{1}{5+5} = \underline{\underline{0.10\ \Omega}}$$

4. Use $V_L = E_q - R_q I_q$.

$$V_L = 129.6 - 0.1 \times 20 = \underline{\underline{127.60}}\ \text{V}$$

5.
$$v_{AVG} = \frac{E_q}{\text{coils/path}} = \frac{129.6}{(Z/2)/2} = \frac{129.6}{10} = \underline{\underline{12.96}}\ \text{V}$$

6. Use the diagram of Fig. 2.10.1 and

$$V_L = R_L I_q \quad \text{or} \quad R_L = \frac{V_L}{I_q}$$

Therefore,

$$R_L = \frac{127.6}{20} = \underline{\underline{6.38\ \Omega}}$$

7. Use $\tau_{em} = -\mathcal{K}\Phi_p I_q$.

$$\tau_{em} = -\frac{40}{\pi} \times 3.24 \times 20 = \underline{\underline{-825}}\ \text{N}\cdot\text{m}$$

The turning effect of τ_{em} opposes the turning effect of the externally applied torque τ_I.

8. Use

$$\text{VR} = \frac{E_q - V_L}{V_L} \times 100$$

Thus

$$\text{VR} = \frac{129.60 - 127.60}{127.60} \times 100 = \underline{\underline{1.57\%}}$$

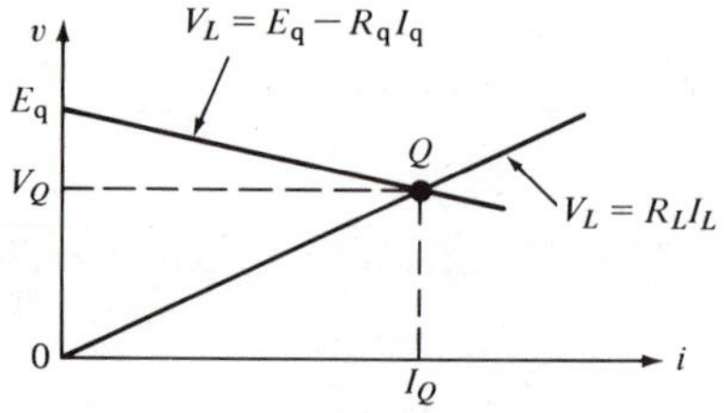

Fig. 2.10.1 Voltage vs. Current Characteristic for the Separately Excited Generator and Load Under Steady State Operation

9. Use

$$G = \frac{p_L}{p_F I_F}$$

Since

$$p_L = V_L I_L = 127.6 \times 20 = \underline{2552}\text{ W}$$

and

$$p_F = E_F I_F$$

Therefore,

$$G = \frac{2552}{?}$$

Note that adequate information is not provided. However, if the field circuit quantities were specified, then the power gain G could be determined.

Shunt-connected dc Generator

As in the case of the shunt motor, the field and rotor windings are connected in parallel and this connection is placed across the load, Fig. 2.36. When compared to the separately excited generator, it is seen that there is no forcing function, E_F, in the field winding of the shunt generator to provide the field that is essential for the voltage generation, $E_q = \mathcal{K}\Phi_p\omega_R$. How then is the voltage generated? The answer to this question is found in the most undesirable (in many instances) properties of ferromagnetic materials. Namely, **retentivity** and **saturation**. Both are utilized here very ingeniously to make the operation of this device possible.

For the condition exhibited in Fig. 2.36, that is, the generator being unloaded and the rotor rotated by a prime mover at a constant speed $\omega_R \equiv \omega_I = k$, what is the terminal voltage V_L?

The describing equations of the dynamic system are

$$v_F(t) = R_F i_F(t) + \frac{d}{dt}[L_F i_F(t)] \tag{2.117}$$

$$v_q(t) = E_q - R_q i_q(t) - \frac{d}{dt}[L_q i_q(t)] \tag{2.118}$$

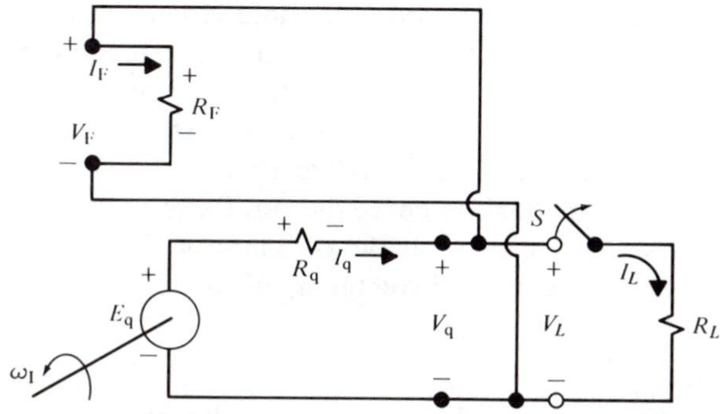

Fig. 2.36 Shunt Generator Under Steady State Operation

Since $R_L = \infty$,

$$i_q(t) = i_F(t) \tag{2.119}$$
$$v_L(t) = v_F(t) = v_q(t) \tag{2.120}$$

Eq. (2.105) yields

$$E_q = \psi[i_F(t)] \tag{2.121}$$

Solution of these equations should give the desired answer. However, since L_F is a nonlinear function of $i_F(t)$ and E_q is nonlinearly related to $i_F(t)$, solution of these equations is cumbersome and is therefore beyond the scope of this text.

To understand the operation of the device, assume that* $R_F \gg R_q$ or $R_q \simeq 0\ \Omega$. Thus for steady state operation, that is, $L_j\, d[i_j(t)]/dt \equiv 0$, $j = \mathrm{F}$, q, Eqs. (2.117), (2.118), and (2.121) in conjunction with Eq. (2.120) become

$$V_F = R_F I_F \equiv V_L \tag{2.122}$$

and

$$V_q = E_q = \mathfrak{M}\omega_R I_F \equiv V_L \tag{2.123}$$

Thus for the shunt generator, which obviously supplies its own excitation current,† V_F versus I_F and E_q versus I_F [i.e., the **no-load magnetization curve** plotted for $\omega_R \equiv \omega_I = k$, Fig. 2.14(b)] can be plotted on the same ordinates, Fig. 2.37.

Fig. 2.37 helps us to understand the manner in which the self-excited shunt generator manages to excite its own field and build a dc voltage V_L across its unloaded terminals.

Assume that the field resistance is $R_F = R_{F2}$ and the prime mover speed is $\omega_R \equiv \omega_I = 0$. Despite the residual magnetism, Fig. 2.14(a), the generated voltage E_q is zero, (i.e., $E_q = 0$). As the speed approaches rated speed, that is, $\omega_R \equiv \omega_I = \omega_R^{RTD}$, the voltage due to residual magnetism and speed increases and becomes $E_q = \mathcal{K}\Phi\omega_R \equiv E_R$. At rated speed this voltage E_R appears across both the rotor and the field circuits. Thus the current in the field circuit is I_R, Fig. 2.37. When I_R flows in the field circuit of the generator, an increase in mmf results due to $I_R N_F$, which aids the residual magnetism in increasing the induced voltage to E_1. Voltage E_1 is now impressed across the field circuit, causing a larger current I_1 to flow in the field circuit, which in turn produces a higher generated voltage E_2.

The process continues until the point Q, where the field resistance line, R_{F2}, crosses the magnetization curve,‡ Fig. 2.37. Here the process stops. The induced voltage produced, when impressed across the field circuit, produces a current flow that in turn produces an induced voltage of the same magnitude, that is, E_q, as shown in Fig. 2.37. The process comes to a stop if a steady state solution exists, that is, if Eqs. (2.122) and (2.123) are simultaneously satisfied. Note that this process *would not have started* if the residual flux were zero and *would not have stopped* if the no-load magnetization curve did not flatten out for large values of I_F. It should be made perfectly clear that the ordinate of the point of intersection of the two curves represents the **no-load voltage** E_q of the self-excited shunt generator.

*Usually $R_F/R_q > 100$.

† E_F is a potential rise across the resistor R_F, but the potential drop is V_F.

‡ Designated as $E_q = \psi(I_F)$.

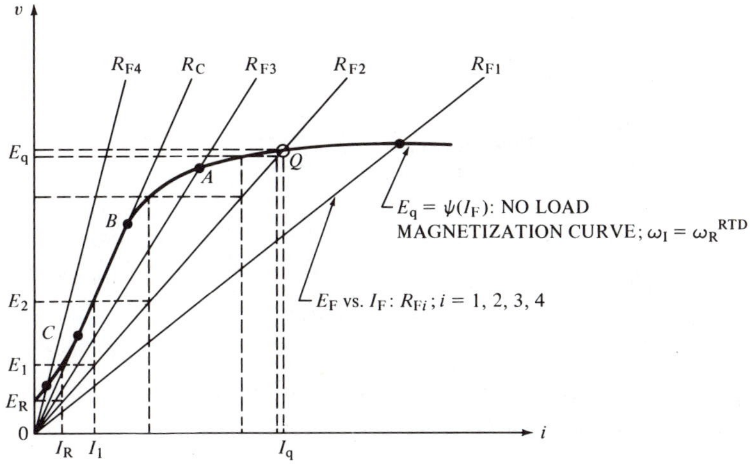

Fig. 2.37 Buildup Characteristic of the Shunt (Self-excited) Generator

The effect of the field resistance on the no-load voltage should be obvious. As the field resistance R_F increases, the no-load voltage decreases (points $A, B, \ldots$) and the device will not operate if the field resistance is greater than the **critical resistance** R_C, which is the slope of the line *just tangent* to the no-load saturation curve.

The perfect match of the field resistance line and the magnetization curve of a magnetic circuit leads to a somewhat unusual application. For a properly designed device with a long linear segment of its magnetization curve and precisely the right resistance, R_C, in the field circuit, the field resistance line lies exactly on top of the straight-line portion of the magnetization curve. Now the shunt generator is said to have a *tuned field*. There is no **single** point of intersection between the two lines, that is, E_F versus I_F and $E_q = \psi(I_F)$, so the generator can theoretically come to equilibrium at **any** point on the common section *BC*, Fig. 2.37. If any slight change is made, in either the field resistance or the saturation curve, a *definite equilibrium point* is established at one end or the other of the straight-line section. This can be made to correspond to a *very large change in power output* for a very small change in power required for the field. In other words, with a tuned field the shunt generator can become a *high-gain power amplifier*. This principle is employed in the design of the rotating amplifiers known as *Rototrol* (Westinghouse) and *Regulex* (Allis Chalmers).

When the generator is loaded and $R_q \neq 0$ the current equation is

$$I_q = I_L + I_F \tag{2.124}$$

Eq. (2.124), in addition to the voltage equation,

$$V_L = E_q - R_q I_q \tag{2.125}$$

Eq. (2.125), are required to describe the device.

Substitution of Eq. (2.124) into Eq. (2.125) yields

$$V_L = E_q - R_q(I_L + I_F) \tag{2.126}$$

Eq. (2.126) is plotted in Fig. 2.38.

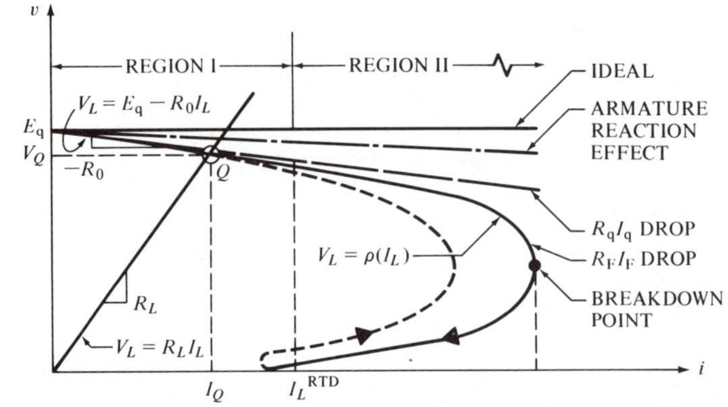

Fig. 2.38 Voltage vs. Current Characteristics of a Shunt (Self-excited) Generator and Load Under Steady State Operation

To obtain data for the V_L versus I_L characteristics,* the following assumptions are considered valid: (1) there is no lack of residual magnetism in the magnetic circuit; (2) the field circuit is properly connected across the rotor circuit; (3) the rotor is rotating in the proper direction (this is essential because the brushes reside at an angle); (4) the field resistance is equal to or less than the critical resistance; (5) the rotor resistance is not too high; (6) the generator is loaded with a variable load, that is, $0<R_L<\infty$ after the buildup process is completed† otherwise, since $R_L \ll R_F$, most of the current would be diverted to R_L and the buildup process would fail.

As R_L is reduced in an attempt to increase the load current I_L, it is observed that the terminal voltage V_L of the generator decreases, Fig. 2.38. The decrease of V_L takes place because of three reasons: (1) the increase of the internal voltage drop $R_q I_q$ (a consequence of the increased I_L and thus I_q); (2) armature reaction; that is, as I_L increases, Φ_P is reduced and distorted (the reduction in Φ_P reduces the generated and terminal voltages, E_q and V_L; thus as the rotor current I_q increases, the effect of the armature reaction is a progressive reduction of Φ_P, E_q, and V_L); and (3) reduction in the field current I_F; that is, because of the armature reaction and the $I_q R_q$-drop, $V_L \equiv V_F$ and thus I_F drops. Reduction of the excitation current I_F results in further reduction of Φ_P, E_q, and V_L. Further reduction of the load causes the generator to reach a breakdown point beyond which further load reduction causes the generator to **unbuild** as it operates on the unsaturated portion of its magnetization curve. This unbuilding process continues until $V_L=0$, at which point the load current is of such magnitude that the $R_q I_q$-drop is equal to E_q, that is, the emf generated on the unsaturated or linear portion of its magnetization curve.

Finally, a continuous increase of the load now will cause the generator to **rebuild** gradually along the dashed line, Fig. 2.38, and never retrace the line it traced during the unbuild process. This phenomenon, of course, is due to hysteresis.

*Designated as $V_L = \rho(I_L)$.

† As far as loading is concerned, the shunt generator should be operated in the saturated portion of the magnetization curve. If it is operated below the knee of the curve, it may unbuild with the application of load.

Plate 2.3 Two 6,000-hp, 40/80 rpm, 700 V, dc Motor Rotors During Assembly for Factory Test; Will be Used as a 12,000-hp, Single-Armature Twin Drive for a Reversing Slabbing Mill. (Courtesy of General Electric Co., Schenectady, N.Y.)

It must be noted in Fig. 2.38 that the external voltage versus current characteristic of the shunt motor remains fairly constant up to its rated load-current value, I_L^{RTD}. Thus the shunt generator is treated as a device that has a fairly constant output voltage, that is, region I. In practice, the device is rarely operated beyond the rated load current, that is, in region II, continuously for any appreciable time. For the useful region, region I, of the external voltage versus current characteristic, the device is treated as a *voltage generator* with a describing equation similar to that of a battery. The describing equation of the linear region is

$$V_L = E_{\mathrm{q}} - R_0 I_L \tag{2.127}$$

where R_0 is the slope of the characteristic, E_{q} is the v-axis intercept, and V_L and I_L are the terminal voltage and current of the device. The circuit analog of Eq. (2.127) is shown in Fig. 2.39.

Example 2.11

A 150-kW, 250-V shunt generator,* Fig. 2.36, has a field resistance of 50 Ω and a rotor resistance of 0.05 Ω. Under steady state operation the rotor speed is measured to be 1800 r/min.

*These values are known as rated values of the generator.

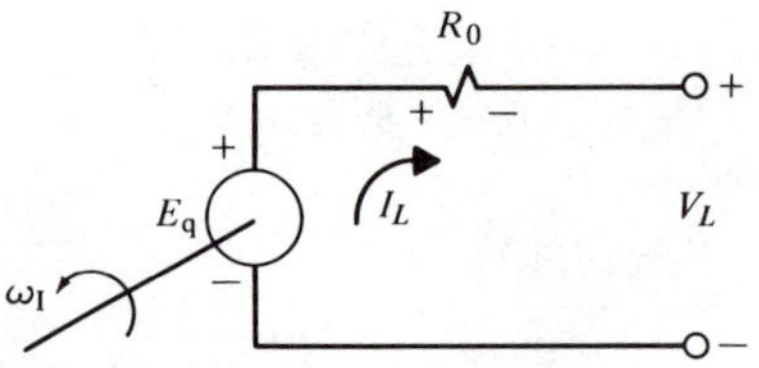

Fig. 2.39 Voltage Source Equivalent Circuit of the Shunt Generator in Region I, Fig. 2.38, Steady State Operation

Part I. Calculate the following quantities: (1) the full-load current L_L flowing into the load, (2) the field current I_F, (3) the rotor current I_q, (4) the full-load generated voltage E_q, (5) the loading resistance for rated conditions of operation, R_L, and (6) the developed electromagnetic torque τ_{em}.

Part II. Neglecting all losses, (1) what is the external torque required for the operation of this device, τ_I, and (2) if the rotor is brought to a halt by some external device, what is the value of E_q? (3) Why?

Solution

Part I.

1. Use $p_O = V_L I_L$. Since $p_O = 150 \times 10^3$ W and $V_L = 250$ V, then

$$I_L = \frac{150 \times 10^3}{250} = \underline{\underline{600}}\text{ A}$$

2. Use $I_F = V_F / R_F$. Since $V_F \equiv V_L = 250$ V and $R_F = 50\ \Omega$, then

$$I_F = \frac{250}{50} = \underline{\underline{5}}\text{ A}$$

3. Use $I_q = I_L + I_F$ (as a generator). Therefore,

$$I_q = I_L + I_F = 600 + 5 = \underline{\underline{605}}\text{ A}$$

4. Use $E_q = R_q I_q + V_L$. Therefore,

$$E_q = 0.05 \times 605 + 250 = \underline{\underline{280.25}}\text{ V}$$

5. Use $p_O = R_L I_L{}^2$. Therefore,

$$R_L = \frac{P_L}{I_L{}^2} = \frac{150 \times 10^3}{600^2} = \underline{\underline{0.417}}\ \Omega$$

6. Use $\tau_{em} = -\mathcal{K}\mathcal{K}_F I_F I_q$. Since $E_q = \mathcal{K}\mathcal{K}_F I_F \omega_R$ and $E_q = 280.25$ V and

$$\omega_R = \frac{2\pi n_R}{60} = \frac{2\pi \times 1800}{60} = \underline{60\pi}\text{ rad/s}$$

then

$$\mathcal{K}\mathcal{K}_F I_F = \frac{E_q}{\omega_R} = \frac{280.25}{60\pi} = \underline{1.49}\text{ MKS units}$$

Therefore,

$$\tau_{em} = -1.49 \times 605 = \underline{\underline{-901}}\text{ N}\cdot\text{m}$$

Direction: Opposing the externally applied torque τ_I.

Part II.

1. $$\tau_I = |\tau_{em}| = \underline{\underline{901}}\ \text{N}\cdot\text{m}$$

2. $$E_q = \mathcal{K}\Phi_P\omega_R|_{\omega_R=0} = \underline{\underline{0}}\ \text{V}$$

3. E_q is zero because the rotor speed on which E_q depends is zero.

Series-connected dc Generator

As in the case of the series motor, the field and rotor windings are connected in series and the combination is connected across the load, Fig. 2.40(a). When compared to the separately excited generator, it is seen that there is no forcing function, E_F, in the field winding to provide the excitation current I_F and thus Φ_P, which is essential for voltage generation, $E_q = \mathcal{K}\Phi_P\omega_R$. However, when compared to Fig. 2.36, that is, the shunt generator, there is a close resemblance between these two devices, at least as far as excitation is concerned, if Fig. 2.40(a) is redrawn as Fig. 2.40(b). Thus we would deduce that the principle for understanding the performance of the series generator is similar to that of the shunt generator. Indeed it is, for $R_L \neq \infty$ and $V_F \to V_F + V_L$, $R_F \to R_F + R_L$, and $R_{Fi} + R_L < R_C$, $i = 1,2,3\ldots$. Thus all the theory of the shunt generator holds if the noted changes to the parameters E_F, R_F, and R_C for the shunt generator are implemented for the series generator as long as $R_L \neq \infty$, that is, there is a load connection. (Why?)

The series voltage generator for no-load connection, that is, $R_L = \infty$ (open circuit), is incapable of building up. This is in contrast to the shunt generator. Thus when the load current is zero, the generated emf $E_q = \phi(I_L)$ is equal to the residual voltage, that is, $E_q = E_R$, Fig. 2.37. If a finite load is connected across the series generator, that is, across the rotor field windings, Fig. 2.40, a common field, rotor, and load current, $I_f = I_q \equiv I_L$, flows through the series field, creating additional mmf, which aids the residual flux, Fig. 2.14(a), to produce a higher generated emf. Automatic buildup continues as long as $R_L + R_{Fi} < R_C$, $i = 1,2,\ldots$, just as it did for the shunt generator.

However, the action of the series generator is somewhat more complex than that of the shunt generator. Here there are **two voltage drops**, which limit the voltage V_L

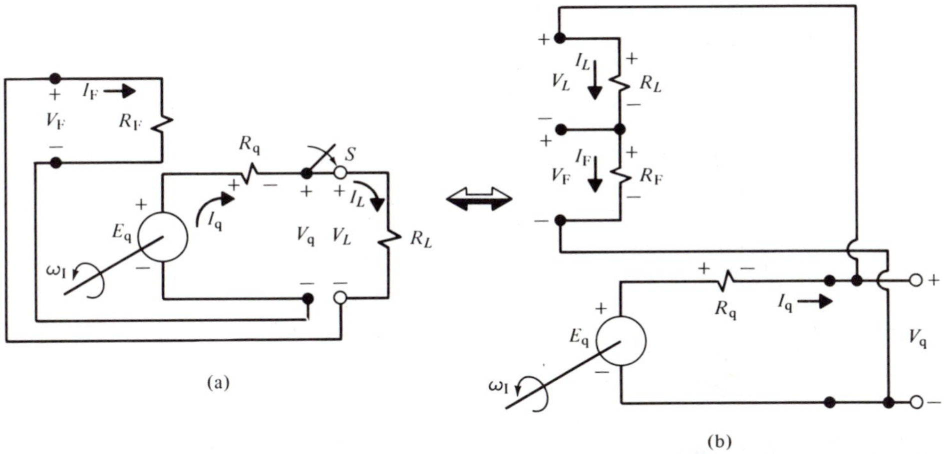

Fig. 2.40 Series Generator Under Steady State Operation

across the load [compare to Eq. (2.126) for the shunt generator], written as

$$V_L = E_q - (R_q + R_F)I_q \tag{2.128}$$

Note that here the generated emf E_q is available in the circuit of Fig. 2.40.

In addition to the $R_F I_q$- and $R_q I_q$-voltage drops, the generated emf E_q is also reduced by the effects of the armature reaction. Thus the voltage across the load, which produces the magnetizing current (excitation current I_F) represents the results of (1) the stated factors, tending to increase the voltage V_L, and (2) the increasing magnetizing current, tending to increase Φ_P and thus E_q. As a consequence, for a given $\omega_R \equiv \omega_I = k$, a maximum voltage E_M, Fig. 2.41, is produced. This voltage represents the critical point at which **buildup** stops and no additional current beyond I_M is automatically produced. At this load current, $I_q = I_M$, the armature reaction drop plus the $R_q I_q$-and $R_F I_q$-voltage drops exactly counterbalance the increased emf by the series field. Thus the terminal voltage remains constant.

Further application of load beyond the critical maximum voltage E_M due to combined factors of increased armature reaction and the voltage drops $R_q I_q$ and $R_F I_q$, will now decrease the load voltage at a faster rate than the rate at which the generated emf E_q is increased by the load current I_q.

Note that the voltage versus current characteristic of Fig. 2.41 has two distinct regions. When the device operates in region I, the rising portion of the characteristic lends itself to *voltage boosting*, used in traction dc transmission lines to compensate for transmission line voltage drops and other similar applications.

Region II enables treatment of the series generator as a *current generator* with a describing equation similar to that of a heavily loaded battery, that is,

$$I_L = I_0 - G_0 V_L \tag{2.129}$$

where G_0 is the slope of the characteristic in region II and I_0 is the i-axis intercept, that is, the current flowing when the loading terminals are short-circuited, and V_L and I_L are the terminal voltage and current of the device. The circuit analog for region II, Eq. (2.129), is as shown in Fig. 2.42.

When operating in region II, the performance of the series generator is measured by its ability to maintain the current it generates constant. This current constancy

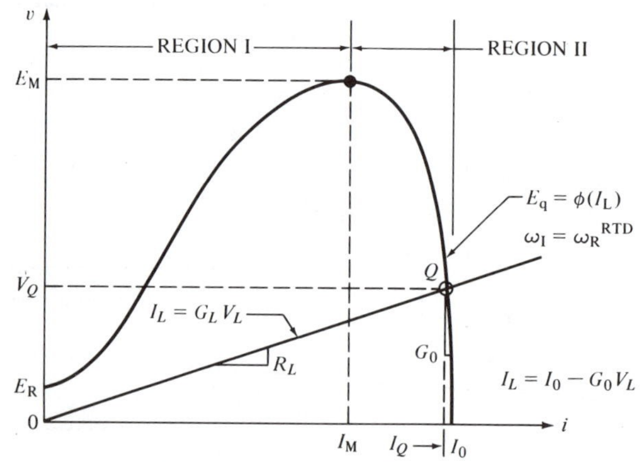

Fig. 2.41 Voltage vs. Current Characteristics of a Series (Self-excited) Generator and Load Under Steady State Operation

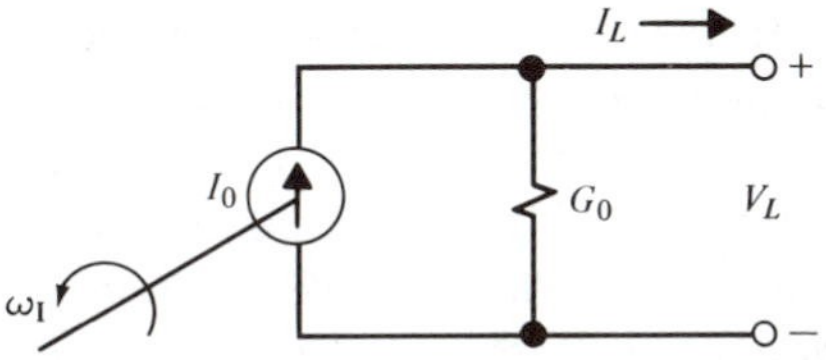

Fig. 2.42 Current Source Equivalent Circuit of the Series Generator in Region II, Fig. 2.41, Steady State Operation

is measured by the percentage **current regulation** (CR), an index of performance of the device, defined as

$$\mathrm{CR} \equiv \frac{I_0 - I_Q}{I_Q} \times 100 = \frac{I_{\text{NO LOAD}} - I_{\text{FULL LOAD}}}{I_{\text{FULL LOAD}}} \times 100 \qquad (2.130)$$

Eq. (2.130) gives a measure of the ability of the series (self-excited) generator to maintain constant terminal current as the loading conditions change from no load to full load.

It should be obvious now that the performance of a device as a dc generator is described by its external voltage versus current characteristics. The external voltage versus current characteristics for the devices that are studied in detail in this section, that is, (1) the separately excited generator, (2) the shunt generator, and (3) the series generator, are exhibited in Fig. 2.43. The scales of the axes are in percentage of rated values in order to simplify comparison between the various characteristics. It is noted that the terminal voltage of a shunt generator decreases faster than it does for the same device operated as a separately excited generator, and the terminal voltage of a series generator increases much faster than it

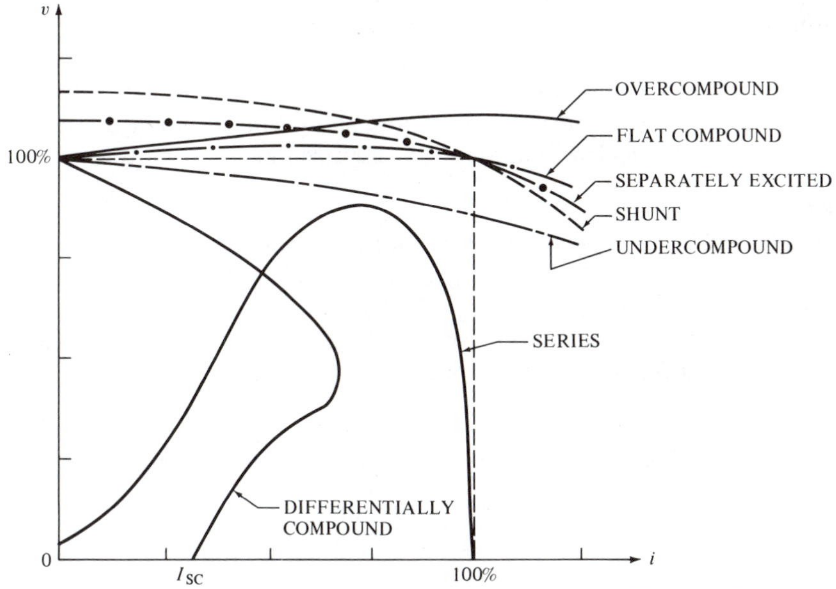

Fig. 2.43 External Voltage vs. Current Characteristics of Various Types of dc Generators, a Comparative Exposé

decreases for the same device operated as a shunt generator. Since *it is usually desirable to operate a voltage generator with a nearly constant voltage at its terminals*, the constancy being measured by a VR $\doteq$ 0%, it can be suspected by looking at the external characteristics of the shunt and series generators that *a generator with two field windings, one in shunt and the other in series with the rotor, may have an external voltage versus current characteristic that is very nearly flat with* I_L. Such a device indeed exists* and is named **compound generator**, Fig. 2.44. The additional $R_S L_S$-winding, when compared to the shunt connection, is connected in series with the $R_q L_q$-winding. Its mmf is either *additive* or *subtractive* to the mmf of the field winding $R_F L_F$ (as in the motor case, Section 2.7).

This new winding gives rise to two new generator types: (1) *cumulatively compound*, that is, the fields of the windings $R_F L_F$ and $R_S L_S$ are *aiding*, and (2) *differentially compound*, that is, the field of the winding $R_S L_S$ *opposes* the field of the $R_F L_F$-winding.

Depending on the relative strength of the mmf's of the $R_F L_F$- and $R_S L_S$-windings, the cumulatively compound generators may be *overcompound*, *flat compound*, or *undercompound*, Fig. 2.43. In the overcompound generator the $R_S L_S$-field is strong enough to give a rising terminal voltage characteristic in the normal load range at constant speed and may thus counteract the effect of a decrease in the prime-mover speed with increasing load. Overcompounding may also be used to compensate for the line *ir*-drop when the load is at a considerable distance from the generator. The flat compound generator maintains a nearly constant voltage from no-load to full-load at the terminals of the device at constant speed; that is, it possesses a nearly perfect, or *zero*, voltage regulation.

The differential compound generator is used in applications where a *wide variation* in load voltage can be tolerated and where the generator may be exposed to load conditions approaching those of a short circuit, Fig. 2.43. Electric power shovels afford an excellent example because the motor supplied by the generator is frequently subjected to loads that produce stalling.

The mathematical analysis of compound generators is not pursued here. However, the external voltage versus current characteristics of the cumulatively compound (*overcompound* and flat *compound*) and differentially compound are exhibited in Fig. 2.43, for comparative purposes, along with the voltage versus current characteristics of the devices that are fully discussed here.

Example 2.12

A 10-kW, 125-V series generator, Fig. 2.40, has a rotor resistance of 0.1 Ω, a field resistance of 0.05 Ω, and is operating at steady state with a rated speed of 1800 rpm. Calculate: (1) the load current I_L, (2) the field current I_F, (3) the rotor current I_q, (4) the generated emf E_q, (5) the loading resistor for rated conditions of operation, R_L, and (6) the developed electromagnetic torque τ_{em}.

Solution

1. Since this device is a series generator, $I_q \equiv I_F \equiv I_L$. Thus from $p_L = V_L I_L$,

$$I_L = \frac{p_L}{V_L} = \frac{10\times 10^3}{125} = \underline{\underline{80}}\text{ A}$$

*It is improbable that any commercially built device is built for either **simple series** or **simple shunt** operation.

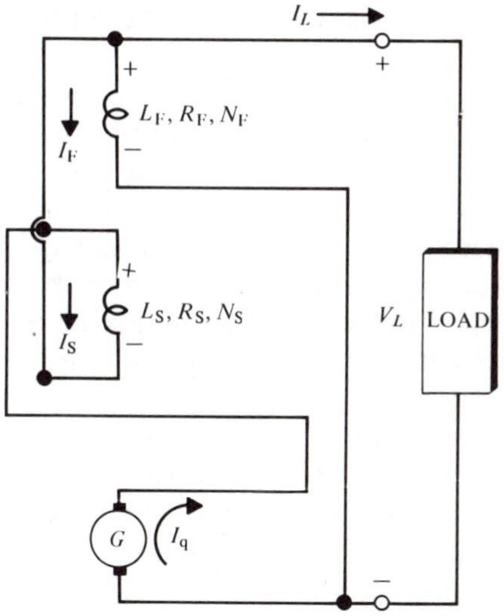

Fig. 2.44 Schematic Diagram for Compound-type Generator; $R_S L_S$- and $R_F L_F$-Windings Are on the Same Stator Axis, but $N_F \gg N_S$: (1) Cumulative Compound, Fields Are Aiding; (2) Differential Compound, Fields Are Opposing

2. and 3. Use $I_F = I_q \equiv I_L$. Therefore,

$$I_F = I_q \equiv I_L = \underline{\underline{80}}\text{ A}$$

4. Use $V_L = E_q - (R_q + R_F)I_q$. Therefore,

$$E_q = V_L + (R_q + R_F)I_q = 125 + (0.15)(80) = \underline{\underline{137}}\text{ V}$$

5. Use $p_L = G_L V_L^2$.

$$G_L = \frac{p_L}{V_L^2} = \frac{10 \times 10^3}{125^2} = \underline{\underline{0.64}}\text{ mho}$$

Therefore,

$$R_L = \frac{1}{G_L} = \underline{\underline{1.56\ \Omega}}$$

Or using $V_L = R_L I_L$,

$$R_L = \frac{V_L}{I_L} = \frac{125}{80} = \underline{\underline{1.56\ \Omega}}$$

6. Use $\tau_{em} = -\mathcal{K}\mathcal{K}_F I_F I_q$. Since $E_q = \mathcal{K}\mathcal{K}_F I_F \omega_R$ and $E_q = 137$ V and

$$\omega_R = \frac{2\pi n_R}{60} = \frac{2\pi \times 1800}{60} = \underline{60\pi}\text{ rad/s}$$

then

$$\mathcal{K}\mathcal{K}_F I_F = \frac{137}{60\pi} = \underline{0.73}\text{ MKS units}$$

Therefore,

$$\tau_{em} = -0.73 \times 80 = \underline{\underline{-58.4}}\text{ N} \cdot \text{m}$$

Direction: Opposes the externally applied torque τ_I.

2.10 APPLICATIONS OF dc ENERGY CONVERTERS

Although through the advances in semiconductor science and technology, semiconductor rectifiers (and, in particular, silicon controlled rectifiers) have dealt a blow to the widespread use of the electromechanical dc energy converter, it still enjoys and will enjoy for some time to come a reputable role in industry and technology. In addition, **if it were not for the overwhelming advantages in the cost of transmission and distribution of the ac type of electrical energy from points of generation to points of consumption, and if electrical energy were to be converted from other forms of energy at the point of consumption, it would be more convenient to use dc-type electrical energy**.

The excellent qualities of the dc motors and generators make them highly desirable in applications where control of speed and voltage or current are essential. Applications of the various types of motors and generators discussed in this chapter follow.

dc Motors

1. The *separately excited motor* is widely used in control systems as a *positioning* device. In this role precise control of its speed and torque over a wide range is often required and achieved by one of the following schemes: (1) **field controlled**, that is, the rotor (armature) winding is connected to a constant current source so that $i_q(t)=k$, and the field (stator) winding is connected to the output of an amplifier that amplifies the controlling variable $i_F(t)$ to a satisfactory level; and (2) **armature controlled**, that is, the field winding is connected to a constant current source, $i_F(t)=k$, and the rotor winding is connected to the output of an amplifier that amplifies the controlling variable $i_q(t)$ to a satisfactory level.

2. The *series motor* is used for drives requiring a very high starting torque, and where adjustable speeds are required. When excited from either dc or a single-phase ac supply, it is used as a drive for electric locomotives. *An ac excitation of the dc series motor is possible because the field and rotor current change direction at the same time*, thus, **the torque remains unidirectional.**

Also, a special design series motor, known as the **universal motor**, operates with either dc or ac excitation. Being very widely used, it is available in sizes of fractional horsepower up to and well beyond 1 hp. Speed ratings range from 2,000 to 12,000 rpm. The universal motor has flooded our households and workshops. It can be found in vacuum cleaners, food mixers, portable hand tools, and other devices.

3. The *shunt motor* is suitable in applications where (1) adjustable speed drive with automatic control features is required, for example, *Ward–Leonard control* (armature voltage control), and (2) nearly constant speed with medium starting torque is required. Some of its applications are rolling mills, paper mills, centrifugal pumps, fans, conveyors, woodworking machines, and machine tools.

dc Generators

1. The *separately excited generator* is used as a main exciter in conjunction with, that is, connected in parallel with, a pilot exciter (a self-excited shunt generator) to provide dc excitation through slip rings to synchronous energy converters (see

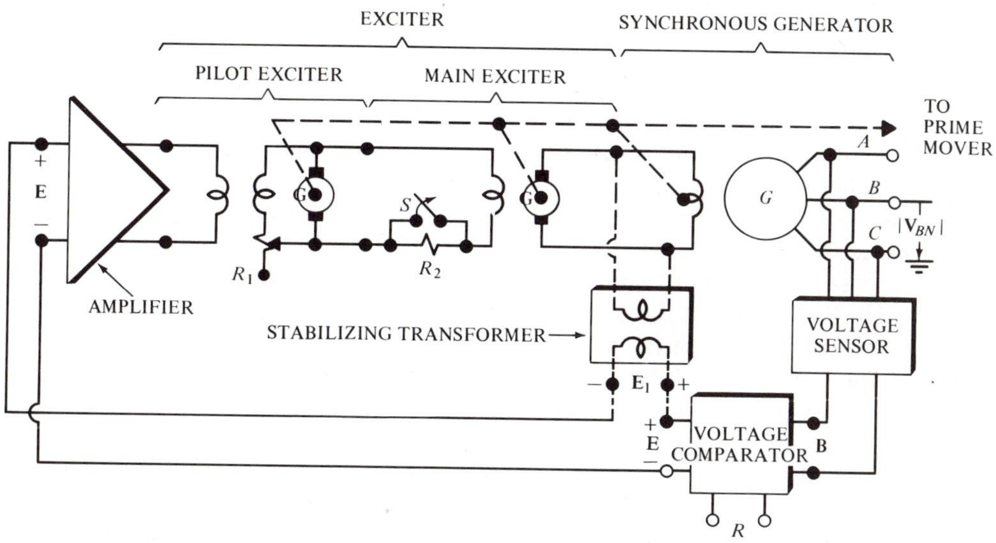

Fig. 2.45 Connection Diagram of an Exciter Within a Generating System

Chapters 3 and 5). Both dc machines are frequently driven from the main rotor shaft, that is, the shaft of the synchronous motor, or that of the generator which is driven by a prime mover. A typical connection diagram for an exciter in place within a generating system is given in Fig. 2.45.

2. The *series generator* is used as a voltage booster in (1) dc distribution systems for some electrified railways and (2) i^2R-loss adder in dc transmission systems. Also, it is used as a constant current source in welding machines.

3. The *shunt generator* is used as a voltage source where absolutely constant voltage is not critical, since its voltage regulation may be as poor as 15 percent.

4. The *cumulative compound generator* is used widely in applications where constant voltage is essential.

5. The *differential compound generator* is used in applications where a large voltage drop is desirable when the current increases, for example, in arc welding.

2.11 DYNAMIC ANALYSIS

Various types of dc generators and motors, along with induction motors and synchronous generators (Chapters 4 and 5), are frequently found in the same interconnected electric power grid system; for example, *the exciter in generating units* is a dc generator and *the positioning device in a control loop* is usually a separately excited motor. As the loads in the power grid system change and shift, the dynamic characteristics of the individual machines taken together will determine whether the system can react in a stable manner to such disturbances. An understanding of the steady state characteristics of such systems alone is not sufficient to provide a useful study of their total performance. After all, if the system of interest fails to achieve steady state operation because of *poor dynamics* either of the entire system or of the individual components, an understanding of steady state performance is an exercise in futility.

Dynamic analysis of electromechanical devices implies knowledge of the behavior of these devices during their transient state. To accomplish this task, we should be able to obtain a closed-form solution of their describing equations. For large interconnected systems we should be able to handle *large blocks* of differential and algebraic equations, which might be *linear*, *nonlinear*, or *combinations* thereof. Therefore, to be able to write the describing equations of a large system, we should be able to handle the describing equations on a component-by-component basis.

The most popular analytical avenues to solutions of the describing equations of any system are (1) **direct solution approach** and (2) the **transfer function approach.** In the first method we seek a closed-form solution of the describing equations in the *t-domain.* In the second method [where the following definition is implied: the transfer function is an *s-domain* mathematical formulation that relates the output variable(s) of a device to its input variable(s)] the describing equations are written, or transformed to the *s*-domain from another domain (e.g., the *t*-domain), utilizing *Laplace transform* techniques, and their solutions are obtained in the *s*-domain. These solutions are then transformed to *t*-domain solutions by utilizing *inverse Laplace transform* techniques.

When a dc device is part of a more extensive system, **only the external** performance of the device and not its internal behavior is of interest. Thus, depending on the situation, we should be able to obtain both (1) solutions of the equations that describe the internal behavior of a component device and (2) a transfer function, that is, a block diagram representation, of the component device. Both of these tasks are demonstrated utilizing the motor and generator modes of operation of a separately excited device, this device being the *most popular* dc device in control systems.

Dynamic Analysis via *s*-Domain Techniques

For this technique we have a separately excited motor in both the field and the armature controlled modes. Fig. 2.17 is redrawn here for convenience as Fig. 2.46, for $\tau_L = 0\ \text{N}\cdot\text{m}$.

To find the transfer function of a given system, the following two steps are essential: (1) we should be able to take the Laplace transform of the describing equations (written in various domains, mainly the *t*-domain) of the given system, and (2) we should be able to take the inverse Laplace transform of the *s*-domain-found solutions to the describing equations of the given system. (Students not familiar with this technique are urged to consult an appropriate text, such as **Network Analysis** by M. E. Van Valkenburg [19], and Appendixes C and D of this text).

FIELD-CONTROLLED MODE, $i_q(t) = \text{constant}$

For $K=0$, a perfectly rigid axis, the transfer function of Fig. 2.46 can be derived as follows.

Utilization of Eq. (2.59) enables writing the Laplace transform, for zero initial conditions, of Eq. (2.69) as

$$(Js^2 + Bs)\Theta_R(s) = \tau_{em}(s) \equiv K_T I_F(s) \tag{2.131}$$

For $E_b = K_V \omega_R(t)$ and initial conditions equal to zero, the Laplace transform of

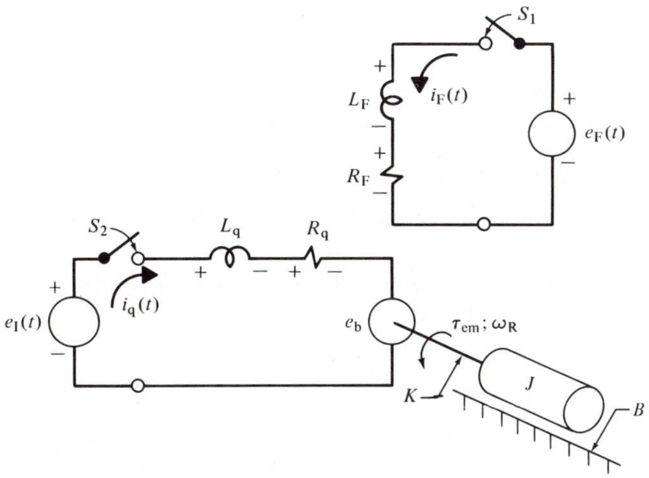

Fig. 2.46 Equivalent Circuit of the Separately Excited Motor with $\tau_L = 0$: (1) Field-controlled Mode $i_F(t) \neq k$ and $i_q(t) = k_1$; (2) Armature-controlled Mode $i_q(t) \neq k$ and $i_F(t) = k_2$

Eq. (2.65) is

$$E_F(s) = (sL_F + R_F)I_F(s) \tag{2.132}$$

Solution of Eq. (2.132) for $I_F(s)$ yields

$$I_F(s) = \frac{E_F(s)}{sL_F + R_F} \tag{2.133}$$

Substitution for $I_F(s)$ from Eq. (2.133) into Eq. (2.131) and the factoring of sB on the left-hand side and R_F on the right-hand side of Eq. (2.131) yields

$$sB\left(s\frac{J}{B} + 1\right)\Theta_R(s) = \left(\frac{K_T}{R_F}\right)\left[\frac{E_F(s)}{s(L_F/R_F) + 1}\right] \tag{2.134}$$

Defining $J/B \equiv \tau_M$ and $L_F/R_F \equiv \tau_F$ as mechanical and electrical time constants, correspondingly, and solving Eq. (2.134) for $\Theta(s)/E_F(s)$ yields

$$\frac{\Theta_R(s)}{E_F(s)} = \frac{K_T/BR_F}{s(s\tau_M + 1)(s\tau_F + 1)} \tag{2.135}$$

Defining $\Theta_R(s)/E_F(s) \equiv G(s)$ and $K_T/(R_F B) \equiv K_N$ enables writing Eq. (2.135) as

$$G(s) \equiv \frac{K_N}{s(s\tau_M + 1)(s\tau_F + 1)} \tag{2.136}$$

where $G(s)$ is defined as the *over-all transfer function* of the field-controlled, separately excited motor and K_N is defined as the *gain constant* of the transfer function $G(s)$.

The transfer function $G(s)$ relates the output angular displacement $\Theta_R(s)$ of the shaft to the controlling input $E_F(s)$, that is, the field voltage of the motor. *The transfer function $G(s)$ is a complex quantity for $s = j\omega$ and is defined as follows*:

$$\mathbf{G}(s)\bigg|_{s=j\omega} \equiv |\mathbf{G}(j\omega)|e^{j\phi} = \frac{K_N}{j\omega(j\omega\tau_M + 1)(j\omega\tau_F + 1)} \tag{2.137}$$

where $|\mathbf{G}(j\omega)|$ and ϕ are the *magnitude* and *phase angle* of the right-hand side of Eq. (2.138).

Utilizing automatic control techniques enables drawing the block diagram of Eq. (2.137) as shown in Fig. 2.47.

ARMATURE-CONTROLLED MODE, $i_F(t)$ = constant

For this mode of operation, the transfer function of Fig. 2.46 can be derived as follows.

For $E_b = K_V\omega_R(t)$ and zero initial conditions, Eqs. (2.66), (2.68), and (2.69) have the following Laplace transforms:

$$E_I(s) = (sL_q + R_q)I_q(s) + K_V s\Theta_R(s) \tag{2.138}$$

and

$$\tau_{em} = \mathfrak{M}I_F(s)I_q(s) = (Js^2 + Bs + K)\Theta_R(s) \tag{2.139}$$

Solution of Eq. (2.138) for $I_q(s)$ yields

$$I_q(s) = \frac{E_i(s) - K_V s\Theta_R(s)}{sL_q + R_q} \tag{2.140}$$

Substitution of Eq. (2.140) into Eq. (2.139) yields

$$\tau_{em} = \mathfrak{M}I_F\left(\frac{E_I(s) - K_V s\Theta_R(s)}{sL_q + R_q}\right) = (s^2J + sB + K)\Theta_R(s) \tag{2.141}$$

Collection of all coefficients of $\Theta_R(s)$ and solution of the resulting equation for the ratio $\Theta_R(s)/E_I(s)$ yields

$$\frac{\Theta_R(s)}{E_I(s)} = \frac{\mathfrak{M}I_F(s)}{(s^2J + sB + K)(sL_q + R_q) + sK_V\mathfrak{M}I_F} \tag{2.142}$$

Division of the numerator and denominator of Eq. (2.142) by $(s^2J + sB + K)(sL_q + R_q)$ yields

$$\frac{\Theta_R(s)}{E_I(s)} = \frac{\mathfrak{M}I_F(s)/(s^2J + sB + K)(sL_q + R_q)}{1 + sK_V\left[\mathfrak{M}I_F(s)/(s^2J + sB + K)(sL_q + R_q)\right]} \tag{2.143}$$

Defining the **forward-path (open loop) transfer function** as

$$G(s) \equiv \frac{\mathfrak{M}I_F(s)}{(s^2J + sB + K)(sL_q + R_q)} \tag{2.144}$$

and the **feedback-path transfer function** as

$$H(s) \equiv sK_V \tag{2.145}$$

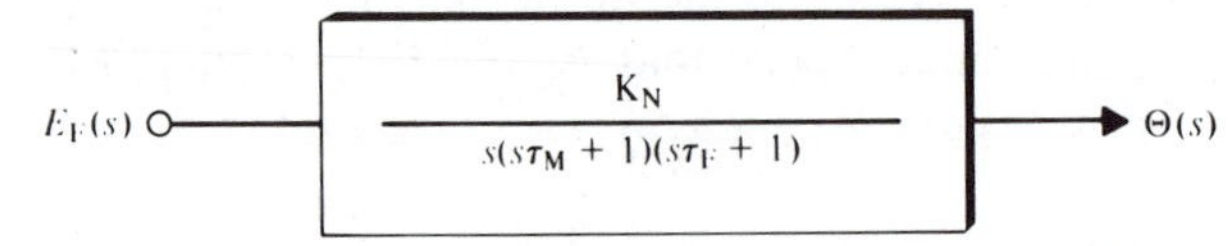

Fig. 2.47 Block Diagram Representation of a Separately Excited Motor, Field-controlled Mode, $K = 0$ N·m/rad and $\tau_L = 0$ N·m

enables writing Eq. (2.143) as

$$\frac{\Theta_R(s)}{E_I(s)} = \frac{G(s)}{1+H(s)G(s)} \equiv M(s) \tag{2.146}$$

where (1) $H(s)G(s)$ is defined as the *loop gain* or *open-loop transfer function* and (2) $M(s)$ is defined as the *control ratio* or *closed-loop transfer function* of the armature-controlled motor. Both have all the properties of $G(s)$ given in Eq. (2.137).

Utilizing automatic control techniques enables us to draw the block diagram of Eq. (2.146) as shown in Fig. 2.48.

There is not much to be gained by representing either controlled mode of the separately excited motor by a block diagram if the system to be analyzed consists only of one component, for example, the separately excited, armature-controlled motor. If this is the case, a direct solution of the describing equation is adequate. The real advantage of a block diagram representation becomes clear when the motor is just *one of the many* components of a given system, for example, Fig. 2.45, and modeling of large systems for *digital, analog, or hybrid computer simulation* becomes absolutely essential.

However, Eqs. (2.136, 2.137, and 2.146), and its counterpart, $\mathbf{M}(j\omega) = \ldots$, are very powerful. They alone provide an abundance of information about the dynamic and the steady state behavior of the motor for the corresponding controlled mode of operation.

The inverse Laplace transform of Eqs. (2.136) and (2.146) yield the time-domain solution for the response of the motor, in the corresponding controlled mode, as a function of the input and time, that is, $\Theta_R(t) = f[e_F(t), t]$ and $\Theta_R(t) = \psi[e_I(t), t]$, (see Appendix C).

In contrast, Eq. (2.137) and its counterpart $\mathbf{M}(j\omega)$ (for $i_F(t) = k$) yield the steady state response of the motor, as a function of the sinusoidal inputs $\mathbf{E}_F(j\omega)$ and $\mathbf{E}_I(j\omega)$. Thus we can perform time- and frequency-domain analyses, as well as design studies, of this device.

Furthermore, with the aid of Eqs. (2.136) and (2.146) we can perform stability studies for the two motor modes of operation by using (1) the Routh–Hurwitz criterion, (2) the Nyquist criterion, (3) the root locus technique, and (4) the Bode plot method. Finally, Eqs. (2.136) and (2.146) make possible the modeling of large systems of which this motor is a component part. Therefore, a mathematical model representing a given large system can be simulated on a digital, analog, or hybrid computer to provide the solutions in numerical data versus time and/or graphical format.

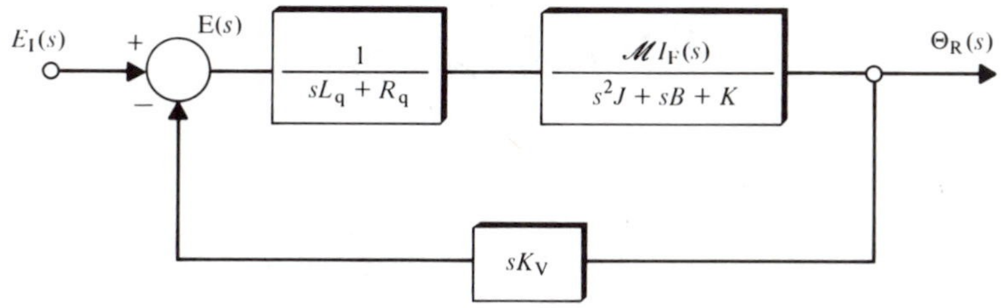

Fig. 2.48 Block Diagram Representation of a Separately Excited Motor, Armature-controlled Mode, $\tau_L = 0$ N·m

Example 2.13

Consider the equivalent circuit of the separately excited generator, Fig. 2.15, and its no-load magnetization curve, Fig. 2.14(b), which can be described by $E_q = K_E i_F$; $K_E = 100$ V/A. Assume that switches S_1 and S_2 are open and the prime mover is rotating at a rated speed of 1800 rpm. At $t=0$, both switches S_1 and S_2 are closed.

Consider $i_q(t)$ to be the output of the generator; then (1) obtain the control ratio that describes the generator, (2) draw the block diagram for this control ratio, and thus of the generator, and (3) utilize s-domain techniques in conjunction with transfer functions and control ratios to calculate: (i) the field current, (ii) the rotor current, and (iii) the developed electromagnetic torque by the generator. Data: $E_F = 100$, $R_F = 40$, $L_F = 8$, $L_L = 2.6$, $L_q = 0.4$, $R_q = 0.5$, and $R_L = 5.5$, all in MKS units.

Solution

1. Utilizing Laplace transform techniques and initial conditions, that is, IC's$\equiv 0$, the describing equations of the field circuit, the rotor circuit, and the generated voltage can be written as

$$(R_F + sL_F)I_F(s) = E_F(s) \tag{1}$$

$$[(R_q + R_L) + s(L_q + L_L)]I_q(s) = E_q(s) \tag{2}$$

$$E_q(s) = K_E I_F(s) \tag{3}$$

Solution of Eq. (1) for $I_F(s)$ yields

$$I_F(s) = \frac{E_F(s)}{R_F + sL_F} \tag{4}$$

Substituting Eq. (4) into Eq. (3) yields

$$E_q(s) = \frac{K_E E_F(s)}{R_F + sL_F} \tag{5}$$

Solution of Eq. (2) for $I_q(s)$ yields

$$I_q(s) = \frac{E_q(s)}{(R_q + R_L) + s(L_q + L_L)} \tag{6}$$

Substituting Eq. (5) into Eq. (6) yields

$$I_q = \left\{ \frac{K_E}{[(R_q + R_L) + s(L_q + L_L)](R_F + sL_F)} \right\} E_F(s) \tag{7}$$

Substitution of available data values into Eq. (7) yields

$$I_q(s) = \left[\frac{100}{(6 + s3)(40 + s8)} \right] E_F(s) \tag{8}$$

or

$$I_q(s) = \left[\frac{100}{8 \times 3(2 + s)(5 + s)} \right] E_F(s) \tag{9}$$

or

$$I_q(s) = \left[\frac{4.16}{(2 + s)(5 + s)} \right] E_F(s) \tag{10}$$

Thus the control ratio becomes

$$M(s) \equiv \frac{I_q(s)}{E_F(s)} = \frac{4.16}{(2+s)(5+s)} \tag{11}$$

Similarly, the transfer function $G_F(s)$ of the field circuit, as obtained from Eq. (4), is

$$G_F(s) \equiv \frac{I_F(s)}{E_F(s)} = \frac{1}{R_F + sL_F} \tag{12}$$

and the transfer function $G_q(s)$ of the rotor circuit, as obtained from Eq. (6), is

$$G_q(s) \equiv \frac{I_q(s)}{E_q(s)} = \frac{1}{(R_q + R_L) + s(L_q + L_L)} \tag{13}$$

2. The block diagram of the generator is drawn, utilizing Eq. (11), as shown in Fig. 2.13.1.
3. Evaluation of the Laplace transformed $I_q(s)$ using Eq. (4) directly, and maintaining zero initial conditions yields

$$I_F(s) = \left[\frac{1}{8(5+s)}\right] E_F(s) = \frac{1}{8(5+s)} \times \frac{100}{s} = \frac{12.5}{s(5+s)} \tag{14}$$

Utilizing partial fraction expansion techniques, Eq. (14) can be written as follows:

$$I_F(s) = \frac{12.5}{s(5+s)} \equiv \frac{K_0}{s} + \frac{K_1}{s+5} \tag{15}$$

In order to evaluate the constants K_0 and K_1,

(a) premultiply Eq. (15) through by s and evaluate at $s=0$, this operation yields

$$K_0|_{s=0} = \frac{12.5}{5} = 2.5 \tag{16}$$

(b) premultiply Eq. (15) through by $(s+5)$ and evaluate the result at $s = -5$ (i.e., set $s+5=0$), this operation yields

$$K_1|_{s=-5} = \frac{12.5}{-5} = -2.5 \tag{17}$$

Thus Eq. (15) can now be written as

$$I_F(s) = \frac{2.5}{s} + \frac{-2.5}{s+5} \tag{18}$$

Taking the inverse Laplace transform of Eq. (18) yields

$$i_F(t) = 2.5(1 - e^{-5t}) \text{ A} \tag{19}$$

Similarly, taking the Laplace transform of Eq. (11), or of Eq. (10) directly, and

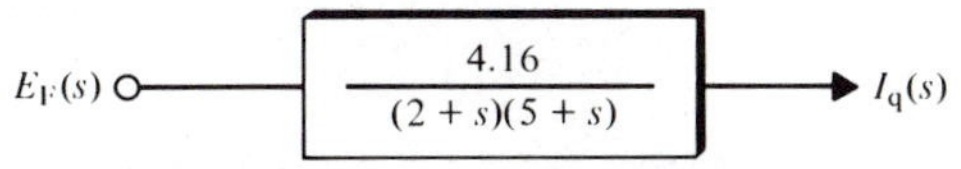

Fig. 2.13.1 Block Diagram Representation of the Separately Excited Generator with Its Excitation $e_F(t)$ and Response $i_q(t)$ Measured Against the Same Interval of Time (with Identical $t=0$)

maintaining the initial conditions at zero yields

$$I_q(s) = \left[\frac{4.16}{(2+s)(5+s)}\right] E_F(s) = \frac{4.17}{(2+s)(5+s)} \times \frac{100}{s} = \frac{417}{s(s+2)(s+5)} \tag{20}$$

Again utilizing partial fraction expansion techniques, Eq. (20) can be written as follows:

$$I_q(s) = \frac{417}{s(2+s)(5+s)} = \frac{K_0}{s} + \frac{K_1}{2+s} + \frac{K_2}{5+s} \tag{21}$$

Applying the techniques used in Eq. (15) yields

$$K_0|_{s=0} = \frac{417}{(5)(2)} = \underline{41.7} \tag{22}$$

$$K_1|_{s=-2} = \frac{417}{(-2)(3)} = \frac{417}{-6} = \underline{-69.5} \tag{23}$$

$$K_3|_{s=-5} = \frac{417}{(-5)(-3)} = \frac{417}{15} = \underline{27.8} \tag{24}$$

Thus Eq. (21) can be written as

$$I_q(s) = \frac{41.7}{s} + \frac{-69.5}{2+s} + \frac{27.8}{5+s} \tag{25}$$

Taking the inverse Laplace transform of Eq. (25) yields

$$i_q(t) = \underline{\underline{41.7 - 69.5e^{-2t} + 27.8e^{-5t}\ \text{A}}} \tag{26}$$

Utilization of Eq. (2.36c), Fig. 2.14(c) [or Fig. 2.14(b) linearized], and $n_R =$ 1800 rpm yields

$$E_q = K_E i_F \tag{27}$$

Eq. (2.36b) gives K_E as

$$K_E = \mathcal{K}\mathcal{K}_F \omega_R \tag{28}$$

Since, as given in the data at $n_R = 1800$ rpm, $E_q = 100$ V, $K_E = 100$, and $i_F = 1.0$ A, then

$$K_E = 100 = \mathcal{K}\mathcal{K}_F \omega_R \tag{29}$$

Thus

$$\mathcal{K}\mathcal{K}_F = \frac{100}{\omega_R} = \frac{100}{2\pi \times n_R/60} = \frac{100}{2\pi \times 1800/60} = \underline{\frac{10}{6\pi}} \tag{30}$$

Eqs. (2.36d) and (2.41) give τ_{em} as

$$\tau_{em} = -\mathcal{M} i_q i_F$$

Thus

$$\tau_{em} = \underline{\underline{-\frac{10}{6\pi} \times (41.7 - 69.5e^{-2t} + 27.8e^{-5t})\left[2.5(1 - e^{-5t})\right]\ \text{N}\cdot\text{m}}} \tag{31}$$

Therefore, for $t \to \infty$,

$$\tau_{em} = -\frac{10}{6\pi} \times 41.7 \times 2.5 = \underline{\underline{-55.3}}\ \text{N}\cdot\text{m}$$

Dynamic Analysis via t-Domain Techniques

In order to observe the total (i.e., internal plus external) behavior of the two devices, namely the separately excited motor and generator, the corresponding t-domain describing equations, Eqs. (2.65) through (2.69) and Eqs. (2.39) through (2.41), must be solved analytically.

Simulation diagram techniques are exploited here rather than pursuing an analytical solution of these equations. No prior knowledge of simulation techniques is essential, as the scope of this presentation is not to teach students the subject of simulation but rather to introduce them to *prepackaged digital computer programs* that provide all the desired information with minimum effort. At the same time, students are introduced to a more sophisticated approach of analysis that can be applied to the study of these devices and large-scale systems.

Solution of Eqs. (2.65) and (2.66), after Eq. (2.61) is substituted into Eq. (2.66) and the defining relationship $e_S(t) - R_S I_q(t) \equiv e_I(t)$ is utilized for the highest derivatives, yields

$$L_F \frac{d}{dt}[i_F(t)] = -R_F i_F(t) + e_F(t) \tag{2.147}$$

and

$$L_q \frac{d}{dt}[i_q(t)] = -R_q i_q(t) - \mathfrak{M} i_F \omega_R + e_I(t) \tag{2.148}$$

Defining $\theta_R(t) \equiv \theta_1(t)$, $d[\theta_R(t)]/dt = d[\theta_1(t)]/dt \equiv \theta_2(t)$ (thus, $d^2[\theta_R(t)]/dt^2 = d^2[\theta_1(t)]/dt^2 = d[\theta_2(t)]/dt$) and dividing through by J enables writing Eq. (2.69) as

$$\frac{d}{dt}[\theta_1(t)] = \theta_2(t) \tag{2.149}$$

$$\frac{d}{dt}[\theta_2(t)] = -\left(\frac{K}{J}\right)\theta_1(t) - \left(\frac{B}{J}\right)\theta_2(t) - \left(\frac{1}{J}\right)(\tau_\ell - \tau_{em}) \tag{2.150}$$

where $\tau_\ell \equiv \pm \tau_L$ and $\tau_{em} = \mathfrak{M} i_F(t) i_q(t)$. Eqs. (2.147) and (2.148) can be written in *matrix* form as

$$\left[\begin{array}{c|c} L_F & 0 \\ \hline 0 & L_q \end{array}\right] \frac{d}{dt}\left[\begin{array}{c} i_F(t) \\ \hline i_q(t) \end{array}\right] = \left[\begin{array}{c|c} -R_F & 0 \\ \hline -\mathfrak{M}\omega_R(t) & -R_q \end{array}\right]\left[\begin{array}{c} i_F(t) \\ \hline i_q(t) \end{array}\right] + \left[\begin{array}{c} e_F(t) \\ \hline e_I(t) \end{array}\right] \tag{2.151}$$

Also, Eqs. (2.149) and (2.150) can be written in matrix form as

$$\frac{d}{dt}\left[\begin{array}{c} \theta_1(t) \\ \hline \theta_2(t) \end{array}\right]\left[\begin{array}{c|c} 0 & 1 \\ \hline -K/J & -B/J \end{array}\right]\left[\begin{array}{c} \theta_1(t) \\ \hline \theta_2(t) \end{array}\right] + \left[\begin{array}{c} 0 \\ \hline 1/J \end{array}\right][\tau_{em} - \tau_\ell] \tag{2.152}$$

where

$$\tau_{em} = [\, i_F(t) | i_q(t) \,]\left[\begin{array}{c|c} 0 & 0 \\ \hline \mathfrak{M} & 0 \end{array}\right]\left[\begin{array}{c} i_F(t) \\ \hline i_q(t) \end{array}\right] \tag{2.153}$$

Since $d[\theta_1(t)]/dt = d[\theta_R(t)]/dt = \omega_R(t)$, the variable $\theta_2(t)$ of Eq. (2.152) (remember that $\theta_2(t) \equiv d[\theta_1(t)]dt$) is affecting the coefficient matrix of the right-hand side of

Eq. (2.153). Also, note the dependence between Eqs. (2.151) and (2.153). This *cross-coupling* between the three equations, that is, between Eqs. (2.151) and (2.152), and (2.151) and (2.153), makes the interrelationship of these equations **nonlinear**. Thus Eq. (2.153) and Eq. (2.152) constitute a *system of nonlinear differential equations with time-varying coefficients*, and Eq. (2.153) constitutes the *quadratic output of the motor* τ_{em}. It might have been guessed by now that an analytical solution of this system of equations is cumbersome. This is true. The task is greatly simplified by utilizing the analog computer to obtain the solution of the nonlinear differential equations.

However, before the solution is pursued further, Eq. (2.156) is solved for the current matrix $d[i]/dt$ by premultiplying the equation through by the *inverse* of the coefficient matrix of the current matrix $d[i]/dt$, namely, Eq. (2.154).

$$\left[\begin{array}{c|c} 1/L_F & 0 \\ \hline 0 & 1/L_q \end{array}\right] \equiv \frac{1}{L_F L_q}\left[\begin{array}{c|c} L_q & 0 \\ \hline 0 & L_F \end{array}\right] \tag{2.154}$$

Thus premultiplication of Eq. (2.151) by Eq. (2.154) and simplification yield

$$\frac{d}{dt}\left[\begin{array}{c} i_F(t) \\ \hline i_q(t) \end{array}\right] = \left[\begin{array}{c|c} -R_F/L_F & 0 \\ \hline -(\mathfrak{M}/L_q)\omega_R(t) & -R_q/L_q \end{array}\right]\left[\begin{array}{c} i_F(t) \\ \hline i_q(t) \end{array}\right] + \left[\begin{array}{c|c} 1/L_F & 0 \\ \hline 0 & 1/L_q \end{array}\right]\left[\begin{array}{c} e_F(t) \\ \hline e_I(t) \end{array}\right] \tag{2.155}$$

It should be noted here that Eq. (2.157), can be written in a compact matrix form as

$$\dot{\mathbf{X}}(t) = \mathbf{A}[\omega_R(t)]\mathbf{X}(t) + \mathbf{Bu}(t) \tag{2.156}$$

where

$$\mathbf{X}(t) \equiv \left[\begin{array}{c} i_F(t) \\ \hline i_q(t) \end{array}\right] \tag{2.157}$$

and

$$\mathbf{u}(t) \equiv \left[\begin{array}{c} e_F(t) \\ \hline e_I(t) \end{array}\right] \tag{2.158}$$

In a similar manner, Eqs. (2.152) and (2.153) may be written as

$$\dot{\boldsymbol{\Theta}}_R(t) = \mathcal{A}\boldsymbol{\Theta}_R(t) + \mathcal{B}\boldsymbol{\nu}(t) \tag{2.159}$$

and

$$\tau_{em} = \mathbf{X}^T(t)\mathbf{G}\mathbf{X}(t) \tag{2.160}$$

These three equations constitute the **state equations** of the separately excited motor. Without going into details here, as this is an advanced topic of study, note

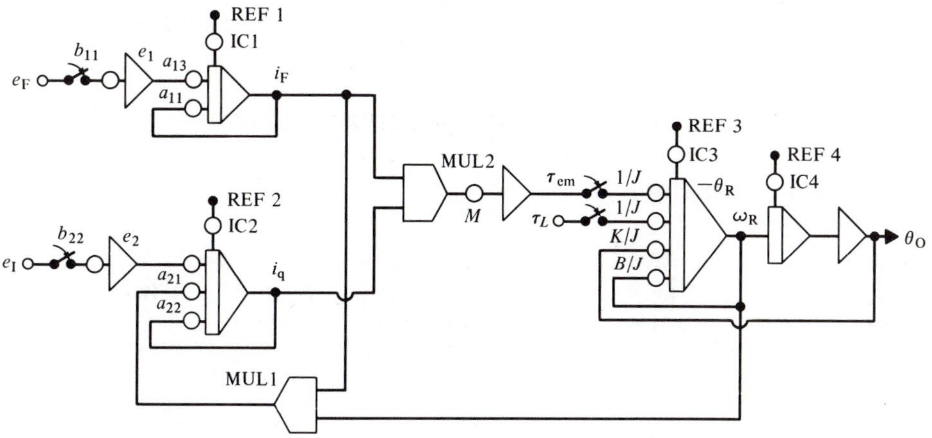

Fig. 2.49 Simulation Diagram for the Separately Excited Motor

that $\dot{\mathbf{X}}(t)$, $\mathbf{X}(t)$, $\mathbf{u}(t)$, $\dot{\mathbf{\Theta}}_R(t)$, $\mathbf{\Theta}_R(t)$, and $\boldsymbol{\nu}(t)$ are *column vectors*, $\mathbf{X}^T(t)$ is a *row vector*, and $\mathbf{A}[\omega_R(t)]$, $\mathbf{B}$, $\mathcal{C}$, $\mathcal{B}$, and $\mathbf{G}$ are rectangular, in fact square, arrays of elements defined as *matrices*. Also, note the one-to-one correspondence between Eqs. (2.156) and (2.155), between Eqs. (2.159) and (2.152), and between Eqs. (2.160) and (2.153).

To obtain solutions for the state variables $\mathbf{X}(t)$, that is, the currents $i_F(t)$ and $i_q(t)$, $\mathbf{\Theta}_R(t)$ [the mechanical variables $\theta_R(t)$ and $\omega_R(t)$], and the electromagnetic torque τ_{em}, the simulation diagram Fig. 2.49 is drawn with the aid of Eqs. (2.155), (2.153), and (2.154), utilizing analog computer simulation techniques.

For example, concentrating our attention on the second row of Eq. (2.157) yields

$$\frac{d}{dt}\left[i_q(t)\right] = -\left(\frac{\mathfrak{M}}{L_q}\right)\omega_R(t)i_F(t) - \left(\frac{R_q}{L_q}\right)i_q(t) + \left(\frac{1}{L_q}\right)e_I(t) \tag{2.161}$$

Solution of Eq. (2.161) for $i_q(t)$ yields

$$i_q(t) = -\frac{1}{\alpha}\int_0^t\left[\left(\frac{\mathfrak{M}}{L_q}\right)\omega_R(t)i_F(t) + \left(\frac{R_q}{L_q}\right)i_q(t) - \left(\frac{1}{L_q}\right)e_I(t)\right]dt + i_q(0) \tag{2.162}$$

Eq. (2.162) is the solution for the current $i_q(t)$, which depends on the following variables: $\omega_R(t)$, $i_F(t)$, $i_q(t)$, $e_I(t)$, $i_q(0)$, and the constant α, which is defined as a **time-scaling factor**. The maximum (or minimum) expected values of all these variables and the time-scaling factor α impose constraints on the use of an analog computer, especially to the beginner. To overcome these constraints, IBM developed a specialized FORTRAN language program that enables analog computer simulation to be performed on a digital computer. The program is known as CSMP (Continuous System Modeling Program). It is this technique that is used here, rather than the analog computer, for the solution of the nonlinear system of Eqs. (2.155), (2.152), and (2.153). Since it is not the stated purpose of this text to teach simulation techniques, the CSMP program is referenced for the convenience of the student.

The CSMP input data consists of three parts (1) initial, (2) dynamic, and (3) timer.* The input data for the dc motor is as follows:

```
                ****CONTINUOUS SYSTEM MODELING PROGRAM****
                           *** VERSION 1.3 ***
TITLE SEPARATELY EXCITED DC MOTOR
INITIAL
PARAMETER        RF=   ,LF=   ,M=   ,LQ=        ,RQ=        ,...
                 EF=   ,EI=   ,K=   ,B=   ,J=        ,TL=
INCON            IC1=  ,IC2=  ,IC3= ,IC4=
DYNAMIC
      A11=RF/LF
      A13=1.
      A21=M/LQ
      A22=RQ/LQ
      A23=1.
      B11=1./LF
      B22=1./LQ
      E1=B11*EF
      E2=B22*EI
      MUL1=OGR*IF
      MUL2=IF*IQ
      IFD=-(A11*IF-A13*E1)
      IF=INTGRL(IC1,IFD)
      IQD=-(A22*IQ+A21*MUL1-A23*E2)
      IQ=INTGRL(IC2,IQD)
      TEM=MUL2*M
      OGRD=-((B/J)*OGR+(K/J)*THETO+(TL/J)-(TEM/J))
      OGR=INTGRL(IC3,OGRD)
      THETA=INTGRL(IC4,OGR)
      THETO=-THETA
TIMER DELT=0.0010,FINTIM=2.0
PRTPLT IF,IQ,TEM,THETA,OGR
END
STOP
```

Upon execution, the program provides *numerical* and *graphical solutions* versus *time* of the above-mentioned electrical and mechanical state variables and of the electromagnetic torque.

The *parameters* that describe the motor to be studied are identified in the PARAMETER portion of the program, and the *initial conditions* to be used are identified in the INCON portion. The DYNAMIC portion contains all the *mathematical equations* that describe the problem. The TIMER portion contains the *integration interval* and the *upper limit* of integration. Finally, the PRTPLT portion is used to specify the variables the user wants to print and plot.

*The first card of the program is: TITLE · · · The program should be preceded by the appropriate CSMP control cards, which should be obtained from your computing center.

Example 2.14

For the separately excited motor, Fig. 2.17, assume that the switches S_1 and S_2 are open, thus yielding IC1 = IC2 = IC3 = IC4 = 0. For $R_F = 104.0$, $L_F = 20.0$, $\mathcal{M} = 0.86$, $L_q = 0.005$, $R_q = 0.11$, $E_F = 125u(t)$, $E_I = 125u(t)$, $K = 0.0$, $B = 0.9$, $J = 10.0$, and $\tau_L = 40.0$, all in MKS units, close the switches S_1 and S_2 at $t = 0$ and study the dynamic behavior of the motor, that is, study the starting dynamics of the motor.

Solution

The provided CSMP program could be used for the study of the dynamic behavior of the motor.

Study carefully the printout in numerical data versus t and the graphical display of the variables $i_F(t)$, $i_q(t)$, $\tau_{em}(t)$, $\omega_R(t)$, and $\theta_R(t)$ exhibited in Fig. 2.50.

Using the same technique as in the motor case, the describing equations of the separately excited dc generator, Eqs. (2.39), (2.40), (2.36e), and (2.56) (as mentioned in Section 2.2, the torque for the generator mode of operation is *negative* because the direction of the response currents in the rotor windings is opposite to that of the impressed currents in the rotor windings for the motor mode of operation), and the describing equation of the series RL-load, which is connected across the rotor winding in place of the input forcing function $e_I(t)$ in the motor mode of operation, enables us to write the describing equations of the separately excited dc generator in matrix form, as follows:

$$\frac{d}{dt}\left[\begin{array}{c} i_F(t) \\ \hline i_q(t) \end{array}\right] = \left[\begin{array}{c|c} -R_F/L_F & 0 \\ \hline -[\mathcal{M}/(L_q+L_L)]\omega_R & (R_q+R_L)/(L_q+L_L) \end{array}\right]\left[\begin{array}{c} i_F(t) \\ \hline i_q(t) \end{array}\right]$$

$$+\left[\begin{array}{c|c} 1/L_F & 0 \\ \hline 0 & 1/(L_q+L_L) \end{array}\right]\left[\begin{array}{c} e_F(t) \\ \hline 0 \end{array}\right] \tag{2.163}$$

$$\tau_{em} = [\,i_F(t) | i_q(t)\,]\left[\begin{array}{c|c} 0 & 0 \\ \hline \mathcal{M} & 0 \end{array}\right]\left[\begin{array}{c} i_F(t) \\ \hline i_q(t) \end{array}\right] \tag{2.164}$$

In Eq. (2.163), ω_R is provided by the prime mover and is assumed to be constant or nearly so, that is, $\omega_R \equiv \omega_I = k$. Thus Eq. (2.163) constitutes a system of *linear differential equations with time-invariant coefficients*. Eq. (2.164) constitutes the *quadratic output of the generator*, that is, the developed electromagnetic torque τ_{em}.

The simulation diagram drawn with the aid of Eqs. (2.163) and (2.164), and analog computer simulation techniques, is exhibited in Fig. 2.51.

As in the case of the separately excited motor, the state equations of the separately excited generator are written in compact matrix form from Eqs. (2.163) and (2.164) as

$$\dot{\mathbf{X}}(t) = \mathbf{AX}(t) + \mathbf{Bu}(t) \tag{2.165}$$

$$\tau_{em} = \mathbf{X}^T(t)\mathbf{GX}(t) \tag{2.166}$$

Note the nondependence on time of the matrix **A** in Eq. (2.165) when compared to the matrix $\mathbf{A}[\omega_R(t)]$ in Eq. (2.156).

MINIMUM IF VERSUS TIME MAXIMUM

0.0 1.2019E+00

TIME	IF
0.0	0.0
2.0000E-02	1.1872E-01
4.0000E-02	2.2571E-01
6.0000E-02	3.2214E-01
8.0000E-02	4.0904E-01
1.0000E-01	4.8735E-01
1.2000E-01	5.5794E-01
1.4000E-01	6.2155E-01
1.6000E-01	6.7887E-01
1.8000E-01	7.3054E-01
2.0000E-01	7.7710E-01
2.2000E-01	8.1906E-01
2.4000E-01	8.5688E-01
2.6000E-01	8.9096E-01
2.8000E-01	9.2167E-01
3.0000E-01	9.4935E-01
3.2000E-01	9.7430E-01
3.4000E-01	9.9679E-01
3.6000E-01	1.0170E+00
3.8000E-01	1.0353E+00
4.0000E-01	1.0518E+00
4.2000E-01	1.0666E+00
4.4000E-01	1.0800E+00
4.6000E-01	1.0920E+00
4.8000E-01	1.1029E+00
5.0000E-01	1.1126E+00
5.2000E-01	1.1215E+00
5.4000E-01	1.1294E+00
5.6000E-01	1.1366E+00
5.8000E-01	1.1430E+00
6.0000E-01	1.1488E+00
6.2000E-01	1.1541E+00
6.4000E-01	1.1588E+00
6.6000E-01	1.1631E+00
6.8000E-01	1.1669E+00
7.0000E-01	1.1704E+00
7.2000E-01	1.1735E+00
7.4000E-01	1.1763E+00
7.6000E-01	1.1788E+00
7.8000E-01	1.1811E+00
8.0000E-01	1.1832E+00
8.2000E-01	1.1850E+00
8.4000E-01	1.1867E+00
8.6000E-01	1.1882E+00
8.8000E-01	1.1895E+00
9.0000E-01	1.1908E+00
9.2000E-01	1.1919E+00
9.4000E-01	1.1929E+00
9.6000E-01	1.1938E+00
9.8000E-01	1.1946E+00
1.0000E+00	1.1953E+00
1.0200E+00	1.1959E+00
1.0400E+00	1.1965E+00
1.0600E+00	1.1971E+00
1.0800E+00	1.1975E+00
1.1000E+00	1.1980E+00
1.1200E+00	1.1984E+00
1.1400E+00	1.1987E+00
1.1600E+00	1.1990E+00
1.1800E+00	1.1993E+00
1.2000E+00	1.1996E+00
1.2200E+00	1.1998E+00
1.2400E+00	1.2000E+00
1.2600E+00	1.2002E+00
1.2800E+00	1.2004E+00
1.3000E+00	1.2005E+00
1.3200E+00	1.2007E+00
1.3400E+00	1.2008E+00
1.3600E+00	1.2009E+00
1.3800E+00	1.2010E+00
1.4000E+00	1.2011E+00
1.4200E+00	1.2012E+00
1.4400E+00	1.2012E+00
1.4600E+00	1.2013E+00
1.4800E+00	1.2014E+00
1.5000E+00	1.2014E+00
1.5200E+00	1.2015E+00
1.5400E+00	1.2015E+00
1.5600E+00	1.2016E+00
1.5800E+00	1.2016E+00
1.6000E+00	1.2016E+00
1.6200E+00	1.2016E+00
1.6400E+00	1.2017E+00
1.6600E+00	1.2017E+00
1.6800E+00	1.2017E+00
1.7000E+00	1.2017E+00
1.7200E+00	1.2018E+00
1.7400E+00	1.2018E+00
1.7600E+00	1.2018E+00
1.7800E+00	1.2018E+00
1.8000E+00	1.2018E+00
1.8200E+00	1.2018E+00
1.8400E+00	1.2018E+00
1.8600E+00	1.2018E+00
1.8800E+00	1.2018E+00
1.9000E+00	1.2019E+00
1.9200E+00	1.2019E+00
1.9400E+00	1.2019E+00
1.9600E+00	1.2019E+00
1.9800E+00	1.2019E+00
2.0000E+00	1.2019E+00

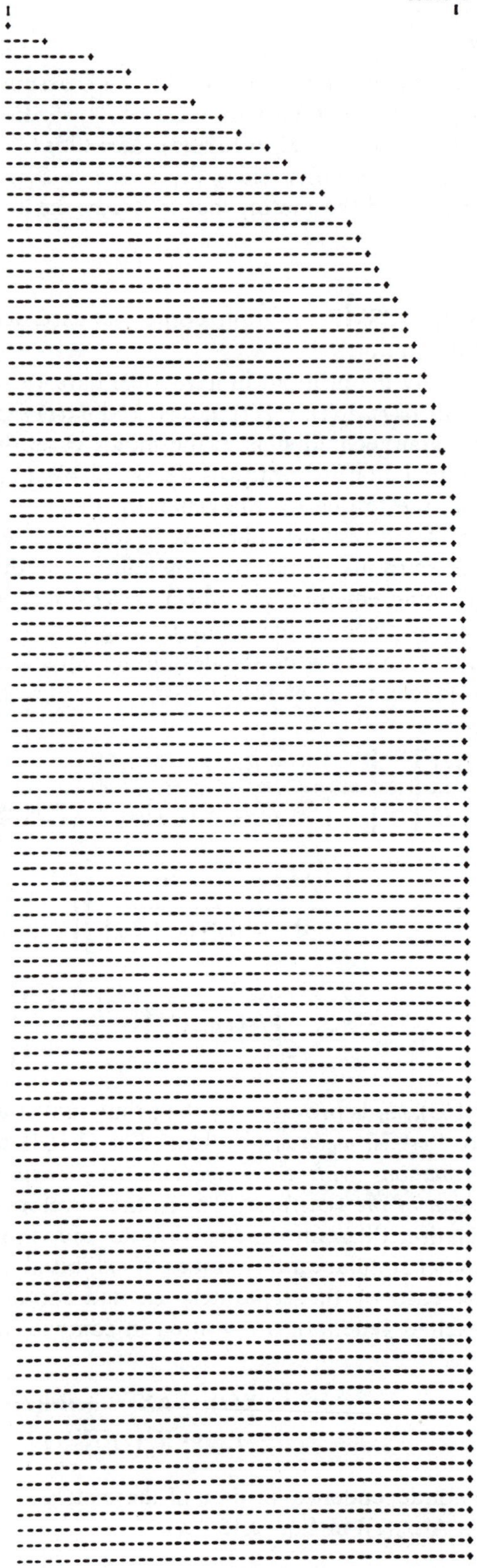

Fig. 2.50 Motor Mode of Operation: Computer Output for Example 2.14

MINIMUM 0.0 IQ VERSUS TIME MAXIMUM 1.0976E+03

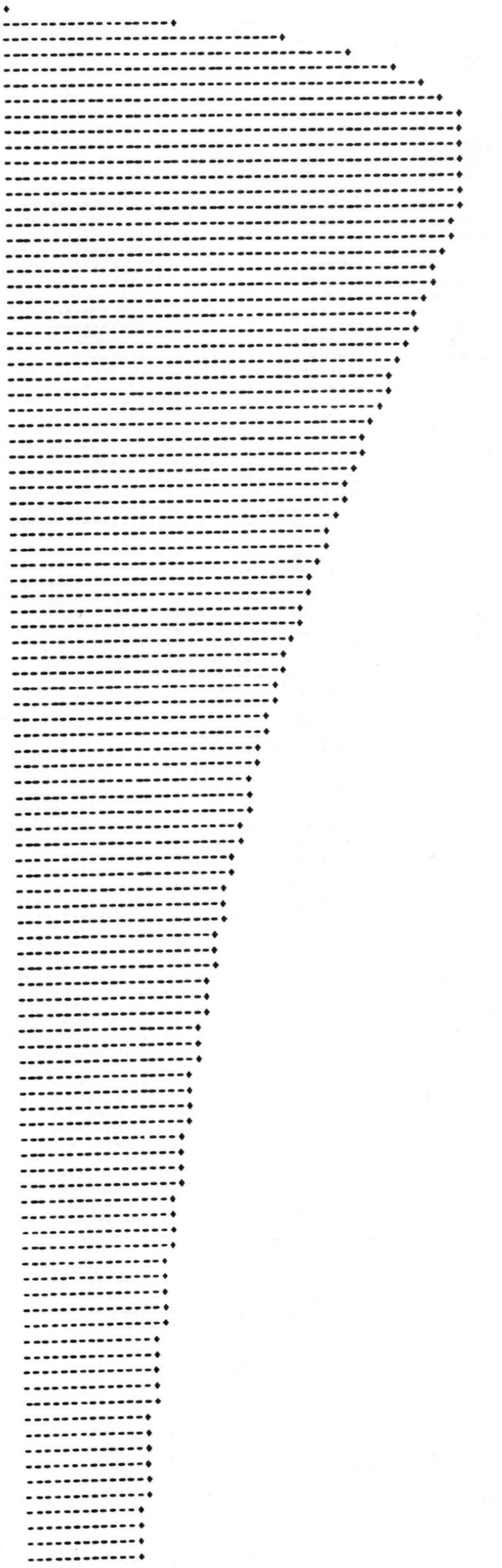

TIME	IQ
0.0	0.0
2.0000E-02	4.0451E+02
4.0000E-02	6.6503E+02
6.0000E-02	8.3267E+02
8.0000E-02	9.4019E+02
1.0000E-01	1.0086E+03
1.2000E-01	1.0513E+03
1.4000E-01	1.0769E+03
1.6000E-01	1.0909E+03
1.8000E-01	1.0971E+03
2.0000E-01	1.0976E+03
2.2000E-01	1.0941E+03
2.4000E-01	1.0876E+03
2.6000E-01	1.0789E+03
2.8000E-01	1.0683E+03
3.0000E-01	1.0564E+03
3.2000E-01	1.0434E+03
3.4000E-01	1.0295E+03
3.6000E-01	1.0149E+03
3.8000E-01	9.9972E+02
4.0000E-01	9.8415E+02
4.2000E-01	9.6828E+02
4.4000E-01	9.5219E+02
4.6000E-01	9.3598E+02
4.8000E-01	9.1970E+02
5.0000E-01	9.0344E+02
5.2000E-01	8.8723E+02
5.4000E-01	8.7113E+02
5.6000E-01	8.5517E+02
5.8000E-01	8.3939E+02
6.0000E-01	8.2381E+02
6.2000E-01	8.0846E+02
6.4000E-01	7.9336E+02
6.6000E-01	7.7852E+02
6.8000E-01	7.6395E+02
7.0000E-01	7.4966E+02
7.2000E-01	7.3566E+02
7.4000E-01	7.2194E+02
7.6000E-01	7.0852E+02
7.8000E-01	6.9539E+02
8.0000E-01	6.8255E+02
8.2000E-01	6.7001E+02
8.4000E-01	6.5774E+02
8.6000E-01	6.4576E+02
8.8000E-01	6.3406E+02
9.0000E-01	6.2263E+02
9.2000E-01	6.1148E+02
9.4000E-01	6.0058E+02
9.6000E-01	5.8994E+02
9.8000E-01	5.7956E+02
1.0000E+00	5.6942E+02
1.0200E+00	5.5952E+02
1.0400E+00	5.4986E+02
1.0600E+00	5.4042E+02
1.0800E+00	5.3121E+02
1.1000E+00	5.2222E+02
1.1200E+00	5.1344E+02
1.1400E+00	5.0486E+02
1.1600E+00	4.9649E+02
1.1800E+00	4.8831E+02
1.2000E+00	4.8033E+02
1.2200E+00	4.7253E+02
1.2400E+00	4.6491E+02
1.2600E+00	4.5747E+02
1.2800E+00	4.5020E+02
1.3000E+00	4.4310E+02
1.3200E+00	4.3617E+02
1.3400E+00	4.2939E+02
1.3600E+00	4.2277E+02
1.3800E+00	4.1630E+02
1.4000E+00	4.0998E+02
1.4200E+00	4.0380E+02
1.4400E+00	3.9776E+02
1.4600E+00	3.9186E+02
1.4800E+00	3.8609E+02
1.5000E+00	3.8046E+02
1.5200E+00	3.7495E+02
1.5400E+00	3.6957E+02
1.5600E+00	3.6430E+02
1.5800E+00	3.5916E+02
1.6000E+00	3.5413E+02
1.6200E+00	3.4921E+02
1.6400E+00	3.4441E+02
1.6600E+00	3.3971E+02
1.6800E+00	3.3511E+02
1.7000E+00	3.3062E+02
1.7200E+00	3.2623E+02
1.7400E+00	3.2194E+02
1.7600E+00	3.1774E+02
1.7800E+00	3.1364E+02
1.8000E+00	3.0963E+02
1.8200E+00	3.0570E+02
1.8400E+00	3.0187E+02
1.8600E+00	2.9811E+02
1.8800E+00	2.9445E+02
1.9000E+00	2.9086E+02
1.9200E+00	2.8735E+02
1.9400E+00	2.8392E+02
1.9600E+00	2.8057E+02
1.9800E+00	2.7729E+02
2.0000E+00	2.7408E+02

Fig. 2.50 (Continued)

MINIMUM 0.0 — TEM VERSUS TIME — MAXIMUM 8.9018E+02

TIME	TEM
0.0	0.0
2.0000E-02	4.1300E+01
4.0000E-02	1.2909E+02
6.0000E-02	2.3068E+02
8.0000E-02	3.3073E+02
1.0000E-01	4.2272E+02
1.2000E-01	5.0442E+02
1.4000E-01	5.7561E+02
1.6000E-01	6.3692E+02
1.8000E-01	6.8924E+02
2.0000E-01	7.3353E+02
2.2000E-01	7.7068E+02
2.4000E-01	8.0148E+02
2.6000E-01	8.2664E+02
2.8000E-01	8.4680E+02
3.0000E-01	8.6250E+02
3.2000E-01	8.7425E+02
3.4000E-01	8.8250E+02
3.6000E-01	8.8767E+02
3.8000E-01	8.9012E+02
4.0000E-01	8.9018E+02
4.2000E-01	8.8818E+02
4.4000E-01	8.8437E+02
4.6000E-01	8.7900E+02
4.8000E-01	8.7231E+02
5.0000E-01	8.6448E+02
5.2000E-01	8.5570E+02
5.4000E-01	8.4612E+02
5.6000E-01	8.3589E+02
5.8000E-01	8.2512E+02
6.0000E-01	8.1393E+02
6.2000E-01	8.0241E+02
6.4000E-01	7.9065E+02
6.6000E-01	7.7873E+02
6.8000E-01	7.6665E+02
7.0000E-01	7.5454E+02
7.2000E-01	7.4242E+02
7.4000E-01	7.3032E+02
7.6000E-01	7.1829E+02
7.8000E-01	7.0634E+02
8.0000E-01	6.9451E+02
8.2000E-01	6.8281E+02
8.4000E-01	6.7126E+02
8.6000E-01	6.5986E+02
8.8000E-01	6.4865E+02
9.0000E-01	6.3761E+02
9.2000E-01	6.2676E+02
9.4000E-01	6.1611E+02
9.6000E-01	6.0565E+02
9.8000E-01	5.9539E+02
1.0000E+00	5.8533E+02
1.0200E+00	5.7547E+02
1.0400E+00	5.6581E+02
1.0600E+00	5.5635E+02
1.0800E+00	5.4709E+02
1.1000E+00	5.3802E+02
1.1200E+00	5.2914E+02
1.1400E+00	5.2046E+02
1.1600E+00	5.1196E+02
1.1800E+00	5.0365E+02
1.2000E+00	4.9552E+02
1.2200E+00	4.8757E+02
1.2400E+00	4.7979E+02
1.2600E+00	4.7219E+02
1.2800E+00	4.6475E+02
1.3000E+00	4.5748E+02
1.3200E+00	4.5037E+02
1.3400E+00	4.4342E+02
1.3600E+00	4.3662E+02
1.3800E+00	4.2998E+02
1.4000E+00	4.2348E+02
1.4200E+00	4.1713E+02
1.4400E+00	4.1091E+02
1.4600E+00	4.0484E+02
1.4800E+00	3.9890E+02
1.5000E+00	3.9310E+02
1.5200E+00	3.8742E+02
1.5400E+00	3.8187E+02
1.5600E+00	3.7645E+02
1.5800E+00	3.7114E+02
1.6000E+00	3.6595E+02
1.6200E+00	3.6088E+02
1.6400E+00	3.5592E+02
1.6600E+00	3.5107E+02
1.6800E+00	3.4633E+02
1.7000E+00	3.4170E+02
1.7200E+00	3.3716E+02
1.7400E+00	3.3273E+02
1.7600E+00	3.2840E+02
1.7800E+00	3.2416E+02
1.8000E+00	3.2002E+02
1.8200E+00	3.1596E+02
1.8400E+00	3.1200E+02
1.8600E+00	3.0813E+02
1.8800E+00	3.0434E+02
1.9000E+00	3.0063E+02
1.9200E+00	2.9701E+02
1.9400E+00	2.9346E+02
1.9600E+00	2.9000E+02
1.9800E+00	2.8661E+02
2.0000E+00	2.8330E+02

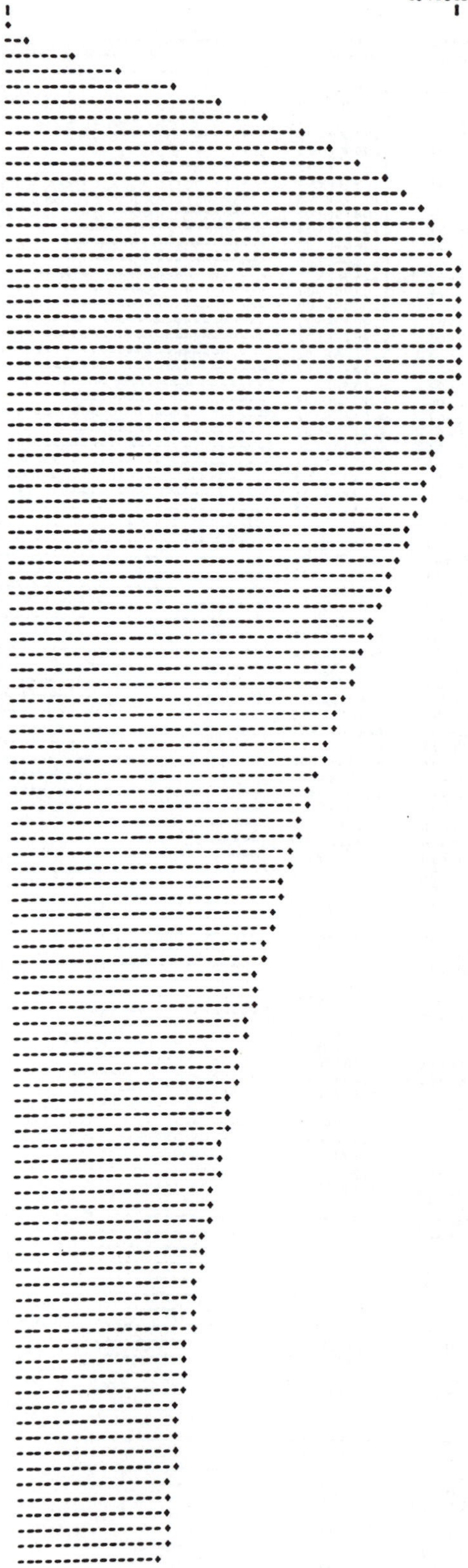

Fig. 2.50 (Continued)

```
                              MINIMUM              OR     VERSUS TIME          MAXIMUM
                             -5.0533E-02            I                         9.2533E+01
 TIME              OR        I                                                I
0.0               0.0        *
2.0000E-02       -5.0533E-02 *
4.0000E-02        3.5547E-02 *
6.0000E-02        3.1545E-01 *
8.0000E-02        7.9586E-01 *
1.0000E-01        1.4685E+00 *
1.2000E-01        2.3144E+00 -*
1.4000E-01        3.3117E+00 -*
1.6000E-01        4.4383E+00 --*
1.8000E-01        5.6765E+00 ---*
2.0000E-01        7.0094E+00 ----*
2.2000E-01        8.4209E+00 -----*
2.4000E-01        9.8975E+00 ------*
2.6000E-01        1.1427E+01 -------*
2.8000E-01        1.3000E+01 --------*
3.0000E-01        1.4605E+01 --------*
3.2000E-01        1.6234E+01 ---------*
3.4000E-01        1.7881E+01 ----------*
3.6000E-01        1.9539E+01 -----------*
3.8000E-01        2.1197E+01 ------------*
4.0000E-01        2.2860E+01 ------------*
4.2000E-01        2.4516E+01 -------------*
4.4000E-01        2.6164E+01 --------------*
4.6000E-01        2.7799E+01 ---------------*
4.8000E-01        2.9419E+01 ----------------*
5.0000E-01        3.1021E+01 -----------------*
5.2000E-01        3.2604E+01 ------------------*
5.4000E-01        3.4166E+01 ------------------*
5.6000E-01        3.5705E+01 -------------------*
5.8000E-01        3.7221E+01 --------------------*
6.0000E-01        3.8711E+01 ---------------------*
6.2000E-01        4.0177E+01 ----------------------*
6.4000E-01        4.1616E+01 -----------------------*
6.6000E-01        4.3029E+01 -----------------------*
6.8000E-01        4.4416E+01 ------------------------*
7.0000E-01        4.5776E+01 -------------------------*
7.2000E-01        4.7109E+01 -------------------------*
7.4000E-01        4.8416E+01 --------------------------*
7.6000E-01        4.9696E+01 ---------------------------*
7.8000E-01        5.0950E+01 ---------------------------*
8.0000E-01        5.2176E+01 ----------------------------*
8.2000E-01        5.3381E+01 ----------------------------*
8.4000E-01        5.4558E+01 -----------------------------*
8.6000E-01        5.5709E+01 ------------------------------*
8.8000E-01        5.6837E+01 ------------------------------*
9.0000E-01        5.7940E+01 -------------------------------*
9.2000E-01        5.9019E+01 -------------------------------*
9.4000E-01        6.0074E+01 --------------------------------*
9.6000E-01        6.1075E+01 --------------------------------*
9.8000E-01        6.2117E+01 ---------------------------------*
1.0000E+00        6.3105E+01 ---------------------------------*
1.0200E+00        6.4071E+01 ----------------------------------*
1.0400E+00        6.5016E+01 ----------------------------------*
1.0600E+00        6.5940E+01 -----------------------------------*
1.0800E+00        6.6844E+01 -----------------------------------*
1.1000E+00        6.7728E+01 ------------------------------------*
1.1200E+00        6.8593E+01 ------------------------------------*
1.1400E+00        6.9438E+01 -------------------------------------*
1.1600E+00        7.0265E+01 -------------------------------------*
1.1800E+00        7.1073E+01 --------------------------------------*
1.2000E+00        7.1866E+01 --------------------------------------*
1.2200E+00        7.2637E+01 ---------------------------------------*
1.2400E+00        7.3392E+01 ---------------------------------------*
1.2600E+00        7.4132E+01 ---------------------------------------*
1.2800E+00        7.4854E+01 ----------------------------------------*
1.3000E+00        7.5561E+01 ----------------------------------------*
1.3200E+00        7.6252E+01 -----------------------------------------*
1.3400E+00        7.6928E+01 -----------------------------------------*
1.3600E+00        7.7589E+01 -----------------------------------------*
1.3800E+00        7.9236E+01 ------------------------------------------*
1.4000E+00        7.8868E+01 ------------------------------------------*
1.4200E+00        7.9486E+01 -------------------------------------------*
1.4400E+00        8.0090E+01 -------------------------------------------*
1.4600E+00        8.0681E+01 -------------------------------------------*
1.4800E+00        8.1259E+01 --------------------------------------------*
1.5000E+00        8.1824E+01 --------------------------------------------*
1.5200E+00        8.2377E+01 --------------------------------------------*
1.5400E+00        8.2917E+01 ---------------------------------------------*
1.5600E+00        8.3445E+01 ---------------------------------------------*
1.5800E+00        8.3963E+01 ---------------------------------------------*
1.6000E+00        8.4468E+01 ----------------------------------------------*
1.6200E+00        8.4963E+01 ----------------------------------------------*
1.6400E+00        8.5446E+01 ----------------------------------------------*
1.6600E+00        8.5919E+01 -----------------------------------------------*
1.6800E+00        8.6381E+01 -----------------------------------------------*
1.7000E+00        8.6833E+01 -----------------------------------------------*
1.7200E+00        8.7275E+01 ------------------------------------------------*
1.7400E+00        8.7708E+01 ------------------------------------------------*
1.7600E+00        8.8130E+01 ------------------------------------------------*
1.7800E+00        8.8544E+01 ------------------------------------------------*
1.8000E+00        8.8948E+01 -------------------------------------------------*
1.8200E+00        8.9344E+01 -------------------------------------------------*
1.8400E+00        8.9731E+01 -------------------------------------------------*
1.8600E+00        9.0104E+01 -------------------------------------------------*
1.8800E+00        9.0487E+01 --------------------------------------------------*
1.9000E+00        9.0840E+01 --------------------------------------------------*
1.9200E+00        9.1194E+01 --------------------------------------------------*
1.9400E+00        9.1543E+01 --------------------------------------------------*
1.9600E+00        9.1877E+01 ---------------------------------------------------*
1.9800E+00        9.2203E+01 ---------------------------------------------------*
2.0000E+00        9.2533E+01 ---------------------------------------------------*
```

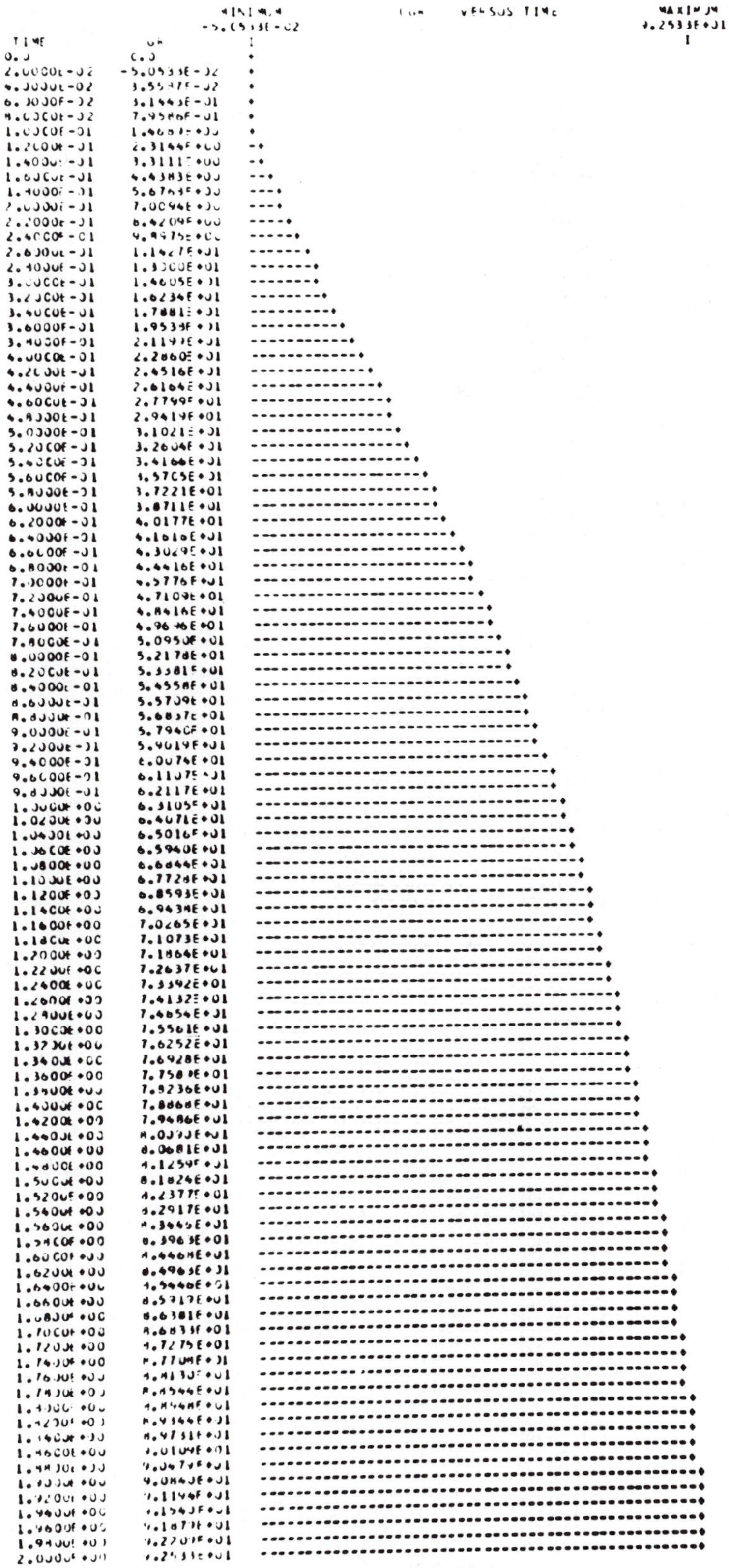

Fig. 2.50 (Continued)

MINIMUM -1.0902E-03 THETA VERSUS TIME MAXIMUM 1.1077E+02

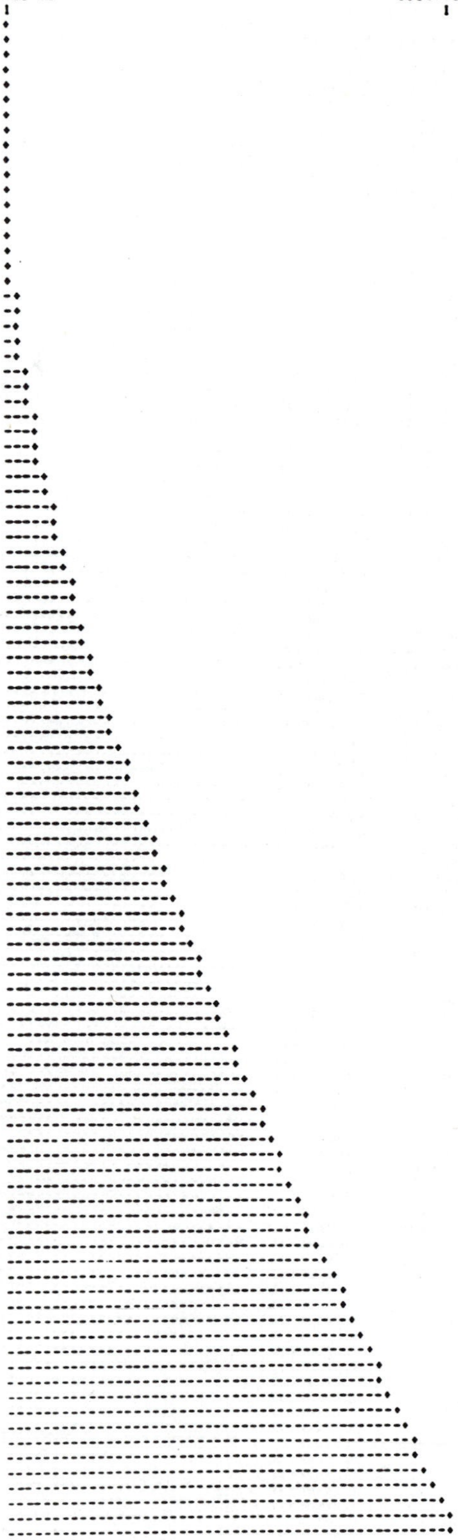

TIME	THETA
0.0	C.0
2.0000E-02	-6.4665E-04
4.0000E-02	-1.0902E-03
6.0000E-02	2.0712E-03
8.0000E-02	1.2841E-02
1.0000E-01	3.5184E-02
1.2000E-01	7.2748E-02
1.4000E-01	1.2877E-01
1.6000E-01	2.0606E-01
1.8000E-01	3.0704E-01
2.0000E-01	4.3376E-01
2.2000E-01	5.8794E-01
4000E-01	7.7103E-01
000E-01	9.8420E-01
000E-01	1.2284E+00
0000E-01	1.5044E+00
3.2000E-01	1.8128E+00
3.4000E-01	2.1539E+00
3.6000E-01	2.5281E+00
3.8000E-01	2.9354E+00
4.0000E-01	3.3760E+00
4.2000E-01	3.8498E+00
4.4000E-01	4.3566E+00
4.6000E-01	4.8963E+00
4.8000E-01	5.4685E+00
5.0000E-01	6.0729E+00
5.2000E-01	6.7092E+00
5.4000E-01	7.3769E+00
5.6000E-01	8.0757E+00
5.8000E-01	8.8050E+00
6.0000E-01	9.5643E+00
6.2000E-01	1.0353E+01
6.4000E-01	1.1171E+01
6.6000E-01	1.2018E+01
6.8000E-01	1.2892E+01
7.0000E-01	1.3794E+01
7.2000E-01	1.4723E+01
7.4000E-01	1.5678E+01
7.6000E-01	1.6660E+01
7.8000E-01	1.7666E+01
8.0000E-01	1.8697E+01
8.2000E-01	1.9753E+01
8.4000E-01	2.0832E+01
8.6000E-01	2.1935E+01
8.8000E-01	2.3061E+01
9.0000E-01	2.4208E+01
9.2000E-01	2.5378E+01
9.4000E-01	2.6569E+01
9.6000E-01	2.7781E+01
9.8000E-01	2.9013E+01
1.0000E+00	3.0265E+01
1.0200E+00	3.1537E+01
1.0400E+00	3.2828E+01
1.0600E+00	3.4137E+01
1.0800E+00	3.5465E+01
1.1000E+00	3.6811E+01
1.1200E+00	3.8174E+01
1.1400E+00	3.9555E+01
1.1600E+00	4.0952E+01
1.1800E+00	4.2365E+01
1.2000E+00	4.3794E+01
1.2200E+00	4.5239E+01
1.2400E+00	4.6700E+01
1.2600E+00	4.8175E+01
1.2800E+00	4.9665E+01
1.3000E+00	5.1169E+01
1.3200E+00	5.2687E+01
1.3400E+00	5.4219E+01
1.3600E+00	5.5764E+01
1.3800E+00	5.7322E+01
1.4000E+00	5.8893E+01
1.4200E+00	6.0477E+01
1.4400E+00	6.2073E+01
1.4600E+00	6.3680E+01
1.4800E+00	6.5300E+01
1.5000E+00	6.6931E+01
1.5200E+00	6.8573E+01
1.5400E+00	7.0226E+01
1.5600E+00	7.1889E+01
1.5800E+00	7.3563E+01
1.6000E+00	7.5248E+01
1.6200E+00	7.6942E+01
1.6400E+00	7.8646E+01
1.6600E+00	8.0360E+01
1.6800E+00	8.2083E+01
1.7000E+00	8.3815E+01
1.7200E+00	8.5556E+01
1.7400E+00	8.7306E+01
1.7600E+00	8.9064E+01
1.7800E+00	9.0831E+01
1.8000E+00	9.2606E+01
1.8200E+00	9.4389E+01
1.8400E+00	9.6179E+01
1.8600E+00	9.7974E+01
1.8800E+00	9.9784E+01
1.9000E+00	1.0160E+02
1.9200E+00	1.0342E+02
1.9400E+00	1.0524E+02
1.9600E+00	1.0708E+02
1.9800E+00	1.0892E+02
2.0000E+00	1.1077E+02

Fig. 2.50 (Continued)

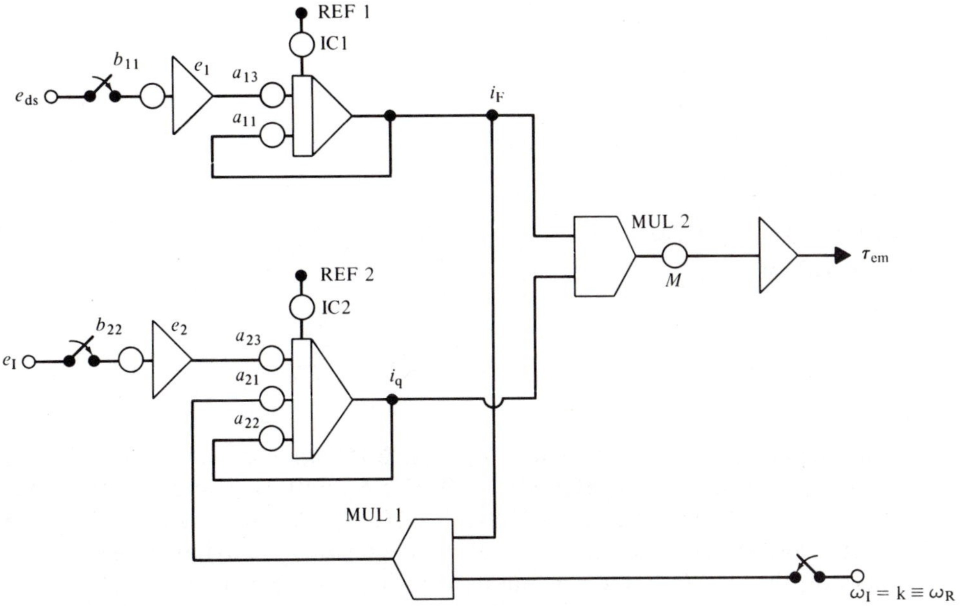

Fig. 2.51 Simulation Diagram for the Separately Excited Generator

The CSMP program* for the separately excited dc generator that provides numerical and graphical results versus time for the state variables $\mathbf{X}(t)$, that is, the currents $i_F(t)$ and $i_q(t)$, and the developed electromagnetic torque τ_{em}, is as follows:

```
              ****CONTINUOUS SYSTEM MODELING PROGRAM****

                          *** VERSION 1.3 ***

TITLE SEPARATELY EXCITED DC GENERATOR
INITIAL
PARAMETER     RF=      ,LF=      ,M=       ,EF=      ,EI=      ,...
              RQ=      ,RL=      ,LQ=      ,LL=      ,FR=
INCON         IC1=     ,IC2=
DYNAMIC
          OGR=6.283185*FR
          RQT=RQ+RL
          LQT=LQ+LL
          A11=RF/LF
          A13=1.
          A21=M/LQT
          A22=RQT/LQT
          A23=1.
          B11=1./LF
          B22=1./LQT
          MUL1=OGR*IF
          MUL2=IF*IQ
```

*Please see footnote on page 136.

```
        E1=B11*EF
        E2=B22*EI
        IFD=-(A11*IF-A13*E1)
        IF=INTGRL(IC1,IFD)
        IQD=-(A22*IQ+A21*MUL1-A23*E2)
        IQ=INTGRL(IC2,IQD)
        TEM=MUL2*M
TIMER DELT=0.0005,FINTIM=1.0
PRTPLT IF,IQ,TEM
END
STOP
```

Example 2.15

For the separately excited dc generator, Fig. 2.15, assume that the switches S_1 and S_2 are open, thus yielding IC1=IC2=0. For $R_F=104.0$, $L_F=20.0$, $\mathcal{M}=0.86$, $R_q=0.11$, $L_q=0.005$, $R_L=4.0$, $L_L=0.001$, $f_R=60$, $E_F=125.0$, and $E_I=0.0$, all in MKS units, close the switches S_1 and S_2 at $t=0$ and study the dynamic behavior of the generator, that is, study the starting dynamics of the generator.

Solution

As before, the provided CSMP program may be used for the study of the dynamic behavior of the generator. Study carefully the printout in numerical data versus t and the graphical display of the variables $i_F(t)$, $i_q(t)$, and $\tau_{em}(t)$ exhibited in Fig. 2.52.

2.12 PARAMETER MEASUREMENT FOR THE EQUIVALENT CIRCUITS OF dc DEVICES

If the equivalent circuit parameters (i.e., $\mathcal{M}$, R_q, L_q, R_F, and L_F, and the rotor's moment of inertia) of an existing, separately excited dc device are not available, they can be obtained experimentally through the following tests: (1) the rotor-winding, *open-circuit test* [Fig. 2.53], (2) the rotor-winding *loaded-circuit test* [Fig. 2.55], and (3) the *motor-starting test* [Fig. 2.57].

During the **open-circuit test**, Fig. 2.53, the field winding is excited from an external dc source of electrical energy, and the rotor is driven, at *rated speed* ω_I with a prime mover, usually a synchronous motor. While rated rotor speed is maintained throughout the test, the field (stator) winding current i_F is varied by means of a rheostat from zero to 120 percent of rated current (i.e., $0<i_F<1.2I_F^{RTD}$). The value of the field current at each setting, along with the corresponding value of the open-circuit voltage across the brushes, designated as e_q^{OC}, is measured, recorded, and plotted, as exhibited in Fig. 2.54.

The plot of these data points, Fig. 2.54—its linear portion described by Eq. (2.36) [i.e., Eq. (2.36a)]—constitutes the **open-circuit characteristic** of the separately excited dc device. This plot exhibits saturation as expected, because the curve constitutes the **magnetization curve** of the device. Note that because the rotor is driven at a known speed, the rated speed of the device $\omega_I \equiv \omega_R^{RTD}$, the slope, $\mathcal{M}\omega_R = \mathcal{K}\,\mathcal{K}_F\omega_R$, of the characteristic is known. Therefore, the design constant

MINIMUM 0.0 — IF VERSUS TIME — MAXIMUM 1.1953E+00

TIME	IF
0.0	0.0
1.0000E-02	6.0903E-02
2.0000E-02	1.1872E-01
3.0000E-02	1.7361E-01
4.0000E-02	2.2571E-01
5.0000E-02	2.7518E-01
6.0000E-02	3.2214E-01
7.0000E-02	3.6672E-01
8.0000E-02	4.0904E-01
9.0000E-02	4.4921E-01
1.0000E-01	4.8736E-01
1.1000E-01	5.2356E-01
1.2000E-01	5.5794E-01
1.3000E-01	5.9057E-01
1.4000E-01	6.2155E-01
1.5000E-01	6.5095E-01
1.6000E-01	6.7887E-01
1.7000E-01	7.0538E-01
1.8000E-01	7.3054E-01
1.9000E-01	7.5442E-01
2.0000E-01	7.7710E-01
2.1000E-01	7.9862E-01
2.2000E-01	8.1906E-01
2.3000E-01	8.3846E-01
2.4000E-01	8.5688E-01
2.5000E-01	8.7436E-01
2.6000E-01	8.9096E-01
2.7000E-01	9.0672E-01
2.8000E-01	9.2167E-01
2.9000E-01	9.3587E-01
3.0000E-01	9.4936E-01
3.1000E-01	9.6215E-01
3.2000E-01	9.7430E-01
3.3000E-01	9.8584E-01
3.4000E-01	9.9679E-01
3.5000E-01	1.0072E+00
3.6000E-01	1.0170E+00
3.7000E-01	1.0264E+00
3.8000E-01	1.0353E+00
3.9000E-01	1.0437E+00
4.0000E-01	1.0518E+00
4.1000E-01	1.0594E+00
4.2000E-01	1.0666E+00
4.3000E-01	1.0734E+00
4.4000E-01	1.0800E+00
4.5000E-01	1.0861E+00
4.6000E-01	1.0920E+00
4.7000E-01	1.0976E+00
4.8000E-01	1.1029E+00
4.9000E-01	1.1079E+00
5.0000E-01	1.1126E+00
5.1000E-01	1.1172E+00
5.2000E-01	1.1215E+00
5.3000E-01	1.1255E+00
5.4000E-01	1.1294E+00
5.5000E-01	1.1331E+00
5.6000E-01	1.1366E+00
5.7000E-01	1.1399E+00
5.8000E-01	1.1430E+00
5.9000E-01	1.1460E+00
6.0000E-01	1.1488E+00
6.1000E-01	1.1515E+00
6.2000E-01	1.1541E+00
6.3000E-01	1.1565E+00
6.4000E-01	1.1588E+00
6.5000E-01	1.1610E+00
6.6000E-01	1.1631E+00
6.7000E-01	1.1650E+00
6.8000E-01	1.1669E+00
6.9000E-01	1.1687E+00
7.0000E-01	1.1704E+00
7.1000E-01	1.1719E+00
7.2000E-01	1.1735E+00
7.3000E-01	1.1749E+00
7.4000E-01	1.1763E+00
7.5000E-01	1.1776E+00
7.6000E-01	1.1788E+00
7.7000E-01	1.1800E+00
7.8000E-01	1.1811E+00
7.9000E-01	1.1821E+00
8.0000E-01	1.1831E+00
8.1000E-01	1.1841E+00
8.2000E-01	1.1850E+00
8.3000E-01	1.1859E+00
8.4000E-01	1.1867E+00
8.5000E-01	1.1874E+00
8.6000E-01	1.1882E+00
8.7000E-01	1.1889E+00
8.8000E-01	1.1895E+00
8.9000E-01	1.1902E+00
9.0000E-01	1.1907E+00
9.1000E-01	1.1913E+00
9.2000E-01	1.1918E+00
9.3000E-01	1.1924E+00
9.4000E-01	1.1928E+00
9.5000E-01	1.1933E+00
9.6000E-01	1.1937E+00
9.7000E-01	1.1942E+00
9.8000E-01	1.1945E+00
9.9000E-01	1.1949E+00
1.0000E+00	1.1953E+00

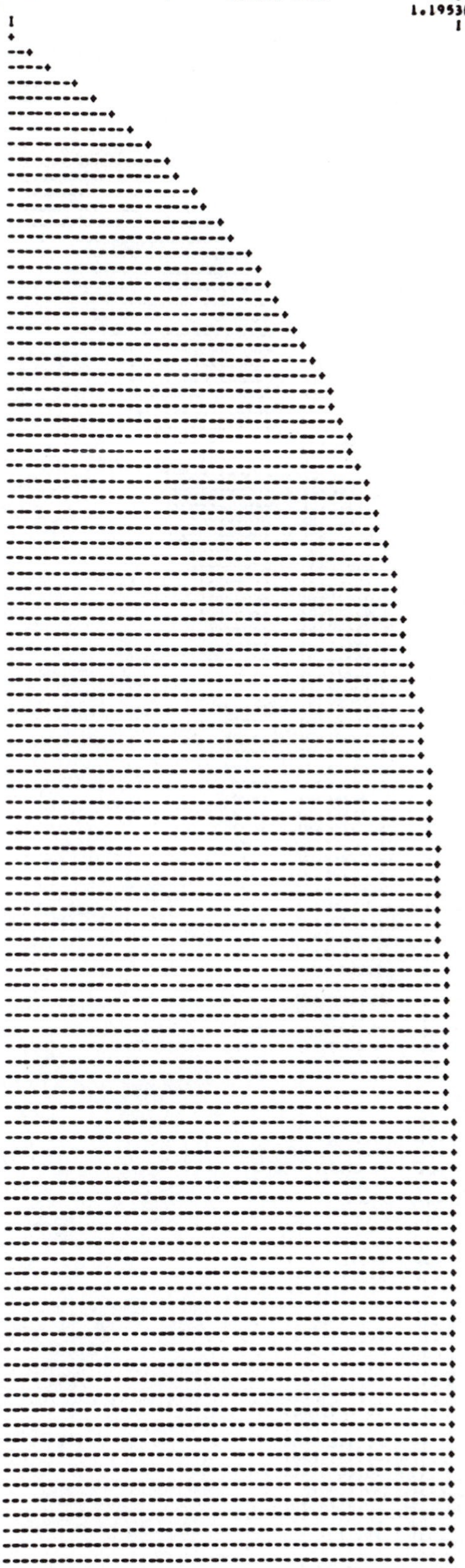

Fig. 2.52 Generator Mode of Operation: Computer Output for Example 2.15

MINIMUM -9.4282E+01 IQ VERSUS TIME MAXIMUM 0.0

TIME	IQ
0.0	0.0
1.0000E-02	-4.1165E+00
2.0000E-02	-8.7114E+00
3.0000E-02	-1.3074E+01
4.0000E-02	-1.7216E+01
5.0000E-02	-2.1148E+01
6.0000E-02	-2.4880E+01
7.0000E-02	-2.8424E+01
8.0000E-02	-3.1788E+01
9.0000E-02	-3.4982E+01
1.0000E-01	-3.8013E+01
1.1000E-01	-4.0891E+01
1.2000E-01	-4.3624E+01
1.3000E-01	-4.6217E+01
1.4000E-01	-4.8680E+01
1.5000E-01	-5.1018E+01
1.6000E-01	-5.3236E+01
1.7000E-01	-5.5343E+01
1.8000E-01	-5.7344E+01
1.9000E-01	-5.9242E+01
2.0000E-01	-6.1044E+01
2.1000E-01	-6.2755E+01
2.2000E-01	-6.4380E+01
2.3000E-01	-6.5922E+01
2.4000E-01	-6.7385E+01
2.5000E-01	-6.8775E+01
2.6000E-01	-7.0095E+01
2.7000E-01	-7.1347E+01
2.8000E-01	-7.2536E+01
2.9000E-01	-7.3666E+01
3.0000E-01	-7.4736E+01
3.1000E-01	-7.5754E+01
3.2000E-01	-7.6719E+01
3.3000E-01	-7.7636E+01
3.4000E-01	-7.8507E+01
3.5000E-01	-7.9334E+01
3.6000E-01	-8.0117E+01
3.7000E-01	-8.0861E+01
3.8000E-01	-8.1568E+01
3.9000E-01	-8.2240E+01
4.0000E-01	-8.2876E+01
4.1000E-01	-8.3481E+01
4.2000E-01	-8.4053E+01
4.3000E-01	-8.4600E+01
4.4000E-01	-8.5118E+01
4.5000E-01	-8.5608E+01
4.6000E-01	-8.6075E+01
4.7000E-01	-8.6518E+01
4.8000E-01	-8.6938E+01
4.9000E-01	-8.7336E+01
5.0000E-01	-8.7715E+01
5.1000E-01	-8.8075E+01
5.2000E-01	-8.8416E+01
5.3000E-01	-8.8741E+01
5.4000E-01	-8.9048E+01
5.5000E-01	-8.9340E+01
5.6000E-01	-8.9616E+01
5.7000E-01	-8.9880E+01
5.8000E-01	-9.0130E+01
5.9000E-01	-9.0367E+01
6.0000E-01	-9.0592E+01
6.1000E-01	-9.0805E+01
6.2000E-01	-9.1009E+01
6.3000E-01	-9.1202E+01
6.4000E-01	-9.1385E+01
6.5000E-01	-9.1558E+01
6.6000E-01	-9.1723E+01
6.7000E-01	-9.1880E+01
6.8000E-01	-9.2028E+01
6.9000E-01	-9.2169E+01
7.0000E-01	-9.2302E+01
7.1000E-01	-9.2428E+01
7.2000E-01	-9.2550E+01
7.3000E-01	-9.2664E+01
7.4000E-01	-9.2772E+01
7.5000E-01	-9.2877E+01
7.6000E-01	-9.2974E+01
7.7000E-01	-9.3068E+01
7.8000E-01	-9.3155E+01
7.9000E-01	-9.3238E+01
8.0000E-01	-9.3319E+01
8.1000E-01	-9.3394E+01
8.2000E-01	-9.3467E+01
8.3000E-01	-9.3535E+01
8.4000E-01	-9.3598E+01
8.5000E-01	-9.3661E+01
8.6000E-01	-9.3719E+01
8.7000E-01	-9.3776E+01
8.8000E-01	-9.3827E+01
8.9000E-01	-9.3876E+01
9.0000E-01	-9.3922E+01
9.1000E-01	-9.3967E+01
9.2000E-01	-9.4011E+01
9.3000E-01	-9.4051E+01
9.4000E-01	-9.4090E+01
9.5000E-01	-9.4126E+01
9.6000E-01	-9.4161E+01
9.7000E-01	-9.4194E+01
9.8000E-01	-9.4226E+01
9.9000E-01	-9.4255E+01
1.0000E+00	-9.4282E+01

Fig. 2.52 (Continued)

MINIMUM -9.6915E+01 TEM VERSUS TIME MAXIMUM 0.0

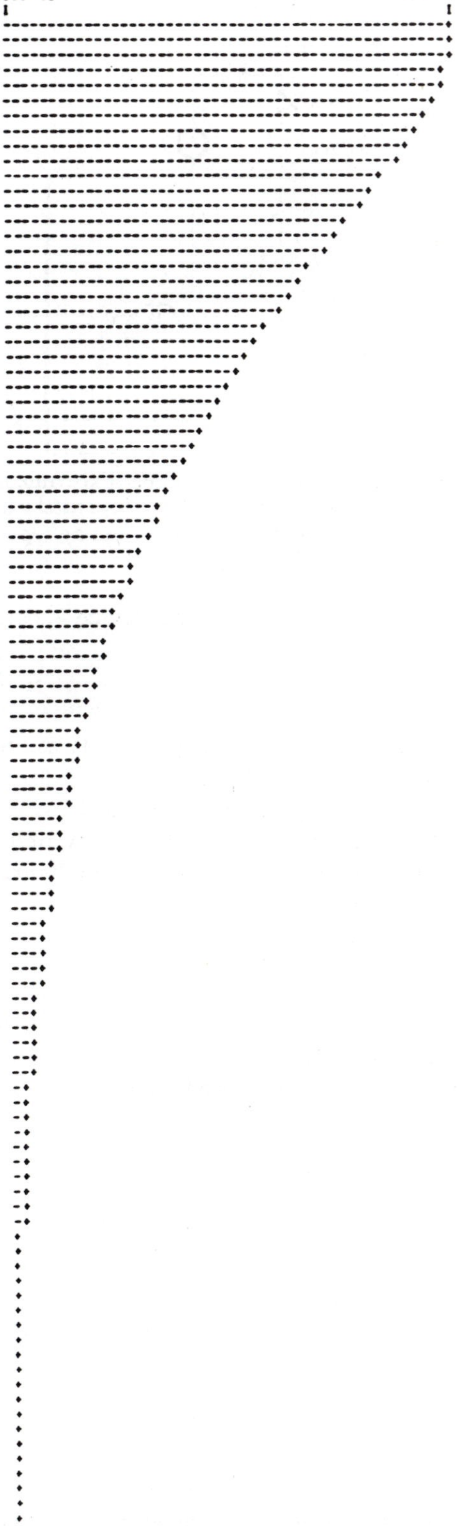

TIME	TEM
0.0	0.0
1.0000E-02	-2.1561E-01
2.0000E-02	-8.8942E-01
3.0000E-02	-1.9520E+00
4.0000E-02	-3.3418E+00
5.0000E-02	-5.0047E+00
6.0000E-02	-6.8928E+00
7.0000E-02	-8.9643E+00
8.0000E-02	-1.1182E+01
9.0000E-02	-1.3514E+01
1.0000E-01	-1.5932E+01
1.1000E-01	-1.8412E+01
1.2000E-01	-2.0932E+01
1.3000E-01	-2.3473E+01
1.4000E-01	-2.6021E+01
1.5000E-01	-2.8561E+01
1.6000E-01	-3.1081E+01
1.7000E-01	-3.3572E+01
1.8000E-01	-3.6027E+01
1.9000E-01	-3.8436E+01
2.0000E-01	-4.0796E+01
2.1000E-01	-4.3101E+01
2.2000E-01	-4.5348E+01
2.3000E-01	-4.7535E+01
2.4000E-01	-4.9657E+01
2.5000E-01	-5.1716E+01
2.6000E-01	-5.3709E+01
2.7000E-01	-5.5635E+01
2.8000E-01	-5.7495E+01
2.9000E-01	-5.9290E+01
3.0000E-01	-6.1018E+01
3.1000E-01	-6.2682E+01
3.2000E-01	-6.4283E+01
3.3000E-01	-6.5821E+01
3.4000E-01	-6.7299E+01
3.5000E-01	-6.8717E+01
3.6000E-01	-7.0075E+01
3.7000E-01	-7.1377E+01
3.8000E-01	-7.2625E+01
3.9000E-01	-7.3820E+01
4.0000E-01	-7.4963E+01
4.1000E-01	-7.6056E+01
4.2000E-01	-7.7100E+01
4.3000E-01	-7.8099E+01
4.4000E-01	-7.9054E+01
4.5000E-01	-7.9964E+01
4.6000E-01	-8.0834E+01
4.7000E-01	-8.1665E+01
4.8000E-01	-8.2456E+01
4.9000E-01	-8.3212E+01
5.0000E-01	-8.3932E+01
5.1000E-01	-8.4619E+01
5.2000E-01	-8.5273E+01
5.3000E-01	-8.5898E+01
5.4000E-01	-8.6491E+01
5.5000E-01	-8.7057E+01
5.6000E-01	-8.7595E+01
5.7000E-01	-8.8109E+01
5.8000E-01	-8.8597E+01
5.9000E-01	-8.9062E+01
6.0000E-01	-8.9505E+01
6.1000E-01	-8.9925E+01
6.2000E-01	-9.0326E+01
6.3000E-01	-9.0708E+01
6.4000E-01	-9.1071E+01
6.5000E-01	-9.1415E+01
6.6000E-01	-9.1744E+01
6.7000E-01	-9.2057E+01
6.8000E-01	-9.2352E+01
6.9000E-01	-9.2634E+01
7.0000E-01	-9.2902E+01
7.1000E-01	-9.3156E+01
7.2000E-01	-9.3400E+01
7.3000E-01	-9.3630E+01
7.4000E-01	-9.3848E+01
7.5000E-01	-9.4057E+01
7.6000E-01	-9.4255E+01
7.7000E-01	-9.4443E+01
7.8000E-01	-9.4622E+01
7.9000E-01	-9.4790E+01
8.0000E-01	-9.4953E+01
8.1000E-01	-9.5105E+01
8.2000E-01	-9.5252E+01
8.3000E-01	-9.5390E+01
8.4000E-01	-9.5520E+01
8.5000E-01	-9.5646E+01
8.6000E-01	-9.5765E+01
8.7000E-01	-9.5879E+01
8.8000E-01	-9.5984E+01
8.9000E-01	-9.6085E+01
9.0000E-01	-9.6180E+01
9.1000E-01	-9.6274E+01
9.2000E-01	-9.6361E+01
9.3000E-01	-9.6442E+01
9.4000E-01	-9.6522E+01
9.5000E-01	-9.6596E+01
9.6000E-01	-9.6667E+01
9.7000E-01	-9.6734E+01
9.8000E-01	-9.6799E+01
9.9000E-01	-9.6859E+01
1.0000E+00	-9.6915E+01

Fig. 2.52 (Continued)

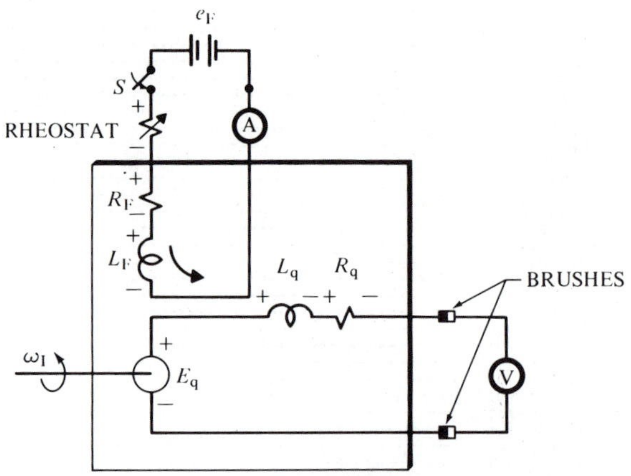

Fig. 2.53 Schematic Diagram for Obtaining Experimentally the Generated Voltage vs. Field Current Characteristic of the Separately Excited dc Device Operating in the Generator Mode, with the Rotor Winding Open-circuited

$\mathfrak{M} = \mathcal{K}\mathcal{K}_F$ of the separately excited dc device is calculated as follows:

$$\mathfrak{M} = \mathcal{K}\mathcal{K}_F = \frac{\Delta e_q^{OC}}{\Delta i_F} \times \frac{1}{\omega_R} \tag{2.167}$$

From the equivalent circuit of the separately excited dc generator, Fig. 2.33, it is clear that the terminal voltage v_L of the generator, that is, the voltage across the brushes, v_q, is equal to the generated voltage E_q if the rotor terminals of the device are open-circuited and the field winding is excited with rated current, $i_F \equiv I_F^{RTD}$. That is, the describing equation of the separately excited dc generator, Eq. (2.104), becomes

$$v_q = (E_q - R_q i_q)\Big|_{\substack{v_q = v_L \\ i_q = 0 \\ i_F = I^{RTD}F}} = E_q \tag{2.168}$$

During the **rotor-winding, loaded-circuit test**, Fig. 2.55, while maintaining the **field current i_F** and the **rotor speed** ω_I constant at *rated values*, ω_R^{RTD}, the load

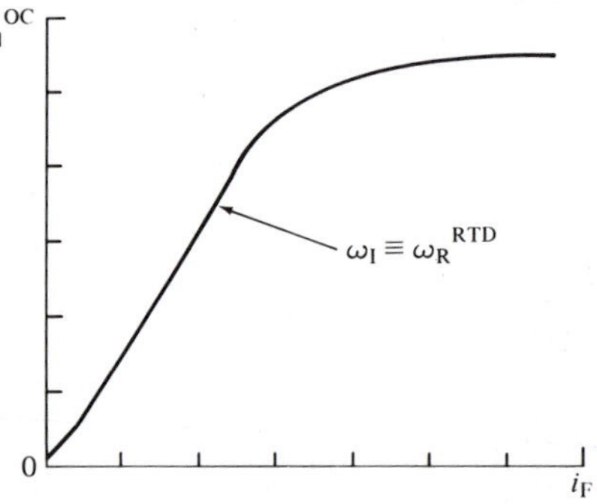

Fig. 2.54 Experimentally Obtained Generated Voltage vs. Field Current Characteristic of the Separately Excited dc Device Operating in the Generator Mode, with the Rotor Winding Open-circuited

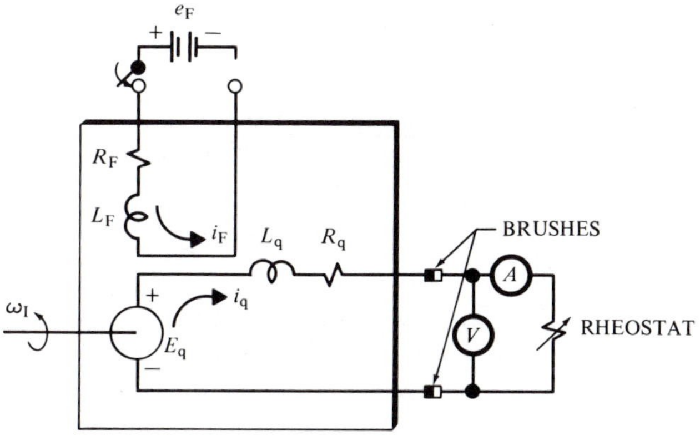

Fig. 2.55 Schematic Diagram for Obtaining Experimentally the Rotor Winding Characteristics (See Fig. 2.56) of the Separately Excited dc Device Operating in the Generator Mode, with the Rotor Winding Loaded

rheostat is *varied* and the following readings are taken: (1) the voltage v_q across the brushes and (2) the current i_q through the brushes, for each setting of the load rheostat. The readings should correspond to a rotor-winding current variation from zero to 120 percent of rated current (i.e., $0<i_q<1.2I_q^{RTD}$). It is essential to note that you should be very careful when taking the readings of the voltage v_q for the corresponding readings of the current i_q, as v_q **decreases slowly** while i_q **increases rapidly**. The recorded data is plotted as in Fig. 2.56(a). The equation of the resulting characteristic is written as

$$v_q = E_q - R_q i_q; \quad v_q = v_L \tag{2.169}$$

It is easy to see that Eq. (2.169) constitutes the steady state describing equation of the dc generator. Therefore, we can read (1) the generated voltage E_q as the v_q-axis intersection of the v_q versus i_q characteristic exhibited in Fig. 2.56(a) and (2) the rotor-winding resistance R_q as the slope of the same characteristic. That is,

$$R_q = \frac{\Delta v_q}{\Delta i_q} \tag{2.170}$$

If the v_q-axis intersection point is not recorded as a data point, the intersection may be established by extrapolating to the left the plotted characteristic, as shown in Fig. 2.56(a).

The inductance L_q of the rotor winding can be calculated easily if we were to set the loading rheostat, Fig. 2.55, to the specific value, say $R_{\text{RHEOSTAT}} = R_L$, and record, or observe on an oscilloscope, the rotor-winding current (designating i_q as i_q^{FL}) as i_q^{FL} versus t, Fig. 2.56(b). Now, use of basic current analysis theory associated with the definition of the term **time constant**, $\tau = L/R$, yields

$$L_q = \tau(R_L + R_q) \tag{2.171}$$

The numerical value of τ is extracted from Fig. 2.56(b); R_L is the set value of the rheostat, Fig. 2.55; R_q is the value of the resistance extracted from Eq. (2.170).

To calculate the moment of inertia J of the rotor (or $J_{EQ} = J +$ loading, if loading is present), one has to run the device in its motor mode of operation, Fig. 2.57. That is, one must excite both the stator and the rotor windings with **rated** dc voltages and record, or observe on an oscilloscope, the currents i_F versus t and i_q

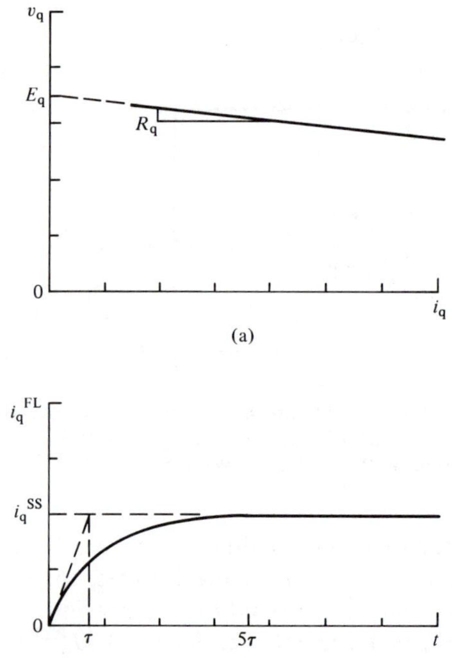

Fig. 2.56 Experimentally Obtained Rotor Winding Characteristics of the Separately Excited dc Device Operating in the Generator Mode: (a) v_q vs. i_q Characteristic, R_L=Variable, and (b) i_q^{FL} vs. t Characteristic, R_L=Constant

versus t and the rotor speed ω_R versus t, as exhibited in Fig. 2.58. The latter objective is achieved through the use of a *tachometer* properly attached to the device, Fig. 2.57. The readings of the two currents at, say, $t=5\tau$ of the plot i_F versus t, Fig. 2.58(a), the utilization of the design constant $\mathcal{M}=\mathcal{K}\mathcal{K}_F$ extracted from Eq. (2.167), and the substitution of these three values into Eq. (2.56) yields

$$\tau_{em} = \mathcal{M} i_F i_q = \mathcal{K}\mathcal{K}_F i_F i_q \tag{2.172}$$

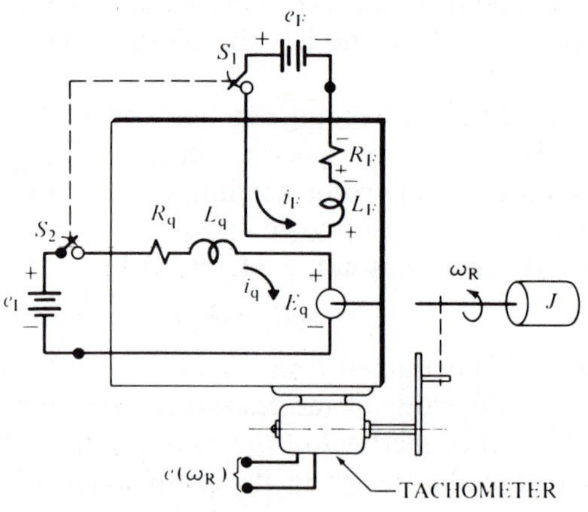

Fig. 2.57 Schematic Diagram for Obtaining Experimentally the Starting Characteristics (See Fig. 2.58) of the Separately Excited Device Operating in the Motor Mode

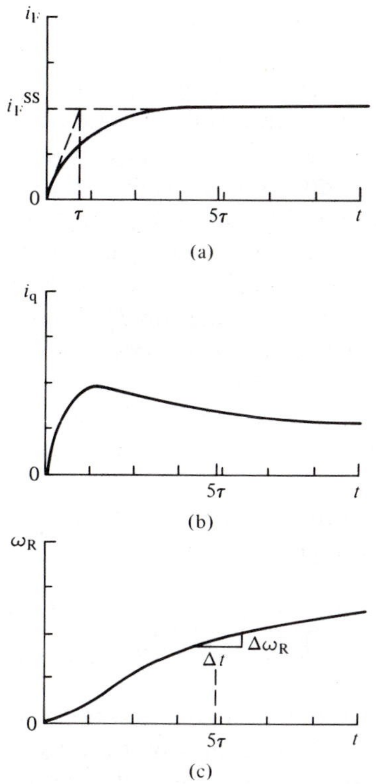

Fig. 2.58 Starting Characteristics of the Separately Excited Device Operating in the Motor Mode: (a) i_F vs. t, (b) i_q vs. t, and (c) ω_R vs. t

Knowledge of the developed electromagnetic torque τ_{em}, Eq. (2.172), enables us to utilize Eq. (2.69), after setting $\tau_L = K = B = 0$, and write

$$\tau_{em} = J\frac{d\omega_R}{dt} \tag{2.173}$$

If we were to calculate the slope of the ω_R versus t plot, Fig. 2.58(c), at $t = 5\tau$ and solve Eq. (2.173) for J, we will obtain

$$J = \frac{\tau_{em}}{(\Delta\omega_R/\Delta t)} = \frac{\mathfrak{M} i_F i_q}{(\Delta\omega_R/\Delta t)}\Bigg|_{\substack{t=5\tau \\ \mathfrak{M}=\mathcal{K}\mathcal{K}F}} \tag{2.174}$$

The circuit parameters R_F and L_F of the field winding are readily obtained. We may either use the rated value of the excitation voltage E_F and the rated, or steady state, current I_F^{RTD} or measure directly the dc value of the field-winding resistance R_F using an ordinary dc ohmmeter. The value of the inductance L_F is calculated through the use of the i_F versus t plot of Fig. 2.58(a) and the basic ideas leading up to and including Eq. (2.171). Thus

$$L_F = \tau R_F \tag{2.175}$$

Basically, this section provides the necessary information for the parameter measurement for the equivalent circuits of the separately excited dc devices. The various connections in which one might find the dc device are mere extensions of the basic separately excited dc device. For example, the circuit parameters for the equivalent circuit of the cumulative compound connection of the dc device can be determined following precisely the procedures above. However, note that in this connection there is a second field winding, Fig. 2.44, connected in series with the rotor winding; therefore, it is excited with the current i_q. This winding causes an additional voltage, designated as $E_q^{CC} = \mathcal{M}'\omega_R i_q$, which, depending on whether the connection is a **cumulative** or **differential compound**, causes an **increase** or **decrease** to the already calculated voltage E_q; that is, $E_q \pm E_q^{CC} = \mathcal{M}\omega_R i_F \pm \mathcal{M}'\omega_R i_q$. Clearly, then, $\mathcal{M}'$ can be calculated from a v_q versus i_q characteristic obtained through the schematic diagram of Fig. 2.55, modified to include the second field winding in the rotor circuit. Then the data is taken while $i_F = I_F^{RTD}$ and $\omega_R = \omega_R^{RTD}$ are maintained and i_q is varied from zero to 120 percent of rated value (i.e., $0 < i_q < 1.2 I_q^{RTD}$).

Before the chapter closes, the following remark should be made about the separately excited dc motor. **The device has an inherent speed runaway capability that is extremely dangerous.** This phenomenon can be seen from the following mathematical development.

The solution of Eq. (2.60) for ω_R after the sequential substitution of Eqs. (2.80) and (2.54) into the resultant equation yields

$$\omega_R = \frac{E_b}{\mathcal{K}\Phi_P} = \frac{E_I - R_q i_q}{\mathcal{K}\Phi_P} = \left(\frac{E_I}{\mathcal{K}\Phi_P} - \frac{R_q \tau_{em}}{[\mathcal{K}\Phi_P]^2} \right) \Bigg|_{\Phi_P = \mathcal{K}_F i_F} \tag{2.176}$$

Although Eq. (2.176) suggests that there are three methods [namely, (1) variation of the impressed voltage E_I, (2) variation of R_q through insertion of an external resistance to augment R_q, and (3) variation of the stator field Φ_P through variation of the field current i_F] for controlling the speed of the dc motor for a fixed mechanical load, it also provides an insight as to what would happen to a dc motor running under **no load** if the field excitation, for any reason, is removed while the device is operating.

The analytical check of Eq. (2.176) is much safer than the experimental investigation of the phenomenon. The reader should verify from Eq. (2.176) that if the field current i_F is to become zero for any reason (i.e., even if the field current i_F is turned off), **the speed of the rotor will tend to increase beyond the structural limit of the dc motor. The motor will explode unless stopped.** Thus if an abnormal *whirr* is heard in the laboratory, or *the rotor speeds abruptly*, the dc supply in the laboratory should be **turned off.** To avoid the possibility of physical injuries, students in the vicinity of the runaway device should **take cover.** However, a gallant attempt should be made **to turn off the dc panic switch**, if such is present in the laboratory.

Example 2.16

Assume that the data you obtained when you performed the tests on the separately excited dc device with the rotor winding open-circuited and loaded yielded the two plots shown in Fig. 2.16.1. Also assume that in addition to these two plots, the following information is available to you: (1) the i_q^{FL} versus t plot for a load comprised of $R_L = 4.00\ \Omega$ and $L_L = 0.684$ H, Fig. 2.52(b); (2) the i_F versus t, i_q versus t, and ω_R versus t plots, Figs. 2.50(a), 2.50(b) and 2.50(d); (3) $E_F = 110.00$ V and $I_F^{RTD} = 1.20$ A; (4) the prime mover used was a two-pole, balanced, three-phase, synchronous motor excited from 110-V, 60-Hz mains. Calculate the circuit parameters $\mathcal{M}$, R_q, L_q, R_F, and L_F for the dc device as well as the rotor moment of inertia J.

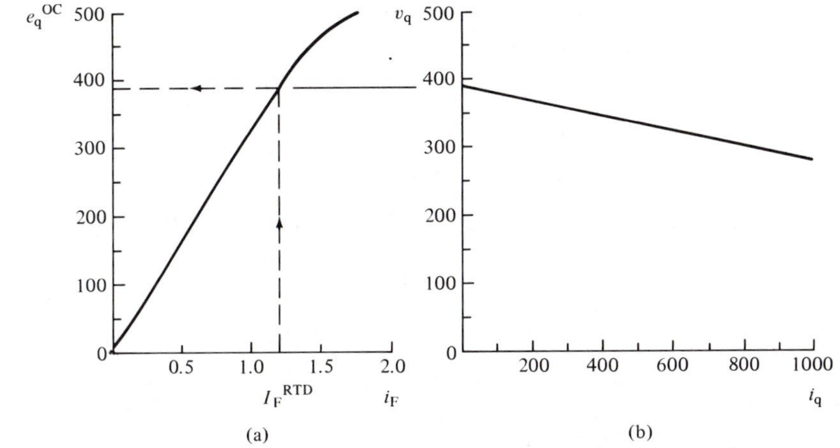

Fig. 2.16.1 Experimentally Obtained Characteristics of a Separately Excited dc Device with (a) the Rotor Winding Open-circuited, Generated Voltage vs. Field Current, and (b) the Rotor Winding Loaded, Brush Voltage vs. Brush Current

Solution

The generated voltage E_q is read directly from the open-circuit characteristic at rated current $I_F^{RTD} = 1.20$ A by direct reflection off this characteristic, Fig. 2.16.1(a). This reading yields $E_q = 389.00$ V.

The design constant, $\mathfrak{M} = \mathcal{K}\mathcal{K}_F$, is calculated with the aid of Fig. 2.16.1(a) and the use of Eq. (2.167). The slope of the open-circuit characteristic, Fig. 2.16.1(a) is

$$\frac{\Delta e_q^{OC}}{\Delta i_F} = \frac{324.00 - 0}{1 - 0} = \underline{\underline{324.00}}\ \text{V/A} \tag{1}$$

The rated speed of the rotor is the speed provided by the two-pole, three-phase synchronous device. Thus from Eq. (2.90)

$$\omega_I = \omega_S = \frac{\omega_E}{(P/2)} = \frac{2\pi \times 60}{(2/2)} = \underline{\underline{376.99}}\ \text{rad/s} \tag{2}$$

Substitution of Eqs. (1) and (2) into Eq. (2.167) yields

$$\mathfrak{M} = \mathcal{K}\mathcal{K}_F = \frac{\Delta e_q^{OC}}{\Delta i_F} \times \frac{1}{\omega_R} = \frac{324.00}{376.99} = \underline{\underline{0.86}}\ \text{V} \cdot \text{s/A} \cdot \text{rad} \tag{3}$$

The resistance of the rotor winding, R_q, is calculated directly from the characteristic obtained with the rotor winding loaded, Fig. 2.16.1(b), in accordance with Eq. (2.170), as

$$R_q = \frac{\Delta R_q^{OC}}{\Delta i_q} = \frac{300.00 - 389.00}{809.00 - 0} = \underline{\underline{-0.11}}\ \Omega \tag{4}$$

The inductance of the rotor winding, L_q, is calculated in accordance with the information leading to, and use of, Eq. (2.171). However, note that since here the load is comprised of a resistor R_L as well as an induction L_L, Eq. (2.171) is modified to encompass L_L and is written as

$$(L_q + L_L) = \tau(R_q + R_L) \tag{5}$$

The numerical values of L_L, R_L, and R_q are known from the given data and Eq. (4).

The numerical value of the time constant τ is read from Fig. 2.52(b) to be $\tau = 0.22$ s. Proper substitution into Eq. (5) yields

$$(L_q + 0.684) = 0.22(0.11 + 4.00) \tag{6}$$

Solution of Eq. (6) for L_q yields

$$L_q = 0.22(0.11 + 4.00) - 0.684 = \underline{\underline{0.22}}\ \text{H} \tag{7}$$

The resistance R_F of the stator winding is calculated with the aid of the input data as follows:

$$R_F = \frac{E_F}{I_F^{RTD}} = \frac{110.00}{1.20} = \underline{\underline{91.67}}\ \Omega \tag{8}$$

The inductance L_F of the stator winding is calculated with the aid of the numerical values of the time constant $\tau = 0.22$ s extracted from Fig. 2.50(a) and of R_F from Eq. (8) and the use of Eq. (2.175):

$$L_F = \tau R_F = 0.22 \times 91.67 = \underline{\underline{20.17}}\ \text{H} \tag{9}$$

Finally, the moment of inertia is calculated from Eq. (2.174) with the aid of the results of Eq. (2.172), Eq. (3), and the use of Figs. 2.50(a), 2.50(b), and 2.50(d), as follows. At $t = 5\tau = 5 \times 0.25 = 1.25$ s, where τ is extracted from Fig. 2.50(a), the numerical values for the currents contained in Eq. (2.172) are read to be $i_F = 1.20$ A and $i_q = 461.20$ A. Substitution of these currents and the results of Eq. (3) into Eq. (2.172) yields

$$\tau_{em} = \mathcal{M} i_F i_q = 0.86 \times 1.20 \times 461.20 = \underline{\underline{475.96}}\ \text{N} \cdot \text{m} \tag{10}$$

The slope $\Delta\omega_R/\Delta t = 36.95$ rad/s^2 required in Eq. (2.174) is calculated at the point $t = 1.25$ s of Fig. 2.50(d). Proper substitution of this slope and the results of Eq. (10) into Eq. (2.174) yields

$$J = \left. \frac{\tau_{em}}{(\Delta\omega_R/\Delta t)} \right|_{\substack{t = 5\tau \\ \mathcal{M} = \mathcal{K}\mathcal{K}_F}} = \frac{475.96}{36.95} = \underline{\underline{12.88}}\ \text{Nms}^2 = 12.88\ \text{kgm}^2 \tag{11}$$

2.13 SUMMARY

Described in this chapter are the operating principles of the separately excited dc energy converter; also explained are the functions of the brushes, the commutator, and the effects of the armature reaction on the motor and generator modes of operation. Corresponding dynamic equivalent circuits for these modes are then derived. From the two circuits, the equivalent circuits that describe the steady state operation of (1) the separately excited, (2) the shunt, and (3) the series **motor** and **generator** modes of operation are derived. Next, the mathematical models are developed that describe the performance of these devices. Their τ_{em} versus ω_R and V_L versus I_L terminal characteristics are also derived. These characteristics are plotted for comparative purposes on common axes, marked on percentage of rated values, along with the τ_{em} versus ω_R and V_L versus I_L terminal characteristics of the compound devices for which only the schematic diagrams are presented.

Next, transfer function or block diagram dynamic modeling techniques are developed, and simulation of the generalized dynamics of the separately excited dc motor and generator (only as an introduction of how such techniques can be extended to simulate the rest of the devices) are set forth; the state-variable approach and simulation via simulation diagrams are introduced.

The chapter closes with a detailed presentation of the experimental procedure used to measure the equivalent circuit parameters for the separately excited dc device operating in both its modes.

Although this chapter is relatively long and fairly complete in its presentation of operating principles and steady state analysis of the separately excited, series and shunt dc motor (and generator), it merely mentions the existence of the compound device. It is not exhaustive in the areas of modeling, simulation, or dynamic analysis of all the devices considered. Neither does it treat the great variety of special types of dc energy converters used in industrial applications such as the amplidyne, metadyne, and others.

Nevertheless, complete mastery of this chapter's content will prove rewarding toward the understanding of the material in the remaining chapters, toward setting the foundations for graduate work, and in the work of the professionally practicing electrical engineer.

PROBLEMS

2.1 For the figure of Example 2.1, assume that (1) $\mathbf{B}=|\mathbf{B}|(-\mathbf{a}_x)$ Tesla, (2) $d\mathbf{L}=dz\,\mathbf{a}_z$ m, (3) $D\rightarrow z_1=-L/2$ m, and (4) $C\rightarrow z_2=L/2$ m. Calculate the electromagnetic force developed on the current-carrying conductor of length $L=2$ m residing within the field $|\mathbf{B}|=1.2$ Tesla.

2.2 For the figure of Example 2.2, assume that (1) friction$=0$ N, (2) $R_{\text{BAR}}=0\ \Omega$, (3) $R_{\text{CONDUCTOR}}=0\ \Omega$, (4) $\mathbf{u}=20\mathbf{a}_y$ m/s, (5) $L=2$ m, and (6) $|\mathbf{B}|=2$ Tesla. Calculate the motional potential v_{21} between bars No. 2 and No. 1.

2.3 Assume that the switch S_1 in Fig. 2.15 has been closed for some time (while the switch S_2 is open) and the field circuit has reached steady state, that is, $L_F\ d/[i_F(t)]/dt\rightarrow 0$. At $t=0$ s, close the switch S_2 and calculate: (1) the response current $i_q(t)$ and (2) the developed electromagnetic torque τ_{em}. What can you say about the turning effect of the torque? DATA: $E_F=200, R_F=25, L_q=1.5, R_q=0.8, R_L=8.5$, all MKS units; for a rated speed of 1800 rpm, the no-load linearized magnetization curve, E_q versus i_F, yields $K_E=100$ V/A, that is, $E_Q=240$ V, $I_Q=1.2$ A.

2.4 Assume that in Fig. 2.2.17 the switch S_2 is in the open position and the switch S_1 has been closed for some time, and the field circuit has reached steady state, that is, $L_F\ d/[i_F(t)]/dt\rightarrow 0$. In addition, assume that $\tau_L=K=B\equiv 0$, that is, the loading torque, the rotational stiffness, and the rotational damping are zero. As the switch S_2 closes, the mechanical coupler C goes to its ON position, that is, the motor is loaded with the moment of inertia J in kilogram meters^2. For $L_q\simeq 0$, that is, $L_q\ d[i_q(t)]/dt \ll R_q i_q(t)$, calculate the speed of the motor as a function of time, $\omega_R(t)$. DATA: $N=1$, $P=2$, $Z=314$, $a=2$, $\Phi_P=5.3\times 10^{-3}$, $E_F=200$, $R_F=25$, $E_S=250$, $R_q=0.8$, $J=2.0$ and $\mathfrak{M}=10/6\pi$, all MKS units.

2.5 The dc motor of Fig. 2.17 is connected in series, to which the commutating or interpole winding (see Fig. 2.32) is also connected in series. The rating of the motor is 80 hp, 1200 V. When tested at 1800 rpm, it exhibits the following results: $I_q=100, R_q=0.8, R_F=0.4, R_C=0.2, P_{CL}=1200, P_{BF}=60, P_{WR}=180$, all MKS units. Calculate: (1) the electromagnetic power p_{em}, (2) the electromagnetic torque τ_{em}, (3) the shaft power P_{SH}, (4) the shaft torque τ_{SH}, (5) the efficiency η, and (6) the shaft power P_{SH} in horsepower.

2.6 The rotor resistance of a separately excited motor, Fig. 2.18, operating at steady state is 0.25 Ω, and its flux per pole is 1.50 W. When connected to a 150-V source, the rotor draws 60 A. For the design constants of the motor, $P=2$ poles, $Z=80$ coils, two turns each, and $a=4$ parallel paths, calculate: (1) the generated back emf E_b, (2) the speed acquired by the rotor in radians per second and in revolutions per minute, and (3) the developed electromagnetic torque τ_{em}.

2.7 For Problem 2.6, (1) obtain a τ_{em} versus ω_R relationship; (2) calculate (i) the no-load speed ω_0 and (ii) the standstill torque τ_0; (3) sketch the τ_{em} versus ω_R characteristic; (4) calculate the speed regulation of the motor $\omega_{RATED}=0.90\omega_0$.

2.8 A 500 V series motor, Fig. 2.23, has a rotor resistance of 0.2 Ω and a field resistance of 0.10 Ω. When the rotor current is 100 A, the speed of the motor is 1000 rpm. Calculate: (1) the generated back emf E_b, (2) the developed electromagnetic torque τ_{em} at this speed, (3) the τ_{em} versus ω_R relationship, and calculate τ_{em} for $\omega_R=0$, 5π, 20π, and ∞ r/s, and (4) the speed when the rotor current is 200 A.

2.9 A 250-V, 50-A shunt motor, Fig. 2.25, has a rotor resistance of 0.25 Ω and a field resistance of 60 Ω. The full-load speed is 1800 rpm. Calculate: (1) the back emf E_b, (2) the torque developed by the motor and its full-load speed, (3) the τ_{em} versus ω_R characteristic and plot it, (4) the speed at half load, (5) the rotor speed at an overload (what does that mean?) of 120%, and (6) the speed when the motor is connected to a 70-A source with R_F decreased to 50 Ω to allow for an increase of flux by 15% (which is used to produce the extra torque).

2.10 A two-pole, separately excited generator, Fig. 2.33, has a rotor containing 50 one-turn coils connected in two parallel paths. The flux per pole is 3.25 W and the steady state speed n_R of the prime mover is 50 rpm. The resistance of each coil side is 0.10 Ω and its current-carrying capacity is 10 A. Calculate: (1) the generated voltage E_q, (2) the full-load current delivered to an external load, I_q, (3) the rotor resistance R_q, (4) the terminal voltage of the generator, V_L, (5) the average voltage generated per coil, v_{AVG}, (6) the loading resistance for the given conditions, R_L, (7) the developed electromagnetic torque τ_{em}, its magnitude and direction, (8) the voltage regulation VR of the generator, and (9) the power gain of the generator.

2.11 A 300-kW, 500-V shunt generator, Fig. 2.36, has a field resistance of 50 Ω and a motor resistance of 0.08 Ω. Under steady state operation the rotor speed is measured to be 1800 rpm.

a. Calculate the following quantities: (i) the full-load current I_L flowing into the load, (ii) the field current I_F, (iii) the rotor current I_q, (iv) the full-load generated voltage E_q, (v) the loading resistance for rated conditions of operation R_L, and (vi) the developed electromagnetic torque τ_{em}.

b. Neglecting all losses, (i) what is the external torque τ_I required for the operation of this device and (ii) if the rotor is brought to a halt by some external device, what is the value of E_q? (iii) Why?

2.12 A 20-kW, 250-V series generator, Fig. 2.40, has a rotor resistance of 0.2 Ω, a field resistance of 0.08 Ω, and is operating at steady state with a rated speed of 1800 rpm. Calculate: (1) the load current I_L, (2) the field current I_F, (3) the rotor current I_q, (4) the generated emf E_q, (5) the loading resistor for rated conditions of operation, R_L, and (6) the developed electromagnetic torque τ_{em}.

2.13 Consider the equivalent circuit of the separately excited generator, Fig. 2.15, and its no-load magnetization curve, Fig. 2.14(b), which can be described by $E_q = K_E i_F, K_E = 125$ V/A. Assume that switches S_1 and S_2 are open and the prime mover is rotating at a rated speed of 1800 rpm. At $t=0$, both switches S_1 and S_2 are closed.

Consider $i_q(t)$ to be the output of the generator; then (1) obtain the control ratio that describes the generator, (2) draw the block diagram for this control ratio, and thus of the generator, and (3) utilize s-domain techniques in conjunction with transfer functions and control ratios to calculate (i) the field current, (ii) the rotor current, and (iii) the electromagnetic torque developed by the generator.

DATA: $E_F = 200, R_F = 50, L_F = 5, L_L = 2.5, L_q = 0.5, R_q = 0.6, R_L = 5.4$, all in MKS units.

2.14 For the separately excited motor, Fig. 2.17, assume that the switches S_1 and S_2 are open, thus yielding IC1 = IC2 = IC3 = IC4 = 0. For $R_F = 100$, $L_F = 25$, $\mathfrak{M} = 0.80$, $L_q = 0.50$, $R_q = 0.20$, $E_F = 250u(t)$, $E_I = 250u(t)$, $K = 0$, $B = 0.50$, $J = 5$, and $\tau_L = 20$, all in MKS units, close the switches S_1 and S_2 at $t=0$ and study the dynamic behavior of the motor, that is, study the starting dynamics of the motor.

2.15 For the separately excited dc generator, Fig. 2.15, assume that the switches S_1 and S_2 are open, thus yielding IC1 = IC2 = 0. For $R_F = 100$, $L_F = 25$, $\mathfrak{M} = 0.80$, $R_q = 0.20$, $L_q = 0.50$, $R_L = 5$, $L_L = 0.20$, $f_R = 60$, $E_F = 250u(t)$, and $E_I = 0$, all in MKS units, close the switches S_1 and S_2 at $t=0$ and study the dynamic behavior of the generator, that is, study the starting dynamics of the generator.

REFERENCES

1. Chapman, C. R., *Electromechanical Energy Conversion*. Wiley, New York, 1965.
2. Del Toro, V., *Electromechanical Devices for Energy Conversion and Control Systems*. Prentice-Hall, Englewood Cliffs, N.J., 1968.
3. Elgerd, O. I., *Electric Energy Systems Theory: An Introduction*. McGraw-Hill, New York, 1971.
4. Fitzgerald, A. E., and Kingsley, C., Jr., *Electric Machinery*, 2d ed. McGraw-Hill, New York, 1961.
5. Gourishankar, V., *Electromechanical Energy Conversion*. Intext, New York, 1965.
6. Hayt, W. H., Jr., *Engineering Electromagnetics*. McGraw-Hill, New York, 1958.
7. Kosow, I. L., *Electric Machinery and Control*. Prentice-Hall, Englewood Cliffs, N.J., 1964.
8. Kron, G., *Tensors for Circuits*. Dover, New York, 1959.
9. Kuo, B. C., *Automatic Control*. Prentice-Hall, Englewood Cliffs, N.J., 1962.
10. Mablekos, V. E., and Grigsby, L. L., "State Equations of Dynamic Systems with Time Varying Coefficients," *IEEE Transactions on Power Apparatus and Systems*, vol. PAS–90, no. 6 (Nov./Dec.), 1971.
11. ———, "An Algorithm for the Unified Solution of the State Equations of Mechanically Driven Systems with Time Varying Coefficients," *IEEE Transactions on Power Apparatus and Systems*, vol. PAS–90, no. 6 (Nov./Dec.), 1971.
12. Mablekos, V. E., and El-Abiad, A. H., "An Algorithm for the Unified Solutions of the State Equations of Electrically Excited Systems with Time Varying Coefficients," *IEEE Transactions on Power Apparatus and Systems*, vol. PAS–90, no. 6 (Nov./Dec.), 1971.

13. Majmudar, H., *Electromechanical Energy Converters*. Allyn & Bacon, Boston, 1965.
14. Matsch, L. W., *Electromagnetic and Electromechanical Machines*. Intext, New York, 1972.
15. Powell, J. E., and Wells, C. P., *Differential Equations*. Wiley, New York, 1950.
16. Rekoff, M. G., Jr., *Analog Computer Programming*. Merrill, Columbus, Ohio, 1967.
17. *System/360 Continuous System Modeling Program* (360*A*-*CX*-16*X*), *User's Manual*, 3d ed. International Business Machines Corporation, White Plains, N.Y., 1968.
18. Thaler, G. J., and Wilcox, M. L., *Electric Machines: Dynamics and Steady State*. Wiley, New York, 1969.
19. Van Valkenburg, M. E., *Network Analysis*, 2d ed. Prentice-Hall, Englewood Cliffs, N.J., 1964.
20. Weedy, B. M., *Electric Power Systems*, 2d ed. Wiley, New York, 1972.

3 THREE-PHASE SYSTEMS THEORY

3.1 INTRODUCTION

In 600 B.C. electricity got its name, when the Greek Thales of Miletos observed the electrostatic effect obtained by rubbing a woolen cloth on amber (electron = ἤλεκτρον). But it was not until the late nineteenth century that electricity had made any significant impact on society.

By 1885, dc electrical energy was generated at 100 or 220 V and was widely used for lighting purposes, as well as to run dc motors. As the demand for electricity increased, so did the need for the generation of larger blocks of electrical energy and its transmission over long distances. In the attempt to meet the needs of the day, engineers were met with great disappointments because of two inherent difficulties in the dc energy systems: (1) the potential level of the generated voltage could not be increased above the potential level of generation and (2) the transmission of electrical energy over long distances proved to be very inefficient.

These difficulties led to further research, which in turn led to the development of (1) the single-phase ac generator and (2) the single-phase transformer. Thus the potential level of the generated voltages could be raised by the transformer and its transmission over long distances could be accomplished efficiently. Therefore, single-phase ac electrical energy was generated and distributed over long distances for lighting purposes, but dc electrical energy was still used to run dc motors.

It was not until 1888, when Tesla invented the three-phase induction motor, that three-phase ac electrical energy proved its superiority. The induction motor was simpler than the dc motor in construction and thus easier to manufacture. Without the elaborate commutator, which necessitated increased labor and material cost, and therefore added economic burdens, the induction motor proved to be rugged and durable, and therefore cheaper, than a comparable dc motor in the long run.

This overall superiority of the induction motor meant a marked reduction in the use of the dc motor and therefore of the generation of dc electrical energy.

By 1890 the superiority of the three-phase ac power system over the dc power system was firmly established. There was little left but to refine the ac power system to its present state of the art: to refine a system's components until the system performed as a balanced three-phase system, and to formulate suitable mathematical processes for its analysis. These events took place in dramatic steps over the years that followed. Perhaps the greatest contribution to power systems analysis was the application of the algebra of complex numbers to power systems analysis, and thus the evolution of the frequency domain, or steady state, per-phase analysis technique of power systems—utilizing the phasor (or sinor) quantities.

Today, regardless of the complexity of the power system, its analysis is carried out, at least as a first approximation, on a per-phase basis through steady-state analysis techniques—on the digital computer. Only in rare occasions does one deal with an unbalanced system—except, of course, in fault analysis.

Thus it is essential for the power systems engineering student to become competent in steady state analysis of balanced (at least as far as this text is concerned) three-phase systems on a per-phase basis, and the ramifications of this analysis. Therefore, this chapter is devoted to this end.

The stated purpose of the chapter is sixfold: (1) to develop the circuit representation of balanced, three-phase, Y- and Δ-connected, ideal alternating current generators, abbreviated as ac generators, and the mathematical relationships between their phase and line voltages and currents; (2) to describe the equivalent circuit representation of balanced three-phase transmission lines for physical lengths ranging from infinitely long lines to short lines; (3) to set forth the circuit representation of balanced, three-phase, Y- and Δ-connected loads; (4) to introduce the two- and three-wattmeter power measurement methods; (5) to present the two-wattmeter, power-factor-angle measurement method; (6) to introduce the student to the properties that unbalanced, three-phase, Y- or Δ-connected loads exhibit when connected, with ideal or nonideal transmission lines, to balanced or unbalanced three-phase Y- or Δ-connected generators.

3.2 SINUSOIDALLY DISTRIBUTED FIELDS

Although the airgap field produced by the concentrated stator winding of the dc energy converter is practically a square waveform, Fig. 2.7, it is approximated by a sinusoidal waveform in the development of the generated speed voltage v_C, Eq. (2.29), and the developed electromagnetic torque τ_{em}, Eq. (2.50), because of the mathematical advantages lended by sinusoidal waveforms. It is possible, however, to generate sinusoidal waveforms directly and thus avoid the approximations of Chapter 2. Thus the waveforms of the fields in induction and synchronous (i.e., alternating current type) energy converters are designed to be sinusoidally distributed in space.

A sinusoidal spatially distributed waveform is generated by properly *distributing dc or ac current-carrying wires in the slots* of the rotor or stator. This current-carrying wire distribution is defined as *current sheet*. Proper rotor current distribution, Fig. 3.1, means that the number of wires per rotor slot increases sinusoidally in

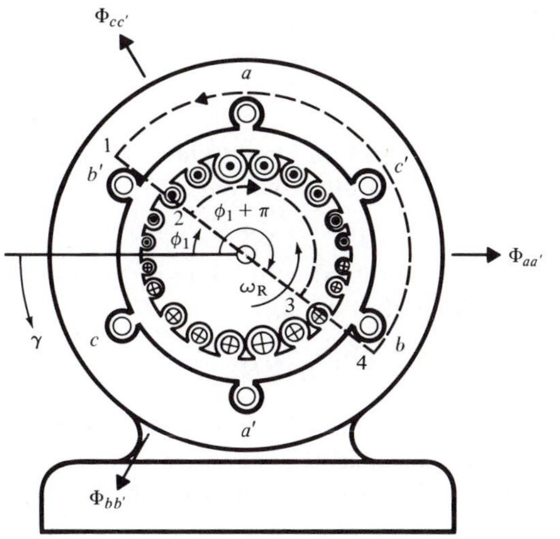

Fig. 3.1 Smooth Airgap Energy Converter with Rotor Winding Producing a Sinusoidally Distributed Field

slots found at $0<\phi<\pi/2$, where ϕ designates the angular displacement on the surface of the rotor, increasing positively in the clockwise direction with zero reference at the negative x-axis. Thus the number of wires *increases sinusoidally* in slots found in the region $0<\phi<o/2$ and *decreases sinusoidally* in the slots found in the region $\pi/2<\phi<\pi$. The cycle repeats itself for the intervals $\pi<\phi<3\pi/2$ and $3\pi/2<\phi<2\pi$.

The change in diameter of the conductor (i.e., *a conductor is considered to be a bundle of wires*) in each slot suggests the increase or decrease of its corresponding number of wires when compared to the conductor in the adjacent slots. The rotor winding is assumed to be the field winding (this is in contrast to the dc energy converters where the field is always in the stator winding). Thus it is assumed that the electromagnetic field $B \equiv |\mathbf{B}|$ generated by the current in the rotor winding is *positive when it exits the rotor,* $\phi=\pi$ rad, and *negative when it enters the rotor,* $\phi=0$ rad. The wires found in the interval $0<\phi<\pi$ carry dc current directed **out** of the page. This direction is assumed to be **positive.** The wires found in the interval $\pi<\phi<2\pi$ carry dc current **into** the page. This direction is assumed to be **negative.** The wires with the positive current constitute a current sheet with a sinusoidally distributed current density $J \equiv |\mathbf{J}|$ (amperes/radian), written as:

$$J = J_M \sin\phi \tag{3.1}$$

where $0 \leqslant \phi \leqslant 2\pi$. Eq. (3.1) is plotted as J versus ϕ in Fig. 3.2.

The mmf produced by the current sheet whose current density distribution is sinusoidal is found (by analogy to mmf's produced by isolated or bundled current-carrying conductors and calculated by $\mathcal{F}=NI$) by integrating Eq. (3.1) and using Ampere's law. The total mmf along the closed path 1–2–3–4–1 is equal to the total current enclosed by the path. This current may be evaluated by integrating the current density between the angles $\phi=\phi_1$ and $\phi=\phi_1+\pi$.

Integrating Eq. (3.1)

$$\mathcal{F}_{1-2-3-4-1} = \int_{\phi_1}^{\phi_1+\pi} J_M \sin\phi \, d\phi \quad \text{(analogous to } \Sigma I) \tag{3.2}$$

which yields

$$\begin{aligned}\mathcal{F}_{1-2-3-4-1} &= -J_M\cos(\phi_1+\pi)+J_M\cos\phi_1\\ &= 2J_M\cos\phi_1\end{aligned} \tag{3.3}$$

The mmf wave is a sinusoid with an amplitude of $2J_M$. The mmf along the entire closed path may be broken down into four parts

$$\mathcal{F} = \mathcal{F}_{1-2-3-4-1} = \mathcal{F}_{1-2}+\mathcal{F}_{2-3}+\mathcal{F}_{3-4}+\mathcal{F}_{4-1} \tag{3.4}$$

The mmf drop may also be expressed as the line integral of the component of the field intensity, H, along the path, that is,

$$\mathcal{F} = \oint H\cdot dL \tag{3.5}$$

The reluctance of the parts of the path that lie in iron is very small compared to the reluctance along the air gap between the inner surface of the rotor, because the permeability of iron is much greater than that of air. Consequently, one may approximate

$$\mathcal{F}_{2-3} = \mathcal{F}_{3-4} = 0 \tag{3.6}$$

Also, the mmf along the remaining two paths is dominated by the mmf across the airgap.

The field intensity along a radial line in the air gap is nearly constant. Let H_ϕ be defined as the magnitude of H along a radial line $\phi=\phi_1=$constant. Then,

$$\mathcal{F}_{1-2} = -H_{\phi_1}g \tag{3.7}$$

and

$$\mathcal{F}_{3-4} = H_{\phi_1+\pi}g \tag{3.8}$$

Recall that H is positive for fields which exit from the rotor. Notice that from the symmetry of field

$$H_{\phi_1+\pi} = -H_{\phi_1} \tag{3.9}$$

Substitution of Eq. (3.9) into Eq. (3.8), and then Eqs. (3.6), (3.7), and (3.8) into Eq. (3.4) yields

$$\mathcal{F} = -2gH_{\phi_1} \tag{3.10}$$

Solving Eq. (3.10) for H_{ϕ_1}, the field intensity in the airgap at ϕ_1, yields

$$H_{\phi_1} = -\mathcal{F}/2g \tag{3.11}$$

It is known that $B=\mu_0H$. Thus the flux density B can be expressed in terms of the current density J, the airgap opening g, and the permeability of free space (i.e., $\mu_0=4\pi\times10^{-7}\ H/\text{m}$), as

$$B = -\mu_0(\mathcal{F}/2g) \tag{3.12}$$

Eqs. (3.3) and (3.12) enable us to write the mathematical expression that describes the waveform of the flux density produced by the winding of the rotor at $\phi_1=\phi$, as

$$B = -\left(\frac{\mu_0}{2g}\right)J_M\cos\phi \tag{3.13}$$

Eq. (3.13) is also plotted in Fig. 3.2 as B versus ϕ. Note the distinct differences between the two waveforms, that is, J versus ϕ and B versus ϕ, and try to correlate

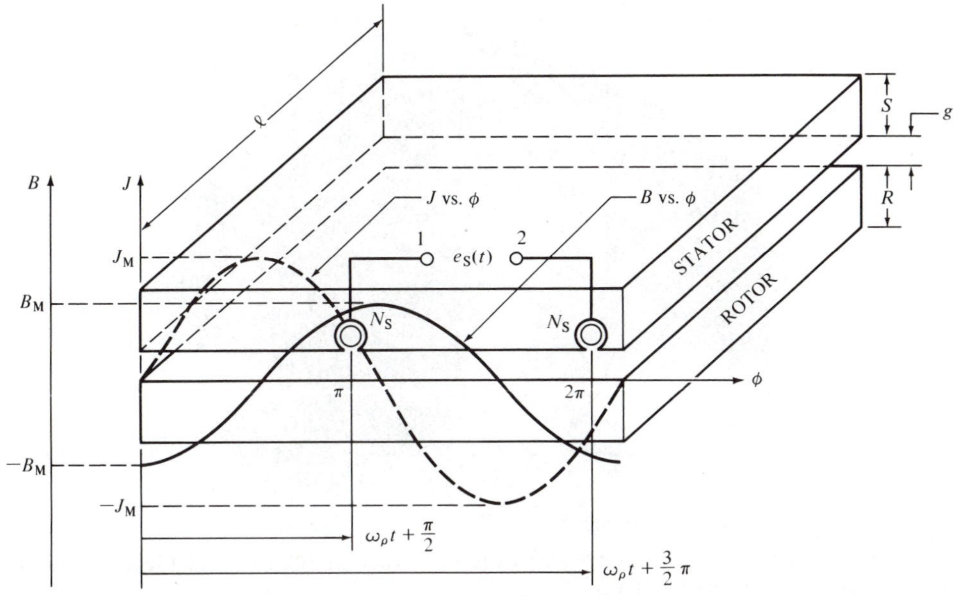

WHERE: $B_M \equiv B_{MAX} = \frac{\mu_0}{2g} J_M$ AND $-B_M \equiv B_{MIN} = \frac{-\mu_0}{2g} J_M$

Fig. 3.2 Rotor Surface Current Density J and Flux Density B Distribution. Background: Pictorial View of Imaginary Rotor-Stator Development Cut Along Negative x-Axis with Lower Half Unfolded in the Direction of γ and Stretched to Place. Note the Physical Dimensions of the Rotor R and ℓ and the Airgap g

them to Fig. 3.1. It will help if we were to **imagine** that both rotor and stator are constructed from a *perfectly elastic* material. If we were to cut with a saw along the negative x-axis from the outside of the stator to the center of the rotor, Fig. 3.1, and allow $\gamma \to \pi$ rad, the center of the rotor will expand (note the small circle drawn at the origin) since our elastic material behaves ideally. The result is two rectangular slabs, Fig. 3.2, whose dimensions are $\ell \times 2\pi R \times R$ and $\ell \times 2\pi R \times S$, where ℓ (meters) is the axial length of the rotor, R (meters) is the radius of the rotor, and S (meters) is the thickness of the stator. It must be noted that $g \ll R$, thus $R+g \to R$, where g (meters) is the airgap opening between the stator and the rotor. The rotor is on the bottom and moves to the right with a velocity* $R\omega_\rho$ for a clockwise rotation of the rotor at an angular velocity ω_ρ radians/second. The front planes of these two slabs are drawn tangent to the plane of the waveforms J versus ϕ and B versus ϕ in Fig. 3.2, for ease of correlation.

The flux density waveform, Eq. (3.23), which is entering the airgap from the rotor and then entering the stator structure, can be written as a vector quantity in the cylindrical coordinate system as

$$\mathbf{B} = -\left(\frac{\mu_0}{2g}\right) J_M \cos\phi \, \bar{a}_R \tag{3.14}$$

It is possible now to calculate *the total flux* Φ_S *linking the stator coil* 1–2, Fig. 3.2, located at $\pi/2+\omega_\rho t$ and $3\pi/2+\omega_\rho t$. Note that the coil spans π rad. Utilization

*Define $\omega_R \equiv -\omega_\rho$.

Plate 3.1 Custom 8000 Prewound Stator Core Construction. (Courtesy of General Electric Co., Schenectady, N.Y.)

of the cylindrical coordinate system enables calculation of Φ_S as

$$\Phi_S = \int_S \mathbf{B}\cdot d\mathbf{S} = \int_{+\omega_\rho t(\pi/2)}^{+\omega_\rho t(3/2)\pi}\left[\left(-\frac{\mu_0}{2g}\right)J_M\cos\phi\,\mathbf{a}_R\right]\cdot \ell R\,d\phi\,\mathbf{a}_R \tag{3.15}$$

Thus

$$\Phi_S = -\left(\frac{\mu_0}{2g}\right)J_M \ell R\sin\phi\Big|_{+\omega_\rho t\,(\pi/2)}^{+\omega_\rho t(3/2)\pi} = \left(\frac{\mu_0}{g}\right)J_M\ell R\cos\omega_\rho t \tag{3.16}$$

The flux linkage of the stator coil λ_S, given by $\lambda_S = N_S\Phi_S$, is

$$\lambda_S = \left(\frac{\mu_0 \ell R}{g}\right)N_S J_M\cos\omega_\rho t \tag{3.17}$$

As the rotor moves to the right (rotates), its field cuts the stator coil. As far as the stator coil is concerned, this (rotor) field changes not only in magnitude but in direction as well, and *when the rotor makes one complete revolution, the stator coil sees one complete reversal of the field.* Note that the field enters the rotor at $\phi=0$ and exits the rotor at $\phi=\pi$, Fig. 3.1. The effect of such changes of the field in coils was studied by Michael Faraday. His observations are presented in the well-known Faraday's law, stated as follows: *The total emf generated in a closed circuit is equal to the negative time rate of change of the flux linkages linking the circuit.* Mathematically the law is stated as

$$e_S(t) = -\frac{d}{dt}\lambda_S \tag{3.18}$$

Substitution for λ_S from Eq. (3.17) into Eq. (3.18) yields

$$e_S(t) = -\frac{d}{dt}\left[\left(\frac{\mu_0 \ell R}{g}\right)N_S J_M\cos\omega_\rho t\right] \tag{3.19}$$

Upon differentiating Eq. (3.19) yields

$$e_S(t) = N_S\omega_\rho\left(\frac{\mu_0 \ell R}{g}\right)J_M \sin\omega_\rho t \tag{3.20}$$

However, note that $\oint \mathbf{H}\cdot d\mathbf{L}$ can also be evaluated in terms of *the dc current* I_F *in the rotor coils* and *the number of the rotor wires designated by* N_R carrying the current I_F out of (or in) the page, Fig. 3.1. Thus

$$\mathcal{F}_{\max} = \mathcal{F}_{\phi_1=0} = \oint \mathbf{H}\cdot d\mathbf{L} = \Sigma I = N_R I_F \tag{3.21}$$

Equating Eqs. (3.4) and (3.21) yields

$$J_M = \frac{N_R I_F}{2} \tag{3.22}$$

Substitution for J_M from Eq. (3.22) into Eq. (3.20) yields the expression for the voltage $e_S(t)$ induced in the stator coil 1–2, Fig. 3.2. Thus

$$e_S(t) = N_S\left(\frac{\mu_0 \ell R N_R}{2g}\right)I_F\omega_\rho \sin\omega_\rho t \tag{3.23}$$

Note the *dependence* of the induced voltage $e_S(t)$ on the *design parameters* ℓ, R, g, N_R, N_S, the rotor speed ω_ρ, and the rotor current I_F.

Classically, it is preferred to perform analyses with devices rotating in the counterclockwise (positive) direction. Thus substituting $\omega_\rho \equiv -\omega_R$ in Eq. (3.23) yields

$$e_S(t) = N_S\left(\frac{\mu_0 \ell R N_R}{2g}\right)I_F\omega_R \sin\omega_R t \tag{3.24}$$

Eq. (3.24), $e_S(t)$ versus $\omega_R t$, is plotted in Fig. 3.3. Note that the voltage reaches e_{SMAX}- and e_{SMIN}-values at $\omega_R t = \pi/2$ and $3\pi/2$, respectively. Now, one might ask the question, "where exactly in the stator is the coil 1–2 physically located?" It can be reasoned out that, for the position of the rotor as shown in Fig. 3.1, *physically the coil* 1–2 *is located in a position such that the coil sides cut zero lines of flux*, that is, the normal to the coil is zero radians away from the origin, or the positive x-axis. The coil 1–2 appears marked as coil sides a–a' in Fig. 3.1.

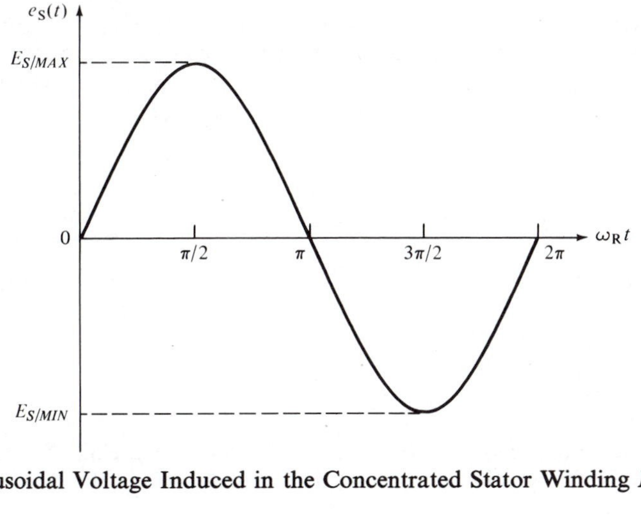

Fig. 3.3 Sinusoidal Voltage Induced in the Concentrated Stator Winding N_S of Fig. 3.2

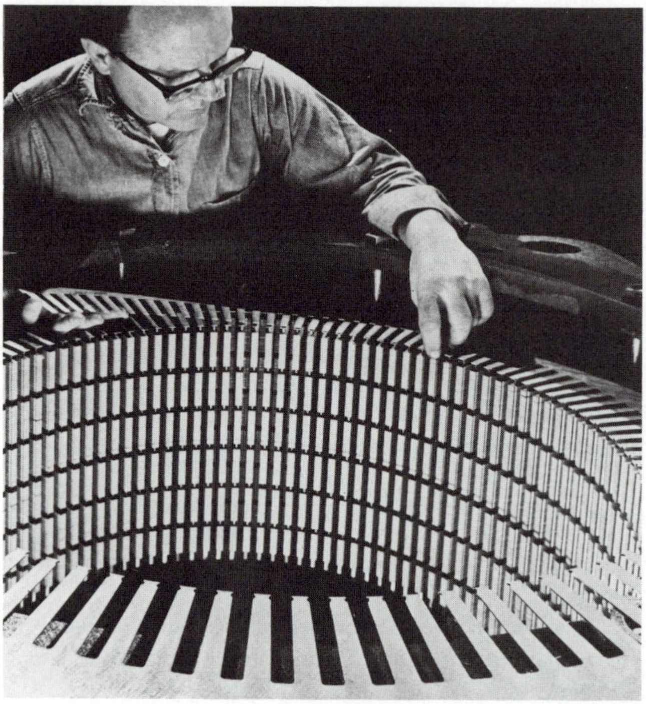

Plate 3.2 Stator Core of a Three-Phase Device Being Completed. The Sections Are Composed of Laminations Held Rigidly in Place Between Two Steel Plates Which Compose the Stator Frame. These Laminations Are Accurately Diecut from Thin, Nonaging, High-Grade, Silicon Sheet Steel Treated with a Monoaluminum Phosphate Compound to Prevent Short-circuiting. (Courtesy of Westinghouse Electric Co., Pittsburgh, Pa.)

Note that the speed voltage in each coil side is zero for the rotor position of Fig. 3.1. Also, note the direction of the potential generating field intensity **E** in the coil sides $a-a'$ and the axis of the flux that would be generated if the current of the coil aa' were allowed to circulate. The direction of this flux, $\Phi_{aa'}$, defines the axis of a *concentrated coil* wound on a core whose axis is along $\Phi_{aa'}$ and produces a flux exactly equal to $\Phi_{aa'}$ for exactly the same current $i(t)_{aa'}$. This concentrated coil is defined as the *equivalent winding*. It is this type of winding to which reference is usually made when one talks about windings. If this winding is wound with ideal

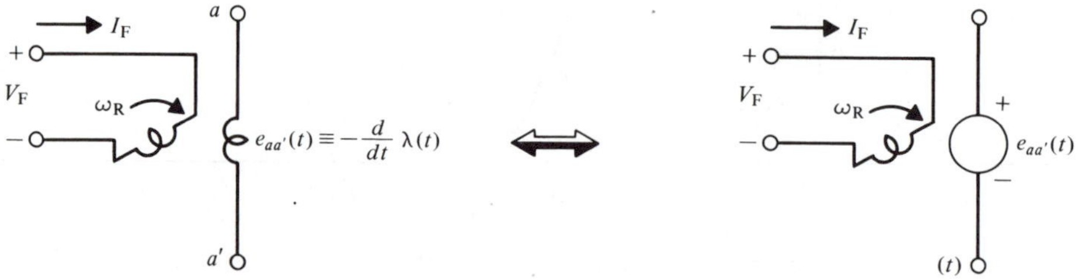

Fig. 3.4 Coil-Ideal Voltage Source Equivalence for the Stator Coil of a Smooth Airgap Energy Converter with Sinusoidally Distributed Field in the Rotor Winding

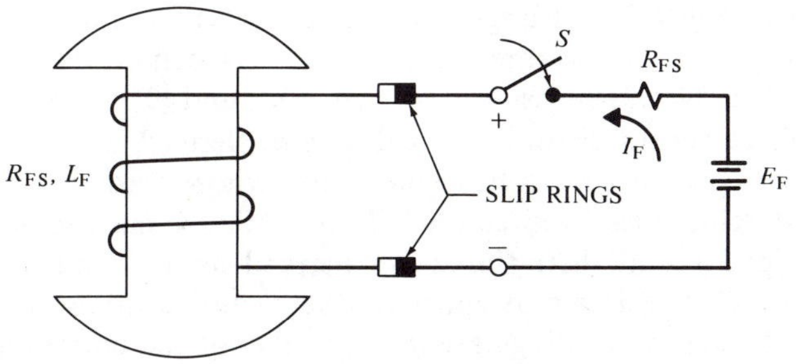

Fig. 3.5 Concentrated dc Winding Wound on a Salient Rotor Shaped to Yield a Sinusoidally Distributed Field

wire, it has no resistance. Thus the voltage that is induced in the concentrated coil, $e_S(t) = -d[\lambda_S]/dt$, can be represented as *the voltage rise of the* **ideal voltage source** in the equivalent circuit of the coil aa', Fig. 3.1. This equivalent circuit is drawn as in Fig. 3.4 and is defined as the **equivalent voltage source of the single phase**, where **phase** designates the circuit comprised of the coil aa' itself when the coil ends are short-circuited.

Before closing this section it should be pointed out that a sinusoidally distributed field in the rotor of a device can also be generated by a concentrated winding, excited with dc current, wound on a **salient rotor** whose poles are **properly shaped** to yield a sinusoidal field. A **salient rotor** with a **concentrated winding** excited with the **dc current** I_F, which is supplied to the winding through **slip rings**, is exhibited in Fig. 3.5.

3.3 THREE-PHASE BALANCED GENERATORS

In a single-phase circuit, Eq. (A.39) [or Eq. (A.54), (A.55), and (A.56)], the instantaneous power to a load is *pulsating*. Even when the voltage and the current are in phase, the instantaneous power is zero twice in each cycle. When the voltage and current are not in phase (e.g., when the power factor is less than unity), the instantaneous power is not only zero four times in each cycle but it is *negative* twice in each cycle. This *pulsating* nature of the instantaneous power in single-phase systems is *highly objectionable* in many applications.

This undesirable phenomenon in loads supplied by single-phase, abbreviated as 1Φ, generators is remedied by the use of polyphase generators. *A polyphase generator consists of two or more single-phases connected in such a way that they provide loads with voltages of equal magnitudes and equal phase differences.* The polyphase generator may be compared to a multicylinder aircraft engine in which the power delivered to the propeller is practically steady. In the *two-phase generator* the two equal voltages differ in phase by 90°, while in the *three-phase generator* the three equal voltages have a phase-angle difference of 120°. In general (except for the special case of two-phase), the electric displacement between phases is equal to $360°/n$, where n is the number of phases. Generators of six, twelve, or even twenty-four phases are sometimes used in conjunction with *polyphase rectifiers* to provide power requirements in the range of kilowatts with low ripple. However, experience has shown that *the most efficient and economic* system for generation, transmission, and utilization of large blocks of power is the three-phase, *abbreviated as* 3Φ, system. The simplest three-phase system is comprised of at least (1) a 3Φ-generator, (2) a 3Φ-transmission line (network), and (3) a 3Φ-load.

The three-phase generator (ideal or real) can be derived from Fig. 3.1 by simply adding two more concentrated coils to the stator, whose axes are displaced by 120° and 240° counterclockwise from coil aa'. These coils are marked as coil sides b–b' and c–c' in Fig. 3.1 and their fluxes are marked as $\Phi_{bb'}$ and $\Phi_{cc'}$, respectively. Although these coils are drawn as concentrated N-turn windings in Fig. 3.1, in the real generator the stator windings are *distributed* over the entire circumference of the stator. Several coils are wound and placed in adjacent slots and then are connected in series. Practical limitations on the design of generators do not allow all the windings to be placed in one slot of the stator. Note, however, that for purposes of analysis, the concept of concentrated windings (or equivalent windings) is sufficient.

The axes of the three equivalent stator coils (and the rotor coil) are drawn parallel to their corresponding fluxes $\Phi_{aa'}$, $\Phi_{bb'}$, and $\Phi_{cc'}$, of Fig. 3.1, in Figure 3.6(a). Also drawn in Fig. 3.6(b) is the corresponding equivalent ideal voltage source representation of each stator coil.

It is shown that Eq. (3.24) describes coil 1–2 of Fig. 3.2. Premultiplication and division of Eq. (3.24) by $\sqrt{2}$ enables writing Eq. (3.24) as

$$e_S(t) = \sqrt{2}\,|\mathbf{E}_{Saa'}|\sin\omega_R t \tag{3.25}$$

where*

$$|\mathbf{E}_{Saa'}| \equiv \left(\frac{N_S}{\sqrt{2}}\right)\left(\frac{\mu_0 \ell R N_R}{2g}\right) I_R \omega_R \tag{3.26}$$

Since coils $C_{aa'}$, $C_{bb'}$, and $C_{cc'}$ are 120° ahead of each other, their corresponding voltages $e_{aa'}$, $e_{bb'}$, and $e_{cc'}$ will reach their maxima one after the other, each maximum delayed by 120°, in a sequence $e_{aa'}$, $e_{bb'}$, and $e_{cc'}$. This sequence is defined as the **sequence a–b–c** or **positive sequence**, whereby each voltage follows the previous one by 120°. Mathematically, these voltages can be written, with the aid of Eqs. (3.25) and (3.26), as

$$e_{aa'} = \sqrt{2}\,|\mathbf{E}_{aa'}|\sin\omega_R t \tag{3.27}$$

$$e_{bb'} = \sqrt{2}\,|\mathbf{E}_{bb'}|\sin(\omega_R t - 120°) \tag{3.28}$$

$$e_{cc'} = \sqrt{2}\,|\mathbf{E}_{cc'}|\sin(\omega_R t - 240°) \tag{3.29}$$

where $|\mathbf{E}_{aa'}|$, $|\mathbf{E}_{bb'}|$, and $|\mathbf{E}_{cc'}|$ represent the RMS phase values $|\mathbf{E}_\phi|$ of the induced voltages $e_{aa'}$, $e_{bb'}$, and $e_{cc'}$.

The waveforms of the sequence $e_{aa'}$, $e_{bb'}$, and $e_{cc'}$, for equal RMS values $|\mathbf{E}_\phi|$, are exhibited in Fig. 3.7. Note that these voltages reach their maxima in the order a–b–c. If the rotation of the rotor, that is, the field rotation, is **reversed**, the sequence of the voltages is reversed. It is defined now as **sequence a–c–b**, or **negative sequence**. The phase sequence can also be *changed* by allowing the rotor to rotate in the *counterclockwise* direction and *interchanging* the letters assigned to windings b and c; the sequence now is sequence a–c–b, rather than sequence a–b–c.

*Henceforth the letter "S" will be deleted from all relevant equations since all voltages of the pertinent equations are stator voltages.

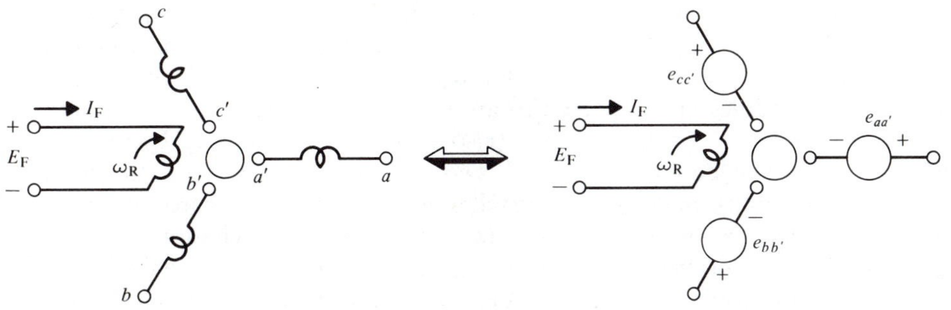

Fig. 3.6 Three-phase Ideal Generator Representation: (a) The Three Concentrated Stator Coils and the Rotor Coil and (b) the Equivalent Ideal Voltage Source Representation of the Three Stator Coils and the Rotor Coil

The three stator windings, Fig. 3.6, of the 3Φ-generator, Fig. 3.1, may be connected, for operation, as a *Y-connected* generator or *Δ-connected* generator. *Commercial generators are mainly Y-connected, with their neutral grounded through a resistor*. The resistor (a typical value is 700 Ω) limits *ground-fault* (short-circuit) *currents* and reduces considerably the amount of possible damage to equipment in the event of ground faults. They have (1) *stationary stator windings*, where the generated voltage is induced; therefore, these windings constitute the *armature* of the generator (contrary to dc devices where the armature was the rotor), and (2) *rotating fields*. The field rotates inside the armature. This arrangement allows solid permanent connection for the high-current-carrying windings in the armature (e.g., a typical rating is 460,000 kVA; stator voltage 24,000 V, stator current 11,066 A, stator voltage frequency 60 Hz, rotor speed 3,600/1,800 rpm, field voltage 500 V, power factor 0.85, and rated hydrogen coolant pressure 60 lb/in^2 gauge). These high-current-carrying windings are located in slots of the *stator* (*armature*). There

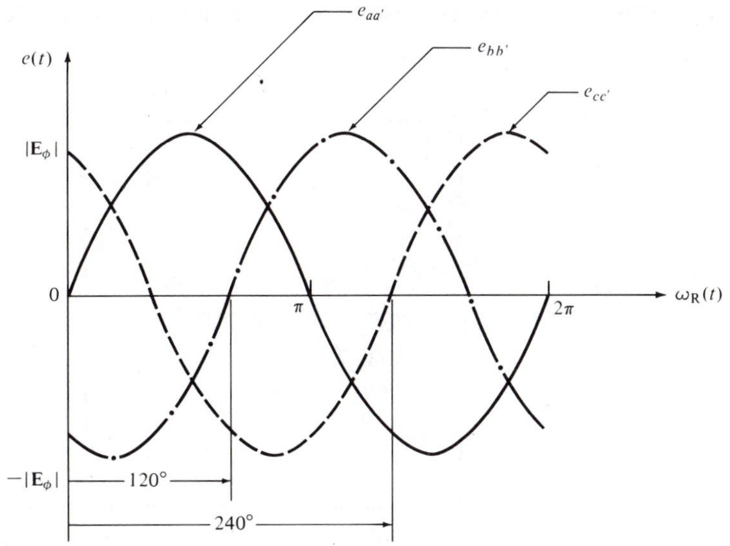

Fig. 3.7 Waveforms of the Stator Coil Voltages in the Sequence *a–b–c* for an Ideal 3Φ Generator

they can be properly *insulated*, since they need heavy insulation, and be properly equipped for easy removal of the generated heat. There are channels in the stator that properly direct *circulating cooled hydrogen* used for cooling purposes. The only stress these windings are subjected to are from magnetic forces on high currents. However, the rotor (field) windings (also cooled by circulating hydrogen gas), which receive their dc current through two slip rings, are subjected to *centrifugal* forces, which limits the types of insulation which may be used on the rotor windings. Since the needed voltage (and current) for excitation is nil when compared to that of the stator, the insulation needed is not as bulky as in the stator. Thus the rotor windings can be easily kept in place, especially when the rotor is the *salient pole* type. Normal speeds of operation for the generator are 1800 or 3,600 rpm.

3.4 Y-CONNECTED GENERATOR

If the coils of Fig. 3.6(a) are connected electrically at points a', b', and c', the device becomes a Y-connected generator, Fig. 3.8. The common point n or N is known as the neutral point. Depending on whether there is or there is not a wire connected to the neutral point, the generator is known as (1) three-phase, four-wire, Y-connected, ideal voltage generator, or (2) three-phase, three-wire, Y-connected, ideal voltage generator. Henceforth, the term "Y- or Δ-connected generator" should be understood to imply "Y- or Δ-connected ideal voltage generator." Before the mathematical analysis of a Y-connected generator proceeds, the following assumptions must be made: (1) the load to which a three-phase generator is to be connected is three-phase and **the load parameters per-phase are of equal magnitude**; (2) the transmission network (known as the transmission line) through which the load is connected to the generator is also three-phase, and **the transmission network parameters/meter of each phase are of equal magnitude**; (3) the **RMS values of the voltages $e_{aa'}$, $e_{bb'}$, and $e_{cc'}$ are equal in magnitude and displaced from each other by a phase angle of 120°**. Thus the response currents of each voltage source, namely,

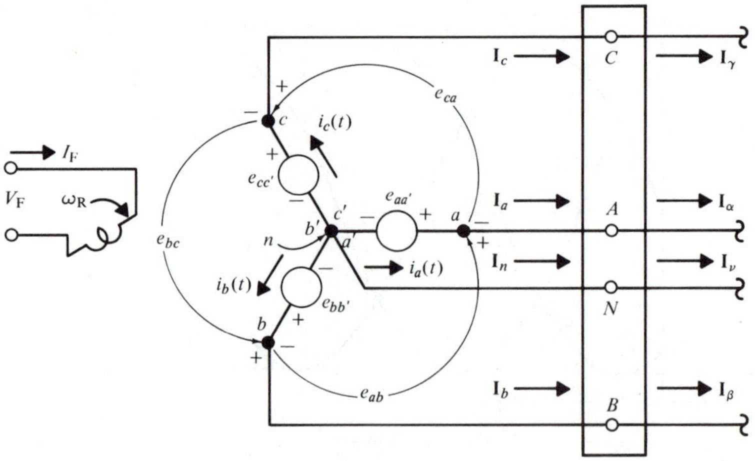

Fig. 3.8 Y-connected Generator Connected to a Transmission Line at Points *A*, *B*, *C*, and *N*

$i_a(t)$, $i_b(t)$, and $i_c(t)$, are of equal magnitude and follow each other by a phase angle of 120°. Under these conditions of operation, the Y-connected generator is defined as a *balanced, three-phase* (3Φ), *Y-connected generator*. However, in practice, when the voltages of a generator are of equal magnitude and displaced by 120°, the generator is referred to as a **balanced** *generator*. Also, the equal-parameter-per-phase transmission lines and loads are known as **balanced transmission lines (networks)** and **loads**, correspondingly. *In a balanced system the total instantaneous power provided by the generator is practically constant.*

The mathematical analysis of a balanced, three-phase, Y-connected generator proceeds as follows: (1) the phasors of the three-phase voltages are extracted from the time equations, that is, Eqs. (3.27), (3.28), and (3.29) (keeping in mind the axis used as a reference, the Im-axis or the Re-axis); (2) the per-phase impedance of the transmission line and the load are obtained; (3) the analysis is carried out in the frequency domain utilizing the techniques of Appendix A; (4) the frequency-domain solution is transformed to the time-domain solution utilizing the technique of inverse transformation set forth in Appendix A.

The per-phase analysis is possible because of the following property of balanced three-phase systems: the current entering or leaving the neutral* n or N (since n and N are connected they are electrically identical) of a balanced, three-phase, Y-connected generator, Fig. 3.8, is equal to zero. This can be proven by utilizing Kirchhoff's current law in the junction n, that is,

$$\sum_{j=0}^{c} I_j = 0 \tag{3.30}$$

Expanding Eq. (3.30) yields

$$\mathbf{I}_n = \mathbf{I}_a + \mathbf{I}_b + \mathbf{I}_c \tag{3.31}$$

For a balanced, three-phase, Y-connected system, the time-domain currents can be written, by analogy, from Eqs. (3.27), (3.28), and (3.29), assuming unity power factor, as

$$i_a(t) = \sqrt{2}\,|\mathbf{I}_a|\sin\omega_\mathrm{R}t \tag{3.32}$$

$$i_b(t) = \sqrt{2}\,|\mathbf{I}_b|\sin(\omega_\mathrm{R}t - 120°) \tag{3.33}$$

$$i_c(t) = \sqrt{2}\,|\mathbf{I}_c|\sin(\omega_\mathrm{R}t - 240°) \tag{3.34}$$

where $|\mathbf{I}_a|$, $|\mathbf{I}_b|$, and $|\mathbf{I}_c|$ represent the RMS phase values $|\mathbf{I}_\phi|$ of the response phase currents $i_a(t)$, $i_b(t)$, and $i_c(t)$. Assuming that the minus sign in Eqs. (3.33) and (3.34) (and similar equations) is always excluded from $\omega_\mathrm{R}t$ [i.e., writing always $(\omega_\mathrm{R}t - \alpha°)$, thus $\alpha°$ always having its own sign], suggesting treatment of 120° and 240° (refer to Figs. 3.6 and 3.7) as negative angles (why?), enables writing of the phasors $\mathbf{I}_a$, $\mathbf{I}_b$, and $\mathbf{I}_c$, utilizing the *Im-axis as reference*, as†

$$\mathbf{I}_a = |\mathbf{I}_a|e^{j0°} \equiv |\mathbf{I}_a|(\cos 0° + j\sin 0°) \tag{3.35}$$

$$\mathbf{I}_b = |\mathbf{I}_b|e^{-j120°} \equiv |\mathbf{I}_b|(\cos 120° - j\sin 120°) \tag{3.36}$$

$$\mathbf{I}_c = |\mathbf{I}_c|e^{-j240°} \equiv |\mathbf{I}_c|(\cos 240° - j\sin 240°) \tag{3.37}$$

*Or the supernode bounding the entire stator of a balanced Δ-connected generator.

†If we were to write these phasors with respect to the Re-axis, we would introduce an angle of −90° to Eqs. (3.35), (3.36), and (3.37).

Plate 3.3 The Three Terminals of the Stator Winding Placed in the Stator Core Are Distinctly Shown. (Courtesy of Westinghouse Electric Co., Pittsburgh, Pa.)

Utilization of the trigonometric properties, $\cos(120°) \equiv \cos(180° - 60°)$, $\cos(240°) \equiv \cos(180° + 60°)$, and so on, and substitution into Eq. (3.31) for $\mathbf{I}_a$, $\mathbf{I}_b$, and $\mathbf{I}_c$, keeping in mind that for balanced systems $|\mathbf{I}_a| = |\mathbf{I}_b| = |\mathbf{I}_c| \equiv |\mathbf{I}_\phi|$ yields

$$\mathbf{I}_n = |\mathbf{I}_\phi|\{(1 + j0) + (-0.5 - j0.866) + (-0.5 + j0.866)\} \tag{3.38}$$

or

$$\mathbf{I}_n = 0 \text{ A} \tag{3.39}$$

Since there is no current entering or leaving the fourth wire connected to the neutral, that is, $\mathbf{I}_n = 0$, the wire may be *removed* altogether, *substituted with ground connections at both ends*, or be retained and *sliced* in three (in one's imagination, of course) to allow analysis of a balanced three-phase system on a **per-phase basis**. The per-phase analysis will become clear shortly.

The phasor diagram of the balanced three-phase stator voltages $\mathbf{E}_{aa'}$, $\mathbf{E}_{bb'}$, and $\mathbf{E}_{cc'}$, of the Y-connected generator, extracted from Eqs. (3.27), (3.28), and (3.29) and drawn in a one-to-one correspondence with the ideal coils of Fig. 3.8, is exhibited in Fig. 3.9. In a real generator with no neutral wire and no accessible ground wire, measurement (i.e., observation) of the phase voltages is not possible. What is possible, however, is the measurement of the line-to-line voltages $|\mathbf{E}_{AB}|$, $|\mathbf{E}_{BC}|$, and $|\mathbf{E}_{CA}|$ of Fig. 3.8. These are the voltages between terminals AB, BC, and CA, at the connection terminals of the generator, the only access that one has to the internal electrical construction of the generator. One end of the transmission line is connected to terminals A, B, and C of the generator, Fig. 3.8, to receive energy to

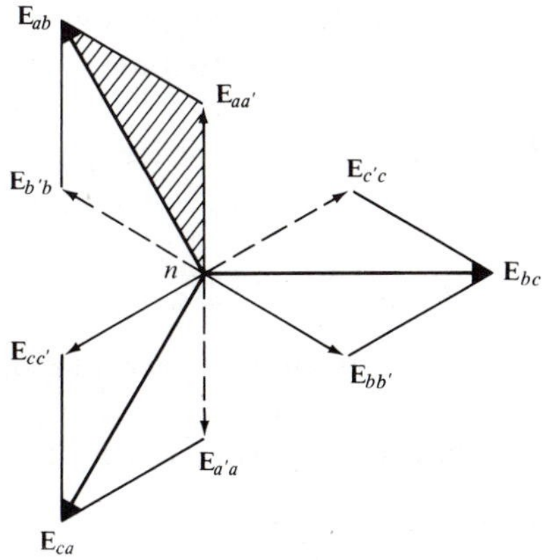

Fig. 3.9 Phasor Diagram of Phase and Terminal Voltages of a Balanced, Three-Phase, Y-connected Generator

be delivered to the load, which is connected to its other end. Since we can measure the line-to-line voltages of Fig. 3.8, then the inaccessible phase voltages can be calculated from the line-to-line voltages by writing three independent equations relating the **three known line-to-line voltages**, designated by $\mathbf{E}_{LL}$, to the **three unknown phase voltages**, designated by $\mathbf{E}_{\phi}$. These equations can be written by utilizing Kirchhoff's voltage law at the terminals of the generator, following the sequence a–b–c and choosing the Kirchhoffian path through the neutral point $n(N)$ each time an independent equation is written. Thus the Kirchhoffian paths A–n–B, B–n–C, and C–n–A of Fig. 3.8 yield

$$+e_{aa'} - e_{bb'} - e_{ab} = 0 \tag{3.40}$$
$$+e_{bb'} - e_{cc'} - e_{bc} = 0 \tag{3.41}$$
$$+e_{cc'} - e_{aa'} - e_{ca} = 0 \tag{3.42}$$

Extraction of the corresponding phasors from Eqs. (3.40), (3.41), and (3.42) yields

$$\mathbf{E}_{aa'} - \mathbf{E}_{bb'} - \mathbf{E}_{ab} = 0 \tag{3.43}$$
$$\mathbf{E}_{bb'} - \mathbf{E}_{cc'} - \mathbf{E}_{bc} = 0 \tag{3.44}$$
$$\mathbf{E}_{cc'} - \mathbf{E}_{aa'} - \mathbf{E}_{ca} = 0 \tag{3.45}$$

Solution of Eqs. (3.43), (3.44), and (3.45) for the voltages $\mathbf{E}_{ab}$, $\mathbf{E}_{bc}$, and $\mathbf{E}_{ca}$ in terms of the voltages $\mathbf{E}_{aa'}$, $\mathbf{E}_{bb'}$, and $\mathbf{E}_{cc'}$ yields

$$\mathbf{E}_{ab} = \mathbf{E}_{aa'} - \mathbf{E}_{bb'} \equiv \mathbf{E}_{aa'} + (-\mathbf{E}_{bb'}) \tag{3.46}$$
$$\mathbf{E}_{bc} = \mathbf{E}_{bb'} - \mathbf{E}_{cc'} \equiv \mathbf{E}_{bb'} + (-\mathbf{E}_{cc'}) \tag{3.47}$$
$$\mathbf{E}_{ca} = \mathbf{E}_{cc'} - \mathbf{E}_{aa'} \equiv \mathbf{E}_{cc'} + (-\mathbf{E}_{aa'}) \tag{3.48}$$

The **negatives** of phasors $\mathbf{E}_{bb'}$, $\mathbf{E}_{cc'}$, and $\mathbf{E}_{aa'}$ are drawn in Fig. 3.9 as $\mathbf{E}_{b'b}$, $\mathbf{E}_{c'c}$, and $\mathbf{E}_{a'a}$, and the additions suggested by Eqs. (3.46), (3.47), and (3.48) are performed graphically. The resulting voltages are $\mathbf{E}_{ab}$, $\mathbf{E}_{bc}$, and $\mathbf{E}_{ca}$. The relationship between these voltages and the phase voltages can be established by observing the geometry of the shaded triangle in Fig. 3.9. We can deduce by inspection that the largest angle of the shaded triangle is 120° and each of the other two angles is 30°. Also, by utilizing plane trigonometry, we can write the magnitude of the phasor $\mathbf{E}_{ab}$, $|\mathbf{E}_{ab}|$, as a function of the magnitude of the phasor $\mathbf{E}_{aa'}$, $|\mathbf{E}_{aa'}|$, and the cosine of the angle defined (bound) by the intersection of these two phasors. Mathematically, this relationship can be written as

$$|\mathbf{E}_{ab}| = 2|\mathbf{E}_{aa'}|\cos 30° = 2|\mathbf{E}_{aa'}|\frac{\sqrt{3}}{2} \tag{3.49}$$

Eq. (3.49) yields

$$|\mathbf{E}_{ab}| = \sqrt{3}\,|\mathbf{E}_{aa'}| \tag{3.50}$$

Eq. (3.50) and the geometric relationship between phasors $\mathbf{E}_{aa'}$ and $\mathbf{E}_{ab}$, Fig. 3.9, suggest the following mathematical relationships among the terminal voltages $\mathbf{E}_{ab}$, $\mathbf{E}_{bc}$, and $\mathbf{E}_{ca}$ and the phase voltages $\mathbf{E}_{aa'}$, $\mathbf{E}_{bb'}$, and $\mathbf{E}_{cc'}$:

$$\mathbf{E}_{ab} = \sqrt{3}\,\mathbf{E}_{aa'}e^{+j30°} \tag{3.51}$$

$$\mathbf{E}_{bc} = \sqrt{3}\,\mathbf{E}_{bb'}e^{+j30°} \tag{3.52}$$

$$\mathbf{E}_{ca} = \sqrt{3}\,\mathbf{E}_{cc'}e^{+j30°} \tag{3.53}$$

Note that the voltages $\mathbf{E}_{ab}$, $\mathbf{E}_{bc}$, and $\mathbf{E}_{ca}$ are of equal magnitude and follow each other by 120°. Thus utilizing the **Im-axis** as **reference**, the voltages $\mathbf{E}_{ab}$, $\mathbf{E}_{bc}$, and $\mathbf{E}_{ca}$ can be written as

$$\mathbf{E}_{ab} = |\mathbf{E}_{ab}|e^{+j30°} \tag{3.54}$$

$$\mathbf{E}_{bc} = |\mathbf{E}_{bc}|e^{-j90°} \tag{3.55}$$

$$\mathbf{E}_{ca} = |\mathbf{E}_{ca}|e^{-j210°} \tag{3.56}$$

However, when the **Re-axis** is used as the **reference**, as is usually the case, the same voltages can be written as

$$\mathbf{E}_{ab} = |\mathbf{E}_{ab}|e^{+j120°} \tag{3.57}$$

$$\mathbf{E}_{bc} = |\mathbf{E}_{bc}|e^{j0°} \tag{3.58}$$

$$\mathbf{E}_{ca} = |\mathbf{E}_{ca}|e^{-j120°} \tag{3.59}$$

When there is no voltage drop between points a and A (why?), b and B, and c and C, Fig. 3.8, the voltages $\mathbf{E}_{ab}$, $\mathbf{E}_{bc}$, and $\mathbf{E}_{ca}$ are identical to the line-to-line voltages $\mathbf{E}_{AB}$, $\mathbf{E}_{BC}$, and $\mathbf{E}_{CA}$. For a balanced, three-phase, Y-connected generator, these voltages are related as follows:

$$\mathbf{E}_{AB} \equiv \mathbf{E}_{ab} \tag{3.60}$$

$$\mathbf{E}_{BC} \equiv \mathbf{E}_{bc} \tag{3.61}$$

$$\mathbf{E}_{CA} \equiv \mathbf{E}_{ca} \tag{3.62}$$

Thus for Fig. 3.8, defining the line-to-line voltages as

$$|\mathbf{E}_{\mathrm{LL}}| \equiv |\mathbf{E}_{AB}| = |\mathbf{E}_{BC}| = |\mathbf{E}_{CA}| \tag{3.63}$$

the phase voltages as

$$|\mathbf{E}_{\phi}| \equiv |\mathbf{E}_{aa'}| = |\mathbf{E}_{bb'}| = |\mathbf{E}_{cc'}| \tag{3.64}$$

the line currents as

$$|\mathbf{I}_{\lambda}| \equiv |\mathbf{I}_{\alpha}| = |\mathbf{I}_{\beta}| = |\mathbf{I}_{\gamma}| \tag{3.65}$$

and the phase currents as

$$|\mathbf{I}_{\phi}| \equiv |\mathbf{I}_{a}| = |\mathbf{I}_{b}| = |\mathbf{I}_{c}| \tag{3.66}$$

enables writing the following **two key equations** *for balanced, three-phase, Y-connected generators:**

$$\mathbf{E}_{\lambda\kappa} = \sqrt{3}\, \mathbf{E}_{\phi} e^{+j30°}, \lambda \neq \kappa = A, B, C, \text{ and } \lambda \equiv \phi = a, b, c \qquad (3.67)$$

and

$$\mathbf{I}_{\lambda} = \mathbf{I}_{\phi}, \lambda \equiv \phi = \alpha, \beta, \gamma, \text{ and } \phi = a, b, c \qquad (3.68)$$

Note the interpretation of these two equations. For a balanced, three-phase, Y-connected generator, (1) the line-to-line voltages are always larger than the phase voltages by a factor of $\sqrt{3}$ and always lead (are ahead of) the (associated) phase voltages by 30°, and (2) the line currents are identical to the phase currents. Also, it should be remembered that in practice usually *the line-to-line voltages* $\mathbf{E}_{\lambda\kappa}$ *are known* and the *phase voltages* $\mathbf{E}_{\phi}$ *are unknown*. Thus Eq. (3.67) can be used with equal flexibility for the evaluation of either line-to-line or phase voltages as a function of the other, once one of them is known. Since the line currents are observable (measurable), the phase currents are automatically known.

3.5 Δ-CONNECTED GENERATOR

If the coils of Fig. 3.6 are shifted parallel to themselves and then are connected electrically, as in Fig. 3.10, the device becomes a Δ-connected generator. Note that there is *no* neutral point in the Δ-connected generator. The absence of the neutral *allows circulation* of currents of both fundamental and higher frequencies, especially of the *third harmonic*, in the windings of the device. Thus *the high-heating effects make the Δ-connected generator undesirable when compared to the Y-connected generator*.

In order to proceed with the mathematical analysis of the Δ-connected generator, it is assumed that all the assumptions made for the balanced, three-phase, Y-connected generator (except that Kirchhoff's current law is applied to the supernode† Δ) apply to the Δ-connected generator. Thus, the analysis of a balanced Δ-connected generator connected to a balanced load through an ideal transmission line is pursued through the use of the *per-phase analysis principle*. The per-phase analysis of such an interconnection is valid because of the following property: The line-to-line (phase) voltages of the Δ-connected generator are impressed directly across the Δ-load and thus the voltage drop across each arm of the Δ-load or Δ-equivalent arm of a Y-connected load is explicitly known. Before proceeding with the per-phase analysis of the balanced, three-phase, Δ-connected generator, the relationship between line-to-line and phase voltages, and line and phase currents, must be set forth, as was done for the Y-connected generator. To accomplish this we draw the phasor diagram of the phase currents $\mathbf{I}_{\phi}$, extracted from Eqs. (3.32), (3.33), and (3.34), as in Fig. 3.11.

In an actual δ-connected generator the phase currents are not observable (i.e., measurable). However, the line currents $\mathbf{I}_{LA} \equiv \mathbf{I}_{\alpha}$, $\mathbf{I}_{LB} \equiv \mathbf{I}_{\beta}$, and $\mathbf{I}_{LC} \equiv \mathbf{I}_{\gamma}$ are observable. Thus it is desirable to set up a set of three independent equations that relate the **unknown phase currents** $\mathbf{I}_a$, $\mathbf{I}_b$, and $\mathbf{I}_c$ to the **known line currents** $\mathbf{I}_{\alpha}$, $\mathbf{I}_{\beta}$, and $\mathbf{I}_{\gamma}$.

*The equation is valid for positive sequence voltages only, that is, $A \rightarrow B \rightarrow C \rightarrow A$.

† A supernode is defined as a collection or set of nodes.

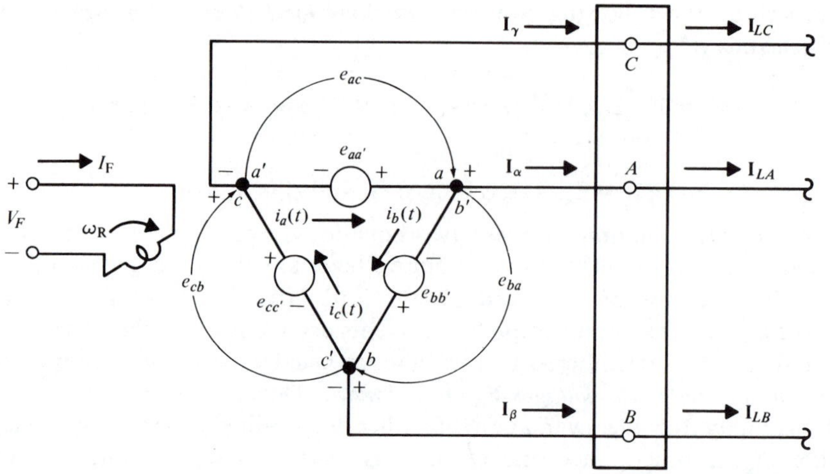

Fig. 3.10 Δ-connected Generator Connected to a Transmission Line at Points A, B, C

These equations can be written in the time domain by utilizing Kirchhoff's current law at the junction points ab', bc', and ca', Fig. 3.10, as follows:

$$i_a(t) - i_b(t) - i_\alpha(t) = 0 \tag{3.69}$$
$$i_b(t) - i_c(t) - i_\beta(t) = 0 \tag{3.70}$$
$$i_c(t) - i_a(t) - i_\gamma(t) = 0 \tag{3.71}$$

Writing Eqs. (3.69), (3.70), and (3.71) in terms of corresponding phasors yields

$$\mathbf{I}_a - \mathbf{I}_b - \mathbf{I}_\alpha = 0 \tag{3.72}$$
$$\mathbf{I}_b - \mathbf{I}_c - \mathbf{I}_\beta = 0 \tag{3.73}$$
$$\mathbf{I}_c - \mathbf{I}_a - \mathbf{I}_\gamma = 0 \tag{3.74}$$

Solving Eqs. (3.72), (3.73), and (3.74) for the currents $\mathbf{I}_\alpha$, $\mathbf{I}_\beta$, and $\mathbf{I}_\gamma$ in terms of the currents $\mathbf{I}_a$, $\mathbf{I}_b$, and $\mathbf{I}_c$ yields

$$\mathbf{I}_\alpha = \mathbf{I}_a - \mathbf{I}_b \equiv \mathbf{I}_a + (-\mathbf{I}_b) \tag{3.75}$$
$$\mathbf{I}_\beta = \mathbf{I}_b - \mathbf{I}_c \equiv \mathbf{I}_b + (-\mathbf{I}_c) \tag{3.76}$$
$$\mathbf{I}_\gamma = \mathbf{I}_c - \mathbf{I}_a \equiv \mathbf{I}_c + (-\mathbf{I}_a) \tag{3.77}$$

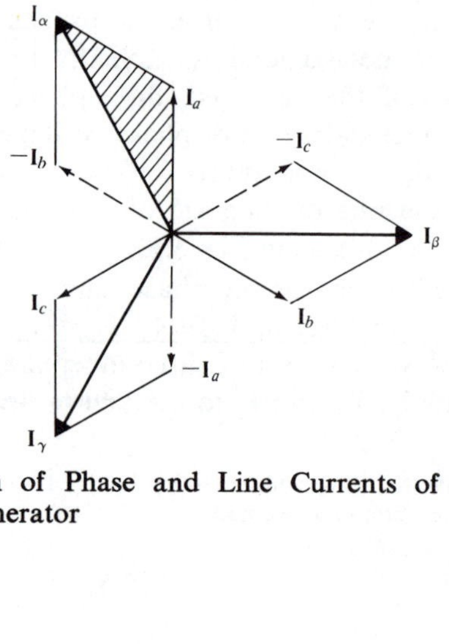

Fig. 3.11 Phasor Diagram of Phase and Line Currents of a Balanced, Three-Phase, Δ-connected Generator

The negatives of phasors $\mathbf{I}_b$, $\mathbf{I}_c$, and $\mathbf{I}_a$ are drawn in Fig. 3.11. Also, the additions suggested by Eqs. (3.75), (3.76), and (3.77) are performed graphically in the figure. The results are $\mathbf{I}_\alpha$, $\mathbf{I}_\beta$, and $\mathbf{I}_\gamma$.

Utilizing the ideas of the paragraph immediately preceding Eq. (3.49) (applied to the balanced, three-phase, Y-connected generators) enables writing the current $\mathbf{I}_\alpha$ of the balanced, three-phase, Δ-connected generator as a function of the current $\mathbf{I}_a$ and the angle defined (bound) by the intersection of these two phasors. Thus

$$|\mathbf{I}_\alpha| = 2|\mathbf{I}_a|\cos 30° = 2|\mathbf{I}_a|\frac{\sqrt{3}}{2} \tag{3.78}$$

Eq. (3.78) yields

$$|\mathbf{I}_\alpha| = \sqrt{3}\,|\mathbf{I}_a| \tag{3.79}$$

Eq. (3.79) and the geometric relationship between phasors $\mathbf{I}_\alpha$ and $\mathbf{I}_a$, Fig. 3.11, suggest the following mathematical relationship between line currents I_α, I_β, and I_γ and phase currents I_a, I_b, and I_c:

$$\mathbf{I}_\alpha = \sqrt{3}\,\mathbf{I}_a e^{+30°} \tag{3.80}$$
$$\mathbf{I}_\beta = \sqrt{3}\,\mathbf{I}_b e^{+j30°} \tag{3.81}$$
$$\mathbf{I}_\gamma = \sqrt{3}\,\mathbf{I}_c e^{+j30°} \tag{3.82}$$

It should be noted that the currents $\mathbf{I}_\alpha$, $\mathbf{I}_\beta$, and $\mathbf{I}_\gamma$ are of equal magnitude and follow each other by 120° When the **Im-axis** is used as the **reference**, these currents can be written as

$$\mathbf{I}_\alpha = |\mathbf{I}_\alpha|e^{j30°} \tag{3.83}$$
$$\mathbf{I}_\beta = |\mathbf{I}_\beta|e^{-j90°} \tag{3.84}$$
$$\mathbf{I}_\gamma = |\mathbf{I}_\gamma|e^{-j210°} \tag{3.85}$$

However, when the **Re-axis** is used as the **reference**, as is usually the case, the same currents can be written as

$$\mathbf{I}_\alpha = |\mathbf{I}_\alpha|e^{+j120°} \tag{3.86}$$
$$\mathbf{I}_\beta = |\mathbf{I}_\beta|e^{j0°} \tag{3.87}$$
$$\mathbf{I}_\gamma = |\mathbf{I}_\gamma|e^{-j120°} \tag{3.88}$$

The line currents are related to the currents $\mathbf{I}_\alpha$, $\mathbf{I}_\beta$, and $\mathbf{I}_\gamma$, Fig. 3.10, as follows:

$$\mathbf{I}_{LA} \equiv \mathbf{I}_\alpha \tag{3.89}$$
$$\mathbf{I}_{LB} \equiv \mathbf{I}_\beta \tag{3.90}$$
$$\mathbf{I}_{LC} \equiv \mathbf{I}_\gamma \tag{3.91}$$

The line-to-line voltages $\mathbf{E}_{AC}$, $\mathbf{E}_{BA}$, and $\mathbf{E}_{CB}$ are identical to the extracted phasors from the voltages e_{ac}, e_{ba}, and e_{cb}, that is, $\mathbf{E}_{ac}$, $\mathbf{E}_{ba}$, and $\mathbf{E}_{cb}$, which are the **phase voltages** of the balanced, three-phase, Δ-connected generator;

$$\mathbf{E}_{AC} = \mathbf{E}_{ac} \tag{3.92}$$
$$\mathbf{E}_{BA} = \mathbf{E}_{ba} \tag{3.93}$$
$$\mathbf{E}_{CB} = \mathbf{E}_{cb} \tag{3.94}$$

Thus, again defining the line-to-line voltages (but for Fig. 3.10) as

$$|\mathbf{E}_{LL}| \equiv |\mathbf{E}_{AC}| = |\mathbf{E}_{BA}| = |\mathbf{E}_{CB}| \tag{3.95}$$

the phase voltages as

$$|\mathbf{E}_\phi| \equiv |\mathbf{E}_{ac}| = |\mathbf{E}_{ba}| = |\mathbf{E}_{cb}| \tag{3.96}$$

the line currents as

$$|\mathbf{I}_L| \equiv |\mathbf{I}_\alpha| = |\mathbf{I}_\beta| = |\mathbf{I}_\gamma| \tag{3.97}$$

and the phase currents as

$$|\mathbf{I}_\phi| \equiv |\mathbf{I}_a| = |\mathbf{I}_b| = |\mathbf{I}_c| \tag{3.98}$$

enables us to write the following **two key** *equations for balanced, three-phase, Δ-connected generators*:

$$\mathbf{I}_{Lj} = \sqrt{3}\,\mathbf{I}_\phi e^{j30°}, j \equiv \phi = A, B, C, \text{ and } \phi = a, b, c \tag{3.99}$$

and

$$\mathbf{E}_{Ljk} = \mathbf{E}_{\phi\mu\nu},\ j \neq k,\ j = A,B,C,\ \ k = C,A,B,\ \ \mu \neq \nu,\ \ \mu = a,b,c,\ \ \nu = c,a,b \tag{3.100}$$

Again note the interpretation of these two equations. For a balanced, three-phase, Δ-connected generator, (1) the line currents are always larger than the phase currents by a factor of $\sqrt{3}$ and always lead (are ahead of) the associated phase currents by 30°, and (2) the line-to-line voltages are identical to the phase voltages. Also, it should be remembered that in practice usually **the line currents $\mathbf{I}_{Lj}$ are known** and **the phase currents $\mathbf{I}_\phi$ are unknown.** Thus Eq. (3.99) can be used with equal flexibility for the evaluation of either current as a function of the other, once one of them is known. Since the line-to-line voltages are observable, the phase voltages are automatically known.

3.6 COMPLEX POWER IN THREE-PHASE BALANCED GENERATORS

In Appendix A, Eq. (A.45), it is shown that the complex power delivered by one source is given by $\mathbf{S} = \mathbf{E}\mathbf{I}^*$. This idea enables writing the complex power expression for *each* voltage source, that is, for each phase, of Figs. 3.8 and 3.10 as

$$\mathbf{S}_{\phi\mu} = \mathbf{E}_{\phi\mu}\mathbf{I}_{\phi\mu}^*,\ \mu = a, b, c \tag{3.101}$$

Thus the *power for each phase* of a balanced, three-phase, Y- or Δ-connected generator can be written as

$$\mathbf{S}_a = \mathbf{E}_a\mathbf{I}_a^* \equiv P_a + jQ_a \tag{3.102}$$

$$\mathbf{S}_b = \mathbf{E}_b\mathbf{I}_b^* \equiv P_b + jQ_b \tag{3.103}$$

$$\mathbf{S}_c = \mathbf{E}_c\mathbf{I}_c^* \equiv P_c + jQ_c \tag{3.104}$$

Since power is a complex quantity, Eqs. (3.102), (3.103), and (3.104) can be added (addition of complex numbers is performed by adding corresponding components). Thus addition of Eqs. (3.102), (3.103), and (3.104) yields

$$\mathbf{S}_{\text{GT}} \equiv P_{\text{GT}} + jQ_{\text{GT}} = \mathbf{S}_a + \mathbf{S}_b + \mathbf{S}_c = (P_a + P_b + P_c) + j(Q_a + Q_b + Q_c) \tag{3.105}$$

where $\mathbf{S}_{\text{GT}}$ designates the *total complex power generated* by a balanced, three-phase, Y- or Δ-connected generator. Since $|\mathbf{E}_{aa'}| \equiv |\mathbf{E}_{bb'}| \equiv |\mathbf{E}_{cc'}|$ and $|\mathbf{I}_a| \equiv |\mathbf{I}_b| \equiv |\mathbf{I}_c|$ and θ_s, where θ_s is the phase angle between two corresponding phasors, is fixed, then

$$P_a = |\mathbf{E}_{aa'}|\,|\mathbf{I}_a|\cos\theta_s = P_b = |\mathbf{E}_{bb'}|\,|\mathbf{I}_b|\cos\theta_s = P_c = |\mathbf{E}_{cc'}|\,|\mathbf{I}_c|\cos\theta_s$$

and

$$Q_a = |\mathbf{E}_{aa'}|\,|\mathbf{I}_a|\sin\theta_s = Q_b = |\mathbf{E}_{bb'}|\,|\mathbf{I}_b|\sin\theta_s = Q_c = |\mathbf{E}_{cc'}|\,|\mathbf{I}_c|\sin\theta_s$$

These identities enable writing Eq. (3.105) as

$$\mathbf{S}_{\mathrm{GT}} = P_{\mathrm{GT}} + jQ_{\mathrm{GT}} \equiv 3\mathbf{S}_{\phi\mu} = 3(P_\phi + jQ_\phi) = 3(\mathbf{E}_{\phi\mathrm{G}}\mathbf{I}_{\phi\mathrm{G}}{}^*)_\mu;\ \mu = a, b, c \quad (3.106)$$

Eq. (3.106) *yields the total complex power generated by a balanced, three-phase, Y- or Δ-connected generator. It can be obtained by multiplying the power generated by any of the phases a, b, or c by a factor of three.*

3.7 TRANSMISSION LINES OR TRANSMISSION NETWORKS

The theory of transmission lines is not undertaken here, as the subject is complex and indeed is beyond the scope of this text. However, a few fundamental concepts about transmission lines are presented in this section in order to explain what transmission lines are and how they are treated in power systems analysis.

A transmission line is defined as an electrical network with two or more conductors, or their equivalence, that is used to transmit electrical energy from one point to another. The energy to be transmitted is of the single-frequency character, such as 60 Hz for ordinary home and commercial use, or it may be of some high-frequency character that is used in communications applications, or it may be of such a character that a band of frequencies is involved, as in telephone transmission. In relation to power and telephone transmission, the lines often become quite long (e.g., a few hundred miles), whereas in transmission of radio frequency energy, the lines are generally short (e.g., a few hundred feet or less). The *physical length* of the transmission line is not generally as important in regard to its electrical behavior as is its *electrical length*. Electrical length is usually expressed in terms of the wavelength for the particular operating frequency. The *wavelength* (λ) of a line, that is, the *velocity of propagation* (v) divided by the operating frequency (f_{E}), never exceeds the *velocity of light* divided by the operating frequency (f_{E}) [6].

In power systems, transmission lines constitute the *arteries* for the transportation of large blocks of electrical energy over large distances. The frequency of the transported energy is 60 Hz. However, the line is also used to send signals to *trip* circuit breakers when faults occur in the system. Thus appropriate parts of the system are removed from service when faults occur and undue damage to the equipment is prevented. Physically, a power system transmission line is comprised of three conductors equilaterally spaced. When the conductors are not equilaterally spaced, the position of the conductors is exchanged at regular intervals along the line so that each conductor occupies the original position of every other conductor over an equal distance. This exchange in position of conductors, taking place at least at switching stations, is known as **transposition**. Its function is to guarantee *balanced* operating conditions for the system by insuring that the inductance per unit length of the transmission line does not vary.

Each conductor of a transmission line is **composite and bare**, that is, the conductor has no insulation whatsoever, except in special cases, and is composed of two or more elements or strands electrically in parallel. In other words, each conductor is composed of strands of wire with alternate layers in opposite directions. The alternation *prevents unwinding* and insures the coincidence of the outer diameter of one layer with the inner diameter of the following layer. Also, stranding *provides flexibility* with large cross-sectional areas. The *tensile strength* of stranded conductors is increased by using *steel* strands in the core of *aluminum* or *copper* strand composite conductor, while the current capacity of the conductor is varied by using various combinations of steel strands and aluminum or copper strands. The number of strands per phase is given (see reference [9]) as $NS = 3x^2 - 3x + 1$, where x is the number of layers including the single center strand.

Transmission lines are described by four parameters, namely, R, L, C, and G, which are measured in MKS units/meter. The standard definition of R and L is applicable. C represents the interwire capacitance, and G, known as the *leakage conductance*, is a measure of the leakage that takes place, due to imperfect insulation, through the air or the insulated supporters.

These parameters enable us to draw the generalized "distributed parameter equivalent circuit" of an infinitesimal length $\Delta\ell$ of a transmission line, on a *per-phase basis*, as in Fig. 3.12. *This equivalent circuit is the most accurate representation of the transmission line, with e and i changing with respect to both time and distance*, the parallel combination constituting an admittance $\Delta\mathbf{Y} = G\Delta\ell + j\omega_E C\,\Delta\ell$ and the series combination constituting an impedance $\Delta\mathbf{Z} = R\Delta\ell + j\omega_E L\,\Delta\ell$. For a transmission line we may define the following two parameters on a *per-unit-length* basis: (1) the *characteristic impedance* of the line $\mathbf{Z}_C \equiv \sqrt{\mathbf{Z}/\mathbf{Y}} \equiv 1/\mathbf{Y}_C$ and (2) the *propagation constant* of the line $\gamma \equiv \sqrt{\mathbf{ZY}}$. With the aid of the parameters $\mathbf{Z}_C$ and γ, we may draw the equivalent circuit of Fig. 3.12 for very long transmission lines as in Fig. 3.13, known as the "long-line equivalent circuit," that is, $150 < \ell < \infty$, where *ℓ is the physical length of the line in miles*. The subscripts S and R designate the

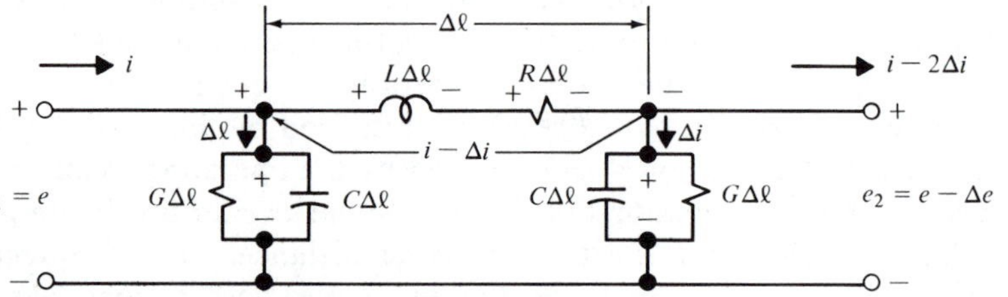

Fig. 3.12 The Distributed Parameter Equivalent Circuit of an Infinitesimal Length of a Single-Phase Transmission Line

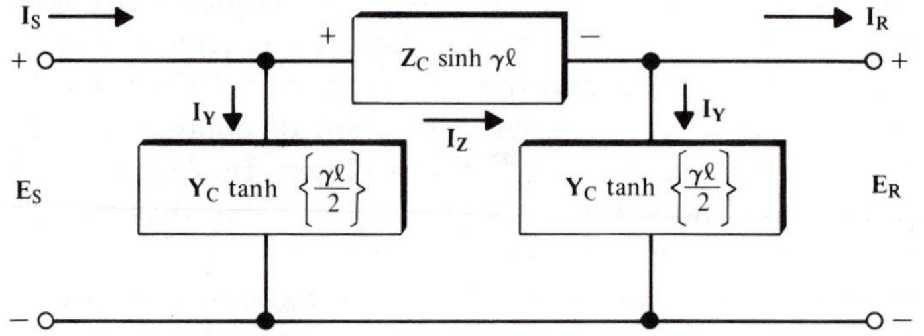

Fig. 3.13 The Long-Line Equivalent Circuit of Single-Phase Transmission Lines, $150 < \ell < \infty$, ℓ in Miles

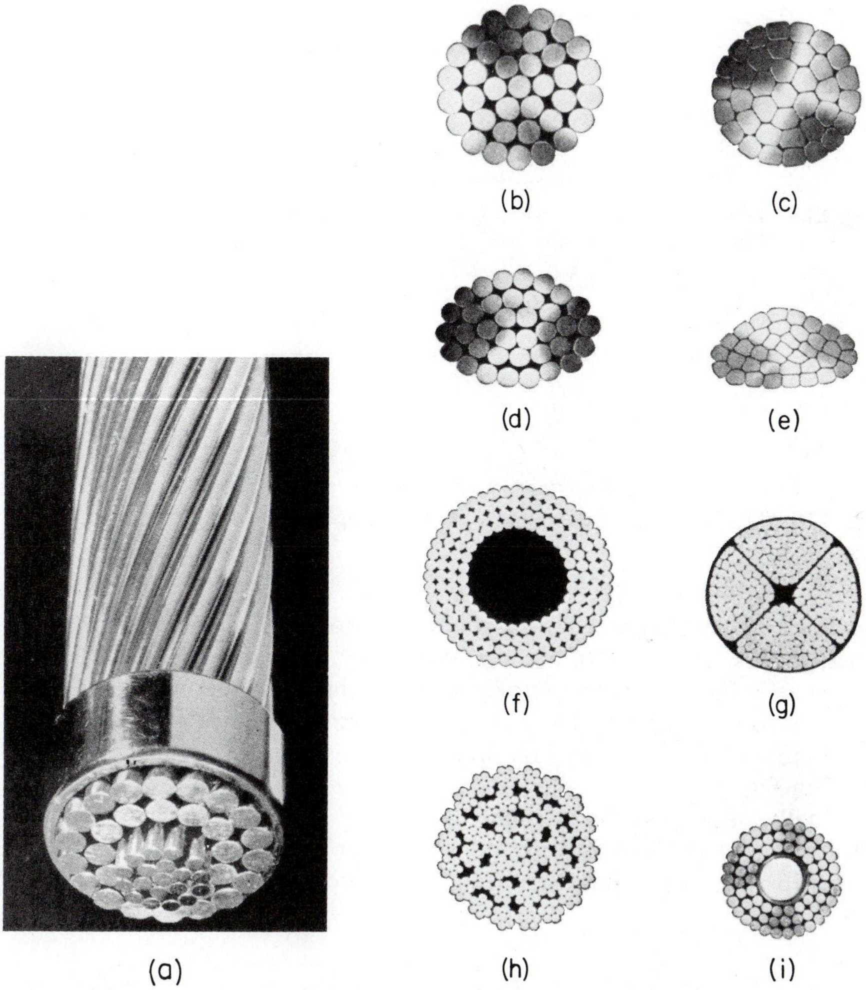

Plate 3.4 Some of the Typical Conductors Used on Overhead Power Lines, Designed Specifically to Reduce Cost and Improve Operating Characteristics: (a) Typical ACSR (Aluminum Cable, Steel Reinforced) Conductor, (b) Standard Concentric Stranded, (c) Compact Round, (d) Noncompact Sector, (e) Compact Sector, (f) Annular Stranded, (g) Segmental, (h) Rope Stranded and (i) Hollow Core. (Reproduced from *Electrical Transmission and Distribution Reference Book*, by Central Station Engineers, Westinghouse Electric Co., Pittsburgh, Pa.)

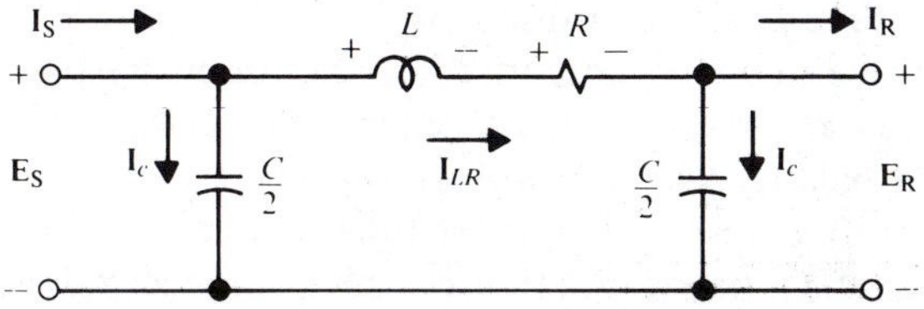

Fig. 3.14 The Medium-Line Equivalent Circuit of Single-Phase Transmission Lines, $50 < \ell \leqslant 150$, ℓ in Miles

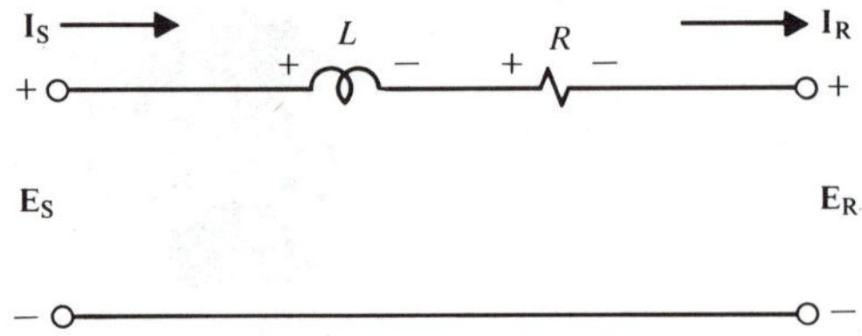

Fig. 3.15 The Short-Line Equivalent Circuit of Single-Phase Transmission Lines, $0<\ell<50$, ℓ in Miles

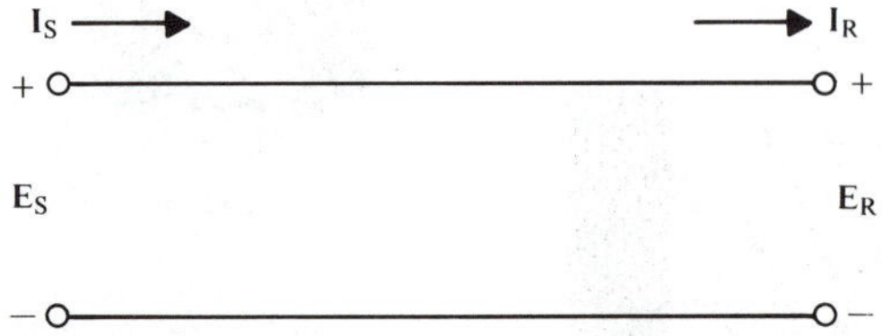

Fig. 3.16 The Ideal-Line Equivalent Circuit of Single-Phase Transmission Lines, L, R, and $G \to 0$

sending and *receiving* ends, correspondingly. However, medium-length transmission lines, that is, $50<\ell<150$, are represented by "medium-line equivalent circuits," drawn as in Fig. 3.14. The approximation is achieved by allowing $G \to 0$ in Fig. 3.12. Short transmission lines, that is, $0<\ell<50$, are represented by a "short-line equivalent circuit," drawn as in Fig. 3.15. The approximation is achieved by allowing $\mathbf{Y} \to 0$ in Fig. 3.12.

Finally, it is possible at times to allow $R \to 0$ and $L \to 0$ in Fig. 3.15. These approximations enable drawing the equivalent circuit of a transmission line as in Fig. 3.16. This is the "ideal-line equivalent circuit" of a single-phase transmission line.

3.8 BALANCED THREE-PHASE LOADS

A generalized, balanced, three-phase load used in conjunction with balanced, three-phase, Y- or Δ-connected generators is comprised of three arms, Fig. 3.17, analogous to the phases of corresponding generators. Each arm is comprised of a *series RLC* network. The parameters in each arm are of *equal* numerical value, and if any of the parameters is missing in one arm, it should be missing in all the others. Thus *for a given frequency of operation* ω_E, *the impedances of all three arms are identical*. Loads with these properties are known as *balanced Y- or Δ-networks*, or *balanced, three-phase Y- or Δ-loads*. When these loads are connected to a generator with a balanced three-phase transmission line, they insure balanced response currents. The resulting network, comprised of the following balanced, three-phase, component units, (1) Y- or Δ-connected generators, (2) transmission lines, and (3) loads, constitutes a balanced *three-phase system*.

The loads of Fig. 3.17 are *equivalent* at their terminals. Thus use of Eq. (1.74), that is, $\mathbf{Z}_\Delta = 3\mathbf{Z}_Y$, enables transformation of any balanced Y-load to an equivalent balanced Δ-load and vice-versa.

The *total complex power dissipated in the load* is calculated by utilizing the ideas preceding Eq. (3.106) and Eq. (A.53). Thus

$$\begin{aligned}\mathbf{S}_{LT} &= P_{LT} + jQ_{LT} \equiv 3\mathbf{S}_{\phi} = 3(P_{\phi} + jQ_{\phi})_{\mu} \\ &= 3(\mathbf{V}_{\phi L}\mathbf{I}_{\phi L}{}^{*})_{\mu} = 3(\mathbf{Z}_{\phi L}|\mathbf{I}_{\phi L}|^2)_{\mu} = 3(|\mathbf{V}_{\phi L}|^2\mathbf{Y}_{\phi L}{}^{*})_{\mu}, \ \mu = a, b \text{ or } c \qquad (3.107)\end{aligned}$$

where the subscripts ϕL designate "phase load" and μ designates the phases a, b, and c.

3.9 POWER MEASUREMENT: THREE- AND TWO-WATTMETER METHODS

To measure the total power drawn by either of the three-phase loads of Fig. 3.17, you might think that you would have to connect the *potential coils* of three wattmeters between points A, B, C and neutral n for the Y-connected load and insert the *current coils* in each phase (proper coil polarities should be taken into consideration). Or you might think that you would have to insert the *current coils* in each phase of the Δ-connected load and connect the appropriate *potential coils* across each phase; note that **maintaining proper polarity of the coils is essential.**

Although the described method is theoretically correct, it is useless because in practice the neutral of a Y-connected load and the phases of a Δ-connected load are *not available*. Actually, only the three terminals of either network (load) are available with 100 percent certainty; the letters designating the sequence of the phases are available only as information on nameplates* of devices and sometimes even they are not available. In reality, then, the situation is that of attempting to measure the power absorbed by a black box, Fig. 3.18.

However, since the terminals of the black box are available, we could interrupt the input and insert the current coils of the three wattmeters. Since two connection points are needed for each potential coil, we could use the *arbitrary reference point*, (i.e., infinity), reference point ρ, Fig. 3.18, as one connection point for each potential coil and the existing three terminals as the other three needed connection points. This idea is exhibited clearly in Fig. 3.18.

The average wattmeter readings for wattmeters W_{L1}, W_{L2}, and W_{L3} are as follows:

$$P[W_{L1}]^{\dagger} = \frac{1}{T}\int_0^T e_{A\rho}(t) i_A(t)\,dt \qquad (3.108)$$

$$P[W_{L2}] = \frac{1}{T}\int_0^T e_{B\rho}(t) i_B(t)\,dt \qquad (3.109)$$

$$P[W_{L3}] = \frac{1}{T}\int_0^T e_{C\rho}(t) i_C(t)\,dt \qquad (3.110)$$

Utilizing Eq. (3.107) enables calculation of the total real power P_{LT} drawn by the network, as measured by the three wattmeters. Thus

$$\begin{aligned}P_{LT} &= P[W_{L1}] + P[W_{L2}] + P[W_{L3}] \\ &= \frac{1}{T}\int_0^T [e_{A\rho}(t) i_A(t) + e_{B\rho}(t) i_B(t) + e_{C\rho}(t) i_C(t)]\,dt \qquad (3.111)\end{aligned}$$

*A nameplate is a small plate attached to a device on which rating and other relevant information about the device is given.

†For example, $P[W_{L1}]$ designates the power measured by wattmeter W_{L1}, Fig. 3.18.

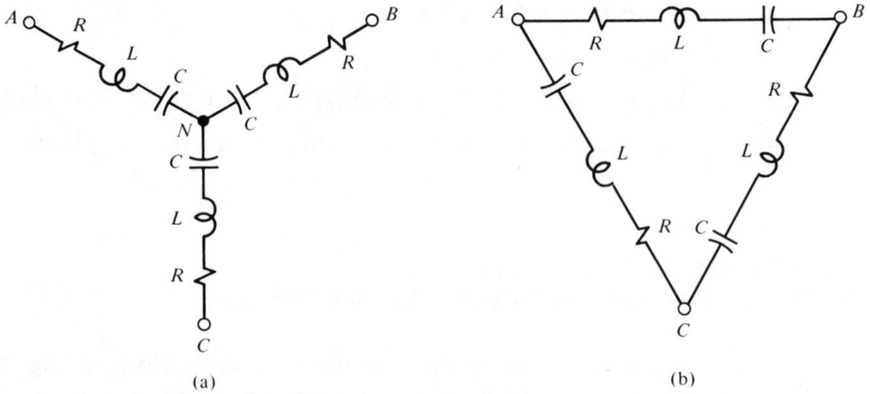

Fig. 3.17 Balanced Three-Phase Loads: (a) Y-connected Load and (b) Δ-connected Load

However, the voltages $e_{A\rho}(t)$, $e_{B\rho}(t)$, and $e_{C\rho}(t)$, Fig. 3.18, can be expressed in terms of the terminal voltages $e_{AR}(t)$, $e_{BR}(t)$, and $e_{CR}(t)$, and $e_{R\rho}(t)$. The point R designates *ground potential*. Thus in a Y-connected load, the point R could designate the neutral point $n(N)$. Thus

$$e_{A\rho}(t) = e_{AR}(t) + e_{R\rho}(t) \tag{3.112}$$

$$e_{B\rho}(t) = e_{BR}(t) + e_{R\rho}(t) \tag{3.113}$$

$$e_{C\rho}(t) = e_{CR}(t) + e_{R\rho}(t) \tag{3.114}$$

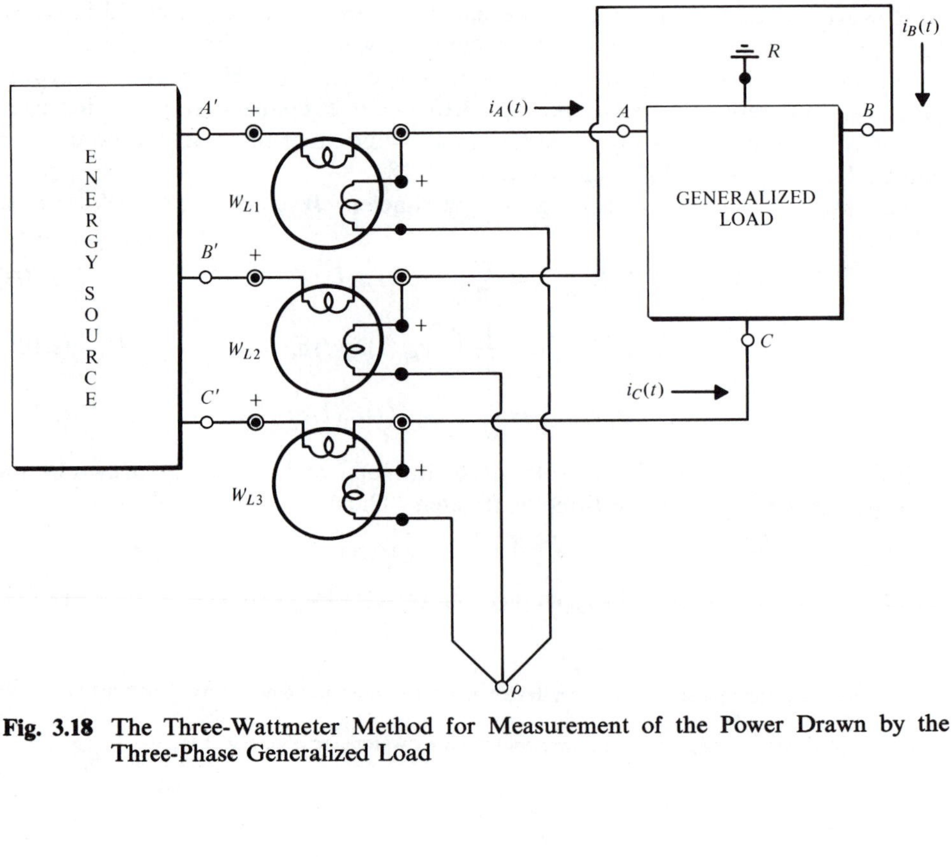

Fig. 3.18 The Three-Wattmeter Method for Measurement of the Power Drawn by the Three-Phase Generalized Load

Substitution of Eqs. (3.112), (3.113), and (3.114) into Eq. (3.111) and factoring yields

$$P_{LT} = \frac{1}{T}\int_0^T [e_{AR}(t)i_A(t) + e_{BR}(t)i_B(t) + e_{CR}(t)i_C(t)]\,dt + \frac{1}{T}\int_0^T e_{R\rho}(t)[i_A(t) + i_B(t) + i_C(t)]\,dt \tag{3.115}$$

Application of Kirchhoff's current law to the supernode, comprised of the entire black box load, yields

$$i_A(t) + i_B(t) + i_C(t) = 0 \tag{3.116}$$

Therefore, Eq. (3.115) becomes

$$P_{LT} = \frac{1}{T}\int_0^T [e_{AR}(t)i_A(t) + e_{BR}(t)i_B(t) + e_{CR}(t)i_C(t)]\,dt \tag{3.117}$$

Since point ρ is an arbitrarily located point in space, it can be moved about and placed in the neighborhood of, say, point A. Now, $e_A(t) = e_{AA}(t) \equiv 0$ and $e_R(t) \rightarrow e_{RA}(t)$. Thus Eq. (3.112) becomes [since $e_{RA}(t) = -e_{AR}(t)$]

$$0 = e_{AR}(t) + e_{R\rho}(t) = e_{AR}(t) + e_{RA}(t) \equiv e_{AR}(t) - e_{AR}(t) = 0 \tag{3.118}$$

This suggests that for this connection wattmeter W_{L1} reads zero. Thus removal of it from the circuit does not change matters at all. As a matter of fact, removal of the wattmeter helps. It allows the total power drawn by the load to be measured by two wattmeters only. This method is known as the *two-wattmeter method* and provides for measurement of the **total real power*** drawn by the black box load. The readings are independent of (1) the balanced or unbalanced nature of the load, (2) the balanced or unbalanced nature of the source of the energy used, (3) the differences in the two wattmeters used, and (4) the waveform of the supplied energy. However, the readings should be *corrected to account* for the power dissipated in the potential coil. The adjustment can be made from the knowledge of the *resistance* of the potential coil, which is usually recorded on the face of the instrument, and the *potential* across it, or other information provided by the manufacturer. It is possible, however, at times to assume that the wattmeter corrections are sufficiently small to be ignored.

3.10 POWER-FACTOR ANGLE MEASUREMENT

In addition to measuring the power drawn by a balanced load, the two-wattmeter method provides a means of determining the *power-factor angle*. Assume the wattmeter connection across the Δ-connected load of Fig. 3.19. Also, assume the phasor diagram of Fig. 3.20, drawn with the information of Fig. 3.11, but in addition to the six currents the three phase voltages $\mathbf{V}_{AC}$, $\mathbf{V}_{BA}$, and $\mathbf{V}_{CB}$ are superimposed.

*In analogy to Eq. (3.111), the **total reactive power** Q_{LT} drawn by the network of Fig. 3.18 can be written in terms of phase (or line) quantities as

$$Q_{LT} = 3|\mathbf{E}_\phi||\mathbf{I}_\phi|\sin\theta \equiv \sqrt{3}\,|\mathbf{E}_{LL}||\mathbf{I}_L|\cos(\theta - 90°) \tag{3.118a}$$

The right-hand side of the identity, Eq. (3.118a), could be implemented if we were to connect (1) the potential coil of the wattmeter between **any two lines** to measure their potential difference and (2) the current coil of the wattmeter in series with a **third line** to measure its current.

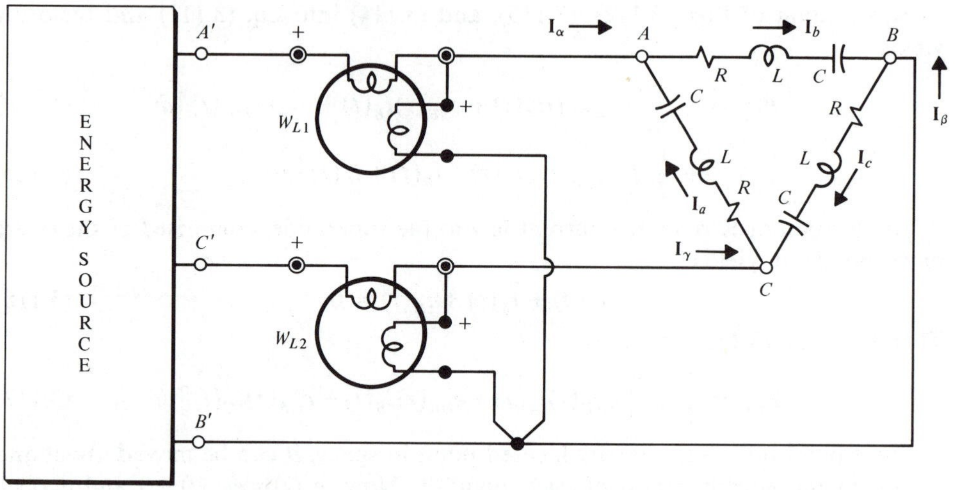

Fig. 3.19 The Two-Wattmeter Method for Measurement of the Power Drawn by a Balanced Δ-connected Load

Note that both the phase angle between the phase and line currents, and phase currents and phase voltages, are measured with the phase current (or Im-axis) as reference. From Fig. 3.19 it is obvious that for a network with an arbitrary lagging power factor $\theta_{\mathbf{Z}}$ (for which $\phi_{\mathbf{V}}+180°+120°+\theta_{\mathbf{Z}}=360°$ yields $\phi_{\mathbf{V}}=60°-\theta_{\mathbf{Z}}$ and $\phi_{\mathbf{I}}=30°$, Fig. 3.20), wattmeters W_{L1} and W_{L2} read as follows:

$$P[W_{L1}] = |\mathbf{V}_{AB}||\mathbf{I}_\alpha|\cos(\phi_{\mathbf{V}}-\phi_{\mathbf{I}})$$
$$= |\mathbf{V}_{AB}||\mathbf{I}_\alpha|\cos[(60°-\theta_{\mathbf{Z}})-30°] = |\mathbf{V}_{AB}||\mathbf{I}_\alpha|\cos(30°-\theta_{\mathbf{Z}}) \quad (3.119)$$

$$P[W_{L2}] = |\mathbf{V}_{CB}||\mathbf{I}_\gamma|\cos(\phi_{\mathbf{V}}-\phi_{\mathbf{I}})$$
$$= |\mathbf{V}_{CB}||\mathbf{I}_\gamma|\cos(30°+\theta_{\mathbf{Z}}) \quad (3.120)$$

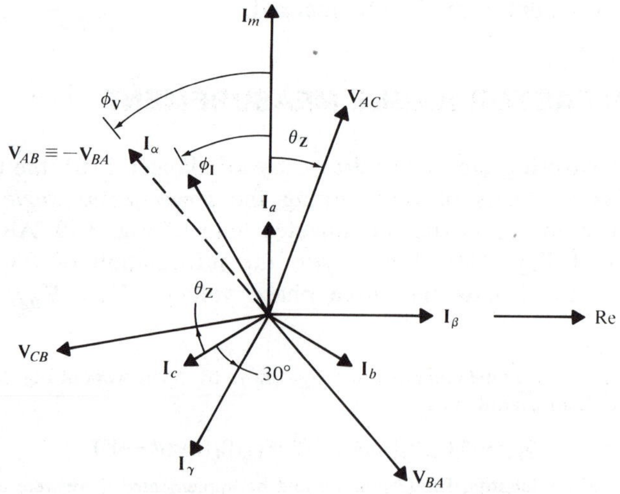

Fig. 3.20 Phasor Diagram of the Δ-connected Load for the Two-Wattmeter, Power-Factor Angle Measurement Method

For the assumed balanced system, $|\mathbf{V}_{AB}|=|\mathbf{V}_{CB}|$ and $|\mathbf{I}_\alpha|=|\mathbf{I}_\beta|$. Thus taking the power ratio of the readings yields

$$\frac{P[W_{L1}]}{P[W_{L2}]}=\frac{\cos(30°-\theta_\mathbf{Z})}{\cos(30°+\theta_\mathbf{Z})} \tag{3.121}$$

Since $\cos(30°\pm\theta_\mathbf{Z})\equiv\cos 30°\cos\theta_\mathbf{Z}\mp\sin 30°\sin\theta_\mathbf{Z}$, upon expansion, proper substitution and collection of terms Eq. (3.121) yields

$$\frac{\sqrt{3}}{2}\{P[W_{L1}]-P[W_{L2}]\}\cos\theta_\mathbf{Z}=0.5\{P[W_{L1}]+P[W_{L2}]\}\sin\theta_\mathbf{Z} \tag{3.122}$$

Taking the ratio of Eq. (3.122) yields

$$\tan\theta_\mathbf{Z}\equiv\frac{\sin\theta_\mathbf{Z}}{\cos\theta_\mathbf{Z}}=\frac{(\sqrt{3}/2)\{P[W_{L1}]-P[W_{L2}]\}}{\frac{1}{2}\{P[W_{L1}]+P[W_{L2}]\}}=\frac{\sqrt{3}\{P[W_{L1}]-P[W_{L2}]\}}{P[W_{L1}]+P[W_{L2}]} \tag{3.123}$$

or

$$\theta_\mathbf{Z}=\tan^{-1}\left\{\frac{\sqrt{3}\{P[W_{L1}]-P[W_{L2}]\}}{P[W_{L1}]+P[W_{L2}]}\right\} \tag{3.124}$$

Thus the relative magnitudes and signs of the wattmeter readings designate the *nature* of the network. For example, equal wattmeter readings, $P[W_{L1}]=P[W_{L2}]$, designate a *unity power-factor* load; equal and opposite readings designate a *purely reactive* load; a reading of $P[W_{L1}]$ that is greater than the reading of $P[W_{L2}]$, that is, $P[W_{L1}]>P[W_{L2}]$, designates an *inductive* load; and a reading of $P[W_{L1}]$ that is less than the reading of $P[W_{L2}]$, that is, $P[W_{L1}]<P[W_{L2}]$, designates a *capacitive* load.

If you are *uncertain* of the results obtained, in particular, the nature (i.e., the sign) of the measured angle, you can check the results by performing the following *test*. Add a high-impedance reactive load across the unknown load and observe the effect the added reactive load has on the calculated angle. The change that has taken place should enable you to deduce the validity of the results. For example, if the added reactive load is a three-phase capacitor and the magnitude of the new angle is smaller than the original angle, the original load is inductive. However, if the magnitude of the new angle is greater than the original angle, the original load is capacitive.

Example 3-1

For the Y-connected ideal voltage generator given in Fig. 3.1.1, the a–b–c sequence line-to-line voltages are $e_{AC}(t)=\sqrt{2}\,(120)\cos\omega_\mathrm{E}t$, $e_{BA}(t)=\sqrt{2}\,(120)\cos(\omega_\mathrm{E}t-120°)$, and $e_{CB}(t)=\sqrt{2}\,(120)\cos(\omega_\mathrm{E}t-240°)$, $\omega_\mathrm{E}=1000$, and the Δ-connected load parameters are $R=30$ and $L=0.03$, all in MKS units.

Calculate: (1) the phase currents $\mathbf{I}_a$, $\mathbf{I}_b$, and $\mathbf{I}_c$, (2) the line currents $\mathbf{I}_\alpha$, $\mathbf{I}_\beta$, and $\mathbf{I}_\gamma$, (3) the complex power delivered to the load, $\mathbf{W}_\mathrm{LT}$. (4) the real power, dissipated in the load P_LT, (5) the reactive power stored in the field Q_LT, (6) the time-domain, phase currents $i_\phi(t)$, (7) the time-domain voltage drop across the resistor, $v_\mathrm{R}(t)$, (8) the time-domain voltage drop across the inductor, $v_\mathrm{L}(t)$, (9) the time-domain line

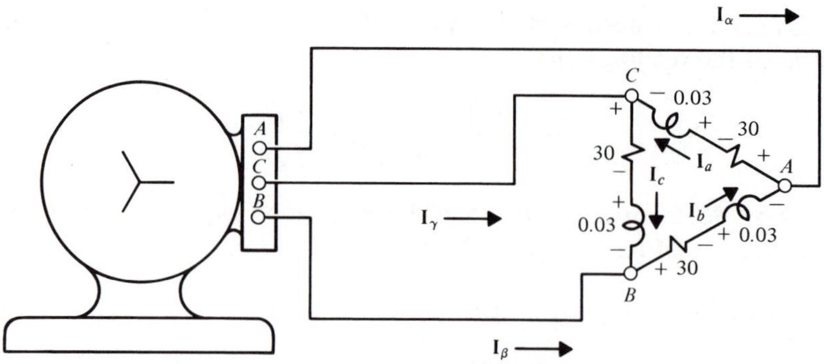

Fig. 3.1.1 Interconnection, through an Ideal Transmission Line, of a Balanced Y-connected Generator to a Balanced Δ-connected Load. Each Leg Comprises a Series *RL* Network

currents $i_L(t)$, (10) the time-domain real power* $p_{LT}(t)$, and (11) the time-domain imaginary power $q_{LT}(t)$.

Solution

1. Using the Re-axis as reference enables writing the phasors of the line voltages as follows:

i. $$\mathbf{E}_{AC} = 120e^{j0°}$$

ii. $$\mathbf{E}_{BA} = 120e^{-j120°}$$

iii. $$\mathbf{E}_{CB} = 120e^{-j240°}$$

Note that in a Δ-connected load, connected to the generator through an ideal transmission line, these voltages are impressed directly across the arms of the Δ-load whose impedances (since the load is balanced) are as follows:

$$\mathbf{Z}_\phi = 30 + j\omega_E L = 30 + j30 = \underline{42.4e^{j45°}\Omega}$$

Thus

i. $$\mathbf{I}_a = \frac{\mathbf{E}_{AC}}{\mathbf{Z}_\phi} = \frac{120e^{j0°}}{42.4e^{j45°}} = \underline{\underline{2.83e^{-j45°}}}\text{ A}$$

ii. $$\mathbf{I}_b = \frac{\mathbf{E}_{BA}}{\mathbf{Z}_\phi} = \frac{120e^{-j120°}}{42.4e^{j45°}} = \underline{\underline{2.83^{-j165°}}}\text{ A}$$

iii. $$\mathbf{I}_c = \frac{\mathbf{E}_{CB}}{\mathbf{Z}_\phi} = \frac{120e^{-j240°}}{42.4e^{j45°}} = \underline{\underline{2.83e^{-j285°}}}\ A$$

2. Use of Kirchhoff's current law at points *A*, *B*, and *C* enables calculation of the line currents as follows:

i. $$\mathbf{I}_\alpha = \mathbf{I}_a - \mathbf{I}_b = 2.0 - j2.0 - (-2.73 - j0.732)$$
$$= 4.73 - j1.27 = \underline{\underline{4.90e^{-j15°}}}\text{ A}$$

*The instantaneous power may be separated into two parts, which we could call "time-domain real power, $p_{LT}(t)$" and "time-domain imaginary power, $q_{LT}(t)$." These are $p_{LT}(t) = \text{Re}(\mathbf{S})[1 - \cos(2\omega_E t)]$, and, $q_{LT}(t) = \text{Im}(\mathbf{S})[-\sin(2\omega_E t)]$

ii. $$\mathbf{I}_\beta = \mathbf{I}_b - \mathbf{I}_c = -2.73 - j0.732 - (0.732 + j2.73)$$
$$= -3.46 - j3.46 = \underline{\underline{4.90e^{j225°}}} \text{ A}$$

iii. $$\mathbf{I}_\gamma = \mathbf{I}_c - \mathbf{I}_a = 0.732 + j2.73 - (2.00 - j2.00)$$
$$= -1.27 + j4.73 = \underline{\underline{4.90e^{j105°}}} \text{ A}$$

3. Use of Eq. (3.107) enables calculation of the power demanded by the load as follows:

$$\mathbf{W}_{\mathrm{LT}} = 3\mathbf{V}_\phi \mathbf{I}_\phi{}^* = 3 \times 120e^{j0°} \times 2.83e^{j45°}$$
$$= \underline{\underline{1.019 \times 10^3 e^{j45°}}} \text{ V-A}$$
$$= \underline{\underline{720 + j720}} \text{ V-A}$$

4. and 5. Using $\mathbf{W}_{\mathrm{LT}} = P_{\mathrm{LT}} + jQ_{\mathrm{LT}} = 720 + j720$ enables writing

i. $$P_{\mathrm{LT}} \equiv \text{power dissipated} = \underline{\underline{720}} \text{ W}$$

ii. $$Q_{\mathrm{LT}} \equiv \text{power stored in the field} = \underline{\underline{720}} \text{ var}$$

6. The time-domain phase currents can be written (using the Re-axis as a reference) from part 1 as

i. $$i_a(t) = \underline{\underline{\sqrt{2}\,(2.83)\cos(\omega_E t - 45°)}} \text{ A}$$

ii. $$i_b(t) = \underline{\underline{\sqrt{2}\,(2.83)\cos(\omega_E t - 165°)}} \text{ A}$$

iii. $$i_c(t) = \underline{\underline{\sqrt{2}\,(2.83)\cos(\omega_E t - 285°)}} \text{ A}$$

7. The time-domain voltage across the resistor is

$$v_{\mathrm{R}}(t) = \mathrm{R}i_a(t)$$
$$= \underline{\underline{\sqrt{2}\,(84.9)\cos(1000t - 45°)}} \text{ V}$$

8. The time-domain voltage across the inductor is

$$v_{\mathrm{L}}(t) = \mathrm{L}\frac{di_a(t)}{dt}$$
$$= \underline{\underline{\sqrt{2}\,(84.9)\cos(1000t + 45°)}} \text{ V}$$

9. The time-domain line currents can be written (using the Re-axis as a reference) from part 2 as

i. $$i_\alpha(t) = \underline{\underline{\sqrt{2}\,(4.90)\cos(\omega_E t - 15°)}} \text{ A}$$

ii. $$i_\beta(t) = \underline{\underline{\sqrt{2}\,(4.90)\cos(\omega_E t + 225°)}} \text{ A}$$

iii. $$i_\gamma(t) = \underline{\underline{\sqrt{2}\,(4.90)\cos(\omega_E t + 105°)}} \text{ A}$$

10. The time domain of the real power can be written with the aid of part 3 and Eq. (A.54) for $\psi° = 0°$ as

$$p_{\mathrm{LT}}(t) = \underline{\underline{720(1 - \cos 2000t)}}\ W$$

11. The time domain of the imaginary power can be written with the aid of part 3 and Eq. (A.55) for $\psi° = 0°$ as

$$q_{\mathrm{LT}}(t) = \underline{\underline{-720 \sin 2000t}} \text{ V-A}$$

Example 3-2

For the Y-connected ideal voltage generator given in Fig. 3.2.1, the a–b–c sequence line-to-line voltages are $e_{AB}(t) = \sqrt{2}\,(120)\cos\omega_E t$, $e_{BC}(t) = \sqrt{2}\,(120)\cos(\omega_E t - 120°)$, and $e_{CA}(t) = \sqrt{2}\,(120)\cos(\omega_E t - 240°)$, $\omega_E = 1000$, and the Y-connected load has parameters $R = 10$ and $L = 0.01$, all in MKS units.

Calculate: (1) the phase currents $\mathbf{I}_a$, $\mathbf{I}_b$, and $\mathbf{I}_c$, (2) the line currents $\mathbf{I}_\alpha$, $\mathbf{I}_\beta$, and $\mathbf{I}_\gamma$, (3) the total power demanded by the load from the generator, and (4) the power factor of the load; (5) utilizing parallel connection correct the system for an operating power factor of 0.866; (6) calculate the power supplied by the generator to the load now, that is, at the new power factor; (7) determine the advantages of power factor correction.

Solution

1. Since the load is Y-connected and the line-to-line voltages are known, the phase voltages must be calculated using Eq. (3.67) and pursue a per-phase analysis. Thus since

$$|\mathbf{E}_{jk}| = \sqrt{3}\,|\mathbf{E}_\phi|$$

$$|\mathbf{E}_\phi| = \frac{120}{\sqrt{3}} = \underline{69.3\ \text{V}}$$

Utilizing the fact that phase voltages lag the line-to-line voltages by 30° enables writing the phase voltages as follows:

$$\mathbf{E}_a = \underline{69.3e^{-j30°}\ \text{V}}$$

$$\mathbf{E}_b = \underline{69.3e^{-j150°}\ \text{V}}$$

$$\mathbf{E}_c = \underline{69.3e^{-j270°}\ \text{V}}$$

Since $\mathbf{Z} = R + j\omega_E L$,

$$\mathbf{Z}_\phi = 10 + j1000(0.01) = 10 + j10 = \underline{14.1e^{j45°}\ \Omega}$$

Since the system is balanced, the phases can be isolated, as drawn in Fig. 3.2.2,

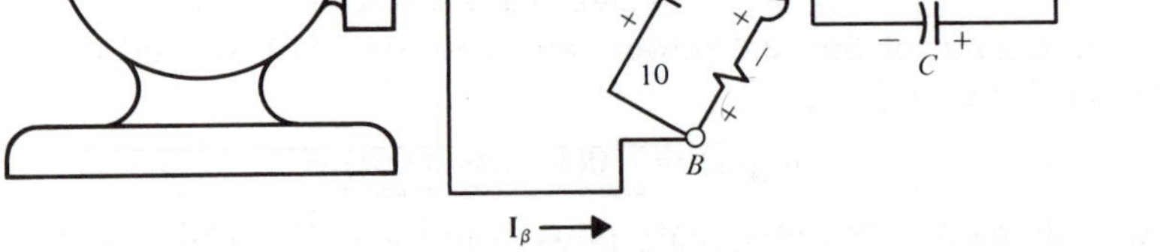

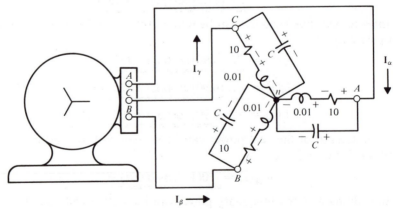

Fig. 3.2.1 Interconnection, through an Ideal Transmission Line, of a Balanced Y-connected Generator to a Balanced Y-connected Load. Each Leg Comprises a Series RL Network Connected in Parallel with a Capacitor C

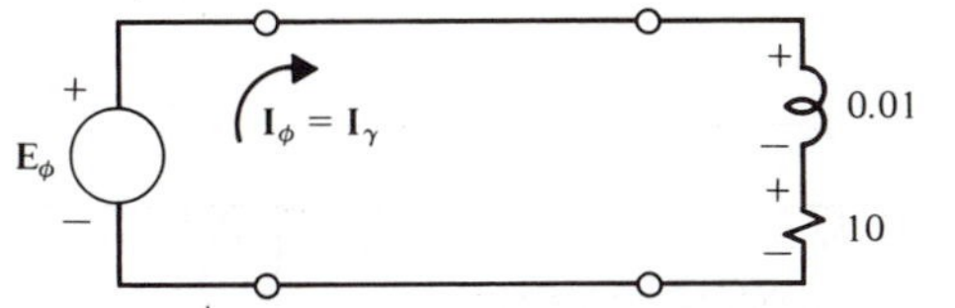

Fig. 3.2.2 The Equivalent Circuit of an Isolated Phase for Fig. 3.2.1. The Capacitor C is not Exhibited

and the analysis proceeds as follows:

$$\mathbf{I}_\alpha = \frac{\mathbf{E}_a}{\mathbf{Z}_\phi} = \frac{69.3e^{-j30°}}{14.1e^{j45°}} = \underline{\underline{4.9e^{-j75°}\text{ A}}}$$

$$\mathbf{I}_\beta = \frac{\mathbf{E}_b}{\mathbf{Z}_\phi} = \frac{69.3e^{-j150°}}{14.1e^{j45°}} = \underline{\underline{4.9e^{-j195°}\text{ A}}}$$

$$\mathbf{I}_\gamma = \frac{\mathbf{E}_c}{\mathbf{Z}_\phi} = \frac{69.3e^{-j270°}}{14.1e^{j45°}} = \underline{\underline{4.9e^{-j315°}\text{ A}}}$$

2. Since the system is Y-connected, the line currents are identical to the phase currents. Thus

$$\mathbf{I}_\alpha = \mathbf{I}_a = \underline{\underline{4.9e^{-j75°}\text{ A}}}$$

$$\mathbf{I}_\beta = \mathbf{I}_b = \underline{\underline{4.9e^{-j195°}\text{ A}}}$$

$$\mathbf{I}_\gamma = \mathbf{I}_c = \underline{\underline{4.9e^{-j315°}\text{ A}}}$$

3. Utilization of Eq. (3.107) yields the total complex power as

$$\begin{aligned}\mathbf{S}_{\mathrm{LT}} &\equiv 3\mathbf{S}_\phi = 3\mathbf{E}_\phi\mathbf{I}_\phi{}^* \\ &= 3(69.3e^{-j30°})(4.9e^{j75°}) \\ &= \underline{\underline{1{,}019e^{j45°}\text{ VA}}}\end{aligned}$$

4. $$\text{PF} = \cos 45° = \underline{\underline{0.707}}$$

5. The angle for the desired power-factor correction, Fig. 3.2.3, is

$$\theta_D = \cos^{-1}(0.866) = 30°$$

$$\mathbf{Y}_\phi = G_\phi - jB_\phi = \frac{1}{\mathbf{Z}_\phi} = \frac{1}{14.2e^{j45°}} = 7.09 \times 10^{-2}e^{-j45°}$$

$$= \underline{5\times 10^{-2} - j5\times 10^{-2}}\text{ mho}$$

Thus the desired susceptance B_{D} is

$$B_{\mathrm{D}} = 5\times 10^{-2}\tan 30° = \underline{2.89\times 10^{-2}}\text{ mho}$$

The desired $\mathbf{Y}_{\mathrm{D}}$ is

$$\mathbf{Y}_D \equiv G_\phi|_{\equiv G_{\mathrm{D}}} - jB_{\mathrm{D}} = \underline{5\times 10^{-2} - j2.89\times 10^{-2}}\text{ mho}$$

and should be obtained from the original $\mathbf{Y}$, designated as $\mathbf{Y}_\phi$, and the corrective element connected in parallel with $\mathbf{Y}_\phi$. Thus

$$\begin{aligned}\mathbf{Y}_D &\equiv Y_\phi + jB_? \\ &= 5\times 10^{-2} - j5\times 10^{-2} + jB_? \\ &= 5\times 10^{-2} + j(-5\times 10^{-2} + B_?) = 5\times 10^{-2} - j2.89\times 10^{-2}\end{aligned}$$

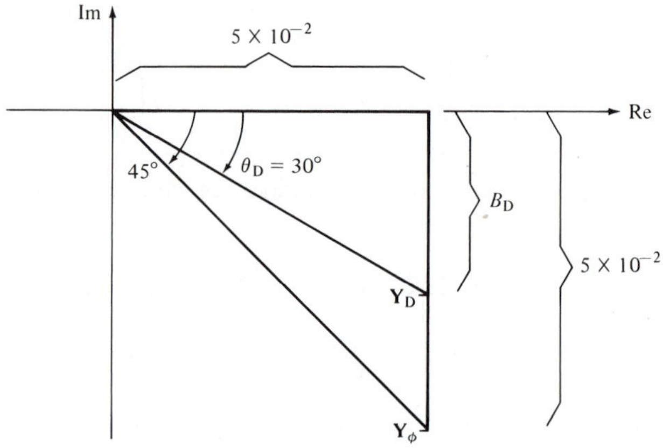

Fig. 3.2.3 The Admittance Plot for an Isolated Phase of the Y-connected Load Exhibited in Fig. 3.2.1

Therefore,

$$-5\times10^{-2}+B_?=-2.89\times10^{-2}$$

and

$$\underline{B_?=2.11\times10^{-2}\text{ mho}}$$

Thus a *capacitive susceptance* of 2.11×10^{-2} mho should be *added in parallel with each arm of the load.* Since $B_?=\omega_E C$ and $B_?=2.11\times10^{-2}$,

$$C=\frac{2.11\times10^{-2}}{10^3}=\underline{\underline{0.211\times10^{-6}\text{ F}}}$$

6. Since $\mathbf{S}_{LT}=3\mathbf{S}_\phi=3\mathbf{E}_\phi\mathbf{I}_\phi{}^*$ and the admittance of each arm has changed, the current $\mathbf{I}_\phi$ has changed. It is now different than the original current, that is, $\mathbf{I}_\phi=4.9e^{-j75°}$, and so on. To find the *new* current $\mathbf{I}_{\phi D}$, use

$$\mathbf{I}_{\phi D}=\mathbf{E}_\phi\mathbf{Y}_D$$

Since

$$\mathbf{Y}_D=5\times10^{-2}-j2.89\times10^{-2}=5.78\times10^{-2}e^{-j30°}$$

$$\mathbf{I}_{\phi D}=(69.3e^{-j30°})(5.78\times10^{-2}e^{-j30°})=\underline{4e^{-j60°}}\text{ A}$$

Therefore,

$$\mathbf{S}_{DT}=3\mathbf{S}_{\phi D}=3\mathbf{E}_\phi\mathbf{I}_{\phi D}{}^*=3(69.3e^{-j30°})(4e^{+j60°})=\underline{\underline{832e^{j30°}}}\text{ V-A}$$

Note: compare the two powers $\mathbf{S}_{LT}$ and $\mathbf{S}_{DI}$. Why are they different?

7. By proper power-factor correction, the reactive power, or part of it, required by the load for satisfactory operation is *supplied by the parallel capacitors* connected across the loads. Thus *the burden of handling this power by the generator and the transmission system is minimized.*

To conclude, the following remarks should be made about *unbalanced Y- and Δ-connected loads* (defined as Y- or Δ-connected networks with *unequal* arm impedances). First, the solution of unbalanced Δ-connected loads consists of computing the phase currents that are responses to balanced impressed line-to-line voltages and then applying Kirchhoff's current law to the load junction points, one at a time, to obtain three current equations that can be solved for the three line currents. Since the phase currents **are not** of equal magnitude and **are not** displaced from each other by 120°, the line currents *are not* equal in magnitude and have phase-angle differences that *differ* from 120°. Second, *in a four-wire, unbalanced, Y-connected load, the neutral conductor* **carries a nonzero current**; that is, the sum of the phase currents, and therefore the line currents, which are responses to balanced line-to-neutral voltages of a Y-connected generator, is nonzero. The phase (line) currents are of unequal magnitudes and do not have a phase difference of 120°. Third, in a three-wire, unbalanced, Y-connected load, with terminals A, B, and C, Fig. 3.21, the phase (line) currents, which are responses to the *unbalanced* voltages $\mathbf{E}_{A\nu}$, $\mathbf{E}_{B\nu}$, and $\mathbf{E}_{C\nu}$, Fig. 3.22, are of **unequal** magnitude and *do not have* a phase difference of 120°. The common point ν of the unbalanced, three-wire, Y-connected network (load) is not at the potential of the neutral $n(N)$ of the balanced Y-connected generator. Thus the voltages across the arms of the impedances $\mathbf{Z}_A \equiv 1/\mathbf{Y}_A$, $\mathbf{Z}_B \equiv 1/\mathbf{Y}_B$, and $\mathbf{Z}_C \equiv 1/\mathbf{Y}_C$ vary from the line-to-neutral voltages of the balanced Y-connected generator by an amount equal to the **displacement neutral voltage** $\mathbf{E}_{\nu n}$. For these impedances and the balanced Y-connected generator with phase (line-to-neutral) voltages $\mathbf{E}_a$, $\mathbf{E}_b$, and $\mathbf{E}_c$, the displacement neutral voltage $\mathbf{E}_{\nu n}$ is found from Fig. 3.21, with the aid of Fig. 3.22, by utilizing the node-voltage method (NVM), which yields

$$\mathbf{E}_{\nu n} = \frac{\mathbf{E}_a\mathbf{Y}_A + \mathbf{E}_b\mathbf{Y}_B + \mathbf{E}_c\mathbf{Y}_C}{\mathbf{Y}_A + \mathbf{Y}_B + \mathbf{Y}_C} \tag{3.125}$$

The displacement neutral voltage $\mathbf{E}_{\nu n}$ enables calculation of the voltages $\mathbf{E}_{A\nu}$, $\mathbf{E}_{B\nu}$, and $\mathbf{E}_{C\nu}$ as follows:

$$\mathbf{E}_{A\nu} = \mathbf{E}_a - \mathbf{E}_{\nu n} \tag{3.126}$$
$$\mathbf{E}_{B\nu} = \mathbf{E}_b - \mathbf{E}_{\nu n} \tag{3.127}$$
$$\mathbf{E}_{C\nu} = \mathbf{E}_c - \mathbf{E}_{\nu n} \tag{3.128}$$

These voltages in turn enable the calculation of the load phase (line) currents.

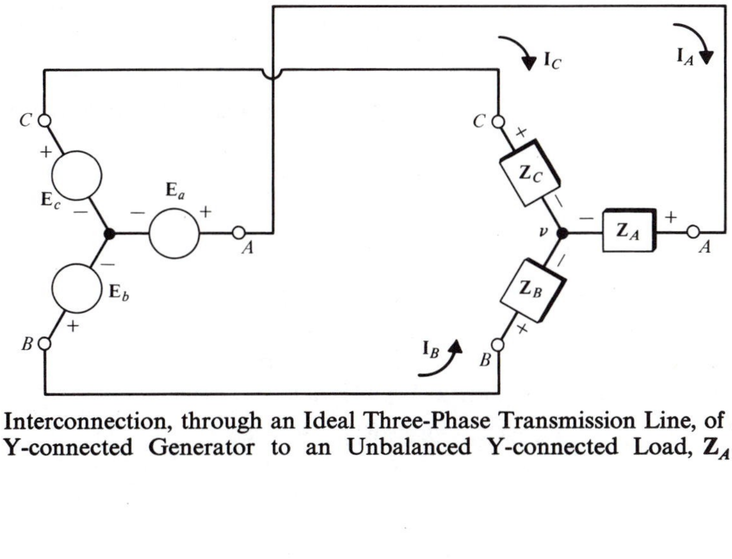

Fig. 3.21 Interconnection, through an Ideal Three-Phase Transmission Line, of a Balanced Y-connected Generator to an Unbalanced Y-connected Load, $\mathbf{Z}_A \neq \mathbf{Z}_B \neq \mathbf{Z}_C$

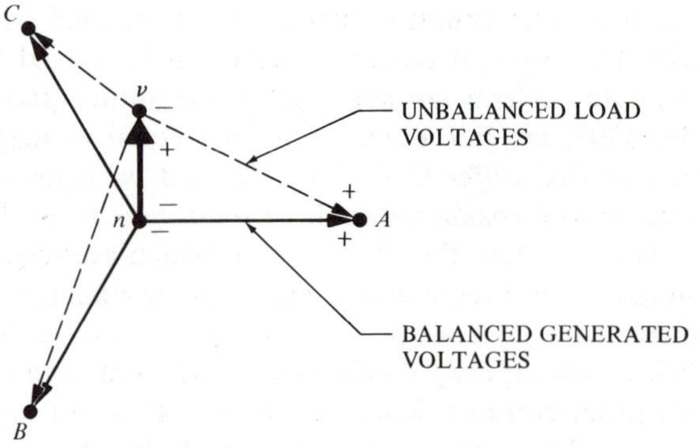

Fig. 3.22 Phasor Diagram Showing the Voltages for the Y-connected Units of Fig. 3.21

Example 3-3

For the network of Fig. 3.21 assume that $e_a(t) = \sqrt{2}\,(100)\cos\omega_E t$, $e_b(t) = \sqrt{2}\,(100)\cos(\omega_E t - 120°)$, and $e_c(t) = \sqrt{2}\,(100)\cos(\omega_E t - 240°)$, and $\mathbf{Z}_A = 5 + j0$, $\mathbf{Z}_B = 0 - j5$, and $\mathbf{Z}_C = 3 + j4$. Calculate the generated currents $i_A(t)$, $i_B(t)$, and $i_C(t)$.

Solution

If we write $\Sigma_{j=A}^{C}\mathbf{I}_j = 0$ for the load,

$$\mathbf{I}_A + \mathbf{I}_B + \mathbf{I}_C = 0 \tag{1}$$

where

$$\mathbf{I}_A = \frac{\mathbf{E}_{Av}}{\mathbf{Z}_A} \tag{2}$$

$\mathbf{E}_{Av}$ is calculated, through the utilization of Fig. 3.22, from the following equation:

$$\mathbf{E}_a - \mathbf{E}_{Av} - \mathbf{E}_{vn} = 0 \tag{3}$$

or

$$\mathbf{E}_{Av} = \mathbf{E}_a - \mathbf{E}_{vn} \tag{4}$$

Substitution of Eq. (4) into Eq. (2) yields

$$\mathbf{I}_A = \frac{\mathbf{E}_a - \mathbf{E}_{vn}}{\mathbf{Z}_A} \tag{5}$$

If we follow a similar procedure, we are able to write similar equations for the currents $\mathbf{I}_B$ and $\mathbf{I}_C$. That is,

$$\mathbf{I}_B = \frac{\mathbf{E}_{Bv}}{\mathbf{Z}_B} = \frac{\mathbf{E}_b - \mathbf{E}_{vn}}{\mathbf{Z}_B} \tag{6}$$

and

$$\mathbf{I}_C = \frac{\mathbf{E}_{Cv}}{\mathbf{Z}_C} = \frac{\mathbf{E}_c - \mathbf{E}_{vn}}{\mathbf{Z}_C} \tag{7}$$

Substitution of Eqs. (6) and (7) into Eq. (1) yields

$$\frac{\mathbf{E}_a - \mathbf{E}_{vn}}{\mathbf{Z}_A} + \frac{\mathbf{E}_b - \mathbf{E}_{vn}}{\mathbf{Z}_B} + \frac{\mathbf{E}_c - \mathbf{E}_{vn}}{\mathbf{Z}_C} = 0 \tag{8}$$

or

$$\mathbf{Y}_A(\mathbf{E}_a - \mathbf{E}_{vn}) + \mathbf{Y}_B(\mathbf{E}_b - \mathbf{E}_{vn}) + \mathbf{Y}_C(\mathbf{E}_c - \mathbf{E}_{vn}) = 0 \tag{9}$$

Proper multiplication of the terms in parentheses, separation of the terms containing the voltage $\mathbf{E}_{vn}$, and solution of the resulting equation for $\mathbf{E}_{vn}$ yield

$$\mathbf{E}_{vn} = \frac{\mathbf{Y}_A\mathbf{E}_a + \mathbf{Y}_B\mathbf{E}_b + \mathbf{Y}_C\mathbf{E}_c}{\mathbf{Y}_A + \mathbf{Y}_B + \mathbf{Y}_C} = \underline{142.48e^{j3.40^\circ}\ \mathrm{V}} \tag{10}$$

Before you proceed, note that Eq. (10) is identical with Eq. (3.125), and Eq. (4) is identical with Eq. (3.126). Therefore, it should be deduced that the numerators of Eqs. (6) and (7) are the terms exhibited in Eqs. (3.127) and (3.128), correspondingly.

Proper substitution from the input data and Eq. (10) into Eqs. (3.126) [or Eq. (4)], (3.127), and (3.128) yields

$$\begin{aligned}\mathbf{E}_{Av} &= 100e^{j0^\circ} - 142.48e^{j3.40^\circ} = (100 + j0) - (142.23 + j8.45)\\ &= -42.23 - j8.45 = \underline{43.07e^{-j168.68^\circ}}\ \mathrm{V}\end{aligned} \tag{11}$$

$$\begin{aligned}\mathbf{E}_{Bv} &= 100e^{-j120^\circ} - 142.48e^{j3.40^\circ} = (-50 - j86.60) - (142.23 + j8.45)\\ &= -192.23 - j95.05 = \underline{214.45e^{j153.69^\circ}}\ \mathrm{V}\end{aligned} \tag{12}$$

$$\begin{aligned}\mathbf{E}_{Cv} &= 100e^{-240^\circ} - 142.48e^{j3.40^\circ} = (-50 + j86.60) - (142.23 + j8.45)\\ &= -192.23 + j78.15 = \underline{207.51e^{-j202.12^\circ}}\ \mathrm{V}\end{aligned} \tag{13}$$

Division of the load phase voltages $\mathbf{E}_{Av}$, $\mathbf{E}_{Bv}$, and $\mathbf{E}_{Cv}$ by the corresponding impedances $\mathbf{Z}_A = 5e^{j0^\circ}$, $\mathbf{Z}_B = 5e^{-j90^\circ}$, and $\mathbf{Z}_C = 5e^{j53.13^\circ}$, given as input data, yields

$$\mathbf{I}_A = \frac{\mathbf{E}_{Av}}{\mathbf{Z}_A} = \frac{43.07e^{-j168.68}}{5e^{j0^\circ}} = \underline{\underline{8.61e^{-j168.68^\circ}}}\ \mathrm{A} \tag{14}$$

$$\mathbf{I}_B = \frac{\mathbf{E}_{Bv}}{\mathbf{Z}_B} = \frac{214.45e^{-j153.69^\circ}}{5e^{-j90^\circ}} = \underline{\underline{42.89e^{-j63.69^\circ}}}\ \mathrm{A} \tag{15}$$

$$\mathbf{I}_C = \frac{\mathbf{E}_{Cv}}{\mathbf{Z}_C} = \frac{207.51e^{-j202.12^\circ}}{5e^{j53.13^\circ}} = \underline{\underline{41.50e^{-j255.25^\circ}}}\ \mathrm{A} \tag{16}$$

Finally, the theory in Appendix A and the results of Eqs. (14), (15), and (16) enable us to write the currents $i_A(t)$, $i_B(t)$, and $i_C(t)$ as

$$i_A(t) = \underline{\underline{\sqrt{2}\,(8.61)\cos(\omega_E t - 168.73^\circ)}}\ \mathrm{A} \tag{17}$$

$$i_B(t) = \underline{\underline{\sqrt{2}\,(42.89)\cos(\omega_E t - 63.69^\circ)}}\ \mathrm{A} \tag{18}$$

$$i_C(t) = \underline{\underline{\sqrt{2}\,(41.50)\cos(\omega_E t - 255.26^\circ)}}\ \mathrm{A} \tag{19}$$

If the transmission line used for the analysis is not ideal, then the line is represented by one of the equivalent circuits of Figs. 3.13, 3.14, and 3.15, and the solution of a given problem becomes much more complex than the three examples presented in this section, in which the transmission line is assumed to be ideal. For Y-to-Y-, Δ-to-Δ-, Y-to-Δ-, or Δ-to-Y-connected systems with **nonideal lines**, the loop- (mesh-) current method (LCM), the node voltage (NVM) and Thevenin's or Norton's theorems become essential, and the *balanced* or *unbalanced* nature of the generator, the load, the transmission line, any combination of two, or all three of

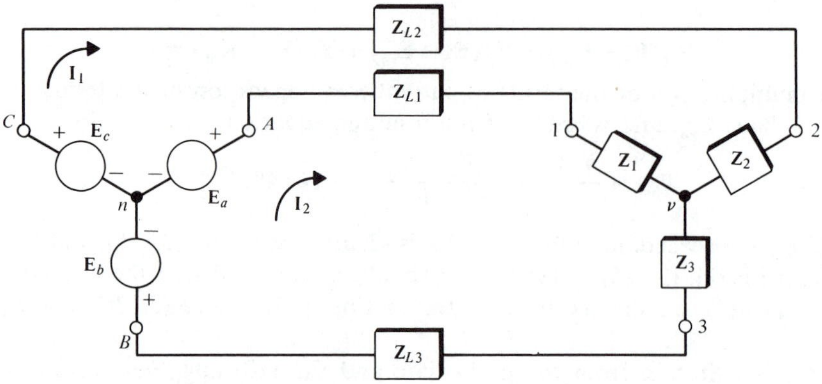

Fig. 3.23 A Generalized Y-connected, Three-Phase, Three-Wire Power System, Short Three-Phase Transmission Line

these networks (i.e., subsystems) are *irrelevant*. Thus for the **generalized** *Y-connected system* of Fig. 3.23, the LCM equations can be written as

$$[\mathbf{Z}_{L2}+\mathbf{Z}_2+\mathbf{Z}_1+\mathbf{Z}_{L1}]\mathbf{I}_1-[\mathbf{Z}_{L1}+\mathbf{Z}_1]\mathbf{I}_2=\mathbf{E}_c-\mathbf{E}_a \tag{3.129}$$

$$-[\mathbf{Z}_1+\mathbf{Z}_{L1}]\mathbf{I}_1+[\mathbf{Z}_{L1}+\mathbf{Z}_1+\mathbf{Z}_3+\mathbf{Z}_{L3}]\mathbf{I}_2=\mathbf{E}_a-\mathbf{E}_b \tag{3.130}$$

and the NVM equations can be written as

$$\mathbf{V}_{vn}\left[\frac{1}{\mathbf{Z}_1+\mathbf{Z}_{L1}}+\frac{1}{\mathbf{Z}_2+\mathbf{Z}_{L2}}+\frac{1}{\mathbf{Z}_3+\mathbf{Z}_{L3}}\right]=\frac{\mathbf{E}_a}{\mathbf{Z}_1+\mathbf{Z}_{L1}}+\frac{\mathbf{E}_b}{\mathbf{Z}_2+\mathbf{Z}_{L2}}+\frac{\mathbf{E}_c}{\mathbf{Z}_3+\mathbf{Z}_{L3}} \tag{3.131}$$

The LCM equations, $\mathbf{Z}_{jk}\mathbf{I}_k=\mathbf{E}_j$, $j=k=1,2,\ldots,m$, and the NVM equations, $\mathbf{Y}_{jk}E_k=\mathbf{I}_j$, $j=k=1,2,\ldots,m$, where m is the order of the given system, can be solved for the corresponding response currents or voltages by digital computer techniques, or determinant or simultaneous equations methods. The methods apply whether the voltages or the impedances are balanced or unbalanced. Since the stated purpose of this text is not that of systems analysis, the subject is not pursued further here.

3.11 SUMMARY

This chapter develops the fundamental concepts of polyphase alternating voltage generation, distribution, and utilization. The principal concerns of this chapter are the balanced, three-phase, Y- and Δ-connected, ideal voltage generator, the balanced, three-phase transmission line, and the balanced, three-phase, Y- and Δ-connected load.

The current sheet concept is used to derive the sinusoidally distributed field in the rotor of a smooth airgap device, which in turn is used to induce alternating voltages to three windings distributed in stator slots and connected in series in a manner such that their voltages are of equal magnitude and exhibit a phase angle of 120°. These voltages are impressed to a balanced three-phase transmission line, which transfers the energy to a balanced, three-phase, Y- or Δ-connected load for utilization.

The equivalent circuits for various physical lengths of transmission lines are obtained from the generalized equivalent circuit of a distributed parameter infinitesimal length of a transmission line. If we were to derive the equivalent circuits presented here and expand on the theory of transmission lines, we would have to use rather extensively the following concepts: (1) velocity of propagation ν (meters per second) of the energy along the line, (2) the wavelength λ (meters per cycle) of the transmission line, (3) the frequency f_E (hertz) of the energy transmitted along the line, (4) the characteristic impedance $\mathbf{Z}_C$ (ohms), and (5) the propagation constant γ (meters^{-1}) of the transmission line.

Also, the methods for measuring the power absorbed by a network (by utilizing three wattmeters and two wattmeters) and the power-factor angle of a network (by the two-wattmeter method) are described in detail.

The concept of the per-phase analysis (involving specific relationships among the *phase* and *line* voltages and currents) for the Y-connected ideal voltage generator, connected, with ideal transmission lines, to Y- and Δ-connected loads, is reinforced with a number of numerical examples.

Finally, we present a small exposé of the properties that unbalanced, three-phase, Y- and Δ-connected loads exhibit when connected to balanced Y- or Δ-connected generators.

It is recommended that the student master the material of this chapter as it constitutes the foundations for power systems analysis and, in particular, the analysis of induction and synchronous devices.

PROBLEMS

3.1 For the Y-connected ideal voltage generator exhibited in Example 3.1, the a–b–c sequence line-to-line voltages are $e_{AB}(t) = \sqrt{2}\,(250)\cos\omega_E t$, $e_{BC}(t) = \sqrt{2}\,(250)\cos(\omega_E t - 120°)$, and $e_{CA}(t) = \sqrt{2}\,(250)\cos(\omega_E t - 240°)$, $\omega_E = 2000$, and the Δ-connected load parameters are $R = 60$ and $L = 0.04$, all in MKS units. Calculate: (1) the phase currents $\mathbf{I}_a$, $\mathbf{I}_b$, and $\mathbf{I}_c$, (2) the line currents $\mathbf{I}_\alpha$, $\mathbf{I}_\beta$, and $\mathbf{I}_\gamma$, (3) the complex power demanded by the load, $\mathbf{S}_{LT}$, (4) the power dissipated in the load, P_{LT}, (5) the power stored in the field, Q_{LT}, (6) the time-domain phase currents $i_\phi(t)$, (7) the time-domain voltage drop across the resistor, $v_R(t)$, (8) the time-domain voltage drop across the inductor, $v_L(t)$, (9) the time-domain line currents $i_L(t)$, (10) the time-domain real power $p_{LT}(t)$, and (11) the time-domain imaginary power $q_{LT}(t)$.

3.2 For the Y-connected ideal voltage generator exhibited in Example 3.2, the a–b–c sequence line-to-line voltages are $e_{AB}(t) = \sqrt{2}\,(250)\cos\omega_E t$, $e_{BC}(t) = \sqrt{2}\,(250)\cos(\omega_E t - 120°)$, and $e_{CA}(t) = \sqrt{2}\,(250)\cos(\omega_E t - 240°)$, $\omega_E = 2000$, and the Y-connected load has parameters $R = 20$ and $L = 0.04$, all in MKS units. Calculate: (1) the phase currents $\mathbf{I}_a$, $\mathbf{I}_b$, and $\mathbf{I}_c$, (2) the line currents $\mathbf{I}_\alpha$, $\mathbf{I}_\beta$, and $\mathbf{I}_\gamma$, (3) the original (i.e., at the original power factor) total power demanded by the load from the generator, $\mathbf{S}_{LT}$, and (4) the power factor of the load; (5) utilize parallel connection and correct the system for an operating factor of 0.90 lag; (6) calculate the final power demanded by the load now, $\mathbf{S}_{DT}$, that is, at the new power factor; (7) discuss the advantages of power-factor correction.

3.3 For the circuit diagram of Problem 3.2, assume that the neutrals are grounded and $R_B = L_C = 0$. Now: (1) calculate the time-domain currents $i_A(t)$, $i_B(t)$, and

$i_C(t)$; (2) sketch the phasor diagram of these currents; (3) calculate the current $\mathbf{I}_N$ that exists in the wire that connects the neutrals to ground; (4) answer the following three questions: (a) Do these currents constitute a balanced or unbalanced set of variables? (b) Is the sequence of these currents positive or negative? (c) Does it matter whether the ground wires are present or not?

3.4 Repeat problem 3.3. However, assume that the two neutrals are ***not grounded***, or ***interconnected***, make the appropriate adjustments in your calculations, and answer the relevant questions.

3.5 For the network of Fig. 3.21 assume that $e_a(t)=\sqrt{2}\,(120)\cos\omega_E t$, $e_b(t)=\sqrt{2}\,(120)\cos(\omega_E t-120°)$, and $e_c(t)=\sqrt{2}\,(120)\cos(\omega_E t-240°)$, and $\mathbf{Z}_A=10+j5$, $\mathbf{Z}_B=5-j10$, and $\mathbf{Z}_C=10+j10$. Calculate: (1) the generated currents $i_A(t)$, $i_B(t)$, and $i_C(t)$; (2) the response voltages $v_A(t)$, $v_B(t)$, and $v_C(t)$; and (3) the complex power components $\mathbf{S}_A$, $\mathbf{S}_B$, and $\mathbf{S}_C$ delivered to the loading impedances $\mathbf{Z}_A$, $\mathbf{Z}_B$, and $\mathbf{Z}_C$.

REFERENCES

1. Boast, W. B., *Principles of Electric and Magnetic Circuits*. Harper & Row, New York, 1950.
2. Del Toro, V., *Electromechanical Devices for Energy Conversion and Control Systems*. Prentice-Hall, Englewood Cliffs, N.J. 1968.
3. Elgerd, O. I., *Electric Energy Systems Theory: An Introduction*. McGraw-Hill, New York, 1971.
4. Fitzgerald, A. E., Higginbotham, D. E., and Grabel, A., *Basic Electrical Engineering*. McGraw-Hill, New York, 1967.
5. Hayt, W. H., Jr., *Engineering Electromagnetics*. McGraw-Hill, New York, 1958.
6. Koehler, G., *Circuits and Networks*. MacMillan, New York, 1955.
7. Majmudar, H., *Electromechanical Energy Converters*. Allyn & Bacon, Boston, 1965.
8. Schmitz, N. L., and Novotny, D. W., *Introductory Electromagnetics*. Ronald Press, New York, 1965.
9. Stevenson, W. D., Jr., *Elements of Power System Analysis*, 2nd ed. McGraw-Hill, New York, 1962.
10. Thomson, H. A., *Alternating Current and Transient Analysis*. McGraw-Hill, New York, 1962.

4 INDUCTION ENERGY CONVERTERS

4.1 INTRODUCTION

An induction device is a rotational electromechanical energy converter that structurally is comprised of a stationary member, **the stator**, with smooth inner surface, and a rotating member, **the rotor**, with smooth outer surface. Both members, for analysis purposes, have sinusoidally distributed three-phase windings with identical resistance, inductance, and capacitance in each phase. The stator windings are *often* excited with a set of balanced three-phase voltages drawn from an external source of electrical energy. The response currents in the stator winding produce a constant-amplitude magnetic field, which rotates at synchronous speed ω_S radians/second in the airgap of the device. This field induces sinusoidal currents in the rotor windings, which in turn produce a second constant-amplitude magnetic field also rotating at a constant speed, proportional to that of the stator field ω_S, in the airgap of the device. The interaction between the airgap stator and rotor fields, which are stationary with respect to each other, creates a torque. This torque may occur in the direction of rotation (motor action) or it may oppose the rotation (generator action). The rotor speed ω_R radians/second, is always lower than the synchronous speed ω_S.

The induction device, whether a three-phase or one of the simpler two-types (two- and single-phase) also available, is noted for its highly desirable torque versus speed characteristic. Therefore, from the day of its inception, the induction device has been used as a motor. The three-phase induction motor has been, and still is, the workhorse of the industrial process throughout the world; the two-phase induction motor is widely used in control systems; and the single-phase induction motor is encountered widely in offices, workshops, and house appliances.

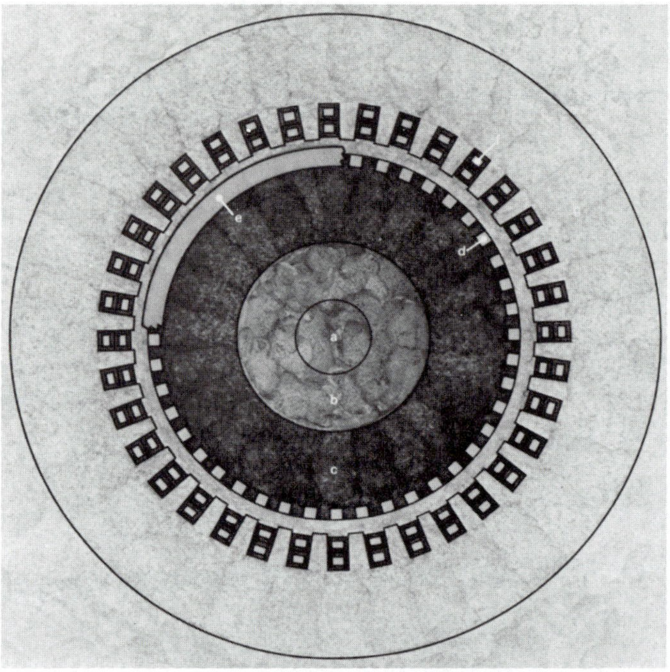

Plate 4.1 Cross Section of an Induction Device: Shaft, Spiker, Rotor-Core Laminations, Rotor Bar, Resistance Ring, Stator Coils, and Stator-Core Laminations. (Courtesy of Westinghouse Electric Co., Pittsburgh, Pa.)

In 1957 the invention of the thyristor* led to the modification of the torque versus speed characteristic of the three-phase induction motor to the point where it, in conjunction with external feedback and thyristor-controlled circuitry, has become the popular choice as a **variable-speed drive**, known as **ac drive**. The lack of brushes in the induction motor makes it an excellent candidate for applications requiring a minimum of routine maintenance and inspection. This asset makes the ac drive highly desirable in difficult environments, such as explosive or radioactive atmospheres and underwater applications.

Thus the objectives of this chapter are sevenfold: (1) to set forth the revolving field theory, which is absolutely essential in ac machine theory, (2) to present the principles of operation of the three-, two-, and single-phase induction devices, (3) to develop mathematical models that describe the steady state performance of the induction devices, (4) to exhibit the terminal characteristics of these devices, (5) to present and expand on the available methods of speed control of the three-phase (mainly) and two-phase induction motors, (6) to present the procedure for the measurement of the equivalent circuit parameters of the three-phase induction device, and (7) to show the reader the most essential applications of the induction devices.

*See Appendix E for pertinent information on this subject.

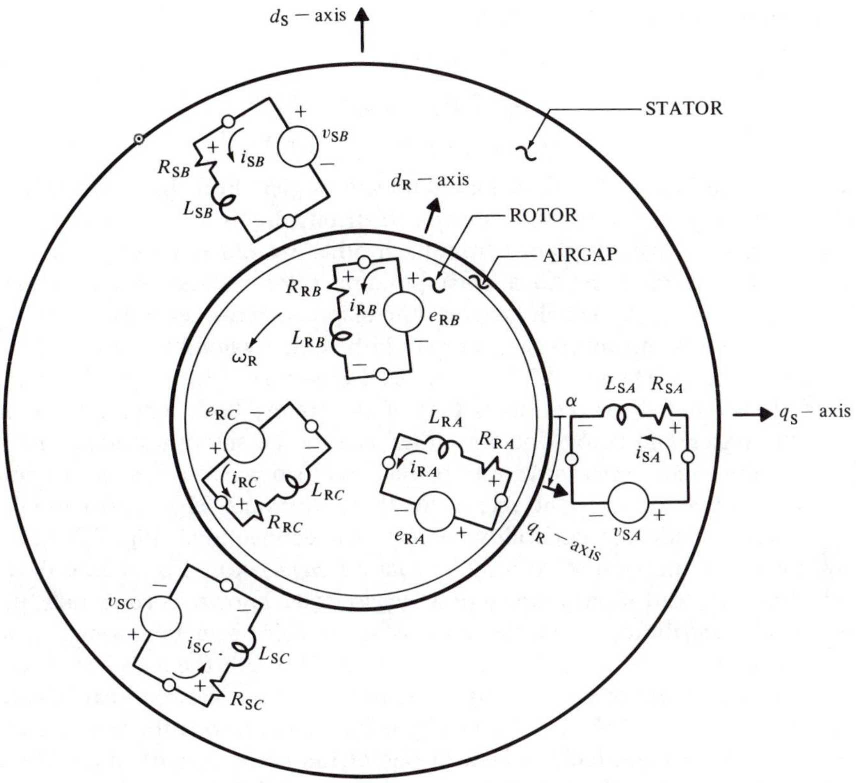

Fig. 4.1 Three-Phase Induction Device, Motor Mode of Operation

4.2 DEVELOPMENT OF THE ROTATING FIELD IN THREE-PHASE, ac BALANCED SYSTEMS

From the analysis point of view, a three-phase induction device is defined as an electromechanical energy converter, Fig. 4.1, which has the following properties. First, it has two or more, but always an even number of, *magnetic poles*, designated by the letter P, in the stator and rotor structures. Second, it has three balanced, sinusoidally distributed, Y- or Δ-connected **stator windings** (balanced windings have their axes displaced from each other by 120 electrical degrees* and have equal resistance R, inductance L, and capacitance C), which are excited from an external, balanced, three-phase *excitation system* defined as $e_{SA}(t) = \sqrt{2}\,|\mathbf{E}_S|\cos[\omega_E t - \alpha]$, $e_{SB}(t) = \sqrt{2}\,|\mathbf{E}_S|\cos[\omega_E t - (\alpha + 120°)]$, and $e_{SC}(t) = \sqrt{2}\,|\mathbf{E}_S|\cos[\omega_E t - (\alpha + 240°)]$ (where ω_E is the radian frequency of excitation in radians/second, $|\mathbf{E}_S|$ is the RMS value of each impressed voltage, and α is its phase angle at $t = 0$), producing three

*It is explained clearly in the text following Eq. (4.18) that the electrical degrees θ_E of a P-pole stator winding are related to its mechanical degrees θ_M by the equation $\theta_M = \theta_E/(P/2)$. Thus, for a two-pole device, $\theta_M \equiv \theta_E$.

balanced *response currents* described as

$$i_{SA}(t) = \sqrt{2}\,|\mathbf{I}_S|\cos\omega_E t \tag{4.1}$$

$$i_{SB}(t) = \sqrt{2}\,|\mathbf{I}_S|\cos(\omega_E t - 120°) \tag{4.2}$$

$$i_{SC}(t) = \sqrt{2}\,|\mathbf{I}_S|\cos(\omega_E t - 240°) \tag{4.3}$$

in the stator windings.* Third, it has a *smooth airgap*. Fourth, it has three (for analysis purposes) balanced, sinusoidally distributed, Y- or Δ-connected **rotor windings**, with their axes displaced from each other by 120 electrical degrees† and in general, each terminating to a corresponding **rotor voltage source** defined as $e_{RA}(t)$, $e_{RB}(t)$, and $e_{RC}(t)$, which generate the response **rotor currents** $i_{RA}(t)$, $i_{RB}(t)$, and $i_{RC}(t)$. In reality, however, the rotor could be either **wound Y-connected** (rarely Δ-connected), Fig. 4.2(a) and 4.2(b), or **squirrel-cage-type**, Fig. 4.2(c). The winding layout of the wound rotor resembles that of the stator; both windings are distributed in the stator and rotor slots on either side of the smooth airgap. The rotor windings of the *wound rotor* have one of their coil sides connected to a common point and the other connected to one of the three slip rings, Figs. 4.2(a) and 4.2(b). The slip rings provide the capability for external connections, Fig. 4.2(b), to the rotor windings of the wound rotor. The *squirrel cage* rotor, Fig. 4.2(c), does not have windings. Instead it has *aluminum* or *copper* rods, known as **rotor rods**, placed in slots evenly distributed along the surface of the rotor structure, which is made from laminated sheets of a highly permeable steel alloy. All the rods are shorted at both ends by aluminum or copper rings. The rotor rods are skewed, that is, spiraled at an angle of *one slot pitch*, as seen in Fig. 4.2(c). The reasons for skewing are (1) to prevent the rotor from locking in with one of the harmonics of the field during starting and thus never achieving the running speed of the motor and (2) to reduce torque variations.

The squirrel cage rotor is **superior** to the wound rotor in that it is simple in construction, is rugged, and is interchangeable between three-, two-, and single-phase devices of similar geometric dimensions, as it has the property to reflect the number of the poles of the stator with which it is associated. However, it is **inferior** to the wound rotor in that it has no capability for external connections to it (the necessity of capability for external connections and its usefulness will become clear as you progress in this chapter). The stator winding is always Y-connected (rarely Δ-connected) with one of the coil sides connected to a common point, Figs. 4.2(a) and 4.2(b), and the other connected to the corresponding phase of a three-phase balanced excitation.

The capability of winding the stator of the three-phase (two- or single-phase) induction device to produce a desired number of magnetic poles P is a very important concept, but a very difficult one to perceive. The following explanation, however, is clear and straightforward, and can be mastered by students if they make a reasonable effort.

In Section 3.2 it is demonstrated how we can, in a two-pole device, generate three fields, the axes of which are 120 electrical degrees from each other. Note that

*Note that Fig. 4.1 exhibits equivalent concentrated circuit analogs of the three stator and the three rotor windings, which, in reality, are distributed correspondingly in slots on the surface of the stator and rotor structures, for example, Figs. 4.2(a), 4.3(a), 4.3(d) through 4.3(h), and 4.25.

†The relationship between the electrical θ_E and mechanical θ_M degrees of the rotor winding is similar to that of the stator winding.

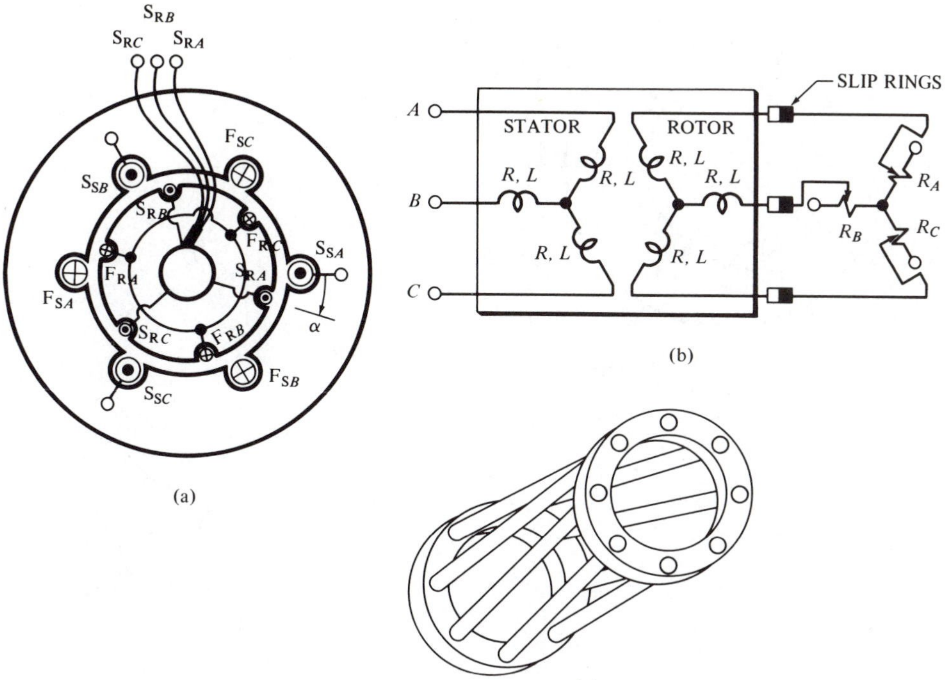

Fig. 4.2 Three-Phase Induction Motor: (a) Three-Phase Wound Rotor Within a Three-Phase Stator, (b) Schematic Diagram of the Three-Phase, Wound-Rotor Induction Motor with a Variable External Resistive Load, and (c) Skewed Rotor Bars and End Rings of a Squirrel-Cage-Rotor Induction Motor

for a two-pole device electrical degrees are equal to mechanical degrees. If, however, we were to design a three-phase, four-pole induction device, we would use the following *rule of thumb*: (1) 3 phases × 4 poles = 12 coil sides; (2) 12 coil sides/3 phases = 4 coil sides/phase ≡ 2⊕coil sides + 2⊙coil sides; (3) 360 mechanical degrees/4 coil sides/phase = 90 mechanical degrees/coil sides/phase. Thus in Fig. 4.3(a), phase *A* provides two coil sides with starting ends, designated as S_A and $S_{A'}$, and two coil sides with finishing ends, designated as F_A and $F_{A'}$. It should be obvious now that these four coil sides constitute two coils with their one end connected to the excitation and their other end connected to a common point for the Y-connected stator. Place the coils that are connected to the phase *A* of the excitation at 0 and 180 mechanical degrees, correspondingly. The other two coil sides that are connected to the common point are located at 90 and 270 mechanical degrees, correspondingly. For a *positive rotation* of the stator's field measure $\frac{2}{3}(90)$ mechanical degrees* counterclockwise from S_A and $S_{A'}$ and locate the coils that are connected to phase *B* of the excitation, S_B and $S_{B'}$, in a similar fashion; measure $\frac{2}{3}(90)$ mechanical degrees counterclockwise from S_B and $S_{B'}$ and locate the coils that are connected to phase *C* of the excitation, S_C and $S_{C'}$, in a similar fashion. Note that the finishing end of each coil leads its starting end by 90 mechanical degrees. Before marking the currents entering and leaving in each coil side, assume

*Two-thirds times the angular displacement between the coil sides of *A* and *A′* (i.e., the computed mechanical degrees/coil sides/phase) yields the angular displacement of the phase *B* with respect to *A*, and *C* with respect to *B*.

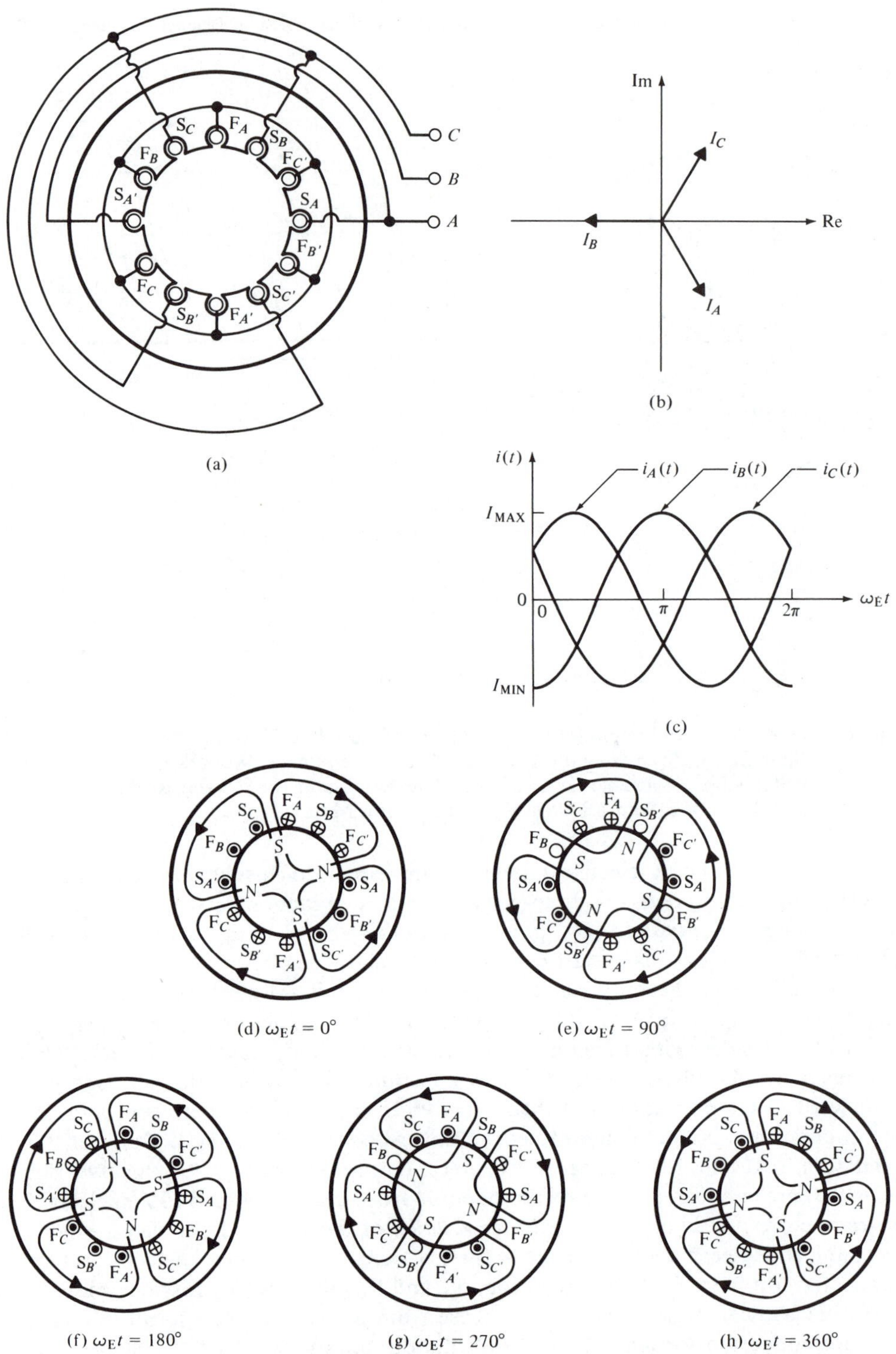

Fig. 4.3 Generation of a Four-Pole Rotating Field in the Stator of a Three-Phase Induction Motor: (a) Three-Phase Winding Distribution, (b) *ABC*-Sequence, Stator Phasor Currents, (c) *ABC*-Sequence, Stator *t*-Domain Currents, and (d) through (h) *t*-Domain Current Orientation in the Concentrated Three-Phase Stator Winding at $\omega_E t = 0°$, 90°, 180°, 270°, and 360° and Thus the Generation of Four Magnetic Poles Rotating Counterclockwise

that the forcing function currents are known, Figs. 4.3(b) and 4.3(c). Figs. 4.3(b) and 4.3(c) reveal that at $t=t_0$ the Re-axis projection of the currents $i_A(\omega_E t_0)$ and $i_C(\omega_E t_0)$ are *positive* (mark these currents as ⊙), and the Re-axis projection of the current $i_B(\omega_E t_0)$ is *negative* (mark this current as ⊕), Fig. 4.3(d). Continue this process for $\omega_E t = 90°, 180°, 270°, 360°, \ldots$, Figs. 4.3(e) through 4.3(h). Note the grouping of the coil sides in accordance with the current they carry and the magnetic field that these currents produce at $\omega_E t = 0°$, 90°, 180°, 270°, and 360°. As time progresses the currents change and the field position changes. In fact, this field is rotating *counterclockwise* with time, Figs. 4.3(d) through 4.3(h). Note that while the currents, Fig. 4.3(c), have gone through one complete cycle, that is, $\omega_E t = \theta_E = 360°$, each pole in Figs. 4.3(d) through 4.3(h) has advanced only half a revolution, that is, $\theta_M = 180°$. This *fact* substantiates the assertion that $\theta_M = \theta_E/(P/2)$.

Also note that for any induction device with a stator field of three phases and four magnetic poles, **two parallel paths of coils per phase** are required. How many **parallel paths of coils per-phase** would you say are required for a stator field of three-phase, six, eight, and so on, magnetic poles? [Answers: (1) $3 \times 6 = 18$; $18/3 = 6 = 3⊙ + 3⊕$,i.e., three parallel paths of coils per phase; (2) $3 \times 8 = 24$, $24/3 = 8 = 4⊙ + 4⊕$, i.e., four parallel paths of coils per phase. Each set of the coil sides (of the three-phase multipole induction device) is 120 electrical degrees [i.e., 2/3(180°) and $\theta_M = \theta_E/(P/2)$] apart, as are those of the three-phase, two-pole induction device.

Example 4.1

Design a three-phase, Y-connected, two-pole induction device and draw the field for $\omega_E t = 0°$, 90°, 180°, 270°, and 360°.

Solution

First, 3 phases×2 poles = 6 coil sides. Second, 6 coil sides/3 phases = 2 coil sides/phase, that is, 1 − ⊕coil side/phase and 1 − ⊙coil side per phase. Third, 360 mechanical degrees/2 coil sides/phase = 180 mechanical degrees/coil side/phase.

Thus phase A requires one coil side with its starting end designated as S_A and one coil side with its finishing end designated as F_A, Fig. 4.1.1. For a positive

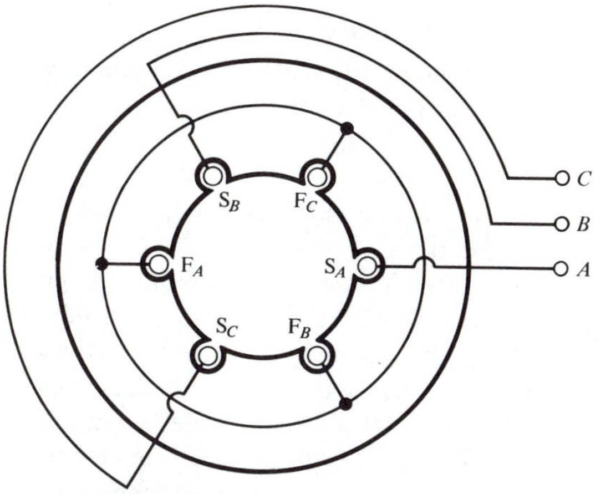

Fig. 4.1.1 Three-Phase Stator Winding Distribution for the Generation of Two Poles in the Stator of an Induction Device

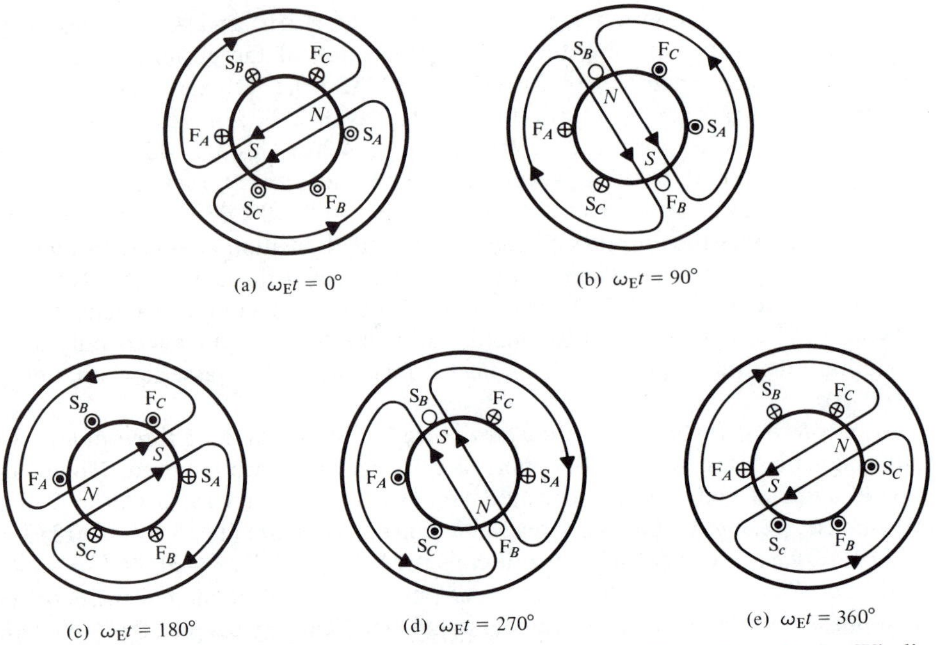

Fig. 4.1.2 t-Domain Current Orientation in the Concentrated Three-Phase Stator Winding at $\omega_E t=0°$, 90°, 180°, 270°, and 360° and Thus the Generation of Two Magnetic Poles Rotating Counterclockwise

sequence the coil sides are laid out in slots along the surface of the stator at 180-mechanical-degree intervals measured counterclockwise. The coil sides of phases B and C are laid out in a similar fashion but with their starting ends displaced from the starting end of coil A, that is, S_A, and from each other by $\frac{2}{3}(180)$ mechanical degrees. Designating positive currents as ⊙ and negative currents as ⊕, and utilizing Fig. 4.3(c), enables drawing the stator field as in Figs. 4.1.2(a) through 4.1.2(e). Note that the relative position of the field, and thus the position of the corresponding poles, is advancing in a counterclockwise direction as t increases.

Example 4.2

Design a three-phase, Y-connected, six-pole induction device and draw the field for $\omega_E t=0°$, 90°, 180°, 270°, and 360°.

Solution

First, 3 phases×6 poles = 18 coil sides. Second, 18 coil sides/3 phases = 6 coil sides/phase, that is, 3 – ⊕coil sides/phase and 3 – ⊙coil sides/phase. Third, 360 mechanical degrees/6 coil sides/phase = 60 mechanical degrees/coil side/phase.

Thus phase A requires three coil sides with starting ends designated as S_A, $S_{A'}$, and $S_{A''}$ and three coil sides with finishing ends designated as F_A, $F_{A'}$, and $F_{A''}$, Fig. 4.2.1. For a *positive rotation* of the stator's field the coil sides are laid out in slots along the surface of the stator at 60-mechanical-degree intervals measured counterclockwise. The coil sides of phases B and C are laid out in a similar fashion but with their starting ends displaced from the starting end of coils A, A', and A'' (i.e., S_A, $S_{A'}$, and $S_{A''}$) and from each other by $\frac{2}{3}(60)$ mechanical degrees. Designating

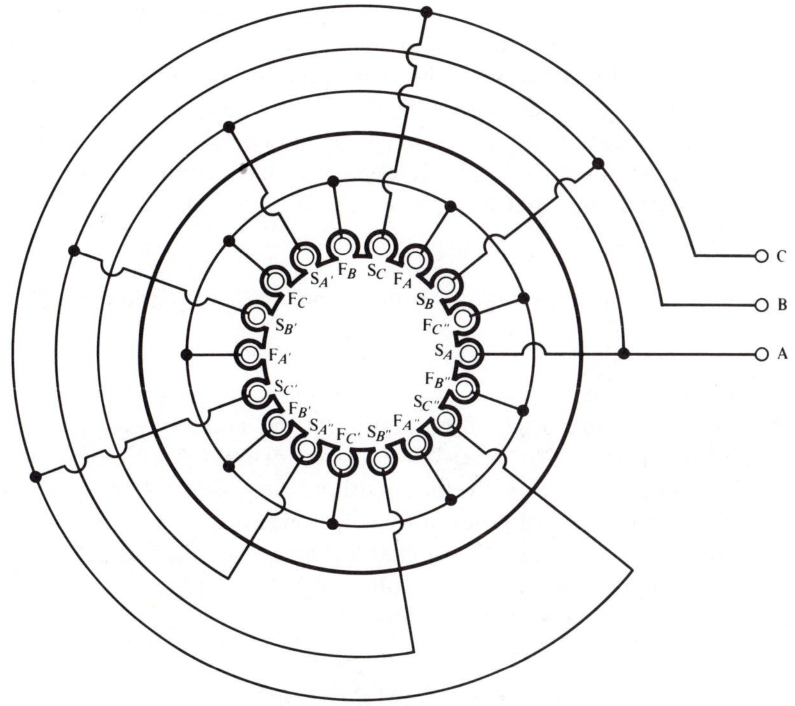

Fig. 4.2.1 Three-Phase Stator Winding Distribution for the Generation of Six Poles in the Stator of an Induction Device

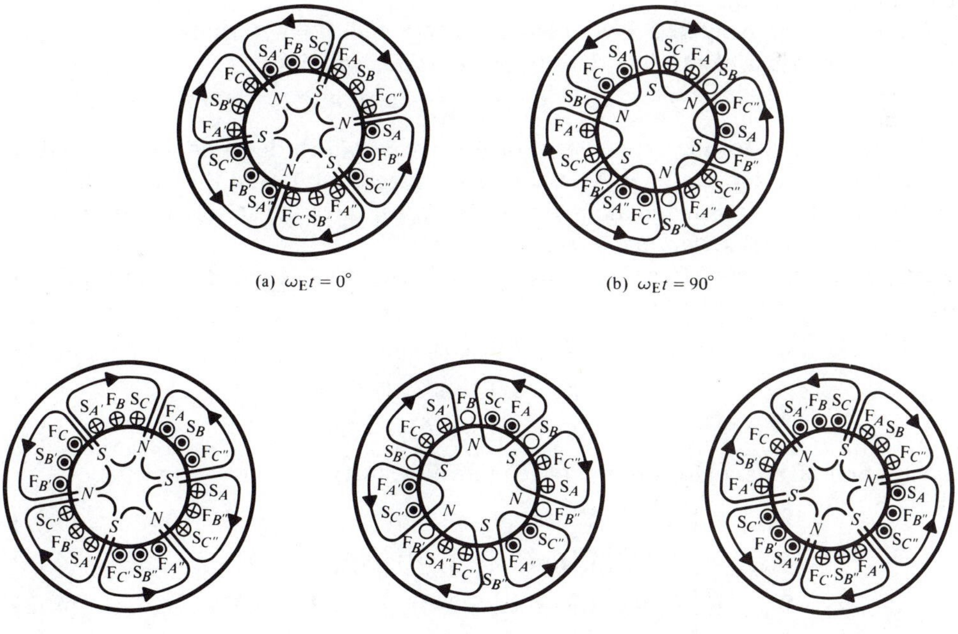

(a) $\omega_E t = 0°$

(b) $\omega_E t = 90°$

(c) $\omega_E t = 180°$

(d) $\omega_E t = 270°$

(e) $\omega_E t = 360°$

Fig. 4.2.2 *t*-Domain Current Orientation in the Concentrated Three-Phase Stator Winding at $\omega_E t = 0°$, 90°, 180°, 270°, and 360° and thus the Generation of Six Magnetic Poles Rotating Counterclockwise

positive currents as ⊙ and negative currents as ⊕, and utilizing Fig. 4.3(c), enables drawing the field as in Figs. 4.2.2(a) through 4.2.2(e). Note that the relative position of the fields and thus the position of the corresponding poles where t increases is advancing in a counterclockwise direction. If the field were plotted in smaller time intervals, we would find that the field sweeps in the counterclockwise direction at the synchronous speed dictated by the frequency of excitation and the number of poles of the stator structure, that is, $\omega_S = \omega_E/(P/2)$.

Example 4.3

What is the physical interpretation of the six-pole field in Example 4.2?

Solution

The six-pole stator field may be thought of as a six-pole natural magnet inserted in a canlike stator structure, Fig. 4.3.1, rotating at the synchronous speed dictated by the equation $\omega_S = \omega_E/(P/2)$. Note: When the stator-field pole count increases or decreases, the number of poles in the fictitious magnet of Fig. 4.3.1 changes correspondingly. The number of poles, however, is **always even.**

Utilization of the ideas of Eq. (3.14) enables writing of the spatial flux density equations for the two-pole (or single-pole pair), three-phase stator windings, *abbreviated* henceforth as 3Φ stator, as follows:*

$$\mathbf{B}_{WSA} = B_{MSA}\cos\beta_S\mathbf{u}_S \tag{4.4}$$
$$\mathbf{B}_{WSB} = B_{MSB}\cos(\beta_S - 120°)\mathbf{u}_S \tag{4.5}$$
$$\mathbf{B}_{WSC} = B_{MSC}\cos(\beta_S - 240°)\mathbf{u}_S \tag{4.6}$$

where β_S is the angular displacement of the stator structure measured positively from the q_S-axis, Fig. 4.1.

Proper use of Ampere's law, $\oint \mathbf{H}\cdot d\mathbf{L} = Ni(t)$ and $\mathbf{B} = \mu\mathbf{H}$ enables writing the magnitude maxima of the flux density of the stator windings as

$$B_{MSj} = K_S\Phi[i_S(t)], \quad j = A, B, C \tag{4.7}$$

where K_S is the stator design constant of the device, which depends on the number of turns of the stator windings N_S, the path of integration ℓ, and the permeability μ (μ_0 and μ_r) of the ferromagnetic material that constitutes the stator structure. Note that in this analysis *saturation is ignored.*

Proper utilization of Eqs. (4.1) through (4.3) enables writing Eq. (4.7) explicitly as

$$B_{MSA} = K_S\sqrt{2}\,|\mathbf{I}_S|\cos\omega_E t \tag{4.8}$$
$$B_{MSB} = K_S\sqrt{2}\,|\mathbf{I}_S|\cos(\omega_E t - 120°) \tag{4.9}$$
$$B_{MSC} = K_S\sqrt{2}\,|\mathbf{I}_S|\cos(\omega_E t - 240°) \tag{4.10}$$

Substitution of Eqs. (4.8) through (4.10) into Eqs. (4.4) through (4.6) yields

$$\mathbf{B}_{WSA} = \left[K_S\sqrt{2}\,|\mathbf{I}_S|\cos\omega_E t\right]\cos\beta_S\mathbf{u}_S \tag{4.11}$$
$$\mathbf{B}_{WSB} = \left[K_S\sqrt{2}\,|\mathbf{I}_S|\cos(\omega_E t - 120°)\right]\cos(\beta_S - 120°)\mathbf{u}_S \tag{4.12}$$
$$\mathbf{B}_{WSC} = \left[K_S\sqrt{2}\,|\mathbf{I}_S|\cos(\omega_E t - 240°)\right]\cos(\beta_S - 240°)\mathbf{u}_S \tag{4.13}$$

*WS*A* designates winding of the stator, phase *A*; MS*A* designates maximum (field) of the stator winding, phase *A*; and $\mathbf{u}_S$ designates the unit vector of the stator field $\mathbf{B}_S$ in the radial direction.

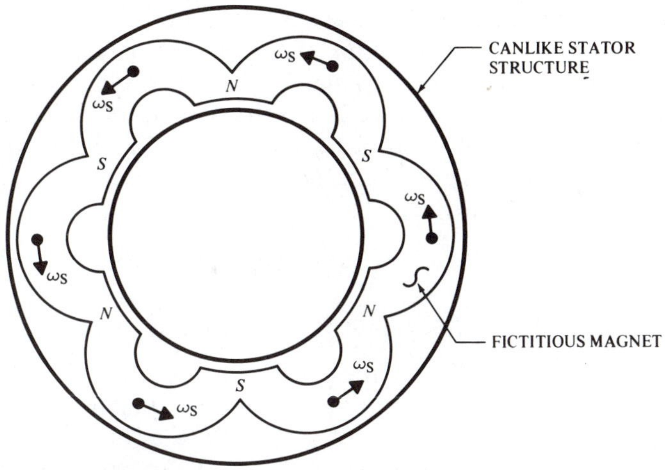

Fig. 4.3.1 Front Cross-sectional View of the Canlike Stator Sturcture, for a Three-Phase, Six-Pole Induction Motor with a Six-Pole Fictitious Magnet Rotating at Synchronous Speed, $\omega_S = \omega_E/(P/2)$

The *total stator magnetic flux density* $\mathbf{B}_{ST}$ at any point T in the airgap due to the currents in the stator windings is the sum of the three stator winding fields. Mathematically, the field $\mathbf{B}_{ST}$ is given by the vector sum of Eqs. (4.11) through (4.13) as

$$\mathbf{B}_{ST} \equiv \mathbf{B}_{WSA} + \mathbf{B}_{WSB} + B_{WSC} = K_S\sqrt{2}\,|\mathbf{I}_S|\big[\cos\omega_E t\cos\beta_S + \cos(\omega_E t - 120°)\cos(\beta_S - 120°) + \cos(\omega_E t - 240°)\cos(\beta_S - 240°)\big]\mathbf{u}_S \tag{4.14}$$

Utilization of the trigonometric identity $\cos X\cos Y = \frac{1}{2}[\cos(X+Y) + \cos(X-Y)]$ enables writing Eq. (4.14) as

$$\mathbf{B}_{ST} = \left(\frac{K_S\sqrt{2}\,|\mathbf{I}_S|}{2}\right)\big[\cos(\omega_E t + \beta_S) + \cos(\omega_E t - \beta_S) + \cos(\omega_E t + \beta_S - 240°) + \cos(\omega_E t - \beta_S) + \cos(\omega_E t + \beta_S - 120°) + \cos(\omega_E t - \beta_S)\big]\,\mathbf{u}_S \tag{4.15}$$

However, since $\cos\chi + \cos(\chi - 120°) + \cos(\chi - 240°) = 0$, where $\chi \equiv \omega_E t + \beta_S$, Eq. (4.15) can be written as

$$\mathbf{B}_{ST} = \left(\frac{3K_S\sqrt{2}\,|\mathbf{I}_S|}{2}\right)\cos(\omega_E t - \beta_S)\mathbf{u}_S \tag{4.16}$$

If, in Eq. (4.16), the quantity $\cos(\omega_E t - \beta_S) = 1$, that is, if and only if

$$\omega_E t = \beta_S \tag{4.17}$$

then $\mathbf{B}_{ST}$ maintains a *constant magnitude,*

$$|\mathbf{B}_{ST}| \equiv \frac{3K_S\sqrt{2}\,|\mathbf{I}_S|}{2} \tag{4.18}$$

This means that $|\mathbf{B}_{ST}|$ is of constant amplitude and travels on the inner surface of the stator a distance of 2π rad counterclockwise with respect to the q_S-axis, Fig. 4.1, around the airgap in $t \equiv 2\pi/\omega_E$ seconds at the speed of ω_E radians/second. When the device has more than two poles, say, four poles [current-carrying coil sides are connected so that the magnetic poles are of alternate polarity, i.e., north, south, north, etc., Figs. 4.3(d) through 4.3(h)], there are two complete wavelengths or cycles in the flux distribution of each stator winding in the airgap; that is, β_S in Eqs. (4.4) through (4.6) becomes $2\beta_M$. When the number of poles increases to six, β_S in Eqs. (4.4) through (4.6) becomes $3\beta_M$, and so on. In general, it can be stated that for a P-pole device an appropriate substitution should be made in Eqs. (4.4) through (4.6) or Eq. (4.16), that is,

$$\beta_S \rightarrow \left(\frac{P}{2}\right)\beta_M \tag{4.19}$$

Substituting for β_S from Eq. (4.19) into Eq. (4.16) yields

$$\mathbf{B}_{ST} = \left(\frac{3K_S\sqrt{2}\,|\mathbf{I}_S|}{2}\right)\cos\left[\omega_E t - \left(\frac{P}{2}\right)\beta_M\right]\mathbf{u}_S \tag{4.20}$$

Dividing both sides of the bracketed expression in Eq. (4.20) by $(P/2)t$ yields

$$\frac{\beta_M}{t} = \frac{\omega_E}{P/2} \equiv \omega_S \tag{4.21}$$

where

$$\omega_S = \frac{\omega_E}{P/2} \text{ radians/second} \tag{4.22}$$

is defined as the *synchronous speed of the total stator magnetic flux density* $|\mathbf{B}_{ST}|$ measured with respect to the q_S-axis, Fig. 4.1. Physically, this means that the magnetic field $|\mathbf{B}_{ST}|$ moves along the inner surface of the stator at a constant angular velocity ω_S radians/second with respect to the stator, maintaining a constant magnitude $|\mathbf{B}_{ST}| = 3K_S\sqrt{2}\,|\mathbf{I}_S|/2$.

Eq. (4.22) can be expressed in revolutions per minute rather than radians per second. Thus premultiplying Eq. (4.22) by 60 s/min yields

$$60\omega_S = \frac{60\omega_E}{P/2} \equiv \frac{2\pi f_E 60}{P/2} \equiv 2\pi n_S \tag{4.23}$$

where n_S is the synchronous speed of the field $|\mathbf{B}_{ST}|$ in revolutions per minute. Thus solving Eq. (4.23) for n_S as a function of the frequency of excitation f_E yields

$$n_S = \frac{(60 \times 2\pi f_E)/(P/2)}{2\pi} = \left(\frac{120}{P}\right) f_E \text{ revolutions/minute} \tag{4.24}$$

Note the significance of Eqs. (4.22) and (4.24). Both give the speed with which the field $|\mathbf{B}_{ST}|$ moves around the inner surface of the stator (i.e., in the smooth airgap of the induction device), either in radians per second (ω_S) or revolutions per minute (n_S), as a function of the frequency of excitation, expressed either in radians per second (ω_E) or hertz (f_E), and the number of poles P in the stator structure.

Example 4.4

If the frequency of excitation of the induction device of Example 4.2 is $f_E = 60$ Hz, what is the synchronous speed of the field in radians/second (ω_S and revolutions/minute (n_S)?

Solution

Utilization of Eqs. (4.22) and (4.24) yields

$$\omega_E = 2\pi f_E = 2\pi \times 60 = \underline{377} \text{ rad/s}$$

$$\omega_S = \frac{\omega_E}{P/2} = \frac{377}{6/2} = \underline{\underline{125.66}} \text{ rad/s}$$

and

$$n_S = \frac{60\omega_S}{2\pi} = \frac{60\times(2\pi f_E)/(P/2)}{2\pi} = \left(\frac{120}{P}\right) f_E = \underline{\underline{1200}} \text{ r/min}$$

4.3 THE INDUCED VOLTAGE IN THE ROTOR WINDINGS

Assume that the rotor windings of Fig. 4.1 are open-circuited, that is, they do not terminate to the suggested rotor voltage sources $e_{RA}(t)$, $e_{RB}(t)$, and $e_{RC}(t)$. The angular velocity of rotor coil, $\omega_C \equiv \omega_{SR}$, Fig. 4.4, can be expressed as the difference between the synchronous speed ω_S of the field $|\mathbf{B}_{ST}|$ rotating counterclockwise and the *rotor speed* ω_R, with the rotor rotating at an angular velocity ω_R radians/second in a counterclockwise direction:

$$\omega_C = \omega_S - \omega_R \tag{4.25}$$

In other words, the *coil velocity* ω_C is the velocity of the synchronous field ω_S relative to the rotor velocity ω_R.

Utilization of the expression $e_R(t) = \oint (\mathbf{u}\times\mathbf{B})\cdot d\mathbf{L}$ enables calculation of the induced voltage in the rotor winding, which, in accordance with Eqs. (2.22) through (2.26), can be written as

$$e_{RC}(t) = N_R 2\ell R\omega_C |\mathbf{B}_{ST}| \cos\left[\left(\frac{P}{2}\right)\omega_C t - \gamma\right] \tag{4.26}$$

In Eq. (4.26), ℓ is the effective length of the rotor wire on which the speed voltage $e_{RC}(t)$ is induced (i.e., ℓ is the length of one coil side of each loop), N_R is the number of wires in each coil side (or the number of turns in each rotor loop), R is

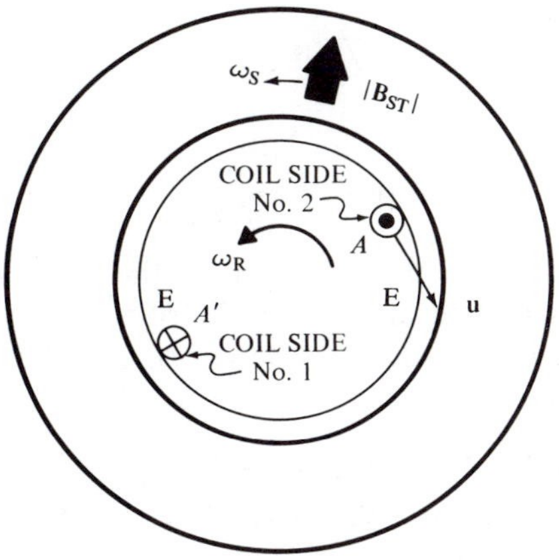

Fig. 4.4 Cross Section of the Induction Energy Converter Exhibiting Only: (1) Concentrated Rotor Winding A, (2) the Stator Field $|\mathbf{B}_{ST}|$, (3) the Synchronous Speed ω_S, and (4) the Rotor Speed ω_R, Coil Speed $\omega_C = \omega_S - \omega_R$

the radius of the rotor, $|\mathbf{B}_{ST}|$ is the amplitude of the stator field that is rotating in the airgap at a synchronous speed ω_S, Eq. (4.22), $(P/2)\omega_C$ is the radian frequency with which the rotating rotor coil cuts the synchronously rotating stator field, and γ is the phase angle of the rotor-coil induced voltage $e_{RC}(t)$ at $t=0$. Lumping all constants of Eq. (4.26) into $\mathcal{K}$ and substituting Eq. (4.25) into Eq. (4.26) yields

$$e_{RC}(t) = \mathcal{K}(\omega_S-\omega_R)\cos\left[\left(\frac{P}{2}\right)(\omega_S-\omega_R)t-\gamma\right] \tag{4.27}$$

Multiplication of each $\omega_C=\omega_S-\omega_R$ in Eq. (4.27) by the ratio ω_S/ω_S yields

$$e_{RC}(t) = \mathcal{K}\omega_S\left[\frac{(\omega_S-\omega_R)}{\omega_S}\right]\cos\left\{\left(\frac{P}{2}\right)\omega_S\left[\frac{(\omega_S-\omega_R)}{\omega_S}\right]t-\gamma\right\} \tag{4.28}$$

The quantity $(\omega_S-\omega_R)/\omega_S$ is defined as *slip*, designated by the lowercase letter s, that is,

$$s \equiv \frac{\omega_S-\omega_R}{\omega_S} \tag{4.29}$$

Note that slip is a dimensionless quantity. It is a measure of the speed of the rotor relative to the synchronous speed of the field $|\mathbf{B}_{ST}|$, which depends on the radian frequency of the excitation voltage sources, $e_{SA}(t)$, $e_{SB}(t)$, and $e_{SC}(t)$. Substitution for s from Eq. (4.29) into Eq. (4.28) yields

$$e_{RC}(t) = \mathcal{K}s\omega_S\cos\left[\left(\frac{P}{2}\right)s\omega_S t-\gamma\right] \tag{4.30}$$

Substitution from Eq. (4.22) for $(P/2)\omega_S=\omega_E$ into Eq. (4.30) yields

$$e_{RC}(t) = \mathcal{K}s\omega_S\cos(s\omega_E t-\gamma) \tag{4.31}$$

The resulting radian frequency in the parentheses of Eq. (4.31), that is, $s\omega_E$, is the frequency of the induced rotor coil voltages e_{Rj}, $j=A,B,C$. It is known as the **slip frequency** of the rotor coils and is defined as

$$\omega_e \equiv s\omega_E \tag{4.32}$$

Utilization of Eq. (4.32) enables writing Eq. (4.31) as follows:

$$e_{RC}(t) = \mathcal{K}s\omega_S\cos(\omega_e t-\gamma) \tag{4.33}$$

or

$$e_{RC}(t) = \sqrt{2}\,|\mathbf{E}_R|\cos(\omega_e t-\gamma) \tag{4.34}$$

where

$$|\mathbf{E}_R| \equiv \frac{\mathcal{K}s\omega_S}{\sqrt{2}} \tag{4.35}$$

Eq. (4.34) enables us to write the equations of the induced voltages in the three balanced rotor windings (which are displaced from each other by 120° and are wound so that the resistance R, the inductance L, and the capacitance C of each rotor coil are identical) as

$$e_{RA}(t) = \sqrt{2}\,|\mathbf{E}_R|\cos[\omega_e t-\gamma] \tag{4.36}$$

$$e_{RB}(t) = \sqrt{2}\,|\mathbf{E}_R|\cos[\omega_e t-(\gamma+120°)] \tag{4.37}$$

$$e_{RC}(t) = \sqrt{2}\,|\mathbf{E}_R|\cos[\omega_e t-(\gamma+240°)] \tag{4.38}$$

When the rotor coils are short-circuited [i.e., the external resistors R_A, R_B, and R_C of Fig. 4.2(b) are replaced by short circuits], balanced currents of slip frequency

ω_e, and γ out of phase with the induced rotor voltages $e_{RA}(t)$, $e_{RB}(t)$, and $e_{RC}(t)$, flow in the rotor windings. These currents can be written as

$$i_{RA}(t) = \sqrt{2}\,|\mathbf{I}_R|\cos\omega_e t \tag{4.39}$$

$$i_{RB}(t) = \sqrt{2}\,|\mathbf{I}_R|\cos(\omega_e t - 120°) \tag{4.40}$$

$$i_{RC}(t) = \sqrt{2}\,|\mathbf{I}_R|\cos(\omega_e t - 240°) \tag{4.41}$$

These currents produce additional fields in the airgap. These fields can be written, with the aid of Eqs. (4.4) through (4.6), as*

$$\mathbf{B}_{WRA} = B_{MRA}\cos\delta_R\mathbf{u}_R \tag{4.42}$$

$$\mathbf{B}_{WRB} = B_{MRB}\cos(\delta_R - 120°)\mathbf{u}_R \tag{4.43}$$

$$\mathbf{B}_{WRC} = B_{MRC}\cos(\delta_R - 240°)\mathbf{u}_R \tag{4.44}$$

where δ_R is the angular displacement on the rotor structure measured (positively) from the q_R-axis, Fig. 4.1.

Again, proper use of Ampere's law, $\oint \mathbf{H}\cdot d\mathbf{L} = Ni(t)$ and $\mathbf{B} = \mu\mathbf{H}$, enables writing the maxima of the flux density of the rotor windings as

$$B_{MAj} = K_R\phi[i_R(t)], \quad j = A, B, C \tag{4.45}$$

Thus for the sinusoidal currents of Eqs. (4.39) through (4.41), $\mathbf{B}_{MRj}$, $j = A, B, C$, can be written as

$$B_{MRA} = K_R\sqrt{2}\,|\mathbf{I}_R|\cos\omega_e t \tag{4.46}$$

$$B_{MRB} = K_R\sqrt{2}\,|\mathbf{I}_R|\cos(\omega_e t - 120°) \tag{4.47}$$

$$B_{MRC} = K_R\sqrt{2}\,|\mathbf{I}_R|\cos(\omega_e t - 240°) \tag{4.48}$$

In Eqs. (4.46) through (4.48), K_R is the rotor design constant that depends on the number of turns of the rotor windings N_R, the path of integration ℓ, and the permeability μ (μ_0 and μ_r) of the ferromagnetic material that constitutes the rotor structure.

Utilization of the ideas of Eqs. (4.11) through (4.19) enables writing the expression of the *total rotor magnetic flux density* $\mathbf{B}_{RT}$ in the airgap, due to the currents in the rotor windings, for a P-pole device as

$$\mathbf{B}_{RT} = \left(\frac{3K_R\sqrt{2}\,|\mathbf{I}_R|}{2}\right)\cos\left[\omega_e t - \left(\frac{P}{2}\right)\delta_M\right]\mathbf{u}_R \tag{4.49}$$

This result is a consequence of the identity $\cos\phi + \cos(\phi - 120°) + \cos(\phi - 240°) \equiv 0$, in which $\phi \equiv \omega_e t + \delta_R$ [reference to Eqs. (4.14) through (4.19) will help].

If in Eq. (4.49) the quantity $\cos[\omega_e t - (P/2)\delta_M] \equiv 1$, that is, if and only if

$$\omega_e t \equiv \left(\frac{P}{2}\right)\delta_M \tag{4.50}$$

then $\mathbf{B}_{RT}$ maintains a constant magnitude, that is,

$$|\mathbf{B}_{RT}| = \frac{3K_R\sqrt{2}\,|\mathbf{I}_R|}{2} \tag{4.51}$$

This means that $\mathbf{B}_{RT}$ is of constant amplitude and travels a distance of 2π rad counterclockwise on the surface of the rotor with respect to the q_R-axis, Fig. 4.1,

*WRA designates winding of the rotor, phase A, and MRA designates maximum (field) of the rotor winding, phase A; and $\mathbf{u}_R$ designates the unit vector of the rotor field $\mathbf{B}_R$ in the radial direction.

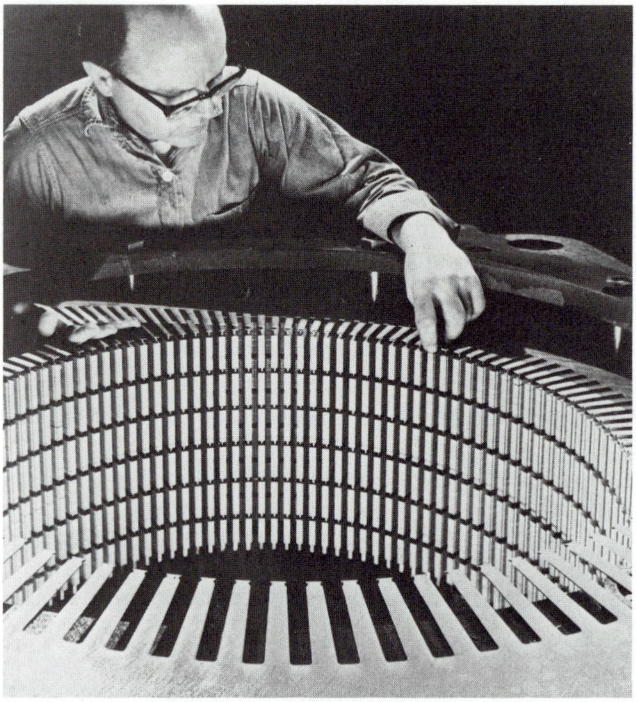

Plate 4.2 Stator of an Induction Motor Being Assembled: A Lamination Sheet Is Clearly Visible, as Are the Resulting Teeth and Slots. (Courtesy of Westinghouse Electric Co., Pittsburgh, Pa.)

around the airgap in $t \equiv 2\pi/\omega_e$ seconds at the speed of ω_e radians/second. Dividing both sides of Eq. (4.50) by $(P/2)t$ yields

$$\frac{\delta_M}{t} = \frac{\omega_e}{P/2} \equiv \omega_s \tag{4.52}$$

where

$$\omega_s = \frac{\omega_e}{P/2} \text{ radians/second} \tag{4.53}$$

is defined as the *synchronous speed of the total rotor magnetic flux density* $|\mathbf{B}_{TR}|$ measured with respect to the q_R-axis. Physically, this means that the magnetic field $|\mathbf{B}_{RT}|$ moves along the surface of the rotor at constant angular velocity ω_s radians/second with respect to the rotor, maintaining a constant magnitude $|\mathbf{B}_{RT}| = 3K_R\sqrt{2}\,|\mathbf{I}_R|/2$. Substitution for ω_e from Eq. (4.32) and for ω_E from Eq. (4.22) into Eq. (4.53) yields

$$\omega_s = \frac{\omega_e}{P/2} = \frac{s\omega_E}{P/2} = \left(\frac{s}{P/2}\right)\omega_S\left(\frac{P}{2}\right) = s\omega_S \tag{4.54}$$

Eq. (4.54) gives the speed with which the rotor field $|\mathbf{B}_{RT}|$ rotates with respect to the rotor as a function of the rotor winding excitation frequency ω_e and the pole-pair $P/2$; or as a function of the stator winding excitation frequency ω_E, the slip s, and the pole-pair $P/2$; or as a function of the slip s and the synchronous speed of the stator field ω_S. The speed with which the stator field $|\mathbf{B}_{ST}|$ rotates, when referred to the rotor, is $\omega_S - \omega_R$. However, Eq. (4.29) yields $\omega_S - \omega_R = s\omega_S$.

Thus the speed of both fields, with respect to the rotor, is $s\omega_S$. On the other hand, the speed of the stator field $|\mathbf{B}_{ST}|$ with respect to the stator is ω_S and the speed of the rotor field $|\mathbf{B}_{RT}|$ when referred to the stator is $\omega_C+\omega_R$. However, Eq. (4.25) yields $\omega_C=\omega_S-\omega_R$. Thus $\omega_C+\omega_R=(\omega_S-\omega_R)+\omega_R=\omega_S$. Also, $\omega_s+\omega_R=s\omega_S+\omega_R=\omega_S$. Again, the speed of both fields, with respect to the stator, is ω_S.

Perhaps it is easier to understand the distinct difference between the two speeds of the fields $|\mathbf{B}_{ST}|$ and $|\mathbf{B}_{RT}|$ when we view the fields through the eyes of a moving observer. If the observer is standing on the inner surface of the stator, he sees both fields, $|\mathbf{B}_{ST}|$ and $|\mathbf{B}_{RT}|$, pass him at a speed ω_S. If the observer is standing on the surface of the rotor, he sees both fields, $|\mathbf{B}_{ST}|$ and $|\mathbf{B}_{RT}|$, pass him at a speed ω_s (or $s\omega_S$). Therefore, whether the observer stands on the inner surface of the stator or on the surface of the rotor, he sees two fields pass him that are stationary with respect to each other. *The interaction of these two fields results in the development of an electromagnetic torque τ_{em} that causes the rotor to rotate at an angular velocity ω_R.*

Example 4.5

When the motor of Example 4.2 delivers rated output horsepower, the slip is found to be $s=0.05$. Determine the following parameters: (1) the radian frequency (ω_E) of the stator currents, (2) the synchronous speed (ω_S) of the revolving field relative to the stator structure, (3) the speed with which the rotor is rotating in radians/second (ω_R) and revolutions/minute (n_R), (4) the radian frequency (ω_e) of the induced rotor currents, (5) the radian speed of the rotor field (ω_s) relative to the rotor structure, (6) the radian speed of the stator field relative to the rotor structure, designated as ω_{SR}, (7) the radian speed of the rotor field relative to the stator structure, designated as ω_{sS}, (8) the speed of the stator field relative to the rotor field, designated as ω_{S-s}, and (9) the speed of the rotor field relative to the stator field, designated as ω_{s-S}. Are the conditions ideal for the development of unidirectional torque?

Solution

Utilization of the theory developed up to now enables calculation of the following parameters.

1. $\omega_E=2\pi f_E=2\pi\times 60=\underline{\underline{377}}\text{ rad/s}$

2. $\omega_S=\dfrac{\omega_E}{P/2}=\dfrac{377}{3}=\underline{\underline{125.66}}\text{ rad/s}$

and

$$n_S=\frac{60\omega_s}{2\pi}=\underline{\underline{1200}}\text{ rpm}$$

3. Use of Eqs. (4.29) and (4.23) yields

i. $\omega_R=\omega_S(1-s)=125.66(1-0.05)=\underline{\underline{119.38}}\text{ rad/s}$

ii. $n_R=\dfrac{60\omega_R}{2\pi}=\underline{\underline{1140.0}}\text{ rpm}$

4. Use of Eq. (4.32) yields

$$\omega_e=s\omega_E=0.05\times 377=\underline{\underline{18.85}}\text{ rad/s}$$

also,

$$f_e = sf_E = s\left(\frac{\omega_E}{2\pi}\right) = \underline{\underline{3}}\ \text{Hz}$$

5. Use of Eq. (4.54) yields

$$\omega_s = s\omega_S = 0.05 \times 125.66 = \underline{\underline{6.28}}\ \text{rad/s}$$

6. The speed of the stator field with respect to the rotor, designated as ω_{SR}, is calculated with the aid of Eq. (4.29) as

$$\omega_{SR} = \omega_S - \omega_R = s\omega_S = \omega_s = 0.05 \times 125.66 = \underline{\underline{6.28}}\ \text{rad/s}$$

7. The speed of the rotor field with respect to the stator, designated as ω_{sS}, is calculated with the aid of Eq. (4.25) as

$$\omega_{sS} = \omega_C + \omega_R = (\omega_S - \omega_R) + \omega_R = \omega_S = \underline{\underline{125.66}}\ \text{rad/s}$$

or utilization of Eq. (4.29) yields

$$\omega_{sS} = \omega_s + \omega_R = s\omega_S + \omega_R = \omega_S = \underline{\underline{125.66}}\ \text{rad/s}$$

8. $\omega_{S-s} = \omega_s - \omega_{sR} = \omega_s - \omega_s = \underline{\underline{0}}\ \text{rad/s}$

9. $\omega_{s-S} = \omega_S - \omega_{sS} = \omega_S - \omega_S = \underline{\underline{0}}\ \text{rad/s}$

Yes; torque is developed and the rotor rotates in the direction of rotation of the stator field.

4.4 THE DEVELOPED ELECTROMAGNETIC TORQUE ON THE ROTOR

To understand the concept of the electromagnetic torque development by such a device, consider Fig. 4.5. For a two-pole device and the field of Eq. (4.4), that is, the field of the stator phase A written as $\mathbf{B}_S \equiv B_{MSA} \cos\beta \mathbf{u}_S$, the voltage induced in the rotor, phase A, $e_{RA}(\beta) = \oint [\mathbf{u} \times \mathbf{B}] \cdot d\mathbf{L}$, can be written as

$$e_{RA}(\beta) = E_{MRA} \cos\beta \tag{4.55}$$

Thus the response current in the corresponding rotor winding, displaced from its

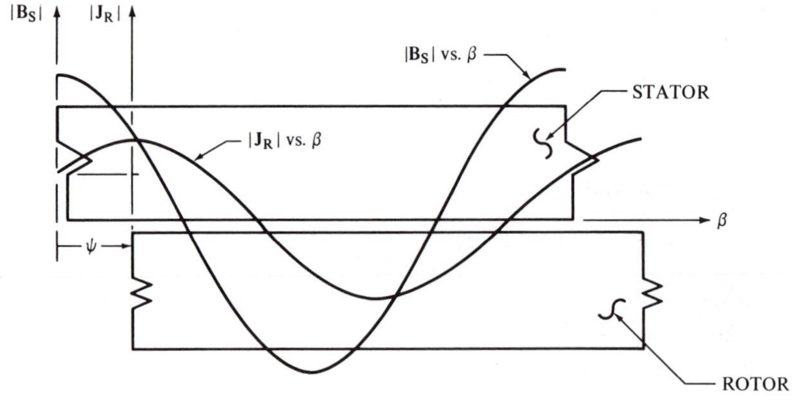

Fig. 4.5 Stator and Rotor Fields Contributing to Electromagnetic Torque Development

generating voltage by angle ψ, can be written in terms of a sinusoidally distributed current density as $i_{RA}(\beta,\psi)=|\mathbf{J}_R|d\beta$, which can be written explicitly as

$$i_{RA}(\beta,\psi) = J_{MRA}\cos(\beta-\psi)\,d\beta \tag{4.56}$$

Reference to Section 3.2 will help in understanding the development of Eq. (4.56). Utilization of Eq. (2.5), $\mathbf{B}_S = B_{MSA}\cos\beta\mathbf{u}_S$, and Eq. (5.46) enables writing the expression of the electromagnetic force developed on each rotor-winding, current-carrying element, that is, $d\mathbf{F}_{Rem} = i\,d\mathbf{L}\times\mathbf{B}\rightarrow|\mathbf{F}_{Rem}| = \ell|\mathbf{B}|i$, as

$$|d\overline{\mathbf{F}}_{Rem}| = \ell(B_{MSA}\cos\beta)[J_{MRA}\cos(\beta-\psi)\,d\beta] \tag{4.57}$$

The total torque developed on the rotor-winding current sheet under one pole of the field distribution is determined by integrating $d\tau_{Rem} = \mathbf{R}\times d\mathbf{F}_{Rem}$ over π electrical radians, which is the span of one pole-pair. Hence, keeping in mind that the vectors $\mathbf{R}$ (the radius of the rotor in meters) and $d\mathbf{F}_{Rem}$ (the force developed on each rotor-winding, current-carrying element in newton meters) are normal to each other, the magnitude of the developed electromagnetic torque/pole on the rotor is

$$\tau_{Rem} = \int_0^{\pi} R\ell B_{MSA}J_{MRA}\cos\beta\cos(\beta-\psi)\,d\beta \tag{4.58}$$

Thus the magnitude of the total torque developed on the rotor due to P stator poles, symbolized at τ_{em} (i.e., when torque is developed in the rotor, it exhibits its presence by rotor rotation, and the presence of R in τ_{em} should always be understood) is

$$\tau_{em} = P\int_0^{\pi} R\ell B_{MSA}J_{MRA}\cos\beta\cos(\beta-\psi)\,d\beta \tag{4.59}$$

Utilization of the trigonometric identity $\cos\beta\cos(\beta-\psi)=\frac{1}{2}[\cos\psi+\cos(2\beta-\psi)]$ enables writing Eq. (4.59) as

$$\tau_{em} = \left(\frac{P}{2}\right)R\ell B_{MSA}J_{MRA}\left[\int_0^{\pi}\cos\psi\,d\beta + \int_0^{\pi}\cos(2\beta-\psi)\,d\beta\right] \tag{4.60}$$

The value of the second integral in the brackets is zero (why?). Thus utilization of the average flux density $B_{AVG}\equiv 2B_{MSA}/\pi\equiv\Phi_P/A_P$, the pole face area $A_P\equiv 2\pi R\ell/P$ (where Φ is the maximum flux per pole in webers, R is the radius of the rotor in meters, and ℓ is the axial length of the rotor in meters), Eq. (3.4) (i.e., $\mathcal{F} = 2J_{MRA}$), and Appendix C of reference [5] enables writing the final expression for the **total developed electromagnetic torque on the rotor of a three-phase induction motor** as

$$\tau_{em} = \left(\frac{P}{2}\right)\pi R\ell B_{MSA}J_{MRA}\cos\psi = 0.177PZ_R K_{WR}\Phi_P I_R\cos\psi \equiv K\Phi_P I_R\cos\psi \tag{4.61}$$

where (1) Φ_P is the maximum flux per pole that links with a rotor coil having Z_R conductors or coil sides (two coil sides constitute one coil turn, designated by N_R) and spanning the *full pitch**, (2) K_{WR} is the **rotor-winding factor** (for further explanation of its meaning see Section 4.5), (3) I_R is the rotor-winding current per phase, and (4) K is a design constant accounting for the three phases of the device, its RMS excitation, and the magnetic field (fundamental component and harmonic components) contribution. Eq. (4.61) is exhibited graphically in Fig. 4.6. Note that

*Full pitch is defined as the 180-electrical-degree displacement between the starting and finishing coil sides of a given coil in any device member.

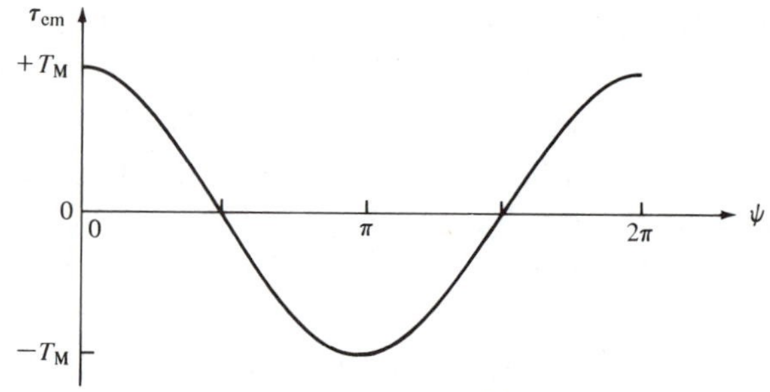

Fig. 4.6 Rotor-developed Electromagnetic Torque τ_{em} as a Function of the Space Displacement Angle ψ (Fig. 4.5) Between the Stator Field Distribution $|\mathbf{B}_S|$ and the Rotor-Current Density Distribution $|\mathbf{J}_R|$ of a Single-Winding-per-Member Induction Device

the developed electromagnetic torque τ_{em} on the rotor is maximum when the space displacement angle ψ, Fig. 4.5, between the stator field $|\mathbf{B}_S|$ and the rotor-current density $|\mathbf{J}_R|$ (note that the rotor field $\mathbf{B}_R$, generated by the current density $\mathbf{J}_R$, lags $\mathbf{J}_R$ by a phase angle of 90° and is of different magnitude, Fig. 3.2) is zero, that is, when the two fields $|\mathbf{B}_S|$ and $|\mathbf{J}_R|$ are perfectly lined up and $\psi = 0°$. Note that the fields $|\mathbf{B}_S|$ and $|\mathbf{B}_R|$ are $\beta = 90°$ out of phase when $\psi = 0°$, Fig. 3.2. This torque is exerted on *both* the stator and rotor coils but in *opposite* directions; that is, *the two fields tend to align themselves* and $\psi \to 0°$. Thus since both the stator and rotor fields travel around the airgap at the same speed (i.e., ω_S or ω_s), **a steady torque is produced by the effort of the rotor field trying to line itself up with the stator field.** It should be emphasized, however, that it is possible for both field and current distributions to be present and still the torque can be equal to zero. This can happen if the field pattern is such that equal and opposite torques are developed by the two rotor conductors, that is, when $\psi = 90°$ or $\psi = 270°$. Also, note that the torque can change direction when the field pattern is such that $90° < \psi < 270°$. Thus when the stator coils of an induction device, Fig. 4.1, are excited from a three-phase excitation system of radian frequency ω_E, and the rotor coils are [refer to Fig. 4.2(b)] (a) short-circuited, (b) loaded through the slip rings with a resistive load, or (c) excited through the slip rings from external sources, rotor currents of slip frequency ω_e circulate. Because of these currents a steady torque, Eq. (4.61) (and Fig. 4.5), is developed that tends to keep the rotor field $|\mathbf{J}_R|$ aligned with the stator field $|\mathbf{B}_S|$ or to keep the rotor field $|\mathbf{B}_R|$ at $\beta = 90°$ out of phase with the stator field $|\mathbf{B}_S|$ at all times. However, the stator field rotates at a synchronous speed $\omega_S = \omega_E/(P/2)$. Ideally the rotor field should lock onto the stator field and the rotor should rotate at an ideal speed $\omega_R \equiv \omega_S$; the space displacement angle ψ between the stator field $|\mathbf{B}_S|$ and the rotor field $|\mathbf{J}_R|$ should be zero, or the rotor field $|\mathbf{B}_R|$ should be at $\beta = 90°$ out of phase with the stator field $|\mathbf{B}_S|$. Although ω_S is the ideal speed of the rotor, it is never achieved because of external loads on the rotor, that is, loads due to (1) moment of inertia, (2) bearing friction, (3) shaft stiffness, (4) externally applied torques, (5) loading torques, and (6) air resistance. Even though it is possible to operate the induction device with no external loading, it is physically impossible to eliminate totally bearing friction and wind resistance. Thus the rotor of an induction device **never** attains its ideal speed, ω_S, the space displacement angle ψ is never zero, and β is never equal to ninety degrees. Therefore, the induction device *always* operates at $\omega_R \neq \omega_S$, $\psi \neq 0°$, and $\beta \neq 90°$.

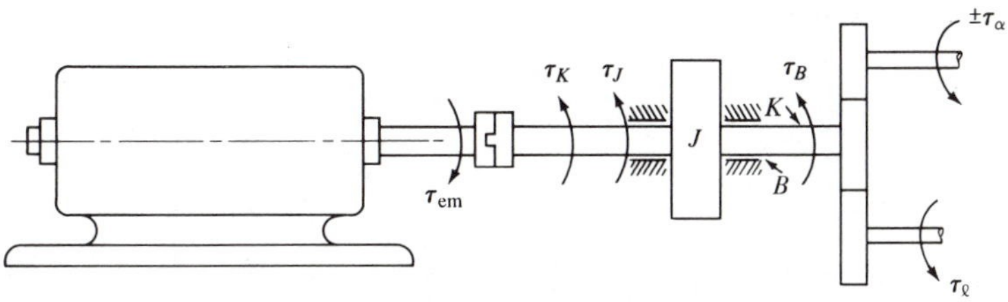

Fig. 4.7 The Relative Directions of the Developed Electromagnetic Torque and the Loading Torques of a Loaded Induction Device

The rotor speed ω_R is the steady state solution of the dynamic equation

$$\tau_{em} = J\frac{d^2}{dt^2}\theta_R(t) + B\frac{d}{dt}\theta_R(t) + K\theta_R(t) + \tau_a \pm \tau_\ell \tag{4.62}$$

for $\omega_R(t) \equiv d\theta_R(t)/dt$, where $\theta_R(t)$ is the angular displacement of the rotor shaft measured counterclockwise from q_S-axis, Fig. 4.1. Eq. (4.62) utilizes D'Alembert's principle, that is, *for any rigid body in rotational motion about a given axis, the algebraic sum of the externally applied torques and the torques resisting this motion is zero.* This equation describes the system of Fig. 4.7 and relates the developed electromagnetic torque τ_{em} by the motor and the loading torques τ_J, τ_B, τ_K, τ_a, and τ_ℓ. These loading torques are due to (1) the moment of inertia J (kilograms meters^2), (2) the coefficient of friction B (Newton meters/radian/second), (3) the stiffness of the shaft K (Newton meters/radian), (4) the externally applied torque τ_a (Newton meters), and (5) the loading torque τ_ℓ (Newton meters). When the windage losses, that is, losses due to air resistance, designated as τ_W, are taken into account, Eq. (4.62) should be adjusted properly by adding to it τ_W. The quantity $\tau_B + \tau_W \equiv \tau_R$ is defined, then, as rotational losses. It should be pointed out, however, that one or more, but not all, loading torques can be equal to zero.

Although Eq. (4.62) is easy to solve, a solution will not be pursued here. For our purposes of analysis, it is sufficient to assume that the rotor of the induction device is rotating at an angular speed ω_R (radians/second) different in magnitude than the synchronous speed ω_S (radians/second) of the stator field and related to it through the slip, Eq. (4.29).

4.5 DEVELOPMENT OF THE EQUIVALENT CIRCUIT OF THE INDUCTION DEVICE UNDER STEADY STATE OPERATION

For the purpose of this analysis, the equivalent stator winding of phase A, Fig. 4.1, is drawn in Fig. 4.8(a) as a concentrated winding with N_{SA}-turns. For the rotor **locked** in place, that is, $\omega_R = 0$, the equivalent rotor winding of phase A, Fig. 4.1, also drawn in Fig. 4.8(a) as a concentrated winding with N_{RA}-turns and maintained open-circuited, the analysis can proceed as follows.

As a first step, the rotating stator field, Eq. (4.16), produced by the application of the three-phase excitation voltages to the stator windings is rewritten here for convenience as*

$$\mathbf{B} = |\mathbf{B}_{ST}|\cos(\omega_E t - \beta_S)\mathbf{u}_S \tag{4.63}$$

*$|\mathbf{B}_{ST}|$ should be interpreted as B_{MAX} in this relationship.

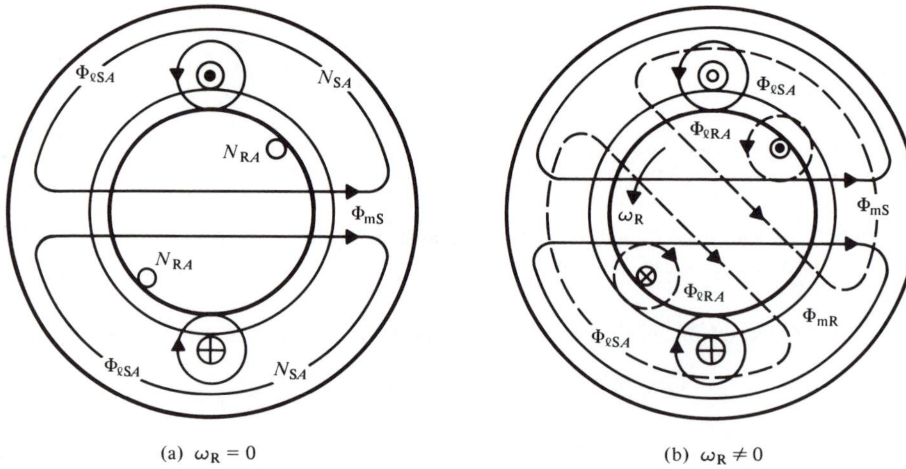

Fig. 4.8 Relative Positions of Concentrated Stator and Rotor Windings of Phase A: (a) Stator Fields $\Phi_{\ell SA}$ and Φ_{mS} for $\omega_R=0$ and (b) Stator Fields $\Phi_{\ell SA}$ and Φ_{mS} and Rotor Fields $\Phi_{\ell RA}$ and Φ_{mR} for $\omega_R \neq 0$

Application of Kirchhoff's voltage law to the stator coil of Fig. 4.8(a) can be written as

$$v_{SA}(t) = R_S i_{SA}(t) + \frac{d}{dt}\lambda_{\ell SA}(t) + e_{SA}(t) \tag{4.64}$$

where

$$\lambda_{\ell SA}(t) = N_{SA}\Phi_{\ell SA}(t) \equiv L_{\ell SA} i_{SA}(t) \tag{4.65}$$

and

$$e_{SA}(t) = \frac{d}{dt}\lambda_m(t) \equiv \oint (\mathbf{u}\times\mathbf{B})\cdot d\mathbf{L} \tag{4.66}$$

The subscripts ℓ and m designate leakage and mutual flux. Since the relative velocity of the stator coil with respect to the revolving field is $\mathbf{u}=R\omega_S\mathbf{a}_T$, $\omega_S=\omega_E/(P/2)$, and $R_{STATOR}\simeq R_{ROTOR}\to R$, $e_{SA}(t)$ can be written as

$$e_{SA}(t) = \left(\frac{N_{SA}2\ell\omega_E R B_{MAX}}{P/2}\right)\cos(\omega_E t - \beta_S) \tag{4.67}$$

The voltage across the rotor coil can be written as

$$e_{RA}(t) = \frac{d}{dt}\lambda_m(t) \tag{4.68}$$

However, $e_{RA}(t)$ is the speed voltage of the rotor winding induced by the rotating stator field $|\mathbf{B}_{ST}|$, which is rotating at the synchronous speed ω_S and is generated by the stator winding excitations. The speed voltage $e_{RA}(t)$ can be written, in accordance with Eqs. (2.22) and (2.26), as

$$e_{RA}(t) = N_{RA}2\ell\omega_S R|\mathbf{B}_{ST}|\cos(\omega_E t - \gamma) \tag{4.69}$$

Utilization of $B_{AVG}=2|\mathbf{B}_{ST}|/\pi$, pole phase area $A_P=2\pi R\ell/P$, and $B_{AVG}=\Phi_P/A_P$ (AVG and P subscripts designate "average" and "per pole" quantities, correspondingly) enables writing Eqs. (4.67) and (4.69) as

$$e_{SA}(t) = N_{SA}\omega_E\Phi_P\cos(\omega_E t - \beta_S) \tag{4.70}$$

and

$$e_{RA}(t) = N_{RA}\omega_E\Phi_P\cos(\omega_E t-\gamma) \tag{4.71}$$

The quantity Φ_P designates the maximum flux per pole that links the stator coil having N_{SA}-turns to the rotor coils having N_{RA}-turns. However, in the practical induction device, the total turns per phase are not concentrated in a single coil, as shown in Fig. 4.8(a). Rather, they are **distributed over one-third of a pole face**, that is, 60 electrical degrees beneath each pole. In addition, the individual coils that make up the N_{SA}-turns and the N_{RA}-turns are intentionally designed to *span only 80 percent to 85 percent of a pole pitch*; that is, they span less than 180 electrical degrees. Such a winding is known as a **distributed winding** with fractional pitch coils. It is used to practically eliminate the harmonics in the flux waveform, while it slightly affects the fundamental component of the flux waveform. This reducing effect in each winding is represented by the winding factors denoted by K_{WS} and K_{WR}, ranging in values as follows: $0.85<K_{WS}$ (or K_{WR})<0.95. The effect of the winding factors K_{WS} and K_{WR} on the voltages $e_{SA}(t)$ across the ideal stator coil and $e_{RA}(t)$ across the ideal rotor coil reflected in Eqs. (4.70) and (4.71), with the $\sqrt{2}$ term included, leads to

$$e_{SA}(t) = \left(\frac{\sqrt{2}\,K_{WS}N_{SA}\omega_E\Phi_P}{\sqrt{2}}\right)\cos(\omega_E t-\beta) \tag{4.72}$$

and

$$e_{RA}(t) = \left(\frac{\sqrt{2}\,K_{WR}N_{RA}\omega_E\Phi_P}{\sqrt{2}}\right)\cos(\omega_E t-\gamma) \tag{4.73}$$

Keeping in mind that in a balanced system $N_{SA}=N_{SB}=N_{SC}\equiv N_S$, $N_{RA}=N_{RB}=N_{RC}\equiv N_R$, $e_{SA}(t)=e_{SB}(t)=e_{SC}(t)\equiv e_S(t)$, and $e_{RA}(t)=e_{RB}(t)=e_{RC}(t)\equiv e_R(t)$, the remaining equations can be written on a **per-phase basis**. The ratio of Eq. (4.72) to Eq. (4.73) yields

$$\frac{e_S(t)}{e_R(t)} = \frac{K_{WS}N_S}{K_{WR}N_R} \equiv \frac{\mathfrak{N}_S}{\mathfrak{N}_R} \tag{4.74}$$

Eq. (4.74) is similar in form and meaning with Eq. (1.47), which describes the ideal transformer. It relates the induced voltage $e_S(t)$ across the stator coil to the induced voltage $e_R(t)$ across the stationary rotor coil through the effective stator and rotor coil turns ratio ($\mathfrak{N}_S/\mathfrak{N}_R\equiv(K_{WS}N_S/K_{WR}N_R)$). The circuit representation of the mathematical analog defined by Eq. (4.74) is given in Fig. 4.9. The similarity

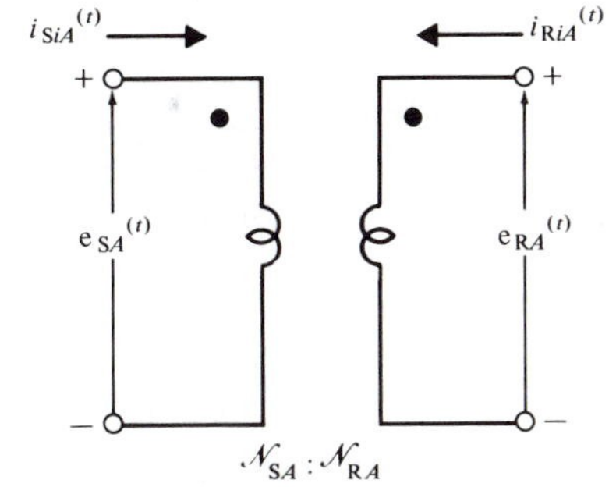

Fig. 4.9 Mutual Coupling Effects Between Stator and Rotor Phase *A* Windings When $\omega_R\equiv 0$

of Fig. 4.9 and Fig. 1.7, and the ideas leading to these circuit representations, enables writing the relationship between the stator current $i_{Si}(t)$ and the rotor current $i_{Ri}(t)$ in terms of the effective turn ratio of the rotor and stator coils as

$$i_{Si}(t) = -\left(\frac{\mathcal{N}_R}{\mathcal{N}_S}\right) i_{Ri}(t) \tag{4.75}$$

Utilization of $\lambda_m = L_m i_m(t)$ enables writing the induced potential rise across the stator coil with N_s-turns, due to the rotating field Φ_m, as

$$e_S(t) = L_m \frac{d}{dt} i_m(t) \tag{4.76}$$

Eq. (4.76), along with Eqs. (4.64), (4.65), (4.74), and (4.75), and Figs. 4.8(a) and 4.9 enable drawing the equivalent circuit of Fig. 4.8(a) as in Fig. 4.10 [note that Fig. 4.8(a) depicts only one stator coil with flux linkages due to $i_S(t)$ current, magnetizing phenomena due to the rotating stator field, a voltage drop due to the resistance of the stator coil, and a nonmoving rotor, i.e., $\omega_R = 0$, with one coil open-circuited but linked by the rotating field].

Application of Kirchhoff's current law to junction 1 of Fig. 4.10 enables relating the magnetizing current $i_m(t)$ through the inductor L_m to the currents $i_S(t)$ and $i_{Si}(t)$ of the ideal transformer, with effective turns ratio $\mathcal{N}_S/\mathcal{N}_R$, as

$$i_m(t) = i_S(t) - i_{Si}(t) \tag{4.77}$$

Substitution of $i_S(t)$ from Eq. (4.75) into Eq. (4.77) yields

$$i_m(t) = i_S(t) + \left(\frac{\mathcal{N}_R}{\mathcal{N}_S}\right) i_R(t)\Bigg|_{i_R(t)=0} = i_S(t) \tag{4.78}$$

For sinusoidal excitation of the stator coil, that is, $v_S(t) = \sqrt{2}\,|\mathbf{E}|\cos(\omega_E t - \alpha)$, the magnetizing field linking the stator coil and the stationary rotor coil, the response currents $i_S(t)$ and $i_m(t)$, and the response voltages $e_S(t)$ and $e_R(t)$ are sinusoidal in nature. Thus for steady state analysis, the describing equation of Fig. 4.10 becomes

$$\mathbf{V}_S = R_R \mathbf{I}_S + j\omega_E L_{\ell S} \mathbf{I}_S + j\omega_E L_m \mathbf{I}_m \tag{4.79}$$

and Eqs. (4.74), (4.75), and (4.78) become

$$\frac{\mathbf{E}_S}{\mathbf{E}_R} = \frac{\mathcal{N}_S}{\mathcal{N}_R} \tag{4.80}$$

$$\frac{\mathbf{I}_{Si}}{\mathbf{I}_{Ri}} = -\frac{\mathcal{N}_R}{\mathcal{N}_S} \tag{4.81}$$

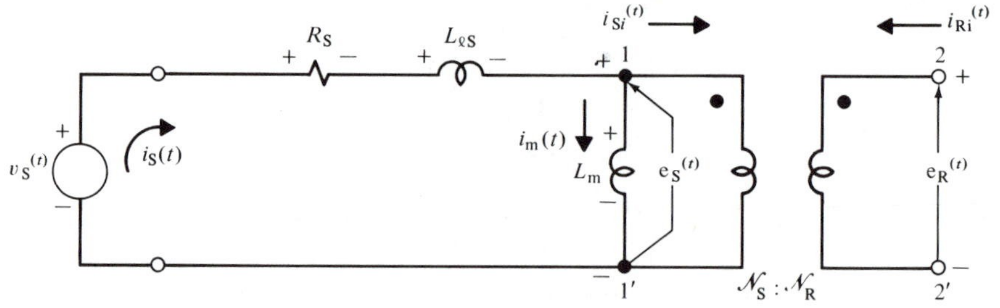

Fig. 4.10 The Per-Phase, t-Domain Equivalent Circuit of the Coupled Stator-Rotor Windings When $\omega_R \equiv 0$ with the Rotor Windings Open Circuited

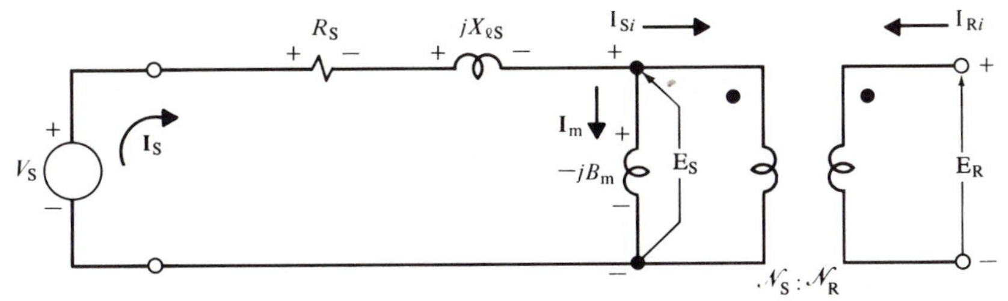

Fig. 4.11 The Per-Phase, $j\omega$-Domain Equivalent Circuit of the Coupled Stator-Rotor Windings When $\omega_R \equiv 0$ with the Rotor Windings Open Circuited

and

$$\mathbf{I}_m = \mathbf{I}_S + \left(\frac{\mathcal{N}_R}{\mathcal{N}_S}\right)\mathbf{I}_R\bigg|_{\mathbf{I}_R=0} = \mathbf{I}_S \tag{4.82}$$

Therefore, for steady state analysis the circuit of Fig. 4.10 can be drawn as in Fig. 4.11.

Now if the rotor windings are short-circuited, Fig. 4.8(b), and the rotor is allowed to rotate at an angular velocity $\omega_R \neq 0$, the induced voltages of Eqs. (4.36), (4.37), and (4.38) cause currents of slip frequency $\omega_e = s\omega_E$ to flow through the rotor coils. These currents are designated $i_R(s,t)$ and cause voltage drops on each rotor coil due to the resistance of the rotor coil R_R and the rotor flux linkage $\Phi_{\ell R}$ linking the rotor coils of N_R-turns. Utilization of $\lambda_{\ell R}(t) = N_R \Phi_{\ell R} \equiv L_{\ell R} i_R(s,t)$ enables writing the describing equation of the rotor coil in motion, Fig. 4.8(b), as

$$se_R(t) = R_R i_R(s,t) + L_{\ell R}\frac{d}{dt} i_R(s,t) \tag{4.83}$$

Since the excitation of the rotor coil is sinusoidal, Eq. (4.83) can be written in terms of the phasors $s\mathbf{E}_R$ and $\boldsymbol{\rho}$* and the slip frequency $\omega_e = s\omega_E$ as

$$s\mathbf{E}_R = \left[R_R \boldsymbol{\rho} + j\omega_e L_{\ell R} \boldsymbol{\rho}\right] \tag{4.84}$$

Solution of Eq. (4.84) for $\mathbf{I}_{Ri}$ yields

$$\boldsymbol{\rho} = \frac{s\mathbf{E}_R}{R_R + j\omega_e L_{\ell R}} \equiv \frac{s\mathbf{E}_R}{R_R + j\chi_{\ell R}} \tag{4.85}$$

Note that the $s\mathbf{E}_R$ and $\chi_{\ell R} \equiv \omega_e L_{\ell R} \equiv s\omega_E L_{\ell R} = sX_{\ell R}$ parameters depend on the rotor motion; that is, they depend on the slip s. Thus they are defined as **slip voltage** and *slip reactance* correspondingly. However, $\mathbf{E}_R$, $X_{\ell R}$, and R_R are speed or *slip independent* and are defined as **standstill parameters**. Eq. (4.85) describes the rotor coil in motion. Thus it should be emphasized that the rotor current phasor $\mathbf{I}_{Ri}$ is of slip frequency ω_e, designated as $\mathbf{I}_{Ri}(\omega_e) \equiv \boldsymbol{\rho}$. The equivalent circuit of Eq. 4.85 is exhibited in Fig. 4.12.

Division of the numerator and denominator of Eq. (4.85) by the slip s and addition and subtraction of the quantity R_R from the denominator of Eq. (4.85) yields

$$\mathbf{I}_R \equiv \mathbf{I}_R(s) = \frac{\mathbf{E}_R}{(R_R/s) + jX_{\ell R}} = \frac{\mathbf{E}_R}{R_R + jX_{\ell R} + (R_R/s)(1-s)} \tag{4.86}$$

*$\boldsymbol{\rho}$ stands for the Greek word ρεύμα, meaning current. It is shown in Eq. (4.85) that $\boldsymbol{\rho}$ is a function of the slip frequency ω_e, where $\rho = \mathbf{I}_{Ri}(\omega_E)$.

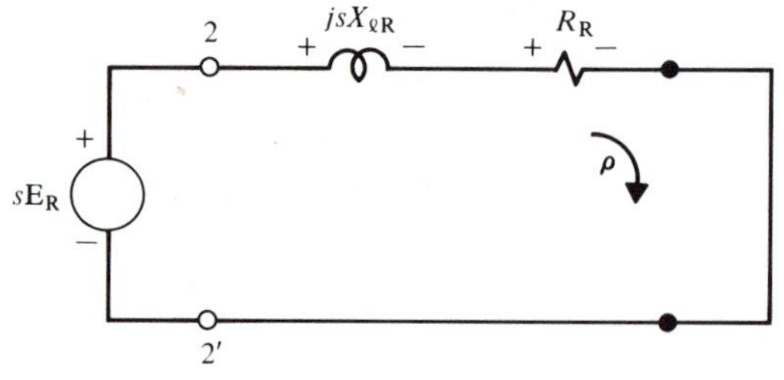

Fig. 4.12 The $j\omega$-Domain Equivalent Circuit of the Rotor Winding on a Per-Phase Basis, with Slip-Dependent, Unreferred Circuit Parameters; $\omega_R \neq 0$

The circuit analog of Eq. (4.86) is exhibited in Fig. 4.13. Note that the rotor winding voltage and reactance [Fig. 4.13 and Eq. (4.86)] are **no longer slip dependent** as they were in Fig. 4.12 and Eq. (4.85). They are in fact **motion independent**; that is, they **possess standstill** values. Also note that the circuit of Fig. 4.13 exhibits a standstill value resistor R_R and a motion-dependent (i.e., slip-dependent) resistor $(R_R/s)(1-s)$, which is drawn with **heavy lines** for distinguishing it from the standstill value resistor. This resistor is defined as the *mechanical load equivalent resistor*, abbreviated henceforth as R_{MLE}, and allows representation of mechanical loads in the equivalent circuit of the induction device. A generalized mechanical load connected on the slip dependent resistor is exhibited in Fig. 4.13. Since the input voltage of Fig. 4.13, $\mathbf{E}_R$, and the output voltage of Fig. 4.11, $\mathbf{E}_R$, represent the standstill value of the voltage across the rotor coil, the two circuits can be joined at the points 2–2′, Fig. 4.14. *Arched connection points* are used at points 2–2′ in order to suggest that the circuit to the right of points 2–2′ represents the rotating rotor windings. However, for analysis purposes (i.e., mathematically) the rotor-winding equivalent circuit is referred to a stationary reference frame fixed on the stator winding; that is, the rotor windings do not move with respect to the stator windings. **The effect of rotor motion is designated by the slip-dependent resistor R_{MLE}, which designates the load of the equivalent circuit of the induction motor.** It must be emphasized that the rotor current in Fig. 4.14 is slip dependent, designated as $\mathbf{I}_R \equiv \mathbf{I}_R(s)$, rather than slip-frequency dependent, designated by $\boldsymbol{\rho} = \mathbf{I}_{Ri}(\omega_e)$, as in Fig. 4.12.

Although the rotor circuit is stationary, Fig. 4.14, the analysis can be much simplified if the rotor circuit is referred to the stator side of the airgap, which is

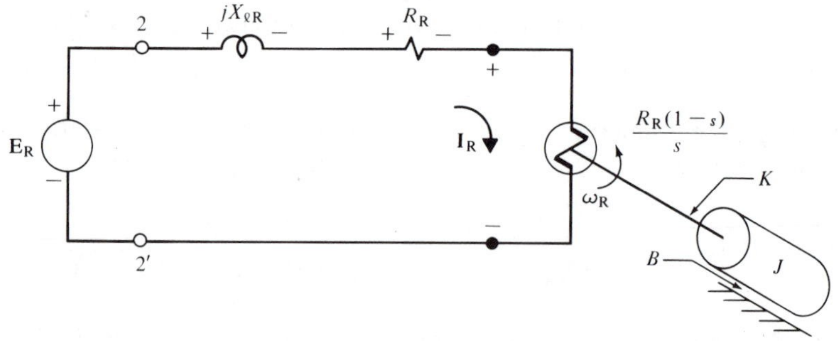

Fig. 4.13 The $j\omega$-Domain Equivalent Circuit of the Rotor Winding on a Per-Phase Basis, with Slip-Independent, Unreferred Circuit Parameters; $\omega_R \neq 0$

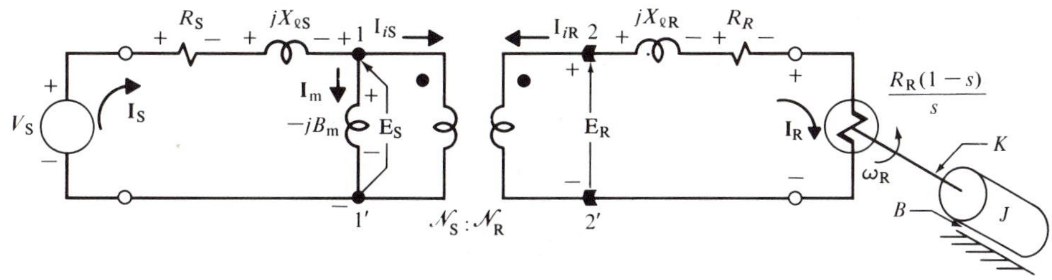

Fig. 4.14 The $j\omega$-Domain Equivalent Circuit of an Induction Device on a Per-Phase Basis, with the Rotor Circuit (on the Rotor Side of the Airgap) Referred to the Stationary Reference Frame of Analysis; $\omega_R \neq 0$

represented by the ideal transformer. To proceed with the transformation of the rotor equivalent circuit to the stator side, the impedance transformation equation is essential. Dividing Eq. (4.74) by Eq. (4.75) yields

$$\mathbf{Z}_S = \left(\frac{\mathcal{N}_S}{\mathcal{N}_R}\right)^2 \mathbf{Z}_R \Big|_{\mathbf{I}_{Ri} = -\mathbf{I}_R} \tag{4.87}$$

Utilization of Eqs. (4.74), (4.75), and (4.87) to refer $\mathbf{E}_R$, $\mathbf{I}_R$, and $\mathbf{Z}_R = R_R + jX_{\ell R} + (R_R/s)(1-s)$ to the stator side of the airgap enables writing the referred-rotor circuit parameters as

$$\mathbf{E}_R{}' = \left(\frac{\mathcal{N}_R}{\mathcal{N}_R}\right)\mathbf{E}_R \tag{4.88}$$

$$\mathbf{I}_R{}' = -\left(\frac{\mathcal{N}_R}{\mathcal{N}_S}\right)\mathbf{I}_R \tag{4.89}$$

and

$$\mathbf{Z}_R{}' = \left(\frac{\mathcal{N}_S}{\mathcal{N}_R}\right)^2 R_R + j\left(\frac{\mathcal{N}_S}{\mathcal{N}_R}\right)^2 X_{\ell R} + \left(\frac{\mathcal{N}_S}{\mathcal{N}_R}\right)^2\left(\frac{R_R}{s}\right)(1-s)$$

$$\equiv R_R{}' + jX_{\ell R}{}' + \left(\frac{R_R{}'}{s}\right)(1-s) \tag{4.90}$$

Utilization of the resistor R_c (or $G_c = 1/R_c$), connected in parallel with the inductor L_m, Fig. 4.10, to represent *eddy current* and *hysteresis losses* in the iron structure of the induction device, as in transformer analysis, enables drawing the complete equivalent circuit of the induction device (with the rotor *referred* to the stator side of the airgap) for steady state analysis as in Fig. 4.15. Note that in order

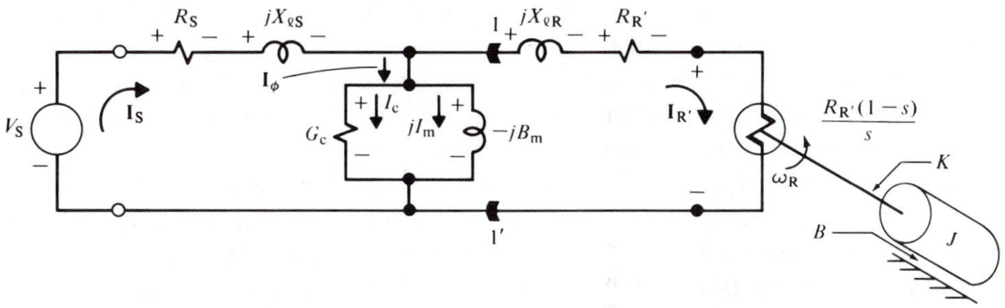

Fig. 4.15 The $j\omega$-Domain Equivalent Circuit of an Induction Device, on a Per-Phase Basis, with the Rotor Circuit Referred to the Stator Side of the Airgap; $\omega_R \neq 0$

to account for the change of the current from $i_m(t)$ in Fig. 4.10 to $i_\phi(t) \equiv i_c(t) + i_m(t)$ in Fig. 4.15, Eq. (4.78) changes to Eq. (4.91), that is

$$\mathbf{I}_\phi = \mathbf{I}_S + \left(\frac{\mathfrak{N}_R}{\mathfrak{N}_S}\right)\mathbf{I}_{Ri}\bigg|_{\mathbf{I}_{Ri} = -\mathbf{I}_R} = \mathbf{I}_S - \mathbf{I}_{R'} \tag{4.91}$$

In phasor notation the current $\mathbf{I}_\phi$ is written as

$$\mathbf{I}_\phi = I_c + jI_m \tag{4.92}$$

$\mathbf{I}_\phi$ is defined as the *exciting current* (with I_c defined as the core-loss component and I_m as the magnetizing component) and enters the magnetizing admittance $\mathbf{Y}_\phi = G_c - jB_m$ ($G_c = 1/R_c$ and $B_m = 1/X_m$). The current $\mathbf{I}_\phi$ is *considerably larger* than the corresponding current for the transformer because of the presence of the airgap—which yields a high reluctance $\mathfrak{R}$—in the magnetic circuit of the induction device. For an induction device the current $|\mathbf{I}_\phi|$ is 25 percent to 40 percent of the rated current, depending upon the size of the induction device. However, for the transformer (depending on the size) the current $|\mathbf{I}_\phi|$ is only 2 percent to 5 percent of the rated current. Also, the leakage reactances of the induction device are *high* when compared to the leakage reactances of the transformer because the windings of the stator and rotor are distributed along the periphery of the airgap rather than being concentrated on a core as in the transformer.

Thus, the equivalent circuit of Fig. 4.15 provides the means for the per-phase analysis of the three-phase induction device, whether its stator windings are Y- or Δ-connected. The parameters of the induction motor's equivalent circuit, however, are *always* given on the basis of its Y-connected stator windings.

4.6 PERFORMANCE CHARACTERISTICS OF THE INDUCTION DEVICE

Before we carry out the mathematical analysis of the induction device, we must have a thorough understanding of how an induction device produces starting torque and, subsequently, running torque. Knowledge of the torque equation $d\tau_R = \mathbf{R} \times d\mathbf{F} = \mathbf{R} \times (i_R d\mathbf{L} \times \mathbf{B}_S)$ makes it clear that **both the presence of the field in the stator and the current in the rotor are essential for its development**. This necessitates the existence of closed paths (i.e., circuits) in the rotor coils. This existence of circuits can be accomplished by (1) shorting the rotor coils, Fig. 4.2(b), (2) forming rotor coil circuits through rotor loading resistors, Fig. 4.2(b), or (3) forming rotor coil circuits through rotor excitation sources, Fig. 4.1. The sense of this torque is always such as *to cause* the rotor to travel in the same direction as the rotating stator field. This becomes evident upon examination of Fig. 4.16.

With the rotor stationary and a counterclockwise rotation of the stator field $|\mathbf{B}_{ST}|$ at synchronous speed ω_S in radians/second, the apparent velocity $\mathbf{u}$ in meters/second of the rotor is oriented in the clockwise direction. The emf-inducing field intensity $\mathcal{E} = \mathbf{u} \times \mathbf{B}_{ST}$ and its direction are indicated appropriately in Fig. 4.16. For the indicated direction of the vector $\mathcal{E}$ and the rotor coil forming a closed circuit, the generated current i_R is out of the page in coil side No. 2. Thus the developed electromagnetic torque in the rotor coil side No. 2, $d\tau_2 = \mathbf{R} \times (i_R d\mathbf{L} \times \mathbf{B}_{ST}) \equiv \mathbf{R} \times d\mathbf{F}_2$, causes a counterclockwise rotation of the rotor. Therefore, the rotor rotates in a direction in which it tries to catch up with the rotating stator field $|\mathbf{B}_{ST}|$.

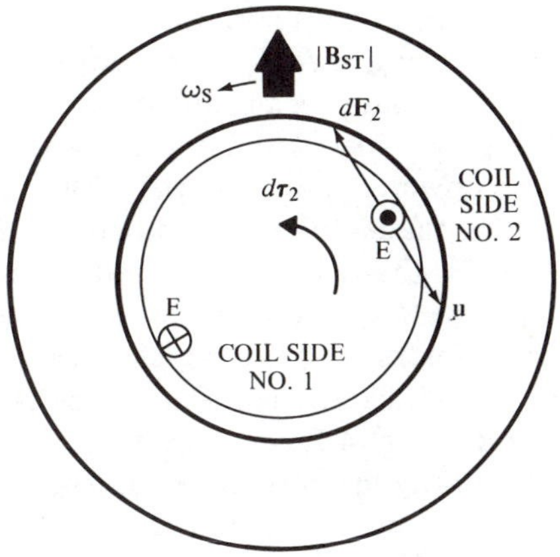

Fig. 4.16 Torque Developed in the Rotor Winding When It Forms a Closed Circuit, Causing the Rotor to Rotate in the Direction of Rotation of the Stator Field $|\mathbf{B}_{ST}|$

As the rotor speed increases, the rate at which the stator field cuts the rotor coils decreases. This reduces the induced emf, $e_R = \oint(\mathbf{u}\times\mathbf{B}_{ST})\cdot d\mathbf{L}$ per phase, which in turn reduces the magnitude of the rotor current per conductor, that is, the rotor current density. This reduction yields less torque. This process continues until the proper rotor speed ω_R is reached, which yields enough emf $e_R(t)$ to produce just the current needed to develop a torque τ_{em} equal to the opposing torque, comprised of the sum of τ_J, τ_B (or* τ_R), τ_K, τ_a, and τ_ℓ. If there is no shaft loading, Fig. 4.7, the opposing torque **consists mainly of frictional losses**. It is emphasized that the reader must understand that as long as there is an opposing torque to overcome, regardless of magnitude and origin, **the rotor speed can never be equal to the synchronous speed**. Since the rotor-winding current is produced by induction, there must always be a difference between the speed ω_S of the stator field $|\mathbf{B}_{ST}|$ and the rotor speed ω_R; that is, *for an induction device with a passive mechanical load,* $\omega_R < \omega_S$ *always.*

Reference to the per-phase equivalent circuit of Fig. 4.15 enables us to understand quantitatively the operation of the induction device. If the device is operating at no-load, that is, only $\tau_R \neq 0$, the slip $s = (\omega_S - \omega_R)/\omega_S$ has a value very close to zero. Since the mechanical equivalent load resistor is a function of s, that is, $R_{MLE} = (R_R/s)(1-s)$, R_{MLE} has a *very large* value, which in turn causes a *small* rotor current $\mathbf{I}_R'$ to flow. The corresponding electromagnetic torque $\tau_{em}(\mathbf{B}_{ST}, i_R)$ merely assumes the value that is required to overcome the rotational losses τ_R, which consist chiefly of friction and windage. If a mechanical load is applied to the shaft as the device is rotating, the initial reaction of the device is a slow down. This reduction in rotor speed ω_R causes an increase in the slip s, Eq. (4.29). The increased slip causes a *reduction* of $R_{MLE} = (R_R/s)(s-1)$ and thus an *increase* of $\mathbf{I}_R'$ to a value that will cause a sufficient electromagnetic torque to develop, which will provide the balance of torque to the load. Thus equilibrium will be established and operation of the device will proceed at a particular value of slip s. In fact, for each loading set, that is, LS$\{\tau_J, \tau_B, \tau_K, \tau_a, \tau_\ell, \tau_R\}$, there is a unique value of slip s. This can be verified from the equivalent circuit of the induction device, Fig. 4.15, which

*Where $\tau_R \equiv \tau_B + \tau_W$; τ_R defines the rotational losses, τ_B defines the frictional losses, and τ_W defines the windage losses, all in Newton meters.

shows that once the slip s is specified, the input power $P_{IN} = \text{Re}\{\mathbf{V}_S\mathbf{I}_S^*\}$, the power losses in the stator windings $P_{SW} = R_S|\mathbf{I}_S|^2$, the airgap power $P_{AG} = \{R_R' + R_R'(1-s)/s\}|\mathbf{I}_R'|^2$ (i.e., the power that is transmitted across the airgap of the device and enters the rotor), the power losses in the rotor windings $P_{RW} = R_R'|\mathbf{I}_R'|^2$, the developed electromagnetic power $P_{em} = (R_R'/s)(1-s)|\mathbf{I}_R'|^2$ (i.e., the power that is utilized when an attempt is made to rotate the rotor), the output shaft power $P_O \equiv P_{SHAFT} \equiv P_{USEFUL} = P_{em} - P_R$ that can be delivered by the shaft to a mechanical load, and the efficiency $\eta = P_O/P_{IN}$ of the device for a given set of core losses P_{CL} (eddy current plus hysteresis losses) and rotational losses P_{RL} (losses due to friction and wind resistance) can all be determined. It must be pointed out that although the resistor R_c (or $G_c = 1/R_c$), Fig. 4.15, representing equivalent core losses in the induction device, is given, usually the core losses are specified in the statement of a problem (e.g., P_{CLT} = XX.XX watts), and the resistor R_c (or G_c) can be omitted from the equivalent circuit of the induction motor, Fig. 4.15, in the ensuing analysis. Rotational losses are specified in a similar manner (e.g., P_{RLT} = YY.YY watts).

Utilization of circuit theory techniques, the theory of Appendix A, and the theory of Chapter 1 enables calculation of the various powers for the **per-phase circuit** of Fig. 4.15 and thus of the induction device by multiplying these answers by a *factor of three*. The procedure is summarized in Fig. 4.17 in the format of a flow diagram. Remembering that power and torque are related by the defining relationship $p = \tau\omega$, we can calculate the electromagnetic torque τ_{em} and the output torque τ_O from the following equations:

$$P_{emT} = \tau_{emT}\omega_R \tag{4.93}$$

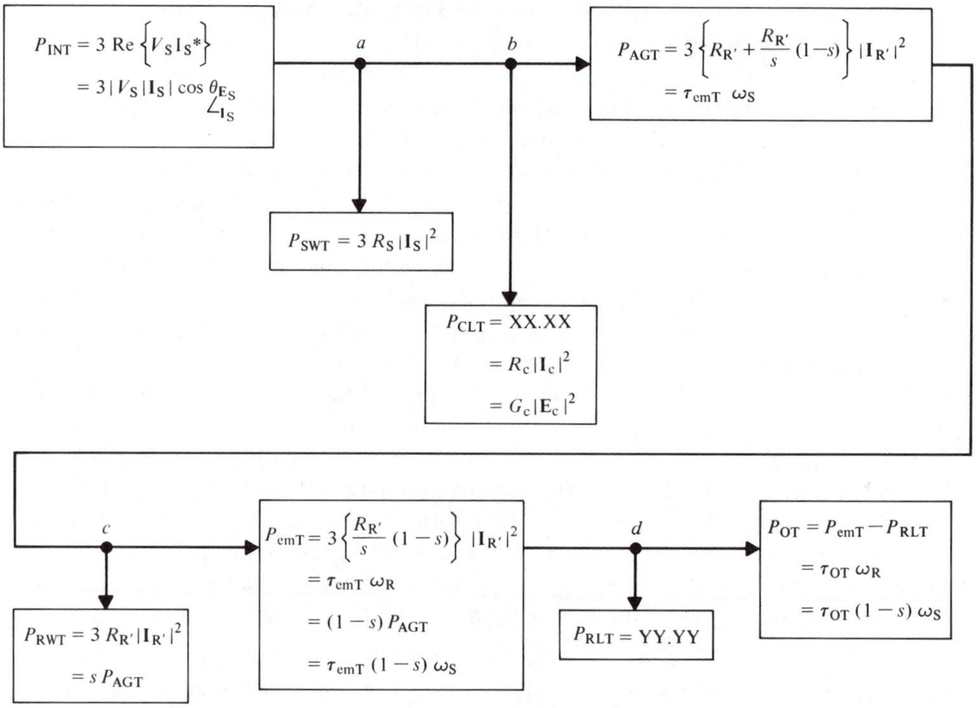

Fig. 4.17 Power Flow Diagram of a Balanced, Three-Phase, *P*-Pole Induction Motor

and

$$P_{OT} = \tau_{OT}\omega_R \tag{4.94}$$

However, utilization of the defining relationship for slip, Eq. (4.29), enables writing ω_R as

$$\omega_R = (1-s)\omega_S \tag{4.95}$$

Substitution for ω_R from Eq. (4.95) into Eq. (4.93) enables writing P_{emT} as

$$P_{emT} = \tau_{emT}(1-s)\omega_S = (1-s)P_{AGT} \tag{4.96}$$

where

$$P_{AGT} \equiv \tau_{emT}\omega_S \tag{4.97}$$

From the power flow diagram, Fig. 4.17, it can be seen, by summing at the junction point c, that

$$P_{AGT} = P_{RWT} + P_{emT} \tag{4.98}$$

Substituting for $P_{emT}=f(s,P_{AGT})$ from Eq. (4.96) into Eq. (4.98) and solving for $P_{RWT}=g(s,P_{AGT})$ yields

$$P_{RWT} = P_{AGT} - (1-s)P_{AGT} = sP_{AGT} \tag{4.99}$$

The performance of the induction device can be judged by examination of two performance indices, (1) the efficiency η and (2) the speed regulation SR, both in percentages. Efficiency of the induction motor is defined as

$$\eta = \frac{P_{OT}}{P_{INT}} = \frac{P_{INT}-\Sigma P_{jT}}{P_{INT}} = 1 - \frac{\Sigma P_{jT}}{P_{INT}}, \quad j = \text{SW, CL, RW, RL} \tag{4.100}$$

where the subscripts refer to: SW = stator winding, CL = core losses, RW = rotor winding, RL = rotational losses, and speed regulation is defined as

$$\text{SR} = \frac{\omega_{NL}-\omega_{FL}}{\omega_{FL}} = \frac{\omega_S-\omega_R}{\omega_R} \tag{4.101}$$

where the subscripts FL and NL designate full load and no load, correspondingly. Note that the ideal speed of the induction motor is the synchronous speed, ω_S. It is achieved only when loading, frictional, and windage losses are zero. **This never occurs in a real device.** At this speed, or at a speed $\omega_R \simeq \omega_S$, the speed regulation is zero. This concurs with the statement made in Section 2.6, that is, the SR (or VR in the case of the transformer) approaches zero as the quality of the device becomes better.

It should be evident now that the power flow diagram of Fig. 4.17 in conjunction with the equivalent circuit of Fig. 4.15 make the computation of the performance characteristics of the induction device an easy and straightforward task. It should be pointed out, however, that before we calculate the performance characteristics of the induction device, the currents $\mathbf{I}_S$, $\mathbf{I}_R'$ and $\mathbf{I}_\phi$ of its equivalent circuit, Fig. 4.15, must be calculated, utilizing pertinent circuit theories.

One final remark that should be made before leaving this section is the following: Considerable simplification of computation can be achieved by moving the magnetizing admittance $\mathbf{Y}_\phi$, Fig. 4.15, to the input terminals of the equivalent circuit. Under these conditions the equivalent circuit of the induction motor becomes that of Fig. 4.18. Although the error introduced is small under some circumstances, this circuit may be quite misleading during start up or heavily

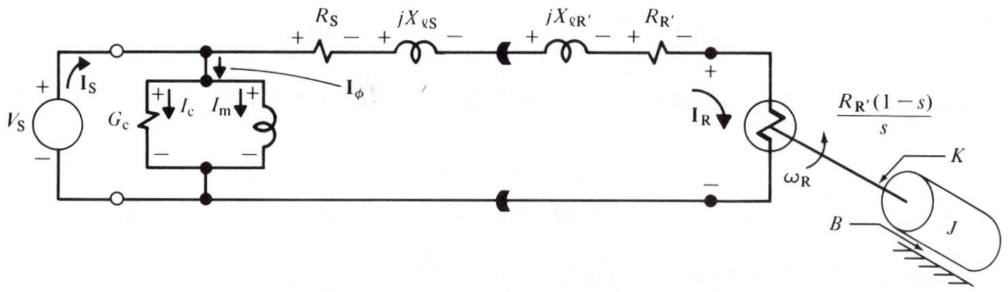

Fig. 4.18 Oversimplified Equivalent Circuit of a Three-Phase Induction Device on a Per-Phase Basis (Inherent Error Is Introduced to the Various Computations)

loaded cases. Therefore, it is recommended that the oversimplified equivalent circuit of Fig. 4.18 should be avoided, *unless* one is seeking rough calculations of the performance characteristics of induction devices.

Example 4.6

For the three-phase, six-pole induction motor of Fig. 4.15, the following parameters are specified: (1) the peak* line-to-line input voltage is 440 V, (2) $f_E = 60$ Hz, (3) $\mathbf{Z}_S = 0.2 + j0.38\ \Omega$, (4) $\mathbf{Z}_{R'} = 0.12/s + j0.38\ \Omega$, (5) $jx_m = j10\ \Omega$, and (6) $P_{LOSSES} = P_{CL} + P_F + P_W = 1600$ W.

If the motor is operating at a slip of 3%, find the following motor parameters: (1) the synchronous speed of the stator field ω_S, (2) the speed of the rotor ω_R, (3) the input power P_{INT}, (4) the stator winding losses P_{SWT}, (5) the airgap power P_{AGT}, (6) the rotor winding losses P_{RWT}, (7) the electromagnetic power P_{emT}, (8) the electromagnetic torque τ_{emT}, (9) the output power P_{OT}, (10) the output torque τ_{OT}, (11) the efficiency η, (12) the speed regulation SR, (13) the frequency of the rotor currents ω_e, (14) the synchronous speed of the rotor ω_s, and (15) determine what would happen if the rotor windings were to be open-circuited while the motor was running (operating).

Solution

Since the total losses are specified, the conductance G_c of the equivalent circuit, Fig. 4.15, must be assumed to be equal to zero. Thus Fig. 4.15 can be drawn as in Fig. 4.6.1.

The input RMS phase voltage $\mathbf{V}_S$ is calculated as

$$\mathbf{V}_S = \frac{440e^{j0°}}{\sqrt{3}\,\sqrt{2}} = \underline{179.63e^{j0°}\text{ V}}$$

and

$$R_{MLE} = \left(\frac{0.12}{s}\right)(1-s) = \left(\frac{0.12}{0.03}\right)(1-0.03) = \underline{3.88}\ \Omega$$

The input impedance $\mathbf{Z}_S$ of the circuit is

$$\mathbf{Z}_S = 0.2 + j0.38 + \left[\frac{j10(4+j0.38)}{4+j(10+0.38)}\right] = \underline{3.97e^{j30.12°}}\ \Omega$$

*Unless otherwise stated, the RMS value of the voltages and currents is given; however, to stress the distinction between RMS and peak values, and their uses, the peak value of the voltage is given here.

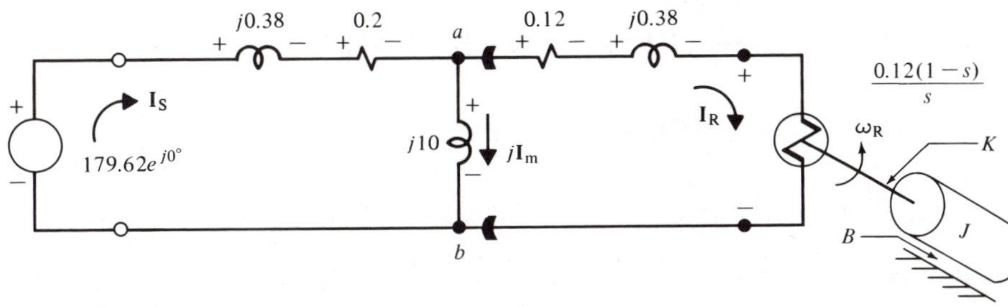

Fig. 4.6.1 The $j\omega$-Domain Equivalent Circuit of the Induction Motor on a Per-Phase Basis; the Rotor Winding Is Referred to Stator Side of the Airgap

Thus the stator phase current $\mathbf{I}_S$ is calculated as

$$\mathbf{I}_S = \frac{\mathbf{V}_S}{\mathbf{Z}_S} = \frac{179.63e^{j0^\circ}}{3.97e^{j30.12^\circ}} = \underline{45.25e^{-j30.12^\circ}}\text{ A}$$

Utilizing the current dividing equation enables calculation of the referred rotor current as

$$\mathbf{I}_R' = \frac{j10}{4+j10.38} \times \mathbf{I}_S = \frac{10e^{j90^\circ}(45.25e^{-j30.12^\circ})}{11.12e^{j68.93^\circ}} = \underline{40.69e^{-9.05^\circ}}\text{ A}$$

Utilizing Eqs. (4.22), (4.29), and Fig. 4.17, and Eqs. (4.32) and (4.53), enables calculation of the following quantities:

1. $\omega_S = \frac{\omega_E}{P/2} = \frac{2\pi \times 60}{6/2} = \underline{\underline{125.7}}\text{ rad/s}$
2. $\omega_R = \omega_S(1-s) = 125.7 \times 0.97 = \underline{\underline{121.93}}\text{ rad/s}$
3. $P_{INT} = 3|\mathbf{V}_S||\mathbf{I}_S|\cos\theta = 3 \times 179.63 \times 45.25\cos(-30.12^\circ) = \underline{\underline{21{,}092.25}}\text{ W}$
4. $P_{SWT} = 3R_S|\mathbf{I}_S|^2 = 3 \times 0.2(45.25)^2 = \underline{\underline{1{,}228.54}}\text{ W}$

Plate 4.3 Fabricated Aluminum Rotor for Induction Motor with Integral Cast—Four or More (Even Number) Poles. (Courtesy of General Electric Co., Schenectady, N.Y.)

5. $P_{AGT}=3\left\{R_R'+\left(\frac{R_R'}{s}\right)(1-s)\right\}|\mathbf{I}_R'|^2=3\left(\frac{R_R'}{s}\right)|\mathbf{I}_R'|^2$

$$=3\times\left(\frac{0.12}{0.03}\right)(40.69)^2=\underline{\underline{19{,}868.11}}\text{ W}$$

6. $P_{RWT}=3R_R'|\mathbf{I}_R'|^2=sP_{AG}=3\times0.12(40.69)^2=0.03\times19{,}868.11=\underline{\underline{596.04}}\text{ W}$

7. $P_{emT}=3\left\{\left(\frac{R_R'}{s}\right)(1-s)\right\}|\mathbf{I}_R'|^2=3\left\{\left(\frac{0.12}{0.03}\right)(1-0.03)\right\}(40.69)^2=\underline{\underline{19{,}272.07}}\text{ W}$

8. $\tau_{emT}=\frac{P_{emT}}{\omega_R}=\frac{19{,}272.07}{121.93}=\underline{\underline{158.06}}\text{ N}\cdot\text{m}$

9. $P_{OT}=P_{emT}-P_{LOSSES}=19{,}272.07-1{,}600=\underline{\underline{17{,}672.07}}\text{ W}$

10. $\tau_{OT}=\frac{P_{OT}}{\omega_R}=\frac{17{,}672.07}{121.93}=\underline{\underline{144.94}}\text{ N}\cdot\text{m}$

11. $\eta=\frac{P_{OT}}{P_{INT}}\times100=\frac{17{,}672.07}{21{,}092.25}\times100=\underline{\underline{83.78\%}}$

12. $SR=\frac{\omega_{NL}-\omega_{FL}}{\omega_{FL}}\times100=\frac{\omega_S-\omega_R}{\omega_R}\times100=\frac{125.70-121.93}{121.93}\times100=\underline{\underline{3.09\%}}$

13. $\omega_e=s\omega_E=0.03\times2\pi\times60=\underline{\underline{11.31}}\text{ rad/s}$

14. $\omega_s=\frac{\omega_e}{P/2}=\frac{11.31}{3}=\underline{\underline{3.77}}\text{ rad/s}$

15. The currents of the rotor coils will cease to exist; the rotor field will vanish and the rotor will be forced to stop due to friction.

4.7 INDUCTION DEVICE TORQUE VERSUS SLIP CHARACTERISTIC

The speed variation of the induction device as related to developed electromagnetic torque or power is of utmost importance to the electrical engineer. To develop the mathematical relationship that leads to τ_{em} versus $s(\omega_R)$ or P_{em} versus $s(\omega_R)$, the Thevenin's equivalent circuit of Fig. 4.15 to the **left** of points 1–1′ is essential. Thus

$$\mathbf{Z}_{TH}\equiv\left.\frac{\mathbf{Z}_S\times\mathbf{Z}_\phi}{\mathbf{Z}_S+\mathbf{Z}_\phi}\right|_{\mathbf{V}_S=0}=\frac{R_S+jX_{\ell S}}{(1+R_SG_c+B_mX_{\ell S})+j(G_cX_{\ell S}-R_SB_m)}$$
$$\equiv R_{TH}+jX_{TH} \tag{4.102}$$

and

$$\mathbf{V}_{TH}\equiv\frac{\mathbf{Z}_\phi}{\mathbf{Z}_\phi+\mathbf{Z}_S}\times\mathbf{V}_S=\frac{1}{(1+R_SG_c+B_mX_{\ell S})+j(G_cX_{\ell S}-R_SB_m)}\times\mathbf{V}_S$$
$$\equiv(\gamma+j\delta)\mathbf{V}_S \tag{4.103}$$

Eqs. (4.102) and (4.103) enable us to draw the equivalent circuit, on a per-phase basis, of the induction motor (given by Fig. 4.15) as exhibited in Fig. 4.19.

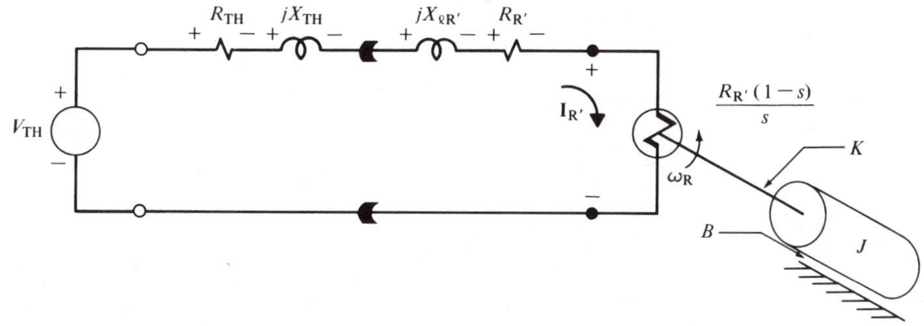

Fig. 4.19 The Equivalent Circuit, on a Per-Phase Basis, of an Induction Motor Simplified by Thevenin's Theorem

The current through the mechanical-loading equivalent resistor $R_{MLE} = R_R'(1-s)/s$ can be calculated with the aid of Fig. 4.19 and Eq. (4.103) as

$$\mathbf{I}_R' = \frac{\mathbf{V}_{TH}}{(R_{TH} + R_R' + R_{MLE}) + j(X_{TH} + X_{\ell R}')} = \frac{\gamma + j\delta}{[R_{TH} + (R_R'/s)] + j(X_{TH} + X_{\ell R}')} \times \mathbf{V}_S \tag{4.104}$$

The **total** developed electromagnetic power P_{emT} by the three-phase induction device is written with the aid of Figs. 4.17 and 4.19 as

$$\begin{aligned} P_{emT} &= 3 \times \left(\frac{R_R'}{s}\right)(1-s)|\mathbf{I}_R'|^2 \\ &= 3 \times \left(\frac{R_R'}{s}\right)(1-s) \times \frac{\gamma^2 + \delta^2}{[R_{TH} + (R_R'/s)]^2 + (X_{TH} + X_{\ell R}')^2} \times |\mathbf{V}_S|^2 \end{aligned} \tag{4.105}$$

Since $P_{emT} = \tau_{emT}\omega_R$ and $\omega_R = (1-s)\omega_S$, the total developed electromagnetic torque by the three-phase induction device is written as

$$\tau_{emT} = \frac{2}{\omega_S} \times \frac{R_R'}{s} \times \frac{\gamma^2 + \delta^2}{[R_{TH} + (R_R'/s)]^2 + (X_{TH} + X_{\ell R}')^2} \times |\mathbf{V}_S|^2 \tag{4.106}$$

The graphical relationship between the total electromagnetic torque (τ_{emT}) and slip (s) or speed (ω_R) is exhibited in Fig. 4.20 for $-\infty < s < \infty$. This curve is simply a sketch here; however, it can be plotted explicitly by utilizing the digital computer if all the parameters of Eq. (4.106) are known and the slip is varied over the range $-\infty < s < \infty$. Such a plot is exhibited in Fig. 4.7.2 of Example 4-7. For the purpose of this discussion, however, Fig. 4.20 will suffice. Since the function of a motor is to produce torque, the region of $s > 0$ for which τ_{emT} versus $s(\omega_R)$ is positive is defined as the **motor region of operation**. However, the region of $s < 0$ for which τ_{emT} versus $s(\omega_R)$ is negative is defined as the *generator region of operation*; that is, the induction device demands electromagnetic torque to function in this region. In addition to these two regions, there is one additional region of operation that is of particular interest to the electrical engineer and four points on the τ_{emT} versus $s(\omega_R)$

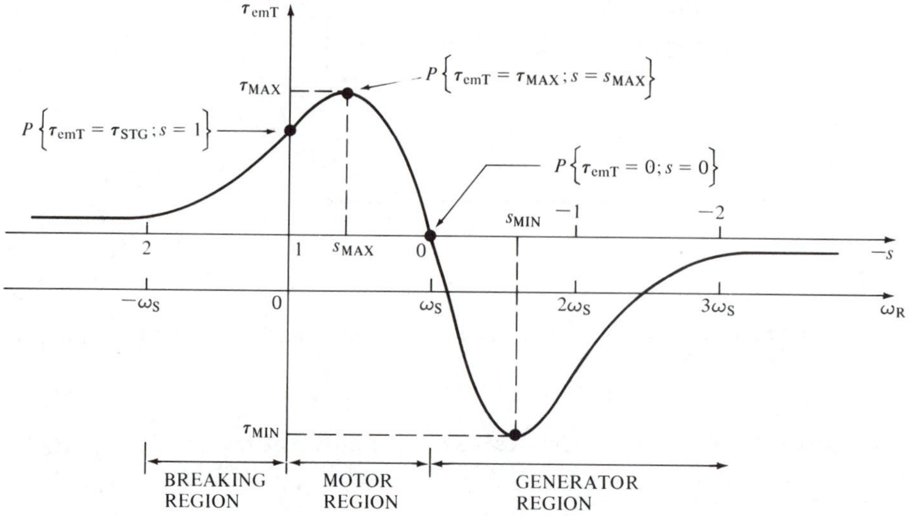

Fig. 4.20 Total Torque vs. Slip (Speed) Characteristic of an Induction Device

characteristic that warrant particular attention. A discussion of these pertinent items follows.

1. Starting point $P\{\tau_{emT}=\tau_{STG}, s=1\}$. When starting the motor, $\omega_R=0$; therefore, $s=(\omega_S-\omega_R)/\omega_S$ and $\omega_S+\omega_E/(P/2)$ leads to $s=1$. This leads to the observation that the τ_{emT} versus $s(\omega_R)$ characteristic has a nonzero value at starting. This torque is designated as τ_{STG} in Fig. 4.20, and its numerical value can be changed by shifting or changing the shape of the τ_{emT} versus $s(\omega_R)$ characteristic. (Such change will be discussed shortly.) The important thing to observe here, however, is that **the three-phase induction motor has an inherent starting torque** and will start to turn as soon as the electrical source is connected unless the mechanical load torque exceeds the starting torque of the motor.

2. Operating points $P\{\tau_{emT}=0, s=0\}$. As the speed of the motor builds up toward the synchronous speed, that is, $\omega_R \to \omega_S$ but $\omega_R \neq \omega_S$, the point $s=0$ is approached. Since there are always rotational losses present, even if there is no other loading present, ω_R is never equal to ω_S. Therefore, *the three-phase induction motor always operates at the region where* $s=0^+$, that is, to the left of the point $P\{\tau_{emT}=0, s=0\}$ but close to it. Typical values of s for the motor mode of operation are $0.03 \leqslant s \leqslant 0.05$.

3. Maximum developed torque point $P\{\tau_{emT}=\tau_{MAX}, s=s_{MAX}\}$. The induction motor can operate near this point for short intervals of time. However, the motor currents are 200 to 300% of the rated value at these high torques. Continuous operation of the device near this point will result in overheating and the destruction of the insulation and ultimately destruction of the device itself. If the load torque exceeds T_{MAX} the motor will stall. The torque τ_{MAX} is defined as the **breakdown** (or maximum) torque of the induction motor. It is a measure of the *reserve capacity* of the motor and permits the motor to operate through momentary peak loads. The value of the slip at which the breakdown torque occurs is defined as s_{MAX} and can be calculated by setting the derivative of torque with respect to slip equal to zero.

Differentiating equation (4.106) yields

$$\frac{\partial \tau_{emT}}{\partial s} = 0 \Rightarrow \left[R_{TH} + \frac{R'}{s_{MAX}}\right]^2 + [X_{TH} + X_{\ell R}']^2 + 2s_{MAX}\left[R_{TH} + \left(\frac{R'}{s_{MAX}}\right)\right]\left[-\frac{R'}{s_{MAX}}\right] = 0 \tag{4.107}$$

or

$$R_{TH}^2 + 2R_{TH}\frac{R'}{s_{MAX}} + \left(\frac{R'}{s_{MAX}}\right)^2 + [X_{TH} + X_{\ell R}']^2 - 2R_{TH}\frac{R'}{s_{MAX}} - 2\left(\frac{R'}{s_{MAX}}\right)^2 = 0 \tag{4.108}$$

Equation (4.108) may be solved for s_{MAX}:

$$s_{MAX} = \frac{R_R'}{\left[R_{TH}^2 + (X_{TH} + X_{\ell R}')^2\right]^{1/2}} \tag{4.109}$$

The breakdown torque τ_{MAX}, as calculated by substituting $s = s_{MAX}$ from Eq. (4.109) and $\omega_R = (1-s)\omega_S$ into Eq. (4.106), is

$$\tau_{MAX} = \frac{3}{\omega_S} \times |V_S|^2 \times \frac{0.5(\gamma^2 + \delta^2)}{R_{TH} + \left[R_{TH}^2 + (X_{TH} + X_{\ell R}')^2\right]^{1/2}} \tag{4.110}$$

or

$$\tau_{MAX} = \frac{3}{\omega_R} \times (\gamma^2 + \delta^2)|\mathbf{V}_S|^2 \times \left\{1 - \frac{R_R'}{\left[R_{TH}^2 + (X_{TH} + X_{\ell R}')^2\right]^{1/2}}\right\} \times \frac{0.5}{R_{TH} + \left[R_{TH}^2 + (X_{TH} + X_{\ell R}')^2\right]^{1/2}} \tag{4.111}$$

Note that τ_{MAX} is a function of the synchronous speed ω_S in Eq. (4.110) and a function of ω_R in Eq. (4.111). Also, note that s_{MAX} in Eq. (4.109) is a linear function of R_R'. Thus an increase or decrease of R_R' will result in a corresponding increase or decrease of s_{MAX}; that is, the point s_{MAX} can be *shifted* left or right from its position in Fig. 4.20. However, Eq. (4.110) suggests that τ_{MAX} is not affected by R_R'. Thus it is obvious from these two observations that by changing R_R' we change the slope of the τ_{emT} versus $s(\omega_R)$ characteristic, Fig. 4.20 [for substantiation see Fig. 4.23(e)], and not its peak value, and therefore we affect the speed of the motor. The subject of speed control will be studied in the following section.

4. Motor region of operation $0 < s < 1$. This region is the defined region for the motor mode of operation. It is divided into two regions: (a) the region $0 < s < s_{MAX}$, which is defined as the **stable** region of operation, with $\doteq 0$ being the normal region of operation, and (b) the region $s_{MAX} < s < 1$, which is defined as the **unstable** region of operation. The motor never operates in this region unless the motor is starting or stopping. If the motor slips into this region of operation, this means that the motor is overloaded. The motor cannot develop enough torque to overcome the load and will come to a full stop.

5. Braking region of operation $1 < s < 2$. The motor cannot physically operate in this region unless the rotor is driven backwards against the direction of the rotating magnetic field $|\mathbf{B}_{ST}|$ by a source of mechanical torque capable of counteracting the developed electromagnetic torque τ_{emT}. The usefulness of this region comes into

play when we want to stop induction motors quickly by utilizing the negative electromagnetic torque developed by the motor itself. The three-phase induction motor *can be stopped* by interchanging any two of the stator leads. When this occurs, the phase sequence of the excitation as far as the motor is concerned changes suddenly, and thus the direction of rotation of the magnetic field $|\mathbf{B}_{ST}|$ changes. The developed electromagnetic torque τ_{emT} has an opposing effect to the direction of rotation of the motor. This torque forces the motor to stop if the forcing function of the motor is disconnected before the motor starts rotating in the opposite direction. This method of stopping a three-phase induction motor is known as **plugging**.

6. Generator region of operation $0 < -s < \infty$. In this region of operation the induction device can perform either as a **frequency converter** or as a **generator of real power**.

Frequency Converter. For this mode of operation, (1) the induction device must have the **wound-type rotor**, Figs. 4.2(a) and 4.2(b), with each of its three windings terminating to a slip ring, and (2) *the rotor must be driven* by an external prime mover at a constant speed ω_R higher than the synchronous speed ω_S of the stator field $|\mathbf{B}_{ST}|$, either in the *same* or the *opposite* direction of the field rotation. The function of the induction device operating in this mode is to generate voltages of a desired frequency ω_e other than the frequency of the forcing function ω_E.

Such frequencies can be used to drive ac motors at higher speeds and efficiencies if silicon-controlled rectifier (SCR) equipment is not available or if mechanical schemes such as gears or belts and pulleys are too inefficient and cumbersome to consider.

This mode of operation is based on the slip equation $s = (\omega_S - \omega_R)/\omega_S$. If the rotor of the device is driven, say, at a speed $\omega_R = 3\omega_S$ (i.e., in the direction in which the stator field rotates), the slip becomes

$$s = \frac{\omega_S - 3\omega_S}{\omega_S} = -2 \tag{4.112}$$

However, if the rotor is driven at a speed $\omega_R = -3\omega_S$ (i.e., in the opposite direction of the stator field rotation), the slip becomes

$$s = \frac{\omega_S + 3\omega_S}{\omega_S} = 4 \tag{4.113}$$

In both instances the voltage extracted from the slip rings is of the frequency

$$\omega_e \equiv s\omega_E = 2\omega_E \quad \text{or} \quad 4\omega_E \tag{4.114}$$

The principal disadvantage of the frequency converter is its poor voltage regulation. This is due to the presence of the airgap in the magnetic circuit, which causes high equivalent circuit reactances and, consequently, high voltage drops across these reactances; thus there are low voltages at the terminals of the device.

The connection and the equivalent circuit of the induction device in this mode of operation are exhibited in Figs. 4.21(a) and 4.21(b). You should consult the literature if you have an interest in this subject (reference [8] is an excellent starting point).

Real Power Generator. For this mode of operation, (1) the windings of the wound-type rotor must be **short-circuited** and (2) the rotor of the induction device *must be driven* by a prime mover, in the **same direction** as that of the rotating stator field $|\mathbf{B}_{ST}|$, at any speed ω_R higher than the synchronous speed ω_S (the speed of the

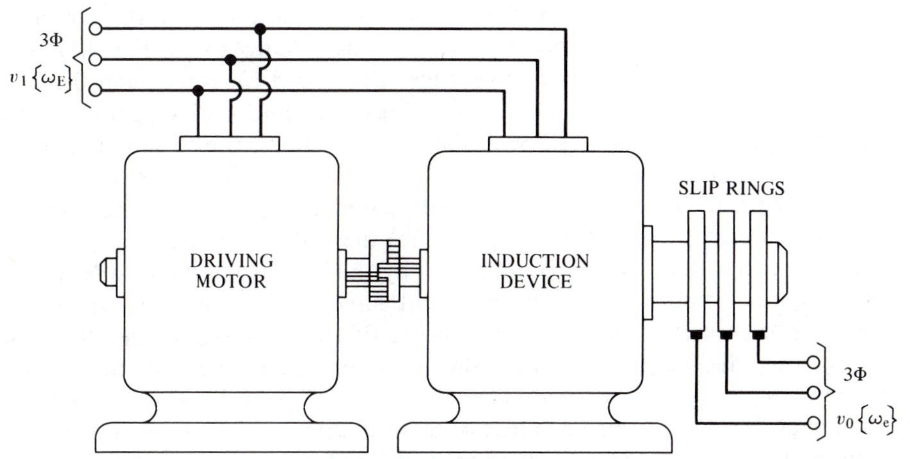

(a)

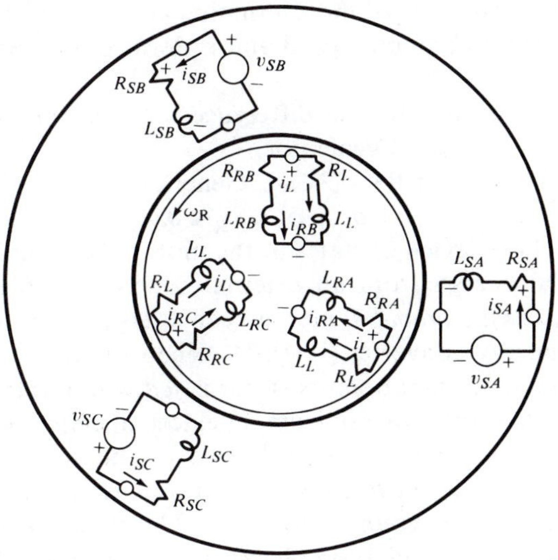

(b)

Fig. 4.21 Induction Device Connection for Its Operation as a Generator. The Input Is of Frequency ω_E rad/s but the Output Is of Frequency $\omega_e = s\omega_E$; $s = (\omega_R - \omega_S)/\omega_S$: (a) Pictorial Diagram and (b) Stator and Rotor-Winding Circuit Representation

synchronously rotating stator field), that is, $\omega_R > \omega_S$. The real power that the device generates is furnished to the supply mains, through which the device receives its excitation, Fig. 4.21(a). Eq. (4.29) reveals that **under these conditions the slip is always negative**. Therefore, the mechanical load equivalent resistance $R_{MLE} = R_R'(1-s)/s$, Fig. 4.15, is negative, and the total developed electromagnetic power P_{emT}, Eq. (4.105) and Fig. 4.17, is negative. This means that **the rotor absorbs negative real power from the stator**—that is, the rotor furnishes the stator with real power. This electric power is a consequence of the transformation of mechanical energy—provided by the prime mover driving the rotor—into electric power. This result also becomes obvious from the following observation: since the slip changes sign, the induced potential in the rotor winding also changes sign, Fig. 4.12, and so does the current in the rotor windings, Figs. 4.12 and 4.13. If we keep the idea of power flow reversal in mind, inspection of Figs. 4.17 and 4.15 suggests that while some of the generated power is lost in the rotor windings, the remaining generated power is delivered to the stator; after the core losses and the stator winding losses are accounted for, the balance is delivered to the supply mains in the form of an alternating current going out of (output from) the stator windings.

Notice that the stator field is not affected by the increase of the rotor's speed but revolves in the same direction as it would have done should the device have operated as an induction motor. Furthermore, although the sign of the slip changes when $\omega_R > \omega_S$, the frequency relationship between the currents in the rotor and stator windings does not change. That is, Eq. (4.32) is still valid; therefore, the frequency of the stator currents is that of the excitation—ω_E radians/second. Since the real power generator is **not self-excited**, the stator windings must be excited by a balanced three-phase system of fixed voltages and frequencies; also, the rotor must be driven at a speed ω_R higher than the synchronous speed ω_S. Precisely for this reason the real power generator is defined as the **asynchronous generator**. In contrast, the classical generator of the alternating current utilized in our homes and industries operates at synchronous speed and is defined as a **synchronous generator** (see Section 5.3).

Example 4.8 illustrates the distinct difference between the induction motor and the asynchronous (or induction) generator.

7. Motor region: $1 \geqslant s \geqslant 0$ or $0 \leqslant \omega_R \leqslant \omega_S$. Generator region: $0 \geqslant s < -\infty$ or $\omega_S \leqslant \omega_R < \infty$. Braking region: $2 \geqslant s \geqslant 1$ or $-\omega_S \leqslant \omega_R \leqslant 0$.

8. When the induction device operates in the motor region of operation, its stator is *fed* with electrical energy from an external source and the rotor, with all its winding currents present, *provides* mechanical energy, that is, the device is an electromechanical energy converter (EMEC). When the induction device operates in the generator region of operation, its stator is *fed* with electrical energy from an external source and its rotor is *driven* from an external source of mechanical energy at a speed **greater** than the synchronous speed of the stator field; consequently voltages of slip frequency ω_e (in radians/second) are available at the slip rings (the induction device must be the wound-rotor type). The device operates in the braking region when any two of the excitation phases are *reversed*, resulting in the reversal of the synchronous field rotation. The opposing rotation of the rotor and the stator fields causes the rotor to slow down, come to a full stop and, if allowed to, reverse its direction of rotation. If the excitation is disconnected before the rotor has come to a complete stop, friction will accelerate the process of stopping, for the rotor. This method of stopping is known as **plugging**.

Example 4.7

For the induction device of Example 4-6: (1) derive the Thevenin's equivalent circuit parameters to the left of terminals *ab* and draw the equivalent circuit of the induction motor on a per-phase basis; (2) derive and plot the total torque versus slip characteristic of the induction device; (3) derive and plot the total power versus slip characteristic of the induction device; (4) obtain the starting torque (power) τ_{STG} (P_{STG}), the breakdown torque (power) τ_{MAX} (P_{MAX}), and the minimum torque (power) τ_{MIN} (P_{MIN}); (5) obtain the slips s_{MAX} and s_{MIN} at which the torque breakdowns occur; (6) obtain the torque (power) readings at $s=\pm 2$; (7) define explicitly the motor region of operation, the generator region, and the braking region; (8) describe the significant difference between these regions.

Solution

1. Utilization of the figure of Example 4.6 and Eqs. (4.102) and (4.103) enables calculation of the parameters for the Thevenin's equivalent circuit of the induction motor as follows:

$$\mathbf{Z}_{TH} \equiv \mathbf{Z}_{ab} = \frac{j10(0.2+j0.38)}{0.2+j(10+0.38)} = \frac{4.29e^{j152.24°}}{10.38e^{j88.9°}}$$

$$= 0.413e^{j63.35°} = \underline{\underline{0.185+j0.369\ \Omega}}$$

$$\mathbf{V}_{TH} \equiv \mathbf{V}_{ab} = \frac{j10}{0.2+j(10+0.38)} \times 179.63e^{j0°}$$

$$= \frac{1796.23e^{j90°}}{10.38e^{j88.9°}} = \underline{\underline{173.05e^{j1.1°}\ \text{V}}}$$

or

$$\mathbf{V}_{TH} \equiv \mathbf{V}_{ab} = \underline{\underline{(0.9998+j0.0192)\times 173.05\ \text{V}}}$$

Now the equivalent circuit of the three-phase induction motor on a per-phase basis can be drawn as in Fig. 4.7.1.

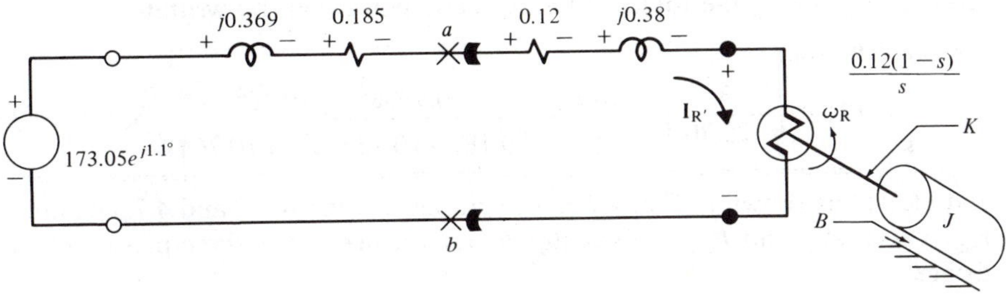

Fig. 4.7.1 The Equivalent Circuit of the Induction Device on a Per-Phase Basis, Drawn Utilizing Thevenin's Theorem to the Left of Terminals *ab*

2. Utilization of Eqs. (4.104) through (4.110) and Fig. 4.1.7 enables calculation of the needed parameters. The current $\mathbf{I}_R'$ as a function of slip s can be calculated by the proper substitution of circuit parameters and variables from Fig. 4.7.1 into Eq. (4.104) as

$$\mathbf{I}_R' = \frac{0.9998+j0.0192}{[0.185+0.12+(0.12/s)(1-s)]+j(0.369+0.38)} \times 173.05$$

The total electromagnetic power P_{emT} developed by the motor is

$$P_{emT} = 3 \times \frac{R_R'}{s} \times (1-s)|\mathbf{I}_R'|^2$$

Since $P_{em} = \omega_R \tau_{em}$ and $\omega_R = \omega_S(1-s)$, the total developed electromagnetic torque

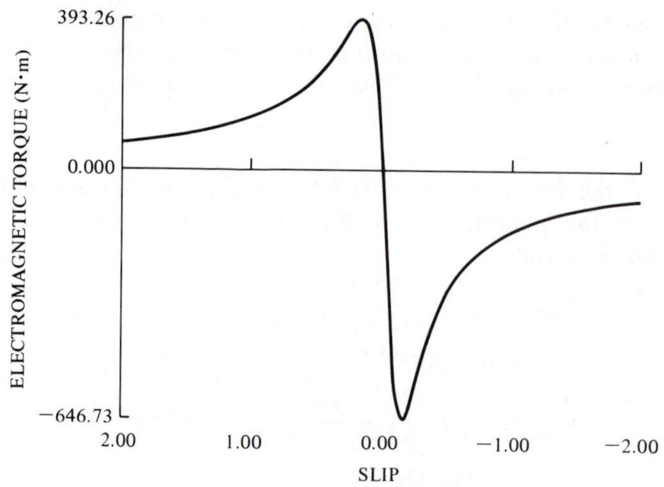

Fig. 4.7.2 τ_{emT} vs. Slip Characteristic of a Balanced, Three-Phase Induction Device

τ_{emT}, as a function of slip s, can be written as

$$\tau_{emT} = \frac{P_{emT}}{\omega_R} = \frac{P_{emT}}{\omega_S(1-s)} = \frac{3R_R'(1-s)|\mathbf{I}_R'|^2}{s\omega_S(1-s)} = \frac{3}{\omega_S}\times\left(\frac{R_R'}{s}\right)|\mathbf{I}_R'|^2 \text{ N}\cdot\text{m}$$

and plotted as shown in Fig. 4.7.2.

3. Since $\omega_R\tau_{em} = P_{em}$, the total electromagnetic power can be written as

$$P_{emT} = \omega_R\tau_{emT}$$
$$= 121.93\left(\frac{3}{125.70}\right)\times\frac{0.12}{s}\times\frac{0.9998^2+0.0192^2}{[0.187+(0.12/s)]^2+(0.764)^2}\times 173.05^2 \text{ W}$$

and plotted as shown in Fig. 4.7.3. The graphs of Figs. 4.7.2 and 4.7.3 exhibit the τ_{emT} versus slip and P_{emT} versus slip characteristics of the three-phase induction device.

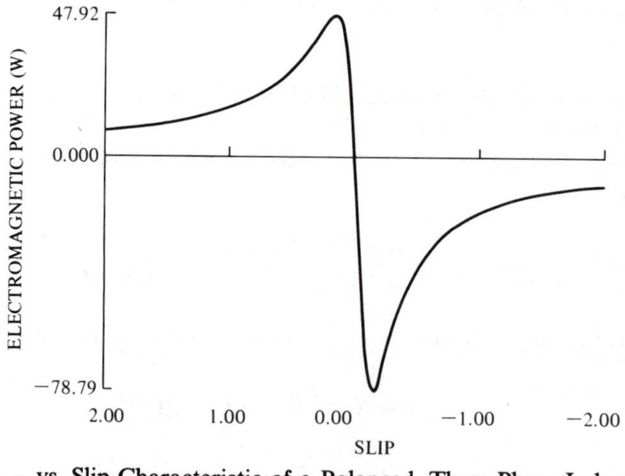

Fig. 4.7.3 P_{emT} vs. Slip Characteristic of a Balanced, Three-Phase Induction Device

The detailed information for parts 4, 5, and 6 is obtained with the aid of a digital computer program.

4. $\tau_{STG} = 135.78$ N·m; $P_{STGT} = 16.54$ kW; $\tau_{MAX} = 393.26$ N·m; $P_{MAX} = 47.91$ kW; $\tau_{MIN} = -646.73$ N·m; $P_{MIN} = -78.79$ kW.
5. $s_{MAX} = 0.152$ and $s_{MIN} = -0.152$.
6. $s = +2$: $\tau_{emT} = 71.44$ N·m, $P_{emT} = 8.704$ KW.
   $s = -2$: $\tau_{emT} = -76.92$ N·m, $P_{emT} = -9.37$ kW.

Example 4.8

The equivalent circuit of the induction motor is given in Fig. 4.15. Assume that a prime mover is used to drive the rotor of a six-pole induction motor excited by 120-V, 60-Hz mains at a speed sufficiently high to yield a slip of -0.05.

Calculate the speed that should be acquired by the rotor in order to produce the given slip and draw (1) the equivalent circuits of the induction device for the **motor** and **generator** modes of operation with everything referred to the rotor side of the equivalent circuit, referring the admittance Y_ϕ, designating excitation, across the mechanical load equivalent resistance R_{MLE}, and (2) the phasor diagrams for the two modes on the same complex plane. Mark clearly all voltages, currents, and angles. Be sure that all quantities that appear in the equivalent circuits of Figs. 4.8.1 and 4.8.2 also appear in Fig. 4.8.3.

Solution

To calculate the speed ω_R, which is required to drive the rotor in order to produce a slip of -0.05, we must first calculate the synchronous speed ω_S. Thus with the aid of Eq. (4.22) and input data, we obtain

$$\omega_S = \frac{\omega_E}{(P/2)} = \frac{2\pi \times 60}{(6/2)} = \underline{40\pi \text{ rad/s}} \tag{1}$$

Substitution for ω_S from Eq. (1) and $s = -0.05$ from the input data into Eq. (4.29) yields

$$\frac{40\pi - \omega_R}{40\pi} = -0.05 \tag{2}$$

Solution of Eq. (2) for ω_R yields

$$\omega_R = 40\pi + 0.05 \times 40\pi = 40\pi + 2\pi = \underline{\underline{42\pi}} \text{ rad/s} \tag{3}$$

Utilization of Eq. (4.23) yields the speed of the rotor n_R in revolutions/minute. Thus

$$n_R = \frac{60\omega_R}{2\pi} = \frac{60 \times 42\pi}{2\pi} = \underline{\underline{1260}} \text{ r/min} \tag{4}$$

The per-phase equivalent circuit of the **induction motor**, with all the quantities and circuit parameters referred from the stator side to the rotor side and the admittance $\mathbf{Y}_\phi$ moved all the way across the slip-dependent resistance $R_R(1-s)/s$, is drawn as exhibited in Fig. 4.8.1. Note the direction of the currents $\mathbf{I}_S'$, $\mathbf{I}_R$, $\mathbf{I}_\phi'$, $\mathbf{I}_c'$, and $\mathbf{I}_m'$. Also note that

$$\mathbf{I}_S' = \mathbf{I}_R + \mathbf{I}_\phi' \tag{5}$$

and

$$\mathbf{I}_\phi' = I_c' + jI_m' \tag{6}$$

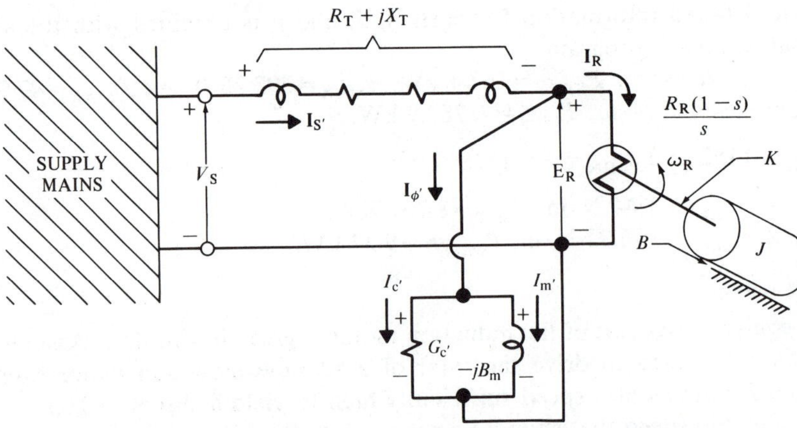

Fig. 4.8.1 The Equivalent Circuit of the Induction Device in Steady State Operation, Referred to the Rotor Side; Motor Mode of Operation ($R_T = R_S' + R_R$ and $X_T = X_{\ell S}' + X_{\ell R}$)

Similarly, the per-phase equivalent circuit of the **asynchronous generator**, with the quantities and circuit parameters referred from the stator side to the rotor side and the admittance $\mathbf{Y}_\phi$ moved all the way across the slip-dependent resistance $R_R(1-s)/s$, is drawn as exhibited in Fig. 4.8.2. Note the direction of the currents $\mathbf{I}_{\mathcal{G}}$, $\mathbf{I}_L'$, $\mathbf{I}_\phi'$, I_c', and I_m'. Also note that

$$\mathbf{I}_{\mathcal{G}} = \mathbf{I}_L' + \mathbf{I}_\phi' \tag{7}$$

and

$$\mathbf{I}_\phi' = I_c' + jI_m' \tag{8}$$

Utilization of the theory in Appendix A, particularly the theory pertinent to Figs. A.9 and A.10, enables us to draw the phasor diagram of Fig. 4.8.3 to the right of the Im-axis. Clearly, this phasor diagram describes Fig. 4.8.1, the motor mode of operation. Similarly drawn, the phasor diagram to the left of the Im-axis, Fig. 4.8.3, describes Fig. 4.8.2, the asynchronous generator mode of operation.

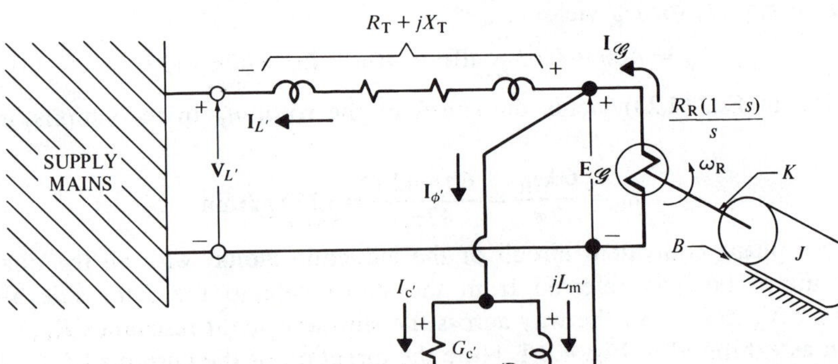

Fig. 4.8.2 The Equivalent Circuit of the Induction Device in Steady State Operation, Referred to the Rotor Side, Asynchronous (or Induction) Generator Mode of Operation ($R_T = R_S' + R_R$ and $X_T = X_{\ell S}' + X_{\ell R}$)

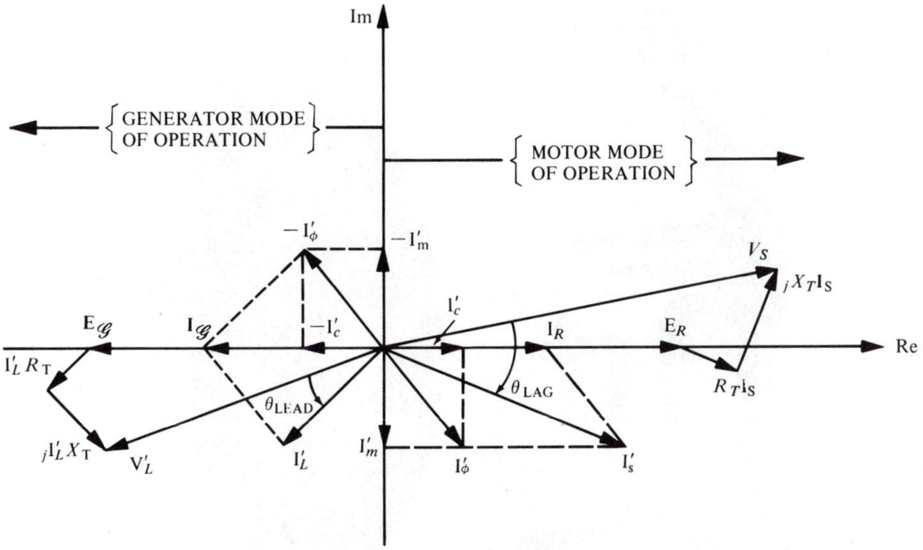

Fig. 4.8.3 Phasor Diagrams Describing the Steady State Operation of the Induction Device: (1) RHP—Motor Mode of Operation and (2) LHP—Asynchronous (or Induction) Generator Mode of Operation

It is essential to note the following pertinent points brought out in this example: (1) the voltage rise $\mathbf{E_S}'$ and the voltage drop $\mathbf{V}_L'$ are out of phase by 180°; (2) the voltage rise $\mathbf{E}_g$ and the voltage drop $\mathbf{V_R}$ are out of phase by 180°; (3) the currents $\mathbf{I_S}'$ and $\mathbf{I}_L'$ are out of phase by 180°; (4) the currents I_c' are in phase with the voltages $\mathbf{V_R}$ and $\mathbf{E}_g$; (5) the currents jI_m' are 90° out of phase with the voltages $\mathbf{V_R}$ and $\mathbf{E}_g$; (6) the response current $\mathbf{I_S}'$ **lags** the impressed voltage $\mathbf{E_S}'$ by an angle θ_{LGG} for the induction motor mode of operation; (7) the generated current $\mathbf{I}_g$ **leads** the generated voltage $\mathbf{E}_g$ by an angle θ_{LDG} for the asynchronous (or induction) generator mode of operation; (8) the magnetizing currents jI_m' is common to both of the phase diagrams that describe these two modes of operation.

It is demonstrated in Fig. 4.8.3 of Example 4.8 that with the excitation terminals of the induction device used as the reference, Fig. 4.8.1, the response current $\mathbf{I_S}'$ **lags** the impressed voltage $\mathbf{E_S}'$ by an angle θ_{LGG} and that the magnetizing current jI_m' lags the voltage $\mathbf{V_R}$ by an angle of 90°. In contrast, with the mechanical load equivalent resistance used as reference, Fig. 4.8.2, the generated current $\mathbf{I}_g$ always **leads** the generated voltage $\mathbf{E}_g$ by an angle θ_{LDG}. Therefore, the magnetizing current jI_m' should be a component of the generated current $\mathbf{I}_g$. However, since the asynchronous generator generates only real power, the magnetizing current jI_m' must be provided from the excitation system to which the device is connected. Thus when the asynchronous generator is connected across a generalized load, the parallel combination should be paralleled by the supply mains or, in the simplest case, by a synchronous generator of sufficient rating to provide the magnetizing current jI_m' for the asynchronous generator as well as the reactive current required by the generalized load. It is emphasized that these parallel connections are essential because the asynchronous generator **generates only real power**. In fact, this generator *requires reactive power* from an external system to furnish its own

Plate 4.4 Stator Core with the Winding in Place. Insulation Is Achieved with Thermoplastic Insulation Materials Composed of Mica Dielectric Material Locked in a Void-free, Stable Elastic Bond to Form a Strong Impervious Barrier That Withstands Prolonged Voltage Stresses, Moisture, Abrasion, Dirt, Thermal Cycling, and Frequent Starting Surges. (Courtesy of Westinghouse Electric Co., Pittsburgh, Pa.)

excitation. Unlike all other types of generators, which have independent field excitations, the asynchronous generator does not have the means to establish an airgap field $|\mathbf{B}_{ST}|$, essential for its operation, when the stator windings are open.

The magnetizing kilovoltamperes could be as low as 12 percent of full-load rating at no-load but never less than 25 percent of full-load rating under no-load.

It is clear, then, that the asynchronous generator does not function *unless* the supply mains—or a synchronous device (motor or generator) connected in parallel—provides the reactive power required for its operation. However, when it does function, at a slip of about -0.05, the asynchronous generator is fully loaded; that is, the rotor windings (or the rotor bars in the squirrel cage type of rotor) and the stator windings carry rated currents. The need for reactive power from external sources, however, restricts this generator's use to a few limited applications.

In contrast to its *disadvantages* (i.e., its lack of self-excitation and its inability to furnish reactive power), the asynchronous generator has three distinct *advantages* that make it desirable in some applications. These advantages are as follows: (1) this device **does not require synchronization** to be put on line (see Section 6.4); (2) it does not **hunt** for the speed of operation after a disturbance, since it does not operate at a fixed speed—such as the synchronous speed ω_S; (3) in case of **short-circuited lines**, it loses its excitation, ceases to function, and stops supplying power to the line faults.

Furthermore, when connected across a load at the end of a transmission line, the asynchronous generator can be used *to improve the power factor of the line.*

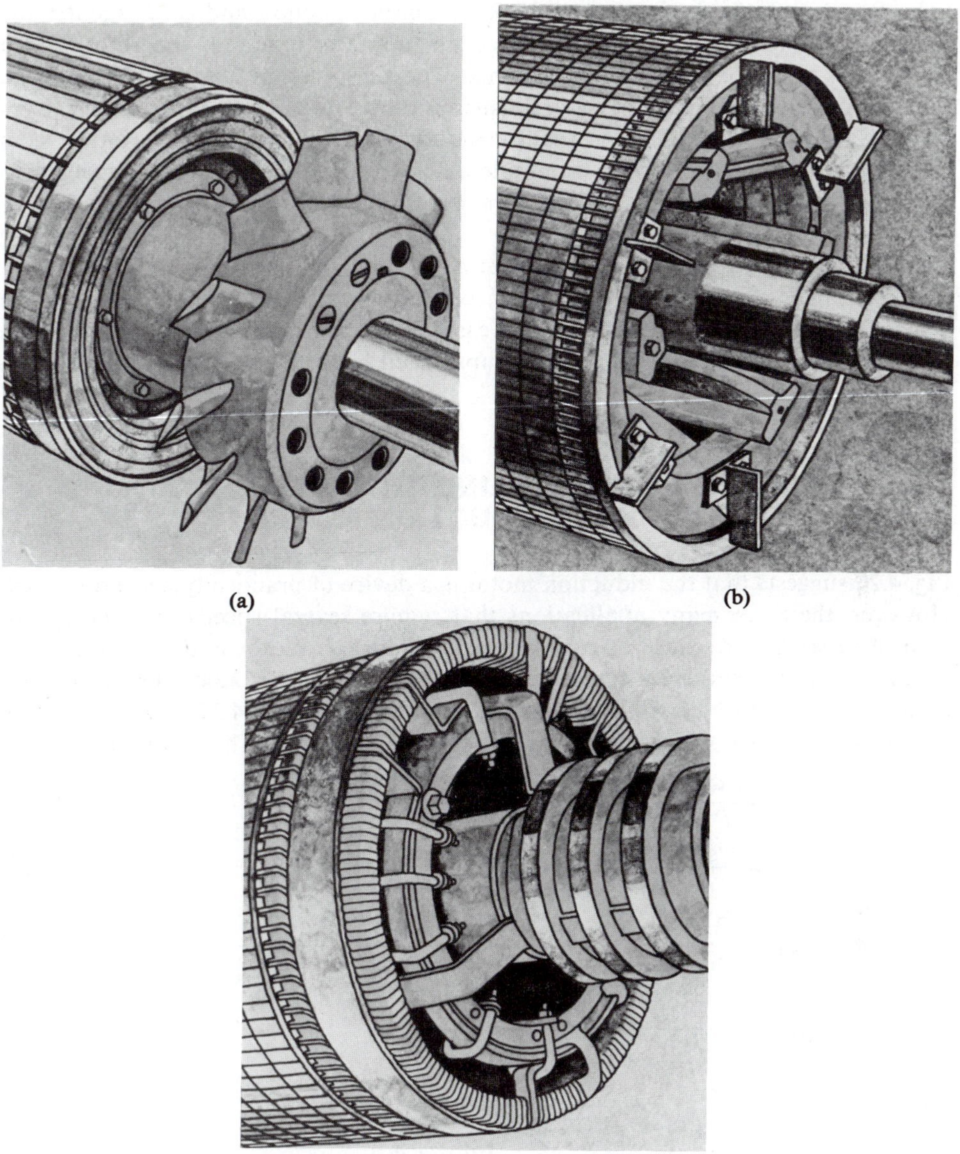

(a) (b) (c)

Plate 4.5 Rotor Construction for Induction Devices: (a) High-Speed, Squirrel Cage Rotor, (b) Low-Speed, Squirrel Cage Rotor, and (c) Wound Rotor. (Courtesy of Westinghouse Electric Co., Pittsburgh, Pa.)

Connected in the same manner, the asynchronous generator-synchronous device (motor or generator) pair can be used *to improve the power factor as well as the voltage regulation of the line*. These schemes, however, are rarely used because of the necessity of a prime mover.

Obviously, either the **wound-rotor** or the **squirrel-cage-rotor induction device** can perform as an asynchronous generator. However, the low magnetizing current requirement, mentioned earlier, necessitates a short airgap, that is, $\ell_g = r_S - r_R \doteq 0$

(where r_S designates the inside radius of the stator structure and r_R designates the radius of the rotor). Also, since **no starting torque** is required, the rotor of the device may be made of very low resistance. These two design specifications suggest that **the squirrel cage type of rotor induction device usually performs as an asynchronous generator**, while **the wound-type rotor induction device performs as a frequency converter**. For the latter mode of operation, the slip rings connected to the rotor windings of Fig. 4.21(a) are essential to extract the slip frequency voltages.

Note that the connection of Fig. 4.21(a) is valid for both modes of operation discussed here, that is, the frequency converter as well as the asynchronous generator modes, *except* that for the case of the asynchronous generator, the wound-type rotor with the slip rings is substituted by an **identical size** squirrel cage type of rotor.

4.8 METHODS FOR CONTROLLING THE SPEED OF THE THREE-PHASE INDUCTION MOTOR

Fig. 4.20 suggests that the induction motor is a device of practically constant speed. However, there are many applications that require several speeds, or even speed-control capability. Because of its nature, the induction motor is the most suitable candidate for a **speed drive whose speed can be continuously adjustable**. Induction motor speed-control capabilities can be studied via Eqs. (4.22), (4.29), (4.106), (4.109), and (4.110), which are rewritten here for simplicity. These equations are

$$\omega_S = \frac{\omega_E}{P/2} \tag{4.115}$$

$$s = \frac{\omega_S - \omega_R}{\omega_S} \tag{4.116}$$

$$\tau_{emT} = \frac{3}{\omega_S} \times \frac{R_R'}{s} \times \frac{\gamma^2 + \delta^2}{\left[R_{TH} + (R_R'/s)\right]^2 + (X_{TH} + X_{\ell R}')^2} \times |\mathbf{V}_S|^2 \tag{4.117}$$

$$s_{MAX} = \frac{R_R'}{\left[R_{TH}^2 + (X_{TH} + X_{\ell R}')^2\right]^{1/2}} \tag{4.118}$$

and

$$\tau_{MAX} = \frac{3}{\omega_S} \times \frac{0.5(\gamma^2 + \delta^2)}{R_{TH} + \left[R_{TH}^2 + (X_{TH} + X_{\ell R}')^2\right]^{1/2}} \times |\mathbf{V}_S|^2 \tag{4.119}$$

Eq. (4.117), which is plotted in Fig. 4.20, yields that τ_{emT} is a function of the following controllable parameters: (1) the synchronous speed ω_S, (2) the impressed voltage $|\mathbf{E}_S|$, (3) the referred rotor reactance $X_{\ell R}'$, and (4) the referred rotor resistance R_R'. In addition, Eq. (4.115) reveals that the synchronous speed ω_S is a function of the following two parameters: (1) the frequency of excitation ω_E and (2) the number of poles P of the device. Therefore, there are four ways of changing the shape of the τ_{emT} versus $s(\omega_R)$ characteristic and therefore the speed of operation ω_{RQ}, which is defined by the point of intersection between τ_{emT} versus $s(\omega_R)$ and τ_L versus $s(\omega_R)$, Fig. 4.22. These methods yield the following **five direct methods of controlling the speed of an induction motor**: (1) variable frequency of excitation ω_E,

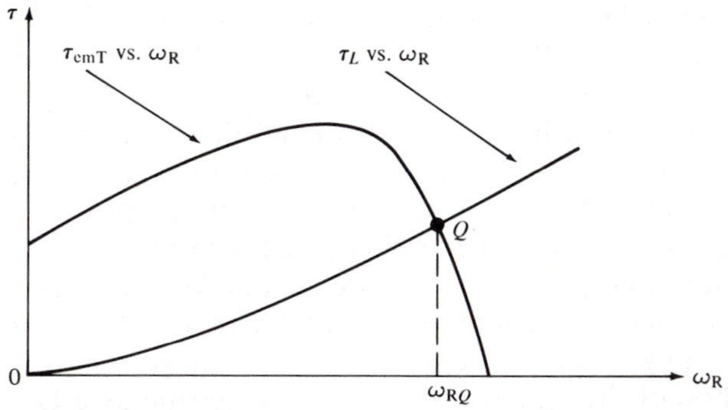

Fig. 4.22 Torque-Speed Characteristics, That Is, τ_{emT} vs. ω_R of an Induction Motor and a Variable Torque Load, That Is, τ_L vs. ω_R. The Point of Intersection Defines the Speed of Operation ω_{RQ}

Fig. 4.23(a); (2) variable number of poles P, Fig. 4.23(b); (3) variable impressed voltage $|\mathbf{E}_S|$, Fig. 4.23(c); (4) variable referred rotor reactance $X_{\ell R}'$, Fig. 4.23(d); and (5) variable referred rotor resistance R_R', Fig. 4.23(e).

The first three methods are adaptable to either the squirrel cage or the wound-rotor induction motor since any control that is possible must take place by manipulating stator variables. In the past, variable-frequency excitation ω_E was very costly and inefficient because it was provided by a motor-generator set with appropriate controls. However, with the advent of silicon-controlled rectifiers (SCR), the supply of variable frequency of excitation is no longer a problem. The method of changing the number of poles P according to requirements is the most popular method for speed control of squirrel cage induction motors. This type of speed control is accomplished by prewiring the stator in such a way as to change

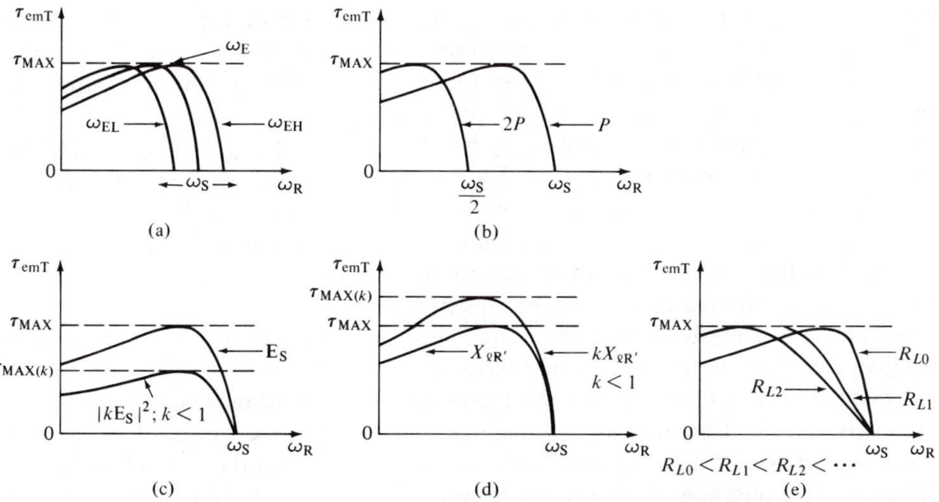

Fig. 4.23 Speed Control of the Induction Motor Is Achieved by Affecting Its Torque vs. Speed Characteristic Through Means of Affecting Changes of: (a) the Frequency of Excitation ω_E, (b) the Number of Stator-Field Poles P, (c) the Magnitude of the Impressed Voltage $\mathbf{E}_S$ at the Stator Windings, (d) the Magnitude of the External Rotor Reactance X_{Lj}, $j=0,1,2,\ldots$, and (e) the Magnitude of the External Rotor Resistance R_{Lj}, $j=0,1,2,\ldots$

the number of poles by a simple throw of a switch and thus change from one synchronous speed to another. Such a device is known as **multispeed motor.** However, speed control is attained in a discrete fashion and speed cannot exceed the range of speeds for which the device is designed. The novelty of this scheme of speed control is that the motor operates at a low slip once a particular speed is selected. As in the case of variable-frequency excitation, speed control by variable impressed voltage $|\mathbf{E}_{\mathrm{S}}|$ suggests the variation of the impressed voltage by a **motor-generator set** with appropriate control schemes or by some other method. Such schemes make the method economically unattractive. In addition, from Eq. (4.117) it is obvious that any reduction of the impressed voltage results in a reduction of the ordinate values of the τ_{emT} versus $s(\omega_{\mathrm{R}})$ characteristic by a factor $|k\mathbf{E}_{\mathrm{S}}|^2$, $k<1$. Thus the reserve capacity of the motor and its starting torque are reduced to the point where the method of variable-impressed-voltage speed control is unattractive.

The last two methods are adaptable to the wound-rotor induction motor only [the squirrel-cage-rotor induction motor is ruled out for referred phase reactance or resistance variation, because $X_{\ell\mathrm{R}}'$ and R_{R}' cannot be changed except by design only, Fig. 4.2(c) (why?)]. Changing the referred reactance per phase, $X_{\ell\mathrm{R}}'$, by introducing external reactance through the slip rings is a method that is rarely, if ever, used **because of its adverse effects on the power factor** of the excitation (which is the transmission line usually) and because of the large size of the added inductor, needed due to slip variation. It should be pointed out, however, that if this method of speed control were chosen, the ordinate values of the τ_{emT} versus $s(\omega_{\mathrm{R}})$ characteristic would change, Fig. 4.23(d). The **most common method** of speed control of the three-phase induction motor is to use external three-phase resistors connected, through the slip rings of the wound rotor [Fig. 4.2(b)], in series with the referred internal resistance or impedance of the rotor winding R_{R}' or $R_{\mathrm{R}}'+jX_{\ell\mathrm{R}}'$, Fig. 4.19. The effects of this added resistance to the speed of the motor becomes obvious by inspection of Eqs. (4.117), (4.118), and (4.119) and Fig. 4.23(e). If the added resistance per phase is R_{Lj} $(j=0,1,2,\ldots)$, the numerator of Eq. (4.118) becomes $R_{\mathrm{R}}'+R_{Lj}'$. As R_{Lj}' increases, s_{MAX} increases; that is, the peak of the τ_{emT} versus $s(\omega_{\mathrm{R}})$ characteristic of Fig. 4.20 moves further and further to the left, while the point $P\{\tau_{\mathrm{emT}}=0, s=0\}$ remains unchanged and τ_{MAX}, the reserve capacity of the motor, remains unchanged, since τ_{MAX} is independent of $R_{\mathrm{R}}'+R_{Lj}'$, Eq. (4.119). The effects of an increased R_{Lj} to Fig. 4.20 are exhibited in Fig. 4.23(e) for $R_{L0}<R_{L1}<R_{L2}<\cdots$. In conjunction with a constant torque load, $\tau_L=k$, a speed control to 50 percent below ω_{S} is acceptable. Beyond this point the efficiency becomes less than 50 percent, an unacceptable level. This drop in efficiency is due to the power dissipated in the external resistors R_{Lj}.

Out of the five methods of speed control just described, only the method of **pole changing** (used in squirrel-cage-rotor induction motors) to achieve a maximum of four distinct speeds of operation, the method of **rotor resistance changing** (used in wound-rotor induction motors) to achieve a speed control range of 50 percent below ω_{S} for a constant torque load, $\tau_L=k$, and the method of **changing the frequency of excitation**, a direct contribution of SCRs, are the most common. However, since a continuous range of speed control, rather than discrete speeds, is desired, the method of speed control by changing the rotor resistances seems more attractive. But since additional external resistance is required to increase the range of operating speeds, the slip s increases, ω_{R} decreases, power is wasted as heat in the added rotor resistance, and the device becomes inefficient. Therefore, a scheme

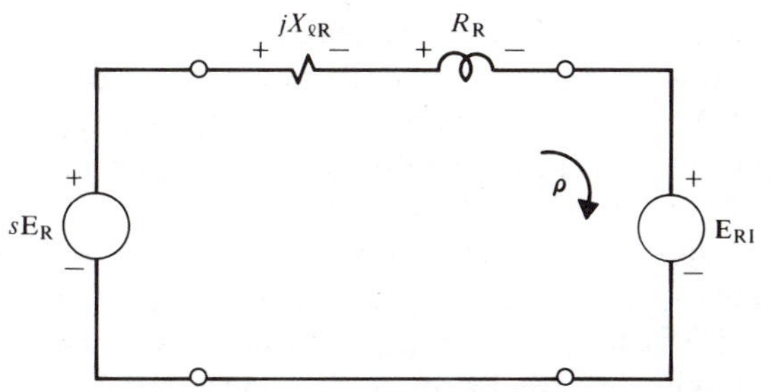

Fig. 4.24 Rotor Equivalent Circuit of an Induction Motor, on a Per-Phase Basis with Rotor Excitation Voltages E_{RI} as Used in Speed Control

of speed control that does not waste power as heat and does not necessitate increase of slip to secure a continuous range of speed control is desired. Inspection of Fig. 4.1 reveals that **speed control can be achieved by controlling the polarity and strength of the rotor excitation voltages** (this scheme is useful only for wound-rotor induction motors). To understand this scheme of speed control, it is helpful to draw the rotor steady state equivalent circuit of Fig. 4.1 as in Fig. 4.24 and designate the rotor excitation phasor voltage as $\mathbf{E}_{RI}$ to distinguish it from the induced rotor phasor voltage E_R. The response current of the unreferred rotor current $\mathcal{I}_R(\omega_e)$ is

$$\rho = \frac{s\mathbf{E}_R + \mathbf{E}_{RI}}{R_R + jsX_{\ell R}} \tag{4.120}$$

To understand this method of speed control, it is assumed that the motor is expected to deliver rated torque to a constant torque load, $\tau_L = k$, at varying speeds. For $\psi = 0$, Eq. (4.61), $\tau_{em} = K\Phi\rho$. Since the design constant K and $\Phi = f(|\mathbf{E}_S|, \omega_E)$ remain constant during the motor operation, τ_{em} can remain constant if and only if ρ is maintained constant. Dependence of ρ on s and $\mathbf{E}_R$ is evident by inspection of Eq. (4.120). Such inspection reveals that **one can affect the magnitude of the current, ρ, Fig. 4.24, in one of five different ways:** (1) if $\mathbf{E}_{RI} = 0$, (2) if the polarity of $\mathbf{E}_{RI}$ opposes the polarity of $s\mathbf{E}_R$ and its strength is smaller than that of $s\mathbf{E}_R$, (3) if $\mathbf{E}_{RI}$ is in phase $s\mathbf{E}_R$ and its strength is smaller than $s\mathbf{E}_R$, (4) if $\mathbf{E}_{RI}$ is in phase with $s\mathbf{E}_R$ and its strength is equal to that of $s\mathbf{E}_R$, and (5) if $\mathbf{E}_{RI}$ is in phase with $s\mathbf{E}_R$ but its magnitude is greater than that of $s\mathbf{E}_R$.

When $\mathbf{E}_R\mathrm{I} = 0$, the rotor windings are short-circuited and the induction motor operates in its normal mode, as described by Fig. 4.15. For the second case, ρ decreases, τ_{em} decreases, speed drops, and s increases until the effect of $\mathbf{E}_{RI}$ on ρ is overcome and the motor has reached equilibrium, that is, when ρ reaches its original value. One should note that although s increases in this scheme of speed control (as in the case of speed control by rotor resistance variation), the rotor excitation voltage sources can **either** usefully utilize the power associated with the drop in speed or **else** return it to the grid network, that is, the stator excitation system. For the third case, ρ increases, τ_{em} increases, the speed increases, and s **decreases** (note that speed control with rotor resistance variation yields only slip **increase**) until the effect of E_{RI} on ρ is overcome and the motor reaches equilibrium, that is, ρ reaches its original value. For the fourth case, $\rho = 0$ and $\tau_{em} = 0$ (i.e., the motor runs at synchronous speed ω_S, Fig. 4.20). This is characteristic of *doubly excited* systems (i.e., both members, stator and rotor, are excited from external energy sources). Finally, for the fifth case, ρ increases, τ_{em} increases, and

ω_R increases constantly until s becomes negative enough so that the effect of $s\mathbf{E}_R$ is adequate to bring $\boldsymbol{\rho}$ to its original value. It should be noted that the addition of the voltages $s\mathbf{E}_R$ and $\mathbf{E}_{RI}$ is only feasible if the voltage $\mathbf{E}_{RI}$ is of the correct slip frequency. Therefore, appropriate equipment of size and rating consistent with the degree of control to be affected is required. These requirements lead to the manufacturing of special units that provide speed control by varying the rotor excitation voltages e_{RA}, e_{RB}, and e_{RC}, Fig. 4.1. Such units are known by the following names: (1) Schrage motor, (2) Leblanc system, and (3) modified Kramer system. All these schemes are large in size, multiunit in nature, and expensive. Therefore, their use for speed control of induction motors over the use of the five methods of speed control mentioned earlier becomes a decision of economics.

Before closing this section it will be wise to explain the τ_{emT} versus $s(\omega_R)$ characteristic of Fig. 4.20 for the region $0 \leqslant s \leqslant 1$ and point out the contribution and physical meaning of s and ψ. Substitution of Eq. (4.120) into Eq. (4.61) and multiplying the result by 3 enables writing Eq. (4.61) as*

$$\tau_{emT} = K\Phi\left(\left|\frac{s\mathbf{E}_R}{R_R + jsX_{\ell R}}\right|\right)\cos\psi \tag{4.121}$$

When the rotor rotates at synchronous speed, $s=0$ and $\tau_{emT}=0$, Eq. (4.121). For this condition, $\phi_R = \tan^{-1}(sX_{\ell R}/R_R) = 0$. Therefore, the angle between the maximum rotor induced voltage $s\mathbf{E}_R$ and the maximum rotor response current $\boldsymbol{\rho}$, Fig. 4.24, is zero. These ideas are depicted in Fig. 4.25(a). Note the geometric relationship between $|\mathbf{B}_{ST}|$, $|\mathcal{F}_{MR}|$, and $|\mathbf{J}_{MR}|$ and the spatial angle ψ between $|\mathbf{B}_{ST}|$ and $|\mathbf{J}_{MR}|$, that is, $\psi=0$. Although this field pattern according to Fig. 4.5 implies the best field pattern for torque development, very little torque is developed at very low slips because the value of the rotor mmf is nearly zero [$\boldsymbol{\rho} = s\mathbf{E}_R/(R + jsX_{\ell R}) \doteq 0$]. As the slip is allowed to increase from zero to about 10 percent of the rotor current, $\boldsymbol{\rho} = s\mathbf{E}_R/(R + jsX_{\ell R})$ increases almost linearly. At the same time, the angle between the maximum rotor induced voltage $s\mathbf{E}_R$ and the maximum rotor response current $\boldsymbol{\rho}$, that is, $\phi_R = \tan^{-1}(sX_{\ell R}/R_R) \neq 0$, leads to the establishment of the fact that $\psi \equiv \phi_R \neq 0$. Thus for the same range of slip variation, $0° < \psi < 15°$. This implies that as s varies, that is, $0 < s < 10\%$, $\cos\psi = k$, and torque increases linearly with slip in this region, Eq. (4.121).

To verify the fact that $\psi \equiv \phi_R$, refer to Fig. 4.25(b) and note the geometric relationship between $|\mathbf{B}_{ST}|$, $|\mathcal{F}_{MR}|$, and $|\mathbf{J}_{MR}|$ and the spatial angle ψ between $|\mathbf{B}_{ST}|$ and $|\mathbf{J}_{MR}|$, that is, $\psi \neq 0$. Also, note that in Fig. 4.25(b) only the current distribution and its relative strength are exhibited in the rotor coils by the appropriate size of dots and crosses. The larger the size of the dot or the cross, the stronger the current is. The voltage that corresponds to each current, although not shown in Fig. 4.25(b), can be drawn ahead of the corresponding currents at an angle ϕ_R. It should be obvious now that in a polyphase induction motor the space displacement between the field $|\mathbf{B}_{ST}|$ and the current density vector $|\mathbf{J}_{MR}|$ is equal to the power-factor angle of the rotor winding ϕ_R, that is, $\psi \equiv \tan^{-1}(sX_{\ell R}/R_R)$.

Returning to the τ_{emT} versus $s(\omega_R)$ characteristic, Fig. 4.20, it is noted that as the slip is increased beyond the region $0 < s < 10\%$, the rotor current continues to increase much more rapidly than at first. This should be obvious from $\boldsymbol{\rho} = s\mathbf{E}_R/(R_R$

*The value of K is assumed to be properly adjusted so that Eqs. (4.121) and (4.117) yield identical results.

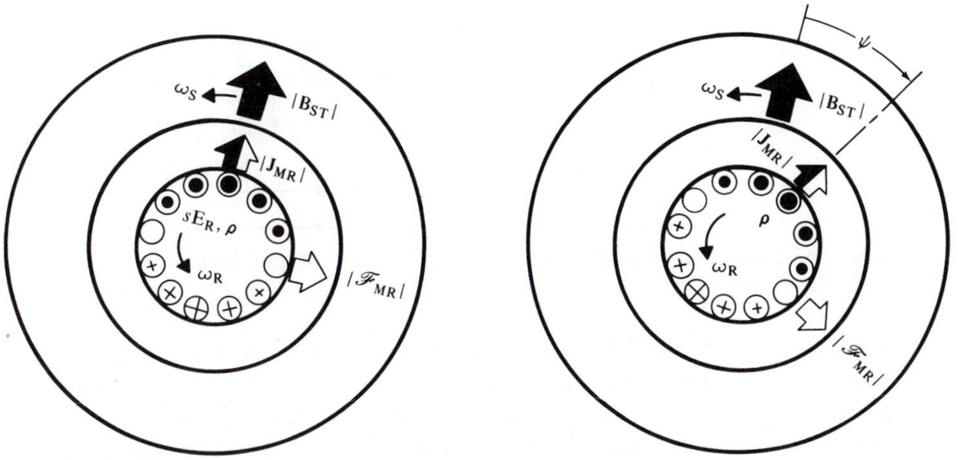

(a) (b)

Fig. 4.25 Waveform Distribution for Electromagnetic Torque Development: (a) $s=0$; $s\mathbf{E_R}$ and ρ in Phase, $\psi=0$, $\tau_{emT}=0$, (b) $s\neq0$, $s\mathbf{E_R}$ (Not Exhibited but Could be Drawn Ahead of ρ by ϕ_R) and ρ Out of Phase by $\phi_R=\tan^{-1}(s\mathbf{X}_{\ell R_R})$, $\phi_R\equiv\psi$, and $\tau_{emT}\neq0$

$+jsX_{\ell R})$. Also, since $\psi=\tan^{-1}(sX_{\ell R}/R_R)$, the space angle ψ begins to increase at a rapid rate. This increase leads to a rapid decrease of the term $\cos\psi$, Eq. (4.121). The rate at which $\cos\psi$ decreases overcomes the rate at which ρ increases, and the developed electromagnetic torque of Eq. (4.121) begins to fall off, Fig. 4.20, point $P\{\tau_{emT}=\tau_{MAX}, s=s_{MAX}\}$. As s increases beyond s_{MAX} (i.e., $s_{MAX}<s<1$), τ_{emT} continues to fall off until the point $P\{\tau_{emT}=\tau_{STG}, s=1\}$, Fig. 4.20, is reached. This point on the τ_{emT} versus $s(\omega_R)$ curve depicts the starting torque developed by a three-phase induction motor, which is always in excess of the rated torque for a typical induction motor. The remaining regions of the τ_{emT} versus $s(\omega_R)$ characteristic can be explained in a similar fashion. It should be pointed out, however, that as s, and thus ψ, increases, the direction of the current in the conductors under the field $|\mathbf{B}_{ST}|$ changes; thus the torque $d\tau=\mathbf{R}\times d\mathbf{F}$, where $d\mathbf{F}=id\mathbf{L}\times\mathbf{B}$, changes direction.

In conclusion, it should be stated that the τ_{emT} versus $s(\omega_R)$ characteristic, Fig. 4.20, describes the performance of the induction device in all its modes of operation. The student should make the effort to understand the meaning of Fig. 4.20 as thoroughly as he understands the voltage versus current characteristic of a battery, for Fig. 4.20 describes the behavior of the induction device in the same way that the voltage versus current characteristic describes the behavior of the battery.

4.9 PARAMETER MEASUREMENT OF THE EQUIVALENT CIRCUIT OF THE THREE-PHASE INDUCTION MOTOR

To compute the performance of a three-phase induction motor or generator, the parameters of its equivalent circuit, Fig. 4.15, must be known. These parameters may be available either from design data or may be obtained from appropriate *tests* that the device may be subjected to. Since the equivalent circuit of the induction

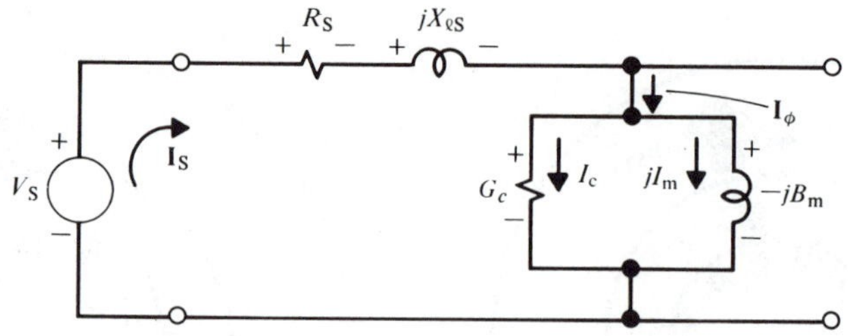

Fig. 4.26 The Equivalent Circuit for the No-Load Test of the Induction Device on a Per-Phase Basis

motor is topologically similar to the equivalent circuit of the transformer, understanding of the principles that govern the open-circuit and short-circuit tests of the transformer leads to the understanding of the principles that govern the equivalent (to the transformer) test applied to the induction device. These two tests are known as (1) the **no-load test** (equivalent to the open-circuit test of the transformer) and (2) the **blocked-rotor test** (equivalent to the short-circuit test of the transformer). To understand these two terms, we must relate the slip expression $s=(\omega_s-\omega_R)/\omega_S$ to the equivalent circuit of the induction device, Fig. 4.15. When the induction device operates *unloaded*, $\omega_R \doteq \omega_S$ and $s \doteq 0$. This leads to a mechanical load equivalent $R_{MLE}=R_R'(1-s)/s \doteq \infty$ and thus to the equivalent circuit of Fig. 4.26. When the induction device operates with its stator coils excited from a three-phase balanced source of frequency, $f_E=60$ Hz, but with the rotor *blocked* by an appropriate locking device so that it does not rotate (i.e., $\omega_R=0$), the slip becomes $s=1$. This leads to a mechanical load equivalent $R_{MLE}=R_R'(1-s)/s=0$ and thus to the equivalent circuit of Fig. 4.27. To make possible the calculation of the parameters of the equivalent circuit of the induction device on a per-phase basis, Fig. 4.26 is redrawn as Fig. 4.28, where $\mathbf{Z}_\phi \equiv 1/\mathbf{Y}_\phi = R_m + jX_m$ is the equivalent series imped-

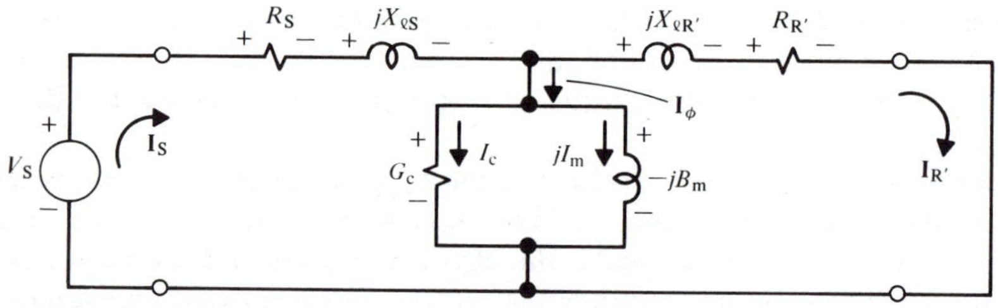

Fig. 4.27 The Equivalent Circuit for the Blocked-Rotor Test of the Induction Device on a Per-Phase Basis

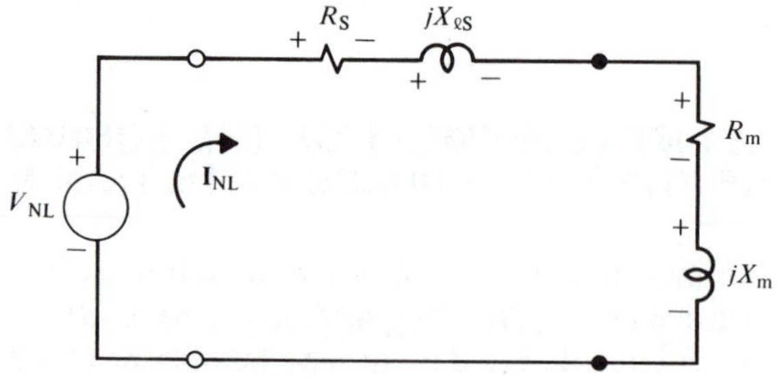

Fig. 4.28 The Equivalent Circuit of the Induction Device Under No-Load Conditions of Operation on a Per-Phase Basis

ance of the magnetizing admittance, $\mathbf{Y}_\phi$, of Fig. 4.26. From practical considerations it can be shown that $R_m \ll |jX_m|$, which leads to $\mathbf{Z}_\phi \doteq jX_m$. Therefore, the two tests can be performed with this assumption being **dominant** throughout this section.

No-Load Test

The no-load test of the induction device, as in the open-circuit test of the transformer, gives information with respect to its exciting current $\mathbf{I}_\phi$ and its no-load losses. Since at no-load the value of the slip is $s \doteq 0$ and $R_{MLE} = \infty$, the no-load currents of the rotor windings are negligible. In addition to these negligible currents, very small currents are produced in the rotor windings mainly due to the harmonics of the flux density waveform and slight nonuniformity of the airgap. These two currents produce enough torque to overcome friction and windage losses. These losses, known as **rotational losses**, at rated voltage and frequency under load usually are considered to be constant and equal to their no-load values. Thus the rotational losses can be estimated accurately by measuring the *no-load power* input to the induction device, P_{NLT}, and subtracting the *copper losses* of the stator windings from it. These ideas are expressed mathematically as

$$P_{RT} = P_{NLT} - 3R_S|\mathbf{I}_{NL}{}^2| \tag{4.122}$$

Since at no-load $s \doteq 0$, the parallel connection of $\mathbf{Z}_X$ and $\mathbf{Z}_R$ yields an equivalent impedance equal to $\mathbf{Z}_\phi = R_m + jX_m$. Thus the equivalent circuit of Fig. 4.28 represents accurately the induction device on a per-phase basis and the measurements of the no-load test should be referred to this circuit. Test data are ordinarily taken at **rated voltage and frequency**, after the motor has been running long enough for proper lubrication of the bearings to be insured. Measurement of the no-load phase voltage $\mathbf{E}_{NL}$, no-load phase current $\mathbf{I}_{NL}$, no-load total power P_{NLT}, and the *dc resistance* of the stator windings,* $R_S = R_{jk}/2$ (where $j \neq k = A, B, C$ are the three terminals of the Y-connected stator winding) at the input stator terminals of a Y-connected motor enables calculation of the following pertinent parameters.

$$|\mathbf{Z}_{NL}| = \frac{|\mathbf{V}_{NL}|/\sqrt{3}}{|\mathbf{I}_{NL}|} \tag{4.123}$$

$$R_{NL} \equiv R_S + R_m = \frac{P_{NLT}/3}{|\mathbf{I}_{NL}|^2} \tag{4.124}$$

and

$$X_{NL} \equiv X_{\ell S} + X_m = \left(|\mathbf{Z}_{NL}|^2 - R_{NL}{}^2\right)^{1/2} \tag{4.125}$$

Close examination of Eqs. (4.122) through (4.125) reveals that additional measurements are required if all the parameters contained in these equations and Fig. 4.15

*More readings than one should be taken for R_{jk} and their average value should be used for R_{jk} in this expression. Because, however, the stator windings are excited with 60-Hz voltages, their dc resistance values should be adjusted to *ac resistance* values, designated as R_S. These dc values, i.e., each R_S, are increased by 10 percent to 25 percent of each dc value. **As a rule of thumb, use 20 percent.** This increase allows for the **skin effect** (i.e., the conduction of an ac current near the surface of its carrying conductor). This phenomenon is caused by the time-varying flux set up within the conductor by the ac current itself. This current concentration near the surface of the conductor creates an **effective** conductor cross sectional area **less** than the **real** conductor cross sectional area. The effect of the effective cross sectional area to the resistance of the conductor is obvious from $R = \rho\ell/A$.

are to be obtained. These measurements are taken with the stator windings excited and the rotor locked, so that it is unable to rotate.

Blocked-Rotor Test

The blocked-rotor test of the induction device, as in the short-circuit test of the transformer, gives information with respect to leakage reactances. For the blocked-rotor test, the speed of the rotor is zero (i.e., $\omega_R = 0$); therefore, the slip is one [i.e., $s = (\omega_S - \omega_R)/\omega_S$] and the mechanical load equivalent is zero [i.e., $R_{MLE} = R_R'(1-s)/s = 0$]. Thus the equivalent circuit of Fig. 4.27, redrawn as in Fig. 4.29, represents accurately the induction device on a per-phase basis, and measurements of the blocked-rotor test should be referred to this circuit. It should be noted, however, that as the speed of the rotor ranges from $\omega_R = 0$ rad/s to a speed almost equal to the synchronous speed of the stator field, that is, $\omega_R = (1-s)\omega_S$, the frequency of the currents in the rotor windings range from $f_e \equiv f_E = 60\,\text{Hz}$ at the time of starting to a few hertz under normal operation of the device. This variation of the frequency of the rotor currents affects the rotor resistance and the leakage reactances, which depend on the magnetic saturation of the paths of the flux leakages $\Phi_{\ell R}$ and $\Phi_{\ell S}$ across the stator and rotor teeth, Fig. 4.8. High currents, Fig. 4.15, due to rated impressed voltages, or high slips saturate the leakage flux paths and yield lower-than-normal values of leakage reactance for the normal range of operating speeds. To avoid the effect of the magnetic saturation on $X_{\ell S}$ and $X'_{\ell R}$, and thus on the performance analysis of the induction motor, the IEEE* in its *test code* recommends that in order to obtain frequency-independent values of the three-phase induction device reactances, the data for the blocked-rotor test should be taken **after** the rotor is blocked so that it cannot rotate and so that the balanced polyphase voltages that are impressed to the stator windings are adjusted to produce approximately rated phase currents and the frequency of these voltages is adjusted to 15 Hz. The blocked-rotor test measurements of the phase voltage $|\mathbf{E}_{BL}|$, the phase current $|\mathbf{I}_{BL}|$, and the total power P_{BLT} at the input terminals of the Y-connected induction device enable calculation of the following parameters:

$$|\mathbf{Z}_{BL}| = \frac{|\mathbf{V}_{BL}|/\sqrt{3}}{|\mathbf{I}_{BL}|} \tag{4.126}$$

$$R_{BL} \equiv R_S + R_{EQ} = \frac{P_{BLT}/3}{|\mathbf{I}_{BL}|^2} \tag{4.127}$$

and

$$X_{BL} = X_{\ell S} + X_{EQ} = \left(|\mathbf{Z}_{BL}|^2 - |R_{BL}|^2\right)^{1/2} \tag{4.128}$$

where $\mathbf{Z}_{EQ} = R_{EQ} + jX_{EQ}$ is the equivalent impedance of the parallel combination of the magnetizing reactance jX_m and the rotor impedance of Fig. 4.29. It can be shown from Fig. 4.29 that

$$\begin{aligned} \mathbf{Z}_{EQ} &= \frac{(jX_m)\times(R_R' + jX_{\ell R}')}{R_R' + j(X_{\ell R}' + X_m)} \\ &= \frac{R_R' X_m^2}{(R_R')^2 + (X_{\ell R}' + X_m)^2} + j\left[\frac{(R_R')^2 X_m + (X_{\ell R}')^2 X_m + X_{\ell R}' X_m^2}{(R_R')^2 + (X_{\ell R}' + X_m)^2}\right] \end{aligned} \tag{4.129}$$

*IEEE Test Code for Polyphase Induction Motors and Generators, no. 112A, 1964 (Institute of Electrical and Electronics Engineers, 345 East 47th Street, New York, N.Y. 10017).

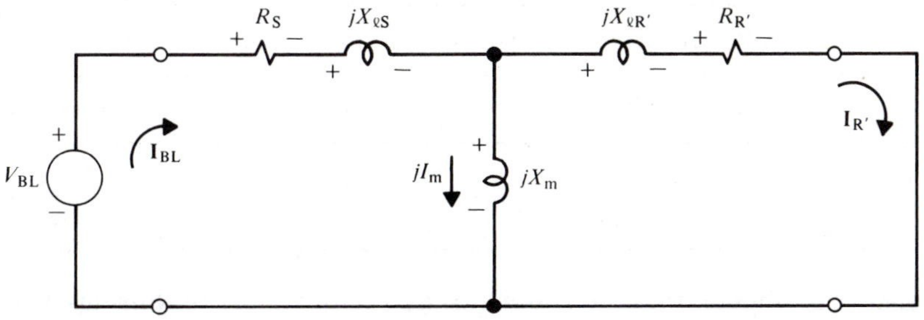

Fig. 4.29 The Equivalent Circuit of the Induction Device Under Blocked-Rotor Conditions of Operation on a Per-Phase Basis

When the rotor is blocked, the exciting current $\mathbf{I}_\phi$ (or I_m if $G_c=0$) is small (why?) when compared to the stator current $\mathbf{I}_S$, Fig. 4.27 (or Fig. 4.29), and the rotor leakage reactance $X_{R\ell}'$ at normal frequency, 60 Hz, is only slightly larger than the equivalent reactance X_{EQ}. Thus the magnitude of the reactance X_{BL}, Eq. (4.128), for the normal frequency of operation, that is, $f_E=60$ Hz, can be computed from the test values (obtained at 15 Hz and rated current) by assuming that it varies proportionally with the frequency, that is,

$$X_{BL(60)} = X_{BL(15)}\left(\frac{60}{15}\right) \tag{4.130}$$

Eq. (4.130) enables writing Eq. (4.128) as

$$X_{BL(60)} \equiv X_{BL(15)}\left(\frac{60}{15}\right) = X_{\ell S(60)} + X'_{\ell R(60)} \tag{4.131}$$

When no design information is available, it is assumed that the reactances $X_{\ell S}$ and $X_{\ell R}'$ are distributed equally between the stator and the rotor, and Eq. (4.131) is written as

$$X_{BL(60)} = X_{\ell R}' + X_{\ell R}' \,\big|_{X_{\ell S}=X_{\ell R}'} = X_{\ell S} + X_{\ell S} \,\big|_{X_{\ell R}'=X_{\ell S}} \tag{4.132}$$

However, if design information is available, the IEEE* in its induction motor test code recommends that the Table 4.1 should be utilized for the appropriate distribution of $X_{\ell S}$ and $X_{\ell R}'$ between the stator and rotor parts of the equivalent circuit of Fig. 4.15. In reference [6] the classification of the induction motors is based on starting torque, Fig. 4.41, starting current, and slip. The breakdown is as follows: (1) Class *A*, normal starting torque and normal starting current; (2) Class *B*, normal starting torque and low starting current; (3) Class *C*, high starting torque and low starting current; (4) Class *D*, high starting torque and high slip.

With this information at hand, that is, $X_{\ell S(60)}$ known, it is possible to calculate the magnetizing reactance X_m from Eq. (4.125) as

$$X_m = X_{NL} - X_{\ell S(60)} \tag{4.133}$$

Since the stator resistance R_S is the measured dc resistance of the winding of each phase of a Y-connected induction device, the resistance of the rotor winding, referred to the stator side, R_R', can be calculated from Eqs. (4.127) and (4.129), which, upon equating of corresponding parts, yield

$$R_{EQ} = R_{BL} - R_S = R_R'\left[\frac{X_m^2}{(R_R')^2+(X_{\ell R}'+X_m)^2}\right] \tag{4.134}$$

*IEEE Test Code, see footnote on page 254.

Table 4.1 STATOR-ROTOR REACTANCE DISTRIBUTION FOR INDUCTION MOTORS

	Class *A*	Class *B*	Class *C*	Class *D*	Wound Rotor
$X_{\ell S}$	0.5	0.4	0.3	0.5	0.5
$X'_{\ell R}$	0.5	0.6	0.7	0.5	0.5

Experience has shown that $|X_{\ell R}'+X_m| \gg R_R'$. Utilization of this approximation enables writing Eq. (4.134) as

$$R_{BL} - R_S = R_R'\left[\frac{X_m^2}{(X_{\ell R}'+X_m)^2}\right] \tag{4.135}$$

where only a 1 percent error is introduced by the approximation. Such error is well within accepted limits of tolerances. Solution of Eq. (4.135) for R_R' yields

$$R_R' = \left[\frac{(X_{\ell R}'+X_m)^2}{X_m^2}\right](R_{BL}-R_S) \tag{4.136}$$

Note that although all the circuit parameters of Fig. 4.15 have been determined and the steady state performance of the induction device under load can be computed, we have to utilize Eqs. (4.87), (4.88), and (4.89) and the ideas of transformer analysis [i.e., Eqs. (1.59) through (1.62)] to obtain the unreferred rotor circuit parameters. Thus the rotor parameters R_R and $X_{\ell R}$ can be calculated as follows:

$$R_R = \left(\frac{\mathcal{N}_R}{\mathcal{N}_S}\right)^2 R_R' \tag{4.137}$$

$$X_{\ell R} = \left(\frac{\mathcal{N}_R}{\mathcal{N}_S}\right)^2 X_{\ell R}' \tag{4.138}$$

Thus the equivalent circuit of the three-phase induction device, on a per-phase basis, with the rotor circuit not referred to the stator side of the airgap, can be drawn as in Fig. 4.30. Note that for Δ-connected motors (which are rarely, if ever, found in practice in the United States) the dc resistance of the stator windings is $R_S = 3/2(R_{jk})$ (where $j \neq k = A, B, C$ are the three terminals of the Δ-connected stator winding). Also, utilization of the theory developed in Chapter 3 enables us to calculate the phase impedance $\mathbf{Z}_\phi$ of the Δ-connected stator winding by taking the ratio of any of the measured line-to-line voltages $|\mathbf{E}_{\lambda\kappa}|$ and phase current $|\mathbf{I}_\phi|$.

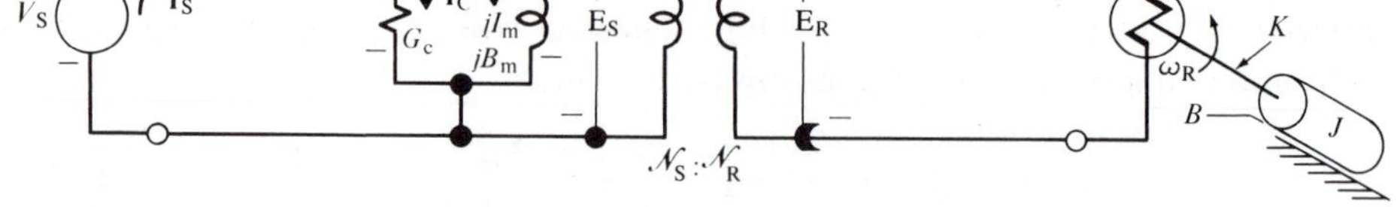

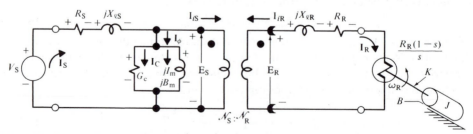

Fig. 4.30 The $j\omega$-Domain Equivalent Circuit of the Three-Phase Induction Device, on a Per-Phase Basis, with the Rotor Circuit Not Referred to the Stator Side of the Airgap, $\omega_R \neq 0$

In conclusion, it should be stated that the effects of frequency to leakage reactances and rotor resistance of induction motors, of rating less than 25 hp, are negligible. Therefore, the blocked-rotor test data on such a device can be measured at 60 Hz.

Also, before leaving this section it should be mentioned that it is customary to measure the blocked-rotor torque and the effects of the rotor position to the blocked-rotor impedance when data for the blocked-rotor test are taken. The effects of rotor position are negligible for squirrel cage rotors (why?).

Example 4.9

For a squirrel cage induction motor rated as Y-connected, three-phase, six-pole, 60 Hz, 440 V (line-to-line RMS voltage), 75 A (line current), and 50 hp, the following test data are obtained:

1. dc test: the dc resistance measured between any two terminals of the stator windings is 0.24 Ω.
2. No-load test: at 60 Hz and rated line-to-line voltage applied, the following measurements are made*:

$$|\mathbf{I}_{\mathrm{NL}}| = 25 \text{ A}$$
$$P_{\mathrm{NLT}} = 1{,}700 \text{ W} \,(P_1 = 900 \text{ W and } P_2 = 800 \text{ W})$$

3. Blocked-rotor test at 15 Hz: for an applied voltage of 40 V (line to line), the following measurements are made:

$$|I_{\mathrm{BL(15)}}| = 75 \text{ A}$$
$$P_{\mathrm{BLT(15)}} = 4{,}250 \text{ W} \,(P_1 = 2{,}000 \text{ W and } P_2 = 2{,}250 \text{ W})$$

4. Blocked-rotor test at 60 Hz: for an applied voltage of 424 V (line to line), the following measurements are made:

$$|\mathbf{I}_{\mathrm{BL(60)}}| = 100 \text{ A}$$
$$P_{\mathrm{BLT(60)}} = 17{,}000 \text{ W} \,(P_1 = 4{,}500 \text{ W and } P_2 = 12{,}500 \text{ W})$$
$$\tau_{\mathrm{STG}} = 95 \text{ N} \cdot \text{m}$$

Compute the rotational losses of the device and the parameters of the equivalent circuit for the normal running conditions (i.e., for a frequency of excitation $f_E = 60$ Hz, $s = 0.03$, and $f_e = 1.8$ Hz). Also, compute the starting torque from the data of part 4. This part of the test is run exclusively for the purpose of calculating the starting torque. The frequency of the rotor currents is never 60 Hz during normal running conditions.

Solution

Utilization of Figs. 4.26 through 4.29 and Eqs. (4.122) to (4.132) enables calculation of all the circuit parameters, based on 60 Hz, for the per-phase equivalent circuit of the induction motor.

From the data of the dc test, it is obvious that for the Y-connected induction device

$$R_S = \frac{R_{\mathrm{LL}}}{2} = \frac{0.24}{2} = \underline{\underline{0.12\ \Omega}}$$

*For information pertinent to the two-wattmeter power measurement, that is, P_1 and P_2 refer to Section 3.10.

Plate 4.6 Four-Pole, 3500-hp, Custom 8000 Induction Motor with Ventilation Louvres to Minimize the Entrance of Liquids and Particles and with Integral Cast Fans on the Rotor to Draw Cooling Air. (Courtesy of General Electric Co., Schenectady, N.Y.)

From the data of the no-load test and Eq. (4.122), the rotational losses of the induction motor are calculated as follows:

$$P_{\mathrm{RT}} = P_{\mathrm{NLT}} - 3R_{\mathrm{S}}|\mathbf{I}_{\mathrm{NL}}{}^2| = 1{,}700 - 3 \times 0.12 \times 25^2 = \underline{\underline{1{,}475}}\ \mathrm{W}$$

Utilization of the same data and Eqs. (4.123), (4.124), and (4.125) enables calculation of

$$|\mathbf{Z}_{\mathrm{NL}}| = \frac{|\mathbf{E}_{\mathrm{NL}}|/\sqrt{3}}{|\mathbf{I}_{\mathrm{NL}}|} = \frac{440/\sqrt{3}}{25} = \underline{10.16}\ \Omega$$

$$R_{\mathrm{NL}} = \frac{P_{\mathrm{NLT}}/3}{|\mathbf{I}_{\mathrm{NL}}|^2} = \frac{1{,}700/3}{25^2} = \underline{0.907}\ \Omega$$

and

$$X_{\mathrm{NL}} = \left(|\mathbf{Z}_{\mathrm{NL}}|^2 - R_{\mathrm{NL}}^2\right)^{1/2} = \underline{10.12}\ \Omega$$

From the blocked-rotor test at reduced frequency and impressed voltage, which is adjusted to yield rated current, and Eqs. (4.126), (4.127), and (4.128), the following quantities are calculated:

$$|\mathbf{Z}_{BL(15)}| = \frac{|\mathbf{V}_{BL}|/\sqrt{3}}{|\mathbf{I}_{BL}|} = \frac{40/\sqrt{3}}{75} = \underline{0.308\ \Omega}$$

$$R_{BL(15)} = \frac{|P_{BLT}|/3}{|\mathbf{I}_{BL}|^2} = \frac{4{,}250/3}{75^2} = \underline{0.252\ \Omega}$$

$$X_{BL(15)} = \left(|\mathbf{Z}_{BL(15)}|^2 - R_{BL(15)}^2\right)^{1/2} = \underline{0.177\ \Omega}$$

Utilization of Eq. (4.130) enables calculation of $X_{BL(60)}$ as

$$X_{BL(60)} = X_{BL(15)}\left(\tfrac{60}{15}\right) = 0.177 \times \tfrac{60}{15} = \underline{0.708\ \Omega}$$

Utilization of Eq. (4.123) (or Table 4.1 if design data are available) enables calculation of $X_{\ell S(60)}$ and $X'_{\ell S(60)}$ as follows:

$$X_{BL(60)} = 2X'_{R\ell(60)} = 2X_{S\ell(60)} = \underline{0.708\ \Omega}$$

Thus

$$X'_{\ell R(60)} \equiv X_{\ell S(60)} = \frac{0.708}{2} = \underline{\underline{0.354\ \Omega}}$$

Use of Eq. (4.133) enables calculation of the magnetizing reactance as

$$X_m = X_{NL} - X_{\ell S(60)} = 10.12 - 0.354 = \underline{\underline{9.766\ \Omega}}$$

Finally, utilization of Eq. (4.136) enables calculation of the referred rotor resistance R_R' as

$$R_R' = \left[\frac{(X'_{\ell R(60)} + X_m)^2}{X_m^{\ 2}}\right](R_{BL(60)} - R_S)\Bigg|_{R_{BL(15)} \equiv R_{BL(60)}}$$

$$= \left[\frac{(0.354 + 9.766)^2}{9.766^2}\right](0.252 - 0.12) = \underline{\underline{0.14\ \Omega}}$$

Use of Eqs. (4.137), (4.138), (4.88), and (4.89) enables calculation of the unreferred rotor circuit parameters as

$$R_R = \left(\frac{\mathcal{N}_R}{\mathcal{N}_S}\right)^2 R_R'$$

$$X_{R\ell} = \left(\frac{\mathcal{N}_R}{\mathcal{N}_S}\right)^2 X_{\ell R}'$$

$$\mathbf{E}_R = \left(\frac{\mathcal{N}_R}{\mathcal{N}_S}\right)\mathbf{E}_R'$$

and

$$I_R = -\left(\frac{\mathcal{N}_S}{\mathcal{N}_R}\right) I_R'$$

Utilization of Fig. 4.17, the data of part 4 of the test, and Eq. (4.22) enables calculation of the airgap power P_{AGT} and the starting torque τ_{STG} as follows:

$$P_{AGT} = P_{INT} - 3R_S|\mathbf{I}_S|^2 = 17{,}000 - 3 \times 0.12 \times (100)^2 = \underline{13{,}400\ \text{W}}$$

$$\omega_S = \frac{\omega_E}{P/2} = \frac{2\pi \times 60}{6/2} = \underline{125.66\ \text{rad/s}}$$

Thus for $\tau_{STG} \equiv \tau_{emT}$,

$$\tau_{STG} = \frac{P_{AGT}}{\omega_S} = \frac{13{,}400}{125.66} = \underline{\underline{106.64 \text{ N} \cdot \text{m}}}$$

The measured value of τ_{STG} is 10.91% less than the calculated value because the calculations do not account for the *stator core* and *stray load losses*.

4.10 THE TWO-PHASE INDUCTION MOTOR: DYNAMIC ANALYSIS VIA *s*-DOMAIN TECHNIQUES

The two-phase induction motor differs from the three-phase induction motor in two main aspects: (1) size, that is, the output of the motor ranges from a fraction of a watt up to a few hundred watts, and (2) structure, that is, *the two-phase induction motor consists of a stator with two sinusoidally distributed windings, displaced from each other spatially* by* 90°, *and a squirrel cage rotor* (the diameter of which is made small in order to reduce the inertia, and thus to improve its acceleration characteristics) *or a drag-cup rotor* (i.e., the rotor is comprised of two parts, the rotating part, which is a canlike metallic part with one end removed and the rotor shaft permanently mounted on the other end, and the stationary part, which is a pluglike iron core inside the cap; the first part is used to reduce the inertia of the rotor, and thus improve acceleration characteristics, the second part is used to complete the magnetic circuit). The latter rotor structure is used in motors whose output is in the range of only a few watts. This rotor structure makes the two-phase induction motor highly desirable in control systems because of its light weight, its ruggedness (a characteristic of squirrel-cage-rotor motors) and the absence of brushes (when compared to dc motors). The schematic diagram of a two-phase induction motor, as it is used in *control systems*, is exhibited in Fig. 4.31. The two windings displaced spatially from voltages $v_R(t)$ and $v_C(t)$ that are *displaced in time* also by 90°. The voltage $v_R(t)$, obtained from a source whose magnitude and frequency ω_E (i.e., $f_E = 60$ Hz or 400 Hz) are maintained constant, is referred to as the **reference voltage**. However, the voltage $v_C(t)$, referred to as **control voltage**, is obtained from the output of an ac amplifier connected at the output of a controller (a device that generates signals from observation of system outputs). The amplitude and polarity of the voltage $v_C(t)$ is a function of the degree of action required of the motor but its frequency is that of the reference voltage (i.e., $f_E = 60$ Hz or 400 Hz). If the variation of the control voltage is slow when compared to the frequency ω_E, the two voltages can be written as

$$v_C(t) = \sqrt{2}\,|\mathbf{V}_C(t)|\sin\omega_E t \tag{4.139}$$

$$v_R(t) = \sqrt{2}\,|\mathbf{V}_R|\cos\omega_E t \tag{4.140}$$

The time quadrature is assured either by introducing the 90° phase shift in the amplifier or by introducing a suitable capacitor in the reference phase (in fact, this

*The angular displacement between the two stator windings is always expressed in electrical degrees θ_E which are related to the mechanical degrees θ_M through the number of poles P of the two-phase motor in accordance with the equation $\theta_M = \theta_E/(P/2)$.

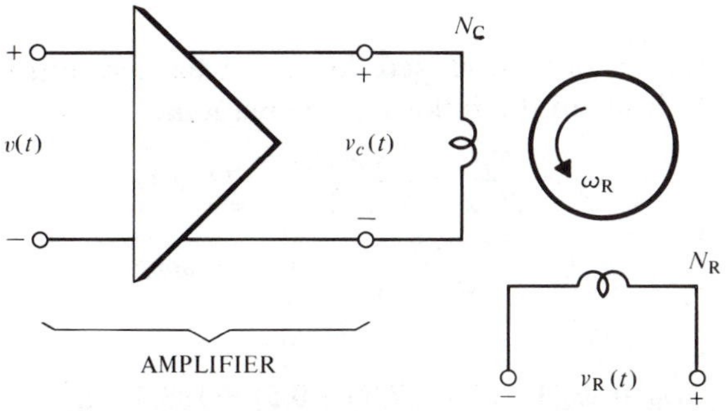

Fig. 4.31 Schematic Diagram of a Two-Phase Induction Motor with Its Control Winding Connected to the Output of an ac Amplifier

is usually the case). For a nonzero $v_R(t)$ leading $v_C(t)$ by 90°, the rotor rotates in the positive direction. However, the direction of rotor rotation changes if the polarity of the control voltage $v_C(t)$ changes. Thus the two-phase induction motor is regarded as a **phase-sensitive** device; the direction of the motor can be changed by reversing the phase of the control winding excitation.

For a steady state analysis of the balanced two-phase induction motor [i.e., when the frequency ω_E and the voltage magnitudes $|\mathbf{V}_C(t)|$ and $|\mathbf{V}_R|$ of the two voltages, Eqs. (4.139) and (4.140), are the same, and the voltage magnitudes are maintained constant, i.e., $|\mathbf{V}_C(t)| \equiv |\mathbf{V}_C| = |\mathbf{V}_R| = k$], the theory of the balanced three-phase induction motor is utilized with these exceptions: (1) the phase voltages in all pertinent equations must be transformed from three- to two-phase excitations, and (2) each equation that exhibits the factor 3, designating "3 phases," should be premultiplied by the factor $\frac{2}{3}$, where 2 designates "2 phases". This multiplication converts these equations automatically to equations describing balanced two-phase induction motors. For example, when the torque equation that describes the three-phase induction motor, Eq. (4.106), is premultiplied by a factor $\frac{2}{3}$, it describes the two-phase induction motor. Similarly, all the rest of the equations that describe the steady state performance of the balanced three-phase induction motor and contain the factor 3 can be used to describe the steady state performance of the balanced two-phase induction motor, simply by premultiplying each by a factor $\frac{2}{3}$.

Example 4.10

The measured, steady state, equivalent circuit parameters of a 5-W, 110-V, 60-Hz, two-pole servomotor are (1) $R_S = 288\ \Omega$, (2) $X_{\ell S} = 60\ \Omega$, (3) $R_R' = 849\ \Omega$, (4) $X_{\ell R}' = 60\ \Omega$, and (5) $X_m = 995\ \Omega$. For 100-V **balanced stator excitation**, $s = 0.5$, and core, friction, and windage losses equal to 0.736 W, calculate the following quantities: (1) the synchronous speed ω_S of the field, (2) the mechanical speed ω_R of the rotor, (3) the frequency of the rotor currents, ω_e, (4) the input current I_S and power P_{INT}, (5) the stalled torque, that is, $\tau_{AGT}(s=1)$, for 100- and 115-V excitation, (6) the developed power* P_{AGT} and torque τ_{AGT}, (7) the output power P_{OT} and torque τ_{OT}, (8) the speed regulation SR, and (9) the efficiency η.

*What is the difference between P_{AGT} and P_{emT} and τ_{AGT} and τ_{emT}?

Solution

Utilization of the theory developed for the balanced three-phase induction motor and Fig. 4.17 leads to the following calculations.

1.
$$\omega_S = \frac{\omega_E}{P/2} = \frac{2\pi \times 60}{2/2} = \underline{\underline{377}} \text{ rad/s}$$

2.
$$s = \frac{\omega_S - \omega_R}{\omega_S} \rightarrow \omega_R = \omega_S - s\omega_S$$

Thus

$$\omega_R = \omega_S(1-s) = 377(1-0.5) = \underline{\underline{188.5}} \text{ rad/s}$$

or

$$\eta_R = \frac{60\omega_R}{2\pi} = \frac{60 \times 180}{2\pi} = \underline{\underline{1718.87}} \text{ rpm}$$

3.
$$f_e = sf_E = 0.5 \times 60 = \underline{\underline{30}} \text{ Hz}$$

and

$$\omega_e = 2\pi f_e = 6.28 \times 30 = \underline{\underline{188.5}} \text{ rad/s}$$

4. Refer to Fig. 4.10.1. Then

$$\mathbf{I_S} = \frac{\mathbf{V_S}}{\mathbf{Z_S} + \mathbf{Z_P}}$$

where

$$\mathbf{Z_P} = \frac{(995e^{j90°})[849 + (849/0.5)(1-0.5) + j60]}{849 + (849/0.5)(1-0.5) + j(60+995)} = \underline{845.68e^{j60.17°}} \ \Omega$$

$$\mathbf{Z_S} + \mathbf{Z_P} = (288 + j60) + (420.67 + j733.63) = 708.67 + j793.63$$

$$= \underline{1063.98e^{j48.24°}} \ \Omega$$

Thus

$$\mathbf{I_S} = \frac{100e^{j0°}}{1063.98e^{j48.24°}} = \underline{\underline{0.094e^{-j48.24°}}} \text{ A}$$

Also,

$$P_{INT} = 2|\mathbf{V_S}||\mathbf{I_S}|\cos\theta = 2 \times 100 \times 0.094 \cos 48.4° = \underline{\underline{12.52}} \text{ W}$$

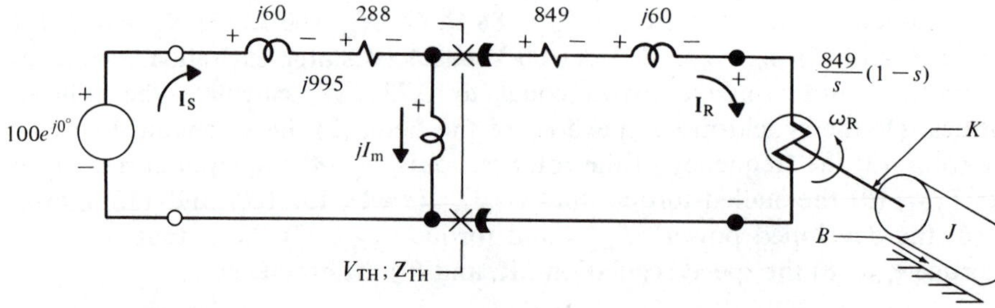

Fig. 4.10.1 The Steady State Equivalent Circuit of a Balanced, Two-Phase Induction Motor on a Per-Phase Basis; the Rotor Is Referred to the Stator Side of the Airgap

5. $$\tau_{AGT} = \frac{1}{\omega_S} \times P_{AGT} = \frac{1}{\omega_S} \times 2 \times \left\{849 + \left(\frac{849}{s}\right)(1-s)\right\} |\mathbf{I_R}'|^2$$

Since $\mathbf{I_R}' = f(s)$, and s is variable, the Thevenin's equivalent of the circuit above is derived as follows:

$$\mathbf{E_{TH}} = \left[\frac{j995}{288 + j(60+995)}\right]\mathbf{V_S} = \left[\frac{995e^{j90°}}{1094e^{j74.73°}}\right]\mathbf{V_S} = \underline{0.91e^{j15.27°}\ \mathbf{V_S}\ \text{V}}$$

$$\mathbf{Z_{TH}} = \frac{(j995)(288+j60)}{288+j(995+60)} = \frac{292{,}713e^{j101.77°}}{1094e^{j74.73°}}$$

$$= \underline{267.56e^{j27.04°}} = \underline{238.31 + j121.64}$$

Thus from Fig. 4.10.1 and the Thevenin parameters given above, the simplified diagram of Fig. 4.10.2 is drawn. At standstill, that is, when the rotor is not rotating, $\omega_R = 0$. Thus $s = (\omega_S - \omega_R)/\omega_S = 1$ and

$$\mathbf{I_R}' = \frac{0.91e^{j15.27°}\mathbf{V_S}}{(238.31+j121.64)+(849+j60)} = \frac{0.91e^{j15.27°}\mathbf{V_S}}{1087.31+j181.64}$$

$$= \frac{0.91e^{j15.24°}\mathbf{V_S}}{1102.38e^{j9.43°}} = \underline{8.25\times10^{-4}e^{j5.84°}\ \mathbf{V_S}}\ \text{A}$$

Thus

$$\mathbf{I}_{R(100)}' = (8.25\times10^{-4}e^{j5.84°})(100e^{j0°}) = \underline{0.0825e^{j5.84°}}\ \text{A}$$

and

$$\mathbf{I}_{R(115)}' = (8.25\times10^{-4}e^{j5.84°})(115e^{j0°}) = \underline{0.09e^{j5.84°}}\ \text{A}$$

Now for $s = 1$,

$$\tau_{AGT} = \tfrac{1}{377}\times 2\times(849)|\mathbf{I_R}'|^2$$

Thus

$$\tau_{AGT(100)} = 4.504\times(0.0825)^2 = \underline{\underline{0.0310}}\ \text{N}\cdot\text{m}$$

and

$$\tau_{AGT(115)} = 4.504\times(0.09)^2 = \underline{\underline{0.036}}\ \text{N}\cdot\text{m}$$

for normal operation, that is, $s = 0.5$ and $\mathbf{V_S} = 100\ e^{j0°}$.

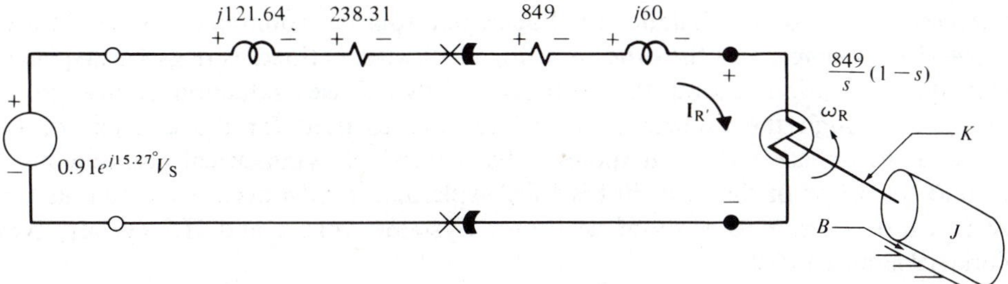

Fig. 4.10.2 The Simplified Circuit of Fig. 4.10.1, Utilizing Thevenin's Theorem

6. $$P_{\text{AGT}} = 2\left\{849 + \left(\frac{849}{0.5}\right)(1-0.5)\right\}|\mathbf{I}_{\text{R}}'|^2 = 2(849+849)|\mathbf{I}_{\text{R}}'|^2 = 3396|\mathbf{I}_{\text{R}}'|^2$$

where

$$\mathbf{I}_{\text{R}}' = \frac{(0.91e^{j15.27^\circ})(100e^{j0^\circ})}{(238.31+j121.64)+[849+(849/0.5)(1-0.5)+j60]} = \frac{91e^{j15.27^\circ}}{1945e^{j5.36^\circ}}$$

$$= \underline{0.0478e^{j9.91^\circ}\ \text{A}}$$

Thus

$$P_{\text{AGT}} = 3396(0.0478)^2 = \underline{\underline{7.76\ \text{W}}}$$

and

$$\tau_{\text{AGT}} = \frac{7.76}{377} = \underline{\underline{0.0206\ \text{N}\cdot\text{m}}}$$

7. $$P_{\text{OT}} = P_{\text{AGT}} - P_{\text{RW}} - P_{\text{LOSSES}} = 7.76 - 2\times 849\times(0.0478)^2 - 0.5 = \underline{\underline{3.380\ \text{W}}}$$

and

$$\tau_{\text{OT}} = \frac{P_{\text{OT}}}{\omega_{\text{R}}} = \frac{3.380}{180} = \underline{\underline{0.0188\ \text{N}\cdot\text{m}}}$$

8. $$\text{SR} = \frac{\omega_{\text{NL}}-\omega_{\text{FL}}}{\omega_{\text{FL}}}\times 100 = \frac{\omega_{\text{S}}-\omega_{\text{R}}}{\omega_{\text{R}}}\times 100 = \frac{377-180}{180}\times 100 = \underline{109.44\%}$$

9. $$\eta = \frac{P_{\text{OT}}}{P_{\text{INT}}}\times 100 = \frac{3.38}{12.52}\times 100 = \underline{\underline{27.00\%}}$$

Note that it is not necessary to find the Thevenin's equivalent in part 5. The stalled torque could have been determined by calculating $\mathbf{I}_{\text{R}}'$ with $s=1$ for the two excitation voltages and using the equation $\tau_{\text{AGT}} = P_{\text{AGT}}/\omega_{\text{S}}$ or $\tau_{\text{emT}} = P_{\text{emT}}/\omega_{\text{R}}$.

In control systems applications the voltage of the control winding $v_{\text{C}}(t)$ is a function of the degree of action required of the two-phase induction motor; for example, $v_{\text{C}}(t)$ may be proportional to the error $\varepsilon(t)$ of a control system. [In Fig. 4.43, for instance, the error $\varepsilon(t)$ depends on the difference between the command signal $\alpha_{\text{G}}(s)$ and the controlled variable $\alpha_{\text{T}}(s)$.] Thus, because the magnitude of the voltage $v_{\text{C}}(t), |\mathbf{V}_{\text{C}}(t)|$, is continuously varying while its frequency ω_{E} remains constant, the voltages of Eqs. (4.139) and (4.140) have the same frequency but different magnitudes, that is, $|\mathbf{V}_{\text{C}}(t)| \neq |\mathbf{V}_{\text{R}}|$. This mode of operation of the two-phase induction motor is known as **unbalances** (because of the unequal magnitude of the two voltages) and necessitates the use of the **symmetrical components** method for the motor's steady state analysis. This method of analysis is based on the assumption that the magnetic structure of the motor is linear. Therefore, the method of superposition is applied to decompose the **one set** of the motor's two unbalanced voltages into **two sets** of balanced voltages, one rotating counterclockwise, defined as **positive sequence**, and the other rotating clockwise, defined as **negative sequence.** This decomposition allows the unbalanced two-phase induction motor to be analyzed through the utilization of the techniques used for the analysis of the balanced two-phase induction motor. The method of symmetrical components is beyond the scope of this text, but is fully explained in advanced texts such as that by E. Clark, **Circuit Analysis of ac Power Systems**, vols. I and II,* (Wiley, New York, 1943 and 1950).

*It should be pointed out, however, that the inventor of the method of symmetrical components for the analysis of unbalanced polyphase systems is C. Z. Fortescue.

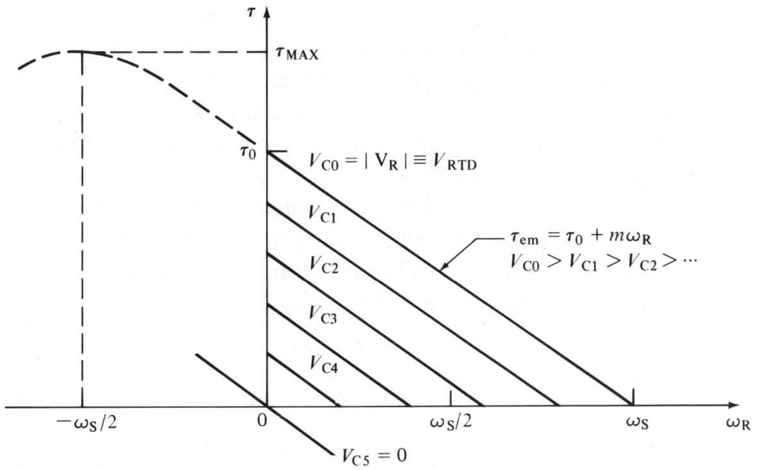

Fig. 4.32 Linearized (Highly Idealized) Torque vs. Speed Characteristics of a Two-Phase Induction Motor

It can be proven, utilizing steady state analysis of the two-phase induction motor with unbalanced excitation, Reference [7], that the torque versus speed characteristic, Fig. 4.20, **as applied to the two-phase induction motor**, is a function of both the control voltage $|\mathbf{V}_C|$ and the slip s. If the control voltage is maintained constant at $|\mathbf{V}_C(t)| = k_j|\mathbf{V}_R|, j = 1,2,3,\ldots$, and for each k_j s is allowed to vary from 0 to 1, a family of characteristics is obtained; Fig. 4.32. The derivation of these characteristics is based on the following **linear** assumptions and approximations: (1) in Fig. 4.20 only the portion of the region $0<\omega_R<\omega_S$ that has the negative slope (a requirement for stable operation, as explained shortly) is utilized; such a characteristic, a straight line, is obtained with the voltages of the control and reference windings of the device set equal to the device's rated voltage, that is, $|\mathbf{V}_R| = |\mathbf{V}_C(t)| \equiv \mathbf{V}_{RTD}$, and (2) a family of characteristics equally spaced for equal increments of $|\mathbf{V}_C(t)|$ and parallel to the characteristic drawn at the rated voltage of the control and reference windings (i.e., $|\mathbf{V}_C(t)| = |\mathbf{V}_R| \equiv \mathbf{V}_{RTD}$) is utilized.

Examination of Fig. 4.32 reveals that: (1) high torques are developed by the two-phase induction motor at $\omega_R = 0\,\text{rad/s}$ and $|\mathbf{V}_C(t)| \neq 0\,\text{V}$, (2) a **zero** torque is developed by this motor at $\omega_R = 0\,\text{rad/s}$ and $|\mathbf{V}_C(t)| = 0\,\text{V}$, and (3) the operating speed of this motor is always **much less** than the synchronous speed ω_S, because the control winding voltage $|\mathbf{V}_C(t)|$ **is always much less** than the rated voltage of the motor. It should be noted that a close approximation of such characteristics **enhanced** the usefulness of this motor in control systems applications.

Manufacturers strive to obtain the characteristics of Fig. 4.32 by designing the two-phase induction motor to (1) develop **maximum** torque at rated voltages $|\mathbf{V}_C(t)| = |\mathbf{V}_R| = \mathbf{V}_{RTD}$ and a reverse speed given by $\omega_R \equiv -\omega_S/2$ rad/s, and (2) exhibit a rotor resistance R_R much **larger** than the rotor reactance, that is, $R_R \gg |jX_{\ell R}|$. This last specification prevents the motor from developing torque when $\omega_R = 0\,\text{rad/s}$ and $|\mathbf{V}_C(t)| = 0\,\text{V}$, a phenomenon known as **single-phasing***, which undermines the usefulness of this motor in control systems.

For control systems analysis, steady state analysis per se, of the two-phase induction motor, is of little importance. However, its dynamic analysis is of great

*For information pertinent to the operation of single-phase induction motors refer to Section 4.11.

value, in the analysis of control systems, of which the two-phase induction motor might be one of the constituent components, is usually carried out in the s-domain. For such analysis, utilization of transfer functions of component subsystems (of large systems) becomes essential. Therefore, should the two-phase induction motor be a control system's power unit, determination of its transfer function would be in order and is, therefore, undertaken here.

The transfer function of the two-phase induction motor operating with the voltage magnitude of the control winding changing, $|\mathbf{V}_C(t)| \equiv \mathbf{V}_{C0}, \mathbf{V}_{C1}, \mathbf{V}_{C2}, \ldots, \mathbf{V}_{Cn}$ (Fig. 4.32) can be derived as follows. The describing equation of the torque versus speed characteristic can be written with the aid of Fig. 4.32 as

$$\tau_{em} = \tau_0 + m\omega_R \tag{4.141}$$

where

$$m = \frac{\Delta\tau}{\Delta\omega} = \frac{\tau_0 - 0}{0 - \omega_S} = -\frac{\tau_0}{\omega_S} \tag{4.142}$$

The torque τ_0 is defined as the **blocked-rotor torque** of the two-phase induction motor. It is the starting torque of the two-phase induction motor and is by design always less than the maximum torque that can be developed by the motor, Fig. 4.32.

If J (in kilogram meters^2) represents the combined moment of inertia of (1) the rotor, (2) the shaft, and (3) the reflected loading moment of inertia J_L, and B (in newton meters/radian/second) represents the viscous friction on the rotor and loading, the dynamic equation of the two-phase induction motor in terms of shaft angular displacement $\theta_R(t)$ (in radians) can be written as

$$\tau_{em} = \tau_0 + m\frac{d\theta_R(t)}{dt} \equiv J\frac{d^2\theta_R(t)}{dt^2} + B\frac{d\theta_R(t)}{dt} \tag{4.143}$$

The blocked-rotor torque is related to the reference or control voltages of the two-phase induction motor in accordance with the equation

$$k \equiv \left.\frac{\tau_0}{v_R(t)}\right|_{v_R(t) = v_C(t)} \tag{4.144}$$

Eq. (4.144) enables writing τ_0 as

$$\tau_0 = kv_R(t) = kv_C(t) \tag{4.145}$$

Substituting from Eq. (4.145) for $\tau_0 = kv_C(t)$ into Eq. (4.143) and taking its *Laplace transform*, assuming that all initial conditions are zero, yields*

$$kV_C(s) + ms\Theta_R(s) = Js^2\Theta_R(s) + Bs\Theta_R(s) \tag{4.146}$$

Separation of the output variable $\Theta(s)$ from the input or control variable $V_C(s)$ [remember that $v_C(t)$ is continuously changing] in Eq. (4.146) yields

$$\left[Js^2 + (B-m)s\right]\Theta_R(s) = kV_C(s) \tag{4.147}$$

Since m is *negative*, Eq. (4.142), Eq. (4.147) shows that the effect of the *negative slope* of the torque versus speed characteristic, Fig. 4.32, is to add more friction to the motor, which does improve the damping of the motor. This added damping effect is referred to as **internal electric damping** of the two-phase induction motor. Note, however, that if m is positive, that is, the slope of the torque versus speed characteristic, Fig. 4.32 [i.e., Eq. (4.20)], is positive, and $m > B$, negative damping

*It should be noted that the "s" associated with this equation, and subsequent equations, is not the slip represented by this symbol previously in this chapter; instead it designates the Laplace operator. The student should use his discretion for the appropriate interpretation of the letter s.

occurs and the motor becomes unstable see Appendix C for stability considerations). These observations justify the remarks that were made earlier with respect to stable regions of the torque versus speed characteristics, Fig. 4.20, of the two-phase induction motor. It should be noted that since the slope of the torque versus speed characteristics of Fig. 4.32 is always negative and the effective damping $(B-m)$, Eq. (4.147), is always positive the motor operation is always stable. There is, however, an idiosyncrasy of the two-phase induction motor performance, such that under certain conditions the motor is capable of running when only the reference phase is excited and the control voltage is zero, a phenomenon known as **single phasing**. This phenomenon is a consequence of the effect of the output impedance of the amplifier, Fig. 4.31, on the internal damping of the motor, which degenerates in such a way as to cause the motor to run as a single-phase induction motor. As mentioned earlier this tendency of the two-phase induction motor can be uprooted by designing the motor to exhibit $R_R \gg |jX_{\ell R}|$.

Division of both sides of Eq. (4.147) by $(B-m)$ and defining

$$\frac{k}{(B-m)} \equiv K_m \tag{4.148}$$

and

$$\frac{J}{(B-m)} \equiv \tau_m \tag{4.149}$$

as the motor *gain constant* and motor *time constant*, respectively, enables writing Eq. (4.147), after factoring s, as

$$s(1+s\tau_m)\Theta_R(s) = K_m V_C(s) \tag{4.150}$$

Solution of Eq. (4.150) for the ratio $\Theta_R(s)/V_C(s)$, which is defined as the *position transfer function*, $G_\Theta(s)$, of the two-phase induction motor, yields

$$G_\Theta(s) = \frac{K_m}{s(1+s\tau_m)} \equiv \frac{K_m/\tau_m}{s[s+(1/\tau_m)]} \tag{4.151}$$

where K_m is defined as the **open-loop gain** and $K_m/\tau_m \equiv \mathcal{K}_m$ is defined as the **static loop sensitivity** of the position transfer function $G_\Theta(s)$. Utilization of automatic control techniques enables drawing the block diagrams of Eq. (4.151) as shown in Fig. 4.33.

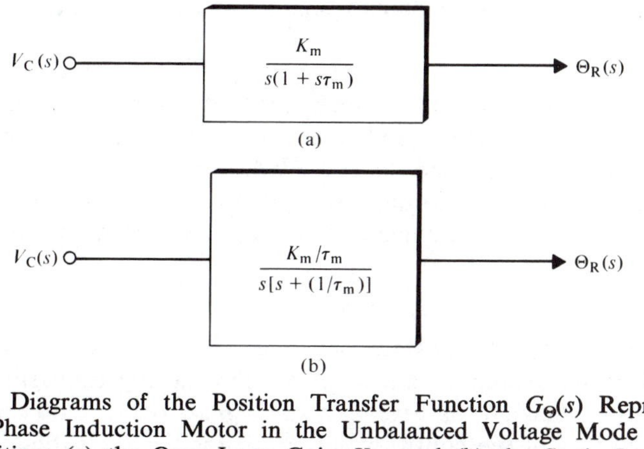

Fig. 4.33 Block Diagrams of the Position Transfer Function $G_\Theta(s)$ Representing the Two-Phase Induction Motor in the Unbalanced Voltage Mode of Operation Exhibiting: (a) the Open-Loop Gain K_m, and (b) the Static Loop Sensitivity $K_m/\tau_m \equiv \mathcal{K}_m$

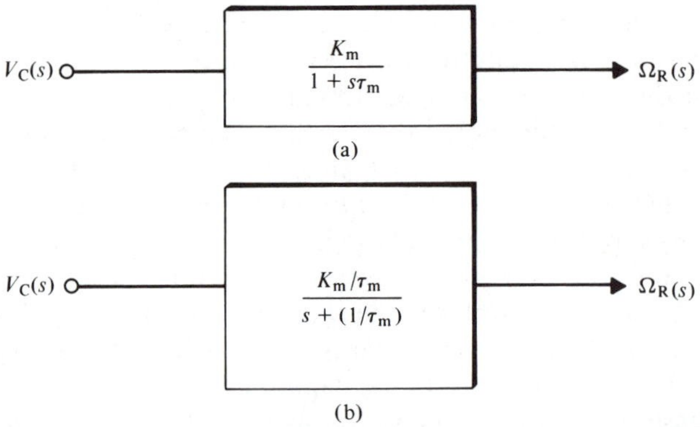

Fig. 4.34 Block Diagrams for the Speed Transfer Function $G_\Omega(s)$ Representing the Two-Phase Induction Motor in the Unbalanced Voltage Mode of Operation Exhibiting: (a) the Open-Loop Gain K_m, and (b) the Static Loop Sensitivity $K_m/\tau_m \equiv \mathcal{K}_m$

Recognizing that the quantity $s\Theta_R(s)$ in Eq. (4.150) defines the Laplace transform of the angular velocity $\Omega_R(s)$ enables us to write $\Omega(s)/V_C(s)$, which is defined as the *speed transfer functions*, $G_\Omega(s)$, of the two-phase induction motor as:

$$G_\Omega(s) = \frac{K_m}{1+s\tau_m} \equiv \frac{K_m/\tau_m}{s+(1/\tau_m)} \tag{4.152}$$

where the definitions of K_m and $\mathcal{K}_m$, pertinent to Eq. (4.151), are valid for the speed transfer function $G_\Omega(s)$ also. As above, the block diagrams of Eq. (4.152) are drawn as shown in Fig. 4.34.

The information presented in Section 2.10 is valid here. Thus it will not be repeated. The t-domain equation for $\omega_R(t)$, obtained from Eq. (4.152) and the defining relationship $G_\Omega(s)=\Omega(s)/V_C(s)$, is

$$\tau_m \frac{d}{dt}\omega_R(t) + \omega_R(t) = v_C(t) \tag{4.153}$$

However, it is emphasized that although one can use the transfer functions to obtain t-domain solutions for $\theta_R(t)$ and $\omega_R(t)$ from Eqs. (4.151) and (4.152), the main objective of obtaining the position and speed transfer functions is their utilization in block diagrams, s-domain simulation of large-scale systems.

Example 4.11

A two-phase servomotor with (1) reference and control-winding voltages, $v_R(t)$ and $v_C(t)$ rated at 115 V, (2) blocked-rotor torque, at rated $|\mathbf{V}_C|$, $\tau_0=3.77$ N·m, (3) combined inertia of rotor and load $J=0.05$ kg·m^2, and (4) viscous friction $B=0.005$ N·m/rad/s, has a torque versus steady state speed curve whose linear approximation is given by (assuming that this curve intersects the speed axis at $n_R=400$ r/min)

$$\tau_{em} = 3.77 - 0.01\omega_R \tag{1}$$

First, assume that the steady state torque equation applies to slow transients and write the differential equation that describes the starting transients of the motor load combination. Second, assume that the signal to the motor consists of the time-varying voltage of the control winding. If this voltage is designated as $v_C(t)$, the torque versus speed equation above is written with the aid of Eqs. (4.144) and

(4.145) as

$$\tau_{\text{em}} = \left(\frac{3.77}{|\mathbf{V}_{\text{R}}|}\right) v_{\text{C}}(t) - 0.01\omega_{\text{R}} \tag{2}$$

Determine the transfer function relating (1) motor shaft position and the voltage of the control winding and (2) motor shaft velocity and the voltage of the control winding. Also, draw their corresponding block diagrams.

Solution

Utilization of Eq. (4.144) and Eq. (1) enables writing the differential equation describing the motor performance as

$$0.05\frac{d^2}{dt^2}\theta_{\text{R}}(t) + 0.005\frac{d}{dt}\theta_{\text{R}}(t) = 3.77 - 0.01\frac{d}{dt}\theta_{\text{R}}(t) \tag{3}$$

or

$$\frac{d^2}{dt^2}\theta_{\text{R}}(t) + \left(\frac{0.005+0.01}{0.05}\right)\frac{d}{dt}\theta_{\text{R}}(t) = \frac{3.77}{0.05} \tag{4}$$

or

$$\frac{d^2}{dt^2}\theta_{\text{R}}(t) + 0.3\frac{d}{dt}\theta_{\text{R}}(t) = 75.4 \tag{5}$$

Since $\omega_{\text{R}}(t) \equiv d\theta_{\text{R}}(t)/dt$, Eq. (5) can be written as

$$\frac{d}{dt}\omega_{\text{R}}(t) + 0.3\omega_{\text{R}}(t) = 75.4 \tag{6}$$

Assume that $\omega_{\text{R}}(t) = \omega_{\text{R}t}(t) + \omega_{\text{R}ss}(t)$, $\omega_{\text{R}t}(t) = Ke^{st}$ and $\omega_{\text{R}ss}(t) = \Omega$. Then for $\omega_{\text{R}t}(t) \equiv \omega_{\text{R}}(t)$, Eq. (6) can be written as

$$s\omega_{Rt}(t) + 0.3\omega_{Rt}(t) = (s+0.3)\omega_{Rt}(t) = 0 \tag{7}$$

Thus $s = -0.3$ and

$$\omega_{\text{R}t}(t) = Ke^{-0.3t} \tag{8}$$

For $\omega_{\text{R}}(t) = \omega_{\text{R}ss}$, Eq. (6) can be written as

$$0 + 0.3\omega_{\text{R}ss} = 75.4 \tag{9}$$

Thus

$$\omega_{\text{R}ss} = 754 \tag{10}$$

and

$$\omega_{\text{R}}(t) = 754 + Ke^{-0.3t} \tag{11}$$

If at $t=0$, $\omega_{\text{R}}(t)=0$, Eq. (11) yields $K = -754$ and Eq. (11) can be written as

$$\omega_{\text{R}}(t) = \underline{754(1-e^{-0.3t})}\ \text{rad/s} \tag{12}$$

Utilization of Eqs. (2) and (4.144), and substitution of $|\mathbf{V}_{\text{R}}| = 115$ in Eq. (2), yields

$$0.05\frac{d^2}{dt^2}\theta_{\text{R}}(t) + 0.005\frac{d}{dt}\theta_{\text{R}}(t) = \left(\frac{3.77}{115}\right)v_{\text{C}}(t) - 0.01\frac{d}{dt}\theta_{\text{R}}(t) \tag{13}$$

Division of Eq. (13) by 0.05 and grouping of appropriate elements yields

$$\frac{d^2}{dt^2}\theta_{\text{R}}(t) + (0.1+0.2)\frac{d}{dt}\theta_{\text{R}} = 0.66v_{\text{C}}(t) \tag{14}$$

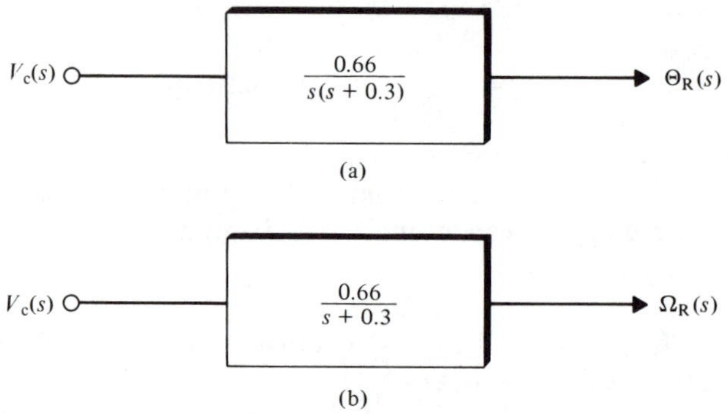

Fig. 4.11.1 Block Diagrams of: (a) the Position and (b) the Velocity Transfer Functions Given by Eqs. (16) and (17), Both Exhibiting the Static Loop Sensitivity $\mathcal{K}_m = 0.66$ Only

Assuming that the initial conditions are zero and taking the Laplace transform of Eq. (14) yields

$$(s^2 + 0.3s)\Theta_R(s) = 0.66\, V_C(s) \tag{15}$$

and

$$G_\Theta(s) \equiv \frac{\Theta_R(s)}{V_C(s)} = \frac{0.66}{s^2 + 0.3s} = \underline{\underline{\frac{0.66}{s(s+0.3)}}} \tag{16}$$

Since $s\Theta_R(s) \equiv \Omega(s)$, Eq. (16) can be written as

$$G_\Omega(s) = \frac{\Omega_R(s)}{V_C(s)} = \underline{\underline{\frac{0.66}{s+0.3}}} \tag{17}$$

Finally, the block diagrams of Eqs. (16) and (17) are drawn as in Fig. 4.11.1.

4.11 THE SINGLE-PHASE INDUCTION MOTOR

Structurally, the single-phase induction motor is a smooth airgap device, Fig. 4.35(a), and is comprised of (1) *one main stator winding* (large circles) appropriately distributed over *two-thirds* of the stator slots and laid to produce a spatially sinusoidal field, (2) *one auxiliary stator winding* (smaller circles) distributed over *one-third* of the stator slots whose axis is 90° out of phase with the main winding (usually used only during starting and is disconnected automatically by a centrifugal switch when the motor has reached about 75 percent of the synchronous speed), and (3) a squirrel cage rotor.

The principle of operation of the single-phase induction motor can be understood best by a close examination of the stator field, which, in accordance with the theory of Section 4.2, can be written (if the space harmonics are neglected) as

$$\mathbf{B} = Ki_S(t)\cos\left[\left(\frac{P}{2}\right)\theta\right]\mathbf{u_B} \tag{4.154}$$

where θ is the **mechanical angle** along the stator airgap surface measured positively from the d-axis of the stator structure, P is the number of poles of the stator field, K is the design constant of the motor, valid only when the magnetic circuit of the

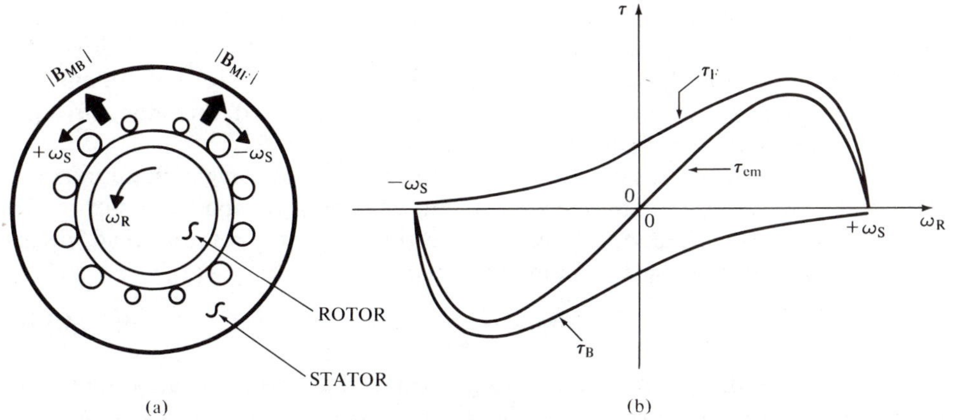

Fig. 4.35 The Single-Phase Induction Motor with the Following Shown: (a) Main (Large Circles) and Auxiliary (Small Circles) Windings and Forward Field $|\mathbf{B}_{MF}|$ with Corresponding Speed $+\omega_S$ and Backward Field $|\mathbf{B}_{MB}|$ with Corresponding Speed $-\omega_S$ and (b) Forward Torque τ_F, Backward Torque τ_B, and Developed Electromagnetic Torque τ_{em}

motor is not saturated, and

$$i(t) = \sqrt{2}\,|\mathbf{I}_S| \cos \omega_E t \tag{4.155}$$

is the current impressed in the stator winding from a single-phase ac source of radian frequency ω_E radians/second. If we substitute Eq. (4.155) into Eq. (4.154), utilize the trigonometric identity $\cos x \cos y \equiv [\cos(x-y)+\cos(x+y)]/2$, and define $\sqrt{2}\,K|\mathbf{I}_S|/2 \equiv |\mathbf{B}_M|$, then Eq. (4.154) becomes

$$\mathbf{B} = |\mathbf{B}_M| \left\{ \cos\left[\omega_E t - \left(\frac{P}{2}\right)(\theta)\right] + \cos\left[\omega_E t + \left(\frac{P}{2}\right)(\theta)\right] \right\} \mathbf{u}_B \tag{4.156}$$

In accordance with Eqs. (4.16) through (4.22), the first term of Eq. (4.156) represents a sinusoidal flux density of amplitude $|\mathbf{B}_M|$ rotating in the *positive* θ-direction with a synchronous speed

$$\omega_S = \frac{\omega_E}{P/2} \tag{4.157}$$

Similarly, the second term of Eq. (4.156) represents a sinusoidal flux density of amplitude $|\mathbf{B}_M|$ rotating in the *negative* θ-direction with a synchronous speed

$$-\omega_S = \frac{\omega_E}{P/2} \tag{4.158}$$

These ideas are exhibited in Fig. 4.35(a). Accordingly, the first term of Eq. (4.156) is known as the **forward rotating field** and is designated as $|\mathbf{B}_{MF}|$, and the second term is known as the **backward rotating field** and is designated as $|\mathbf{B}_{MB}|$. In the case of symmetrical polyphase windings, with balanced excitation, the total backward rotating field due to various phases does not exist; it cancels out. Thus *the balanced three-phase motor does not have a backward-rotating field.*

This interpretation of the field decomposition of the single-phase induction motor into a forward rotating field $|\mathbf{B}_{MF}|$ rotating at a synchronous speed ω_S radians/second, and a backward rotating field $|\mathbf{B}_{MB}|$, rotating at a synchronous speed $-\omega_S$ radians/second, suggests that the single-phase induction motor may be analyzed by applying the *principle of superposition* on two three-phase induction motors. The rotor of the *first* motor rotates in the positive (i.e., forward) direction, having a torque versus slip characteristic designated as τ_F in Fig. 4.35(b), and the rotor of the *second* motor rotates in the negative (i.e., backward) direction, having a torque versus slip characteristic designated as τ_B in Fig. 4.35(b).

On the basis of this analysis, assume that the rotor of the single-phase motor is rotating at a speed ω_R radians/second in the positive direction. The slip of the rotor rotating in the direction of the positive (i.e., forward) rotating field, designated as s_F, can be expressed in terms of the speeds ω_S and ω_R as

$$s_F = \frac{\omega_S - \omega_R}{\omega_S} \tag{4.159}$$

On the other hand, the slip of the single-phase rotor rotating at a speed ω_R radians/second in the direction of the negative (i.e., backward) rotating field, designated as s_B, can be expressed in terms of the speeds $-\omega_S$ and ω_R as

$$s_B = \frac{-\omega_S - \omega_R}{-\omega_S} \tag{4.160}$$

Substitution of ω_R from Eq. (4.159) into Eq. (4.160) yields

$$s_B = 2 - s_F \tag{4.161}$$

Eqs. (4.159) and (4.161) provide an insight to the events taking place in the rotor of the single-phase motor. The field $|\mathbf{B}_{MF}|$ induces rotor currents of frequency $\omega_{eF} = s_F\omega_E$ and the field $|\mathbf{B}_{MB}|$ induces rotor currents of frequency $\omega_{eB} = s_B\omega_E = (2 - s_F)\omega_E$. These currents produce corresponding rotor fields that interact with their corresponding stator fields to produce the forward and backward torques, τ_F and τ_B. The resultant torque developed by the single-phase induction motor is the algebraic sum of τ_F and τ_B, designated as τ_{em} in Fig. 4.35(b). Inspection of Fig. 4.35(b) reveals that there is no torque developed by the single-phase induction motor when the squirrel cage rotor is at standstill. However, the same figure reveals that a net torque is developed by the single-phase induction motor when the rotor is running at speeds ω_R close to either $+\omega_S$ or $-\omega_S$. This suggests that if the rotor were to be brought up to such speeds by some auxiliary means, it will continue to run at a speed $\omega_R \cong +\omega_S$ or $\omega_R \cong -\omega_S$.

Utilization of the theory of the three-phase induction motor, and of the fact that the stator field is divided into one forward and one backward component, whose magnitudes are half the magnitude of the total stator field, Eq. (4.156), enables us to draw the equivalent circuit of the single-phase induction motor as in Fig. 4.36, valid *only* for steady state analysis.

The steady state analysis of the single-phase induction motor follows precisely that of the three-phase induction motor. However, in this analysis we should pay particular attention to the two slips, s_F and s_B, that we will encounter in the analysis of the single-phase induction motor. For example, the mechanical load equivalent due to the forward field is

$$R_{MLEF} = \frac{0.5R_R'(1 - s_F)}{s_F} \tag{4.162}$$

and the mechanical load equivalent due to the backward field is

$$R_{MLEB} = \frac{0.5R_R'(1 - s_B)}{s_B} = \frac{0.5R_R'(s_F - 1)}{2 - s_F} \tag{4.163}$$

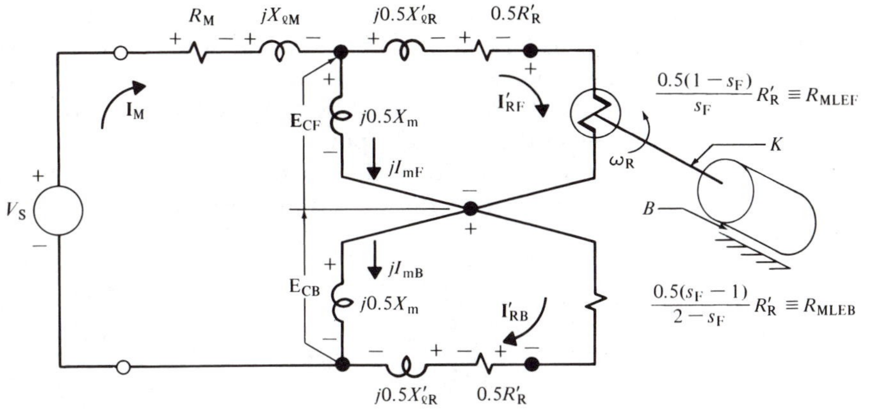

Fig. 4.36 The Equivalent Circuit for the Steady State Analysis of the Single-Phase Induction Motor

Utilization of the theory of steady state analysis for the three-phase induction motor and Fig. 4.36 enable us to draw the power flow diagram of Fig. 4.37. With the aid of circuit analysis for the calculation of the rotor current I'_{RF} due to the forward field and the rotor current I'_{RB} due to the backward field, and the rotational losses P_R known, the following power quantities can be calculated from Fig. 4.37: (1) the input power to the motor P_I, (2) the copper losses in the main winding, P_{CM}, (3) the airgap power due to the forward field, P_{AGF}, (4) the copper

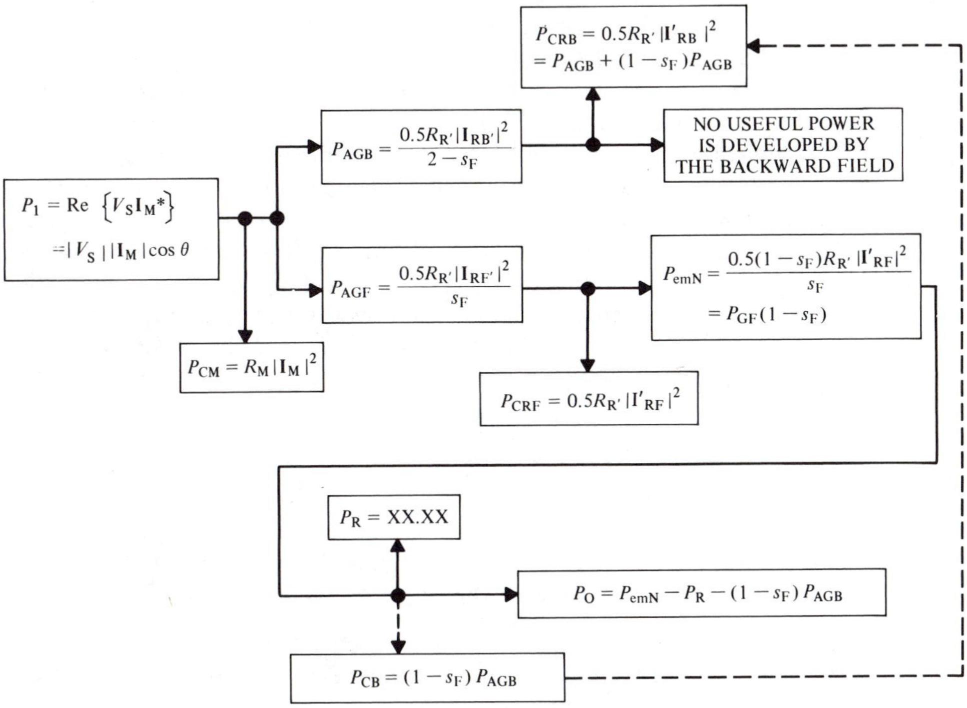

Fig. 4.37 Power Flow Diagram for the Single-Phase Induction Motor

losses in the rotor winding due to the forward field, P_{CRF}, (5) the net developed power due to the forward field,* (P_{emN}), (6) the output power due to the forward field, P_o, (7) the airgap power due to the backward field, P_{AGB}, and (8) the copper losses in the rotor winding due to the backward field, P_{CRB}. Since the rotor travels in the direction of the forward field, only power due to the forward field is developed. The airgap power due to the backward field is consumed entirely as rotor copper losses. This can be seen by utilizing Fig. 4.37 to calculate

$$P_{AGB} = \frac{0.5R_R'|\mathbf{I}_{RB}'|^2}{2-s_F} \tag{4.164}$$

and

$$P_{CRB} = 0.5R_R'|\mathbf{I}_{RB}'|^2 \tag{4.165}$$

Rearranging Eq. (4.164) yields

$$0.5R_R'|\mathbf{I}_{RB}'|^2 = (2-s_F)P_{AGB} \equiv P_{AGB} + (1-s_F)P_{AGB} = P_{CRB} \tag{4.166}$$

Comparison of Eq. (4.166) and Eq. (4.165) reveals that the copper losses, due to the backward current $|\mathbf{I}_{RB}'|$, in the rotor winding, Eq. (4.165), are comprised of two components: (1) the airgap power P_{AGB} that is fed directly to the rotor winding through the backward field and (2) the portion $P_{AGB}(1-s_F)$ that is supplied mechanically by the rotor in response to the forward field. That is, this power, the opposing action of the backward field, must be overcome by the rotor before any useful power developed by the forward field is delivered to the load. Although this might seem strange at first, an examination of Fig. 4.36 reveals that for normal slips of operation, that is, $0 \leqslant s_F \leqslant 0.15$, $0.5[R_R'/(2-s_F)] \doteq 0.25R_R'$ and $|j0.5X_m| \gg 0.5|\{R_R'/(2-s_F)\} + jX_{\ell R}'|$. Thus elimination of $j0.5X_m$ from the circuit of Fig. 4.36 and utilization of the approximation $0.25R_R'$ enable us to draw the equivalent circuit of Fig. 4.36 as in Fig. 4.38. This makes obvious the fact that $P_{AGF} \gg P_{AGB}$ (i.e., $\tau_F \gg \tau_B$). If the motor is started in the backward direction, the upper loop of Fig. 4.36 is simplified and the lower loop (i.e., the loop with the current $\mathbf{I}_{RB}'$ is retained. Topologically, the simplified equivalent circuit for backward operation is identical to that of Fig. 4.38 with only the numerical value of the parameters appropriately changed. Thus it is obvious that the power P_{AGB} consumed in the rotor winding is a very small quantity. It should be mentioned that comparison of P_{emN} and P_{AGF}, Fig. 4.37, enables us to write $P_{emN} = f(P_{AGF})$. Before concluding this topic, it should be pointed out that (1) the power-torque relationship, $P = \omega\tau$, enables calculation of the electromagnetic torque τ_{emN} and the output torque τ_o from the following two equations:

$$P_{emN} = \tau_{emN}\omega_R \tag{4.167}$$

and

$$P_o = \tau_o\omega_R \tag{4.168}$$

and (2) the definitions of efficiency and speed regulation set forth by Eqs. (4.110) and (4.101) are valid for the analysis of the single-phase induction motor also.

*The letter "N" in the subscripts designates *net* since it can result either from the forward or backward field.

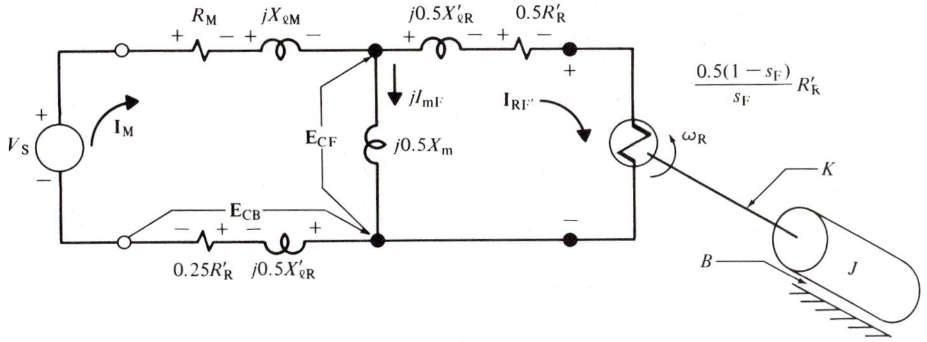

Fig. 4.38 Simplified Equivalent Circuit for Steady State Analysis and $0 \leqslant s_F \leqslant 0.15$ of the Single-Phase Induction Motor in the Forward-Motion Mode of Operation

Example 4.12

The parameters of the equivalent circuit of a 115-V, 50-Hz, four-pole, $\frac{1}{4}$-hp, single-phase induction motor measured at $n_R = 1750$ are $R_M = 1.6\ \Omega$, $X_{\ell M} = 2.07\ \Omega$, $X_m = 51.4\ \Omega$, $R_R' = 2.31\ \Omega$, $X_R' = 2.07\ \Omega$, $P_{LOSSES} = 33$ W (friction, windage, and core losses). Calculate the following quantities: (1) the input current $\mathbf{I}_M$, (2) the input power P_{IN}, (3) the developed electromagnetic power (torque) P_{emN} (τ_{emN}), (4) the output power (torque) P_o (τ_o), (5) the efficiency η, and (6) the speed regulation SR.

Solution

Utilization of Eqs. (4.157) through (4.168) enables calculation of the parameters of the equivalent circuit of the single-phase induction motor, Fig. 4.36, as follows:

$$\omega_S = \frac{\omega_E}{P/2} = \frac{2\pi \times 60}{4/2} = \frac{377}{2} = \underline{188.5} \text{ rad/s}$$

or

$$n_S = \frac{60 \times \omega_S}{2\pi} = \frac{60 \times 188.5}{2\pi} = \underline{1800}$$

and

$$\omega_R = \frac{2\pi n_R}{60} = \frac{2\pi \times 1750}{60} = \underline{183.26} \text{ rad/s}$$

The slip s_F then is

$$s_F = \frac{\omega_S - \omega_R}{\omega_S} \equiv \frac{n_S - n_R}{n_S} = \frac{1800 - 1750}{1800} = \underline{0.0278}$$

and

$$R_{MLEF} = \frac{0.5(1-s_F)R_R'}{s_F} = \frac{0.5 \times (1-0.0278) \times 2.31}{0.0278} = \underline{40.4}\ \Omega$$

$$R_{MLEB} = \frac{0.5(s_F - 1)R_R'}{2 - s_F} = \frac{0.5 \times (0.0278 - 1) \times 2.31}{2 - 0.0278} = \underline{-0.569}\ \Omega$$

$$0.5R_R' = 0.5 \times 2.31 = \underline{1.155}\ \Omega$$

$$0.5X_{\ell R}' = 0.5 \times 2.07 = \underline{1.035}\ \Omega$$

$$0.5X_m = 0.5 \times 51.4 = \underline{25.70}\ \Omega$$

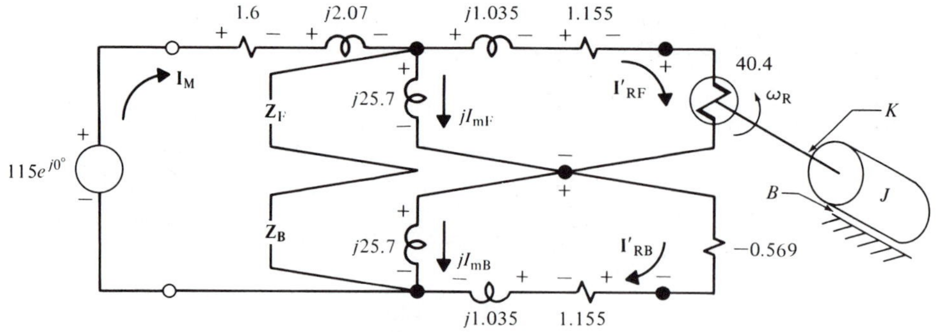

Fig. 4.12.1 The Steady State Equivalent Circuit of the Single-Phase Induction Motor; Power Is Developed by the Stator's Forward Field $\mathbf{B}_{MF}$

The equivalent circuit of the single-phase induction motor is drawn as in Fig. 4.12.1. The forward and backward impedances of the equivalent circuit are

$$\mathbf{Z}_F = \frac{(j25.70)(40.4+1.155+j1.035)}{40.4+1.155+j(25.70+1.035)} = \frac{(j25.70)(41.555+j1.035)}{41.555+j26.735}$$

$$= \frac{(25.7e^{j90°})(41.57e^{j1.43°})}{49.41e^{j32.76°}} = \underline{21.62e^{j58.67°}} = \underline{11.24+j18.47}\ \Omega$$

$$\mathbf{Z}_B = \frac{(j25.70)(1.155-0.569+j1.035)}{1.155-0.569+j(25.70+1.035)} = \frac{(j25.70)(0.586+j1.035)}{(0.586+j26.735)}$$

$$= \frac{(25.70e^{j90°})(1.19e^{j60.48°})}{26.74e^{j88.74°}} = \underline{1.144e^{j61.74°}} = \underline{0.54+j1.0}\ \Omega$$

The input impedance of the circuit is

$$\mathbf{Z}_{IN} = R_M + jX_{\ell M} + \mathbf{Z}_F + \mathbf{Z}_B = 1.60 + j2.07 + 11.24 + j18.47 + 0.54 + j1.0$$

$$= \underline{13.37+j21.54} = \underline{25.35e^{j58.17°}}\ \Omega$$

and the input current is

$$\mathbf{I}_M = \frac{\mathbf{V}_S}{\mathbf{Z}_{IN}} = \frac{115e^{j0°}}{25.35e^{j58.17°}} = \underline{\underline{4.54e^{-j58.17°}}}\ \text{A}$$

The input power is

$$P_{IN} = |\mathbf{E}_S|\,|\mathbf{I}_M|\cos\theta = 115 \times 4.54 \times \cos 58.17° = \underline{\underline{275.36}}\ \text{W}$$

Utilizing Fig. 4.38 enables calculation of the following pertinent quantities:

$$P_{AGF} = \frac{0.5R_R'|\mathbf{I}_{RF}'|^2}{s_F}$$

and

$$P_{AGB} = \frac{0.5R_R'|\mathbf{I}_{RB}'|^2}{2-s_F}$$

where

$$\mathbf{I}_{RF}' = \frac{j25.70}{40.4+1.155+j(1.035+25.70)} \times 4.54e^{-j58.17°}$$

$$= \frac{25.70e^{j90°}}{49.41e^{j32.76°}} \times 4.54e^{-j58.17°} = \underline{2.36e^{-j0.93°}}\ \text{A}$$

and

$$\mathbf{I}_{RB}' = \frac{j25.70}{1.155 - 0.569 + j(25.70 + 1.035)} \times 4.54e^{-j56.54°}$$

$$= \frac{j25.70}{0.586 + j26.735} \times 4.54e^{-j58.17°} = \frac{(25.70e^{j90°})(4.54e^{-j58.17°})}{26.74e^{j88.74°}}$$

$$= \underline{4.36e^{-j56.91°}} \text{ A}$$

Thus

$$P_{AGF} = \frac{0.5 \times 2.31 \times (2.36)^2}{0.0278} = \underline{231.40} \text{ W}$$

and

$$P_{AGB} = \frac{0.5 \times 2.31 \times (4.36)^2}{2 - 0.0278} = \underline{11.13} \text{ W}$$

The net developed electromagnetic power and torque are calculated as follows:

$$P_{emN} = \frac{0.5(1 - s_F)R_R'|\mathbf{I}_{RF}'|^2}{s_F} = P_{AGF}(1 - s_F)$$

$$= \frac{0.5 \times (1 - 0.0278) \times 2.31 \times (2.36)^2}{0.0278} = \underline{\underline{224.97}} \text{ W}$$

and

$$\tau_{emN} = \frac{P_{emN}}{\omega_R} = \frac{224.97}{183.26} = \underline{\underline{1.23}} \text{ N} \cdot \text{m}$$

Finally, the output power and torque are

$$P_o = P_{emN} - P_{LOSSES} - (1 - s_F)P_{AGB} = 224.97 - 33 - 10.82 = \underline{\underline{181.15}} \text{ W}$$

and

$$\tau_o = \frac{P_o}{\omega_R} = \frac{181.15}{183.26} = \underline{\underline{0.988}} \text{ N} \cdot \text{m}$$

The efficiency and speed regulation of the motor are

$$\eta = \frac{P_o}{P_{IN}} \times 100 = \frac{181.15}{275.36} \times 100 = \underline{\underline{65.79\%}}$$

and

$$\text{SR} = \frac{\omega_{NL} - \omega_{FL}}{\omega_{FL}} \times 100 = \frac{\omega_S - \omega_R}{\omega_R} \times 100 = \frac{188.5 - 183.26}{183.26} \times 100 = \underline{\underline{2.86\%}}$$

Examination of the equivalent circuit of Fig. 4.36 enables us to draw the following conclusion about the operation of the single-phase induction motor: When the motor is running at a small slip ($s_F \to 0$), the resistance of the forward field circuit, that is, $0.5R_{R'}/s_F$, increases while the resistance of the backward field circuit, that is, $0.5R_{R'}/(2 - s_F)$, decreases. These changes in the impedances result in an increase of the forward field and a reduction in the backward field (forward rotation of the single-phase induction motor is defined to be the direction of the rotor at the moment of discussion); however, the sum of these two fields remains constant as it must produce a counter emf ($\mathbf{V}_C \equiv \mathbf{V}_{CF} + \mathbf{V}_{CB}$), which should remain constant if $\mathbf{Z}_M\mathbf{I}_M \doteq 0$. Thus when the rotor runs at small slips, the forward field is

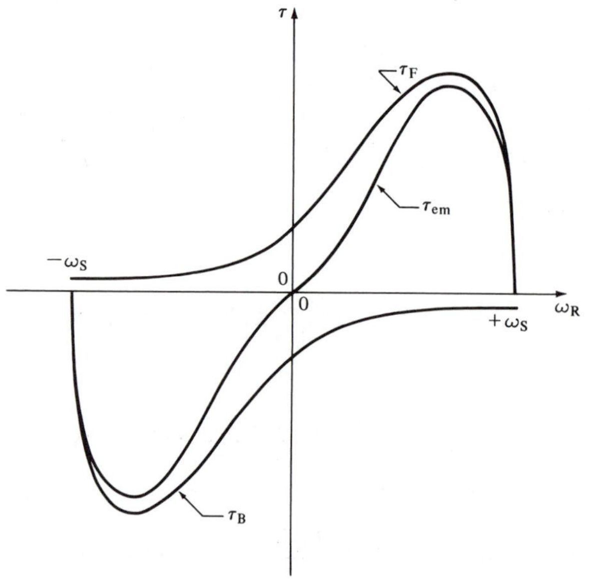

Fig. 4.39 Torque vs. Speed Characteristic of the Single-Phase Induction Motor with the Effects of the Forward and Backward Fields, $\mathbf{B}_{MF}$ and $\mathbf{B}_{MB}$, Exhibited as τ_F and τ_B

several times greater than the backward field. In a way it resembles the revolving field of a three-phase induction motor. Accordingly, for normal speeds of operation, the torque versus speed characteristics of a single-phase induction motor and a three-phase induction motor, with the same squirrel cage rotor and equally strong fields, are practically identical. These ideas lead to the updating of the torque versus speed characteristic of the single-phase induction motor in order to account for the changes, with rotation, in $|\mathbf{B}_{MF}|$ and $|\mathbf{B}_{MB}|$. This more realistic torque versus speed characteristic is exhibited in Fig. 4.39.

It is mentioned earlier in this section that **unless the single-phase induction motor is brought to approximately 75 percent of synchronous speed, the motor is not functional**. Thus an understanding of the auxiliary methods by which this task is accomplished is essential.

For starting purposes the single-phase induction motors are designed to produce a rotating magnetic field at standstill and are classified in accordance with the method of starting and referred to by descriptive names. The most widely used methods are (1) split-phase motor, Figs. 4.40(a) through 4.40(c), (2) capacitor-start motor, Figs. 4.40(d) through 4.40(f), (3) permanent capacitor-split motor, Figs. 4.40(g) through 4.40(i), (4) two-value-capacitor motor, Figs. 4.40(j) through 4.40(l), and (5) shaded-pole motor, Figs. 4.40(m) and 4.40(n). Fig. 4.40 exhibits the schematic, the **Z**-plane plots, when possible, and the torque versus speed characteristics of these five motors. The main stator winding of the single-phase induction motor (which occupies *two-thirds* of the stator slots), Fig. 4.35(a), is designed to have a winding leakage reactance $X_{\ell M}$ *higher* than its resistance, that is, $|jX_{\ell M}| \gg R_M$. On the other hand, the auxiliary winding (which occupies *one-third* of the stator slots) is designed to have a winding leakage reactance $X_{\ell A}$ *lower* than its resistance, that is, $|jX_{\ell A}| \ll R_A$. These reactance-to-resistance ratios lead to a time

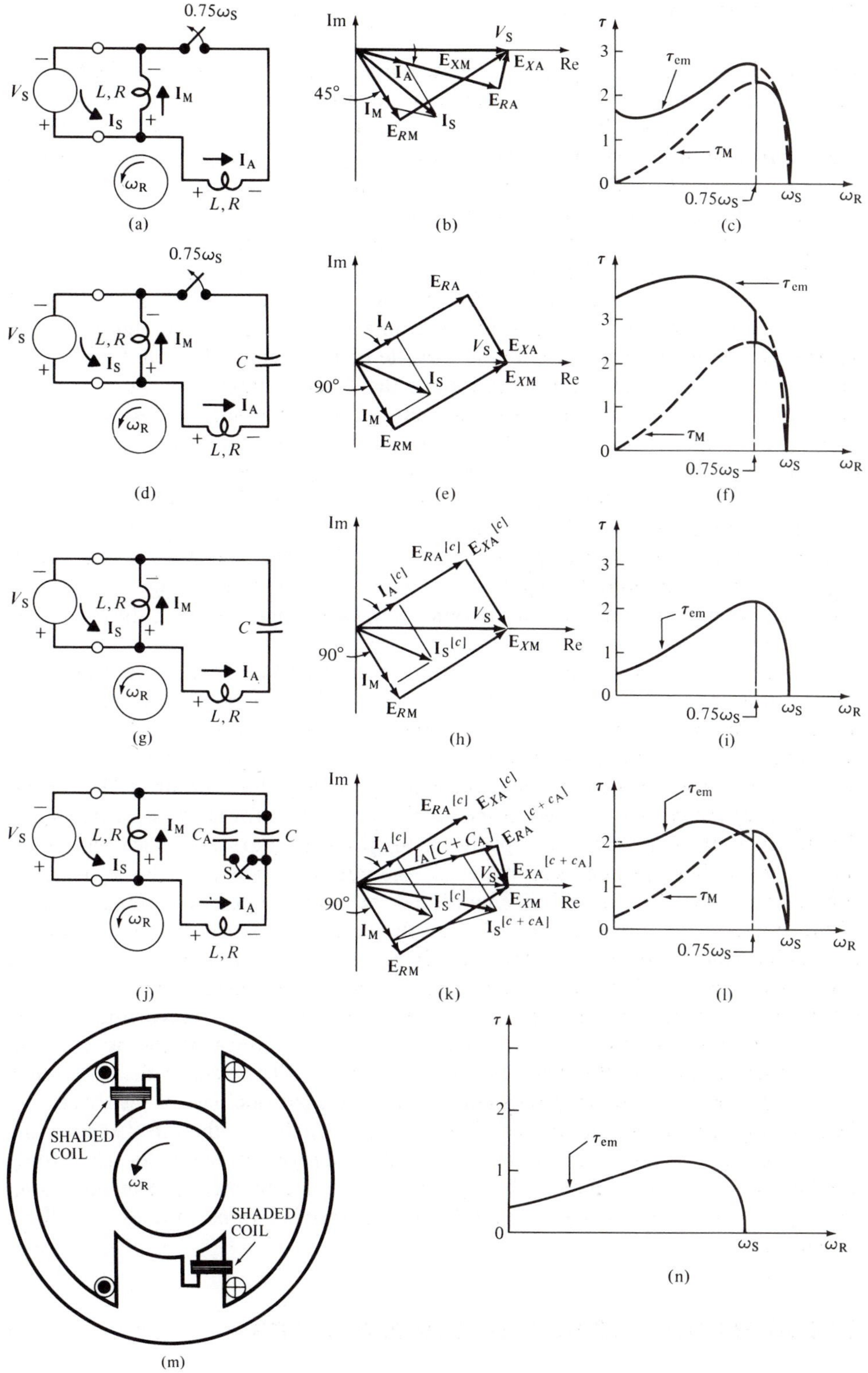

Fig. 4.40 Schematic Diagram, *Z*-Plane Plots (when possible), and Torque vs. Speed Characteristics of the Following Types of Single-Phase Motors: (a)–(c) Split-Phase Motor, (d)–(f) Capacitor-Start Motor, (g)–(i) Permanent Capacitor-Split Motor, (j)–(l) Two-Value-Capacitor Motor, and (m)–(n) Shaded-Pole Motor; Torque Scale Is Calibrated in Percentage of Rated Torque Units

phase angle between the current of the auxiliary winding $\mathbf{I}_A$ and the current of the main winding $\mathbf{I}_M$. Note (1) the magnitude of the angles between these two currents in the various motors, Fig. 4.40, and (2) the spatial position of the two windings in the various motor schematics. The phasor diagrams of Figs. 4.40(b), 4.40(e), 4.40(h), and 4.40(k) reveal that the current of the auxiliary winding $\mathbf{I}_A$ leads the current of the main winding $\mathbf{I}_M$. This suggests that the field becomes maximum along the auxiliary winding first and then along the main winding. The net result is the rotation of the stator field in the positive direction, which causes the rotor to rotate in the positive direction also. As the rotor reaches a speed in the vicinity of $0.75\omega_S$, a centrifugal switch opens and disconnects the auxiliary winding from the source. However, the main winding is capable of producing adequate torque now and the rotor will continue to rotate until the forcing function is disconnected. Note that the torque versus speed characteristics of the single-phase induction motors, Figs. 4.40(c), 4.40(f), 4.40(i), 4.40(l), and 4.40(n), exhibit the torque developed by the main winding, designated as the τ_M curve, and/or the torque developed by both windings, designated as the τ_{em} curve. The motors which are equipped with a disconnecting switch operate along the solid portions of the τ_{em} and τ_M curves; that is, along the τ_{em} curve, when the speed is in the range $0 < \omega_R < 0.75\omega_S$ and along the τ_M curve when the speed is in the range $0.75\omega_S < \omega_R < \omega_S$. Motors not equipped with a disconnecting switch operate along the τ_{em} curve all the time.

Before closing this section the following remarks should be made about the single-phase induction motor: (1) the resistance-to-leakage reactance ratio of the auxiliary winding is higher than the same ratio of the main winding (this is achieved by using a smaller wire size and fewer turns in the main winding); (2) the leakage reactance can be reduced somewhat by placing the winding in the top of the slots; (3) the capacitors used in the auxiliary windings for starting purposes are the **dry** type of ac electrolytic capacitors; (4) the capacitors used for continuous service are **oil-impregnated paper** type; (5) the cutout switch could be a potential source of trouble; (6) economics demand that the phase angle between the currents of the auxiliary winding and main winding should always be less than 90°. Finally, the shaded-pole motor provides the ruggedness that the other types of motors lack. Fig. 4.40(m) should illustrate the fact that except for overheating nothing can go wrong with this motor. Since it has no auxiliary winding, the revolving field is obtained as follows. As the flux changes due to the change of the ac current, currents are induced in the shaded coils (copper rings). The field of these currents always opposes their inducing (changing) field. Thus for one part of the cycle, the field is concentrated on the unshaded portion of the pole; for the other part of the cycle, the field is concentrated on the shaded portion of the pole. Therefore, the field appears to be moving across the pole phase in the positive direction, thereby producing a starting torque (about 50 percent of rated torque) as well as running torque.

4.12 APPLICATIONS OF INDUCTION MOTORS

The user of electric motors is concerned with the choice of one motor, from the many available, that is most suited for a particular application and whose characteristics must be known explicitly. The objective then becomes one of matching

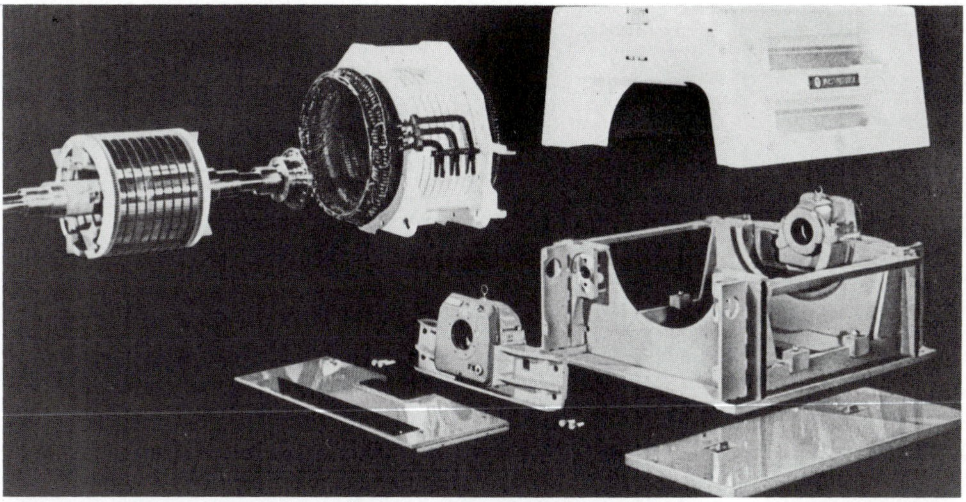

Plate 4.7 Easily Disassembled and Assembled Self-ventilated Induction Motor. (Courtesy of Westinghouse Electric Co., Pittsburgh, Pa.)

characteristics of a given motor to a given load. These characteristics include (1) the horsepower requirements of the motor, (2) the starting torque capabilities of the motor, (3) the starting current capacity of the motor, (4) the accelerating capability of the motor, (5) the speed variation requirements of the motor, (6) the duty cycle of the load, and (7) the environment in which the motor is to operate. Once the capabilities of a given motor are matched to the demands of a load, economic considerations come into focus. That is, the problem becomes one of finding a motor that is technically the best suited for the job to be done but is the least expensive available.

Although it is prohibitive to devote space and time here to give a complete account of what type of motor should be used for a particular job (the interested reader should refer to the literature for such information; good starting points are pp. 206–207 of [5], pp. 389–402 of [11], and pp. 386–387 of [5]), the following comments, of a general nature, can be made about the widely used three- and single-phase induction motors.

For the three-phase induction motor, the torque versus speed characteristics of Fig. 4.41 (with SC-A interpreted to mean squirrel-cage-rotor, three-phase, induction motor, Class A, etc., and WR_0 interpreted to mean wound-rotor, three-phase induction motor with external resistance R_{L0}, etc.) gives a comprehensive account of the torque and speed capabilities of the available three-phase induction motors. Note that the distinct difference between the squirrel-cage-rotor and the wound-rotor motors is the constancy of the operating speed for the squirrel-cage-rotor motors versus the capability of approximately 50 percent of speed control for the wound-rotor motors. Another set of unique differences between the two classes of motors is the following: *the squirrel-cage-rotor motors are rugged and inexpensive while the wound-rotor motors require maintenance for reliable service and are more expensive*.

For applications that require motors of fractional horsepower rating, the single-phase induction motor becomes important. Inspection of Figs. 4.40(a) through 4.40(n) indicates that as the principle of operation for the single-phase induction

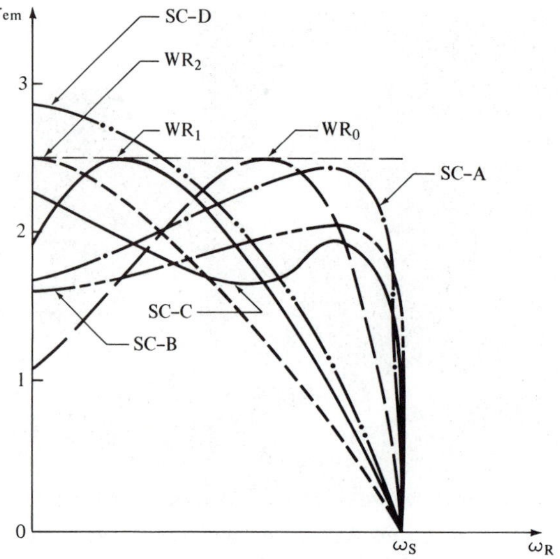

Fig. 4.41 Torque vs. Speed Characteristics for Four Classes of Squirrel-Cage (Designated as SC-j, $j = A, B, \ldots$) and Wound-Rotor (Designated as WR$_i$, $i = 0, 1, 2, \ldots$) Induction Motors with Variable External Resistance; Torque Scale Is in Percentage of Full-Load Torque

motor approaches that of the two-phase induction motor, Figs. 4.31 and 4.35, the starting torque drops; also, the audible **noise level** drops (a consequence of vibrations, a unique characteristic of single-phase induction motors, due to high harmonics in the current of the rotor). Thus one has to compromise between starting torque and audible noise level when one chooses a single-phase induction motor for a particular application. The *quietest* and *most rugged* single-phase induction motor is the shaded-pole motor. However, its *starting torque is the lowest* in the class of single-phase induction motors and its rating lies in the lowest end of the fractional horsepower-rating spectrum. On the other hand, the *noisiest* single-phase induction motor is the capacitor-start motor, but its *starting torque is the highest*; its rating lies in the upper end of the fractional horsepower-rating spectrum.

Of the many tasks that can be performed by the three classes of motors discussed in this chapter, the most important applications are discussed in the following sections.

Three-Phase Induction Motors

The three-phase induction motor is used extensively in industrial processes, for example, machinery tools, shears, punch presses, die stamping, crushers, compressors, pumps, agitators, fans, conveyors, hoists, and woodworking tools. In its generator mode of operation, the induction device performs three very important tasks, which causes the induction device to be associated with the following industrial names: (1) frequency converter, (2) real power generator, (3) voltage regulator, and (4) three-phase power selsyn. The frequency conversion and real power generation principles are explained in Section 4.7. The induction *voltage*

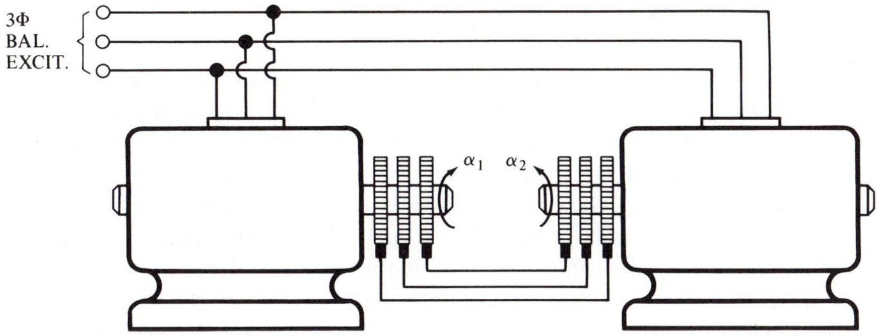

Fig. 4.42 Three-Phase Power Selsyn for Heavy-Torque Transmission

regulator uses a wound-rotor type of induction device with its rotor speed maintained at zero, that is, $\omega_R \equiv 0$, and its position (with respect to the stator) adjusted automatically. Thus by automatically varying the stator-to-rotor turns ratio, the voltages at the rotor output are maintained constant. The voltages to be regulated are fed in the stator windings. The voltage regulator is used in ac distribution feeders and in loads that require constant ac voltages. The three-phase *power selsyn*, Fig. 4.42, is used for position control, that is, to cause remotely the angular position,* α_2, of one shaft to follow the angular position, α_1, of another shaft as closely as possible. One such application is the maintenance of synchronism between the two hoist motors that raise the two ends of large lift bridges. The three-phase power selsyn is comprised of two three-phase, wound-rotor induction motors (they can be of fractional horsepower rating) with their Y-connected stator windings excited from the same balanced three-phase excitation systems and their Y-connected rotor windings connected together through the slip rings, Fig. 4.42. Corresponding windings should be connected in parallel. Now if the shaft of motor No. 1 is driven mechanically to a position α_1, the shaft of motor No. 2 will follow in synchronism with the shaft of motor No. 1, as if the two shafts were connected mechanically. Thus the shaft of motor No. 2 follows the position of the shaft of motor No. 1 until $\alpha_2 \equiv \alpha_1$. Therefore, torque is transmitted remotely from point No. 1 to point No. 2 [for further explanation, please refer to the light torque transmitter, the material associated with Fig. 4.48 and Eq. (4.171)].

Two-Phase Induction Motors

The two-phase induction motor is used primarily as the basic power unit, that is, to position and drive shafts, in low-power (power ranges from a fraction of a watt to a few hundred watts), ac *feedback control systems* [systems comprising one or more feedback loops which compare the controlled signal $c(t)$ with the command signal $r(t)$; the error $\varepsilon(t) = c(t) - r(t)$ is used to drive $c(t)$ into correspondence with $r(t)$] or *servomechanisms* (feedback controlled systems in which one or more system signals represent mechanical motion†, Fig. 4.44); an instrument-type

*It must be kept in mind that the angles α_2 and α_1 exhibited here, and α_G, α_T, and α_D, exhibited later in this section, are always functions of time.

†For information pertinent to Fig. 4.44 refer to p. 498.

servomechanism, for example, requires two-phase motors of power rating 50 W or less.

In addition to the use described above, the two-phase induction device with drag-cup rotor can operate in a **generator-like** mode known as **ac tachometer**, Fig. 4.43. This device is used to **obtain a measure of the angular velocity of a shaft in the form of an ac voltage of constant frequency**. The ac tachometer has a reference winding designated by $e_{RT}(t)$, displaced spatially by 90° from the control winding, designated by $e_{CT}(t)$, which is used for a pick-off of the output voltage $e_O(t) \equiv e_{CT}(t)$. This voltage is nonzero if and only if the speed of the rotor is nonzero. The speed for the ac tachometer is provided by a gear train from the shaft whose speed is to be measured. To set forth the functional differences of the two-phase induction device performing the tasks of (1) positioning a load (ac motor) and (2) measuring the angular velocity of the load shaft (ac tachometer), observe their application in Fig. 4.43. Note the distinct difference in the winding connections of the two devices. While both devices have a **reference winding** and a **control winding** each, **spatially** 90° apart, and in both devices the control winding is connected to an amplifier, there is a distinct difference between the two devices. The control winding of the two-phase induction motor is connected to the **low-impedance output of the amplifier** and always (except when the load has reached the desired position, preset by the command signal) carries a current that sets up the field of the control winding of the two-phase motor. The fields of the reference and control windings cause the rotor of the two-phase induction motor to rotate counterclockwise or clockwise, depending on the polarity of the voltages across these two windings. However, the control (output) winding of the ac tachometer is connected to the **high-impedance input of the amplifier**, and therefore it can be considered as open-circuited with no current flowing through it. Thus the field that links the control winding is produced by the reference winding only (which is excited by an

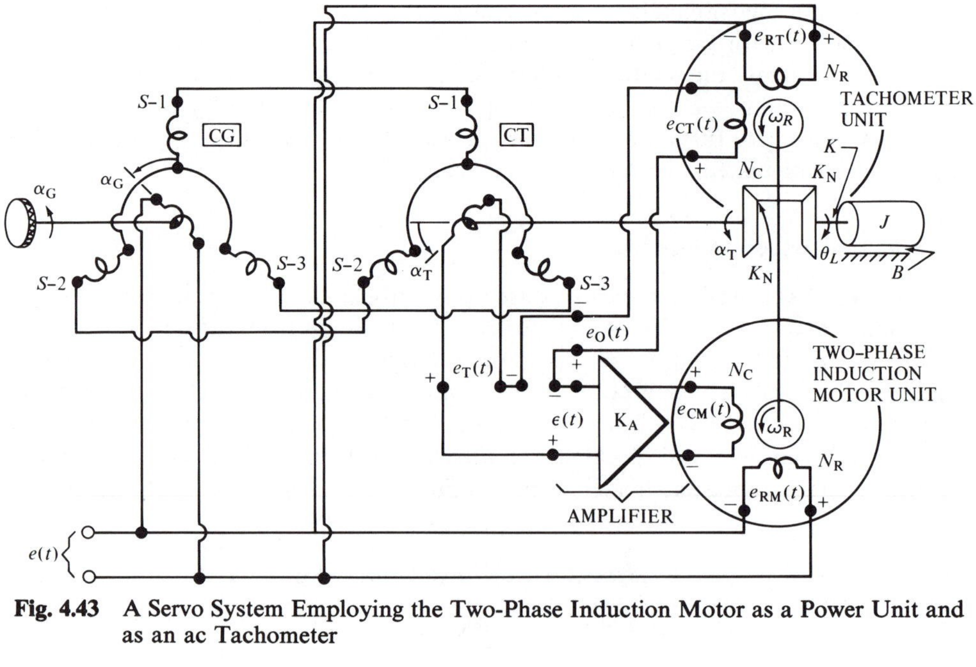

Fig. 4.43 A Servo System Employing the Two-Phase Induction Motor as a Power Unit and as an ac Tachometer

ac voltage of constant magnitude and frequency). This field, which depends on the externally applied speed of the tachometer rotor, induces the output voltage $e_O(t) \equiv e_{CT}(t)$ of the same frequency as the input or reference voltage.

The understanding of the voltage induction across the control (output) winding is based on the *two-revolving-field theory* governing the operation of the single-phase induction motor. Note that, as viewed from the reference winding [across $e_{RT}(t)$], the ac tachometer is equivalent to a single-phase induction motor, whose theory of analysis is applicable here. The two revolving fields induce [in addition to the counter emf $e_C(t) = e_{CB}(t) + e_{CF}(t)$ across the reference winding, Fig. 4.36] the voltage $e_{CT}(t)$ across the control winding. By proper design (turns ratio, Figs. 4.31, 4.43, etc.) the voltage $e_{CT}(t)$ can be made to be a function of the voltages $e_{CF}(t)$ and $e_{CB}(t)$, Fig. 4.36. At standstill the two fields $\mathbf{B}_{MF}$ and $\mathbf{B}_{MB}$ are identical and no voltages are generated. However, as the speed of the rotor approaches the synchronous speed ω_S, Fig. 4.39, the voltage $e_{CF}(t)$ *increases* and the voltage $e_{CB}(t)$ *decreases*. Since the difference between these voltages is a function of the rotor speed, the voltage $e_{CT}(t)$ is also a function of the rotor speed. Clearly, as the direction of rotation of the rotor reverses, Figs. 4.36, 4.38, and 4.39, the phase of the voltage $e_{CT}(t)$ reverses.

By proper design of the motor (the rotor self-reactance-to-resistance ratio is a critical design criterion) the magnitude of the voltage $e_{CT}(t)$ is made to be a linear function of the tachometer rotor speed ω_R (radians/second), written as

$$e_{CT}(t) = K_T \frac{d\theta_R(t)}{dt} \equiv K_T \omega_R(t) \tag{4.169}$$

where K_T is defined as the *sensitivity* of the tachometer in volts/radians/second (or volts/revolution/minute) and $\theta_R(t)$ is the angular displacement of the tachometer shaft. The phase of the voltage $e_{CT}(t)$ should be fixed with respect to the applied voltage $e_{RT}(t) \equiv e(t)$. Taking the Laplace transform of Eq. (4.169), with the initial conditions assumed to be zero, and defining the ratio of the output voltage $E_{CT}(s)$ to the input angular displacement of the tachometer shaft $\Theta_R(s)$ as the transfer function of the ac tachometer, Eq. (4.169) yields

$$G_T(s) \equiv \frac{E_{CT}(s)}{\Theta_R(s)} = sK_T \tag{4.170}$$

Eq. (4.170) provides the means for a block diagram representation of the ac tachometer, useful in the s-domain analysis of automatic control (servomechanism) systems. For example, utilizing automatic control theory, the block diagram of Fig. 4.43 can be drawn as in Fig. 4.44, utilizing Eqs. (4.151), (4.170), and* (4.182), where K_T is the sensitivity of the tachometer, K_A is the gain of the amplifier, K_N is the gain of the gear train, and K_S is the sensitivity of the synchro error detector. It can be shown that the tachometer feedback improves the response of the system; that is, it *improves* the stability of the system. In passing it should be mentioned that the task that the ac tachometer performs in Fig. 4.43 can be performed instead by the dc tachometer if a modulator is used to convert its dc output voltage into an ac voltage. A dc tachometer is a permanent magnet dc motor, whose rotor is driven by the shaft, the speed of which is to be measured. It can be shown that the transfer function of the dc tachometer is also given by Eq. (4.170).

*For the discussion pertinent to Eq. (4.182) refer to p. 294.

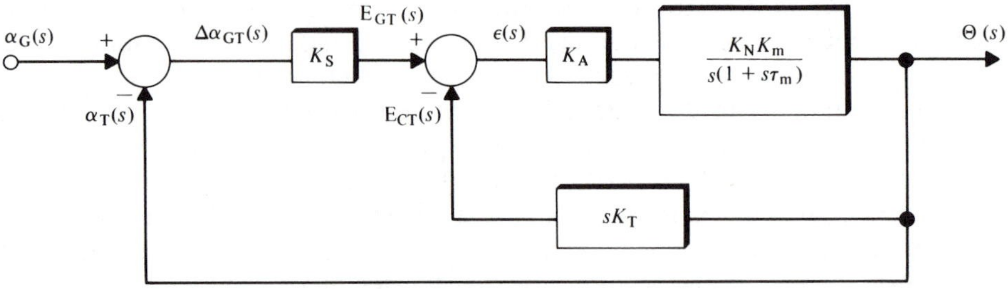

Fig. 4.44 Block Diagram of the Servo System Exhibited in Fig. 4.43

Single-Phase Induction Motors

Although the single-phase induction motor in its classical connection is instrumental in food preparation, washing machines, hair dryers, office appliances, home workshops, and toys, in its nonclassical connection it has become essential in the most ambitious engineering achievements. In a unique connection known as **synchro**, it is used to synchronize the motion of two shafts without requiring mechanical coupling. Such combinations are used for (1) remote positioning applications, if the loads are light (for heavy loads see power selsyns, Section 4.12), (2) indicating instruments for remotely detected signals that can be translated into the angular position of a shaft (e.g., aircraft compass headings, position of control surfaces in aircraft, remote indication of water level in reservoirs), and (3) **error detectors**, the most essential components of servomechanisms. The function of an error detector is **to produce a signal whose magnitude and phase angle represent the amount and direction of displacement of an output shaft as compared to the displacement of one or two input (command) shafts** (signals).

There are four basic types of synchros: (1) the *control generator*, denoted by CG, Fig. 4.45, (2) the *control receiver*, denoted by CR, Fig. 4.45, (3) the *control transformer*, denoted by CT, Fig. 4.46, and (4) the *control differential*, denoted by CD, Fig. 4.47. The stator of all four synchros has a *smooth* inner surface which has *skewed* slots and is designed to accommodate sinusoidally distributed, Y-connected, three-phase windings. However, **it should be pointed out that although three-phase windings are involved, these four units deal with single-phase voltages**. Also, it should be pointed out that structurally the stator units are identical except that the stator

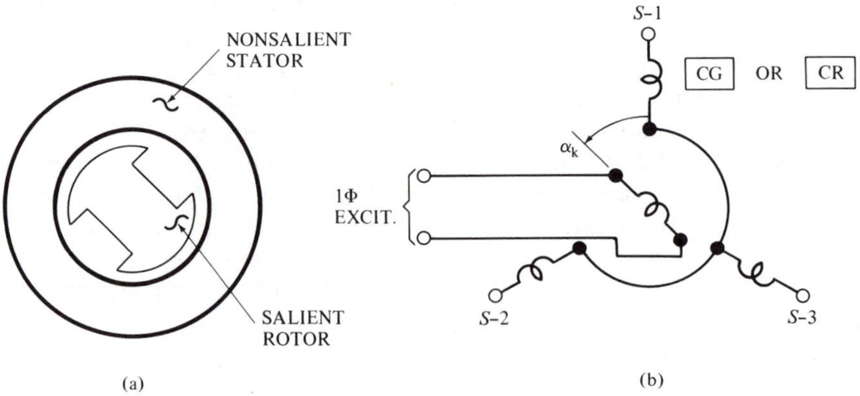

Fig. 4.45 Control Generator or Control Receiver: (a) Structure and (b) Schematic Diagram with Rotor Winding α_k (k=G, R) Degrees Off Its Electrical Zero Position

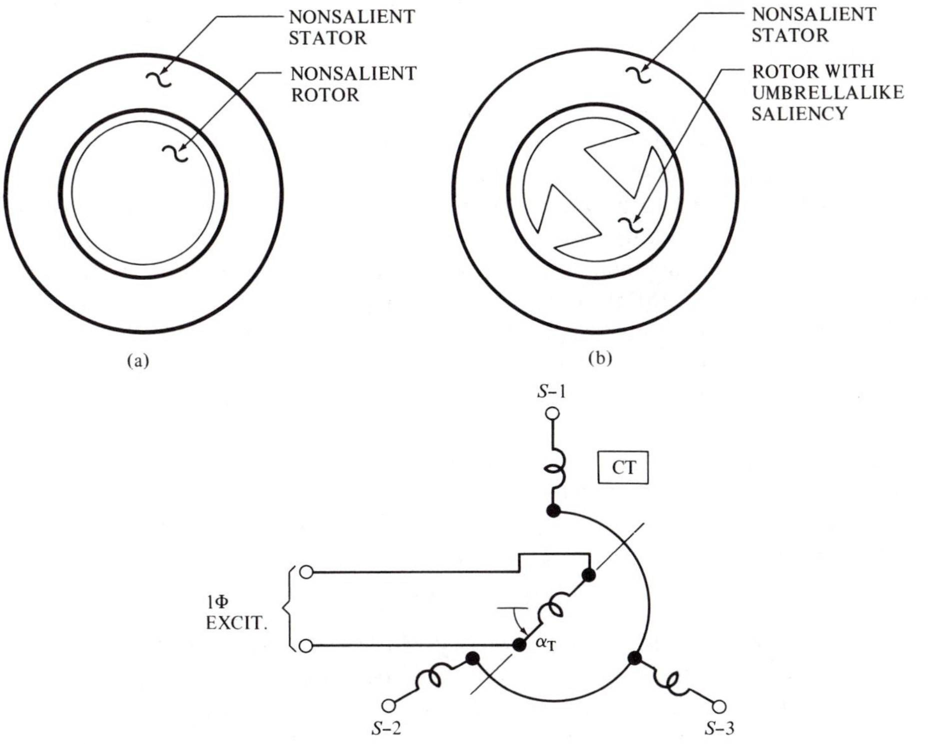

Fig. 4.46 Control Transformer: (a) and (b) Smooth Airgap Structures and (c) Schematic Diagram with Rotor Winding α_T Degrees Off Its Electrical Zero Position

winding of the control transformer (CT) has a higher per-phase impedance in order to allow several control transformers to be fed by the same source (the control generator will be described shortly). As seen in Figs. 4.45 through 4.47, the rotor structure of the four synchros differs from unit to unit but all have the appropriate number of slip rings for energy transfer to and from the rotor winding. For the control generator and receiver (CG or CR), the rotor is of the *salient-pole* or

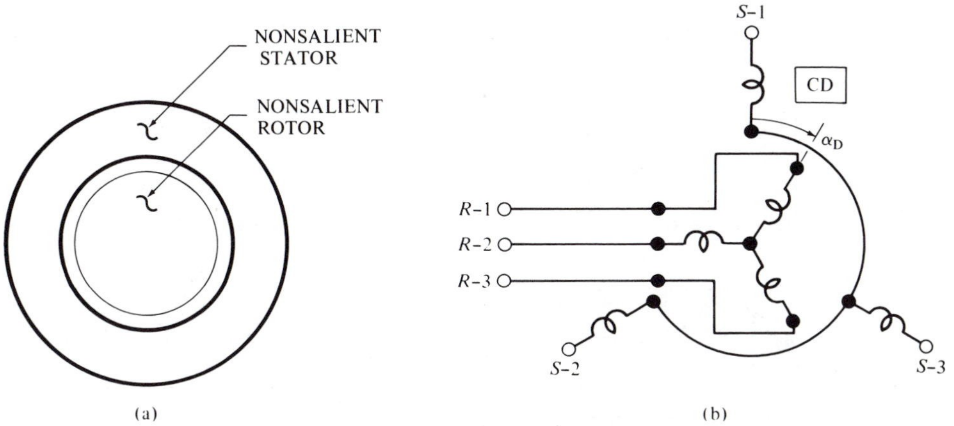

Fig. 4.47 Control Differential: (a) Structure and (b) Schematic Diagram with Rotor Winding α_D Degrees Off Its Electrical Zero Position

dumbbell type, Fig. 4.45(a), with many turns of wire wrapped around the stem, and its electrical zero position defined to be the position of *maximum coupling* between the rotor winding and the $S-1$ stator winding. For the control transformer (CT), Fig. 4.46, the rotor is of (1) the *cylindrical type*, Fig. 4.46(a), with a distributed single-phase winding, or (2) the *umbrellalike* saliency type, Fig. 4.46(b), with the single-phase winding wrapped around the stem. The electrical zero position of either rotor is defined to be the position of *minimum coupling* (i.e., zero coupling) between the rotor winding and the $S-1$ stator winding. The uniform airgap structure minimizes the magnetizing current drawn by the control transformer (CT). For the control differential (CD), Fig. 4.47, the rotor is *cylindrically shaped* and has balanced, Y-connected, three-phase distributed windings with its electrical zero position defined to be the position of *maximum coupling* between windings $S-1$ and $R-1$, and so on.

For remote positioning and indicating instrument applications, the generator (a position-to-voltage transducer) is connected to a receiver (voltage-to-position transducer) as shown in Fig. 4.48. This connection is known as *CG-CR synchro* or *light torque transmitter*. Note that the stator coils are connected in *parallel* and the rotor coils are excited by the *common* forcing function

$$v(t) = \sqrt{2}\,|\mathbf{V}|\sin\omega_E t \tag{4.171}$$

where ω_E (radian/second) is the frequency of the excitation source.

The operation of the torque transmitter or position indicator, Fig. 4.48, can be explained as follows. The generator rotor, acting as a transformer primary, induces voltages in the generator stator coils acting as transformer secondaries. The CG-stator-coil voltages, measured from terminal-to-terminal because the junction point of the Y-connected stator windings is not accessible, depend upon the angle at which the magnetic field of the rotor coil cuts the stator coil turns. Thus the voltage induced in any of the three secondaries is determined by the physical position of the CG rotor and is sinusoidal in nature. These terminal-to-terminal, CG stator voltages are applied across the corresponding stator terminals of the CR. The CR stator windings now generate a composite magnetic field in the airgap of the control receiver (CR), whose strength and direction are determined by the

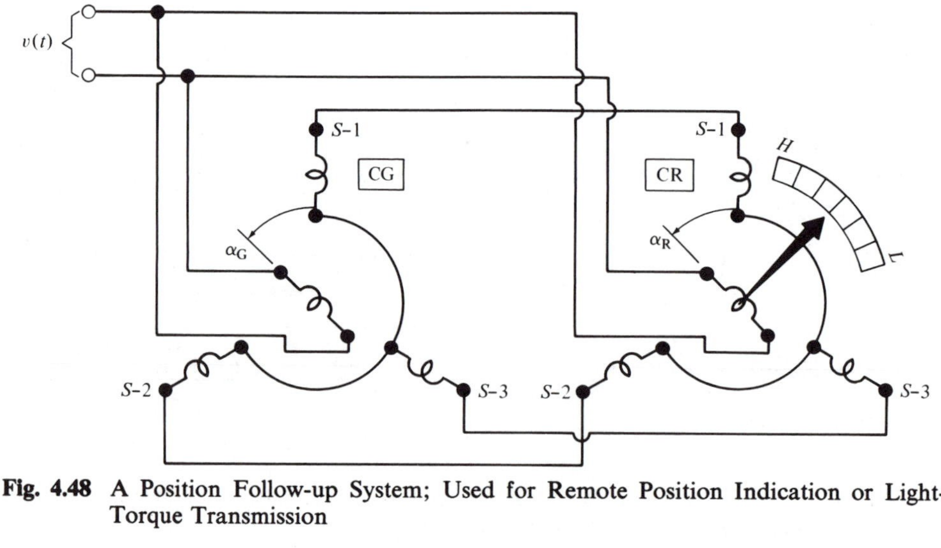

Fig. 4.48 A Position Follow-up System; Used for Remote Position Indication or Light-Torque Transmission

applied terminal-to-terminal, CR stator voltages. The rotor of the CR, connected to the same (with the CG) excitation source, also generates a magnetic field. The two CR fields tend to line up along the stable position of their axes. But as the rotor of the CR rotates to line up with the CR stator field, it acts as a transformer primary. Thus it induces voltages in the CR stator windings, which depend on the CR rotor-to-stator position. The CR rotor position continues to rotate until it reaches the position in which the voltages it induces in the CR stator windings cancel exactly the voltages applied to the CR stator windings by the CG stator windings. Now the net voltage in each of both sets of the stator windings is zero—that is, one set of voltages opposes the other—and the stator currents become zero. Thus the CR stator coils generate no magnetic field and the CR rotor stops at a position that corresponds to the position of the CG rotor. If the CG rotor is moved to a new position, the CR rotor will rotate until it duplicates the position of the CG rotor. However, if the rotor of the CG is turned continuously, the rotor of the CR will attempt to keep up with it. In the process, depending on the momentum of the CR rotor, the rotor (of the CR) might overshoot the position of the CG rotor and oscillate (or hunt) about the correct position until friction causes it to stop or it (the CR rotor) might even *spin* like the rotor of a motor. To avoid this undesirable behavior, or even to smooth the motion of the CR rotor, the CR is equipped with a mechanical damper. The presence of this *damper* is the only difference in structure between the CG and the CR.

Although the torque produced by the torque transmitter, that is, the CG-CR synchro, is enough to position light loads of the order of 0.07 to 1.75 in·oz/degree, such as the pointer of an instrument, it is not enough, because of the horsepower rating of the devices used, to position heavy loads such as radar antennas, antiaircraft guns, and so on.

Since positioning of such heavy loads to prescribed, or continuously changing positions (e.g., tracking a moving target), is essential, the *CG-CT synchro*, Fig. 4.49, is employed in conjunction with the powerful two-phase induction motors (a

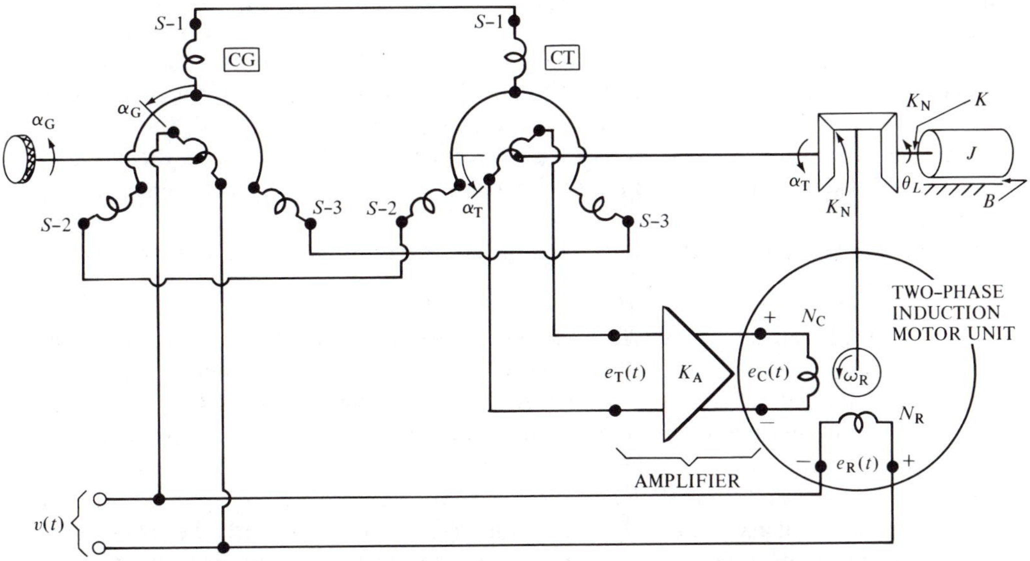

Fig. 4.49 Generator-Transformer Synchro Used as Error Detector in Positional Servomechanisms When the Position Is a Function of One-Command Input

servomechanism application) to do the positioning. The CT is designed to produce a sinusoidal signal $e_T(t)$ at its rotor winding the **magnitude** of which is defined as

$$|\mathbf{E}_T| = K_S \sin(\alpha_G - \alpha_T) \tag{4.172}$$

where α_T is the CT rotor position in degrees, α_G is the CG rotor position in degrees, and K_S, defined as the *sensitivity* of the CG-CT synchro, is the maximum RMS voltage per degree induced on the CT rotor winding, by the CT's airgap flux, when the CT rotor-to-stator winding coupling is *maximum*. Note that this voltage is a measure of lack of correspondence between the CT and CG rotor positions. This signal is fed into the amplifier, Fig. 4.49, while the CT rotor shaft is mechanically coupled to the load to be positioned. The operation of the *CG-CT synchro*, known as a **GT error detector**, can be explained as follows. Assume that the CG rotor (in this case representing the command signal) is set to the desired angular position α_G that the load is expected to assume. The field of the CG rotor, as in the case of the CG-CR synchro, will induce terminal-to-terminal voltages that are proportional to the CG rotor position. These voltages are impressed, from terminal-to-terminal, across the stator windings of the CT. The resulting currents in the stator windings of the CT generate a composite magnetic field whose magnitude depends upon the magnitude and phase of the applied terminal-to-terminal CT stator voltages and, therefore, the CG terminal-to-terminal stator voltages. Thus the direction of the composite field established in the airgap of the CT duplicates the CG rotor position. If this field position is not in correspondence with the CT rotor position, a voltage will be induced in the rotor winding of the CT in accordance with Eq. (4.172). This voltage is amplified and is applied to the control winding of the two-phase induction motor. The motor will cause a rotation of its rotor in such a direction as to rotate the rotor of the CT in a direction that minimizes the induced voltage $|\mathbf{E}_T|$. If $|\mathbf{E}_T|$ becomes zero, the angle of the rotor of the CT, α_T, is equal to that of the CG (i.e., the angle of the command shaft α_G); the load is in the desired position and the rotor of the two-phase motor stops. If the desired position α_G that the load must follow is again changed, the procedure is repeated. If the desired position to be taken by the load changes continuously, the voltage $\mathbf{E}_T$ of the rotor of the CT changes continuously. Thus the two-phase induction motor turns continuously in such a direction as to try to maintain $\mathbf{E}_T = 0$ and thus maintain an equality between the positions of the CG and CT rotors. Since the rotor position of the CG-rotor is constantly followed by the rotor position of the CT, the angular difference $\Delta\alpha_{GT} \equiv \alpha_G - \alpha_T$ of Eq. (4.172) is kept to small values and the voltage magnitude $|\mathbf{E}_T|$ is very nearly linear with respect to the angular difference $\Delta\alpha_{GT}$, Fig. 4.50.

The instantaneous voltage induced in the rotor winding of the CT, expressed as

$$e_T(t) = \sqrt{2}\, |\mathbf{E}_T| \sin \omega_E t \tag{4.173}$$

[where ω_E (radians/second) is the frequency of excitation of the CG rotor and the reference phase, i.e., $e_R(t)$, of the two-phase induction motor] may be either in phase or 180° out of phase with the excitation applied to the CG rotor. This depends on the relative magnitudes of the angles α_G and α_T *measured positively* in the positive or counterclockwise direction when viewed from the shaft end of the synchros. When the CG shaft leads the CT shaft, that is, $\alpha_G > \alpha_T$, $|\mathbf{E}_T|$ is positive and $e_T(t)$ is in phase with the excitation $v(t)$ [and $e_R(t)$] and the two-phase induction motor provides rotation in the counterclockwise direction, tending to bring the rotor of the CT in correspondence with the rotor of the CG. When the

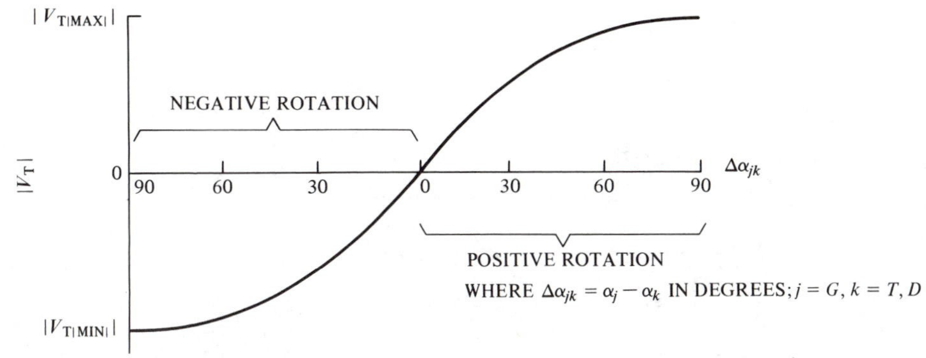

Fig. 4.50 Magnitude of the Voltage $|\mathbf{E}_T|$ Induced in the Rotor Winding of an Error Detector as a Function of the Angle Difference $\Delta\alpha_{jk}$ Between the CG and CT, or CG and CD, Synchro Shafts

CG shaft lags the CT shaft, that is, $\alpha_G < \alpha_T$, $|\mathbf{E}_T|$ is negative and $e_T(t)$ is out of phase with the excitation $v(t)$ [and $e_R(t)$] and the two-phase induction motor provides rotation in the clockwise direction, again tending to bring the rotor of the CT in correspondence with the rotor of the CG. It should be obvious now that there are **two** positions in the CT rotor position for which $e_T(t)$ is zero. They are 180° out of phase and represent the rotor positions in which the CT rotor-winding loops are parallel to the composite airgap field of the CT, which is parallel to the CG rotor position. Thus the CT has **two zero-output positions**, 180° apart. It is essential, then, that when the CT is connected as a component of a servomechanism, as in Fig. 4.49, the correct zero-output position must be selected a priori, *otherwise* the system will tend to increase rather than decrease the amount of error between correspondence of the CG rotor position α_G (command signal) and the CT rotor position α_T (load position). For the correct method of determining the proper zero-output position of the CT rotor position, one should refer to the synchro literature.

It should be obvious that it is more economical to use wiring for remote transformation of information rather than mechanical linkage. Thus the system of Fig. 4.49 can be connected as in Fig. 4.51. Please study these two figures and note the differences. Note that the principles of operation are the same in both systems.

In certain servomechanism applications it is desirable that the load position itself according to the difference or sum of *two* command signals. For example, in antiaircraft firing systems, the firing gun can position itself in accordance with the tracking signal α_G that can be provided by a radar or optical tracker and the lead angle of the target α_D. Thus it is essential that the error signal $e_T(t)$ of Fig. 4.49 is the difference or sum of two mechanical inputs (shaft positions). This can be achieved by connecting the CG synchro to the CT synchro through a CD synchro, as shown in Fig. 4.52. The *CG-CD-CT synchro* of Fig. 4.52, known as a GDT **error detector**, is designed so that the magnitude of the induced voltage in the rotor winding of the CT is

$$|\mathbf{E}_T| = K_D \sin(\alpha_G - \alpha_D) \tag{4.174}$$

where α_G is the angle of the CG rotor in degrees, α_D is the angle of the CD rotor in degrees, and K_D, defined as the *sensitivity* of the GDT synchro, is the maximum RMS voltage per degree induced on the CT rotor winding by the airgap flux when

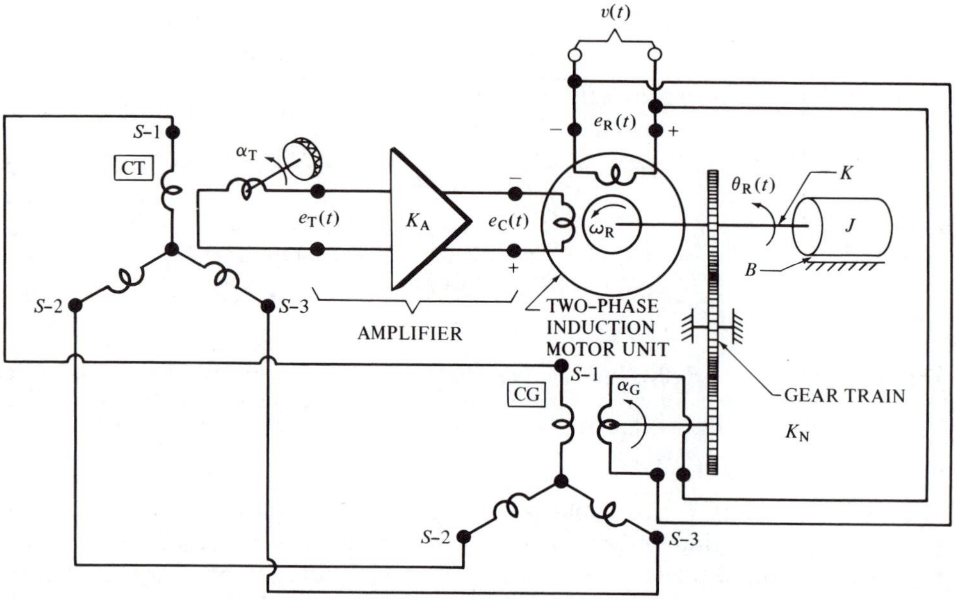

Fig. 4.51 Transformer-Generator Synchro Used as Error Detector in Positional Servomechanisms When the Position Is a Function of One-Command Input

the CT rotor-to-stator coupling is *maximum*. Note that for a counterclockwise rotation of the CG rotor, the CT rotor voltage magnitude $|\mathbf{E}_T|$ is positive if the CD rotor is rotated also counterclockwise but lags the CG rotor. Thus the voltage $e_T(t)$ is in phase with the excitation voltage $e(t)$. The two-phase induction motor would provide a counterclockwise rotation if this synchro system were inserted in the servomechanism of Fig. 4.49 in place of the GT synchro. If, on the other hand, the CD rotor rotates clockwise, the voltage $\mathbf{E}_T$ is again positive and the results are the same as above. You might want to investigate what would happen if the CG rotor rotates counterclockwise and the CD rotor leads the CG rotor and (1) rotates counterclockwise, (2) clockwise, and so on.

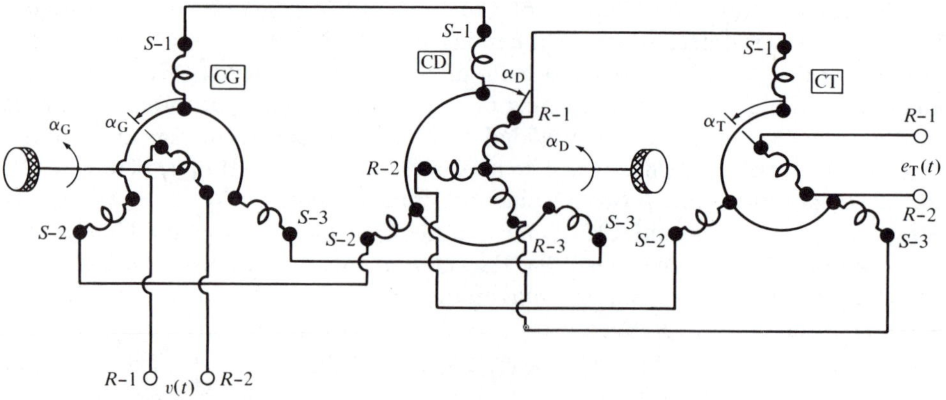

Fig. 4.52 Generator-Differential-Transformer Synchro Used as Error Detector in Positional Servomechanisms When the Position Is a Function of the Two Command Inputs α_G and α_D

If the CT in Fig. 4.52 is substituted with a CR, the system becomes a *torque transmitter* and is used for indication of the **difference or the sum** of two shaft positions. In this connection the rotor position of the CR is affected by the angular **difference** in the rotor position of the CG and CD synchros in accordance with the equation

$$\alpha_R = \alpha_G - \alpha_D \equiv \Delta\alpha_{GD^-} \tag{4.175}$$

If the interconnection of the GDT error detector of Fig. 4.52 (with CT→CR) is affected as follows, $S-1_{CG}$ to $S-1_{CD}$, $S-2_{CG}$ to $S-3_{CD}$, $S-3_{CG}$ to $S-2_{CD}$ and $R-1_{CD}$ to $S-1_{CT}$, $R-2_{CD}$ to $S-3_{CT}$, $R-3_{CD}$ to $S-2_{CT}$, then Eqs. (4.174) and (4.175) respectively become

$$|\mathbf{E}_T| = K_D \sin(\alpha_G + \alpha_D) \tag{4.176}$$

and

$$\alpha_R = \alpha_G + \alpha_D \equiv \Delta\alpha_{GD^+} \tag{4.177}$$

Thus the signal $e_T(t)$ at the CT rotor output or the angle α_R of the CR rotor (if the CT of Fig. 4.52 were substituted by a CR, e.g., in angle-measuring devices or light power selsyns) is a measure of the *sum* of the angular positions of the CG and CD rotors.

Utilization of the following two facts, (1) the dependence of the angles α_G, α_T, and α_D on time and (2) the fact that for small values of $\Delta\alpha_{jk}$ ($j=\text{G}$, $k=\text{T,D}$), that is, $0 \leqslant \Delta\alpha_{jk} \leqslant 15°$ (3.26 rad), $\sin\Delta\alpha_{jk} \doteq \Delta\alpha_{jk}$, enables us to write Eq. (4.173) [after proper substitution for $|\mathbf{E}_T|$ from Eq. (4.172)] as

$$v_{T[GT]}(t) = e_{T[GT]}(t)\sin\omega_E t \tag{4.178}$$

and

$$v_{T[GD]}(t) = e_{T[GD]}(t)\sin\omega_E t \tag{4.179}$$

where

$$e_{T[GT]}(t) \equiv K_S \Delta\alpha_{GT}(t) \equiv K_S[\alpha_G(t) - \alpha_T(t)] \tag{4.180}$$

and

$$e_{T[G_D]}(t) \equiv K_D \Delta\alpha_{GD}(t) \equiv K_D[\alpha_G(t) - \alpha_D(t)] \tag{4.181}$$

are time-dependent magnitudes of the output voltage of the GT error detector, Fig. 4.49, designated as $e_{T[GT]}(t)$, and the output voltage of the GDT error detector, Fig. 4.52, designated as $e_{T[GD]}(t)$. A typical variation of $\Delta\alpha_{jk}$ [Eqs. (4.180) and (4.181)] is exhibited in Fig. 4.53(a). The effect of $\Delta\alpha_{jk}$, and thus of Fig. 4.53(a), on the impressed voltage $v(t)=\sqrt{2}\,|\mathbf{V}|\sin\omega_E t$ (Figs. 4.43, 4.49, and 4.51) exhibited in Fig. 4.53(b) is described mathematically by Eqs. (4.178) and (4.179) and is exhibited in Fig. 4.53(c) as $e_{T[jk]}(t)$, $j=\text{G}$, $k=\text{T,D}$.

As seen in Figs. 4.49 and 4.51, $e_{T[jk]}(t)$ is amplified and impressed in the control winding of the two-phase induction motor used as the positioning device in the ac positional control system. Since, as pointed out earlier in this section, the transient behavior of the two-phase induction motor is determined by the time-varying amplitude of the control voltage $e_{T[jk]}(t)$, Figs. 4.49 and 4.51, rather than the voltage $v_{T[jk]}(t)$ itself, the amplitudes $e_{T[jk]}(t)$ of the voltages [i.e., Eqs. (4.180) and (4.181)] are of **greater interest** than the voltages $v_{T[jk]}(t)$ [i.e., Eqs. (4.178) and (4.179)] themselves. Consequently, the transfer functions of the GT and GDT error detectors used extensively in control systems theory are found, by taking the Laplace transform of Eqs. (4.180) and (4.181) (with their initial conditions assumed

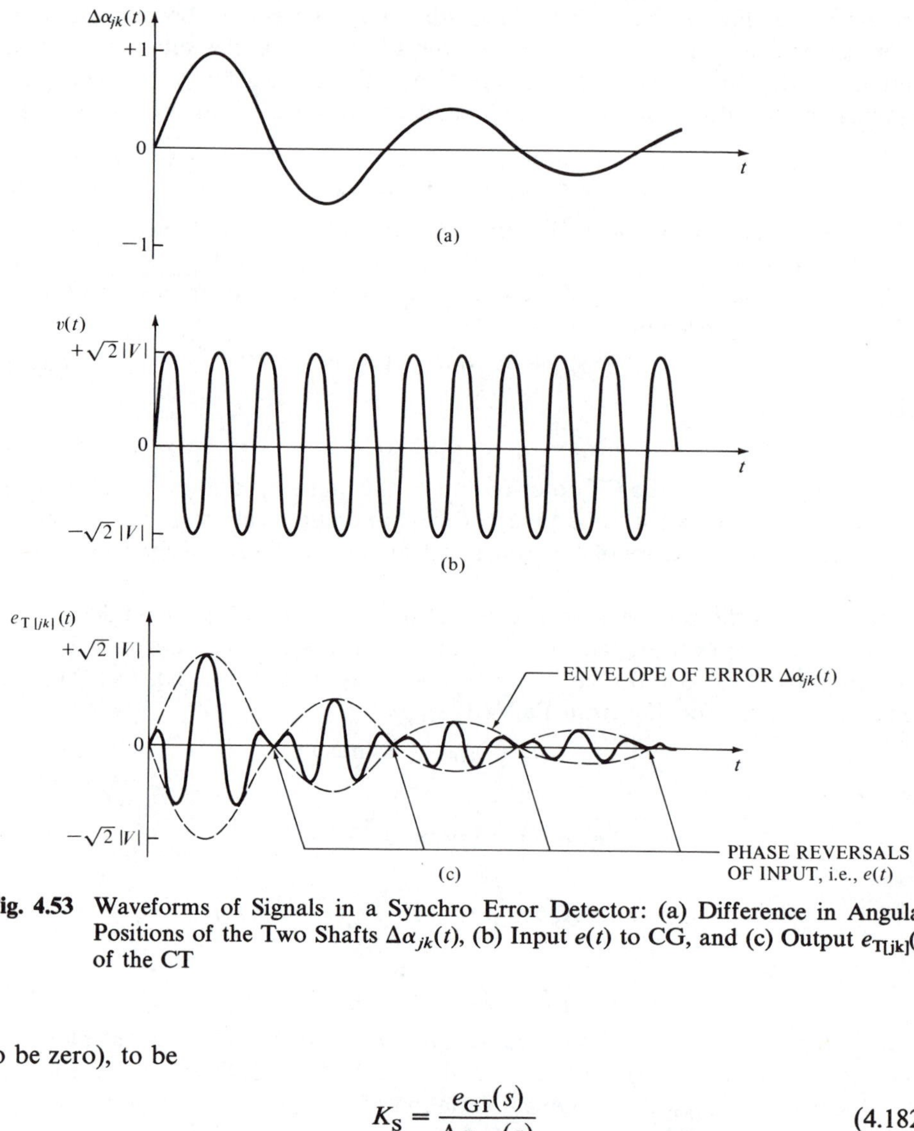

Fig. 4.53 Waveforms of Signals in a Synchro Error Detector: (a) Difference in Angular Positions of the Two Shafts $\Delta\alpha_{jk}(t)$, (b) Input $e(t)$ to CG, and (c) Output $e_{T[jk]}(t)$ of the CT

to be zero), to be

$$K_S = \frac{e_{GT}(s)}{\Delta\alpha_{GT}(s)} \tag{4.182}$$

$$K_D = \frac{e_{GD}(s)}{\Delta\alpha_{GD}(s)} \tag{4.183}$$

A practical application of these ideas is exhibited in Fig. 4.44 where the GT synchro error detector of Fig. 4.43 (or of Fig. 4.49 for that matter) is shown in place.

Before closing, it should be pointed out that the theory that supports Eqs. (4.179) and (4.181) is still valid when Eq. (4.176) is to be considered. In fact, one can define $\Delta\alpha_{jk} \equiv [\alpha_G(t) \pm \alpha_D(t)]$ (in the immediately previous theory) at the very outset and make the development valid for $\Delta\alpha_{jk} = [\alpha_G(t) - \alpha_T(t)]$, defining **negative feedback**, and $\Delta\alpha_{jk} = [\alpha_G(t) + \alpha_T(t)]$, defining **positive feedback**, Fig. 4.44. Note that only negative feedback is exhibited in Fig. 4.44. If the sign of $\alpha_T(s)$ were to become plus, Fig. 4.44 would designate a system with positive feedback.

4.13 SUMMARY

In this chapter we present the operating principles of the balanced three-phase, balanced and unbalanced (voltages) two-phase, and single-phase induction motors. Corresponding equivalent circuits and describing equations are developed. Also, the terminal characteristics, τ_{em} versus ω_R, of the devices discussed are presented. In addition, for the three-phase induction device, the three distinct regions of operation for the τ_{emT} versus s (or ω_R) characteristic, that is, motor, generator, and braking regions of operation, are discussed in great detail. Moreover, the available methods of speed control for the three-phase induction motor are discussed extensively: and the procedure for measuring the equivalent circuit parameters of this device is presented. In addition to the steady state performance analysis of the balanced two-phase induction motor, its dynamic analysis via s-domain techniques as used in automatic control theory is presented; in this method of analysis the excitation of the two-phase induction motor is comprised of a set of two unbalanced voltages. The steady state performance analysis of the single-phase induction motor is based on the equivalent circuit, which is derived with the aid of the forward and backward rotating fields. Finally, a good account, although not complete, of the applications of the induction devices is presented.

Although this chapter is relatively long and fairly complete in its presentation of the operating principles and steady state analysis of the three-, two-, and single-phase induction motors, it merely mentions the dynamic analysis of the balanced three-phase induction devices and makes no mention of the analysis of the unbalanced three-phase induction device; it simply mentions the existence of unbalanced operating conditions for the two-phase induction motor.

The reader is urged to master the material of this chapter, because in addition to the fact that all industrial-type and home (and office) appliance motors utilize these concepts, the revolving field theory set forth here is essential for the operating principles of the three-phase synchronous device, that is, the synchronous generator and motor.

PROBLEMS

4.1 Design a three-phase, Y-connected, four-pole induction device and draw the field for $\omega_E t = 0°$, 90°, 180°, 270°, and 360°.

4.2 Design a three-phase, Y-connected, eight-pole induction device and draw the field for $\omega_E t = 0°$, 90°, 180°, 270°, and 360°.

4.3 What is the concept of the eight-pole field in Problem 4.2?

4.4 If the frequency of excitation of the induction device of Problem 4.2 is $f_E = 60$ Hz, what is the synchronous speed of the field in radians/second (ω_S) and revolutions/minute (n_S)?

4.5 When the motor of Problem 4.2 delivers rated output horsepower, the slip is found to be $s = 0.03$. Determine the following parameters: (1) the radian frequency (ω_E) of the stator currents, (2) the synchronous speed (ω_S) of the revolving field relative to the stator structure, (3) the speed with which the rotor is rotating in radians/second (ω_R) and revolutions/minute (n_R), (4) the radian frequency (ω_e) of the induced rotor currents, (5) the radian speed of the rotor field (ω_s) relative to the rotor structure, (6) the radian speed of the stator field relative to the rotor structure, designated as ω_{SR}, (7) the radian speed of the rotor field relative to the stator structure, designated as ω_{sS}, (8) the speed of the stator field relative to the rotor field, designated as ω_{S-s}, and (9) the speed of the rotor field relative to the stator field, designated as ω_{s-S}. Are the conditions ideal for the development of unidirectional torque?

4.6 For the three-phase, eight-pole induction motor of Fig. 4.15, the following parameters are specified: (1) the RMS line-to-line input voltage is 220 V, (2) $f_E = 60$ Hz, (3) $\mathbf{Z}_S = 0.5 + j0.4\ \Omega$, (4) $\mathbf{Z}_{R'} = 0.2/s + j0.3\ \Omega$, (5) $j\mathrm{X}_m = j20\ \Omega$, and (6) $P_{LOSSES} = P_{CL} + P_F + P_W = 1000$ W. If the motor is operating at a slip of 2%, find the following motor parameters: (1) the synchronous speed of the stator field, ω_S, (2) the speed of the rotor, ω_R, (3) the input power P_{INT}, (4) the stator winding losses P_{SWT}, (5) the airgap power P_{AGT}, (6) the rotor winding losses P_{RWT}, (7) the electromagnetic power P_{emT}, (8) the electromagnetic torque τ_{emT}, (9) the output power P_{OT}, (10) the output torque τ_{OT}, (11) the efficiency η, (12) the speed regulation SR, (13) the frequency of the rotor currents, ω_e, (14) the synchronous speed of the rotor, ω_s, and (15) determine what would happen if the rotor windings were to be open-circuited while the motor was running (operating).

4.7 For the induction device of Problem 4.6, (1) derive the Thevenin's equivalent circuit parameters to the left of the airgap terminals and draw the equivalent circuit of the induction motor on a per-phase basis; (2) derive and plot the total torque versus slip characteristic of the induction device; (3) derive and plot the total power versus slip characteristic of the induction device; (4) obtain the starting torque (power) τ_{STG} (P_{STG}), the breakdown torque (power) τ_{MAX} (P_{MAX}), and the minimum torque (power), τ_{MIN} (P_{MIN}); (5) obtain the slips s_{MAX} and s_{MIN} at which the torque breakdowns occur; (6) obtain the torque (power) readings at $s = \pm 2$; (7) define explicitly the motor region of operation, the generator region, and the braking region; (8) determine the significant difference between these regions.

4.8 For a squirrel cage induction motor rated as Y-connected, three-phase, eight-pole, 60-Hz, 500-V (line-to-line RMS voltage), 70-A (line current), and 60-hp, the following test data are obtained:

a. dc test: the dc resistance measured between any two terminals of the stator windings is 0.30 Ω.

b. No-load test: at 60 Hz and rated line-to-line voltage applied, the following measurements are made:

$$|\mathbf{I}_{NL}| = 30\ \mathrm{A}$$
$$P_{NLT} = 2{,}000\ \mathrm{W}(P[W_1] = 1{,}200\ \mathrm{W}\ \text{and}\ P[W_2] = 800\ \mathrm{W})$$

c. Blocked-rotor test at 15 Hz: for an applied voltage of 50 V (line-to-line), the following measurements are made:

$$|\mathbf{I}_{BL(15)}| = 80\ \mathrm{A}$$
$$P_{BLT(15)} = 5{,}000\ \mathrm{W}(P[W_1] = 2{,}000\ \mathrm{W}\ \text{and}\ P[W_2] = 3{,}000\ \mathrm{W})$$

d. Blocked-rotor test at 60 Hz: for an applied voltage of 450 V (line-to-line), the following measurements are made:

$$|\mathbf{I}_{BL(60)}| = 120\ \mathrm{A}$$
$$P_{BLT(60)} = 20{,}000\ \mathrm{W}(P[W_1] = 4{,}500\ \mathrm{W}\ \text{and}\ P[W_2] = 15{,}500\ \mathrm{W})$$
$$\tau_{STG} = 98\ \mathrm{N \cdot m}$$

Compute the rotational losses of the device and the parameters of the equivalent circuit for the normal running conditions (i.e., for a frequency of excitation $f_E = 60$ Hz and $s = 0.02$ and $f_e = 1.2$ Hz). Also, compute the starting torque from the data of part d. This part of the test is run exclusively for the purpose of calculating the starting torque. The frequency of the rotor currents is never 60 Hz during normal running conditions.

4.9 The measured steady state equivalent circuit parameters of a 10-W, 120-V, 60-Hz, two-pole servomotor are (1) $R_S = 300\ \Omega$, (2) $X_{\ell S} = 80\ \Omega$, (3) $R_R' = 900\ \Omega$, (4) $X_{\ell R}' = 80\ \Omega$, and (5) $X_m = 1000\ \Omega$. For 100-V **balanced stator excitation**, $s = 0.4$, and core friction, and windage losses equal to 0.85 W, calculate the following quantities: (1) the synchronous speed ω_S of the field, (2) the mechanical speed ω_R of the rotor, (3) the frequency of the rotor currents, ω_e, (4) the input current $\mathbf{I}_S$ and power P_{INT}, (5) the stalled torque τ_{AGT} ($s = 1$) for 100- and 120-V excitation, (6) the electromagnetic power P_{emT} and torque τ_{emT}, (7) the developed power P_{AGT} and torque τ_{AGT}, (8) the output power P_{OT} and torque τ_{OT}, (9) the speed regulation SR, and (10) the efficiency η.

4.10 A two-phase servomotor with (1) reference and control winding voltages $v_R(t)$ and $v_C(t)$, respectively, rated as 120 V, (2) blocked-rotor torque, at rated $|\mathbf{V}_C|$, $\tau_0 = 7.54$ N·m, (3) combined inertia of rotor and load $J = 0.10$ kg·m², and (4) viscous friction $B = 0.01$ N·m/rad/s has a torque versus steady state speed curve whose linear approximation is given by (assuming that this curve intersects the speed axis at $n_R = 420$ rpm)

$$\tau_{em} = 7.54 - 0.02\omega_R$$

Assume that (1) the steady state torque equation applies to slow transients, and write the differential equation that describes the starting transients of the motor load combination, and (2) the signal to the motor consists of the time-varying voltage of the control winding. If this voltage is designated as $v_C(t)$, the preceding torque versus speed equation is written as

$$\tau_{em} = \left(\frac{7.54}{|\mathbf{V}_R|}\right) v_C(t) - 0.02\omega_R$$

Determine the transfer function relating (1) motor shaft position and the voltage of the control winding and (2) motor shaft velocity and the voltage of the control winding. Also, draw their corresponding block diagrams.

4.11 The parameters of the equivalent circuit of a 120-V, 60-Hz, six-pole, $\frac{1}{2}$-hp, single-phase induction motor measured at $n_R = 1800$ rpm are $R_M = 2.0\ \Omega$, $X_{\ell M} = 2.47\ \Omega$, $X_m = 51.8\ \Omega$, $R_R' = 2.71\ \Omega$, $X_{\ell R}' = 2.47\ \Omega$, $P_{LOSSES} = 40$ W (friction, windage, and core losses). Calculate the following quantities: (1) the input current $\mathbf{I}_M$, (2) the input power P_{IN}, (3) the developed electromagnetic power (torque) P_{emN} (τ_{emN}), (4) the output power (torque) P_O (τ_O), (5) the efficiency η, and (6) the speed regulation SR.

REFERENCES

1. Buckstein, E., *Basic Servomechanisms*. Holt, Rinehart and Winston, New York, 1963.
2. Chapman, C. R., *Electromechanical Energy Conversion*. Wiley, New York, 1965.
3. Davis, S. A., *Outline of Servomechanisms*. Regents, New York, 1966.

4. D'azzo, J. J., and Houpis, C. H., *Feedback Control System Analysis*. McGraw-Hill, New York, 1960.
5. Del Toro, V., *Electromechanical Devices for Energy Conversion and Control Systems*. Prentice-Hall, Englewood Cliffs, N.J., 1968.
6. Fitzgerald, A. E, and Kingsley, C., Jr., *Electric Machinery*, 2nd ed. McGraw-Hill, New York, 1961.
7. Gourishankar, V., *Electromechanical Energy Conversion*. Intext, New York, 1965.
8. Heller, S., *Frequency Changers-Rotating Type*. Tab Books, Blue Ridge Summit, Pa., 1968.
9. Koscow, I. L., *Electric Machinery and Control*. Prentice-Hall, Englewood Cliffs, N.J., 1964.
10. Kuo, B. C., *Automatic Control Systems*. Prentice-Hall, Englewood Cliffs, N.J., 1962.
11. Majmudar, H., *Electromechanical Energy Converters*. Allyn & Bacon, Boston, 1965.
12. Matsch, L. W., *Electromagnetic and Electromechanical Machines*. Intext, New York, 1972.
13. Thaler, G. J., and Brown, R. G., *Analysis and Design of Feedback Control Systems*. McGraw-Hill, New York, 1960.

5 SYNCHRONOUS ENERGY CONVERTERS — ELEMENTARY CONCEPTS

5.1 INTRODUCTION

The synchronous device is defined as the electromechanical energy converter that is comprised of a stationary member, **the stator**, with smooth inner surface, and a rotating member, **the rotor**, with either a smooth or indented (yielding saliencies) surface.

For the balanced three-phase synchronous device, the stator has a three-phase sinusoidally distributed winding with identical resistance, inductance, and capacitance (if any) in each phase. The winding is designed either to receive balanced three-phase ac currents from an external source of electrical energy or to send balanced three-phase ac currents to a three-phase load. The rotor, depending on whether it is smooth or salient, has either a sinusoidally distributed winding or a concentrated winding, known as the field winding, which is **always** excited from a dc source of electrical energy through a set of **slip rings**.

Some synchronous machines have a rotating armature configuration, that is, the field winding is on the stator and the three phase (armature) windings are on the rotor. There are three slip rings which carry the three phase currents. The rotating field configuration is most often used for large machines because of the smaller slip ring currents. The analysis of synchronous machines is identical whether it has a rotating field or a rotating armature configuration. Therefore, in the remainder of the text, a rotating field configuration is assumed.

As in the case of the induction motor, when three-phase ac currents are present in the stator winding of the synchronous device, they produce a constant-amplitude, sinusoidally distributed magnetic field in the airgap of the device. There, this field rotates at synchronous speed ω_S radians/second. The direction in which the field rotates depends on the sequence of the excitation currents. Similarly, the sinusoidal distribution of the rotor winding (when the rotor has a smooth outer surface) or the proper shaping of the magnetic pole faces (when the rotor has saliencies) guarantees the presence of a second sinusoidally distributed magnetic field in the airgap of the device.

If these two fields are stationary with respect to each other, they interact in a manner such that they tend to align themselves and therefore develop an electromagnetic torque. If the torque is to be unidirectional, the two fields must rotate at the same speed and in the same direction. That is, the rotor must rotate precisely at the same speed and direction as the stator field does. The synchronous device is distinguished by its torque versus torque angle (the angular displacement between the terminal and generated voltages) characteristic. It shows that the same device can operate either in the **generator mode**, in which case the stator windings provide electrical energy in the form of three-phase currents, or the **motor mode**, in which case the stator windings draw electrical energy in the form of three-phase currents. In the generator mode, the developed electromagnetic torque opposes rotation, and mechanical torque must be applied from the prime mover in order to sustain the rotation. As the developed electromagnetic torque increases, so does the mechanical torque applied from the prime mover. Therefore, this developed electromagnetic torque is the mechanism through which the electric power output from the generator is related to its mechanical power input. In the motor mode, the developed torque is in the direction of rotation and balances the opposing torque, which results mainly from the motor's mechanical load.

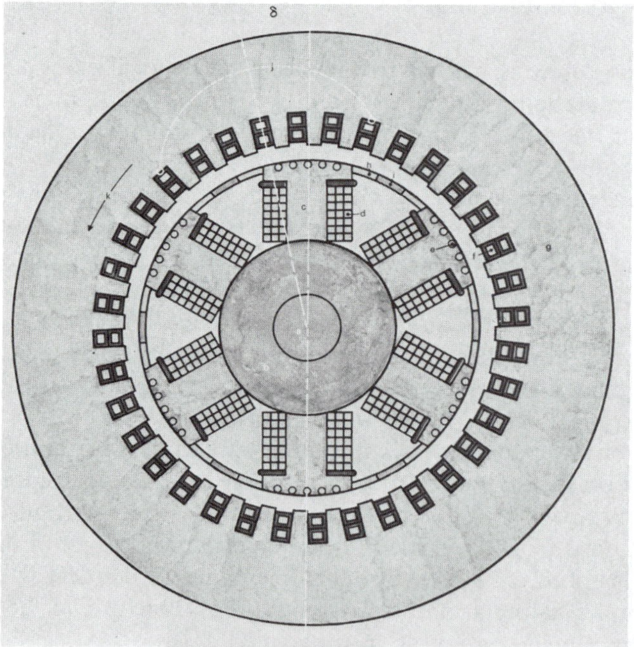

Plate 5.1 Ross Section of a Salient-Pole Synchronous Device: a = Shaft, b = Spider, c = Field Pole, d = Field Winding, e = Damper Winding, f = Stator Coils, g = Stator Core Laminations, h = Short-Circuiting Segments, iu = Short-Circuiting Connectors, j = Stator Magnetic Pole, k = Direction of Stator Flux Rotation, and j = $S\delta$ = Torque Angle. (Courtesy of Westinghouse Electric Co., Pittsburgh, Pa.)

An inherent characteristic of the synchronous device's motor mode of operation (regardless of whether the device is a three-, two-, or single-phase synchronous device) is that it *does not have* a starting torque. Therefore, it will not start by itself, but it must be brought up to synchronous speed by auxiliary means before it can produce a net torque of its own. This means that **the synchronous motor runs only at synchronous speed**—thus the name of the device. Overloading of the rotor to the point where the maximum torque developed by the motor is exceeded by the motor's mechanical load causes the rotor to slow down and lose synchronism. Under such conditions the motor is usually disconnected from its excitation by automatic circuit breakers.

The synchronous device is perhaps the greatest invention in the field of electrical engineering. The **abundance of electrical energy** has been the key to industrial expansion. Nearly all ac power generation is provided by one device—the synchronous generator—regardless of what fuel (fossil fuel, nuclear energy, etc.) is used in the prime mover. As a generator the synchronous device can be found operating singly, but most commonly it is found connected in parallel and operating in synchronism with other synchronous generators in large interconnected power system grids.

As a motor it is used in such a manner as to capitalize on its inherent property: **to provide constant speed regardless of load constancy**. Therefore, it is found in timing systems, pulsating loads that need to be driven at constant speeds, and dc generator prime movers. Perhaps the most important application of the three-phase synchronous motor is that of its use in large power system grids **to generate required reactive power**.

The quantitative analysis of the three-phase synchronous device is pursued here. Thus the objectives of this chapter are fivefold: (1) to present the principles of operation of the balanced, three-phase, two-phase, and single-phase synchronous devices, and to discuss the structural differences between these devices as well as those between the synchronous generator and motor, (2) to develop mathematical models that describe the steady state performance of these devices, (3) to exhibit the terminal characteristics of these devices, (4) to introduce the procedure for the measurement of the equivalent circuit parameters of the balanced three-phase synchronous devices, and (5) to show the reader the most essential applications of the balanced three-phase and the single-phase synchronous devices.

5.2 STRUCTURAL THEORY OF THREE-PHASE SYNCHRONOUS ENERGY CONVERTERS

The synchronous energy converter functions both as a generator, Fig. 5.1, and as a motor, Fig. 5.2. Inspection of these two figures reveals distinct structural differences between the two devices. Thus they can generally be distinguished from each other by their physical appearance. For example, the high-speed synchronous generator, Fig. 5.1, is comprised of (1) a rotor that has (i) a smooth (round) structure with a winding designed to produce a sinusoidal field (it is usually excited by an external dc source through two slip rings), (ii) a long axial length, (iii) a small diameter, (iv) two magnetic poles (or at most four magnetic poles), and (2) a smooth stator structure with three windings, with their axes displaced from each other by 120 electrical degrees, in the positive direction, distributed sinusoidally in

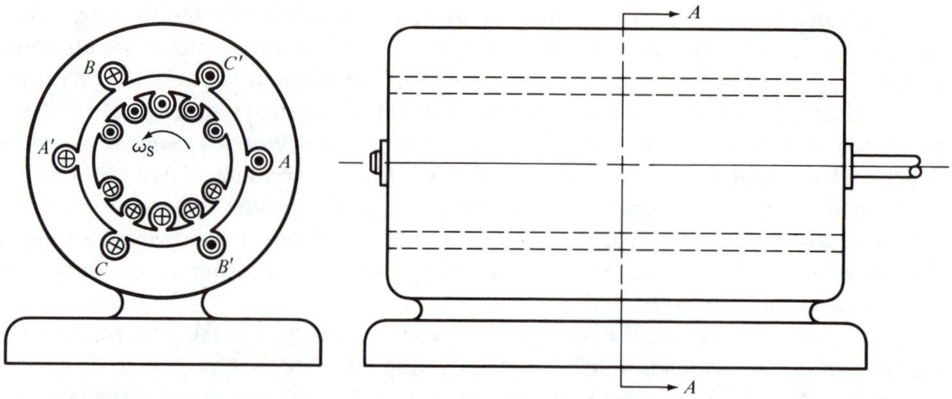

Fig. 5.1 Synchronous Generator Characterized by a Long Axial Length, a Smooth Rotor with Small Diameter, Low Number of Poles and High Speed, and Smooth Stator Structure

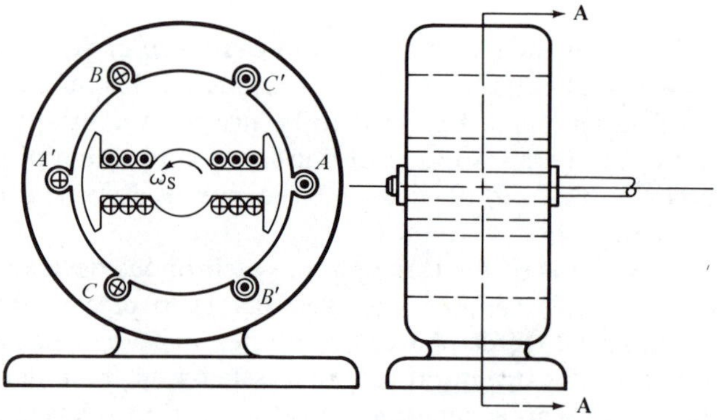

Fig. 5.2 Synchronous Motor (or Generator) Characterized by Short Axial Length, a Salient Rotor with Large Diameter, Large Number of Poles and Low Speed, and Smooth Stator Structure

slots along the surface of the stator. The rotor of the generator is always driven by a prime mover at a high constant, or nearly so, speed, that is, at a synchronous speed designated by ω_S.* High-speed rotational motion is provided to the rotor by a steam-driven (or gas-driven) turbine. The steam is generated in a power plant that may, depending on fuel resources, employ a coal-fired, or oil-fired, boiler or a nuclear plant. When a low-speed hydroturbine is used as the prime mover of a synchronous generator (hydroelectric power plants utilize the potential energy of water collected in river dams or pump-and-storage dams), a large number of poles is required in order to generate a voltage of a desired frequency f_E. The frequency of the induced voltages f_E (in hertz) is related to the speed of the generator's rotor n_S (in revolutions/minute) and the generator's number of poles P, in accordance with Eq. (4.24). Eq. (4.24) is solved for f_E and is rewritten here for simplicity,

*Throughout this chapter it should be understood that for the generator mode of operation, $\omega_S \equiv \omega_I$ radians/second and $n_S \equiv n_I$ rpm, where "S" designates synchronous and "I" designates input.

because of repeated use, as

$$f_E = \frac{n_S P}{120} \tag{5.1}$$

Example 5.1

A synchronous generator has four poles and operates at a speed of 1800 rpm. Calculate: (1) the frequency of the generated voltage and (2) the prime-mover speed required to generate voltages whose frequencies are 60 Hz and 30 Hz.

Solution

The required quantities can be calculated utilizing Eq. (5.1). Thus Eq. (5.1) used intact yields

$$f_E = \frac{n_S \times P}{120} = \frac{1800 \times 4}{120} = \underline{\underline{60}} \text{ Hz} \tag{1}$$

However, rewriting Eq. (5.1) as $n_S = (120 \times f_E)/P$ yields the prime-mover speed required to generate voltages of frequencies 60 and 30 Hz. Thus

$$n_{S60} = \frac{120 \times f_E}{P} = \frac{120 \times 60}{4} = \underline{\underline{1800}} \text{ rpm} \tag{2}$$

and

$$n_{S30} = \frac{120 \times f_E}{P} = \frac{120 \times 30}{4} = \underline{\underline{900}} \text{ rpm} \tag{3}$$

To obtain a large number of poles, a large rotor diameter is used. Thus for low-speed prime movers, a salient-pole-rotor synchronous generator, Fig. 5.2, is used. However, the salient-pole-rotor device is used primarily as a motor; its structural description follows. The synchronous motor, Fig. 5.2, is comprised of (1) a rotor that has (i) a salient-multipole (the number of poles is always **even**) structure with windings excited by an external dc source through two slip rings, (ii) a short axial length, (iii) a large diameter with a large number of protruding magnetic poles (this structure gives rise to relatively high mechanical stresses at high speeds; however this rotor structure is simple and economical to manufacture), and (2) a smooth stator structure with three windings, with their axes displaced from each other by 120 electrical degrees, in the positive direction, distributed in slots along the surface of the stator. When this device functions in its generator mode, the rotor must be driven at a *low*, constant (or nearly constant), synchronous speed ω_S, as defined by design specifications. Such speeds could be provided by hydroturbines, gasoline, diesel, gas, or steam engines. Whether the synchronous device operates as a motor or generator, the stator should be wound so that the number of its stator poles is identical to the number of the poles in the rotor structure, whether the rotor structure is smooth or salient.

The winding distribution of a smooth rotor, whether the rotor is a two- or four-pole structure, is similar to that of the induction device. That is, the winding is designed to produce a sinusoidal field. The airgap of this device is of uniform length except for the slots in the rotor and stator structures. One end of the multiple-turn rotor coil is connected to the slip ring that is fed by the positive brush and the other end of this coil is connected to the slip ring that is fed by the negative brush of the dc excitation. For the salient-pole rotor, the windings are concentrated on the stem of the rotor protrusions. The ends of these windings are connected to

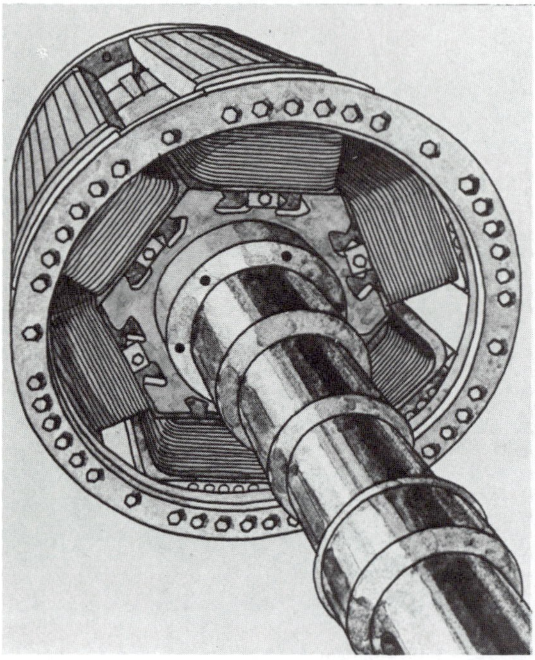

(a)

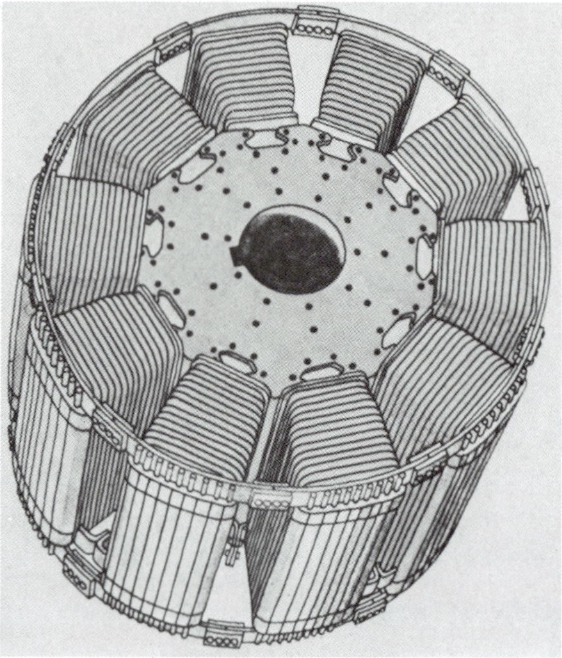

(b)

(c)

Plate 5.2 Rotor Construction for Synchronous Devices: (a) High-Speed Rotor, (b) Medium-Speed Rotor, and (c) Low-Speed Rotor. (Courtesy of Westinghouse Electric Co., Pittsburgh, Pa.)

Plate 5.3 Side View of an Assembled Low-Speed Synchronous Device with Its Multipole Rotor Structure Clearly Visible. (Courtesy of Westinghouse Electric Co., Pittsburgh, Pa.)

the positively and negatively fed slip rings in such a manner as to have the poles of the rotor alternate in a sequence $N, S, N, S, \ldots$ (remember that the number of these poles is always even). To assure the sinusoidal nature of the field generated by the rotor windings, the pole shoes* of the salient rotor, Fig. 5.2, are shaped so that **the airgap is smaller under the center of the poles and larger under the tips of the poles.**

To produce a perfect sinusoidal voltage generation, particular attention is paid to the layout of the stator windings in the stator slots. The stator windings are diamond-shaped preformed coils, Fig. 5.3, which are properly inserted in the stator

*Pole shoes are defined to be the two extensions of each rotor pole, Fig. 5.2.

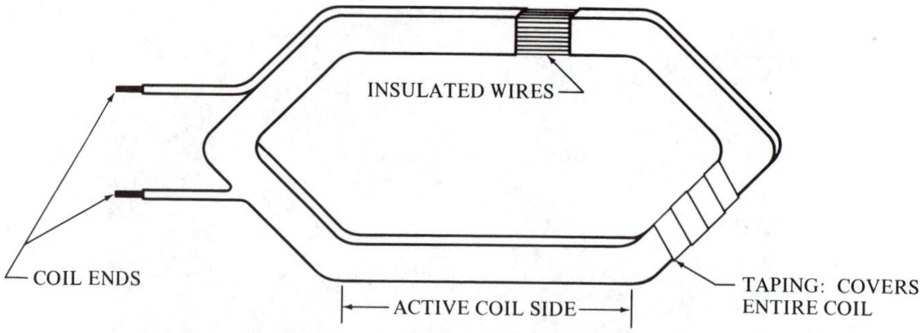

Fig. 5.3 Preformed Coil Used in Stator Windings of Synchronous (and Induction) Devices: Consists of Many Turns of Fine Silk-Covered (or Cotton-Covered, or Enamel-Covered) Wire, Individually Taped and Lacquer-Dipped

slots and are interconnected so that their electric and magnetic effects are cumulative in such a manner as to produce three distinct windings with their axes displaced from each other by 120 positive electrical degrees. Each coil consists of many turns of fine silk-covered (or cotton-covered, or enamel-covered) wire individually taped, lacquer-dipped, and insulated from the stator slot. In general, stator coils span 180 electrical degrees (or 180 mechanical degrees for a two-pole

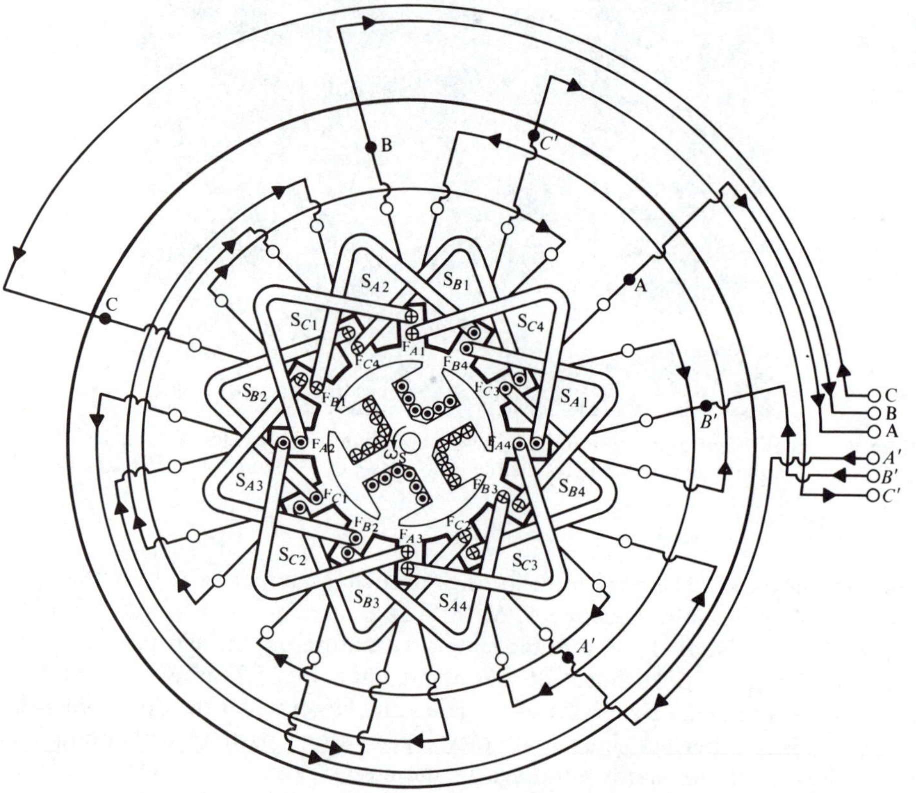

Fig. 5.4 Three-Phase, Full-Pitch, Double-Layer, Integral-Slot, Series-Connected, Concentrated Winding as Used in ac Synchronous Generators and Motors (and Induction Motors), the Winding Shown Here Is Wound to Generate Four Poles

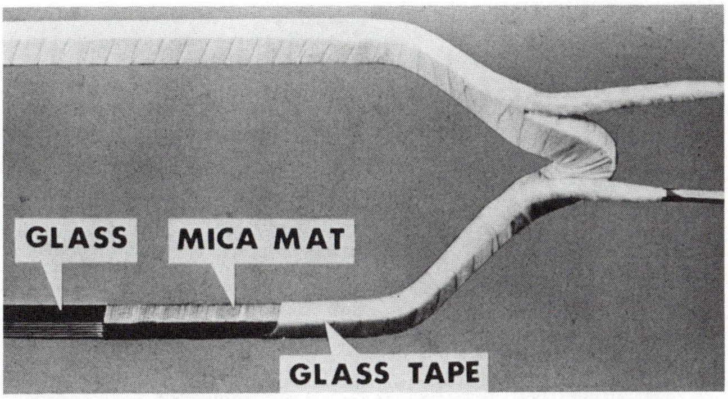

Plate 5.4 Insulation Materials Used in High-Voltage Stator Coils. (Courtesy of General Electric Co., Schenectady, N.Y.)

synchronous device); such coils are defined as **full-pitch coils**. The coils that constitute each winding are distributed over several slots per phase and pole. However, the same characteristics can be achieved if the coils span an angle $120 \leqslant \theta_S \leqslant 180$ electrical degrees, with much less copper wasted since the front end and back end of each coil are inactive. The ratio of the span angle to 180°, that is, $p \equiv \theta_S/180°$, is defined as **pitch factor**; such a coil is defined as a **fractional-pitch coil**, and the windings wound by fractional-pitch coils are defined as **chorded windings**. Most stator windings of ac polyphase devices are two-layer windings; that is, there are two coil sides per slot, Fig. 5.4 [2.12(a)]. This type of winding allows the use of fractional-pitch windings. Experience has shown that a **double-layer, fractional-pitch winding** provides the following advantages: (1) savings in copper and thus reduction of winding resistance without a proportional decrease in the winding's flux linkage, (2) reduction of mmf harmonics generated in the stator winding, and (3) reduction of the emf harmonics induced in the stator windings. The fundamental components in the mmf and emf waveforms generated in the stator windings remain unchanged. This property makes the double-layer, fractional-pitch winding universally attractive to synchronous (and induction) devices. For descriptive purposes a full-pitch, double-layer, four-pole, three-phase winding is laid along the stator slots of Fig. 5.4, utilizing the rule of thumb established in Chapter 4 (for the development of the induction motor's windings). Thus (1) 3 phases × 4 poles = 12 coil sides; (2) 12 coil sides/3 phases = 4 coil sides/phase [or 2 coils per phase]; (3) 360 mechanical degrees/4 coil sides/phase = 90 mechanical degrees/coil side/phase. Thus phases A, B, and C are comprised of four* coils each laid in a full-pitch (i.e., each coil spans 180 electrical degrees), double-layer (i.e., two coil sides/slot) winding as follows. Phase A is comprised of four full-pitch coils, Fig. 5.4, with starting ends designated as S_{A1}, S_{A2}, S_{A3}, and S_{A4} and finishing ends designated as F_{A1}, F_{A2}, F_{A3}, and F_{A4}. The coil sides that constitute a coil are placed in slots separated 90 mechanical degrees and are always paired. Each pair is comprised of the starting end of a given coil, which is placed in the bottom of the slot, and the finishing end of the previous coil placed on the top of the slot. Phase B is comprised of a similar number of coils. Their starting ends are designated as

*If the winding were to be a single-layer winding, only two coils would have been used. The coils S_{A2}–F_{A2} and S_{A4}–F_{A4} would have been omitted.

Plate 5.5 After the Assembly of the Stator Core for a Synchronous Device, Slot Dimensions Verified. The Core Is Held Tightly in Place by the Use of Through-Bolts. (Courtesy of Westinghouse Electric Co., Pittsburgh, Pa.)

Plate 5.6 Coils Being Placed into the Stator Slots of a Synchronous Device. (Courtesy of Westinghouse Electric Co., Pittsburgh, Pa.)

Plate 5.7 The Stator Winding of a Synchronous Device Is Being Secured in Place—Note the Enormous Size of This Stator Structure. (Courtesy of Westinghouse Electric Co., Pittsburgh, Pa.)

S_{B1}, S_{B2}, S_{B3}, and S_{B4} and their finishing ends are designated as F_{B1}, F_{B2}, F_{B3}, and F_{B4}. The coil side S_{B1} is located at $\frac{2}{3}(90)$ mechanical degrees ahead of S_{A1} in the positive direction. The rest of the coil sides are located at 90-mechanical-degree intervals in a fashion similar to that of phase A. Similarly, the starting and finishing ends of the coil sides of phase C, designated as S_{C1}, S_{C2}, S_{C3}, and S_{C4} and F_{C1}, F_{C2}, F_{C3}, and F_{C4}, are laid in a similar fashion to phases A and B, with the coil side S_{C1} located $\frac{2}{3}(90)$ mechanical degrees ahead of S_{B1} in the positive direction. The four coils that constitute each phase may be connected in **series** to form a single-circuit winding or may be connected in **parallel** to form a four-circuit winding. Note that in Fig. 5.4 the four coils that constitute each phase are connected in series. (However, they could have been connected in parallel.) The corresponding load of the generator can be connected between points $A-A'$, $B-B'$, and $C-C'$. Note that the exiting (positive) and the entering (negative) currents which are assumed to be present in the sides of the coils that constitute phases A, B, and C are designated as $\odot$ and $\oplus$, correspondingly. The signs of these currents are read from Fig. 5.5(b) for $\omega_E t=0$ and $e(t)\rightarrow i(t)$. If points A', B', and C', Fig. 5.5(a), are connected to a common point, and if this point is grounded internally, the generator is known as a **three-phase, three-wire**, balanced, Y-connected, synchronous generator. However, if this common point is connected to a wire that is brought outside the device (i.e., if the wire becomes accessible), the device is known as a **three-phase, four-wire**, balanced, Y-connected, synchronous generator. However, if the coil ends are connected A to B', B to C', and C to A', the device becomes a **three-phase, three-wire, balanced, Δ-connected**, synchronous generator. It

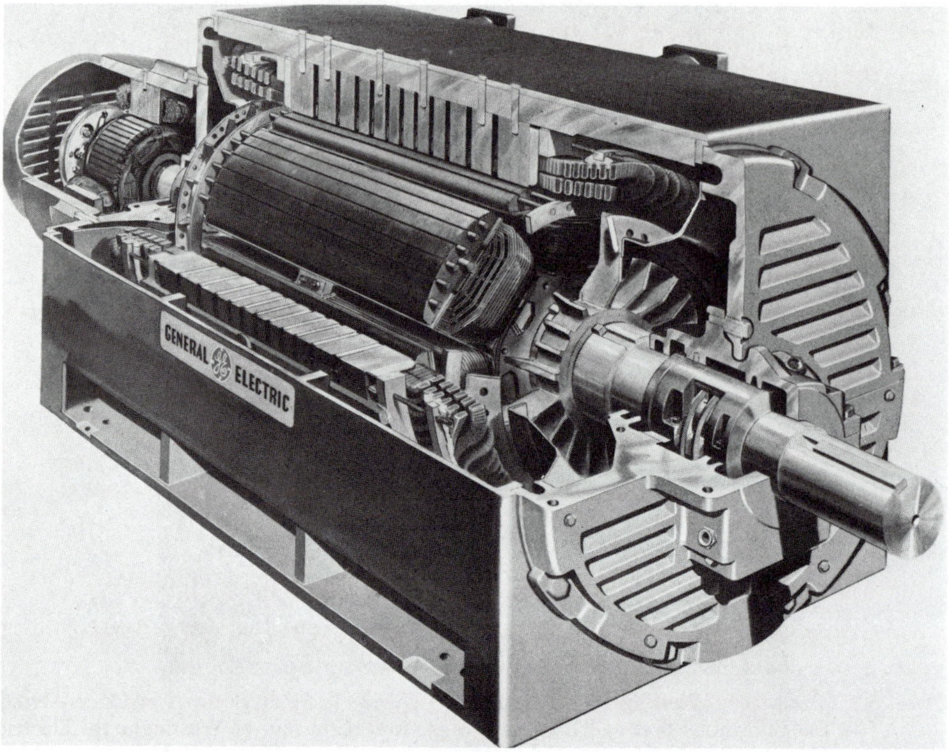

Plate 5.8 Cutaway View of an Eight-Pole, 2000-hp, Custom 8000 Synchronous Motor with Exciter, with Ventilation Louvres to Minimize the Entrance of Liquids and Particles, and with Integral Cast Fans on the Rotor to Draw Cooling Air. (Courtesy of General Electric Co., Schenectady, N.Y.)

should be noted at the outset, however, that similar connections are valid for the **motor mode of operation** of the synchronous device.

If the poles of the salient-pole rotor are designed, Fig. 5.4, to occupy an arc in the rotor circumference that overlaps an arc in the stator circumference that bounds three slots of the stator winding (i.e., one slot/phase) per pole, the winding is known as an **integral-slot winding**. If the poles of the salient-pole rotor are designed to occupy only part (fraction) of the stator sector that bounds the three stator winding slots, the winding is known as a **fractional-slot winding**. Although the analysis of integral-slot windings is simpler in bringing out the basic principles regarding the mmf's and inductances in the stator windings, the fractional-slot windings (i.e., each coil spans less than 180 but seldom less than 120 electrical degrees) have two distinct advantages: (1) it is possible to use the same stator frame (resulting in a lower investment of manufacturing processes) for machines with different numbers of poles provided that a salient-pole-rotor structure is used, and (2) the contribution toward good mmf and emf waveform is equivalent to that of an integral-slot winding with a larger number of slots per pole. (Fractional-slot windings are found to some extent in induction motors.) Inspection of Fig. 5.4 reveals that the series connected coils that constitute the winding of each phase progress from rotor pole to rotor pole, are concentrated, have two coils sides per slot, and have three slots (i.e., one slot/phase) per pole. Thus these windings are

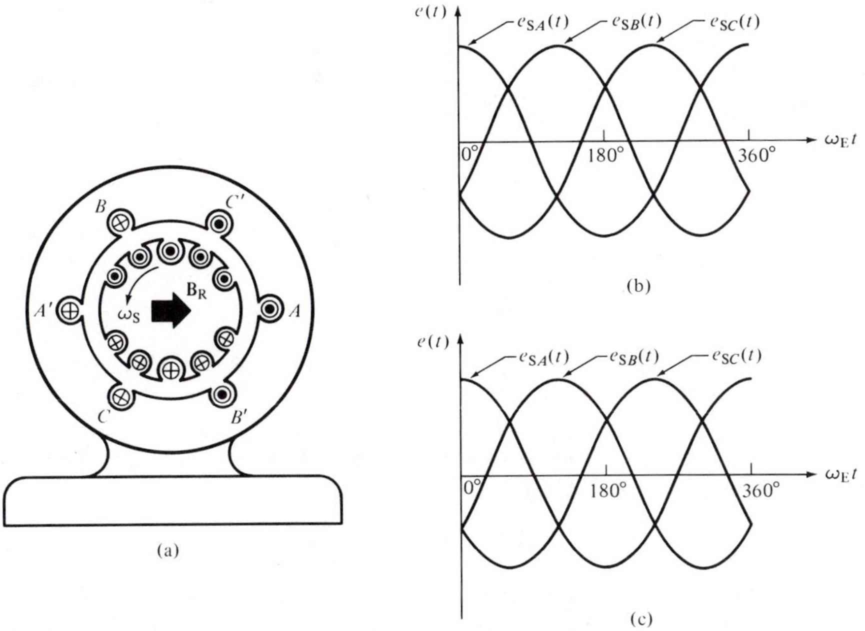

Fig. 5.5 (a) Cross Section of Three-Phase, Single-Layer, Full-Pitch, Concentrated Winding and Two-Pole, Smooth-Rotor, Synchronous Generator, (b) Positive Sequence of the Induced Voltages, $e_{SA}(t), e_{SB}(t), e_{SC}(t), \ldots$, and (c) Negative Sequence of the Induced Voltages $e_{SA}(t), e_{SC}(t), e_{SB}(t), \ldots$, Plotted for $\alpha = 0$

defined as full-pitch, double-layer, integral-slot, series-connected, concentrated windings. Note, however, that these windings could have been connected differently and be defined as fractional-pitch, single-layer, fractional-slot, concentrated (or distributed) windings. Before pursuing a mathematical analysis, it will be to our advantage to gain a qualitative understanding of how the synchronous generator functions. This can be accomplished by a careful study of Fig. 5.5. The device depicted in Fig. 5.5 is a synchronous generator with a three-phase, single-layer, full-pitch, concentrated stator winding and a two-pole smooth rotor with a distributed dc winding. This device is chosen for its relative simplicity when in a diagram form [compare Fig. 5.5(a) to Fig. 5.4].

As the rotor, driven by a prime mover, is rotating counterclockwise at synchronous speed ω_S, the conductors A, C', B, A', C, and B' move against the field $\mathbf{B}_R$ with a tangential velocity $\mathbf{u}$. For the present position of the rotor, that is, the rotor pole N under the stator conductor A and the rotor pole S under the stator conductor A', this motion leads to the presence of the voltage-producing, speed field intensities $\mathcal{E}_{RA}$ and $\mathcal{E}_{RA'}$ in the conductors A and A'. Mathematically, these speed field intensities are written as

$$\mathcal{E}_{RA} = \mathbf{u} \times \mathbf{B}_R \tag{5.2}$$

and

$$\mathcal{E}_{RA'} = \mathbf{u} \times \mathbf{B}_R \tag{5.3}$$

where the directions of $\mathbf{u}$ and $\mathbf{B}_R$ are considered at the corresponding coil sides A and A' (also B, B', C, and C').

Similarly, the voltage-producing, speed field intensities $\mathcal{E}_{RB}$, $\mathcal{E}_{RB'}$, $\mathcal{E}_{RC}$, and $\mathcal{E}_{RC'}$ induced in the conductors B, B', C, and C' (when the rotor pole N lies under the stator conductors B and C sequentially and the rotor pole S lies under the stator conductors B' and C' sequentially) can be written as

$$\mathcal{E}_{RB} = \mathbf{u} \times \mathbf{B}_{R} \tag{5.4}$$
$$\mathcal{E}_{RB'} = \mathbf{u} \times \mathbf{B}_{R} \tag{5.5}$$
$$\mathcal{E}_{RC} = \mathbf{u} \times \mathbf{B}_{R} \tag{5.6}$$

and

$$\mathcal{E}_{RC'} = \mathbf{u} \times \mathbf{B}_{R} \tag{5.7}$$

Utilization of Eqs. (5.2) through (5.7), $\mathbf{u}=\omega_S \times \mathbf{R}$, where $\mathbf{R}$ is the vector radius of the rotor at the center of the conductor, and ω_S the angular speed of the rotor (at the center of the conductor), normal to $\mathbf{R}$, $\omega_S = \omega_E/(P/2)$, and $e=\oint \mathcal{E} \cdot d\ell$ [see Eqs. (2.9), (2.10), and (2.23)], enables us to write the generalized equation of the induced voltages in phases A, B, and C as

$$e_{S_i}(t) = \phi\{|\mathbf{B}_R|, R, P, \omega_E t, \alpha\}, i = A, B, C \tag{5.8}$$

where $|\mathbf{B}_R|$ (in webers/meter^2) is the sinusoidally distributed field generated by the rotating smooth rotor excited by a dc source through two slip rings, R (in meters) is the radius of the rotor, P is the number of poles of the stator (always an even number), ω_E (in radians/second) is the radian frequency of the induced voltages in the stator windings, $\omega_E t$ (in radians) is the angular displacement of the induced voltages in the stator windings, and α (in radians or degrees) is the phase angle of the voltages at $t=0$. Qualitatively, for a counterclockwise rotation of the rotor, the voltages of Eq. (5.8) are plotted, for $\alpha \equiv 0$, in Fig. 5.5(b). These plots can be verified by following the rotor position as the rotor moves in the counterclockwise direction in 120-degree intervals, considering only the conductors in which current exits (i.e., coil sides A, B, and C) or the conductors in which current enters (i.e., coil sides A', B', and C'), or by following the rotor position as the rotor moves in the counterclockwise direction in 60-degree intervals, considering the voltages of the exiting conductors (i.e., coil sides A, B, and C) as positive and the voltages of the entering conductors (i.e., coil sides A', B', and C') as negative. Note that if the direction of rotation of the rotor is reversed, the voltages are plotted as in Fig. 5.5(c). The distinct difference between the plots of Figs. 5.5(b) and 5.5(c) is the sequence in which the maxima of the voltages in the phases A, B, and C are reached. The voltage sequence of Fig. 5.5(b), that is, $ABCABCAB\ldots$, is defined as a **positive sequence**. On the other hand, the voltage sequence of Fig. 5.5(c), $ACBACBAC\ldots$, is defined as a **negative sequence**.

Before closing it should be pointed out that the describing equations for the induced voltages in the coils of Fig. 5.5(a), which are plotted in Figs. 5.5(b) and 5.5(c) for $\alpha \equiv 0°$, can be written, if and only if $|\mathbf{E}_{SA}| = |\mathbf{E}_{SB}| = |\mathbf{E}_{SC}| \equiv |\mathbf{E}_G|$, as follows

1. **Positive sequence:**

$$e_{SA}(t) = \sqrt{2}\, |\mathbf{E}_G| \cos[\omega_E t - \alpha] \tag{5.9}$$
$$e_{SB}(t) = \sqrt{2}\, |\mathbf{E}_G| \cos[\omega_E t - (\alpha + 120°)] \tag{5.10}$$
$$e_{SC}(t) = \sqrt{2}\, |\mathbf{E}_G| \cos[\omega_E t - (\alpha + 240°)] \tag{5.11}$$

2. **Negative sequence:**

$$e_{SA}(t) = \sqrt{2}\,|\mathbf{E}_G|\cos[\omega_E t - \alpha] \tag{5.12}$$

$$e_{SC}(t) = \sqrt{2}\,|\mathbf{E}_G|\cos[\omega_E t - (\alpha + 120°)] \tag{5.13}$$

$$e_{SB}(t) = \sqrt{2}\,|\mathbf{E}_G|\cos[\omega_E t - (\alpha + 240°)] \tag{5.14}$$

5.3 THE EQUIVALENT CIRCUIT OF THE SYNCHRONOUS GENERATOR (INDUCED VOLTAGES IN THE STATOR WINDINGS): STEADY STATE OPERATION

The stator winding of the synchronous generator (as in the case of the rotor winding of the dc generator) is the source of all the ac energy utilized by the industry and by the household appliances of a modern society. Since (1) the synchronous generator is universally used by the electric power industry for supplying three-phase, as well as single-phase, power to its customers (the single phase power that is brought to our homes, shops, and offices originates from one-phase of a three-phase synchronous generator, with commercial loads assigned to each phase in order to keep the phases balanced), and (2) the synchronous motor is used widely in industry (used extensively for power-factor correction, etc.), it is essential that equivalent circuits are developed to provide the tools for the mathematical analysis of the synchronous devices themselves, as well as complex systems that incorporate these devices as components.

For the purposes of this chapter there is no distinction made between the smooth-rotor and salient-rotor synchronous devices in the derivation of the equivalent circuits for both the synchronous generator and motor and their related theory of analysis. It should be emphasized, however, that the saliency of the rotor will be taken into account in Section 6.6. There the appropriate describing equations and the equivalent circuits of the salient-rotor synchronous device operating in both modes are derived and other theories relevant to the salient-rotor synchronous device are developed. The theory of analysis developed in this chapter is based on the following two key assumptions: (1) the rotor of a synchronous energy converter is a smooth cylinder (or a multiple salient-rotor, which approximates a smooth cylinder), as in the case of turbine generators, and (2) saturation and hysteresis are neglected. Figure 5.6 exhibits some of the structural details of a three-phase, two-pole, N-turn winding synchronous generator with a sinusoidally distributed field. Figure 5.6(a) exhibits one, out of three, concentrated stator winding with N_S turns and pitch p ($0.666 \leq p \leq 1.000$) linking the sinusoidally distributed two-pole rotor field $\mathbf{B}_R$ and the development of the stator structure with its angular measurements all expressed in electrical degrees. The direction of the flux density $\mathbf{B}_R$ is assumed to be radial and positive when it exits the N pole of the **stator** structure.

In Fig. 5.6(a) the rotor poles (labeled N and S) and the rotor field plot, $\mathbf{B}_R$, move to the right while the stator windings (labeled 1, 2, and 3) are stationary. The relative angular position of the rotor with respect to the stator winding is ξ. In Fig. 5.6(b) ℓ (in meters) is the effective axial length of the stator, R (in meters) is the inner radius of the stator (neglecting slots), and $0° \leq \theta_S \leq 360°$ is the mechanical

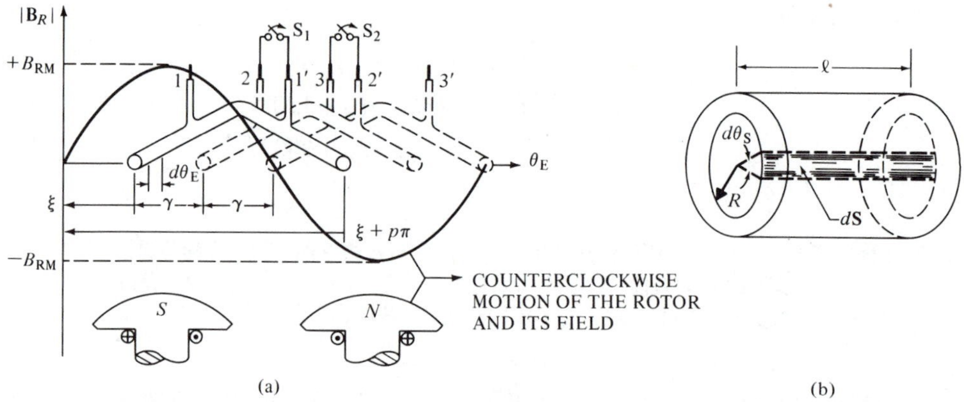

Fig. 5.6 (a) One N-Turn Concentrated Winding of a Three-Phase Synchronous Generator in a Two-Pole, Sinusoidally Distributed Rotor Field and (b) Stator Structure with Pertinent Dimensions

measure of the stator's inside circumference. This measure is related to the electrical measure θ_E of the field $\mathbf{B_R}$ by the equation

$$\omega_S = \frac{\omega_E}{P/2} \rightarrow \theta_S = \frac{\theta_E}{P/2} \tag{5.15}$$

The flux density at a given point θ_E is given by

$$\mathbf{B_R} = B_{RM} \sin\theta_E \mathbf{a_R} \tag{5.16}$$

where B_{RM} is the amplitude of the flux density, which is assumed to be time independent. The flux $d\phi_C$ (in webers) through the incremental area $d\mathbf{S}$, Fig. 5.6(b), that is,

$$d\mathbf{S} = \ell R\, d\theta_S \mathbf{a_R} \tag{5.17}$$

of the stator coil is given by

$$d\phi_C = \mathbf{B}\cdot d\mathbf{S} = (B_{RM}\sin\theta_E \mathbf{a_R})\cdot(\ell R\, d\theta_S \mathbf{a_R}) = \ell R B_{RM} \sin\theta_E\, d\theta_S \tag{5.18}$$

Substitution for θ_S from Eq. (5.15) into Eq. (5.18) yields the incremental flux $d\phi_C$ through $d\theta_E$ of the stator coil, as a function of the number of poles, of a P-pole device (note that p designates the pitch factor of the coil):

$$\int_0^{\Phi_C} d\phi_C = \left(\frac{2R\ell}{P}\right) B_{RM} \int_{\xi}^{\xi+p\pi} \sin\theta_E\, d\theta_E \tag{5.19}$$

Regarding the coil sides of Fig. 5.6(a) as filaments, integration of Eq. (5.19) over the limits $0 \leqslant \phi_C \leqslant \Phi_C$ and $\xi \leqslant \theta_E \leqslant (\xi + p\pi)$ yields the total flux that links the stator coil, written as

$$\Phi_C = -\left(\frac{2R\ell}{P}\right) B_{RM}[\cos(\xi + p\pi) - \cos\xi] \tag{5.20}$$

Utilization of the trigonometric identity $\cos x - \cos y = -2\sin[(x+y)/2]\sin[(y-x)/2]$ enables rewriting Eq. (5.20) as

$$\Phi_C = \left[\left(\frac{4R\ell}{P}\right) B_{RM} \sin p\left(\frac{\pi}{2}\right)\right] \sin\left[\xi + p\left(\frac{\pi}{2}\right)\right] \tag{5.21}$$

Since $\lambda = N\Phi$, the flux linkages of the stator coil are

$$\lambda_C = \left[\left(\frac{4R\ell}{P}\right) N_S B_{RM} \sin p\left(\frac{\pi}{2}\right)\right] \sin\left[\xi + p\left(\frac{\pi}{2}\right)\right] \qquad (5.22)$$

When the generator is driven by a prime mover at an angular velocity $\omega_S = \omega_E/(P/2)$, the angle ξ in Fig. 5.6(a) decreases. Thus setting $\xi \equiv -\omega_E t$ in Eq. (5.22) and extracting the minus sign of $\omega_E t$ from the parentheses yields

$$\lambda_C = -\left[\left(\frac{4R\ell}{P}\right) N_S B_{RM} \sin p\left(\frac{\pi}{2}\right)\right] \sin\left[\omega_E t - p\left(\frac{\pi}{2}\right)\right] \qquad (5.23)$$

Since $e_C(t) \equiv -d\lambda_C(t)/dt$, the voltage induced in the stator coil becomes

$$e_C(t) = \omega_E N_S \left(\frac{4R\ell}{P}\right) B_{RM} \sin p\left(\frac{\pi}{2}\right) \cos\left[\omega_E t - p\left(\frac{\pi}{2}\right)\right] \qquad (5.24)$$

Recognizing that the resultant flux per magnetic pole, Φ_P (in webers), can be written as

$$\Phi_P = (B_{AVG})(\underline{\underline{\text{pole area}}}) \qquad (5.25)$$

and that (1) $B_{AVG} = (2/\pi)B_{RM}$ and (2) $\underline{\underline{\text{pole area}}} = 2\pi R\ell/P$, the flux per pole, Eq. (5.25), can be written as

$$\Phi_P = \left(\frac{2}{\pi}\right)(B_{RM})\left(\frac{2\pi R\ell}{P}\right) = \left(\frac{4R\ell}{P}\right)(B_{RM})\left(\frac{\pi}{\pi}\right) = \left(\frac{4R\ell}{P}\right)(B_{RM}) \qquad (5.26)$$

Eq. (5.26) enables us to write Eq. (5.24) as

$$e_C(t) = \left[\omega_E N_S \Phi_P \sin p\left(\frac{\pi}{2}\right)\right] \cos\left[\omega_E t - p\left(\frac{\pi}{2}\right)\right] \qquad (5.27)$$

Defining $\sin p(\pi/2)$ in Eq. (5.27) as the **pitch factor** of the winding and designating it by k_p yields

$$k_p = \sin p\left(\frac{\pi}{2}\right) \qquad (5.28)$$

Example 5.2

A 90-slot, six-pole, three-phase, synchronous generator is wound with coils having $\frac{13}{16}$ fractional pitch. Calculate the pitch factor k_p.

Solution

Utilization of Eq. (5.28) yields

$$k_p = \sin p\left(\frac{\pi}{2}\right) = \sin\left(\frac{13}{16} \times \frac{180}{2}\right) = \sin 73.12° = \underline{\underline{0.9569}}$$

Utilization of Eq. (5.28) and rewriting $\omega_E = 2\pi f_E$ enables us to write Eq. (5.27) in terms of RMS magnitudes as

$$e_C(t) = \sqrt{2}\left(\frac{2\pi f_E N_S k_p \Phi_P}{\sqrt{2}}\right) \cos\left[\omega_E t - p\left(\frac{\pi}{2}\right)\right] \qquad (5.29)$$

Substitution for f_E from Eq. (5.1) and simplification enables writing Eq. (5.29) in terms of design parameters as

$$e_C(t) = \sqrt{2}\left(\frac{4.44 n_S P N_S k_p \Phi_P}{120}\right) \cos\left[\omega_E t - p\left(\frac{\pi}{2}\right)\right] \qquad (5.30)$$

Eq. (5.30) yields the voltage induced in a concentrated coil that constitutes one phase (of the three-phase synchronous generator), occupies one slot per pole, and has a pitch factor $k_p \leqslant 1$. In many devices, however, the winding that constitutes one phase is distributed, that is, it occupies more than one slot per pole; the individual coils that occupy the corresponding slots may be connected in series, in parallel, or in a series-parallel connection, depending on the number of poles and the general arrangement of the stator winding. However, the connections of all phases must be identical. If the switches S_1 and S_2 in Fig. 5.6(a) are closed, the three coils are connected in series and constitute one phase with ends 1–3′. For such a connection, however, the flux linkages of the coils occupying the six slots under a pair of poles differ since the angle ξ of Eq. (5.22) must have different values for the different coils that are displaced from each other, and from the first coil, by an angle γ. As a result, the flux linkages of the three coils and the corresponding induced voltages are not in time phase with each other; they lag each other by an angle γ. Thus for the flux density $\mathbf{B}_R$, that is sinusoidally distributed, the magnitude of the resultant flux linkage λ_C, or voltage E_C, of the three coils that constitute one phase, connected in series, Fig. 5.6(a), can be found by means of a corresponding phasor addition, as exhibited in Figs. 5.7(a) and 5.7(b) for an arbitrary location of the first coil and the remaining coils lagging by an angle γ. Note that the voltages in Fig. 5.7(b) are drawn 90 electrical degrees behind the flux linkages of Fig. 5.7(a) in accordance with Eq. (5.23) ($\sin[\omega_E t - p(\pi/2)]$) and Eq. (5.30) ($\cos[\omega_E t - p(\pi/2)]$).

The ratio of the magnitude of the vector sum of the voltages $|\mathbf{E}_C|$ to the arithmetic sum of the voltages $3|\mathbf{E}|$ ($|\mathbf{E}_1| = |\mathbf{E}_2| = |\mathbf{E}_3| \equiv |\mathbf{E}|$) is defined as the **breadth factor**, designated by k_b, that is, $k_b \equiv |\mathbf{E}_C|/3|\mathbf{E}|$. For n coils constituting the winding of each phase under each pole, the breadth factor is defined as

$$k_b \equiv \frac{|\mathbf{E}_C|}{n|\mathbf{E}|} \tag{5.31}$$

The breadth factor k_b can be written in terms of design parameters, that is, the

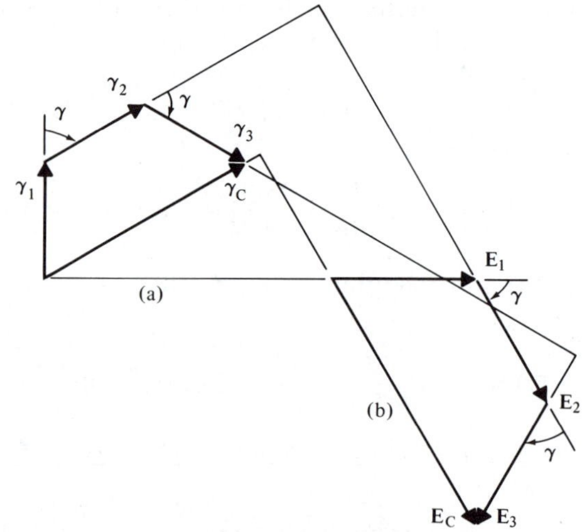

Fig. 5.7 Phasor Diagrams of Flux Linkages and Induced Voltages in the Individual Coils and the Total-Phase Distributed Winding of Fig. 5.6(a) When the Switches S_1 and S_2 Are Closed

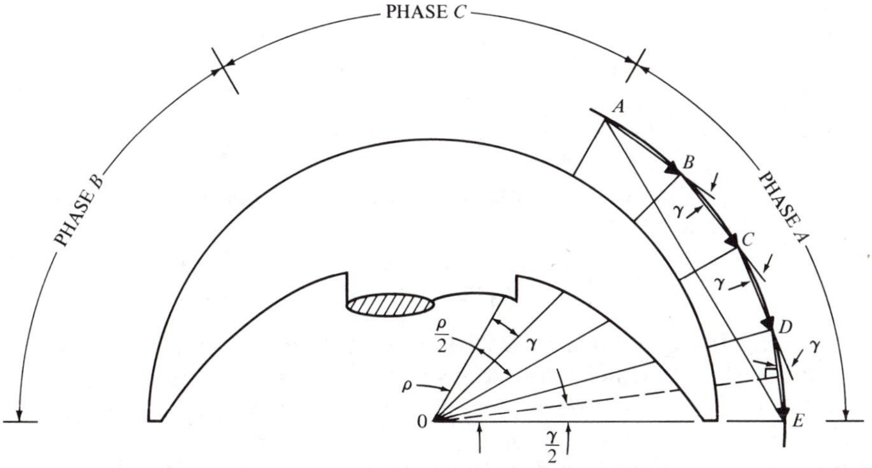

Fig. 5.8 Voltage Phasor Diagram of a Stator Phase Winding Distributed over Five Slots per Pole Used to Calculate the Breadth Factor, ρ Is the Angle Occupied by One Phase Under Each Pole and γ Is the Angle Between Adjacent Slots (the Pole Shape and Size Are Exaggerated to Achieve Drawing Clarity (compare to Fig. 5.4)

angle occupied by one phase under one pole ρ (in electrical degrees), Fig. 5.8, the angle between the adjacent slots γ (in electrical degrees), and the number of slots per pole per phase, n. Reference and correspondence of Eq. (5.31) to the geometry of Fig. 5.8, for five slots per pole per phase, lead to

$$k_b = \frac{|\mathbf{E}_C|}{n|\mathbf{E}|} = \frac{AE}{nAB} = \frac{2[OA\sin(\rho/2)]}{2[nOA\sin(\gamma/2)]}\bigg|_{n=5} = \frac{\sin(\rho/2)}{5\sin(\gamma/2)} \tag{5.32}$$

Example 5.3

From pertinent data given in Example 5.2, calculate the breadth factor k_b.

Solution

Utilization of Eq. (5.32), the theory thereof, and Fig. 5.8 enables calculation of

$$\rho = \frac{180 \text{ degrees}}{3 \text{ phases}} = \underline{60 \text{ degrees/phase}} \tag{1}$$

$$n = \frac{90 \text{ slots}}{6 \text{ poles} \times 3 \text{ phases}} = \frac{90 \text{ slots}}{18 \text{ pole-phase}} = \underline{5 \text{ slots/pole-phase}} \tag{2}$$

Thus

$$\gamma = \frac{\rho}{n} = \frac{60 \text{ degrees/phase}}{5 \text{ slots/pole phase}} = \underline{12 \text{ degrees/slot pole}} \tag{3}$$

and

$$k_b = \frac{\sin(60/2)}{5\sin(12/2)} = \frac{\sin 30^\circ}{5\sin 6^\circ} = \frac{0.5000}{5\times 0.1045} = \underline{\underline{0.9568}} \tag{4}$$

Eq. (5.32) is utilized to adjust Eq. (5.30), by multiplying Eq. (5.30) by Eq. (5.32), so that Eq. (5.30) yields practically the actual induced voltage in a distributed stator winding. Thus when we take into account both the pitch and breadth factors, we write Eq. (5.30) as

$$e_C(t) = \sqrt{2}\left(\frac{4.44 n_S P N_S k_p k_b \Phi_P}{120}\right)\cos\left[\omega_E t - p\left(\frac{\pi}{2}\right)\right] \tag{5.33}$$

If the coils of the stator winding are connected in parallel (instead of series, as in Fig. 5.4), forming **"a" parallel paths per phase**, the generalized equation of the induced voltage per phase of the synchronous generator, taking all refinements into account, is written as

$$e_C(t) = \sqrt{2}\left(\frac{4.44 n_S P N_S k_p k_b \Phi_P}{120a}\right)\cos\left[\omega_E t - p\left(\frac{\pi}{2}\right)\right] \tag{5.34}$$

Generalizing from Fig. 5.8, it can be stated that the coils that constitute each phase occupy one kth (for a k-phase device) of the slots under each pole. These coils may be connected in **series**, in **parallel**, or in **series-parallel**, depending on the design requirements.

In practical devices the flux density waveform is not an exact sinusoid. Thus it contains harmonics that may or may not be negligible. It can be shown (see reference [11]) that if we were to start with a flux density waveform whose harmonic content is indicated by the letter χ, that is, $B_\chi = B_{RM\chi}\sin(\chi\theta_E + \delta_\chi)$, we would be able to prove that the angles of the pitch and breadth factors, Eqs. (5.28) and (5.32), are multiplied by the factor χ; in fact, they become $k_{p\chi} = \sin[\chi p(\pi/2)]$ and $k_{b\chi} = [\sin(\chi\rho/2)]/[n\sin(\chi\gamma/2)]$ correspondingly (where $\chi = 3, 5, 7, \ldots$). Even harmonics are absent from the flux density waveform of the symmetry of the magnetic circuit. The harmonic components of the flux can be calculated by modifying Eq. (5.26) to read as $\Phi_{P\chi} = (4\ell R/\chi P)B_{RM\chi}$. Then appropriate substitution of $k_{p\chi}$, $k_{b\chi}$, $\Phi_{P\chi}$, $\omega_E t \rightarrow \chi\omega_E t + \gamma_\chi$, and $p(\pi/2) \rightarrow \chi P(\pi/2)$ into Eq. (5.33) or Eq. (5.34) enables calculation of the harmonic components of the coil voltage $e_C(t)$. The RMS value of the coil voltage $e_C(t)$, including the fundamental and its harmonics, is calculated from $|\mathbf{E}_{RMS}| = (|\mathbf{E}_1|^2 + |\mathbf{E}_3|^2 + |\mathbf{E}_5|^2 + \cdots + |\mathbf{E}_j|^2)^{1/2}$, where $j \equiv$ odd. Experience has shown (see reference [17]) that **distributed windings with fractional pitch** reduce the fundamental component of the voltage by about 8 percent and its harmonics by a greater percentage. Thus these two winding qualities tend to purify the induced voltage waveform, even though the flux density waveform might not be a perfect sinusoid. The necessity for a perfect sinusoidal voltage in the coils of a synchronous generator should be obvious.

The influence of the harmonics on the generated voltages of the balanced three-phase synchronous generator is examined here in some detail. The third harmonic and its multiple components, even though they might be appreciable in the phase voltages, become negligible in the line-to-line voltages. In the case of the Y-connected synchronous generators, the line-to-line voltage is the phasor difference between the corresponding phase voltages, for example, $\mathbf{E}_{LL(AB)} = \mathbf{E}_{AO} - \mathbf{E}_{BO}$, Eq. (3.46). This difference yields a line-to-line voltage $e_{LL(AB)}(t) = |\mathbf{E}_1|\{\sin\omega_E t - \sin(\omega_E t - 120°)\} + |\mathbf{E}_3|\{\sin 3\omega_E t - \sin 3(\omega_E t - 120°)\}$, which is independent of the third harmonic. (Why?) However, the rest of the harmonics, that is, all harmonics except the third and its multiples, if present in the line-to-line voltages are larger than they are in the phase voltages by a factor of $\sqrt{3}$, Eq. (3.50). In the case of Δ-connected synchronous generators, the third harmonic and its multiples (and all other harmonics), if present, cause corresponding currents to circulate in the windings. Their effects (heating, etc.) should be obvious. It is clear, then, that the

Y-connection is far more common for balanced synchronous generators than is the Δ-connection.

Comparison of Eq. (5.33) [or (5.34)] with Eq. (5.9) yields

$$|\mathbf{E}_G| \equiv \frac{4.44 n_S P N_S k_p k_b \Phi_P}{120a}\Bigg|_{\substack{a=1;\ \text{Eq. (5.33)}\\ a\neq 1;\ \text{Eq. (5.34)}}} \tag{5.35}$$

and

$$\alpha \equiv p\left(\frac{\pi}{2}\right) \tag{5.36}$$

Thus the generated voltage of phase A can be written as

$$\mathbf{E}_G = \frac{4.44 n_S P N_S k_p k_b \Phi_P}{120a} e^{-jp(\pi/2)} \tag{5.37}$$

The generated voltages of phases B and C have the same magnitude as phase A but the corresponding angles are $-[p(\pi/2)+120°]$ and $-[p(\pi/2)+240°]$.

If the coils of the stator windings, Fig. 5.5(a), are loaded with an arbitrary balanced load, sinusoidal currents, similar in nature to the voltages of Figs. 5.5(b) or 5.5(c) but with a phase angle ψ, exist throughout the stator coils. These currents are written (if and only if $|\mathbf{I}_{SA}|=|\mathbf{I}_{SB}|=|\mathbf{I}_{SC}|\equiv|\mathbf{I}_G|$), with the aid of Eqs. (5.9) through (5.14), as follows:

1. **Positive sequence:**

$$i_{SA}(t) = \sqrt{2}\,|\mathbf{I}_G|\cos[\omega_E t - \psi] \tag{5.38}$$

$$i_{SB}(t) = \sqrt{2}\,|\mathbf{I}_G|\cos[\omega_E t - (\psi + 120°)] \tag{5.39}$$

$$i_{SC}(t) = \sqrt{2}\,|\mathbf{I}_G|\cos[\omega_E t - (\psi + 240°)] \tag{5.40}$$

2. **Negative sequence:**

$$i_{SA}(t) = \sqrt{2}\,|\mathbf{I}_G|\cos[\omega_E t - \psi] \tag{5.41}$$

$$i_{SC}(t) = \sqrt{2}\,|\mathbf{I}_G|\cos[\omega_E t - (\psi + 120°)] \tag{5.42}$$

$$i_{SB}(t) = \sqrt{2}\,|\mathbf{I}_G|\cos[\omega_E t - (\psi + 240°)] \tag{5.43}$$

In Fig. 5.9(a) the current $i_{SA}(t)$ is arbitrarily shown to lag the voltage $e_{SA}(t)$ by an angle ψ. Note that the current is drawn to flow through a separate winding. This

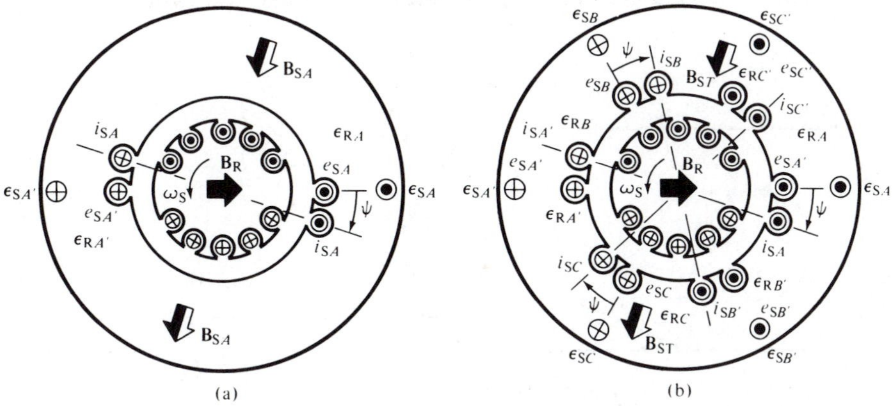

Fig. 5.9 Two-Pole, Stator-Field Generation of a Three-Phase Synchronous Generator: (a) Contribution of Phase A Only and (b) Contribution of Phases A, B, and C

is not the case in a real device, however, because the windings of each phase are distributed. Thus the speed voltages and their corresponding currents appear in different conductors of the same winding. The field $\mathbf{B}_{SA}$ caused by the current $i_{SA}(t)$ is normal to the current loop $i_{SA}(t)-i_{SA'}(t)$ and its direction, for the given current, is established by the right-hand rule. This field can be written arbitrarily as

$$\mathbf{B}_{SA} = B_{MA}\cos\beta_S \mathbf{u}_B \tag{5.44}$$

where β_S is the angular displacement of the stator phase-A field measured in electrical degrees, and B_{MA} is its maximum amplitude. In accordance with Eqs. (4.4) through (4.6), the field equations for phase B and phase C can be written as

$$\mathbf{B}_{SB} = B_{MB}\cos(\beta_S - 120°)\mathbf{u}_B \tag{5.45}$$

$$\mathbf{B}_{SC} = B_{MC}\cos(\beta_S - 240°)\mathbf{u}_B \tag{5.46}$$

Similarly, the total stator field due to currents $i_{SA}(t)$, $i_{SB}(t)$, and $i_{SC}(t)$ can be written in accordance with Eq. (4.16) as

$$\mathbf{B}_{ST} = \left(\frac{3K_S\sqrt{2}\,|\mathbf{I}_G|}{2}\right)\cos\left[\omega_E t-(\psi+\beta_S)\right]\mathbf{u}_B \tag{5.47}$$

This field is exhibited in Fig. 5.9(b). This figure represents the structure of a three-phase, two-pole, synchronous generator with the currents $i_{SA}(t)$, $i_{SB}(t)$, and $i_{SC}(t)$ [their amplitudes are obtained from Fig. 5.5(b) at $\omega_E t=0$ for $e(t)\to i(t)$ and $\alpha\to\psi$; *positive currents are designated by dotted circles* and *negative currents are designated by crossed circles*] lagging the corresponding voltages $e_{SA}(t)$, $e_{SB}(t)$, and $e_{SC}(t)$ by an arbitrary angle ψ. Note that Eq. (5.47) represents a P-pole generator if $(\psi+\beta_S)\to(p/2)(\psi_M+\beta_M)$.

As the field $\mathbf{B}_{ST}$ rotates at synchronous speed ω_S, in the direction of rotation, following the field $\mathbf{B}_R$, it induces a new voltage-producing speed field intensity designated by $\mathcal{E}_{Sk}=\mathbf{u}\times\mathbf{B}_{ST}$ $(k=A,B,C)$ in each of the stator windings. These new speed field intensities are drawn as separate circles with appropriate dots and crosses outside each coil side, Fig. 5.9(b). In reality, however, the speed field intensities $\mathcal{E}_{Rk}$ and $\mathcal{E}_{Sk}$ coexist within the same coil side and cause currents to appear to be flowing in the same direction through each coil once a closed path is established.

These ideas can be described best by referring to Fig. 5.10. Here the kth phase, $k=A$, B, and C, is exhibited connected to a load through the switch S. When the

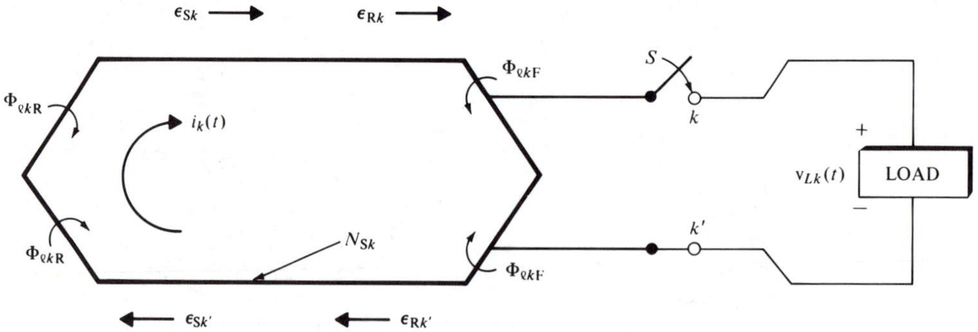

Fig. 5.10 Effects of the Two Voltage-Producing Total Field Intensities $\mathcal{E}_{RAT}$, and $\mathcal{E}_{SAT}$ on Phase A of a Three-Phase Synchronous Generator (Where, in General, $\mathcal{E}_{RkT}=\mathcal{E}_{Rk}+\mathcal{E}_{Rk}'$ and $\mathcal{E}_{SkT}=\mathcal{E}_{Sk}+\mathcal{E}_{Sk}'$; $k=k'=A,B,C$)

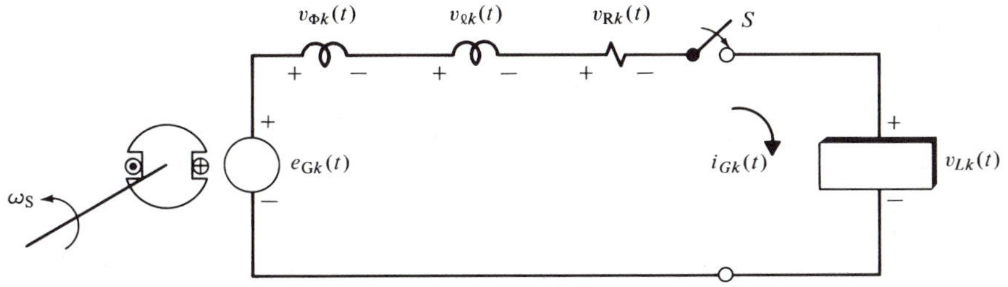

Fig. 5.11 Circuit Analog of the Cause and Effect Phenomena of Fig. 5.10

switch is open and the fields $\mathbf{B}_R$ and $\mathbf{B}_{ST}$, Fig. 5.9(b), are rotating counterclockwise, the speed field intensities $\mathcal{E}_{Rk}$ and $\mathcal{E}_{Sk}$ are oriented along the effective coil sides of the coil, as shown in Fig. 5.10. As the switch closes, the current $i_k(t)$ is established throughout the coil and the load. This current causes the following effects: (1) a voltage drop along the coil due to the resistance of the coil R_{Sk}, designated as $v_{Rk}(t)$, and (2) a flux leakage $\Phi_{\ell k}$ comprised of two components, the flux leakages at the *front* of the coil, designated as $\Phi_{\ell kF}$, and the flux leakages at the *rear* of the coil, designated as $\Phi_{\ell kR}$. These flux leakages multiplied by the number of turns of the stator coil, N_{Sk}, yield the coil's flux linkages $\lambda_{\ell k}=N_{Sk}\Phi_{\ell k}$. The time derivative of these flux linkages constitutes a voltage drop along the coil, designated as $v_{\ell k}=d\lambda_{\ell k}/dt$. In addition, the current causes a voltage drop across the load, designated as $v_{Lk}(t)$.

If we designate the effects of the speed field intensity $\mathcal{E}_{Rk}$ in the coil, that is, the current generation, by a voltage generator, designated by $e_{Gk}(t)$, and the effects of $\mathcal{E}_{Sk}$ as the tendency to aid the current $i_{Gk}(t)$—and thus to cause an increased voltage drop along the coil, designated by $v_{\phi k}(t)[\lambda_{\phi k}=L_{\phi k}i_k(t)]$—then we can view the effects of the two fields on the kth phase, and its response to the two fields, from the circuit analog point of view, as exhibited in Fig. 5.11. Since, for the steady state operation of a balanced three-phase synchronous generator, all voltages and currents are sinusoidal and of equal magnitudes, and displaced from each other by 120 positive electrical degrees, we can utilize the principles just set forth and draw the circuit analog of Fig. 5.9(b) **for the steady state operation of a balanced synchronous generator,** as exhibited in Fig. 5.12. For balanced operation all the circuit parameters and voltages and current magnitudes of Fig. 5.12 are identical; the phasor voltages, currents, and the coil axes are displaced from each other by 120 positive electrical degrees.

Utilizing the theory given in Chapters 3 and 4, we can view the performance of the balanced, three-phase, P-pole, synchronous generator through the single-phase diagram of Fig. 5.13, where (1) $\mathbf{E}_G$ is defined as the *generated phase voltage* [it is given in terms of design parameters in Eq. (5.37)], (2) $\mathbf{I}_G$ is defined as the *generated phase current*, (3) R_S is the *resistance of the winding* that constitutes each stator phase, (4) jX_ℓ is the *leakage reactance of the winding* that constitutes the stator phase, and (5) jX_ϕ is the **armature reaction reactance** *of the effective length of the winding* that constitutes each stator phase. *It (i.e., jX_ϕ) is a measure of the reaction of the stator (also known as armature*) coils to the stator's own synchronously rotating field $\mathbf{B}_{ST}$. The quantity $j(X_\ell+X_\phi)$ is defined as the* **synchronous reactance** *of*

*Armature is defined to be the member of the dc or ac generator in which voltages are induced.

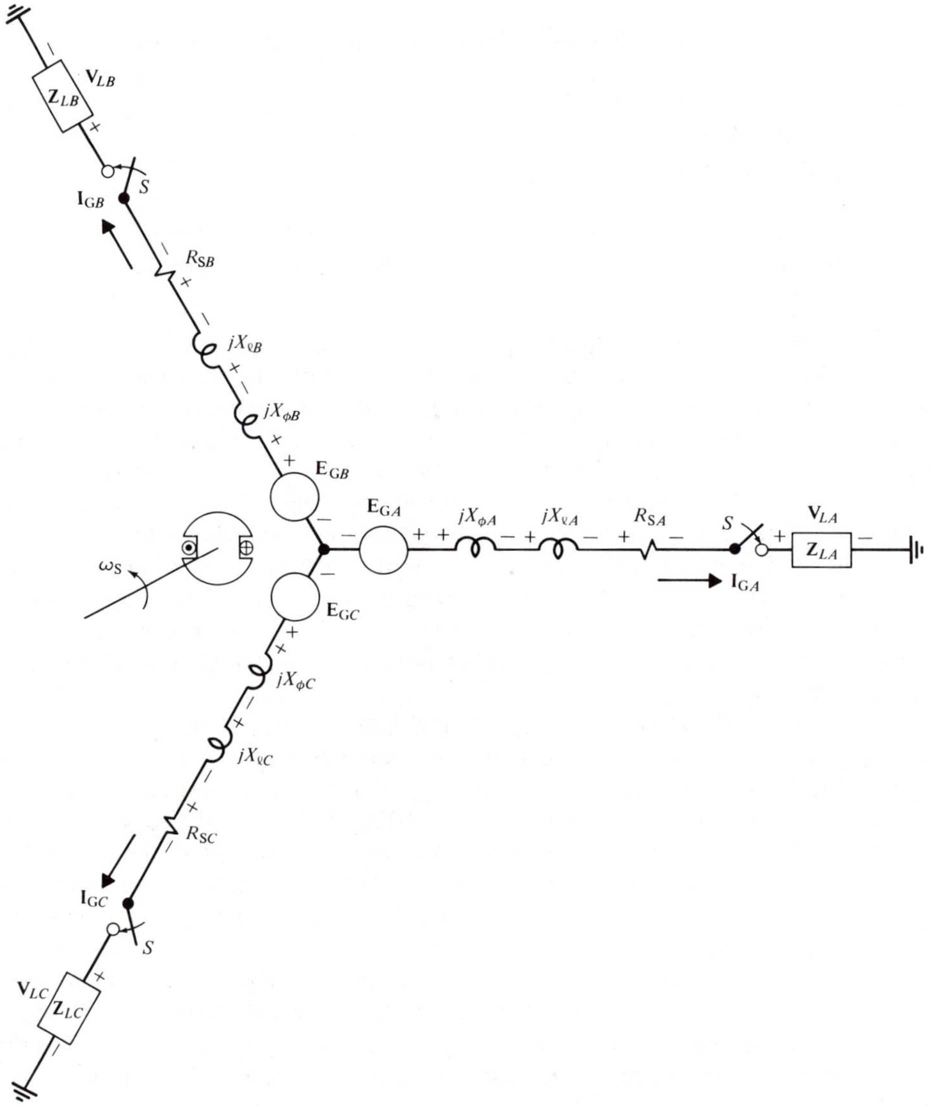

Fig. 5.12 Circuit Analog of Fig. 5.9(b), Depicting the Steady State Operation of a Balanced, Three-Phase, *P*-Pole, Synchronous Generator

the synchronous generator (and motor); it is designated by jX_S, *and its defining mathematical equation is*

$$jX_S \equiv j(X_\ell + X_\phi) \tag{5.48}$$

The quantity $R_S + jX_S$ is defined as the **synchronous impedance** of the synchronous generator (and motor); it is designated by Z_S, and its defining mathematical equation is

$$Z_S \equiv R_S + jX_S \tag{5.49}$$

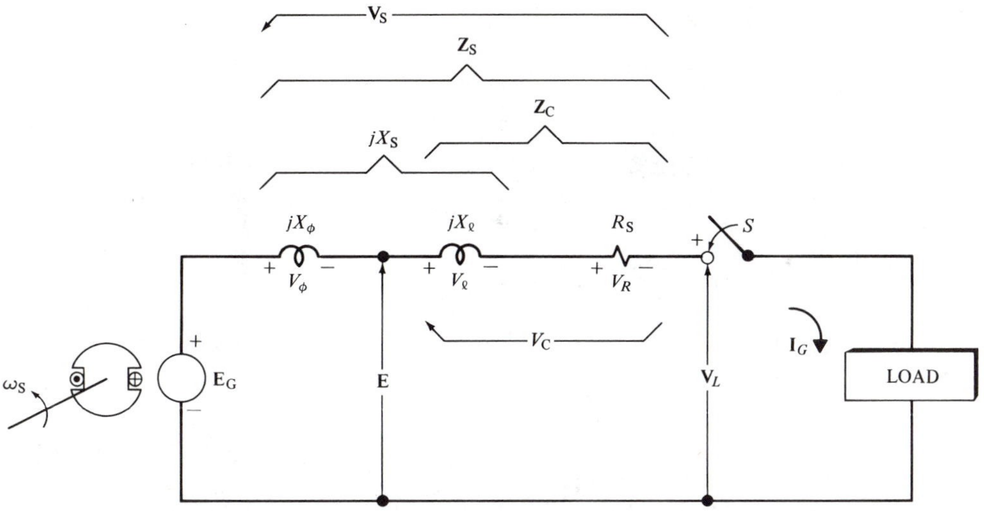

Fig. 5.13 The Single-Phase, Steady State, Equivalent Circuit of a Balanced, Three-Phase, *P*-Pole, Synchronous Generator as Utilized in Its Per-Phase Analysis; All Pertinent Circuit Parameters and Voltages Are Explicitly Defined

However, the quantity $R_S + jX_\ell$ is defined as the **stator-coil impedance**; it is designated by Z_C, and its defining mathematical equation is

$$Z_C = R_S + jX_\ell \tag{5.50}$$

It should be understood here that the synchronous reactance, jX_S, is a fictitious quantity that represents, in terms of phase circuit parameters, the effects of (1) the flux leakage $\Phi_{\ell k}$ ($k = A$, B, and C) and (2) the speed field intensity $\mathcal{E}_{Sk}$, due to the rotating stator field $\mathbf{B}_{ST}$, on each stator phase.

Before closing this section the following observations should be made with respect to Fig. 5.9(b): (1) as the rotor rotates at a synchronous speed ω_S, the field $\mathbf{B}_R$ is also rotating at a synchronous speed ω_S; (2) the field $\mathbf{B}_{ST}$ is also rotating at a synchronous speed ω_S, lagging the field $\mathbf{B}_R$ by an angle $90° + \psi$. These two fields react upon each other and produce a torque opposing the direction of rotation. This torque is designated by a negative sign. In addition, the relative position of the stator field $\mathbf{B}_{ST}$ to the rotor field $\mathbf{B}_R$ results in the following three overall effects on the device: (1) demagnetizing* effect, that is, the angle between the two fields is greater than 90°, Fig. 5.14(a); thus the component $\mathbf{B}_{ST}^{d-}$ of the field $\mathbf{B}_{ST}$ is inhibiting $\mathbf{B}_R$ [contrary to aiding it as does $\mathbf{B}_{ST}^{d+}$, Fig. 5.14(c)]; (2) neutral, that is, the angle between the two fields $\mathbf{B}_R$ and $\mathbf{B}_{ST}$ is exactly 90°, Fig. 5.14(b); and (3) magnetizing, that is, the angle between the two fields is less than 90°, Fig. 5.14(c). Thus the component $\mathbf{B}_{ST}^{d+}$ of the field $\mathbf{B}_{ST}$ is aiding $\mathbf{B}_R$ [contrary to inhibiting it as does $\mathbf{B}_{ST}^{d-}$ in Fig. 5.14(a)]. It should be obvious from Fig. 5.9(b) that the relative position of the two fields $\mathbf{B}_R$ and $\mathbf{B}_{ST}$ depends on the angle ψ, that is, the angle between the speed-dependent phase voltage $e_j(t)$ and the response phase current $i_j(t)$, $j = A$, B, and C.

*The terms "magnetizing" and "demagnetizing" refer to the net increase or decrease of the rotor field $\mathbf{B}_R$ due to a component of the stator field $\mathbf{B}_{ST}$ in the direction of the rotor field $\mathbf{B}_R$, Fig. 5.14.

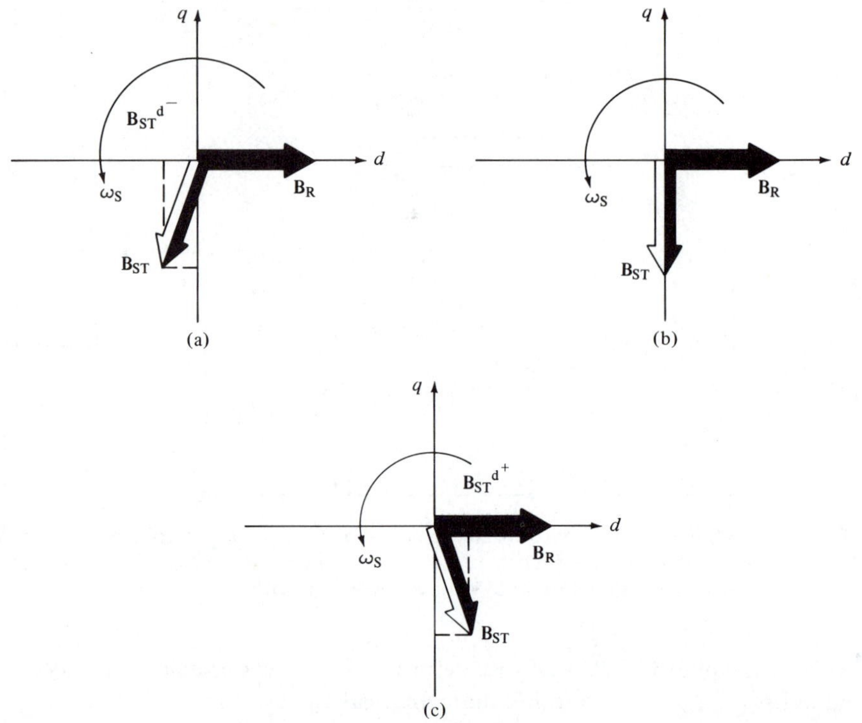

Fig. 5.14 The Rotating Fields in a Synchronous Generator and Their Corresponding Effects on the Device: (a) Demagnetizing, (b) Neutral, and (c) Magnetizing

5.4 GENERATOR MODE OF OPERATION: PHASOR DIAGRAM ANALYSIS

For the balanced, three-phase, P-pole, synchronous generator (i.e., generated phase voltages of equal amplitude but out of phase by 120 electrical degrees; equal phase resistances, and loads; and response currents of equal amplitude but out of phase by 120 electrical degrees), the mathematical analysis can be performed on a per-phase basis, utilizing the theories given in Appendix A and in Chapters 3 and 4. Thus the circuit analog of a given phase of the balanced, three-phase, P-pole, synchronous generator is adequately represented by the per-phase equivalent circuit of Fig. 5.13 if the load is defined explicitly, that is, $\mathbf{Z}_L = R_L + jX_L$, or is defined in terms of its voltampere rating and its power factor, $\cos\theta_L$.

If the circuit parameters $\mathbf{Z}_S$ and $\mathbf{Z}_L$ and the phase-generated voltage $\mathbf{E}_G$ of Fig. 5.13 are known explicitly, the mathematical analysis of the synchronous generator is a straightforward process. The generated current, $\mathbf{I}_G$, can be calculated, utilizing circuit analysis theory, as follows:

$$\mathbf{I}_G = \frac{\mathbf{E}_G}{\mathbf{Z}_S + \mathbf{Z}_L} = \frac{\mathbf{E}_G}{(R_S + R_L) + j(X_S + X_L)} \tag{5.51}$$

Once the current $\mathbf{I}_G$ is known, the voltages $\mathbf{V}_\phi$, $\mathbf{V}_\ell$,$\mathbf{V}_R$, $\mathbf{V}_C$, $\mathbf{V}_S$, $\mathbf{E}_{\mathfrak{R}}$ ($\mathbf{E}_{\mathfrak{R}}$, defined as the **airgap potential rise**, designates the total speed-dependent voltage along the active length of the stator coil, due to the synchronously rotating fields $\mathbf{B}_R$ and $\mathbf{B}_{ST}$,

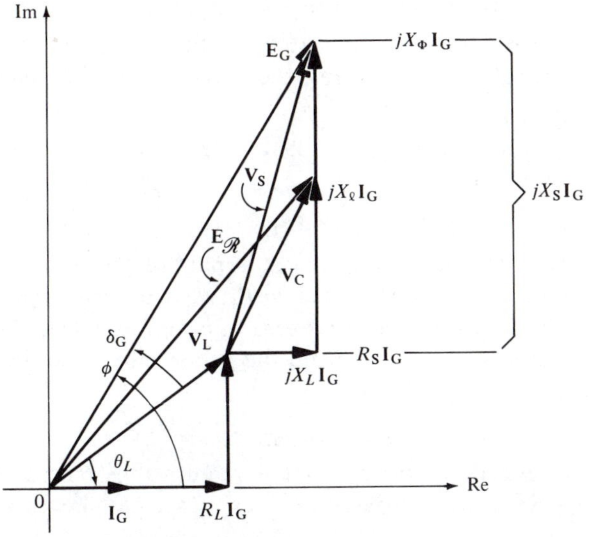

Fig. 5.15 Phasor Diagram Exhibiting the Relative Position of the Variables of Eqs. (5.52) Through (5.58); the Loading of the Generator Is Explicitly Defined

for the generator mode of operation), and $\mathbf{V}_L$ are calculated utilizing Eqs. (5.52) through (5.58). Thus

$$\mathbf{V}_\phi = jX_\phi \mathbf{I}_G \tag{5.52}$$

$$\mathbf{V}_\ell = jX_\ell \mathbf{I}_G \tag{5.53}$$

$$\mathbf{V}_R = R_S \mathbf{I}_G \tag{5.54}$$

$$\mathbf{V}_C = Z_c \mathbf{I}_G = (R_S + jX_\ell)\mathbf{I}_G \tag{5.55}$$

$$\mathbf{V}_S = \{R_S + j(X_\phi + X_\ell)\}\mathbf{I}_G \equiv \{R_S + jX_S\}\mathbf{I}_G \tag{5.56}$$

$$\mathbf{E}_\mathscr{R} = \mathbf{E}_G - \mathbf{V}_\phi = \mathrm{Re}(E_G) + j[\mathrm{Im}(E_G) - X_\phi \mathbf{I}_G] \tag{5.57}$$

and

$$\mathbf{V}_L = \mathbf{Z}_L \mathbf{I}_G = (R_L + jX_L)\mathbf{I}_G \tag{5.58}$$

The relationships among the various voltages of Eqs. (5.52) through (5.58), and their relative positions in the complex plane, are exhibited in Fig. 5.15. Also, utilization of the current of Eq. (5.51) enables calculation of the phase power generated by the generator, designated by $\mathbf{S}_G$, the phase power required by the stator winding, designated by $\mathbf{S}_S$, and the phase power demanded by the load, designated by $\mathbf{S}_D$. The power $\mathbf{S}_D$ is identical to the output power of the generator, $\mathbf{S}_O$, defined as $\mathbf{S}_O \equiv \mathbf{V}_L \mathbf{I}_G^*$, $\mathbf{I}_G \equiv \mathbf{I}_L$. Thus*

$$\mathbf{S}_G = P_G + jQ_G \equiv \mathbf{E}_G \mathbf{I}_G^* = |\mathbf{S}_G|\cos\phi + j|\mathbf{S}_G|\sin\phi \tag{5.59}$$

$$\mathbf{S}_S = P_S + jQ_S \equiv \mathbf{Z}_S |\mathbf{I}_G|^2 = (R_S + jX_S)|\mathbf{I}_G|^2 = \{R_S + j(X_\ell + X_\phi)\}|\mathbf{I}_G|^2 \tag{5.60}$$

and

$$\mathbf{S}_D \equiv P_D + jQ_D = \mathbf{V}_L \mathbf{I}_G^* = \mathbf{Z}_L |\mathbf{I}_G|^2 = |\mathbf{S}_D|\cos\theta_L + j|\mathbf{S}_D|\sin\theta_L \tag{5.61}$$

***ϕ_C was used earlier in this chapter to designate the flux linking the stator coils. However, this use was very limited. Here the unsubscripted ϕ designates the angle of the voltage $\mathbf{E}_G$, and it is used extensively with this meaning. Please refer to Figs. 5.15 through 5.18 and Figs. 5.28 through 5.32.**

Once the phase powers $\mathbf{S}_G$, $\mathbf{S}_S$, and $\mathbf{S}_D$ are calculated, the total generated power, $\mathbf{S}_{GT}$, the total power required by the stator winding, $\mathbf{S}_{ST}$, and the total demanded power, $\mathbf{S}_{DT}$, can be calculated utilizing the theory given in Chapters 3 and 4 as follows:

$$\mathbf{S}_{GT} = P_{GT} + jQ_{GT} \equiv 3\mathbf{S}_G = 3P_G + j3Q_G = \cdots \tag{5.62}$$

$$\mathbf{S}_{ST} = P_{DT} + jQ_{ST} \equiv 3\mathbf{S}_S = 3P_S + j3Q_S = \cdots \tag{5.63}$$

$$\mathbf{S}_{DT} = P_{DT} + jQ_{DT} \equiv 3\mathbf{S}_D = 3P_D + j3Q_D = \cdots \tag{5.64}$$

The real part of the total generated power is part of the total mechanical power provided by the prime mover, designated as P_{INM}. If the *total windage power losses*, P_{WLT}, the *total bearing-friction power losses*, P_{BFT}, and *total core power losses*, P_{CLT}, are known, then the power required from the prime mover can be calculated from the following equation:

$$P_{GT} = P_{INM} - (P_{WLT} + P_{BFT} + P_{CLT}) \tag{5.65}$$

Since power is related to torque by the relationship $P = \tau\omega$, the total torque provided by the prime mover, designated as τ_{INM}, can be calculated from the following equation:

$$P_{INM} = \tau_{INM}\omega_S \tag{5.66}$$

The efficiency, η, and voltage regulation, VR, of the synchronous generator, for loading conditions where the loading impedance is explicitly defined, are calculated as follows:

$$\eta = \frac{P_{DT}}{P_{INM}} \times 100 \tag{5.67}$$

and

$$\mathrm{VR} = \frac{|\mathbf{V}_{NL}| - |\mathbf{V}_{FL}|}{|\mathbf{V}_{FL}|} \times 100 \Bigg|_{|\mathbf{V}_{NL}| \equiv |\mathbf{E}_G|;\ |\mathbf{V}_{FL}| \equiv |\mathbf{Z}_L||\mathbf{I}_G|} \tag{5.68}$$

where $\mathbf{V}_{NL}$ and $\mathbf{V}_{FL}$ are defined as (1) the voltage drop across the output terminals of the generator when the loading terminals are open-circuited, thus yielding $\mathbf{V}_{NL} = \mathbf{E}_G$, and (2) the voltage drop across the output terminals of the generator when the full load, $\mathbf{Z}_L$, is connected to it, thus yielding $\mathbf{V}_{FL} = \mathbf{Z}_L\mathbf{I}_G$.

Example 5.4

The per-phase parameters of a synchronous generator are as follows: $\mathbf{E}_G = 14{,}800e^{j62.42°}$ V, $X_\phi = 3.04\ \Omega$, $X_\ell = 0.36\ \Omega$, and $R_S = 0.20\ \Omega$. If the loading impedance is $\mathbf{Z}_L = 2.53 + j1.826 = 3.12e^{j35.82°}\ \Omega$, and the windage, bearing friction, and core losses are negligible, calculate the following per-phase quantities: (1) the synchronous reactance X_S, (2) the synchronous impedance $\mathbf{Z}_S$, (3) the stator coil impedance $\mathbf{Z}_C$, (4) the voltages $\mathbf{V}_\phi$, $\mathbf{V}_\ell$, $\mathbf{V}_R$, $\mathbf{V}_C$, $\mathbf{V}_S$, and $\mathbf{V}_L$ and the airgap potential rise $\mathbf{E}_{\mathfrak{R}}$, (5) the generated power $\mathbf{S}_G$, (6) the power required by the stator winding, $\mathbf{S}_S$, (7) the power demanded by the load, $\mathbf{S}_D$, (8) the total powers $\mathbf{S}_{GT}$, $\mathbf{S}_{ST}$, and $\mathbf{S}_{DT}$, (9) the required input power from the prime mover P_{INM}, (10) the efficiency η, and (11) the voltage regulation VR.

Solution

Utilization of Eq. (5.48) yields the synchronous reactance X_S as

$$jX_S = j(X_\phi + X_\ell) = j(3.04 + 0.36) = \underline{\underline{j3.40\ \Omega}} \tag{1}$$

In turn, Eq. (5.49) yields the synchronous impedance $\mathbf{Z}_S$ as

$$\mathbf{Z}_S = R_S + jX_S = 0.20 + j3.40 = \underline{\underline{3.41e^{j86.63^\circ}\ \Omega}} \tag{2}$$

and Eq. (5.50) yields the stator coil impedance $\mathbf{Z}_C$ as

$$\mathbf{Z}_C = R_S + jX_\ell = 0.20 + j0.36 = \underline{\underline{0.412e^{j60.95^\circ}\ \Omega}} \tag{3}$$

Utilization of Eq. (5.51) yields the phase-generated current $\mathbf{I}_G$ as

$$\begin{aligned} \mathbf{I}_G &= \frac{\mathbf{E}_G}{\mathbf{Z}_S + \mathbf{Z}_L} = \frac{14{,}800e^{j62.42^\circ}}{(0.20 + j3.40) + (2.53 + j1.826)} \\ &= \frac{14{,}800e^{j62.42^\circ}}{2.73 + j5.226} = \frac{14{,}800e^{j62.42^\circ}}{5.896e^{j62.42^\circ}} = \underline{2{,}510.18e^{j0^\circ}\ \text{A}} \end{aligned} \tag{4}$$

Now utilization of the magnitude of the generated current $|\mathbf{I}_G| = 2{,}510.18$ and Eqs. (5.52) through (5.58) (also, see Fig. 5.15) yields

$$\mathbf{V}_\phi = jX_\phi \mathbf{I}_G = j3.04 \times 2{,}510.18 = \underline{\underline{j7{,}630.93}}\ \text{V} \tag{5}$$

$$\mathbf{V}_\ell = jX_\ell \mathbf{I}_G = j0.36 \times 2{,}510.18 = \underline{\underline{j903.66}}\ \text{V} \tag{6}$$

$$\mathbf{V}_R = R_S \mathbf{I}_G = 0.20 \times 2{,}510.18 = \underline{\underline{502.04}}\ \text{V} \tag{7}$$

$$\mathbf{V}_C = Z_C \mathbf{I}_G = (0.412e^{j60.95^\circ}) \times 2{,}510.18 = \underline{\underline{1{,}034.19e^{j60.95^\circ}}}\ \text{V} \tag{8}$$

$$\mathbf{V}_S = \mathbf{Z}_S \mathbf{I}_G = (3.40e^{j86.63^\circ}) \times 2{,}510.18 = \underline{\underline{8{,}534.61e^{j86.63^\circ}}}\ \text{V} \tag{9}$$

$$\mathbf{V}_L = \mathbf{Z}_L \mathbf{I}_G = (3.12e^{j35.82^\circ}) \times 2{,}510.18 = \underline{\underline{7{,}831.76e^{j35.82^\circ}\ \text{V}}} \tag{10}$$

$$\begin{aligned} \mathbf{E}_{\mathcal{R}} &= \mathbf{E}_G - \mathbf{V}_\phi = 14{,}800e^{j62.42^\circ} - j7{,}630.95 \\ &= 6{,}852.20 + j13{,}118.21 - j7{,}630.95 \\ &= 6{,}852.20 + j5{,}487.26 = \underline{\underline{8{,}778.53e^{j38.69^\circ}}}\ \text{V} \end{aligned} \tag{11}$$

Once the voltages $\mathbf{V}_L$ and $\mathbf{E}_G$ are explicitly known, that is, Eq. (10) and input data, the phase angle between these voltages, defined as the power angle and designated as δ_G, is calculated as

$$\delta_G = \angle \mathbf{E}_G - \angle \mathbf{V}_L = 62.42^\circ - 35.82^\circ = \underline{26.60^\circ} \tag{12}$$

The generated power, $\mathbf{S}_G$, is calculated with the aid of Eq. (5.59) as follows:

$$\begin{aligned} \mathbf{S}_G &= \mathbf{E}_G \mathbf{I}_G{}^* = (14{,}800e^{j62.42^\circ}) \times 2{,}510.18 = 37{,}150{,}664.00e^{j62.42^\circ} \\ &= \underline{\underline{17{,}200{,}261.97 + j32{,}929{,}057.44\ \text{VA}}} \end{aligned} \tag{13}$$

The power required by the stator winding, $\mathbf{S}_S$, is calculated with the aid of Eq. (5.60) as follows:

$$\begin{aligned} \mathbf{S}_S &= \mathbf{Z}_S |\mathbf{I}_G|^2 = (3.41e^{j86.63^\circ}) \times 2{,}510.18^2 \\ &= \underline{\underline{21{,}486{,}422.39e^{j86.64^\circ} = 1{,}2630{,}51.57 + j21{,}449{,}266.86}}\ \text{VA} \end{aligned} \tag{14}$$

Similarly, the power demanded by the load, $\mathbf{S}_D$, is calculated with the aid of Eq. (5.61) as

$$\begin{aligned} \mathbf{S}_D &= \mathbf{V}_L \mathbf{I}_G{}^* = \mathbf{Z}_L |\mathbf{I}_G|^2 = (2.53 + j1.826) \times (2{,}510.18)^2 \\ &= \underline{\underline{15{,}941{,}539.19 + j11{,}505{,}632.63 = 19{,}659{,}914.9e^{j35.82^\circ}\ \text{VA}}} \end{aligned} \tag{15}$$

Utilization of Eqs. (5.62) through (5.64) enables calculation of the total powers, the power generated, $\mathbf{S}_{GT}$, the power required by the stator winding, $\mathbf{S}_{ST}$, and the power demanded by the load, $\mathbf{S}_{DT}$, as follows:

$$\mathbf{S}_{GT} = 3\mathbf{S}_G = 3 \times 37{,}150{,}664.00e^{j62.42°} = \underline{\underline{111{,}451{,}992e^{j62.42°}\text{ VA}}} \quad (16)$$

$$\mathbf{S}_{ST} = 3\mathbf{S}_S = 3 \times 21{,}486{,}422.39e^{j86.63°} = \underline{\underline{6{,}449{,}267.17e^{j86.63°}\text{ VA}}} \quad (17)$$

$$\mathbf{S}_{DT} = 3\mathbf{S}_D = 3 \times 19{,}659{,}914.9e^{j35.82°} = \underline{\underline{58{,}979{,}744.7e^{j35.82°}\text{ VA}}} \quad (18)$$

For zero total losses (i.e., $P_{WLT} + P_{BFT} + P_{CLT} = 0$), Eq. (5.65), in conjunction with Eq. (17), yields the total input power from the prime mover, P_{INM}, as

$$P_{INM} = P_{GT} = \text{Re}(\mathbf{S}_{GT}) = \text{Re}(111{,}451{,}992e^{j62.42°}) = \underline{\underline{51{,}600{,}785.92\text{ W}}} \quad (19)$$

The efficiency of the generator, as calculated using Eq. (5.67) is

$$\eta = \frac{P_{DT}}{P_{INM}} \times 100 = \frac{\text{Re}(\mathbf{S}_{DT})}{\text{Re}(\mathbf{S}_{GT})} \times 100 = \frac{47{,}824{,}291.07}{51{,}600{,}785.92} \times 100 = \underline{\underline{92.68\%}} \quad (20)$$

Finally, the voltage regulation, VR, of the generator is calculated with the aid of Eq. (5.68) as follows:

$$\text{VR} = \frac{|\mathbf{V}_{NL}| - |\mathbf{V}_{FL}|}{|\mathbf{V}_{FL}|}\Bigg|_{|\mathbf{V}_{FL}| = |\mathbf{Z}_L||\mathbf{I}_G|;\ |\mathbf{V}_{NL}| = |\mathbf{E}_G|} \times 100$$

$$= \left(\frac{14{,}800 - 7{,}831.76}{7{,}831.76}\right) \times 100 = \frac{6{,}968.24}{7{,}831.76} \times 100 = \underline{\underline{88.97\%}} \quad (21)$$

If $\mathbf{Z}_S$ and $\mathbf{Z}_C$ are explicitly defined, and if the load is defined in terms of its voltampere rating and its power factor, $\cos\theta_L$, the mathematical analysis of the synchronous generator, that is, the calculation of $\mathbf{E}_G$, $\mathbf{S}_G$, and so on, becomes more challenging. This happens because the power factor can take values that yield three distinct cases of mathematical analysis: (1) unity power-factor analysis, (2) lagging power-factor analysis, and (3) leading power-factor analysis. These three analyses are pursued in accordance with the theory of Appendix A.

Unity Power-Factor (Designated by $\cos\theta_L \equiv 1.0$) Analysis

Since a unity power factor implies the load current $\mathbf{I}_L \equiv \mathbf{I}_G$, is in phase with the load voltage $\mathbf{V}_L$, that is, $\theta_L = 0$, the phasors $\mathbf{V}_L$ and $\mathbf{I}_G$ (for the circuit exhibited in Fig. 5.13, with unity power-factor loading, i.e., the generator is loaded with a **resistively behaving load**) are drawn as in Fig. 5.16. However, in order to complete the phasor diagram of Fig. 5.16, and thus to determine the desired phase-generated voltage $\mathbf{E}_G$, the describing equation of Fig. 5.13 is written, utilizing Kirchhoff's voltage law, as

$$\mathbf{E}_G = R_S\mathbf{I}_G + j(X_\ell + X_\phi)\mathbf{I}_G + \mathbf{V}_L \quad (5.69)$$

or,

$$\mathbf{E}_G = \mathbf{Z}_S\mathbf{I}_G + \mathbf{V}_L \equiv (\mathbf{V}_L + R_S\mathbf{I}_G) + jX_S\mathbf{I}_G \equiv |\mathbf{E}_G|e^{j\phi} \quad (5.70)$$

where, as mentioned earlier and repeated here for clarity,

$$\mathbf{Z}_S = R_S + jX_S \quad (5.71)$$

and

$$X_S = X_\phi + X_\ell \quad (5.72)$$

The relative positions of the phasors of Eq. (5.70) are depicted in Fig. 5.16. Utilization of Fig. 5.16 also enables us to write the equations for the voltage drops $\mathbf{V}_S$ across the synchronous impedance $\mathbf{Z}_S$ and $\mathbf{V}_C$ across the stator coil impedance

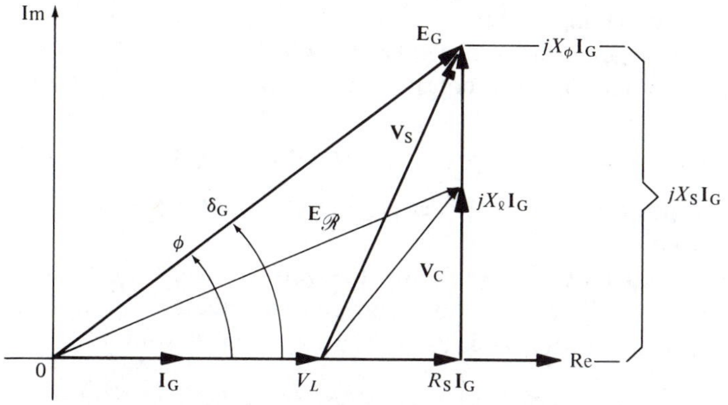

Fig. 5.16 Phasor Diagram for the Per-Phase Analysis of Synchronous Generators with Unity Power-Factor Loading, PF$=k$

$\mathbf{Z}_C$ and the airgap potential rise, $\mathbf{E}_{\mathcal{R}}$, as follows:

$$\mathbf{V}_S = \mathbf{Z}_S\mathbf{I}_G = (R_S + jX_S)\mathbf{I}_G \tag{5.73}$$

$$\mathbf{V}_C = \mathbf{Z}_C\mathbf{I}_G = (R_S + jX_\ell)\mathbf{I}_G \tag{5.74}$$

and

$$\mathbf{E}_{\mathcal{R}} = \mathbf{V}_L + \mathbf{V}_C = \mathbf{V}_L + (R_S + jX_\ell)\mathbf{I}_G \tag{5.75}$$

Lagging Power-Factor Analysis

Since a lagging power factor implies that the load current $\mathbf{I}_L \equiv \mathbf{I}_G$ lags the load voltage $\mathbf{V}_L$ by an angle $\theta_L = \cos^{-1}(k_{LGG})$, where $\cos\theta_L = k_{LGG}$ and $\theta_L > 0$, the phasors $\mathbf{V}_L$ and $\mathbf{I}_G$ (for the circuit exhibited in Fig. 5.13, with lagging power-factor loading, i.e., the generator is loaded with an **inductively behaving load**) are drawn as in Fig. 5.17. Completion of this phasor diagram, however, necessitates the use of

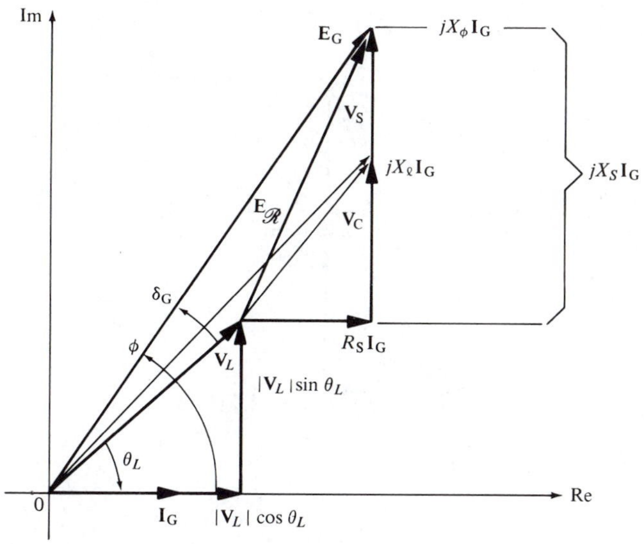

Fig. 5.17 Phasor Diagram for the Per-Phase Analysis of Synchronous Generators with Lagging Power-Factor Loading, PF$=k_{LGG}$

the generalized describing equation of the circuit of Fig. 5.13, that is, Eq. (5.69). The lagging current $\mathbf{I}_G$, which is set along the Re-axis, Fig. 5.17, and the decomposition of the voltages $\mathbf{V}_L$ into components along the Re- and Im-axes, allows us to write Eq. (5.69) as

$$\mathbf{E}_G = (|\mathbf{V}_L|\cos\theta_L + R_S\mathbf{I}_G) + j(|\mathbf{V}_L|\sin\theta_L + X_S\mathbf{I}_G) \equiv |\mathbf{E}_G|e^{j\phi} \qquad (5.76)$$

The relative positions of the phasors (and their corresponding components) of Eq. (5.76) are depicted in Fig. 5.17.

As in the case of unity power factor, the voltage drops $\mathbf{V}_S$, across the synchronous impedance $\mathbf{Z}_S$, $\mathbf{V}_C$, across the stator coil impedance $\mathbf{Z}_C$, and $\mathbf{E}_{\mathscr{R}}$, the airgap potential rise can be written with the aid of Fig. 5.17 as follows:

$$\mathbf{V}_S = \mathbf{Z}_S\mathbf{I}_G = (R_S + jX_S)\mathbf{I}_G \qquad (5.77)$$

$$\mathbf{V}_C = \mathbf{Z}_C\mathbf{I}_G = (R_S + jX_\ell)\mathbf{I}_G \qquad (5.78)$$

and

$$\mathbf{E}_{\mathscr{R}} = \mathbf{V}_L + \mathbf{V}_C = (|\mathbf{V}_L|\cos\theta_L + R_S\mathbf{I}_G) + j(|\mathbf{V}_L|\sin\theta_L + X_\ell\mathbf{I}_G) \qquad (5.79)$$

Leading Power-Factor Analysis

Since a leading power factor implies that the load current, $\mathbf{I}_L \equiv \mathbf{I}_G$, leads the load voltage $\mathbf{V}_L$ by an angle $\theta_L = \cos^{-1}(k_{LDG})$, where $\cos\theta_L = k_{LDG}$ and $\theta_L < 0$, the phasors $\mathbf{V}_L$ and $\mathbf{I}_G$ (for the circuit exhibited in Fig. 5.13, with leading power-factor excitation, i.e., the generator is loaded with a **capacitively behaving load**) are drawn as in Fig. 5.18. As in the previous case, completion of this phasor diagram necessitates the use of the generalized describing equation of the circuit of Fig. 5.13, that is, Eq. (5.69). The leading current $\mathbf{I}_G$, which is set along the Re-axis, Fig. 5.18, and the decomposition of the voltage $\mathbf{V}_L$ into components along the Re- and Im-axes allows us to write Eq. (5.69) as

$$\mathbf{E}_G = (|\mathbf{V}_L|\cos\theta_L + R_S\mathbf{I}_S) + j(|\mathbf{V}_L|\sin\theta_L + X_S\mathbf{I}_G) \equiv |\mathbf{E}_G|e^{j\phi} \qquad (5.80)$$

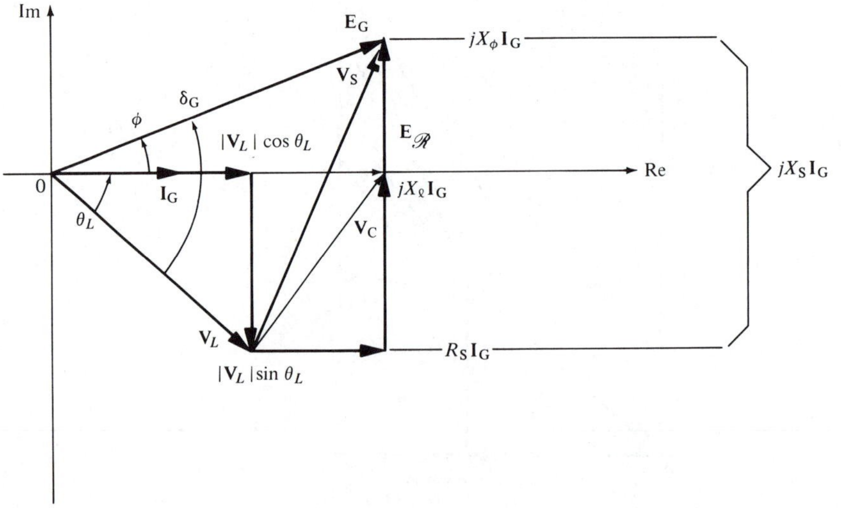

Fig. 5.18 Phasor Diagram for the Per-Phase Analysis of Synchronous Generators with Leading Power-Factor Loading, PF = k_{LDG}

The relative positions of the phasors (and their corresponding components) of Eq. (5.80) are depicted in Fig. 5.18.

Again, as in the cases of unity and lagging power factors, the voltage drops $\mathbf{V}_S$ across the synchronous impedance $\mathbf{Z}_S$ and $\mathbf{V}_C$ across the stator coil impedance $\mathbf{Z}_C$, and the airgap potential rise, $\mathbf{E}_{\mathcal{R}}$, can be written with the aid of Fig. 5.18 as follows:

$$\mathbf{V}_S = \mathbf{Z}_S|\mathbf{I}_G| = (R_S + jX_S)\mathbf{I}_G \tag{5.81}$$

$$\mathbf{V}_C = \mathbf{Z}_C|\mathbf{I}_G| = (R_S + jX_\ell)\mathbf{I}_G \tag{5.82}$$

and

$$\mathbf{E}_{\mathcal{R}} = \mathbf{V}_L + \mathbf{V}_C = (|\mathbf{V}_L|\cos\theta_L + R_S\mathbf{I}_G) + j(|\mathbf{V}_L|\sin\theta_L + X_\ell\mathbf{I}_G) \tag{5.83}$$

Example 5.5

A 1000-kVA (magnitude of total demanded power, $\mathbf{S}_{TD}$), 4600-line-to-line-V, three-phase, Y-connected, synchronous generator has a synchronous resistance $R_S = 2\ \Omega$ and a synchronous reactance $X_S = 20\ \Omega$ per phase. Find: (1) the full-load generated voltage $\mathbf{E}_G$, per phase, at PF $= k_{LGG} = 0.4$, 0.8, PF $= 1$, and PF $= k_{LDG} =$ 0.8, 0.4; (2) the corresponding power angle δ_G; (3) the corresponding voltage regulations, VR.

Solution

Since the analysis of the synchronous generator is performed on a per-phase basis, the phase full-load voltage and phase current should be calculated first. Thus use of Eq. (3.67) yields

$$|\mathbf{V}_L| = \frac{|\mathbf{V}_{LL}|}{\sqrt{3}} = \frac{4600}{1.732} = \underline{2655.89\ \text{V}} \tag{1}$$

Recognizing the fact that the magnitude of the total loading apparent power is $|\mathbf{S}_{TD}| = 3|\mathbf{V}_L|\,|\mathbf{I}_G^*| = 1000$ kVA, calculation of the generated current is as follows:

$$|\mathbf{I}_G| = \frac{1000 \times \text{kVA}}{3|\mathbf{V}_L|} = \frac{1000 \times 1000}{3 \times 2655.89} = \underline{125.50\ \text{A}} \tag{2}$$

Since, for analysis purposes, the current $\mathbf{I}_G$ is placed along the Re-axis, Figs. 5.16, 5.17, and 5.18, the voltage drop across the synchronous impedance $\mathbf{Z}_S$, Fig. 5.19, is calculated as follows:

$$\mathbf{Z}_S\mathbf{I}_G = (2 + j20) \times 125.50 = \underline{251 + j2510}\ \text{V} \tag{3}$$

Eq. (3) in conjunction with Figs. 5.10, 5.17, and 5.18 and Eqs. (5.76), (5.70), and (5.80), and $\delta_G = \phi - \theta_L$ enables us to calculate the full-load generated voltage $\mathbf{E}_G$, per phase, and the corresponding power angle δ_G at the suggested power factors, that is, PF $= \cos\theta_L$, as follows:

1. $k_{LGG} = 0.4$; $\theta_L = 66.42°$; Eq. (5.76), and Fig. 5.17.

$$\begin{aligned}\mathbf{E}_G &= (|\mathbf{V}_L|\cos\theta_L + R_S\mathbf{I}_G) + j(|\mathbf{V}_L|\sin\theta_L + X_S\mathbf{I}_G)\\ &= (1062.36 + 251) + j(2434.16 + 2510)\\ &= 1313.36 + j4944.16 = \underline{\underline{5115.63e^{j75.12°}}}\ \text{V}\end{aligned} \tag{4}$$

$$\delta_G = \angle\mathbf{E}_G - \angle\mathbf{V}_L = 75.12° - 66.42° = \underline{\underline{8.70°}} \tag{5}$$

2. $k_{LGG}=0.8$; $\theta_L=36.87°$; Eq. (5.76), and Fig. 5.17.

$$\begin{aligned}\mathbf{E}_G &= (|\mathbf{V}_L|\cos\theta_L + R_S|\mathbf{I}_G|) + j(|\mathbf{V}_L|\sin\theta_L + X_S|\mathbf{I}_G|) \\ &= (2124.71+251) + j(1593.53+2510) \\ &= 2375.71 + j4103.53 = \underline{\underline{4741.62e^{j59.93°}}}\ \text{V}\end{aligned} \tag{6}$$

$$\delta_G = \angle\mathbf{E}_G - \angle\mathbf{V}_L = 59.93° - 36.87° = \underline{\underline{23.06°}} \tag{7}$$

3. PF = 1; $\theta_L=0$; Eq. (5.70), and Fig. 5.16.

$$\begin{aligned}\mathbf{E}_G &= (|\mathbf{V}_L| + R_S\mathbf{I}_G) + jX_S\mathbf{I}_G = (2655.89+251) + j2510 \\ &= 2906.89 + j2510 = \underline{\underline{3840.59e^{j40.81°}}}\ \text{V}\end{aligned} \tag{8}$$

$$\delta_G = \angle\mathbf{E}_G - \angle\mathbf{V}_L = 40.81° - 0.00° = \underline{\underline{40.81°}} \tag{9}$$

4. $k_{LDG}=0.8$; $\theta_L=-36.87°$; Eq. (5.80), and Fig. 5.18.

$$\begin{aligned}\mathbf{E}_G &= (|\mathbf{V}_L|\cos\theta_L + R_S\mathbf{I}_G) + j(|\mathbf{V}_L|\sin\theta_L + X_S\mathbf{I}_G) \\ &= (2124.71+251) + j(-1593.53+2510) \\ &= 2375.71 + j916.47 = \underline{\underline{2546.35e^{j21.10°}}}\ \text{V}\end{aligned} \tag{10}$$

$$\delta_G = \angle\mathbf{E}_G - \angle\mathbf{V}_G = 21.10° - (-36.87°) = \underline{\underline{57.97°}} \tag{11}$$

5. $k_{LDG}=0.4$; $\theta_L=-66.42°$; Eq. (5.80), and Fig. 5.18.

$$\begin{aligned}\mathbf{E}_G &= (|\mathbf{V}_L|\cos\theta_L + R_S\mathbf{I}_G) + j(|\mathbf{V}_L|\sin\theta_L + X_S\mathbf{I}_G) \\ &= (1062.36+251) + j(-2434.16+2510) \\ &= 1313.36 + j75.84 = \underline{\underline{1315.55e^{j3.31°}}}\ \text{V}\end{aligned} \tag{12}$$

$$\delta_G = \angle\mathbf{E}_G - \angle\mathbf{V}_L = 3.31° - (-66.42°) = \underline{\underline{69.73°}} \tag{13}$$

It is essential to note that the generated voltage $\mathbf{E}_G$ **leads** the loading (i.e., the terminal) voltage $\mathbf{V}_L$ in all cases.

The voltage regulation, for the corresponding power factor, is calculated with aid of Eq. (5.68), rewritten here for clarity,

$$\text{VR} = \frac{|\mathbf{V}_{NL}| - |\mathbf{V}_{FL}|}{|\mathbf{V}_{FL}|} \times 100 \Bigg|_{|\mathbf{V}_{FL}|=|\mathbf{Z}|\mathbf{I}_G||;\ |\mathbf{V}_{NL}|=|\mathbf{E}_G|;\ \cos\theta_L=?}$$

as follows:

1. $$\text{VR} = \frac{5115.63-2655.89}{2655.89} \times 100 \Bigg|_{k_{LGG}=0.4} = \underline{\underline{92.61\%}} \tag{14}$$

2. $$\text{VR} = \frac{4741.63-2655.89}{2655.89} \times 100 \Bigg|_{k_{LGG}=0.8} = \underline{\underline{78.53\%}} \tag{15}$$

3. $$\text{VR} = \frac{3840.59-2655.89}{2655.89} \times 100 \Bigg|_{P_F=1} = \underline{\underline{44.61\%}} \tag{16}$$

4. $$\text{VR} = \frac{2546.35-2655.89}{2655.89} \times 100 \Bigg|_{k_{LDG}=0.8} = \underline{\underline{-4.12\%}} \tag{17}$$

5. $$\text{VR} = \frac{1315.54-2655.89}{2655.89} \times 100 \Bigg|_{k_{LDG}=0.4} = \underline{\underline{-50.47\%}} \tag{18}$$

Note from Table 5.1 that as the power factor changes (ascending through lagging values, going through unity power factor, and descending through leading values), the values of the power angle δ_G increase monotonically while the values of the magnitude of the generated voltage, $\mathbf{E}_G$, and the voltage regulation, VR, decrease monotonically. What do these changes suggest?

Table 5.1 TABULATION OF ANSWERS: EXAMPLE 5.5

			Pertinent Quantities, MKS Units		
			$\mathbf{E}_G$ (V)	δ_G (degrees)	VR (%)
Power Factor	k_{LGG}	0.4	$5115.62e^{j75.12°}$	8.70	92.61
		0.8	$4741.62e^{j59.93°}$	23.06	78.53
	k	1.0	$3840.59e^{j40.81°}$	40.81	44.61
	k_{LDG}	0.8	$2546.35e^{j21.10°}$	57.97	−4.12
		0.4	$1315.54e^{j3.31°}$	69.73	−50.47

As a rule only the synchronous impedance of the synchronous generator is usually defined explicitly (i.e., $\mathbf{Z}_S = R_S + jX_S$); therefore, the circuit of Fig. 5.13 can be simplified and drawn as in Fig. 5.19. Under these conditions only the quantities designated with *heavy lines* in Figs. 5.16, 5.17, and 5.18 along with Eqs. (5.70), (5.76), and (5.80) are essential for the complete analysis of synchronous generators that are loaded with loads of unity, lagging, and leading power factors.

A comparison among Figs. 5.16, 5.17, and 5.18 reveals that (1) the generated voltage per-phase, $\mathbf{E}_G$, is seemingly larger than the loading voltage, $\mathbf{V}_L$, and (2) the generated voltage $\mathbf{E}_G$ always leads the response (i.e., the terminal) voltage $\mathbf{V}_L$ by an angle δ_G. This angle is defined as the **power angle**, or the **torque angle** (or **displacement angle**), of the synchronous generator and is defined as $\delta_G \equiv \angle \mathbf{E}_G - \angle \mathbf{V}_L$. It is always *positive* when the synchronous device functions in the *generator mode*, and it represents the measure of the angular displacement, in the counterclockwise direction, of the generated voltage $\mathbf{E}_G$ from the response (i.e., the terminal) voltage $\mathbf{V}_L$. (See Section 5.6 for the definition of the power angle δ_M of the synchronous motor.) More will be said about this angle in Section 5.7 and later sections.

In this type of analysis—that is, when the synchronous impedance $\mathbf{Z}_S$ of the synchronous generator and both the voltampere rating (i.e., $|\mathbf{V}_L|$ and $|\mathbf{I}_L| \equiv |\mathbf{I}_G|$) and the power factor (i.e., $\cos\theta_L$) of the generator's load are explicitly defined—the generated phase voltages $\mathbf{E}_G$ are easily calculated with the aid of Eqs. (5.70), (5.76), and (5.80), and the relevant theory of Chapter 3. With this information at hand, we

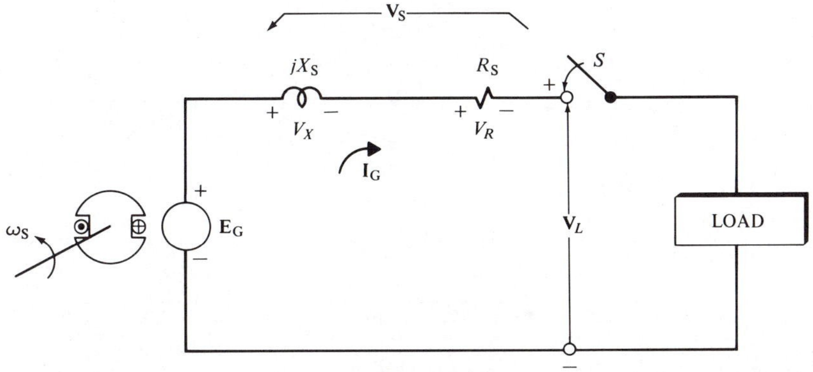

Fig. 5.19 Simplified Single-Phase, Steady State, Equivalent Circuit of a Balanced, Three-Phase, *P*-Pole, Synchronous Generator Utilized in Its Per-Phase Analysis

can calculate the per-phase generated power $\mathbf{S}_G$ as

$$\mathbf{S}_G = P_G + jQ_G \equiv \mathbf{E}_G\mathbf{I}_G^* \tag{5.84}$$

Similarly, the per-phase power required by the stator winding (its effects represented by $\mathbf{Z}_S$), $\mathbf{S}_S$, and the phase power demanded by the load, $\mathbf{S}_D$, can be calculated readily from the input data as

$$\mathbf{S}_S = P_S + jQ_S \equiv \mathbf{Z}_S|\mathbf{I}_G|^2 = \{R_S + j(X_\ell + X_\phi)\}|\mathbf{I}_G|^2 \tag{5.85}$$

and

$$\mathbf{W}_D = P_D + jQ_D = |\mathbf{V}_L||\mathbf{I}_G|\cos\theta_L + j|\mathbf{V}_L||\mathbf{I}_G|\sin\theta_L \tag{5.86}$$

Once the phase powers, $\mathbf{S}_G$, $\mathbf{S}_S$, and $\mathbf{S}_D$, are known, the total generated power, $\mathbf{S}_{GT}$, the total power required by the stator winding, $\mathbf{S}_{ST}$, and the total power demanded by the load, $\mathbf{S}_{DT}$, can be calculated with the aid of Eqs. (5.62), (5.63), and (5.64). With this information at hand, the total power, P_{INM}, and total torque, τ_{INM}, provided by the prime mover can be calculated with the aid of Eqs. (5.65) and (5.66). With this information available, the efficiency η of the synchronous generator can be calculated by utilizing Eq. (5.67). In addition, since the generated voltage, $\mathbf{E}_G$, constitutes the no-load voltage for the synchronous generator, that is, $|\mathbf{V}_{NL}| = |\mathbf{E}_G|$, and the loading voltage constitutes the full-load voltage for the synchronous generator, that is, $|\mathbf{V}_{FL}| = |\mathbf{V}_L|$, the synchronous generator's voltage regulation, VR, can be calculated utilizing Eq. (5.68).

For comparative analysis, Fig. 5.20 displays the $v = f(i)$ characteristics for the power factors PF $= k_{LGG} = 0.4$, 0.8; PF $= 1$, and PF $= k_{LDG} = 0.8$, 0.4, as calculated in Example 5.5, $Q(\mathbf{V}_L)$, $(\mathbf{I}_L) = (2655.89, 125.50)$. The graphs of Fig. 5.20 illustrate how the nature of a load on a synchronous generator affects the magnitude of its phase-generated voltage $\mathbf{E}_G$. $|\mathbf{E}_G|$ is the v-axis intercept for the various power-factor characteristics.

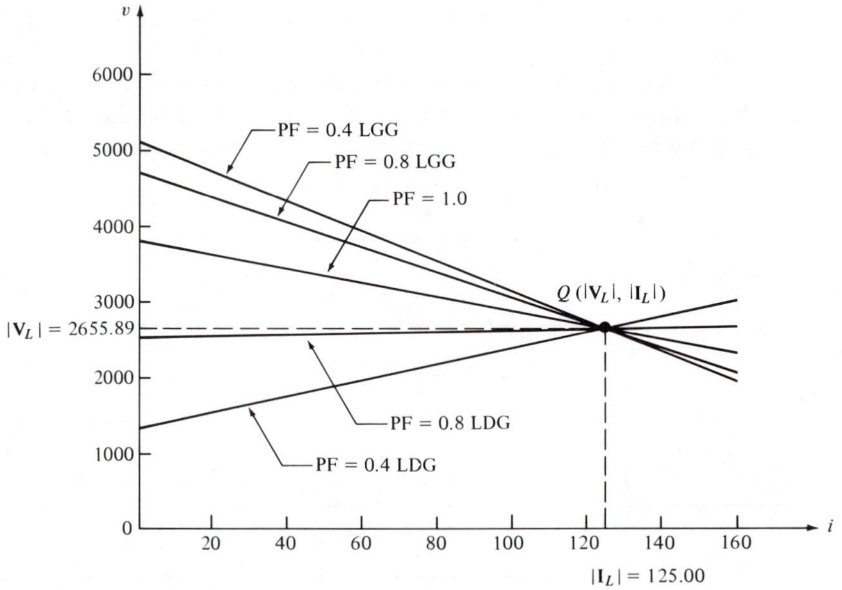

Fig. 5.20 Effects of the Loading of a Synchronous Generator on the Generated Phase Voltage $\mathbf{E}_G$

Inspection of this figure reveals that (1) for loads with **leading power factors**, as the power factor decreases, the generated voltage $|\mathbf{E}_G|$ decreases; (2) for loads with **lagging power factors**, as the power factor decreases, the generated voltage $|\mathbf{E}_G|$ increases; (3) for loads with **unity power factor**, the generated voltage $|\mathbf{E}_G|$ is always greater than the loading voltage $|\mathbf{V}_L|$; (4) for the same rated loading voltage $|V_L|$, loads with lagging power factors require greater generated voltages $|E_G|$ than do loads with leading power factors.

Fig. 5.20 also shows the effects of the armature reaction to the synchronous generator. As the power factor of the load is varied, the angle ψ between the induced phase voltages and response currents, Fig. 5.9(b), changes. The angle ψ determines the position of $\mathbf{B}_{ST}$ with respect to $\mathbf{B}_R$, Fig. 5.14, and thus the magnetizing or demagnetizing effects of $\mathbf{B}_{ST}$ on $\mathbf{B}_R$. Recall that the magnitude of $\mathbf{E}_G$ depends upon $\mathbf{B}_R$, Eq. (5.24). For loads with lagging power factors, the armature reaction, jX_ϕ, is demagnetizing, Fig. 5.14(a), and its effect in reducing the generated voltage $\mathbf{E}_G$, coupled with the voltage drop across the impedance of the stator coil, $\mathbf{Z}_C$, results in a decrease of the loading voltage $\mathbf{V}_L$ as load is applied. For loads with leading power factors, the armature reaction, jX_ϕ, has magnetizing effects, Fig. 5.14(c), and tends to increase the generated voltage $\mathbf{E}_G$ as load is applied. Note that at one specific leading power factor, the generated voltage $\mathbf{E}_G$ and the terminal voltage $\mathbf{V}_L$ of the synchronous generator are identical and the voltage regulation is zero.

Example 5.6

It is known that the per-phase generated voltage of a six-pole, three-phase, Y-connected, synchronous generator is $\mathbf{E}_G = 3{,}845e^{j40.64°}$ V per phase, its frequency is 60 Hz, the per-phase synchronous impedance of the winding is $\mathbf{Z}_S = 2 + j20$, and the per-phase loading impedance is $\mathbf{Z}_L = 21.3 + j0$. What are the minimum prime-mover requirements in speed and torque so that the generated voltage maintains its present magnitude and frequency if all losses are 10 kW (i.e., $P_{WLT} + P_{BFT} + P_{CLT} = 10{,}000$ W)?

Solution

For the calculation of the minimum speed requirements, we utilize Eq. (5.1), rewritten here for convenience as

$$f_E = \frac{n_S \times P}{120} = \frac{n_S \times 6}{120} = \frac{n_S}{\underline{20}} \text{ Hz} \tag{1}$$

Solution of Eq. (1) for n_S yields the required speed of the prime mover, in revolutions per minute (rpm);

$$n_S = 20 \times f_E = 20 \times 60 = \underline{\underline{1{,}200}} \text{ rpm} \tag{2}$$

Division of Eq. (2) by 60 s/min yields the speed of the prime mover in revolutions per second (rps). Thus

$$f_S = \frac{n_S \text{ rpm}}{60 \text{ s/min}} = \frac{1{,}200 \text{ rpm}}{60 \text{ s/min}} = \underline{20} \text{ rps} \tag{3}$$

Finally, the required speed of the prime mover in MKS units, that is, in radians/second, is calculated as follows:

$$\omega_S = 2\pi \times f_S = 2\pi \times 20 = 40\pi \text{ rad/s} = \underline{\underline{125.66}} \text{ rad/s} \tag{4}$$

For the calculation of the minimum torque requirements, we can proceed, by utilizing Eq. (5.51), to calculate the generated phase current $\mathbf{I}_G$ as follows:

$$\mathbf{I}_G = \frac{\mathbf{E}_G}{\mathbf{Z}_S + \mathbf{Z}_L} = \frac{3{,}845e^{j40.64°}}{(2+j20)+(21.3+j0)} = \frac{3{,}845e^{j40.64°}}{23.3+j20}$$
$$= \frac{3{,}845e^{j40.64°}}{30.71e^{j40.64°}} = \underline{125.20e^{j0°}} \text{ A} \qquad (5)$$

The per-phase generated power $\mathbf{S}_G$ is calculated, utilizing Eq. (5.59), as

$$\mathbf{S}_G = \mathbf{E}_G\mathbf{I}_G{}^* = P_G + jQ_G = 3{,}845e^{j40.64°} \times 125.20e^{j0°}$$
$$= \underline{481{,}394.000e^{j40.64°} = 365{,}289.85 + j313{,}533.90} \text{ VA} \qquad (6)$$

Now, the total generated power is calculated, with the aid of Eq. (5.62) and Eq. (6), as follows:

$$\mathbf{S}_{GT} = 3\mathbf{S}_G = P_{GT} + jQ_{GT} = \underline{1{,}095{,}869.55 + j940{,}601.70} \text{ VA} \qquad (7)$$

Utilization of Eq. (7) and Eq. (5.65) enables us to calculate the required power, P_{INM}, from the prime mover as follows:

$$P_{INM} = P_{GT} + (P_{WLT} + P_{BFT} + P_{CLT}) = 1{,}095{,}869.55 + 10{,}000 = \underline{1{,}105{,}869.55} \text{ W} \qquad (8)$$

Finally, the total required torque from the prime mover is calculated with the aid of Eqs. (5.66) and (4) as follows:

$$\tau_{INM} = \frac{P_{INM}}{\omega_S} = \frac{1{,}105{,}869.55}{125.66} = \underline{\underline{8{,}800.49}} \text{ N} \cdot \text{m} \qquad (9)$$

Before closing this section it should be mentioned that since most commercial electrical loads are inherently of lagging power factor, the loading voltage $\mathbf{V}_L$, that is, the full-load voltage, is reduced for full-load conditions by the voltage drops across the stator coil impedance, $\mathbf{Z}_C$, and the armature reaction reactance, jX_ϕ. Thus the synchronous generator has a poor voltage regulation. This is primarily due to the demagnetizing effects of the stator field component $\mathbf{B}_{ST}^{d-}$, Fig. 5.14(a). Although many attempts have been made to devise a compensating scheme to offset armature reaction at various power factors, none has been fully successful. Therefore, the inherent poor voltage regulation of the synchronous generator is ignored in practice, but the generator's output voltage is maintained constant by means of **external voltage regulators** that automatically increase or decrease the field excitation from the exciter (usually a dc generator) with changes in the electrical load and its power factor. As mentioned in Chapter 2, the exciter is usually connected on the same shaft as the prime mover and the rotor of the synchronous generator. Thus its characteristics are closely related to the voltage regulation of the synchronous generator; that is, if the exciter is to maintain a constant voltage over a wide range of load, the limits of field current, power, and exciter rating depend on the amount of the field current required by the synchronous generator to maintain good voltage regulation.

5.5 THE EQUIVALENT CIRCUIT OF THE SYNCHRONOUS MOTOR: STEADY STATE OPERATION

Although the synchronous motor is structurally identical to the synchronous generator, there is one distinct difference between the two. **The synchronous motor is doubly excited**, while the **synchronous generator is singly excited**. The rotor winding of the motor is excited with a dc voltage derived from the exciter (a dc generator that is driven by the shaft of the synchronous motor), and the stator coils are excited by the external voltages

$$v_{SA}(t) = \sqrt{2}\,|\mathbf{E}_S| \cos \omega_E t \tag{5.87}$$

$$v_{SB}(t) = \sqrt{2}\,|\mathbf{E}_S| \cos(\omega_E t - 120°) \tag{5.88}$$

$$v_{SC}(t) = \sqrt{2}\,|\mathbf{E}_S| \cos(\omega_E t - 240°) \tag{5.89}$$

These voltages, in accordance with the theory of Section 4.2, cause the total stator field $\mathbf{B}_{ST}$ to rotate in the airgap at the synchronous speed

$$\omega_S = \frac{\omega_E}{P/2} \tag{5.90}$$

If the rotor is at rest and an attempt is made to start the synchronous motor, the following events will take place. With the field of the rotor $\mathbf{B}_R$ established and the field of the stator $\mathbf{B}_{ST}$ rotating in the airgap at a synchronous speed ω_S, when the two north poles, N_R and N_S, of the fields come in proximity, a repulsion between the poles takes place, and the rotor attempts to move. This motion, however, is very slow because of the inertia of the rotor or of the load on it. By this time the south pole, S_S, of the stator field $\mathbf{B}_{ST}$ has come in proximity with the north pole, N_R, of the rotor field, and an attraction between the two fields develops. This **continuous attraction and repulsion** of the rotor field $\mathbf{B}_R$ by the rotating stator field $\mathbf{B}_{ST}$ results in the **inability** of the rotor to start rotation. It should be obvious now that *the synchronous motor is not self-starting*, as in one instant the north pole of the rotor field, N_R, is repelled by the approaching north pole of the stator field, N_S, producing torque in the counterclockwise direction; in the next instant the same rotor field pole, N_R, is attracted in the opposite direction by the passing south pole of the stator field, S_S, producing torque in the clockwise direction. The resultant torque produced in a given time Δt is zero since the rotor has, in effect, been pushed alternately counterclockwise and clockwise an equal number of times over Δt seconds.

If by some means, however, the rotor of a two-pole synchronous motor is moving counterclockwise at some speed near or at synchronous speed ω_S, then the north pole of the rotor field, N_R, is locked in synchronism with the south pole of the resultant stator field, S_S, rotating at synchronous speed in the counterclockwise direction. The same phenomenon takes place between the south pole of the rotor field, S_R, and the north pole of the resultant stator field, N_S. Thus for a polyphase synchronous motor, the north rotor poles, N_R, are locked in synchronism with the south stator poles, S_S, and vice versa, both rotating in the counterclockwise direction at synchronous speed ω_S. If a load is placed on the shaft of the synchronous motor, the countertorque created by the load, designated by τ_L, will cause the rotor to drop back momentarily but to continue to rotate at the synchronous speed ω_S as long as the countertorque does not exceed the **pull-in** torque of the motor. [**Pull-in** torque of the synchronous motor is defined to be the torque at which the rotor of a synchronous motor pulls into synchronism with the field of the stator $\mathbf{B}_{ST}$ when the synchronous motor is operating as an induction motor (see the discussion of the methods of starting a synchronous motor that follows.)] This phenomenon is in no way similar to the phenomenon of the induction motor, where the rotor **always** runs at a speed less than the synchronous speed ω_S of the stator field $\mathbf{B}_{ST}$. If the countertorque (i.e., the applied load) ever exceeds the **pull-out** torque of the motor, (**pull-out** torque of a synchronous motor is defined to be the maximum torque the motor can deliver and still stay in synchronism), the rotor **slips** out of synchronism and the synchronous motor stops. Note that as the countertorque, τ_L, increases, the rotor falls back. The rotor, or mutual airgap flux, reduces slightly, Fig. 5.21, because of the increased airgap reluctance, and the rotor slips out of synchronism and begins to slow down. As the rotor slows down, the rotating field of the stator slips by the rotor field poles so rapidly that it is unable to **mesh**, that is, lock synchronously with the rotating stator field. Thus the rotor comes to full stop eventually. **The synchronous motor, therefore, either runs at synchronous speed or does not run at all, regardless of how strong the excitations of the stator and rotor coils are.**

There are **four methods** to bring the rotor of the synchronous motor up to synchronous speed. They are as follows: (1) use of a dc motor coupled to the shaft of the synchronous motor; (2) use of the field exciter as a dc motor; (3) use of a small induction motor whose number of poles is less than the number of poles of the synchronous motor by at least 2; (4) the use of a squirrel cage structure at the faces of the rotor poles. The rotor copper bars and the shorting strips are exhibited in Fig. 5.22. Note the four connecting bolts at the ends of the shorting strips. These bolts are used to connect sets of copper bars and shorting strips attached to the faces of adjacent magnetic poles. Bolting from one set to the next enables the formation of a complete squirrel cage winding, similar to that of the induction

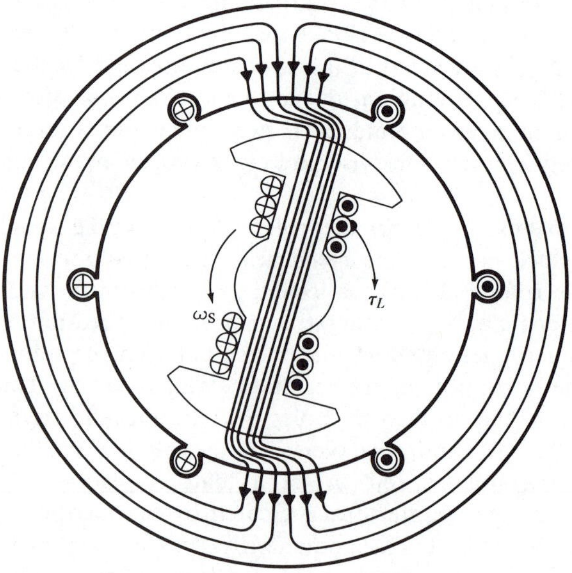

Fig. 5.21 Effects of Overloading the Rotor of a Two-Pole, Three-Phase, Synchronous Motor

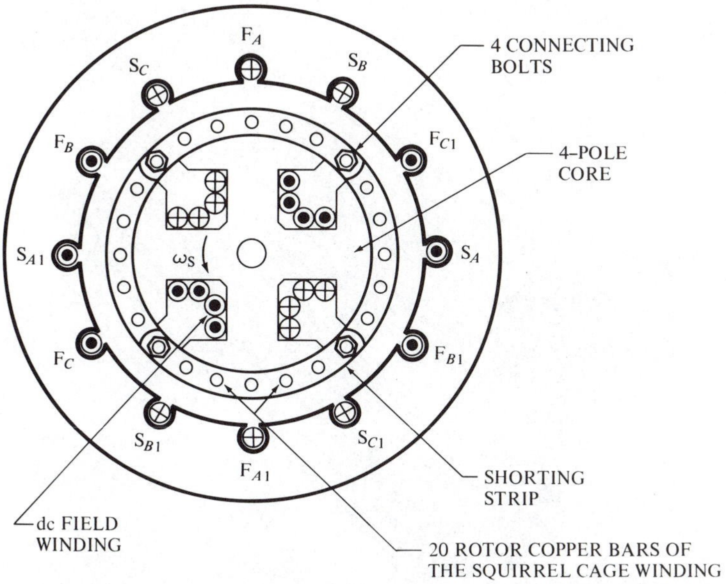

Fig. 5.22 The Rotor Structure of a Three-Phase Synchronous Motor with Squirrel-Cage-Type Damping Winding in Addition to the dc Field Winding, the Stator Winding Is Three-Phase, Full-Pitch, Single-Layer, Four-Pole, Integral-Slot, and Series-Connected

motor; that is, the rotor of the synchronous motor has two sets of windings, the dc winding and the squirrel cage winding. (The squirrel cage winding of the synchronous motor is also known as a *damper*, an *amortisseur*, or a *starting winding*.) Sometimes it is better to use a wound-rotor winding rather than a squirrel cage winding in the rotor pole face of the synchronous motor. This winding is called a *phase-wound damper winding*. Such a rotor has five slip rings, two for the dc field and three for the ac, Y-connected, wound-rotor winding. This rotor structure, known as a **simplex rotor**, has the starting characteristics of the wound-rotor induction motor. The first method of the four starting methods is used with synchronous motors not equipped with damping windings. The second method is actually the same as the first, except that the exciter, a dc shunt generator, Fig. 2.45, is operated as a motor. For the third method the induction motor has at least one less pair of poles than the synchronous motor. This lack of poles is required to compensate for the loss of the induction motor speed due to slip. The third method provides (1) the capability of self-starting for the synchronous motor and (2) the capability of damping out speed oscillations of the rotor due to pulsating load torques.

Synchronous motors are generally started and synchronized by means of automatic control equipment. Since it is practically impossible to start a synchronous motor with its dc winding energized, or deenergized and open-circuited, the dc winding is short-circuited through a discharge resistor during the starting period. The induced voltages and their corresponding currents in the rotor winding aid in the production of induction motor action. Thus the induction motor action causes the rotor of the synchronous motor to start rotating (self-start). As the speed of the rotor approaches the synchronous speed of the stator field $\mathbf{B}_{ST}$, the dc winding is connected to the dc excitation through a polarized relay, the short circuit is

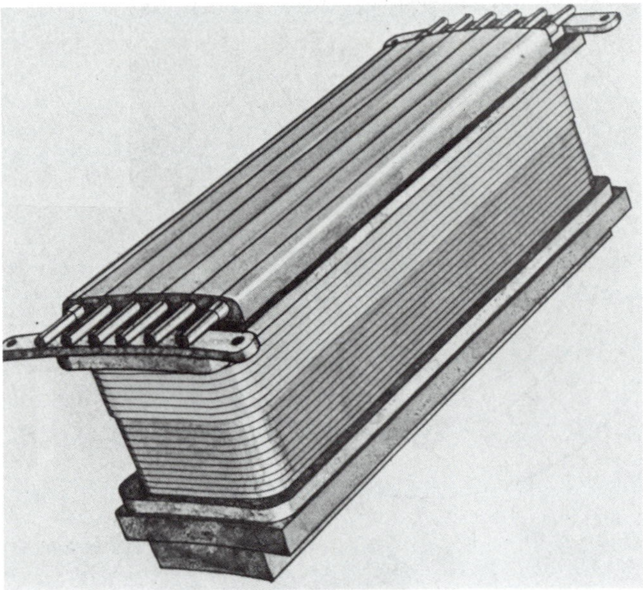

Plate 5.9 Field Winding Pole with the Bars of the Squirrel-Cage-Type Damper Winding Clearly Visible. (Courtesy of Westinghouse Electric Co., Pittsburgh, Pa.)

removed from the dc winding, and the rotor pulls into synchronism with the rotating stator field. The speed of the rotor now is $\omega_R \equiv \omega_S = \omega_E/(P/2)$.

A common phenomenon that is occasionally encountered when the dc winding is energized is the sudden inrush of alternating current and a consequent **thump** indicating a sudden line disturbance. It occurs when the field emf produces poles on the rotor that are directly under the rotating field of the same polarity. That is, for a P-pole synchronous motor, the stator and rotor poles line up as N_S to N_R, S_S to S_R, N_S to N_R, and so on. This causes a sudden reduction in the airgap flux, which reduces the generated emf and suddenly increases the armature current. The consequent reduction in torque causes the rotor to slip back one pole, that is, 180 electrical degrees, where it operates now in synchronism with the stator field; that is, the poles line up as N_S to S_R, S_S to N_R, N_S to S_R, and so on, and the line current is restored to normal.

The methods above necessitate the following constraints for successful starting: (1) for methods 1 through 3, the load on the synchronous motor should be zero or light and the capacity of the starting motors (dc or ac) should be 90 percent to 95 percent of the synchronous motor coupled to it; (2) for method 4, that is, the squirrel cage winding (whose characteristics are **low resistance** and **high reactance**), the starting torque is 30 percent to 50 percent of full-load torque. For loads that are functions of speeds (i.e., fans, compressors, etc.), the synchronous motor with squirrel cage windings can start and come up to synchronous speed and lock into synchronism with such a load on its shaft.

The growing popularity of the synchronous motor created a demand for motors that produce starting torques ranging from full-load up to 30 percent of full load. This starting torque development can be accomplished by one of two methods: (1) use of high-resistivity alloys for the bars of the squirrel-cage-rotor windings or (2) use of wound-rotor windings. Although the high-resistivity bars provide high

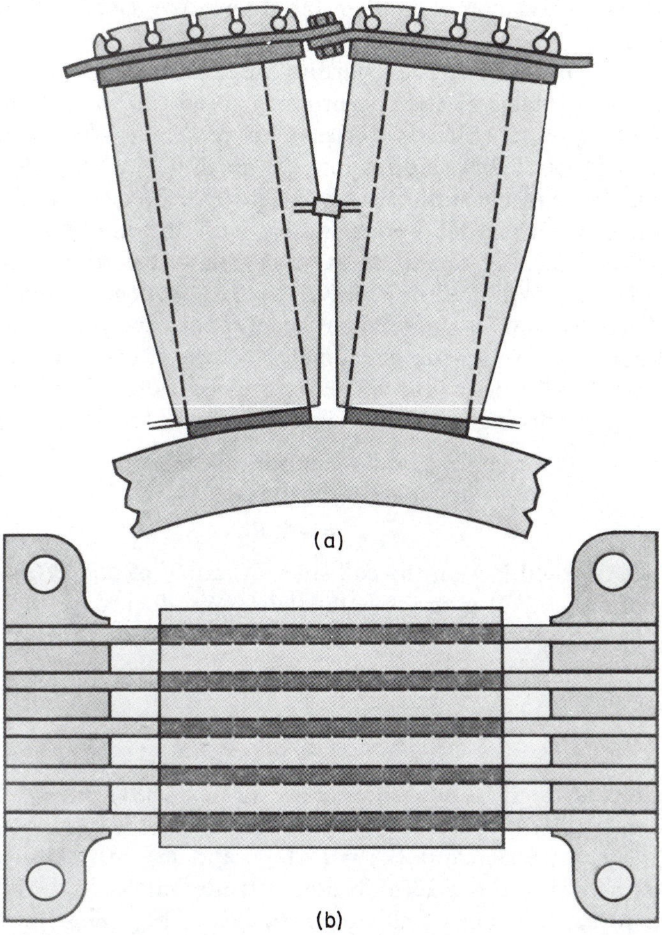

Plate 5.10 Field and Damper-Winding Structure: (a) Damper Bar Segments Are Bolted Together Between Field Winding Poles to Form a Complete Squirrel Cage Damper Winding and (b) Top View of Field Winding Pole with Damper Winding in Place. (Courtesy of Westinghouse Electric Co., Pittsburgh, Pa.)

starting torques, they do not allow the rotor to come as close to synchronous speed as low-resistivity squirrel cage bars. This is due to an increase in slip because of high winding resistance. However, the rotor can be locked into synchronism simply by replacing the discharge resistors with short circuits for an instant just before it is excited by the dc source. This technique causes the rotor to surge forward sufficiently so that it can lock in synchronism with the stator field.

The most effective technique of starting synchronous motors with high loads, however, is to use a wound-rotor winding rather than a squirrel-cage-rotor winding and to vary the resistance of the rotor winding through externally varying, Y-connected, resistance banks connected to the wound-rotor winding. The starting performance of the **simplex-rotor** synchronous motor is exactly identical to that of the wound-rotor induction motor; that is, the external resistance is employed to improve the starting torque. The motor is started with full external resistance per phase and the dc winding **open**. The motor approaches synchronous speed as starting resistance is reduced and, when the dc winding is excited, the rotor pulls into synchronism.

Regardless of which of the four starting methods is employed, the rotor of a synchronous motor rotates at the synchronous speed ω_S, and follows the field of the stator winding, $\mathbf{B}_{ST}$, which is designed to rotate at synchronous speed ω_S, (usually) in the counterclockwise direction; the north pole of the rotor field $\mathbf{B}_R$, N_R, follows the south pole of the stator field $\mathbf{B}_{ST}$, S_S, Fig. 5.23(b), always.

For the smooth-rotor synchronous motor, with the impressed stator current distribution of Fig. 5.23(a), operating at steady state, the relative position of the fields $\mathbf{B}_R$ and $\mathbf{B}_{ST}$ are exhibited in Fig. 5.23(b). The sequential effects of the north poles of the fields $\mathbf{B}_R$ and $\mathbf{B}_{ST}$, rotating at synchronous speed ω_S, on each winding, Fig. 5.9(b), leads to the following phenomena: (1) the effect of the field $\mathbf{B}_R$ on the coil sides B and B' of coil B (as well as on the coil sides of coils C and A) is to cause the voltage-producing speed field intensities $\mathcal{E}_{RB}$ and $\mathcal{E}_{RB'}$, given as

$$\mathcal{E}_{RB} = \mathbf{u} \times \mathbf{B}_R \tag{5.91}$$

and

$$\mathcal{E}_{RB'} = \mathbf{u} \times \mathbf{B}_R \tag{5.92}$$

(2) the effect of the field $\mathbf{B}_{ST}$ on the coil sides B and B' of coil B (as well as on the coil sides of coils C and A) is to cause the voltage-producing speed field intensities $\mathcal{E}_{SB}$ and $\mathcal{E}_{SB'}$, given as

$$\mathcal{E}_{SB} = \mathbf{u} \times \mathbf{B}_{ST} \tag{5.93}$$

and

$$\mathcal{E}_{SB'} = \mathbf{u} \times \mathbf{B}_{ST} \tag{5.94}$$

The directions of the field intensities given by Eqs. (5.91) through (5.94) (as well as the field intensities in the coil sides of coils C and A, obtained utilizing $\mathcal{E}_{Rk} = \mathbf{u} \times \mathbf{B}_R$, $\mathcal{E}_{Rk'} = \mathbf{u} \times \mathbf{B}_R$, and $\mathcal{E}_{Sk} = \mathbf{u} \times \mathbf{B}_{ST}$, and $\mathcal{E}_{Sk'} = \mathbf{u} \times \mathbf{B}_{ST}$, $k = A$, C and $k' = A'$, C') are shown in Fig. 5.23(b) as the outermost circles in the stator structure. In reality, however, the speed field intensities $\mathcal{E}_{Rk}$, $\mathcal{E}_{Sk}$, $\mathcal{E}_{Rk'}$, and $\mathcal{E}_{Sk'}$ coexist within the same coil side and cause currents, in addition to the impressed currents, to flow in opposite directions through each coil.

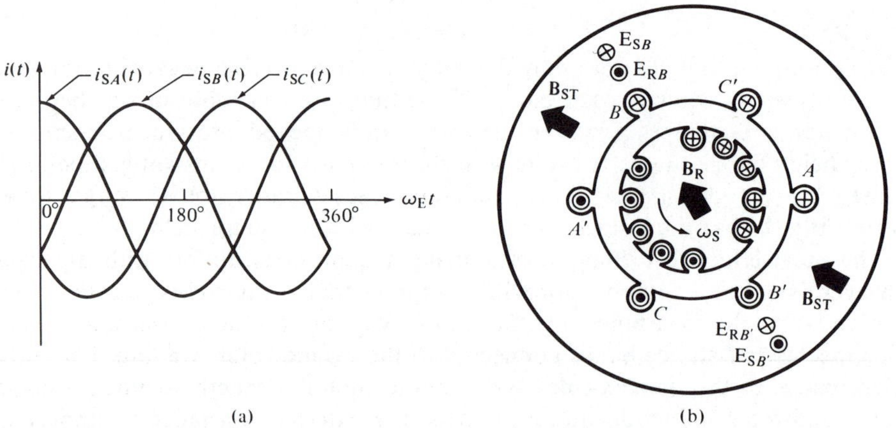

Fig. 5.23 Two Pole Synchronous Motor: (a) Stator Winding Currents, and (b) Stator Field Orientation

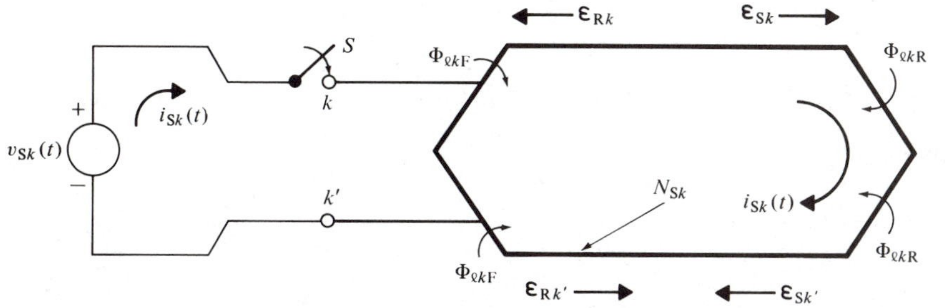

Fig. 5.24 Effects of the Voltage-Producing Total Field Intensities $\mathcal{E}_{RkT}$ and $\mathcal{E}_{SkT}$ on the kth Phase of a Three-Phase Synchronous Motor

These ideas can be described best by referring to Fig. 5.24. Here the kth phase is connected to a source $v_{Sk}(t)$, $k=A$, B, and C, through the switch S. When the switch is closed, the response currents $i_{Sk}(t)$, $k=A$, B, and C, are established, which cause the following effects: (1) a voltage drop along the coil due to the resistance of the entire winding R_{Sk}, designated as $v_{Rk}(t)$, and (2) a flux leakage $\Phi_{\ell k}$ comprised of two components, that flux leakage at the *front* of the coil, designated as $\Phi_{\ell kF}$, and the flux leakage at the *rear* of the coil, designated as $\Phi_{\ell kR}$. These flux leakages multiplied by the number of turns of the stator coil, N_{Sk}, yield the coils flux linkages $\lambda_{\ell k}=N_{Sk}\Phi_{\ell k}$. The time derivative of these flux linkages constitutes a voltage drop along the coil, designated as $v_{\ell k}(t)=d\lambda_{\ell k}(t)/dt$.

If we designate (1) the effects of the total stator field intensity, $\mathcal{E}_{SkT}\equiv\mathcal{E}_{Sk}+\mathcal{E}_{Sk'}$, by the tendency to increase the response currents $i_{Sk}(t)$, Figs. 5.23(b) and 5.24, and thus to cause an increased voltage drop along the winding designated by $v_{\phi k}(t)$; (2) the effects of the rotor field intensity, $\mathcal{E}_{RkT}\equiv\mathcal{E}_{Rk}+\mathcal{E}_{Rk'}$, by the tendency to generate a current in the winding, opposing the response currents $i_{Sk}(t)$; (3) the effects of the total stator field $\mathcal{E}_{SkT}$, designated by the voltage generator $e_{Gk}(t)$, with appropriate polarity, then we can view the effects described above on the kth phase from the circuit analog point of view, as exhibited in Fig. 5.25. Since for the steady state operation of a balanced three-phase synchronous motor, all voltages and currents are sinusoidal and of equal amplitudes, and displaced from each other by 120 positive electrical degrees, we can utilize the principles just set forth and draw the circuit analog of Fig. 5.23(b) **for the steady state operation of the three-phase, *P*-pole, balanced synchronous motor** as exhibited in Fig. 5.26. For balanced operation all the circuit parameters and voltage and

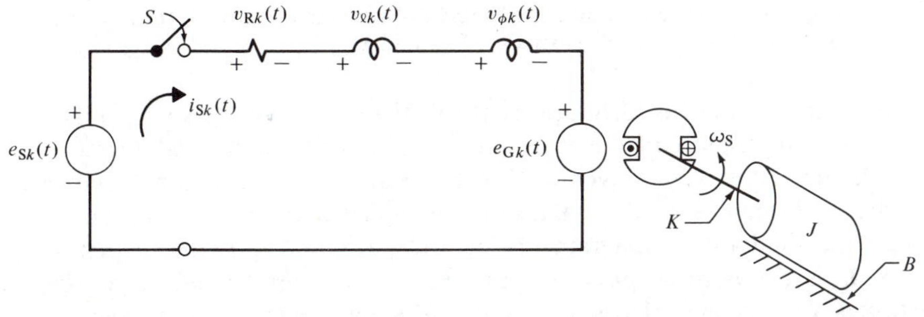

Fig. 5.25 Circuit Analog of the Cause and Effect Phenomena of Fig. 5.24

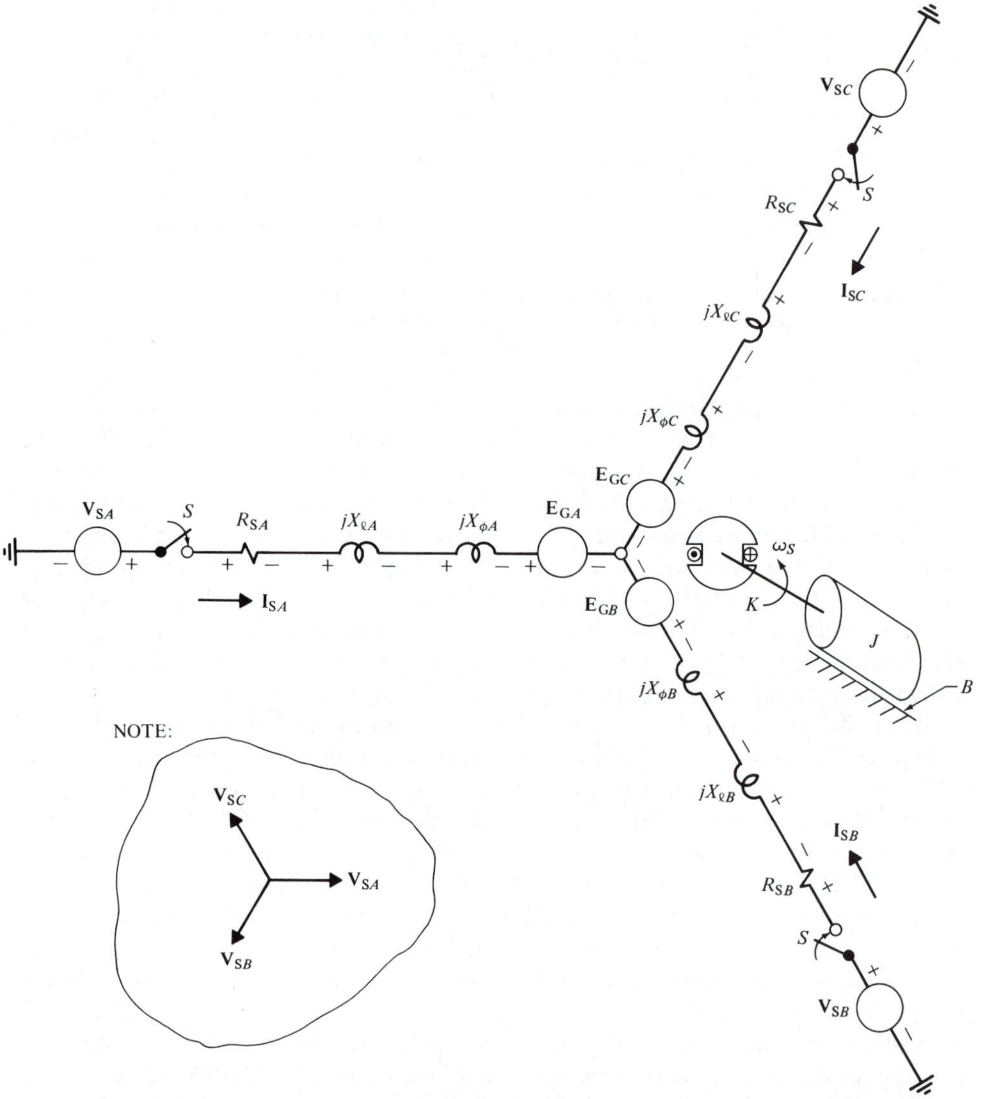

Fig. 5.26 Circuit Analog of Fig. 5.24(b) Depicting the Steady State Operation of a Three-Phase, *P*-Pole, Balanced Synchronous Motor. Note That When Compared to Fig. 5.12, Each Phase Is Seemingly Rotated 180° About a Vertical Axis on the Page. This Allows Placement of the Load (Rotor) on the Right-hand Side of the Page for Consistency and Also Maintains a Positive Sequence (i.e., Counterclockwise Rotation of the Phasors $\mathbf{E}_{SA}$, $\mathbf{E}_{SB}$ and $\mathbf{E}_{SC}$)

current magnitudes of Fig. 5.26 are identical; the phasor voltages and currents and the coil axes are displaced from each other by 120 positive electrical degrees.

Utilization of the theory given in Chapters 3 and 4 and Section 5.3 enables us to view the performance of the balanced, three-phase, *P*-pole, synchronous motor through the single-phase diagram of Fig. 5.27, where $\mathbf{V}_{ST}$ is the *impressed phase voltage*, $\mathbf{I}_S$ is the *response phase current*, R_S is the *resistance of the winding* that constitutes each stator phase, jX_ℓ is the *leakage reactance of the winding* that constitutes each stator phase, jX_ϕ is the **armature reaction reactance** *of the effective*

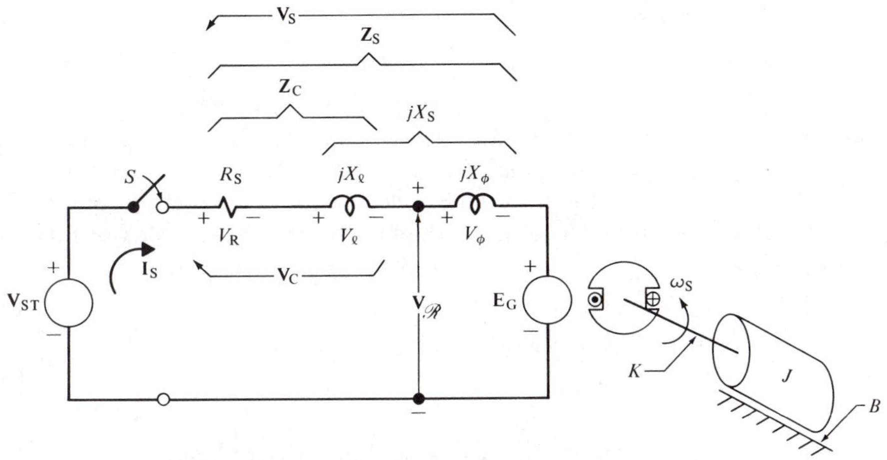

Fig. 5.27 The Single-Phase, Steady State, Equivalent Circuit of a Balanced, Three-Phase, *P*-Pole, Synchronous Motor as Utilized in Its Per-Phase Analysis; All pertinent Parameters and Voltages Are Explicitly Defined

length of the winding that constitutes each stator phase (the latter is the measure of the reaction of the stator coils to the stator's own synchronously rotating field $\mathbf{B}_{ST}$), and $\mathbf{E}_G$ is defined as the *developed phase voltage* (*back emf*) of the synchronous motor. The voltage $\mathbf{E}_G$ is the measure of the reaction of the stator coils to the rotor's synchronously rotating field $\mathbf{B}_R$. The parameters jX_S, $\mathbf{Z}_S$, and $\mathbf{Z}_C$ are defined by Eqs. (5.48), (5.49), and (5.50).

5.6 MOTOR MODE OF OPERATION: PHASOR DIAGRAM ANALYSIS

For the balanced, three-phase, *P*-pole, synchronous motor (i.e., impressed phase voltages of equal amplitude but out of phase by 120 positive electrical degrees; equal phase resistances and reactances; response phase currents of equal amplitude but out of phase by 120 positive electrical degrees; and developed phase voltages of equal amplitude but out of phase by 120 positive electrical degrees), the mathematical analysis can be performed on a per-phase basis, utilizing the theory given in Appendix A, in Chapters 3 and 4, and in Section 5.4. Thus the circuit analog of a given phase of the balanced three-phase, *P*-pole, synchronous motor is adequately represented by the per-phase equivalent circuit of Fig. 5.27 if and only if (1) both the impressed and developed phase voltages $\mathbf{V}_{ST}$ and $\mathbf{E}_G$ are explicitly defined or (2) the excitation is explicitly defined in terms of its voltampere rating (i.e., $|\mathbf{E}_S|$ and $|\mathbf{I}_S|$) and its power factor (i.e., $\cos\theta_W$).

If the circuit parameters R_S, jX_ℓ, and jX_ϕ (or $jX_S=j(X_\ell+X_\phi)$ and the voltages $\mathbf{V}_{ST}$ and $\mathbf{E}_G$ are explicitly defined (i.e., $\mathbf{V}_{ST}=|\mathbf{V}_{ST}|e^{j\theta_S}$ and $\mathbf{E}_G=|\mathbf{E}_G|e^{j\phi}$), the mathematical analysis of the synchronous motor is a straightforward process. The response current $\mathbf{I}_S$, Fig. 5.27, can be calculated utilizing circuit analysis theory as follows. Kirchhoff's voltage law yields

$$\mathbf{V}_{ST}=\left[R_S+j(X_\ell+X_\phi)\right]\mathbf{I}_S+E_G\equiv\mathbf{Z}_S\mathbf{I}_S+\mathbf{E}_G \tag{5.95}$$

Solution of Eq. (5.95) for $\mathbf{I}_S$ yields

$$\mathbf{I}_S = \frac{\mathbf{V}_{ST} - \mathbf{E}_G}{\mathbf{Z}_S} \tag{5.96}$$

Once the current $\mathbf{I}_S$ is known, the voltages $\mathbf{V}_\phi$, $\mathbf{V}_\ell$, $\mathbf{V}_R$, $\mathbf{V}_C$, $\mathbf{V}_S$, and $\mathbf{V}_{\mathcal{R}}$ ($\mathbf{V}_{\mathcal{R}}$, defined as the **airgap potential drop**, designates the total speed-dependent voltage along the active length of the stator coils, due to the synchronously rotating fields $\mathbf{B}_R$ and $\mathbf{B}_{ST}$, **for the motor mode of operation**) are calculated utilizing Eqs. (5.97) through (5.102):

$$\mathbf{V}_R = R_R\mathbf{I}_S \tag{5.97}$$
$$\mathbf{V}_\ell = jX_\ell\mathbf{I}_S \tag{5.98}$$
$$\mathbf{V}_\phi = jX_\phi\mathbf{I}_S \tag{5.99}$$
$$\mathbf{V}_C = \mathbf{Z}_C\mathbf{I}_S = (R_S + jX_\ell)\mathbf{I}_S \tag{5.100}$$
$$\mathbf{V}_S = Z_R\mathbf{I}_S = (R_S + jX_S)\mathbf{I}_S \equiv \{R_S + j(X_\ell + X_\phi)\}\mathbf{I}_S \tag{5.101}$$

and

$$\mathbf{V}_{\mathcal{R}} = \mathbf{V}_\phi + \mathbf{E}_G \tag{5.102}$$

The relationships among the various voltages, as calculated with the aid of Eqs. (5.97) through (5.102), and their relative positions in the complex plane are exhibited in Fig. 5.28.

Also, utilization of the current of Eq. (5.96) enables calculation of the per-phase electrical powers, that is, (1) the **impressed power**, designated by $\mathbf{S}_{IN}$, (2) the **power required** by the stator winding, designated by $\mathbf{S}_S$, and (3) the **power developed** by the motor, designated by $\mathbf{S}_G$. Thus

$$\mathbf{S}_{IN} = P_{IN} + jQ_{IN} \equiv \mathbf{E}_S\mathbf{I}_S{}^* = |\mathbf{S}_{IN}|\cos\theta_S + j|\mathbf{S}_{IN}|\sin\theta_S \tag{5.103}$$
$$\mathbf{S}_S = P_S + jQ_S = Z_S|\mathbf{I}_S|^2 = (R_S + jX_S)|\mathbf{I}_S|^2 \tag{5.104}$$

and

$$\mathbf{S}_G = P_G + jQ_G \equiv \mathbf{E}_G\mathbf{I}_S{}^* = |\mathbf{S}_G|\cos\phi + j|\mathbf{S}_G|\sin\phi \tag{5.105}$$

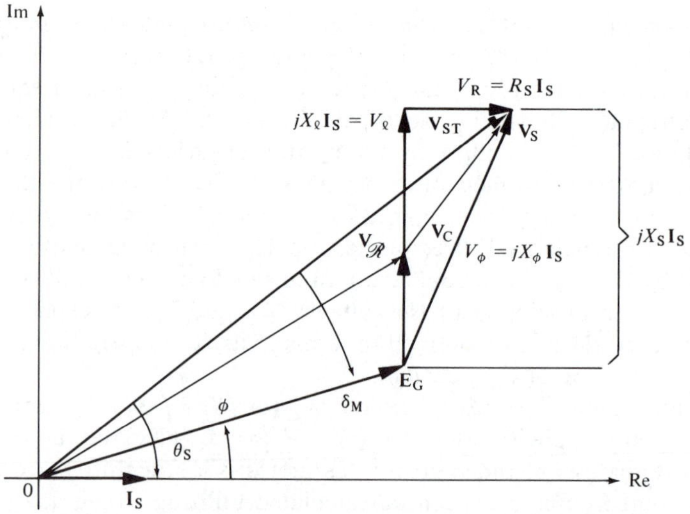

Fig. 5.28 Phasor Diagram Exhibiting the Relative Position of the Variables of Eqs. (5.97) through (5.102); the Excitation Is Explicitly Defined, That Is, PF$=k_{LGG}$; the Developed Phase Voltage and Response Phase Voltages Are Explictly Shown

Once the phase electrical powers $\mathbf{S}_{IN}$, $\mathbf{S}_S$, and $\mathbf{S}_G$ are calculated, the electrical powers, that is, (1) the **total impressed power**, $\mathbf{S}_{INT}$, (2) the **total required power** by the stator windings, $\mathbf{S}_{ST}$, and (3) the **total developed power**, $\mathbf{S}_{GT}$, can be calculated utilizing the theory given in Chapters 3 and 4 and Section 5.4, as follows:

$$\mathbf{S}_{INT} = P_{INT} + jQ_{INT} \equiv 3\mathbf{S}_{IN} = 3P_{IN} + j3Q_{IN} = \cdots \tag{5.106}$$

$$\mathbf{S}_{ST} = P_{ST} + jQ_{ST} \equiv 3\mathbf{S}_S = 3P_S + j3Q_S = \cdots \tag{5.107}$$

$$\mathbf{S}_{GT} = P_{GT} + jQ_{GT} \equiv 3\mathbf{S}_G = 3P_G + j3Q_G = \cdots \tag{5.108}$$

If the *total windage power losses*, P_{WLT}, the *total bearing friction power losses*, P_{BFT}, and the *total core power losses*, P_{CLT}, are known, then the mechanical power delivered by the motor to the load, designated by P_{OM}, is calculated as

$$P_{OM} = P_{GT} - (P_{WLT} + P_{BFT} + P_{CL}) \tag{5.109}$$

Remembering again that power and torque are related by the defining relationship $P = \tau\omega$, we can calculate the **total developed torque**, designated by τ_{GT}, and the **total output torque**, designated by τ_{OM}, as

$$P_{GT} = \tau_{GT}\omega_S \tag{5.110}$$

and

$$P_{OM} = \tau_{OM}\omega_S \tag{5.111}$$

Eqs. (5.106) through (5.111) designate the various power levels of the synchronous motor as derived from the per-phase equivalent circuit of such a motor as exhibited in Fig. 5.27. The multiplication of the phase powers by a factor of 3 converts the synchronous motor's phase power levels to total power levels. This power-level calculation procedure is summarized in Fig. 5.29 in the format of a flow diagram. Note that this flow diagram represents the flow of power from the ac source to the shaft of the synchronous motor only. The ac source does not supply the rotor-winding copper losses (as it does for the induction motor, Fig. 4.17). It

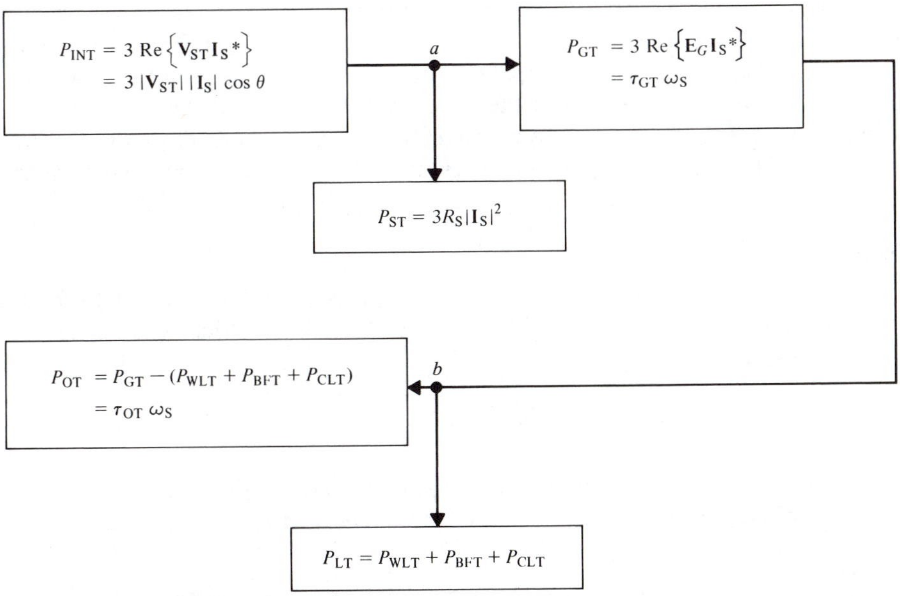

Fig. 5.29 Stator-Winding-Power Flow Diagram of a Balanced, Three-Phase, *P*-Pole, Synchronous Motor

does not have to do this in this case because the rotor is rotating at synchronous speed ω_S (contrary to the rotor of the induction motor, which rotates at a speed $\omega_R < \omega_S$). It must be emphasized, however, that the rotor losses exist, but they are supplied by the dc excitation of the rotor winding. Section 5.7 deals with the mechanism by which the synchronous motor senses the presence of the shaft load and with the process by which the electrical energy ac source responds to the shaft load and therefore proceeds to provide electrical energy for continuous shaft rotation—or fails to do so and the shaft stops rotating.

As in the cases of the induction and dc motors, the performance of the balanced, three-phase, P-pole, synchronous motor can be judged by the examination of two performance indices: (1) the efficiency η and (2) the speed regulation SR, both in percentages. The efficiency of the synchronous motor is defined as

$$\eta \equiv \frac{P_{OM}}{P_{INT}} = \frac{P_{INT} - \Sigma P_j}{P_{INT}} = 1 - \frac{\Sigma P_j}{P}, \quad j = \text{ST, WLT, BFT, and CLT} \tag{5.112}$$

The speed regulation is defined as

$$\text{SR} \equiv \frac{\omega_{NL} - \omega_{FL}}{\omega_{FL}} = \frac{\omega_S - \omega_R}{\omega_R} \tag{5.113}$$

where the subscripts NL and FL designate *no-load* and *full-load*, correspondingly (T designates *total*) and the subscript R designates rotor speed. It is established earlier in Section 5.5 that the synchronous motor runs at synchronous speed ω_S or it does not run at all. Therefore, $\omega_R \equiv \omega_S$, and Eq. (5.113) implies that *the speed regulation of the synchronous motor is zero.*

Example 5.7

A 60-Hz, four-pole, three-phase, Y-connected, synchronous motor has the following per-phase parameters: $R_S = 0.5$, $X_\phi = 3.6$, $X_\ell = 0.4$, all in ohms; excitation voltage $\mathbf{V}_{ST} = 2{,}886.75e^{j36.86°}$ V; generated voltage $\mathbf{E}_G = 2{,}519.39e^{j27.36°}$ V; and total losses are 68,540.37 W. Calculate: (1) the power angle δ_M, (2) the response current $\mathbf{I}_S$, (3) the voltages $\mathbf{V}_R$, $\mathbf{V}_\ell$, $\mathbf{V}_\phi$, $\mathbf{V}_C$, $\mathbf{V}_S$, and $\mathbf{V}_{\mathfrak{R}}$, (4) the total input power to the motor, P_{INT}, (5) the power lost in the stator winding, P_{ST}, (6) the power developed by the motor, P_{GT}, (7) the power delivered by the motor, P_{OM}, (8) the torque delivered at the shaft, τ_{OM}, (9) the efficiency of the motor η, and (10) the speed regulation SR of the synchronous motor.

Solution

Before proceeding with a per-phase analysis, it is essential to calculate the speed ω_S of the motor. From Eq. (5.90) the speed of the rotor, which is the speed of the field of the stator windings, is calculated as

$$\omega_S = \frac{\omega_E}{P/2} = \frac{2\pi \times f_E}{P/2} = \frac{2\pi \times 60}{4/2} = \underline{\underline{188.50}} \text{ rad/s} \tag{1}$$

Since the voltages $\mathbf{V}_{ST}$ and $\mathbf{E}_G$ are explicitly known, the angle between these voltages, defined as the power angle and designated as $\delta_M \equiv \angle\mathbf{E}_G - \angle\mathbf{V}_{ST}$, can be calculated as

$$\delta_M \equiv \angle\mathbf{E}_G - \angle\mathbf{V}_{ST} = 27.36° - 36.86° = \underline{\underline{-9.50°}} \tag{2}$$

Reference to Fig. 5.27 enables us to write the describing equation of the motor as [also see Eqs. (5.95) and (5.96)]

$$\mathbf{V}_{ST} = \left[R_S + j(X_\phi + X_\ell)\right]\mathbf{I}_S + \mathbf{E}_G = (R_S + jX_S)\mathbf{I}_S + \mathbf{E}_G \tag{3}$$

Since the voltages $\mathbf{V}_{ST}$ and $\mathbf{E}_G$, and the synchronous impedance X_S, are defined explicitly, Eq. (3) can be solved for the current $\mathbf{I}_S$:

$$\begin{aligned}\mathbf{I}_S &= \frac{\mathbf{V}_{ST} - \mathbf{E}_G}{R_S + jX_S} = \frac{2{,}886.75e^{j36.86°} - 2{,}519.39e^{j27.36°}}{0.5 + j4}\\ &= \frac{(2{,}309.70 - 2{,}237.56) + j(1{,}731.65 - 1{,}157.86)}{0.5 + j4}\\ &= \frac{72.14 + j573.79}{0.5 + j4} = \frac{578.31e^{j82.87°}}{4.03e^{j82.87°}} = \underline{\underline{143.5e^{j0°}}}\text{ A}\end{aligned} \tag{4}$$

Now utilization of Eqs. (5.97) through (5.102) enables calculation of the suggested voltages as follows:

$$\mathbf{V}_R = R_S\mathbf{I}_S = 0.5 \times 143.5 = \underline{\underline{71.75}}\text{ V} \tag{5}$$

$$\mathbf{V}_\ell = jX_\ell\mathbf{I}_S = j0.4 \times 143.5 = \underline{\underline{j57.40}}\text{ V} \tag{6}$$

$$\mathbf{V}_\phi = jX_\phi\mathbf{I}_S = j3.6 \times 143.5 = \underline{\underline{j516.60}}\text{ V} \tag{7}$$

$$\begin{aligned}\mathbf{V}_C &= (R_S + jX_\ell)\mathbf{I}_S = (0.5 + j0.4) \times 143.5 = 71.75 + j57.40\\ &= \underline{\underline{91.88e^{j38.66°}}}\text{ V}\end{aligned} \tag{8}$$

$$\begin{aligned}\mathbf{V}_S &= (R_S + jX_S)\mathbf{I}_S = (0.5 + j4.0) \times 143.5 = 71.75 + j574.00\\ &= \underline{\underline{578.47e^{j82.87°}}}\text{ V}\end{aligned} \tag{9}$$

and

$$\begin{aligned}\mathbf{V}_{\mathfrak{R}} &= \mathbf{V}_\phi + \mathbf{E}_G = j516.60 + 2{,}519.39e^{j27.36°}\\ &= j516.60 + 2{,}237.56 + j1{,}157.86\\ &= 2{,}237.56 + j1{,}674.46 = \underline{\underline{2{,}794.73e^{j36.81°}}}\text{ V}\end{aligned} \tag{10}$$

Utilization of Eqs. (5.103), (5.106), and (4), or Fig. 5.30 and Eq. (4), enables calculation of the total input power to the motor as

$$\begin{aligned}P_{INT} &= \text{Re}(3\mathbf{V}_{ST}\mathbf{I}_S{}^*) = \text{Re}\left[3 \times 2{,}886.75e^{j36.86°} \times 143.5\right]\\ &= \text{Re}(1{,}242{,}745.88e^{j36.86°}) = \underline{\underline{994{,}325.5}}\text{ W}\end{aligned} \tag{11}$$

Similarly, utilization of Eqs. (5.104), (5.107), and (4), or Fig. 5.30 and Eq. (4), enables calculation of the total power lost in the resistance of the stator winding as

$$P_{ST} = \text{Re}(3\mathbf{S}_S) = 3R_S|\mathbf{I}_S|^2 = 3 \times 0.5 \times (143.5)^2 = \underline{\underline{30{,}888.38}}\text{ W} \tag{12}$$

The power developed by the motor, P_{GT}, is calculated with the aid of Eqs. (5.105), (5.108), and (4), or Fig. 5.30 and Eq. (4), as

$$\begin{aligned}P_{GT} &= \text{Re}(3\mathbf{E}_G\mathbf{I}_S{}^*) = \text{Re}\left[3 \times (2{,}519.39e^{j27.36°}) \times 143.5\right]\\ &= \text{Re}\{1{,}084{,}597.40e^{j27.36°}\} = \underline{\underline{963{,}270.48}}\text{ W}\end{aligned} \tag{13}$$

The power delivered by the motor, P_{OM}, is calculated with the aid of Eqs. (5.109) and (13), or Fig. 5.30 and Eq. (13), as

$$P_{OM} = P_{GT} - (P_{WLT} + P_{BFT} + P_{CLT}) = 963{,}270.48 - 68{,}540.37 = \underline{\underline{894{,}730.11 \text{ W}}} \tag{14}$$

The torque delivered at the shaft of the synchronous motor is calculated with the aid of Eqs. (5.111), (14), and (1), or Fig. 5.30 and Eqs. (14) and (1), as

$$\tau_{OM} = \frac{P_{OM}}{\omega_S} = \frac{894{,}730.10}{188.50} = \underline{\underline{4{,}746.58 \text{ N}\cdot\text{m}}} \tag{15}$$

Now the efficiency η of the synchronous motor is calculated with the aid of Eqs. (5.112), (14), and (11) as

$$\eta = \frac{P_{OM}}{P_{INT}} \times 100 = \frac{894{,}730.11}{994{,}325.50} \times 100 = \underline{\underline{89.98\ \%}} \tag{16}$$

Finally, the speed regulation of the synchronous motor, SR, is calculated with the aid of Eqs. (5.113) and (1) as

$$\text{SR} = \frac{\omega_{NL} - \omega_{FL}}{\omega_{FL}} = \left.\frac{\omega_S - \omega_R}{\omega_R}\right|_{\omega_R = \omega_S} = \underline{\underline{0}} \tag{17}$$

If Z_S (and Z_C) are explicitly defined, and if the excitation is defined in terms of its voltampere rating (i.e., $|\mathbf{V}_{ST}|$ and $|\mathbf{I}_S|$) and its power factor (i.e., $\cos\theta_S$), the mathematical analysis of the synchronous motor, that is, the calculation of $\mathbf{E}_G$, $\mathbf{S}_G$, η, and so on, becomes more challenging. This occurs, as in the case of the three-phase, P-pole, synchronous generator, because the power factor can take values that yield three distinct cases of mathematical analysis: (1) unity power-factor analysis, (2) lagging power-factor analysis, and (3) leading power-factor analysis. These analyses are pursued in accordance with the theory given in Appendix A and Section 5.4.

Unity Power-Factor (Designated by $\cos\theta_S \equiv 1.0$) Analysis

Since a unity power factor implies that the impressed voltage, $\mathbf{E}_S$, and the response current, $\mathbf{I}_S$, Fig. 5.27, are in phase, that is,* $\cos\theta_S = 1$ and $\theta_S = 0$. The phasors $\mathbf{V}_{ST}$ and $\mathbf{I}_S$ (for the current exhibited in Fig. 5.27 with unity power-factor excitation) are drawn as in Fig. 5.30. However, in order to complete the phasor diagram of Fig. 5.30, and thus to solve for the desired developed voltage $\mathbf{E}_G$, the describing equation of Fig. 5.27 is written, utilizing Kirchhoff's voltage law, as

$$\mathbf{V}_{ST} = R_S\mathbf{I}_S + j(X_\ell + X_\phi)\mathbf{I}_S + \mathbf{E}_G \tag{5.114}$$

The current $\mathbf{I}_S$ and the voltage $\mathbf{E}_S$ allows us to write the solution of Eq. (5.114) for $\mathbf{E}_G$ as

$$\mathbf{E}_G = (\mathbf{E}_S - R_S\mathbf{I}_S) - j(X_\ell + X_\phi)\mathbf{I}_S \equiv |\mathbf{E}_G|e^{j\phi} \tag{5.115}$$

The relative positions of the phasors of Eq. (5.115) are depicted in Fig. 5.30. Utilization of Fig. 5.30 also enables us to write the equations for the voltage drops

*θ_S was used earlier in this chapter, Eq. (5.15), to designate the angular position of the total stator field $\mathbf{B}_{ST}$. However, this use was very limited. Here θ_S designates the angle of the voltage $\mathbf{V}_{ST}$, and it is **extensively** used with this meaning.

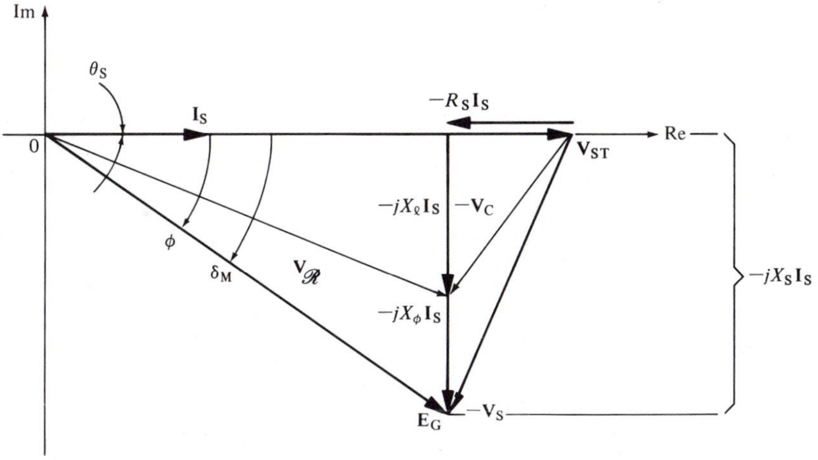

Fig. 5.30 Phasor Diagram for the Per-Phase Analysis of a Balanced, Three-Phase, *P*-Pole, Synchronous Motor with Unity Power-Factor Excitation, $\cos\theta_S = k$

$\mathbf{V}_S$ across the synchronous impedance $\mathbf{Z}_S$ and $\mathbf{V}_C$ across the stator coil impedance $\mathbf{Z}_C$ and the airgap potential drop, $\mathbf{V}_{\mathfrak{R}}$, as

$$-\mathbf{V}_S = -\mathbf{Z}_S\mathbf{I}_S = -R_S\mathbf{I}_S - j(X_\ell + X_\phi)\mathbf{I}_S \tag{5.116}$$

$$-\mathbf{V}_C = -\mathbf{Z}_C\mathbf{I}_S = -R_S\mathbf{I}_S - jX_\ell\mathbf{I}_S \tag{5.117}$$

and

$$\mathbf{V}_{\mathfrak{R}} = (\mathbf{V}_{ST} - R_S\mathbf{I}_S) - jX_\ell\mathbf{I}_S \tag{5.118}$$

Lagging Power-Factor Analysis

Since a lagging power factor implies that the response current, $\mathbf{I}_S$, Fig. 5.27, lags the impressed voltage, $\mathbf{V}_{ST}$, by an angle $\theta_S = \cos^{-1}(k_{LGG})$, where $\cos\theta_S = k_{LGG}$ and $\theta_S > 0$, the phasors $\mathbf{V}_{ST}$ and $\mathbf{I}_S$ (for the circuit exhibited in Fig. 5.27, with lagging power-factor excitation) are drawn as in Fig. 5.31. Completion of this phasor diagram, however, necessitates the use of the generalized describing equation of the circuit of Fig. 5.27, that is, Eq. (5.113). The lagging current, $\mathbf{I}_S$, which is set along the Re-axis, Fig. 5.31, and the decomposition of the voltage $\mathbf{V}_{ST}$ into components along the Re- and Im-axes allows us to write the solution of Eq. (5.114) for $\mathbf{E}_G$ as

$$\mathbf{E}_G = (|\mathbf{V}_{ST}|\cos\theta_S - R_S\mathbf{I}_S) + j\left\{|\mathbf{V}_{ST}|\sin\theta_S - (X_\ell + X_\phi)\mathbf{I}_S\right\} \equiv |\mathbf{E}_G|e^{j\phi} \tag{5.119}$$

The relative positions of the phasors (and their corresponding components) of Eq. (5.119) are depicted in Fig. 5.31. Utilization of Fig. 5.31 also enables writing of the equations for the voltage drops $\mathbf{V}_S$ across the synchronous impedance $\mathbf{Z}_S$ and $\mathbf{V}_C$ across the stator coil impedance $\mathbf{Z}_C$ and the airgap potential drop, $\mathbf{V}_{\mathfrak{R}}$, as

$$-\mathbf{V}_S = -\mathbf{Z}_S\mathbf{I}_S = -R_S\mathbf{I}_S - j(X_\ell + X_\phi)\mathbf{I}_S \tag{5.120}$$

$$-\mathbf{V}_C = -\mathbf{Z}_C\mathbf{I}_S = -R_S\mathbf{I}_S - jX_\ell\mathbf{I}_S \tag{5.121}$$

and

$$\mathbf{V}_{\mathfrak{R}} = (|\mathbf{V}_{ST}|\cos\theta_S - R_S\mathbf{I}_S) + j(|\mathbf{V}_{ST}|\sin\theta_S - X_\ell\mathbf{I}_S) \tag{5.122}$$

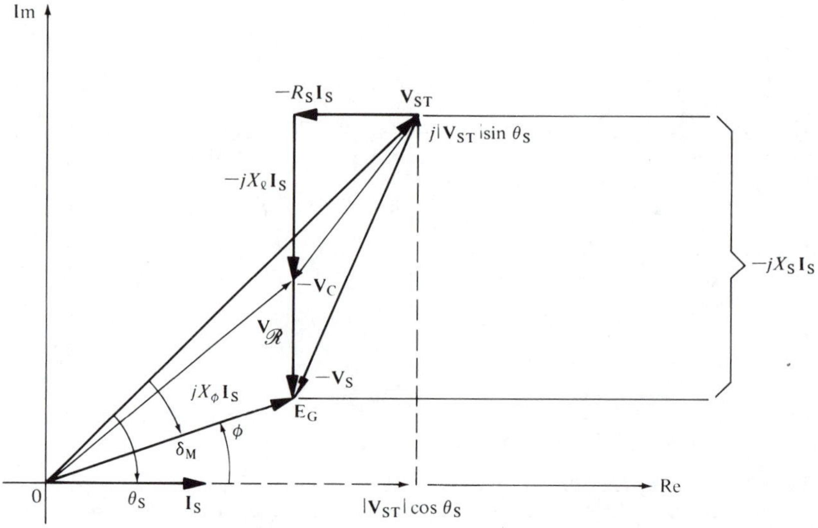

Fig. 5.31 Phasor Diagram for the Per-Phase Analysis of a Balanced, Three-Phase, P-Pole, Synchronous Motor, with Lagging Power-Factor Excitation, $\cos\theta_S = k_{LGG}$

Leading Power-Factor Analysis

Since a leading power factor implies that the response current, $\mathbf{I}_S$, Fig. 5.27, leads the impressed voltage, $\mathbf{E}_S$, by an angle $\theta_S = \cos^{-1}(k_{LDG})$, where $\cos\theta_S = k_{LDG}$ and $\theta_S < 0$, the phasors $\mathbf{E}_S$ and $\mathbf{I}_S$ (for the circuit exhibited in Fig. 5.27, with leading power-factor excitation) are drawn as in Fig. 5.32. As in the previous case, completion of this phasor diagram necessitates the use of the generalized describing equation of the circuit of Fig. 5.27, that is, Eq. (5.114). The leading current $\mathbf{I}_S$, which is set along the Re-axis, Fig. 5.32, and the decomposition of the voltage $\mathbf{V}_{ST}$ to components along the Re- and Im-axes allows us to write the solution of Eq.

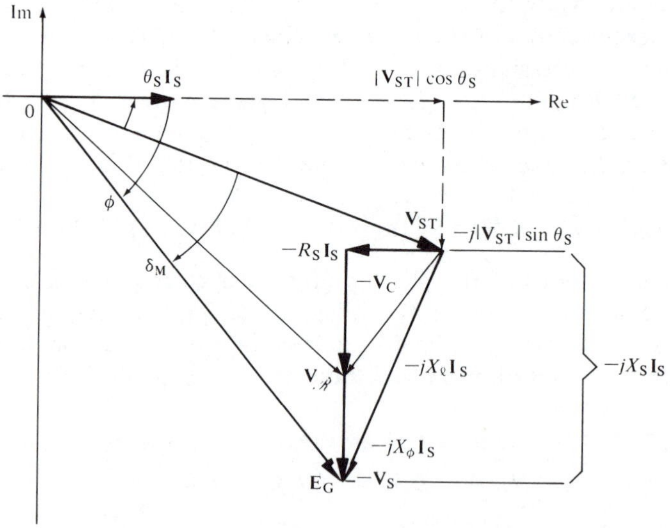

Fig. 5.32 Phasor Diagram for the Per-Phase Analysis of a Balanced, Three-Phase, P-Pole, Synchronous Motor with Leading Power-Factor Excitation, $\cos\theta_S = k_{LDG}$

(5.114) for $\mathbf{E}_G$ as

$$\mathbf{E}_G = (|\mathbf{V}_{ST}|\cos\theta_S - R_S\mathbf{I}_S) + j\left(|\mathbf{V}_{ST}|\sin\theta_S - (X_\ell + X_\phi)\mathbf{I}_S\right) = |\mathbf{E}_G|e^{j\phi} \quad (5.123)$$

The relative positions of the phasors (and their corresponding components) of Eq. (5.123) are depicted in Fig. 5.32. Utilization of Fig. 5.32 also enables writing of the equations for the voltage drops $\mathbf{V}_S$ across the synchronous impedance $\mathbf{Z}_S$ and $\mathbf{V}_C$ across the stator coil impedance $\mathbf{Z}_C$ and the airgap potential drop, $\mathbf{V}_{\mathfrak{R}}$, as

$$-\mathbf{V}_S = -\mathbf{Z}_S\mathbf{I}_S = -R_S\mathbf{I}_S - j(X_\ell + X_\phi)\mathbf{I}_S \quad (5.124)$$

$$-\mathbf{V}_C = -Z_C\mathbf{I}_S = -R_S\mathbf{I}_S - jX_\ell\mathbf{I}_S \quad (5.125)$$

and

$$\mathbf{V}_{\mathfrak{R}} = (|\mathbf{V}_{ST}|\cos\theta_S - R_S\mathbf{I}_S) + j(|\mathbf{V}_{ST}|\sin\theta_S - X_\ell\mathbf{I}_S) \quad (5.126)$$

As mentioned in Section 5.4, as a rule only the synchronous impedance of the synchronous motor is usually defined explicitly (i.e., $\mathbf{Z}_S = R_S + jX_S$). Therefore, the circuit of Fig. 5.27 can be simplified and drawn as in Fig. 5.33. Under these conditions only the quantities designated with *heavy lines* in Figs. 5.30, 5.31, and 5.32 along with Eqs. (5.115), (5.119), and (5.123) are essential for the complete analysis of synchronous motors whose excitation is of unity, lagging, or leading power factors. Inspection of Figs. 5.30, 5.31, and 5.32 reveals that (1) the developed voltage, $\mathbf{E}_G$, per phase, is seemingly larger than the excitation voltage, $\mathbf{V}_{ST}$, for leading and unity power factors but is seemingly smaller than $\mathbf{V}_{ST}$ for lagging power factors, and (2) the developed voltage, $\mathbf{E}_G$, always lags the excitation (i.e., terminal) voltage $\mathbf{V}_{ST}$. This angle is defined as the **power angle** or the **torque angle** (or the **displacement angle**) of the synchronous motor and is designated by $\delta_M \equiv \angle\mathbf{E}_G - \angle\mathbf{V}_{ST}$. It is always negative when the synchronous device functions as a motor, and it represents the measure of the angular displacement, in the clockwise direction, of the developed phase voltage $\mathbf{E}_G$ measured from the impressed (i.e., the terminal) voltage $\mathbf{V}_{ST}$. (See Section 5.4 for the definition of the power angle δ_G of synchronous generators.) More will be said about this angle in Section 5.7 and later sections.

The magnetizing, neutral, and demagnetizing effects of the stator field $\mathbf{B}_{ST}$ on the rotor field $\mathbf{B}_R$ for the motor mode of operation of the synchronous device are applicable as they were for the generator mode of operation. A comparison of Fig. 5.33 with Figs. 5.9 and 5.14 demonstrates these three effects.

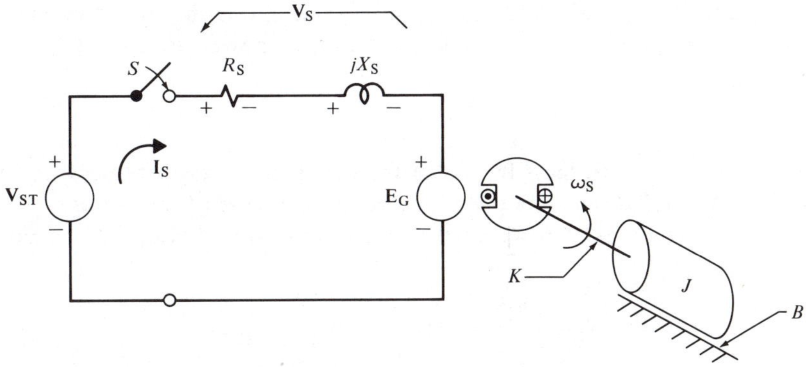

Fig. 5.33 Simplified, Single-Phase, Steady State, Equivalent Circuit of a Balanced, Three-Phase, *P*-Pole, Synchronous Motor Utilized in its Per-Phase Analysis

In this type of analysis—that is, when the synchronous impedance $\mathbf{Z}_S$ of the synchronous motor and both the voltampere rating (i.e., $|\mathbf{V}_{ST}|$ and $|\mathbf{I}_S|$) and the power factor (i.e., $\cos\theta_S$) of the motor's excitation are explicitly defined—the generated phase voltages $\mathbf{E}_G$ are easily calculated with the aid of Eqs. (5.115), (5.119), and (5.123), and the relevant theory from Chapter 3. With this information at hand, we can calculate the per-phase generated power $\mathbf{S}_G$ as

$$\mathbf{S}_G = P_G + jQ_G = \mathbf{E}_G\mathbf{I}_S{}^* = |\mathbf{S}_G|\cos\phi + j|\mathbf{S}_G|\sin\phi \tag{5.127}$$

Similarly, the per-phase power required by the stator winding, designated by $\mathbf{S}_S$, and the per-phase input power, designated by $\mathbf{S}_{IN}$, can be calculated readily from the input data as

$$\mathbf{S}_S = P_S + jQ_S = \mathbf{Z}_S|\mathbf{I}_S|^2 = (R_S + jX_S)|\mathbf{I}_S|^2 \tag{5.128}$$

and

$$\mathbf{S}_{IN} = P_{IN} + jQ_{IN} = |\mathbf{E}_S||\mathbf{I}_S|\cos\theta_S + j|\mathbf{E}_S||\mathbf{I}_S|\sin\theta_S \tag{5.129}$$

Once the phase powers, $\mathbf{S}_G$, $\mathbf{S}_S$, and $\mathbf{S}_{IN}$, are calculated, the total developed power, $\mathbf{S}_{GT}$, the total power required by the stator windings, $\mathbf{S}_{ST}$, and the total impressed power, $\mathbf{S}_{INT}$, can be calculated utilizing Eqs. (5.108), (5.107), and (5.106). If this information is available and if the total windage power losses, P_{WLT}, the total bearing friction power losses, P_{BFT}, and the total core power losses, P_{CLT}, are explicitly known, then the mechanical power delivered by the motor, designated by P_{OM}, the total developed torque, designated by τ_{GT}, and the total output torque, designated by τ_{OM}, can be calculated with the aid of Eqs. (5.109), (5.110), and (5.111). With all this information at hand, the efficiency η and the speed regulation of the balanced, three-phase, P-pole, synchronous motor can be calculated utilizing Eqs. (5.112) and (5.113).

Example 5.8

A 60-Hz, four-pole, three-phase, Y-connected, 1,200-hp, 5,000-V, line-to-line synchronous motor has a per-phase synchronous impedance $\mathbf{Z}_S = 0.5 + j4 = 4.03e^{j82.87°}$ Ω. Its efficiency, at rated load and PF = 1, is $\eta = 90\%$. Neglecting the field losses due to dc excitation, calculate: (1) the developed voltage $\mathbf{E}_G$, (2) the power angle δ_M, (3) the total developed power P_{GT}, (4) the total developed torque τ_{GT}, (5) the total output torque τ_{OM}, and (6) the total losses, $P_{LT} = P_{WLT} + P_{BFT} + P_{CLT}$. Repeat calculations (1) through (5) and in addition calculate (a) the efficiency η, and (b) the speed regulation of the synchronous motor for power factors $k_{LGG} = 0.8$ and $k_{LDG} = 0.8$. (Note: The total output power $P_{OM} = 1{,}200$ hp and 1 hp = 746 W.)

Solution

Before proceeding with the solution of the problem, we must remember that this analysis is to be carried out on a per-phase basis. Thus the phase voltage, $|\mathbf{V}_{ST}|$, Fig. 5.34, should be calculated first from the given line-to-line voltage. Now Eq. (3.67) leads to

$$|\mathbf{V}_{ST}| = \frac{|\mathbf{V}_{LL}|}{\sqrt{3}} = \frac{5{,}000}{\sqrt{3}} = \underline{2{,}886.75 \text{ V}} \tag{1}$$

Similarly, the synchronous speed of the motor is calculated with the aid of Eq.

(5.90). Thus

$$\omega_S = \frac{\omega_E}{P/2} = \frac{2\pi f_E}{P/2} = \frac{2\pi \times 60}{4/2} = \underline{188.50} \text{ rad/s} \tag{2}$$

1. PF = 1.0; $\theta_L = \cos^{-1}(1)$, $\theta_L = 0°$. For unity power factor, correlation of Fig. 5.30 with the given data suggests that the voltage equation, Eq. (1), for unity power factor can be written explicitly as

$$\mathbf{V}_{ST} = |\mathbf{V}_{ST}|e^{j\theta_S} = \underline{2{,}886.75e^{j0°}} \text{ V} \tag{3}$$

Since $\eta \equiv P_{OM}/P_{INT} = 0.90$ and $P_{OM} = 1{,}200$ hp, it follows that P_{INT} can be calculated readily, in MKS units, as follows:

$$P_{INT} = \frac{P_{OM}}{\eta} = \frac{1{,}200 \text{ hp} \times 746 \text{ W/hp}}{0.90} = \underline{994{,}666.67} \text{ W} \tag{4}$$

Also, since $P_{INT} = \text{Re}(\mathbf{S}_{INT})$, we are able to calculate the current of the stator winding from Eq. (5.103), written as

$$\mathbf{S}_{INT} = P_{INT} + jQ_{INT} = 3\mathbf{V}_{ST}\mathbf{I}_S{}^* = 3|\mathbf{V}_{ST}||\mathbf{I}_S|\cos\theta_S + j3|\mathbf{V}_{ST}||\mathbf{I}_S|\sin\theta_S \tag{5}$$

by equating the real part of Eq. (5) with Eq. (4). Thus

$$3|\mathbf{V}_{ST}||\mathbf{I}_S|\cos\theta_S = \underline{994{,}666.67} \tag{6}$$

Utilization of Eq. (6), **but adjusting it to reflect unity power factor**, yields

$$3|\mathbf{V}_{ST}||\mathbf{I}_S| \times 1 = \underline{994{,}666.67} \tag{7}$$

Solution of Eq. (7) for $|\mathbf{I}_S|$ yields

$$|\mathbf{I}_S| = \frac{994{,}666.67}{3 \times 2{,}886.75 \times 1} = \underline{114.85} \text{ A} \tag{8}$$

Utilization of Eq. (5.115) in conjunction with Fig. 5.30 and Eq. (8) yields

$$\begin{aligned} \mathbf{E}_G &= (\mathbf{V}_{ST} - R_S\mathbf{I}_S) - jX_S\mathbf{I}_S = (2{,}886.75 - 0.5 \times 114.85) - j4 \times 114.85 \\ &= 2{,}829.33 - j459.40 = \underline{\underline{2{,}866.38e^{-j9.22°}}} \text{ V} \end{aligned} \tag{9}$$

From Fig. 5.30 and Eqs. (3) and (9), the torque angle δ_M is calculated as

$$\delta_M = \angle\mathbf{E}_G - \angle\mathbf{V}_{ST} = \phi - \theta_S = -9.22° - 0.00° = \underline{\underline{-9.22°}} \tag{10}$$

Note that the developed voltage $\mathbf{E}_G$ is **lagging** the impressed (i.e., terminal) voltage $\mathbf{V}_{ST}$ by an angle of 9.22°.

The developed power P_{GT} is calculated with the aid of Eqs. (5.105), (5.106), (9), and (8) as follows:

$$\begin{aligned} P_{GT} &= \text{Re}(3 \times \mathbf{S}_G) = \text{Re}(3 \times \mathbf{E}_G\mathbf{I}_S{}^*) = \text{Re}(3 \times 2{,}866.38e^{-j9.22°} \times 114.85e^{j0°}) \\ &= \text{Re}(987{,}611.23e^{-j9.22°}) = \underline{\underline{974{,}851.68}} \text{ W} \end{aligned} \tag{11}$$

Utilization of Eqs. (5.110), (11), and (2) yields the total developed torque τ_{GT} as

$$\tau_{GT} = \frac{P_{GT}}{\omega_S} = \frac{974{,}851.68}{188.50} = \underline{\underline{5{,}171.63}} \text{ N} \cdot \text{m} \tag{12}$$

Eqs. (5.111) and (2), and the input data for P_{OM}, enable calculation of the total output torque τ_{OM} as

$$\tau_{OM} = \frac{P_{OM}}{\omega_S} = \frac{1{,}200 \times 746}{188.50} = \frac{895{,}200}{188.50} = \underline{\underline{4{,}749.07}} \text{ N} \cdot \text{m} \tag{13}$$

Similarly, Eq. (5.109) in conjunction with Eq. (11) and the input data for P_{OM} enable calculation of the total losses as

$$P_{LT} = P_{WLT} + P_{BFT} + P_{CLT} = P_{GT} - P_{OM} = 974{,}851.68 - (1{,}200 \times 746) = \underline{\underline{79{,}651.68}} \text{ W} \tag{14}$$

2. $PF = k_{LGG} = 0.8$; $\theta_L = \cos^{-1}(0.8)$, $\theta_L = 36.87°$. For lagging power factor, correlation of Fig. 5.31 with the given data suggests that the voltage equation, Eq. (1), for lagging power factor can be written explicitly as

$$\mathbf{V}_{ST} = |\mathbf{V}_{ST}|e^{j\theta_S} = \underline{2{,}886.75e^{j36.87°}} \text{ V} \tag{15}$$

Again, utilization of Eq. (6), but adjusting it to reflect a lagging power factor of $k_{LGG} = 0.8$, yields

$$3|\mathbf{V}_{ST}||\mathbf{I}_S|\cos\theta_S = 994{,}666.67 \tag{16}$$

Solution of Eq. (16) for $|\mathbf{I}_S|$ yields

$$|\mathbf{I}_S| = \frac{994{,}666.67}{3 \times 2{,}886.75 \times 0.8} = \underline{143.57} \text{ A} \tag{17}$$

Utilization of Eq. (5.119) in conjunction with Fig. 5.31 and Eq. (17) yields

$$\begin{aligned}\mathbf{E}_G &= (|\mathbf{V}_{ST}|\cos\theta_S - R_S\mathbf{I}_S) + j(|\mathbf{V}_{ST}|\sin\theta_S - X_S\mathbf{I}_S) \\ &= (2{,}886.75 \times 0.8 - 0.5 \times 143.57) + j(2{,}886.75 \times 0.6 - 4.0 \times 143.57) \\ &= \underline{\underline{2{,}237.62 + j1{,}157.77}} = \underline{\underline{2{,}519.40e^{j27.36°}}} \text{ V}\end{aligned} \tag{18}$$

From Fig. 5.31 and Eqs. (18) and (15), it is obvious that the power angle δ_M is

$$\delta_M = \angle\mathbf{E}_G - \angle\mathbf{V}_{ST} = \phi - \theta_S = 27.36° - 36.87° = \underline{\underline{-9.51°}} \tag{19}$$

Note that the developed voltage $\mathbf{E}_G$ is lagging the impressed voltage $\mathbf{V}_{ST}$ by an angle of 9.51°.

The total developed power P_{GT} is calculated with the aid of Eqs. (5.105), (5.108), (18), and (17) as follows:

$$\begin{aligned}P_{GT} &= \text{Re}(3 \times \mathbf{S}_G) = \text{Re}(3 \times \mathbf{E}_G\mathbf{I}_s^*) = \text{Re}\left[3 \times (2{,}519.40e^{27.36°}) \times 143.57e^{j0°}\right] \\ &= \text{Re}(1{,}085{,}130.77e^{j27.36°}) = \underline{\underline{963{,}744.19}} \text{ W}\end{aligned} \tag{20}$$

The total developed torque τ_{GT} is calculated with the aid of Eqs. (5.110), (2), and (10) as follows:

$$\tau_{GT} = \frac{P_{GT}}{\omega_S} = \frac{963{,}744.19}{188.50} = \underline{\underline{5{,}112.70}} \text{ N} \cdot \text{m} \tag{21}$$

The total output power P_{OM} is calculated using Eqs. (5.109), or Fig. 5.29, and Eqs. (14) and (20) as

$$P_{OM} = P_{GT} - P_{LT} = 963{,}744.19 - 79{,}651.68 = \underline{\underline{884{,}092.51}} \text{ W} \tag{22}$$

The total output torque τ_{OM} is calculated with the aid of Eq. (5.111), or Fig. 5.29, and Eqs. (22) and (2) as

$$\tau_{OM} = \frac{P_{OM}}{\omega_S} = \frac{884{,}092.51}{188.50} = \underline{\underline{4{,}690.15}} \text{ N} \cdot \text{m} \tag{23}$$

The efficiency η can be calculated using Eqs. (5.112), (22), and (24) as

$$\eta = \frac{P_{OM}}{P_{INT}} \times 100\% = \frac{884{,}092.51}{994{,}666.67} \times 100\% = \underline{\underline{88.88}} \% \tag{24}$$

Finally, the speed regulation of the synchronous motor is calculated utilizing Eq. (5.113), and the theory following that equation, as

$$\mathrm{SR} = \frac{\omega_{\mathrm{NL}} - \omega_{\mathrm{FL}}}{\omega_{\mathrm{FL}}} = \frac{\omega_{\mathrm{S}} - \omega_{\mathrm{R}}}{\omega_{\mathrm{R}}}\bigg|_{\omega_{\mathrm{R}} \equiv \omega_{\mathrm{S}}} = \underline{\underline{0}} \tag{25}$$

3. $\mathrm{PF} = k_{\mathrm{LDG}} = 0.8$; $\theta_L = \cos^{-1}(0.8)$, $\theta_L = -36.87°$. For a leading power factor, correlation of Fig. 5.32 with the given data suggests that the voltage equation, Eq. (1), for leading power factor can be written explicitly as

$$\mathbf{V}_{\mathrm{ST}} = |\mathbf{V}_{\mathrm{ST}}|e^{j\theta_{\mathrm{S}}} = \underline{\underline{2{,}886.75e^{-j36.87°}}} \text{ V} \tag{26}$$

Utilization of Eq. (6), but adjusting it to reflect lagging power factor, yields

$$3|\mathbf{E}_{\mathrm{S}}||\mathbf{I}_{\mathrm{S}}|\cos\theta_{\mathrm{S}} = 994{,}666.67 \tag{27}$$

Solution of Eq. (28) for $|\mathbf{I}_{\mathrm{S}}|$ yields

$$|\mathbf{I}_{\mathrm{S}}| = \frac{994{,}666.67}{3 \times 2{,}886.75 \times 0.8} = \underline{\underline{143.57}} \text{ A} \tag{28}$$

Utilization of Eq. (5.123), in conjunction with Fig. 5.32 and Eq. (29), yields

$$\begin{aligned}\mathbf{E}_{\mathrm{G}} &= (|\mathbf{V}_{\mathrm{ST}}|\cos\theta_{\mathrm{S}} - R_{\mathrm{S}}\mathbf{I}_{\mathrm{S}}) + j(|\mathbf{V}_{\mathrm{ST}}|\sin\theta_{\mathrm{S}} - X_{\mathrm{S}}\mathbf{I}_{\mathrm{S}}) \\ &= (2{,}886.75 \times 0.8 - 0.5 \times 143.57) + j(-2{,}886.75 \times 0.6 - 4 \times 143.57) \\ &= 2{,}237.62 - j2{,}306.33 = \underline{\underline{3{,}213.42e^{-j45.87°}}} \text{ V}\end{aligned} \tag{29}$$

From Fig. 5.32, and Eqs. (27) and (30), the torque angle δ_{M} is calculated as

$$\delta_{\mathrm{M}} = \angle\mathbf{E}_{\mathrm{G}} - \angle\mathbf{V}_{\mathrm{ST}} = \phi - \theta_{\mathrm{S}} = -45.87° - (-36.87°) = \underline{\underline{-9.00°}} \tag{30}$$

Note again that the developed voltage $\mathbf{E}_{\mathrm{G}}$ is lagging the impressed (i.e., terminal) voltage $\mathbf{V}_{\mathrm{ST}}$ by an angle of 9.00°.

Utilization of Eqs. (5.105), (5.108), (30), and (29) yields the total developed power P_{GT} as follows:

$$\begin{aligned}P_{\mathrm{GT}} &= \mathrm{Re}(3 \times \mathbf{S}_{\mathrm{G}}) = \mathrm{Re}(3 \times \mathbf{E}_{\mathrm{G}}\mathbf{I}_{\mathrm{S}}^*) = \mathrm{Re}(3 \times 3{,}213.42e^{-j45.87°} \times 143.57) \\ &= \mathrm{Re}(1{,}384{,}052.13e^{-j45.87°}) = \underline{\underline{963{,}699.87}} \text{ W}\end{aligned} \tag{31}$$

The total developed torque τ_{GT} is calculated with the aid of Eqs. (5.110), (32), and (2) as follows:

$$\tau_{\mathrm{GT}} = \frac{P_{\mathrm{GT}}}{\omega_{\mathrm{S}}} = \frac{963{,}699.87}{188.50} = \underline{\underline{5{,}112.47}} \text{ N} \cdot \text{m} \tag{32}$$

The total output power P_{OM} is calculated with the aid of Eq. (5.109), or Fig. 5.29, and Eqs. (32) and (14) as

$$P_{\mathrm{OM}} = P_{\mathrm{GT}} - P_{\mathrm{LT}} = 963{,}699.87 - 79{,}651.68 = \underline{\underline{884{,}048.19}} \text{ W} \tag{33}$$

The total output torque τ_{OM} is calculated with the aid of Eq. (5.111), or Fig. 5.29, and Eqs. (34) and (2) as

$$\tau_{\mathrm{OM}} = \frac{P_{\mathrm{OM}}}{\omega_{\mathrm{S}}} = \frac{884{,}048.19}{188.50} = \underline{\underline{4{,}689.91}} \text{ N} \cdot \text{m} \tag{34}$$

The efficiency of η of the synchronous motor can be calculated with the aid of Eqs. (5.112), (34), and (36) as

$$\eta = \frac{P_{\mathrm{OM}}}{P_{\mathrm{INT}}} \times 100 = \frac{884{,}048.19}{994{,}666.67} \times 100\% = \underline{\underline{88.88}}\ \% \tag{35}$$

Table 5.2 TABULATION OF ANSWERS: EXAMPLE 5.8

Quantity†	Power Factor (PF) $k_{LGG}=0.8$	$k=1$	$k_{LDG}=0.8$
$\mathbf{V}_{ST}$ (V)	$2{,}886.75e^{j36.87°}$	$2{,}886.75e^{j0°}$	$2{,}886.75e^{-j36.87°}$
ω_S (rad/s)	188.50	188.50	188.50
$\mathbf{I}_S$ (A)	$143.57e^{j0°}$	$114.85e^{j0°}$	$143.57e^{j0°}$
$\mathbf{E}_G$ (V)	$2{,}519.40e^{j24.36°}$	$2{,}866.38e^{-j9.22°}$	$3{,}213.42e^{-j45.87°}$
δ_M (degrees)	−9.51°	−9.22°	−9.00°
P_{GT} (W)	963,744.19	974,851.68	963,699.87
τ_{GT} (N·m)	5,112.70	5,171.63	5,112.47
P_{OM} (W)	884,092.51	895,200.00	884,048.19
τ_{OM} (N·m)	4,690.15	4,749.07	4,689.91
P_{INT} (W)	994,666.67	994,666.67	994,666.67
η (%)	88.88	90.00	88.88
SR (%)	0.00	0.00	0.00

†Pertinent Quantities, MKS Units

Finally, the speed regulation of the synchronous motor is calculated with the aid of Eq. (5.113), and the theory following it, as

$$\text{SR} = \frac{\omega_{NL}-\omega_{FL}}{\omega_{FL}} = \frac{\omega_S-\omega_R}{\omega_R}\bigg|_{\omega_R\equiv\omega_S} \underline{\underline{=0}} \tag{36}$$

The behavior of the motor for the various power factors becomes clear when you study Table 5.2 carefully. Note that the magnitude of the developed voltage $|\mathbf{E}_G|$ and the power angle δ_M increase monotonically. In contrast, the efficiency drops by equal increments for the equal lagging and leading power factors, when compared to the given efficiency at unity power factor.

5.7 REAL AND REACTIVE POWER VERSUS POWER-ANGLE CHARACTERISTICS OF THE SYNCHRONOUS DEVICES AND REGIONS OF OPERATION

In a synchronous generator, mechanical energy is changed to electrical energy. In a synchronous motor, the converse is true. The power required to effect the conversion is the rate of expenditure of energy or of doing work. Work is done when a torque acts through an angle θ_S, that is, $dw=\tau\, d\theta_S$.

Sufficient torque is required to cause the rotor of a synchronous generator, Fig. 5.12, to run at synchronous speed to overcome the **countertorque** caused by **windage** and **bearing friction**. If excitation is applied to the field, additional torque is required because of the **core losses. If, in addition, the stator circuit is closed through**

an external circuit, a power output results if and only if additional torque is supplied to the rotor by the prime mover. The transfer of energy from the rotor circuit to the stator circuit must occur through the flux common to both circuits.

In the case of the synchronous motor, Fig. 5.26, the electrical power input, except for the **stator winding, windage, bearing friction,** and **core losses**, is transferred to the rotor through the flux common to both circuits, and it supplies the torque demands of the load attached to the shaft.

To establish criteria for the two modes of operation, and thus to derive the terminal characteristics of the synchronous device, the mathematical analysis of the single-phase equivalent circuit of the three-phase, *P*-pole, smooth-rotor, balanced (1) synchronous generator, Fig. 5.19, and (2) synchronous motor, Fig. 5.33, is pursued.

The circuit of the generator is redrawn here for simplicity and realistic depiction of its phasor diagram, Figs. 5.34(a), 5.34(b), and 5.34(c). By design, the ratio of the synchronous reactance X_S to the synchronous resistance R_S for synchronous devices is set to be $(20:1) < (X_S : R_S) < (100:1)$, depending on the size of the device

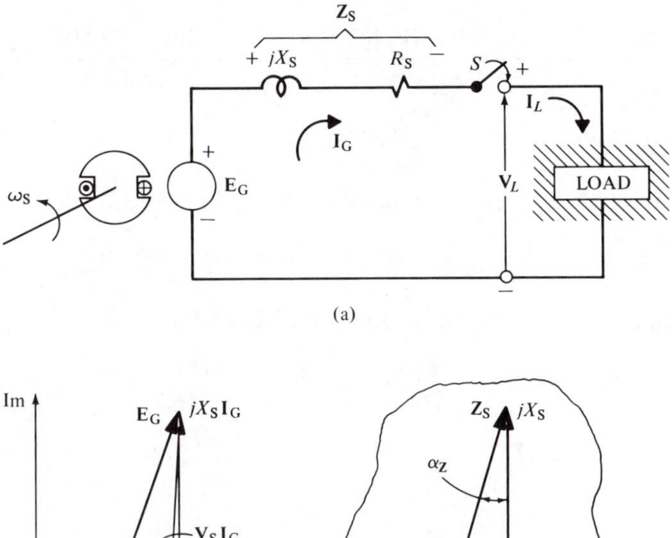

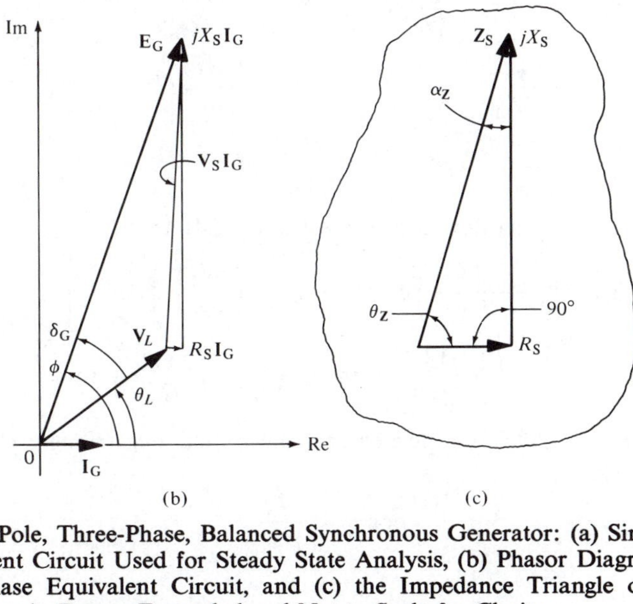

Fig. 5.34 *P*-Pole, Three-Phase, Balanced Synchronous Generator: (a) Single-Phase Equivalent Circuit Used for Steady State Analysis, (b) Phasor Diagram of the Single-Phase Equivalent Circuit, and (c) the Impedance Triangle of the Equivalent Circuit, Drawn Expanded and Not to Scale for Clarity

and as well as other design considerations. Since the angles of the impedance triangle, Fig. 5.34(b), whose vertical and horizontal sides differ so much, cannot be drawn in detail, the same impedance triangle is drawn expanded, and to scale, as in Fig. 5.34(c).

The electric power demanded by the load from the synchronous generator, Fig. 5.34(a), is written as

$$\mathbf{S}_D = P_D + jQ_D = \mathbf{V}_L \mathbf{I}_G^* \tag{5.130}$$

This is the electric power that the synchronous device (generator) delivers to the outside world. In power systems analysis, generation of electrical energy (power) is defined to be *positive*, that is, $\mathbf{S}_O \equiv +\mathbf{V}_L\mathbf{I}_G^*$. The quantity $\mathbf{S}_O$ is the electric output power of the generator and is identical to the power demanded by the load, $\mathbf{S}_D$; thus $\mathbf{S}_D \equiv \mathbf{S}_O$, and $\mathbf{S}_{DT} \equiv \mathbf{S}_{OT}$.

Utilization of Kirchhoff's voltage law enables us to write the describing equation of Fig. 5.34(a) as

$$\mathbf{E}_G - \mathbf{Z}_S\mathbf{I}_G - \mathbf{V}_L = 0 \tag{5.131}$$

Solution of Eq. (5.131) for the current $\mathbf{I}_G$, and utilization of Fig. 5.34(b), enable us to write†

$$\begin{aligned}\mathbf{I}_G &= \frac{\mathbf{E}_G - \mathbf{V}_L}{\mathbf{Z}_S} = \left.\frac{|\mathbf{E}_G|e^{j\phi} - |\mathbf{V}_L|e^{j\theta_L}}{|\mathbf{Z}_S|e^{j\theta_Z}}\right|_{\phi=\theta_L+\delta_G} = \frac{|\mathbf{E}_G|e^{j(\theta_L+\delta_G)}}{|\mathbf{Z}_S|e^{j\theta_Z}} - \frac{|\mathbf{V}_L|e^{j\theta_L}}{|\mathbf{Z}_S|e^{j\theta_Z}} \\ &= \frac{|\mathbf{E}_G|}{|\mathbf{Z}_S|}e^{j(\theta_L+\delta_G-\theta_Z)} - \frac{|\mathbf{V}_L|}{|\mathbf{Z}_S|}e^{j(\theta_L-\theta_Z)}\end{aligned} \tag{5.132}$$

The conjugate of the current $\mathbf{I}_G$ is obtained from Eq. (5.132) and is written as

$$\mathbf{I}_G^* = \frac{|\mathbf{E}_G|}{|\mathbf{Z}_S|}e^{-j(\theta_L+\delta_G-\theta_Z)} - \frac{|\mathbf{V}_L|}{|\mathbf{Z}_S|}e^{-j(\theta_L-\theta_Z)} \tag{5.133}$$

Substitution for $\mathbf{I}_G^*$ from Eq. (5.133) into Eq. (5.130) yields

$$\begin{aligned}\mathbf{S}_D &= |\mathbf{V}_L|e^{j\theta_L}\left\{\frac{|\mathbf{E}_G|}{|\mathbf{Z}_S|}e^{-j(\theta_L+\delta_G-\theta_Z)} - \frac{|\mathbf{V}_L|}{|\mathbf{Z}_S|}e^{-j(\theta_L-\theta_Z)}\right\} \\ &= \frac{|\mathbf{V}_L||\mathbf{E}_G|}{|\mathbf{Z}_S|}e^{j(\theta_L-\theta_L-\delta_G+\theta_Z)} - \frac{|\mathbf{V}_L|^2}{|\mathbf{Z}_S|}e^{j(\theta_L-\theta_L+\theta_Z)} \\ &= \frac{|\mathbf{V}_L||\mathbf{E}_G|}{|\mathbf{Z}_S|}e^{j(\theta_Z-\delta_G)} - \frac{|\mathbf{V}_L|^2}{|\mathbf{Z}_S|}e^{j\theta_Z}\end{aligned} \tag{5.134}$$

Use of Euler's identity enables us to write Eq. (5.134) as

$$\begin{aligned}\mathbf{S}_D &= \left[\left(\frac{|\mathbf{V}_L||\mathbf{E}_G|}{|\mathbf{Z}_S|}\right)\cos(\theta_Z-\delta_G) - \left(\frac{|\mathbf{V}_L|^2}{|\mathbf{Z}_S|}\right)\cos\theta_Z\right] \\ &\quad + j\left[\left(\frac{|\mathbf{V}_L||\mathbf{E}_G|}{|\mathbf{Z}_S|}\right)\sin(\theta_Z-\delta_G) - \left(\frac{|\mathbf{V}_L|^2}{|\mathbf{Z}_S|}\right)\sin\theta_Z\right]\end{aligned} \tag{5.135}$$

Utilization of trigonometric properties and the impedance triangle of Fig. 5.34(c) enables us to write

$$\cos\theta_Z = \frac{R_S}{|\mathbf{Z}_S|} \tag{5.136}$$

†Note that θ_L is a measure of the angle of departure of the voltage $\mathbf{V}_L$ from the Re-axis.

and

$$\sin\theta_Z = \frac{X_S}{|\mathbf{Z}_S|} \tag{5.137}$$

Also, utilization of geometric properties and Fig. 5.34(c) enables us to write $\theta_Z + \alpha_Z + 90 = 180$ (all in degrees), which yields

$$\theta_Z = 90 - \alpha_Z \tag{5.138}$$

Eq. (5.138) enables writing the quantities $\cos(\theta_Z - \delta_G)$ and $\sin(\theta_Z - \delta_G)$ of Eq. (5.135) as

$$\cos(\theta_Z - \delta_G) = \cos[90 - (\alpha_Z + \delta_G)] \equiv \sin(\alpha_Z + \delta_G) \tag{5.139}$$

and

$$\sin(\theta_Z - \delta_G) = \sin[90 - (\alpha_Z + \delta_G)] \equiv \cos(\alpha_Z + \delta_G) \tag{5.140}$$

Substitution for $\cos(\theta_Z - \delta_G)$, $\cos\theta_Z$, $\sin(\theta_Z - \delta_G)$, and $\sin\theta_Z$ from Eqs. (5.139), (5.136), (5.140), and (5.137) into Eq. (5.135) yields

$$\begin{aligned}\mathbf{W}_D = &\left[\left(\frac{|\mathbf{V}_L||\mathbf{E}_G|}{|\mathbf{Z}_S|}\right)\sin(\alpha_Z + \delta_G) - \left(\frac{|\mathbf{V}_L|^2}{|\mathbf{Z}_S|^2}\right)R_S\right] \\ &+ j\left[\left(\frac{|\mathbf{V}_L||\mathbf{E}_G|}{|\mathbf{Z}_S|}\right)\cos(\alpha_Z + \delta_G) - \left(\frac{|\mathbf{V}_L|^2}{|\mathbf{Z}_S|^2}\right)X_S\right]\end{aligned} \tag{5.141}$$

Since $(20:1) < (X_S : R_S) < (100:1)$ we may state that in practical synchronous generators, $X_S \gg R_S$. Then the following three conditions are valid for Eq. (5.141): (1) the resistance R_S can be neglected, (2) the angle α_Z of the real impedance triangle, Fig. 5.34(b), becomes

$$\alpha_Z = \tan^{-1}\left(\frac{R_S}{X_S}\right) \simeq 0° \tag{5.142}$$

and thus it can be neglected, and (3) the impedance $\mathbf{Z}_S$ is considered to be equal to the synchronous reactance X_S, that is, $|\mathbf{Z}_S| \simeq |jX_S| = X_S$. Under these conditions Eq. (5.141) can be written as

$$\begin{aligned}\mathbf{S}_D &= P_D + jQ_D \\ &= \left(\frac{|\mathbf{V}_L||\mathbf{E}_G|}{X_S}\sin\delta_G\right) \\ &\quad + j\left(\frac{|\mathbf{V}_L||\mathbf{E}_G|}{X_S}\cos\delta_G - \frac{|\mathbf{V}_L|^2}{X_S}\right)\end{aligned} \tag{5.143}$$

Utilization of Eq. (5.64) enables us to write the total real power P_{DT} and the total reactive power Q_{DT} demanded by the load and supplied by the synchronous generator as

$$P_{DT} \equiv P_{OT} = 3\left(\frac{|\mathbf{V}_L||\mathbf{E}_G|}{X_S}\right)\sin\delta_G \tag{5.144}$$

and

$$Q_{DT} \equiv Q_{OT} = 3\left(\frac{|\mathbf{V}_L||\mathbf{E}_G|}{X_S}\right)\cos\delta_G - 3\left(\frac{|\mathbf{V}_L|^2}{X_S}\right) \tag{5.145}$$

Inspection of Fig. 5.34(a) reveals that $|\mathbf{V}_L| < |\mathbf{E}_G|$. This idea, in conjunction with the fact that δ_G always has positive values, enables drawing $P_{DT} \equiv P_{OT} = f(\delta_G)$ and

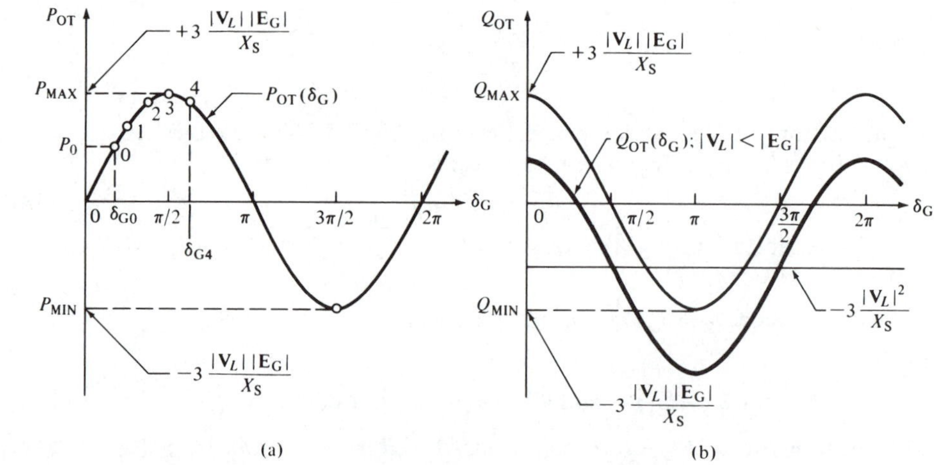

Fig. 5.35 Power-Angle Characteristics of a Smooth-Rotor Synchronous Generator: (a) Real Electric Output Power vs. Power Angle and (b) Reactive Output Power vs. Power Angle

$Q_{DT}=Q_{OT}=g(\delta_G)$, that is, Eqs. (5.144) and (5.145), for constant values of $|\mathbf{E}_G|$, $|\mathbf{V}_L|$, and X_S as exhibited in Figs. 5.35(a) and 5.35(b), respectively. The constancy of the voltage $\mathbf{V}_L$ is not assured at all points in the transmission network. However, the voltage at the output bus of a power station (or substation) is maintained constant at a predetermined value by means of automatic voltage regulators.

The variation of P_{OT} and Q_{OT} as functions of δ_G, Fig. 5.35, although clear mathematically, needs additional explanation; however, the physical significance of δ_G and how δ_G varies must be explained first. Consider the synchronous generator of Fig. 5.36(a) operating at unity power factor with respect to the generated voltages; that is, the generated voltages $\mathbf{E}_{GA}$, $\mathbf{E}_{GB}$, and $\mathbf{E}_{GC}$ of phases A, B, and C and their corresponding response currents $\mathbf{I}_{GA}$, $\mathbf{I}_{GB}$, and $\mathbf{I}_{GC}$ are in phase, Fig. 5.36(b). For the given rotor field $\mathbf{B}_R$ and the counterclockwise rotation of the rotor at synchronous speed ω_S, the speed-dependent field intensity $\mathcal{E}_j=\mathbf{u}\times\mathbf{B}_R$ ($j=A, C', B, A', C$, and B')* yields the current notation of Fig. 5.36(a). These currents produce the stator field $\mathbf{B}_{ST}$, which rotates counterclockwise also at synchronous speed ω_S. We then can draw the phasor and flux density diagrams, as in Figs. 5.36(c) and 5.36(d), respectively, if we make the following three classical assumptions: (1) $R_S\simeq 0$; (2) saturation is neglected, that is, superposition of flux densities is valid; and (3) while $\mathbf{B}_R$ and $\mathbf{B}_{ST}$ produce potential rises, the stator field $-\mathbf{B}_{ST}$ produces a potential drop.

If we employ the principles involved in the drawing of Fig. 5.7, we can draw Figs. 5.36(c) and 5.36(d) as in Fig. 5.37. By use of plane geometry theorems† on

*A designates the coil side S_A, A' designates the coil side F_A. The same principle is valid for coil sides S_B, F_B, S_C, and F_C.

†Ἂν αἱ πλευραὶ μιᾶζ γωνίαζ εἶναι, μία πρὸζ μίαν, κάθετοι ἐπὶ τὰζ πλευρὰζ ἄλληζ γωνιάζ, αἱ γωνιάι αὗται εἶναι μὲν ἴσαι, ἂν ἀμφότεραι εἶναι ὀξεῖαι ἢ (ἀμφότεραι) ἀμβλεῖαι, παραπληρωματικαὶ δὲ, ἂν ἡ μία εἶναι ὀξεῖα καὶ ἡ ἄλλη ἀμβλεῖα. The translation is: (If the sides of an angle are perpendicular, one-to-one, to the sides of another angle, then the angles are equal if both angles are acute or if both are obtuse. The angles are complementary if one angle is acute and the other obtuse.)

N. D. Nikolaou, *Theoretiki Geometria* (Athens, Greece: OESB Publishing House, 1950), p. 86.

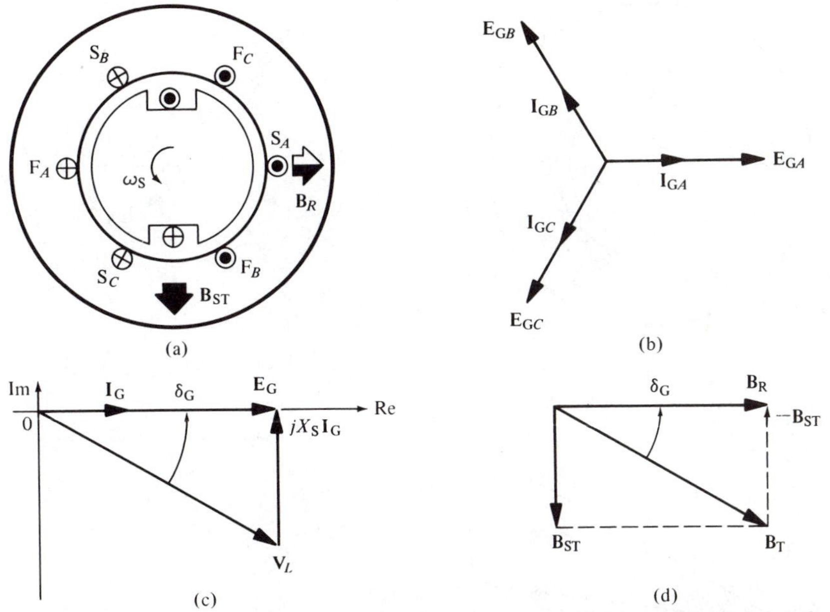

Fig. 5.36 The Power Angle δ_G and Its Relationship to the Phasor and Magnetic Field Diagrams for a Two-Pole, Three-Phase, Balanced, Synchronous Generator Operating at PF $= k$ with Respect to the Generated Phase Voltage $\mathbf{E}_G$

Fig. 5.37 and the ideas of Figs. 5.16 through 5.18, we can conclude that $\angle \frac{\mathbf{V}_L}{\mathbf{E}_G} \equiv \angle \frac{\mathbf{B}_T}{\mathbf{B}_R}$. Thus the physical significance of δ_G is this: **δ_G is a measure of the angular displacement of the rotor field $\mathbf{B}_R$ and thus of the rotor axis from the resultant field $\mathbf{B}_T$ of the generator**. Note that since the fields $\mathbf{B}_R$ and $\mathbf{B}_{ST}$ produce the voltages $\mathbf{E}_G$ and $jX_S\mathbf{I}_G$, respectively, we can conclude that the field $\mathbf{B}_T$ produces the loading voltage $\mathbf{V}_L$. Thus with respect to the angle δ_G, we may mentally interchange these fields with their corresponding voltages and vice versa.

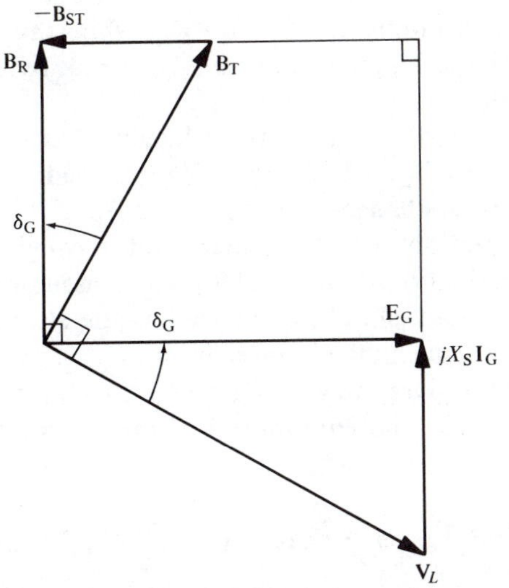

Fig. 5.37 Relative position of the Phasor and Field Diagrams of a Two-Pole, Three-Phase, Balanced, Synchronous Generator Operating in Steady State

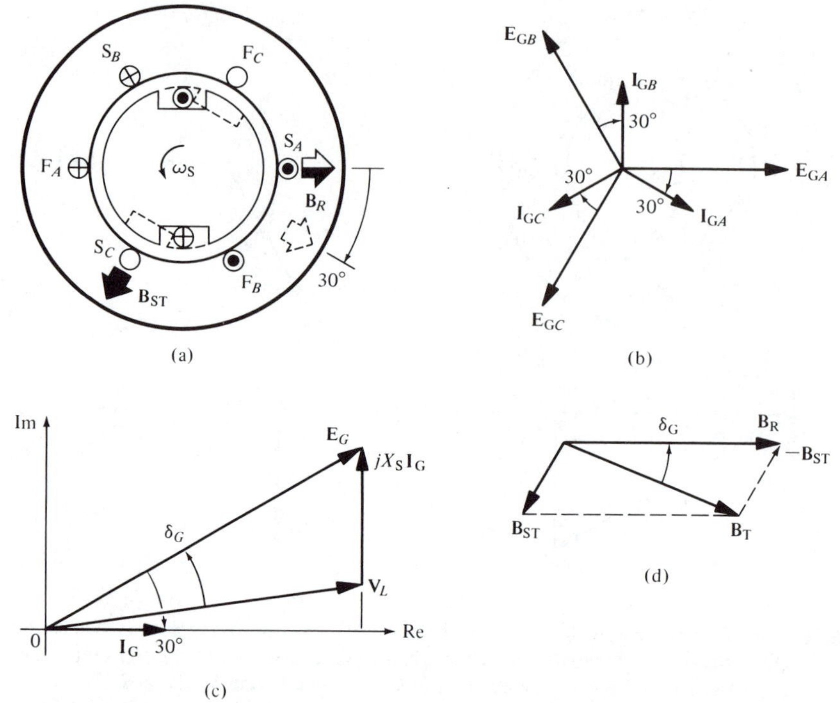

Fig. 5.38 The Power Angle δ_G and Its Relationship to the Phasor and Magnetic Field Diagrams for a Two-Pole, Three-Phase, Balanced, Synchronous Generator Operating at $\mathrm{PF}=k_{\mathrm{LGG}}$ with Respect to the Generated Phase Voltage $\mathbf{E}_{\mathrm{G}}$

The reasoning used to obtain Fig. 5.37 from Fig. 5.36 could also be used to obtain figures analogous to Fig. 5.37 from Figs. 5.38 and 5.39. Such figures would substantiate the findings with respect to the physical significance of the power angle δ_{G}. The effects of lagging and leading are represented by the dotted rotor position in each of these two figures. The real rotor position designates the location of the speed-induced field intensities $\mathcal{E}_j = \mathbf{u}\times\mathbf{B}_{\mathrm{R}}$, while the dotted position of the rotor designates that of the lagging or leading currents by 30° and consequently that of the stator field $\mathbf{B}_{\mathrm{ST}}$.

Examination of Fig. 5.38(d) reveals that for lagging power factors the stator field $\mathbf{B}_{\mathrm{ST}}$ **inhibits** the rotor field $\mathbf{B}_{\mathrm{R}}$, while Fig. 5.39(d) reveals that for leading power factors the stator field $\mathbf{B}_{\mathrm{ST}}$ **enhances** the rotor field $\mathbf{B}_{\mathrm{R}}$.

The power-angle variation can be understood through examination of Fig. 5.35(a). Since for the derivation of Eq. (5.144), R_{S} is assumed to be zero, P_{DT}, the power demanded by the load, should be viewed as the electric power provided by the generator. For a given mechanical power input, P_{INM}, by the prime mover (with **all losses neglected**) and a given power angle δ_{G0}, the electric output power of the generator is P_0, Fig. 5.35(a). The equilibrium of the generator is described by Eq. (5.146):

$$P_0 \equiv P_{\mathrm{INM0}} \equiv P_{\mathrm{OT0}} = \left(\frac{|\mathbf{V}_L||\mathbf{E}_{\mathrm{G}}|}{X_{\mathrm{S}}}\right)\sin\delta_{\mathrm{G0}} \tag{5.146}$$

When, by means of the prime mover, added power (or torque $\tau_{\mathrm{INM}}\omega_{\mathrm{S}} = P_{\mathrm{INM}}$) is applied to the generator, the input mechanical power to the generator will exceed

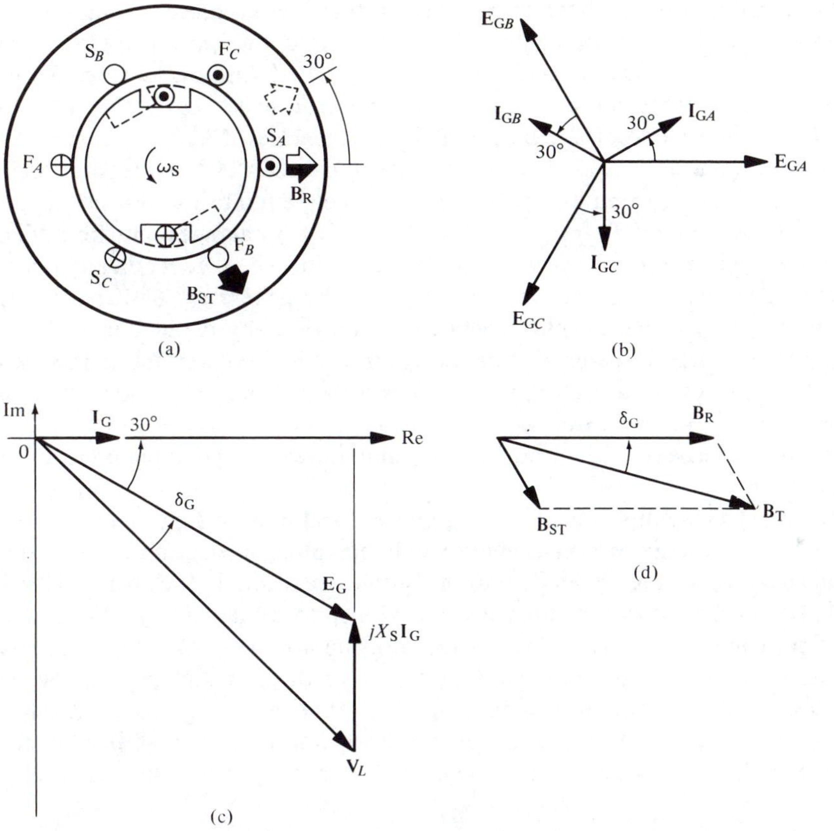

Fig. 5.39 The Power Angle δ_G and Its Relationship to the Phasor Diagram and the Magnetic Fields of a Two-Pole, Three-Phase, Balanced, Synchronous Generator Operating at $\text{PF} = k_{\text{LDG}}$ with Respect to the Generated Phase Voltage $\mathbf{E}_G$

the electric output power by the generator, that is, $P_{\text{INM1}} > P_0$. Thus the rotor will accelerate. This means that the magnetic field of the rotor $\mathbf{B}_R$, which corresponds to the generated voltage $\mathbf{E}_G$, will begin to rotate at a speed ω_R slightly greater than the synchronous speed ω_S, that is, $\omega_R > \omega_S$. However, the resultant field of the generator $\mathbf{B}_T$, which corresponds to the voltage $\mathbf{V}_L$ (throughout this analysis it is assumed that the generator is connected to an infinitely strong network rather than an isolated load), must remain constant in magnitude and must continue to rotate at the synchronous speed ω_S. Thus the departure of $\mathbf{B}_R$ from $\mathbf{B}_T$ or $\mathbf{E}_G$ from $\mathbf{V}_L$ suggests that the power angle δ_G will increase. This process will continue until δ_{G0} reaches a new value designated by $\delta_{GSS} = \delta_{G0} + \Delta\delta_G$, Fig. 5.40, corresponding to a new power level designated by $P_1 = P_0 + \Delta P_{\text{OT}}$, which satisfies the condition

$$P_1 \equiv P_{\text{INM1}} \equiv P_{\text{OT1}} = P_0 + \Delta P_{\text{OT}} = \left(\frac{|\mathbf{V}_L||\mathbf{E}_G|}{X_S} \right) \sin \delta_{GSS} \qquad (5.147)$$

There is an upper limit on $\Delta P_{\text{OT}} \equiv \Delta P_{\text{INM}}$ that the generator can tolerate and still remain stable. If this limit is exceeded inadvertently, the peak value of the power angle δ_G may exceed 90° and the generator rotor will accelerate beyond the synchronous speed.

Here, at $\delta_G = \delta_{GSS}$, there is no accelerating power present and the generator operates at a steady state again. The new value of the power angle δ_G for which steady state operation is achieved is not reached **instantaneously**. Rather, it is reached, depending on the design of the generator, usually after the rotor oscillates about the final value of δ_G, that is, δ_{GSS}, a few times, Fig. 5.40, curve 1, with the constraint that the peak value of δ_G never exceeds 90°. Note that the normal behavior of the rotor as it strives to reach a new equilibrium power angle—where the generated power $P_k = P_{k-1} + \Delta P_{OT}$ $(k = 1, 2, \ldots)$ balances out the added power from the prime mover—resembles the behavior of a generalized second-order system. That is, depending on the design of the generator, the final value of the power angle δ_{GSS} **might** (1) **be reached** by following any of the curves 1 through 4, Fig. 5.40; (2) **not be reached**, but the rotor would oscillate about this value with equal departures from it at equal time intervals, curve 5, Fig. 5.40; or (3) **never be reached**, but now the rotor would oscillate about this value with the departures becoming larger and larger and with the time intervals remaining constant, curve 6, Fig. 5.40.

Curves 1 through 4, inclusive, designate a **stable operating mode** of the generator; curve 5 designates a **marginally stable operating mode**, and curve 6 designates an **unstable operating mode**. The term "stable operating mode" (or **stability**) means that the synchronous generator has the ability to remain in synchronism with its terminal voltage (i.e., the voltage of the transmission network—the infinitely strong network—known as the bus voltage) whenever there is a change in the operating conditions of the generator. The term "unstable operating mode" (or **instability**) means that the synchronous generator does not have the ability to remain in synchronism with its terminal voltage whenever there is a change in the operating

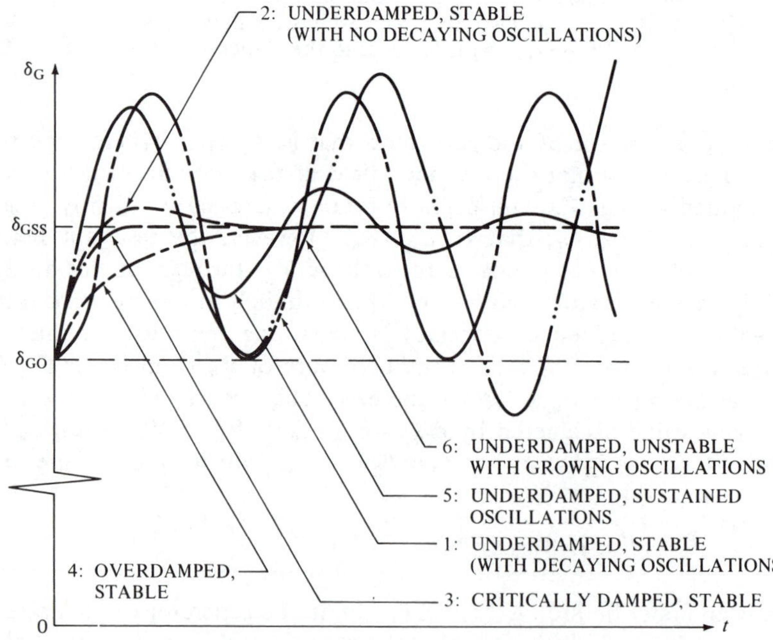

Fig. 5.40 Variation of the Power Angle δ_G of a Synchronous Generator When Prime-Mover Input Increases Suddenly

conditions of the generator. Finally, the term "marginally stable operating mode" (or **marginal stability**) means that the synchronous generator loses synchronism with its terminal voltage whenever there is a slight increase in the input from the prime mover.

This process whereby the generator acquires a new power angle δ_{GSS} for a new prime-mover input can continue through points $1,2,\ldots$, Fig. 5.35(a), until the input power from the prime mover P_{INM3} is equal to the maximum power P_{MAX} that the generator can deliver to the load, that is, $P_{MAX} \equiv P_{INM3} \equiv P_{OT3}$. At this point $\delta_G = \pi/2$ rad (90°).

If the input power from the prime mover is now increased again to a value $P_{INMX} > P_{MAX}$, the rotor (and thus $\mathbf{E}_G$ or $\mathbf{B}_R$) will accelerate while $\mathbf{V}_L$ or $\mathbf{B}_T$ will continue to move at the synchronous speed ω_S. This will cause δ_G to increase beyond $\pi/2$ rad (90°), say to δ_{G4}. However, even though δ_G increases, the power generated by the generator decreases. Thus $P_{INMX} > P_{O4}$; that is, the synchronous generator is unable to convert the input mechanical power to electric power. Thus the excess mechanical power is absorbed by the rotor, and the rotor continues to accelerate with no equilibrium point in sight. The rotor of the generator will begin to rotate at speeds much higher than the synchronous speed; thus it steps out of synchronism and "skips poles." Soon the power angle can become greater than π rad (180°), and the generator will begin to absorb electric power, if and only if such is available, rather than generate it. Then the synchronous generator will behave as a motor and will continue to accelerate. If it is left alone, it would destroy itself mechanically as a result of very high electromechanical forces, that is, fatigue due to torque oscillations. Overheating is also detrimental when the synchronous generator operates at power angles of $\delta_G > \pi/2$.

It is clear then from Fig. 5.35(a) that the phenomenon just described (i.e., the rotor of a synchronous generator accelerates and pulls out of synchronism) will occur whether the initial power angle is equal to or greater than $\pi/2$ rad. It is concluded, therefore, that the synchronous generator can be operated in synchronism with the terminal voltage $\mathbf{V}_L$ for power angles in the range $0 \leqslant \delta_G < \pi/2$.

As a consequence of what was just described, the maximum power the synchronous generator can deliver to the load, Fig. 5.34(a), before it pulls out of synchronism is defined as the **generator pull-out power** and is designated as P_{GPO}. Mathematically, it is defined as

$$P_{GPO} \equiv P_{MAX} = 3\left(\frac{|\mathbf{V}_L||\mathbf{E}_G|}{X_S}\right) \tag{5.148}$$

Thus the point $P(P_{MAX}, \pi/2)$ on the power versus power angle graph, Fig. 5.35(a), is defined as the **steady state stability limit** of the synchronous generator and is designated by $SSSL_G$.

The preceding explanation suffices for describing the behavior of the synchronous generator. Thus any comments pertaining to Fig. 5.35(b) are postponed for the time being. This subject is considered again in Section 5.8.

Example 5.9

A balanced, three-phase, 20-pole, Y-connected, 60-Hz, 12,000-line-to-line-V, 10-MW synchronous generator is connected to an infinite bus. Such a network has the property of maintaining its node voltages at constant magnitude, phase angle, and frequency. If the per-phase synchronous reactance is $jX_S = j10\ \Omega$ and neglecting the per-phase stator winding resistance, and if the field (rotor winding) dc excitation is maintained constant to a level that assures power delivery to the network at a lagging power factor of 0.8, calculate: (1) the generated voltage $\mathbf{E}_G$, (2) the power angle δ_G, (3) the power, P_{INM}, required from the prime mover for steady state operation, (4) the maximum power the generator can deliver, that is, the **generator pull-out power**, $P_{GPO} = P_{MAX}$ (5) the reactive power, Q_{OT}, the generator delivers or absorbs to or from the network at rated load, (6) the maximum reactive power, Q_{OTMAX}, the generator can deliver or absorb, (7) the minimum reactive power, Q_{OTMIN}, the generator can deliver or absorb, (8) the **steady state stability limit** ($SSSL_G$).

Solution

From the input data, the per-phase power delivered to the loading network is calculated from Eq. (5.64) as

$$P_D = \frac{P_{DT}}{3} = \frac{10{,}000{,}000}{3} = \underline{3{,}333{,}333.33}\ \text{W} \tag{1}$$

and the phase loading voltage $\mathbf{V}_L$ is calculated, utilizing Eq. (3.67), as

$$|\mathbf{V}_L| = \frac{|\mathbf{V}_{L(L-L)}|}{\sqrt{3}} = \frac{12{,}000}{\sqrt{3}} = \underline{6928.20}\ \text{V} \tag{2}$$

Because the power factor is defined as $\cos\theta_L = 0.8$ (LGG), the angle θ_L between the phase voltage $\mathbf{V}_L$ and the phase current $\mathbf{I}_G$, Figs. 5.34 and 5.17, is

$$\theta_L = \cos^{-1}(0.8) = 36.87° \tag{3}$$

Thus if the current is set along the Re-axis, the voltage is written as

$$\mathbf{V}_L = |\mathbf{V}_L| e^{j\theta_L} = \underline{6{,}928.20 e^{j36.87°}}\ \text{V} \tag{4}$$

Now the magnitude of the generated current can be calculated from Eq. (5.61) through direct substitution from Eqs. (1), (2), and (3) as

$$\mathbf{I}_G = \frac{P_D}{|\mathbf{V}_L| \cos\theta_L} = \frac{3{,}333{,}333.33}{(6{,}928.20) \times (0.8)} = \underline{601.41}\ \text{A} \tag{5}$$

Finally, the voltage drop across the synchronous reactance jX_S is calculated as

$$\mathbf{V}_S = jX_S \mathbf{I}_G = j10 \times 601.41 = \underline{j6{,}014.10}\ \text{V} \tag{6}$$

Application of Kirchhoff's law to Fig. 5.34 and substitution from Eqs. (6) and (4) (utilizing Euler's identity) yields

$$\begin{aligned} \mathbf{E}_G &= \mathbf{V}_S + \mathbf{V}_L = j6{,}014.10 + 6{,}928.20 e^{j36.87°} \\ &= 5{,}542.55 + j4{,}156.93 + j6{,}014.10 \\ &= 5{,}542.55 + j10{,}171.03 = \underline{\underline{11{,}583.17 e^{j61.41°}}}\ \text{V} \end{aligned} \tag{7}$$

The power angle δ_G, defined earlier as $\delta_G \equiv \measuredangle \mathbf{E}_G - \measuredangle \mathbf{V}_L$, can be calculated by substituting directly the angles of Eqs. (7) and (4). Thus

$$\delta_G = 61.41° - 36.87° = \underline{\underline{24.54°}} \tag{8}$$

The power input from the prime mover, P_{INM}, when all losses are ignored, is calculated by direct substitution from Eqs. (4), (7), and (8) and the input data into

Eq. (5.144). Thus

$$P_{\text{INM}} \equiv P_{\text{OT}} = 3\left(\frac{|\mathbf{V}_L||\mathbf{E}_{\text{G}}|}{X_{\text{S}}}\right)\sin\delta_{\text{G}}$$
$$= P_{\text{DT}} = 10{,}000 \text{ kW} \tag{9}$$

The maximum power the generator can deliver, that is, the **generator pull-out power**, is calculated from Eq. (5.144) by allowing $\delta_{\text{G}}=24.53°\rightarrow\delta_{\text{G}}=90°$, or from Eq. (5.148). Thus

$$P_{\text{GPO}} = 3\left(\frac{|\mathbf{V}_L||\mathbf{E}_{\text{G}}|}{X_{\text{S}}}\right)\sin(90°)$$
$$= \frac{3\times(6{,}928.20)\times(11{,}583.17)\times 1.0}{10} = \underline{\underline{24{,}075{,}155.52 \text{ W}}} \tag{10}$$

The reactive power that the generator delivers or absorbs at rated load, that is, when $\delta_{\text{G}}=24.54°$, can be calculated from Eq. (5.145) by direct substitution from Eqs. (4), (7), and (8) and input data. Thus

$$Q_{\text{OT}}(24.54°) \equiv 3\left(\frac{|\mathbf{V}_L||\mathbf{E}_{\text{G}}|}{X_{\text{S}}}\right)\cos\delta_{\text{G}} - 3\left(\frac{|\mathbf{V}_L|^2}{X_{\text{S}}}\right)$$
$$= \left[\frac{3\times(6{,}928.20)\times(11{,}583.16)}{10}\right]\cos(24.54°) - \frac{3\times(6{,}928.20)^2}{10}$$
$$= 21{,}900{,}483.76 - 14{,}399{,}986.57 = \underline{\underline{7{,}500{,}497.19}} \text{ var} \tag{11}$$

Since $Q_{\text{OT}}>0$, the generator *delivers reactive power* to the network and thus its behavior is **capacitive in nature.**

The maximum reactive power the generator can deliver or absorb is calculated from Eq. (5.145) by directly substituting from Eqs. (4) and (7) and from the input data, and by setting $\delta_{\text{G}}=0°$ [or from Eq. (11) by letting cos(24.54°)→cos(0°)]. Thus

$$Q_{\text{OTMAX}} = 3\left(\frac{|\mathbf{V}_L||\mathbf{E}_{\text{G}}|}{X_{\text{S}}}\right)\cos(0°) - 3\left(\frac{|\mathbf{V}_L|^2}{X_{\text{S}}}\right)$$
$$= \frac{3\times(6{,}928.20)\times(11{,}583.17)}{10} - \frac{3\times(6{,}928.20)^2}{10}$$
$$= 24{,}075{,}155.52 - 14{,}399{,}986.57 = \underline{\underline{9{,}675{,}168.95}} \text{ var} \tag{12}$$

Again, since $Q_{\text{OTMAX}}>0$, the generator *delivers reactive power* to the network and thus its behavior is **capacitive in nature.**

The minimum reactive power that the generator can deliver or absorb is calculated from Eq. (5.145) by directly substituting from Eqs. (4) and (7) and from the input data, and by setting $\delta_{\text{G}}\equiv 90°$ [or from Eq. (11) by letting cos(24.54°)→cos(90°)]. Thus

$$Q_{\text{OTMIN}} = 3\left(\frac{|\mathbf{V}_L||\mathbf{E}_{\text{G}}|}{X_{\text{S}}}\right)\cos(90°) - 3\left(\frac{|\mathbf{V}_L|^2}{X_{\text{S}}}\right)$$
$$= 0 - \frac{3\times(6{,}928.20)^2}{10} = \underline{\underline{-14{,}399{,}986.57}} \text{ var} \tag{13}$$

Since $Q_{\text{OTMIN}}<0$, the generator *absorbs reactive power* when $\delta_{\text{G}}=90°$ and thus its behavior is **inductive in nature**. However, the device never operates for long intervals of time at $\delta_{\text{G}}=90°$ since the real power output would exceed the design capabilities. It is the intention of this calculation to show that the device has the capability of producing positive and negative vars as δ_G varies and $|\mathbf{E}_G|$, $|\mathbf{V}_L|$, and X_S remain constant. It is shown in the next section that the same effects on the output reactive power can be achieved by varying the field (rotor-winding) excitation.

The steady state stability limit of the generator is calculated by direct substitution from Eq. (10) into Eq. (6.6). Thus

$$\text{SSSL}_G \equiv P_{\text{GPO}} = 3\left(\frac{|\mathbf{E}_G||\mathbf{V}_L|}{X_S}\right) = \underline{\underline{24{,}075{,}134.74 \text{ W}}} \tag{14}$$

As in the case of the synchronous generator, the circuit of the synchronous motor is redrawn here for simplicity and realistic depiction of its phasor diagram, Figs. 5.41(a), 5.41(b), and 5.41(c).

The electric power utilized (absorbed) by the synchronous motor, Fig. 5.41(a), as defined in Eq. (5.103) is $\mathbf{S}_{\text{IN}} \equiv \mathbf{V}_{\text{ST}}\mathbf{I}_S{}^*$. However, it is our intention in this section to use the synchronous device in a dual mode of operation (generator or motor).

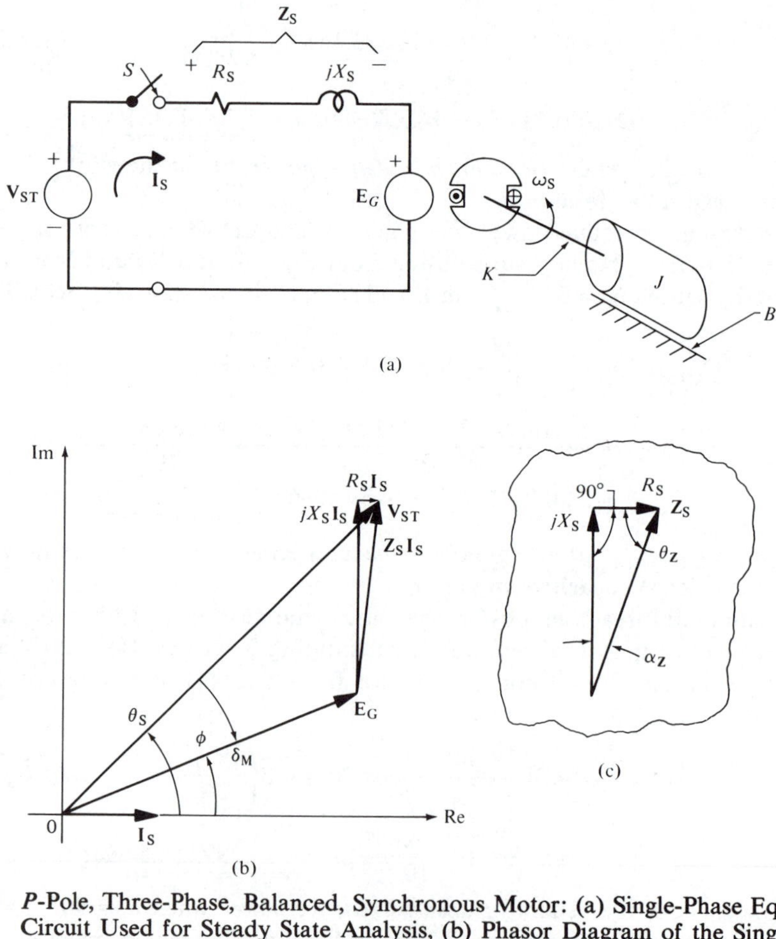

Fig. 5.41 *P*-Pole, Three-Phase, Balanced, Synchronous Motor: (a) Single-Phase Equivalent Circuit Used for Steady State Analysis, (b) Phasor Diagram of the Single-Phase Equivalent Circuit, and (c) the Impedance Triangle of the Equivalent Circuit, Drawn Expanded and Not to Scale for Clarity

Therefore, the electric output power of the motor is the quantity of interest. This electric output power flows in the opposite direction of the electric input power. Thus **the electric output power of the synchronous device** is defined as the **negative** of the input power and is written as follows:

$$\mathbf{S}_\mathrm{O} = P_\mathrm{O} + jQ_\mathrm{O} \equiv -\mathbf{S}_\mathrm{IN} = -\mathbf{V}_\mathrm{ST}\mathbf{I}_\mathrm{S}^* \tag{5.149}$$

Utilization of Kirchhoff's voltage law enables us to write the describing equation of Fig. 5.41(a) as

$$\mathbf{V}_\mathrm{ST} - \mathbf{Z}_\mathrm{S}\mathbf{I}_\mathrm{S} - \mathbf{E}_\mathrm{G} = 0 \tag{5.150}$$

Solution of Eq. (5.150) for the current $\mathbf{I}_\mathrm{S}$, and utilization of Fig. 5.41(b), enables us to write

$$\begin{aligned}\mathbf{I}_\mathrm{S} &= \frac{\mathbf{V}_\mathrm{ST} - \mathbf{E}_\mathrm{G}}{\mathbf{Z}_\mathrm{S}} = \left.\frac{|\mathbf{V}_\mathrm{ST}|e^{j\theta_\mathrm{S}} - |\mathbf{E}_\mathrm{G}|e^{j\phi}}{|\mathbf{Z}_\mathrm{S}|e^{j\theta_\mathrm{Z}}}\right|_{\phi=\theta_\mathrm{S}+\delta_\mathrm{M}} \\ &= \frac{|\mathbf{V}_\mathrm{ST}|e^{j(\theta_\mathrm{S}-\theta_\mathrm{Z})}}{|\mathbf{Z}_\mathrm{S}|} - \frac{|\mathbf{E}_\mathrm{G}|e^{j(\theta_\mathrm{S}+\delta_\mathrm{M}-\theta_\mathrm{Z})}}{|\mathbf{Z}_\mathrm{S}|}\end{aligned} \tag{5.151}$$

The conjugate of the current $\mathbf{I}_\mathrm{S}$ is obtained from Eq. (5.151) and is written as

$$\mathbf{I}_\mathrm{S}^* = \frac{|\mathbf{V}_\mathrm{ST}|e^{-j(\theta_\mathrm{S}-\theta_\mathrm{Z})}}{|\mathbf{Z}_\mathrm{S}|} - \frac{|\mathbf{E}_\mathrm{G}|e^{-j(\theta_\mathrm{S}+\delta_\mathrm{M}-\theta_\mathrm{Z})}}{|\mathbf{Z}_\mathrm{S}|} \tag{5.152}$$

Substitution for $\mathbf{I}^*_\mathrm{S}$ from Eq. (5.152) into Eq. (5.149) yields

$$\begin{aligned}\mathbf{S}_\mathrm{O} &= -|\mathbf{V}_\mathrm{ST}|e^{j\theta_\mathrm{S}}\left[\frac{|\mathbf{V}_\mathrm{ST}|e^{-j(\theta_\mathrm{S}-\theta_\mathrm{Z})}}{|\mathbf{Z}_\mathrm{S}|} - \frac{|\mathbf{E}_\mathrm{G}|e^{-j(\theta_\mathrm{S}+\delta_\mathrm{M}-\theta_\mathrm{Z})}}{|\mathbf{Z}_\mathrm{S}|}\right] \\ &= -\left(\frac{|\mathbf{V}_\mathrm{ST}|^2}{|\mathbf{Z}_\mathrm{S}|}\right)e^{-j(\theta_\mathrm{S}-\theta_\mathrm{Z}-\theta_\mathrm{S})} + \left(\frac{|\mathbf{V}_\mathrm{ST}||\mathbf{E}_\mathrm{G}|}{|\mathbf{Z}_\mathrm{S}|}\right)e^{-j(\theta_\mathrm{S}+\delta_\mathrm{M}-\theta_\mathrm{Z}-\theta_\mathrm{S})} \\ &= -\left(\frac{|\mathbf{V}_\mathrm{ST}|^2}{|\mathbf{Z}_\mathrm{S}|}\right)e^{-j(-\theta_\mathrm{Z})} + \left(\frac{|\mathbf{V}_\mathrm{ST}||\mathbf{E}_\mathrm{G}|}{|\mathbf{Z}_\mathrm{S}|}\right)e^{-j(\delta_\mathrm{M}-\theta_\mathrm{Z})} \\ &= \left(\frac{|\mathbf{V}_\mathrm{ST}||\mathbf{E}_\mathrm{G}|}{|\mathbf{Z}_\mathrm{S}|}\right)e^{j(\theta_\mathrm{Z}-\delta_\mathrm{M})} - \left(\frac{|\mathbf{V}_\mathrm{ST}|^2}{|\mathbf{Z}_\mathrm{S}|}\right)e^{j\theta_\mathrm{Z}}\end{aligned} \tag{5.153}$$

Again, utilization of Euler's identity enables us to write Eq. (5.153) as

$$\begin{aligned}\mathbf{S}_\mathrm{O} &= \left[\left(\frac{|\mathbf{V}_\mathrm{ST}||\mathbf{E}_\mathrm{G}|}{|\mathbf{Z}_\mathrm{S}|}\right)\cos(\theta_\mathrm{Z}-\delta_\mathrm{M}) - \left(\frac{|\mathbf{V}_\mathrm{ST}|^2}{|\mathbf{Z}_\mathrm{S}|}\right)\cos\theta_\mathrm{Z}\right] \\ &\quad + j\left[\left(\frac{|\mathbf{V}_\mathrm{ST}||\mathbf{E}_\mathrm{G}|}{|\mathbf{Z}_\mathrm{S}|}\right)\sin(\theta_\mathrm{Z}-\delta_\mathrm{M}) - \left(\frac{|\mathbf{V}_\mathrm{ST}|^2}{|\mathbf{Z}_\mathrm{S}|}\right)\sin\theta_\mathrm{Z}\right]\end{aligned} \tag{5.154}$$

Similarly, utilization of trigonometric properties and the impedance triangle of Fig. 5.41(c) enables us to write

$$\cos\theta_\mathrm{Z} = \frac{R_\mathrm{S}}{|\mathbf{Z}_\mathrm{S}|} \tag{5.155}$$

and

$$\sin\theta_\mathrm{Z} = \frac{X_\mathrm{S}}{|\mathbf{Z}_\mathrm{S}|} \tag{5.156}$$

As in the case of the synchronous generator, utilization of geometric properties and Fig. 5.41(c) enables us to write $\theta_\mathrm{Z} + \alpha_\mathrm{Z} + 90 = 180$ (all in degrees), which yields

$$\theta_\mathrm{Z} = 90 - \alpha_\mathrm{Z} \tag{5.157}$$

Eq. (5.157) enables writing the quantities $\cos(\theta_Z-\delta_M)$ and $\sin(\theta_Z-\delta_M)$ of Eq. (5.154) as

$$\cos(\theta_Z-\delta_M) = \cos[90-(\alpha_Z+\delta_M)] \equiv \sin(\alpha_Z+\delta_M) \tag{5.158}$$

and

$$\sin(\theta_Z-\delta_M) = \sin[90-(\alpha_Z+\delta_M)] \equiv \cos(\alpha_Z+\delta_M) \tag{5.159}$$

Substitution for $\cos(\theta_Z-\delta_M)$, $\cos\theta_Z$, $\sin(\theta_Z-\delta_M)$, and $\sin\theta_Z$ from Eqs. (5.158), (5.155), (5.159), and (5.156) into Eq. (5.154) yields

$$\mathbf{S}_O = \left[\left(\frac{|\mathbf{V}_{ST}||\mathbf{E}_G|}{|\mathbf{Z}_S|}\right)\sin(\alpha_Z+\delta_M) - \left(\frac{|\mathbf{V}_{ST}|^2}{|\mathbf{Z}_S|^2}\right)R_S\right] + j\left[\left(\frac{|\mathbf{V}_{ST}||\mathbf{E}_G|}{|\mathbf{Z}_S|}\right)\cos(\alpha_Z+\delta_M) - \left(\frac{|\mathbf{V}_{ST}|^2}{|\mathbf{Z}_S|^2}\right)X_S\right] \tag{5.160}$$

Since for synchronous devices $(20:1)<(X_S:R_S)<(100:1)$ then for the practical synchronous motor, $X_S \gg R_S$. As in the case of the synchronous generator, the following three conditions are valid for Eq. (5.160): (1) the resistance R_S can be neglected, (2) the angle α_Z of the real impedance triangle, Fig. 5.41(b), becomes

$$\alpha_Z = \tan^{-1}\left(\frac{R_S}{X_S}\right) \simeq 0° \tag{5.161}$$

and thus it can be neglected, and (3) the impedance $\mathbf{Z}_S$ is considered to be equal to the synchronous reactance X_S, that is, $|\mathbf{Z}_S| \simeq |jX_S| = X_S$. Under these conditions Eq. (5.160) can be written as

$$\mathbf{S}_O = P_O + jQ_O = \left[\left(\frac{|\mathbf{V}_{ST}||\mathbf{E}_G|}{|X_S|}\right)\sin\delta_M\right] + j\left[\left(\frac{|\mathbf{V}_{ST}||\mathbf{E}_G|}{|X_S|}\right)\cos\delta_M - \frac{|\mathbf{V}_{ST}|^2}{X_S}\right] \tag{5.162}$$

Utilization of Eq. (5.108) enables us to write the total real power P_{OT} and the total reactive power Q_{OT} exiting the motor as

$$P_{OT} = 3\left(\frac{|\mathbf{V}_{ST}||\mathbf{E}_G|}{|X_S|}\right)\sin\delta_M \tag{5.163}$$

and

$$Q_{OT} = 3\left(\frac{|\mathbf{V}_{ST}||\mathbf{E}_G|}{|X_S|}\right)\cos\delta_M - 3\left(\frac{|\mathbf{V}_{ST}|^2}{X_S}\right) \tag{5.164}$$

Inspection of Fig. 5.41(a) reveals that $|\mathbf{E}_G|<|\mathbf{V}_{ST}|$, not considering the effects of various excitations that will be considered shortly. This idea, in conjunction with the fact that δ_M always has negative values, enables us to draw $P_{OT}=\phi(\delta_M)$ and $Q_{OT}=\gamma(\delta_M)$, that is, Eqs. (5.163) and (5.164), for constant values of $|\mathbf{V}_{ST}|$, $|\mathbf{E}_G|$, and X_S as exhibited in Figs. 5.42(a) and 5.42(b), respectively.

The variation of P_{OT} and Q_{OT} as functions of δ_M, Fig. 5.42, can be explained clearly if the physical significance of δ_M and how δ_M varies are understood first. Consider the synchronous motor of Fig. 5.43(b) operating at unity power factor with respect to the impressed voltages; that is, the impressed voltages $\mathbf{V}_{SA}$, $\mathbf{V}_{SB}$, and $\mathbf{V}_{SC}$ of phases A, B, and C and the excitation currents $\mathbf{I}_{SA}$, $\mathbf{I}_{SB}$, and $\mathbf{I}_{SC}$ are in

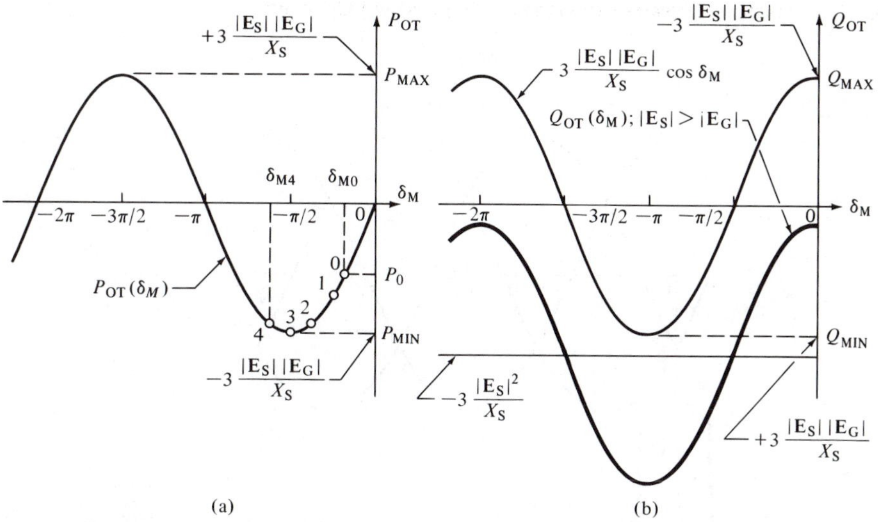

Fig. 5.42 Power-Angle Characteristics of a Smooth-Rotor Synchronous Motor: (a) Real Electric Output Power vs. Power Angle and (b) Reactive Output Power vs. Power Angle

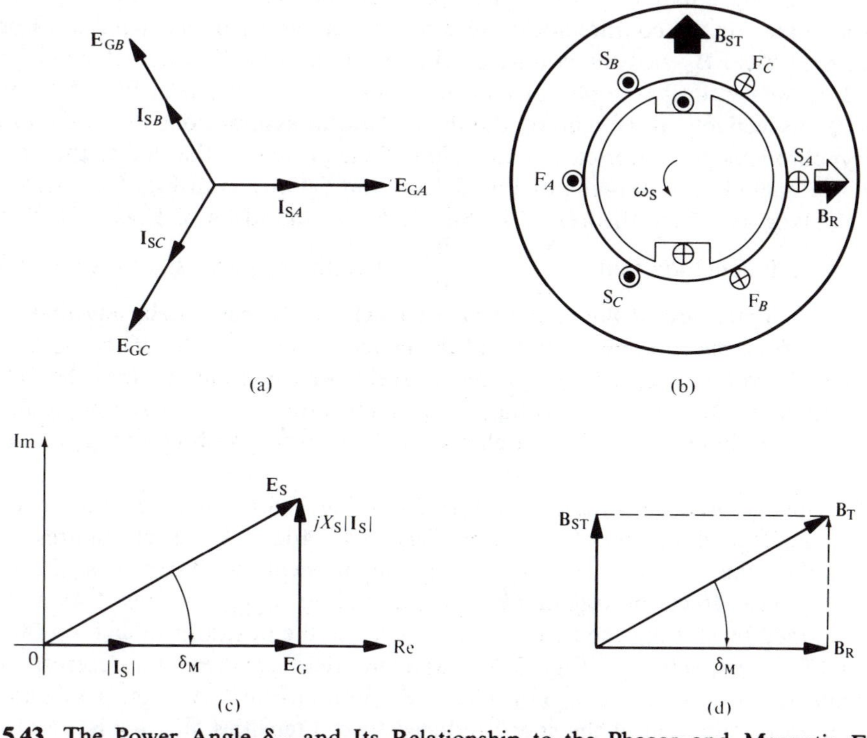

Fig. 5.43 The Power Angle δ_M and Its Relationship to the Phasor and Magnetic Field Diagrams for a Two-Pole, Three-Phase, Balanced, Synchronous Motor Operating at PF $= k$ with Respect to the Impressed Phase Voltage $\mathbf{E_S}$

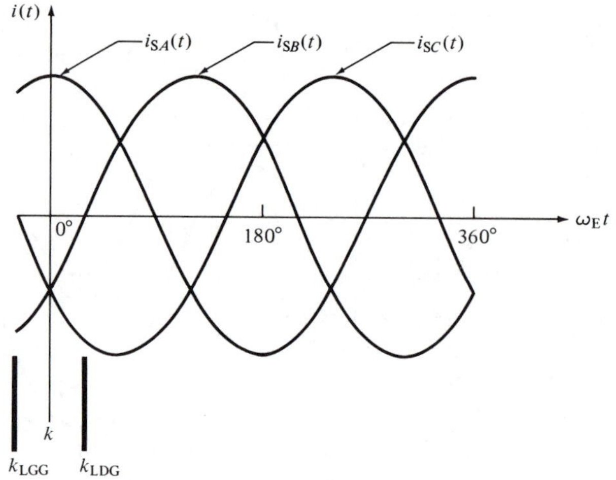

Fig. 5.44 Positive Sequence of Excitation Currents; Zero Axis Position for (1) PF $= k_{LGG}$ (2) PF $= k$, and (3) PF $= k_{LDG}$

phase, Fig. 5.43(a). When the stator coils are excited with the currents of Fig. 5.44 measured at $\omega_E t = 0$ (note that negative currents are designated by ⊕'s and positive currents are designated by ⊙'s), they produce the stator field $\mathbf{B}_{ST}$, Fig. 5.43(b), which rotates in the counterclockwise direction at synchronous speed ω_S. For the given rotor field $\mathbf{B}_R$ and the counterclockwise rotation of the rotor at synchronous speed ω_S, we can draw the phasor and flux density diagrams, as in Figs. 5.43(c) and 5.43(d), respectively, if we employ the three classical assumptions that were used in the synchronous generator case. Utilization of the principles that led to the drawing of Fig. 5.7 allows the drawing of Fig. 5.43(c) and 5.43(d) as in Fig. 5.45. Again, by use of plane geometry theorems* in Fig. 5.45 and the ideas of Figs. 5.30 through 5.32, we can conclude that $\measuredangle \frac{\mathbf{V}_{ST}}{\mathbf{E}_G} \equiv \measuredangle \frac{\mathbf{B}_T}{\mathbf{B}_R}$. Thus the physical significance of δ_M is this: **δ_M is a measure of the angular displacement of the rotor field and thus of the rotor axis from the resultant field $\mathbf{B}_T$ of the motor.** Note again that since $\mathbf{B}_R$ and $\mathbf{B}_{ST}$ produce the voltages $\mathbf{E}_G$ and $jX_S\mathbf{I}_S$, respectively, we can conclude that the field $\mathbf{B}_T$ corresponds to the excitation voltage $\mathbf{V}_{ST}$. Thus with respect to the angle δ_M, we may mentally interchange these fields with their corresponding voltages and vice versa.

The theory used to obtain Fig. 5.45 from Fig. 5.43 could be used to obtain figures analogous to Fig. 5.45 from Figs. 5.46 and 5.47. Such figures would substantiate the findings with respect to the physical significance of the power angle δ_M. The effects of lagging (k_{LGG}) and leading (k_{LDG}) in Figs. 5.46 and 5.47 are obtained by shifting the zero of the $\omega_E t$-axis to the positions of $\omega_E t = -30°$ and $\omega_E t = +30°$, respectively (Fig. 5.44), and by reading there the corresponding currents $i_{SA}(t)$, $i_{SB}(t)$, and $i_{SC}(t)$. The real rotor position in Figs. 5.46 and 5.47 designates the position of the speed-induced field intensities $\mathcal{E}_j = \mathbf{u} \times \mathbf{B}_R$. Examination of Fig. 5.46(d) reveals that for lagging power factors the stator field **enhances**

***Please see theorem in footnote on page 362.**

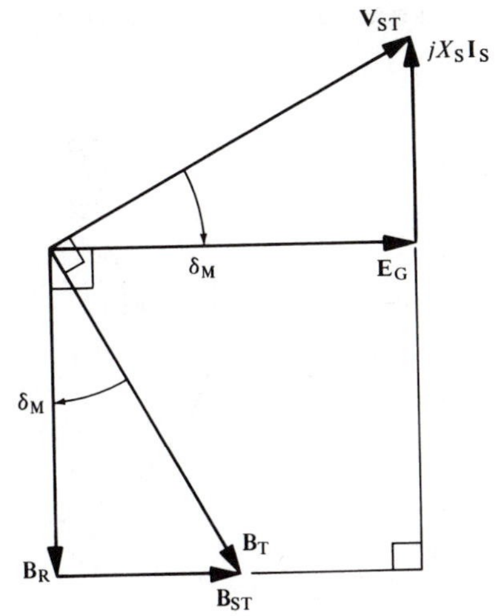

Fig. 5.45 Relative Position of the Phasor and Field Diagrams of a Two-Pole, Three-Phase, Balanced, Synchronous Motor Operating in Steady State

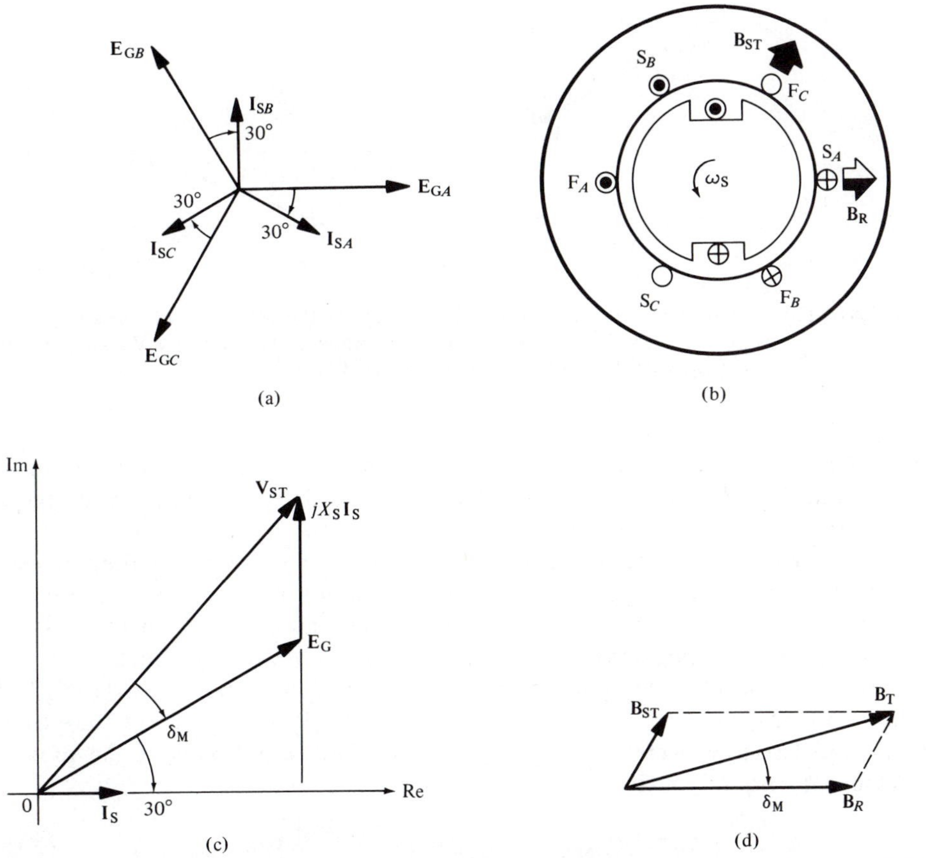

Fig. 5.46 The Power Angle δ_M and Its Relationship to the Phasor and Magnetic Field Diagrams for a Two-Pole, Three-Phase, Balanced, Synchronous Motor Operating at PF $= k_{LGG}$ with Respect to the Impressed Phase Voltage $\mathbf{V}_{ST}$

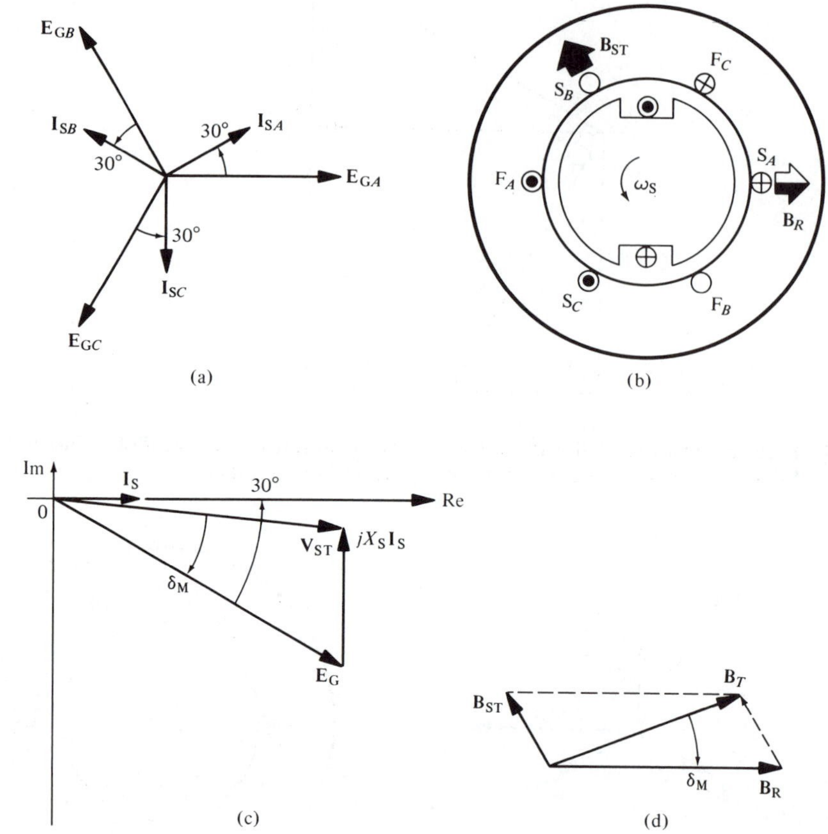

Fig. 5.47 The Power Angle δ_M and Its Relationship to the Phasor and Magnetic Field Diagrams for a Two-Pole, Three-Phase, Balanced, Synchronous Motor Operating at PF $= k_{LDG}$ with Respect to the Impressed Phase Voltage $\mathbf{V}_{ST}$

the rotor field $\mathbf{B}_R$, while Fig. 5.47(d) reveals that for leading power factors the stator field $\mathbf{B}_{ST}$ **inhibits** the rotor field $\mathbf{B}_R$. Note that the power-angle variation can be understood through the examination of Fig. 5.42(a).

The mechanical load, henceforth referred to as the mechanical loading power, is the power (or torque, $P_{OM} = \tau_{OM}\omega_S$) that the motor must develop in order to overcome the load and maintain the speed of the load constant. In the following discussion, **only the magnitudes** of the various powers are considered.

For a given mechanical loading power P_{OM} and a given power angle $|\delta_{MO}|$, the electric power developed [i.e., $\text{Re}(-\mathbf{E}_G\mathbf{I}_S^*) = \text{Re}(-\mathbf{V}_{ST}\mathbf{I}_S^*) = P_{OT}$, Eq. (5.163)] by the motor **(all losses neglected)** is P_O, Fig. 5.42(a). The equilibrium of the motor is described by Eq. 5.165, that is,

$$|P_O| \equiv |P_{OMO}| \equiv |P_{OTO}| = \left(\frac{|\mathbf{V}_{ST}||\mathbf{E}_G|}{X_S}\right)\sin|\delta_{MO}| \tag{5.165}$$

When the mechanical loading is increased, the loading power on the motor will exceed the developed power of the motor, that is, $|P_{OM1}| > |P_O|$. Thus the rotor will decelerate. This means that the magnetic field of the rotor, which corresponds to the generated voltage $\mathbf{E}_G$, will begin to rotate at speeds ω_R slightly lower than the synchronous speed ω_S, that is, $\omega_R < \omega_S$. However, the resultant field of the motor, $\mathbf{B}_T$, which corresponds to the voltage $\mathbf{V}_{ST}$, must remain constant in magnitude and must continue to rotate at synchronous speed ω_S. Thus the departure of $\mathbf{B}_R$ from $\mathbf{B}_T$, or $\mathbf{E}_G$ from $\mathbf{V}_{ST}$, suggests that the power angle $|\delta_M|$ will increase. This process will continue until $|\delta_{MO}|$ reaches a new value, designated by $|\delta_{MSS}| = |\delta_{MO}| + |\Delta\delta_M|$, Fig. 5.48, corresponding to a new power level designated by $|P_1| = |P_O| + |\Delta P_{OT}|$, which satisfies the condition

$$|P_1| \equiv |P_{OM1}| \equiv |P_{OT1}| \equiv |P_O| + |\Delta P_{OT}| = \left(\frac{|\mathbf{V}_{ST}||\mathbf{E}_G|}{X_S}\right)\sin|\delta_{MSS}| \qquad (5.166)$$

There is an upper limit on $|\Delta P_{OT}| \equiv \Delta P_{OM}$ that the motor can tolerate and still remain stable. If this limit is exceeded inadvertently, the peak value of the power angle δ_M may exceed $-90°$ and the motor will eventually stop. Here there is no decelerating power present and the motor operates at steady state again. The new value of the power angle $|\delta_M|$ for which steady state operation is achieved is **not reached instantaneously**. Rather, it is reached, depending on the design of the motor, usually after the rotor oscillates about the final value of $|\delta_M|$, that is, $|\delta_{MSS}|$, a few times, Fig. 5.48, curve 1, with the constraint that the negative peak value of δ_M never exceeds $-90°$. Note that the normal behavior of the rotor as it strives to reach a new equilibrium power angle—where the power developed by the motor, $|P_k| = |P_{k-1}| + |\Delta P_{OT}|$, $k = 1, 2, \ldots$, balances out the added loading power—resembles the behavior of a generalized second-order system. That is, depending on the design of the motor, the final value of the power angle $|\delta_{MSS}|$ might (1) **be reached** by following any of the curves 1 through 4, Fig. 5.48; (2) **not be reached**, but the

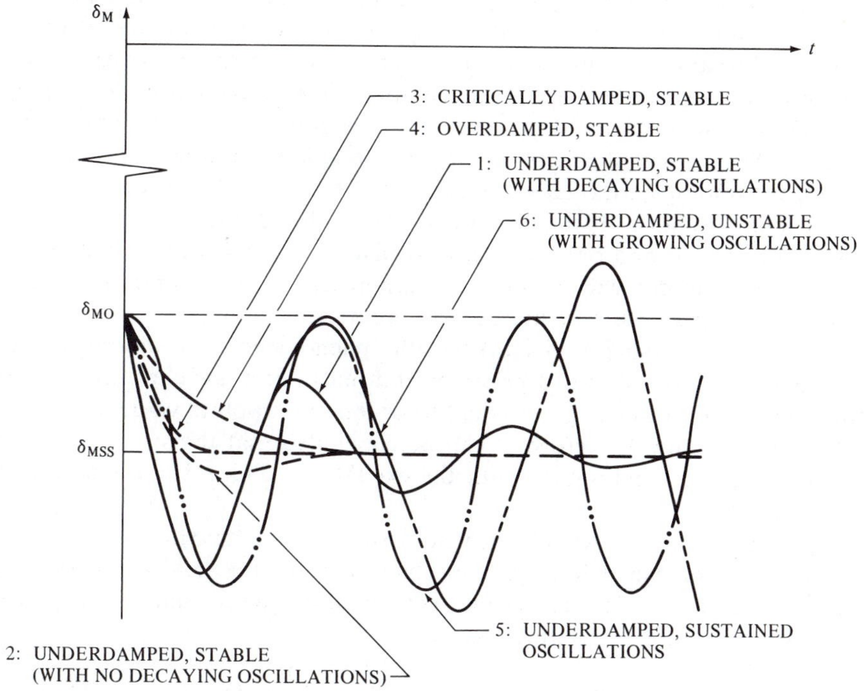

Fig. 5.48 Variation of the Power Angle δ_M of a Synchronous Motor When Loading Increases Suddenly

rotor would oscillate about this value with equal departures from it at equal time intervals, curve 5, Fig. 5.48; or (3) **never be reached**, but now the rotor would oscillate about this value, with the departures becoming larger and larger and the time intervals remaining constant, curve 6, Fig. 5.48.

Again, curves 1 through 4, inclusive, designate a **stable operating mode** of the motor, curve 5 designates a **marginally stable operating mode**, and curve 6 designates an **unstable operating mode**. The term "stable operating mode" (or **stability**) means that the synchronous motor has the ability to remain in synchronism with its excitation voltage (i.e., the voltage of the transmission network—the infinitely strong network—known as an infinite bus, or the voltage of an ideal voltage source) whenever there is a change in the operating conditions of the motor. The term "unstable operating mode" (or **instability**) means that the synchronous motor does not have the ability to remain in synchronism with its excitation voltage whenever there is a change in the operating conditions of the motor. Finally, the term "marginally stable operating mode" (or **marginal stability**) means that the synchronous motor loses synchronism with its excitation voltage whenever there is a slight increase in the loading conditions of the motor.

This process can continue through points 1,2, ..., Fig. 5.42(a), until the external loading power P_{OM3} is equal to the maximum power $|P_{MIN}|$ that can be developed by the motor, that is, $|P_{MIN}| \equiv |P_{OM3}| \equiv |P_{OT3}|$, at this point, $\delta_M = -\pi/2$ rad$(-90°)$.

If the applied mechanical loading power is now increased again to a new value $|P_{OMX}| > |P_{MIN}|$, the rotor (and thus $\mathbf{E}_G$ or $\mathbf{B}_R$) will decelerate while $\mathbf{V}_{ST}$ or $\mathbf{B}_T$ will continue to rotate at the synchronous speed ω_S. This will cause $|\delta_M|$ to increase beyond $\pi/2$ rad (i.e., $\delta_M < -\pi/2$), say to δ_{M4}. However, even though $|\delta_M|$ increases, the power developed by the motor decreases. Thus $|P_{OMX}| > |P_{OT4}|$; that is, the motor is unable to develop enough power to overcome the loading power. Thus the excess loading power is overwhelming and the rotor continues to decelerate with no equilibrium point in sight. The rotor of the motor continues to rotate at speeds continuously decreasing (i.e., much lower than the synchronous speed); thus it steps out of synchronism, "skips poles," and eventually stops at a power angle of $-\pi$ rad $(-180°)$, unless it is kept moving by induction-motor action resulting from the damper windings that may be present, Fig. 5.22. When the power angle becomes more negative than $-\pi$ rad $(-180°)$, the motor will begin to absorb mechanical power, if and only if such is available, rather than to develop it. Then the synchronous motor will behave as a generator and will continue to generate electric power if the necessary mechanical input is available.

It is clear, then, from Fig. 5.42(a) that the phenomenon just described (i.e., the rotor of a synchronous motor decelerates and pulls out of synchronism) will occur whether the initial power angle is equal to or less (in absolute value) than $\pi/2$ rad (i.e., $\delta_M < -\pi/2$ rad). Therefore, it can be concluded that the synchronous motor can be operated in synchronism with the excitation voltage $\mathbf{V}_{ST}$ for power angles $-\pi/2 \leqslant \delta_M < 0$.

As a consequence of what was just described, the maximum power that a synchronous motor can deliver to a mechanical load, Fig. 5.41(a), before it pulls out of synchronism is defined as the **motor pull-out power** and is designated as P_{MPO}. Mathematically it is defined as

$$P_{MPO} \equiv P_{MIN} = -3\left(\frac{|\mathbf{V}_{ST}||\mathbf{E}_G|}{X_S}\right) \tag{5.167}$$

Thus the point $P(P_{MIN}, -\pi/2)$ on the power versus power angle graph, Fig. 5.42(a), is defined as the **steady state stability limit** of the synchronous motor and is designated as $SSSL_M$.

The above explanation suffices for describing the behavior of the synchronous motor. Thus any comments pertaining to Fig. 5.42(b) will be considered in Section 6.2.

Example 5.10

A three-phase, 30-pole, Y-connected, balanced synchronous motor is excited from an infinitely strong network (remember that a motor can also be excited from an independent three-phase voltage source) whose line-to-line voltage is 2,500 V with a frequency of 60 Hz. The motor delivers 2,000 hp to a mechanical load when its phase parameters are $R_S = 0\ \Omega$ and $jX_S = j1.9\ \Omega$, and when the field (rotor-winding) dc excitation I_R is adjusted to yield a power factor equal to unity (i.e., $\cos\theta_S = 1$) at rated load.

Assume all losses to be zero and calculate: (1) the maximum mechanical power that the motor can deliver, that is, the **motor pull-out power**, P_{MPO}, (2) the maximum torque the motor can deliver, that is, the **motor pull-out torque**, τ_{MPO}, (3) the reactive power that the motor delivers or absorbs to or from the network at rated load, Q_{OT}, (4) the maximum reactive power, Q_{OTMAX}, that the motor can deliver or absorb, (5) the minimum reactive power, Q_{OTMIN}, that the motor can deliver or absorb, (6) the steady state stability limit of the motor ($SSSL_M$).

Solution

Since all losses are neglected, the per-phase output power P_O of the motor, in MKS units, is calculated through the total output power as follows:

$$P_{OMT} = (2{,}000 \text{ hp}) \times (746 \text{ W/hp}) = \underline{1{,}492{,}000.00} \text{ W} \tag{1}$$

and the per-phase output power is

$$P_{OM} = \frac{P_{OMT}}{3} = \frac{1{,}492{,}000.00}{3} = \underline{497{,}333.33 \text{ W}} \tag{2}$$

The phase excitation voltage $\mathbf{E_S}$ of the motor is calculated utilizing Eq. (3.67) as follows:

$$|\mathbf{V}_{ST}| = \frac{|\mathbf{V}_{ST(L-L)}|}{\sqrt{3}} = \frac{2{,}500}{\sqrt{3}} = \underline{1{,}443.38 \text{ V}} \tag{3}$$

Since for zero losses, $P_{OM} \equiv P_G \equiv P_{IN} \equiv \text{Re}(\mathbf{V}_{ST}\mathbf{I}_S^*) = |\mathbf{V}_{ST}||\mathbf{I}_S|\cos\theta_S$ and $\cos\theta_S \equiv 1$ (i.e., $\theta_S = 0$), the phase excitation voltage $\mathbf{V}_{ST}$ can be written from Eq. (3) and Fig. 5.30 as

$$\mathbf{V}_{ST} = \underline{1{,}443.37 e^{j0°} \text{ V}} \tag{4}$$

Now the phase stator current $\mathbf{I}_S$ is calculated utilizing Eqs. (5.103), (2), and (3) as follows:

$$\mathbf{I}_S = \frac{P_{IN}}{|\mathbf{V}_{ST}|\cos\theta_S} = \frac{P_{OM}}{|\mathbf{V}_{ST}|\cos\theta_S} = \frac{497{,}333}{(1{,}443.37)\times(1.00)} = \underline{344.56} \text{ A} \tag{5}$$

It is assumed that the current $\mathbf{I}_S$ is set along the Re-axis of the complex plane, Fig. 5.30. Then ($\cos\theta_S \equiv 1$ leads to the fact that $\measuredangle_{\mathbf{I}_S}^{\mathbf{V}_{ST}} \equiv 0$), Eq. (5) enables us to calculate the voltage drop across the reactance jX_S, Fig. 5.41(a), as follows:

$$\mathbf{V}_S = jX_S\mathbf{I}_S = j1.90 \times (344.56) = \underline{j654.67 \text{ V}} \tag{6}$$

Now the developed voltage $\mathbf{E}_G$, Fig. 5.41(a), can be calculated with the aid of Fig. 5.30 (for $R_S = 0$), or Fig. 5.41(a) and Kirchhoff's voltage law, and substitution from Eqs. (4) (utilizing Euler's identity) and (6) as follows:

$$\mathbf{E}_G = \mathbf{V}_{ST} - \mathbf{V}_S = 1{,}443.37 - j654.67 = \underline{1{,}584.90e^{-j24.40°}} \text{ V} \tag{7}$$

Inspection of Fig. 5.30 enables us to calculate (deduce) the power angle $\delta_M = \measuredangle \mathbf{E}_G - \measuredangle \mathbf{V}_{ST}$ as follows:

$$\delta_M = -24.40° - 0° \equiv \phi = \underline{-24.40°} \tag{8}$$

The maximum mechanical power P_{OMT} that the motor can deliver is equal to the absolute value of the **motor pull-out power,*** that is, $|P_{MPO}|$, which can be calculated by substituting for $|\mathbf{V}_{ST}|$ and $\mathbf{E}_G$ from Eqs. (4) and (7) and $\delta_M = -90°$ to Eq. (5.163) [or substitute for $|\mathbf{V}_{ST}|$ and $|\mathbf{E}_G|$ from Eqs. (4) and (7) directly into Eq. (5.167)]. Thus

$$P_{OMTMAX} \equiv |P_{MPO}| = 3\left(\frac{|\mathbf{V}_{ST}||\mathbf{E}_G|}{X_S}\right)\sin|(-90)|$$

$$= \frac{3 \times (1{,}443.38) \times (1{,}584.91)(1.0)}{1.9} = \underline{\underline{3{,}612{,}043.26 \text{ W}}} \tag{9}$$

As a check, we can calculate the power that the motor delivers at the power angle $\delta_M = -24.40°$ [Eq. (8)] and compare it to the rated power of 2,000 hp. Substitution in Eq. (5.163) of $\sin(-24.40°)$, instead of $\sin(-90°)$, yields

$$P_{OMT} \equiv P_{GT} \equiv P_{INT} = 3\left(\frac{|\mathbf{V}_{ST}||\mathbf{E}_G|}{X_S}\right)\sin|(-24.40°)|$$

$$= \frac{3 \times (1{,}443.38) \times (1{,}584.91) \times \sin(24.4)}{1.9} = \underline{1{,}492{,}151.07 \text{ W}} \tag{10}$$

Division of Eq. (10) by 746 W/hp yields

$$P_{OMT} = \frac{1{,}492{,}151.07}{746} = 2{,}000.20 \simeq \underline{2{,}000} \text{ hp} \tag{11}$$

The maximum torque, also known as the **motor pull-out torque**, can be calculated from the well-known relationship $P = \tau\omega_S$ and appropriate substitution from Eqs. (10) and (5.15) and input data. Thus

$$|\tau_{MPO}| = \frac{|P_{MPO}|}{\omega_S} = \frac{3{,}612{,}043.26}{2\pi \times 60/(30/2)} = \underline{\underline{143{,}718.63 \text{ N} \cdot \text{m}}} \tag{12}$$

The reactive power that the motor delivers or absorbs at $\delta_M = -24.40°$ is calculated from Eq. (5.164) by direct substitution from Eqs. (4), (7), and (8) and the

*The absolute value of P_{MPO}, that is, the **electric power delivered by the motor to the outside world** is defined to be the negative of the input, that is, $\mathbf{S}_O = -\mathbf{S}_{IN} = -\mathbf{V}_{ST}\mathbf{I}_S$, while **the mechanical power P_{OMT} provided by the motor to the outside world is defined to be positive.**

input data. Thus

$$Q_{OT}(-24.40°) = 3\left(\frac{|\mathbf{V}_{ST}||\mathbf{E}_G|}{X_S}\right)\cos\delta_M - 3\left(\frac{|\mathbf{V}_{ST}|^2}{X_S}\right)$$

$$= \left(\frac{3\times(1{,}443.38)\times(1{,}584.91)}{1.9}\right)\cos(-24.40°) - \frac{3\times(1{,}443.38)^2}{1.9} = \underline{\underline{-64.63}} \text{ var} \tag{13}$$

Since $Q_{OT} < 0$, the motor *absorbs reactive power* from the infinitely strong network and thus, its behavior is **inductive in nature**.

The maximum reactive power that the motor can deliver or absorb occurs at $\delta_M = 0°$, Fig. 5.42(b). Proper substitution to Eq. (5.164) [or Eq. (13)] yields

$$Q_{OTMAX} = 3\left(\frac{|\mathbf{V}_{ST}||\mathbf{E}_G|}{X_S}\right) - 3\left(\frac{|\mathbf{V}_{ST}|^2}{X_S}\right)$$

$$= \frac{3\times(1{,}443.38)\times(1{,}584.91)}{1.9} - \frac{3\times(1{,}443.38)^2}{1.9} = \underline{\underline{322{,}549.84}} \text{ var} \tag{14}$$

Since $Q_{OTMAX} > 0$, the motor *delivers reactive power* to the infinitely strong network and thus, its behavior is **capacitive in nature**.

The minimum reactive power that the motor can deliver or absorb occurs at $\delta_M = -90°$, Fig. 5.42(b). Proper substitution to Eq. (5.164) [or Eq. (13)] yields

$$Q_{OTMIN} = -3\left(\frac{|\mathbf{V}_{ST}|^2}{X_S}\right) = \frac{-3\times(1{,}443.38)^2}{1.9} = \underline{\underline{-3{,}289{,}493.41}} \text{ var} \tag{15}$$

Again, since $Q_{OTMIN} < 0$, the motor *absorbs reactive power* from the network and thus, its behavior is **inductive in nature**. However, the device never operates for long intervals of time at $\delta_M = -90°$ since the mechanical power output would exceed the design capabilities. It is the intention of this calculation to show that the device has the capability of producing positive and negative vars as δ_M varies and $|\mathbf{E}_G|$, $|\mathbf{V}_{ST}|$, and X_S remain constant. It is shown in Section 6.2 that the same effects on the output reactive power can be achieved by varying the field (rotor-winding) excitation.

The steady state stability limit of the motor is from Eq. (9) (the absolute value symbol deleted)

$$\text{SSSL}_M \equiv P_{MPO} = -3\left(\frac{|\mathbf{V}_{ST}||\mathbf{E}_G|}{X_S}\right)\Bigg|_{\mathbf{E}_S \equiv \mathbf{E}_T} = \underline{\underline{-3{,}612{,}043.26}} \text{ W} \tag{16}$$

5.8 THE BALANCED, TWO-PHASE, SYNCHRONOUS DEVICE

The balanced, two-phase, synchronous device is identical to the balanced, three-phase, synchronous device with **one exception.** *The two-phase device has* **two,** instead of three, *sinusoidally distributed stator windings*. The axes of these two windings are displaced from each other by 90 electrical degrees in space and time. The cross section of such a device is exhibited in Fig. 5.49(a). Note that **the connections for the motor and generator modes of operation** of this device are exhibited in Figs. 5.49(b) and 5.49(c), respectively. Also note that in Figs. 5.49(b) and 5.49(c) the *equivalent windings*, rather than the real ones, are exhibited.

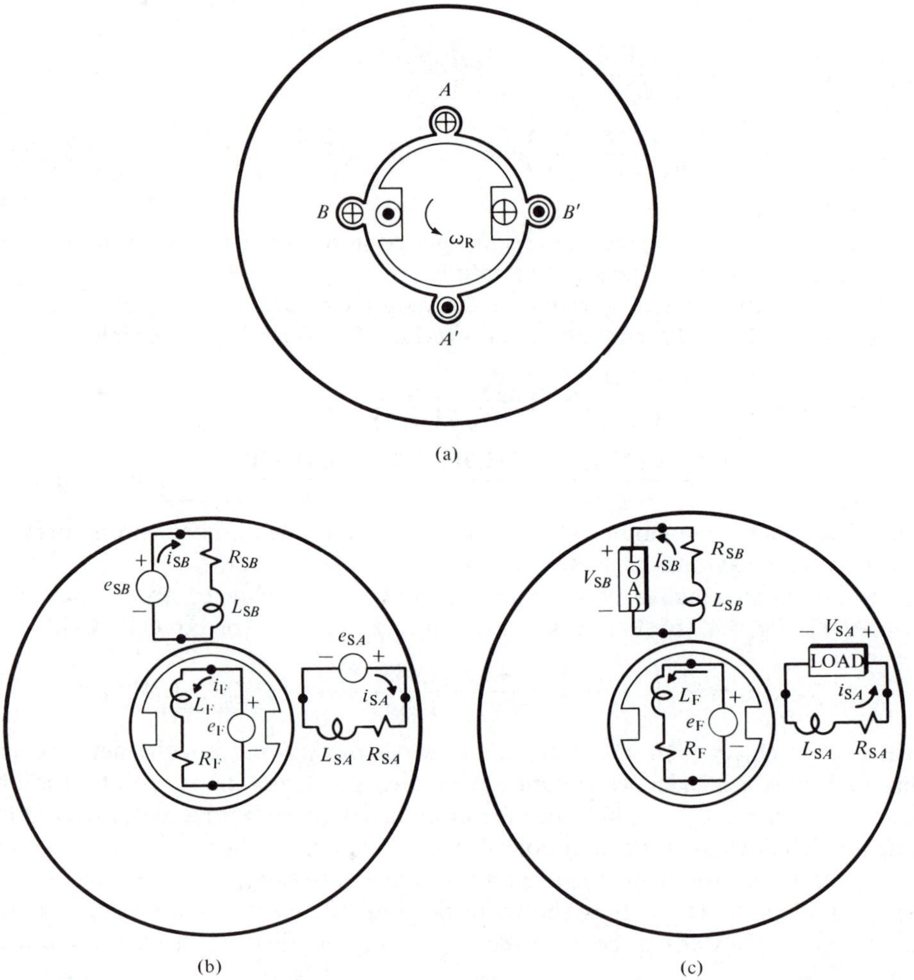

Fig. 5.49 Cross Section of the Balanced, Two-Pole, Two-Phase, Synchronous Device: (a) The Windings Are Exhibited as Concentrated—in the Real Device the Windings Are Distributed, (b) Motor Mode of Operation, and (c) Generator Mode of Operation

In practice, the two phases of a synchronous device generally are not connected internally. The four stator wires are brought out of the frame separately. For this reason the device is known as a *two-phase, four-wire*, synchronous device. However, if one end of phase A is connected to one end of phase B, and the common point is brought out of the frame through a common wire, the device is known as a *two-phase, three-wire*, synchronous device.

For the motor mode of operation, Fig. 5.49(b), the two stator windings are excited with balanced sinusoidal voltages that are displaced from each other, in space and time, by 90 electrical degrees. These **stator-winding excitations** are defined as

$$v_{SA}(t) = \sqrt{2}\,|\mathbf{V}_{ST}|\cos\omega_E t \tag{5.168}$$

and

$$v_{SB}(t) = \sqrt{2}\,|\mathbf{V}_{ST}| \sin \omega_E t \tag{5.169}$$

The excitation of the rotor winding, always dc, is defined as

$$e_F(t) = E_F \tag{5.170}$$

The two stator windings produce, in the airgap of the device, a stator field that is rotating in the counterclockwise direction at a synchronous speed—defined by Eqs. (4.22) and (5.15) and repeated here for clarity—given as

$$\omega_S = \frac{\omega_E}{(P/2)} \tag{5.171}$$

As in the case of the three-phase synchronous motor, the rotor field $\mathbf{B}_R$ of the two-phase synchronous device locks into synchronism with the rotating stator field $\mathbf{B}_{ST}$, and the rotor rotates at a speed ω_R identical to the synchronous speed. That is,

$$\omega_R \equiv \omega_S = \frac{\omega_E}{(P/2)} \tag{5.172}$$

It can be shown (see reference [14]) that (as in the case of the balanced, three-phase, synchronous device) the electromagnetic torque τ_{em} developed by the two-phase synchronous motor, because of the interaction of the rotating (ac-excited) stator field $\mathbf{B}_{ST}$ and the (dc-excited) rotor field $\mathbf{B}_R$, is given by

$$\tau_{em} = \left(\frac{K_T |\mathbf{V}_{ST}| I_F}{\omega_S} \right) \sin \delta \Big|_{\delta<0} \equiv \tau_{MAX} \sin \delta \Big|_{\delta<0} \tag{5.173}$$

Here K_T is a design constant, $|\mathbf{V}_{ST}| \equiv |\mathbf{V}_{SA}| = |\mathbf{V}_{SB}|$ is the RMS value of the stator winding excitation, I_F is the response current in the rotor winding, ω_S is the synchronous speed of the rotor, and δ, defined as the **torque or power angle** in electrical degrees, is the angle between the airgap field $\mathbf{B}_T$, that is, $\mathbf{B}_T = \mathbf{B}_R + \mathbf{B}_{ST}$, and the rotor field $\mathbf{B}_R$, Fig. 5.43(d). The torque angle δ is a measure of the angular displacement of the rotor field $\mathbf{B}_R$ from the airgap field $\mathbf{B}_T$. This angle is assigned **negative values when $\mathbf{B}_R$ lags $\mathbf{B}_T$**, and therefore for the motor mode of operation, and **positive values when $\mathbf{B}_R$ leads $\mathbf{B}_T$**, and therefore for the generator mode of operation. Examination of Eq. (5.173) reveals that since $|\mathbf{V}_{ST}|$ and I_F are controlled externally, for a given set of a design constant K_T and a synchronous speed ω_S, the device has a maximum torque capability defined as τ_{MAX}, and any changes in load are accommodated by changes in the torque angle δ. If an attempt were made to load the motor so that the externally applied load plus the internal frictional losses exceed the motor's maximum torque capability τ_{MAX}, the torque angle δ would increase to an angle greater than $|-90°|$; the developed torque $|\tau_{em}|$ would decrease, the rotor would slow down, and the rotor field would no longer be in synchronism with the airgap field $\mathbf{B}_T$. The motor is then said to have **lost synchronism** or to have **pulled out of synchronism.**

If instead of exciting the two stator windings of the balanced, two-phase, synchronous device, as well as the rotor winding, we were to (1) excite the rotor winding with a dc voltage, (2) apply a mechanical torque to the rotor, and (3) load the stator coils with identical loads, Fig. 5.49(c), we would have a *balanced, two-phase, synchronous generator*. The torque versus torque angle characteristics of

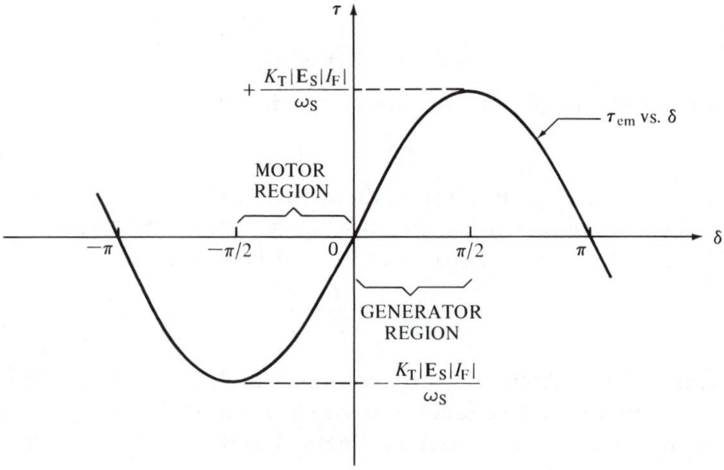

Fig. 5.50 Torque vs. Torque Angle Characteristic of the Balanced, Two-Phase, Synchronous Device

the device can be shown to be

$$\tau_{em} = \left(\frac{K_T|\mathbf{V}_{ST}|I_F}{\omega_S} \right) \sin\delta \Bigg|_{\delta>0} \equiv \tau_{MAX} \sin\delta \Bigg|_{\delta>0} \tag{5.174}$$

Fig. 5.50 exhibits the torque versus torque angle characteristic of the balanced, two-phase, synchronous device with the regions of operation for the motor and generator modes designated explicitly.

Comparison of Eqs. (5.173) and (5.174) with Eqs. (5.163) and (5.144), respectively, as well as Fig. 5.50 with Figs. 5.42(a) and 5.35(a), suggests that the analysis of the two-phase synchronous device could be carried out as if it were a balanced, three-phase, synchronous device. That is, the analysis would be carried out on a **per-phase basis**. However, the analyst should keep in mind that the phase voltages and the phase currents of the balanced, two-phase, synchronous device are out of phase by 90 electrical degrees and that the total power of the balanced, two-phase, synchronous device is **twice** its phase power. Therefore, the power equations for this device are written in analogy with Eqs. (5.62) through (5.64) and (5.106) through (5.108), except that the multiplier 3 is substituted by 2.

The presentation of this device in this text is only of **academic interest** as there is only limited use of the balanced, two-phase, synchronous device. Therefore, the subject is not pursued any further. For additional information, the interested reader is referred to the literature [for example see K. Y. Tang, **Alternating Current Circuits**, (Intext, New York, 1948)].

5.9 THE SINGLE-PHASE SYNCHRONOUS DEVICE

The operation of the single-phase synchronous device depends on either the **variable-reluctance principle**, which yields the reluctance-type synchronous device, or the **hysteresis principle**, which yields the hysteresis-type synchronous device.

Reluctance-type Synchronous Device

The analysis of this device, the reluctance of which depends on the space variable θ_R, is based on the application of the **virtual displacement* principle** to the moving member, the rotor, of the device. That is, for an incremental displacement $\Delta\theta_R$ of such a device's rotating member, and the added constraint that the device's flux is maintained constant, the principle of conservation of energy, $dw_E = dw_F + dw_M$, Fig. 2.4, can be written (with the aid of $dw_E \equiv ei\,dt \equiv i\,d\lambda \equiv i\,d(N\phi) \equiv Ni\,d\phi \equiv \mathcal{F}\,d\phi$) as

$$dw_M = \mathcal{F}\,d\phi|_{d\phi=0} - dw_F \tag{5.175}$$

If, however, we were to express in Eq. (5.175) the work done dw_M on the rotating member of the given device as $dw_M = \tau_{em}\,d\theta_R$ and solve for the electromagnetic torque τ_{em}, we can write the resulting equation as

$$\tau_{em} = -\frac{d}{d\theta_R} w_F \tag{5.176}$$

In addition, if we were to **assume linearity in the λ versus *i* characteristic** of the device being studied, **even though the device's magnetic circuit may be saturated**, then we would be able to express the mmf $\mathcal{F}$ of the device's magnetic circuit as a function of its reluctance $\mathcal{R}(\theta_R)$ and its flux ϕ. Under these conditions, the field energy w_F of the device can be written as

$$w_F\{\phi,\theta_R\} \equiv \int_0^i \lambda(i)\,di \equiv \tfrac{1}{2}\lambda i \equiv \tfrac{1}{2}\phi\mathcal{F} \equiv \tfrac{1}{2}\phi^2\mathcal{R}(\theta_R) \tag{5.177}$$

Substitution of Eq. (5.177) into Eq. (5.176) yields

$$\tau_{em} = -\tfrac{1}{2}\phi^2 \frac{\partial}{\partial\theta_R}\mathcal{R}(\theta_R) \tag{5.178}$$

The developed electromagnetic torque τ_{em} is, therefore, equal to the rate of decrease of the field energy $w_F\{\phi,\theta_R\}$ and acts upon the moving member of the device in such a direction as to tend to decrease the reluctance of the device's magnetic circuit.

The most popular single-phase device with variable reluctance is exhibited in Fig. 5.51(a). The essential feature of this device is that **both** the reluctance of the magnetic circuit and the excitation vary sinusoidally. Therefore, the rotor is shaped so that the reluctance of the device's magnetic circuit depends sinusoidally on the angular displacement θ_R of the rotor which is measured in the positive direction using as reference the *d*-axis through the stator poles, Fig. 5.51(a). The device is designed so that the sinusoidal dependence of the reluctance $\mathcal{R}$ of its magnetic circuit on the rotor's angular displacement θ_R is expressed as

$$\mathcal{R}(\theta_R) = \tfrac{1}{2}(\mathcal{R}_q + \mathcal{R}_d) - \tfrac{1}{2}(\mathcal{R}_q - \mathcal{R}_d)\cos 2\theta_R \tag{5.179}$$

Inspection of Eq. (5.182) and Fig. 5.51(a) reveals that the reluctance† $\mathcal{R}(\theta_R)$ has a **miminum value** $\mathcal{R}_d$, defined as direct-axis (*d*-axis) reluctance, when the axis of the rotor is directly in line with the *d*-axis through the stator poles (i.e., $\theta_R = 0, \pi, 2\pi, \ldots$)

*The term "virtual displacement" is derived from the mathematical process followed when the motion of a device's rotor is to be described by $\theta_R = \theta_{RO} + \Delta\theta_R$, and $\Delta\theta_R$ is allowed to approach zero in the limit—which means that there is no displacement.

†Remember that the mathematical expression for the reluctance is $\mathcal{R} = \ell/(\mu A)$.

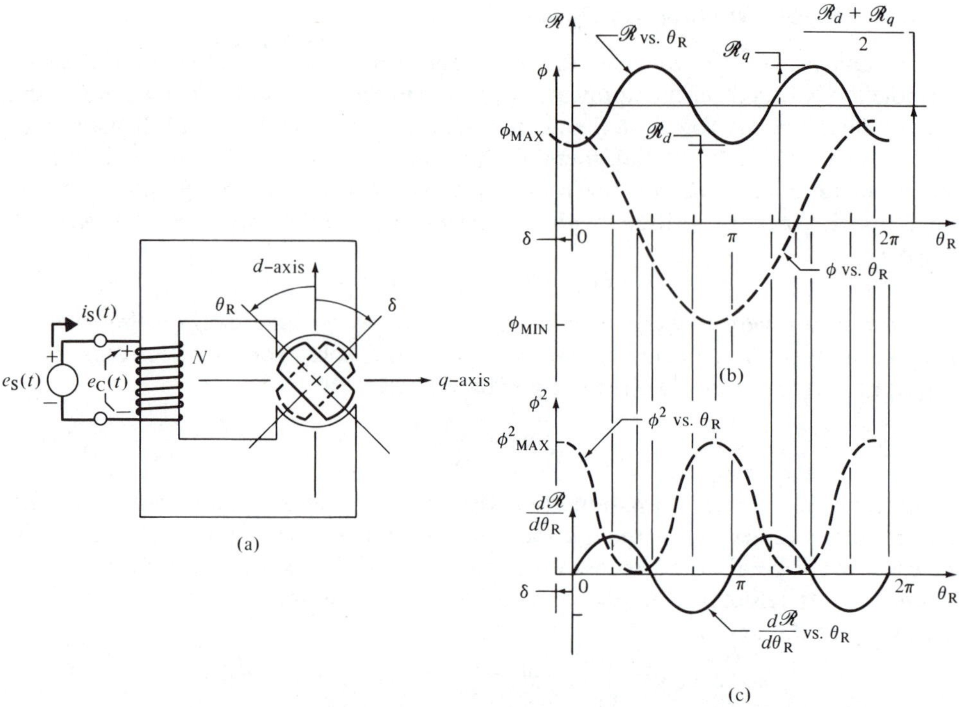

Fig. 5.51 Single-Phase, Two-Pole, Synchronous (Reluctance) Device: (a) Structure, (b) Flux and Reluctance vs. Rotor Angular Displacement Variations, and (c) Flux Squared and Reluctance Derivative vs. Rotor Angular Displacement Variations (Scale: $\phi_{MIN} = 1.30$ Units and $\phi_{MAX} = 1.69$ Units)

and a **maximum value** $\mathscr{R}_q$, defined as quadrature-axis (q-axis) reluctance, when the rotor axis is at right angles to the axis of the stator poles (i.e., $\theta_R = \pi/2, 3\pi/2, \ldots$).

The exciting or stator winding is connected to a source of sinusoidal voltage $e_S(t)$ designed to produce the cosinusoidal flux ϕ given by Eq. (5.180), that is,

$$\phi = \phi_{MAX} \cos \omega_E t \tag{5.180}$$

The plots for Eqs. (5.179) and (5.180) are exhibited in Fig. 5.51(b).

Practically all electromechanical energy-conversion devices are designed to have the resistance of their windings as close to zero as possible. Thus if we were to assume that (1) the voltage drop across the resistance of the windings of Fig. 5.51(a) is negligible when compared to the impressed voltage $v_S(t)$ and (2) the voltage source is ideal, when the flux ϕ of Eq. (5.180) must vary in such a manner as to induce a counter emf $e_C(t)$ across the coil of Fig. 5.51(a) equal to the impressed voltage $e_S(t)$. This suggests that, for the flux of Eq. (5.180), this counter emf $e_C(t)$ becomes

$$e_C(t) = -\frac{d\lambda(t)}{dt} = -\frac{d[N\phi(t)]}{dt} = N\phi_{MAX}\omega_E \sin \omega_E t \tag{5.181}$$

Therefore, the impressed voltage $v_S(t)$, Fig. 5.51(a), should be

$$v_S(t) = \sqrt{2}\, |\mathbf{V}_{ST}| \sin \omega_E t \tag{5.182}$$

Since $e_C(t)$ should be equal to $v_S(t)$, if Kirchhoff's voltage law pertaining to Fig.

5.51(a) is to be satisfied, the RMS values of Eqs. (5.181) and (5.182) would be related as follows:

$$\sqrt{2}\,|\mathbf{V}_{ST}| = N\Phi_{MAX}\omega_E \tag{5.183}$$

Solution of Eq. (5.183) for ϕ_{MAX} yields

$$\phi_{MAX} = \frac{\sqrt{2}\,|\mathbf{V}_{ST}|}{N\omega_E} \tag{5.184}$$

In order to substitute ϕ and $\mathcal{R}(\theta_R)$ into Eq. (5.178), we must first square the flux ϕ and take the derivative of $\mathcal{R}(\theta_R)$ with respect to the space variable θ_R. These two mathematical operations yield

$$\phi^2 = \tfrac{1}{2}\phi^2_{MAX}(1+\cos 2\omega_E t) \tag{5.185}$$

and

$$\frac{d}{d\theta_R}\mathcal{R}(\theta_R) = (\mathcal{R}_q - \mathcal{R}_d)\sin 2\theta_R \tag{5.186}$$

The plots of Eqs. (5.185) and (5.186) are exhibited in Fig. 5.51(c).

Substitution of Eqs. (5.185) and (5.186) into Eq. (5.178) yields

$$\tau_{em} = -\tfrac{1}{4}\phi^2_{MAX}(\mathcal{R}_q - \mathcal{R}_d)(\sin 2\theta_R + \sin 2\theta_R \cos 2\omega_E t) \tag{5.187}$$

Utilization of the trigonometric identity $\sin X \cos Y = \frac{1}{2}[\sin(X+Y) + \sin(X-Y)]$ enables us to write Eq. (5.187) as

$$\tau_{em} = -\tfrac{1}{4}\phi^2_{MAX}(\mathcal{R}_q - \mathcal{R}_d)\left[\sin 2\theta_R + \tfrac{1}{2}\sin(2\theta_R + 2\omega_E t) + \tfrac{1}{2}\sin(2\theta_R - 2\omega_E t)\right] \tag{5.188}$$

Clearly, if the torque τ_{em} is to have a nonzero average value, the following conditions must be satisfied:

$$\theta_R = \omega_R t - \delta \tag{5.189}$$

$$\theta_R = -\omega_R t - \delta \tag{5.190}$$

$$\theta_R = -\delta \tag{5.191}$$

Substitution of Eq. (5.189) into Eq. (5.188) yields [please note that $\pm\omega_R$, Eqs. (5.189) and (5.190), designates a counterclockwise or clockwise rotor rotation; for additional explanation consult the text following Eq. (5.198)]

$$\tau_{em} = -\tfrac{1}{4}\phi^2_{MAX}(\mathcal{R}_q - \mathcal{R}_d)\left\{\sin(2\omega_R t - 2\delta) + \tfrac{1}{2}\sin[2(\omega_R + \omega_E)t - 2\delta] + \tfrac{1}{2}\sin[2(\omega_R - \omega_E)t - 2\delta]\right\} \tag{5.192}$$

Now the average torque over one complete revolution of the rotor, Fig. 5.51(a), that is, over one period of time T, is calculated as follows:

$$\begin{aligned}\tau_{AVG} &= \frac{1}{T}\int_0^T \tau_{em}\,dt \\ &= -\left[\frac{\phi^2_{MAX}(\mathcal{R}_q - \mathcal{R}_d)}{4T}\right]\int_0^T \left\{\sin(2\omega_R t - 2\delta) + \tfrac{1}{2}\sin[2(\omega_R + \omega_E)t - 2\delta] \right. \\ &\quad \left. + \tfrac{1}{2}\sin[2(\omega_R - \omega_E)t - 2\delta]\right\} dt\end{aligned} \tag{5.193}$$

The value of the integral in Eq. (5.193) is zero (why?) unless $\omega_R \equiv \omega_E$. This condition sets forth the constraint that *the rotor of the single-phase synchronous (reluctance) device must rotate at a radian speed exactly equal to the radian frequency of the excitation source*. That is,

$$\omega_R \equiv \omega_E \tag{5.194}$$

Under this condition the first and second terms of the integrand, Eq. (5.193), yield **zero average value**. These two terms represent pulsating components of the torque at a frequency **two** and **four times** the excitation frequency ω_E. The last term of the integrand, however, is independent of time. It therefore yields an average value of the electromagnetic torque, given as

$$\tau_{AVG} = \tfrac{1}{8}\phi_{MAX}^2(\mathcal{R}_q - \mathcal{R}_d)\sin 2\delta \tag{5.195}$$

It could be shown that substitution of either Eqs. (5.190) or (5.191) into Eq. (5.188) also yields the average torque given by Eq. (5.195). This suggests that this device has a built-in capability to start rotating either **counterclockwise** or **clockwise**. Therefore, the designer must provide a built-in discriminator that will allow the rotor to rotate **only** in the desired direction. For example, self-starting electric clock motors are designed to start electrically as shaded-pole induction motors, Fig. 4.40(e), and then to operate synchronously as reluctance motors.

The angular displacement θ_R in Eqs. (5.189) through (5.191) designates the instantaneous position of the rotor, Fig. 5.51(a); the angle δ designates the position of the rotor at $t=0$ when the flux ϕ is passing through its maximum value ϕ_{MAX}, Fig. 5.51(b). When $\theta_R = -\delta$ and δ is positive constant, the rotor is in the position designated by dashed lines in Fig. 5.51(a). According to Eq. (5.195), τ_{AVG} is **positive** for $0<|\delta|\leqslant\pi/2$ rad. Therefore, it acts in a direction so as to **increase** the angular displacement of the rotor θ_R. A **negative** value of τ_{AVG} would suggest that τ_{AVG} acts in the direction to **decrease** the angular displacement of the rotor θ_R. That is, **a negative torque would force the rotor to assume such a position as to minimize the reluctance of the magnetic circuit**; the rotor axis would line up with the *d*-axis, a position defined as **stable**. It is clear, then, that the average value τ_{AVG}, of the torque τ_{em} depends on the initial displacement angle δ of the rotor, that is, the phase relation between the sinusoidaly varying waveforms of the reluctance $\mathcal{R}(\theta_R)$ and the flux ϕ, as well as the speed of the rotor ω_R. Inspection of Eq. (5.192) reveals that the instantaneous torque τ_{em} is not constant but contains second-harmonic functions of ω_E; it therefore cannot contribute average torque unless the time rate of change of θ_R, that is, ω_R, is constant and equal in absolute value to ω_E, Eq. (5.194). This constancy of ω_R is maintained by the inertias of the device's rotor and its load. Thus if the torque angle δ and the rotor speed ω_R are such that the average value of the waveform of the flux squared, Fig. 5.51(c), is greater while the reluctance is decreasing than it is while the reluctance is increasing, we can, without solving Eq. (5.193), deduce that τ_{AVG} of Eq. (5.178) is positive. **Therefore, τ_{AVG} acts in such a way so as to move the rotor in the direction of increasing θ_R and thus to keep the rotor of Fig. 5.51(a) rotating.**

These remarks suggest that, **except for the negligible effect of the harmonics**, no average torque is produced at speeds other than $\omega_R \equiv \omega_E$. Therefore, in the motor mode of operation, the device is **not self-starting** and must be started to rotate in the desired direction by auxiliary means, such as auxiliary windings that produce induction motor action, Fig. 4.40. The rotating field of the induction motor created to start the reluctance device is created by the exciting or stator winding, usually referred to as the **main winding** of the reluctance device, and an **auxiliary winding** displaced from the main winding, electrically and spatially, by 90 degrees. The torque versus speed characteristic of a typical single-phase reluctance motor is exhibited in Fig. 5.52. Note that the motor starts anywhere from 300 percent to 400 percent of its full-load torque—depending on the position of the rotor axis with respect to the d-axis at $t=0$, Fig. 5.51(a)—and follows the curve of the two-phase induction motor action, that is, the τ_{em} versus ω_R main-and-auxiliary-winding curve, Fig. 5.52. At about 75% ω_S, a centrifugal switch opens the auxiliary winding; the motor continues to develop single-phase torque produced by the main winding only. As the rotor approaches synchronous speed, the reluctance torque developed by the motor is sufficient to pull, that is, **to snap**, the rotor into synchronism with the pulsating single-phase field $|\mathbf{B}_{MF}|$ or $|\mathbf{B}_{MB}|$, Fig. 4.35. The device operates as a constant speed, single-phase, singly excited (i.e., there is no rotor excitation), synchronous motor up to approximately 200 percent of its full-load torque. When the reluctance motor is loaded with an extremely light load (e.g., a clock motor), the motor could be started manually—or by means of a gear rack that engages a pinion in the clock's gear train—by spinning the rotor in the **proper direction** above synchronous speed. As the rotor approaches synchronous speed from above, it is **pulled into step** and continues to run at the synchronous speed, that is, at $\omega_R = \pm\omega_S$ radians/second, Fig. 4.35.

Inspection of Eqs. (5.195) and (5.184) enables us to draw the following conclusions. If the RMS value of the applied voltage $|\mathbf{V}_{ST}|$ and its frequency ω_E are maintained constant, the flux ϕ_{MAX} is maintained constant. Since the reluctances $\mathcal{R}_q$ and $\mathcal{R}_d$ depend on the configuration of the magnetic circuit, they are also constant. Therefore, the torque angle δ is the only variable in the right-hand side of Eq. (5.195).

For a given load connected to the shaft of the motor, the torque angle δ adjusts itself so that the torque τ_{AVG} equals the torque required to drive the given load plus the rotational losses in the motor. If the load increases, the rotor momentarily slows down, thereby increasing the torque angle δ until sufficient torque τ_{AVG} is developed to carry the increased load. After the brief time required for the adjustment of the torque angle δ, during which $\omega_R \neq \omega_S$, operation is resumed at constant speed, that is, $\omega_R \equiv \omega_E$.

Note that the maximum torque $\tau_{AVG(MAX)}$ that can be developed by the motor is

$$\tau_{AVG(MAX)} = \tfrac{1}{8}\phi_{MAX}^2(\mathcal{R}_q - \mathcal{R}_d) \tag{5.196}$$

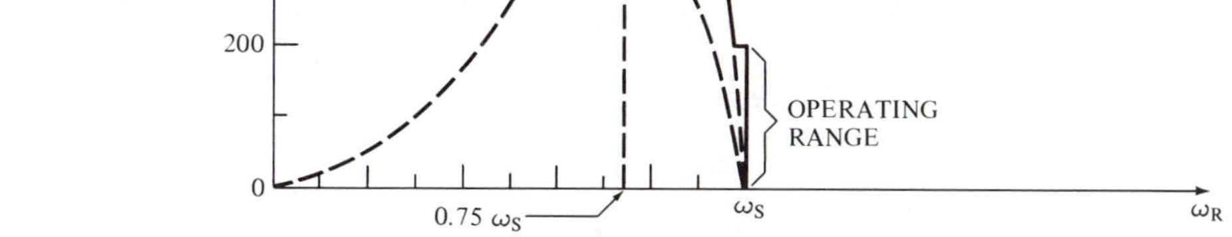

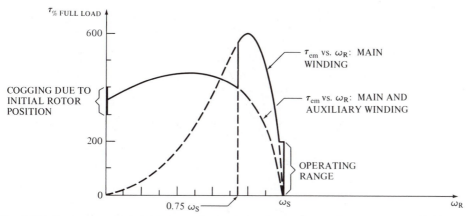

Fig. 5.52 Torque vs. Speed Characteristic of the Single-Phase Synchronous (Reluctance) Motor

This maximum occurs at $|\delta| = \pi/4$ rad or 45°. If the applied load plus the rotational losses exceed the torque given by Eq. (5.196), the motor will lose synchronism but will continue to operate as a single-phase induction motor up to 600 percent of its full load; but eventually it will stall, Fig. 5.52. Notice the high values of the induction motor torque in Fig. 5.52. The reason for their presence is as follows. To obtain satisfactory synchronous motor performance, it has been found necessary to build reluctance-type synchronous motors on frames that would be suitable for induction motor performance of two to three times the synchronous motor rating.

If we were to supply torque to the shaft of the rotor, Fig. 5.51(a), the rotor would advance in phase. That is, **δ would change sign**, and the average electromagnetic torque given by Eq. (5.195) would change sign. Therefore, it would represent power absorbed by the device and converted into electrical energy. As with the motor action, there is an upper limit to the mechanical torque that can be absorbed by the device and can be converted to electrical energy. This limit would occur at $|\delta| = \pi/2$ rad or 45°. If the driving torque were to increase beyond the upper limit specified by Eq. (5.195), it would cause overspeeding and therefore loss of synchronism.

If we were to designate the angle δ measured from the *d*-axis to the rotor axis as (1) **positive**, when measured in the counterclockwise direction, and (2) **negative**, when measured in the clockwise direction, we would be able to draw the torque versus torque angle characteristic of the single-phase synchronous (reluctance) device discussed here as exhibited in Fig. 5.53. Note that the regions of operation for the motor and generator modes are designated explicitly. Also note the similarity of the curves in Figs. 5.53, 5.50, and 5.35(a) and 5.42(a).

Furthermore, reluctance motors can be started as induction motors by making use of the squirrel cage rotor from which *teeth have been removed* in locations diametrically opposite. Removal of such teeth leaves the rotor bars and the end rings intact, resulting in a two-pole, modified, squirrel cage rotor, Fig. 5.54. Thus the motor can be started as a single-phase induction motor, by use of any of the

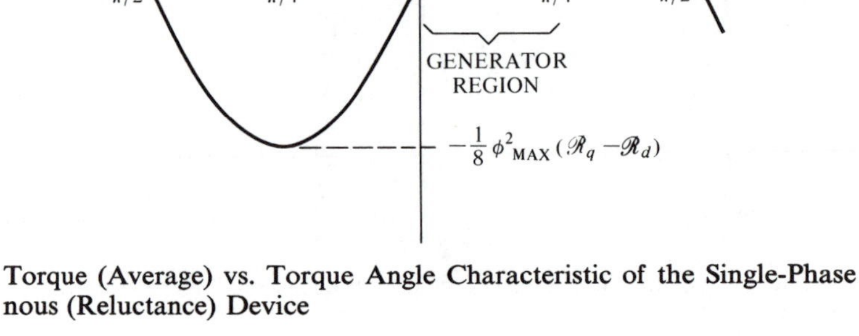

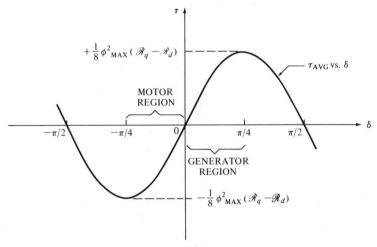

Fig. 5.53 Torque (Average) vs. Torque Angle Characteristic of the Single-Phase Synchronous (Reluctance) Device

schemes exhibited in Fig. 4.40, and can then accelerate towards the synchronous speed $\omega_R < \omega_S$. The tendency of the rotor, however, to align itself in the **minimum reluctance position** with respect to, say, the forward synchronously revolving field $|\mathbf{B}_{MF}|$, Fig. 4.35, produces the reluctance torque τ_{em}. If the moments of inertia of the rotor and its load are not excessive, the developed reluctance torque accelerates the rotor up to synchronous speed. The rotor **pulls into step** with the revolving field and continues to rotate at synchronous speed $\omega_R \equiv \omega_S$. However, in this case the backward rotating field $|\mathbf{B}_{MB}|$, Fig. 4.35, of the single-phase induction motor action causes a backward torque to be reflected on the rotor shaft. This torque affects the performance of the reluctance motor in the same way that an additional shaft load would. That is, its maximum value of the τ_{AVG} is reduced.

In passing, it is simply mentioned here that polyphase reluctance motors are also available. Such devices can be created if one were to remove teeth from a squirrel cage rotor of a polyphase induction motor, Figs. 5.54(b) and 5.54(c), in such a way as to produce the same number of salient poles in the rotor structure as in the stator field (structure) of a given device. The removal of the teeth should be such as to leave the rotor bars and the end rings intact, so that the device can start readily as an induction motor, its rotor then able to accelerate towards synchronous speed,

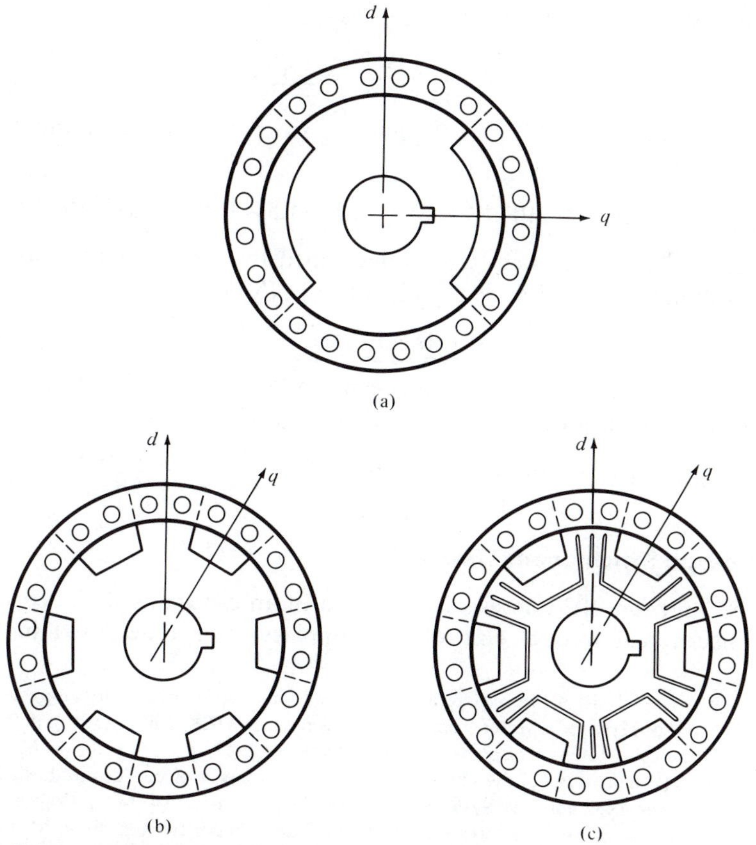

Fig. 5.54 Cross Sections of Rotors for Reluctance Motors: (a) Two-Pole, Modified, Squirrel Cage Rotor, (b) Six-Pole, Modified, Squirrel Cage Rotor, and (c) Six-Pole, Modified, Squirrel Cage Slitted Rotor

i.e., $\omega_R \simeq \omega_S$. The rotor then *pulls into synchronism* with the aid of the reluctance torque developed at the salient poles of the rotor structure, which have lower reluctance airgaps. The slits in the rotor structure of Fig. 5.54(c) are used *to direct the field* along more effective paths to produce a more abrupt change in reluctance (i.e., a greater $\partial\mathcal{R}(\theta_R)/\partial\theta_R$ in the transition from the *d*-axis to the *q*-axis), which results in a higher value of maximum torque.

Example 5.11

When the rotor of the reluctance motor exhibited in Fig. 5.52(a) is in the *d*-axis position, the reluctance of the magnetic circuit is $\mathcal{R}_d = 1.5 \times 10^6$ At/Wb. When the rotor is in the *q*-axis position, the reluctance is $\mathcal{R}_q = 2.5 \times 10^6$ At/Wb. If the stator winding has 2000 turns and is excited by the source $v_S(t) = \sqrt{2}\,(155.56)\sin 377t$, (1) calculate the maximum average torque $\tau_{AVG(MAX)}$ that can be developed by the reluctance motor, (2) write the general expression for the average torque τ_{AVG} developed by this motor, and (3) calculate the torque developed by this motor when $\delta = 15°$

Solution

To calculate the maximum average torque from Eq. (5.196), we must first calculate the maximum flux ϕ_{MAX}. Thus substitution into Eq. (5.184) for $|\mathbf{V}_{ST}|$, N, and ω_E from the input data yields

$$\phi_{MAX} = \frac{\sqrt{2} \times 155.56}{2000 \times 377} = \underline{2.92 \times 10^{-4}\ \text{W}} \tag{1}$$

Substituting into Eq. (5.199) for $\mathcal{R}_d$ and $\mathcal{R}_q$ from the input data and ϕ_{MAX} from Eq. (1) yields

$$\tau_{AVG(MAX)} = \tfrac{1}{8}(2.92 \times 10^{-4})^2 \times (2.5 \times 10^6 - 1.5 \times 10^6) = \underline{\underline{1.07 \times 10^{-2}}}\ \text{N}\cdot\text{m} \tag{2}$$

The expression for the average torque developed by the reluctance motor can be written, by direct substitution of Eq. (2) into Eq. (5.195), as

$$\tau_{AVG} = \underline{\underline{1.07 \times 10^{-2} \sin 2\delta}}\ \text{N}\cdot\text{m} \tag{3}$$

The torque developed by the reluctance motor when $\delta = -15°$ is calculated by direct substitution for δ (from the input data) into Eq. (3) as

$$\tau_{AVG(-15°)} = 1.07 \times 10^{-2} \sin 2(-15°) = \underline{\underline{-0.535 \times 10^{-2}}}\ \text{N}\cdot\text{m} \tag{4}$$

Hysteresis-type Synchronous Device

The phenomenon of hysteresis is exploited in an especially designed device* to develop mechanical torque. One of the simplest rotor types used in hysteresis

*For more detailed analysis of the hysteresis motor, the following references are suggested: (1) B. R. Teare, Jr., "Theory of Hysteresis Motor Torque," **Transactions of AIEE**, vol. 59 (1940), p. 907; (2) H. C. Roters, "The Hysteresis Motor," **Transactions of AIEE**, vol. 66 (1947), pp. 1419–30; (3) M. A. Copeland and G. R. Slemon, "An Analysis of the Hysteresis Motor: I—Analysis of the Idealized Machine," **IEEE Transactions on Power Apparatus and Systems**, vol. 32 (1963), pp. 34–42; (4) M. A. Copeland and G. R. Slemon, "An Analysis of the Hysteresis Motor: II—The Circumferential Flux Machine," **IEEE Transactions**, vol. 33 (1964), pp. 619–25; and (5) G. R. Slemon, R. D. Jackson, and M. A. Rachman, "Performance Predictions for Large Hysteresis Motors," Paper F 77 063-1 (IEEE Power Engineering Society, Winter Meeting, New York, N.Y., January 30–February 4, 1977).

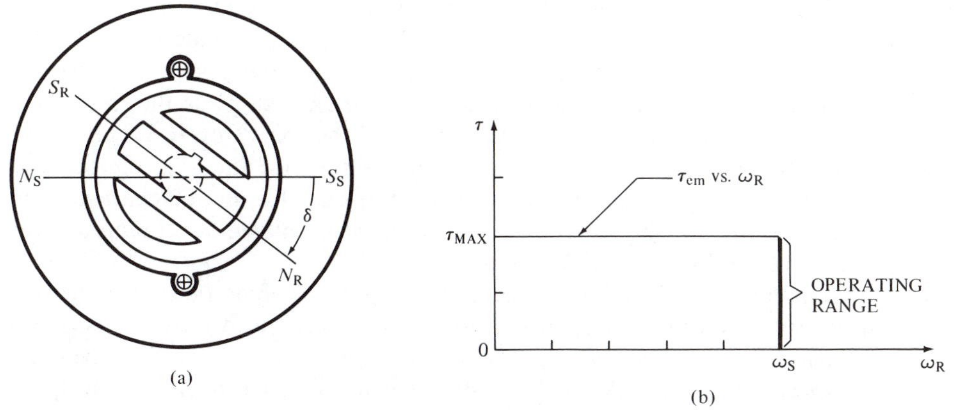

Fig. 5.55 Single-Phase, Two-Pole, Synchronous (Hysteresis) Motor: (a) Cross-Sectional Front View and (b) Idealized Torque vs. Speed Characteristic

motors is exhibited in Fig. 5.55(a). The laminations are made of hardened, high-retentivity steel. The stator has a distributed winding designed to produce a sinusoidal field. This winding is a permanent capacitor-split type, Fig. 4.40(c), with the capacitor chosen so that the stator behaves as if the device were a balanced two-phase induction motor. Therefore, the stator produces a rotating field $\mathbf{B}_{ST}$ of constant amplitude that revolves at the synchronous speed ω_S. As a result of this rotating stator field, eddy currents, which travel across the two **bar paths** of the rotor, Fig. 5.55(a), are induced in the steel structure of the rotor. These eddy currents produce a field $\mathbf{B}_R$ in the rotor structure, which causes the rotor to rotate. As the rotor approaches synchronous speed, the frequency of the eddy current reversal in the cross bars decreases and the rotor becomes permanently **magnetized** in one direction, a result of the high retentivity of the rotor steel. Thus the rotor runs as a hysteresis motor on hysteresis torque **because the rotor is permanently magnetized**.

The torque developed by the hysteresis motor at synchronous speed is a function of torque angle δ, **the angle between the magnetic axes of the stator and the rotor**. Fig. 5.55(a) exhibits an elementary two-pole hysteresis motor with the stator magnetic axis passing through the stator poles N_S and S_S, produced by the stator windings, and the rotor magnetic axis passing through the rotor poles S_R and N_R, produced by the high retentivity of the rotor steel. *The hysteresis torque developed is extremely steady in amplitude and phase, in spite of fluctuations of the supply voltage.* An idealized torque versus speed characteristic of the hysteresis motor is exhibited in Fig. 5.55(b). The high starting torque is produced as a result of the high rotor resistance, proportional to the hysteresis loss.

If the rotor is stationary, starting torque proportional to $\sin\delta$ is developed. That is,*

$$\tau_{em} = \mathcal{K}_T|\mathbf{B}_R|\sin\delta \equiv \tau_{MAX}\sin\delta \tag{5.197}$$

If the loading torque plus the frictional losses of the motor are less than the developed torque of the motor, the rotor accelerates. As long as the rotor turns at a

*Note that the format of Eq. (5.197) is identical to that of Eqs. (5.173) and (5.174). Therefore, the plot of Fig. 5.50 could be adjusted to represent Eq. (5.197).

speed less than the slip frequency, the rotor structure is subjected to a repetitive hysteresis cycle of slip frequency. While the rotor accelerates, δ remains constant if the rotor field is constant, because δ depends on the hysteresis loop of the rotor steel rather than the frequency at which the loop is traversed. Thus the torque τ_{em} developed by the hysteresis motor is constant for all the speeds up to synchronous speed ω_S, and the acceleration of the driven load is smooth; the load inertia only affects the time taken by the rotor to reach synchronism. This unique characteristic of the hysteresis motor makes the device an **excellent candidate** for timing devices and record player and tape recorder motors.

Note that in contrast to the reluctance motor, Fig. 5.52, which must **snap** its load into synchronism from an induction motor torque versus speed characteristic, *the hysteresis motor, Fig. 5.55(b), can synchronize any load that it can accelerate, no matter how great is its inertia*. After the rotor has reached synchronism, it continues to run at synchronous speed and adjusts its torque angle so as to develop the torque required by the load. However, since reluctance torque can be produced more cheaply than hysteresis torque for the same rating motors, *high-torque hysteresis motors of good quality are more expensive than reluctance motors of the same rating*.

5.10 APPLICATIONS OF SYNCHRONOUS DEVICES

The applications of the synchronous devices are many. The choice of a synchronous device for a given task depends on the characteristics of the task the device is to perform, as well as the dollar value per unit power of the device's capability compared to that of other available devices. The qualifications and capabilities of each of the types of available synchronous devices are discussed in the following sections.

The Balanced, Three-Phase, Synchronous Device

This device has distinguished itself in both the generator and the motor modes of operation.

As a generator, known as an **ac generator** or **alternator**, it provides more than 99 percent of the electric energy used by people. However, it should be mentioned in passing that new types of generators, namely, fuel cell, thermionic, thermoelectric, and magnetohydrodynamic (abbreviated as MHD), are under research and development. In addition, there is a tremendous amount of research activity in the areas of solar energy, wind energy, and seawater (wave) energy.

As a motor, the balanced, three-phase, synchronous device has a wide range of applications because of several inherent unique capabilities. These applications are (1) power-factor control, (2) constant-speed rotation, and (3) high starting torque and high operating efficiency.

POWER-FACTOR CONTROL

The balanced, three-phase, synchronous motor operating at a **maximum** and **constant** excitation, a state known as **overexcitation**,* behaves as a capacitor. Thus

*Refer to Section 6.2.

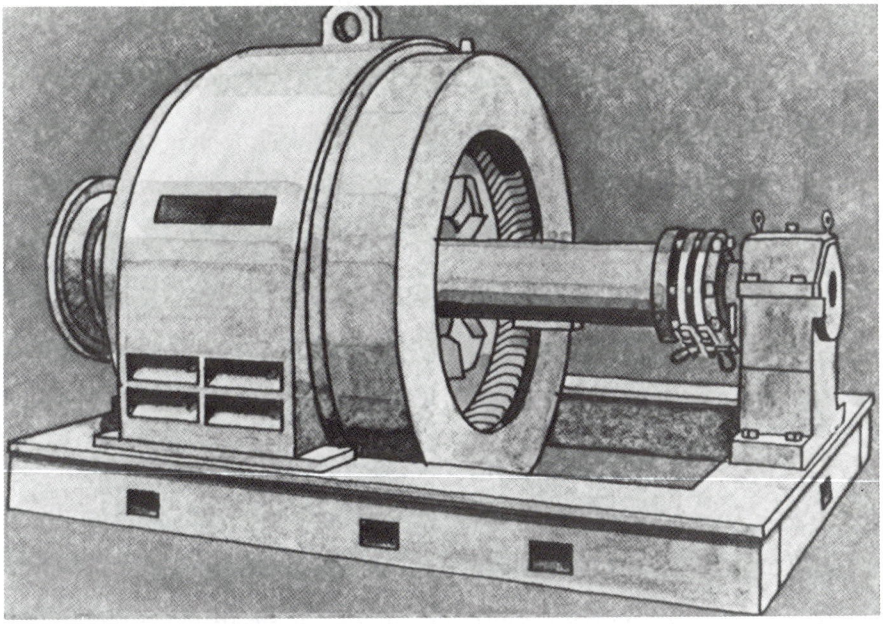

(a)

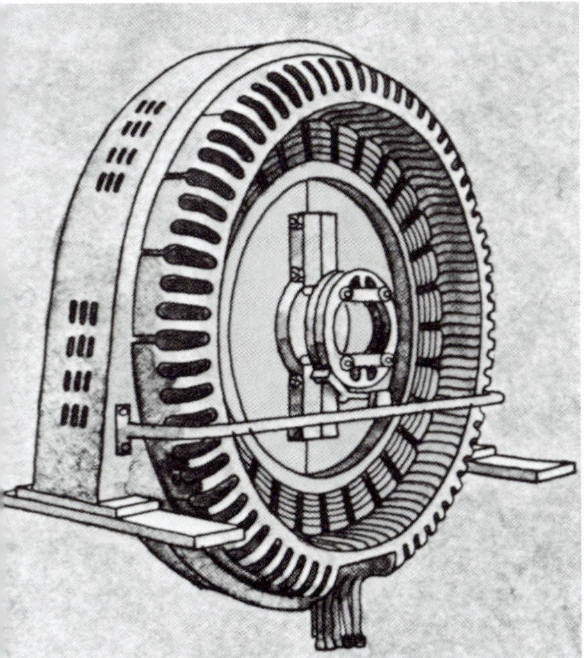

(b)

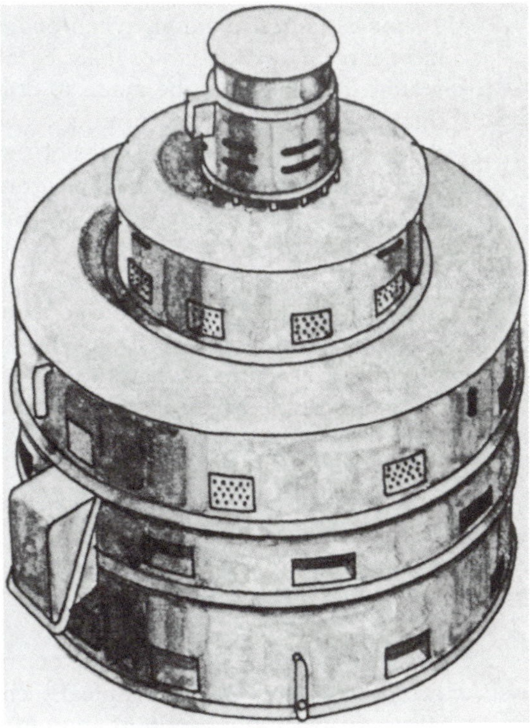

(c)

Plate 5.11 Synchronous Motor Types: (a) Pedestal-Bearing Design-Type (Provides for Low Shaft Height and Stator Shift), (b) Engine-type Low-Speed, and (c) Vertical Type. (Courtesy of Westinghouse Electric Co., Pittsburgh, Pa.)

it is known as a **synchronous condenser** and is used for power-factor correction. As a synchronous condenser, it generates lagging reactive kilovoltamperes; therefore, it is used to relieve the power source of the necessity of supplying this component of power. Power companies use such devices across transmission lines at the receiving ends, that is, at the load points, to (1) improve the power factors of the loading systems or (2) reduce line voltage drops and thereby improve or control (stabilize) the voltages at these points. Thus they improve the transmission capacity of their systems.

From the standpoint of the utility companies' supplying power, it is of great importance that the system power factor be maintained at a high value. This is so because fewer generated kilovoltamperes are needed to supply a given demand at a high power factor than are needed at a low power factor. Because of this fact, **power-factor penalty** and **bonus** clauses are often written into power contracts to encourage industrial consumers to raise the power factor of their equipment.

When a synchronous motor is used for power-factor correction purposes, it supplies that function in addition to acting as a prime mover. Its kilovoltampere capacity is then determined by the combination of these two functions. As a prime mover the synchronous motor requires a power component of current, and as a power-factor corrector it requires a current component in quadrature leading. When multiplied into the loading voltage, the vector sum of these current components gives the kilovoltamperes taken by the synchronous motor, and, obviously, it should not exceed the rating of the device.

For a similar function, synchronous motors are often applied to a system that has a large induction motor load. By adjusting the dc excitation of the synchronous motor, the motor can be made to draw a leading current that would compensate for some of the lagging currents drawn by the induction motors and that would, therefore, raise the power factor of the entire system.

If the dc excitation of the synchronous motor is not maintained constant at the maximum value, as it is for the synchronous condenser, but is made (with the aid of a voltage regulator in a feedback loop) to reduce as a function of the loading current $\mathbf{I}_L$ and the voltage $\mathbf{E}_\mathrm{R}$ at the receiving end (i.e., the **load end**) of the transmission line, it is possible to maintain the voltage $\mathbf{E}_\mathrm{R}$, at the receiving end of the transmission line, **constant** by underexciting the synchronous motor. The synchornous motor is then known to behave as a **synchronous reactor** (inductor) rather than as a synchronous condenser (capacitor).

CONSTANT-SPEED ROTATION

Because of its constant-speed characteristics, the balanced, three-phase, synchronous motor may be used as a prime mover of (1) dc shunt generators that maintain a fairly constant dc voltage from no-load to full-load, Fig. 2.43, and (2) synchronous generators of various frequencies that maintain a constant generated voltage $\mathbf{E}_\mathrm{G}$ at a constant frequency ω_G. A synchronous motor-generator set in which a change of frequency $f_j, j = \mathrm{M, G}$, occurs is referred to as a **frequency changer**. Since two such devices are mechanically coupled, they are both operating at the same speeds in revolutions/minute, that is, $n_\mathrm{M} = 120 f_\mathrm{M}/P_\mathrm{M}$ and $n_\mathrm{G} = 120 f_\mathrm{G}/P_\mathrm{G}$. Therefore, the equality of speeds, $n_\mathrm{M} = n_\mathrm{G}$, yields

$$\frac{120 f_\mathrm{M}}{P_\mathrm{M}} = \frac{120 f_\mathrm{G}}{P_\mathrm{G}} \tag{5.198}$$

or

$$\frac{f_M}{f_G} = \frac{P_M}{P_G} \tag{5.199}$$

where f_M and f_G designate the respective frequencies in hertz for the synchronous motor and generator, and P_M and P_G designate the respective number of poles for the motor and generator.

Eq. (5.199) defines the desired pole combination of the synchronous motor-generator set that yields the desired frequency conversion.

HIGH STARTING TORQUE AND HIGH OPERATING EFFICIENCY

Balanced, three-phase, synchronous motors are rated at the mechanical power they can deliver and at a power factor other than unity, usually leading. This means that at rated load and excitation, the synchronous motor supplies reactive power (kilovolt-amperes) equal to a high percentage of its mechanical power rating. Since the torque is a function of both the excitation voltage $\mathbf{V}_{ST}$ and the generated voltage $\mathbf{E}_G$, Eq. (5.164), the synchronous motor with the *nonunity* power-factor rating has greater* pull-in and pull-out torques than the synchronous motor with the **unity** power-factor rating, for the same mechanical power rating. The former motor is therefore capable of meeting larger peak loads (which are characteristic of ball mills and crushers).

Synchronous motors are never used below 50 hp and medium speeds because of (1) initial high cost when compared to induction motors, (2) required dc excitation, and (3) starting and control device expenses. However, synchronous motors are particularly suitable for speeds less than 500 rpm when directly connected to loads (i.e., no reduction gears are necessary) such as compressors, grinders, and mixers, especially in ratings of 100 hp or more. At these low-speed ratings, the synchronous motor is less expensive than an induction motor of the same rating and has the distinct advantage of being a source of reactive power simultaneously.

There is an optimum combination of horsepower and speeds for which the disadvantage of high initial cost of the synchronous motor, when compared to the induction motor, vanishes and its use becomes advantageous. Combinations of horsepower and speeds for which the synchronous motor (1) is **inferior** to the induction motor, (2) **matches** the induction motor, and (3) is **superior** to the induction motor are given in Fig. 5.56, (see reference [1]). Note that at **low speeds** and **high horsepower**, the induction motor is no longer cheaper to use. Here large amounts of iron are needed if the flux density is not to exceed airgap flux densities of the order of 0.7 Wb/m^2. However, for the synchronous device this is not a problem because an airgap flux density of 1.4 (i.e., 2×0.7) Wb/m^2 is permissible, because of the separate excitation.

The most important characteristics of the synchronous motor along with some typical applications are exhibited in Table 5.3. However, in addition to the applications listed in Table 5.3, the synchronous motor is found in applications characterized by operation at **low speeds** and **high horsepower**. Such typical applications are (1) large, low-head pumps, (2) flour-mill line shafts, (3) rubber mills and

*Application of Eq. (5.167) to Example 5.8 verifies this statement. In addition, one can verify that the **motor pull-out power** increases as the power factor goes from lagging through unity to leading.

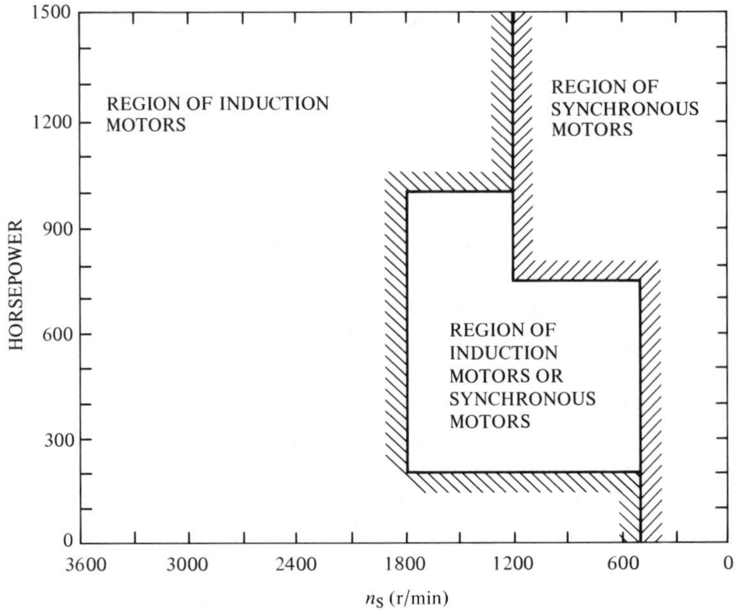

Fig. 5.56 Horsepower vs. Speed Regions, Designating Regions of Application of (1) Synchronous Motors, (2) Induction Motors, and (3) Synchronous or Induction Motors

mixers, (4) crushers, (5) shippers, (6) pulp grinders and jordans, and (7) refiners used in the papermaking industry.

It should be noted that **synchronous motors have no inherent starting torque** and thus cannot be used to start high inertia loads.

The Balanced, Two-Phase, Synchronous Device

The two-phase synchronous device is essentially no longer in use; therefore, no attempt is made here to discuss its applications.

THE SINGLE-PHASE SYNCHRONOUS DEVICE

The single-phase, singly excited (i.e., no dc field excitation is essential for its operation), synchronous device, in its **motor mode of operation**, is well suited to enhance the comfort and pleasure afforded by our environment. A person encounters this device in almost every step he or she makes in the course of the day. For example, the **reluctance motor** is found in such devices as electric clocks, electric shavers, electric clippers, vibrator sandpapering machines, and engraving tools, while the **hysteresis motor** is a popular choice for driving high-quality record players and tape recorders. Such uses are feasible because of the special qualities of these two devices.

For example, the reluctance motor has the characteristic ability to operate at fixed speed when supplied with voltages of fixed frequency from an ac power system. Such systems operate at a nearly constant frequency. Checked constantly for accuracy, their day-to-day average value is determined through comparison of time indications of an electric clock with international standard time signals. Other

Table 5.3 SYNCHRONOUS MOTOR CHARACTERISTICS AND TYPICAL APPLICATIONS*

	Synchronous Speed	
	$n_S > 500$ rpm	$n_S < 500$ rpm
Starting Torque (% of normal)	25 up to several thousand	Usually above 25 up to several thousand
Pull-in Torque (% of Normal)	Up to 120	Low 40
Pull-out Torque	Up to 200	Up to 180
Starting Current	800–700	200–350
Slip	0	0
Power Factor	High, but varies with (1) load and (2) excitation	High, but varies with excitation
Efficiency	92–96%; the highest of all motors	92–96%; the highest of all motors
Typical Applications	(1) power-factor correction, (2) constant-speed generations, (3) frequency changing, (4) fans, (5) blowers, (6) dc generators, (7) line shafts, (8) centrifugal pumps and compressors, (9) reciprocating pumps and compressors	(1) power-factor control, (2) constant-speed generation, (3) flywheel used for pulsating loads, (4) low-speed, direct-connected loads such as reciprocating compressors when started unloaded, (5) dc generators, (6) rolling mills, (7) band mills, (8) ball mills, (9) pumps

*By permission from M. Liwschitz-Garik and C. C. Whipple, **Electric Machinery**, vol. II (New York: Van Nostrand, 1946).

Plate 5.12 Easily Disassembled and Assembled, Self-Ventilated, Synchronous Device. (Courtesy of Westinghouse Electric Co., Pittsburgh, Pa.)

variations of the singly excited reluctance device include small motors that employ cam-operated contacts to close a coil circuit—when the space rate of change of reluctance is favorable to torque production—and to interrupt the circuit at other times. Such motors can be found in electric shavers. An additional variation of the singly excited reluctance device is the linear motion motor, which is capable of synchronous operation. Such a device could have the outlay of the figure exhibited in Example 1.4, whereby the bar would be suspended by a stiff spring and would be tuned to mechanical resonance with a sinusoidal excitation. The force equations that would describe such a system could then be written from Eq. (5.176) if we were to allow $\tau \to f$ and $d\theta_R \to dx$ and to follow the theory given there. In such a system mechanical energy would be stored in the spring during motion in one direction and later would be released to be expended to do useful work. Examples of devices utilizing this type ot motor are electric clippers, vibrators, sandpapering machines, and engraving tools.

Similarly, the hysteresis motor has been used in electric clocks and other timing devices. Its most important application, however, is that it is the preferred device for driving high-quality record players and tape recorders. The device is favored for these applications because of its quietness—a consequence of its smooth rotor. It also has the unique characteristic of synchronizing heavy-inertia loads (which affect only the time required by the rotor to reach synchronous speed) with smooth acceleration.

Because of their available low ratings (hysteresis motor rating: 0 to 1 hp; reluctance motor rating: 0 to 100 hp), the applications of the reluctance motor and the hysteresis motor are limited. Nevertheless, the service they provide for the improvement of our environment is of great value.

5.11 SUMMARY

In this chapter the basic theory of the smooth-rotor, balanced, three-phase, synchronous device is developed. Then the equivalent circuits for the steady state analysis of the device's generator and motor modes of operation are developed. These equivalent circuits are used for the modeling and detailed analysis of synchronous generators or motors connected to loads or forcing function with (1) unity, (2) lagging, or (3) leading power factors.

Furthermore, the real and reactive power versus the power-angle characteristics of the synchronous device are derived, and the distinct regions of operation for the generator and motor modes are defined and discussed in detail. Also included is the introduction to the development of the detailed theory pertaining to the measurement of the circuit parameters for the equivalent circuits of the smooth-rotor synchronous device. The bulk of the work, however, is presented in Section 7.2. This is done to avoid duplication, as the same information is essential for the measurement of the equivalent circuit parameters for the salient-rotor synchronous device, presented in Section 6.6, as well as the measurement of associated time constants.

The chapter continues with the presentation of the principle of operation of the balanced, two-phase, synchronous device; however, no mathematical analysis is undertaken, for the device is essentially no longer in use. The chapter also sets forth in detail the theory describing the operation of the single-phase synchronous device. Of the two types available, the reluctance motor is given an extensive mathematical treatment, while the hysteresis motor is treated descriptively.

Finally, the chapter concludes with a thorough account of the applications of the synchronous devices in use.

PROBLEMS

5.1 A synchronous generator has 8 poles and operates at a speed of 1800 rpm. Calculate: (1) the frequency of the generated voltage and (2) the prime-mover speed required to generate voltages whose frequencies are 60 Hz and 50 Hz.

5.2 A 96-slot, eight-pole, three-phase, synchronous generator is wound with coils having $\frac{12}{14}$ fractional pitch. Calculate the pitch factor k_p.

5.3 From pertinent data given in Problem 5.2, calculate the breadth factor k_b.

5.4 The per-phase parameters of a synchronous generator are as follows: $\mathbf{E}_G = 15{,}000e^{j60°}$ V, $X_\phi = 4.00\ \Omega$, $X_\ell = 0.50\ \Omega$, and $R_S = 0.34\ \Omega$. If the loading impedance is $\mathbf{Z}_L = 2.70 + j1.98\ \Omega$, and the windage, bearing friction, and core losses are negligible, calculate the following per-phase quantities: (1) the synchronous reactance X_S, (2) the synchronous impedance $\mathbf{Z}_S$, (3) the stator coil impedance $\mathbf{Z}_C$, (4) the voltages $\mathbf{V}_\phi$, $\mathbf{V}_\ell$, $\mathbf{V}_R$, $\mathbf{V}_C$, $\mathbf{V}_S$, and $\mathbf{V}_L$ and the airgap potential rise $\mathbf{E}_{\mathscr{R}}$, (5) the generated power $\mathbf{S}_G$, (6) the power required by the stator winding, $\mathbf{S}_S$, (7) the power demanded by the load, $\mathbf{S}_D$, (8) the total powers $\mathbf{S}_{GT}$, $\mathbf{S}_{ST}$, and $\mathbf{S}_{DT}$, (9) the required input power from the prime mover P_{INM}, (10) the efficiency η, and (11) the voltage regulation VR.

5.5 A 1200-kVA (magnitude of total demanded power, $\mathbf{S}_D$), 5000-line-to-line-V, three-phase, Y-connected, synchronous generator has a synchronous resistance $R_S = 2.50\ \Omega$ and a synchronous reactance $X_S = 25\ \Omega$ per phase. Find: (1) the full-load generated voltage $\mathbf{E}_G$, per phase, at PF $= k_{LGG} = 0.35$, 0.75, PF $= 1$, and PF $= k_{LDG} = 0.75$, 0.35, (2) the corresponding power angle δ_G, and (3) the corresponding voltage regulations VR.

5.6 It is known that the per-phase generated voltage of an eight-pole, three-phase, Y-connected, synchronous generator is $\mathbf{E}_G = 3{,}800e^{j45°}$ V and its frequency is 60 Hz, the per-phase synchronous impedance of the winding is $\mathbf{Z}_S = 2.5 + j18$, and the per-phase loading impedance is $\mathbf{Z}_L = 21.5 + j22.5$. What are the minimum prime-mover requirements in speed and torque so that the generated voltage maintains its present magnitude and frequency if all losses are 10 kW (i.e., $P_{WLT}+P_{BFT}+P_{CLT}=10{,}000$ W)?

5.7 A 60-Hz, eight-pole, three-phase, Y-connected, synchronous motor has the following per-phase parameters: $R_S=0.8, X_\phi=3.4, X_\ell=0.6$, all in ohms; excitation voltage $\mathbf{V}_{ST}=3500e^{j40°}$ V; generated voltage $\mathbf{E}_G=2200e^{j20°}$ V; and total losses of 6500 W. Calculate: (1) the power angle δ_M, (2) the response current $\mathbf{I}_S$, (3) the voltages V_R, V_ℓ, V_ϕ, $\mathbf{V}_C$, $\mathbf{V}_S$ and $\mathbf{V}_{\mathfrak{R}}$, (4) the total input power to the motor, P_{INT}, (5) the power lost in the stator winding, P_{ST}, (6) the power developed by the motor, P_{GT}, (7) the power delivered by the motor, P_{OM}, (8) the torque delivered at the shaft, τ_{OM}, (9) the efficiency of the motor, η, and (10) the speed regulation SR of the synchronous motor.

5.8 A 60-Hz, eight-pole, three-phase, Y-connected, 1500-hp, 5000-V-line-to-line, synchronous motor has a per-phase synchronous impedance $\mathbf{Z}_S=0.8+j4.8$ Ω. Its efficiency, at rated load and PF = 1, is $\eta=95\%$. Neglecting the field losses due to dc excitation, calculate: (1) the developed voltage $\mathbf{E}_G$, (2) the power angle δ_M, (3) the total developed power P_{GT}, (4) the total developed torque τ_{GT}, (5) the total output torque τ_{OM}, and (6) the total losses, $P_{LT}=P_{WLT}+P_{BFT}+P_{CLT}$. Repeat calculations (1) through (5) and in addition calculate (a) the total input power to the stator winding, (b) the efficiency η, and (c) the speed regulation of the synchronous motor for power factors $k_{LGG}=0.85$ and $k_{LDG}=0.85$. (Note: The total output power $P_{OM}=1500$ hp and 1 hp = 746 W.)

5.9 A balanced, three-phase, 24-pole, Y-connected, 60-Hz, 14,000-line-to-line V, 20-MW, synchronous generator is connected to an infinitely strong network. Such a network has the property of maintaining its node voltages at constant magnitude, phase angle, and frequency. If the per-phase synchronous reactance $X_S=12$ Ω and the per-phase, stator-winding resistance $R_S=0.5$ Ω, and if the field (rotor-winding) dc excitation is maintained constant to a level that assures power delivery to the network at a lagging power factor of 0.85, calculate: (1) the generated voltage $\mathbf{E}_G$, (2) the power angle δ_G, (3) the power, P_{INM}, required from the prime mover for steady state operation, (4) the maximum power the generator can deliver, that is, **the generator pull-out power**, $P_{GPO}=P_{MAX}$, (5) the reactive power, Q_{OT}, the generator delivers or absorbs to or from the network at rated load, (6) the maximum reactive power, Q_{OTMAX}, the generator can deliver or absorb, (7) the minimum reactive power, Q_{OTMIN}, the generator can deliver or absorb, (8) **the steady state stability limit** ($SSSL_G$).

5.10 A three-phase, 26-pole, Y-connected, balanced synchronous motor is excited from an infinitely strong network (remember that a motor can also be excited from a three-phase voltage source) whose line-to-line voltage is 5000 V with a frequency of 60 Hz. The motor delivers 4000 hp to a mechanical load when its phase parameters are $R_S=0.5$ Ω and $X_S=4.0$ Ω and when the field (rotor-winding) dc excitation I_F is adjusted to yield a power factor equal to unity (i.e., $\cos\theta_S=1$) at rated load. Assume all losses to be zero and calculate: (1) the maximum mechanical power that the motor can deliver, that is, the **motor pull-out power**, P_{MPO}, (2) the maximum torque the motor can deliver, that is, the **motor pull-out torque**, τ_{MPO}, (3) the reactive power that the motor delivers or absorbs to or from the network at rated load, Q_{OT}, (4) the maximum reactive power, Q_{OTMAX}, that the motor can deliver or absorb, (5) the minimum reactive power, Q_{OTMIN}, that the motor can deliver or absorb, (6) the steady state stability limit of the motor ($SSSL_M$).

5.11 When the rotor of the reluctance motor exhibited in Fig. 5.51(a) is in the d-axis position, the reluctance of the magnetic circuit is $\mathcal{R}_d = 2.2 \times 10^6$ At/W. When the rotor is in the q-axis position, the reluctance is $\mathcal{R}_q = 3.0 \times 10^6$ At/W. If the stator winding has 3000 turns and is excited by the source $v_S(t) = \sqrt{2}\,(200)\sin 377t$, (1) calculate the maximum average torque $\tau_{AVG(MAX)}$ that can be developed by the reluctance motor, (2) write the general expression for the average torque τ_{AVG} developed by this motor, and (3) calculate the torque developed by this motor when $\delta = -20°$.

REFERENCES

1. Del Toro, V., *Electromechanical Devices for Energy Conservation and Control Systems*. Prentice-Hall, Englewood Cliffs, N.J., 1968.
2. Elgerd, O. I., *Electric Energy Systems Theory: An Introduction*. McGraw-Hill, New York, 1971.
3. Fitzgerald, A. E., and Kingsley, C., Jr., *Electric Machinery*, 2d ed., McGraw-Hill, New York, 1961.
4. Gourishankar, V., and Kelly, D. H., *Electromechanical Energy Conversion*, 2d ed., Intext, New York, 1973.
5. Kimbark, E. W., *Power System Stability*, vol. I. Wiley, New York, 1948.
6. Kimbark, E. W., *Power Systems Stability*, vol. III. Wiley, New York, 1956.
7. Kosow, I. L., *Electric Machinery and Control*. Prentice-Hall, Englewood Cliffs, N.J., 1964.
8. Kuo, B. C., *Automatic Control Systems*. Prentice-Hall, Englewood Cliffs, N.J., 1962.
9. Lansdorf, A. S., *Theory of Alternating Current Machinery*. McGraw-Hill, New York, 1955.
10. Mablekos, Van E. and Grigsby, L. L., "State Equations of Dynamic Systems with Time Varying Coefficients," *IEEE Transactions on Power Apparatus and Systems*, vol. PAS-90, no. 6 (Nov./Dec., 1971), pp. 2589–98.
11. Mablekos, Van E. and Grigsby, L. L., "An Algorithm for the Unified Solutions of the State Equations of Mechanically Driven Systems with Time Varying Coefficients," *IEEE Transactions*.
12. Mablekos, Van E. and El-Abiad, A. H., "An Algorithm for the Unified Solutions of the State Equations of Electrically Excited Systems with Time Varying Coefficients," *IEEE Transactions*.
13. Majmudar, H., *Electromechanical Energy Converters*. Allyn & Bacon, Boston, 1965.
14. Matsch, L. W., *Electromagnetic and Electromechanical Machines*. Intext, New York, 1972.
15. McFarland, T. C., *Alternating Current Machines*. Van Nostrand, New York, 1948.
16. O'Kelly, D., and Simons, S., *Introduction to Generalized Electrical Machine Theory*. McGraw-Hill, London, 1968.
17. Schmitz, N. L., and Novatny, D. W., *Introductory Electromechanics*. Ronald Press, New York, 1965.
18. Thaler, G. J., and Wilcox, M. L., *Electric Machines: Dynamics and Steady State*. Wiley, New York, 1966.
19. Thompson, H. A., *Alternating Current and Transient Circuit Analysis*. McGraw-Hill, New York, 1955.
20. Weedy, B. M., *Electric Power Systems*. Wiley, London, 1972.

6 SYNCHRONOUS ENERGY CONVERTERS—SYSTEM CONSIDERATIONS

6.1 INTRODUCTION

Power companies want to provide their customers with continuity of service and economic power production. To achieve these objectives the power industry recognized the need to coordinate the operation of many generating stations and their associated networks. Thus prevailing customer demand could be constantly fulfilled by the establishment and maintenance of power interchange with neighboring generating stations and their associated networks. With this capability at hand, power companies can, by taking advantage of load diversity or availability of lower cost capacity, (1) lower overall operating cost and possibly defer capital investment for new generating plants, (2) stagger scheduled outages for maintenance purposes, and (3) share spinning reserve capacity during emergencies and thereby guarantee the continuity of service.

Consequently, the utility industry in the United States joined into five huge networks known as **interconnections** or **pools**. They are as follows: (1) the Interconnected Systems Group (ISG)–Canadian, U.S. Eastern Interconnection (CANUSE)–Pennsylvania–New Jersey–Maryland (PJM) Pool, stretching from the Rocky Mountains to the Gulf Coast to the Canadian border to the eastern seaboard; (2) Northwest Power Pool–Rocky Mountain Pool; (3) Pacific Southwest Interconnected System–New Mexico Power Pool; (4) Southwest Publish Service System; and (5) Texas Pool.*

Each generating station has at least one synchronous generator and usually two connected in parallel. Obviously, in a power grid the synchronous device

*Cohn, N., **Control Generation and Power Flow on Interconnected Systems** (New York: Wiley 1971), p. 3.

operating as a generator (or synchronous condenser) can be viewed **as one of many** synchronous generators operating in parallel, all supplying an assigned load (comprised of combinations of synchronous motors, induction motors, lighting systems, etc.)—these generators and their assigned loads constitute an **area**—and maintained in synchronism with the synchronous generators and the corresponding loads of all other areas. When all these areas are joined together, they constitute an interconnection.

Consequently, whether the synchronous devices function as generators or as motors or as capacitors (rarely as inductors) in an interconnection, they should perform their assigned tasks precisely. As generators, they should be reliable sources of electrical energy; as capacitors, they should be reliable in maintaining assigned voltage levels at their terminals; as motors, they should be reliable in delivering mechanical power at the proper speeds. If all these devices perform their assigned tasks precisely, the network (i.e., the power system) to which they are connected functions within its design limits and it is said that the power system is **stable**. Conversely, **instability** denotes a condition involving failure of some or all of the power system's synchronous devices to perform their assigned tasks.

The cause of instability (usually a **fault** in the system, causing wide variation of system voltages and therefore bad effects on customer loads) must be detected and removed from the system immediately. This is usually accomplished by identifying the part that contains the cause of instability and disconnecting it from the parent system, usually through relay-circuit-breaker sets. When the remaining system returns to its normal mode of operation, the removed synchronous device(s) should be resynchronized and reconnected with the parent system.

Clearly, then, two aspects of each synchronous device are essential: (1) the individual behavior of a given synchronous device in a stable or unstable mode of operation and (2) the effect of a given synchronous device on the parent power system. For the purpose of studying the behavior of interconnected systems, then, we can draw an **equivalent** of the parent system to which a given synchronous device is to be connected and then study the cross effects of the interconnection.

It is possible, therefore, (1) to represent the parent system as an infinitely strong network and identify it as a source of electrical energy with **fixed voltage amplitude, phase angle, and frequency**, and therefore investigate the behavior of the total system as if it were comprised of only one synchronous device connected across this fictitious source, and (2) to represent the parent system as an **equivalent synchronous device**, and therefore investigate the system behaviour as if it were comprised of two synchronous devices connected in parallel.

With such equivalents, we can study the effect of a synchronous device on a given system, or vice versa, through the study of two synchronous devices connected in parallel or through stability studies of a synchronous device connected to an infinite bus or to another synchronous device—thus the classical problem, **stability of the two-machine power system.**

To understand the function of the equivalents just described, consider the interconnection of two large power systems for the purpose of exchanging power to obtain economy in generation or to provide reserve capacity. For example, how would we go about studying the effects on the system of a poorly designed *interconnecting tie line*, capable of transmitting only part of the generating capacity of either system, if a severe fault occurred in the system? (A study through these equivalents would reveal that the synchronizing power that the interconnecting tie line is capable of transmitting is not enough to maintain system stability and that therefore the system would fail. Does the problem above remind you of anything that happened in New York in 1965 and 1977?)

To answer such questions, power engineers must have the capability of **modeling a power system**. That is, power engineers must be able to draw on equivalent circuit of the system and write its describing equations. To do this they must have the circuit parameters of all the system's components and for all contingencies. Therefore, it is mandatory **to have knowledge** of the effects of the various types of excitation on the synchronous device under various constraints of operation, **to know** the circuit parameters of the synchronous device during a balanced three-phase fault operation, immediately after the fault and long after the fault, and **to be able to draw** the corresponding equivalent circuits and so on.

The qualitative analysis of relevant subjects is undertaken here. The objectives of this chapter are fivefold: (1) to study the behavior of the synchronous device under various modes and constraints of operation when it is connected to an infinite bus; (2) to familiarize the reader with excitation and prime-mover systems used in conjunction with three-phase synchronous devices; (3) to introduce the principles of parallel operation of three-phase synchronous devices; (4) to present two methods of stability analysis; and (5) to acquaint the reader with the analysis of balanced three-phase faults on synchronous generators and thus derive their simplified equivalent circuits for their normal, transient, and subtransient analysis.

6.2 SYNCHRONOUS DEVICE CONNECTED TO INFINITE BUS

It was concluded in Section 5.7 that for the interval $0 \leqslant \delta_G < \pi/2$, Fig. 5.35(a) describes the acceptable region of operation for the synchronous generator, Fig. 5.34(a); and for the interval $-\pi/2 < \delta_M \leqslant 0$, Fig. 5.42(a) describes the acceptable region of operation for the synchronous motor, Fig. 5.41(a). These two figures, Figs. 5.35(a) and 5.42(a), and their counterparts, Figs. 5.35(b) and 5.42(b), can be plotted on a common set of axes, P versus δ and Q versus δ if and only if (1) $|P| \equiv |P_{DT}| \equiv |P_{OT}|$, (2) $|Q| \equiv |Q_{DT}| \equiv |Q_{OT}|$, (3) $|\mathbf{E}_T| \equiv |\mathbf{V}_L| \equiv |\mathbf{E}_S|$, (4) $|\mathbf{E}_G| > |\mathbf{E}_T|$, and (5) $-\pi/2 < \delta < \pi/2$, where $0 \leqslant \delta \equiv \delta_G < \pi/2$ and $-\pi/2 < \delta \equiv \delta_M \leqslant 0$. This can be achieved by redrawing Figs. 5.34(a) and 5.41(a) as in Fig. 6.1(a), and redrawing Figs. 5.35(a) and 5.42(a) as in Fig. 6.1(b), and redrawing Figs. 5.35(b) and 5.42(b) as in Fig. 6.1(c).

Here the one port (to the right of the terminals *ab*) described by the terminal variables $\mathbf{E}_T$ and $\mathbf{I}_T$ represents an **infinitely strong network** defined as an **infinite bus.*** Such a network has the following properties: (1) the phase voltage(s) $\mathbf{E}_j$, $j = 1, 2, 3, \ldots$ is (are) of constant magnitude, that is, $|\mathbf{E}_j| = k_j$, (2) the phase angle(s) of the voltage(s) is (are) constant, that is, $\theta_m = k_m$, $m = 1, 2, 3, \ldots$, and (3) the frequency of the voltage(s) is constant, that is, $f_E = k$.

*__Bus__ is defined as the nodal point of a transmission network. However, an **infinite bus** is defined as a source of voltage constant in magnitude, phase angle, and frequency and is not affected by the amount of current drawn from it.

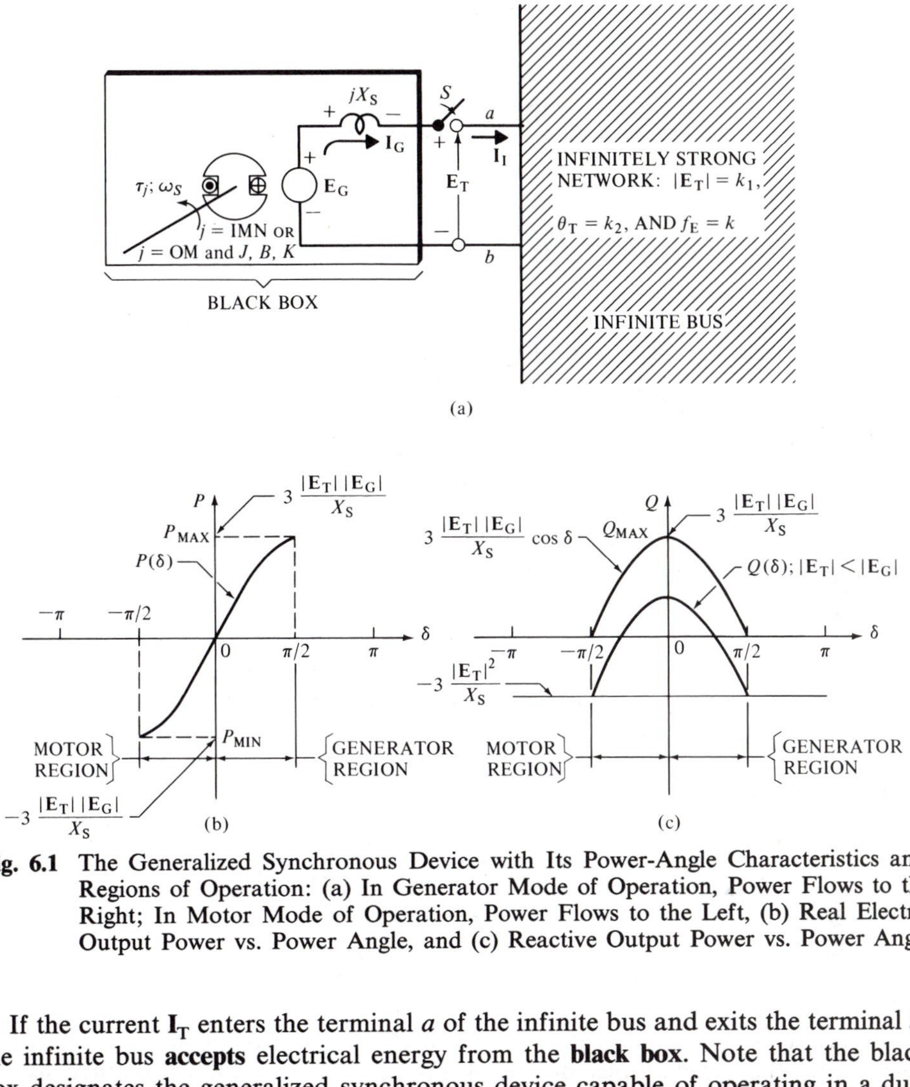

Fig. 6.1 The Generalized Synchronous Device with Its Power-Angle Characteristics and Regions of Operation: (a) In Generator Mode of Operation, Power Flows to the Right; In Motor Mode of Operation, Power Flows to the Left, (b) Real Electric Output Power vs. Power Angle, and (c) Reactive Output Power vs. Power Angle

If the current $\mathbf{I}_T$ enters the terminal *a* of the infinite bus and exits the terminal *b*, the infinite bus **accepts** electrical energy from the **black box**. Note that the black box designates the generalized synchronous device capable of operating in a dual mode (i.e., as a generator or as a motor). Consequently, the black box functions as a synchronous generator, that is, $\mathbf{I}_T \equiv \mathbf{I}_G$ and $\mathbf{E}_T \equiv \mathbf{V}_L$, and can be described by Eqs. (5.144) and (5.145) if and only if $|\mathbf{V}_L| \to |\mathbf{E}_T|$. Therefore, for $0 \leqslant \delta < \pi/2$, $\delta_G \to \delta$, Eqs. (5.144) and (dT5.145) become

$$P_{DT} \to P = 3\left(\frac{|\mathbf{E}_T||\mathbf{E}_G|}{X_S}\right)\sin\delta, \quad 0 \leqslant \delta < \pi/2 \tag{6.1}$$

and

$$Q_{DT} \to Q = 3\left(\frac{|\mathbf{E}_T||\mathbf{E}_G|}{X_S}\right)\cos\delta - 3\left(\frac{|\mathbf{E}_T|^2}{X_S}\right), \quad 0 \leqslant \delta < \pi/2 \tag{6.2}$$

However, if the current $\mathbf{I}_T$ *exits* the terminal *a* of the infinite bus and enters the terminal *b*, the infinite bus **delivers** electrical energy to the black box. Thus the black box functions as a synchronous motor, that is, $\mathbf{I}_T \equiv -\mathbf{I}_S$ and $\mathbf{E}_T \equiv \mathbf{E}_S$; then it can be described by Eqs. (5.163) and (5.164) if and only if $|E_S| \to |E_T|$. Therefore, for $-\pi/2 < \delta \leqslant 0$, $\delta_M \to \delta$, Eqs. (5.163) and (5.164) become

$$P_{OT} \to P = 3\left(\frac{|\mathbf{E}_T||\mathbf{E}_G|}{X_S}\right)\sin\delta, \quad -\pi/2 < \delta \leqslant 0 \tag{6.3}$$

and

$$Q_{OT} \to Q = 3\left(\frac{|\mathbf{E}_T||\mathbf{E}_G|}{X_S}\right)\cos\delta - 3\left(\frac{|E_T|^2}{X_S}\right), \quad -\pi/2 < \delta \leqslant 0 \tag{6.4}$$

It is important that the student fully understands the power versus power angle mechanism just described and depicted in Fig. 6.1 because it is basic to the operation of power systems. **Whether a synchronous device behaves as a generator or motor depends entirely on the direction of power flow**, that is, **into or out of the generalized synchronous device**.

Before closing this section, however, the following observations are in order.

Since the curve of Fig. 6.1(b) is symmetrical about the origin, the *pull-out* power for the generator and motor modes of operation, Eqs. (5.148) and (5.167), are identical (for this connection), that is,

$$|P_{GPO}| \equiv |P_{MPO}| \equiv 3\left(\frac{|\mathbf{E}_T||\mathbf{E}_G|}{X_S}\right)\Bigg|_{\mathbf{E}_T = \mathbf{V}_L, \mathbf{E}_S} \tag{6.5}$$

Thus the *steady state stability limits* that define the generator and motor modes of operation of the synchronous device, Fig. 6.1(a), designated as SSSL_G and SSSL_M, respectively, are defined by the corresponding pull-out powers [Eqs. (5.148) and (5.167), when $|\mathbf{V}_L| \to |\mathbf{E}_T|$ and $|\mathbf{E}_S| \to |\mathbf{E}_T|$] as

$$\text{SSSL}_G \equiv P_{GPO} = +3\left(\frac{|\mathbf{E}_T||\mathbf{E}_G|}{X_S}\right) \tag{6.6}$$

and

$$\text{SSSL}_M \equiv P_{MPO} = -3\left(\frac{|\mathbf{E}_T||\mathbf{E}_G|}{X_S}\right) \tag{6.7}$$

Inspection of Fig. 6.1(b) yields information with respect to the rate of power delivery by the synchronous device of Fig. 6.1(a). This rate is defined as the **stiffness** of the synchronous device, designated by K_S (in watts/degree or watts/radian) and is obtained by differentiating Eqs. (6.1) or (6.3) with respect to the power angle δ. Thus

$$\mathrm{K}_S \equiv \frac{dP}{d\delta} = 3\left(\frac{|\mathbf{E}_T||\mathbf{E}_G|}{X_S}\right)\cos\delta \tag{6.8}$$

Note that the stiffness can be controlled only by controlling $|\mathbf{E}_G|$ and the synchronous reactance, since $|\mathbf{E}_T|$ is defined to be constant. Since X_S is a design parameter, only $|\mathbf{E}_G|$ can be controlled by varying the field (rotor-winding) dc excitation I_F. The dependence of $|\mathbf{E}_G|$ on I_F, derived through the utilization of Eqs. (5.35), (5.26), (3.22), and (3.13), is

$$|\mathbf{E}_G| = \left(\frac{4.44 n_S k_p k_b \mu_0 N_S N_R}{120a}\right) I_F = \left(\frac{1.11 \omega_S k_p k_b \mu_0 N_S N_R}{\pi a}\right) I_F \tag{6.9}$$

Also, it is essential to note that the stiffness at the pull-out points is zero.

Finally, examination of Eqs. (6.2) and (6.4), and thus of Fig. 6.1(c), reveals that because $\cos\delta$ is an even function of δ, if

$$|\mathbf{E}_G|\cos\delta > |\mathbf{E}_T| \qquad (6.10)$$

then

$$Q > 0 \qquad (6.11)$$

Thus the synchronous device, whether it functions in its generator or its motor mode, **produces** reactive power; that is, it acts, from the circuit point of view, as a *shunt capacitor* connected across the infinite bus. Eq. (6.10) evidently depends upon δ, that is, upon P, Fig. 6.1(b). In general, however, the inequality holds for *high* values of $|\mathbf{E}_G|$, that is, for **strong excitation**, referred to as **overexcitation**. (Please see explanation associated with Figs. 6.3 through 6.6.)

If, on the other hand,

$$|\mathbf{E}_G|\cos\delta < \mathbf{E}_T \qquad (6.12)$$

then

$$Q < 0 \qquad (6.13)$$

Thus the synchronous device, again whether it functions in its generator or its motor mode, **absorbs** reactive power; that is, it acts, from the circuit point of view, as a *shunt inductor* connected across the infinite bus. In general, this inequality holds for **low** values of $|\mathbf{E}_G|$, that is, for **weak excitation**, referred to as **underexcitation.**

From the preceding discussion we can formulate this very important rule:

An overexcited synchronous device (operated either as a generator or motor) acts as a shunt capacitor connected across the infinitely strong network (i.e., the infinite bus) and provides reactive power to it, while an underexcited synchronous device (operated as a generator or motor) acts as a shunt inductor connected across the infinitely strong network (i.e., the infinite bus) and absorbs reactive power from it.

The ideas of this rule can be explained as follows. Consider the power, $\mathbf{S}_O$, **provided** from a synchronous device to the infinite bus. The equation for output power is $\mathbf{S}_O = \mathbf{E}_T\mathbf{I}_T^*$, Fig. 6.1(a). If the device functions as a generator, $\mathbf{I}_T = \mathbf{I}_G$ and $P_O > 0$. If the device functions as a motor, $\mathbf{I}_T = -\mathbf{I}_S$ so that $P_O < 0$. On the other hand, the power absorbed by a synchronous device is designated as $\mathbf{S}_{IN} \equiv -\mathbf{E}_T\mathbf{I}_T^*$. If the device functions as a generator, $\mathbf{I}_T = \mathbf{I}_G$, thus $P_{IN} < 0$. If the device functions as a motor, $\mathbf{I}_T = -\mathbf{I}_S$, thus $P_{IN} > 0$. The explanation above can be simplified by stating that P_O **for a generator is positive** and P_{IN} **for a motor is positive.**

Now consider the power absorbed, $\mathbf{S}_{IN}$, by a system (or a synchronous device) that can either be capacitive in nature, that is, $\mathbf{Y}_C = G + jB_C \rightarrow \mathbf{Y}_C^* = G - jB_C$, or inductive in nature, that is, $\mathbf{Y}_L = G - jB_L \rightarrow \mathbf{Y}_L^* = G + jB_L$. The concepts of admittance as set forth in Appendix A, Figs. A.10 and A.11, yield $\mathbf{I} = \mathbf{YE}$, Eq. (A.48), where $\mathbf{I}$ is the current that **enters** the positive terminal of the element. Since $\mathbf{I}_T$ in Fig. 6.1(a) **leaves** the positive terminal, $\mathbf{I}_T = -\mathbf{I}$, and thus $\mathbf{I}_T = -\mathbf{YE}_T$ and $\mathbf{I}_T^* = -\mathbf{E}_T\mathbf{Y}^*$. Therefore, the power absorbed by the system, $\mathbf{S} = \mathbf{EI}^*$, Eq. (A.45), becomes $\mathbf{S}_{IN} = -\mathbf{E}_T\mathbf{I}_T^* \rightarrow \mathbf{S}_{IN} = -\mathbf{E}_T(-\mathbf{E}_T^*\mathbf{Y}^*) \rightarrow \mathbf{S}_{IN} = |\mathbf{E}_T|^2\mathbf{Y}^*$.

Next consider the output power, $\mathbf{S}_O$, of a system (or synchronous device) that can either be capacitive or inductive in nature, that is, $\mathbf{Y}_C = G + jB_C$ or $\mathbf{Y}_L = G - jB_L$. Following the same reasoning as above, $\mathbf{I}_T = -\mathbf{YE}_T$ and $\mathbf{I}_T^* = -\mathbf{E}_T^*\mathbf{Y}^*$. Thus the power output is $\mathbf{S}_O=\mathbf{EI}^*\rightarrow\mathbf{S}_O=\mathbf{E}_T\mathbf{I}_T^*\rightarrow\mathbf{S}_O=\mathbf{E}_T(-\mathbf{E}_T^*\mathbf{Y}^*)\rightarrow\mathbf{S}_O=-|\mathbf{E}_T|^2\mathbf{Y}^*$. This is consistent with the result obtained above, that is, $\mathbf{S}_{IN}=|\mathbf{E}_T|^2\mathbf{Y}^*$, since $\mathbf{S}_O=-\mathbf{S}_{IN}$.

Thus the output power of capacitive and inductive systems can be written as

$$\mathbf{S}_{OC} = -|\mathbf{E}_T|^2 Y_C^* = -|\mathbf{E}_T|^2(G-jB_C) = -|\mathbf{E}_T|^2 G + j|\mathbf{E}_T|^2 B_C = -P_C + jQ_C \equiv \mathcal{S}_{LDG} = \mathcal{P}_{LDG} + j\mathcal{Q}_{LDG} \tag{6.14}$$

and

$$\mathbf{S}_{OL} = -|\mathbf{E}_T|^2 \mathbf{Y}_L^* = -|\mathbf{E}_T|^2(G+jB_L) = -|\mathbf{E}_T|^2 G - j|\mathbf{E}_T|^2 B_L = -P_L - jQ_L \equiv \mathcal{S}_{LGG} \equiv \mathcal{P}_{LGG} - j\mathcal{Q}_{LGG} \tag{6.15}$$

Eq. (6.14) describes a **capacitive** system whose power factor is **leading. The real power is absorbed (negative)** but the **reactive power is generated (positive)**, that is, $\mathcal{P}_{LDG}\equiv -P_C$ and $\mathcal{Q}_{LDG}\equiv +Q_C$. In contrast, Eq. (6.15) describes an **inductive** system whose power factor is **lagging.** In this case **both the real and the reactive powers are absorbed (negative),** that is $\mathcal{P}_{LDG}\equiv -P_L$ and $\mathcal{Q}_{LDG}\equiv -Q_L$.

Also note that in Eqs. (6.14) and (6.15) the script variables are introduced, with the correspondence as follows: $\mathbf{S}\rightarrow\mathcal{S}$, $-P\rightarrow+\mathcal{P}$, and $Q\rightarrow\mathcal{Q}$. When $\mathcal{P}>0$, the system **absorbs** real power; thus $\mathcal{P}$ is the input real power. But when $\mathcal{Q}>0$, the system **generates** reactive power; thus $\mathcal{Q}$ is the output reactive power. The reason for this choice of correspondence between regular and script variables is apparent when the output complex power $\mathcal{S}$ is plotted. Eqs. (6.14) and (6.15) are plotted on the complex output power plane as in Fig. 6.2. Now compare Fig. 6.2 with Fig. A.11, the admittance plane. When the quantities **I** of Fig. A.11 and $\mathbf{S}_{LDG}$ of Fig. 6.2 are in the first quadrant, they both designate systems that are *capacitive* in nature and thus have *leading* power factors. In contrast, when **I** and $\mathcal{S}_{LGG}$ of Figs. A.11 and 6.2, respectively are in the fourth quadrant, they both designate systems that are *inductive* in nature and thus have *lagging* power factors. This is in contrast to the impedance plane, Fig. A.9, where location of the quantities **E** and **S** in the first

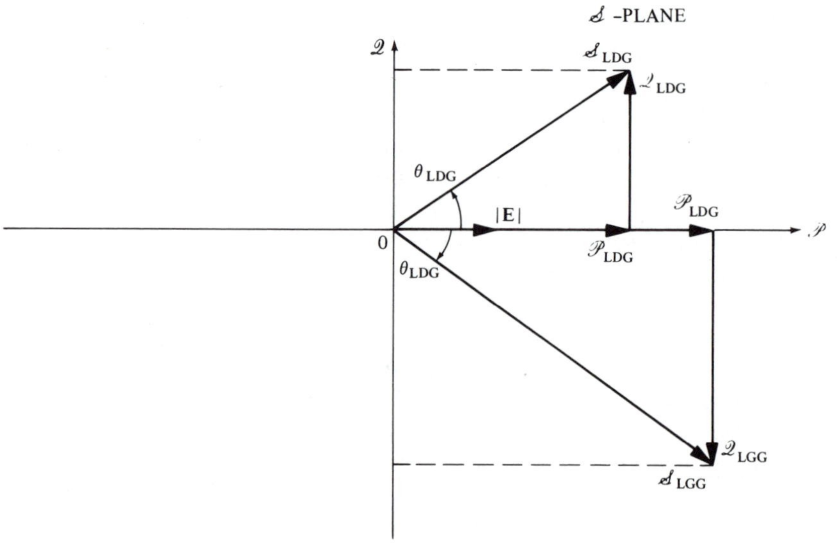

Fig. 6.2 Output Complex Power Plots Exhibiting Pertinent Information Associated with the Leading and Lagging Power Factors of Power Systems—See Fig. A. 11.

quadrant designates systems with lagging power factors and location of the quantities **E** and **S** in the fourth quadrant designates systems with leading power factors. The usefulness of the $\mathbb{S}$- and **Y**-planes is apparent when the synchronous device is used as a shunt capacitor or inductor for power-factor correction (see Examples 6.1 and 6.2).

When the synchronous device operates in one or the other of these two modes, it is known as a **synchronous condenser** or **synchronous inductor** (or **phase modifier)** and is used either **to correct an undesirable power factor** or **to control both the power factor and voltage** on long transmission line systems. It should be clear from the material covered in Appendix A that the synchronous condenser mode has a greater number of applications. In this mode of operation, the synchronous device, usually of salient-pole type, has neither a prime mover nor a mechanical load. It is, in effect, a synchronous motor running **idle** with an overexcited field, and it is permitted to **float** on the line simply for the purposes of power-factor improvement. Can the synchronous generator be used for the same function? When the total losses and the stator resistance of the synchronous condenser are neglected, its real power is zero, Fig. 6.1(b), and consequently the power angle $\delta = 0$.

Under these conditions Eq. (6.2) becomes

$$Q = 3\left(\frac{|\mathbf{E}_{\mathrm{T}}||\mathbf{E}_{\mathrm{G}}| - |\mathbf{E}_{\mathrm{T}}|^2}{X_{\mathrm{S}}}\right) \tag{6.16}$$

Note that in the vicinity of $\delta \simeq 0$, Fig. 6.1(c), and, in fact, for the region $-30° \leqslant \delta \leqslant 30°$, $\cos\delta$ is rather *insensitive* to changes in δ; therefore, the variation in Q is insignificant.

It should be pointed out, before closing, that the attractive merits of the synchronous condenser when compared to commercial fixed capacitors (i.e., capacitor banks) of extremely high ratings and voltages are (1) the capability of *simple and continuous control* of Q via $|\mathbf{E}_{\mathrm{G}}|$, which can be varied by controlling the field (rotor-winding) dc excitation I_{F} of the synchronous condenser, and (2) the capability to manufacture synchronous condensers at a *lower cost*. The primary disadvantage is the higher maintenance cost.

In summary, the behavior of the synchronous device, designated as a black-box in Fig. 6.3, is studied from the systems point of view through the dependence of the **output variables** P, Q, δ, and K_{S} on the **input variables** I_{F} and τ_j ($j = $ INM, OM). The dependence of the outputs on the inputs is designated by the arrowheads connecting the inputs to the outputs. Solid lines designate *strong coupling* (i.e., strong dependence); dotted lines designate *weak coupling* (i.e., weak dependence).

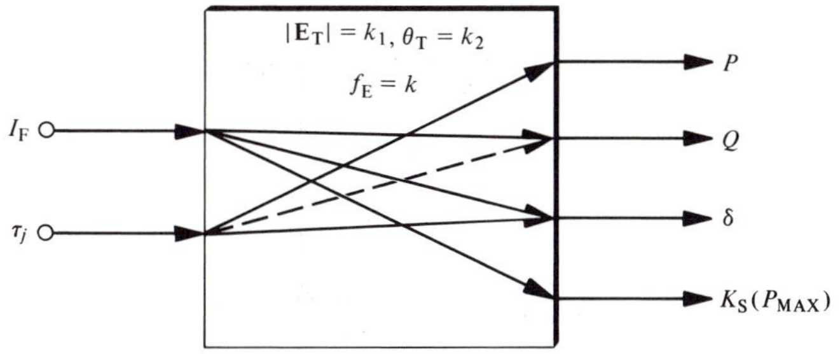

Fig. 6.3 Black Box Representation of the Synchronous Device, $j = $ INM, OM

For example, a reduction in the field (rotor-winding) dc excitation $|I_F|$ reduces $|\mathbf{E}_G|$ and thus P_{MAX} and the stiffness S of the device, Fig. 6.1(b), and consequently the magnitude of the power angle δ, even possibly to the point of pull-out, for a given level of power.

Example 6.1

A plant draws 2 MW at 10,000 V from an infinitely strong transmission network (i.e., an infinite bus) at a power factor of 0.6 lagging. It is desired that a synchronous condenser should be used to correct the overall power factor to unity. Assume that the losses of the synchronous condenser are 275 kW. Calculate: (1) the original reactive power of the plant, (2) the reactive power required to bring the power factor to unity, and (3) the kilovoltampere rating of the synchronous condenser and its power factor. Repeat all the calculations for a power-factor correction to 0.85 lagging and comment on the feasibility of the results. This unique capability of the synchronous motor to operate at various power factors can be achieved by (1) *underexcitation*, (2) *normal excitation*, and (3) *overexcitation*.

Solution

Because power-factor correction is always accomplished by connecting the required corrective component—that is, capacitors (condensers) or inductors, **conventional** or **synchronous**—in parallel with (i.e., across) the given plant, the $\mathbb{S}$-plane is utilized. The basic concepts of power-factor correction are discussed in Appendix A and in Eqs. (6.14) and (6.15) and the related theory. The key equations are repeated here for convenience.

$$\mathbb{S}_{LDG} = \mathcal{P}_{LDG} + j\mathcal{Q}_{LDG} \tag{1}$$

and

$$\mathbb{S}_{LGG} = \mathcal{P}_{LGG} - j\mathcal{Q}_{LGG} \tag{2}$$

Recalling that $\mathcal{P}$ is the input real power and $\mathcal{Q}$ is the output reactive power, we can draw the power triangle as shown in Fig. 6.1.1. Now the following calculations are

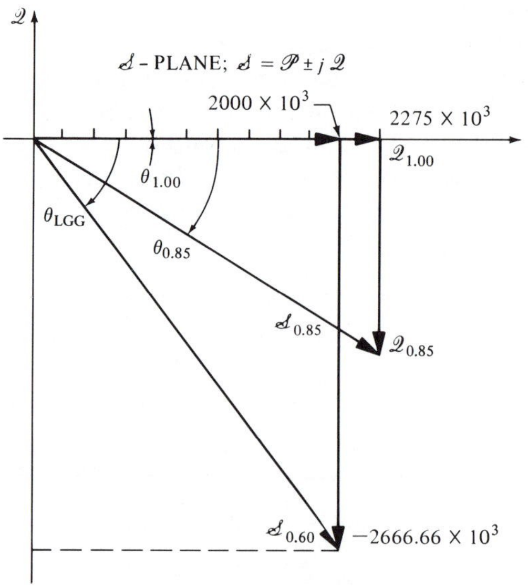

Fig. 6.1.1 Output Complex Power Diagram Exhibiting Complex Powers Directly Related to Corresponding Power Factors: (1) 0.60 LGG, (2) 0.85 LGG and (3) 1.00

pertinent:

$$\theta_{LGG} = \cos^{-1}(0.6) = \underline{-53.13°} \tag{3}$$

and

$$\mathcal{Q}_{LGG} = \mathcal{P}_{LGG} \tan\theta_{LGG} = (2\times10^6)\times(-1.33) = \underline{-2666.66\times10^3 \text{ var}} \tag{4}$$

Thus the complex power that defines the lagging load, designated as $\mathcal{S}_{LGG} \equiv \mathcal{P}_{LGG} + j\mathcal{Q}_{LGG}$, is written as

$$\mathcal{S}_{LGG} = \underline{\underline{2\times10^6 - j2.66\times10^6}} = \underline{\underline{3.33\times10^6 e^{-j53.13°}}} \text{ VA} \tag{5}$$

The describing equation of the total system comprised of the plant $\mathcal{S}_{LGG} = \mathcal{P}_{LGG} + j\mathcal{Q}_{LGG}$ and the synchronous device $\mathcal{S}_{SD} = \mathcal{P}_{SD} + j\mathcal{Q}_{SD}$, designated as $\mathcal{S}_H$ (H≡HAVE), is written as

$$\mathcal{S}_H = \mathcal{S}_{LGG} + \mathcal{S}_{SD} = (\mathcal{P}_{LGG} + \mathcal{P}_{SD}) + j(\mathcal{Q}_{LGG} + \mathcal{Q}_{SD}) \tag{6}$$

Proper substitution from Eq. (5) and the given data into Eq. (6) yields

$$\mathcal{S}_H = 2275\times10^3 + j(-2666.66\times10^3 + \mathcal{Q}_{SD}) \tag{7}$$

To correct the plant for unity power factor, the describing equation of the total system, designated as $\mathcal{S}_{W(1.00)}$ (W≡WANT), should become

$$\mathcal{S}_{W(1.00)} \equiv \mathcal{P}_W + j\mathcal{Q}_{1.00} \tag{8}$$

where, from Fig. 6.1.1, the following calculations are pertinent:

$$\theta_{1.00} = \cos^{-1}(1.00) = \underline{0.00°} \tag{9}$$

and

$$\mathcal{Q}_{1.00} = \mathcal{P}_W \tan\theta_{1.00} = \underline{0.00} \text{ var} \tag{10}$$

Substitution from Eq. (10) into Eq. (8) yields

$$\mathcal{S}_{W(1.00)} = \mathcal{P}_W + j0.00 \tag{11}$$

Thus equating corresponding parts of the "have" and "want" complex powers, that is, Eqs. (7) and (11) with $\mathcal{Q}_{SD}$ adjusted to designate unity power-factor correction, $\mathcal{Q}_{SD(1.00)}$, yields

$$\mathcal{P}_W \equiv \underline{2275\times10^3} \text{ W} \tag{12}$$

and

$$0.00 \equiv -2666.66\times10^3 + \mathcal{Q}_{SD(1.00)} \tag{13}$$

Solution of Eq. (13) for $\mathcal{Q}_{SD(1.00)}$ yields

$$\mathcal{Q}_{SD(1.00)} = \underline{\underline{2666.66\times10^3}} \text{ var} \tag{14}$$

Since $\mathcal{Q}_{SD(1.00)} > 0$, Eq. (14) suggests that the synchronous device is *capacitive* in nature.

Once $\mathcal{Q}_{SD(1.00)}$ is calculated, the describing equation of the synchronous condenser that yields unity power-factor correction is written as

$$\mathcal{S}_{SD(1.00)} = \mathcal{P}_{SD(1.00)} + j\mathcal{Q}_{SD(1.00)} = \underline{\underline{275\times10^3 + j2666.66\times10^3}}$$
$$= \underline{\underline{2680.80\times10^3 e^{j84.11°}}} \text{ VA} \tag{15}$$

Now utilization of Eq. (15) enables us to calculate the power factor of the synchronous condenser as follows:

$$\cos\theta_{SD(1.00)} = \cos(84.11°) = \underline{\underline{0.103 \text{ LDG}}} \tag{16}$$

Therefore, the kilovoltampere rating and the power factor of the synchronous condenser required to correct the given plant to unity power factor, as written from Eqs. (15) and (16), correspondingly, are 2680.80 kVA and 0.103 LDG.

To correct the plant for a power factor of 0.85 LGG, the describing equation for the total system, written in accordance with Eq. (8) and Fig. 6.1.1, becomes

$$\mathcal{S}_W = \mathcal{P}_W + j\mathcal{Q}_{0.85} \tag{17}$$

where, from Fig. 6.1.1 and Eqs. (9) and (10), the following calculations are pertinent:

$$\theta_{0.85} = \cos^{-1}(0.85) = -31.79° \tag{18}$$

and

$$\mathcal{Q}_{0.85} = \mathcal{P}_W \tan\theta_{0.85} = (2{,}275\times 10^3)\times(-0.62) = \underline{\underline{-1409.92\times 10^3 \text{ var}}} \tag{19}$$

Substitution from Eq. (19) into Eq. (17) yields

$$\mathcal{S}_W = \mathcal{P}_W + j(-1409.92\times 10^3) \tag{20}$$

Now equating **real** and **imaginary** parts of Eqs. (20) and (7) with $\mathcal{Q}_{SD}$ adjusted to designate 0.85 power-factor correction, $\mathcal{Q}_{SD(0.85)}$, yields

$$\mathcal{P}_W = \underline{2.275\times 10^3} \text{ W} \tag{21}$$

and

$$-1409.92\times 10^3 = -2666.66\times 10^3 + \mathcal{Q}_{SD(0.85)} \tag{22}$$

Solution of Eq. (22) for $\mathcal{Q}_{SD(0.85)}$ yields

$$\mathcal{Q}_{SD(0.85)} = -1409.92\times 10^3 + 2666.66\times 10^3 = 1256.74\times 10^3 \text{ var} \tag{23}$$

Since $\mathcal{Q}_{SD(0.85)} > 0$, Eq. (23) suggests that the synchronous device is again *capacitive* in nature.

Now the describing equation of the synchronous condenser that yields a power-factor correction of 0.85 LGG is written as

$$\begin{aligned}\mathcal{S}_{SD(0.85)} &= \mathcal{P}_{SD(0.85)} + j\mathcal{Q}_{SD(0.85)} = \underline{\underline{275\times 10^3 + j1256.74\times 10^3}} \\ &= \underline{\underline{1286.48\times 10^3 e^{j77.66°}}} \text{ VA}\end{aligned} \tag{24}$$

Thus Eq. (24) leads to

$$\cos\theta_{SD(0.85)} = \cos(77.66°) = \underline{\underline{0.21 \text{ LDG}}} \tag{25}$$

Now, the kilovoltampere rating and the power factor of the synchronous condenser required to correct the given plant to a power factor of 0.85 LGG, as written from Eqs. (24) and (25), correspondingly, are 1286.47 kVA and 0.21 LDG.

Note that substitution for $\mathcal{P}_W$ from Eq. (21) into Eq. (20) yields

$$\begin{aligned}\mathcal{S}_{W(0.85)} &= \underline{\underline{2275\times 10^3 - j1409.92\times 10^3}} \\ &= \underline{\underline{2676.47\times 10^3 e^{-j31.78°}}} \text{ VA}\end{aligned} \tag{26}$$

Thus correcting the system for 1.00 and 0.85 LGG power factors leads to a kilovoltampere net improvement, designated by $|\mathcal{S}_{NI(j)}|$, $j = k_{LDG}$, k and k_{LGG}, calculated from Eqs. (5), (11) [after proper substitution for Eq. (12)] and Eq. (26) as follows:

$$\begin{aligned}|\mathcal{S}_{NI(1.00)}| &= |\mathcal{S}_{LGG}| - |\mathcal{S}_{W(1.00)}| = 3333.33\times 10^3 - 2275\times 10^3 \\ &= \underline{\underline{1058.33}} \text{ kVA}\end{aligned} \tag{27}$$

and

$$|\mathbb{S}_{NI(0.85)}| = |\mathbb{S}_{LGG}| - |\mathbb{S}_{W(0.85)}| = 3333.33\times 10^3 - 2676.47\times 10^3 = \underline{\underline{656.86\ \text{kVA}}} \tag{28}$$

From Eqs. (15) and (24), respectively, it is seen that these kilovoltampere improvements are accomplished at the expense of synchronous condensers that are capable of delivering 2666.66 kvar and 1286.74 kvar correspondingly. For the **equipment cost** of a given dollar per kilovar rate, it is clear that the extra dollars spent for the unity power-factor correction are more than the dollars spent for 0.85 lagging power-factor correction, by a factor of 2.12. This *extra expense* has to be compared to the *additional savings* that the user would realize from the power company by reducing the penalty of the **demand charge policy** that power companies impose on their customers for the peak kilovoltamperes drawn. This is a reasonable charge by the power companies, because for a poor power-factor load, they must have the capability of supplying the reactive voltamperes (i.e., *useless power*) in addition to the actual real voltamperes (i.e., *useful power*) which they must supply to the same load operating at unity power factor. Moreover, the delivery of these additional reactive voltamperes causes higher current in the transmission lines and consequently greater i^2R (real) transmission losses.

The behavior of synchronous motors and generators as capacitors (synchronous condensers) or inductors (synchronous inductors) is explained in the remainder of this section.

Synchronous Motor

The describing equation of the simplified equivalent circuit of the synchronous device, Fig. 6.1(a), describing the motor mode of operation (i.e., $\mathbf{E}_T \equiv \mathbf{E}_S$ and $-\mathbf{I}_T \equiv \mathbf{I}_S$) can be written as

$$\mathbf{E}_S - jX_S\mathbf{I}_S - \mathbf{E}_G \equiv \mathbf{E}_S - \mathbf{V}_{SM} - \mathbf{E}_G = 0 \tag{6.17}$$

Solution of Eq. (6.17) for $\mathbf{V}_{SM}$ yields

$$\mathbf{V}_{SM} \equiv jX_S\mathbf{I}_S = \mathbf{E}_S - \mathbf{E}_G \equiv |\mathbf{V}_{SM}|e^{j\sigma_M} \tag{6.18}$$

Finally, solution of Eq. (6.18) for $\mathbf{I}_S$ yields

$$\mathbf{I}_S = \frac{|\mathbf{V}_{SM}|e^{j\sigma_M}}{jX_S} \equiv \frac{\mathbf{E}_S - \mathbf{E}_G}{jX_S} \equiv \frac{\mathbf{E}_S + (-\mathbf{E}_G)}{jX_S} \tag{6.19}$$

Close examination of Eq. (6.19) provides the following information: (1) the stator winding current $\mathbf{I}_S$ depends on the magnitude of the voltage $\mathbf{V}_{SM}$ and its phase angle σ_M, and thus on the voltages $\mathbf{E}_S$ and $\mathbf{E}_G$, and (2) the stator winding current $\mathbf{I}_S$ is **always** 90° behind the voltage $\mathbf{V}_{SM}$.

For $\theta \to \theta_S$ this current is utilized to calculate the *input* power to the motor, P_{INT}, given as

$$P_{INT} \equiv 3\,\text{Re}(\mathbf{E}_S\mathbf{I}_S^*) = 3|\mathbf{E}_S|\,|\mathbf{I}_S|\cos\theta_S \tag{6.20}$$

Since the winding resistance R_S is assumed to be zero, for $\delta \to \delta_M$ the *developed* power by the motor, that is,

$$P_{GT} \equiv 3\,\text{Re}(\mathbf{E}_G\mathbf{I}_S^*) = 3|\mathbf{E}_G|\,|\mathbf{I}_S|\cos\phi \equiv \left(\frac{3|\mathbf{E}_S|\,|\mathbf{E}_G|}{X_S}\right)\sin\delta_M \tag{6.21}$$

is identical to the *output* (loading) power, P_{OM}, applied to the motor shaft if and only if all the losses are assumed to be zero [otherwise the losses should be accounted for, as is stated earlier, Eq. (5.109)], which in turn is identical to the *input* power to the motor given by Eq. (6.20). Thus

$$P_{INT} \equiv P_{GT} \equiv P_{OM} = 3|\mathbf{E}_S||\mathbf{I}_S|\cos\theta_S \equiv 3\left(\frac{|\mathbf{E}_S||\mathbf{E}_G|}{X_S}\right)\sin\delta_M \tag{6.22}$$

Since the developed voltage $\mathbf{E}_G$ is a function of the field (rotor-winding) dc excitation I_F, Eq. (6.9), the last term of Eq. (6.22) is a function of I_F. It is, therefore, the stated purpose of this section to show what the effects of the **motor field (rotor-winding) dc excitation** I_F are and which of the variables $\mathbf{I}_S$, θ, δ, P_{INT}, and Q_{INT} are affected by its changes.

NORMAL EXCITATION: CONSTANT MECHANICAL LOADING P_{OM} AND STATOR WINDING FORCING FUNCTION $\mathbf{E}_S$

Assume that a synchronous motor is operating at **normal excitation**, that is, unity power factor, with respect to the stator winding forcing function $\mathbf{E}_S$ for a given mechanical load P_{OM}, Fig. 6.4(b). In Fig. 6.4(b), which is drawn in accordance with the information provided by Eq. (6.19), the developed voltage $\mathbf{E}_G$ has been *adjusted* [by a controlled variation of the field (rotor-winding) dc excitation I_F, Eq. (6.9)] *for unity power factor* for the given load P_{OM} on the motor shaft. The adjustment of the developed voltage $\mathbf{E}_G$ is accomplished as follows: for a given stator winding forcing function $\mathbf{E}_S$, I_F (a measurable quantity) is adjusted until the developed voltage $\mathbf{E}_G$ (a calculated quantity) becomes almost equal in magnitude to $\mathbf{E}_S$, that is, $|\text{Re}(-\mathbf{E}_G)| = |\mathbf{E}_S|$. These two voltages, $\mathbf{E}_G$ lagging $\mathbf{E}_S$ by an angle δ_{M1}, drawn in accordance with Eq. (6.19), yield the voltage $\mathbf{V}_{SM}$ and 90° behind it the current $\mathbf{I}_S$ in phase with the stator winding forcing function $\mathbf{E}_S$, Fig. 6.4(b). The power necessary to move the load is calculated, from the information of Fig. 6.4(b), to be

$$P_{OM} = P_{INT} \equiv 3|\mathbf{E}_S||\mathbf{I}_S|\cos 0 = 3|\mathbf{E}_S||\mathbf{I}_S| \tag{6.23}$$

Also, from the information provided by Fig. 6.4(b), the power developed by the motor is calculated to be

$$P_{GT} = 3\left(\frac{|\mathbf{E}_S||\mathbf{E}_G|}{X_S}\right)\sin\delta_{M1} \tag{6.24}$$

Since (for all losses neglected) the *electric power* taken from the bus (infinitely strong network) must always be equal to the *mechanical power* delivered to the rotor of the synchronous motor (the synchronous motor would *not* continue to run continuous state synchronous speed ω_S if these two powers were not equal), *the* **electric power** *of the synchronous motor is completely determined by its* **mechanical** load torque requirement at synchronous speed. Also, since the magnitude of the developed voltage $\mathbf{E}_G$ is a function of the field (rotor-winding) dc excitation I_F, a synchronous motor can be made to operate at a *constant* value of mechanical power P_{OM} but with *different* values of developed voltages $\mathbf{E}_G$. Variation (or control) of $\mathbf{E}_G$ would have an effect on the various quantities of interest, such as stator winding current $\mathbf{I}_S$, power factor $\cos\theta$, power angle δ, and reactive power Q_{INT}. These variations are studied now.

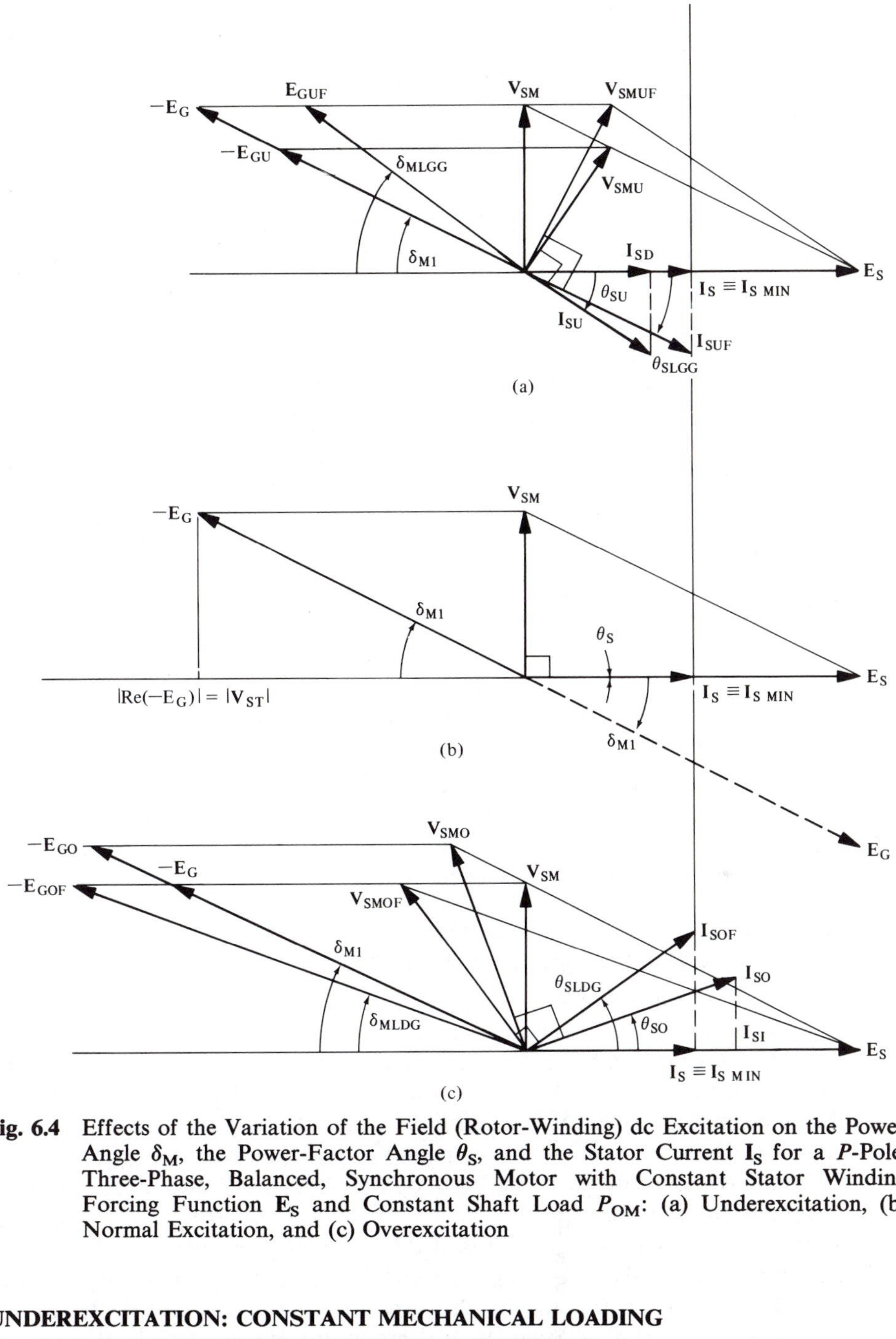

Fig. 6.4 Effects of the Variation of the Field (Rotor-Winding) dc Excitation on the Power Angle δ_M, the Power-Factor Angle θ_S, and the Stator Current $\mathbf{I}_S$ for a *P*-Pole, Three-Phase, Balanced, Synchronous Motor with Constant Stator Winding Forcing Function $\mathbf{E}_S$ and Constant Shaft Load P_{OM}: (a) Underexcitation, (b) Normal Excitation, and (c) Overexcitation

UNDEREXCITATION: CONSTANT MECHANICAL LOADING P_{OM} AND STATOR WINDING FORCING FUNCTION $\mathbf{E}_S$

Study carefully Fig. 6.4(a). It exhibits Fig. 6.4(b) but superimposed on it are (1) the initial effects of a suddenly *decreased* field (rotor-winding) dc excitation I_F, designated by the letter "U," for example, $\mathbf{E}_{GU}$, $\mathbf{V}_{SMU}$, $\mathbf{I}_{SU}$, and θ_{SU}, and (2) the *final value* of these variables, that is, the value which they will evolve to after the ensuing transient settles out, designated by the letter "F." For example, $\mathbf{E}_{GUF}$, $\mathbf{V}_{SMUF}$, $\mathbf{I}_{SUF}$, and θ_{SUF}. Note that the simple *decrease* of excitation E_F(i.e., I_F), designated by $\mathbf{E}_{GU}$, causes the *clockwise rotation* of the voltage $\mathbf{V}_{SM}$, designated by $\mathbf{V}_{SMU}$, which in turn causes the *new* current $\mathbf{I}_{SU}$ to **lag** $\mathbf{V}_{SMU}$ by an angle of 90° and thus to **lag** the stator winding forcing function $\mathbf{E}_S$ by an angle θ_{SU} (degrees). Note that the component $|\mathbf{I}_{SD}|$ of the current $\mathbf{I}_{SU}$, that is, $|\mathbf{I}_{SD}| = |\mathbf{I}_{SU}| \cos\theta_{SU}$ is less than $|\mathbf{I}_S|$.

These ideas lead to the following two conclusions: (1) the initial value of the input power immediately after I_F is adjusted $|\mathbf{E}_S|\,|\mathbf{I}_{SU}|\cos\theta_{SU} < |\mathbf{E}_S|\,|\mathbf{I}_S|\cos 0$ and (2) this *new* developed power $(|\mathbf{E}_S|\,|\mathbf{E}_{GU}|/X_S)\sin\delta_{M1} < (|\mathbf{E}_S|\,|\mathbf{E}_G|/X_S)\sin\delta_{M1}$. Since, the mechanical load P_{OM} is maintained constant, the rotor must be **retarded** from δ_{M1} to a new angle δ_{MLGG} whose magnitude is less than δ_{M1} in order to develop enough power so that the rotor continues to rotate at synchronous speed ω_S. This rotor retardation causes the developed voltage to occur at a new angle. The rotated developed voltage is designated as $\mathbf{E}_{GUF}$. This rotation in turn causes the final current $\mathbf{I}_{SUF}$ to **lag** $\mathbf{V}_{SMUF}$ by an angle of 90° and to **lag** the voltage $\mathbf{E}_S$ by an angle θ_{SLGG} (degrees). The magnitude of this current is such that $|\mathbf{I}_{SUF}|\cos\theta_{SLGG} \equiv |\mathbf{I}_S|$. Now the rotor rotates at synchronous speed ω_S and the describing equations become

$$P_{GT} = 3\left(\frac{|\mathbf{E}_S|\,|\mathbf{E}_G|}{X_S}\right)\sin\delta_{M1} \equiv 3\left(\frac{|\mathbf{E}_S|\,|\mathbf{E}_{GUF}|}{X_S}\right)\sin\delta_{MLGG} \qquad (6.25)$$

and

$$P_{INT} = 3|\mathbf{E}_S|\,|\mathbf{I}_S|\cos 0 = 3|\mathbf{E}_S|\,|\mathbf{I}_{SUF}|\cos\theta_{SLGG} \qquad (6.26)$$

These ideas in conjunction with information from Fig. 6.4(a) lead to the following conclusions: (1) $\delta_{MLGG} < \delta_{M1}$, (2) $\theta_{SLGG} < 0$, and (3) $|\mathbf{I}_{SUF}| > |\mathbf{I}_S|$. In other words, the *power angle* δ and the *power factor angle* θ_S by which the stator current $\mathbf{I}_S$ can be made to **lag** the stator winding forcing function $\mathbf{E}_S$ *can be controlled* by adjusting the field (rotor-winding) dc excitation I_F to a desired value.

OVEREXCITATION: CONSTANT MECHANICAL LOADING P_{OM} AND STATOR WINDING FORCING FUNCTION $\mathbf{E}_S$

Contrary to Fig. 6.4(a), Fig. 6.4(c) exhibits Fig. 6.4(b) but superimposed on it are (1) the effects of an *increased* field (rotor-winding) dc excitation I_F, designated by the letter "O," for example, $\mathbf{E}_{GO}$, $\mathbf{V}_{SMO}$, $\mathbf{I}_{SO}$, and θ_{SO} and (2) the *final* value of these variables, designated by the letter "F," for example, $\mathbf{E}_{GOF}$, $\mathbf{V}_{SMOF}$, $\mathbf{I}_{SOF}$, and θ_{SOF}. Note that the simple *increase* of excitation E_F (i.e., I_F), designated by $\mathbf{E}_{GO}$, causes the *counter-clockwise rotation* of the voltage $\mathbf{V}_{SM}$, designated by $\mathbf{V}_{SMO}$, which in turn causes the *new* current $\mathbf{I}_{SO}$ to **lag** $\mathbf{V}_{SMO}$ by an angle of 90° and thus to **lead** the stator winding forcing function $\mathbf{E}_S$ by an angle θ_{SO} (degrees). Note that the component $|\mathbf{I}_{SI}|$ of the current $\mathbf{I}_{SO}$, that is, $|\mathbf{I}_{SI}| = |\mathbf{I}_{SO}|\cos\theta_{SO}$, is greater than $|\mathbf{I}_S|$.

These ideas lead to the following two conclusions: (1) the *new* input power $|\mathbf{E}_S||\mathbf{I}_{SO}|\cos\theta_{SO} > |\mathbf{E}_S||\mathbf{I}_S|\cos 0$ and (2) the **new** developed power $(|\mathbf{E}_S||\mathbf{E}_{GO}|/X_S)\sin\delta_{M1} > (|\mathbf{E}_S||\mathbf{E}_G|/X_S)\sin\delta_{M1}$. Since by definition the mechanical load is maintained constant, the rotor must be **advanced** from δ_{M1} to a new angle δ_{MLDG} whose magnitude is related to δ_{M1} by the expression $\delta_{MLDG} > \delta_{M1}|$ in order to develop the necessary power requirements so that the rotor continues to rotate at synchronous speed ω_S. This rotor advance causes a *new* developed voltage, designated as $\mathbf{E}_{GOF}$, which in turn causes the **newest** current $\mathbf{I}_{SOF}$ to **lag** $\mathbf{V}_{SMOF}$ by an angle of 90° and to

lead the voltage $\mathbf{E}_S$ by an angle θ_{SLDG} (degrees). The magnitude of this current is such that $|\mathbf{I}_{SOF}|\cos\theta_{SLDG} \equiv |\mathbf{I}_S|$. Now the rotor rotates at synchronous speed ω_S and the describing equations become

$$P_{GT} = 3\left(\frac{|\mathbf{E}_S||\mathbf{E}_G|}{X_S}\right)\sin\delta_{M1} = 3\left(\frac{|\mathbf{E}_S||\mathbf{E}_{GOF}|}{X_S}\right)\sin\delta_{MLDG} \tag{6.27}$$

and

$$P_{INT} = 3|\mathbf{E}_S||\mathbf{I}_S|\cos 0 = 3|\mathbf{E}_S||\mathbf{I}_{SOF}|\cos\theta_{SLDG} \tag{6.28}$$

These ideas in conjunction with information from Fig. 6.4(c) lead to the following conclusions: (1) $\delta_{MLDG} > \delta_{M1}$, (2) $\theta_{SLDG} > 0$, and (3) $|\mathbf{I}_{SOF}| > |\mathbf{I}_S|$. In other words, the *power angle* δ and the *power-factor angle* θ_S by which the stator current $\mathbf{I}_S$ can be made to **lead** the stator winding forcing function $\mathbf{E}_S$ *can be controlled* by adjusting the field (rotor-winding) dc excitation I_F to a desired value.

Note that the process of *reducing excitation* can continue until the limiting case is reached, that is, $|\text{Re}(-\mathbf{E}_{GUF})| \equiv 0$ for the particular condition of loading. Similarly, the process of *increasing excitation* can continue until the limiting case is reached, that is, $|\text{Im}(-\mathbf{E}_{GOF})| \equiv 0$ for the particular condition of loading.

In summary, inspection of Figs. 6.4(a), 6.4(b), and 6.4(c) and utilization of the observations above lead to the following conclusions: (1) when the synchronous motor is **underexcited**, the stator current $\mathbf{I}_{SUF}$ is *large*, that is, $|\mathbf{I}_{SUF}| > |\mathbf{I}_S|$, the power factor $\cos\theta$, $\cos\theta_{SLGG}$, is *lagging* in nature, and the power angle δ, δ_{MLGG}, is *small*, that is, $\delta_{MLGG} < \delta_{M1}$; (2) when the synchronous motor is **normally excited** [$|\text{Re}(-\mathbf{E}_G)| = |\mathbf{E}_S|$ and $\measuredangle \mathbf{V}_{SM} = 90°$], the stator current $\mathbf{I}_S$ is at its *minimum* value $|\mathbf{I}_S| \equiv |\mathbf{I}_S^{MIN}|$ (but not zero), the power factor $\cos 0°$ is *unity*, and the power angle δ has a modest value, designated as δ_{M1}, that is, $\delta_{M1} > \delta_{MLGG}$; and (3) when the synchronous motor is **overexcited**, the stator current $\mathbf{I}_{SOF}$ is *large*, the power factor $\cos\theta$, $\cos\theta_{SLDG}$, is *leading* in nature, and the power angle δ, δ_{MLDG}, is *large*, that is, $\delta_{MLDG} > \delta_{M1}$. In general, however, $\delta_{MLDG} > \delta_{M1} > \delta_{MLGG}$. These observations suggest that for a *constant stator winding forcing function* $|\mathbf{E}_S|$ and *constant shaft load* P_{OM}, we may **plot** the stator winding current $|\mathbf{I}_S|$ as a function of the field (rotor-winding) dc excitation I_F. As the field (rotor-winding) dc excitation I_F increases: (1) the stator current starts at high values, $|\mathbf{I}_{SUF}|$, and decreases monotonically, reaches a minimum value $|\mathbf{I}_S| \equiv |\mathbf{I}_S^{MIN}|$, and increase again monotonically, $|\mathbf{I}_{SOF}|$, (2) the power factor $\cos\theta_S$ goes from **lagging** $\cos\theta_{SLGG}$, to **unity** $\cos 0°$, to **leading** $\cos\theta_{SLDG}$, and (3) the power angle δ **increases** from $-90°$ to $0°$. The lower limit of δ is defined by **instability** and its upper limit by **field coil destruction** due to very high I_F. The practical limitation of $|\mathbf{E}_G|$ occurs before the overheating of the field coil due to the magnetic saturation of the motor iron. Such plots, which can also be obtained experimentally in the laboratory, for various shaft loads, that is, P_{OM0}, P_{OM1}, $P_{OM2}, \ldots$ (where $P_{OM2} > P_{OM1} > P_{OM0}$) and *constant* stator winding forcing function $|\mathbf{E}_S|$ are exhibited in Fig. 6.5. Their general shape resembles the shape of the English script letter $\mathcal{V}$; therefore, they are known as *V-curves of the synchronous motor*.

Inspection of Fig. 6.4(c) and Fig. 6.4 suggest that the synchronous motor described by the no-load V-curve, Fig. 6.5 at **leading power factors** and **large power angles**, that is, $-30° < \delta < 0°$, represents the **synchronous condenser** *mode of operation* of the synchronous motor, while the same V-curve at **lagging power factor** and **small power angles**, that is, $-90° < \delta < -30°$, represents the **synchronous inductor** *mode of operation* of the synchronous motor.

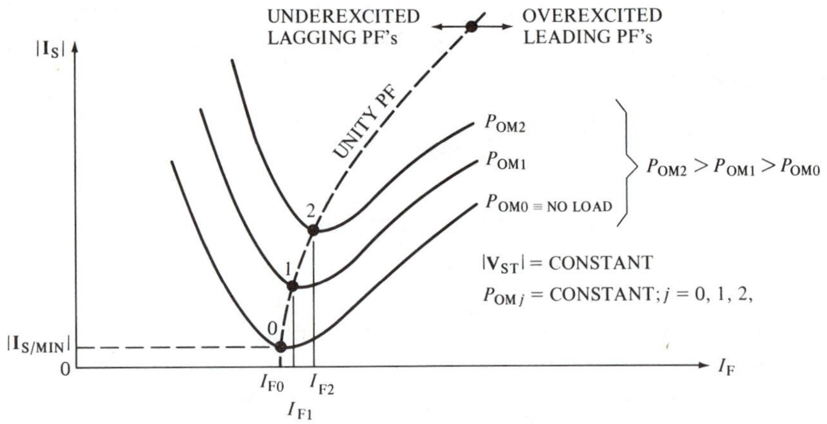

Fig. 6.5 *V*-Curves: Relationship Between the Stator Current $|\mathbf{I}_S|$ and the Field (Rotor-Winding) dc Excitation I_F for a *P*-Pole, Three-Phase, Balanced, Synchronous Motor

Note that at no-load the stator current at unity power-factor point 0, Fig. 6.5, is *not* zero but has some *small* value per phase. *This current contributes to the development of enough torque by the synchronous motor to overcome the rotational losses*. Also, note that the *V*-curves show that a slightly increased field current, $I_{F0} < I_{F1} < I_{F2}, \ldots$, Fig. 6.5, is required to produce normal excitation as the mechanical load increases, points 0, 1, 2, and so on. This is due to saturation.

Synchronous Generator

The describing equation of the simplified equivalent circuit of the synchronous device, Fig. 6.1(a), describing the generator mode of operation (i.e., $\mathbf{E}_T \equiv \mathbf{V}_L$ and $\mathbf{I}_T \equiv \mathbf{I}_G \equiv I_L$) can be written as

$$\mathbf{E}_G - jX_S\mathbf{I}_G - \mathbf{V}_L \equiv \mathbf{E}_G - \mathbf{V}_{SG} - \mathbf{V}_L = 0 \tag{6.29}$$

Solution of Eq. (6.29) for $\mathbf{V}_{SG}$ yields

$$\mathbf{V}_{SG} \equiv jX_S\mathbf{I}_G = \mathbf{E}_G - \mathbf{V}_L \equiv |\mathbf{V}_{SG}|e^{j\sigma_G} \tag{6.30}$$

Finally, solution of Eq. (6.30) for $\mathbf{I}_G$ yields

$$\mathbf{I}_G = \frac{|\mathbf{V}_{SG}|e^{j\sigma_G}}{jX_S} = \frac{\mathbf{E}_G - \mathbf{V}_L}{jX_S} = \frac{\mathbf{E}_G + (-\mathbf{V}_L)}{jX_S} \tag{6.31}$$

Examination of Eq. (6.31) provides the following information: (1) the stator winding current $\mathbf{I}_G$ depends on the magnitude of the voltage $\mathbf{V}_{SG}$ and its phase angle σ_G, and thus on the generated voltage $\mathbf{E}_G$ and the stator winding loading voltage $\mathbf{V}_L$, and (2) the stator current $\mathbf{I}_G$ is *always* 90° behind the voltage $\mathbf{V}_{SG}$.

For $\theta \to \theta_L$ this current is utilized to calculate the power *demanded* by the load, P_{DT}, as

$$P_{DT} = 3\,\mathrm{Re}(\mathbf{V}_L\mathbf{I}_G^*) = 3|\mathbf{V}_L||\mathbf{I}_L|\cos\theta_L \tag{6.32}$$

Since the stator winding resistance R_S is assumed to be zero, for $\delta \to \delta_G$ the power *generated* by the generator, that is,

$$P_{GT} = 3\,\mathrm{Re}(\mathbf{E}_G\mathbf{I}_G^*) = 3|\mathbf{E}_G||\mathbf{I}_G|\cos\phi = 3\left(\frac{|\mathbf{E}_G||\mathbf{V}_L|}{X_S}\right)\sin\delta_G \tag{6.33}$$

is identical to the *input* power P_{INM}, provided by the prime mover, if and only if all the losses are assumed to be zero [otherwise the losses should be accounted for, as stated earlier in Eq. (5.65)], which in turn is identical to the power *demanaed* by tne load, given by Eq. (6.32). Thus

$$P_{\text{INM}} \equiv P_{\text{GT}} \equiv P_{\text{DT}} = 3|\mathbf{V}_L||\mathbf{I}_L|\cos\theta_L = 3\left(\frac{|\mathbf{E}_G||\mathbf{V}_L|}{X_S}\right)\sin\delta_G \tag{6.34}$$

Since the generated voltage $\mathbf{E}_G$ is a function of the field (rotor-winding) dc excitation I_F, Eq. 6.6, the last term of Eq. 6.34 is a function of I_F. It is therefore the stated purpose of this section to show what the effects of the generator field (rotor-winding) dc excitation I_F are and which of the variables $\mathbf{I}_G$, θ, δ, P_{DT}, and Q_{DT} are affected by its changes.

NORMAL EXCITATION: CONSTANT MECHANICAL INPUT P_{INM} AND STATOR WINDING LOADING VOLTAGE $\mathbf{V}_L$

Assume that a synchronous generator is operating at normal excitation, that is, unity power factor with respect to the stator winding loading voltage $\mathbf{V}_L$, for a given mechanical input P_{INM}, Fig. 6.6(b). In Fig. 6.6(b), which is drawn in accordance with the information provided by Eq. (6.31), the developed voltage $\mathbf{E}_G$ has been *adjusted* [by a controlled variation of the field (rotor-winding) dc excitation I_F, Eq. (6.9)] *for unity power factor*, for the given load to the generator, designated by $\mathbf{V}_L$. The adjustment is accomplished as follows: for a given stator winding loading voltage $\mathbf{V}_L$, I_F (a measurable quantity) is adjusted until the generated voltage $\mathbf{E}_G$ (a calculated quantity) becomes almost equal in magnitude to the voltage $\mathbf{V}_L$, that is, $|\text{Re}(\mathbf{E}_G)| = |-\mathbf{V}_L|$. These two voltages, drawn in accordance with Eq. (6.31), yield the voltage $\mathbf{V}_{SG}$ and 90° behind it the current $\mathbf{I}_G$ in phase with the loading voltage $\mathbf{V}_L$, Fig. 6.6(b). The power demanded by the load is calculated from the information of Fig. 6.6(b) to be

$$P_{\text{DT}} = 3|\mathbf{V}_L||\mathbf{I}_G|\cos 0 = 3|\mathbf{V}_L||\mathbf{I}_G| \tag{6.35}$$

From the information provided by Fig. 6.6(b), the power generated by the generator is calculated to be

$$P_{\text{GT}} = 3\left(\frac{|\mathbf{E}_G||\mathbf{V}_L|}{X_S}\right)\sin\delta_{G1} \tag{6.36}$$

Since (for all losses neglected) the *electric power* delivered to the bus must always be equal to the *mechanical power* supplied by the rotor (through the prime mover) of the synchronous generator (the synchronous generator would not run at steady state synchronous speed ω_S if these two powers were not equal), *the* **electric power** *of a synchronous generator is completely determined by its mechanical power*. Also, since the magnitude of the generated voltage $\mathbf{E}_G$ is a function of the field (rotor-winding) dc excitation I_F, a synchronous generator can be made to operate at a *constant* value of mechanical power P_{INM} but with *different* values of generated voltages $\mathbf{E}_G$. Variation (or control) of the generated voltage $\mathbf{E}_G$ would have an effect on the various quantities of interest, such as stator winding current $\mathbf{I}_G$, power factor angle θ, power angle δ, and reactive power Q_{DT}. These variations can be studied through Figs. 6.6(a) and 6.6(c). The process of extracting information from these two figures is identical to the process followed in Figs. 6.4(a) and 6.4(c). Therefore, only the describing equations that define steady state operation and relevant information will be written for underexcitation and overexcitation of the synchronous generator.

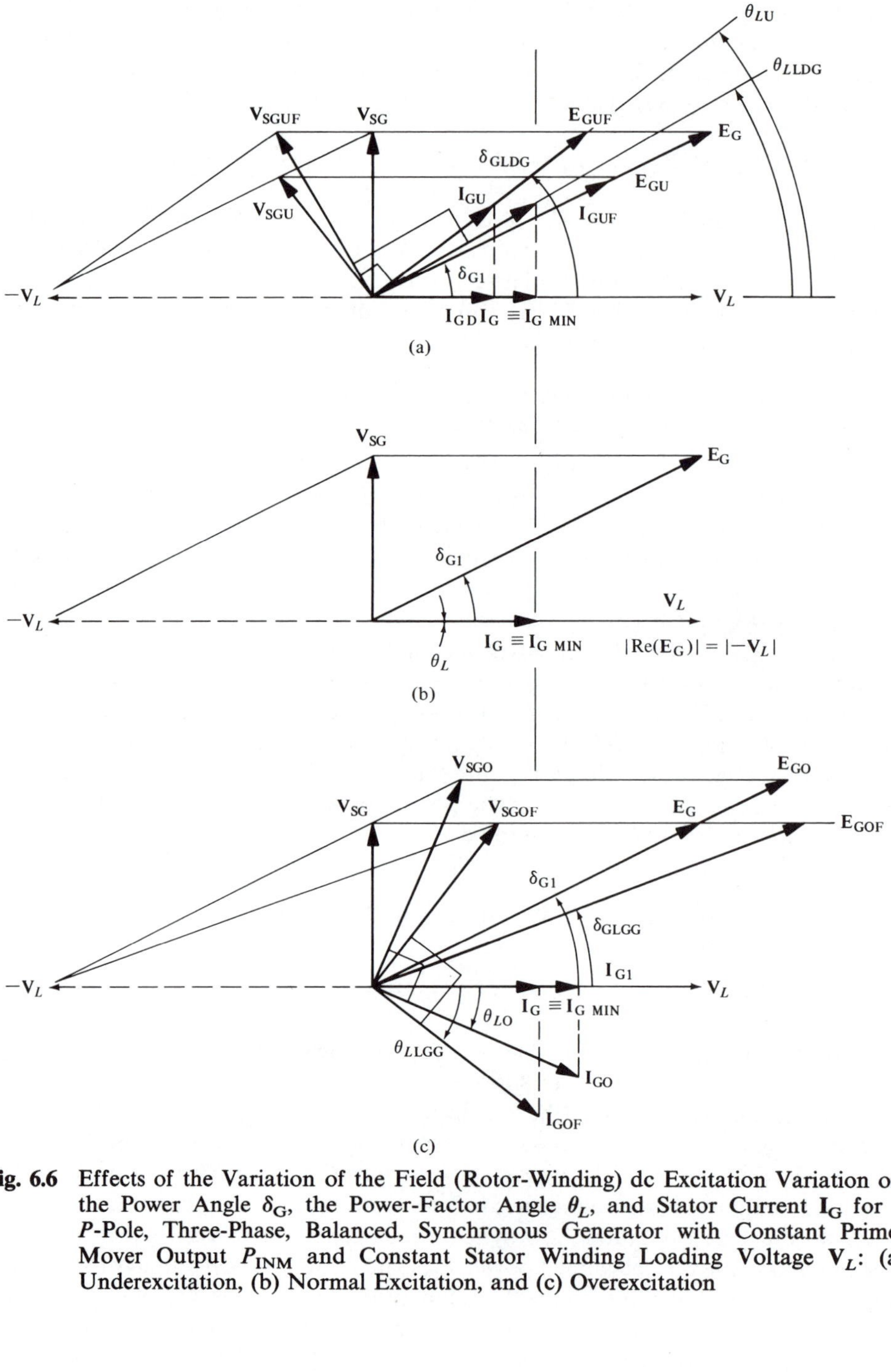

Fig. 6.6 Effects of the Variation of the Field (Rotor-Winding) dc Excitation Variation on the Power Angle δ_G, the Power-Factor Angle θ_L, and Stator Current $\mathbf{I}_G$ for a P-Pole, Three-Phase, Balanced, Synchronous Generator with Constant Prime-Mover Output P_{INM} and Constant Stator Winding Loading Voltage $\mathbf{V}_L$: (a) Underexcitation, (b) Normal Excitation, and (c) Overexcitation

UNDEREXCITATION: CONSTANT MECHANICAL INPUT P_{INM} AND STATOR WINDING LOADING VOLTAGE $\mathbf{V}_L$

When the synchronous generator settles to steady state operation (please go over the section on underexcitation for the synchronous motor now), the describing equations written, with the aid of Fig. 6.6(a), become

$$P_{\mathrm{GT}} = 3\left(\frac{|\mathbf{E}_{\mathrm{G}}||\mathbf{V}_L|}{X_{\mathrm{S}}}\right)\sin\delta_{\mathrm{G1}} = 3\left(\frac{|\mathbf{E}_{\mathrm{GUF}}||\mathbf{V}_L|}{X_{\mathrm{S}}}\right)\sin\delta_{\mathrm{GLDG}} \tag{6.37}$$

$$P_{\mathrm{DT}} = 3|\mathbf{V}_L||\mathbf{I}_{\mathrm{G}}|\cos 0 = 3|\mathbf{V}_L||\mathbf{I}_{\mathrm{GUF}}|\cos\theta_{L\mathrm{LDG}} \tag{6.38}$$

$$\delta_{\mathrm{GLDG}} > \delta_{\mathrm{G1}} \tag{6.39}$$

$$\theta_{L\mathrm{LDG}} > 0 \tag{6.40}$$

and

$$|I_{\mathrm{GUF}}| > |I_{\mathrm{G}}| \tag{6.41}$$

That is, the *power angle* δ and the *power factor angle* θ_L by which the stator current $\mathbf{I}_{\mathrm{G}}$ can be made to **lead** the stator winding loading voltage $\mathbf{V}_L$ *can be controlled* by adjusting the field (rotor-winding) dc excitation I_{F} to a desired value.

OVEREXCITATION: CONSTANT MECHANICAL INPUT P_{INM} AND STATOR WINDING LOADING VOLTAGE $\mathbf{V}_L$

When the synchronous generator settles to steady state operation (please go over the section on overexcitation for synchronous motors now), the describing equations, written with the aid of Fig. 6.6(c), become

$$P_{\mathrm{GT}} = 3\left(\frac{|\mathbf{E}_{\mathrm{G}}||\mathbf{V}_L|}{X_{\mathrm{S}}}\right)\sin\delta_{\mathrm{G1}} = 3\left(\frac{|\mathbf{E}_{\mathrm{GOF}}||\mathbf{V}_L|}{X_{\mathrm{S}}}\right)\sin\delta_{\mathrm{GLGG}} \tag{6.42}$$

$$P_{\mathrm{DT}} = 3|\mathbf{V}_L||\mathbf{I}_{\mathrm{G}}|\cos 0 = 3|\mathbf{V}_L||\mathbf{I}_{\mathrm{GOF}}|\cos\theta_{L\mathrm{LGG}} \tag{6.43}$$

$$\delta_{\mathrm{GLGG}} < \delta_{\mathrm{G1}} \tag{6.44}$$

$$\theta_{L\mathrm{LGG}} < 0 \tag{6.45}$$

and

$$|I_{\mathrm{GOF}}| > |I_{\mathrm{G}}| \tag{6.46}$$

That is, the power angle δ and the power factor angle θ_L by which the stator current $\mathbf{I}_{\mathrm{G}}$ can be made to **lag** the stator winding loading voltage $\mathbf{V}_L$ can be *controlled* by adjusting the field (rotor-winding) dc excitation I_{F} to a desired value.

Again, note that the process of *reducing excitation* can continue until the limiting case is reached, that is, $|\mathrm{Re}(\mathbf{E}_{\mathrm{GUF}})| \equiv 0$, for the particular conditions of loading. Similarly, the process of *increasing excitation* can continue until the limiting case is reached, that is, $|\mathrm{Im}(\mathbf{E}_{\mathrm{GOF}})| \equiv 0$, for the particular condition of loading.

In summary, inspection of Figs. 6.6(a), 6.6(b), and 6.6(c) and utilization of the observations above lead to the following conclusions: (1) when the synchronous generator is **underexcited**, the stator current $\mathbf{I}_{\mathrm{GUF}}$ is *large*, that is, $|\mathbf{I}_{\mathrm{GUF}}| > |\mathbf{I}_{\mathrm{G}}|$, the power factor $\cos\theta$, $\cos\theta_{L\mathrm{LDG}}$, is *leading* in nature, and the power angle δ, δ_{GLDG}, is *large*, that is, $\delta_{\mathrm{GLDG}} > \delta_{\mathrm{G1}}$; (2) when the synchronous generator is **normally excited** [$|\mathrm{Re}(\mathbf{E}_{\mathrm{G}})| = |-\mathbf{V}_L|$ and $\measuredangle \mathbf{V}_{\mathrm{SG}} = 90°$], the stator current $\mathbf{I}_{\mathrm{G}}$ is at its minimum value $|\mathbf{I}_{\mathrm{G}}| \equiv |\mathbf{I}_{\mathrm{G}}^{\mathrm{MIN}}|$ (but not zero), the power factor $\cos 0°$ is *unity*, and the power angle δ has a modest value, designated as δ_{G1} (that is, $\delta_{\mathrm{G1}} < \delta_{\mathrm{GLDG}}$); and (3) when the synchronous generator is **overexcited**, the stator current $\mathbf{I}_{\mathrm{GOF}}$ is *large*, the power factor $\cos\theta_L$, $\cos\theta_{\mathrm{GLGG}}$, is *lagging* in nature, and the power angle δ, δ_{GLGG}, is *small*, that is, $\delta_{GLGG} < \delta_{G1}$. In general, however, $\delta_{GLGG} < \delta_{G1} < \delta_{GLDG}$. These observations suggest that for a *constant stator winding loading voltage* $|\mathbf{V}_L|$ and a *constant mechanical input* P_{INM}, we may **plot** the stator winding current $|\mathbf{I}_G|$ as a function of the field (rotor-winding) dc excitation I_F. As the field (rotor-winding) dc excitation I_F increases; (1) the stator current starts at high values, $|\mathbf{I}_{GUF}|$, and decreases monotonically, reaches a minimum value $|\mathbf{I}_G| \equiv |\mathbf{I}_G^{MIN}|$, and increases again monotonically, $|\mathbf{I}_{GOF}|$, (2) the power factor $\cos\theta_L$ goes from **leading** $\cos\theta_{LLDG}$, to **unity** $\cos 0°$, to **lagging** $\cos\theta_{LLGG}$, and (3) the power angle δ **decreases** from 90° to 0°. The upper limit of δ is now defined by **instability** and its lower limit by the **field coil destruction** due to very high I_F or saturation of the iron. Such plots, also obtainable experimentally, for various inputs from the prime mover, that is, P_{INM0}, P_{INM1}, $P_{INM2}, \ldots$ (where $P_{INM2} > P_{INM1} > P_{INM0}$), and a *constant* loading voltage $|\mathbf{V}_L|$ are exhibited in Fig. 6.7. They resemble the plots of Fig. 6.5; thus they are known as the *V-curves of the synchronous generator*.

Inspection of Fig. 6.1(c) and Fig. 6.6 suggest that the synchronous generator described by the low-input *V*-curve, Fig. 6.7, at **lagging power factors** and **small power angles**, that is, $0° < \delta < 30°$, represents the **synchronous condenser** *mode of operation* of the synchronous generator, while the same *V*-curve at **leading power factors** and **large power angles**, that is, $30° < \delta < 90°$, represents the **synchronous inductor** *mode of operation* of the synchronous generator.

Note that at low input the stator current at unity power factor, point 0, Fig. 6.7, is *not* zero but has some small value per phase. *This current contributes to the development of enough countertorque to equalize the torque required by the prime mover to overcome the windage, bearing friction, and core losses.* Also, note that the *V*-curve show that a slightly increased field, $I_{F0} < I_{F1} < I_{F2}, \ldots$, Fig. 6.7, is required to produce normal excitation as the mechanical input increases, points 0, 1, 2, and so on due to magnetic saturation.

Finally, the dotted lines in Figs. 6.5 and 6.7 are only one (in each figure) of many such lines that can be drawn through points of the same power-factor point in each *V*-curve. The resulting dotted lines (if drawn) would be the loci of constant power factors. These lines are known as **compounding curves**.

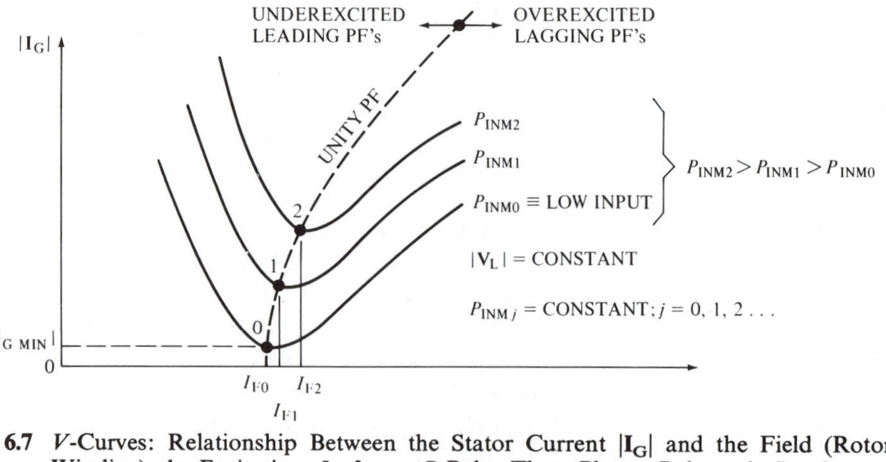

Fig. 6.7 *V*-Curves: Relationship Between the Stator Current $|\mathbf{I}_G|$ and the Field (Rotor-Winding) dc Excitation I_F for a *P*-Pole, Three-Phase, Balanced, Synchronous Generator

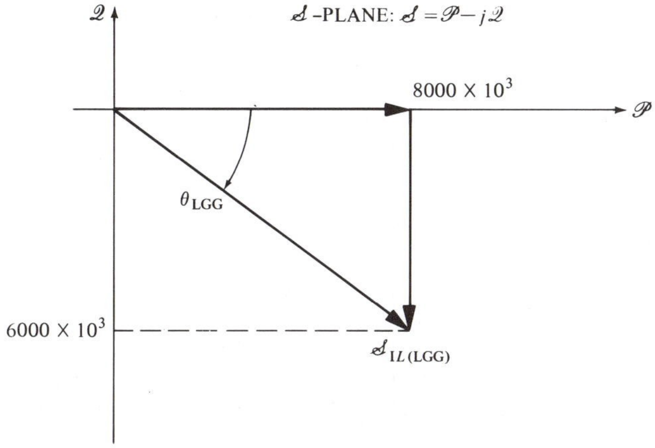

Fig. 6.2.1 Output Complex Power Plane for an Industrial Process Operating at a Power Factor 0.8 LGG

Example 6.2

An industrial process has an installed load rated at 10,000 kVA* and a power factor of 0.8 LGG. An expansion of the process is planned whereby the added load, a large press, is rated at 1,500 kW. A choice is to be made whether to drive the process with (1) an **induction motor** of power factor 0.80 LGG or (2) with a **synchronous motor** capable of having its dc field (rotor-winding) excitation adjusted to give a motor power factor 0.50 LDG. Which motor should one choose and why?

Solution

The two given power factors yield the following corresponding angles:

$$\theta_{\text{LGG}} = \cos^{-1}(0.8) = \underline{-36.87°} \tag{1}$$

and

$$\theta_{\text{LDG}} = \cos^{-1}(0.50) = \underline{60.00°} \tag{2}$$

From the given data, the *installed load* of the industrial process, designated as $\mathbb{S}_{IL(\text{LGG})}$, can be written, utilizing the $\mathbb{S}$-plane, Fig. 6.2.1, that is, $_{\text{LGG}} = \mathcal{P}_{\text{LGG}} - j\mathcal{R}_{\text{LGG}}$, as

$$\mathbf{I}_{L(\text{LGG})} = 10{,}000e^{-j36.87°} = \underline{8{,}000 \times 10^3 - j6{,}000 \times 10^3}\text{VA} \tag{3}$$

Also, from the given data, the *added load* to the processes, designated as $\mathbb{S}_{AL(\text{LGG})}$, when using the **induction motor**, is written with the aid of Fig. 6.2.2 as

$$\mathbb{S}_{AL(\text{LGG})} = 1{,}500 \times 10^3 + jQ_{AL}\,\text{VA} \tag{4}$$

Now the reactive power of the added load $Q_{AL(\text{LGG})}$ can be calculated with the aid of Fig. 6.2.2 as

$$Q_{AL(\text{LGG})} = P_{AL}\tan\theta_{\text{LGG}} = (1{,}500 \times 10^3) \times \tan(-36.87°) = -1{,}125.00 \times 10^3\,\text{var} \tag{5}$$

Substituting for $Q_{AL(\text{LGG})}$ from Eq. (5) into Eq. (4) yields

$$\mathbb{S}_{AL(\text{LGG})} = \underline{1{,}500 \times 10^3 - j1{,}125 \times 10^3}\,\text{VA} \tag{6}$$

*Note that rating of a load can be given in voltamperes, that is, |S|, or in watts, that is, *P*.

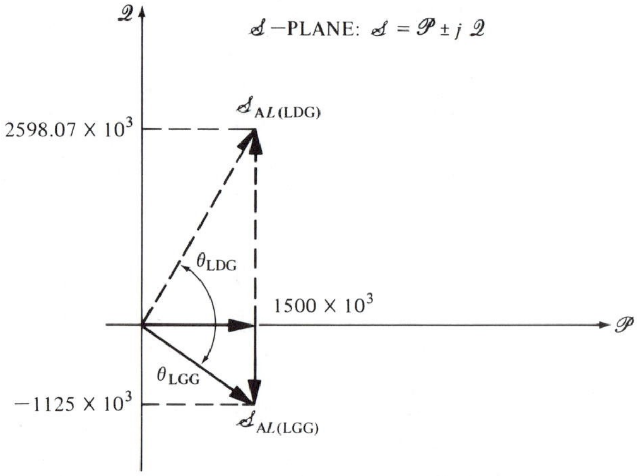

Fig. 6.2.2 Output Complex Power Plane of a Process Operating at Power Factors of (1) 0.80 LGG and (2) 0.50 LDG

Now the total industrial process comprised of the loads of Eqs. (3) and (6) when the induction motor is used—that is, the motor is operating at a power factor of 0.80 LGG—is written as

$$\mathbb{S}_{\mathrm{TIM}} = \mathbb{S}_{IL(\mathrm{LGG})} + \mathbb{S}_{AL(\mathrm{LGG})} = 9{,}500 \times 10^3 - j7{,}125 \times 10^3$$
$$= \underline{11{,}875 \times 10^3 e^{-j36.87^\circ}} \mathrm{VA} \quad (7)$$

Thus the kilovoltampere rating and the power factor of the total process, when the induction motor is used, are

$$|\mathbb{S}_{\mathrm{TIM}}| = \underline{\underline{11{,}875.00}} \mathrm{kVA} \quad (8)$$

and

$$\cos\theta_{\mathrm{TIM}} = \cos(-36.87^\circ) = \underline{\underline{0.80 \text{ LGG}}} \quad (9)$$

However, when the synchronous motor is utilized (instead of the induction motor), the plant is still described by Eq. (3) but the added load, designated as $\mathbb{S}_{AL(\mathrm{LDG})}$, is written, with the aid of Fig. 6.2.2 for power factor 0.50 LDG, as

$$\mathbb{S}_{AL(\mathrm{LDG})} = 1{,}500 \times 10^3 + jQ_{AL(\mathrm{LDG})} \quad (10)$$

The reactive power $Q_{AL(\mathrm{LDG})}$ can be calculated with the aid of Fig. 6.2.2 as

$$Q_{AL(\mathrm{LDG})} = P_{AL} \tan\theta_{\mathrm{LDG}} = (1{,}500 \times 10^3) \times \tan 60.00^\circ = \underline{2{,}598.07 \times 10^3} \mathrm{var} \quad (11)$$

Substituting for $Q_{AL(\mathrm{LGG})}$ from Eq. (11) into Eq. (10) yields

$$\mathbb{S}_{AL(\mathrm{LDG})} = 1{,}500 \times 10^3 + j2{,}598.07 \times 10^3 = \underline{3{,}000.00 \times 10^3 e^{j60.00^\circ}} \mathrm{VA} \quad (12)$$

Now the total industrial process comprised of the loads of Eqs. (3) and (12), as when the synchronous motor is utilized—that is, the motor is operating at a power factor of 0.5 LDG—is written as

$$\mathbb{S}_{\mathrm{TSM}} = \mathbb{S}_{IL(\mathrm{LGG})} + \mathbb{S}_{AL(\mathrm{LDG})} = 9{,}500 \times 10^3 - j3{,}401.93 \times 10^3$$
$$= \underline{10{,}090.74 \times 10^3 e^{-j19.70^\circ}} \mathrm{VA} \quad (13)$$

Thus the kilovoltampere rating and power factor of the total process, when the synchronous motor is utilized, are

$$|\mathbb{S}_{TSM}| = \underline{\underline{10{,}090.74\,\text{kVA}}} \tag{14}$$

and

$$\cos\theta_{TSM} = \cos(-19.70°) = \underline{\underline{0.94\ \text{LGG}}} \tag{15}$$

A comparison between Eqs. (8) and (14) reveals that use of the synchronous motor results in a net improvement of kilovoltamperes, designated by $\mathbb{S}_{NI}$, given by

$$|\mathbb{S}_{NI}| = |\mathbb{S}_{TIM}| - |\mathbb{S}_{TSM}| = 11{,}875.00 - 10{,}090.74 = \underline{\underline{1{,}784.2\,\text{kVA}}} \tag{16}$$

The decision as to whether to use the induction or the synchronous motor depends strictly on **economics**. The difference in cost between the two motors must be compared to the *savings* resulting from the lower kilovoltamperes and the demand charges when the synchronous motor is used.

Before leaving this section is should be **pointed out** that the concepts covered here are valid only when the synchronous device is connected to an **infinitely strong network** (i.e., an **infinite bus**).

It should be made clear, however, that although the **isolated synchronous motor supplying its own load behaves as if it were connected to an infinite bus, the situation is quite different when we consider an isolated generator supplying its own load.** In such a connection there are no other parallel synchronous generators to compensate for changes in the field (rotor-winding) dc excitation, or in the prime mover output, in order to maintain the terminal voltage and the frequency of the **synchronous generator** constant.

If the isolated synchronous generator is driven at "constant speed" (i.e., constant frequency $\tau_{INM} = k_1$, and the field (rotor-winding) dc excitation **is increased**, the terminal voltage $\mathbf{V}_L$ **increases**, which, in general, is accompanied by an **increase** in the **real** (P_{DT}) and **reactive** (Q_{DT}) **powers** delivered to the isolated load. If, for the same prime-mover conditions, the field (rotor-winding) dc excitation is decreased, the **converse is true**. If, on the other hand, the field (rotor-winding) dc excitation **is maintained constant**, that is, $I_F = k_2$, and the prime mover output **is increased**, the **frequency** f_E of the stator currents ($\mathbf{I}_G \equiv \mathbf{I}_S$) and the **terminal voltage** $\mathbf{V}_L$ **increase**. This voltage increase is, in general, accompanied by an **increase** in the **real** (P_{DT}) and **reactive** (Q_{DT}) powers delivered to the isolated load. If, for the same field (rotor-winding) dc excitation conditions, the prime-mover output **is decreased**, the **converse is true**.

Finally, it should be pointed out that the power factor of the isolated synchronous generator is determined by the load connected to the generator, rather than by the field (rotor-winding) dc excitation I_F.

6.3 EXCITATION SYSTEMS FOR SYNCHRONOUS DEVICES

An exciter is defined to be a system comprised of **one** or **more** dc generators [Figs. 6.9(a) and 6.9(b)] or ac generators in conjunction with rectifying systems [Fig. 6.10] designed to provide dc **current** to the rotor winding, known also as the field winding, of the synchronous device operating either in the **generator** or the **motor** mode. The magnitude of this dc current may be (1) maintained constant or (2) continuously changing in accordance with the system. These changes may be manual or automatic, using feedback control schemes, depending upon the complexity and the requirements of the system to which the device is connected.

The necessity of the auxiliary exciter is unquestionable in its use in conjunction with the synchronous generator because of the role it plays in (1) the generation of the voltage $\mathbf{E}_G$ and (2) the control on the power factor of the generator. However, it is a disadvantage, especially in the case of synchronous motors of small size. (Remember that no excitation system is needed for the induction motor.) In large-size motors, however, this disadvantage is counterbalanced by such distinctive operating characteristics as constancy of average speed and the possibility of power-factor correction of the system to which the synchronous motor is connected.

Excitation systems are commonly designed for 125 V, though 250 V instead are commonly used, with ratings of up to 50 kW. The most common excitation system for either the synchronous generator or the synchronous motor is comprised of (1) **the main exciter** and (2) **the pilot exciter**, Figs. 6.8(a) and 6.8(b).

The main exciter is normally a separately excited dc generator. This type of generator is chosen because of its (1) **quick response to field current variations** (a feature that is very important in the case of synchronous generators when there are disturbances on the system to which the generator is connected) and (2) **stable**

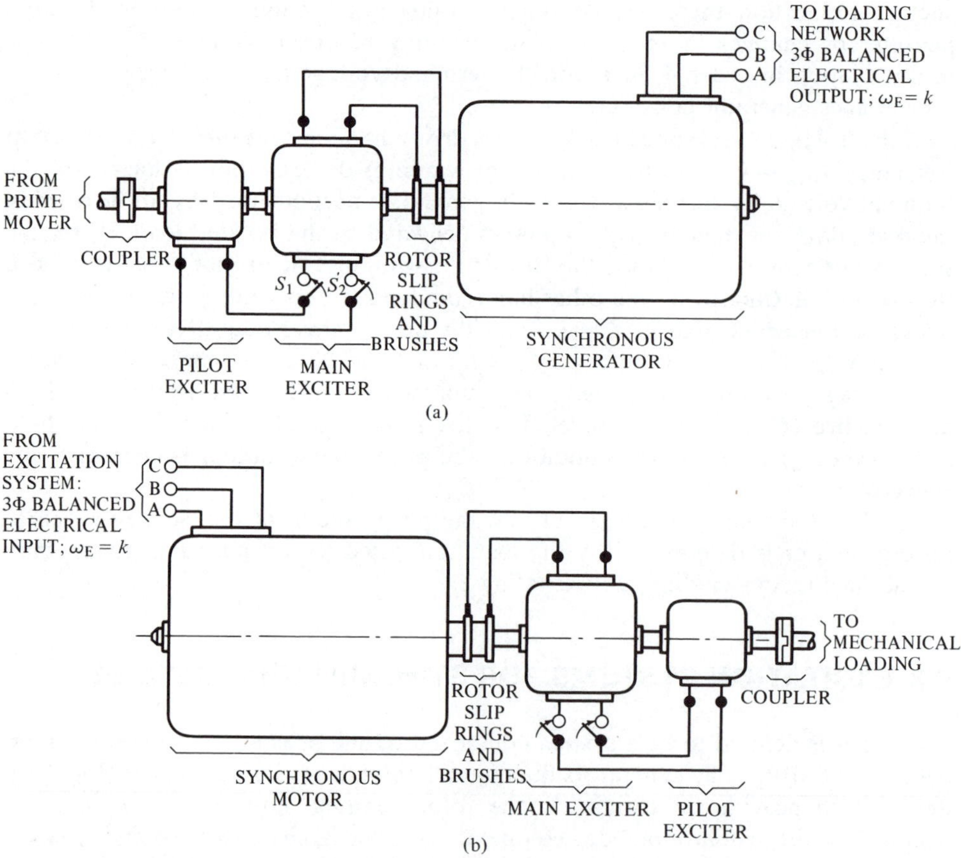

Fig. 6.8 Physical Arrangement of Shaft-Mounted Conventional Excitation Systems: (a) Synchronous Generator and (b) Synchronous Motor

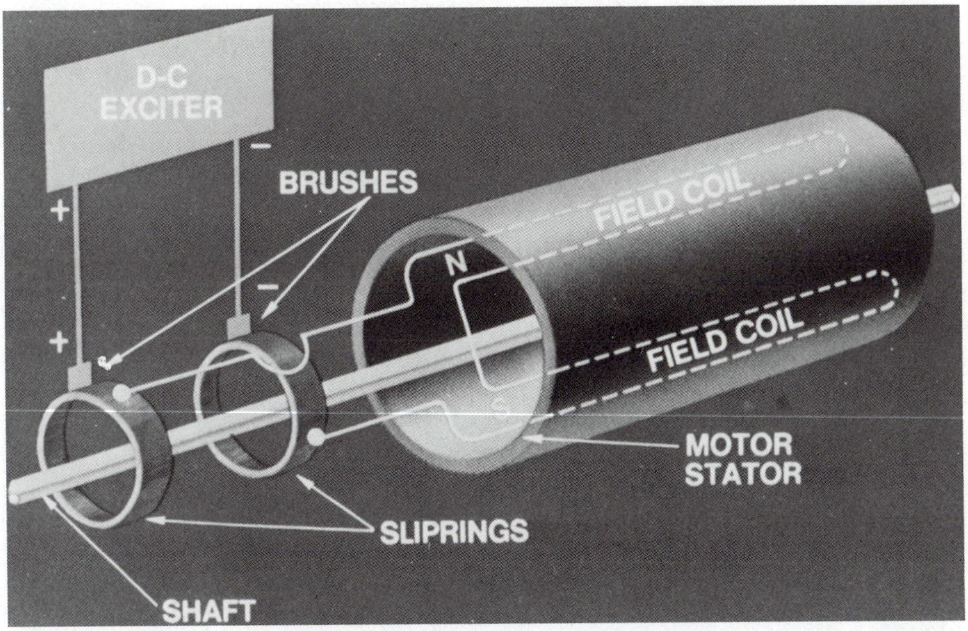

Plate 6.1 Schematic Diagram of the Normal Excitation System for Synchronous Motors. (Courtesy of Westinghouse Electric Co., Pittsburgh, Pa.)

operation at low voltages (a feature used to implement the underexcitation of synchronous condensers). However, a combination of the separate excitation with self- and series excitation have enhanced the properties of the main exciter. On the other hand, **the pilot exciter is primarily a self-excited dc generator**, with the unique property of uniform response to the control of voltage regulators. It is operated at constant voltage. Thus the output of the pilot exciter, that is, the excitation to the rotor winding of the synchronous generator I_F, is controlled by variation of the field rheostat of the pilot exciter or the field rheostat of the main exciter, if any. Contrary to the main excitation—which utilizes only separately excited dc generators—or variations of it, pilot excitation, in addition to self-excited dc generators, utilizes rotating amplifiers operating at varying voltages as pilot exciters. Examples of rotating amplifiers are (1) the *Amplidyne* (General Electric Co.), (2) the *Rototrol* (Westinghouse Electric Co.), and (3) the *Regulex* (Allis-Chalmers Manufacturing Co.). A **rotating amplifier** is a dc generator especially designed as a power amplifier, that is, *a large output may be controlled by a very small input*, which can be supplied from a voltage regulator. In some cases, where more power gain is required than is provided by the voltage regulator, a magnetic amplifier can be inserted between the voltage regulator and the rotating amplifier, Fig. 6.9(b).

The excitation systems for units installed since 1965 are primarily either AC systems using thyristers such as G.E.'s "ALTERREX" or brushless lash-ups involving rotating armature configuration exciters connected to rotating field configuration main generators by way of rotating diodes. The latter are harder to control but reduce maintenance due to the lack of brushes.

The main and pilot exciters may be *mounted* on the shaft of the synchronous devices in one of two ways: (1) **directly (direct-connected exciters)**, that is, the speed of the shafts of both exciters is identical to the speed of the synchronous device to

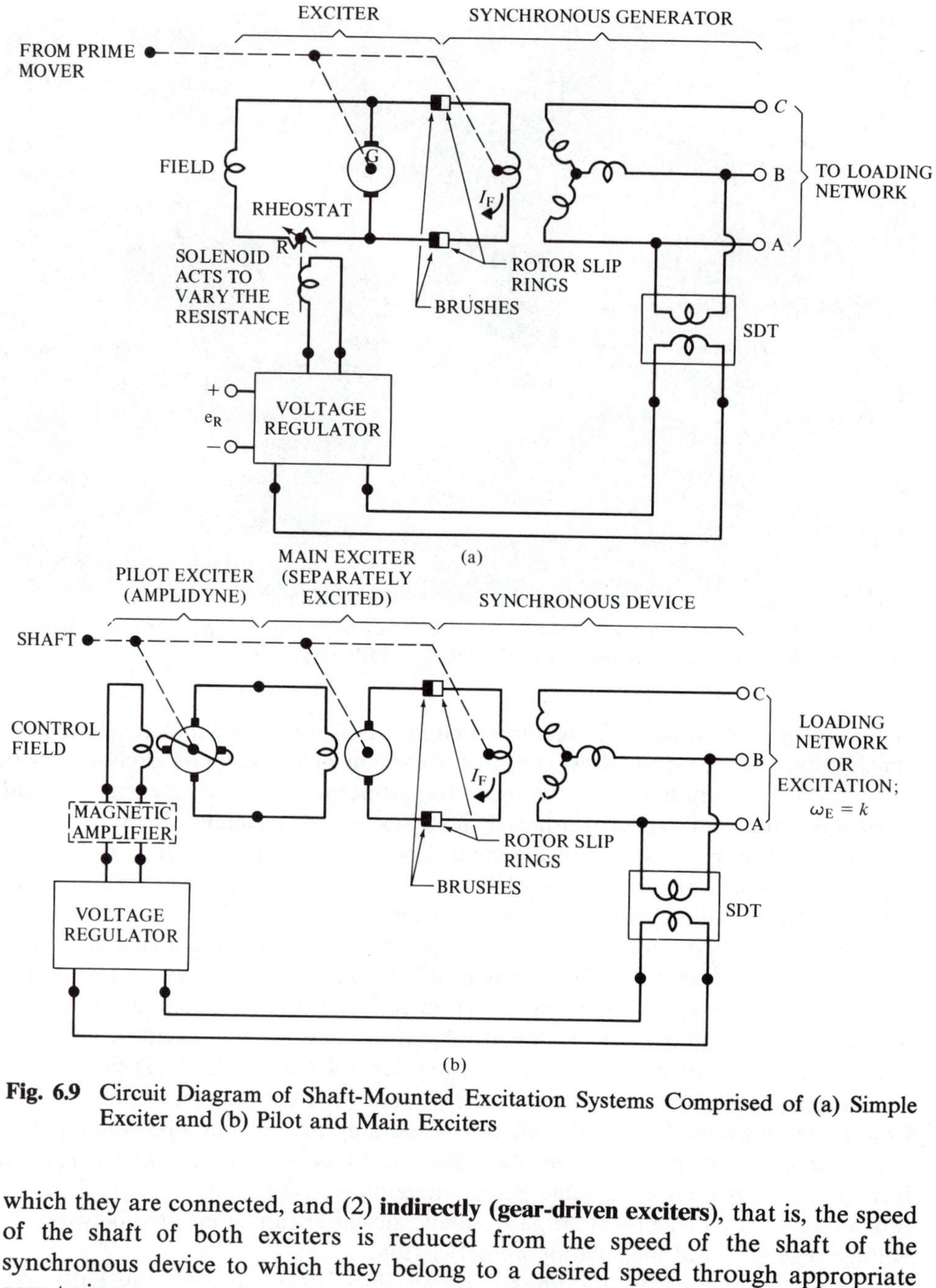

Fig. 6.9 Circuit Diagram of Shaft-Mounted Excitation Systems Comprised of (a) Simple Exciter and (b) Pilot and Main Exciters

which they are connected, and (2) **indirectly (gear-driven exciters)**, that is, the speed of the shaft of both exciters is reduced from the speed of the shaft of the synchronous device to which they belong to a desired speed through appropriate gear trains.

In some installations the exciter, usually comprised of the pilot exciter only, is placed in a less expensive and more convenient location than on the shaft of the synchronous device and is driven by an induction motor. This motor is designed to have (1) maximum torque of about five times its full-load torque and (2) a large flywheel in its axis, designed to carry the exciter through short periods (12 to 15 cycles) of severely reduced driving voltage (i.e., forcing function of the induction motor) due to severe faults. Thus since this **motor-driven exciter** requires a reliable source of electric power, in some stations the motor may be excited from a house generator, which also supplies power to other station auxiliaries. In other stations the motor may be fed through a transformer from the **main synchronous generators** or the **main bus**. In the latter two cases, the induction motor's source of electric energy (i.e., its forcing function) is subject to reductions or interruptions during system disturbances.

The motor-driven exciter system yields two distinct advantages: (1) there is freedom in the choice of speed, and (2) maintenance on the exciter can be carried out by switching to an auxiliary exciter, if available, without shutting down the entire system, that is, the synchronous generator (and perhaps the power plant) or the synchronous motor (and perhaps the plant to which it is assigned).

The type of excitations exhibited in Figs. 6.8(a) and 6.8(b), known as the **unit-exciter scheme** (i.e., **each synchronous generator and motor** has its own excitation system, in contrast to the method used in practice, especially when it concerns generating units, where each station has a number of exciters operating in parallel feeding an **excitation bus**, which in turn is used to feed all synchronous devices in the station), has three distinct advantages that should be **stressed**: (1) great reliability, that is, trouble on one exciter affects only the device it serves, rather than all devices present, (2) great adaptability to automatic control equipment for synchronous device field (rotor-winding) current, I_F, control, and (3) elimination of switchgear associated with the excitation bus and cumbersome and inefficient main-field rheostats, used for synchronous device, field (rotor-winding) current, I_F, control.

An excitation system, commonly used on the smaller synchronous generators, in which a conventional **dc shunt generator** mounted on the shaft of the synchronous generators furnishes the field (rotor-winding) dc excitation I_F is shown in Fig. 6.9(a). The field (rotor-winding) dc excitation current I_F of the synchronous generator is adjusted by continuously comparing the rectified line-to-line voltage $\mathbf{V}_{AB}$, detected by the step-down transformer (SDT), to the reference voltage e_R of the voltage regulator. The reference voltage represents a means of setting the magnitude of the output voltage at the desired value. The difference between these two voltages is applied to the solenoid, which is designed to act on the rheostat of the self-excited dc generator to vary its resistance R in such way as to have a neutralizing effect on the departure of the line-to-line voltage $\mathbf{V}_{AB}$ from the desired output voltage.

A somewhat more complex system that makes use of the pilot exciter (in this instance the **rotating amplifier** Amplidyne), whose output excites the field of the main exciter (in this instance, a simple **separately excited dc generator**), is shown in Fig. 6.9(b). Note that both exciters are mounted on the shaft of the synchronous device. Here the field (rotor-winding) dc excitation current I_F of the synchronous generator is adjusted by continuously comparing the rectified line-to-line voltage $\mathbf{V}_{AB}$, detected by the step-down transformer, to the reference voltage of the voltage regulator. The difference between these two voltages is applied to the control field winding of the rotating amplifier. The amplifier is designed to produce a current whose effect on the separately excited dc generator would be such as to have a neutralizing effect on the departure of the line-to-line voltage $\mathbf{V}_{AB}$ from the desired output voltage.

The **reliability** of large synchronous devices, particularly that of synchronous generators in power generation plants, depends heavily on the excitation system, and especially on the compounded problems of the natural wear, and thus on the

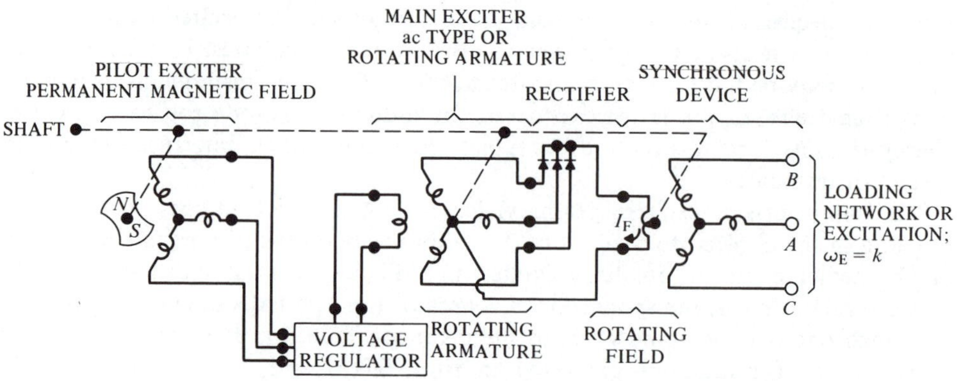

Fig. 6.10 Circuit Diagram of a Shaft-Mounted, Brushless Excitation System

maintenance of brushes, commutators, and slip rings of the various devices that constitute the excitation system. It is clear, then, that elimination of such components from the excitation system, a phenomenon that leads to the **brushless excitation**, enhances the overall system's reliability.

One such arrangement in which (1) **a permanent magnetic field plot exciter**, (2) an ac- **(synchronous device) type main exciter**, known as a rotating-armature exciter, and (3) **a rotating rectifier** constitute the brushless excitation system for synchronous devices is exhibited in Fig. 6.10. Note that (1) the permanent-magnet rotor of the pilot exciter, (2) the armature (stator windings) of the main exciter, (3) the rectifier, and (4) the field (rotor-winding) of the synchronous device are mounted on the shaft of the synchronous device (motor or generator modes of operation). The rotating permanent magnetic field of the pilot exciter induces a balanced, three-phase, 420-Hz signal, which is fed to the voltage regulator which in turn supplies a regulated dc signal to the **stationary field winding** of the main exciter. The three-phase balanced output of the rotating-armature main exciter is rectified

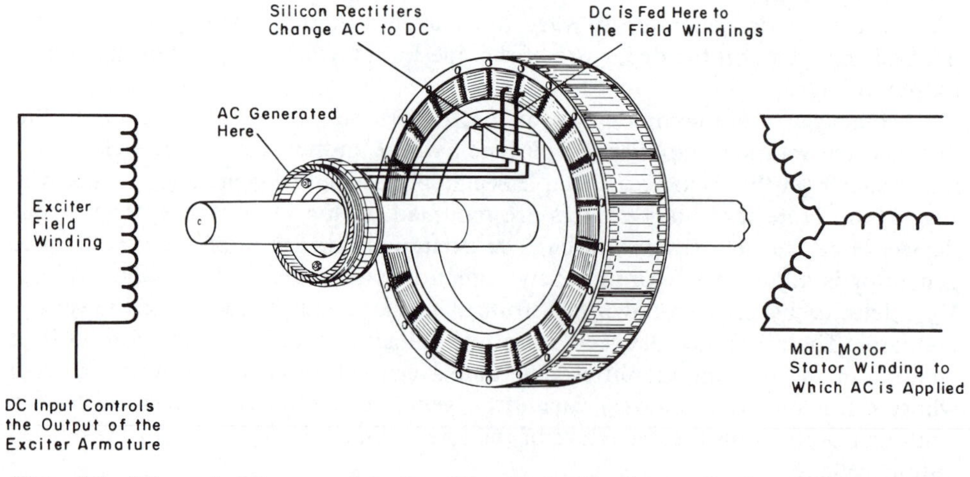

Plate 6.2 Schematic Diagram of the Brushless Excitation System for Synchronous Motors. (Courtesy of General Electric Co., Schenectady, N.Y.)

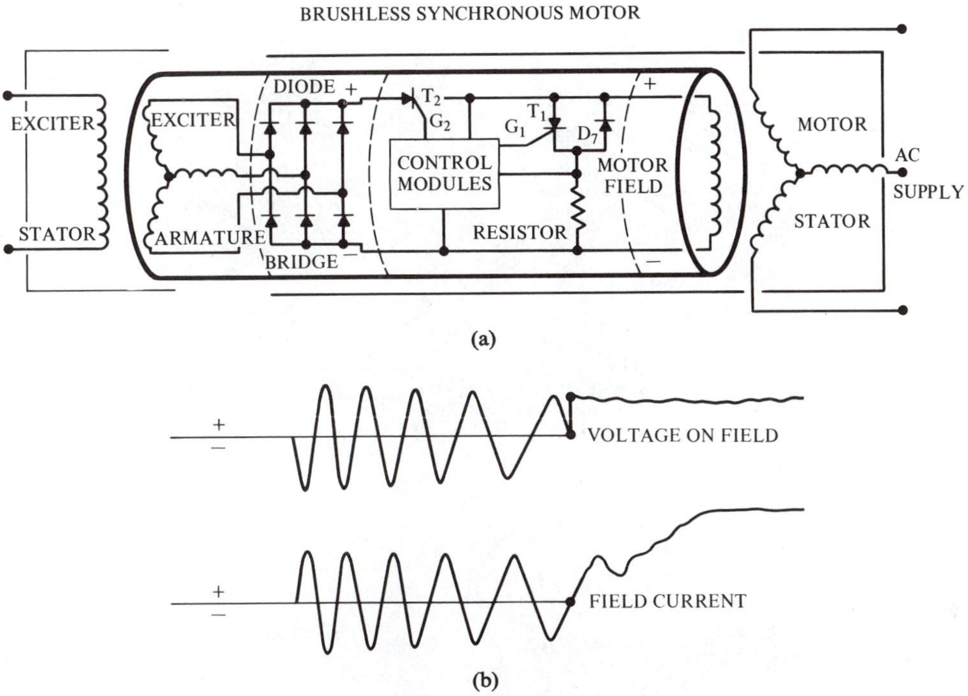

Plate 6.3 Brushless Excitation: (a) Device and (b) Field-Winding Variables. (Courtesy of Westinghouse Electric Co., Pittsburgh, Pa.)

Plate 6.4 Brushless Exciter Armature and Heat Sink Attached to Rotor: (a) Exciter Armature and (b) Rectifiers and Heat Sink. (Courtesy of General Electric Co., Schenectady, N.Y.)

Plate 6.5 Brushless Exciter for Generator Application; Rectifier Assembly Uses Full-Wave Bridge Circuit with GE A-90 Type of Diodes and Varistor Protection Circuit. (Courtesy of General Electric Co., Schenectady, N.Y.)

by solid-state diodes and is fed as I_F to the field (rotor winding) of the synchronous device. The merits of this exciter, Fig. 6.10, when compared to the excitation systems exhibited in Figs. 6.9(a) and 6.9(b) should be obvious.

6.4 PARALLEL OPERATION OF SYNCHRONOUS GENERATORS

To secure **reliable** and **economic** operation of power systems, the power industry utilizes extensive interconnection of utility power systems into large groupings known as **power pools**; power pools are interconnected into still larger groupings known as **regional power pools**; regional power pools are interconnected into **a nationwide power grid system.**

A utility system usually consists of several **generating stations, all interconnected electrically in parallel**. It is common for **each generating station to be comprised of two or more synchronous generators connected electrically in parallel.** Under normal operating conditions, all the synchronous generators and motors in an electrically interconnected system operate in **synchronism** with each other, that is, the frequencies of all the synchronous devices are **exactly equal** except during momentary changes in load or excitation. One or more, of the large synchronous devices may pull out of synchronism from the rest of the system, as a consequence of a severe disturbance. This is called **system instability**, a condition that may result in a complete shutdown unless remedial steps are taken immediately to rectify the problem, that is, reinstate synchronism to the unsynchronized generators or take the unsynchronized generators out of the system.

Electric parallel interconnection of generating stations (as well as units within a generating station; i.e., it is preferable to replace single large units by several smaller units of equivalent capacity) provides several advantages from the military and economic points of view, of which the most important are as follows. First, if a single large unit becomes inoperative, regardless of reason, the generating station ceases to be functional. However, if one of several units that constitute an equivalent capacity becomes inoperative, regardless of the reason, the remaining smaller units remain functional and provide service as needed. Second, for maximum efficiency each unit must be loaded to capacity; thus it is advantageous to have several small units operating in parallel (in each generating station) that can be added in to, or removed from, service as the demand fluctuates with each unit operating near its capacity. Third, for purposes of repair or maintenance, it is preferable to have several units of equivalent capacity rather than a single unit; while some units become inoperative, the remaining units remain functional and provide needed service. Fourth, as the demand from each generating station increases, additional units may be purchased and added to the system; thus capital is invested efficiently. Finally, multiunit (connected electrically in parallel) stations are imposed by the physical and economic limitations of the possible capacity of each unit. That is, no single unit can be constructed to supply the output of the largest stations, for example, 1,000,000 MW. However, units of approximately 1,000 MW are available.

It is clear, from these reasons, that electrical parallel operation of synchronous generators is essential. Generally, the policy of each utility is to utilize one large unit to handle the minimum demand and to add additional smaller units as the station's demand increases over a 24-h period. When the demand requires the generating station to continuously employ several smaller units to handle the minimum demand, a large unit of equivalent capacity is then added. This process continues as the demand increases.

The electric parallel interconnection of n synchronous generators to a three-phase bus, feeding a three-phase balanced load, is exhibited in Fig. 6.11. Since circuit analysis theorems dictate that the same terminal voltage exist across all elements connected in parallel, the line-to-line voltages across the n synchronous generators and the load, and consequently the generated phase voltages of Fig. 6.11, are identical. (This is true if and only if the voltage drops along the lines are neglected and all pertinent components are balanced.) Thus any synchronous generator to be added to the system depicted in Fig. 6.11, commonly referred to as **to be put on line**, must satisfy the following requirements: (1) the voltage waveforms of the incoming generator must have the same shape as the voltage waveforms of the interconnected three-phase system, (2) the effective values of the voltages of the incoming synchronous generator must be the same as the effective values of the voltages of the interconnected three-phase system, (3) the phase sequence of the voltages of the incoming generator must be the same as the phase sequence of the voltages of the interconnected three-phase system*, (4) the

*Correct phase-sequencing is essential for the proper connection of motors, generators, transformers, watthour meters, instruments, and relays. Utilities realize considerable savings in core losses and other losses when distribution transformers are connected properly. Time and money can be saved if rotating machinery is connected properly the first time.

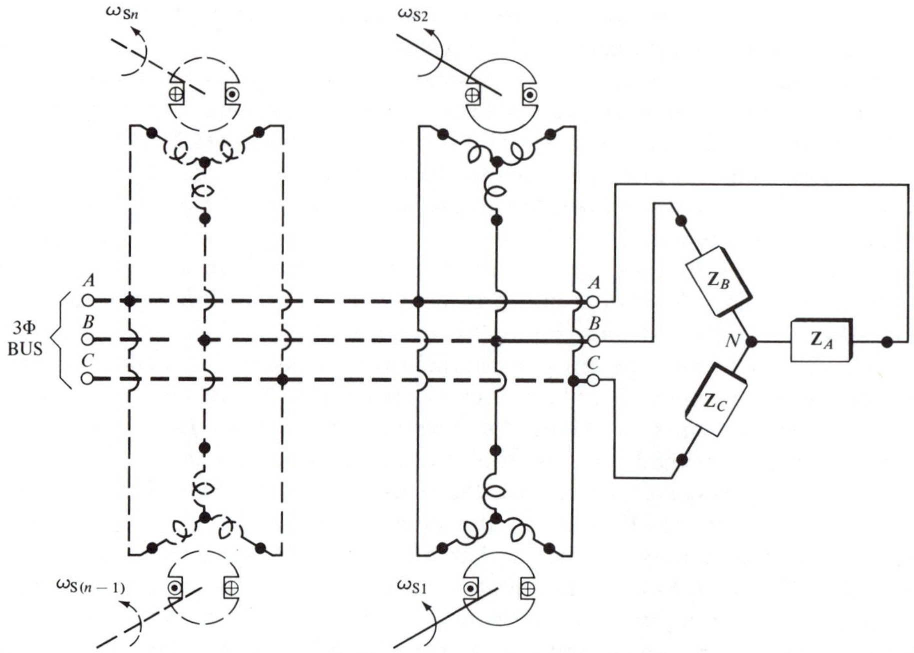

Fig. 6.11 Electric Parallel Interconnection of n Identical Synchronous Generators to a Three-Phase Bus, Feeding a Three-Phase Balanced Load

phase voltages of the incoming generator must be in phase with the phase voltages of the interconnected three-phase system, and (5) the frequency ω_E (or f_E) of the voltages of the incoming generator must be equal to the frequency of the voltages of the interconnected three-phase system.

These requirements can be checked with various instruments before closing the switching system to put the incoming generator on line. Most of the measurements are quite obvious. However, two of the measurements need particular attention.

First, the phase sequence of the voltages can be checked with the aid of an instrument known as **phase-sequence inductor.** The basic circuit of this instrument is comprised of **two identical neon lamps**, N_1 and N_3, and a capacitor C_2, each connected in series with the corresponding resistors R_1, R_2, and R_3. The resulting three branches are designed to yield unequal branch impedances which are interconnected to a Y-connected unbalanced three-phase load. In addition, these three branch resistors are designed to yield potential rises above and below the ignition potential of the neon lamps such that the following two conditions are satisfied. For the phase sequence 3-2-1 (i.e., C-B-A), the potential rise $\mathbf{E}_{N1}$ across the lamp N_1 is related to the potential rise $\mathbf{E}_{N2}$ across the lamp N_2 as $|\mathbf{E}_{N1}| > |\mathbf{E}_{N2}|$; this causes lamp N_1 to illuminate the lamp N_2 to be extinguished. However, for the phase sequence 1-2-3 (i.e., A-B-C) the potential rise $\mathbf{E}_{N1}$ across the lamp N_1 is related to the potential rise $\mathbf{E}_{N2}$ across the lamp N_2 as $|\mathbf{E}_{N1}| < |\mathbf{E}_{N2}|$; this causes lamp N_1 to be extinguished while it causes lamp N_2 to illuminate. A **small induction motor** accomplishes the same task, however. If one were to connect this motor to the terminals of the interconnected three-phase system where the incoming synchronous generator is to be connected, observe the direction of rotation of the shaft of the induction motor, and then find the terminals of the incoming generator that

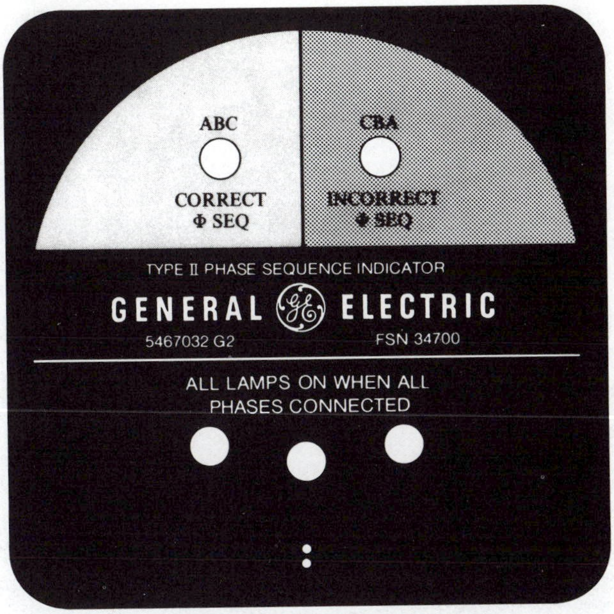

(a)

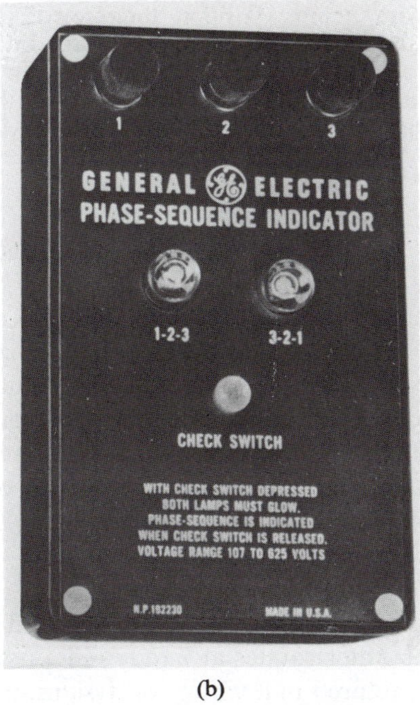

(b)

Plate 6.6 Phase Sequence Indicator: (a) Panel Mounted Model and (b) Portable Model. (Courtesy of General Electric Co., Schenectady, N.Y.)

Plate 6.7 Synchroscope with Pointer Position Interpreted: (a) Vertical Position—the Frequencies of the Incoming Generator and the Infinite Bus Are Identical, (b) SLOW—the Frequency of the Incoming Generator Is Lower than that of the Infinite Bus, and (c) FAST—the Frequency of the Incoming Generator Is Higher than that of the Infinite Bus. (Courtesy of General Electric Co., Schenectady, N.Y.)

yield the same rotation of the motor's shaft, then one has the correct phase sequence.

Secondly, the frequency (or speed) of the incoming generator can be compared to the frequency (or speed) of the interconnected three-phase system with the aid of an instrument known as a **synchroscope**. This instrument is designed to compare the frequency of the voltage of one phase of the incoming generator with the frequency of the voltage of the corresponding phase of the interconnected infinite bus. When the two frequencies are identical, the pointer of the synchroscope **locks to a vertical position**. However, if the radian frequency of the incoming generator is 380 rad/s and that of the interconnected three-phase system is 377 rad/s, the pointer rotates at 3 rad/s in a direction marked "FAST." Conversely, if the radian frequency of the incoming generator is 370 rad/s and that of the three-phase system is 377 rad/s, the pointer rotates at 7 rad/s in a direction marked "SLOW." Synchroscopes are manufactured in a variety of designs, namely the polarized-vane type, the moving iron type and the crossed-coil types. The synchroscope is designed to operate on single-phase circuits and therefore may be used to synchronize single-phase as well as polyphase synchronous generators. The pointer of the instrument is mounted on the instrument's rotor whose winding is polarized (magnetized) to the frequency of the infinite bus through its connection across any two lines of the infinite bus. However, the instrument's stator winding is comprised of **two** coils, i.e., R_1L_1 and $R_2L_2C_2$ connected in parallel across any two lines of the incoming generator, which are distributed in the instrument's circumference in the same manner as the single-phase, capacitor-split, induction motor, Fig. 4.40(g). The operation of the instrument can be explained as follows. The field of its **stator** winding rotates at the frequency of the **incoming generator,** while the **pointer** is polarized (magnetized) at the frequency of the **infinite bus**. Therefore, when the frequencies of the two windings are identical the stator field is at right angles to the pointer; the pointer is unmagnetized, and aligns itself in the fixed vertical position. However, when the pointer is slightly magnetized due to a difference in the frequency of the voltages in the two windings, the pointer would move slowly **clockwise** (FAST) or **counterclockwise** (SLOW) at a speed (in radians per second) equal to the difference between the voltage frequencies of the infinite bus and the incoming generator.

To put a synchronous generator, say the jth, on line, the following procedure must be completed. First, the synchronous generator, with its field (rotor) winding deenergized, is brought up to speed by means of the prime mover. Second, once up to speed, the field (rotor) winding of the incoming synchronous generator is energized and its dc excitation I_F is adjusted until the generated voltages $|\mathbf{E}_{GA}|$, $|\mathbf{E}_{GB}|$, and $|\mathbf{E}_{GC}|$ of the incoming synchronous generator are equal to the terminal voltages $|\mathbf{V}_{LA}|$, $|\mathbf{V}_{LB}|$, and $|\mathbf{V}_{LC}|$ of the interconnected three-phase system, that is, $|\mathbf{E}_{Gk}|=|\mathbf{V}_{Lk}|$; $k=A,B,C$. Under these conditions the jth synchronous generator **neither delivers nor draws current from the three-phase system**; it is said that this generator is **floating** across the interconnected three-phase system.

If, however, the field (rotor) winding dc excitation I_F of the incoming synchronous generator is **increased**, but **no adjustments are made** on its prime-mover input, the phase voltages $|\mathbf{E}_{GA}|$, $|\mathbf{E}_{GB}|$, and $|\mathbf{E}_{GC}|$ exceed the terminal voltage $|\mathbf{V}_{Lk}|$, and the synchronous generator delivers current to the interconnected three-phase system at $\delta_G \neq 0$. This mode of operation clearly designates **generation** or **generator action**. If, however, the field (rotor) winding dc excitation I_F of the incoming synchronous generator is **decreased, but no adjustments are made on the prime-mover input**, the phase voltages $|\mathbf{E}_{GA}|$, $|\mathbf{E}_{GB}|$, and $|\mathbf{E}_{GC}|$ are less than the terminal voltage $|\mathbf{V}_L|$, and the synchronous generator draws current from the interconnected three-phase system at $\delta_G \neq 0$. This mode of operation designates **motoring** or **motor action**. Consequently, any synchronous generator in parallel with an infinite bus whose dc excitation I_F is reduced so that the phase-generated voltages are less than the terminal bus voltage is operating as a synchronous motor. **Such a synchronous generator is said to be motorized**. This property turns out to be **beneficial rather than detrimental**, as is shown shortly.

When a synchronous motor (e.g., a synchronous condenser) is to be put on line, the procedure is similar to putting a synchronous generator on line. That is, the motor is started with the field (rotor) winding deenergized. Large motors are started by smaller auxiliary motors (induction or dc motors), while smaller synchronous motors are started as induction motors by means of their own squirrel cage damping windings. As the synchronous motor is brought close to synchronous speed ω_S, the dc excitation current I_F is applied to the field (rotor) winding and the rotor pulls into synchronism with the three-phase system—that is, if the loading is not excessive.

The behavior of synchronous generators connected in parallel across an infinite bus are better demonstrated through the mathematical analysis of the per-phase diagram of Fig. 6.11, which is exhibited in Fig. 6.12. The describing equation of

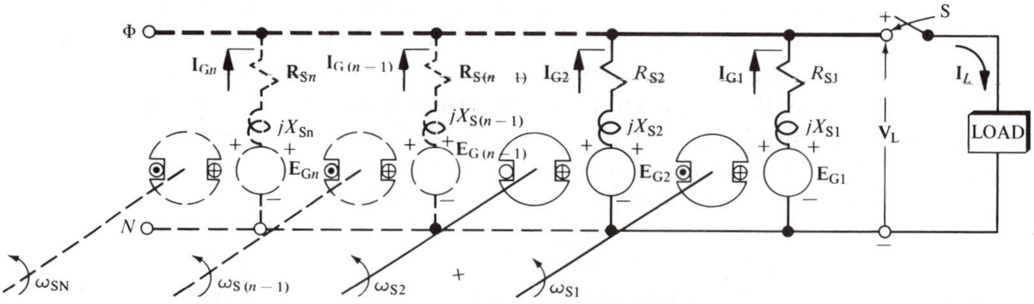

Fig. 6.12 Per-Phase Circuit Diagram of the Three-Phase System Exhibited in Fig. 6.11, Where $\Phi = A$, B, or C and N = Neutral (or Ground)

Fig. 6.12 is written, utilizing Kirchhoff's voltage law and circuit theory, as

$$\mathbf{E}_{G1} - \mathbf{Z}_{S1}\mathbf{I}_{G1} = \mathbf{E}_{G2} - \mathbf{Z}_{S2}\mathbf{I}_{G2} = \cdots = \mathbf{E}_{G(n-1)} - \mathbf{Z}_{S(n-1)}\mathbf{I}_{G(n-1)}$$
$$= \mathbf{E}_{Gn} - \mathbf{Z}_{Sn}\mathbf{I}_{Gn} = \mathbf{V}_L \tag{6.47}$$

It is not necessary that each of the sources, $\mathbf{E}_{Gj}$, $j = 1, 2, \ldots, n$, Fig. 6.12, **produce the same emf or deliver the same current to supply the load.** In order, however, for any unit to serve as a source of emf, as indicated by Eq. (6.47), it is necessary that the emf generated by that unit exceed the terminal voltage $\mathbf{V}_L$ in order for the unit to deliver current to the infinite bus. Eq. (6.47) also points out that regardless of the division of real and reactive power loads between the generators, the terminal voltages of all generators are equal. However, the phase currents of the interconnected three-phase system (all generators and loads) must satisfy Kirchhoff's current law. That is,

$$\mathbf{I}_L = \mathbf{I}_{G1} + \mathbf{I}_{G2} + \cdots + \mathbf{I}_{G(n-1)} + \mathbf{I}_{Gn} \tag{6.48}$$

It should be emphasized, however, that with respect to a load on the interconnected three-phase system, all generators should have the same instantaneous polarity or phase relation if circulating currents between the generators are to be avoided. With respect to each other, generators operating in parallel must have their voltages in phase opposition; that is, in the loop containing generators No. 1 and No. 2, Fig. 6.12, $\mathbf{E}_{G1}$ and $\mathbf{E}_{G2}$ oppose each other.

Example 6.3

Two synchronous generators, having identical design characteristics and parameters of $R_S = 47.5\ \Omega$ and $jX_S = j475.0\ \Omega$, are connected to a 15,822.70-V bus, as per Fig. 6.12. The generated voltages are adjusted to be $\mathbf{E}_{G1} = 18{,}200.00e^{j0°}$ V and $\mathbf{E}_{G2} = 13{,}445.41e^{j0°}$ V. For each generator calculate: (1) the current delivered to or drawn from the bus, (2) the power generated or accepted, and (3) the power drawn from or delivered to the bus. In addition, calculate: (4) the load current $\mathbf{I}_L$ and (5) the current circulating to the local loop comprised of the two generators.

Solution

Utilization of Eq. (6.47) and the input data enables us to calculate the currents $\mathbf{I}_{G1}$ and $\mathbf{I}_{G2}$, by using the equations $\mathbf{E}_{G1} - \mathbf{Z}_{S1}\mathbf{I}_{G1} = \mathbf{V}_L$ and $\mathbf{E}_{G2} - \mathbf{Z}_{S2}\mathbf{I}_{G2} = \mathbf{V}_L$. Thus,

$$\mathbf{I}_{G1} = \frac{\mathbf{E}_{G1} - \mathbf{V}_L}{\mathbf{Z}_{S1}} = \frac{18{,}200.00e^{j0°} - 15{,}822.70e^{j0°}}{47.5 + j475} = \frac{2{,}377.30e^{j0°}}{477.37e^{j84.29°}}$$
$$= \underline{\underline{4.98e^{-j84.29°}\text{A}}} \tag{1}$$

and

$$\mathbf{I}_{G2} = \frac{\mathbf{E}_{G2} - \mathbf{V}_L}{\mathbf{Z}_{S2}} = \frac{13{,}445.41e^{j0°} - 15{,}822.71e^{j0°}}{47.5 + j475} = \frac{-2{,}377.30e^{j0°}}{477.37e^{j84.29°}}$$

$$= \underline{\underline{-4.98e^{-j84.29°}\text{A}}} \tag{2}$$

Eqs. (1) and (2) suggest that generator No. 1 **delivers** $4.98e^{-j84.29°}$ A to the bus and generator No. 2 **draws** $4.98e^{-j84.29°}$ A from the bus.

The powers generated by each generator are calculated with the aid of the equation $P_{GTj} = 3\,\text{Re}(\mathbf{E}_{Gj}\mathbf{I}_{Gj}{}^*)$, $j = 1, 2$, input data, and Eqs. (1) and (2) as follows:

$$P_{GT1} = 3\,\text{Re}(\mathbf{E}_{G1}\mathbf{I}_{G1}{}^*) = 3\,\text{Re}(18{,}200.00 \times 4.98e^{j84.29°})$$
$$= \underline{\underline{27{,}055.88\text{ W}}} \tag{3}$$

and

$$P_{GT1} = 3\,\text{Re}\big(\mathbf{E}_{G2}\mathbf{I}_{G2}{}^* = 3\,\text{Re}\big[13{,}445.41 \times (-4.98e^{j84.29°})\big]$$
$$= \underline{\underline{-19{,}987.77\text{ W}}} \tag{4}$$

Eqs. (3) and (4) suggest that generator No. 1 generates 27,055.88 W and generator No. 2 accepts 19,987.77 W in the same direction in which it is driven by its prime mover.

The powers delivered to or drawn from the bus are calculated with the aid of the equation $P_{DTj} = 3\,\text{Re}(\mathbf{V}_L\mathbf{I}_{Gj}{}^*)$, $j = 1, 2$, input data, and Eqs. (1) and (2) as follows:

$$P_{DT1} = 3\,\text{Re}(\mathbf{V}_L\mathbf{I}_{G1}{}^*) = 3\,\text{Re}(15{,}822.70 \times 4.98e^{j84.29°})$$
$$= \underline{\underline{23{,}521.82\text{ W}}} \tag{5}$$

and

$$P_{DT2} = 3\,\text{Re}(\mathbf{V}_L\mathbf{I}_{G2}{}^*) = 3\,\text{Re}\big[15{,}822.70 \times (-4.98e^{j84.29°})\big]$$
$$= \underline{\underline{-23{,}521.82\text{ W}}} \tag{6}$$

Eqs. (5) and (6) suggest that generator No. 1 delivers 23,521.82 W to the bus and generator No. 2 draws 23,521.82 W from the bus.

The load current is calculated with the aid of the equation $\mathbf{I}_L = \mathbf{I}_{G1} + \mathbf{I}_{G2}$, Eq. 6.48, and Eqs. (1) and (2). Thus

$$\mathbf{I}_L = \mathbf{I}_{G1} + \mathbf{I}_{G2} = 4.98e^{-j84.29°} + (-4.98e^{-j84.29°}) = \underline{\underline{0.00}}\text{ A} \tag{7}$$

Finally, since $\mathbf{V}_L = \mathbf{E}_{G1} - \mathbf{Z}_{S1}\mathbf{I}_{G1} = \mathbf{E}_{G2} - \mathbf{Z}_{S2}\mathbf{I}_{G2}$, Eq. 6.47, we can write

$$\mathbf{E}_{G1} - \mathbf{Z}_{S1}\mathbf{I}_{G1} = \mathbf{E}_{G2} - \mathbf{Z}_{S2}\mathbf{I}_{G2} \tag{8}$$

Eq. (8) can be written as

$$\mathbf{E}_{G1} - \mathbf{E}_{G2} = \mathbf{Z}_{S1}\mathbf{I}_{G1} - \mathbf{Z}_{S2}\mathbf{I}_{G2} = 2\mathbf{Z}_{S1}\left[\frac{(\mathbf{I}_{G1} - \mathbf{I}_{G2})}{2}\right]\Bigg|_{\mathbf{Z}_{S1} = \mathbf{Z}_{S2}} \tag{9}$$

Defining the term $(\mathbf{I}_{G1} - \mathbf{I}_{G2})/2$ of Eq. (9) as $\mathbf{I}_\sigma$ enables writing Eq. (9) as

$$\mathbf{I}_\sigma = \frac{\mathbf{I}_{G1} - \mathbf{I}_{G2}}{2} = \frac{\mathbf{E}_{G1} - \mathbf{E}_{G2}}{2\mathbf{Z}_S} \tag{10}$$

Utilization of input data and Eqs. (1) and (2) yields

$$\mathbf{I}_\sigma = \frac{4.98e^{-j84.29°} - (-4.98e^{-j84.29°})}{2} = \frac{18{,}200.00e^{j0°} - 13{,}445.41e^{j0°}}{2 \times (47.5 + j475)}$$

$$= \underline{\underline{4.98e^{-j84.29°}\text{ A}}} \tag{11}$$

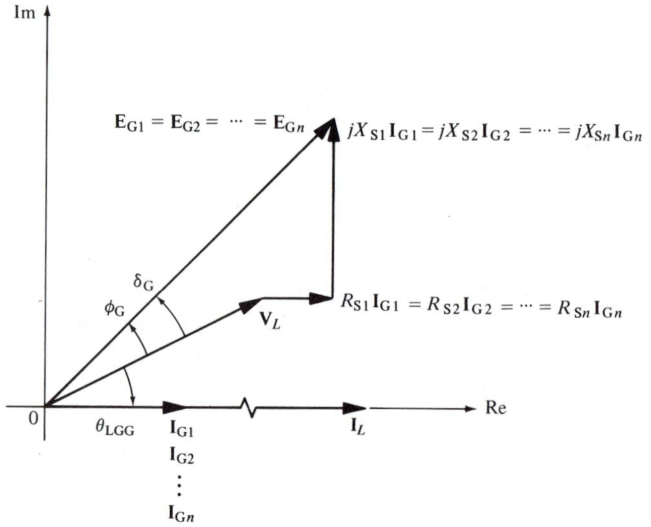

Fig. 6.13 Phasor Diagram of the Per-Phase Circuit Diagram Exhibited in Fig. 6.11, Note That the Synchronous Generators Are Identical and the Power Factor of the Chosen Load Is Lagging

This current, defined as **synchronizing current** later in this section, flows in the counterclockwise direction in the local loop comprised of the two generators.

The voltage and current relationship of Eqs. 6.47 and 6.48 for **n** identical synchronous generators and a load whose power factor is lagging is exhibited in Fig. 6.13. Note that since the synchronous generators are identical, all internal voltages may be added to the terminal voltage $\mathbf{V}_L$ to yield the generated voltages $\mathbf{E}_{Gj}$, $j = 1, 2, \ldots, n$, which are equal and bear the same relationship to the load current $\mathbf{I}_L$.

Perhaps the mathematical analysis of the synchronous generators No. 1 and No. 2, with generator No. 1 already on line and generator No. 2 coming on line, can demonstrate the **behavior of synchronous generators operating in parallel** as well as their effects on (1) bus voltage variation (i.e., how the bus voltage can be raised or lowered) and (2) transfer of load between generators. Phasor analysis is used here to describe the electrical behavior of the system, although the mechanical behavior of the generators is dynamic in nature. Use of phasors is possible because of long time constants of the mechanical system. The following analysis assumes that the mechanical system is frozen at a given time t_1. The electrical quantities are then calculated for this time. As time passes to a new time t_2, the dynamics of the mechanical system change very slowly, and the electrical quantities must be recalculated.

Effects of Underexcitation on the Synchronization of Synchronous Generators Connected in Parallel to an Infinite Bus

Consider generator No. 2 being connected in parallel with generator No. 1, of identical design characteristics, Fig. 6.14. At the instant of closing switch S_2, (1) the generated voltage $\mathbf{E}_{G2}$ is exactly equal to, and in phase with, the generated voltage $\mathbf{E}_{G1}$ of generator No. 1 with respect to the load* and (2) the frequency ω_{E2} is

*The points of generators that are connected together have identical polarity.

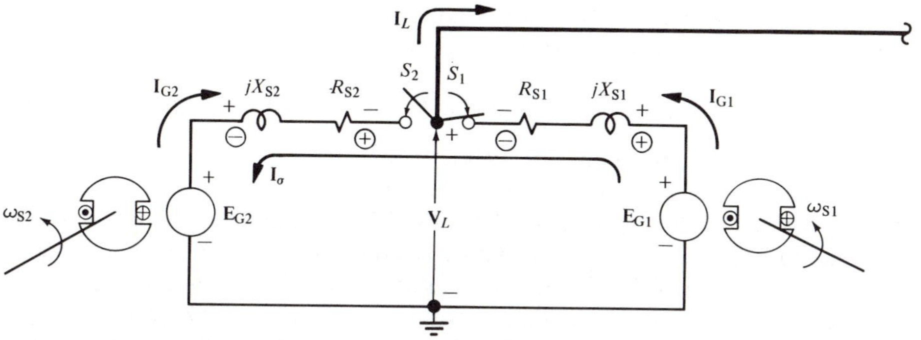

Fig. 6.14 Per-Phase Circuit Diagram of Two Identical Synchronous Generators Feeding a Load with Lagging Power Factor Through an Infinite Bus

exactly equal to the frequency ω_{E1} of generator No. 1. The load (not shown in Fig. 6.14) has a lagging power factor and is fed through an infinite bus. Before switch S_2 is closed, generator No. 1 supplies all the load current; Eq. 6.48 is satisfied for $j=1$, that is $\mathbf{I}_L = \mathbf{I}_{G1}$. As soon as switch S_2 is closed, Eq. 6.48 is satisfied for $j=2$, that is, $\mathbf{I}_L = \mathbf{I}_{G1} + \mathbf{I}_{G2}$, and Fig. 6.13 (for $j=2$) depicts the behavior, from the load point of view, of Fig. 6.14. The phasor diagram of Fig. 6.15, where generator No. 1 is used as the reference, shows the relationship of the pertinent quantities of the two generators under load. Note that all voltages and currents of the synchronous generators are equal in magnitude but are out of phase with respect to each other. This becomes obvious by examining the **local loop** created by the two generators, Fig. 6.14. When generator No. 2, Fig. 6.14, is underexcited with respect to generator No. 1, that is, $|\mathbf{E}_{G1}| > |\mathbf{E}_{G2}|$, a current designated as $\mathbf{I}_\sigma$ begins to flow in the counterclockwise direction, causing voltage drops designated by the signs ⊕

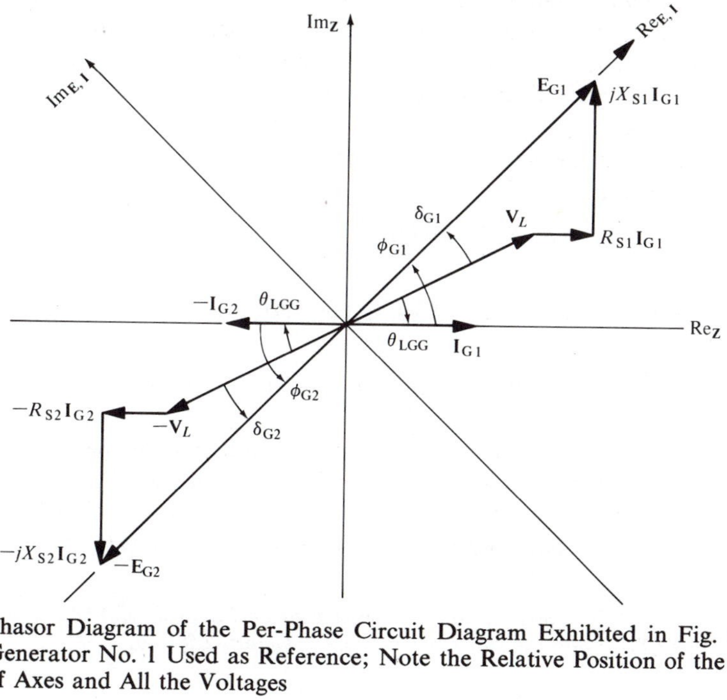

Fig. 6.15 Phasor Diagram of the Per-Phase Circuit Diagram Exhibited in Fig. 6.14 with Generator No. 1 Used as Reference; Note the Relative Position of the Two Sets of Axes and All the Voltages

and Θ. Now utilization of Kirchhoff's voltage law enables writing the describing equation of the **local loop** comprised of the two synchronous generators as

$$\mathbf{E}_{G1} - \mathbf{E}_{G2} \equiv \mathbf{E}_\sigma \equiv |\mathbf{E}_\sigma| e^{j\phi_E} = (\mathbf{Z}_{S1} + \mathbf{Z}_{S2})\mathbf{I}_\sigma \tag{6.49}$$

where the following definitions are valid: (1) $\mathbf{E}_\sigma$ is the **phasor difference** between the generated voltages of the synchronous generators No. 1 and No. 2 and **is defined as the synchronizing voltage of the local loop** (sometimes the bus voltage $\mathbf{V}_L$ is used instead of $\mathbf{E}_{G1}$), and (2) $\mathbf{I}_\sigma$ **is defined as the synchronizing current, circulating in the local loop formed by the two generators, or the incoming generator and the infinite bus.** Note that if the incoming generator is perfectly synchronized, $\mathbf{E}_{G2} = \mathbf{E}_{G1}$. Thus $\mathbf{E}_\sigma = 0$ and consequently $\mathbf{I}_\sigma = 0$. If, however, $\mathbf{E}_{G2} \neq \mathbf{E}_{G1}$, then $\mathbf{E}_\sigma \neq 0$ and $\mathbf{I}_\sigma \neq 0$.

The effects that the synchronizing current $\mathbf{I}_\sigma$ has on the two generators can be understood by solving Eq. (6.49) for $\mathbf{I}_\sigma$. Thus

$$\mathbf{I}_\sigma \equiv |\mathbf{I}_\sigma| e^{j\phi_I} = \frac{\mathbf{E}_{G1} - \mathbf{E}_{G2}}{(R_{S1} + jX_{S1}) + (R_{S2} + jX_{S2})}$$
$$= \frac{|\mathbf{E}_\sigma| e^{j\phi_E}}{|\mathbf{Z}_{ST}| e^{j\phi_Z}} \tag{6.50}$$

where

$$\phi_Z = \tan^{-1}\left(\frac{X_{S1} + X_{S2}}{R_{S1} + R_{S2}}\right) \tag{6.51}$$

The voltages $\mathbf{E}_{G1}$, $\mathbf{E}_{G2}$, and $\mathbf{E}_\sigma$ and the current $\mathbf{I}_\sigma$ of Eq. (6.50) are drawn on the **E**, **I**-plane of Fig. 6.15 as exhibited in Fig. 6.16. Close examination of Fig. 6.16 and of Eq. (6.50) reveals that the synchronizing current $\mathbf{I}_\sigma$ always lags the voltage $\mathbf{E}_\sigma$ by an angle $\phi_\sigma \equiv \phi_E - \phi_I = \phi_E - (\phi_E - \phi_Z) = \phi_Z < 90°$ (remember that $R_S \ll |jX_S|$). This angle plays a very important role in the behavior of synchronous generators connected in parallel to an infinite bus. In fact, it makes possible the synchronization of the two generators being discussed (see if you can deduce this contribution of $\phi_\sigma \equiv \phi_Z$ as you read through the remainder of this section). Now Fig. 6.16 enables us to write the expressions for the developed powers, designated as $P_{\sigma T1}$ and $P_{\sigma T2}$, generated by generators No. 1 and No. 2 of Fig. 6.14, due to the synchronizing current $\mathbf{I}_\sigma$, as†

$$P_{\sigma T1} = 3\,\mathrm{Re}(\mathbf{E}_{G1}\mathbf{I}_\sigma^*) = 3|\mathbf{E}_{G1}||\mathbf{I}_\sigma|\cos\phi_\sigma \tag{6.52}$$

and

$$P_{\sigma T2} \equiv 3\,\mathrm{Re}(-\mathbf{E}_{G2}\mathbf{I}_\sigma^*) = 3|-\mathbf{E}_{G2}||\mathbf{I}_\sigma|\cos(180° - \phi_\sigma) = -3|-\mathbf{E}_{G2}||\mathbf{I}_\sigma|\cos\phi_\sigma \tag{6.53}$$

These powers are **in addition** to the powers delivered to the load by the two generators.

Note the **signs** of the developed powers $P_{\sigma T1}$ and $P_{\sigma T2}$ of the generators No. 1 and No. 2 operating in parallel. In fact, $P_{\sigma T1}$ is **positive** and $P_{\sigma T2}$ is **negative**. This suggests that generator No. 1 operates in the **generator mode**, that is, it generates power, while generator No. 2 operates in the **motor mode**, that is, it seems to absorb

†In general, the angle $\measuredangle_{\mathbf{I}_\sigma}^{\mathbf{E}_{G1}}$ is used. However, in this subsection Eqs. (6.52) to (6.58), $\measuredangle \mathbf{E}_{G1} = \measuredangle \mathbf{E}_{G2} = \phi_E = 0°$ and $\measuredangle_{\mathbf{I}_\sigma}^{\mathbf{E}_{G1}} = \measuredangle \mathbf{E}_{G1} - \phi_I = -\phi_I$. Since $\phi_\sigma = \phi_E - \phi_I = -\phi_I$, then $\measuredangle_{\mathbf{I}_\sigma}^{\mathbf{E}_{G1}} = \phi_\sigma$.

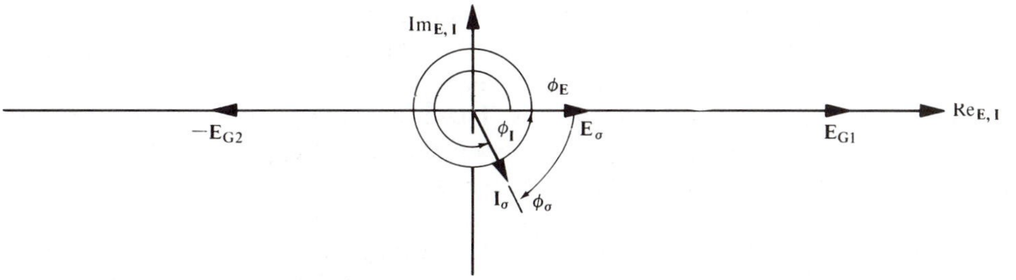

Fig. 6.16 Generator of Synchronizing Current $\mathbf{I}_\sigma$ Due to Overexcitation of Generator No. 1, $|\mathbf{E}_{G1}|>|\mathbf{E}_{G2}|$, Fig. 6.14

power from generator No. 1. The power flow from the overexcited generator No. 1, that is, $|\mathbf{E}_{G1}|>|-\mathbf{E}_{G2}|$, to generator No. 2—in fact, the power flow between synchronous generators during such a transient process—is defined as **synchronizing power**. This seemingly absorbed power, or the synchronizing power $P_{\sigma T2}$, causes generator No. 2 to tend to run as a motor in the same direction as it is being driven by its own prime mover. However, since generator No. 2, driven by its own prime mover, generates power and supplies its own losses in addition to providing power to the load, it cannot absorb the synchronizing power $P_{\sigma T2}$ supplied by generator No. 1. Thus the synchronizing power disturbs the power balance at both generators causing generator No. 2 to generate less power than it receives from its prime mover and generator No. 1 to generate more power (momentarily) than it receives from its prime mover.

The power **dissipated** in the coils of both synchronous generators that constitute the local loop can be written as

$$P_{\sigma LT} = 3|\mathbf{E}_\sigma||\mathbf{I}_\sigma|\cos(\phi_E-\phi_I)|_{\phi_\sigma=\phi_E-\phi_I} = 3(R_{S1}+R_{S2})|\mathbf{I}_\sigma|^2 \tag{6.54}$$

Use of the defining voltage relationship of the numerator of the right-hand side of Eq. (6.50), Fig. 6.16, the theory of Appendix A, and trigonometric identities, enables us to write Eq. (6.54) as

$$P_{\sigma LT} = 3|\mathbf{E}_{G1}||\mathbf{I}_\sigma|\cos\phi_\sigma - 3|\mathbf{E}_{G2}||\mathbf{I}_\sigma|\cos\phi_\sigma = 3(R_{S1}+R_{S2})|\mathbf{I}_\sigma|^2 \tag{6.55}$$

Eq. (6.55) can be written, with the aid of Eqs. (6.52) and (6.53), as

$$P_{\sigma LT} = P_{\sigma T2} + P_{\sigma T1} = 3(R_{S1}+R_{S2})|\mathbf{I}_\sigma|^2 \tag{6.56}$$

or

$$P_{\sigma T1} = -P_{\sigma T2} + P_{\sigma LT} = 3|-\mathbf{E}_{G2}||\mathbf{I}_\sigma|\cos\phi_\sigma + 3(R_{S1}+R_{S2})|\mathbf{I}_\sigma|^2 \tag{6.57}$$

That is, the overexcited generator (No. 1 with respect to No. 2) must supply both (1) the synchronizing power to generator No. 2, that is, the standby power at the bus for use by the load over and above the powers generated by both generators, P_{GT2}, and (2) the heat losses for both synchronous generators, that is, $P_{\sigma LT}$, Eqs. (6.53) and (6.56).

As soon as the synchronizing power $P_{\sigma T1}$ is generated (due to $|\mathbf{E}_{G1}|>|-\mathbf{E}_{G2}|$, Fig. 6.16, or due to a change in any of the generated voltages $\mathbf{E}_{Gj}$, $j=2,3,\ldots,n$, of various generators of a multigenerator system, in parallel with an inifinite bus), it effects the following changes: (1) generator No. 1 generating the synchronizing power **drops back** in phase, or tends to slow down, as a result of the increased load and the developed electromagnetic countertorque and (2) generator No. 2 receiving the synchronizing power **advances** in phase, or tends to speed up, because of the motor action produced in it, acting in the same direction as its prime mover. That is, power imbalance at each machine is made up by changes in the mechanical kinetic energy stored in the rotor inertias. These two changes lead to the redrawing of Fig. 6.16 as exhibited in Fig. 6.17. Note that the **dropping back** in phase of generator No. 1 and the **advancing** in phase of generator No. 2 bring the voltages $\mathbf{E}_{G1}$ and $-\mathbf{E}_{G2}$ closer together. Their vectorial summation yields a *new* position of the synchronizing voltage $\mathbf{E}_\sigma$ followed by the new synchronizing current $\mathbf{I}_\sigma$ by the angle $\phi_\sigma < 90°$. The process discussed earlier, and described by Eqs. (6.50) through (6.56), continues with the two (or more) generators continuously adjusting their positions, that is, **dropping back or advancing until an equilibrium is reached where no synchronizing power or motor action is produced by either generator**, that is, $\measuredangle_{\mathbf{I}_{\sigma F}}^{-\mathbf{E}_{G2F}}| \equiv |\measuredangle_{\mathbf{I}_{\sigma F}}^{\mathbf{E}_{G1F}}| < 90°$. The final (equilibrium) values of the pertinent quantities, designated as $\mathbf{E}_{G1F}$, $-\mathbf{E}_{G2F}$, $\mathbf{E}_{\sigma F}$, and $\mathbf{I}_{\sigma F}$, are not shown in Fig. 6.17. This figure is drawn to exaggerate the effects of advancing and dropping back of $\mathbf{E}_{G1}$ and $-\mathbf{E}_{G2}$. In reality, the final values lie closer to the real axis. Under such conditions, the only power that is produced is the power lost as heat in the stator windings of **both** generators. **This power is produced by both generators** and is written, with the aid of Eq. (6.56), as

$$\begin{aligned} P_{\sigma LT} &= 3(R_{S1} + R_{S2})|\mathbf{I}_{\sigma F}|^2 \\ &= 3|\mathbf{E}_{G1F}||\mathbf{I}_{\sigma F}|\cos\left(\measuredangle_{\mathbf{I}_{\sigma F}}^{\mathbf{E}_{G1F}}\right) + 3|-\mathbf{E}_{G2F}||\mathbf{I}_{\sigma F}|\cos\left(\measuredangle_{\mathbf{I}_{\sigma F}}^{-\mathbf{E}_{G2F}}\right) \end{aligned} \tag{6.58}$$

It should be noted that, since the magnitudes of the final generated voltages are equal in magnitude to the initial generated voltages, that is, $|\mathbf{E}_{G1F}| = |\mathbf{E}_{G1}|$, $|-\mathbf{E}_{G2F}| = |-\mathbf{E}_{G2}|$, and $\mathbf{E}_{G1F}$ and $-\mathbf{E}_{G2F}$ are closer in phase than $\mathbf{E}_{G1}$ and $-\mathbf{E}_{G2}$, then the resultant synchronizing voltage $\mathbf{E}_\sigma$ and consequently the synchronizing current $\mathbf{I}_\sigma$ **are both increased**, that is, $|\mathbf{E}_{\sigma F}| > |\mathbf{E}_\sigma|$ and $|\mathbf{I}_{\sigma F}| > |\mathbf{I}_\sigma|$. Thus the synchronizing currents used in Eqs. (6.58) and (6.56) are not the same. In fact, the $3(R_{S1} + R_{S2})|\mathbf{I}_{\sigma F}|^2$ losses given by Eq. (6.58) are higher than the losses given by Eq. (6.56). However, those of Eq. (6.58) are provided by both generators instead of being provided only by generator No. 1, as in Eq. (6.56).

If we were to change the excitation of generator No. 2 so that its generated voltage $|\mathbf{E}_{G2}| > |\mathbf{E}_{G1}$, then the synchronizing power would be supplied by generator

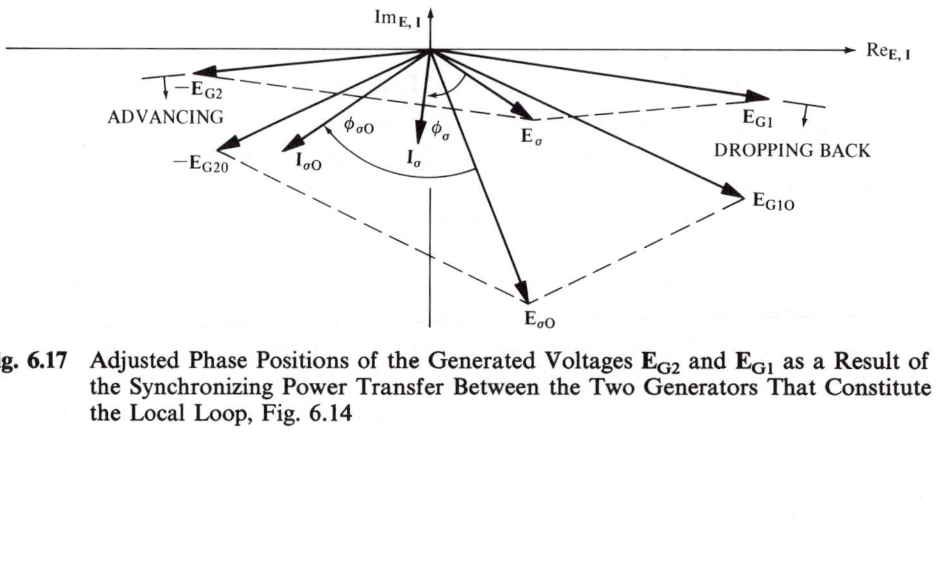

Fig. 6.17 Adjusted Phase Positions of the Generated Voltages $\mathbf{E}_{G2}$ and $\mathbf{E}_{G1}$ as a Result of the Synchronizing Power Transfer Between the Two Generators That Constitute the Local Loop, Fig. 6.14

No. 2. **That is, the overexcited generator always supplies the synchronizing power, which eventually leads to equilibrium.** In both instances, however, the terminal voltage $\mathbf{V}_L$ of both (or all) generators is the same. The reason is simple. The two (or more) generators are connected in parallel. Since the synchronizing current $\mathbf{I}_\sigma$ is all internal to the local loop, and since whatever occurs internally does not change the loading voltage, changing the resistance of the field (rotor) winding and the dc excitation of any synchronous generator in parallel with other synchronous generators **does not affect the division of load between these generators.**

Returning to the case when $|\mathbf{E}_{G1}|>|\mathbf{E}_{G2}|$, inspection of Fig. 6.14 leads to the following observation. The synchronizing current $\mathbf{I}_\sigma$ of the **local loop** affects the load current contribution of the two generators as follows: (1) the net current contribution by generator No. 1 is **increased**, that is, $\mathbf{I}_{G1N}=\mathbf{I}_{G1}+\mathbf{I}_\sigma$, Fig. 6.18, and (2) the net current contribution by generator No. 2 is **decreased**, that is, $\mathbf{I}_{G2N}=\mathbf{I}_{G2}-\mathbf{I}_\sigma$, Fig. 6.18. These changes are more obvious when one superimposes the relevant information of Fig. 6.15, that is, the E, I-plane, the generated voltages $\mathbf{E}_{G1}$ and $-\mathbf{E}_{G2}$, the generated currents $\mathbf{I}_{G1}$ and $\mathbf{I}_{G2}$ with Fig. 6.16 but allowing $\phi_\sigma \rightarrow 90°$ and $|\mathbf{E}_{G1}|>|-\mathbf{E}_{G2}|$, as exhibited in Fig. 6.18. Fig. 6.18 reveals the following striking pieces of information: (1) the net current $|\mathbf{I}_{G2N}|$ of the underexcited generator is **decreased,** while the net current $|\mathbf{I}_{G1N}|$ of the other generator is **increased** and (2) the power-factor angle of the underexcited generator becomes **less lagging**, that is, **decreases or improves**, while the power factor-angle of the other generator becomes **more lagging**, that is, **increases or worsens**. The consequences of these phenomena are (1) inhibition of the capacity of generator No. 2 to deliver useful current to the load and (2) improvement of the capacity of generator No. 1 to deliver useful current to the load. In addition to the observations above, two more observations are relevant and equally important: (1) since $\mathbf{I}_\sigma$ leads $-\mathbf{E}_{G2}$, $\mathbf{I}_\sigma$ causes a demagnetizing effect on generator No. 2 and (2) since $\mathbf{I}_\sigma$ lags $\mathbf{E}_{G1}$, $\mathbf{I}_\sigma$ causes a magnetizing effect on generator No. 1. (Magnetization and demagnetization phenomena become clear when one remembers that the armature reaction voltage, i.e., the voltage $\mathbf{E}_\sigma$ caused by the synchronizing current $\mathbf{I}_\sigma$, leads the current $\mathbf{I}_\sigma$ by an angle of 90°.)

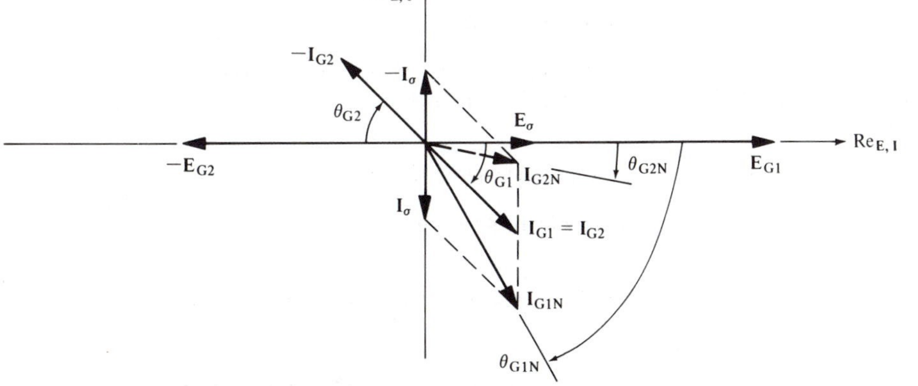

Fig. 6.18 The Effects of the Synchronizing Current $\mathbf{I}_\sigma$ on the Power Factors of the Generators That Constitute the Local Loop of Fig. 6.14. Generator No. 1 Is Overexcited, That Is, $|\mathbf{E}_{G1}|>|-\mathbf{E}_{G2}|$

Example 6.4

Two synchronous generators, having identical design characteristics and parameters of $R_S=47.5\ \Omega$ and $jX_S=j475.0\ \Omega$, are synchronized so that the phase voltages of the incoming generator No. 2 are in phase opposition with respect to the phase voltages of the on-line generator No. 1. The generated voltages are adjusted to be $\mathbf{E}_{G1}=18{,}200.00e^{j0°}$ V and $\mathbf{E}_{G2}=13{,}445.41e^{j0°}$ V. Calculate the following quantities at the instant after the synchronizing switch is closed to parallel the two generators: (1) the synchronizing current $\mathbf{I}_\sigma$, (2) the synchronizing power generated by generator No. 1 (i.e., generator No. 1 is generating), (3) the synchronizing power seemingly absorbed by, or delivered to, generator No. 2 (i.e., generator No. 2 is motoring), (4) the power loss in both stator windings, and (5) the terminal voltage $\mathbf{V}_L$ of both generators.

Solution

The synchronizing voltage $\mathbf{E}_\sigma$ with respect to $\mathbf{E}_{G1}$ is calculated with the aid of Fig. 6.16 and Eq. 6.49 as

$$\mathbf{E}_\sigma = 18{,}200.00e^{j0°} - 13{,}445.41e^{j0°} = \underline{4{,}754.59e^{j0°}\ \text{V}} \tag{1}$$

The impedance of the windings of each generator is

$$\mathbf{Z}_S = R_S + jX_S = 47.5 + j475.0 = \underline{477.37e^{j84.29°}\ \Omega} \tag{2}$$

and that of both generators is

$$\mathbf{Z}_{ST} = (47.5+j475.0) + (47.5+j475.0) = \underline{954.75e^{j84.29°}\ \Omega} \tag{3}$$

Thus, the synchronizing current with respect to $\mathbf{E}_{G1}$, calculated with the aid of Eqs. (6.50), (1), and (3) is

$$\mathbf{I}_\sigma = \frac{\mathbf{E}_\sigma}{\mathbf{Z}_\sigma} = \frac{4{,}754.59e^{j0°}}{954.75e^{j84.29°}} = \underline{\underline{4.98e^{-j84.29°}\ \text{A}}} \tag{4}$$

and

$$\phi_\sigma = \phi_{\mathbf{E}} - \phi_{\mathbf{I}} = 0° - (-84.29°) = \underline{84.29°} \tag{5}$$

Now the synchronizing power generated by generator No. 1, designated as $P_{\sigma T1}$, is calculated with the aid of input data and Eqs. (6.52), (4), and (5) as

$$P_{\sigma T1} = 3 \times (18{,}200.00) \times (4.98) \times \cos(84.29°) = \underline{27{,}055.82\ \text{W}} \tag{6}$$

The synchronizing power seemingly absorbed by, or delivered to, generator No. 2, designated as $P_{\sigma T2}$, is calculated with the aid of input data and Eqs. (6.53), (4), and (5) as

$$P_{\sigma T2} = -3 \times (13{,}445.41) \times (4.98) \times \cos(84.29°) = \underline{\underline{-19{,}987.73\ \text{W}}} \tag{7}$$

The power loss in both stator windings is calculated with the aid of Eqs. (6.54), (1), (4), (5), and (3) as

$$P_{\sigma LT} = 3 \times (4{,}754.59) \times (4.98)\cos(84.29°) = \underline{\underline{7{,}068.10\ \text{W}}} \tag{8}$$

or with the aid of Eq. (6.55) as

$$P_{\sigma LT} = 3 \times (47.5+47.5) \times (4.98)^2 = \underline{\underline{7{,}068.10\ \text{W}}} \tag{8a}$$

or with the aid of Eq. (6.56) as

$$P_{\sigma LT} = 27{,}055.82 - 19{,}987.73 = \underline{\underline{7{,}068.10\ \text{W}}} \tag{8b}$$

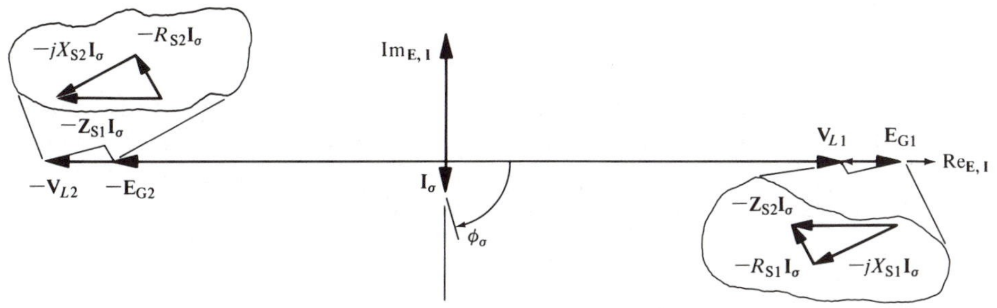

Fig. 6.4.1 Phasor Diagram of Two Identical Synchronous Generators Synchronized to Have the Phase Voltages of the Incoming Generator No. 2 in Phase Opposition with Respect to the Phase Voltages of the On-Line Generator No. 1

Finally, the terminal voltage $\mathbf{V}_L$ of each generator, designated as $\mathbf{V}_{L1}$ and $\mathbf{V}_{L2}$, respectively, can be found with the aid of input data and Eq. (4), and the utilization of Kirchhoff's voltage law on Fig. 6.14, which enables us to write the describing equations of the two generators as (1) $\mathbf{E}_{G1}-\mathbf{Z}_{S1}\mathbf{I}_\sigma-\mathbf{V}_{L1}=0$ and (2) $\mathbf{V}_{L2}-\mathbf{Z}_{S2}\mathbf{I}_\sigma-\mathbf{E}_{G2}=0$. Thus the terminal voltage $\mathbf{V}_{L1}$ of generator No. 1, which is generating, is

$$\begin{aligned}\mathbf{V}_{L1} &= \mathbf{E}_{G1}-\mathbf{Z}_{S1}\mathbf{I}_\sigma = 18{,}200.00e^{j0^\circ}-(477.34e^{j84.29^\circ})(4.98e^{-j84.29^\circ}) \\ &= 18{,}200.00-(2{,}377.30) = \underline{\underline{15{,}822.70e^{j0^\circ}\ \mathrm{V}}}\end{aligned} \tag{9}$$

and the terminal voltage $\mathbf{V}_{L2}$ of the generator No. 2, which is motoring, is

$$\begin{aligned}\mathbf{V}_{L2} &= \mathbf{E}_{G2}+\mathbf{Z}_{S2}\mathbf{I}_\sigma = 13{,}445.41e^{j0^\circ}+(477.34e^{j84.28^\circ})(4.98e^{-j84.28^\circ}) \\ &= 13{,}445.41+(2{,}377.30) = \underline{\underline{15{,}822.71e^{j0^\circ}\ \mathrm{V}}}\end{aligned} \tag{10}$$

Finally, the phasor diagram of the local loop voltages with respect to generator No. 1 can be drawn as shown in Fig. 6.4.1.

Effects of Speed Variation on the Synchronization of Synchronous Generators Connected in Parallel to an Infinite Bus

If $|\mathbf{E}_{G1}|=|\mathbf{E}_{G2}|$ and the speed of the prime mover of generator No. 1, Figs. 6.14 and 6.15, increases, its generated voltage $\mathbf{E}_{G1}$ advances in phase, as exhibited in Fig. 6.19. The resultant synchronizing voltage $\mathbf{E}_\sigma$, Eq. (6.50), causes the synchronizing current $\mathbf{I}_\sigma$ to circulate counterclockwise in the local loop, Fig. 6.14, comprised of the two generators. This current causes the synchronizing power $P_{\sigma T1}$ to be

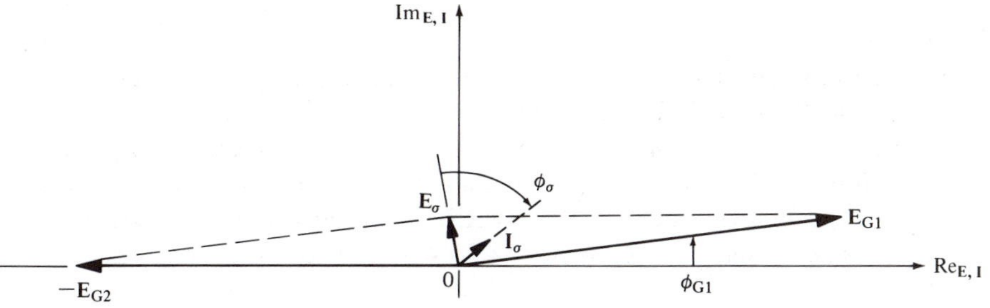

Fig. 6.19 Effects of Overdriving Generator No. 1 of Fig. 6.14 on the Synchronizing Voltage $\mathbf{E}_\sigma$, the Synchronizing Current $\mathbf{I}_\sigma$, and Thus on the Synchronizing Power $P_{\sigma T1}$

generated by generator No. 1, since $\measuredangle_{\mathbf{I}_\sigma}^{\mathbf{E}_{G1}} < 90°$, and is given as

$$P_{\sigma T1} = 3|\mathbf{E}_{G1}||\mathbf{I}_\sigma|\cos\left(\measuredangle_{\mathbf{I}_\sigma}^{\mathbf{E}_{G1}}\right) \tag{6.59}$$

The synchronizing power $P_{\sigma T1}$ contains the stator winding losses, that is, $(R_{S1}+R_{S2})|\mathbf{I}_\sigma|^2$, and the synchronizing power component $P_{\sigma T2}$ transferred to generator No. 2 ($\measuredangle_{\mathbf{I}_\sigma}^{-\mathbf{E}_{G2}} > 90°$), causing it to function in the motor mode. These two powers are written [with the aid of Eqs. (6.54) and (6.52)] as

$$P_{\sigma LT} = 3|\mathbf{E}_\sigma||\mathbf{I}_\sigma|\cos\left(\measuredangle_{\mathbf{I}_\sigma}^{\mathbf{E}_\sigma}\right) = 3(R_{S1}+R_{S2})|\mathbf{I}_\sigma|^2 \tag{6.60}$$

and

$$P_{\sigma T2} = 3|-\mathbf{E}_{G2}||\mathbf{I}_\sigma|\cos\left(\measuredangle_{\mathbf{I}_\sigma}^{-\mathbf{E}_{G2}}\right) \tag{6.61}$$

As in the previous case, since generator No. 1 is more loaded, as a result of the additional power it generates, it tends to **drop back** both in phase and speed, while generator No. 2, as a result of motor action, tends to **advance** both in phase and speed. Thus, as in the case of underexcitation, *the synchronizing current acts in such a way as to keep the generators in synchronism.* When in equilibrium, both generators **share equally** (1) the $(R_{S1}+R_{S2})|\mathbf{I}_\sigma|^2$ losses of the two synchronous generators and (2) a given load if their speed versus load characteristics of their prime movers are drooping and identical.*

Note that the synchronizing power received by generator No. 2 depends on the angle $\phi_\sigma = \phi_Z$ of the synchronous impedance $\mathbf{Z}_{S1}+\mathbf{Z}_{S2}$, Eq. (6.51) and Fig. 6.19. For a given angle of advance ϕ_{G1}, Fig. 6.19, of generator No. 1, the angle $\measuredangle_{\mathbf{I}_\sigma}^{\mathbf{E}_{G1}}$ depends on the angle ϕ_σ. If ϕ_σ is small, $\measuredangle_{\mathbf{I}_\sigma}^{\mathbf{E}_{G1}}$ is large and its $\cos(\measuredangle_{\mathbf{I}_\sigma}^{\mathbf{E}_{G1}})$ has a **small** value. Thus in order to develop the same synchronizing power in generator No. 1, more synchronizing current would be required for a small ϕ_σ than would be required for a large ϕ_σ. Similarly, an extremely high value of synchronous impedance would reduce the synchronizing current faster than the reduction of the angle $\measuredangle_{\mathbf{I}_\sigma}^{\mathbf{E}_{G1}}$. Therefore, a reasonably high ratio of synchronous reactance to synchronous resistance would produce a rapid and sufficient synchronizing power to assure successful parallel operation. Although such synchronous generators have poor voltage regulation, they have been proven to operate more successfully when connected in parallel than those that have good voltage regulation. In summary, then, it can be stated that regarding changes in prime-mover speed (or sudden application or removal of load, for maximum stability), synchronous generators must have (1) a **high ratio** of $|jX_S|$ to R_S, and thus $\phi_\sigma \simeq 90°$, and (2) sufficiently low synchronous impedances so that small changes in their phase angle ϕ_σ will affect the advance angle ϕ_{G1} of the overdriven generator in such a way as to produce large values of synchronizing current and power.

Example 6.5

The generators of Example 6.3 have their generated voltages adjusted to be $|E_{G1}| = |E_{G2}| = 18{,}200.00$ V but generator No. 2 is in phase opposition with respect to generator No. 1. If the prime mover of generator No. 1 drives $\mathbf{E}_{G1}$ ahead of its correct position by 20°, that is, $\mathbf{E}_{G1} = 18{,}200e^{j20°}$, calculate the following quantities at the instant after the synchronizing switch is closed for the two given values of coil impedances, that is, $\mathbf{Z}_{S1} = \mathbf{Z}_{S2} = 47.5 + j475.0\ \Omega$ and $\mathbf{Z}_{S1} = \mathbf{Z}_{S2} = 915.96 +$

*For detailed discussion on load division please read information pertaining to Figs. 6.22 and 6.23.

$j1{,}091.60\ \Omega$: (1) the synchronizing current $\mathbf{I}_\sigma$, (2) the synchronizing power generated by generator No. 1, (3) the synchronizing power received by generator No. 2, (4) the power loss in both stator windings, and (5) the terminal voltage $\mathbf{V}_L$ of both generators.

Solution

The impedance of each generator is

$$\mathbf{Z}_{SA} = 47.5 + j475.0 = \underline{477.37e^{j84.29^\circ}}\ \Omega \tag{1}$$

and

$$\mathbf{Z}_{SB} = 915.96 + j1{,}091.60 = \underline{1{,}424.98e^{j50.00^\circ}}\ \Omega \tag{2}$$

and that of both generators is

$$\mathbf{Z}_{SAT} = 2\mathbf{Z}_{SA} = \underline{954.74e^{j84.28^\circ}}\ \Omega \tag{3}$$

and

$$\mathbf{Z}_{SBT} = 2\mathbf{Z}_{SB} = \underline{2{,}849.96e^{j50.00^\circ}}\ \Omega \tag{4}$$

The synchronizing voltage with respect to generator No. 1 is calculated with the aid of Fig. 6.16 and Eq. (6.49) as

$$\begin{aligned}\mathbf{E}_\sigma &= \mathbf{E}_{G1} - \mathbf{E}_{G2} = 18{,}200.00e^{j20.00^\circ} - 18{,}200.00e^{j0^\circ} \\ &= 17{,}102.41 + j6{,}224.77 - 18{,}200.00 - j0.00 \\ &= -1{,}097\ 59 + j6{,}224.77 = \underline{6{,}320.79e^{j100.00^\circ}}\ \mathrm{V}\end{aligned} \tag{5}$$

Thus the synchronizing currents calculated with the aid of Eqs. (6.50), (5), (3), and (4) are

$$\mathbf{I}_{\sigma A} = \frac{6{,}320.79e^{j100.00^\circ}}{954.74e^{j84.29^\circ}} = \underline{6.62e^{j15.71^\circ}}\ \mathrm{A} \tag{6}$$

and

$$\mathbf{I}_{\sigma B} = \frac{6{,}320.79e^{j100.00^\circ}}{2{,}849.96e^{j50.00^\circ}} = \underline{2.22e^{j50.00^\circ}}\ \mathrm{A} \tag{7}$$

Now the synchronizing powers generated by generator No. 1, designated as $P_{\sigma T1A}$ and $P_{\sigma T1B}$, respectively, are calculated with the aid of input data and Eqs. (6.50), (6), and (7) as

$$\begin{aligned}P_{\sigma T1A} &= 3 \times (18{,}200.00) \times (6.62) \times \cos(20.00^\circ - 15.71^\circ) \\ &= \underline{\underline{360{,}463.91\ \mathrm{W}}}\end{aligned} \tag{8}$$

and

$$\begin{aligned}P_{\sigma T1B} &= 3 \times (18{,}200.00) \times (2.22) \times \cos(20.00^\circ - 50.00^\circ) \\ &= \underline{\underline{104{,}871.07\ \mathrm{W}}}\end{aligned} \tag{9}$$

Similarly, the synchronizing powers seemingly absorbed by, or delivered to, generator No. 2, designated as $P_{\sigma T2A}$ and $P_{\sigma T2B}$, respectively, are calculated with the aid of input data and Eqs. (6.61), (6), and (7) as

$$\begin{aligned}P_{\sigma T2A} &= -3 \times (18{,}200.00) \times (6.62) \times \cos(0.00^\circ - 15.71^\circ) \\ &= \underline{\underline{-347{,}972.27\ \mathrm{W}}}\end{aligned} \tag{10}$$

and

$$P_{\sigma T2B} = -3 \times (18{,}200.00) \times (2.22) \times \cos(0.00° - 50.00°) = \underline{\underline{-77{,}838.19 \text{ W}}} \qquad (11)$$

The power losses in both stator windings, designated as $P_{\sigma LTA}$ and $P_{\sigma LTB}$, respectively, calculated with the aid of input data and Eqs. (6.60) [or (6.56)], (5), (6), and (7) are

$$P_{\sigma LTA} = 3 \times (6{,}320.79) \times (6.62) \times \cos(100.00° - 15.71°) = \underline{\underline{12{,}491.64 \text{ W}}} \qquad (12)$$

or

$$P_{\sigma LTA} = P_{\sigma T1A} + P_{\sigma T2A} = 360{,}463.91 - 347{,}972.27 = \underline{\underline{12{,}491.64 \text{ W}}} \qquad (13)$$

or

$$P_{\sigma LTA} = 3 \times (47.5 + 47.5) \times (6.62)^2 = \underline{\underline{12{,}489.95 \text{ W}}} \qquad (14)$$

and

$$P_{\sigma LTB} = 3 \times (6{,}320.79) \times (2.22) \times \cos(100.00° - 50.00°) = \underline{\underline{27{,}032.88 \text{ W}}} \qquad (15)$$

or

$$P_{\sigma LTB} = P_{\sigma T1B} + P_{\sigma T2B} = 104{,}871.07 - 77{,}838.19 = \underline{\underline{27{,}032.88 \text{ W}}} \qquad (16)$$

or

$$P_{\sigma LTB} = 3 \times (915.96 + 915.96) \times (2.22)^2 = \underline{\underline{27{,}085.30 \text{ W}}} \qquad (17)$$

Finally, the terminal voltages, designated as $\mathbf{V}_{L1A}$, $\mathbf{V}_{L1B}$, $\mathbf{V}_{L2A}$, and $\mathbf{V}_{L2B}$, of the generators, as calculated with the aid of input data, Eqs. (1), (2), (6), and (7), and application of Kirchhoff's voltage law to Fig. 6.14 with the synchronizing current $\mathbf{I}_\sigma$ flowing in the counterclockwise direction, are

$$\begin{aligned}\mathbf{V}_{L1A} &= \mathbf{E}_{G1} - \mathbf{Z}_{SA}\mathbf{I}_{\sigma A} = 18{,}200.00e^{j20°} - (477.37e^{j84.29°}) \times (6.62e^{j15.71°}) \\ &= 17{,}102.41 + j6{,}224.77 - (-548.80 + j3{,}112.38) \\ &= 17{,}651.21 + j3{,}112.39 = \underline{\underline{17{,}923.51e^{j10.00°} \text{ V}}}\end{aligned} \qquad (18)$$

or

$$\begin{aligned}\mathbf{V}_{L1B} &= \mathbf{E}_{G1} - \mathbf{Z}_{SB}\mathbf{I}_{\sigma B} = 18{,}200.00e^{j20°} - (1{,}424.98e^{j50°}) \times (2.22e^{j50.00°}) \\ &= 17{,}102.41 + j6{,}224.77 - (-548.80 + j3{,}112.38) \\ &= 17{,}651.21 + j3{,}112.39 = \underline{\underline{17{,}923.51e^{j10.00°} \text{ V}}}\end{aligned} \qquad (19)$$

and

$$\begin{aligned}\mathbf{V}_{L2A} &= \mathbf{E}_{G2} + \mathbf{Z}_{SA}\mathbf{I}_{\sigma A} = 18{,}200.00 + (477.37e^{j84.29°}) \times (6.62e^{j15.71°}) \\ &= 18{,}200.00 + (-548.80 + j3{,}112.38) = 17{,}651.20 + j3{,}112.38 \\ &= \underline{\underline{17{,}923.50e^{j10.00°} \text{ V}}}\end{aligned} \qquad (20)$$

or

$$\begin{aligned}\mathbf{V}_{L2B} &= \mathbf{E}_{G2} + \mathbf{Z}_{SB}\mathbf{I}_{\sigma B} = 18{,}200.00 + (1{,}424.98e^{j50.00°}) \times (2.22e^{j50.00°}) \\ &= 18{,}200.00 + (-549.33 + j3{,}115.40) = 17{,}650.67 + j3{,}115.40 \\ &= \underline{\underline{17{,}923.50e^{j10.00°} \text{ V}}}\end{aligned} \qquad (21)$$

Note that the use of higher impedance at a lower angle, or $|jX_{SB}|/R_{SB}=1.19$, Eq. (2), versus a lower impedance at a higher angle, or $|jX_{SA}|/R_{SA}=10$, Eq. (1), results in (1) reduction in synchronizing power, Eqs. (8) and (10) versus Eqs. (9) and (11), and (2) an increase in losses, Eqs. (12), (13), and (14) versus Eqs. (15), (16), and (17), despite the reduction in synchronizing current. **This is one of the reasons why the high ratio of synchronous reactance to synchronous resistance is preferred in synchronous generators regardless of its adverse effect on regulation.** Also note that the terminal voltages of the two generators remain constant regardless of the variation in the identical synchronous impedances.

The Effects of Variable Prime-Mover Output and dc Excitation on Synchronous Generators Connected in Parallel

Generally speaking, we can summarize the information of this section up to this point as follows: (1) variation of the dc excitation of a synchronous generator **does not affect** its output when other generators, connected in parallel, maintain $|\mathbf{V}_L|=k$ but results in the **production** of synchronizing current, which is undesirable because of its heating effects, (2) the synchronizing current $\mathbf{I}_\sigma$ **enhances** the power factor of the underexcited generator and **adversely affects** the power factor of the other generator, and (3) the synchronizing power produced because of the jX_S's **forces an equilibrium** to be reached by the two generators; that is, when a change in the dc excitation or in the prime-mover output of one generator is made, corrective changes must be made in the other generator in order to maintain a constant bus voltage $\mathbf{V}_L$ and constant frequency ω_E.

To understand the process of load sharing between the two generators, it is essential to recollect that the power output of any synchronous generator cannot exceed the difference between the mechanical input power (i.e., the mechanical output of the prime mover) and its own internal losses, that is, $P_{DT}=P_{INM}-P_L$. Thus since the excitation current affects the internal losses only to a minor extent, *the power output of any synchronous generator becomes* $P_{DT}=f(P_{INM})$ *and can be varied only by controlling the input to the prime mover*. This is accomplished by *manipulating the valves* controlling the steam supply to an engine- or turbine-driven generator or the water supply of a hydroelectric unit. Thus the speed versus load characteristic of each prime mover determines the portion of the load that each synchronous generator connected in parallel to an infinite bus assumes. The speed versus load characteristics, and thus the load division, can be controlled only by the *valve settings* on each prime mover. In turn, these settings are controlled by a *centrifugal force sensitive governor*.* A basic governor is a mechanical device connected to a prime-mover shaft through a gear train that uses the shaft speed to control the throttle-valve opening in such a manner as to maintain the shaft speed constant. For example, if the shaft speed increases, the governor would close the throttle valve to reduce the speed. Thus *throttle valve openings are prime-mover speed sensitive*. For stable operation and equal division of loads among synchronous generators connected in parallel to an infinite bus, the prime-mover speed versus load characteristics must be (1) dropping (i.e., have negative slopes) and (2) **identical** [i.e., the slopes of the two (all) prime movers must be identical]. If they

*The simplest type of governor available is explained here in order to bring forth the principle of operation. More sophisticated types do exist. The interested reader can refer to the literature.

have dissimilar speed versus load characteristics they will divide the load in proportion to their ratings only at the speed at which their characteristics intersect each other. At any other speed, the division of load departs from this proportion.

With these ideas in mind, the following phenomena are examined in detail: (1) how the generated voltages $\mathbf{E}_{G1}$ and $\mathbf{E}_{G2}$, the generated currents $\mathbf{I}_{G1}$ and $\mathbf{I}_{G2}$, and the reactive powers Q_{DTG1} and Q_{DTG2} vary in order to maintain the real powers P_{DTG1} and P_{DTG2}, the loading voltage $\mathbf{V}_L$, the loading current $\mathbf{I}_L$, and the frequency ω_E of the two synchronous generators constant; (2) how the generated voltages $\mathbf{E}_{G1}$ and $\mathbf{E}_{G2}$, the generated currents $\mathbf{I}_{G1}$ and $\mathbf{I}_{G2}$, and the reactive powers Q_{DTG1} and Q_{DTG2} of the two generators vary in order to maintain the real power P_{DTG1} and P_{DTG2} variable and the loading voltage $\mathbf{V}_L$, the loading current $\mathbf{I}_L$ and the frequency ω_E of the generated voltages of the two synchronous generators constant; and (3) how the speed of operation ω_S and thus the frequency ω_E of the generated voltages are maintained constant while the real powers P_{DTG1} and P_{DTG2} of the two generators vary.

If the two synchronous generators of Fig. 6.14 have (1) identical design parameters, (2) equal excitation, and (3) their prime movers adjusted so that each generator carries **half load**, that is, half the external loading current, then Eq. (6.48) is written, for $n=2$, as

$$\mathbf{I}_L = 2\mathbf{I}_{G1} = 2\mathbf{I}_{G2} = \frac{\mathbf{V}_L}{\mathbf{Z}_L} \tag{6.62}$$

For $n=2$ the phasor diagram of Fig. 6.13 is valid for describing these two generators. However, since it is already established that $|jX_S| \gg R_S$, the phasor diagram of Fig. 6.13 can be simplified if we were to let $R_{S1}|\mathbf{I}_{G1}| = R_{S2}|\mathbf{I}_{G2}| \to 0$ and still describe rather accurately the two synchronous generators. This simplified phasor diagram is exhibited in Fig. 6.20. Note that under this condition of operation, the two synchronous generators deliver equal real and reactive powers, that is, $\mathbf{S}_{DTG1} \equiv \mathbf{S}_{DTG2} = P_{DTG1} + jQ_{DTG1} = P_{DTG2} + jQ_{DTG2}$. Note, however, the phenomena that take place when the power output of each generator is maintained constant, by continuously adjusting the valve settings of the prime movers while the generated voltages $\mathbf{E}_{G1}$ and $\mathbf{E}_{G2}$ have simultaneously changed, by continuously adjusting the field (rotor-winding) dc excitations to the new values designated as $\mathbf{E}_{G1F}$ and $\mathbf{E}_{G2F}$, Fig. 6.20, in a manner such that the frequency ω_E, the terminal voltage $\mathbf{V}_L$, and the loading current $\mathbf{I}_L$ remain constant. The output phase power of generator No. 1 has changed from

$$\begin{aligned}\mathbf{S}_{DG1} = P_{DG1} + jQ_{DG1} &= |\mathbf{V}_L||\mathbf{I}_{G1}|\cos\theta_{LGG} \\ &+ j|\mathbf{V}_L||\mathbf{I}_{G1}|\sin\theta_{LGG}\end{aligned} \tag{6.63}$$

to

$$\begin{aligned}\mathbf{S}_{DG1F} = P_{DG1F} + jQ_{DG1F} &= |\mathbf{V}_L||\mathbf{I}_{G1F}|\cos\theta_{LGGF} \\ &+ j|\mathbf{V}_L||\mathbf{I}_{G1F}|\sin\theta_{LGGF}\end{aligned} \tag{6.64}$$

and that of generator No. 2 has changed from

$$\begin{aligned}\mathbf{S}_{DG2} = P_{DG2} + jQ_{DG2} &= |\mathbf{V}_L||\mathbf{I}_{G2}|\cos\theta_{LGG} \\ &+ j|\mathbf{V}_L||\mathbf{I}_{G2}|\sin\theta_{LGG}\end{aligned} \tag{6.65}$$

to

$$\begin{aligned}\mathbf{S}_{DG2F} = P_{DG2F} + jQ_{DG2F} &= |\mathbf{V}_L||\mathbf{I}_{G2F}|\cos\theta_{LDGF} \\ &+ j|\mathbf{V}_L||\mathbf{I}_{G2F}|\sin\theta_{LDGF}\end{aligned} \tag{6.66}$$

where $P_{DG1} = P_{DG1F}$ and $P_{DG2} = P_{DG2F}$.

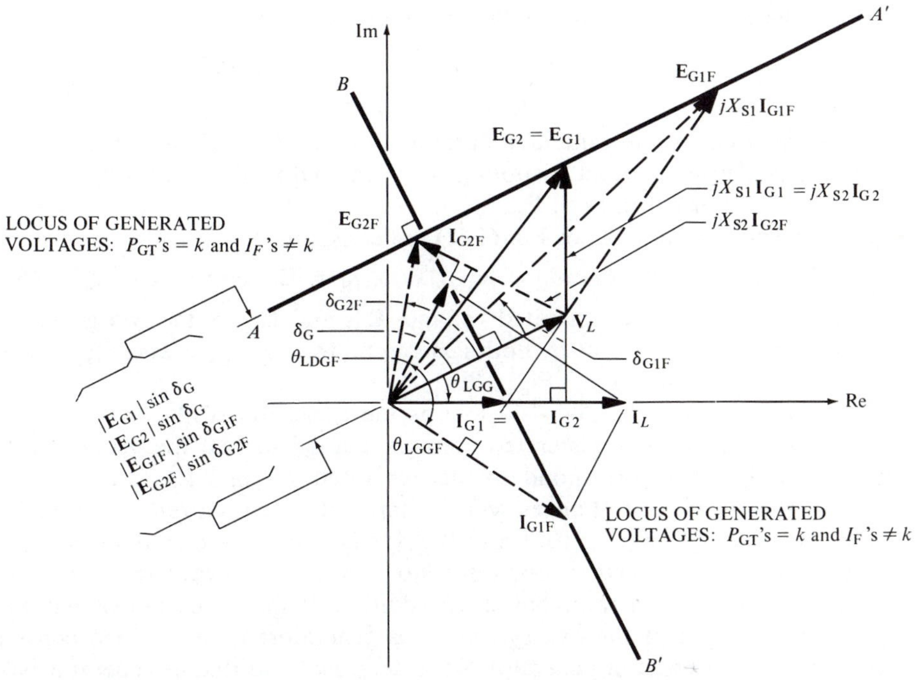

Fig. 6.20 Effects of Excitation Variation of the Generators of Fig. 6.14 (with Real Power, Frequency, and Terminal Voltage Maintained Constant) on the Generated Voltages $\mathbf{E}_{G1}$ and $\mathbf{E}_{G2}$, the Generated Currents $\mathbf{I}_{G1}$ and $\mathbf{I}_{G2}$, and Thus the Reactive Powers of Both Generators

Since, for all internal losses of the two generators discarded, the output power of each generator comes directly from its prime mover, the power angle δ_{Gj} must assume such values as to satisfy the equation

$$P_{DTGj} \equiv P_{INMj} = 3\left(\frac{|\mathbf{V}_L||\mathbf{E}_{Gj}|}{X_S}\right)\sin\delta_{Gj}, \qquad j = 1,2 \tag{6.67}$$

for any given operating condition defined by the generated voltages $\mathbf{E}_{G1}$, $\mathbf{E}_{G2}$, $\mathbf{E}_{G1F}$, and $\mathbf{E}_{G2F}$, Fig. 6.20. It should be obvious now that for the output power of the generators maintained constant, Eq. (6.67) as applied to Figs. 6.14 and 6.20 can be written as

$$\frac{P_{INM}}{3} \equiv P_D\text{'s} = \left(\frac{|\mathbf{V}_L||\mathbf{E}_{G1}|}{X_S}\right)\sin\delta_G = \left(\frac{|\mathbf{V}_L||\mathbf{E}_{G2}|}{X_S}\right)\sin\delta_G$$
$$= \left(\frac{|\mathbf{V}_L||\mathbf{E}_{G1F}|}{X_S}\right)\sin\delta_{G1F} = \left(\frac{|\mathbf{V}_L||\mathbf{E}_{G2F}|}{X_S}\right)\sin\delta_{G2F} \tag{6.68}$$

Note that since the output power from each generator is defined to be constant, Eqs. (6.63) and (6.64), and (6.65) and (6.66), yield

$$|\mathbf{V}_L||\mathbf{I}_{G1}|\cos\theta_{LGG} = |\mathbf{V}_L||\mathbf{I}_{G1F}|\cos\theta_{LGGF} = k_1 \tag{6.69}$$

and

$$|\mathbf{V}_L||\mathbf{I}_{G2}|\cos\theta_{LGG} = |\mathbf{V}_L||\mathbf{I}_{G2F}|\cos\theta_{LDGF} = k_2 \tag{6.70}$$

However, since $|\mathbf{V}_L|$ is constant, and the power developed by each generator remains unchanged at the value* of P_D's given by Eq. (6.68),

$$|\mathbf{I}_{G1}|\cos\theta_{LGG} = |\mathbf{I}_{G1F}|\cos\theta_{LGGF} = |\mathbf{I}_{G2}|\cos\theta_{LGG} = |\mathbf{I}_{G2F}|\cos\theta_{LDGF} = k_3 \tag{6.71}$$

Therefore, the locus of the generated currents $\mathbf{I}_{G1F}$ and $\mathbf{I}_{G2F}$ of the two generators, for constant real-power output by both generators, that is, P_{DTG2} and P_{DTG1}, is the straight line BB' normal to $|\mathbf{V}_L|$, Fig. 6.20.

Similarly, since $\mathbf{V}_L$ and X_S of Eq. (6.68) are constant,

$$|\mathbf{E}_{G1}|\sin\delta_G = |\mathbf{E}_{G2}|\sin\delta_G = |\mathbf{E}_{G1F}|\sin\delta_{G1F} = |\mathbf{E}_{G2F}|\sin\delta_{G2F} = k_4 \tag{6.72}$$

Therefore, the locus of the generated voltages $\mathbf{E}_{G1}$ and $\mathbf{E}_{G2}$ of the two generators, for constant real-power output by both generators, that is, P_{DTG1} and P_{DTG2}, is the straight line AA' parallel to $\mathbf{V}_L$, Fig. 6.20.

It is essential to note in Fig. 6.20 that as the excitation of the two generators changes, the corresponding generated voltages change so that the generated currents $\mathbf{I}_{G1}$ and $\mathbf{I}_{G2}$ that correspond to the generated voltages $\mathbf{E}_{G1}$ and $\mathbf{E}_{G2}$ have undergone two prominent changes when compared to the currents of Fig. 6.13. These changes are (1) $|\mathbf{I}_{G1F}|>|\mathbf{I}_{G1}|$ and $|\mathbf{I}_{G2F}|>|\mathbf{I}_{G2}|$ and (2) $\mathbf{I}_{G1F}$ is *lagging*, thus causing *demagnetizing* effects in generator No. 1, and I_{G2F} is *leading*, thus causing *magnetizing* effects in generator No. 2. In addition, it should be noted that while the real-power output of the two synchronous generators is maintained constant, the receiver-power output of generator No. 1 decreases and that of generator No. 2 increases.

It should be emphasized that if the three conditions mentioned here, that is, (1) $|\mathbf{V}_L|=k_A$ (and thus $|\mathbf{I}_L|=k_B$), (2) P_D's$=k_C$, and (3) $\omega_E=k_D$, are to hold true, the changes of $\mathbf{E}_{G1}$ to $\mathbf{E}_{G1F}$ and $\mathbf{E}_{G2}$ to $\mathbf{E}_{G2F}$ are not random but must satisfy the following constraint derived from Fig. 6.20:

$$\mathbf{E}_{G1F} = \mathbf{V}_L + jX_{S1}\mathbf{I}_{G1F} \tag{6.73}$$

and

$$\mathbf{E}_{G2F} = \mathbf{V}_L + jX_{S2}\mathbf{I}_{G2F} \tag{6.74}$$

Proper addition of Eqs. (6.73) and (6.74) and utilization of the **constraint**

$$\mathbf{I}_L = \mathbf{I}_{G1F} + \mathbf{I}_{G2F} = \frac{\mathbf{V}_L}{\mathbf{Z}_L} \tag{6.75}$$

leads to

$$\mathbf{E}_{G1F} + \mathbf{E}_{G2F} = (2\mathbf{Z}_L + jX_S)\mathbf{I}_L \equiv k_4 \tag{6.76}$$

Eq. (6.76), therefore, suggests that if $\mathbf{E}_{G2}$ is changed to $\mathbf{E}_{G2F}$ without at the same time changing $\mathbf{E}_{G1}$ to a new value as dictated by Eq. (6.76), the terminal voltage $\mathbf{V}_L$ will change to a new value, which will in turn affect the current $\mathbf{I}_L$ and thereby change the magnitude of the load, $P_L=\text{Re}(\mathbf{V}_L\mathbf{I}_L^*)=P_{INM}/3$. However, if these changes occur without a change in the **governor settings** of the prime movers, the changed load, if it decreases, will cause the speed of the two generators to increase, Fig. 6.23†; if on the other hand, the changed load increases, the speed of the two generators will decrease, Fig. 6.23. In either case, the frequency of the bus voltage

*The same conclusion can be reached if we were to compare Eq. (6.69) with Eq. (6.70) since $|\mathbf{I}_{G1}|=|\mathbf{I}_{G2}|$, Fig. 6.20.

†Fig. 6.23 is exhibited later in order to accompany the material explaining it.

will be affected and this will affect the load if the load contains inductance, capacitance, or both. What would happen, however, if the load is comprised of induction motors and others? It is suggested, therefore, that the parallel operation of two identical synchronous generators presents an interesting problem, to say the least.

In contrast with the phenomena exhibited in Fig. 6.20, if the generated voltages of the two synchronous generators $\mathbf{E}_{G1}$ and $\mathbf{E}_{G2}$ are maintained constant [by continuously adjusting the field (rotor-winding) dc excitations of the two generators independently] while it is allowed to vary (i.e., increase P_{INMG2} and decrease P_{INMG1}) the output power of each synchronous generator (by continuously adjusting the valve setting of the prime movers independently) in such a way as to maintain the frequency ω_E, the terminal voltage $\mathbf{V}_L$, and the loading current $\mathbf{I}_L$ constant, the phenomena that take place are as exhibited in Fig. 6.21. If the synchronous generators are to supply power to the infinite bus, the value of the power angle δ_{Gj} must change sufficiently so that the output of each generator

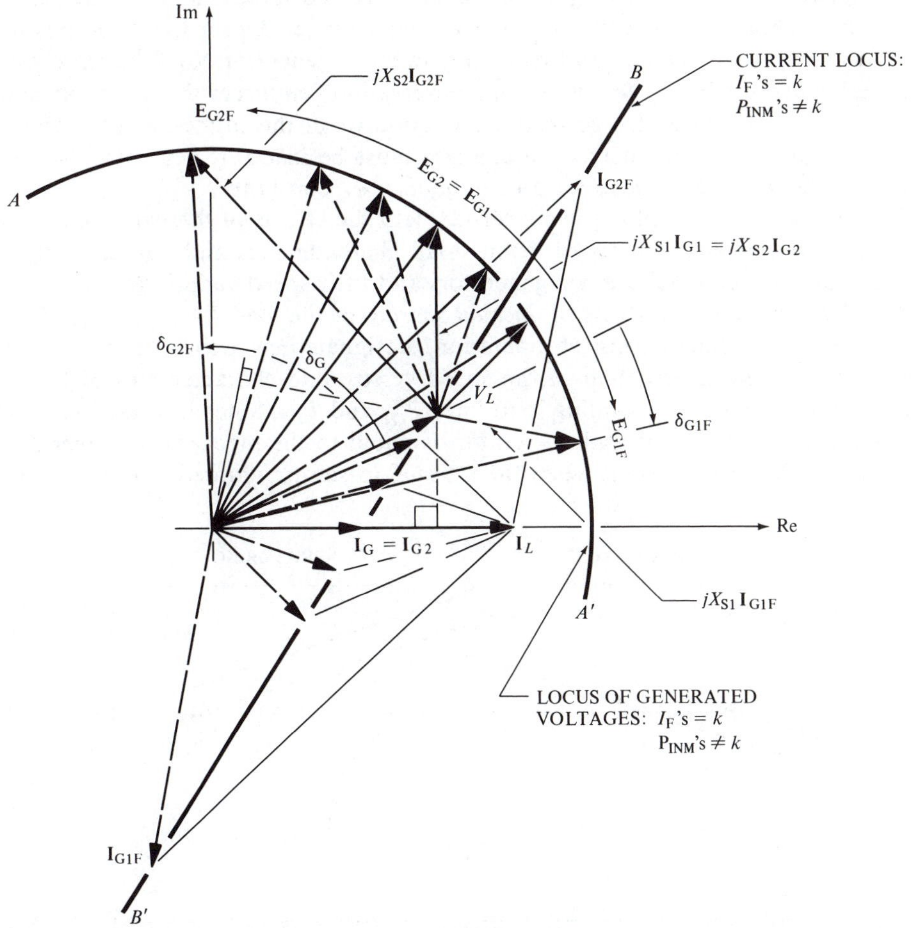

Fig. 6.21 Effects of Prime-Mover Inputs to the Generators of Fig. 6.14 (with Field Current, Frequency, and Terminal Voltage Maintained Constant) on the Generated Voltages $\mathbf{E}_{G1}$ and $\mathbf{E}_{G2}$, the Generated Currents $\mathbf{I}_{G1}$ and $\mathbf{I}_{G2}$, and Thus the Generated Real and Reactive Powers of Both Generators

matches the output of the corresponding prime mover. Thus

$$P_{\mathrm{DTG}j} \equiv P_{\mathrm{INM}j} = 3\left(\frac{|\mathbf{V}_L||\mathbf{E}_{\mathrm{G}j}|}{X_{\mathrm{S}}}\right)\sin\delta_{\mathrm{G}j}, \qquad j = 1,2 \tag{6.77}$$

Since $P_{\mathrm{INM}j}$ is varied and $|\mathbf{V}_L|$, $|\mathbf{E}_{\mathrm{G}j}|$, and X_{S} are maintained constant, only $\delta_{\mathrm{G}j}$ can change by adjusting the valve settings of the two prime movers. Thus for $\mathbf{E}_{\mathrm{G1}}$ moving clockwise and $\mathbf{E}_{\mathrm{G2}}$ moving counterclockwise from the position designated by $\mathbf{E}_{\mathrm{G1}}=\mathbf{E}_{\mathrm{G2}}$, Fig. 6.21, and both voltages maintaining constant magnitudes, that is, $|\mathbf{E}_{\mathrm{G1}}|=|\mathbf{E}_{\mathrm{G2}}|$, it is obvious that the locus of the generated voltages is the circle sector AA' of radius $|\mathbf{E}_{\mathrm{G1}}|=|\mathbf{E}_{\mathrm{G2}}|$ with center at the origin, Fig. 6.21. A plot of $\mathbf{E}_{\mathrm{G1}}$ for three values of the angle δ_{G}, that is, $|\delta_{\mathrm{G}}|>|\delta_{\mathrm{G1}}|>|\delta_{\mathrm{G1F}}|$, and for $\mathbf{E}_{\mathrm{G2}}$ also for three values of the angle δ_{G}, that is, $|\delta_{\mathrm{G}}|<|\delta_{\mathrm{G2}}|<|\delta_{\mathrm{G2F}}|$, yields that the locus of the two currents $\mathbf{I}_{\mathrm{G1}}$ and $\mathbf{I}_{\mathrm{G2}}$ is the straight line BB', Fig. 6.21.

Inspection of Fig. 6.21 reveals that as $\mathbf{E}_{\mathrm{G2}}$ moves counterclockwise, its corresponding **current $\mathbf{I}_{\mathrm{G2}}$ increases in magnitude continuously as it becomes more leading.** However, as $\mathbf{E}_{\mathrm{G1}}$ moves clockwise, its corresponding **current $\mathbf{I}_{\mathrm{G1}}$ becomes more lagging but its magnitude goes through a minimum value and then increases.** In addition, it should be pointed out that as the voltages depart from the position $\mathbf{E}_{\mathrm{G2}}=\mathbf{E}_{\mathrm{G1}}$, Fig. 6.21, the real and reactive powers of generator No. 2 **increase** while those of generator No. 1 **decrease**. This observation reinforces the assertion made earlier; that is, when a change in the excitation or in the prime-mover output of generator No. 2 is made, corrective changes must be made by generator No. 1 to maintain the terminal voltage $\mathbf{V}_L$ and frequency ω_{E} constant.

The discussion that follows is concerned with the effects of the valve settings of the prime movers on their speed versus load characteristics and consequently on the load division between the two generators and their speed variation.

Fig. 6.22 exhibits the effects of *unequal sharing* of the load $P=P_{\mathrm{G1}}+P_{\mathrm{G2}}, P_{\mathrm{G1}}>P_{\mathrm{G2}}$. The equilibrium points of operation, designated as points A and B, are determined by the intersection of the speed versus load characteristics of the two generators and the corresponding load lines P_{G1} and P_{G2}. Since the electrical load, that is, real power P_{DT} of each generator, is equal to the mechanical power P_{INM} supplied by the prime mover, when the internal losses are ignored, the load can be

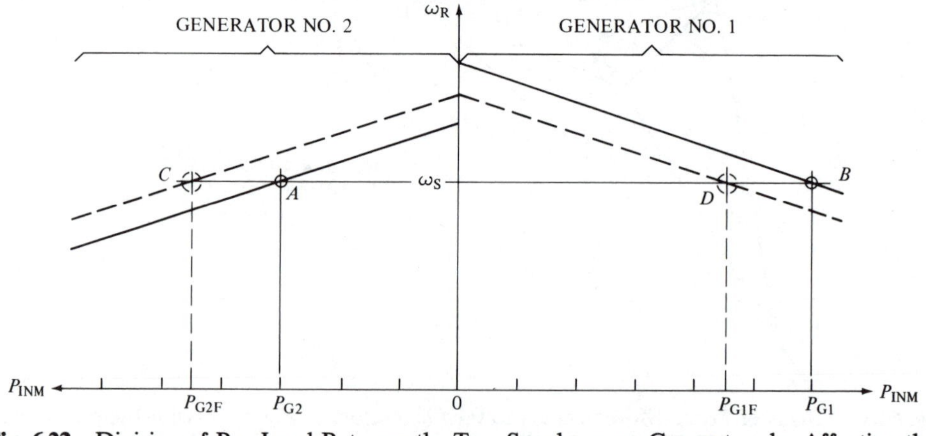

Fig. 6.22 Division of Bus Load Between the Two Synchronous Generators by Affecting the Speed vs. Load Characteristics of the Prime Movers Through Changes in Governor Settings

shared equally between the two generators by **decreasing** the governor setting, that is, **closing** the throttle, of prime mover No. 1 and **increasing** the governor setting, that is, **opening** the throttle, of prime mover No. 2. The effects are to lower the speed versus load characteristic of generator No. 1 and raise that of generator No. 2. The new characteristics are shown by the dashed lines of Fig. 6.22. The new operating points, designated as points C and D, are determined by the intersection of the extended line AB and the newly acquired speed versus load characteristics. Note that the two generators continue to run at the same speed ω_S and thus produce voltages of frequency ω_E. The two ω's are related by Eq. (5.15), that is, $\omega_S = \omega_E/(P/2)$.

Fig. 6.23 exhibits the effects of increasing or decreasing the load on the two generators, which are assumed to be *sharing equally* the load $P = P_{G1} + P_{G2}, P_{G1} = P_{G2}$. The equilibrium points of operation, designated as points E and F, are determined by the intersection of the speed versus load characteristics of the two generators and the corresponding load lines $P_{G1} = P_{G2}$. If the governor settings are not changed, an extra load or a deficiency of load is shared equally by the two generators. The operating points, designated as points G and H, and I and J, respectively, are determined by the intersection of the load lines P_{G1H} and P_{G2H}, and P_{G1L} and P_{G2L}, and the speed versus load characteristics, Fig. 6.23. However, for these points of operation, the speed of operation of the generators ω_S, and thus the frequency of their voltages ω_E, either **decreases** (intersection of line GH with the speed axis, i.e., ω_L) or **increases** (intersection of line IJ with the speed axis, i.e., ω_H). Since industrial applications demand voltages of **absolutely constant frequency**, the frequency of the bus voltage, supplied by the parallel generators, must be maintained constant. This is accomplished by use of **automatic frequency regulators**. Their function is to control the governor settings to maintain the speed of operation ω_S, and thus the frequency of the generated voltages ω_E, constant. In Fig. 6.23 the governor settings are either automatically **increased**, that is, the throttles are **opened** to shift the characteristics upward, or are automatically **decreased**, that is, the throttles are **closed** to shift the characteristics downward to secure constant frequency. The operating points are determined by the intersection of (1) the

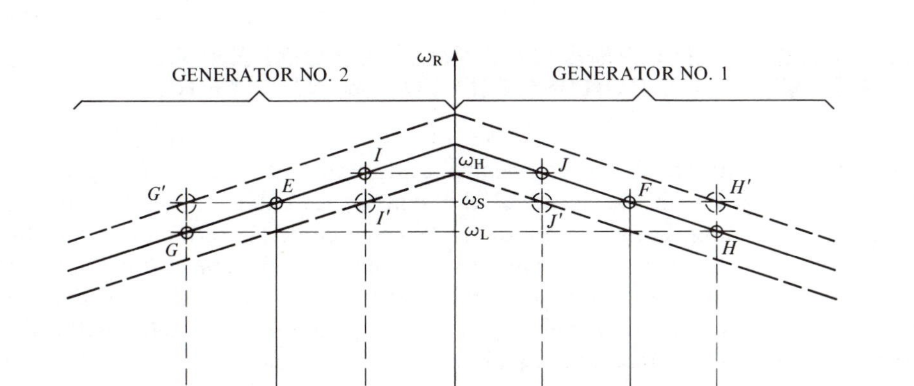

Fig. 6.23 Effects of Bus Load Variation on the Speed of the Prime Movers and Thus on the Governor Settings and Thereby the Speed vs. Load Characteristics of the Prime Movers

extended line EF and the newly acquired speed versus load characteristics, designated as the dashed lines, that are **shifted upward for an increased load** or (2) the line EF and the newly acquired speed versus load characteristics, designated as long-short-long dashed lines, that are **shifted downward for a reduced load**. The corresponding points of operation are designated as points G' and H', and I' and J', respectively. In order to assure the constancy of frequency within a tolerance of $\Delta f_{\mathrm{E}} = \pm 0.05$ Hz, an electric *clock* supplied from the bus whose frequency is to be controlled is compared with a *standard clock* and the frequency is regulated by automatic control schemes so that practically no discrepancy exists between the two clocks over long intervals of time. Thus when an isolated bus is fed from several generators connected in parallel, the usual practice is to run all generators **except one** with fixed governor settings and **put one generator on automatic frequency control** to take care of minor variations of the load and maintain the frequency of the bus voltage constant. On the other hand, in the case of interconnected power systems, which give rise to a quasi-infinite bus at each generating station of the grid network, **a few stations are jointly assigned the task of frequency control** to take care of minor variations of the load and maintain the frequency of the bus voltage constant. In order to assure that the voltage of a given bus is within prescribed limits, **automatic voltage regulators** are connected to the field circuits of the synchronous generators. However, regulating devices such as regulating transformers are also necessary to control the system voltages and reactive-power distribution.

It is clear, then, that the only way to raise or lower the terminal or bus voltage $\mathbf{V}_L$ without affecting the power output and the power factors of the individual synchronous generators connected in parallel to an infinite bus is to increase or decrease the generated voltage of all parallel generators simultaneously. In contrast, however, the terminal or bus voltage of the isolated or quasi-infinite bus can be changed by changing the excitations of one of the synchronous generators or the load connected to such a bus. Therefore, the changes in excitations of generators connected to such buses affect the terminal voltage as well as the reactive-power distribution.

6.5 DYNAMIC ANALYSIS OF SYNCHRONOUS ENERGY CONVERTERS CONNECTED TO AN INFINITE BUS

The dynamic analysis of power systems deals with the effects of large sudden system disturbances on the speed of the synchronous devices whether they are generators or motors. Large system disturbances include (1) sudden application or removal of loads, (2) sudden switching of lines, or (3) line faults. Successful operation of such devices demands equality of the mechanical speed of the rotors ω_{R} and the speed of the stator fields ω_{S}. If such equality is disturbed, synchronizing forces that tend to restore this equality of speeds come into play and cause a behavior of the system that can only be described by differential equations of associated electrical and mechanical variables. For example, if a large load is suddenly applied to the shaft of a synchronous motor connected to an infinitely strong network, that is, to an infinite bus, whose $\mathbf{E}_{\mathrm{T}}$ and ω_{E} are maintained constant, and $\mathbf{E}_{\mathrm{G}}$ remains invariant, the motor must slow down, at least momentarily, in order that the power angle δ (i.e., $|\delta_{\mathrm{M}}|$) may assume the increased value required by the motor to develop the necessary power in order to overcome the added load. In fact, until the new torque angle $|\delta_{MSS}|$, Fig. 5.48, is reached, an appreciable portion of the energy supplied to the load comes from energy stored in the rotating mass of the rotor as it slows down. When the newly acquired value of the torque angle $|\delta_{MSS}|$, Fig. 5.48, is first reached, equilibrium is not yet attained because the mechanical speed of the rotor ω_R is then below the synchronous speed ω_S, that is, $\omega_R < \omega_S$, because of the reduction in its kinetic energy. This causes a further increase in the torque angle $|\delta_M|$. As $|\delta_M|$ increases beyond $|\delta_{MSS}|$, the power developed exceeds the loading power and the motor accelerates until the rotor obtains synchronous speed at an angle $|\delta_{M(MAX)}|$, after which further acceleration causes the torque angle $|\delta_M|$ to decrease. In the absence of damping, or losses, the motor speed fluctuates with the torque angle oscillating about $|\delta_{MSS}|$ in a manner such that $|\delta_{M(MIN)}| < |\delta_M| < |\delta_{M(MAX)}|$, unless the suddenly applied load is so excessive as to cause $|\delta_M|$ to increase indefinitely, in which case both the (1) steady state stability limit and (2) **transient stability limit*** (i.e., the definite upper limit to the load which the motor will carry without pulling out of synchronism) are exceeded. In real devices damping, caused by **damping** or **amortisseur windings**, Fig. 5.22, causes the oscillations to die out gradually and the torque angle reaches its final value of $|\delta_{MSS}|$ at which the torque developed by the motor, designated† as $\tau_{GT}(t)$, matches the opposing shaft loading torque. Shaft loading torque is comprised of (1) the shaft torque, designated as $\tau_{SH}(t)$, (2) the torque due to the moment of inertia J, designated as $\tau_J(t)$ and (3) the torque due to the coefficient of friction B, designated as $\tau_B(t)$. In addition to the torques mentioned, there is one more torque that affects the motion of the rotor of the synchronous motor as well as that of the synchronous generator. This torque is caused by currents circulating in the **damping** or **amortisseur windings** of the rotor and is designated as $\tau_D(t)$. This torque *aids* (i.e., $\tau_D(t) < 0$) rotation when $\omega_R < \omega_S$ and *inhibits* (i.e., $\tau_D(t) > 0$) rotation when $\omega_R > \omega_S$.

The relative directions of these torques are exhibited in Fig. 6.24. These torques are related by

$$\tau_{GT}(t) = \tau_J(t) + \tau_B(t) + \tau_D(t) + \tau_{SH}(t) \tag{6.78}$$

where (1) $\tau_{GT}(t)$ is the developed electromagnetic torque by the motor; (2) $\tau_J(t) = Jd^2\theta_R(t)/dt^2, \tau_B(t) = Bd\theta_R(t)/dt$, are the loading torques due to the moment of inertia $J(\text{kg}\cdot\text{m}^2)$ and the coefficient of friction $B(\text{N}\cdot\text{m/rad/s})$ respectively; and (3) $\tau_{SH}(t)$ is the shaft torque.

Use of these ideas enables us to write Eq. (6.78) as

$$J\frac{d^2}{dt^2}\theta_R(t) + B\frac{d}{dt}\theta_R(t) + \tau_D(t) + \tau_{SH}(t) = \tau_{GT}(t) \tag{6.79}$$

Although Eq. (6.79) as derived describes the dynamic behavior of the synchronous motor excited from an infinite bus, that is, an infinitely strong network, the same equation can describe the dynamic behavior of the synchronous generator driven by a prime mover if the positions of the torques $\tau_{SH}(t)$ and $\tau_{GT}(t)$ in Eq. (6.76) are interchanged. The concept of the interchange can be comprehended from

*A different definition of the **transient stability limit** in terms of the maximum value of the power angle δ and energy stored and given up by the rotor mass is given later in this section.

†Since, as a rule, lowercase letters always designate time-varying quantities, "(t)" is omitted in the drawings.

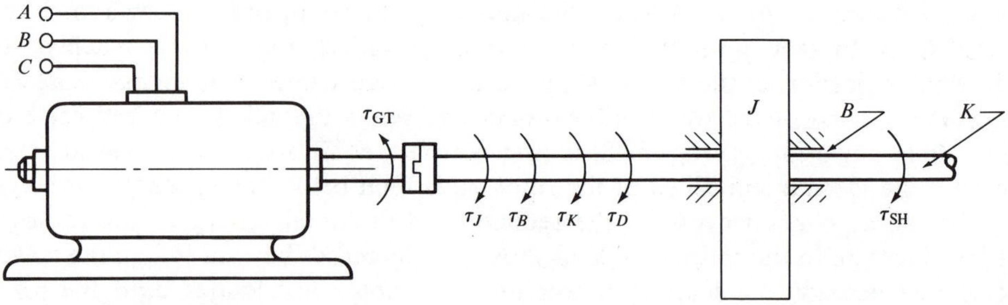

Fig. 6.24 Torque Relationship in a Round-Rotor, Balanced, *P*-Pole, Three-Phase, Synchronous Motor

the close examination of Figs. 6.25(b) and 6.25(a). Thus with the positions of the torques $\tau_{GT}(t)$ and $\tau_{SH}(t)$ in Eq. (6.79) interchanged, the describing equation of the synchronous generator is written as

$$J\frac{d^2}{dt^2}\theta_R(t) + B\frac{d}{dt}\theta_R(t) + \tau_D(t) + \tau_{GT}(t) = \tau_{SH}(t) \tag{6.80}$$

Premultiplication of Eq. (6.80) by the rotor speed $\omega_R(t)$ yields the describing equation of the synchronous generator in terms of power. [Note that $\tau(t)\omega_R(t) = p(t)$.] Thus

$$\omega_R(t)J\frac{d^2}{dt^2}\theta_R(t) + \omega_R(t)B\frac{d}{dt}\theta_R(t) + \omega_R(t)\tau_D(t) + \omega_R(t)\tau_{GT}(t) = \omega_R(t)\tau_{SH}(t) \tag{6.81}$$

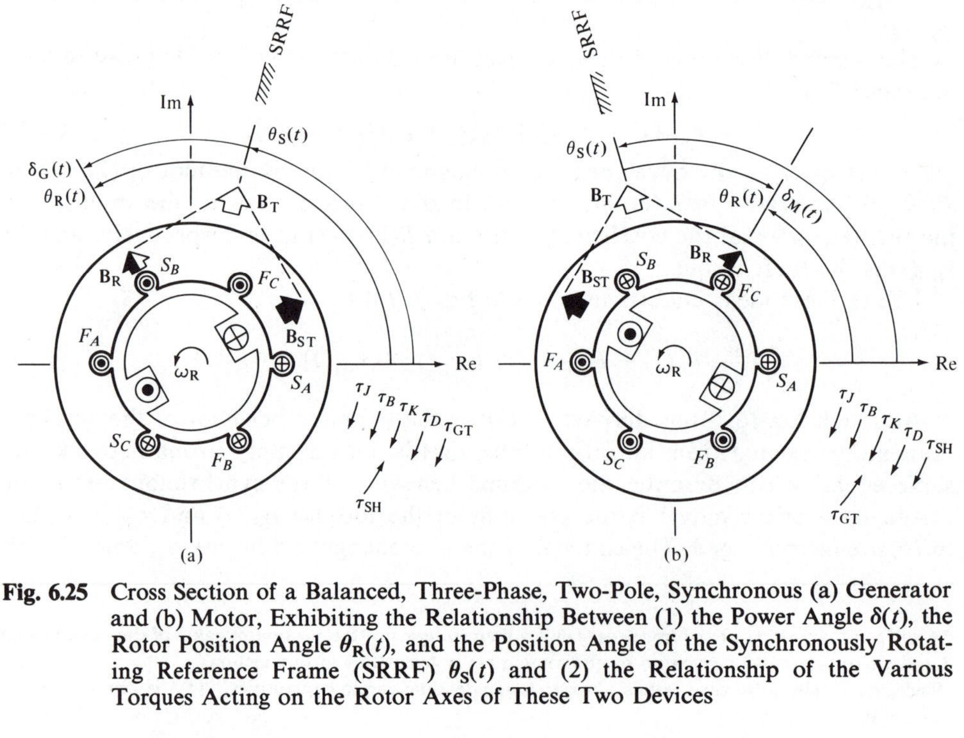

Fig. 6.25 Cross Section of a Balanced, Three-Phase, Two-Pole, Synchronous (a) Generator and (b) Motor, Exhibiting the Relationship Between (1) the Power Angle $\delta(t)$, the Rotor Position Angle $\theta_R(t)$, and the Position Angle of the Synchronously Rotating Reference Frame (SRRF) $\theta_S(t)$ and (2) the Relationship of the Various Torques Acting on the Rotor Axes of These Two Devices

If for the synchronous generator, Fig. 6.25(a), we were to express the angular displacement of the rotor $\theta_R(t)$ as a function of the angular displacement $\theta_S(t)$ of the total field $\mathbf{B}_T$, and thus of the generated voltages rotating at synchronous speed ω_S (i.e., the frequency of the terminal voltages of the synchronous generator is locked on the frequency of the infinitely large network or infinite bus, designated by the synchronously rotating reference frame, abbreviated as SRRF, Fig. 6.25), and the power angle $\delta_G(t)$, Fig. 6.25(a), we would be able to write

$$\theta_R(t) = \theta_S(t) + \delta_G(t) \tag{6.82}$$

Similarly, from Fig. 6.25(b) we would be able to write

$$\theta_R(t) = \theta_S(t) + \delta_M(t) \tag{6.83}$$

However, we may generalize these two equations into Eq. (6.84),

$$\theta_R(t) = \theta_S(t) + \delta_R(t) \tag{6.84}$$

and thus be able to describe both the synchronous generator if $0<\delta_R\leqslant\pi/2$ or the synchronous motor if $-\pi/2\leqslant\delta_R<0$. Taking the first and second derivatives of Eq. (6.84) and keeping in mind that (1) **the synchronous speed $\omega_S \equiv d\theta_S(t)/dt$ is constant** and (2) $p(t)=\omega_R(t)\tau(t)$, we can write

$$\frac{d}{dt}\theta_R(t) = \frac{d}{dt}\theta_S(t) + \frac{d}{dt}\delta_R(t) = \omega_S + \frac{d}{dt}\delta_R(t) \tag{6.85}$$

$$\frac{d^2}{dt^2}\theta_R(t) = \frac{d^2}{dt^2}\delta_R(t) \tag{6.86}$$

$$p_D(t) = \omega_R(t)\tau_D(t) \tag{6.87}$$

$$p_{GT}(t) = \omega_R(t)\tau_{GT}(t) \tag{6.88}$$

and

$$p_{SH}(t) = \omega_R(t)\tau_{SH}(t) \tag{6.89}$$

Substitution of Eqs. (6.86), (6.85), (6.84), (6.87), (6.88), and (6.89) into Eq. (6.81) yields

$$\omega_R(t)J\frac{d^2}{dt^2}\delta_R(t) + \omega_R(t)B\left[\omega_S + \frac{d}{dt}\delta_R(t)\right] + p_D(t) + p_{GT}(t) = p_{SH}(t) \tag{6.90}$$

In power systems analysis and design, the design parameter B is very small, and thus its power contribution to the left-hand side of Eq. (6.90) is negligible when compared to the rest of the terms. Therefore, we can let $p_B(t)=0$ and define $J\omega_R(t)$ as $M(t)\equiv J\omega_R(t)$ and write Eq. (6.90) as

$$M(t)\frac{d^2}{dt^2}\delta_R(t) + p_D(t) + p_{GT}(t) = p_{SH}(t) \tag{6.91}$$

where $M(t)$ **is defined as the angular momentum**, whose MKS unit is watts per unit acceleration, that is, watts per mechanical radian per second per second or joules seconds per mechanical radian. This unit asserts that $\delta_R(t)$ is measured in mechanical radians. Usually, however, $\delta(t)$ is measured in electrical degrees. If such is the case, M **firstly, must be multiplied by the factor $2\pi/360$ rad/degree* and be expressed in watts per mechanical degree per second per second** or **joules seconds per mechanical degree and, secondly, the mechanical degrees must be converted to**

***Remember that 2π rad = 360 degrees → 1 degree = $2\pi/360$ rad**

electrical degrees as

$$\delta_R = \frac{\delta_E}{(P/2)} \tag{6.92}$$

The angular momentum $M(t)$ is not strictly constant because the rotor speed $\omega_R(t)$ of Eq. (5.258), $M(t) \equiv J\omega_R(t)$, is not constant but varies somewhat during the oscillations that follow a disturbance. However, in practical cases the change in rotor speed $\omega_R(t)$, before synchronism is lost, is so small when compared to synchronous speed ω_S that very little error is introduced to Eq. (6.91) by allowing $\omega_R(t) \to \omega_S$ in the first term. In addition, utilization of Eqs. (6.92) and (4.23) enables the writing of the first term of Eq. (6.91) as

$$\begin{aligned} M(t)\frac{d^2}{dt^2}\delta_R(t) &= J\omega_R(t)\frac{d^2}{dt^2}\left[\frac{\delta_E(t)}{P/2}\right] = J\omega_S\left(\frac{2}{P}\right)\frac{d^2}{dt^2}\delta_E(t) \\ &= J\left(\frac{2}{P}\right) \times \left(\frac{2\pi n_S}{60}\right)\frac{d^2}{dt^2}\delta_E(t) \equiv M\frac{d^2}{dt^2}\delta_E(t) \end{aligned} \tag{6.93}$$

Thus $M(t)$, defined in Eq. (6.91), is replaced by $M \equiv (J \times 2 \times 2\pi n_S)/(P \times 60)$, **the value of the angular momentum at synchronous speed**, in Eq. (6.93). Note that $M \neq M(t)$. This value of M is defined as the **momentum constant** of the synchronous generator (or motor). Substitution of $M d^2\delta_E(t)/dt^2$ for $M(t)d^2\delta_R(t)/dt^2$ in Eq. (6.91), as in Eq. (6.93), allows us to write the describing equation of the synchronous generator in terms of the electrical angle $\delta_E(t)$ and its derivatives as

$$M\frac{d^2}{dt^2}\delta_E(t) + p_D(t) + p_{GT}(t) = p_{SH}(t) \tag{6.94}$$

Eq. (6.94) henceforth is referred to as the **swing equation** for the synchronous generator.

From the theory of induction devices, it should be obvious that currents exist in the **damping** or **amortisseur windings** only when $\omega_R(t) \neq \omega_S$. Thus the damping power $p_D(t)$ is present as long as $\omega_R(t) \neq \omega_S$, that is, as long as $d\delta_E(t)/dt \neq 0$. Therefore, as a first approximation, the damping power is considered to vary linearly with $d\delta_E(t)/dt$, written as

$$p_D(t) = P_D\frac{d}{dt}\delta_E(t) \tag{6.95}$$

where P_D **is the damping power per unit speed**, that is, watts per electrical radian per second or watts per electrical degree per second.

The last term of the left-hand side of Eq. (6.94) depends on the nature of the network to which the synchronous generator is connected. If the generator is connected to an infinitely large network (infinite bus), such as in Fig. 6.1(a), and the synchronous generator (or motor) has a smooth rotor, then $p_{GT}(t)$ is written with the aid of Eq. (6.1) as

$$p_{GT}(t) = 3\left(\frac{|\mathbf{V}_{ST}||\mathbf{E}_G|}{X_S}\right)\sin\delta_E(t) \tag{6.96}$$

Elimination of the subscript "E" from the power angle $\delta_E(t)$, Eqs. (6.94) through (6.96), but always keeping in mind that $\delta(t)$ *is a measure of the power angle in electrical degrees, and therefore* $P/2$ *is accounted for*, and substitution of Eqs. (6.95) and (6.96) into Eq. (6.94) yield the final form of the swing equation for a balanced, three-phase, P-pole, smooth-rotor, synchronous generator connected to an in-

Plate 6.8 Power Circuit Breakers Connected in the Power Grid. (Courtesy of General Electric Co., Schenectady, N.Y.)

finitely large network, that is, an infinite bus, written as

$$M\frac{d^2}{dt^2}\delta(t) + P_D\frac{d}{dt}\delta(t) + 3\left(\frac{|\mathbf{V}_{ST}||\mathbf{E}_G|}{X_S}\right)\sin\delta(t) = p_{SH}(t) \tag{6.97}$$

where (1) **positive values of $\delta(t)$ designate generator action,** (2) **positive values of $p_{SH}(t)$ designate mechanical power input to the rotor shaft from the prime mover,** (3) **positive values of $d\delta(t)/dt$ designate speeds above the synchronous speed*** ω_S, **and** (4) **positive values of $d^2\delta(t)/dt^2$ designate acceleration of the rotor.** However, a reverse convention may be employed and Eq. (6.97) may describe the dynamic behavior of a balanced, three-phase, P-pole, smooth-rotor, synchronous motor connected to an infinitely large network, that is, an infinite bus. For such convention the following definitions apply in conjunction with Eq. (6.97): (1) **positive values of $\delta(t)$ designate motor action,** (2) **positive values of $p_{SH}(t)$ designate mechanical power output from the shaft,** (3) **positive values of $d\delta(t)/dt$ designate speeds below the synchronous speed ω_S, and** (4) **positive values of $d^2\delta(t)/dt^2$ designate deceleration of the rotor.**

Eq. (6.97) is a second-order nonlinear differential equation with constant coefficients and nonzero forcing function. A variety of simulation techniques may be used to solve this equation for $\delta(t)$. **The plot of δ versus t is known as the swing**

*This becomes obvious when one differentiates Eq. (6.84) and examines $d\delta_R(t)/dt = \omega_R(t) - \omega_S$.

curve of the given synchronous device and reveals information relevant to the stability, instability, or oscillatory operation of the given device.

The method to be used for the solution of Eq. (6.97) depends on the nature of the oscillations of the system under study. If the oscillations are (or assumed to be) small, the **linear analysis** is used. However, if the oscillations are (or are assumed to be) large, the **nonlinear analysis** is used.

Although either of these methods can be applied to systems of equations of the type of Eq. (6.97), the methods are utilized here for the solution of one equation, that is, one synchronous device connected to an infinitely large network (i.e., an infinite bus).

Linear Analysis

Analytically, Eq. (6.94) can be solved by *linearizing* the last term of the left-hand side of Eq. (6.97). Linearization is based on the assumption that the variation of the torque angle $\delta(t)$ is small. Then either we replace this term by a first-order equation through the operating point of the power versus power angle curve, Figs. 5.35 or 5.42 or 6.1(b), or when $-\pi/6<\delta<\pi/6$ electrical radians, we replace* $\sin\delta$ by δ in Eq. (6.97), that is, $(3|\mathbf{V}_{\mathrm{ST}}||\mathbf{E}_{\mathrm{G}}|/X_{\mathrm{S}})\sin\delta \rightarrow P_{\mathrm{G}}\delta$, where $P_{\mathrm{G}}=3|\mathbf{V}_{\mathrm{ST}}||\mathbf{E}_{\mathrm{G}}|/X_{\mathrm{S}}$ is the slope of the power versus power angle curve, evaluated at the origin.

The substitution of $\sin\delta$ by δ allows writing Eq. (6.97) as

$$M\frac{d^2}{dt^2}\delta(t) + P_{\mathrm{D}}\frac{d}{dt}\delta(t) + 3\left(\frac{|\mathbf{V}_{\mathrm{ST}}||\mathbf{E}_{\mathrm{G}}|}{X_{\mathrm{S}}}\right)\delta(t) = p_{\mathrm{SH}}(t) \qquad (6.98)$$

Division of Eq. (6.98) by M yields

$$\frac{d^2}{dt^2}\delta(t) + \left(\frac{P_{\mathrm{D}}}{M}\right)\frac{d}{dt}\delta(t) + 3\left(\frac{|\mathbf{V}_{\mathrm{ST}}||\mathbf{E}_{\mathrm{G}}|}{MX_{\mathrm{S}}}\right)\delta(t) = \frac{p_{\mathrm{SH}}(t)}{M} \qquad (6.99)$$

Comparison of Eq. (6.99) to Eq. (5.18) of reference [13] or Eq. (6.93) of reference [29], rewritten here for convenience as

$$\frac{d^2}{dt^2}\delta(t) + 2\zeta\omega_0\frac{d}{dt}\delta(t) + \omega_0^2\delta(t) = p(t) \qquad (6.100)$$

yields the following two pertinent equations:

$$\omega_0^2 = 3\left(\frac{|\mathbf{V}_{\mathrm{ST}}||\mathbf{E}_{\mathrm{G}}|}{MX_{\mathrm{S}}}\right) \qquad (6.101)$$

and

$$2\zeta\omega_0 = \frac{P_{\mathrm{D}}}{M} \qquad (6.102)$$

Solution of Eq. (6.101) for the **natural**, or **undamped, radian frequency of oscillation** ω_0 (radians/second) yields

$$\omega_0 = \left[3\left(\frac{|\mathbf{V}_{\mathrm{ST}}||\mathbf{E}_{\mathrm{G}}|}{MX_{\mathrm{S}}}\right)\right]^{1/2} \qquad (6.103)$$

Now substitution for ω_0 from Eq. (6.103) into Eq. (6.102) and solution of the

*$\sin\delta = \delta - \dfrac{\delta^3}{3!} + \dfrac{\delta^5}{5!} - \cdots + (-1)^{n+1}\left[\dfrac{\delta^{2n-1}}{(2n-1)!}\right] + \cdots$

Plate 6.9 Oil-immersed Equipment: (a) Short-Circuit, Current-Limiting Reactor Connected in Series with (b) a Power Circuit Breaker. (Courtesy of General Electric Co., Schenectady, N.Y.)

resulting equation for the **damping ratio** ζ (dimensionless) yield

$$\zeta = \frac{P_D}{2\omega_0 M} = \frac{X_S{}^{1/2} P_D}{(12M|\mathbf{V}_{ST}||\mathbf{E}_G|)^{1/2}} \tag{6.104}$$

Utilization of the theory of differential equations enables us to write the solution of Eq. (6.99) for the power angle $\delta(t)$ as a function of (1) $\delta_{ss}(t)$, **the steady state response**, and (2) $\delta_{tr}(t)$, **the transient response of the system**, as

$$\delta(t) = \delta_{ss}(t) + \delta_{tr}(t) \tag{6.105}$$

where

$$\delta_{ss}(t) \equiv g\left[\frac{p_{SH}(t)}{M}\right] \tag{6.106}$$

and

$$\delta_{tr}(t) \equiv Ke^{st} \tag{6.107}$$

with the constants K and s of Eq. (6.107), and the function $g[\ ]$ of Eq. (6.106) still to be evaluated. Their evaluation as explained in Appendix C necessitates the knowledge of the initial conditions of the differential equation.

Explicit expressions for $\delta_{ss}(t)$ and $\delta_{tr}(t)$ can be found by (1) substituting directly $\delta_{ss}(t)$ and its derivatives into Eq. (6.99) after allowing only $\delta(t) \rightarrow \delta_{ss}(t)$ and (2) substituting directly $\delta_{tr}(t)$ and its derivatives into Eq. (6.99) after allowing $\delta(t) \rightarrow \delta_{tr}(t)$ and $[p_{SH}(t)/M] \rightarrow 0$.

EVALUATION OF $\delta_{ss}(t)$

Taking the first and second derivatives of Eq. (6.106), after assuming that $p_{SH}(t)$ is constant, and thus defining $g[p_{SH}(t)/M]$ as Δ, yields

$$\delta_{ss}(t) = g\left[\frac{p_{SH}(t)}{M}\right] \equiv \Delta \tag{6.108}$$

$$\frac{d}{dt}\delta_{ss}(t) = 0 \tag{6.109}$$

and

$$\frac{d^2}{dt^2}\delta_{ss}(t) = 0 \tag{6.110}$$

Substitution of Eqs. (6.108), (6.109), and (6.110) into Eq. (6.99), after letting $\delta(t) \rightarrow \delta_{ss}(t)$, yields

$$0 + \left(\frac{P_D}{M}\right)(0) + 3\left(\frac{|\mathbf{V}_{ST}||\mathbf{E}_G|}{MX_S}\right)(\Delta) = \frac{p_{SH}(t)}{M} \tag{6.111}$$

Solution of Eq. (6.111) for Δ, and thus for $\delta_{ss}(t)$, yields

$$\Delta = \frac{p_{SH}(t)X_S}{3|\mathbf{V}_{ST}||\mathbf{E}_G|} \equiv \delta_{ss}(t) \tag{6.112}$$

It should be obvious that the case when $p_{SH}(t)$ is constant is the simplest case possible. As $p_{SH}(t)$ becomes a more and more complex function of time, the evaluation of $\delta_{ss}(t)$ becomes more and more tedious but not impossible. It should be noted, however, that one should become extremely competent in the field of differential equations. This subject is well documented in reference [23].

EVALUATION OF $\delta_{tr}(t)$

Taking the first and second derivatives of Eq. (6.107) yields

$$\frac{d}{dt}\delta_{tr}(t) = sKe^{st} \tag{6.113}$$

and

$$\frac{d^2}{dt^2}\delta_{tr}(t) = s^2Ke^{st} \tag{6.114}$$

Substitution of Eqs. (6.114), (6.113), and (6.107) into Eq. (6.99), after letting $\delta(t) \rightarrow \delta_{tr}(t)$ and $[p_{SH}(t)/M] \rightarrow 0$, and extracting the common factor Ke^{st}, yields

$$\left[s^2 + \left(\frac{P_D}{M}\right)s + 3\left(\frac{|\mathbf{E}_T||\mathbf{E}_G|}{MX_S}\right)\right]Ke^{st} = 0 \tag{6.115}$$

Since $\delta_{tr}(t)$ is chosen to be nonzero, Eq. (6.107), only the brackets of Eq. (6.115) can be equal to zero. This idea enables us to write the characteristic equation* of the synchronous device under study as

$$s^2 + \left(\frac{P_D}{M}\right)s + 3\left(\frac{|\mathbf{V}_{ST}||\mathbf{E}_G|}{MX_S}\right) = 0 \tag{6.116}$$

*Please read Appendix C if you are not familiar with the theory of differential equations.

Utilization of the well-known techniques for the solution of the quadratic equation in s yields the two roots of the characteristic equation as

$$s_{1,2} = \frac{1}{2}\left\{-\frac{P_{\mathrm{D}}}{M} \pm \left[\left(\frac{P_{\mathrm{D}}}{M}\right)^2 - 4\times 3\left(\frac{|\mathbf{V}_{\mathrm{ST}}||\mathbf{E}_{\mathrm{G}}|}{MX_{\mathrm{S}}}\right)\right]^{1/2}\right\}$$

$$= -\frac{P_{\mathrm{D}}}{2M} \pm j\left\{\left[\sqrt{3\left(\frac{|\mathbf{V}_{\mathrm{ST}}||\mathbf{E}_{\mathrm{G}}|}{MX_{\mathrm{S}}}\right)}\,\right]^2 - \left[\frac{P_{\mathrm{D}}}{2M}\right]^2\right\}^{1/2} \qquad (6.117)$$

Automatic control or systems techniques enable us to identify the roots s_1 and s_2 as **complex frequencies** and to write Eq. (6.117) as

$$s_{1,2} = -\alpha \pm j\omega_{\mathrm{d}} \equiv -\alpha \pm j(\omega_0{}^2 - \alpha^2)^{1/2} \qquad (6.118)$$

and therefore identify the components of these complex frequencies s_1 and s_2 as (1) the **damping coefficient or neper frequency** α (nepers/second) and (2) the **damped radian frequency** ω_{d} (radians/second) of the synchronous device under study, where

$$\alpha = \frac{P_{\mathrm{D}}}{2M} \qquad (6.119)$$

and*

$$\omega_{\mathrm{d}} = (\omega_0{}^2 - \alpha^2)^{1/2} \quad \text{or} \quad \omega_0 = (\alpha^2 + \omega_{\mathrm{d}}{}^2)^{1/2} \qquad (6.120)$$

Since there are two values of complex frequencies s, that is,

$$s_1 = -\alpha + j\omega_{\mathrm{d}} \qquad (6.121)$$

and

$$s_2 = -\alpha - j\omega_{\mathrm{d}} \qquad (6.122)$$

it follows, from the theory of differential equations, that for a second-order system, Eq. (6.107) can be written as

$$\delta_{tr}(t) = K_1 e^{s_1 t} + K_2 e^{s_2 t} \qquad (6.123)$$

where K_1 and K_2 are arbitrary constants and $K_2 e^{s_2 t}$ is linearly independent from $K_1 e^{s_1 t}$. Depending on the parameters of the power system under study, Eq. (6.117), the following conditions might arise: (1) $s_1 \equiv s_2 = \mathrm{Re}$; (2) $s_1 = \mathrm{Re}$ and $s_1 \neq s_2 = \mathrm{Re}$; (3) $s_1 = -\alpha + j\omega_{\mathrm{d}}$ and $s_2 = -\alpha - j\omega_{\mathrm{d}}$; and (4) $s_1 = j\omega_0$ and $s_2 = -j\omega_0$. These four conditions lead to four distinct expressions of Eq. (6.123), written as

$$\delta_{tr}(t) = K_1 e^{s_1 t} + K_2 t e^{s_1 t} \qquad (6.124)$$

$$\delta_{tr}(t) = K_1 e^{s_1 t} + K_2 e^{s_2 t} \qquad (6.125)$$

$$\delta_{tr}(t) = \Gamma_3 e^{-\alpha t}\cos(\omega_{\mathrm{d}} t - \gamma_3) \equiv \Gamma_3 e^{-\alpha t}\sin(\omega_{\mathrm{d}} t - \gamma_3 + 90°) \qquad (6.126)$$

and

$$\delta_{tr}(t) = \Gamma_4 \cos(\omega_0 t - \gamma_4) \equiv \Gamma_4 \sin(\omega_0 t - \gamma_4 + 90°) \qquad (6.127)$$

It must be emphasized here that the values of the roots s_1 and s_2 of Eqs. (6.124) and (6.125) and the damping coefficient α of Eq. (6.126) might be **positive** or **negative.** These two sets of values describe systems that are either **unstable** or **stable**. (Why? If you do not understand why, please study carefully Appendix C.)

*Note the dependence of ω_{d} on ω_0, Eq. (6.103), as well as the geometric relationship between the vectors ω_0, α, and ω_{d} and thus the application of Pythagoras's theorem.

The linear equations for the machine rotor dynamics will always be stable provided the parameters P_D, M, and X_s are all positive in Eq. (6.98). It is possible for a synchronous energy converter connected to an infinite bus to go unstable. However, the nonlinear nature of the electric power term, that is, $\sin\delta$, must not be ignored if the stability of the machine is to be analyzed.

Once $\delta_{ss}(t)$ and $\delta_{tr}(t)$ are explicitly known, Eqs. (6.124) through (6.127) and Eq. (6.112), the solution of Eq. (6.99) for $\delta(t)$ can be written by direct substitution from the appropriate equations into Eq. (6.105). Thus if, for example, the roots of the power system under study are complex conjugates, the solution for $\delta(t)$ is written by direct substitution of Eqs. (6.112) and (6.126) into Eq. (6.105). Thus

$$\delta(t) = \delta_{ss}(t) + \delta_{tr}(t) = \Delta + \Gamma_3 e^{-\alpha t}\cos(\omega_d t - \gamma_3) \tag{6.128}$$

If in addition to the parameters of Eq. (6.99) the state of the power system is defined explicitly for a given time designated as t_k, or can be evaluated somehow, that is, if $\delta(t_k)$ and $d\delta(t_k)/dt$ are known explicitly, then the arbitrary constants Γ_3 and γ_3 of Eq. (6.128) can be evaluated with great ease, utilizing Eq. (6.128) and its first derivative.

Example 6.6

A 250-hp, 2,500-V, three-phase, 60-Hz, 30-pole, synchronous motor is directly connected to an infinitely large network (i.e., to an infinite bus) and has the following parameters: (1) $J = 450$ kg·m², (2) $P_D = 80$ W/electrical degree/s, (3) $P_G = 11{,}000$ W/electrical degree. Assume that the motor is unloaded and is operating at steady state. With all internal losses neglected, $\delta(0^+) = d\delta(0^+)/dt = 0$. Suddenly the motor is loaded with rated mechanical load. Define the switching time of the loading at zero, that is, $t = 0$ and calculate: (1) the synchronous speed of the motor ω_S and n_S, and (2) the momentum constant of the motor M. Also, write the power balance equation that describes the electrodynamic behavior of the synchronous motor when loaded and calculate: (3) the natural or undamped frequency of the motor ω_0, (4) the damping ratio ζ, (5) the damping factor α, (6) the damped frequency ω_d, and (7) the solution for the power angle $\delta(t)$.

Solution

The synchronous speed ω_S in radians/second of the motor can be calculated with the aid of Eq. (5.15) as

$$\omega_S = \frac{\omega_E}{P/2} = \frac{2\pi \times f_E}{P/2} = \frac{2\pi \times 60}{30/2} = \underline{\underline{25.13}} \text{ rad/sec} \tag{1}$$

Similarly, the synchronous speed n_S in revolutions per minute is calculated with the aid of Eq. (4.24) as

$$n_S = \frac{\omega_E}{P/2} \times \frac{60}{2\pi} = \frac{f_E}{P/2} \times 60 = \frac{60 \times 60}{30/2} = \underline{\underline{240}} \text{ rpm} \tag{2}$$

Now the momentum constant M (in watts/electrical degree/second²) of the motor can be calculated with the aid of Eq. (6.93), input data, and Eq. (1) or (2) as

$$M = J\omega_S \times \frac{2}{P} \times \frac{2\pi}{360°} = J\left(\frac{2\pi}{P/2} \times \frac{n_S}{60}\right) \times \frac{2\pi}{360°} = 450 \times 25.13 \times \frac{2}{30} \times \frac{2\pi}{360°}$$

$$= 450 \times \left(\frac{2\pi}{30/2} \times \frac{240}{60}\right) \times \frac{2\pi}{360°} = \underline{\underline{13.16}} \text{ W/electrical degree/sec}^2 \tag{3}$$

Now with the aid of Eq. (3) and input data, the describing equation of the synchronous motor can be written by direct substitution into Eq. (6.98). Thus

$$13.16\frac{d^2}{dt^2}\delta(t) + 80\frac{d}{dt}\delta(t) + 11{,}000\delta(t) = p_{SH}(t) \tag{4}$$

where

$$p_{SH}(t) = 250 \text{ hp} \times 746 \text{ W/hp} = 186{,}500 \tag{5}$$

Substitution of Eq. (5) into Eq. (4) and division of the resulting equation by $M = 13.16$ yields

$$\frac{d^2}{dt^2}\delta(t) + 6.08\frac{d}{dt}\delta(t) + 836.0\delta(t) = 14{,}170.0 \tag{6}$$

Equating the coefficients of Eq. (6) and Eq. (7) [i.e., Eq. (6.100)]

$$\frac{d^2}{dt^2}\delta(t) + 2\zeta\omega_0\frac{d}{dt}\delta(t) + \omega_0^2\delta(t) = p(t) \tag{7}$$

yields

$$\omega_0^2 = 836.0 \tag{8}$$

and

$$2\zeta\omega_0 = 6.08 \tag{9}$$

Solution of Eq. (8) for ω_0 yields

$$\omega_0 = (836.0)^{1/2} = \underline{\underline{28.91}} \text{ rad/sec} \tag{10}$$

Solution of Eq. (9) for ζ (dimensionless) and substitution for ω_0 from Eq. (10) yields

$$\zeta = \frac{6.08}{2\omega_0} = \frac{6.08}{(2)\times(28.91)} = \underline{\underline{0.11}} \tag{11}$$

The damping factor α (nepers/second or Np/s) is calculated with the aid of Eq. (6.119) as

$$\alpha = \frac{P_D}{2M} = \frac{80}{2\times 13.16} = \underline{\underline{3.04}} \text{ Np/sec} \tag{12}$$

The damped frequency ω_d is calculated with the aid of Eq. (6.120) as

$$\omega_d = (\omega_0^2 - \alpha^2)^{1/2} = (836.0 - 9.24)^{1/2} = \underline{\underline{28.75}} \text{ rad/sec} \tag{13}$$

The solution for the power angle $\delta(t)$ is calculated as follows. Utilization of the theory of Section 6.5, that is, Eqs. (6.108) through (6.112), suggests that since $p_{SH}(t)/M = 14{,}170.0u(t)$ in Eq. (6),

$$\delta(t) = \delta_{ss}(t) \equiv \Delta \tag{14}$$

Then

$$\frac{d\delta(t)}{dt} = 0 \tag{15}$$

and

$$\frac{d^2\delta(t)}{dt^2} = 0 \tag{16}$$

Substitution of Eqs. (16), (15), and (14) into Eq. (6) yields

$$0 + 6.08(0) + 835.87(\Delta) = 14{,}170.0 \tag{17}$$

Solution of Eq. (17) for Δ yields

$$\Delta = \frac{14{,}170.0}{836.0} = \underline{16.95^\circ} \tag{18}$$

Substitution for Δ from Eq. (18) into Eq. (14) yields

$$\delta_{ss}(t) = \underline{16.95^\circ} \tag{19}$$

Similarly, utilization of the theory of Section 6.5, that is, Eqs. (6.113) through (6.123), suggests that in Eq. (6) if

$$\frac{p_{SH}(t)}{M} = 14{,}170.0 \to 0 \tag{20}$$

and

$$\delta(t) \equiv \delta_{tr}(t) = Ke^{st} \tag{21}$$

then

$$\frac{d\delta(t)}{dt} = Kse^{st} \tag{22}$$

and

$$\frac{d^2\delta(t)}{dt^2} = Ks^2e^{st} \tag{23}$$

Substitution of Eqs. (23), (22), (21), and (20) into Eq. (6) and extraction of the common factor Ke^{st} yield

$$(s^2 + 6.08s + 836.0)Ke^{st} = 0 \tag{24}$$

Since, as mentioned earlier, $Ke^{st} \neq 0$, Eq. (24) yields

$$s^2 + 6.08s + 836.0 = 0 \tag{25}$$

Thus

$$\begin{aligned} s_{1,2} &= \tfrac{1}{2}\{-6.08 \pm [6.08^2 - 4\times 836.0]^{1/2}\} \\ &= -3.04 \pm (9.24 - 836.0)^{1/2} = \underline{-3.04 \pm j28.75} \end{aligned} \tag{26}$$

Substituting for s_1 and s_2 from Eq. (26) into Eq. (6.123), utilizing algebraic theorems that dictate exponentiation properties and Eulers rule, yields

$$\begin{aligned} \delta_{tr}(t) &= K_1e^{(-3.04+j28.75)t} + K_2e^{(-3.04-j28.75)t} \\ &= e^{-3.04t}[K_1(\cos 28.75t + j\sin 28.75t) \\ &\quad + K_2(\cos 28.75t - j\sin 28.75t)] \end{aligned} \tag{27}$$

or

$$\delta_{tr}(t) = \underline{e^{-3.04t}(A\cos 28.75t + B\sin 28.75t)^\circ} \tag{28}$$

where

$$A = K_1 + K_2 \tag{29}$$

and

$$B = j(K_1 - K_2) \tag{30}$$

Note that Eq. (26) verifies Eqs. (12), (13), and (6.120), that is,

$$\omega_0 = \left(\alpha^2 + \omega_d{}^2\right)^{1/2} = (3.04^2 + 28.75^2) = \underline{\underline{28.91}} \text{ rad/sec} \tag{31}$$

Now substitution of Eqs. (19) and (28) into Eq. (6.105) yields

$$\delta(t) = \delta_{ss}(t) + \delta_{tr}(t) = 16.95° + e^{-3.04t}(A\cos 28.75t + B\sin 28.75t) \tag{32}$$

The first derivative of Eq. (32) becomes

$$\frac{d}{dt}\delta(t) = -3.04e^{-3.04t}(A\cos 28.75t + B\sin 28.75t)$$
$$+ e^{-3.04t}(-28.75A\sin 28.75t + 28.75B\cos 28.75t) \tag{33}$$

Since the initial conditions on $\delta(t)$ and $d\delta(t)/dt$ are defined to be zero, that is, $t=0$, $\delta(0^+) = d\delta(0^+)/dt = 0$, Eqs. (32) and (33) become, for the appropriate substitution of $t=0$, $\delta(0^+)$, and $d\delta(0^+)/dt$,

$$0 = 16.95 + 1(A) + 0(B) \tag{34}$$

and

$$0 = -3.04(A) + 28.75(B) \tag{35}$$

Solution of Eq. (34) for A yields

$$A = -16.95° \tag{36}$$

Substituting for A from Eq. (36) into Eq. (35) yields

$$B = \frac{3.04 \times A}{28.75} = \frac{3.04 \times (-16.95)}{28.75} = -1.79° \tag{37}$$

Substituting for A and B from Eqs. (36) and (37) into Eq. (32), and taking initial condition effects into account, yields

$$\delta(t) = 16.95° - e^{-3.04t}(16.95°\cos 28.75t + 1.79°\sin 28.75t) \tag{38}$$

Utilization of trigonometric properties enables us to calculate the magnitude Γ_3 and the phase angle γ_3 of the parentheses of Eq. (38) as

$$\Gamma_3 = \left[(16.95)^2 + (1.79)^2\right]^{1/2} = 17.04° \tag{39}$$

and

$$\gamma_3 = \tan^{-1}\left(\frac{1.79}{16.95}\right) = 6.03° \tag{40}$$

These two quantities enable writing Eq. (38) as

$$\delta(t) = 16.95° - 17.04° e^{-3.04t}\cos(28.75t - 6.03°) \tag{41}$$

or

$$\delta(t) = 16.95° - 17.04° e^{-3.04t}\sin(28.75t + 83.97°) \tag{42}$$

or

$$\delta(t) = 16.95°\left[1 - 1.01e^{-3.04t}\sin(28.75t + 83.97°)\right] \tag{43}$$

Eq. (43) is exhibited graphically in Fig. 6.6.1.

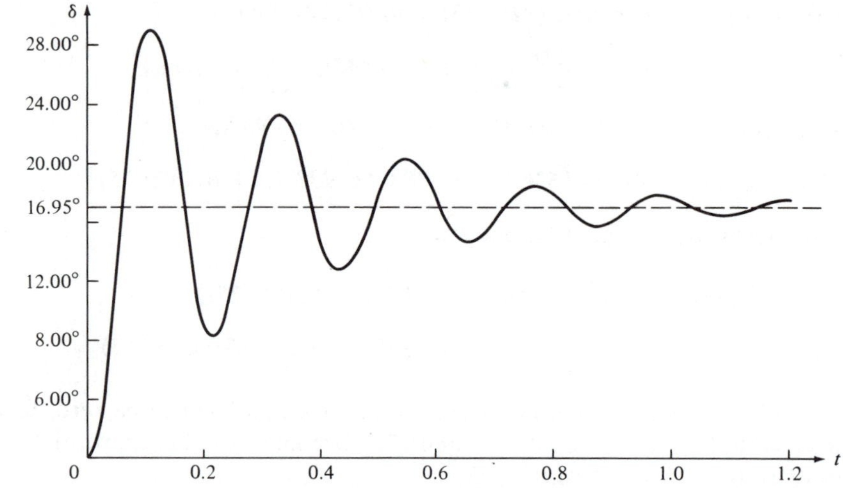

Fig. 6.6.1 Graphical Display of the Solution of Eq. (43); $\alpha = 3.04$Np/s; $\omega_d = 28.75$ rad/s, $\zeta = 0.11$, $\delta(0^+)$, and $d\delta(0^+)/dt = 0$

Nonlinear Analysis (Case 1)

In most of the serious dynamic problems, the oscillations are of such magnitude that the foregoing linearization is not permissible. The describing equations of motion, which are of the form of Eq. (6.97), must be retained in *nonlinear* form, and use of the analog, digital, or hybrid computer must be made to aid in their solution. Regardless of which computer is used and what method is employed for the simulation of such a dynamic equation, the objective of the study is usually to find whether or not synchronism is maintained. That is, whether or not the power angle $\delta(t)$ of a given synchronous device settles down to a steady state value δ_{GSS} or δ_{MSS}, Fig. 5.40 or Fig. 5.48, after the given power system has been subjected to a sizable disturbance. Such a disturbance might result from (1) *line faults*, that is, *short circuits*, (2) *line switching*, or (3) *sudden load changes, anywhere in the given power system.* A discussion of line faults is deferred to later in this section.

Although power systems are comprised of many synchronous devices and are described by a corresponding number of swing equations, no attempt is made here to model such a system or to obtain the solution of the describing swing equations. Instead, the case of one synchronous generator connected to an infinite bus, Fig. 6.1(a), is considered in detail. The describing equation of such a system is given by Eq. (6.97). Its formal solution necessitates the use of **elliptic integrals**, or **perturbation techniques**. The plot of the solution δ (in electrical degrees) versus time (in seconds) gives an indication as to whether the rotor of the synchronous generator relative to the synchronously rotating reference frame (SRRF, Fig. 6.25) (1) **remains in synchronism (is stable)**, (2) **goes out of synchronism (becomes unstable)**, or (3) **oscillates around the equilibrium point with constant amplitude (is marginally stable).**

It is possible to determine which of the three forms the solution has without having to calculate the swing curves using the solution of the swing equation, but rather through a more or less intuitive approach known as the **equal-area criterion.** Although *not applicable* to systems with more than two synchronous generators or motors, this method helps in understanding how certain factors affect the stability of any power system.

The derivation of the equal-area criterion is valid for the one synchronous generator connected to an infinite bus, Fig. 6.1(a). However, the method can be adapted to two-generator systems as well. The total output electrical power of the synchronous generator P_{OT} as given by Eq. (6.1), is rewritten here for convenience as

$$P_{OT} = 3\left(\frac{|\mathbf{V}_{ST}||\mathbf{E}_G|}{X_S}\right)\sin\delta \tag{6.129}$$

Most disturbances of critical nature (i.e., short circuits, line switching, and sudden load changes) entail the sudden change of electrical output power P_{OT} during a disturbance, while the mechanical input power P_{SH} from the prime mover remains relatively constant. However, the equal-area criterion is best demonstrated if the mechanical power input [$p_{SH}(t) = P_{SH}u(t)$, i.e., the prime-mover output] is suddenly increased from P_{SH0} to P_{SH1}, Fig. 6.26, as the first disturbance is noted, while the infinite bus absorbs any change in the electrical output power and maintains $\mathbf{V}_{ST}$ and f_E constant. The graph of Fig. 6.1(b) is replotted here as in Fig. 6.27. For $P_D = 0$ Eq. (6.97) is written as

$$\begin{aligned} M\frac{d^2}{dt^2}\delta(t) &= P_{SH} - 3\left(\frac{|\mathbf{V}_{ST}||\mathbf{E}_G|}{X_S}\right)\sin\delta(t) \\ &\equiv P_{SH} - P_{MAX}\sin\delta(t) \end{aligned} \tag{6.130}$$

where $M\,d^2\delta(t)/dt^2$ designates the *accelerating power*, designated as $p_\alpha(t)$ [e.g., $F = m\,d^2x(t)/dt^2$], that is, power that stores kinetic energy in the rotating mass of the rotor. Thus if the initial value of the rotor shaft power, designated as P_{SH0} with the frictional (and all other) losses neglected is superimposed in Fig. 6.27, the intersection of the graphs P_{OT} and P_{SH0} defines the equilibrium point O where $\delta = \delta_O$. When the shaft power P_{SH}, that is, the prime mover output, is increased to the new value, designated as P_{SH1}, Fig. 6.27, there is momentarily more shaft power input P_{SH1} than electrical power output P_{OT}. Thus there is an excess of accelerating

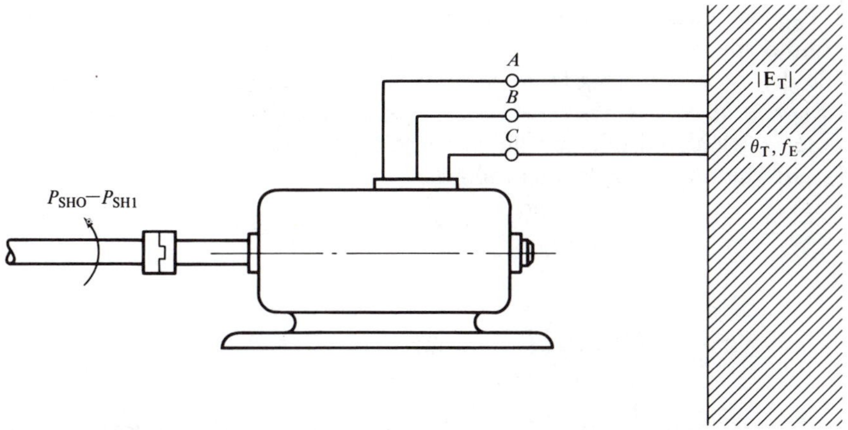

Fig. 6.26 A Round-Rotor, Balanced, P-Pole, Three-Phase, Synchronous Generator Driven by an Adjustable Output Power P_{SH} ($P_{SH0} \rightarrow P_{SH1}$) Prime Mover Feeding an Infintely Strong Network, That Is, an Infinite Bus: $|\mathbf{E}_T| = k_1$, $\theta_T = k_2$, and $f_E = k$

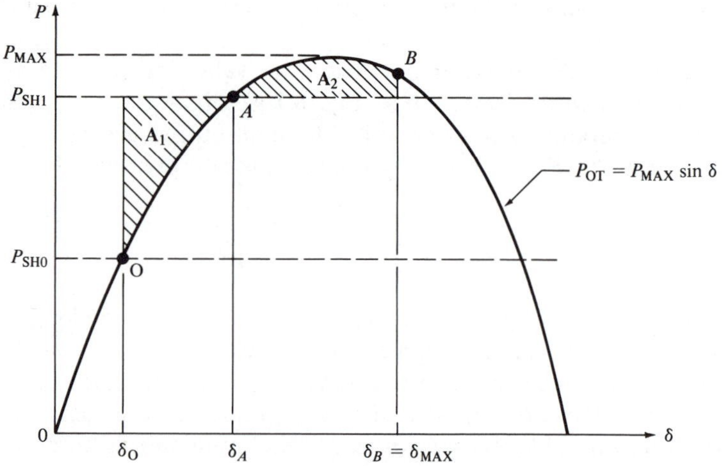

Fig. 6.27 Power-Angle Characteristics for a Round-Rotor, Balanced, Three-Phase, *P*-Pole, Synchronous Generator, with Prime-Mover Output Power $P_{SH}(P_{SH0} \rightarrow P_{SH1})$ Superimposed

power, that is, $P_{SH1} > P_{OT}$, Eq. (6.130), which causes the rotor to accelerate to point A where $\delta = \delta_A$ and $P_{SH1} = P_{OT}$. At this point the rotor stops accelerating but its speed ω_R is slightly higher than the synchronous speed, that is, $\omega_R > \omega_S$. Thus the power angle continues to increase, that is, $\delta > \delta_A$, point A is overshot, and $P_{OT} > P_{SH1}$. Since the generator output power P_{OT} now exceeds the input shaft power P_{SH1}, the rotor returns its extra stored kinetic energy to the infinite bus and decelerates until it reaches some maximum point B where its speed $\omega_R = \omega_S$ and $\delta_B = \delta_{MAX}$. Having returned all its extra kinetic energy back to the network, the rotor continues to decelerate, falling through point A and back toward point 0. In the absence of losses or damping, the rotor speed fluctuates with the power angle, oscillating between δ_O and δ_{MAX}. In practical synchronous generators, however, damping causes the oscillations to gradually die out and the power angle δ to reach its final value $\delta \equiv \delta_A \equiv \delta_{GSS}$, Figs. 6.27 and 5.40, correspondingly, at which the input power P_{SH} matches the output power P_{OT}, and the generator is running at synchronous speed ω_S.

While the rotor is moving from point O to point A, Fig. 6.27, *work is being done on the rotor in increasing the kinetic energy of the revolving mass*, since $P_{SH1} > P_{OT}$. This work is calculated utilizing the accelerating torque $\tau_\alpha(t), p_\alpha(t) = \tau_\alpha(t)\omega_R(t)$, Eq. (6.139), and the power angle δ as follows:

$$\int_{W_O}^{W_A} dw = \int_{\delta_O}^{\delta_A} \tau_\alpha(t)\, d\delta = W_A - W_O \equiv \Delta W_{AO} \tag{6.131}$$

or

$$\Delta W_{AO} = \int_{\delta_O}^{\delta_A} \frac{p_\alpha(t)}{\omega_R(t)}\bigg|_{\omega_R(t) \rightarrow \omega_S} d\delta$$

$$= \frac{1}{\omega_S}\left\{ \int_{\delta_O}^{\delta_A} (P_{SH1} - P_{MAX} \sin\delta)\bigg|_{P_{SH1} > P_{OT}} d\delta \right\} = k\mathbf{A}_1 \tag{6.132}$$

where $k \equiv 1/\omega_S$.

While the rotor is moving from point A to B, Fig. 6.27, *energy is returned from the rotor mass to the infinite bus*, since $P_{OT} > P_{SH1}$. When the rotor reaches point B, the rotor is again at synchronous speed ω_S and has given back all the kinetic energy that was stored in its mass. This energy is again calculated utilizing the accelerating torque $\tau_\alpha(t)$ and the power angle δ as follows:

$$\int_{W_A}^{W_B} dw = \int_{\delta_A}^{\delta_B} \tau_\alpha(t)\, d\delta = W_B - W_A = \Delta W_{BA} \tag{6.133}$$

or

$$\begin{aligned}\Delta W_{BA} &= \int_{\delta_A}^{\delta_B} \frac{P_\alpha(t)}{\omega_R(t)}\bigg|_{\omega_R(t)\to\omega_S} d\delta \\ &= \frac{1}{\omega_S}\left\{\int_{\delta_A}^{\delta_B} (P_{SH1} - P_{MAX}\sin\delta)\Big|_{P_{SH1}<P_{OT}} d\delta\right\} \\ &= -k\mathbf{A}_2\end{aligned} \tag{6.134}$$

where, again, $k \equiv 1/\omega_S$.

The integrals of Eqs. (6.132) and (6.134) represent the *increase* and *decrease* in the stored kinetic energy in the rotor mass. Since the speed at points δ_O and δ_B is synchronous ω_S, the resultant change in the stored kinetic energy, as the power angle takes the values $\delta_O \leqslant \delta \leqslant \delta_B$, must be zero. Mathematically, this idea is expressed as

$$\Delta W_{AO} + \Delta W_{BA} = k\mathbf{A}_1 - k\mathbf{A}_2 = 0 \tag{6.135}$$

Cancellation of the constant k from Eq. (6.135) yields

$$\mathbf{A}_1 = \mathbf{A}_2 \tag{6.136}$$

Eq. (6.136) suggests that if the energies of Eqs. (6.132) and (6.134) are to be equal, their equality can be checked through the equality of the *areas* $\mathbf{A}_1$ and $\mathbf{A}_2$. Thus sufficient area must be available between the curves P_{SH1} and P_{OT}, Fig. (6.27), *to satisfy the equal-area criterion*, that is, $\mathbf{A}_1 = \mathbf{A}_2$. If the available area $\mathbf{A}_2 < \mathbf{A}_1$, then the excess stored kinetic energy in the rotor mass will cause δ to continue to increase in an effort by the motor to return power back to the infinite bus. This, however, is impossible, since P_{OT} is decreasing with an increased δ beyond $\delta_B \doteq 90°$. It should be noted, however, that it is permissible for the rotor to oscillate past the point where $\delta_B = 90°$, as long as the equal-area criterion, that is, $\mathbf{A}_1 = \mathbf{A}_2$, is satisfied. The equal-area criterion defines the maximum power angle $\delta_B \equiv \delta_{MAX}$, Fig. 6.27, to which δ can swing without the rotor losing synchronism. This point is defined as the **transient stability limit** and is **less restrictive** than the **steady state stability limit**, that is, $\delta_{MAX} \leqslant 90°$. In other words, the **transient stability limit** is defined as the **limiting value** of the mechanical power input to a synchronous generator, or the mechanical load to a synchronous motor, below which the synchronous device is **stable** and above which it is **unstable**.

Although the equal-area criterion can be applied only for **one synchronous device connected to an infinitely strong network, that is, an infinite bus,** or the case of **two synchronous devices connected in parallel,** it is a very useful means for beginning to see what happens when a fault occurs to a given power system. For very large power systems, however, the digital or the hybrid computer is the only practical approach to determine their stability status. However, **because the equal-area criterion is so basic in helping one to understand the transient stability intricacies of**

power systems comprised of a maximum of two synchronous devices, it is the only method that is introduced in this text.

Usually a synchronous generator supplies power to an **infinite bus represented by an $\mathbf{E}_B$ and jX_B Thevenin equivalent**, Figs. 6.28(a) and 6.28(b), through two parallel transmission lines connected to two paralleling buses with **relay-circuit-breaker sets** at both ends of each line [or relay-circuit-breaker transformer sets at both ends of each line and relay-circuit-breaker sets between the paralleling bus and the synchronous generator on the left-hand side and the paralleling bus and the infinite bus on the right-hand side, Fig. 6.8.1(a)]. A **relay** is a device that, when energized by appropriate system quantities, indicates an abnormal condition. When the relay contacts close, the associated **circuit-breaker trip circuits** are energized and the breaker contacts open, thus isolating the faulty component from the system, that is, line AB, Fig. 6.28(a) and 6.28(b), when a fault occurs at its far end or its midpoint.

Plate 6.10 242-kV Oil Circuit Breakers Allow High-Voltage Power to be Transmitted into Heavily Populated Areas and Interrupted Safely. (Courtesy of General Electric Co., Schenectady, N.Y.)

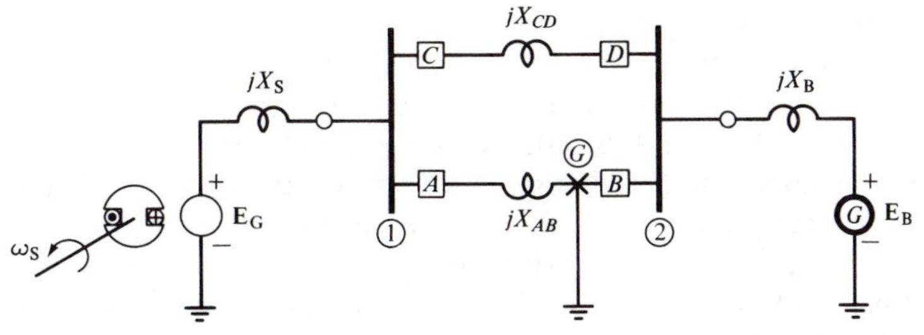

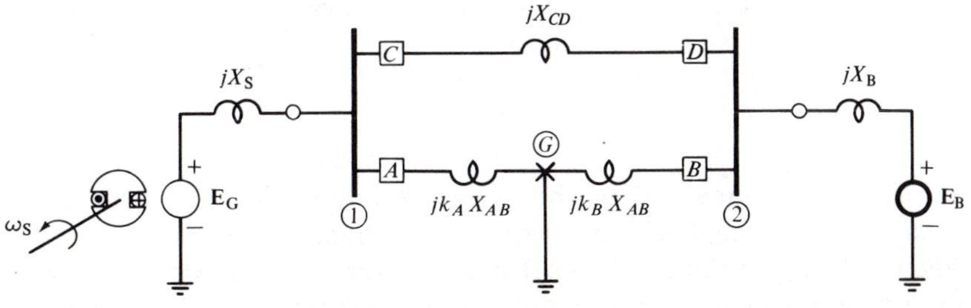

Fig. 6.28 A Synchronous Generator Feeding an Infinite Bus, $\mathbf{E}_B$ and jX_B, Through Two Parallel Transmission Lines Connected to the Paralleling Buses No. 1 and No. 2 Through Relay-Circuit-Breaker Sets: (a) Fault at the Far End of the line AB and (b) Fault at the Midpoint of the Line AB

Relay-circuit-breaker sets adjacent to a fault on both sides are arranged to open simultaneously. Their standard interrupting times are 2, 3, 5, or 8 cycles after a fault occurs, and thus breaker speeds may be specified. Removing one line from the system by opening the relay circuit breakers A and B at both ends of the line may cause the generator to lose synchronism, that is, to become unstable, even though the load could be supplied over the remaining line under steady state conditions. **If a three-phase short circuit occurs on a paralleling bus to which two parallel lines are connected, no power can be transmitted over either line** [Fig. 6.31(b)]. However, if the fault occurs at either end or along the given line AB, the opening of the relay circuit breakers A and B at both ends of this line isolates the fault from the system and allows power to flow to the bus through the unfaulted second parallel line [Fig. 6.31(c)].

When the three-phase fault occurs at **some point** along one of the lines, rather than on the paralleling buses or at the extreme ends of either line, there is some nonzero impedance between the connecting buses and the fault [Figs. 6.33(b) and 6.33(c)]. Thus there is power still transmitted while the system is still faulted. Actually, regardless of the location, short-circuit faults not involving all three phases allow the transmission of some power, because the faults are represented by connecting some nonzero impedance, rather than a short circuit, between the fault point and the reference bus in the one-line impedance diagram of the system. **A one-line diagram is a per-phase representation of a balanced three-phase system in which the components, that is, generators, transformers, buses, lines, relays, loads, and so on, are represented in a simplified form.** It should be obvious, then, that the power transmitted to the infinite bus during a fault is a function of the net impedance connected between the synchronous generator and the infinite bus.

The power transmitted to an infinite bus, through the transmission system, shown in Figs. 6.28(a) and 6.28(b), can be calculated after representing the reactance (impedance) of (1) the synchronous generator jX_S, (2) the effective transmission network jX_{12}, and (3) the infinite bus jX_B, by an equivalent reactance designated by jX_{EQ} connected between the two voltage sources $\mathbf{E}_G$ and $\mathbf{V}_B$, Fig. 6.29 (remember that $|jX_{EQ}| \ggg R_{EQ}$; thus $\mathbf{Z}_{EQ} \doteq j\mathbf{X}_{EQ}$). Thus Fig. 6.29 exhibits the common per-phase diagram used for the per-phase analysis of a synchronous generator and an infinite bus connected in parallel, as used in Fig. 6.1(a). The real power transmitted to the infinite bus of Fig. 6.29 is then written, by analogy from Eq. (6.1), as

$$P_{OT} = 3\left(\frac{|\mathbf{V}_B||\mathbf{E}_G|}{X_{EQ}}\right)\sin\delta \tag{6.137}$$

Thus, if the fault analysis of the system exhibited in Fig. 6.28(b) is to be performed, the power-angle characteristics to be utilized are those exhibited in Fig. 6.30, where (1) P_{SH} is the mechanical input power to the shaft of the generator from the prime mover with all functional losses neglected, (2) $P_{OT}^{I} = k_1 P_{MAX} \sin\delta$, $k = 1.00$ and $k_1 > k_3 > k_2$, is the electric power that the generator can deliver to the infinite bus before the fault occurs [Fig. 6.33(a)], (3) $P_{OT}^{II} = k_2 P_{MAX} \sin\delta$ is the electric power that the generator can deliver to the infinite bus during the fault [Fig. 6.33(b) and 6.33(c)], and (4) $P_{OT}^{III} = k_3 P_{MAX} \sin\delta$ is the electric power that the generator can deliver to the infinite bus after the fault is cleared [Fig. 6.33(d)] by switching the faulty line off service at the instant when $\delta = \delta_C$. The power angle δ_C is defined as the **critical clearing angle**. It is the largest possible value the power angle δ can acquire before clearing the fault without the value of the transmitted power exceeding the transient stability limit. Examination of Fig. 6.30 indicates that if clearing were to occur at a value of δ greater than δ_C, the area $\mathbf{A}_2$ between the horizontal line P_{SH} and the curve P_{OT}^{III} would be smaller than the area $\mathbf{A}_1$ between the horizontal line P_{SH} and the curve P_{OT}^{II}. This *excess* of stored kinetic energy designated by the inequality $\mathbf{A}_1 > \mathbf{A}_2$ suggests that the input mechanical power provided by prime mover P_{SH} is *greater* than the electric power P_{OT} developed by the generator. Thus the speed of the rotor increases, that is, $\omega_R > \omega_S$, the power angle δ continues to increase beyond δ_{MAX}, and the generator loses synchronism, that is, becomes **unstable**. It is obvious from Fig. 6.30 that the amount of power transmitted to the infinite bus during the fault affects the value of kinetic energy represented by the area $\mathbf{A}_1$ for a given clearing angle δ_C. Thus

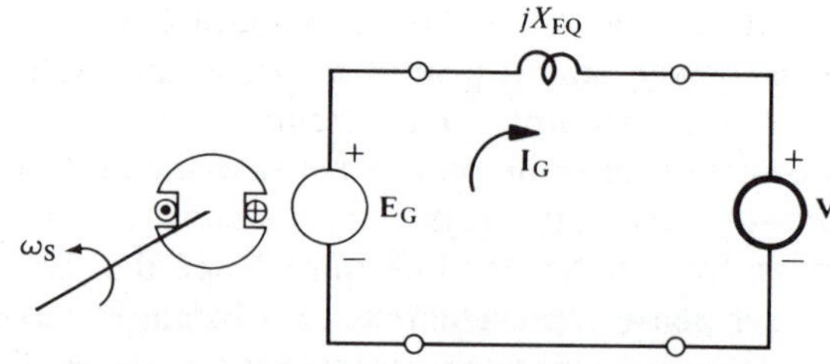

Fig. 6.29 Per Phase Diagram Exhibiting the Equivalent Reactance Connected Between the Voltage Sources Representing the Synchronous Generator and the Infinite Bus

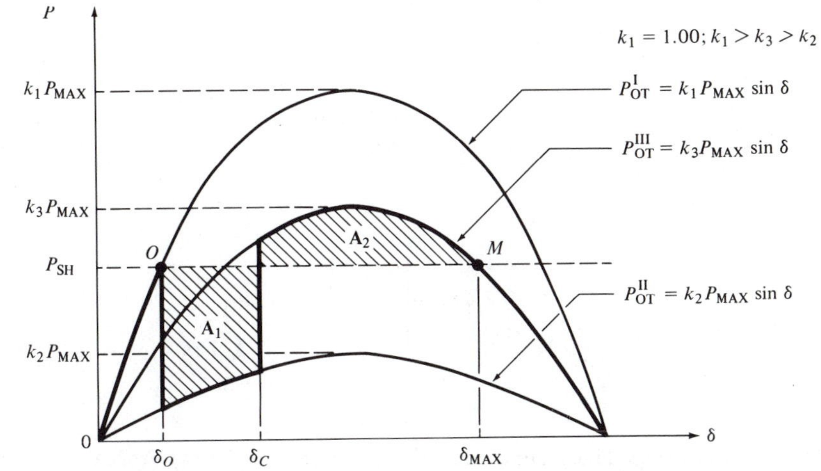

Fig. 6.30 Power-Angle Characteristics of a Synchronous Generator—Infinite Bus Interconnection with Powers Transmitted to the Infinite Bus at the (a) Prefaulted Status, P_{OT}^{I}, (b) Faulted Status, P_{OT}^{II}, and (c) Postfaulted Status, P_{OT}^{III}

smaller values of k_2 designate greater disturbances to the system, since low k_2 means low power transmitted to the infinite bus during the fault and greater $\mathbf{A}_1$. In order of **increasing severity**, that is, decreasing $k_2 P_{MAX} \sin \delta$, the various faults are (1) **single-line-to-ground**, (2) **line-to-line**, (3) **double-line-to-ground**, and (4) **three-phase faults**. Of these, the **single-line-to-ground faults occur most frequently**. They constitute approximately 75 percent of all faults that occur, and are mostly a result of insulator flashover during electrical storms. The **three-phase faults occur least frequently**. They constitute approximately 5 percent of all faults that occur.

In studying the effects of faults upon generator stability, we are often looking for the **critical clearing time** in cycles (the standard number of cycles is 2, 3, 5, or 8 after the fault occurs) or seconds, during which time the circuit breakers must isolate the faulted line from the system. If the circuit breakers fail to open within this period, the equal-area criterion for stability would not be met. The equal-area criterion will not of itself determine this critical clearing time, but it does offer a means for determining the critical clearing angle corresponding to this time. The most severe fault and the easiest to handle mathematically, to be considered when designing a system for complete reliability, is the three-phase fault at the worst locations. It might be permissible, however, from the economic point of view, to extend the clearing time to accommodate a less severe fault. Thus the usual criterion used for transient stability studies is the *double-line-to-ground fault*. However, because of calculation simplicity, **only** three-phase faults are considered in this text (in conjunction with one synchronous generator connected to an infinite bus through a two-parallel-transmission-line network connected through relay-circuit-breaker sets to two paralleling buses). The analysis of systems with more than two synchronous machines is *beyond* the scope of this text.

For the purpose of carrying out comprehensive fault analysis, consider the one-line impedance diagram of Fig. 6.28 with the fault occurring at (1) the far end of the transmission line *AB*, Fig. 6.28(a), or (2) at the midpoint of the transmission line *AB*, Fig. 6.28(b). Three distinct states of operation characterize the system as the system goes through the cycle of (1) prefaulted, (2) faulted, and (3) postfaulted

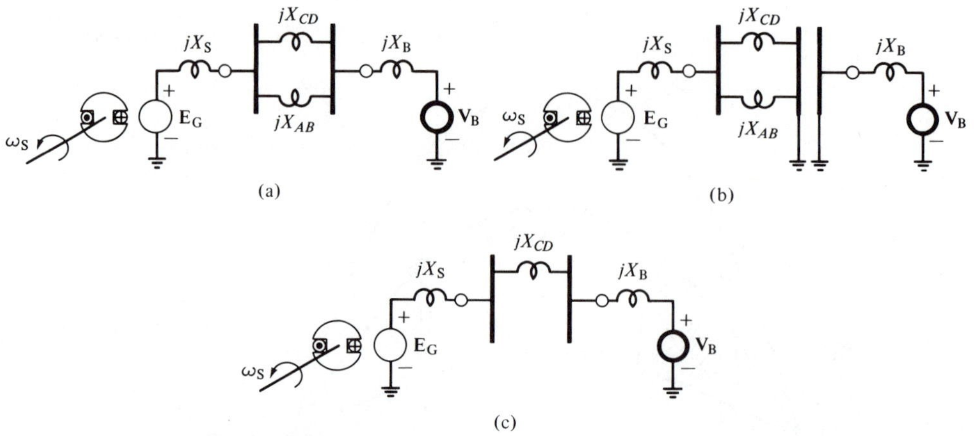

Fig. 6.31 One-Line Impedance Diagrams of Fig. 6.28(a) for the (a) Prefaulted Status, (b) Faulted Status, and (c) Postfaulted Status

status. These states are described by (1) Figs. 6.31(a), 6.31(b), 6.31(c), and 6.32, and (2) Figs. 6.33(a), 6.33(b), 6.33(c), 6.33(d), and 6.34.

When the far end of the transmission line AB, Fig. 6.28(a), is faulted, the generated power P_{OT}^{I}, Fig. 6.32 and Fig. 6.31(a), delivered from the synchronous generator to the infinite bus, drops suddenly to zero P_{OT}^{II}, Fig. 6.32 and Fig. 6.31(b). At a preset time relay circuit breakers A and B open and thus remove the transmission line AB from the system and isolate the fault. The generated power P_{OT} delivered to the infinite bus now picks up to some new value P_{OT}^{III}, Fig. 6.32 and Fig. 6.31(c), as the synchronous generator transfers power to the infinite bus over the remaining transmission line CD. The status of the system, Fig. 6.28(a), as it goes through the cycle of (1) prefaulted, (2) faulted, and (3) postfaulted operation is described by the single-line impedance diagrams of Figs. 6.31(a), 6.31(b), 6.31(c), and 6.32 and is analyzed as described in the following sections.

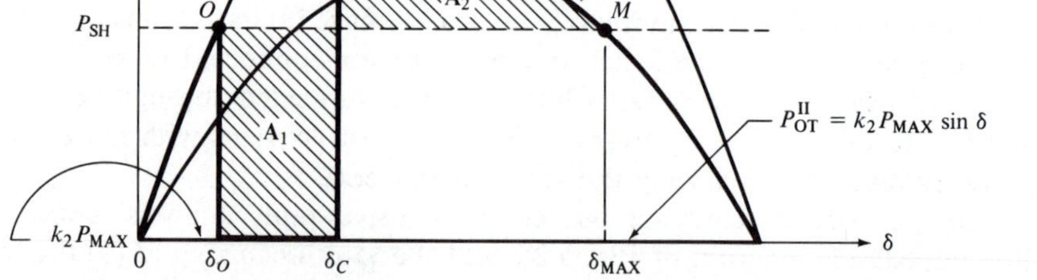

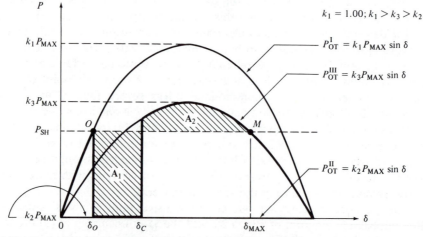

Fig. 6.32 Power-Angle Characteristics for the System of Fig. 6.28(a) with Powers Transmitted to the Infinite Bus at the (a) Prefaulted Status, Fig. 6.31(a), P_{OT}^{I}, (b) Faulted Status, Fig. 6.31(b), P_{OT}^{II}, and (c) Postfaulted Status, Fig. 6.31(c), P_{OT}^{III}

PREFAULTED STATUS, FIG. 6.31(a)

The system is operating in a normal mode defined by the system requirements. The system operates at the equilibrium point O, that is, the intersection of line P_{SH} and curve $P_{\mathrm{OT}}^{\mathrm{I}}$, Fig. 6.32, where the power angle value is δ_{O}, $P_{\mathrm{SH}} = P_{\mathrm{OT}}^{\mathrm{I}}$, and the rotor rotates at synchronous speed, that is, $\omega_{\mathrm{R}} = \omega_{\mathrm{S}}$. The equivalent reactance of the system, $X_{\mathrm{EQ}}^{\mathrm{I}}$, is calculated from Fig. 6.31(a) as

$$X_{\mathrm{EQ}}^{\mathrm{I}} = X_{\mathrm{S}} + \frac{(X_{AB})\times(X_{CD})}{X_{AB}+X_{CD}} + X_{\mathrm{B}} \tag{6.138}$$

The generated power $P_{\mathrm{OT}}^{\mathrm{I}}$, Fig. 6.32, transmitted to the infinite bus is calculated with the aid of Eqs. (6.137) and (6.138) as

$$P_{\mathrm{OT}}^{\mathrm{I}} = 3\left(\frac{|\mathbf{V}_{\mathrm{B}}||\mathbf{E}_{\mathrm{G}}|}{X_{\mathrm{EQ}}^{\mathrm{I}}}\right)\sin\delta \tag{6.139}$$

Now utilization of the constraint $P_{\mathrm{SH}} = P_{\mathrm{OT}}^{\mathrm{I}}$ that prevails at the equilibrium point O, Fig. 6.32, enables us to calculate the power angle $\delta = \delta_{\mathrm{O}}$ by equating the generated power $P_{\mathrm{OT}}^{\mathrm{I}}$, Eq. (6.139), to the power P_{SH} input to the generator (neglecting frictional and all other losses) from the prime mover. Thus

$$P_{\mathrm{SH}} \equiv 3\left(\frac{|\mathbf{V}_{\mathrm{B}}||\mathbf{E}_{\mathrm{G}}|}{X_{\mathrm{EQ}}^{\mathrm{I}}}\right)\sin\delta_0 \tag{6.140}$$

Solution of Eq. (6.140) for the power angle δ_{O} at the equilibrium point O, Fig. 6.32, yields

$$\delta_{\mathrm{O}} = \sin^{-1}\left\{\frac{P_{\mathrm{SH}}}{3(|\mathbf{V}_{\mathrm{B}}||\mathbf{E}_{\mathrm{G}}|/X_{\mathrm{EQ}}^{\mathrm{I}})}\right\} \tag{6.141}$$

FAULTED STATUS, FIG. 6.31(b)

The fault at the far end of the transmission line AB, Fig. 6.28(a), has occurred. The system is disturbed, that is, the synchronous generator and the two transmission lines are isolated from the infinite bus, Fig. 6.31(b). Thus the generated power $P_{\mathrm{OT}}^{\mathrm{I}}$, Fig. 6.32, transmitted to the infinite bus, falls suddenly to zero as the fault occurs. That is, the curve $P_{\mathrm{OT}}^{\mathrm{II}}$, Fig. 6.32, coincides with the δ-axis. Thus

$$P_{\mathrm{OT}}^{\mathrm{II}} = k_2 P_{\mathrm{MAX}} \sin\delta \equiv 0 \tag{6.142}$$

The area $\mathbf{A}_1$, Fig. 6.32, is proportional to the kinetic energy stored in the rotor mass as it accelerates striving for the (critical) power angle δ_{C}, an *unknown* quantity still to be calculated.

POSTFAULTED STATUS, FIG. 6.31(c)

The relay circuit breakers A and B have opened simultaneously and have removed the transmission line AB, Fig. 6.28(a), from the system. Thus they have removed the fault from the system, Fig. 6.31(c). The status of the system is improved; that is, the synchronous generator is connected to the infinite bus now through the transmission line CD and power flows from the synchronous generator to the infinite bus. The equivalent reactance of the system, $X_{\mathrm{EQ}}^{\mathrm{III}}$, is calculated from Fig. 6.31(c) as

$$X_{\mathrm{EQ}}^{\mathrm{III}} = X_{\mathrm{S}} + X_{CD} + X_{\mathrm{B}} \tag{6.143}$$

The generated power $P_{\rm OT}^{\rm III}$, Fig. 6.32, transmitted to the infinite bus, is calculated with the aid of Eqs. (6.137) and (6.143) as

$$P_{\rm OT}^{\rm III} = 3\left(\frac{|\mathbf{V}_{\rm B}||\mathbf{E}_{\rm G}|}{X_{\rm EQ}^{\rm III}}\right)\sin\delta \tag{6.144}$$

After the faulted transmission line AB is removed from the system, the rotor overshoots the (critical) power angle $\delta_{\rm C}$. Since power is absorbed by the infinite bus, the rotor decelerates until it reaches the new equilibrium point M, that is, the intersection of the line $P_{\rm SH}$ and the curve $P_{\rm OT}^{\rm III}$, Fig. 6.32, but the power angle increases to a new value $\delta_{\rm MAX}$, whereupon $P_{\rm SH} = P_{\rm OT}^{\rm III}$, and the rotor is again at synchronous speed, that is, $\omega_{\rm R} = \omega_{\rm S}$. At this point all the kinetic energy stored in the rotor mass, designated by the area $\mathbf{A}_1$, Fig. 6.32, has been returned to the infinite bus in the form of electrical energy, designated by the area $\mathbf{A}_2$, Fig. 6.32.

Again, utilization of the constraint $P_{\rm SH} = P_{\rm OT}^{\rm III}$ that prevails at the equilibrium point M, Fig. 6.32, enables us to calculate the power angle $\delta = \delta_{\rm MAX} \equiv 180° - \delta_M$, $90° < \delta_{\rm MAX} < 180°$, by equating the generated power $P_{\rm OT}^{\rm III}$, Eq. (6.144), to the power $P_{\rm SH}$ input to the generator (neglecting frictional losses, etc.) from the prime mover. Thus

$$P_{\rm SH} \equiv 3\left(\frac{|\mathbf{V}_{\rm B}||\mathbf{E}_{\rm G}|}{X^{\rm III}X_{\rm EQ}^{\rm III}}\right)\sin\delta_M \tag{6.145}$$

Solution of Eq. (6.145) for the power angle δ_M, at the equilibrium point M, Fig. 6.32, yields

$$\delta_M = \sin^{-1}\left\{\frac{P_{\rm SH}}{3(|\mathbf{V}_{\rm B}||\mathbf{E}_{\rm G}|/X_{\rm EQ}^{\rm III})}\right\} \tag{6.146}$$

Finally, the power angle at the equilibrium point M, Fig. 6.32, is calculated by subtracting δ_M, Eq. (6.146), from 180° (why?). Thus

$$\delta_{\rm MAX} = 180° - \delta_M = 180° - \sin^{-1}\left\{\frac{P_{\rm SH}}{3(|\mathbf{V}_{\rm B}||\mathbf{E}_{\rm G}|/X_{\rm EQ}^{\rm III})}\right\} \tag{6.147}$$

The area $\mathbf{A}_1$, Fig. 6.32, is calculated, utilizing Eqs. (6.132) and (6.142) and ignoring the constant k of Eq. (6.135) because it cancels out in further mathematical operations, Eq. (6.136). Thus

$$\mathbf{A}_1 = \int_{\delta_{\rm O}}^{\delta_{\rm C}}(P_{\rm SH} - P_{\rm OT}^{\rm II})\,d\delta = \int_{\delta_{\rm O}}^{\delta_{\rm C}}(P_{\rm SH} - 0)\,d\delta = P_{\rm SH}(\delta_{\rm C} - \delta_{\rm O}) \tag{6.148}$$

Similarly, the area $\mathbf{A}_2$, Fig. 6.32, is calculated utilizing Eqs. (6.134) and (6.144), and again ignoring the constant k, as

$$\begin{aligned}\mathbf{A}_2 &= -\int_{\delta_{\rm C}}^{\delta_{\rm MAX}}(P_{\rm SH} - P_{\rm OT}^{\rm III})\,d\delta = -\int_{\delta_{\rm C}}^{\delta_{\rm MAX}}(P_{\rm SH} - k_3P_{\rm MAX}\sin\delta)\,d\delta \\ &= -(P_{\rm SH}\delta + k_3P_{\rm MAX}\cos\delta)\Big|_{\delta_{\rm C}}^{\delta_{\rm MAX}} \\ &= -P_{\rm SH}(\delta_{\rm MAX} - \delta_{\rm C}) - k_3P_{\rm MAX}(\cos\delta_{\rm MAX} - \cos\delta_{\rm C})\end{aligned} \tag{6.149}$$

The critical clearing angle $\delta_{\rm C}$ can now be calculated by equating the values of the areas $\mathbf{A}_1$ **and** $\mathbf{A}_2$, **that is, by applying the equal-area criterion and solving the resulting equation for** $\delta_{\rm C}$. Since

$$\mathbf{A}_2 = -[+P_{\rm SH}(\delta_{\rm MAX} - \delta_{\rm C}) + k_3P_{\rm MAX}(\cos\delta_{\rm MAX} - \cos\delta_{\rm C})] \tag{6.150}$$

then the application of the equal-area criterion, that is, $\mathbf{A}_1 = \mathbf{A}_2$, to Eqs. (6.148) and (6.150) and simplification yield

$$k_3 P_{MAX} \cos\delta_C = P_{SH}(\delta_{MAX} - \delta_O) + k_3 P_{MAX} \cos\delta_{MAX} \tag{6.151}$$

Keeping in mind that (1) $90° < \delta_{MAX} < 180°$ always and (2) the angles in the nontrigonometric terms must always be expressed in **radians**, we can solve Eq. (6.151) for the critical clearing angle δ_C as

$$\delta_C = \cos^{-1}\left[\left(\frac{P_{SH}}{k_3 P_{MAX}}\right)(\delta_{MAX} - \delta_O) + \cos\delta_{MAX}\right] \tag{6.152}$$

Note that all quantities of the right-hand side of Eq. (6.152) are known either directly, through input data, or indirectly through calculations, for example, Eqs. (6.141) and (6.147).

Example 6.7

The per-phase single-line diagram of Fig. 6.7.1 shows a generator connected through two parallel high-voltage transmission lines, connected between buses No. 1 and No. 2, to a large metropolitan system considered as an infinite bus. The relay-circuit-breaker sets A and B, and C and D, are adjusted to clear simultaneously. [In practice, each of the four relay-circuit-breaker sets is connected to each of the paralleling buses* through a transformer-relay-circuit-breaker set. Usually the synchronous generator and the infinite bus are also connected to the corresponding buses through relay-circuit-breaker sets (see Example 6.8). However, for the purpose of this example, the circuit diagram of Fig. 6.7.1 suffices.] Assume that a balanced three-phase fault is applied to the far end of the line AB near the circuit breaker B, as shown in Fig. 6.7.1. Calculate the **critical clearing angle** δ_C for clearing the fault with simultaneous opening of the circuit breakers A and B if the synchronous generator is delivering the prime-mover input (neglecting frictional losses and so on) 300.0 MW at the instant preceding the fault.

Solution

1. Prefault status: The reactance of the system of Fig. 6.7.1, exhibited in detail in Fig. 6.31(a), is calculated utilizing Eq. (6.138) and input data as

$$X_{EQ}^{I} = X_S + \frac{(X_{AB}) \times (X_{CD})}{X_{AB} + X_{CD}} + X_B = 23.00 + \frac{(60) \times (50)}{60 + 50} + 16 = \underline{66.27\ \Omega} \tag{1}$$

The generated power P_{OT}^{I} transmitted to the infinite bus is calculated with the aid of Eq. (6.139) and input data as

$$P_{OT}^{I} = \left(\frac{3|\mathbf{V}_B||\mathbf{E}_G|}{\mathbf{X}_{EQ}^{I}}\right)\sin\delta = \left[\frac{3(100{,}000)(125{,}000)}{66.27}\right]\sin\delta = \underline{565.9\,\sin\delta}\ \text{MW} \tag{2}$$

2. Faulted status: When the fault is applied to the system of Fig. 6.7.1, exhibited in detail in Fig. 6.31(b), there is no power transmitted to the infinite bus. Therefore, from Eq. (6.142),

$$P_{OT}^{II} = \underline{0}\ \text{W} \tag{3}$$

*A nodal point of a transmission network is defined to be a "bus."

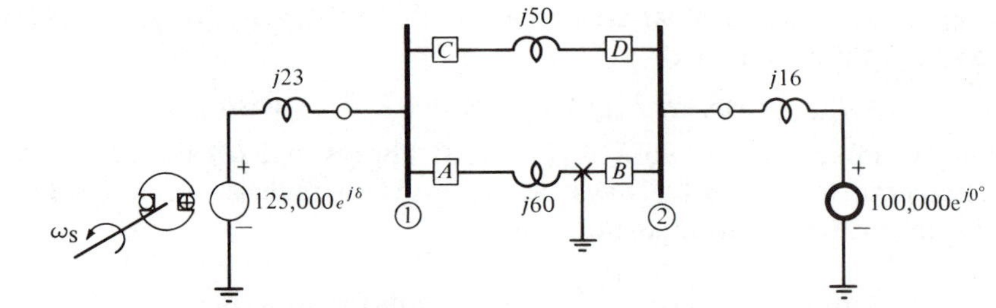

Fig. 6.7.1 A Synchronous Generator Connected Through Two Parallel High-Voltage Transmission Lines and the Paralleling Buses No. 1 and No. 2 to a Large Metropolitan System Considered as an Infinite Bus: Note the Number and Position of the Relay-Circuit-Breaker Sets and the Position of the Fault

3. Postfaulted status: When the relay-circuit-breaker sets A and B open and remove the transmission line AB from the system of Fig. 6.7.1, exhibited in detail in Fig. 6.31(c), the reactance X_{EQ}^{III} of the system is calculated utilizing Eq. (6.143) and input data as

$$X_{EQ}^{III} = X_S + X_{CD} + X_B = 23 + 50 + 16 = \underline{89}\ \Omega \tag{4}$$

The generated power P_{OT}^{III} transmitted to the infinite bus is calculated with the aid of Eq. (6.144) and input data as

$$P_{OT}^{III} = \left(\frac{3|\mathbf{V}_B||\mathbf{E}_G|}{\mathbf{X}_{EQ}^{III}}\right)\sin\delta = \left[\frac{3(100{,}000)(125{,}000)}{89}\right]\sin\delta$$
$$= \underline{421.3 \sin\delta \text{ MW}} \tag{5}$$

Now the fault power angle $\delta = \delta_O$ can be calculated by equating the powers P_{SH} and P_{OT}^{I} at the first equilibrium point of Fig. 6.32, that is, point O. Thus equating the power P_{SH} of 300.0 MW delivered from the prime mover, neglecting frictional losses, and so on, to the generator, with the power developed at the fault status, that is, Eq. (2), yields

$$P_{SH} \equiv P_{OT}^{I} \tag{6}$$

Proper substitution from the input data and Eq. (2) into Eq. (6) yields

$$300.0 \times 10^6 = 565.9 \times 10^6 \sin\delta|_{\delta=\delta_O} \tag{7}$$

Solution of Eq. (7) for the power angle δ_O, at the prefault equilibrium point O, yields

$$\delta_O = \sin^{-1}\left(\frac{300.0}{565.9}\right) = \underline{32.01°} \equiv \underline{0.5588} \text{ rad} \tag{8}$$

The maximum allowable (or postfault) power angle $\delta_{MAX} = 180° - \delta_M$ can be calculated by equating the powers P_{SH} and P_{OT}^{III} at the second equilibrium point of Fig. 6.32, that is, point M, and calculating the angle δ_M. Thus equating the power of 300.0 MW delivered from the prime mover, neglecting frictional losses and so on, to the generator, with the power calculated at the postfault status, that is, Eq. (5), yields

$$P_{SH} \equiv P_{OT}^{III} \tag{9}$$

Proper substitution from the input data and Eq. (5) into Eq. (9) yields

$$300.0 \times 10^6 = 421.3 \times 10^6 \sin\delta|_{\delta=\delta_M} \tag{10}$$

Plate 6.11 Vacuum Circuit Breaker with Front Plate Removed and Polyurethane Shield Added to Demonstrate Contact Erosion Indicator Check. Vacuum Is a Nearly Perfect Medium for Arc Interruption. One Set of Contacts Performs the Function of Both Main and Arcing Contacts—The Vacuum Environment Guarantees No Contact Maintenance. (Courtesy of General Electric Co., Schenectady, N.Y.)

Solution of Eq. (10) for the power angle δ_M (i.e., the transient stability limit) at the second equilibrium point of Fig. 6.32, that is, point M, yields

$$\delta_M = \sin^{-1}\left(\frac{300.0}{421.3}\right) = \underline{45.40°} \tag{11}$$

Finally, utilization of Eq. (6.147) enables the calculation of the maximum allowable angle, designated as δ_{MAX} rather than δ_M (why?) as

$$\delta_{MAX} = 180° - \delta_M = 180.00° - 45.40° = \underline{134.6°} \equiv \underline{2.349} \text{ rad} \tag{12}$$

The kinetic energy stored in the mass of the rotor during acceleration, viewed in terms of the area **A**, can be calculated with the aid of Fig. 6.32 and Eqs. (6.148) and (6.147) and input data as

$$\mathbf{A}_1 = \int_{\delta_O=0.5588}^{\delta_C} (P_{SH} - P_{OT}^{II})\, d\delta = (300.0\times10^6 - 0)\times(\delta_C - 0.5588)$$
$$= \underline{300.0\delta_C - 167.6 \text{ MW}\cdot\text{rad}} \tag{13}$$

Similarly, the kinetic energy delivered to the infinite bus from the mass of the rotor during deceleration, viewed in terms of the area $\mathbf{A}_2$, can be calculated with the aid of Fig. 6.32 and Eqs. (6.149) and (6.144) and input data as

$$\mathbf{A}_2 = -\int_{\delta_C}^{\delta_{MAX}=2.349} (P_{SH} - P_{OT}^{III})\,d\delta$$

$$= -\int_{\delta_C}^{\delta_{MAX}=2.349} (300.0 - 421.3\sin\delta) \times 10^6 d\delta$$

$$= -(300.0\delta + 421.3\cos\delta) \times 10^6 \Big|_{\delta_C}^{2.349}$$

$$= -300.0(2.349) - 421.3\cos(2.349) + 300.0\delta_C + 421.3\cos\delta_C \quad \text{MW}\cdot\text{rad}$$

$$= \underline{-704.7 + 295.8 + 300.0\delta_C + 421.3\cos\delta_C} \quad \text{MW}\cdot\text{rad} \tag{14}$$

$$\mathbf{A}_2 = \underline{(-408.9 + 300.0\delta_C + 421.3\cos\delta_C)} \quad \text{MW}\cdot\text{rad} \tag{15}$$

The equal-area criterion states that $\mathbf{A}_1 = \mathbf{A}_2$. Thus, equating the right-hand sides of Eqs. (13) and (15) yields

$$300.0\delta_C - 167.6 = -408.9 + 300.0\delta_C + 421.3\cos\delta_C \tag{16}$$

Simplification of Eq. (16) yields

$$421.3\cos\delta_C = 241.3 \tag{17}$$

Finally, solution of Eq. (17) for the **critical clearing angle** δ_C yields

$$\delta_C = \cos^{-1}\left(\frac{241.3}{421.3}\right) = \underline{\underline{55.06°}} \tag{18}$$

Nonlinear Analysis (Case 2)

When the midpoint of the transmission line AB, Fig. 6.28(b), is to be faulted, Fig. 6.33(a), the generated power P_{OT}^{I}, Fig. 6.34, delivered by the synchronous generator to the infinite bus, suddenly reduces to a nonzero value P_{OT}^{II}, Fig. 6.34 and Fig. 6.33(b) and 6.33(c). At a preset time the relay circuit breakers A and B open, and thus they remove the transmission line AB from the system. This action isolates the fault. The generated power delivered to the infinite bus picks up to some new value P_{OT}^{III}, Fig. 6.34 and Fig. 6.33(d), as the synchronous generator transfers power to the infinite bus over the remaining transmission line CD. The status of the system, Fig. 6.28(b), as it goes through the cycle of (1) prefaulted, (2) faulted, and (3) postfaulted operation is described by the single-line diagrams of Figs. 6.33(a), 6.33(b), 6.33(c), 6.33(d), and Fig. 6.34, and is analyzed as described in the following sections.

PREFAULTED STATUS, FIG. 6.33(a)

The system is operating at its normal status. It operates at the equilibrium point O, that is, the intersection of line P_{SH} and curve P_{OT}^{I}, Fig. 6.34, where the power angle value is δ_0, $P_{SH} = P_{OT}^{I}$, and the rotor rotates at synchronous speed, that is, $\omega_R = \omega_S$. The equivalent reactance of the system X_{EQ}^{I} is calculated from Fig. 6.33(a) as*

$$X_{EQ}^{I} = X_S + \frac{(k_A X_{AB} + k_B X_{AB}) \times (X_{CD})}{(k_A X_{AB} + k_B X_{AB}) + X_{CD}} + X_B \tag{6.153}$$

The generated power transmitted to the infinite bus is calculated with the aid of

*The values of k_A and k_B should be such as to satisfy the equation $k_A X_{AB} + k_B X_{AB} = X_{AB}$. In this case consider $k_A = k_B = 0.50$.

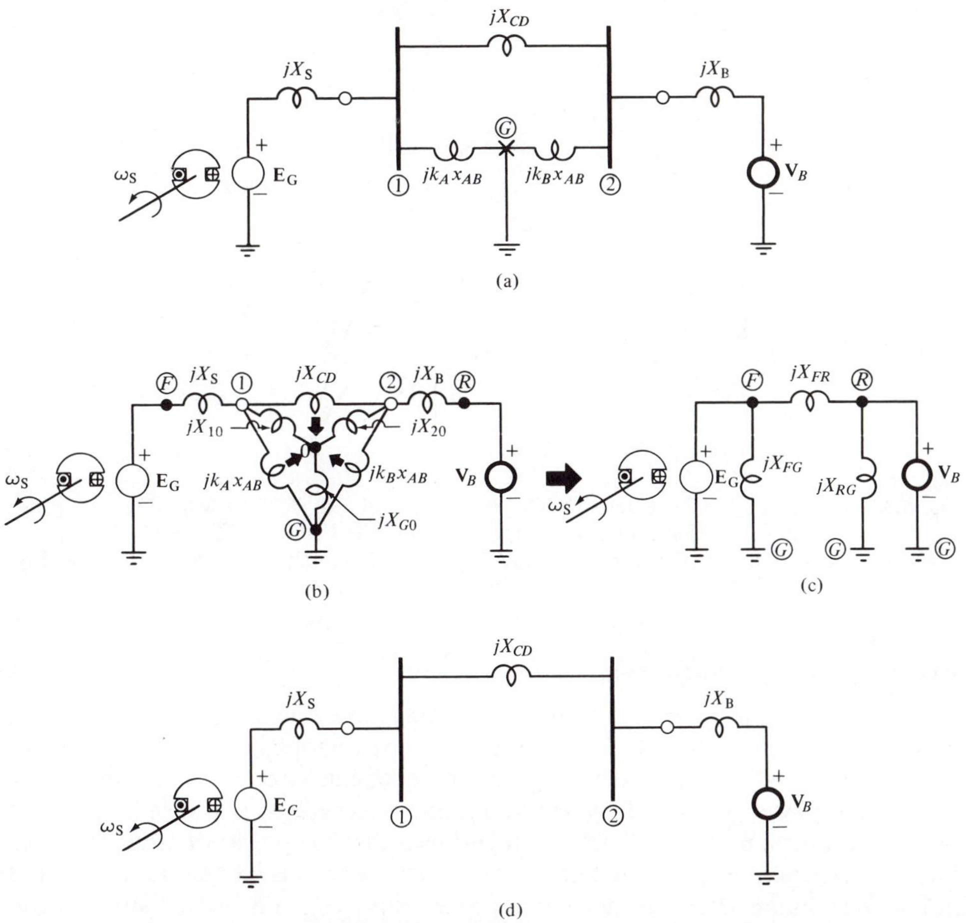

Fig. 6.33 One-line Impedance Diagram of Fig. 6.28(b) for (a) the Immediately Prefaulted Status, (b) the Faulted Status [the $\Delta \rightarrow Y$-Equivalent Transformation Is Designated by the Three Thick Arrowheads], (c) the Faulted Status [Note the Δ-Equivalent of the Modified Y-Equivalent], and (d) the Postfaulted Status

Eqs. (6.137) and (6.153) as

$$P_{OT}^{I} = 3\left(\frac{|\mathbf{V}_B||\mathbf{E}_G|}{X_{EQ}^{I}}\right)\sin\delta \tag{6.154}$$

As in the previous case [Eqs. (6.139) through (6.141)], the power angle at the equilibrium point O, Fig. 6.34, is calculated by equating the generating power P_{OT}^{I}, Eq. (6.154), to the input power P_{SH} to the generator (neglecting frictional losses, etc.) from the prime mover. Thus

$$P_{SH} \equiv 3\left(\frac{|\mathbf{V}_B||\mathbf{E}_G|}{X_{EQ}^{I}}\right)\sin\delta_O \tag{6.155}$$

Solution of Eq. (6.155) for the power angle δ_O at the equilibrium point O, Fig. 6.34, yields

$$\delta_O = \sin^{-1}\left\{\frac{P_{SH}}{3(|\mathbf{V}_B||\mathbf{E}_G|/X_{EQ}^{I})}\right\} \tag{6.156}$$

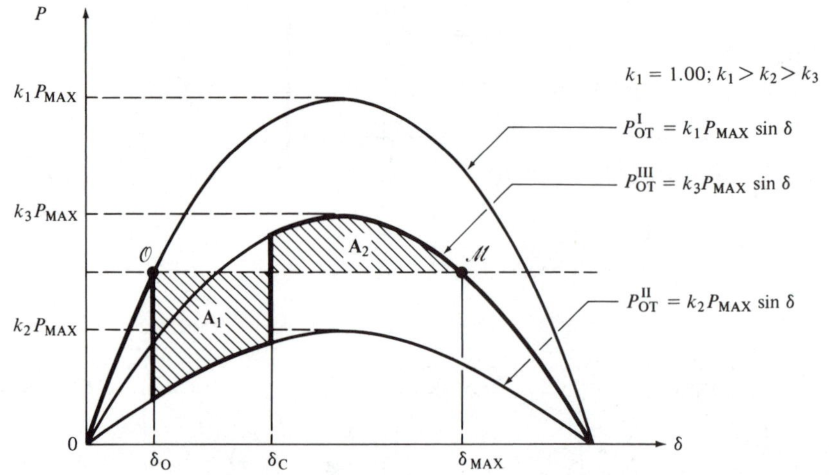

Fig. 6.34 Power-Angle Characteristics for the System of Fig. 6.28(b) with Powers Transmitted to the Infinite Bus at the (a) Prefaulted Status, Fig. 6.33(a), P^{I}_{OT}, (b) Faulted Status, Figs. 6.33(b) and 6.33(c), P^{II}_{OT}, and (c) Postfaulted Status, Fig. 6.33(d), P^{III}_{OT}

FAULTED STATUS, FIG. 6.33(b)

The fault at the midpoint G of the transmission line AB, Fig. 6.28(b), has occurred. The system is disturbed, that is, the synchronous generator loses part of its generated power to ground through the short circuit at point G. However, now part of the power generated by the synchronous generator is transmitted to the infinite bus, Figs. 6.33(b) and 6.33(c). It is shown that the power transmitted to the bus is a function of the reactance between the terminals of the generator and infinite bus, Fig. 6.33(c). The generated power, curve P^{I}_{OT}, Fig. 6.34, transmitted to the infinite bus, falls suddenly to a new value designated by curve P^{II}_{OT}, Fig. 6.34. This power is calculated with the aid of Eq. (6.137) and the equivalent reactance X^{II}_{EQ}, Eq. 6.168, of the faulted system. Because of the complexity of the faulted network [compare Fig. 6.33(b) with Fig. 6.31(b)], we must undertake a series of $\Delta \rightarrow Y$, and modified $Y \rightarrow \Delta$, transformations of the faulted network of Fig. 6.33(b) in order to be able to calculate its equivalent reactance X^{II}_{EQ}. Thus the Δ-network [defined by the terminals 1, 2, and G, Fig. 6.33(b)] comprised of the elements X_{CD}, $k_A X_{AB}$, and $k_B X_{AB}$ is converted to a Y-equivalent network utilizing the following equations:

$$\Sigma\{\Delta\} = X_{CD} + k_A X_{AB} + k_B X_{AB} \tag{6.157}$$

$$X_{10} = \frac{(X_{CD}) \times (k_A X_{AB})}{\Sigma\{\Delta\}} \tag{6.158}$$

$$X_{20} = \frac{(X_{CD}) \times (k_B X_{AB})}{\Sigma\{\Delta\}} \tag{6.159}$$

and

$$X_{G0} = \frac{(k_A X_{AB}) \times (k_B X_{AB})}{\Sigma\{\Delta\}} \tag{6.160}$$

It is obvious that the elements of the calculated Y-equivalent network that are connected to the terminals 1 and 2, Fig. 6.33(b), should be increased by the reactances X_S and X_B, respectively. Thus this **modified Y-equivalent** is comprised from the elements given by the following three equations:

$$X_{FO} = X_S + X_{10} \tag{6.161}$$

$$X_{RO} = X_B + X_{20} \tag{6.162}$$

and [the element X_{GO} of Eq. (6.160) remains unchanged; why?]

$$X_{GO} = X_{GO} \tag{6.163}$$

Finally, the Δ-equivalent of the modified Y-equivalent of Fig. 6.33(b), that is, Fig. 6.33(c), is calculated utilizing the following four equations:

$$\Sigma\{\Pi\} = (X_{FO}) \times (X_{RO}) + (X_{RO}) \times (X_{GO}) + (X_{GO}) \times (X_{FO}) \tag{6.164}$$

$$X_{FR} = \frac{\Sigma\{\Pi\}}{X_{OG}} \tag{6.165}$$

$$X_{FG} = \frac{\Sigma\{\Pi\}}{X_{RO}} \tag{6.166}$$

and

$$X_{RG} = \frac{\Sigma\{\Pi\}}{X_{FO}} \tag{6.167}$$

Note that since the elements X_{FG} and X_{RG} of Fig. 6.33(c) are (1) connected across the voltage sources $\mathbf{E}_G$ and $\mathbf{E}_B$ representing the synchronous generator and the infinite bus, respectively, and (2) purely reactive and do not absorb real power, they are not utilized in any calculations. Thus X_{FG} and X_{RG} *need not* be calculated at all. It is obvious, therefore, that the equivalent reactance X_{EQ}^{II} of Fig. 6.33(b) is equal to the reactance X_{FR} of the Δ-equivalent, Fig. 6.33(c). That is,

$$\mathrm{X}_{EQ}^{II} \equiv X_{FR} \tag{6.168}$$

Now the generated power P_{OT}^{II}, Fig. 6.34, transmitted to the infinite bus, Fig. 6.33(b), is calculated with the aid of Eqs. (6.137) and (6.168) as

$$P_{OT}^{II} = 3\left(\frac{|\mathbf{V}_B||\mathbf{E}_G|}{X_{FR}}\right)\Bigg|_{X_{FR}=\mathrm{X}_{EQ}^{II}} \sin\delta \tag{6.169}$$

The area $\mathbf{A}_1$, Fig. 6.34, is proportional to the kinetic energy stored in the rotor mass as it accelerates striving for the power angle δ_C. Note that the area $\mathbf{A}_1 \neq \mathbf{A}_1$, Figs. 6.34 and 6.32.

POSTFAULTED STATUS, FIG. 6.33(d)

The relay circuit breakers *A* and *B*, Fig. 6.33(a), have opened simultaneously and have removed the transmission line *AB* from the system. Thus they have removed the fault from the system, Fig. 6.33(d). The status of the system is improved, that is, the synchronous generator is connected to the infinite bus now directly through the transmission line *CD* and power flows from the synchronous generator to the infinite bus. The equivalent reactance X_{EQ}^{II} of the system is calculated from Fig. 6.33(d) as

$$\mathrm{X}_{EQ}^{II} = X_S + X_{CD} + X_B \tag{6.170}$$

Finally, the generated power P_{OT}^{III}, Fig. 6.34, transmitted to the infinite bus, is calculated with the aid of Eqs. (6.137) and (6.170) as

$$P_{OT}^{III} = 3\left(\frac{|\mathbf{V}_B||\mathbf{E}_G|}{\mathrm{X}_{EQ}^{II}}\right)\sin\delta \tag{6.171}$$

Again, after the faulted transmission line AB is removed from the system, the rotor overshoots the power angle $\delta_{\mathcal{C}}$. Since power is absorbed by the infinite bus, the rotor decelerates until it reaches the new equilibrium point $\mathcal{M}$, that is, the intersection of line P_{SH} and the curve P_{OT}^{III}, Fig. 6.34, but the power angle increases to a new value δ_{MAX} whereupon $P_{SH}=P_{OT}^{III}$, and the rotor is again at synchronous speed, that is, $\omega_R=\omega_S$. Now all the kinetic energy stored in the rotor mass, designated by the area $\mathbf{A}_1$, Fig. 6.34, has been returned to the infinite bus in the form of electrical energy, designated by the area $\mathbf{A}_2$, Fig. 6.34.

Again, as in the previous case [Eqs. (6.145) through (6.147)], the power angle at the equilibrium point $\mathcal{M}$, Fig. 6.34, $\delta=\delta_{MAX}\equiv 180°-\delta_{\mathcal{M}}$, $90°<\delta_{MAX}<180°$, is found by equating the generated power P_{OT}^{III}, Eq. (6.171), to the power P_{SH} input to the generator (neglecting frictional losses, etc.) from the prime mover. Thus

$$P_{SH} \equiv 3\left(\frac{|\mathbf{V}_B||\mathbf{E}_G|}{\mathbf{X}_{EQ}^{II}}\right)\sin\delta_{\mathcal{M}} \tag{6.172}$$

Solution of Eq. (6.172) for the power angle $\delta_{\mathcal{M}}$, at the equilibrium point $\mathcal{M}$, Fig. 6.34, yields

$$\delta_{\mathcal{M}} = \sin^{-1}\left\{\frac{P_{SH}}{3(|\mathbf{V}_B||\mathbf{E}_G|/\mathbf{X}_{EQ}^{II})}\right\} \tag{6.173}$$

Finally, the power angle δ_{MAX} at the equilibrium point $\mathcal{M}$, Fig. 6.34, is calculated by subtracting $\delta_{\mathcal{M}}$, Eq. (6.173), from 180° (why?). Thus

$$\delta_{MAX} = 180° - \delta_{\mathcal{M}} = 180° - \sin^{-1}\left\{\frac{P_{SH}}{3(|\mathbf{V}_B||\mathbf{E}_G|/\mathbf{X}_{EQ}^{III})}\right\} \tag{6.174}$$

As in the previous case, ignoring the constant k of Eq. (6.136) because of its cancellation in further mathematical operations, the area $\mathbf{A}_1$ is calculated utilizing Eqs. (6.132) and (6.169) as

$$\begin{aligned}\mathbf{A}_1 &= \int_{\delta_O}^{\delta_C}(P_{SH}-P_{OT}^{II})\,d\delta = \int_{\delta_O}^{\delta_C}(P_{SH}-k_2P_{MAX}\sin\delta)\,d\delta \\ &= (P_{SH}\delta + k_2P_{MAX}\cos\delta)\Big|_{\delta_O}^{\delta_C} \\ &= P_{SH}(\delta_C-\delta_O) + k_2P_{MAX}(\cos\delta_C - \cos\delta_O)\end{aligned} \tag{6.175}$$

Similarly, the area $\mathbf{A}_2$ is calculated, utilizing Eqs. (6.134) and (6.171), and again ignoring the constant k, as

$$\begin{aligned}\mathbf{A}_2 &= -\int_{\delta_C}^{\delta_{MAX}}(P_{SH}-P_{OT}^{III})\,d\delta = -\int_{\delta_C}^{\delta_{MAX}}(P_{SH}-k_3P_{MAX}\sin\delta)\,d\delta \\ &= -(P_{SH}\delta + k_3P_{MAX}\cos\delta)\Big|_{\delta_C}^{\delta_{MAX}} \\ &= -P_{SH}(\delta_{MAX}-\delta_C) - k_3P_{MAX}(\cos\delta_{MAX} - \cos\delta_C)\end{aligned} \tag{6.176}$$

Again, the critical clearing angle δ_C can be calculated by equating the areas $\mathbf{A}_1$ and $-\mathbf{A}_2$, that is, by applying the **equal-area criterion** and solving the resulting equation for δ_C.

The area $\mathbf{A}_2$ is calculated from Eq. (6.176) as follows:

$$\mathbf{A}_2 = -P_{SH}(\delta_{MAX}-\delta_C) - k_3P_{MAX}(\cos\delta_{MAX} - \cos\delta_C) \tag{6.177}$$

Now the application of the equal-area criterion, to Eqs. (6.175) and (6.177) and simplification yield

$$(k_2P_{MAX} - k_3P_{MAX})\cos\delta_C = P_{SH}(\delta_O - \delta_{MAX}) + k_2P_{MAX}\cos\delta_O - k_3P_{MAX}\cos\delta_{MAX} \tag{6.178}$$

Again, keeping in mind that (1) $90° < \delta_{MAX} < 180°$ always and (2) the angles in the nontrigonometric terms must be expressed in **radians**, we can solve Eq. (6.178) for the critical clearing angle δ_C as

$$\delta_C = \cos^{-1}\left\{\left(\frac{1}{k_2 P_{MAX} - k_3 P_{MAX}}\right)\left[P_{SH}(\delta_O - \delta_{MAX}) + k_2 P_{MAX}\cos\delta_O - k_3 P_{MAX}\cos\delta_{MAX}\right]\right\} \tag{6.179}$$

Note again that all the quantities of the right side of Eq. (6.179) are known either indirectly, through input data, or directly through calculations, for example, Eqs. (6.156) and (6.174).

Example 6.8

The per-phase, single-line diagram of Fig. 6.8.1 shows a generator connected through two parallel high-voltage transmission lines, connected between buses No. 1 and No. 2, to a large metropolitan system considered as an infinite bus. This system is identical to the system of Example 6.7 except that the transmission lines of reactance $j24.0\ \Omega$ are connected to the paralleling buses through transformers of reactance $j13.0\ \Omega$ and relay-circuit-breaker sets on both sides of each transformer. In addition, the synchronous generator and the infinite bus are connected to the paralleling buses through relay-circuit-breaker sets also. The relay-circuit-breaker sets *A* and *B*, and *C* and *D*, are adjusted to open or close simultaneously.

Assume that a balanced three-phase fault is applied to the far end of the relay-circuit-breaker-transformer line unit near the relay circuit breaker *G* as

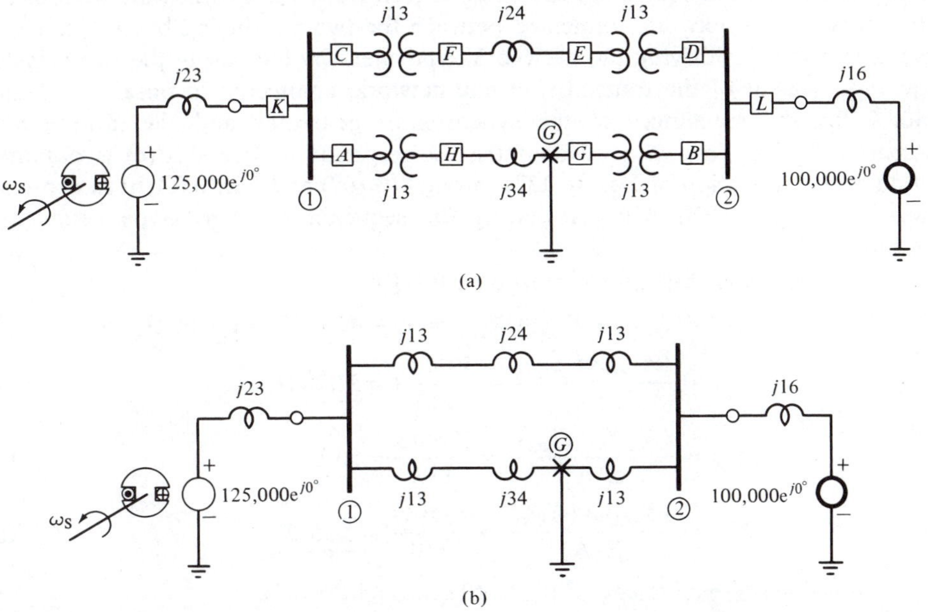

Fig. 6.8.1 A Synchronous Generator Connected to a Large Metropolitan System Considered as an Infinite Bus. Note Through Two Parallel High-Voltage Transmission Lines, Four Coupling Transformers, and the Paralleling Buses No. 1 and No. 2. (a) the Number and Position of the Relay-Circuit-Breaker Sets, the Coupling Transformers, and the Position of the Fault; and (b) the Equivalent Circuit of the System Exhibited in Part (b)

shown in Fig. 6.8.1. Calculate the **critical clearing angle** δ_C for clearing the fault with simultaneous opening of the relay circuit breakers A and B if the synchronous generator is delivering the prime-mover input, neglecting frictional losses and so on, of 300.0 MW at the instant preceding the fault.

Solution

1. Prefaulted status: The reactance of the system of Fig. 6.8.1 [(a) or (b)], exhibited in detail in Fig. 6.33(a), is calculated utilizing Eq. (6.153) and input data as

$$\mathbf{X}_{\mathrm{EQ}}^{\mathrm{I}} = X_{\mathrm{S}} + \frac{(k_A X_{AB} + k_B X_{AB}) \times (X_{CD})}{k_A X_{AB} + k_B X_{AB} + X_{CD}} + X_{\mathrm{B}}$$

$$= 23.00 + \frac{(13+34+13)\times(13+24+13)}{13+34+13+13+24+13} + 16 = \underline{66.27\ \Omega} \tag{1}$$

The generated power $P_{\mathrm{OT}}^{\mathrm{I}}$ transmitted to the infinite bus is calculated with the aid of Eq. (6.154) and input data as

$$P_{\mathrm{OT}}^{\mathrm{I}} = \left(\frac{3|\mathbf{V}_{\mathrm{B}}||\mathbf{E}_{\mathrm{G}}|}{\mathbf{X}_{\mathrm{EQ}}^{\mathrm{I}}}\right)\sin\delta = \left[\frac{3(100{,}000)(125{,}000)}{66.27}\right]\sin\delta$$

$$= \underline{565.9 \sin\delta\ \mathrm{MW}} \tag{2}$$

2. Faulted status: When the fault is applied to the system of Fig. 6.8.1 [(a) or (b)], exhibited in detail in Figs. 6.33(b) and 6.33(c), the network between the terminals front (F) and rear (R) to which the voltage sources, which designate the synchronous generator and the infinite bus, are connected can be viewed as a Δ-equivalent. This Δ-equivalent has been derived from a Δ-Y-transformation of the transmission line network connected between the two paralleling buses No. 1 and No. 2 and then modifying the derived Y-equivalent by increasing the two arms of the Y-equivalent of the transmission line network, connected to buses No. 1 and No. 2, by the reactances of the synchronous generator and the infinite bus, respectively. The final Δ-equivalent that corresponds to Fig. 6.33(c) is obtained through the utilization of Eqs. (6.157) through (6.167) and Fig. 6.8.1(b) in conjunction with Fig. 6.33(b). Thus, following the sequence of steps given below one obtains:

i. Y-equivalent; Eqs. (6.157) through (6.160):

$$\Sigma\{\Delta\} = X_{CD} + X_{AG} + X_{GB} = 50 + 47 + 13 = \underline{110.00\ \Omega} \tag{3}$$

$$X_{10} = \frac{(X_{12})\times(X_{1\mathrm{G}})}{\Sigma\{\Delta\}} = \frac{50\times 47}{110} = \underline{21.36\ \Omega} \tag{4}$$

$$X_{20} = \frac{(X_{21})\times(X_{2\mathrm{G}})}{\Sigma\{\Delta\}} = \frac{50\times 13}{110} = \underline{5.91\ \Omega} \tag{5}$$

$$X_{\mathrm{G0}} = \frac{(X_{\mathrm{G2}})\times(X_{\mathrm{G1}})}{\Sigma\{\Delta\}} = \frac{13\times 47}{110} = \underline{5.55\ \Omega} \tag{6}$$

ii. Modified Y-equivalent; Eqs. (6.161) through (6.163):

$$X_{\mathrm{F0}} = X_{\mathrm{F1}} + X_{10} = 23.00 + 21.36 = \underline{44.36\ \Omega} \tag{7}$$

$$X_{\mathrm{R0}} = X_{\mathrm{R2}} + X_{20} = 16.00 + 5.91 = \underline{21.91\ \Omega} \tag{8}$$

$$X_{\mathrm{G0}} = \underline{5.55\ \Omega} \tag{9}$$

iii. Δ-equivalent; Eqs. (6.164) through (6.167):

$$\Sigma\{\Pi\} = (X_{FO})\times(X_{RO}) + (X_{RO})\times(X_{GO}) + (X_{GO})\times(X_{FO})$$
$$= (44.36)(21.91) + (21.91)(5.55) + (5.55)(44.36)$$
$$= \underline{1340.0\ \Omega} \quad (10)$$

$$X_{FR} = \frac{\Sigma(\Pi)}{X_{GO}} = \frac{1340.0}{5.55} = \underline{241.44\ \Omega} \quad (11)$$

$$X_{FG} = \frac{\Sigma(\Pi)}{X_{RO}} = \frac{1340.0}{21.91} = \underline{61.2}\ \Omega \quad (12)$$

$$X_{RG} = \frac{\Sigma(\Pi)}{X_{FO}} = \frac{1340.0}{44.36} = \underline{30.3}\ \Omega \quad (13)$$

The generated power transmitted to the infinite bus is calculated with the aid of Eqs. (6.169) and (6.168) and input data as

$$P_{OT}^{II} = \frac{3|\mathbf{V}_B||\mathbf{E}_G|}{X_{FR}}\sin\delta = \left[\frac{3(100{,}000)(125{,}000)}{241.39}\right]\sin\delta$$
$$= \underline{155.4\sin\delta\ \text{MW}} \quad (14)$$

3. Postfaulted status: When the relay-circuit-breaker sets A and B open and remove the transmission line AB from the system of Fig. 6.8.1 [(a) or (b)], exhibited in detail in Fig. 6.33(d), the reactance X_{EQ}^{III} of the system is calculated utilizing Eq. (6.170) and input data as

$$X_{EQ}^{III} = X_S + X_{CD} + X_B = 23 + (13+24+13) + 16 = \underline{\underline{89}}\ \Omega \quad (15)$$

The generated power P_{OT}^{III} transmitted to the infinite bus is calculated with the aid of Eq. (6.171) and input data as

$$P_{OT}^{III} = \left(\frac{3|\mathbf{V}_B||\mathbf{E}_G|}{X_{EQ}^{III}}\right)\sin\delta = \left[\frac{3(100{,}000)(125{,}000)}{89}\right]\sin\delta$$
$$= \underline{421.3\sin\delta\ \text{MW}} \quad (16)$$

Now the prefault power angle $\delta = \delta_O$ can be calculated by equating the powers P_{SH} and P_{OT}^{III} at the first equilibrium point of Fig. 6.34, that is, point O. Thus equating the generated power P_{SH} of 300.0 MW delivered from the prime mover (neglecting frictional losses, etc.) to the generator, with the power developed at the prefault status, that is, Eq. (2), yields

$$P_{SH} \equiv P_{OT}^{I} \quad (17)$$

Proper substitution from the input data and Eq. (2) into Eq. (17) yields

$$300.0\times10^6 = 565.9\times10^6\sin\delta|_{\delta=\delta_O} \quad (18)$$

Solution of Eq. (18) for the power angle δ_O, at the prefault equilibrium point, yields

$$\delta_O = \sin^{-1}\left(\frac{300.0}{565.9}\right) = \underline{32.01°} \equiv \underline{0.5588}\ \text{rad} \quad (19)$$

The maximum allowable (or postfault) power angle $\delta_{MAX} = 180° - \delta_{\mathfrak{M}}$ can be calculated by equating the powers P_{SH} and P_{OT}^{III} at the second equilibrium point of Fig. 6.34, that is, point $\mathfrak{M}$, and calculating the angle $\delta_{\mathfrak{M}}$. Thus equating the power of 300.0 MW delivered from the prime mover (neglecting frictional losses, etc.) to the generator, with the power calculated at the postfaulted status, that is, Eq. (16),

Plate 6.12 13.8-kV Metal-clad Switchgear. Interruption Principle: Air Magnetic Power Circuit Breakers. (Courtesy of General Electric Co., Schenectady, N.Y.)

yields

$$P_{SH} = P_{OT}^{III} \tag{20}$$

Proper substitution from the input data and Eq. (16) into Eq. (20) yields

$$300.0 \times 10^6 = 421.3 \times 10^6 \sin\delta|_{\delta=\delta_{\mathfrak{M}}} \tag{21}$$

Solution of Eq. (21) for the power angle $\delta_{\mathfrak{M}}$ at the second equilibrium point of Fig. 6.34, that is, point $\mathfrak{M}$, yields

$$\delta_{\mathfrak{M}} = \sin^{-1}\left(\frac{300.0}{421.3}\right) = \underline{45.40°} \tag{22}$$

Finally, the utilization of Eq. (6.174) enables the calculation of the maximum allowable angle, designated as δ_{MAX} rather than $\delta_{\mathfrak{M}}$ (why?), as

$$\delta_{MAX} = 180° - \delta_{\mathfrak{M}} = 180.00° - 45.40° = \underline{134.6°} \equiv \underline{2.349}\text{ rad} \tag{23}$$

The kinetic energy stored in the mass of the rotor during acceleration, viewed in terms of the area $\mathbf{A}_1$, Fig. 6.34, can be calculated with the aid of Fig. 6.34 and Eqs.

(6.175) and (6.169) and input data as

$$\mathbf{A}_1 = \int_{\delta_O=0.5588}^{\delta_C} (P_{SH} - P_{OT}^{II})\,d\delta$$
$$= \int_{\delta_O=0.5588}^{\delta_C} (300.0 - 155.4\sin\delta) \times 10^6\,d\delta$$
$$= (300.0\delta + 155.4\cos\delta) \times 10^6 \Big|_{0.5588}^{\delta_C}$$
$$= 300.0\delta_C + 155.4\cos\delta_C - 300.0(0.5588) - 155.4\cos(0.5588)\ \text{MW}\cdot\text{rad}$$
$$= \underline{155.4\cos\delta_C + 300.0\delta_C - 299.4\ \text{MW}\cdot\text{rad}} \tag{24}$$

Similarly, the kinetic energy delivered to the infinite bus from the mass of the rotor during deceleration, viewed in terms of the area $\mathbf{A}_2$, Fig. 6.34, can be calculated with the aid of Fig. 6.34 and Eqs. (6.177) and (6.166) and input data as

$$\mathbf{A}_2 = -\int_{\delta_C}^{\delta_{MAX}=2.349} (P_{SH} - P_{OT}^{III})\,d\delta$$
$$= -\int_{\delta_C}^{\delta_{MAX}=2.349} (300.0 - 421.3\sin\delta) \times 10^6\,d\delta$$
$$= -(300.0\delta + 421.3\cos\delta) \times 10^6 \Big|_{\delta_C}^{2.349}$$
$$= -300.0(2.349) - 421.3\cos(2.349) + 300.0\delta_C + 421.3\cos\delta_C \quad \text{MW}\cdot\text{rad}$$
$$= \underline{-704.7 + 295.8 + 300.0\delta_C + 421.3\cos\delta_C} \quad \text{MW}\cdot\text{rad} \tag{25}$$

The equal-area criterion states that $\mathbf{A}_1 \equiv \mathbf{A}_2$. Thus equating Eqs. (24) and (25) yields

$$155.4\cos\delta_C + 300.0\delta_C - 299.4 = 421.3\cos\delta_C + 300.0\delta_C - 408.9 \tag{26}$$

Simplification of Eq. (26) yields

$$265.9\cos\delta_C = 109.5 \tag{27}$$

Finally, solution of Eq. (27) for the **critical clearing angle δ_C** yields

$$\delta_C = \cos^{-1}\left(\frac{109.5}{265.9}\right) = \underline{\underline{65.68^\circ}} \tag{28}$$

It can be summarized that the object of the nonlinear analysis is fivefold. That is; (1) to determine whether or not synchronism—stability—is maintained, that is, whether or not the power angle δ settles down to a steady state operating value, after the synchronous device has been subjected to a sizable disturbance, (2) to calculate the precise plot of the power angle versus time, that is, the **swing curve**, (3) to calculate the **critical clearing angle δ_C, the largest possible value of δ for fault clearing to occur without the system losing synchronism, that is, becoming unstable**, (4) to calculate the **critical clearing time t_C, the time required by a synchronous device to swing from its original position to its critical clearing angle δ_C**, and (5) to estimate the **relay-circuit-breaker interrupting time** k_{RCB} (in cycles) or t_{RCB} (in seconds) after a fault occurs.

Most of this information is achieved by the *graphical* interpretation of (1) the energy stored and (2) the energy given up by the rotating mass of the synchronous device's rotor as areas $\mathbf{A}_1$ and $\mathbf{A}_2$, Figs. 6.32 and 6.34, respectively. The relative magnitudes of these areas are used as an aid in determining (1) the maximum angle of swing (and thus, the loss, preservation, or marginal preservation of synchronism, i.e., instability, stability, or marginal stability) and (2) the critical clearing angle δ_C.

Thus if $\mathbf{A}_j > \mathbf{A}_k$, $j=1$, 3, and 5 and $k=2$, 4, and 6, the accelerating momentum of the synchronous generator's rotor can never be overcome; the power angle versus time curve follows the course of any of the curves designated as No. 1, Fig. 6.35, and synchronism is lost, that is, the power angle δ increases without bound, and the system is said to have become **unstable.** If, on the other hand, $\mathbf{A}_1 < \mathbf{A}_2$, the accelerating momentum of the synchronous generator's rotor is overcome, the power angle versus time curve follows the course of the curve No. 2, Fig. 6.35, and synchronism is maintained with a margin indicated by the difference in areas. The power angle δ is bounded, that is, $0 \leqslant \delta \leqslant \delta_{ST}$, $\delta_{ST} < \delta_{MAX}$, and the system is said to be **stable.** Equality of the areas, however, that is, $\mathbf{A}_1 = \mathbf{A}_2$, yields the borderline case of marginal stability; the power angle versus time curve follows the course of curve No. 3, Fig. 6.35, and the system is said to be **marginally stable.**

In order to assure oneself of the stability of a given system, one must plot the swing curve of the seemingly stable system, for example, curve No. 2, Fig. 6.35, over a period of time **long enough** to guarantee that the power angle δ does not increase again without returning to a low value. **Sustained oscillations of the power angle δ, as long as $\delta_{ST} < \delta_{MAX}$, Fig. 6.35, designate stable systems.**

In addition to the stability information provided by comparing the relative magnitudes of the areas $\mathbf{A}_1$ and $\mathbf{A}_2$, the equality of areas $\mathbf{A}_1$ and $\mathbf{A}_2$, that is, the application of the equal-area criterion, enables the determination of the critical fault clearing angle δ_C. If we were to plot the power angle versus time utilizing Eq. (6.130), we could then read from this plot the **critical clearing time t_C,** as the time that corresponds to the critical clearing angle δ_C, calculated with the aid of Eqs. (6.152) and (6.179).

The critical clearing time t_C is essential in the calculation of the relay-circuit-breaker-set interrupting time, designated as t_{RCB} (seconds). However, usually it is

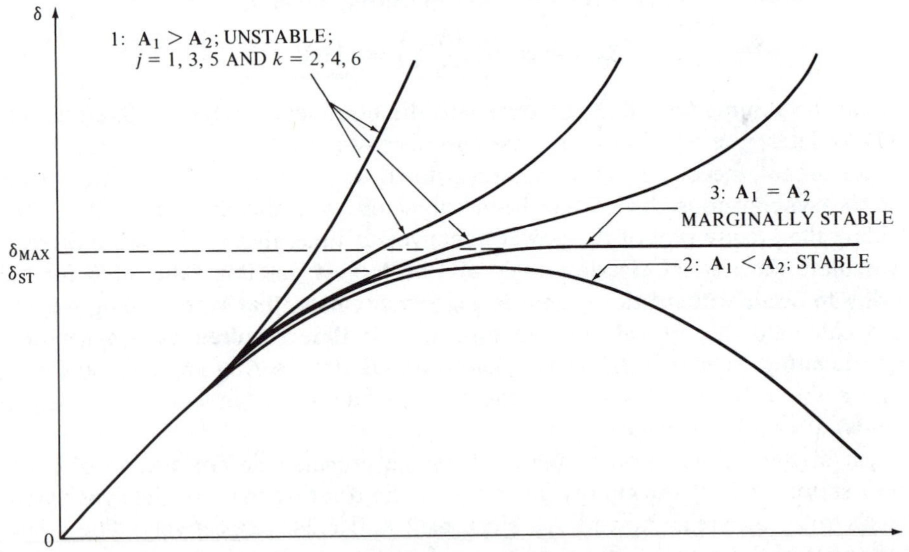

Fig. 6.35 Swing Curves for a Round-Rotor, Balanced, *P*-Pole, Three-Phase, Synchronous Generator Power System, Fig. 6.28(a), with a Balanced, Three-Phase Fault at the End of Line *AB*, the Various Curves Are the Results of Corresponding Critical Clearing Times of the Relay Circuit Breakers *A* and *B*

expressed in cycles instead and is designated as k_{RCB} (cycles). The relay-circuit-breaker time t_{RCB} must be within the critical clearing time t_C calculated. Standard interrupting time for relay-circuit-breaker sets is commonly 2, 3, 5, or 8 cycles. Mathematically, these quantities are related to each other and to the frequency of excitation f_E (hertz) or its corresponding period T_E (seconds) as

$$t_C = k_{RCB} T_E = \frac{k_{RCB}}{f_E}, \qquad t_{RCB} \leqslant t_C \tag{6.180}$$

Example 6.9

Which of the standard relay-circuit-breaker sets (2, 3, 5, or 8 cycles) would be used in the system of the Example 6.7 operating at 60 Hz if it were necessary to clear the fault in $t_C \leqslant 0.12$ sec?

Solution

Utilization of Eq. (6.180) yields

$$0.12 = k_{RCB} T_E = \frac{k_{RCB}}{60} \tag{1}$$

Solution of Eq. (1) for the relay-circuit-breaker critical clearing time k_{RCB} in cycles yields

$$k_{RCB} = (0.12) \times (60) = \underline{\underline{7.20}} \text{ cycles} \tag{2}$$

From the standard interrupting times of 2, 3, 5, or 8 cycles of available relay-circuit-breaker sets, **we would choose a 5-cycle set**. If we were to choose an 8-cycle set, we would *run the risk* of forcing the system to go unstable, because the fault would not be cleared quickly enough.

6.6 THE SOLUTION OF THE SWING EQUATION

In order to plot the swing curve, that is, δ vs. t, we must have the explicit solution of Eq. (6.130) in a format that would allow the point by point plot of such a curve. The task of deriving such a solution of $\delta(t)$ is undertaken beginning with Eq. (6.130) rewritten here for convenience, with "(t)" omitted and P_A used to define the **accelerating power** of the system. Thus,

$$M \frac{d^2}{dt^2} \delta = P_{SH} - P_{MAX} \sin\delta \equiv P_A \tag{6.181}$$

Eq. (6.181) can be rewritten as

$$\frac{d}{dt}\left(\frac{d}{dt}\delta\right) = \frac{d}{dt}\omega = \frac{P_A}{M}; \quad \omega \equiv \frac{d}{dt}\delta \tag{6.182}$$

or

$$d\omega = \frac{P_A}{M} dt \tag{6.183}$$

Integration of Eq. (6.183) yields

$$\int_{\omega_0}^{\omega} d\omega = \int_{t_0}^{t} \frac{P_A}{M} dt \tag{6.184}$$

Plate 6.13 Switchgear, Front View; Left-to-Right Sections Are ac Entry, Three SCRs, Control, Relay, and Breaker. Air Entry at Bottom Front; Exhaust at the Top. (Courtesy of General Electric Co., Schenectady, N.Y.)

or

$$\omega = \omega_0 + \frac{P_A}{M} t; \qquad t_0 \equiv 0 \tag{6.185}$$

Use of the defining relationship of Eq. (6.182) enables us to write Eq. (6.185) as

$$\frac{d}{dt}\delta = \omega_0 + \frac{P_A}{M} t \tag{6.186}$$

or

$$d\delta = \left(\omega_0 + \frac{P_A}{M} t\right) dt \tag{6.187}$$

Integration of Eq. (6.187) yields

$$\int_{\delta_0}^{\delta} d\delta = \int_{t_0}^{t}\left(\omega_0 + \frac{P_A}{M} t\right) dt \tag{6.188}$$

or

$$\delta = \delta_0 + \omega_0 t + \frac{P_A}{2M} t^2; \qquad t_0 \equiv 0 \tag{6.189}$$

It should be clear that in Eqs. (6.182) through (6.189), δ and ω designate the angular displacement and the angular velocity, respectively, of the rotor's d-axis with respect to the synchronously rotating reference frame, (SRRF), Fig. 6.25.

Note that Eqs. (6.185) and (6.189) yield the values of ω and δ at the end of the time interval $\Delta t = t - t_0$, if we designate the interval over which we integrate as the

kth interval, then allow (1) $t \to t_k$ and $t_0 \to t_{k-1}$, and (2) $\omega \to \omega_k$, $\omega_0 \to \omega_{k-1}$, $\delta \to \delta_k$, and $\delta_0 \to \delta_{k-1}$; where $k-1$ designates the beginning of the kth time interval and k designates the end of the same time interval, Eqs. (6.185) and (6.189). It is clear then that P_A in Eqs. (6.185) and (6.189) designates the accelerating power existing at the beginning of the kth time interval. Thus, if we allow $P_A \to P_{A(k-1)}$ Eqs. (6.185) and (6.189) can be written as

$$\omega_k = \omega_{k-1} + \frac{\Delta t}{M} P_{A(k-1)}; \qquad \Delta t \equiv t_k - t_{k-1} \tag{6.190}$$

and

$$\delta_k = \delta_{k-1} + \omega_{k-1}\Delta t + \frac{(\Delta t)^2}{2M} P_{A(k-1)}; \qquad \Delta t = t_k - t_{k-1} \tag{6.191}$$

These two equations can be solved for the speed and angle intervals as

$$\Delta\omega_k \equiv \omega_k - \omega_{k-1} = \frac{\Delta t}{M} P_{A(k-1)} \tag{6.192}$$

and

$$\Delta\delta_k \equiv \delta_k - \delta_{k-1} = \omega_{k-1}\Delta t + \frac{(\Delta t)^2}{2M} P_{A(k-1)} \tag{6.193}$$

In order to make δ_k of Eq. (6.191) independent of ω_{k-1}—for the plotting of the swing curve—we write Eq. (6.191) for the preceding time interval as

$$\delta_{k-1} = \delta_{k-2} + \omega_{k-2}\Delta t + \frac{(\Delta t)^2}{2M} P_{A(k-2)} \tag{6.194}$$

Subtraction of Eq. (6.194) from Eq. (6.191) yields

$$(\delta_k - \delta_{k-1}) = (\delta_{k-1} - \delta_{k-2}) + (\omega_{k-1} - \omega_{k-2})\Delta t + \frac{(\Delta t)^2}{M}\left\{\tfrac{1}{2}\left[P_{A(k-1)} - P_{A(k-2)}\right]\right\} \tag{6.195}$$

When the first three parentheses of Eq. (6.195) are defined as

$$\Delta\delta_k \equiv \delta_k - \delta_{k-1} \tag{6.196}$$

$$\Delta\delta_{k-1} \equiv \delta_{k-1} - \delta_{k-2} \tag{6.197}$$

and

$$\Delta\omega_{k-1} \equiv \omega_{k-1} - \omega_{k-2} \tag{6.198}$$

Eq. (6.195) is written as

$$\Delta\delta_k = \Delta\delta_{k-1} + \Delta\omega_{k-1}\Delta t + \frac{(\Delta t)^2}{M}\left\{\tfrac{1}{2}\left[P_{A(k-1)} - P_{A(k-2)}\right]\right\} \tag{6.199}$$

In order to eliminate $\Delta\omega_{k-1}$ from Eq. (6.199) we must rewrite Eq. (6.199) for the previous time interval first. This yields

$$\Delta\omega_{k-1} \equiv \omega_{k-1} - \omega_{k-2} = \frac{\Delta t}{M} P_{A(k-2)} \tag{6.200}$$

Now, substitution of Eq. (6.200) into Eq. (6.199) and proper collection of terms yield

$$\Delta\delta_k = \Delta\delta_{k-1} + \frac{(\Delta t)^2}{M}\left\{\tfrac{1}{2}\left[P_{A(k-1)} + P_{A(k-2)}\right]\right\} \tag{6.201}$$

Substitution of Eq. (6.201) into Eq. (6.196) and solution of the resulting equation for δ_k yield

$$\delta_k = \delta_{k-1} + \Delta\delta_{k-1} + \frac{(\Delta t)^2}{M}\left\{\tfrac{1}{2}\left[P_{A(k-1)} + P_{A(k-2)}\right]\right\} \tag{6.202}$$

Inspection of Eq. (6.202) yields the following pertinent information. The torque angle at the end of a given time interval is a function of the same angle at the end of the previous time interval, the increment in angle during the previous time interval, the momentum constant M, the average value of the accelerating power applied at the beginning of the two previous time intervals, and the time interval Δt squared. The **heavy dependence** of δ_k on Δt that is, on $(\Delta t)^2$, suggests that the time increment Δt should be chosen wisely so that the required accuracy of the swing curve is obtained but not so short as to unduly increase the number of points to be computed for a given swing curve.

Eq. (6.202) is understood better if we inspect Fig. 6.36(a), which designates the positive descending portion of a sinusoidal acceleration waveform. Note the physical significance of the terms $P_{A(k-2)}$, $P_{A(k-1)}$, and P_{Ak} and the effects these terms would have on calculations that take place on the corresponding time intervals, if the system's true accelerating power designated by the continuous-line curve, is compared to the assumed accelerating power designated by the dashed-line curve (i.e., curve 1, 2, 3, 4, 5, 6, 7). Since the accelerating power during each time interval is assumed constant and at a value equal to its value at the beginning of the time interval, the **assumed acceleration** $M\alpha \equiv P_A$, Eq. (6.181), over this time interval is always greater than the **true acceleration.** Consequently the acceleration, α, yields a calculated speed, $\omega = \omega_0 + \int_{t_0}^{t} \alpha dt$, which becomes progressively higher than the true speed, ω, which in turn yields a calculated displacement $\delta = \delta_0 + \int_{t_0}^{t} \omega dt$, which is greater than the true displacement. The second half of the sinusoidal oscillation, Fig. 6.36(a), therefore, begins with an amplitude greater than the true amplitude and continues either with this amplitude, or with an always increasing amplitude, if the phenomena of the first half of the oscillation repeat themselves in the second half of the oscillation. These phenomena would cause the calculated amplitude of the swing curve and its period of oscillation to increase with each swing, and thus, the plot would yield *erroneous* results.

It should be clear now that such results can be improved if we were to use the accelerating power at the **middle** of the time interval considered rather than the accelerating power at the interval's beginning. This becomes clear if we shift the dashed-line curve of Fig. 6.36(a) to the left, a distance equal to half of the interval Δt—note the thick continuous-line curve (i.e., curve a, b, c, d, e, f, g, h). Close examination of this curve yields that the accelerating power at the middle of the time interval is nearly equal to the **average** accelerating power during the time interval Δt. This is verified by the near equality of the areas of the triangles $3cm$ and $3dn$, for example. These ideas enable us to write the underlying principle under which the solution of the swing curve is obtained as: **the accelerating power as calculated at the beginning of a particular time interval is assumed to remain constant from the middle of the preceding time interval to the middle of the time interval being considered.**

With these ideas in mind then, let us calculate the angular displacement at the kth interval which begins at $t = (k-1)\Delta t$ and ends at $t = k\Delta t$. At this moment the

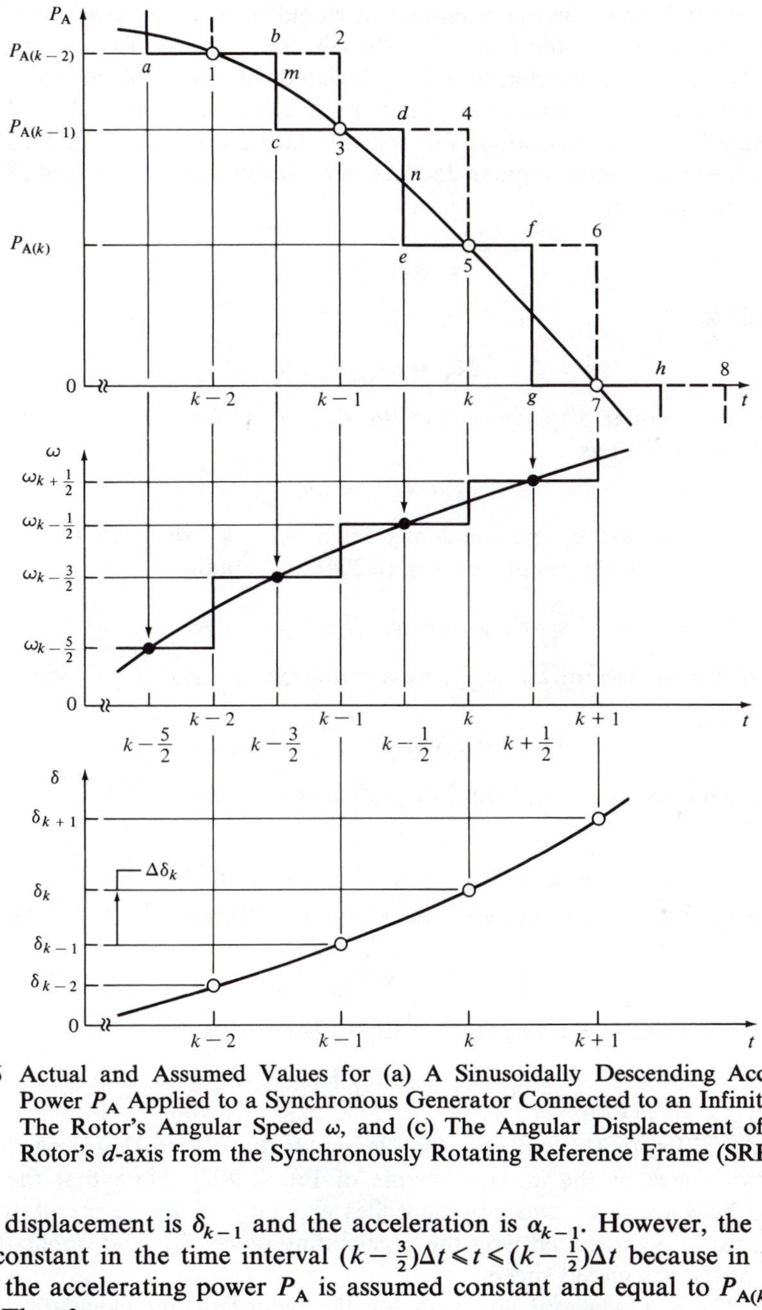

Fig. 6.36 Actual and Assumed Values for (a) A Sinusoidally Descending Accelerating Power P_A Applied to a Synchronous Generator Connected to an Infinite Bus, (b) The Rotor's Angular Speed ω, and (c) The Angular Displacement of δ of the Rotor's d-axis from the Synchronously Rotating Reference Frame (SRRF)

angular displacement is δ_{k-1} and the acceleration is α_{k-1}. However, the acceleration is constant in the time interval $(k-\frac{3}{2})\Delta t \leqslant t \leqslant (k-\frac{1}{2})\Delta t$ because in this time interval the accelerating power P_A is assumed constant and equal to $P_{A(k-1)}$, Fig. 5.36(a). Therefore,

$$\alpha_{k-1} \equiv \frac{\Delta\omega_{k-(1/2)}}{\Delta t}\bigg|_{\omega_{k-(3/2)}<\omega_{k-1}<\omega_{k-(1/2)}} = \frac{P_{A(k-1)}}{M} \tag{6.203}$$

where, from Fig. 5.36(b), the change in speed is given as

$$\Delta\omega_{k-(1/2)} \equiv \omega_{k-(1/2)} - \omega_{k-(3/2)} \tag{6.204}$$

Although the change in speed would occur linearly in time, as an outcome of the assumptions made regarding acceleration, it would be assumed that this change in speed occurs as a step at the middle of the kth time interval. That is, at the point $t=(k-1)\Delta t$ at which acceleration is calculated. Between steps, the speed is assumed constant. Therefore, throughout the kth interval, that is, $(k-1)\Delta t \leqslant t \leqslant k\Delta t$, the speed ω is assumed to remain constant at the value $\omega_{k-(1/2)}$. This speed is related to the incremental angular displacement during the kth interval of time by the following equation

$$\omega_{k-(1/2)} = \frac{\Delta\delta_k}{\Delta t} \tag{6.205}$$

which leads to

$$\Delta\delta_k = \omega_{-k(1/2)}\Delta t \tag{6.206}$$

Therefore, the angular displacement at the end of the kth interval is written with the aid of Fig. 5.36(c) as

$$\delta_k = \delta_{k-1} + \Delta\delta_k \tag{6.207}$$

In order to express δ_k independently from $\omega_{k-(1/2)}$ we solve Eq. (6.203) for $\omega_{k-(1/2)}\Delta t$, substitute the result into Eq. (6.204) and obtain

$$\frac{\Delta t}{M} P_{A(k-1)} = \omega_{k-(1/2)} - \omega_{k-(3/2)} \tag{6.208}$$

Solution of Eq. (6.208) for $\omega_{k-(1/2)}$ and substitution into Eq. (6.206) yields

$$\Delta\delta_k = \omega_{k-(3/2)}\Delta t + \frac{(\Delta t)^2}{M} P_{A(k-1)} \tag{6.209}$$

In order to eliminate $\omega_{k-(3/2)}$ from Eq. (6.209) we write Eq. (6.206) for the previous interval as

$$\Delta\delta_{k-1} = \omega_{k-(1/2)-1}\Delta t \equiv \omega_{k-(3/2)}\Delta t \tag{6.210}$$

Now, substitution of the right hand side of Eq. (6.210) into Eq. (6.209) yields

$$\Delta\delta_k = \Delta\delta_{k-1} + \frac{(\Delta t)^2}{M} P_{A(k-1)} \tag{6.211}$$

Finally, substitution of Eq. (6.211) into Eq. (6.209) yields

$$\delta_k = \delta_{k-1} + \Delta\delta_{k-1} + \frac{(\Delta t)^2}{M} P_{A(k-1)} \tag{6.212}$$

Comparison of Eq. (6.212) with Eq. (6.202) reveals that Eq. (6.212) is much simpler to use and is free of the inherent errors of Eq. (6.202). Note that the distinct difference between these two equations lies explicitly in the terms of these two equations that are associated with the accelerating power P_A, and implicitly in the accuracy of the calculated speed.

Eq. (6.212) is a **powerful** equation for the **point-by-point** calculation for the change, as a function of time, in the angular position δ of the synchronous device's rotor d-axis, with respect to the synchronously rotating reference frame (SRRF), during abrupt changes in the generator's electric output power.

It is important to remember that Eq. (6.212) is valid under the following two constraints: (1) the accelerating power P_A computed at the beginning of a time interval is constant from the middle of the proceeding time interval to the middle of the time interval considered, and (2) the speed (angular velocity) of the rotor is constant, throughout any time interval, and of value equal to that computed for the middle of the time interval considered. It should be noted, however, that neither of these principles are true as δ is changing continuously and both P_A and ω are functions of δ. As the time interval is decreased, however, the point-by-point computed swing curve approaches the true curve. Experience has shown that when a $\Delta t = 0.05$ s is used in Eq. (6.212), the results are excellent.

It is mentioned earlier that when faults occur or clear, or when switching operations take place, the accelerating power P_A exhibits discontinuities (Figs. 6.30, 6.32 and 6.34). As a result, three accelerating power discontinuities occur. They are as follows: A discontinuity may occur at (1) the beginning of a time interval, (2) the middle of a time interval, and (3) at a point other than the beginning or the middle of a time interval. These three cases are discussed at some length.

First, if the discontinuity occurs at the beginning of a time interval—a case encountered when we make an attempt to compute the increment of the angle δ occurring during the first time interval after a fault is applied to a system at $t = 0$ s —the accelerating power of Eq. (6.212) is written as

$$P_{A(k-1)} = \tfrac{1}{2}\left[P^-_{A\,(k-1)} + P^+_{A\,(k-1)}\right] \equiv \tfrac{1}{2}\left(P_{A(0^-)} + P_{A(0^+)}\right) \tag{6.213}$$

where $P^-_{A(k-1)}$ designates the accelerating power immediately before the fault is applied, and $P^+_{A(k-1)}$ designates the accelerating power immediately after the fault is applied. Because the system is operated in steady state, $P_{A(0^-)} \equiv 0$, and $P_{A(k-1)} \rightarrow (1/2)P_{A(0^+)}$, Eq. (6.213). For the same conditions of system operation the angular displacement interval $\Delta\delta$ in Eq. (6.212), for the time interval previous to that considered, is zero, that is, $\Delta\delta_{k-1} \rightarrow \Delta\delta_0 = 0$. If the fault is cleared at the beginning of the ℓth increment, the accelerating power in Eq. (6.212) is written as

$$P_{A(\ell-1)} = \tfrac{1}{2}\left[P^-_{A\,(\ell-1)} + P^+_{A\,(\ell-1)}\right] \tag{6.214}$$

where the components $P^-_{A(\ell-1)}$ and $P^+_{A(\ell+1)}$ designate the accelerating power immediately before and immediately after the fault is cleared, and $\Delta\delta_{k-1} \neq 0$.

Second, if the discontinuity occurs at the middle of a time interval, say the λth, no special procedure is applied and the accelerating power term of Eq. (6.212) is written as

$$P_{A(k-1)} \equiv P_{A(\lambda-1)} \tag{6.215}$$

and $\Delta\delta_{k-1} \neq 0$. The definition applied to $P_{A(k-1)}$ also applies to $P_{A(\lambda-1)}$.

Third, if the discontinuity occurs at some time other than the beginning or the middle of a time interval, a "weighted average" of the values of $\mathrm{P}_{A(k-1)}$ immediately before and immediately after the discontinuity should be used. Because, however, the time intervals for these computations are so short, these computations are performed with the discontinuity assumed to occur either at the beginning or the middle of the time interval with no substantial error introduced. As in the previous two cases, $\Delta\delta_{k-1} \neq 0$ in Eq. (6.212).

The momentum constant M of Eq. (6.212) was expressed in watts-seconds squared per electrical degree, earlier in the passage associated with Eq. (6.92). If the per unit system is to be used in the analysis of a system, the value of M that appears in Eq. (6.212) must be divided by the base power $|\mathrm{S}_{\mathcal{B}}|$ in VA, (as the entire equation is divided by $|\mathrm{S}_{\mathcal{B}}|$) to convert the equation to pu units. Now, M is expressed in units of **unit power seconds squared per electrical degree**, a value defined as the **pu value of** M.

In practice, however, M is not given in the units mentioned, but rather in terms of the **inertia constant** H of the synchronous device's rotor, which is defined as the stored kinetic energy, KE, at rated speed divided by the rated power, $|\mathbf{S}_R|$ in VA (or kVA or MVA), of the synchronous device. Mathematically these ideas are expressed as

$$H \equiv \frac{KE}{|\mathbf{S}_R|} \frac{\text{joules}}{\text{voltamps}} \tag{6.216}$$

The relationship of the inertia constant H to the momentum constant M is derived through the expression for the kinetic energy, KE, of a rotating body, written in terms of the moment of inertia J(kg·m^2) of the rotating body at an angular speed ω_R (rad/s) as

$$KE = \tfrac{1}{2} J\omega_R^2 \equiv \tfrac{1}{2} JM\omega_R \tag{6.217}$$

where, the expression for the angular momentum M(J-s per rad) is written as

$$M = J\omega_R \tag{6.218}$$

Solution of Eq. (6.217) for M, and use of Eq. (5.15), yields

$$M = \frac{2KE}{\omega_R}\bigg|_{\omega_R \equiv \omega_S = \frac{\omega_E}{(P/2)}} = \left(\frac{P}{2}\right) \times \left(\frac{2KE}{\omega_E}\right) \tag{6.219}$$

However, use of the expressions 2π rad$=360$ deg.$\rightarrow 1$ rad$=(360/2\pi)$ deg. and $\omega_E = 2\pi f_E$ enables us to write Eq. (6.219) as

$$M = \left(\frac{P}{2}\right) \times \left[\frac{2KE}{2\pi f_E\left(\dfrac{360}{2\pi}\right)}\right] = \left(\frac{P}{2}\right) \times \left(\frac{KE}{180 f_E}\right) \tag{6.220}$$

Substitution for KE into Eq. (6.220) from Eq. (6.216) yields

$$M = \left(\frac{P}{2}\right) \times \left(\frac{H|\mathbf{S}_R|}{180 f_E}\right) \frac{\text{joules}}{\text{elec. degree}} \tag{6.221}$$

or

$$M = \frac{H|\mathbf{S}_R|}{180 f_E} \frac{\text{joules}}{\text{mech. degree}} \tag{6.222}$$

Division of Eqs. (6.221) and (6.222) by the base power, $|\mathbf{S}_{\mathfrak{B}}|$ in VA, yields the pu value* of M, designated by $M^{[\text{pu}]}$, as

$$M^{[\text{pu}]} \equiv \frac{M}{|\mathbf{S}_{\mathfrak{B}}|} = \left(\frac{P}{2}\right) \times \left(\frac{H|\mathbf{S}_R|}{180 f_E |\mathbf{S}_{\mathfrak{B}}|}\right) = \frac{H|\mathbf{S}_R|}{180 f_R |\mathbf{S}_{\mathfrak{B}}|} \text{pu} \tag{6.223}$$

*In systems theory if one were to know the maximum value x_{MAX} of the variable x, then one can obtain the normalized variable $[x/x_{MAX}]$ whose ratio does not exceed ± 1. Similarly, in power systems engineering, if one were to define a common voltampere throughout a system and a voltage as base values designated as $|\mathbf{S}_{\mathfrak{B}}|$ and $|\mathbf{V}_{\mathfrak{B}}|$ correspondingly, then one can fix the base values of all other quantities used to describe a given system. The corresponding ratio of a given quantity q to its base value, designated as q_B (which is usually the rated design value for a given device) is written as $[q/q_B]$, which is expressed in **per unit** (voltamperes, volts, amperes, ohms or mhos) and is **abbreviated as pu (VA, V, A, Ω, or Ω^{-1})**. The per-unit quantities are the working quantities in power systems engineering. They are extensively used because they provide two distinct advantages: (1) the parameters of components lie in a very narrow numerical range in the neighborhood of one, and (2) one does not have to refer circuit variables to one side or the other of a given transformer. One can draw the equivalent circuit of the transformer in place using per unit values of all the variables associated with the given transformer. Those who have an interest in the per-unit analysis are referred to reference [7] pages 543–46, reference [22] pages 55–60, and reference [27] pages 128–45.

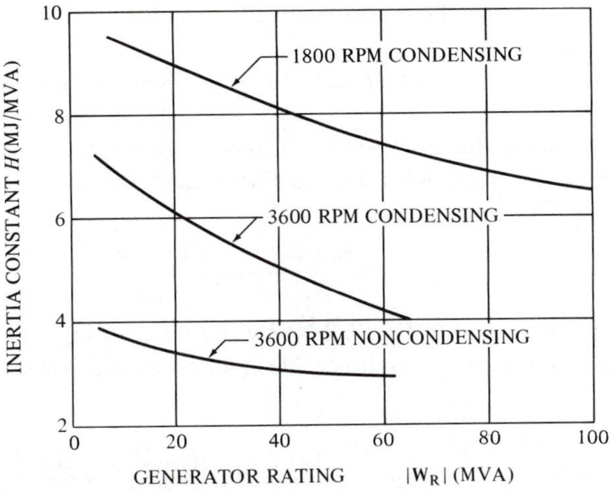

Fig. 6.37 Inertia Constant vs. Rating of Large Steam Turboalternators, Turbine Included

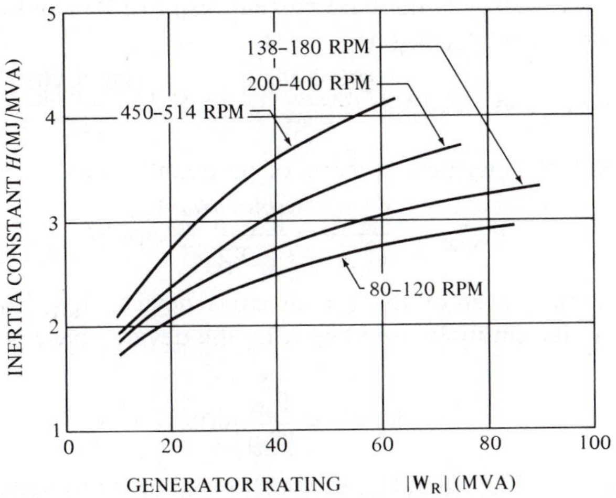

Fig. 6.38 Inertia Constant vs. Rating of Large Vertical-Type Water Wheel Generators, Including Allowance of 15% for Water Wheels

It should be noted that *the inertia constant H is usually used in practice rather than the momentum constant M*, because it has the desirable property that its value, unlike that of the momentum constant M, does not vary greatly with the rated kVA and the speed of the synchronous device, but has a characteristic value, or set of values, for each class of synchronous devices. From this point of view H is similar to the pu reactance of synchronous devices. Thus, in the absence of more definite information, a characteristic value of H, extracted from curves,* similar to those given in Figs. 6.37 and 6.38, may be used. The characteristics exhibited in these two figures describe large steam turbogenerators and large vertical-shaft water wheel generators with the inertia of the prime movers included—30 to 60% of the total inertia of a steam turbogenerator unit is that of the prime mover, while only 4 to

*By permission of E. W. Kimbark, *Power System Stability*, Vol. I New York, N.Y., John Wiley & Sons.

15% of the total inertia of the hydroelectric generator unit is that of the water wheel, including water.

Values of the inertia constant H for the other devices may be obtained from Table 6.1.

In order to calculate the coefficient of the output electric power P_{MAX}, Eq. (6.212), in pu units we start with Eq. (6.130) which we modify to describe the system of Fig. 6.29. Thus,

$$P_{MAX} = 3\frac{|\mathbf{E}_G||\mathbf{E}_T|}{X_S} \to 3\frac{|\mathbf{E}_G||\mathbf{V}_B|}{X_{EQ}} \tag{6.224}$$

Division of the two voltages of Eq. (6.224) by the base voltage, designated as $|\mathbf{E}_{\mathcal{B}}|$, and the reactance by the base reactance, designated as $X_{\mathcal{B}}$, yields

$$\mathrm{P}_{MAX} = 3\frac{\left(|\mathbf{E}_{\mathcal{B}}|\dfrac{|\mathbf{E}_G|}{|\mathbf{E}_{\mathcal{B}}|}\right)\left(|\mathbf{E}_{\mathcal{B}}|\dfrac{|\mathbf{V}_B|}{|\mathbf{E}_{\mathcal{B}}|}\right)}{\mathrm{X}_{\mathcal{B}}\left(\dfrac{X_{EQ}}{X_{\mathcal{B}}}\right)} \equiv 3\frac{|\mathbf{E}_{\mathcal{B}}||\mathbf{E}_{\mathcal{B}}|}{X_{\mathcal{B}}} \times \frac{|\mathbf{E}_G^{[pu]}||\mathbf{V}_B^{[pu]}|}{X_{EQ}^{[pu]}} \tag{6.225}$$

If we define $|\mathbf{E}_{\mathcal{B}}|/X_{\mathcal{B}} \equiv |\mathbf{I}_{\mathcal{B}}|$ as the base current, Eq. (6.225) can be written in terms of the base power $|\mathbf{S}_{\mathcal{B}}| = 3|\mathbf{E}_{\mathcal{B}}||\mathbf{I}_{\mathcal{B}}|$ as

$$P_{MAX} = 3|\mathbf{E}_{\mathcal{B}}||\mathbf{I}_{\mathcal{B}}|\frac{|\mathbf{E}_G^{[pu]}||\mathbf{V}_B^{[pu]}|}{X_{EQ}^{[pu]}} \equiv |\mathbf{S}_{\mathcal{B}}|\frac{|\mathbf{E}_G^{[pu]}||\mathbf{V}_B^{[pu]}|}{X_{EQ}^{[pu]}} \tag{6.226}$$

Therefore, Eq. (6.224) is written in terms of pu quantities as

$$\mathrm{P}_{MAX}^{[pu]} \equiv \frac{P_{MAX}}{|\mathbf{S}_{\mathcal{B}}|} = \frac{|\mathbf{E}_G^{[pu]}||\mathbf{V}_B^{[pu]}|}{X_{EQ}^{[pu]}}\,\mathrm{puW} \tag{6.227}$$

Finally, the shaft power of the synchronous device, P_{SH}, Eq. (6.181), can be expressed in pu units simply by dividing it by the device's base power, $|\mathbf{S}_{\mathcal{B}}|$, in VA. Thus,

$$\mathrm{P}_{SH}^{[pu]} = \frac{\mathrm{P}_{SH}}{|\mathbf{S}_{\mathcal{B}}|}\,\mathrm{puW} \tag{6.228}$$

Now use of Eqs. (6.223), (6.227), and (6.228) enables us to write Eq. (6.212), with all pertinent quantities expressed in pu units as

$$\delta_k = \delta_{k-1} + \Delta\delta_{k-1} + \frac{(\Delta t)^2}{M^{[pu]}}\left(P_{SH}^{[pu]} - P_{MAX}^{[pu]}\sin\delta_k\right), \qquad k = 1,2,3,\ldots \tag{6.229}$$

Table 6.1 AVERAGE VALUES OF STORED ENERGY IN ROTATING MACHINES*

Type of Machine	H—Stored Energy at Rated Speed (Megajoules per megavolt—ampere = sec)
Synchronous Motors	2.00
Synchronous Condensers:	
(Hydrogen Cooled, 25% Less)	
Large	1.25
Small	1.00
Induction Motors	0.50

*By permission of E. W. Kimbark, *Power System Stability*, Vol. I New York, N.Y., John Wiley & Sons.

Note that, in Eq. (6.229) Δt is usually set equal to 0.05 s (i.e., $\Delta t = 0.05$ s). However, one may deviate from this value of Δt if one chooses.

It must be recalled here that the speed of the synchronous device's rotor is calculated by use of the increment of the rotor's angular displacement $\Delta\delta_k$ and the time interval Δt. The pertinent mathematical equation is derived earlier as Eq. (6.205) but is rewritten here for emphasis. Thus,

$$\omega_{k-(1/2)} = \frac{\Delta\delta_k}{\Delta t} \frac{\text{elec. degrees}}{\text{sec}} \tag{6.230}$$

Inspection of Eq. (6.181) reveals that in order to obtain its solution, which is given by Eq. (6.212), it was tacitly assumed that the accelerating power P_A is known at the beginning of each time interval. However, because P_A is defined as a function of P_{SH} and P_{OT}, both the shaft power, P_{SH}, and the output electric power, P_{OT}, must be known. In the determination of these two powers the following assumptions are made: (1) the shaft power, P_{SH}, remains constant during the entire period of the swing curve, (2) the damping or asynchronous power is negligible, [see passage prior to Eq. (6.130)], (3) the output electric power is calculated from the steady-state solution of the network to which the synchronous device is connected, Eq. (6.1), and (4) each synchronous device in the network is represented by a Thevenin's equivalent comprised of a voltage source $\mathbf{E}_G$ connected in series with the reactance jX_S, which becomes the ***d*-axis transient reactance** jX'_d if the synchronous device is of the salient rotor type [see Fig. 6.48(b)]. In this analysis the angular displacement δ of the synchronous device's rotor designates the angle of departure of each rotor's *d*-axis from the synchronously rotating reference frame (SRRF), Fig. 6.25.

The constancy of the shaft power P_{SH} can be justified by the following reasoning. The mechanical input to the synchronous generator is controlled by the governor of its prime mover. This governor does not act until the speed change exceeds a **preset** value, usually 1% of rated speed, and even then, there is a time lag before the governor changes the input. Experience has shown that during swings of synchronous devices the percent change in speed is very small until after synchronism is lost. Therefore, since governor action does not come into play until synchronism is really lost, governor action is **neglected** in stability analysis via the swing curves and P_{SH} **is assumed to remain constant throughout this analysis** (i.e., **during the swing curve's period**).

The assumed steady state performance of the system's network Eq. (6.1), is not truly accurate both because of sudden changes, such as occurrence or removal of faults within the network, and because of the gradual change in the phase angles of the voltage generators due to the swinging of the rotors. In reality, the angular displacement δ of the rotor's *d*-axis coincides with the phase angle of the generator's voltage. Therefore as one angle changes, the other changes also. Because, however, the time constants of the network are relatively *short* when compared to the periods of oscillation of the synchronous devices (these periods are of the order of **1** s), the network may be assumed, without significant error, to operate in steady state at all times.

Example 6.10

For the system exhibited in Fig. 6.10.1(a) assume that the inertia constant $H = 2.7$ MJ/MVA and the pu values of the pertinent quantities, all measured on a

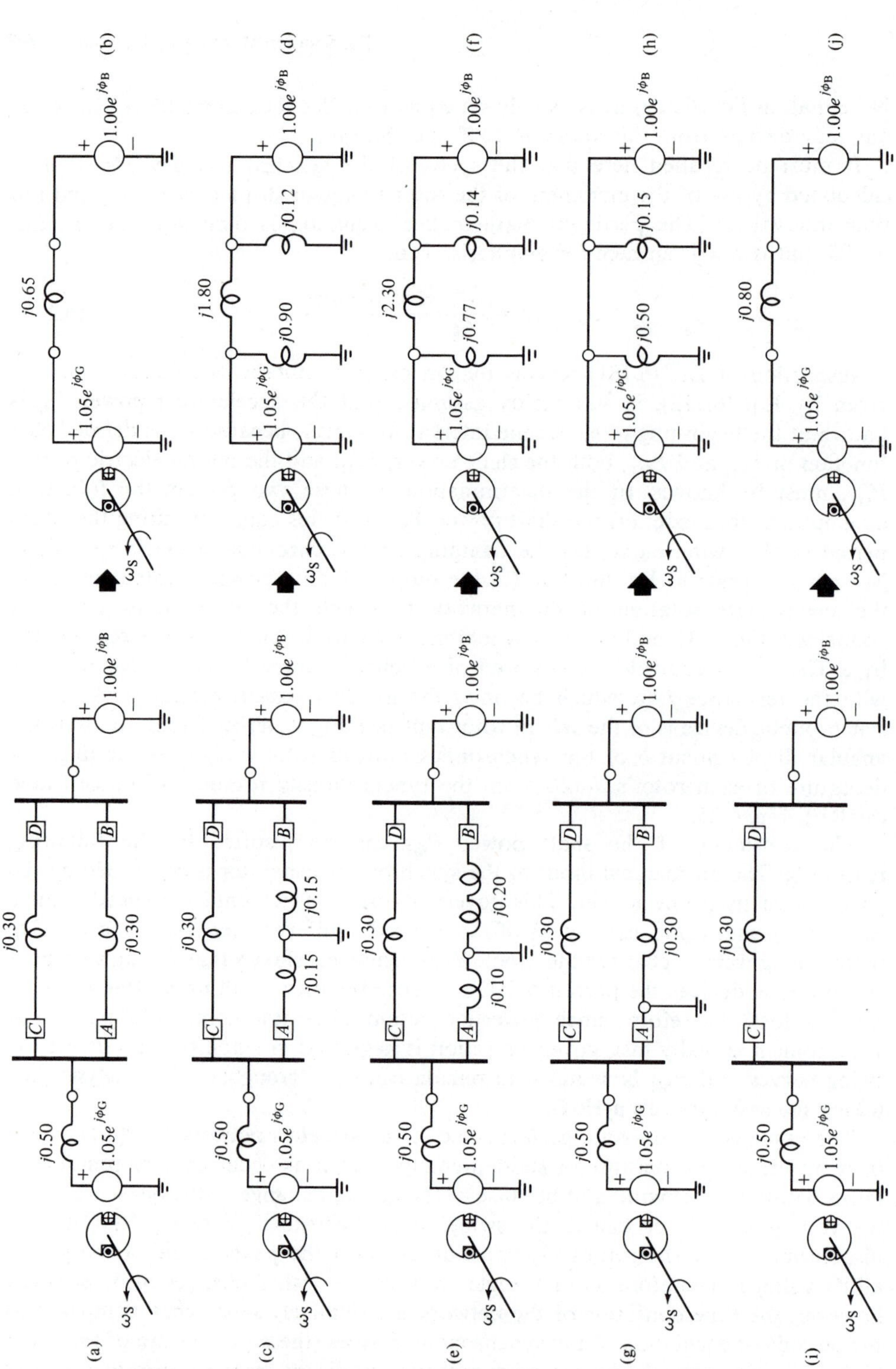

Fig. 6.10.1 Synchronous Generator Connected to a Metropolitan Area Through Two Parallel Transmission Lines Connected at Two Paralleling Buses: (a) and (b) Normal System Operation; (b) and (c) System Faulted at $k_A = 0.50$; (d) and (e) System Faulted at $k_A = 0.33$; (g) and (h) System Faulted at $k_A = 0.00$; and (i) and (j) System with Fault at $k_A = 0.50$ Cleared at $t_C = 0.45$, 0.425, 0.40 and 0.35 Seconds, Respectively

120 MVA base, are $\mathbf{E}_G = 1.05e^{j\phi_G}$ puV, $jX_S \equiv jX'_d = j0.50$ puΩ, $R_{AB} = R_{CD} = R_S = R_B = 0$ puΩ, and $jX_{AB} = jX_{CD} = j0.30$ puΩ. Under normal operating conditions, the 120 MVA, 2 pole 60 Hz synchronous generator delivers 75 MW over a two parallel conductor transmission line to a large metropolitan area which may be regarded as an infinite bus designated as $\mathbf{E}_B = 1.00e^{j\phi_B}$ and $jX_B = 0$ puΩ.

If a three-phase fault were to occur at $k_A = 0.50$, 0.33, and 0.00, respectively, along the transmission line AB away from the generator side paralleling bus (1) plot the swing curves, that is, δ vs. t, for $t = 1$ s for all three cases, and comment on the stability status of the system operating under the constraints of the three sustained faults, (2) for the faulted status for which $k_A = 0.5$ determine the approximate critical clearing time t_C, and the approximate critical angle δ_C, (3) specify the circuit breakers that can be used at both ends of the transmission lines to isolate the faults, and (4) determine whether or not a governor with a sensitivity 1% of normal speed would have any effects on the stability status of the system at hand.

Solution

The solution of this example is based on the solution curve of Eq. (6.229), known as swing curve. Therefore, the plotting of the swing curves for all specified cases is undertaken as well as the plotting of an additional number of such curves deemed necessary to determine the critical clearing time t_C and the critical angle δ_C.

The maxima of the output electric power for the prefaulted status $P^{\mathrm{I}}_{\mathrm{MAX}}$, faulted status $P^{\mathrm{II}}_{\mathrm{MAX}}$, and the postfaulted status $P^{\mathrm{III}}_{\mathrm{MAX}}$ are based on the network configuration during each mode of the system's operation. Thus, use of the basic network theory associated with series and parallel connection of network impedances as well as $Y \rightarrow \Delta$ transformations, Eqs. (6.164) through (6.167), enable us to calculate the reactances $\mathrm{X}^{\mathrm{I}}_{\mathrm{EQ}}$, $\mathrm{X}^{\mathrm{II}}_{\mathrm{EQ}}$, and $\mathrm{X}^{\mathrm{III}}_{\mathrm{EQ}}$ for $k_A = 0.50$, 0.33, and $\mathrm{X}^{\mathrm{I}}_{\mathrm{EQ}}$, $\mathrm{X}^{\mathrm{II}}_{\mathrm{EQ}}$, and $\mathrm{X}^{\mathrm{III}}_{\mathrm{EQ}}$ for $k_A = 0.00$, utilizing Eqs. (6.153), (6.164) through (6.167), (6.168), (6.170), (6.138), and (6.143) as:

$$k_A = \underline{0.50}\ \underline{\text{Figs. } 6.10.1(a) \rightarrow 6.10.1(b),\ 6.10.1(c) \rightarrow 6.10.1(d), \text{ and } 6.10.1(i) \rightarrow 6.10.1(j)}$$

$$\mathrm{X}^{\mathrm{I}}_{\mathrm{EQ}} = j0.5 + \frac{(j0.3) \times (j0.3)}{j0.3 + j0.3} = \underline{j0.65}\ \text{pu}\Omega \tag{1}$$

$$\mathrm{X}^{\mathrm{II}}_{\mathrm{EQ}} = \frac{(j0.5) \times (j0.3) + (j0.3) \times (j0.15) + (j0.15) \times (j0.5)}{j0.15} = \underline{j1.8}\ \text{pu}\Omega \tag{2}$$

and

$$\mathrm{X}^{\mathrm{III}}_{\mathrm{EQ}} = j0.5 + j0.3 = \underline{j0.8}\ \text{pu}\Omega \tag{3}$$

$$k_A = \underline{0.33}: \underline{\text{Figs. } 6.10.1(a) \rightarrow 6.10.1(b),\ 6.10.1(e) \rightarrow 6.10.1(f), \text{ and } 6.10.1(i) \rightarrow 6.10.1(j)}$$

$$\mathrm{X}^{\mathrm{I}}_{\mathrm{EQ}}\big|_{k_A = 0.33} \equiv \mathrm{X}^{\mathrm{I}}_{\mathrm{EQ}}\big|_{k_A = 0.50} = \underline{j0.65}\ \text{pu}\Omega \tag{4}$$

$$\mathrm{X}^{\mathrm{II}}_{\mathrm{EQ}} = \frac{(j0.5) \times (j0.3) + (j0.3) \times (j0.1) + (j0.1) \times (j0.5)}{j0.1} = \underline{j2.3}\ \text{pu}\Omega \tag{5}$$

and

$$\mathbf{X}_{\mathrm{EQ}}^{\mathrm{III}}\Big|_{k_A=0.33} \equiv \mathbf{X}_{\mathrm{EQ}}^{\mathrm{III}}\Big|_{k_A=0.50} = \underline{j0.8}\ \mathrm{pu}\Omega \tag{6}$$

$k_A = \underline{0.00}$: $\underline{\text{Figs. 6.10.1}(a)\rightarrow 6.10.1(b),\ 6.10.1(g)\rightarrow 6.10.1(h),\ \text{and } 6.10.1(i)\rightarrow 6.10.1(j)}$

$$\mathbf{X}_{\mathrm{EQ}}^{\mathrm{I}}\Big|_{k_A=0.00} \equiv \mathbf{X}_{\mathrm{EQ}}^{\mathrm{I}}\Big|_{k_A=0.50} = \underline{j0.65}\ \mathrm{pu}\Omega \tag{7}$$

$$\mathbf{X}_{\mathrm{EQ}}^{\mathrm{II}} = \frac{(j0.5)\times(j0.3)+(j0.3)\times(j0.0)+(j0.0)\times(j0.5)}{j0.0} \rightarrow j\infty \tag{8}$$

and

$$\mathbf{X}_{\mathrm{EQ}}^{\mathrm{III}}\Big|_{k_A=0.00} \equiv \mathbf{X}_{\mathrm{EQ}}^{\mathrm{III}}\Big|_{k_A=0.50} = \underline{j0.8}\ \mathrm{pu}\Omega \tag{9}$$

Now, the use of Eqs. (6.154), (6.169), (6.171), (6.139), (6.142), and (6.144) in conjunction with Eq. (6.229) and Figs. 6.10.1(b), 6.10.1(d), and 6.10.1(j); 6.10.1(b), 6.10.1(f), and 6.10.1(j); and 6.10.1(b), 6.10.1(h), and 6.10.1(j) enables us to calculate $P_{\mathrm{MAX}}^{[\mathrm{pu}]}$ of Eq. (6.229) for the three cases as

$$k_A = \underline{0.50}:$$

$$P_{\mathrm{MAX}}^{[\mathrm{pu}]} = \frac{(1.05)\times(1.0)}{0.65} = \underline{1.62}\ \mathrm{puW} \tag{10}$$

$$P_{\mathrm{MAX}}^{[\mathrm{pu}]} = \frac{(1.05)\times(1.0)}{1.8} = \underline{0.58}\ \mathrm{puW} \tag{11}$$

and

$$P_{\mathrm{MAX}}^{[\mathrm{pu}]} = \frac{(1.05)\times(1.0)}{0.8} = \underline{1.31}\ \mathrm{puW} \tag{12}$$

$$k_A = \underline{0.33}:$$

$$P_{\mathrm{MAX}}^{[\mathrm{pu}]}\Big|_{k_A=0.33} = P_{\mathrm{MAX}}^{[\mathrm{pu}]}\Big|_{k_A=0.50} = \frac{(1.05)\times(1.0)}{0.65} = \underline{1.62}\ \mathrm{puW} \tag{13}$$

$$P_{\mathrm{MAX}}^{[\mathrm{pu}]} = \frac{(1.05)\times(1.0)}{2.3} = \underline{0.46}\ \mathrm{puW} \tag{14}$$

and

$$P_{\mathrm{MAX}}^{[\mathrm{pu}]}\Big|_{k_A=0.33} = P_{\mathrm{MAX}}^{[\mathrm{pu}]}\Big|_{k_A=0.50} = \frac{(1.05)\times(1.0)}{0.8} = \underline{1.31}\ \mathrm{puW} \tag{15}$$

$$k_A = \underline{0.00}:$$

$$P_{\mathrm{MAX}}^{[\mathrm{pu}]}\Big|_{k_A=0.00} = P_{\mathrm{MAX}}^{[\mathrm{pu}]}\Big|_{k_A=0.50} = \frac{(1.05)\times(1.0)}{0.65} = \underline{1.62}\ \mathrm{puW} \tag{16}$$

$$P_{\mathrm{MAX}}^{[\mathrm{pu}]} = \frac{(1.05)\times(1.0)}{\infty} = \underline{0.00}\ \mathrm{puW} \tag{17}$$

$$P_{\mathrm{MAX}^2}^{[\mathrm{pu}]}\Big|_{k_A=0.00} = P_{\mathrm{MAX}}^{[\mathrm{pu}]}\Big|_{k_A=0.50} = \frac{(1.05)\times(1.0)}{0.8} = \underline{1.31}\ \mathrm{puW} \tag{18}$$

Note that the reactances that are connected in parallel with the voltage sources, Figs. 6.10.1(d), 6.10.1(f), and 6.10.1(h), have no effect on the output electric power of the sources.

The momentum constant of Eq. (6.229) is calculated with the aid of input data and Eq. (6.223) in which $|\mathbf{S}_{\mathrm{R}}| \equiv |\mathbf{S}_{\mathfrak{B}}| = 120\times 10^6$ VA (this is **not true** always) as

$$M^{[\mathrm{pu}]} = \left(\frac{2}{2}\right)\times\left(\frac{120\times 10^6\times 2.7}{180\times 60}\right)\times\left(\frac{1}{120\times 10^6}\right) = \underline{2.5\times 10^{-4}\ \frac{\mathrm{Ws}}{\mathrm{elec.\ degree}}} \tag{19}$$

Initial conditions:

In the prefaulted status of the system, the swing equation is written as

$$M^{[pu]}\frac{d^2}{dt^2}\delta = P_{SH}^{[pu]} - P_{MAX}^{[pu]}\sin\delta \equiv 0 \tag{20}$$

This is true because the system operates in its steady-state mode and the acceleration is zero. Now, Eq. (20) enables us to calculate the initial angle δ_0 as follows

$$P_{SH}^{[pu]} = P_{MAX}^{[pu]}\sin\delta \rightarrow P_{H}^{[pu]} = P_{MAX}^{[pu]}\sin\delta_0 \tag{21}$$

where, from Eq. (6.228)

$$P_{SH}^{[pu]} = \frac{P_{SH}}{|\mathbf{S}_{\mathcal{B}}|} = \frac{75\times10^6}{120\times10^6} = \underline{0.625}\ \text{puW} \tag{22}$$

Substitution from Eqs. (22) and (10) into Eq. (21) and solution for δ_0 yields

$$\delta_0 = \sin^{-1}\left(\frac{0.625}{1.62}\right) = \underline{22.69°} \tag{23}$$

Immediately before the fault, the output electric power is written with the aid of Eqs. (10), (21), and (23) as

$$P_{OT(0^-)} = P_{MAX}\sin\delta_0 = 1.62\sin(22.69°) = \underline{0.625}\ \text{puW} \tag{24}$$

Immediately after the fault, the angular position has not changed, however the output electric power can be written with the aid of Eqs. (11), (21), and (23) as

$$P_{OT(0^+)} = P_{MAX}\sin\delta_0 = 0.58\sin(22.69°) = \underline{0.225}\ \text{puW} \tag{25}$$

Substitution of each of these two equations into the $P_{A(k-1)}$ term of Eq. (6.229) [i.e., the right hand side of Eq. (6.181) expressed in pu units] and of the results into Eq. (6.213) yields

$$P_{A(0_{AVG})} = \tfrac{1}{2}\left[P_{A(0^-)} + P_{A(0^+)}\right] = \tfrac{1}{2}\left[0+0.40\right] = \underline{0.20}\ \text{puW} \tag{26}$$

Note the entries of Eqs. (10), (11), (24), and (25) in the second and third line of Table 6.10.1 and of Eqs. (23) and (26) in the table's fourth line. Now follow the 0_{AVG} line of this table from left to right for the entries $\Delta t = 0.05$ s, and $M^{[pu]} = 2.5\times10^{-4}$, Eq. (19), and observe that

$$\Delta\delta_0 \equiv \Delta\delta_- + \frac{(\Delta t)^2}{M^{[pu]}}P_{A(0-1)}^{[pu]} = 0 + 2 = \underline{2°} \tag{27}$$

Continue along this same line and observe that the angle δ_k, $k \equiv 1$ is

$$\delta_1 \equiv \Delta\delta_0 + \delta_0 = 2.00 + 22.69 = \underline{24.69°} \tag{28}$$

Also, note that with the aid of Eq. (6.230) the spread $\omega_{0-(1/2)}$ is calculated as

$$\omega_{0-(1/2)} = \frac{\Delta\delta_0}{\Delta t} = \frac{0.20}{0.05} = \underline{40.00}\,\frac{\text{elec. degrees}}{\text{sec}} \tag{29}$$

Now, pay close attention to line $k=1$ of this table and follow the pattern. Note that δ_1 allows us to calculate all the quantities of the row $k=1$ as well as δ_2. Once the line $k=0$ is completed, however, one may use the relevant entries of line $k=0$, the angle δ_1, the calculated value of $M^{[pu]}$, Eq. (19), and the specified value of $\Delta t = 0.05$ s to initial and run the following program* on a modern calculator.

*This program was prepared by Mr. B. A. Echeverri, a 1978 EE-449 student. The program was written to run on an HP 25 Calculator.

CALCULATOR PROGRAM FOR THE SOLUTION OF THE SWING EQUATION
ALL PERTINENT QUANTITIES ARE EXPRESSED IN PU UNITS

$$\delta_k = \delta_{k-1} + \Delta\delta_{k-1} + \frac{(\Delta t)^2}{M^{[pu]}}\left(P_{SH}^{[pu]} - P_{MAX}^{[pu]}\sin\delta_k\right)$$

INSTRUCTIONS AND PROGRAM CODING: (HEWLETT-PACKARD CODES)

1. SET SWITCH ON PRGM MODE
2. READ IN PROGRAM

```
RCL 1
f SIN
R/S
RCL 2
×
R/S
CHS
RCL 3
+
R/S
RCL 4
gx²
RCL 5
÷
×
R/S
RCL 6
+
R/S
STO 6
RCL 1
+
R/S
STO 1
RCL 6
RCL 4
÷
GTO 00
```

3. SET SWITCH ON RUN MODE
4. SET INITIAL VALUES*:

A. δ_k	STO	1
B. $P_{MAX}^{[pu]}$	STO	2
C. $P_{SH}^{[pu]}$	STO	3
D. Δt	STO	4
E. $M^{[pu]}$	STO	5
F. $\Delta\delta_{k-1}$	STO	6

5. PRESS KEYS f AND PRGM
6. PRESS KEY R/S SEQUENTIALLY TO READ TABLE CONTENT

*All initial values are expressed in the decimal system except $M^{[pu]}$ which is usually expressed in the exponential form. Therefore, to enter the number 2.5×10^{-4} follow the following instructions: (1) enter 2.5, (2) press key EE, (3) enter 4, and (5) press key CHS.

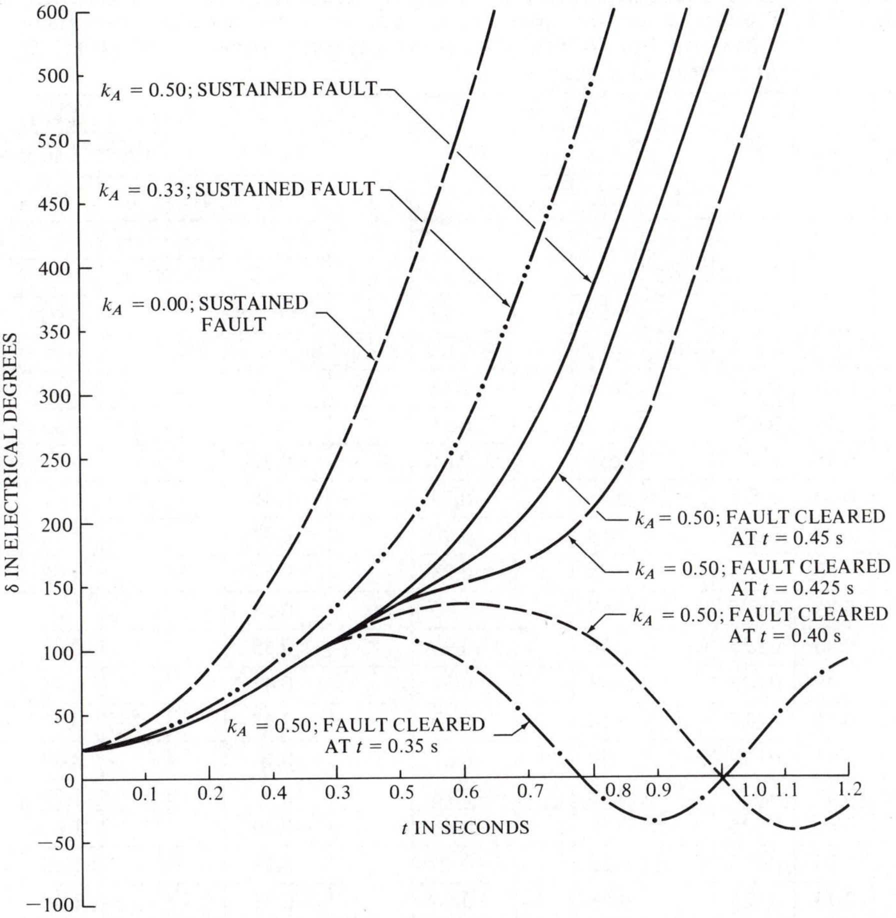

Fig. 6.10.2 The Swing Curves of Example 6.10: Sustained Three-Phase Fault at $k_A=0.50$, 0.33, and 0.00; and Three-Phase Fault at $k_A=0.50$ Cleared at $t_C=0.45$, 0.425, 0.40, and 0.35 Seconds, Respectively

The results of the point-by-point solution of Eq. (6.229) as applied to Example 6.10 are exhibited in Table 6.10.1. The swing curve for these data are exhibited in Fig. 6.10.2 as the curve $k_A=0.50$; Sustained Fault. Note that both the data of Table 6.10.1 and the corresponding curve in Fig. 6.10.2 describe the system of Fig. 6.10.1(a) faulted at the middle of the transmission line AB, with the fault sustained, Figs. 6.10.1(c) and 6.10.1(d). The pertinent data used to initiate and execute the solution of Eq. (6.229) is contained in the **shaded** portion of Table 6.10.1.

Similar data are obtained using the same program and pertinent data, for sustained faults at $k_A=0.33$ and $k_A=0.00$, respectively. However, these data are not exhibited here. The swing curves for these two cases are also exhibited in Fig. 6.10.2. Inspection of swing curves for the sustained faults at $k_A=0.50$, 0.33, and 0.00 reveals that the angular displacement of the synchronous device's rotor d-axis from the synchronously rotating reference frame (SRRF) increases with time. In addition, the curves reveal that the closer the sustained fault to the generator is the more pronounced the departure is; that is, the faster the departure of the rotor's d-axis from the SRRF is (Fig. 6.25). All three curves designate **unstable** operation.

Table 6.10.1 TABULATION OF DATA ASSOCIATED WITH THE SWING EQUATION DESCRIBING THE SYSTEM OF FIG. 6.10.1(a) WITH A SUSTAINED THREE-PHASE FAULT AT THE MIDPOINT OF LINE AB*

k	t	$P_{SH}^{[pu]}$	$P_{MAX}^{[pu]}$			$\sin\delta_k$	$P_{MAX}^{[pu]}\sin\delta$			$P_{A(k-1)}^{[pu]}$ $P_{SH}^{[pu]} - P_{MAX}^{[pu]}\sin\delta_k$		
			I	II	III		I	II	III	I	II	III
—	—	—	—	—	—	—	—	—	—	—	—	
	0^-	0.625	1.62			0.39	0.625			0		
	0^+	0.625		0.58		0.39		0.225			0.40	
0	0_{AVG}	0.625		0.58								
1	0.05	0.625		0.58		0.42		0.24			0.38	
2	0.10	0.625		0.58		0.51		0.29			0.33	
3	0.15	0.625		0.58		0.64		0.37			0.25	
4	0.20	0.625		0.58		0.78		0.45			0.17	
5	0.25	0.625		0.58		0.90		0.52			0.10	
6	0.30	0.625		0.58		0.98		0.57			0.06	
7	0.35	0.625		0.58		1.00		0.58			0.05	
8	0.40	0.625		0.58		0.94		0.55			0.08	
9	0.45	0.625		0.58		0.81		0.47			0.15	
10	0.50	0.625		0.58		0.59		0.35			0.28	
11	0.55	0.625		0.58		0.27		0.16			0.47	
12	0.60	0.625		0.58		−0.16		−0.09			0.72	
13	0.65	0.625		0.58		−0.66		−0.39			1.01	
14	0.70	0.625		0.58		−0.99		−0.58			1.20	
15	0.75	0.625		0.58		−0.66		−0.38			1.01	
16	0.80	0.625		0.58		0.39		0.23			0.40	
17	0.85	0.625		0.58		1.00		0.58			0.05	
18	0.90	0.625		0.58		0.33		0.19			0.43	
19	0.95	0.625		0.58		−0.81		−0.47			1.10	
20	1.00	0.625		0.58		−0.66		−0.38			1.01	

*System exhibited in Figs. 6.10.1(c) and 6.10.1(d).

Table 16.10.1 (*Continued*)

Avg $P^{[pu]}_{A(k-1)}$ at Discontinuity $\frac{1}{2}[P^{[pu]-}_{A(k-1)} P^{[pu]+}_{A(k-1)}]$ I	II	III	$\frac{(\Delta t)^2}{M^{[pu]}}$	$\frac{(\Delta t)^2}{M^{[pu]}} P^{[pu]}_{A(k-1)}$	$\Delta\delta_k$ elec. deg.	δ_k elec. deg.	$\omega_{k-(1/2)}$ el. deg./s
—	—	—	—	—	0	—	—
	0.20		10	2.00	2.00	22.69	40.00
			10	3.83	5.83	24.69	116.55
			10	3.30	9.13	30.52	182.64
			10	2.55	11.68	39.65	233.62
			10	1.72	13.40	51.33	268.05
			10	1.00	14.41	64.73	288.15
			10	0.55	14.96	79.14	299.23
			10	0.46	15.43	94.10	308.53
			10	0.78	16.21	109.53	324.20
			10	1.54	17.75	125.74	355.04
			10	2.80	20.55	143.49	411.03
			10	4.66	25.21	164.04	504.14
			10	7.18	32.39	189.25	647.78
			10	10.10	42.49	221.64	849.86
			10	12.02	54.51	264.13	1090.25
			10	10.08	54.59	318.64	1291.89
			10	3.96	68.56	383.24	1371.13
			10	0.45	69.01	451.79	1380.18
			10	4.34	73.35	520.80	1467.04
			10	10.95	84.30	594.16	1686.07
			10	10.10	94.40	678.46	1888.00
						772.86	

In Table 6.10.2 are exhibited data for the plot of the curve $k_A = 0.50$, Fault Cleared at 0.40 seconds. The missing data preceding $t = 0.35$ s, data for $t = 0.35$ s inclusive, are identical to that of Table 6.10.1 (i.e., the data of Table 6.10.1 for $0 \leqslant t \leqslant 0.35$ are also valid in Table 6.10.2. Note that for $t = 0.40$ sec, $\rightarrow k = 8$. Thus, the δ_8 and $\Delta\delta_7$ required to initiate the solution consitutes only part of the data of Eq. (6.229) valid only in the postfault mode. Also, note that since the switching occurs at $t = 0.40$ s, $P_{A(7^-)}^{[pu]}$ and $P_{A(7^+)}^{[pu]}$, and $P_{A(7_{AVG})}^{[pu]}$ are calculated with the aid of Eqs. (24) and (25) for which $P_{MAX}^{[pu]}$ is set equal to values given by Eqs. (11) and (12), respectively, while $\delta_0 \equiv \delta_8 = 109.53°$. Note that the information used to initiate the solution of Eq. (6.229) is again **shaded** in Table 6.10.2. The data for the curves: $k_A = 0.50$; Fault Cleared at $t = 0.45$, 0.425, and 0.35 s, are obtained following the procedure used to obtain the data for clearing the fault at $t = 0.40$ s. Although not exhibited, it can be obtained through the use of the same program, the appropriate initial conditions, and pertinent data. The estimate for clearing time $t_C = 0.45$ s is made on the basis that at $t = 0.45$ s, the swing curve for $k_A = 0.50$; Sustained Fault, begins to depart from the straight line portion of the curve. Several clearing times, smaller than $t = 0.45$ s, are tried until two clearing times, differing slightly, are found—for one, the system is **stable** and $t \rightarrow t_{CS}$, for the other, the system is **unstable** and $t \rightarrow t_{CU}$. The critical clearing time t_C is chosen to lie between the stable clearing time, t_{CS}, and the unstable clearing time, t_{CU}, as defined by these two swing curves. That is, $t_{CS} \leqslant t_C < t_{CU}$. It is concluded from inspection of Fig. 6.10.2 that 0.40 s $\leqslant t_C < 0.425$ s and $109.53° < \delta_C < 117.64°$, with the lower bounds on both t_C and δ_C chosen as the preferred values. Application of these ideas to Eq. (6.180) suggests that

$$k_{RCB} = k_C f_E \rightarrow t_{CS} f_E \tag{30}$$

or

$$k_{RCB} = t_{CS} f_E = (0.40) \times (60) = \underline{24} \text{ cycles} \tag{31}$$

From the standard interrupting times of 2, 3, 5, or 8 cycles of available relay-circuit-breaker-sets, we can choose only the 8-cycle set, which definitely assures stable operation.

To determine whether or not the governor of the system has an effect on the system's stability status we check the speed columns of Tables 6.10.1 and 6.10.2 to see whether or not the speed of the rotor exceeds $\pm 1\%$ of the normal speed corresponding to the frequency of 60 Hz. Since 2π rad $= 360$ elec. deg.$\rightarrow 1$ rad $=$ $(360/2\pi)$ elec. deg., the normal speed of the synchronous device is

$$\omega_E = 2\pi f_E = 2\pi \times 60 \times \frac{360}{2\pi} = 21{,}600 \frac{\text{elec. deg.}}{\text{sec}} \tag{32}$$

and

$$1\% \, \omega_E = 216 \frac{\text{elec. deg.}}{\text{sec}} \tag{33}$$

The inspection of Tables 6.10.1 and 6.10.2 during the stable operation of the system yields maximum speed deviation of 308.53 and -639.09 elec. deg./s at $t = 0.35$ and 0.90 s, respectively. These speeds correspond to 1.43 and 2.96% speed deviations. Both speeds are greater than 216 elec. deg./s, or the speed deviations are greater than 1% of the normal speed. It is concluded, therefore, that the governor **has** a definite effect on the stability status of the systems at hand.

Finally, with respect to stability note that as the clearing time decreases, the system becomes more stable. Note that in Fig. 6.10.2 the curves $k_A = 0.50$ and

$t_C = 0.40$ s, and $k_A = 0.50$ and $t_C = 0.35$ s designate **stable** system operation. The remaining curves designate **unstable** system operation. **The swing curve that designates marginally stable system operation is not exhibited**; however, it is anticipated that a clearing time in the neighborhood of 0.4125 s would yield such a curve.

It is clear then, that aside from the three-phase faults and their location, the stability of power systems is a function of the acceleration of the synchronous generators in a given system. For a system with only one synchronous generator, Fig. 6.29 and Eq. (6.130), the acceleration is $d^2\delta/dt^2$. The greater the acceleration is, the shorter the fault clearing time must be if stability of the system is to be maintained. From Eq. (6.130) adapted to Fig. 6.29 it is evident that for a given mechanical input, P_{SH}, and an electrical load, $|\mathbf{E}_B|$, the acceleration of the system at hand is a function of: (1) the momentum constant M^{-1}, (2) the phase voltage $|\mathbf{E}_G|$ of the synchronous generator, and (3) the network reactance $[X_{EQ}]^{-1}$. The effect of the momentum constant M on the system's performance becomes more evident from the inspection of Eq. (6.212), which reveals that an increase in the momentum constant of the synchronous generator reduces the torque angle through which the rotor swings during any time interval, thus allowing a longer time for circuit-breaker operation to isolate a given fault before the rotor exceeds its critical clearing angle. Thus, although an increase in M offers a means to improve system stability, the method has not been used, strictly for economic reasons.

Since stability is improved by the decrease of the accelerating power P_A, for constant P_{SH}, $|\mathbf{E}_T|$, and M, the accelerating power P_A can be reduced by increasing P_{MAX} either through the **increase** of $|\mathbf{E}_G|$ or the reduction of X_{EQ}. Increasing P_{MAX} results in raised curves, Figs. 6.30, 6.32, and 6.34, therefore a lowered δ_0 and an increased δ_{MAX}, and thus, a greater difference between δ_0 and the clearing angle δ_C. This of course implies that the rotor of the synchronous device is allowed to swing through a larger angle from its original position, that is, it has an increased clearing time before the system becomes unstable. Consequently, an increased P_{MAX} yields an improved probability of maintaining system stability.

Reduction of X_{EQ} can be achieved by either **compensating** for the reactance of transmission lines with series capacitors or **increasing** the number of parallel transmission lines between two paralleling buses. It should be understood that multiple parallel transmission lines allow energy to be transmitted between the two paralleling buses even if a three-phase fault were to occur on one of the lines, unless the fault occurs on one of the paralleling buses. Note that for any type of fault anywhere along **one** line (but **not** on any of the buses), more and more power is transferred during the fault between the paralleling buses, as the number of the parallel transmission lines between these buses becomes two or more, than is transferred over a single faulted line. For an increased P_{MAX} and constant P_{SH} and M, the accelerating power P_A is decreased and so is the acceleration of the synchronous generator's rotor. Therefore, again the probability of maintaining system stability is increased.

It should be noted that when faults occur in a power system, they create a disturbance in the system causing the system to tend toward instability. To maintain stability the fault has to be cleared within a critical clearing time. This requirement led to the development of high-speed circuit–breakers which function, practically within any critical clearing time, to clear faults. If, however, clearing of faults is delayed, or a large load is suddenly lost from a system, stability can be maintained by connecting a large resistive load, defined as a **braking resistor**, at, or

Table 6.10.2 TABULATION OF DATA ASSOCIATED WITH SWING EQUATION DESCRIBING THE SYSTEM OF FIG. 6.10.1(a) WITH THE THREE-PHASE FAULT AT THE MIDPOINT OF LINE AB CLEARED AT $t_C = 0.40$ S [EXHIBITED IN FIGS. 6.10.1(i) AND 6.10.1(j).

k	t	$P_{SH}^{[pu]}$	$P_{MAX}^{[pu]}$			$\sin\delta_k$	$P_{MAX}^{[pu]}\sin\delta_k$			$P_{A(k-1)}^{[pu]}$ $P_{SH}^{[pu]} - P_{MAX}^{[pu]}\sin\delta_k$		
			I	II	III		I	II	III	I	II	III
0												
1												
2												
3												
4												
5												
6												
7	0.35											
	0.40^-			0.58		0.94		0.55			0.08	
	0.40^+				1.31	0.94			1.23			−0.60
8	0.40_{AVG}	0.625			1.31							
9	0.45				1.31	0.84			1.11			−0.48
10	0.50				1.31	0.76			1.00			−0.37
11	0.55				1.31	0.71			0.93			−0.31
12	0.60				1.31	0.70			0.91			−0.29
13	0.65				1.31	0.72			0.94			−0.32
14	0.70				1.31	0.77			1.01			−0.39
15	0.75				1.31	0.86			1.13			−0.50
16	0.80				1.31	0.96			1.25			−0.63
17	0.85				1.31	1.00			1.31			−0.68
18	0.90				1.31	0.87			1.13			−0.51
19	0.95				1.31	0.47			0.61			0.01
20	1.00				1.31	−0.07			−0.09			0.71

Table 6.10.2 (*Continued*)

Avg $P_{A(k-1)}^{[pu]}$ at Discontinuity $\frac{1}{2}[P_{A(k-1)}^{[pu]-} P_{A(k-1)}^{[pu]}]$ I	II	III	$\dfrac{(\Delta t)^2}{M^{[pu]}}$	$\dfrac{(\Delta t)^2}{M^{[pu]}} P_{A(k-1)}^{[pu]}$	$\Delta\delta_k$ elec. deg.	δ_k elec. deg.	$\omega_{k-(1/2)}$ el. deg. / s
					15.43		
		−0.26		−2.61	12.82	109.53	256.40
				−4.82	8.00	122.35	160.06
				−3.73	4.27	130.35	85.40
				−3.07	1.20	134.62	23.93
				−2.88	−1.68	135.82	−33.67
				−3.15	−4.84	134.14	−96.70
				−3.89	−8.72	129.30	−174.44
				−5.03	−13.75	120.58	−275.01
				−6.29	−20.04	106.83	−400.79
				−6.83	−26.87	86.79	−537.38
				−5.09	−31.95	59.92	−639.09
				0.11	−31.85	27.97	−636.96
				7.14	−24.71	−3.88	−494.22
						−28.59	

near, the generator bus, Fig. 6.31(a). The function of such load is to compensate for at least part of the load on the generator, thus, to reduce the acceleration and improve the probability of maintaining system stability.

Stability can also be maintained through the implementation of the following procedure. We sense the difference between the mechanical input, P_{SH}, and the electrical output, P_{OT}, caused by a fault, and use this difference to initiate the closing of a set of certain turbine valves which cause a reduction in the mechanical power input, P_{SH}, to the generator. This method of maintaining system stability is known by the code name **fast valving**.

When the power system at hand is comprised of n synchronous devices, rather than one synchronous device and an infinite bus, the system is described by n simultaneous differential equations written with the aid of Eq. (6.97) as

$$\left.\begin{array}{llll} M_1\dfrac{d^2}{dt^2}\delta_1 & +P_D^1\dfrac{d}{dt}\delta_1 & +P_{OT}^1(\delta_1,\delta_2,\delta_3,\ldots,\delta_n) & =P_{SH}^1 \\ M_2\dfrac{d^2}{dt^2}\delta_2 & +P_D^2\dfrac{d}{dt}\delta_2 & +P_{OT}^2(\delta_1,\delta_2,\delta_3,\ldots,\delta_n) & =P_{SH}^2 \\ \cdot & \cdot & \cdot & \cdot \\ \cdot & \cdot & \cdot & \cdot \\ \cdot & \cdot & \cdot & \cdot \\ M_n\dfrac{d^2}{dt^2}\delta_n & +P_D^n\dfrac{d}{dt}\delta_n & +P_{OT}^n(\delta_1,\delta_2,\delta_3,\ldots,\delta_n) & =P_{SH}^n \end{array}\right\} \quad (6.231)$$

Solution of this system of equations is beyond the scope of this book.* It should be pointed out in passing, however, that the digital computer is used for a point-by-point solution of Eq. (6.231). To accomplish this task a typical **stability program**, similar to the program used earlier in this section, is interfaced with a **load-flow program**, which supplies the initial values for the calculations. (These calculations begin with a disturbance due to a fault or a switching operation.) The solution can be refined by incorporating simulations for each (1) voltage regulator, (2) field excitation, (3) governor action, and (4) saturation of the flux path in each synchronous device as excitation is changed, while the transient analysis is in progress for either the smooth rotor or salient rotor synchronous devices.

6.7 ANALYSIS OF BALANCED, THREE-PHASE, SALIENT-ROTOR, SYNCHRONOUS DEVICES

In contrast to the cylindrical-rotor synchronous device, which has a smooth airgap, the salient-pole synchronous device has a highly nonuniform airgap because of the protruding pole structure. Compare Fig. 5.5(a) to 6.39(a).

In Fig. 6.39(c) the stator flux component that flows through the two protruding poles of the rotor—that is, it traces the d-axis of the rotor structure—is identified as the d-axis component of the stator flux Φ_{ds}. Since all the mmf drop occurs across

*Those interested in the solution of Eq. (6.231) should refer to "Transient Stability Regions of Multimachine Power Systems," by A. H. El-Abiad and K. Nagappan, *IEEE Transactions on Power Apparatus and Systems*, Vol. PAS 85, No. 2, February 1966.

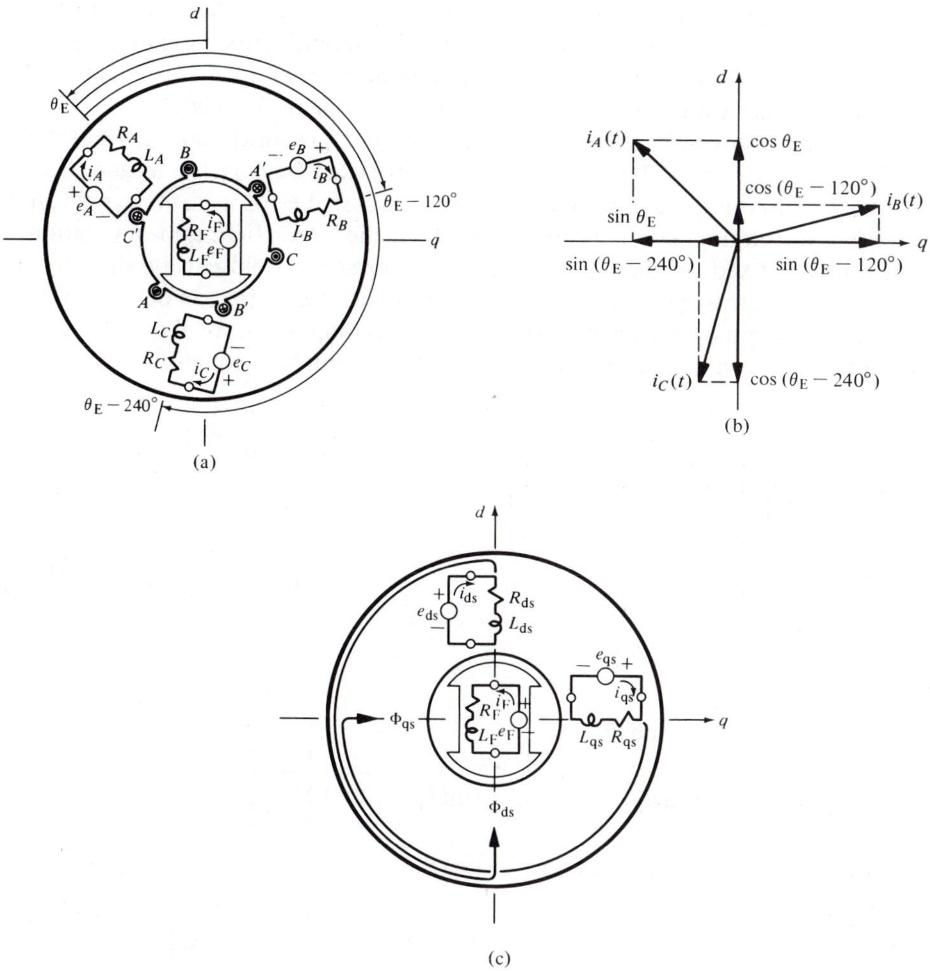

Fig. 6.39 Three-Phase, Two-Pole, Salient-Rotor, Synchronous Device: (a) Physical Winding Representation by Concentrated Equivalent Ficticious Windings, (b) Decompositon of Phase Variables into *dq*-Variables, and (c) *dq*-Reference-Frame Equivalent Windings of the Three Stator Windings

the two small airgaps ℓ_d with effective area A^d_{EFF}, this path is the path of minimum reluctance, given as $\mathcal{R}_d = 2\ell_d/\mu_0 A^d_{EFF}$. On the other hand, the stator flux component that flows through the two large airgaps—that is, it traces the *q*-axis of the rotor structure—is identified as the *q*-axis component of the stator flux Φ_{qs}. Since all the mmf drop occurs across the two large airgaps ℓ_q, this path is the path of maximum reluctance, given as $\mathcal{R}_q = 2\ell_q/\mu_0 A^q_{EFF}$. It should be noted that $\mathcal{R}_q \gg \mathcal{R}_d$.

In the normal operation of synchronous devices, the field of the stator winding $\mathbf{B}_{ST}$ is sinusoidally distributed within the airgap with its peak value located somewhere between the *d*-, and *q*-axes, Fig. 6.25(a) and 6.25(b). Accordingly, it produces significantly different effects in each axis, when compared to the smooth airgap synchronous device, because of the considerable difference between the reluctance of the two paths, that is, $\Phi = \mathcal{F}/\mathcal{R}$ and $\mathcal{R}_q \gg \mathcal{R}_d$. Consequently, the cylindrical-rotor theory developed in the preceding sections cannot be used in its present form for the analysis of salient-pole synchronous devices, unless it is modified properly *to account for the saliency*. This modification is explained in the remainder of this section through use of a minimum amount of mathematics. The detailed mathematical development is beyond the scope of this text.*

The balanced three-phase currents $i_A(t)$, $i_B(t)$, and $i_C(t)$ in the stator winding of a synchronous device produce a sinusoidally distributed field in the airgap, Section 4.2, which is rotating at synchronous speed ω_S. Since the rotor of the synchronous device is also rotating at synchronous speed ω_S, the stator field appears stationary to an observer fixed on the rotor. This stationary stator field can be decomposed into two components: (1) one along the rotor d-axis, designated as $\mathbf{B}_{ds}$, and (2) one along the rotor q-axis, designated as $\mathbf{B}_{qs}$. This decomposition of the stator field is accomplished automatically by applying the transformation equations

$$\begin{bmatrix} g_{ds}(t) \\ g_{qs}(t) \\ g_{0s}(t) \\ g_F(t) \end{bmatrix} = \sqrt{\frac{2}{3}} \begin{bmatrix} \cos\theta_E & \cos(\theta_E - 120°) & \cos(\theta_E - 240°) & 0 \\ -\sin\theta_E & -\sin(\theta_E - 120°) & -\sin(\theta_E - 240°) & 0 \\ \frac{1}{\sqrt{2}} & \frac{1}{\sqrt{2}} & \frac{1}{\sqrt{2}} & 0 \\ 0 & 0 & 0 & \sqrt{\frac{3}{2}} \end{bmatrix} \begin{bmatrix} f_A(t) \\ f_B(t) \\ f_C(t) \\ f_F(t) \end{bmatrix} \tag{6.232}$$

and

$$\begin{bmatrix} f_A(t) \\ f_B(t) \\ f_C(t) \\ f_F(t) \end{bmatrix} = \sqrt{\frac{2}{3}} \begin{bmatrix} \cos\theta_E & -\sin\theta_E & \frac{1}{\sqrt{2}} & 0 \\ \cos(\theta_E - 120°) & -\sin(\theta_E - 120°) & \frac{1}{\sqrt{2}} & 0 \\ \cos(\theta_E - 240°) & -\sin(\theta_E - 240°) & \frac{1}{\sqrt{2}} & 0 \\ 0 & 0 & 0 & \sqrt{\frac{3}{2}} \end{bmatrix} \begin{bmatrix} g_{ds}(t) \\ g_{qs}(t) \\ g_{0s}(t) \\ g_F(t) \end{bmatrix} \tag{6.233}$$

to the currents and voltages of the synchronous device. The variables $f(\ )$ in Eqs. (6.232) and (6.233) are in the ABC-reference frame while the variables $g(\ )$ are in the dq-reference frame. The subscripts A, B, C, and F designate phase A, B, C, and rotor excitation winding components, respectively, while the subscripts ds, qs, and 0s designate the stator variables projected onto the d-axis, q-axis and 0-axis, respectively. Note that f_F and g_F, the rotor components, remain invariant. The 4×4 *coefficient matrix* of Eq. (6.233) is henceforth abbreviated as $\mathbf{C}[\theta_E(t)]$. Similarly, the 4×4 *coefficient matrix* of Eq. (6.232) is abbreviated as $\mathbf{C}^T[\theta_E(t)]$. It is possible to prove, by employing the power invariance principle, that

$$\mathbf{C}^T[\theta_E(t)] = \mathbf{C}^{-1}[\theta_E(t)] \tag{6.234}$$

Examination of Fig. 6.39(a) yields that the physical windings (AA', BB', and CC') are represented by equivalent RL coils located along the axes of the fluxes

*V. E. Mablekos and L. L. Grigsby, "State Equations of Dynamic Systems with Time Varying Coefficients," *IEEE Transactions on Power Apparatus and Systems*, vol. PAS-90, no. 6 (Nov./Dec., 1971), pp. 2589–98.

produced by the physical coils. For example, coils AA' and $R_A L_A$ produce identical fluxes and have identical voltages and currents, and so on. Therefore, the describing equations of the equivalent windings of the synchronous device, Fig. 6.39(a), can be written in compact form as

$$e_j(t) = R_{jk}i_k(t) + \frac{d}{dt}\lambda_j(t) \equiv R_{jk}i_k(t) + \frac{d}{dt}\{\mathfrak{L}[\theta_E(t)]_{jk}i_k(t)\}, \quad j = k = A, B, C, \mathrm{F} \tag{6.235}$$

In the motor mode of operation, **applied** voltages and currents **to** the windings are **positive**, and positive currents produce **positive flux linkages**. For the motor mode of operation, the following definitions are pertinent for Fig. 6.39(a); (1) $e_j(t)$ constitutes a 4×1 *column matrix* whose elements are the impressed voltages across the windings; (2) $i_k(t)$ constitutes a 4×1 *column matrix* whose elements are the currents **into** the windings; (3) $R_{jk}\equiv\mathbf{R}$ constitutes a 4×4 *diagonal matrix* whose elements are the resistances of the windings; (4) $\lambda_j(t)$ designates the appropriate *flux linkages* of the windings; (5) $\mathfrak{L}[\theta_E(t)]_{jk}\equiv[\theta_E(t)]$ constitutes a 4×4 *matrix* whose elements are the time-dependent inductances of the four windings comprised of self- and mutual-inductance terms; and (6) for a P-pole synchronous device, the mechanical angular displacement θ_R is related to the electrical angular displacement θ_E in accordance with the equation

$$\theta_R = \frac{\theta_E}{(P/2)} \tag{6.236}$$

For a two-pole synchronous device, $\theta_R\equiv\theta_E$. However, for a P-pole synchronous device, we can perform a per-pole-pair analysis and thus we can still work with (1) the electrical angular displacement θ_E, (2) the electrical angular frequency ω_E, that is, $\theta_E=\omega_E t$, and (3) the inductance $[\theta_E(t)]$.

If we were to apply Eq. (6.233) to currents, we would set* $f_k(t)=i_k(t)\equiv\mathbf{i}(t)$, $k=A,B,C,\mathrm{F}$, and $g_\nu(t)=\hat{\mathbf{i}}_\nu(t)\equiv\hat{\mathbf{i}}(t)$, $\nu=$ ds, qs, 0s, F, and be able to write Eq. (6.233) in terms of the **current vectors** $i(t)$ and $\hat{\mathbf{i}}(t)$ and the matrix coefficient $\mathbf{C}[\theta_E(t)]$ as

$$\mathbf{i}(t) = \mathbf{C}[\theta_E(t)]\hat{\mathbf{i}}(t) \tag{6.237}$$

Note that script variables, for example, $\hat{\mathbf{i}}(t)$ or $\hat{\mathbf{e}}(t)$, designate dq-reference-frame variables.

Similarly, if we were to apply Eq. (6.232) to voltages, we would set $g_\eta(t)=\hat{\mathbf{e}}_\eta(t)$ $\equiv\hat{\mathbf{e}}(t)$, $\eta=$ ds, qs, 0s, F, and $f_j(t)=e_j(t)\equiv\mathbf{e}(t)$, $j=A,B,C,\mathrm{F}$, and be able to write Eq. (6.232) in terms of the **voltage vectors** $\hat{\mathbf{e}}(t)$ and $e(t)$ and the matrix coefficient $\mathbf{C}^T[\theta_E(t)]$ as

$$\hat{\mathbf{e}}(t) = \mathbf{C}^T[\theta_E(t)]\mathbf{e}(t) \tag{6.238}$$

Substitution for $\mathbf{i}(t)$ from Eq. (6.237) into Eq. (6.235), premultiplication of the resulting equation by $\mathbf{C}^T[\theta_E(t)]$, and equating the newly derived equation to Eq. (6.238) yield

$$\hat{\mathbf{e}}(t) = \mathbf{C}^T[\theta_E(t)]\mathbf{R}\mathbf{C}[\theta_E(t)]\hat{\mathbf{i}}(t) + \mathbf{C}^T[\theta_E(t)]\frac{d}{dt}\{\mathfrak{L}[\theta_E(t)]\mathbf{C}[\theta_E(t)]\hat{\mathbf{i}}(t)\} \tag{6.239}$$

If we were to keep in mind the time dependence of $\mathfrak{L}[\theta_E(t)]$, $\mathbf{C}[\theta_E(t)]$, and $\hat{\mathbf{i}}(t)$ in

*It should be understood here that the unsubscripted variables, $\mathbf{e}(t)$, $\mathbf{i}(t)$, $\hat{\mathbf{e}}(t)$, and $\hat{\mathbf{i}}(t)$, designate vectors.

Eq. (6.239) and perform the suggested mathematical operations, we could write Eq. (6.239) as

$$\hat{\mathbf{e}}(t) = \Big\{\mathbf{C}^{\mathrm{T}}[\theta_{\mathrm{E}}(t)]\mathbf{R}\mathbf{C}[\theta_{\mathrm{E}}(t)] + \mathbf{C}^{\mathrm{T}}[\theta_{\mathrm{E}}(t)]\frac{d}{dt}\mathfrak{L}[\theta_{\mathrm{E}}(t)]\mathbf{C}[\theta_{\mathrm{E}}(t)][\theta_{\mathrm{E}}(t)] + \mathbf{C}^{\mathrm{T}}[\theta_{\mathrm{E}}(t)]\mathfrak{L}\frac{d}{dt}\mathbf{C}[\theta_{\mathrm{E}}(t)]\Big\}\hat{\mathbf{i}}(t) + \mathbf{C}^{\mathrm{T}}[\theta_{\mathrm{E}}(t)]\mathfrak{L}[\theta_{\mathrm{E}}(t)]\mathbf{C}[\theta_{\mathrm{E}}(t)]\frac{d}{dt}\hat{\mathbf{i}}(t) \tag{6.240}$$

Eq. (6.240) describes the three-phase synchronous motor with all the variables reflected on the rotor *dq*-reference frame, which is considered to be stationary in space, Fig. 6.39(c). If we substitute the device parameters into Eq. (6.240), recognizing the fact that in a balanced, three-phase synchronous motor, (1) $R_S \equiv R_A = R_B \neq R_{\mathrm{F}}$, (2) $L_{\mathrm{d}} = f_1(L_{jk})$ and $L_{\mathrm{q}} = f_2(L_{jk})$, and (3) $d\theta_{\mathrm{E}}(t)/dt = \omega_{\mathrm{E}}(t)$, and simplify properly, we obtain four equations that describe the balanced three-phase synchronous motor referred to the *dq*-reference frame. These equations are

$$\hat{\mathbf{e}}_{\mathrm{ds}}(t) = R_S\hat{\mathbf{i}}_{\mathrm{ds}}(t) - L_{\mathrm{q}}\hat{\mathbf{i}}_{\mathrm{qs}}(t)\omega_{\mathrm{E}} + \frac{d}{dt}\lambda_{\mathrm{ds}}(t) \tag{6.241}$$

$$\hat{\mathbf{e}}_{\mathrm{qs}}(t) = R_S\hat{\mathbf{i}}_{\mathrm{qs}}(t) + L_{\mathrm{d}}\hat{\mathbf{i}}_{\mathrm{ds}}(t)\omega_{\mathrm{E}} + L_{A\mathrm{F}}\hat{\mathbf{i}}_{\mathrm{F}}(t)\omega_{\mathrm{E}} + \frac{d}{dt}\lambda_{\mathrm{qs}}(t) \tag{6.242}$$

$$\hat{\mathbf{e}}_{0\mathrm{s}}(t) = R_S\hat{\mathbf{i}}_{0\mathrm{s}}(t) + \frac{d}{dt}\lambda_{0\mathrm{s}}(t) \tag{6.243}$$

$$\hat{\mathbf{e}}_{\mathrm{F}}(t) = R_{\mathrm{F}}\hat{\mathbf{i}}_{\mathrm{F}}(t) + \frac{d}{dt}\lambda_{\mathrm{F}}(t) \tag{6.244}$$

If the stator currents were defined to be the balanced sinusoidal currents of Eq. (6.245), with the arbitrary phase angle β at $t=0$,

$$\begin{bmatrix} i_A(t) \\ \hline i_B(t) \\ \hline i_C(t) \\ \hline i_{\mathrm{F}}(t) \end{bmatrix} = \begin{bmatrix} \sqrt{2}\,|\mathbf{I}|\cos[\omega_{\mathrm{E}}t + \beta] \\ \hline \sqrt{2}\,|\mathbf{I}|\cos[\omega_{\mathrm{E}}t + (\beta - 120^\circ)] \\ \hline \sqrt{2}\,|\mathbf{I}|\cos[\omega_{\mathrm{E}}t + (\beta - 240^\circ)] \\ \hline i_{\mathrm{F}}(t) \end{bmatrix} \tag{6.245}$$

and were substituted for $f_k(t) \equiv i_k(t)$, $k = A, B, C, \mathrm{F}$, into Eq. (6.233), simplification would yield the currents $\hat{\mathbf{i}}_\nu(t) = \hat{\mathbf{i}}_{\mathrm{ds}}(t), \hat{\mathbf{i}}_{\mathrm{qs}}(t), \hat{\mathbf{i}}_{0\mathrm{s}}, \hat{\mathbf{i}}_{\mathrm{F}}(t)$ written in matrix form as

$$\begin{bmatrix} \hat{\mathbf{i}}_{\mathrm{ds}}(t) \\ \hline \hat{\mathbf{i}}_{\mathrm{qs}}(t) \\ \hline \hat{\mathbf{i}}_{0\mathrm{s}}(t) \\ \hline \hat{\mathbf{i}}_{\mathrm{F}}(t) \end{bmatrix} = \begin{bmatrix} \sqrt{2}\,|\mathbf{I}|\cos\beta \\ \hline \sqrt{2}\,|\mathbf{I}|\sin\beta \\ \hline 0 \\ \hline i_{\mathrm{F}}(t) \end{bmatrix} \tag{6.246}$$

Therefore, since the stator currents i_{ds} and i_{qs} are constant, the fields Φ_{ds} and Φ_{qs}, Fig. 6.39(c), can be looked upon as steady fields with respect to the rotor just as if they were produced by the currents i_{ds} and i_{qs} in the stator equivalent $R_{\mathrm{ds}}L_{\mathrm{ds}}$ and $R_{\mathrm{qs}}L_{\mathrm{qs}}$ coils located on the rotor *d*- and *q*-axes, respectively. Therefore, the terms containing $\lambda_{\mathrm{ds}}(t)$, $\lambda_{\mathrm{qs}}(t)$, $\lambda_{0\mathrm{s}}(t)$, $\lambda_{\mathrm{F}}(t)$, and $i_{0\mathrm{s}}(t)$ in Eqs. (6.241) through (6.244) drop out. It should be noted in passing that Eq. (6.243) is decoupled from the rest of the equations. Thus it can be treated separately. For the pole-pair analysis of the synchronous motor, Eqs. (6.241) through (6.244) can be written in matrix form as

$$\begin{bmatrix} \hat{\mathbf{e}}_{\rm ds}(t) \\ \hline \hat{\mathbf{e}}_{\rm qs}(t) \\ \hline \hat{\mathbf{e}}_{\rm 0s}(t) \\ \hline \hat{\mathbf{e}}_{\rm F}(t) \end{bmatrix} = \begin{bmatrix} R_{\rm S}\hat{\mathbf{i}}_{\rm ds}(t) - X_{\rm q}\hat{\mathbf{i}}_{\rm qs}(t) \\ \hline \omega_{\rm E}L_{\rm SF}\hat{\mathbf{i}}_{\rm F}(t) + {\rm R}_{\rm S}\hat{\mathbf{i}}_{\rm qs}(t) + X_{\rm d}\hat{\mathbf{i}}_{\rm ds}(t) \\ \hline 0 \\ \hline R_{\rm F}\hat{\mathbf{i}}_{\rm F}(t) \end{bmatrix} \tag{6.247}$$

where

$$X_{\rm d} \equiv \omega_{\rm E}L_{\rm ds} \tag{6.248}$$

$$X_{\rm q} \equiv \omega_{\rm E}L_{\rm qs} \tag{6.249}$$

are defined as the *d*- **and** *q*-**axes synchronous reactances**, respectively.

Eqs. (6.248) and (6.249) enable us to account for the inductive effects created by the stator field $\mathbf{B}_{\rm ST}$ by computing the effects separately on the *d*-axis and *q*-axis components of the field $\mathbf{B}_{\rm ST}$. In balanced steady state, then, the stator time-varying variables $i_k(t)$ and $e_j(t)$, Eq. (6.235), appear as *dq*-reference-frame components $\hat{\mathbf{i}}(t)$ and $\hat{\mathbf{e}}(t)$, Eqs. (6.240), (6.246), and (6.247), which are time-invariant variables when viewed from the stationary rotor *dq*-reference frame.

This idea of transformation was initiated by André Blondel of France and was refined by R. Doherty, C. Nickle, R. Park, G. Kron, E. Clark, and others in the United States. The significance of the variable $g_{\rm 0s}(t)$ in Eq. (6.232) can be explained by considering the current $\hat{\mathbf{i}}_{\rm 0s}(t) \equiv g_{\rm 0s}(t)$. Since the original set of stator phase currents is comprised of three currents, namely, $f_k(t) = i_k(t)$, $k = A$, B, and C, Fig. 6.39(b), the new set of stator currents must also be comprised of three currents, that is, the set of currents $g_\nu(t) = \hat{\mathbf{i}}_\nu(t)$, $\nu = $ ds, qs, and 0s. However, since the currents $\hat{\mathbf{i}}_{\rm ds}(t)$ and $\hat{\mathbf{i}}_{\rm qs}(t)$, Fig. 6.39(b), produce the correct magnetic field, the third variable $\hat{\mathbf{i}}_{\rm 0s}$ must produce no space fundamental field in the airgap. Blondel and his associates showed that this condition is met by the variable $\hat{\mathbf{i}}_{\rm 0s}(t)$ when defined in terms of the stator currents $i_A(t)$, $i_B(t)$, and $i_C(t)$ as*

$$\hat{\mathbf{i}}_{\rm 0s}(t) \equiv \frac{k}{3}\left[i_A(t) + i_B(t) + i_C(t) \right] \tag{6.250}$$

When the phase currents are constrained so that their sum is zero, as in the case of the ungrounded Y-connection, or under any balanced three-phase conditions, $\hat{\mathbf{i}}_{\rm 0s}(t)$ must be zero. This is the case with which this text is concerned. The multiplying factor $k/3$ is chosen arbitrarily to simplify the numerical coefficients of the transformed equations.

Substitution for $g_\eta(t) = \hat{\mathbf{e}}_\eta(t)$, $\eta = \rm ds, qs, 0s, F$, into Eq. (6.233) from Eq. (6.247) and solution for the impressed voltage of phase *A*, designated as $e_j(t) \equiv e_A(t)$, for the balanced, three-phase, synchronous motor yield

$$\begin{aligned} e_A(t) &= \left[R_{\rm S}\hat{\mathbf{i}}_{\rm ds}(t) - X_{\rm q}\hat{\mathbf{i}}_{\rm qs}(t) \right]\sqrt{\tfrac{2}{3}}\,\cos\omega_{\rm E}t \\ &\quad - \left[\omega_{\rm E}L_{\rm SF}\hat{\mathbf{i}}_{\rm F}(t) + R_{\rm S}\hat{\mathbf{i}}_{\rm qs}(t) + X_{\rm d}\hat{\mathbf{i}}_{\rm ds}(t) \right]\sqrt{\tfrac{2}{3}}\,\sin\omega_{\rm E}t \end{aligned} \tag{6.251}$$

*Power engineering students will come to know this current (for $k = 1$) as a **zero-sequence** current in advanced studies.

Utilization of phasor properties and trigonometric identities enables writing Eq. (6.251), for an assumed phase impressed voltage $e_A(t)=|\mathbf{E}_A|\cos(\omega_E t+\alpha)$, as

$$\mathrm{Re}\left\{\sqrt{2}\left[\left(\frac{|\mathbf{E}_A|}{\sqrt{2}}\right)e^{j\alpha}\right]e^{j\omega_E t}\right\} = \mathrm{Re}\left\{\sqrt{2}\left[\left(\frac{R_S\hat{\mathbf{i}}_{ds}(t)-X_q\hat{\mathbf{i}}_{qs}(t)}{\sqrt{2}}\right)e^{j0^\circ}\right]\sqrt{\tfrac{2}{3}}\;e^{j\omega_E t}\right\}$$
$$+\mathrm{Re}\left\{\sqrt{2}\left[\left(\frac{\omega_E L_{SF}\hat{\mathbf{i}}_F(t)+R_S\hat{\mathbf{i}}_{qs}(t)+X_d\hat{\mathbf{i}}_{ds}(t)}{\sqrt{2}}\right)e^{+j90^\circ}\right]\sqrt{\tfrac{2}{3}}\;e^{j\omega_E t}\right\} \tag{6.252}$$

Similarly, substitution for $g_\nu(t)=\hat{\mathbf{i}}_\nu(t)$, $\nu=\mathrm{ds,qs,0s,F}$, into Eq. (6.233) from Eq. (6.246), where $\hat{\mathbf{i}}_{ds}(t)\to\hat{\mathbf{i}}_{ds}$ and $\hat{\mathbf{i}}_{qs}(t)\to\hat{\mathbf{i}}_{qs}$, and solution for the impressed current of phase A, designated as $i_A(t)$, for the balanced synchronous motor yields

$$i_A(t) = \hat{\mathbf{i}}_{ds}\sqrt{\tfrac{2}{3}}\,\cos\omega_E t-\hat{\mathbf{i}}_{qs}\sqrt{\tfrac{2}{3}}\,\sin\omega_E t \tag{6.253}$$

Again, utilization of phasor properties and trigonometric identities enables us to write Eq. (6.253), for an assumed impressed current $i_A(t)=|\mathbf{I}_A|\cos(\omega_E t+\beta)$, as

$$\mathrm{Re}\left\{\sqrt{2}\left[\left(\frac{|\mathbf{I}_A|}{\sqrt{2}}\right)e^{j\beta}\right]e^{j\omega_E t}\right\} = \mathrm{Re}\left\{\sqrt{2}\left[\frac{\hat{\mathbf{i}}_{ds}}{\sqrt{2}}\right]\sqrt{\tfrac{2}{3}}\;e^{j\omega_E t}\right\}$$
$$-\mathrm{Re}\left\{\sqrt{2}\left[\left(\frac{\hat{\mathbf{i}}_{qs}}{\sqrt{2}}\right)e^{-j90^\circ}\right]\sqrt{\tfrac{2}{3}}\;e^{j\omega_E t}\right\} \tag{6.254}$$

It should be clear from Eq. (6.252) that the impressed voltage of phase A can be written in phasor form as

$$\mathbf{E}_A=\sqrt{\frac{2}{3}}\left\{R_S\left(\frac{i_{ds}}{\sqrt{2}}\right)-X_q\left(\frac{i_{qs}}{\sqrt{2}}\right)\right\}$$
$$+j\sqrt{\frac{2}{3}}\left\{\frac{L_{SF}\omega_E i_F}{\sqrt{2}}+R_S\left(\frac{i_{qs}}{\sqrt{2}}\right)+X_d\left(\frac{i_{ds}}{\sqrt{2}}\right)\right\} \tag{6.255}$$

If we were to define

$$\frac{i_{ds}}{\sqrt{3}}\equiv jI_d \tag{6.256}$$

and

$$\frac{i_{qs}}{\sqrt{3}}\equiv I_q \tag{6.257}$$

which yield

$$I_d=-j\left(\frac{i_{ds}}{\sqrt{3}}\right) \tag{6.258}$$

$$I_q=\frac{i_{qs}}{\sqrt{3}} \tag{6.259}$$

and

$$E_{GA}\equiv\left.\frac{L_{SF}\omega_E\hat{\mathbf{i}}_F(t)}{\sqrt{3}}\right|_{\hat{\mathbf{i}}_F(t)=i_F(t)} \tag{6.260}$$

and substitute Eqs. (6.256), (6.257), and (6.260) into Eq. (6.255), simplify, and group appropriate terms, we would be able to write Eq. (6.255) as

$$\mathbf{E}_A = jR_S(I_q + I_d) - X_d I_d - X_q I_q + j\mathbf{E}_{GA} \tag{6.261}$$

The voltage E_{GA}, Eq. (6.260), designates the RMS value of the stator phase voltage $e_{GA}(t) = \omega_E L_{AF}\hat{\mathbf{i}}_F(t)$ induced by the rotor-winding current alone, that is, when $\hat{\mathbf{i}}_F(t) \neq 0$ and $\hat{\mathbf{i}}_{ds}(t) \equiv \hat{\mathbf{i}}_{qs}(t) = 0$. Similarly, from Eq. (6.254) the impressed current of phase A can be written in phasor form with the aid of Eqs. (6.256) and (6.257) as

$$\mathbf{I}_A = j(I_q + I_d) \tag{6.262}$$

Inspection of Eqs. (6.256) and (6.257) reveals that the working quadrant of these equations is quadrant I of the *dq*-reference frame, Fig. 6.39(b), where the correspondence between the axes of the *dq*-reference frame and the complex reference frame is as follows: (1) the *q*-axis↔Re-axis and (2) the *d*-axis↔Im-axis. However, the working quadrant of the *dq*-reference frame has been established in literature to be quadrant IV of Fig. 6.39(b). The reason becomes apparent in the passage following Eq. (6.224).

Under these conditions Eqs. (6.261) and (6.262) can be rotated from quadrant I to quadrant IV by *rotating* all vectors 90° in the clockwise direction. This rotation is effected automatically by multiplying Eqs. (6.261) and (6.262) by $-j$.

Thus multiplication of Eqs. (6.261) and (6.262) by $-j$, allowing $E_{GA} \to \mathbf{E}_G$, and simplification yields

$$\mathbf{V}_{ST} = R_S(I_q + I_d) + jX_d I_d + jX_q I_q + \mathbf{E}_G \tag{6.263}$$

and

$$I_S = I_q + I_d \tag{6.264}$$

where

$$\mathbf{E}_S \equiv -jE_A \tag{6.265}$$

and

$$\mathbf{I}_S \equiv -jI_A \tag{6.266}$$

designate the phase A (or the **per-phase**) *impressed* phasor voltage $\mathbf{V}_{ST}$ and the *response* current $\mathbf{I}_S$ of Fig. 6.39(a), correspondingly, referred to quadrant IV. Finally, utilization of the equations $\lambda = Li = N\Phi$, $\Phi = Ni/\mathfrak{R}$, and $\mathfrak{R} = \ell/\mu A$ in conjunction with $\ell_d \ll \ell_q$, as set forth earlier in this section, yields that in Eq. (6.263) the following relationship is valid:

$$X_d > X_q \tag{6.267}$$

The theory developed in this section, and thus Eq. (6.263) in accordance with the principles set forth earlier, describes the *motor mode of operation* of the salient-rotor synchronous device. Since the synchronous device is used **mostly** in its generator mode, the corresponding describing equation should be derived. This is easy, however, since the distinct difference between the two modes of operation is the direction of the stator winding currents $\mathbf{I}_S = -\mathbf{I}_G$ and their components I_d and I_q. Note that $\mathbf{I}_S$ is the current **into** the stator windings for the motor mode of operation and $\mathbf{I}_G$ is the current **out** of the stator windings for the generator mode of operation. Thus if all the currents of Eq. (6.263) are **negated** and $\mathbf{V}_{ST} \to \mathbf{V}_L$, the equation is written as

$$\mathbf{E}_G = \mathbf{V}_L + R_S(I_q + I_d) + jX_d I_d + jX_q I_q \tag{6.268}$$

where

$$\mathbf{I}_G \equiv I_q + I_d \tag{6.269}$$

Eq. (6.268) describes the **generator mode of operation** of the salient-rotor synchronous generator.

Utilization of Eq. (6.268) enables us to draw the phasor diagram that describes the salient-rotor synchronous generator as in Fig. 6.40. Note that in Fig. 6.40, the quantities that bear a j are drawn 90° ahead (counterclockwise) of the defining axes. Thus, jI_d is 90° ahead of I_d, and so on. The reason why the Im- and Re-axes do not coincide with the d- and q-axes becomes apparent shortly.

Use of the phasor diagram of Fig. 6.40 necessitates decomposition of the generated current $\mathbf{I}_G$, as per Eq. (6.269), into d- and q-axes components. However, this resolution presumes knowledge of the power-factor angle θ_L and the power angle δ. Knowledge of both these angles is not always available. Usually the power angle δ is not known. Thus the position of the current $\mathbf{I}_G$ with respect to the d- and q-axes is **unknown** and must be established before one proceeds with the analysis of the problem at hand. **Note that we cannot draw an equivalent circuit of the salient-rotor synchronous generator yet.**

These difficulties, however, are circumvented by modifying the phasor diagram of Fig. 6.40 as shown in Fig. 6.41. Study carefully the *heavy-line* phasor diagram of Fig. 6.41 and compare it with Fig. 6.40 and Fig. 5.17. Note that the heavy-line phasor diagram of Fig. 6.41, derived from Fig. 6.40 by rotating the Re- and Im-axes $|\theta_L|+|\delta|$ degrees counterclockwise, is identical in format to the phasor diagram of Fig. 5.17. **Therefore, we can now draw the equivalent circuit of the salient-rotor synchronous generator as in Fig. 6.42.** Inspection of Fig. 6.41 enables us to write the equation for the *approximate* generated voltage $\mathbf{E}_{\mathcal{G}}$ as

$$\mathbf{E}_{\mathcal{G}} \equiv |\mathbf{E}_{\mathcal{G}}|e^{j\phi_{\mathcal{G}}} = \mathbf{V}_L + (R_S + jX_d)\mathbf{I}_G \tag{6.270}$$

However, if we wish to calculate the *actual* generated voltage $\mathbf{E}_G$, again with the aid of Fig. 6.41, we can write Eqs. (6.271) and (6.275) for $\mathbf{E}_G$ and $\mathbf{E}$. Thus

$$\mathbf{E}_G = \mathbf{E} + [X_d - X_q]|I_d|e^{j\phi} \equiv |\mathbf{E}_{\mathcal{G}}|\cos\Delta\phi e^{j\phi}, \phi = \measuredangle \mathbf{E}_G \equiv \measuredangle \mathbf{E} \tag{6.271}$$

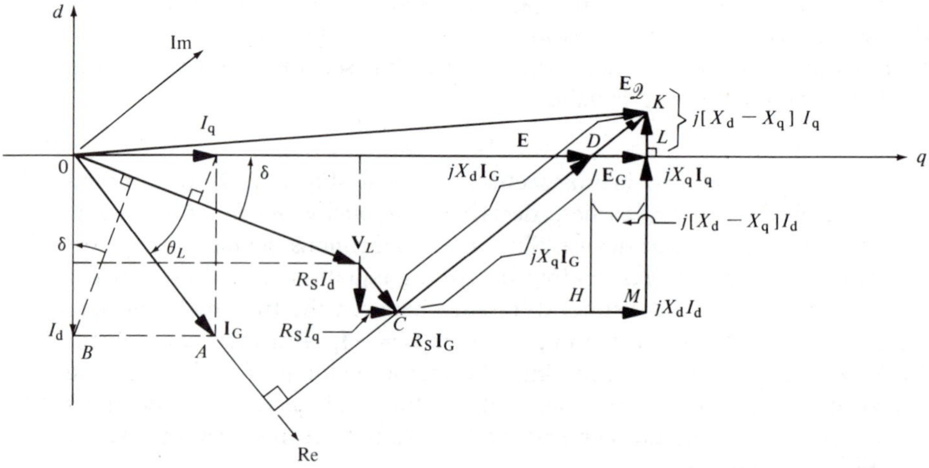

Fig. 6.40 The Phasor Diagram of a Balanced, Three-Phase, *P*-Pole, Salient-Rotor, Synchronous Generator

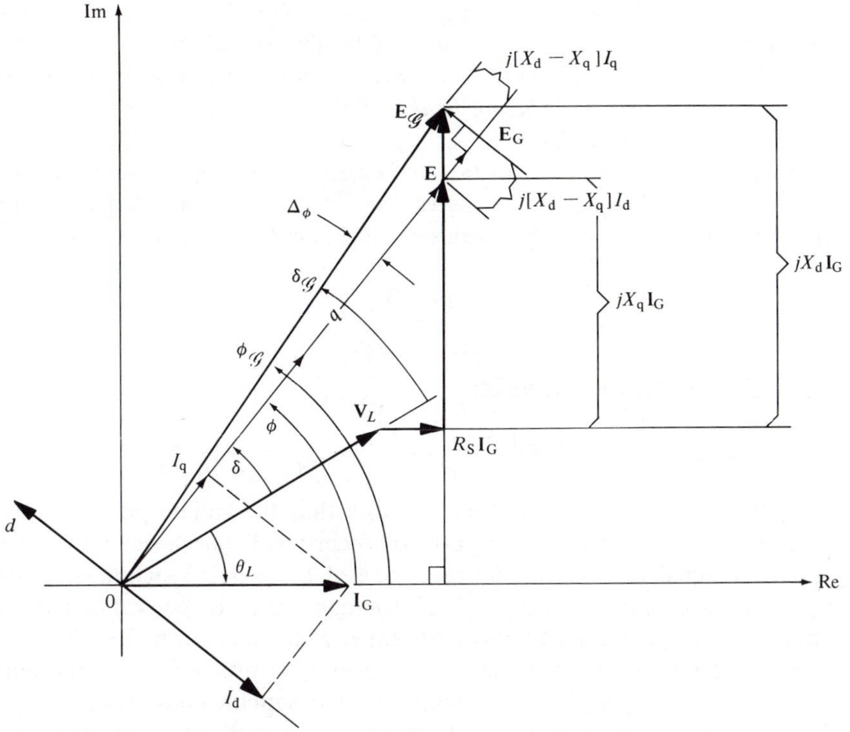

Fig. 6.41 Modified Phasor Diagram of a Three-Phase, Salient-Rotor, Synchronous Generator

where

$$\Delta\phi = |\phi_{\mathscr{G}}| - |\phi| \equiv |\delta_{\mathscr{G}}| - |\delta| \tag{6.272}$$
$$|I_d| = |\mathbf{I}_G|\sin(|\theta_L| + |\delta|) \equiv |\mathbf{I}_G|\sin\phi \tag{6.273}$$
$$|I_q| = |\mathbf{I}_G|\cos(|\theta_L| + |\delta|) \equiv |\mathbf{I}_G|\cos\phi \tag{6.274}$$

and

$$\mathbf{E} \equiv |\mathbf{E}|e^{j\phi} = \mathbf{V}_L + (R_S + jX_q)\mathbf{I}_G \tag{6.275}$$

It is clear that Eq. (6.275) locates the angular position of the generated voltage $\mathbf{E}_G$ and therefore the position of the d- and q-axes with respect to the Re-axis. Note

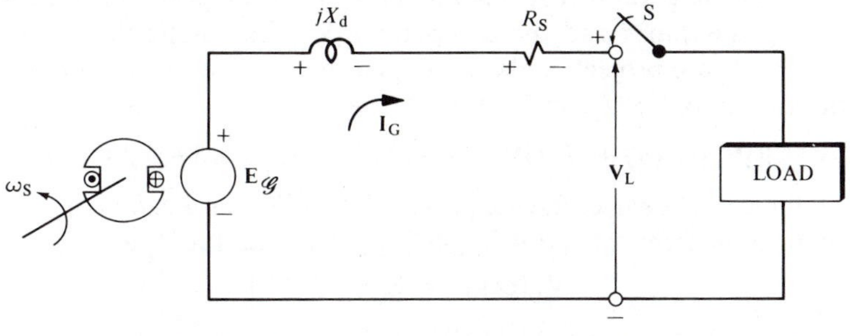

Fig. 6.42 Approximate Single-Phase, Steady State, Equivalent Circuit of a Balanced, Three-Phase, P-Pole, Salient-Rotor, Synchronous Generator Utilized in Its Per-Phase Analysis

that the generated voltage $\mathbf{E}_G$ is along the q-axis. This is the reason for the 90° clockwise rotations of Eqs. (6.261) and (6.262). Therefore, the generating field is along the d-axis as it should be in a normal synchronous generator; that is, *the generated voltage* $\mathbf{E}_G$ *lags its inducing field by 90°*—note the alignment of the rotor with the d-axis in Fig. 6.39(a).

Eqs. (6.270) through (6.275) can be verified from Fig. 6.41 with the aid of Fig. 6.40 if we observe the similar triangles that are formed and equate ratios of mutually perpendicular sides. For example, the similar triangles OAB and DKL yield

$$\frac{\overline{KL}}{\overline{AB}} = \frac{\overline{DL}}{\overline{OB}} \tag{6.276}$$

Solution of Eq. (6.276) for $\overline{KL}$ yields

$$\overline{KL} = \frac{\overline{DL}}{\overline{OB}} \times \overline{AB} = \frac{[X_d - X_q]|I_d|}{|I_d|} \times |I_q| = [X_d - X_q]|I_q| \tag{6.277}$$

The equivalent circuit of Fig. 6.42 suggests that the salient-pole synchronous device can be treated using cylindrical-rotor theory **as if the device had a synchronous reactance equal in value to the reactance of the d-axis reactance X_d, that is,** $X_S \equiv X_d$. Since the side $[X_d - X_q]|I_q|$ of triangle OKL is perpendicular to the generated voltage $\mathbf{E}_G$, Figs. 6.40 and 6.41, experience has shown that there is *little difference* in magnitude between the actual generated voltage $\mathbf{E}_G$ and the approximate generated voltage $\mathbf{E}_g$ for a normally excited synchronous device. However, there is *considerable difference* in the phase angles ϕ and ϕ_g of the voltages $\mathbf{E}_G$ and $\mathbf{E}_g$, respectively. This difference $\Delta\phi$ is caused by the effects of reluctance torque* in a salient-pole synchronous device.

Thus, as far as interrelations among stator-winding terminal voltage and current, and excitation over the normal operating range, are concerned, the effects of saliency are usually of minor importance, and the characteristics of the salient-pole device can be computed with satisfactory accuracy by employing the cylindrical-rotor theory developed in this chapter. It should be noted, however, that (1) for small excitations the differences between cylindrical-rotor and salient-rotor theory become important, and (2) the total power developed P_{DT} by the salient-rotor synchronous device is distinctly different than that developed by its smooth-rotor counterpart.

If the generator of Fig. 6.1(a) is of the salient-rotor type, the total power, $P_{DT} = 3P_D$, delivered to the infinite bus by the generator can be calculated with the aid of the phasor diagram of Fig. 6.40 if (1) $\mathbf{V}_L$ is decomposed into the component $|\mathbf{V}_L|\sin\delta$ along the d-axis and the component $|\mathbf{V}_L|\cos\delta$ along the q-axis or (2) the currents I_d and I_q are projected along the phasor $\mathbf{V}_L$. This current projection yields a total current $|I_q|\cos\delta + |I_d|\sin\delta$. Thus

$$P_D = |I_q|(|\mathbf{V}_L|\cos\delta) + |I_d|(|\mathbf{V}_L|\sin\delta) \equiv |\mathbf{V}_L|(|I_q|\cos\delta + |I_d|\sin\delta) \tag{6.278}$$

Since in the actual device, $R_S \simeq 0$, $|\mathbf{V}_L|\cos\delta$ and $|\mathbf{V}_L|\sin\delta$ can be written, with the aid of Fig. 6.40, in terms of the voltages $|\mathbf{E}_G|$, $X_d I_d$, and $X_q I_q$ as

$$|\mathbf{V}_L|\cos\delta = |\mathbf{E}_G| - X_d|I_d| \tag{6.279}$$

*Inspection of Eq. (6.284) and Fig. 6.43 enables us to interpret the significance of reluctance torque. Please see Section 5.10 for additional information.

and

$$|\mathbf{V}_L|\sin\delta = X_q|I_q| \tag{6.280}$$

Solution of Eq. (6.279) for $|I_d|$ yields

$$|I_d| = \frac{|\mathbf{E}_G| - |\mathbf{V}_L|\cos\delta}{X_d} \tag{6.281}$$

Similarly, solution of Eq. (6.280) for $|I_q|$ yields

$$|I_q| = \frac{|\mathbf{V}_L|\sin\delta}{X_q} \tag{6.282}$$

Substitution for $|I_d|$ and $|I_q|$ from Eqs. (6.281) and (6.282) into Eq. (6.278), and letting $|\mathbf{V}_L| \to |\mathbf{V}_{ST}|$, yields

$$P_D = |\mathbf{V}_{ST}|\left\{\frac{|\mathbf{E}_G| - |\mathbf{V}_{ST}|\cos\delta}{X_d}\right\}\sin\delta + |\mathbf{V}_{ST}|\left\{\frac{|\mathbf{V}_{ST}|\sin\delta}{X_q}\right\}\cos\delta \tag{6.283}$$

Expansion of Eq. (6.283), factoring out the common term $|\mathbf{V}_{ST}|^2$, utilization of the trigonometric identity $\mathbf{2\cos\delta\sin\delta \equiv \sin 2\delta}$, putting all relevant terms under a common denominator, and multiplication by a factor of 3 yields the total power developed by the three-phase, salient-rotor, synchronous device, that is, $P_{DT} \to P$, as

$$P_{DT} \to P = 3\left(\frac{|\mathbf{V}_{ST}||\mathbf{E}_G|}{X_d}\right)\sin\delta + 3|\mathbf{V}_{ST}|^2\left\{\frac{X_d - X_q}{2X_dX_q}\right\}\sin 2\delta \tag{6.284}$$

Plotting of Eq. (6.284) yields the power versus power-angle characteristic, Fig. 6.43, of the salient-rotor synchronous device for (1) the generator mode of operation, that is, $\delta \geqslant 0$, and (2) the motor mode of operation, that is, $\delta \leqslant 0$.

Inspection of Eq. (6.284) reveals that the first term is the same as the expression obtained for the cylindrical-rotor synchronous device, Eq. 6.1. The second term, a *second-harmonic* of the power angle δ, introduces the effect of the salient poles. It represents the fact that *the airgap field creates power (i.e., torque), tending to align the field poles in the position of minimum reluctance*. The term is independent of the field excitation and is the power that corresponds to what is known as the **saliency effect** or **reluctance power**. When $X_d = X_q$, as in the uniform airgap synchronous device, the second term of Eq. (6.284) becomes zero. That is, Eq. (6.284) reduces to the power equation for the cylindrical-rotor synchronous device whose synchronous reactance is X_d, that is, Eq. (6.1) for $X_S \equiv X_d$.

Inspection of Fig. 6.43 yields that, for the effects of the rotor resistance neglected, the generator and motor regions of operation are symmetrical. The difference between the two modes of operation is given by the sign of the power angle δ and its physical significance. That is, for the generator mode of operation, $\mathbf{E}_G$ *leads* $\mathbf{V}_{ST} \to \mathbf{V}_L$, Figs. 6.40 and 6.41, and for the motor mode of operation, $\mathbf{E}_G$ *lags* $\mathbf{V}_{ST} \to \mathbf{V}_L$, Figs. 6.40 and 6.41. As in the case of the smooth-rotor synchronous device, Fig. 6.1(b), steady state operation of the salient synchronous device is *stable* over the range of the power angle where the slope of the power versus power-angle characteristic is *positive*. It should be obvious from the resultant curve, Fig. 6.43, that because of the reluctance power, *a salient-pole synchronous device is stiffer* [please see Eq. (6.8) for stiffness] *than one with a cylindrical rotor*. That is, for equal voltages and equal values of reactance, that is, $X_d = X_S$, the salient-pole synchronous device when compared to the smooth-airgap synchronous device develops (1)

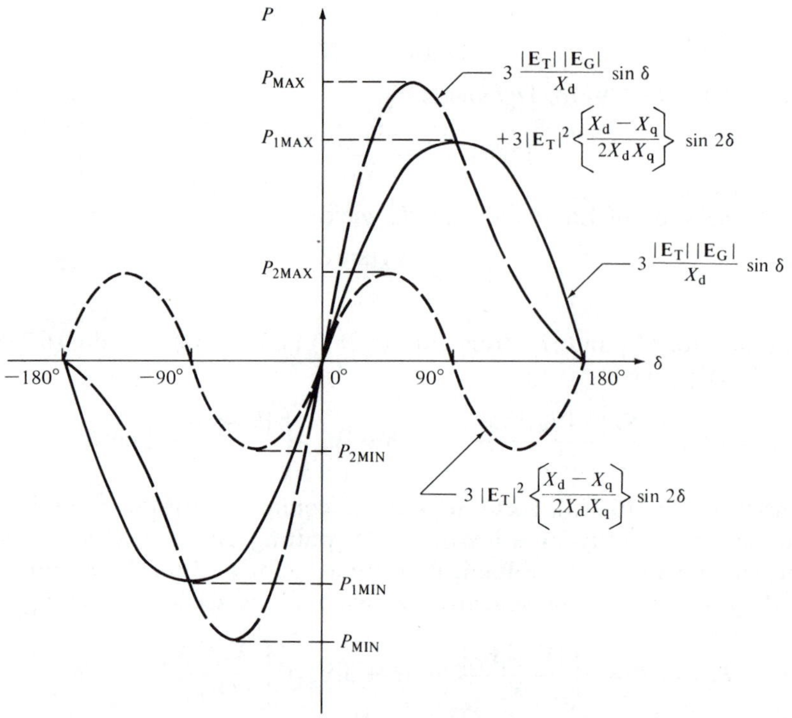

Fig. 6.43 Power vs. Power-Angle Characteristics of a Balanced, Three-Phase, *P*-Pole, Salient-Rotor, Synchronous Device Showing the Fundamental Component Due to Field Excitation and the Second-Harmonic Component Due to Reluctance Power

a given power at a smaller value of power angle δ and (2) a greater maximum power at $|\delta|<90°$. For most salient-pole synchronous devices, the power angle at which maximum power occurs is $65° < \delta_{MAX} < 75°$.

The power angle at which maximum power occurs, designated as δ_{MAX}, can be found by differentiating Eq. (6.284) with respect to δ and setting $dP_{DT}/d\delta = 0$. Thus, with $|\mathbf{E}_G|$ and $|\mathbf{V}_{ST}|$ maintained constant,

$$\frac{dP_{DT}}{d\delta} = 3\left(\frac{|\mathbf{V}_{ST}||\mathbf{E}_G|}{X_d}\right)\cos\delta + 6|\mathbf{V}_{ST}|^2\left\{\frac{X_d - X_q}{X_d X_q}\right\}\cos 2\delta = 0 \tag{6.285}$$

Utilization of the trigonometric identity $\mathbf{\cos 2\delta \equiv 2\cos^2\delta - 1}$, putting both terms of Eq. (6.285) over the same denominator, and simplification yields

$$4|\mathbf{V}_{ST}|^2(X_d - X_q)\cos^2\delta_{MAX} + |\mathbf{V}_{ST}||\mathbf{E}_G|X_q\cos\delta_{MAX} - 2|\mathbf{V}_{ST}|^2(X_d - X_q) = 0 \tag{6.286}$$

Solution of the quadratic equation of $\cos\delta_{MAX}$, Eq. (6.286), for δ_{MAX} yields

$$\delta_{MAX} = \pm\cos^{-1}\left\{\frac{-|\mathbf{E}_G|X_q \pm \left[(|\mathbf{E}_G|X_q)^2 + 32|\mathbf{V}_{ST}|^2(X_d - X_q)^2\right]^{1/2}}{8|\mathbf{V}_{ST}|(X_d - X_q)}\right\} \tag{6.287}$$

It should be noted that discretion should be used as to which of the four values of δ_{MAX} should be chosen.

Example 6.11

The reactances of a salient-pole synchronous generator are $X_d = 1.00$ pu Ω and $X_q = 0.80$ pu Ω. The stator winding resistance R_S is negligible. If the generator delivers rated per-unit voltamperes at rated per-unit volts and lagging current at a power factor of 0.82, calculate: (1) the approximate generated voltage $\mathbf{E}_g$, (2) the approximate power angle δ_g, (3) the approximate generated power P_g, (4) the actual generated voltage $\mathbf{E}_G$, (5) the actual power angle δ, and (6) the actual generated power P.

Compare the calculated quantities and draw conclusions with respect to the accuracy of the approximate equivalent circuit of the synchronous device, Figs. 6.41 and 6.42.

Solution

Utilization of Fig. 6.41, Eq. (6.270), and input data [where $\cos^{-1}(0.82) = 35°$] enables the calculation of the approximate generated voltage $\mathbf{E}_g$ as

$$\mathbf{E}_g = |\mathbf{E}_g|e^{j\phi_g} = \mathbf{V}_L + (R_S + jX_d)\mathbf{I}_G = 1.00\, e^{j35°} + j1.00 \times 1.00$$
$$= 0.82 + j0.57 + j1.00 = \underline{\underline{1.77e^{j62.42°}}} \text{ pu V} \qquad (1)$$

The approximate power angle δ_g, defined as $\delta_g \equiv \measuredangle \mathbf{E}_g - \measuredangle \mathbf{V}_L$, is calculated as

$$\delta_g = \phi_g - \theta_L = 62.42° - 35.00° = \underline{\underline{27.42°}} \qquad (2)$$

The approximate power P_g is calculated with the aid of Eq. (6.284) if we let $|\mathbf{E}_G| \rightarrow |\mathbf{E}_g|$ and $\delta \rightarrow \delta_g$. Thus proper substitution from the input data and Eqs. (1) and (2) into Eq. (6.284) yields

$$P_g = 3\left(\frac{|\mathbf{E}_T||\mathbf{E}_g|}{X_d}\right)\sin\delta_g + 3|\mathbf{E}_T|^2\left(\frac{X_d - X_q}{2X_dX_q}\right)\sin 2\delta_g$$
$$= \left(\frac{3 \times 1.00 \times 1.77}{1.00}\right)\sin 27.42° + 3 \times (1.00)^2\left(\frac{1.00 - 0.80}{2 \times 1.00 \times 0.80}\right)\sin 54.84°$$
$$= 5.31 \times 0.46 + 0.38 \times 0.82 = 2.44 + 0.31 = \underline{\underline{2.75}} \text{ pu W} \qquad (3)$$

Similarly, utilization of Fig. 6.41, Eqs. (6.275), (6.273), and (6.271) and input data enables us to calculate the actual generated voltage as follows:

$$|I_d| = |I_G|$$
$$\mathbf{E} = |\mathbf{E}|e^{j\phi} = \mathbf{V}_L + jX_q\mathbf{I}_G = 1e^{j35°} + j0.8 \times 1.0 = 0.82 + j0.57 + j0.80$$
$$= 0.82 + j1.37 = \underline{\underline{1.60e^{j59.10°}}} \text{ pu V} \qquad (4)$$
$$|I_d| = |\mathbf{I}_G|\sin\phi = 1.0 \sin 59.10° = \underline{\underline{0.86}} \text{ pu A} \qquad (5)$$

and (recognizing $\phi = \measuredangle \mathbf{E}_G$)

$$\mathbf{E}_G = |\mathbf{E}_G|e^{j\phi} = \mathbf{E} + (X_d - X_q)|I_d|e^{j\phi}$$
$$= 1.60e^{j59.10°} + [1.0 - 0.8] \times 0.86e^{j59.10°} = \underline{\underline{1.77e^{j59.10°}}} \text{ pu V} \qquad (6)$$

Or, by substituting $|\mathbf{E}_g|$, ϕ_g, and ϕ from Eqs. (1) and (4) into Eqs. (6.271) and (6.272),

$$\mathbf{E}_G = |\mathbf{E}_g|\cos\Delta\phi e^{j\phi} = 1.77\cos(62.42° - 59.10°)e^{j59.10°}$$
$$= 1.77\cos 3.32°\, e^{j59.10°} = \underline{\underline{1.77e^{j59.10°}}} \qquad (7)$$

Table 6.11.1 TABULATION OF THE ANSWERS OF EXAMPLE 6.11

	$\lvert \mathbf{E}_G \rvert$	ϕ	δ	P
Actual Values	1.77	59.10°	24.10°	2.45
Percentage Differences	0.00	+5.62	+13.79	+11.79
Approximate Values	1.77	62.42°	27.42°	2.75
	$\lvert \mathbf{E}_{\mathcal{g}} \rvert$	$\phi_{\mathcal{g}}$	$\delta_{\mathcal{g}}$	$P_{\mathcal{g}}$

The real power angle δ, defined as $\delta \equiv \measuredangle \mathbf{E}_G - \measuredangle \mathbf{V}_L$, is calculated as

$$\delta = |\phi| - |\theta_L| = 59.10° - 35.00° = \underline{\underline{24.10°}} \tag{8}$$

The real power P, calculated by substituting properly from the input data and Eqs. (6) or (7) and (8) into Eq. (6.284), is

$$P = 3\left(\frac{|\mathbf{V}_{ST}||\mathbf{E}_G|}{X_d}\right)\sin\delta + 3|\mathbf{V}_{ST}|^2\left(\frac{X_d - X_q}{2X_dX_q}\right)\sin 2\delta$$

$$= \left(\frac{3\times 1.00\times 1.77}{1.00}\right)\sin 24.10° + 3\times(1.00)^2\left(\frac{1.00-0.80}{2\times 1.00\times 0.80}\right)\sin 48.20° = 2.45$$

Note that for the same terminal voltage and current, the salient-pole theory and the cylindrical-rotor theory yield the same generated voltages, that is, $|\mathbf{E}_{\mathcal{g}}| = |\mathbf{E}_G|$. However, the phase angles of the generated voltages, that is, ϕ and $\phi_{\mathcal{g}}$, differ considerably. Consequently so do the power angles δ and $\delta_{\mathcal{g}}$ and the powers P and $P_{\mathcal{g}}$. In conclusion, it should be pointed out, however, that the percentage differences between the tabulated values are not that unreasonable. Therefore, the cylindrical-rotor theory could be used for the analysis of the salient-rotor synchronous devices as a first approximation.

6.8 BALANCED THREE-PHASE FAULTS ON SYNCHRONOUS GENERATORS

It is clear from the preceding discussion that the analysis of the salient-rotor synchronous device is more complex than the analysis of the smooth-rotor synchronous device. The complexity arises from the different reluctances that the *d*- and *q*-axes of the rotor present to the corresponding stator field components Φ_{ds} and Φ_{qs}. The *unity* power-factor current component I_q, of the stator current $\mathbf{I}_G$, with respect to the voltage $\mathbf{E}_G$, Fig. 6.40, should always be identified with the *q*-axis reactance X_q since this current component sets up the stator flux component Φ_{qs} in quadrature with the main field poles, Fig. 6.39(c). Similarly, the *lagging* (or *leading*) zero power-factor current component I_d, of the stator current $\mathbf{I}_G$, with respect to the voltage $\mathbf{E}_G$, Fig. 6.37, should always be identified with the *d*-axis reactance X_d, since this current sets up the stator field component Φ_{ds} in line with the main poles, Fig. 6.39(c).

The theory of analysis of the salient-rotor synchronous device, because of the two reactances $X_d \neq X_q$, is referred to as the **two-reactance theory**. In contrast, the analysis of the cylindrical-rotor synchronous device, because $X_d = X_q \equiv X_S$, is referred to as the **single-reactance theory**. For balanced three-phase fault (short-circuit) studies, the short-circuit currents, in general, feed through transmission lines and transformers where the $|jX|$'s$\gg R$'s. Thus the power factors approach zero as currents go lagging; the I_q component of the current $\mathbf{I}_G$ can be neglected, and therefore $|\mathbf{I}_G| \to I_d$, Fig. 6.40. (The stator field then follows the normal steady state path of minimum reluctance $\mathcal{R}_d$, meaning that X_d for steady state operation is high.) Thus for $I_q \to 0$, the phasor diagram of Fig. 6.40 for the salient-rotor synchronous device approaches that of Fig. 6.44(a), redrawn in the classical position as in Fig. 6.44(b). Therefore, although the two-reactance theory is more accurate than the single-reactance theory, it so happens that because $|I_G| \to I_d$ for fault analysis, the single-reactance theory can be employed. For systems comprised of salient-rotor synchronous devices, jX_d merely replaces jX_S used in cylindrical-rotor synchronous device theory, and the same steady state analysis theory is used in balanced three-phase faults with $\mathbf{E}_G$, X_d, and $\mathbf{I}_G$ taking three distinct values, designated as $\mathbf{E}_{\mathcal{V}}$; $\mathcal{V} = G'', G', G$, $X_\nu = X_d'', X_d', X_d$, and $\mathbf{I}_V = I_F'', I_F', I_F$, corresponding to *three distinct intervals of time* following a balanced three-phase fault. Since the main component of large-scale power systems is the synchronous generator, a detailed study is carried out here for a loaded synchronous generator *faulted* either well into its loading network, or at its terminals.

When a synchronous generator is operating with no load as shown in Fig. 6.45(a), that is, its terminals are open-circuited, and it is generating sinusoidal voltages, Fig. 6.45(b), it can be shown that the value of the current $i_V(t)$ through the **short circuit** at the terminals of the generator following a balanced three-phase fault, that is, for $t > 0^+$, Fig. 6.45(c), depends to a great extent upon the instant in the cycle, and thus on the phase angle α, Fig. 6.45(b), at which the short circuit occurs. Since it is

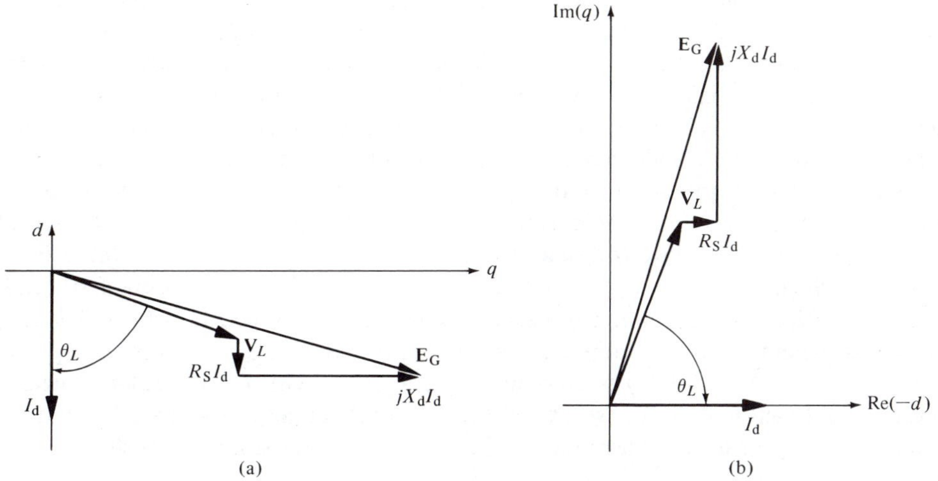

Fig. 6.44 Per-Phase Phasor Diagrams of Three-Phase, *P*-Pole, Synchronous Generators Used in Balanced Fault Studies: (a) *dq*-Reference-Frame and (b) Classical Reference Frame, That Is, the Phasor Diagram of Part (a) is Rotated 90° in the Counterclockwise Direction

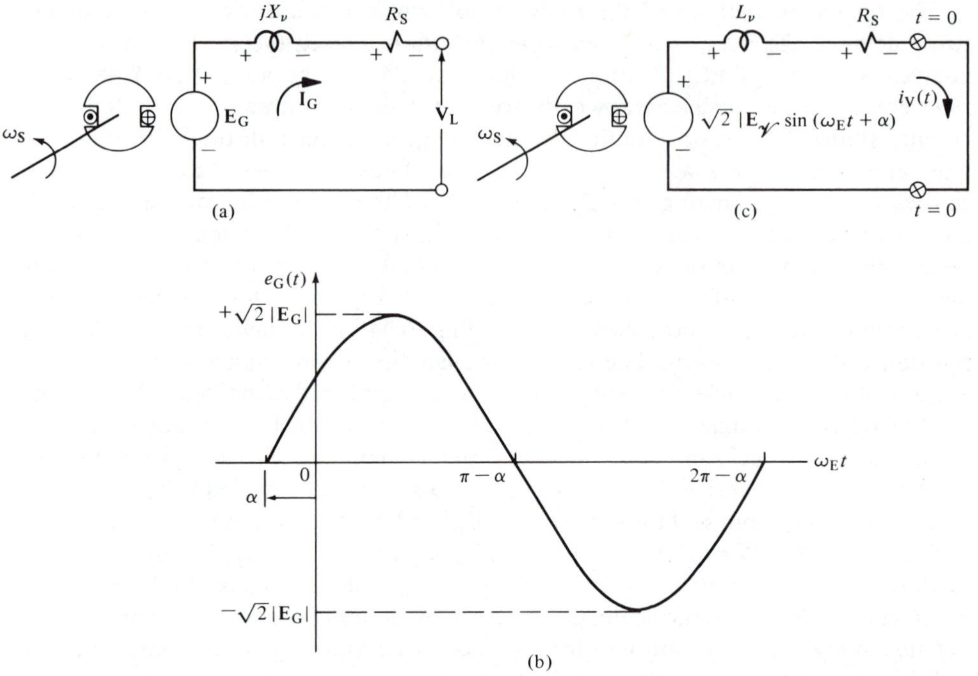

Fig. 6.45 Per-Phase Representation of a Balanced Three-Phase, Synchronous Generator: (a) Unloaded Generator, (b) a Balanced, Three-Phase Fault is Applied to the Generator at $t_F = 0$, and (c) Time-Domain Generated Voltage with an Initial Phase Angle α

impossible to predict at which instant a generator fault might occur, we should be prepared to deal with the **worst case**. In order to obtain such information, assume that the generator of Fig. 5.12 is initially open-circuited and a **balanced three-phase fault** is placed upon the terminals of the generator, that is, *all three terminals are shorted simultaneously* (this short could, or could not, be grounded) at $t = 0$.

To understand the behavior of the initially unloaded synchronous generator, Fig. 5.12—when a balanced three-phase fault is placed upon the terminals of the generator, that is, all three terminals are short-circuited simultaneously—the following procedure is followed on a per-phase basis. First, **the constant-reactance technique**, in which the synchronous generator is considered to be a normal series *RL*-circuit, that is, the response current $i_V(t)$ of Fig. 6.45(c) is studied in great detail when (a) **the rotor structure is absent**, (b) there is constant resistance R_S, (c) there is **constant inductance** L_v, and (d) it is suddenly excited with an ac source of strength $\sqrt{2}\,|\mathbf{E}_{\gamma}|\sin(\omega_E t + \alpha)$. Second, **the variable-reactance technique**, in which the synchronous generator is considered to be in its normal mode, that is, the response current $i_V(t)$ of Fig. 6.45(c) is studied in great detail with (a) **the rotor structure present and functional**, (b) constant resistance R_S, (c) **variable inductance** L_v, and (d) sudden faulting of all the terminals of the stator windings simultaneously.

Constant-Reactance Technique

For the first case, as the source $\sqrt{2}\,|E_{\gamma}|\sin(\omega_E t + \alpha)$ is suddenly applied to the series *RL*-circuit by applying the short circuit at $t = 0$, Fig. 6.45(c), the phase angle

α of the voltage, Fig. 6.45(b), determines the magnitude of the voltage when the short circuit is applied. If the instantaneous voltage is zero and increasing in a positive direction when it is applied, α is equal to zero. If the instantaneous voltage is at its maximum value, α is equal to $\pi/2$.

The time-domain describing equation, for $t>0$, for this case is written as

$$L_\nu \frac{d}{dt} i_V(t) + R_S i_V(t) = \sqrt{2}\,|E_{\mathcal{V}}| \sin(\omega_E t + \alpha) \tag{6.288}$$

Utilization of the impedance of the series RL-circuit, Eq. (6.289),

$$\mathbf{Z}_\nu = R_S + j\omega_E L_\nu \equiv |\mathbf{Z}_\nu| e^{j\theta_Z}, \theta_Z = \tan^{-1}\left(\frac{\omega_E L_\nu}{R_S}\right) \tag{6.289}$$

Fig. 6.45(c), and Laplace transformation theory, enables us to write the time-domain solution of Eq. (6.288) as

$$i_V(t) = \sqrt{2}\,|\mathbf{I}_V| \{\sin[\omega_E t + (\alpha - \theta_Z)] - \sin(\alpha - \theta_Z) e^{-(R_S/L_\nu)t}\} \tag{6.290}$$

where

$$|\mathbf{I}_V| \equiv \frac{|\mathbf{E}_{\mathcal{V}}|}{|\mathbf{Z}_\nu|} \tag{6.291}$$

It is essential to examine carefully Eq. (6.290) and verify the assertion set forth earlier. That is, to note the *dependence* of the current $i_V(t)$ on the phase angle α of the voltage waveform, Fig. 6.45(b), at $t=0$, that is, the instant the short circuit is applied. Note that the first term of Eq. (6.290) varies sinusoidally with time. It is called the **steady state term** and is designated as $i_{AC}(t)$. The second term of Eq. (6.290) is nonperiodic and decays exponentially with a time constant $\tau = L_\nu / R_S$. This nonperiodic term is referred to as the **transient term** or **dc component** and is designated as $i_{DC}(t)$, $0^+ \leqslant t < 5\tau$. If the value of the steady state term is not zero when $t=0$, the dc component is present in the solution in order to satisfy the physical condition of zero current $i_V(t)$ at the instant the short circuit is applied. Note that if (1) $\alpha - \theta_Z = 0$ or (2) $\alpha - \theta_Z = \pi$, the dc component does not exist. The dependence of $i_V(t)$ on $\alpha - \theta_Z$, and thus upon the phase angle α of the voltage waveform, Fig. 6.45(b), becomes more obvious if we were to plot Eq. (6.290) for (1) $\alpha - \theta_Z = -90°$ and (2) $\alpha - \theta_Z = 0°, 180°$.

For the first case the current is plotted as in Fig. 6.46(a), and for the second case the current is plotted as in Fig. 6.46(b). Note the existence of the transient or dc component $i_{DC}(t)$ in Fig. 6.46(a). Also, note that its maximum magnitude, designated as i_{DCMAX}, as extracted from the second term of Eq. (6.290) by setting $\alpha - \theta_Z = -90°$ at $t=0$, is as large as the amplitude of the steady state current in Fig. 6.46(b). That is,

$$i_{DCMAX} = \sqrt{2}\,|\mathbf{I}_V| = \sqrt{2}\left(\frac{|\mathbf{E}_{\mathcal{V}}|}{|\mathbf{Z}_\nu|}\right) \tag{6.292}$$

Therefore, Fig. 6.46(a) represents the **worst** possible transient situation. The maximum amplitude of the current $i_V(t)$, that is, i_{VMAX}, as extracted from Eq. (6.290) by setting $\alpha - \theta_Z = -90°$ and $\omega_E t = 180°$, is **twice** the amplitude of the steady state value, also extracted from Eq. (6.290) by setting $\alpha - \theta_Z = -90°$, $\omega_E t = 180°$, and $5\tau \leqslant t < \infty, \tau = L_\nu / R_S$. That is, in Fig. 6.46(a), letting i_{VMAX} designate the maximum amplitude of the short-circuit current $i_V(t)$ occurring at $t \doteq 0$ and i_{VMIN} designate the minimum amplitude of the short-circuit current $i_V(t)$, $t \geqslant 5\tau$, which equals the

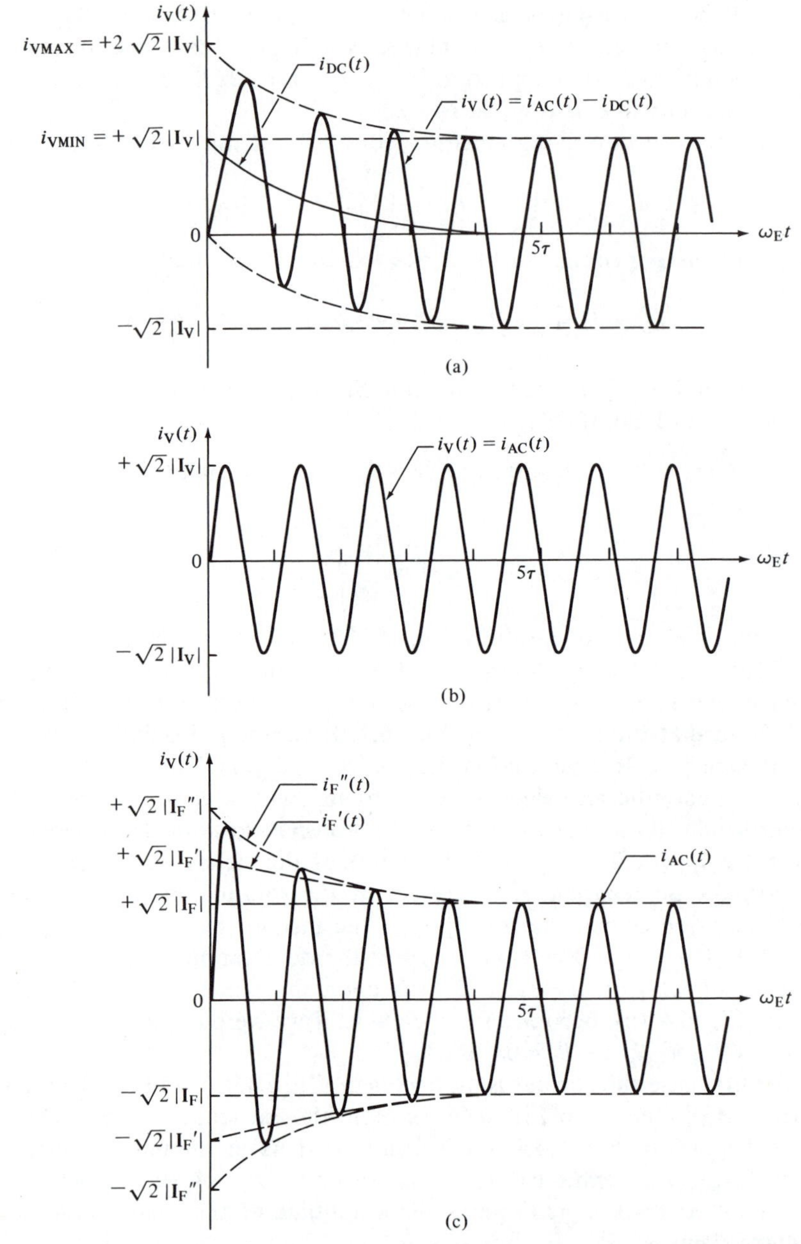

Fig. 6.46 Comparison of Short-Circuit Current Waveform for Fig. 6.42(c): (1) Constant Reactance Technique Assuming Constant Parameter Series *RL*-Circuit Excited Suddenly with a Sinusoidal Forcing Function when (a) $\alpha - \theta_Z = -90°$ and (b) $\alpha - \theta_Z = 0$, and (2) Variable Reactance Technique Assuming a Synchronous Generator with Variable Inductance (c) $X_d'' < X_d' < X_d$. Note: Frequency Is Not Drawn to Scale

amplitude of the steady state current of Fig. 6.46(b), we obtain

$$i_{V\text{MAX}} = 2i_{\text{DCMAX}} = 2\sqrt{2}\,|\mathbf{I}_V| = 2\sqrt{2}\left(\frac{|\mathbf{E}_{q_V}|}{|\mathbf{Z}_v|}\right) \tag{6.293}$$

and

$$i_{V\text{MIN}} = i_{\text{DCMAX}} = \sqrt{2}\,|\mathbf{I}_V| = \sqrt{2}\left(\frac{|\mathbf{E}_{q_V}|}{|\mathbf{Z}_v|}\right) \tag{6.294}$$

This is in contrast with Fig. 6.46(b), which represents the **best** possible case where the current follows steady state from the outset, that is, $0^+ \leqslant t < \infty$. The maximum amplitude of the current $i_V(t)$ in this case, as extracted from Eq. (6.290) by setting $\alpha - \theta_Z = 0°$, 180°, is

$$i_{\text{ACMAX}} = \sqrt{2}\,|\mathbf{I}_V| = \sqrt{2}\left(\frac{|\mathbf{E}_{q_V}|}{|\mathbf{Z}_v|}\right) \tag{6.295}$$

Note that Fig. 6.46(a) exhibits the *troublesome* unidirectional transient current component, that is, $i_{\text{DC}}(t)$, which is so very essential. It should be obvious now that the transient or dc component $i_{\text{DC}}(t)$ of the current $i_V(t)$ may have any value from 0 to $\pm\sqrt{2}\,|\mathbf{E}_{q_V}|/|\mathbf{Z}_v|$ depending on the instantaneous value of the voltage of a given phase when the short circuit is applied and on the power factor angle θ_Z of the circuit, and thus on the angles α, θ_Z, and $\alpha - \theta_Z = \pm\pi/2$. Also, note that the transient or dc component $i_{\text{DC}}(t)$ takes on different values for each phase since the three phase voltages are displaced 120 electrical degrees from each other and thus α varies correspondingly. Observe, however, that at the instant of application of the short circuit, the dc and steady state components, that is, $i_{\text{DC}}(t)$ and $i_{\text{AC}}(t)$, always have the same magnitudes but are opposite in sign. This assures that the value of the current $i_V(t)$ is zero the instant the short circuit is applied.

Variable-Reactance Technique

In contrast to the first case in which the rotor structure is absent and the parameters of the series *RL*-circuit are constant, the second case represents a synchronous generator with a balanced three-phase fault with **the rotor structure present and functional**. In the first case the current $i_V(t)$, Fig. 6.46(a), resembles the current $i_V(t)$ of the second case, Fig. 6.46(c). However, there are **striking differences** resulting from the effect of the stator winding current $i_V(t)$ on the field of the rotor structure, designated as Φ_R.

The best way to analyze the effect of a balanced three-phase fault at the terminals of a previously **unloaded** synchronous generator is to take an oscillogram of the current $i_V(t)$ in one of the phases upon the occurrence of a balanced three-phase fault. Since the voltages generated in the three-phases of the balanced three-phase synchronous generator are displaced 120 electrical degrees from each other, the short circuit occurs at **different points** on the voltage waveform of each phase. For this reason the transient or dc component in the generated current of each phase is **different**. However, if this transient or dc component is eliminated from the oscillograms of the three phase currents, that is, $i_A(t)$, $i_B(t)$, and $i_C(t)$, by using proper instrumentation, the resulting plot of the one phase current $i_V(t)$, that is, $i_V(t)$ versus t—designating a per-phase analysis—is exhibited in Fig. 6.46(c).

Comparison of Figs. 6.46(b) and 6.46(c) exemplifies the difference between applying a voltage to an ordinary series *RL*-circuit **suddenly** and applying a

balanced fault (i.e., a short circuit) to a synchronous generator. Note that the phase angle α of the voltage is adjusted so that $\alpha-\theta_Z=0,\ \pi$ in both cases. That is, the transient or dc component of the phase current $i_V(t)$ in both these figures is zero. The waveform of Fig. 6.46(c) differs from that of Fig. 6.46(b) for the following reasons. For the synchronous generator, at the instant the short circuit occurs, the flux across the airgap is comprised totally of the flux of the rotor (or field) winding Φ_R and is much higher than it is after steady state is reached, that is, $t \geqslant 5\tau, \tau = L_\nu / R_S$. This reduction of flux is caused totally by the effects of the currents in the stator windings, an effect known as **armature reaction.** As this flux diminishes, the current in the stator winding decreases because the magnitude of the generated voltage $e_G(t)$, which is a function of field flux Φ_R, decreases and eventually reaches a steady state RMS value designated as $\mathbf{E}_G$.

The plot of Fig. 6.46(c) is a very important one. It allows one to neglect the transient or dc component in the calculations by avoiding time-domain analysis. Instead, a *steady state analysis is performed over three distinct time intervals*, in the time interval $0^+ \leqslant t < 5\tau$, $\tau = L_\nu / R_S$, following a balanced three-phase fault utilizing the changing reactance concept. That is, **three distinct values of reactance** X_ν, designated by $\nu \equiv d'', d'$, and d and ordered as $X_d'' < X_d' < X_d$, are used to perform steady state analysis of the given system over three distinct intervals of time referred to the period T of the frequency of excitation, 60 Hz: (1) X_d'' $(0 < t \leqslant 2T)$, (2) X_d' $(2T < t \leqslant 60T)$, and (3) X_d $(t > 60T)$. These time intervals by no means should be taken as absolute. They simply represent a rule of thumb, and each utility or manufacturer uses its own rule of thumb.

The steady state analysis in the **first range** enables us to calculate the fault current $i_V(t)$, $0 < t \leqslant 2T$, **immediately after the fault has occurred.** This **postfault current** is defined as the **subtransient current**, is designated as $\mathbf{I}_F''$, and is calculated with the aid of Eq. (6.291) by substituting $\mathbf{I}_F''$ for $\mathbf{I}_V$, $\mathbf{E}_G''$ for $\mathbf{E}_{\mathcal{V}}$, and $\mathbf{Z}_d''$ for $\mathbf{Z}_\nu$. Thus

$$\mathbf{I}_F'' \equiv \frac{\mathbf{E}_G''}{\mathbf{Z}_d''} = \frac{\mathbf{E}_G''}{R_S + jX_d''} \tag{6.296}$$

The physical significance of the subtransient current $\mathbf{I}_F''$ becomes evident if we inspect Fig. 6.46(c). Note that $|\mathbf{I}_F''|$ is the RMS value of the current envelope of the first time interval at the current axis $i_V(t)$. It is the initial current in a power system just after a fault occurs. Since the fastest relay-circuit-breaker sets open in approximately 2 cycles, they **never detect** the subtransient current $\mathbf{I}_F''$.

The steady state analysis in the **second range** enables us to calculate the fault current $i_V(t)$, $2T < t \leqslant 60T$, **after the fault has occurred but before steady state is reached.** This **postfault current** is defined as the **transient current**, is designated as $\mathbf{I}_F'$, and is calculated with the aid of Eq. (6.291) by substituting $\mathbf{I}_F'$ for $\mathbf{I}_V$, $\mathbf{E}_G'$ for $\mathbf{E}_{\mathcal{V}}$, and $\mathbf{Z}_d'$ for $\mathbf{Z}_\nu$. Thus,

$$\mathbf{I}_F' \equiv \frac{\mathbf{E}_G'}{\mathbf{Z}_d'} = \frac{\mathbf{E}_G'}{R_S + jX_d'} \tag{6.297}$$

The physical significance of the transient current $\mathbf{I}_F'$ becomes evident if we inspect Fig. 6.46(c). Note that as in the case of subtransient current $\mathbf{I}_F''$, the transient current $|\mathbf{I}_F'|$ is the RMS value of the current envelope of the second time interval at the current axis $i_V(t)$. Since this current exists in the system after approximately the second cycle, it is *detected* by at least the fastest (2 cycles) relay-circuit-breaker sets.

The steady state analysis in the third range enables us to calculate the fault current $i_V(t)$, $t > 60T$, **after the faulted system has reached steady state.** This **postfault current** is defined as **the steady state current**, is designated as $\mathbf{I}_F$, and is calculated with the aid of Eq. (6.291) by substituting $\mathbf{I}_F$ for $\mathbf{I}_V$, $\mathbf{E}_G$ for $\mathbf{E}_{\mathcal{V}}$, and $\mathbf{Z}_d$ for $\mathbf{Z}_\nu$. Thus

$$\mathbf{I}_F = \frac{\mathbf{E}_G}{\mathbf{Z}_d} = \frac{\mathbf{E}_G}{R_S + jX_d} \tag{6.298}$$

Note that the current of Eq. (6.298) is identical to the steady state current calculated by Eq. (6.268) allowing $\mathbf{V}_L = I_q = 0$.

The physical significance of the steady state current $\mathbf{I}_F$ as well as the differences between it and (1) the subtransient current $\mathbf{I}_F''$ and (2) the transient current $\mathbf{I}_F'$ become evident if we inspect Fig. 6.46(c). As in the case of subtransient $\mathbf{I}_F''$ and transient $\mathbf{I}_F'$ currents, the steady state current $|\mathbf{I}_F|$ is the RMS value of the current envelope of the third time interval at the current axis $i_V(t)$.

Upon completion of the postfault subtransient current $\mathbf{I}_F''$ calculation, the effect of the unidirectional transient or dc current component $i_{DC}(t)$ is added by applying a *multiplying factor*, designated as $\sqrt{2}\, k_{DC}$, to the findings for $\mathbf{I}_F''$. This multiplying factor has the range $1.0 \leqslant k_{DC} \leqslant 1.6$, depending on the type of the circuit breaker to be used and its speed of operation. [Note that $\sqrt{2}\, k_{DC} \neq 2\sqrt{2}$, Eq. (6.293).] Even though $i_{DC}(t)$ is different for each phase, the effect of the *worst possible case* of the transient or dc currents is accounted for by the multiplication of $\mathbf{I}_F''$ of each phase by $\sqrt{2}\, k_{DC}$.

The changing-reactance concept during transient conditions can be explained as follows. First, as the balanced three-phase fault (short circuit) occurs, the stator currents of the previously unloaded synchronous generator increase very rapidly, and so does the field of the stator windings. This causes the damper windings to carry induced currents. The field of these induced currents not only opposes the flow of the stator field, designated as Φ_{ST}'', Fig. 6.47(a), through the rotor structure, but forces it out of the rotor-pole structure to a great extent; it forces the flux to complete its path through air rather than the highly permeable material of the rotor structure. This causes an apparently increased reluctance, designated as $\mathcal{R}_d''$, in the path of the flux Φ_{ST}'', and thus an apparently decreased reactance, since $X_d'' = \omega_E L_d'' = (N^2/\mathcal{R}_d'')\omega_E$. This reactance is defined as the **subtransient reactance** of the salient-pole synchronous generator (device). Second, as the stator field makes its way through the rotor-pole structure, it encounters a second obstacle caused by the field of the induced current in the rotor excitation winding. This new obstacle causes an opposition smaller than the previous opposition to the stator field through the rotor structure, designated as Φ_{ST}', Fig. 6.47(b). Thus it causes an apparently reduced reluctance, designated as $\mathcal{R}_d' < \mathcal{R}_d''$, and thereby an apparently increased reactance since $X_d' = \omega_E L_d' = (N^2/\mathcal{R}_d')\omega_E$. This reactance is defined as the **transient reactance** of the salient-pole synchronous generator (device) and is related to the subtransient reactance as $X_d' > X_d''$. Third, as the currents of the stator windings approach steady state the stator field through the rotor structure, designated as Φ_{ST}, Fig. 6.47(c), becomes constant. The rotor locks on the rotating stator field and the induced fields of the damper and rotor excitation windings cease to exist. Thus the stator field encounters no opposition, Fig. 6.47(c), and the reluctance of the field's path, designated as $\mathcal{R}_d < \mathcal{R}_d'$, becomes minimum, and thereby the reactance of the salient-pole synchronous generator (device) becomes

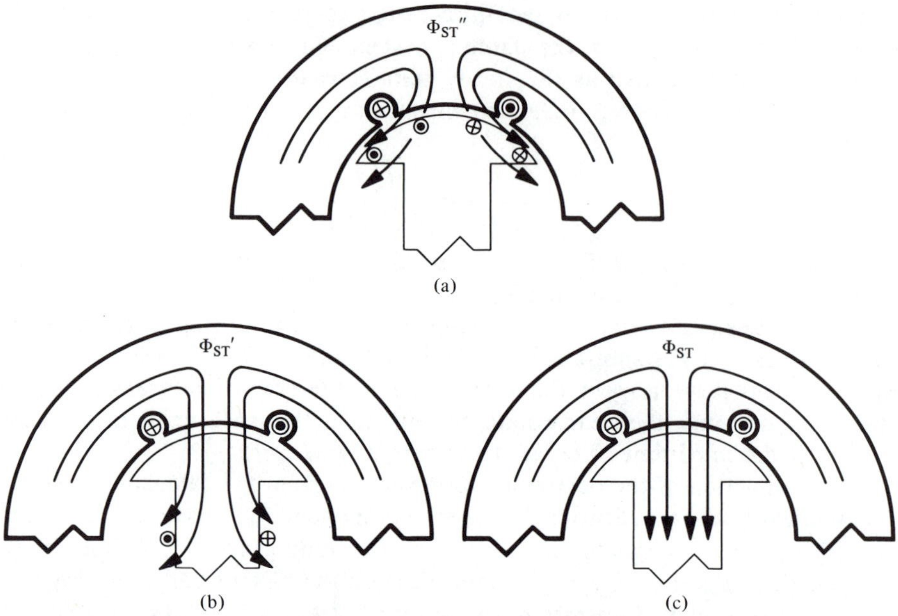

Fig. 6.47 Effects of the Rotor Windings to the Stator Winding Field Crossing the Airgap (i.e., the Armature Reaction Field): (a) Subtransient Effect of the Damper Windings, (b) Transient Effect of the Field Winding, and (c) Steady State—That Is, Unimpeded—Path of the Stator Field

maximum, that is, $X_d = L_d\omega_E = (N^2/\mathcal{R}_d)\omega_E$. This reactance is defined as the synchronous (steady state) reactance and is related to the transient reactance as $X_d > X_d'$. From the explanation above, the relative magnitudes of the (1) **subtransient** X_d'', (2) **transient** X_d', and (3) **synchronous** X_d **reactances** are given as

$$X_d'' < X_d' < X_d \tag{6.299}$$

The relative magnitudes of these reactances depend on the design characteristics of the individual device. However, for an existing synchronous device, these parameters can be measured by performing appropriate tests. (Please see Chapter 7.)

The phasor diagram and the corresponding subtransient, transient, and synchronous reactance equivalent per-phase circuits for approximating balanced postfault operating conditions of synchronous devices are exhibited in Figs. 6.48(a) and 6.48(b.i), 6.48(b.ii) and 6.48(b.iii), respectively. The choice of X_d'', X_d', or X_d depends upon the *time after the fault* at which the study is to take place. It is clear from Eq. (6.299) that (1) X_d'' represents the first and **worst** condition, (2) X_d' represents the **intermediate** condition, and (3) X_d represents the last and **best** condition.

The use of these parameters can be summarized as follows: (1) **the initial current when a fault occurs in a given system** is calculated with the aid of the subtransient reactance X_d'' of synchronous generators and motors, (2) **the interrupting capacity of circuit breakers**, except those that open instantaneously, is calculated with the aid of the subtransient reactance X_d'' of synchronous *generators* and the transient reactance X_d' of synchronous *motors*, (3) **stability studies of power systems** are

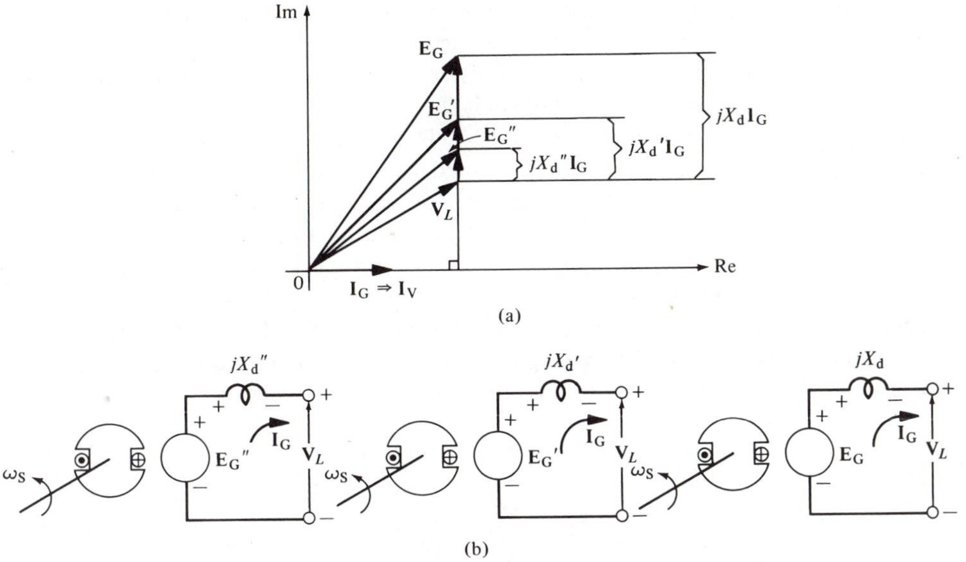

Fig. 6.48 Phasor Diagram and Corresponding Per-Phase, Equivalent Circuits of the Balanced, Three-Phase, *P*-Pole, Salient-Rotor, Synchronous Generator Faulted with a Balanced, Three-Phase, Short Circuit Depicting (a) the Generalized Phasor Diagram and (b) the Equivalent Circuits for Postfault (i) Subtransient, (ii) Transient, and (iii) Steady State Analysis

performed with the aid of transient reactances X_d' of synchronous devices, and (4) **steady state analyses of power systems** are performed with the aid of the synchronous reactance X_d of synchronous devices.

In addition to these four uses of the (1) subtransient X_d'', (2) transient X_d', and (3) synchronous X_d reactances, we may use these parameters in conjunction with Eqs. (6.290) and (6.291) to approximate the *maximum amplitude* of the short-circuit current $i_{V\text{MAX}}$, Eq. (6.293), and the *minimum amplitude* of the short-circuit current $i_{V\text{MIN}}$, Eq. (6.294). These are only rough approximations and can be deduced by inspection of Figs. 6.46(a) and 6.46(c) and drawing appropriate conclusions about their similarities. (Which of the values X_d'', X_d', and X_d would you use for the calculation of $i_{V\text{MAX}}$ and which would you use for the calculation of $i_{V\text{MIN}}$?) Note that the multiplying factor $\sqrt{2}\, k_{\text{DC}}$ is used only to account for the effects of the transient or dc component and not to calculate $i_{V\text{MAX}}$.

Utilization of the concepts set forth here enables us to perform the fault analysis of the *three distinct cases*, designated as (1) loaded generator with the fault applied either within the loading network or at the input terminals of the loading network, that is, the terminals of the generator, (2) unloaded generator faulted at the far end of a loading network, for example, faulted at the far end of an unloaded transmission line, or output of an unloaded transformer, and (3) unloaded generator faulted at its terminals.

Case 1. If the **prefault current I_G**, Fig. 6.48, **is known and is nonzero**, that is, the generator is *loaded* prior to the occurrence of the fault, and the balanced three-phase fault occurs at points other than the terminals of the synchronous generator, the principle of variable reactance is still valid. Under these conditions the currents change rapidly from the prefault value, designated $i_G(t)$, to much higher postfault values, designated as $i_V(t)$. Utilization of the principle of variable reactance enables us to employ the phasor diagram and the corresponding equivalent circuits of Fig. 6.48 and perform the postfault (1) subtransient, (2) transient, and (3) steady state analysis. Before proceeding with the analysis, however, the following definitions pertaining to Fig. 6.48 are relevant: (1) $\mathbf{E}_G''$ is defined as the voltage behind the subtransient reactance X_d'', (2) $\mathbf{E}_G'$ is defined as the voltage behind the transient reactance X_d', and (3) $\mathbf{E}_G$ is defined as the voltage behind the synchronous reactance X_d. The proper choice of the voltage source strengths, that is, $\mathbf{E}_G''$, $\mathbf{E}_G'$, and $\mathbf{E}_G$, depends upon the *time elapsed* since the fault has occurred. These voltages, the equivalent circuits of which are drawn in Figs. 6.48(b.i), 6.48(b.ii), and 6.48(b.iii), respectively, are expressed with the aid of the phasor diagram of Fig. 6.48(a), and the **prefault loading voltage** $\mathbf{V}_L$, the **prefault loading current** $\mathbf{I}_G$, and the **subtransient** X_d'', **transient** X_d', and **synchronous** X_d **reactances** as

$$\mathbf{E}_G'' = \mathbf{V}_L + jX_d''\mathbf{I}_G \tag{6.300}$$

$$\mathbf{E}_G' = \mathbf{V}_L + jX_d'\mathbf{I}_G \tag{6.301}$$

$$\mathbf{E}_G = \mathbf{V}_L + jX_d\mathbf{I}_G \tag{6.302}$$

where $\mathbf{V}_L = k_1 e^{j\theta_1}$ and $\mathbf{I}_G = k_2$.

Once the voltages $\mathbf{E}_G''$, $\mathbf{E}_G'$, and $\mathbf{E}_G$ are calculated with the aid of Eqs. (6.300) through (6.302), the postfault **subtransient** $\mathbf{I}_F''$, **transient** $\mathbf{I}_F'$, and **steady state** $\mathbf{I}_F$ currents can be calculated by utilizing the equivalent circuits of Figs. 6.48(b.i), 6.48(b.ii), and 6.48(b.iii) as Thevenin's equivalent forcing functions to the faulted networks, described by $\mathcal{V}_L''\{\mathbf{Z}_{FN}, \mathbf{I}_F''\}$, $\mathcal{V}_L'\{\mathbf{Z}_{FN}, \mathbf{I}_F'\}$, and $\mathcal{V}_L\{\mathbf{Z}_{FN}, \mathbf{I}_F\}$, where $\mathbf{Z}_{FN}$ designates the impedance of the corresponding faulted network, by writing corresponding describing equations and solving them for the respective **postfault** currents $\mathbf{I}_F''$, $\mathbf{I}_F'$, and $\mathbf{I}_F$. That is,

$$\mathbf{E}_G'' = \mathcal{V}_L''\{\mathbf{Z}_{FN}, \mathbf{I}_F''\} + jX_d''\mathbf{I}_F'' \tag{6.303}$$

$$\mathbf{E}_G' = \mathcal{V}_L'\{\mathbf{Z}_{FN}, \mathbf{I}_F'\} + jX_d'\mathbf{I}_F' \tag{6.304}$$

$$\mathbf{E}_G = \mathcal{V}_L\{\mathbf{Z}_{FN}, \mathbf{I}_F\} + jX_d\mathbf{I}_F \tag{6.305}$$

where $\mathcal{V}_L^-\{\ \}$ designates the loading (terminal) voltage of the faulted network.

Perhaps these ideas can be clarified through Example 6.12.

Example 6.12

The per-phase impedance diagram of a synchronous generator connected to a Y-Y-transformer with all parameters expressed in per-unit values is exhibited in Fig. 6.12.1. If the subtransient reactance is $X_d'' = 0.15$ pu Ω, the transformer reactance is $X_T = 0.12$ pu Ω, and the load is such that the terminal voltage of the generator is $\mathbf{V}_L = 1.0e^{j42.61°}$ pu V and the loading current is $|\mathbf{I}_G| = 0.74$ pu A at a power factor 0.74 LGG, calculate: (1) the voltage $\mathbf{E}_G''$ behind the subtransient reactance X_d'', (2) the subtransient current $\mathbf{I}_F''$ when a balanced three-phase fault is placed across the load at $t=0$, (3) the maximum amplitude i_{DCMAX} of the transient or dc current component $i_{DC}(t)$ of the fault current $i_F''(t)$, (4) the maximum possible value i_{FMAX}'', of the fault current $i_F''(t)$, utilizing time-domain analysis and empirical formulas, and (5) the minimum possible amplitude i_{FMIN}'', of the fault current, that is, the amplitude of the steady state current $i_{AC}(t)$.

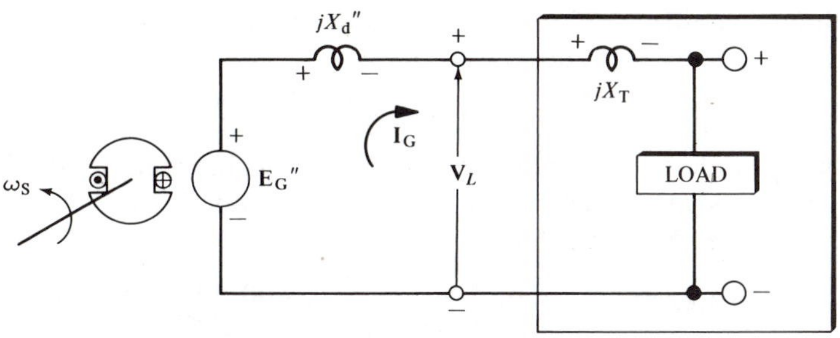

Fig. 6.12.1 Single-Phase, Steady State, Equivalent Circuit Representation of a Balanced, Three-Phase, *P*-Pole, Salient-Rotor, Synchronous Generator with the Load at the Output of the transformer

Solution

The voltage $\mathbf{E}_G''$ behind the subtransient reactance is calculated by substituting properly from the input data into Eq. (6.300). Thus

$$\mathbf{E}_G'' = \mathbf{V}_L + jX_d''|\mathbf{I}_G| = 1e^{j42.61°} + j0.15 \times 0.74 = 0.74 + j0.68 + j0.11$$
$$= 0.74 + j0.79 = \underline{\underline{1.08e^{j47°}}} \text{ pu V} \quad (1)$$

The Thevenin equivalent forcing function, that is, $\mathbf{E}_G''$ in series with jX_d'', is connected across the faulted network as in Fig. 6.11.2. Note the distinct difference between Figs. 6.12.1 and 6.12.2. Now the describing equation of Fig. 6.12.2 is written with the aid of Eq. (6.303) as

$$\mathbf{E}_G'' = \mathcal{V}_L''\{\mathbf{Z}_{FN}, \mathbf{I}_F''\} + jX_d''\mathbf{I}_F'' = \mathbf{Z}_{FN}\mathbf{I}_F'' + jX_d''\mathbf{I}_F'' = (\mathbf{Z}_{FN} + jX_d'')\mathbf{I}_F'' \quad (2)$$

Solution of Eq. (2) for the subtransient current $\mathbf{I}_F''$ and proper substitution from the input data and Eq. (1) yields

$$\mathbf{I}_F'' = \frac{\mathbf{E}_G''}{\mathbf{Z}_{FN} + jX_d''} = \frac{\mathbf{E}_G''}{jX_T + jX_d''} = \frac{1.08e^{j47°}}{j0.12 + j0.15} = \frac{1.08e^{j47°}}{0.27e^{j90°}}$$
$$= \underline{\underline{4.0\, e^{-j43°}}} \text{ pu A} \quad (3)$$

The maximum amplitude i_{DCMAX} of the transient or dc current component $i_{DC}(t)$ of the fault current $i_F''(t)$, is calculated with the aid of Eqs. (6.292) and (3) as

$$i_{DCMAX} = \sqrt{2}\, |\mathbf{I}_F''| = \sqrt{2} \times 4.00 = \underline{\underline{5.66}} \text{ pu A} \quad (4)$$

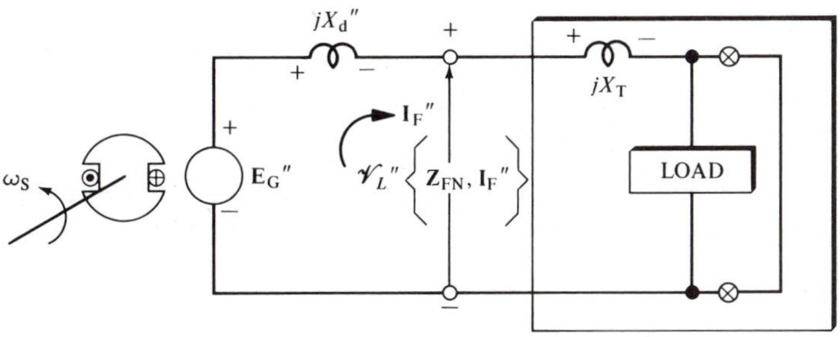

Fig. 6.12.2 The Synchronous Generator of Fig. 6.11.1 Designating Postfault Subtransient Operation

The maximum possible amplitude of the fault current $i_V(t) = i''_F(t)$, is calculated with the aid of Eqs. (6.293) and (4) as

$$i_{V\text{MAX}} = 2i_{\text{DCMAX}} = 2 \times 5.66 = \underline{\underline{11.32}} \text{ pu A} \tag{5}$$

Utilizing the empirical formula, the maximum possible value of the fault current $i_V(t)$, and $t=0$, designated as $i_{V\text{MAX}}$, is calculated with the aid of input data, the multiplying constant k_{DC}, and Eq. (3) as

$$i_{V\text{MAX}} = 1.6\sqrt{2}\,|\mathbf{I}_\text{F}''| = 1.6\sqrt{2} \times 4.00 = \underline{\underline{9.05}} \text{ pu A} \tag{6}$$

Note the difference between Eqs. (5) and (6). We can draw the conclusion that Eq. (6.292) yields a better safety margin than the empirical formula.

The minimum possible value $i_{V\text{MIN}}$, of the fault current $i_V(t), t \geqslant 5\tau$, is the amplitude of the sinusoidal current of the type exhibited in Fig. 6.45(b), the amplitude of which is equal to i_{DCMAX}, as given by Eq. (6.294). Thus, with the aid of Eq. (4), this current is written as

$$i_{V\text{MIN}} = \sqrt{2}\,|\mathbf{I}_\text{F}''| = \sqrt{2} \times 4.00 = \underline{\underline{5.66}} \text{ pu A} \tag{7}$$

Note that if the synchronous reactance X_d were given, the value of i_{FMIN}, would have been calculated to be **less** than the value calculated in Eq. (7). It should be emphasized that if $\mathbf{I}_\text{F}$ were used in place of $\mathbf{I}_\text{F}''$, then Eq. (4) would not yield i_{DCMAX}.

Case 2. When the synchronous generator is *unloaded* prior to the application of the fault, the **prefault current** is zero, that is, $\mathbf{I}_\text{G} = 0$, in Eqs. (6.300) through (6.302). Thus, Eqs. (6.300) through (6.302) can be written as

$$\mathbf{E}_\text{G}'' = \mathbf{V}_L, \mathbf{I}_\text{G} = 0 \tag{6.306}$$
$$\mathbf{E}_\text{G}' = \mathbf{V}_L, \mathbf{I}_\text{G} = 0 \tag{6.307}$$
$$\mathbf{E}_\text{G} = \mathbf{V}_L, \mathbf{I}_\text{G} = 0 \tag{6.308}$$

It is clear from these equations that for the **unloaded** generator the voltages above are equal to each other. That is, they are equal to the RMS value of the **no-load, line-to-neutral voltage** of the synchronous generator, that is,

$$\mathbf{E}_\text{G} \equiv \mathbf{E}_\text{G}' \equiv \mathbf{E}_\text{G}'' \tag{6.309}$$

Thus when a balanced three-phase fault (short circuit) is applied to a loading network of a synchronous generator that maintains zero prefault current, that is, $\mathbf{I}_\text{G} = 0$, as in an unloaded transmission line or transformer, Fig. 6.48, the postfault subtransient, transient, and steady state performances of the interconnected system are described by Eqs. (6.252) through (6.305) modified, with the aid of Eq. (6.309), to read as

$$\mathbf{E}_\text{G} = \mathcal{V}_L''\{\mathbf{Z}_{\text{FN}}, \mathbf{I}_\text{F}''\} + jX_\text{d}''\mathbf{I}_\text{F}'' \tag{6.310}$$
$$\mathbf{E}_\text{G}' = \mathcal{V}_L'\{\mathbf{Z}_{\text{FN}}, \mathbf{I}_\text{F}'\} + jX_\text{d}'\mathbf{I}_\text{F}' \tag{6.311}$$
$$\mathbf{E}_\text{G} = \mathcal{V}_L\{\mathbf{Z}_{\text{FN}}, \mathbf{I}_\text{F}\} + jX_\text{d}\mathbf{I}_\text{F} \tag{6.312}$$

The solution of Eqs. (6.310) through (6.312) yields the postfault subtransient current $\mathbf{I}_\text{F}''$, the transient current $\mathbf{I}_\text{F}'$, and the steady state current $\mathbf{I}_\text{F}$.

Example 6.13

The per-phase impedance diagram of a synchronous generator connected to a Y-Y-transformer, with all quantities expressed in per-unit values, is exhibited in Fig. 6.13.1. If the generator is unloaded, the generated voltage is $\mathbf{E}_\text{G} = 1.00e^{j0°}$

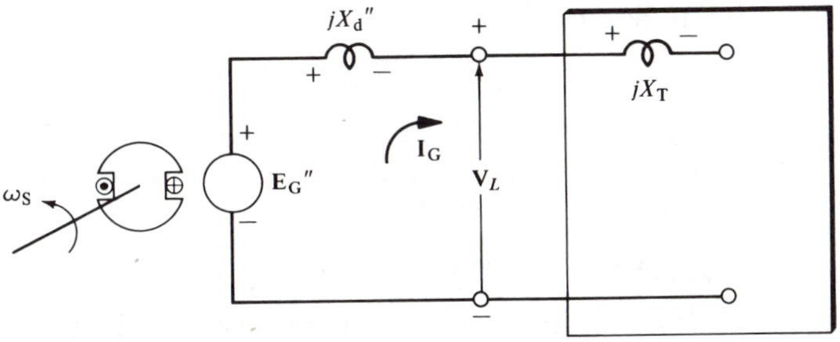

Fig. 6.13.1 Single-Phase, Steady State, Equivalent Circuit Representation of a Balanced, Three-Phase, *P*-Pole, Salent-Rotor, Synchronous Generator Unloaded

pu V, the subtransient reactance is $X_d''=0.15$ pu Ω, and the transformer reactance is $X_T=0.12$ pu Ω when a balanced three-phase fault is placed suddenly at all output terminals of the transformer, calculate: (1) the subtransient current I_F'', (2) the maximum value of the transient or dc component of the current $i_V(t)$, (3) the maximum possible value i_{VMAX}, of the fault current $i_V(t)$, $V=F''$, and (4) the minimum possible amplitude i_{VMIN}, of the fault current $i_V(t)$.

Solution

Since the voltage behind the subtransient reactance is known, Eq. (6.309), the Thevenin equivalent forcing function, that is, $\mathbf{E}_G$ in series with jX_d'', is connected directly across the faulted network as in Fig. 6.13.2. The describing equation of the faulted network is written with the aid of Eq. (6.310) as

$$\mathbf{E}_G = \mathbf{Z}_{FN}\mathbf{I}_F'' + jX_d''\mathbf{I}_F'' = (\mathbf{Z}_{FN} + jX_d'')\mathbf{I}_F'' \tag{1}$$

Solution of Eq. (1) for the subtransient current $\mathbf{I}_F''$ and proper substitution from the input data yields

$$\mathbf{I}_F'' = \frac{\mathbf{E}_G}{\mathbf{Z}_{FN}+jX_d''} = \frac{\mathbf{E}_G}{jX_T+jX_d''} = \frac{1.00e^{j0°}}{j0.12+j0.15} = \frac{1.00e^{j0°}}{0.27e^{j90°}}$$
$$= \underline{\underline{3.70e^{-j90°}}} \text{ pu A} \tag{2}$$

The maximum value of the transient or dc component of the current $i_V(t)$, designated as i_{DCMAX}, is calculated with the aid of Eq. (6.292) as

$$i_{DCMAX} = \sqrt{2}\,|\mathbf{I}_F''| = \sqrt{2} \times 3.70 = \underline{\underline{5.24}} \text{ pu A} \tag{3}$$

Finally, the maximum possible value, i_{VMAX}, of the fault current $i_V(t)$ is

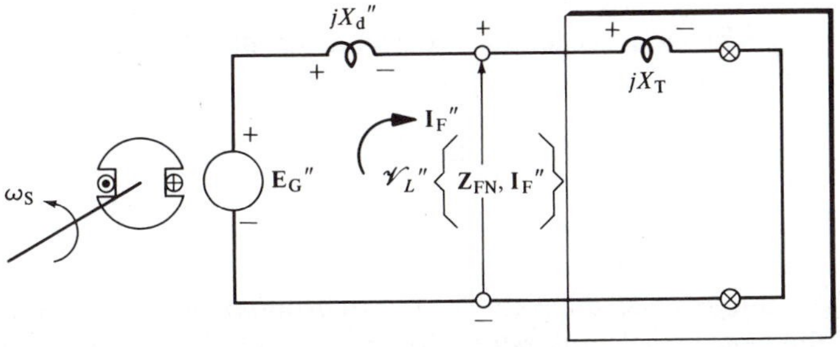

Fig. 6.13.2 The Synchronous Generator of Fig. 6.12.1 Designating Postfault Subtransient Operation

calculated with the aid of Eq. (6.293) as

$$i_{V\text{MAX}} = 2i_{\text{DCMAX}} = 2 \times 5.24 = \underline{\underline{10.48}} \text{ pu A} \tag{4}$$

In the limit, that is, $t \geqslant 5\tau$, the fault current $i_V(t)$, is sinusoidal of the type exhibited in Fig. 6.46(b). The amplitude of this current is equal to i_{DCMAX}, as given by Eqs. (6.294) and (6.292). Thus

$$i_{V\text{MIN}} = \sqrt{2}\,|\mathbf{I}_F''| = \sqrt{2} \times 3.70 = \underline{\underline{5.24}} \text{ pu A} \tag{5}$$

With the aid of Eq. (5), this current is written in the time domain as

$$i_V(t) = \underline{\underline{5.24 \sin 377t}};\ t \geqslant 5\tau \tag{6}$$

Case 3. When the synchronous generator is unloaded prior to the application of the fault, that is, $\mathbf{I}_G = 0$, Eqs. (6.300) through (6.302) become Eqs. (6.306) through (6.308). Under such conditions a balanced three-phase fault (short-circuit) applied directly at the *terminals* of the synchronous generator yields, $\mathcal{V}_L''\{\mathbf{Z}_{FN}, \mathbf{I}_F''\} = \mathcal{V}_L'\{\mathbf{Z}_{FN}, \mathbf{I}_F'\} = \mathcal{V}_L\{\mathbf{Z}_{FN}, \mathbf{I}_F\} = 0$. Therefore, Eqs. (6.310) through (6.312) are written as

$$\mathbf{E}_G = jX_d''\mathbf{I}_F'' \tag{6.313}$$

$$\mathbf{E}_G = jX_d'\mathbf{I}_F' \tag{6.314}$$

$$\mathbf{E}_G = jX_d\mathbf{I}_F \tag{6.315}$$

The solutions of Eqs. (6.313) through (6.315) yield the postfault subtransient current $\mathbf{I}_F''$, the transient current $\mathbf{I}_F'$, and the steady state current $\mathbf{I}_F$. Note that Eqs. (6.313) through (6.315) concur with Eqs. (6.296) through (6.298) if both Eq. (6.309) and the simplification $|R_S| \ll |jX_\nu|, \nu = d'', d', d$, are applied to Eqs. (6.296) through (6.298).

It is easy now to see that it is possible to **define the three reactances of the synchronous device** in terms of the RMS values of the measurable quantities (1) the no-load, line-to-neutral voltage $|\mathbf{E}_G|$, Eq. (6.309), and (2) the subtransient, transient, and steady state currents $|\mathbf{I}_F''|$, $|\mathbf{I}_F'|$, and $|\mathbf{I}_F|$, respectively, Fig. 6.46(c).

Thus the subtransient reactance X_d'', designating the effects of the damper winding, is defined as

$$X_d'' \equiv \frac{|\mathbf{E}_G|}{|\mathbf{I}_F''|} \tag{6.316}$$

Similarly, the transient reactance X_d', designating the effects of the field winding, is defined as

$$X_d' \equiv \frac{|\mathbf{E}_G|}{|\mathbf{I}_F'|} \tag{6.317}$$

Finally, the synchronous reactance, designating steady state phenomena, is defined as

$$X_d \equiv \frac{|\mathbf{E}_G|}{|\mathbf{I}_F|} \tag{6.318}$$

Example 6.14

The per-phase impedance diagram of a synchronous generator is exhibited in Fig. 6.14.1. If the generator is unloaded, the generated voltage is $\mathbf{E}_G = 1.0e^{j0°}$ pu V, the subtransient reactance is $X_d'' = 0.15$ pu Ω, the transient reactance is $X_d' =$

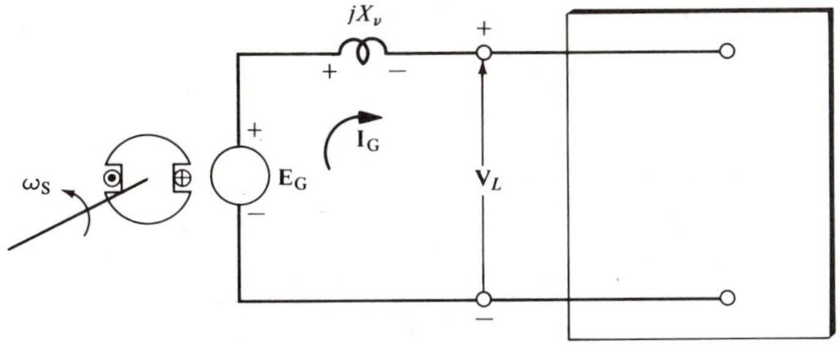

Fig. 6.14.1 Single-Phase, Steady State, Equivalent Circuit Representation of a Balanced, Three-Phase, P-Pole, Salient-Rotor, Synchronous Generator Unloaded

0.20 pu Ω, and the steady state reactance is $X_d = 0.25$ pu Ω. If a balanced three-phase fault is placed suddenly at all terminals of the generator, calculate: (1) the subtransient current $\mathbf{I}_F''$, (2) the transient current $\mathbf{I}_F'$, (3) the postfault steady state current $\mathbf{I}_F$, (4) the minimum amplitude i''_{FMIN}, utilizing the constant-reactance technique, and (5) the minimum amplitude i_{FMIN}.

What is the difference between i''_{FMIN} and i_{FMIN} as calculated in parts (4) and (5)?

Solution

The subtransient current $\mathbf{I}_F''$ is calculated with the aid of Fig. 6.14.2 for $X_\nu = X_d''$ and Eq. (6.313), utilizing the input data, as

$$\mathbf{I}_F'' = \frac{\mathbf{E}_G}{jX_d''} = \frac{1.00e^{j0°}}{0.15e^{j90°}} = \underline{\underline{6.67e^{-j90°}}} \text{ pu A} \tag{1}$$

The transient current $\mathbf{I}_F'$ is calculated with the aid of Fig. 6.14.2 for $X_\nu = X_d'$ and Eq. (6.314), utilizing the input data, as

$$\mathbf{I}_F' = \frac{\mathbf{E}_G}{jX_d'} = \frac{1.00e^{j0°}}{0.20e^{j90°}} = \underline{\underline{5.00e^{-j90°}}} \text{ pu A} \tag{2}$$

The postfault steady state current $\mathbf{I}_F$ is calculated with the aid of Fig. 6.14.2 for $X_\nu = X_d$ and Eq. (6.315), utilizing the input data, as

$$\mathbf{I}_F = \frac{\mathbf{E}_G}{jX_d} = \frac{1.00e^{j0°}}{0.25e^{j90°}} = \underline{\underline{4.00e^{-j90°}}} \text{ pu A} \tag{3}$$

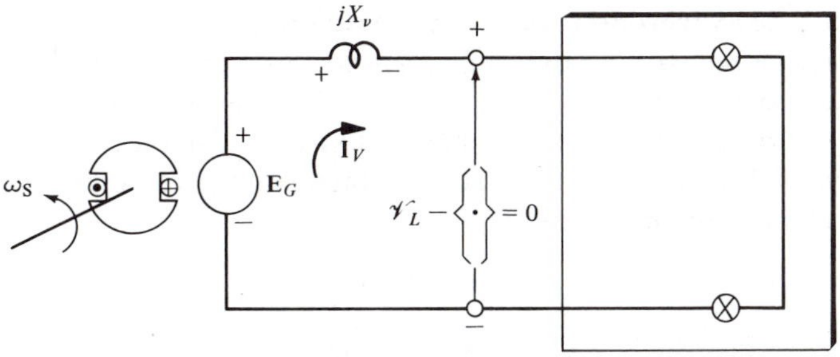

Fig. 6.14.2 The Synchronous Generator of Fig. 6.13.1 Designating Postfault (a) Subtransient, (b) Transient, and (c) Steady State Operation. Note: $X_\nu = X_d''$, X_d', X_d, and $\mathbf{I}_V = \mathbf{I}_F''$, $\mathbf{I}_F'$, $\mathbf{I}_F$.

The minimum amplitude of the fault current i_{FMIN} as calculated using Eqs. (1) and (6.294), is

$$i''_{FMIN} = \sqrt{2}\,|\mathbf{I}_F''| = \sqrt{2} \times 6.67 = \underline{\underline{9.43}} \text{ pu A} \tag{4}$$

Similarly, the minimum amplitude of the fault current i_{FMIN} as calculated utilizing Eqs. (3) and (6.294), is

$$i_{FMIN} = \sqrt{2}\,|\mathbf{I}_F| = \sqrt{2} \times 4.00 = \underline{\underline{5.66}} \text{ pu A} \tag{5}$$

The difference between i''_{FMIN} and i_{FMIN} is explained as follows. The current i''_{FMIN} is the amplitude of the steady state fault current when the synchronous device is modeled using the **constant-reactance technique**. Therefore, i''_{FMIN} is the amplitude of the steady state solution to the differential equation shown in Eq. (6.288). However, i_{FMIN} is the amplitude of the steady state fault current in a **physical** synchronous device after all transients have died out. This current (i_{FMIN}) has been calculated utilizing the **variable-reactance technique**. Note that since $i''_{FMIN} > i_{FMIN}$, a **design safety factor** is realized when one calculates the steady state fault current by using the constant-reactance technique.

6.9 SUMMARY

In this chapter the theory pertaining to the connection of the synchronous device to an infinite bus is developed, and the behavior of the synchronous device, during underexcitation, normal excitation, and overexcitation, is studied. Of particular importance is the **capacitive** and **inductive** behavior of the synchronous device.

Next, the most common excitation systems of synchronous devices are presented. Also, the parallel operation of synchronous devices is presented in great detail. Considerable effort is made to bring forth the effects of underexcitation, speed variation, and prime-mover output on the synchronization of synchronous generators connected in parallel to an infinite bus.

Furthermore, the theory describing the dynamic behavior, and thus the stability conditions, of the synchronous device connected to an infinite bus is set forth in detail. Linear and nonlinear model behavior, with particular emphasis to the (1) prefaulted, (2) faulted, and (3) postfaulted status of the system, is studied.

In addition, the reader is introduced to the *dq*-reference frame of analysis for the synchronous device, through which the exact and approximate steady state models of the salient-pole synchronous device are developed.

The chapter closes with a study of balanced three-phase faults on synchronous generators, and the subtransient X_d'', transient X_d', and steady state X_d reactances are defined. Also derived are the equivalent circuits (of the faulted synchronous generator), valid for postfault analysis. Pertinent numerical examples are used to support the theory set forth here.

PROBLEMS

6.1 A plant draws 4 MW at 10,000 V from an infinitely strong transmission network at a power factor of 0.5 LGG. It is desired that a synchronous condenser should be used to correct the overall power factor to unity. Assume that the losses of the synchronous condenser are 250 kW. Calculate: (1) the original reactive power of the plant, (2) the reactive power required to bring the power factor to unity, and (3) the kilovoltamperes rating of the synchronous condenser and its power factor. Repeat all the calculations above for a power-factor correction to 0.80 LGG, and comment on the feasibility of the results. This unique capability of the synchronous motor to operate at various power factors can be achieved by (1) underexcitation, (2) normal excitation, and (3) overexcitation.

6.2 An industrial process has an installed load rated at 15,000 kVA and a power factor of 0.85 LGG. An expansion of the process is planned whereby the added load, a large press, is rated at 1,500 kW. A choice is to be made whether to drive the process with (1) an **induction motor** of power factor 0.85 LGG or (2) a **synchronous motor** capable of having its dc field (rotor-winding) excitation adjusted to give a motor power factor 0.55 LDG. Which motor should one choose and why?

6.3 Two synchronous generators, having identical design characteristics and parameters of $R_S = 50\ \Omega$ and $X_S = 450\ \Omega$, are connected to a 15,000-V bus, as per Fig. 6.12. The generated voltages are adjusted to be $\mathbf{E}_{G1} = 20{,}000e^{j0°}$ V and $\mathbf{E}_{G2} = 10{,}000e^{j0°}$ V. For each generator calculate: (1) the current delivered to or drawn from the bus, (2) the power generated or accepted, and (3) the power drawn from or delivered to the bus. In addition, calculate: (4) the load current $\mathbf{I}_L$ and (5) the current circulating to the local loop comprised of the two generators.

6.4 Two synchronous generators, having identical design characteristics and parameters of $R_S = 50\ \Omega$ and $X_S = 450\ \Omega$, are synchronized so that the phase voltages of the incoming generator No. 2 are in phase opposition with respect to the phase voltages of the on-line generator No. 1. The generated voltages are adjusted to be $\mathbf{E}_{G1} = 20{,}000e^{j0°}$ V and $\mathbf{E}_{G2} = 10{,}000e^{j0°}$ V. Calculate the following quantities at the instant after the synchronizing switch is closed to parallel the two generators: (1) the synchronizing current $\mathbf{I}_\sigma$, (2) the synchronizing power generated by generator No. 1 (i.e., generator No. 1 is generating), (3) the synchronizing power seemingly absorbed by, or delivered to, generator No. 2 (i.e., generator No. 2 motoring), (4) the power loss in both stator windings, and (5) the terminal voltage $\mathbf{V}_L$ of both generators.

6.5 The generators of Problem 6.3 have their generated voltages adjusted to be $|\mathbf{E}_{G1}| = |\mathbf{E}_{G2}| = 20{,}000$ V but generator No. 2 is in phase opposition with respect to generator No. 1. If the prime mover of generator No. 1 drives $\mathbf{E}_{G1}$ ahead of its correct position by 20°, that is, $\mathbf{E}_{G1} = 20{,}000e^{j20°}$, calculate the following quantities at the instant after the synchronizing switch is closed for the two given values of coil impedances, (a) $\mathbf{Z}_{S1} = \mathbf{Z}_{S2} = 50 + j450\ \Omega$ and (b) $\mathbf{Z}_{S1} = \mathbf{Z}_{S2} = 900 + j1{,}200\ \Omega$: (1) the synchronizing current $\mathbf{I}_\sigma$, (2) the synchronizing power generated by generator No. 1, (3) the synchronizing power received by generator No. 2, (4) the power loss in both stator windings, and (5) the terminal voltage $\mathbf{V}_L$ of both generators.

6.6 A 250-hp, 2,500-V, three-phase, 60-Hz, 26-pole, synchronous motor is directly connected to an infinitely strong network, that is, to an infinite bus, and has the following parameters: (1) $J = 400\ \text{kg}\cdot\text{m}^2$, (2) $P_D = 100$ W/electrical degree/sec, (3) $P_G = 12{,}000$ W/electrical degree. Assume that the motor is unloaded and is operating at steady state. With all internal losses neglected, $\delta(0^+) = d\delta(0^+)/dt = 0$. Suddenly the motor is loaded with rated mechanical load. Define the switching time of the loading as zero, that is, $t=0$, and calculate: (1) the synchronous speed of the motor ω_S and n_S, and (2) the inertia constant of the motor M. Write the power balance equation that describes the electrodynamic behavior of the synchronous motor when loaded and calculate: (3) the natural or undamped frequency of the motor ω_0, (4) the damping ratio ζ, (5) the damping factor α, (6) the damped frequency ω_d, and (7) the solution for the power angle $\delta(t)$.

6.7 The per-phase single-line diagram of Fig. 6.7.1 shows a generator connected through two, parallel, high-voltage transmission lines, connected between buses No. 1 and No. 2, to a large metropolitan system considered as an infinite bus. The relay-circuit-breaker sets A and B, and C and D, are adjusted to clear simultaneously. [In practice, each of the four relay-circuit-breaker sets is connected to each of the paralleling buses through a transformer-relay-circuit-breaker set. Usually the synchronous generator and the infinite bus are also connected to the corresponding buses through relay-circuit-breaker sets (see Example 6.8). However, for the purpose of this problem, the circuit diagram of Fig. 6.7.1 suffices.] Assume that a balanced three-phase fault is applied to the far end of the line AB near the circuit breaker B, as shown in Fig. 6.7.1. Calculate the **critical clearing angle** δ_C for clearing the fault with simultaneous opening of the circuit breakers A and B if the synchronous generator is delivering the prime-mover input (neglecting frictional losses, etc.) 400 MW at the instant preceding the fault.

6.8 The per-phase single-line diagram of Fig. 6.8.1 shows a generator connected through two parallel, high-voltage transmission lines, connected between buses No. 1 and No. 2, to a large metropolitan system considered as an infinite bus. This system is identical to the system of Problem 6.7 except that the transmission lines of reactance $j0.50\ \Omega$ are connected to the paralleling buses through transformers of reactance $j0.20\ \Omega$ and relay-circuit-breaker sets on both sides of each transformer. In addition, the synchronous generator and the infinite bus are connected to the paralleling buses through relay-circuit-breaker sets also. The relay-circuit-breaker sets A and B, and C and D, are adjusted to open or close simultaneously. Assume that a balanced three-phase fault is applied to the far end of the relay-circuit-breaker, transformer-line unit near the relay circuit breaker G as shown in Fig. 6.8.1. Calculate the **critical clearing angle** δ_C for clearing the fault with simultaneous opening of the relay circuit breakers A and B if the synchronous generator is delivering the prime-mover input (neglecting frictional losses, etc.) of 500 MW at the instant preceding the fault.

6.9 Which of the standard relay-circuit-breaker sets (2, 3, 5, or 8 cycles) would be used in the system of Problem 6.7 operating at 60 Hz if it were necessary to clear the fault in $t_C \leqslant 0.20$ sec?

6.10 For the system exhibited in Fig. 6.10.1(a) assume that the inertia constant $H=2.7$ MJ/MVA and the pu values of the pertinent quantities, all measured on a 120 MVA base, are $E_G=1.10\ e^{j\phi_G}$ pu V, $jX_S \equiv jX'_d = j0.50$ pu Ω, $R_{AB} = R_{CD} = R_S = R_B = 0$ pu Ω, and $jX_{AB} = jX_{CD} = j0.40$ pu Ω. Under normal operating conditions, the 120 MVA, 6-pole 60 Hz synchronous generator delivers 75 MW over a two parallel conductor transmission line to a large metropolitan area which may be regarded as an infinite bus designated as $E_B = 1.00e^{j0°}$ and $jX_B = 0$ pu Ω. If a three-phase fault were to occur at $k_A = 0.25$, 0.50, and 1.00, respectively, along the transmission line AB away from the generator side paralleling bus (1) plot the swing curves, that is, δ vs. t, for $t=1$ sec for all three cases, and comment on the stability status of the system operating under the constraints of the three sustained faults, (2) for the faulted status for which $k_A=0.25$ determine the approximate critical clearing time t_C, and the approximate critical angle δ_C, (3) specify the circuit–breakers that can be used at both ends of the transmission lines to isolate the faults and (4) determine whether or not a governor with a sensitivity 1% of normal speed would have any effects on the stability status of the system at hand.

6.11 The reactances of a salient-pole synchronous generator are $X_d=1.00$ pu Ω and $X_q=0.75$ pu Ω. The stator-winding resistance R_S is negligible. If the generator delivers rated per-unit voltamperes at rated per-unit volts and lagging current at a power factor of 0.85 LGG, calculate: (1) the approximate generated voltage $\mathbf{E}_G$, (2) the approximate power angle δ_g, (3) the approximate generated power P_g, (4) the actual generated voltage $\mathbf{E}_G$, (5) the actual power angle δ, and (6) the actual generated power P. Compare the calculated quantities and draw conclusions with respect to the accuracy of the approximate equivalent circuit of the synchronous device, Figs. 6.41 and 6.42.

6.12 The per-phase impedance diagram of a synchronous generator connected to a Y-Y-transformer, with all parameters expressed in per-unit values, is exhibited in Fig. 6.12.1. If the subtransient reactance is $X_d''=0.20$ pu Ω, the transformer reactance is $X_T=0.15$ pu Ω, and the load is such that the terminal voltage of the generator is $\mathbf{V}_L=1.0e^{j45°}$ pu V and the loading current is $\mathbf{I}_G=0.80$ pu A at a power factor 0.80 LGG, calculate: (1) the voltage $\mathbf{E}_G''$ behind the subtransient reactance X_d'', (2) the subtransient current $\mathbf{I}_F''$ when a balanced three-phase fault is placed across the load at $t=0$, (3) the maximum amplitude i_{DCMAX} of the transient or dc current component $i_{DC}(t)$ of the fault current $i_F''(t)$, (4) the maximum possible value i_{FMAX}'', of the fault current $i_F''(t)$, utilizing time-domain analysis and empirical formulas, and (5) the minimum possible amplitude i_{FMIN}'', of the fault current, that is, the amplitude of the steady state current $i_{AC}(t)$.

6.13 The per-phase impedance diagram of a synchronous generator connected to a Y-Y-transformer, with all quantities expressed in per-unit values, is exhibited in Fig. 6.13.1. If the generator is unloaded, the generated voltage is $\mathbf{E}_G=1.00e^{j0°}$ pu V, the subtransient reactance is $X_d''=0.20$ pu Ω, and the transformer reactance is $X_T=0.15$ pu Ω when a balanced three-phase fault is placed suddenly at all output terminals of the transformer, calculate: (1) the subtransient current $\mathbf{I}_F''$, (2) the maximum value of the transient or dc component of the current $i_F''(t)$, (3) the maximum possible value i_{FMAX}'', of the fault current $i_F''(t)$, and (4) the minimum possible amplitude i_{FMAX}'', of the fault current $i_F''(t)$.

6.14 The per-phase impedance diagram of a synchronous generator is exhibited in Fig. 6.14.1. If the generator is unloaded, the generated voltage is $\mathbf{E}_G=1.0e^{j0°}$ pu V, the subtransient reactance is $X_d''=0.20$ pu Ω, the transient reactance is $X_d'=0.25$ pu Ω, and the steady state reactance is $X_d=0.30$ pu Ω. If a balanced three-phase fault is placed suddenly at all terminals of the generator, calculate: (1) the subtransient current $\mathbf{I}_F''$, (2) the transient current $\mathbf{I}_F'$, (3) the postfault steady state current $\mathbf{I}_F$, (4) the minimum amplitude i_{FMIN}'', utilizing the constant-reactance technique, and (5) the minimum amplitude i_{FMIN}. What is the difference between i_{FMIN}'' and i_{FMIN} as calculated in parts (4) and (5)?

REFERENCES

1. Balabanian, N., *Fundamentals of Circuit Theory*. Allyn & Bacon, Boston, 1961.
2. Cheng, D. K., *Analysis of Linear Systems*. Addison-Wesley, Reading, Mass., 1961.
3. D'Azzo, J. J., and Houpis, C. H., *Feedback Control System Analysis and Synthesis*, 3d ed. McGraw-Hill, New York, 1960.
4. Del Toro, V., *Electromechanical Devices for Energy Conversion and Control Systems*. Prentice-Hall, Englewood Cliffs, N.J., 1968.
5. Elgerd, O. I., *Electric Energy Systems Theory: An Introduction*. McGraw-Hill, New York, 1971.
6. Eveleigh, V. W., *Introduction to Control Systems Design*. McGraw-Hill, New York, 1972.
7. Fitzgerald, A. E., and Kingsley, C., Jr., *Electric Machinery*, 2d ed. McGraw-Hill, New York, 1961.
8. Gourishankar, V., and Kelly, D. H., *Electromechanical Energy Conversion*, 2d ed. Intext, New York, 1973.
9. Hoover, D. B., "The Brushless Excitation System for Large AC Generators," *Westinghouse Engineer*, September 1964, pp. 141–45.
10. Kimbark, E. W., *Power System Stability*, vol. I. Wiley, New York, 1948.
11. Kimbark, E. W., *Power Systems Stability*, vol. III. Wiley, New York, 1956.
12. Kosow, I. L., *Electric Machinery and Control*. Prentice-Hall, Englewood Cliffs, N.J., 1964.
13. Kuo, B. C., *Automatic Control Systems*. Prentice-Hall, Englewood Cliffs, N.J., 1962.
14. Langill, A. W., *Automatic Control Systems Engineering*. vol. I. Prentice-Hall, Englewood Cliffs, N.J., 1965.
15. Lansdorf, A. S., *Theory of Alternating Current Machinery*. McGraw-Hill, New York, 1955.
16. Majmudar, H., *Electromechanical Energy Converters*. Allyn & Bacon, Boston, 1965.
17. Matsch, L. W., *Electromagnetic and Electromechanical Machines*. Intext, New York, 1972.
18. McFarland, T. C., *Alternating Current Machines*. Van Nostrand, New York, 1948.
19. Miller, D. H., and Rubenstein, A. S., "Excitation Systems for Small Generators," *Electrical Engineering*, June 1962, pp. 434–40.
20. Myers, E. H., "Rotating Rectifier Exciters for Large Turbine ac Generators," *Proceedings of the American Power Conference*, vol. XXVII (1965), pp. 1130–37.
21. Neuenswander, J. W., *Modern Power Systems*. Intext, New York, 1971.
22. O'Kelly, D., and Simons, S., *Introduction to Generalized Electrical Machine Theory*. McGraw-Hill, London, 1968.
23. Powell, J. E., and Wells, C. P., *Differential Equations*. Wiley, New York, 1950.
24. Schmitz, N. L., and Novatny, D. W., *Introductory Electromechanics*. Ronald Press, New York, 1965.
25. Smail, L. L., *Analytical Geometry and Calculus*. Prentice-Hall, Englewood Cliffs, N.J., 1953.
26. Stevenson, W. D., Jr., *Elements of Power Systems Analysis*, 3d ed. McGraw-Hill, New York, 1975.
27. Thaler, G. J., and Wilcox, M. L., *Electric Machines: Dynamics and Steady State*. Wiley, New York, 1966.
28. Thompson, H. A., *Alternating Current and Transient Circuit Analysis*. McGraw-Hill, New York, 1955.
29. Van Valkenburg, M. E., *Network Analysis*, 2d ed. Prentice-Hall, Englewood Cliffs, N.J., 1964.
30. Weedy, B. M., *Electric Power Systems*. Wiley, London, 1972.

7 PARAMETER MEASUREMENT OF THREE-PHASE SYNCHRONOUS DEVICES

7.1 INTRODUCTION

It is essential that the systems analyst be able to write the mathematical equations, based on scientific laws, for example, Kirchhoff's voltage law, which describe the behavior of the system at hand, and then solve these equations to get information about the **state** of the system. Such equations can be written with great ease if the analyst has a circuit analog of the system he or she is to analyze. To draw an accurate circuit analog, however, necessitates knowledge of the parameters that define it and, therefore, the system it represents. If these parameters are not available, the analyst should be able to measure them experimentally.

It is, therefore, the stated purpose of this chapter to arm the student with the capability to obtain the circuit parameters for the smooth- and the salient-rotor synchronous device, that is, R_S, X_S, X_d, X_d', X_d'', X_q, X_q', and X_q'', as well as the pertinent time constants, τ_d, τ_d', τ_d'', and τ_a, which are associated with the dynamic modeling of the synchronous device.

7.2 MEASUREMENT OF THE STATOR COIL REACTANCES X_S AND X_d

If the reactance X_S of the smooth-rotor synchronous device or the reactance X_d (d-axis) of the salient-rotor synchronous device are not available, they can be obtained experimentally by performing (1) the **open-circuit test** (Fig. 7.1), (2) the **short-circuit test** (Fig. 7.3), and (3) the **zero power-factor test** (Figs. 7.4 or 7.5).

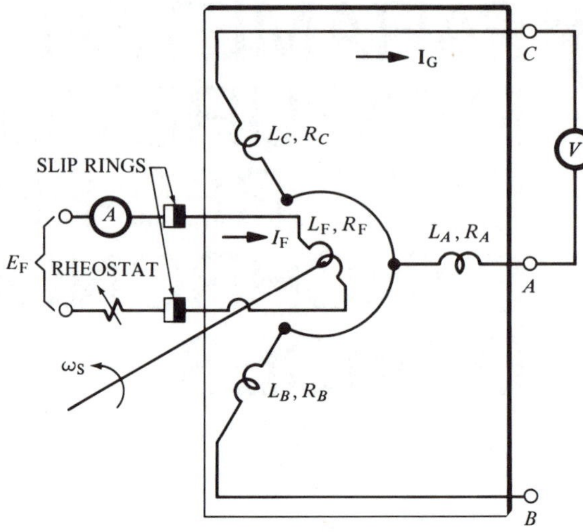

Fig. 7.1 Schematic Diagram for Obtaining the Open-Circuit or No-Load Characteristic of the Synchronous Device Operating in the Generator Mode

During the open-circuit test, Fig. 7.1, the rotor is driven at synchronous speed ω_S throughout the test by a constant-speed prime mover. The field- (rotor-) winding current I_F is varied by means of a rheostat and at each setting the corresponding open-circuit, line-to-neutral voltage $|\mathbf{V}_{LN}| = |\mathbf{V}_{LL}|/\sqrt{3} \equiv |\mathbf{V}|/\sqrt{3} \equiv |\mathbf{V}_L|$, designated as the $|\mathbf{V}_{L(OC)}|$-axis in Fig. 7.2, is measured. From the equivalent circuits of the synchronous generator, Figs. 5.19 and 6.42, it is clear that the terminal voltage $\mathbf{V}_L$ equals the excitation voltages $\mathbf{E}_G$ or $\mathbf{E}_{\mathcal{G}}$ if the terminals of the synchronous generator are **open**. That is, the describing equations of the synchronous generator, Eqs. (5.69) and (6.268) written in a compact form as

$$\mathbf{E}_\xi = \mathbf{V}_L + \{R_S + jX_\sigma\}\mathbf{I}_G,\ \xi = G, \mathcal{G} \text{ and } \sigma = \text{S, d} \tag{7.1}$$

become

$$\mathbf{E}_\xi = \mathbf{V}_L \tag{7.2}$$

The plot of these data points constitutes the voltage **open-circuit** or **no-load characteristic**, designated as curve **OCC** or **NLC** in Fig. 7.2. This curve is the **magnetization curve** of the synchronous device tested and exhibits saturation as expected. The straight-line portion of this curve, along with its extrapolation, is defined as the **airgap characteristic** and is designated as **AGC-curve** in Fig. 7.2.

During the short-circuit test, Fig. 7.3, where, except for small synchronous devices, the ammeters are connected at the secondaries of current transformers, designated as CT's (see Plate B.12), the rotor is again driven at synchronous speed ω_S throughout the test by a constant-speed prime mover. The field- (rotor-) winding current I_F is varied by a rheostat and at each setting the corresponding stator-winding, short-circuit current, designated as the $|\mathbf{I}_{G(SC)}|$-axis in Fig. 7.2, is measured. The plot of these data points constitutes the **short-circuit characteristic**, designated as **SCC-curve** in Fig. 7.2. Note that this characteristic is a straight line. Thus although the saturation of the magnetic material is clearly evident in the voltage open-circuit characteristic, it fails to show up in the current short-circuit characteristic. The reason for this is as follows: Fig. 5.38 shows that the stator magnetic field $\mathbf{B}_{ST}$

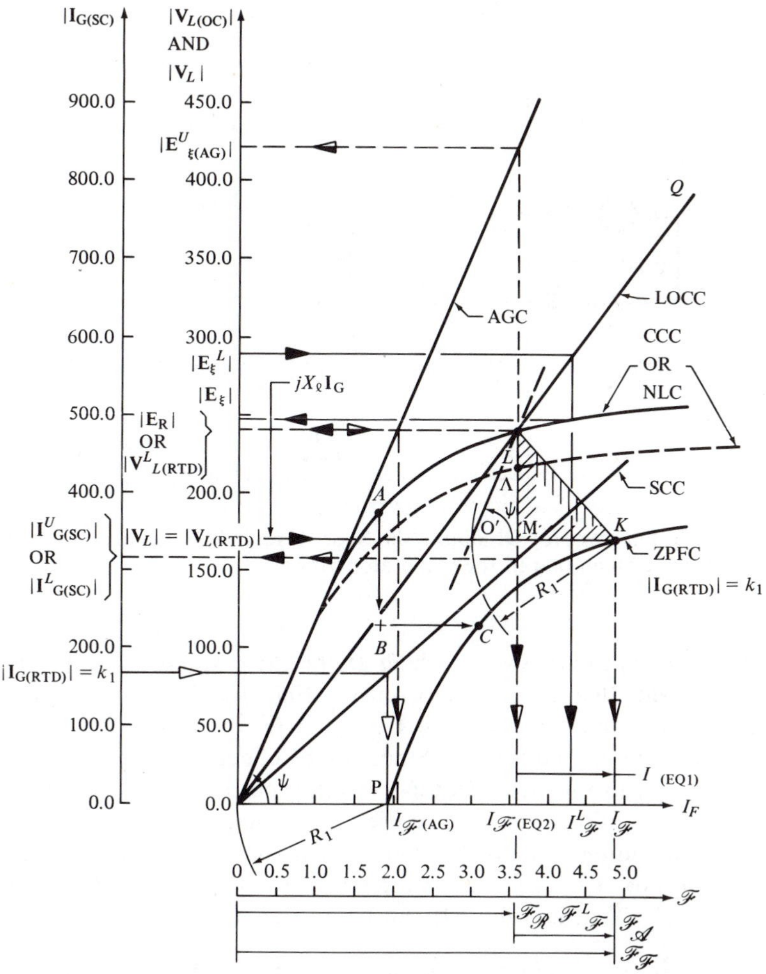

Fig. 7.2 Experimentally Obtainable Open-Circuit or No-Load, Short-Circuit, and Zero Power-Factor Characteristics for a Three-Phase, Y-Connected, Synchronous Device

partly opposes that of the rotor field $\mathbf{B}_R$ when the stator-winding current $\mathbf{I}_G$ lags with respect to the excitation voltage $\mathbf{E}_G$. From the equivalent circuits of the synchronous generator, Figs. 5.19 and 6.42, it is clear that for the terminals of the synchronous generator shorted, the terminal voltage becomes zero, that is, $\mathbf{V}_L = 0$, and the describing equations of the synchronous generators, Eqs. (5.69) and (6.268), written in a compact form as

$$\mathbf{E}_\xi = \mathbf{V}_L|_{\mathbf{V}_L=0} + (R_S + jX_\sigma)\mathbf{I}_G \tag{7.3}$$

become

$$\mathbf{E}_\xi = (R_S + jX_\sigma)\mathbf{I}_G,\ \xi = \mathrm{G}, \mathcal{G} \text{ and } \sigma = \mathrm{S}, \mathrm{d} \tag{7.4}$$

However, since $R_S \ll |jX_\sigma|$, $\sigma = \mathrm{S}, \mathrm{d}$, the stator coil current $\mathbf{I}_G$ lags the generated voltage $\mathbf{E}_\xi$, $\xi = \mathrm{G}, \mathcal{G}$, by nearly 90°. Thus the stator field $\mathbf{B}_{ST}$ is nearly in opposition with the rotor field $\mathbf{B}_R$. Consequently, the magnetic circuit is **not saturated** during the short-circuit test, even at exceedingly high values of field current I_F that

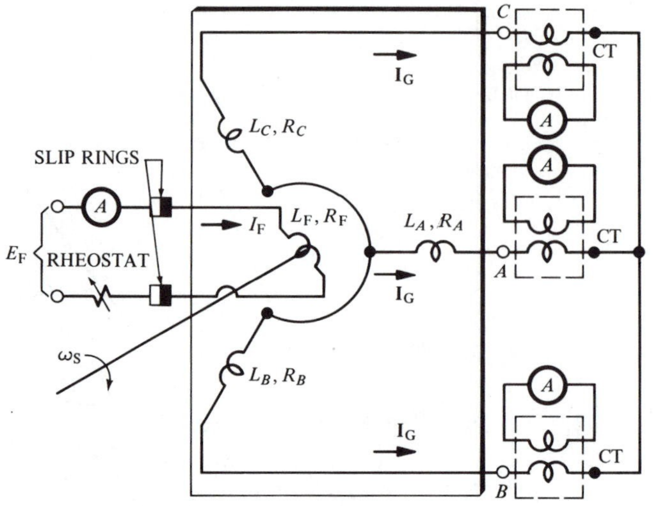

Fig. 7.3 Schematic Diagram for Obtaining the Short-Circuit Characteristic of the Synchronous Device Operating in the Generator Mode

produce stator currents even twice as high as the rated value of $\mathbf{I}_G$. Therefore, the **airgap characteristic**, that is, the AGC-curve, Fig. 7.2, should be taken as the relation between the generated voltage $\mathbf{E}^U_{\xi(AG)}$, $\xi = G, \mathcal{G}$ and $U \equiv$ Unsaturated, and the field- (rotor-) winding current I_F. Since the SCC-curve relates the short-circuit stator current $\mathbf{I}_{G(SC)}$ to the field- (rotor-) winding current I_F, the field current I_F determined from the AGC-curve can be used in conjunction with the SCC-curve to obtain the corresponding value of the short-circuit stator current $\mathbf{I}_{G(SC)}$. Therefore, the generated voltage $\mathbf{E}^U_{\xi(AG)}$ can be expressed as a function of the short-circuit stator current $\mathbf{I}^U_{G(SC)}$. That is, Eq. (7.4) can be written as

$$\mathbf{E}^U_{\xi\,(AG)} = (R_S + jX_\sigma)\mathbf{I}^U_{G(SC)},\ \xi = G, \mathcal{G} \text{ and } \sigma = S, d \qquad (7.5)$$

If the excitation current I_F of the field (rotor) winding is set at a value equal to $I_{\mathcal{F}(EQ2)}$, that is, $I_F \equiv I_{\mathcal{F}(EQ2)}$, Fig. 7.2, the variables $|\mathbf{E}^U_{\xi(AG)}|$ and $|\mathbf{I}^U_{G(SC)}|$ of Eq. (7.5) are located on the corresponding axes $|\mathbf{V}_{L(OC)}|$ and $|\mathbf{I}_{G(SC)}|$ of Fig. 7.2 by reflecting off the AGC- and SCC-curves. Note that the values of $|\mathbf{E}^U_{\xi(AG)}|$ and $|\mathbf{I}^U_{G(SC)}|$ can be obtained by following the dashed lines identified by the **three half-shaded arrowheads** resulting from the dashed line originating at $I_F \equiv I_{\mathcal{F}(EQ2)}$ and reflected off the SCC- and AGC-curves.

Solution of Eq. (7.5) for jX_σ, $\sigma = S, d$, for the condition $|R_S| \ll |jX_\sigma|$ yields the unsaturated value of **the synchronous reactance**, designated as $X_{\sigma U}$, $U \equiv$ Unsaturated, for either the smooth-rotor synchronous device, designated by $\sigma \equiv S$, or the salient-rotor synchronous device, designated by $\sigma \equiv d$. Therefore,

$$X_{\sigma U} = \frac{|\mathbf{E}^U_{\xi\,(AG)}|}{|\mathbf{I}^U_{G(SC)}|},\ \sigma = S, d \qquad (7.6)$$

If $|\mathbf{E}^U_{\xi(AG)}|$ and $|\mathbf{I}^U_{G(SC)}|$ are expressed in volts per phase and amperes per phase, respectively, the synchronous reactance $X_{\sigma U}$ is expressed in ohms per phase. However, if $|\mathbf{E}^U_{\xi(AG)}|$ and $|\mathbf{I}^U_{G(SC)}|$ are expressed in per-unit volts and amperes, respectively, the synchronous reactance $X_{\sigma U}$ is expressed in per-unit ohms.

The zero power-factor test (Figs. 7.4 and 7.5) yields the **zero power-factor characteristic**, designated as the ZPFC-curve in Fig. 7.2. When the rating of the synchronous generator at hand is 50 A or less (i.e., for small- or medium-size generators), an almost zero power-factor load can be obtained using three low resistance inductive coils, with variable inductance, connected in a Y-configuration, Fig. 7.4. Initially, the loading coils are set for zero reactances and the field- (rotor-) winding current I_F is adjusted to yield rated stator coil currents, designated as $|\mathbf{I}_{G(RTD)}| = k_1$, Fig. 7.2. The coil reactances are then increased, which causes a drop in the rated stator currents $|\mathbf{I}_{G(RTD)}|$. However, the field- (rotor-) winding current I_F is increased until the rated stator coil currents $|\mathbf{I}_{G(RTD)}|$ are reestablished. Both the terminal voltage $|\mathbf{V}_L| \equiv |\mathbf{V}|/\sqrt{3} \equiv |\mathbf{V}_{LL}|/\sqrt{3} = |\mathbf{V}_{LN}|$ of the synchronous device and the field- (rotor-) winding current I_F are recorded for the rated stator coil currents $|\mathbf{I}_{G(RTD)}|$. The coil reactances are again increased, which again causes a drop in the rated stator coil currents $|\mathbf{I}_{G(RTD)}|$. Then the field- (rotor-) winding current I_F is increased again until the rated stator current $|\mathbf{I}_{G(RTD)}|$ is reestablished. Again both the terminal voltage and the field- (rotor-) winding current I_F are recorded for rated stator coil currents $|\mathbf{I}_{G(RTD)}|$. The process is continued until the terminal voltages $|\mathbf{V}_L|$ across the zero power-factor load **exceed their rated value** by a factor of 1.5 to 2.0, while the rated, full-load, stator coil currents $|\mathbf{I}_{G(RTD)}|$ are maintained constant. The plot of these data points constitutes the zero power-factor characteristic, that is, the ZPFC-curve, in Fig. 7.2.

When the rating of the synchronous generator is greater than 50 A, an unloaded synchronous motor of approximately the same kilovoltampere rating as that of the synchronous generator is used in place of the purely inductive, Y-connected loading network, Fig. 7.5. The per-phase equivalent circuit of Fig. 7.5(a), exhibited in Fig. 7.5(b), is described by Eq. 7.7. That is,

$$\mathbf{I}_G = \frac{\mathbf{E}_{GG} - \mathbf{E}_{GM}}{\mathbf{Z}_{SG} + \mathbf{Z}_{SM}} = \frac{\mathbf{E}_{GG} - \mathbf{E}_{GM}}{(R_{SG} + R_{SM}) + j(X_{SG} + X_{SM})} \tag{7.7}$$

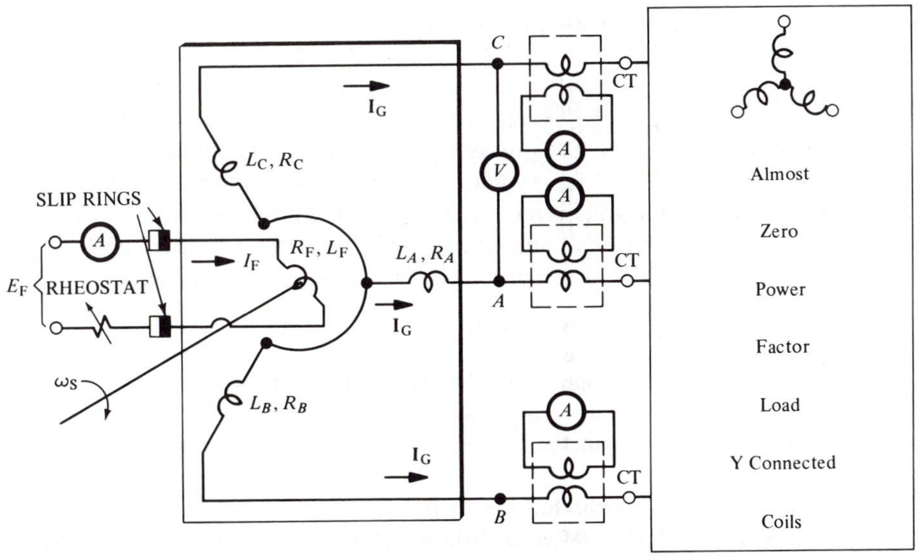

Fig. 7.4 Schematic Diagram for Obtaining the Zero Power-Factor Characteristic of the Synchronous Generator, Utilizing a Y-Connected, Purely Inductive Loading Network

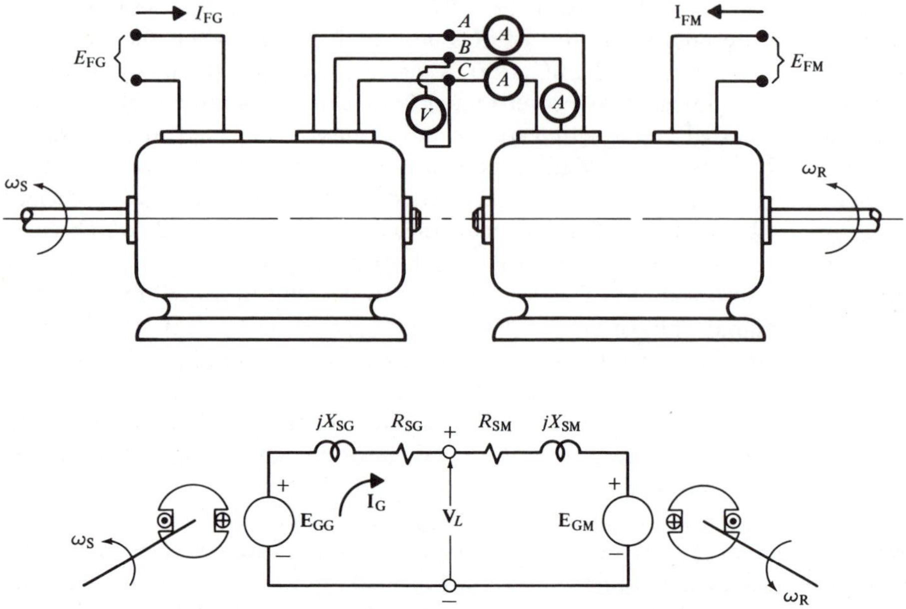

Fig. 7.5 Device Setup for Obtaining the Zero Power-Factor Characteristic of the Synchronous Generator, Utilizing a Y-Connected Synchronous Motor as a Load: (a) Two Device Interconnection and (b) Equivalent Circuit Interconnection of the Two Synchronous Devices

If (1) the generator is driven at synchronous speed ω_S throughout the test by a constant-speed prime mover and (2) the two excitation voltages $\mathbf{E}_{GG}$ and $\mathbf{E}_{GM}$, Fig. 7.5(b), are adjusted so that the generated current $\mathbf{I}_G$ is maintained at its rated full-load value $|\mathbf{I}_{G(RTD)}|$, data can be obtained for a curve exhibiting the terminal voltage $|\mathbf{V}_L|$ as a function of the generator field- (rotor-) winding current I_F. This curve constitutes the **zero power-factor characteristic** of the synchronous generator at rated generated current $|\mathbf{I}_{G(RTD)}|$. It is identical to the test curve obtained from Fig. 7.4 and is designated as ZPFC in Fig. 7.2. If the generator is overexcited while the motor is underexcited, the power factor is lagging, and the curve is similar to that obtained with the aid of the purely inductive Y-connected network.

The zero power factor of the synchronous generator-motor set is explained as follows. Since the synchronous motor is unloaded, the power required by the motor during this test is only the small amount needed to provide the required losses, Fig. 5.29. That is, from Fig. 7.5(b) the input power to the motor is given as

$$P_{INT} = 3\,\mathrm{Re}(\mathbf{V}_L\mathbf{I}_G^*) \equiv P_{ST} + P_{LT} \doteq 0 \tag{7.8}$$

Therefore, for $|jX_{SG}| \gg R_{SG}$ and $|jX_{SM}| \gg R_{SM}$, the power factor of the generator, Fig. 7.5(b), is nearly zero, since $P_{GT} = 3|\mathbf{E}_{GG}|\,|\mathbf{I}_G|\cos\phi = P_{INT} \doteq 0$. That is, the input power to the synchronous generator is very small, in fact it is almost equal to zero watts.

Inspection of Fig. 7.2 yields that the ZPFC-curve is similar in shape to the OCC- or NLC-curve but is shifted **down** a distance $\overline{AB}$ and to the **right** a distance $\overline{BC}$. This shape of ZPFC-curve can be explained through the phasor diagrams of Fig. 7.6 and the OCC- or NLC-curve of Fig. 7.2. From Fig. 7.6(a) we can write the

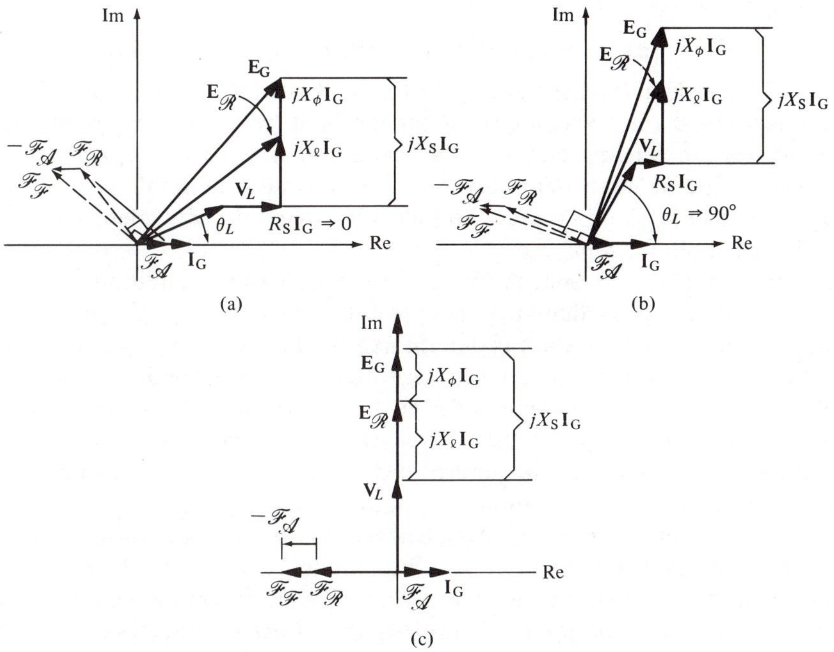

Fig. 7.6 Phasor Diagrams of a System Whose Power Factor Is Lagging: (a) $R_S \gg 0$, (b) $R_S \doteq 0$, and (c) $R_S = 0\ \Omega$

following pertinent equations for the generated voltage $\mathbf{E}_G$ and the airgap voltage $\mathbf{E}_{\mathcal{R}}$:

$$\mathbf{E}_G = \mathbf{V}_L + (R_S + jX_S)|\mathbf{I}_G|,\ \xi = G \tag{7.9}$$

$$\mathbf{E}_{\mathcal{R}} = \mathbf{V}_L + (R_S + jX_\ell)|\mathbf{I}_G| \tag{7.10}$$

and

$$\mathcal{F}_{\mathcal{R}} = \mathcal{F}_{\mathcal{F}} + \mathcal{F}_{\mathcal{A}} \tag{7.11}$$

where (1) $\mathcal{F}_{\mathcal{F}}$ designates the mmf due to the field (rotor) winding, (2) $\mathcal{F}_{\mathcal{A}}$ designates the mmf due to the stator windings, and (3) $\mathcal{F}_{\mathcal{R}}$ designates the resultant mmf, that is, the geometric sum of the mmf's $\mathcal{F}_{\mathcal{F}}$ plus $\mathcal{F}_{\mathcal{A}}$. As the power factor of the system described by Fig. 7.6(a) approaches zero, that is, as the system becomes more and more inductive, $|jX_S| \gg R_S$ and $\theta_L \to 90°$, the phasor diagram of Fig. 7.6(a) approaches that of Fig. 7.6(c) through that of Fig. 7.6(b). Thus all voltages, $\mathbf{V}_L, jX_\ell\,\mathbf{I}_G$, $\mathbf{E}_{\mathcal{R}}, jX_\phi\mathbf{I}_G, jX_S\mathbf{I}_G$, and $\mathbf{E}_G$, become colinear as do all the mmf's, that is, $\mathcal{F}_{\mathcal{F}}, -\mathcal{F}_{\mathcal{A}}$, and $\mathcal{F}_{\mathcal{R}}$. Therefore, we can use Fig. 7.6(c) and rewrite Eqs. (7.9) through (7.11) as

$$|\mathbf{E}_G| = |\mathbf{V}_L| + |jX_S\mathbf{I}_G|,\ \xi = G \tag{7.12}$$

$$|\mathbf{E}_{\mathcal{R}}| = |\mathbf{V}_L| + |jX_\ell\,\mathbf{I}_G| \tag{7.13}$$

and

$$|\mathcal{F}_{\mathcal{F}}| = |\mathcal{F}_{\mathcal{R}}| + |-\mathcal{F}_{\mathcal{A}}| \tag{7.14}$$

Note that in order to plot all mmf's on the I_F-axis of Fig. 7.2, they must be expressed in **equivalent field- (rotor-)** winding amperes. That is,

$$|\mathcal{F}_{\mathcal{F}}| = K_{WR}N_R I_F \equiv \mathcal{N}_R I_{\mathcal{F}} \tag{7.15}$$

$$|-\mathcal{F}_{\mathcal{A}}| = K_{WS}N_S\mathbf{I}_G \equiv \mathcal{N}_R I_{\mathcal{F}(EQ1)} \tag{7.16}$$

and

$$|\mathcal{F}_{\mathcal{R}}| = |\mathcal{F}_{\mathcal{F}}| - |-\mathcal{F}_{\mathcal{A}}| = \mathcal{N}_{\mathbf{R}} I_{\mathcal{F}} - \mathcal{N}_{\mathbf{R}} I_{\mathcal{F}(\mathrm{EQ1})} \equiv \mathcal{N}_{\mathbf{R}} I_{\mathcal{F}(\mathrm{EQ2})} \tag{7.17}$$

where (1) K_{WR} designates the **winding factor** of the rotor winding, (2) $\mathcal{N}_{\mathbf{R}}$ designates the **equivalent rotor winding turns for the field (rotor) winding** mmf $|\mathcal{F}_{\mathcal{F}}|$, (3) K_{WS} designates the **winding factor of the stator winding,** (4) $I_{\mathcal{F}(\mathrm{EQ1})}$ designates the field- (rotor-) winding equivalent current that constitutes the mmf $\mathcal{F}_{\mathcal{A}}$ in the stator winding, and (5) $I_{\mathcal{F}(\mathrm{EQ2})}$ designates the field rotor-winding equivalent current that constitutes the resultant mmf $\mathcal{F}_{\mathcal{R}}$.

The mmf's of Eq. 7.17 specify the magnetic operating conditions of a given synchronous device. In particular, to simplify the analysis of synchronous devices, it is assumed that their saturation is determined by the resultant airgap flux $\Phi_{\mathcal{R}}$ and (in addition) it is assumed that the resultant mmf $\mathcal{F}_{\mathcal{R}}$ corresponding to a specified value of airgap flux $\Phi_{\mathcal{R}}$ is the same **under load** as **under no-load.** The open-circuit or no-load characteristic, that is, the OCC- or NLC-curve, Fig. 7.2, of a given synchronous device can then be interpreted as the relation between the airgap voltage $\mathbf{E}_{\mathcal{R}}$ and the resultant mmf $\mathcal{F}_{\mathcal{R}}$, that is, $|\mathbf{E}_{\mathcal{R}}| = f(\mathcal{F}_{\mathcal{R}})$. This assumption neglects (1) the effects of stator-winding leakage flux on the saturation of the stator iron, (2) the changes of the field- (rotor-) winding leakage flux on the saturation of the rotor iron, (3) the effect of the stator winding mmf $\mathcal{F}_{\mathcal{A}}$ on the waveform of the synchronously rotating airgap flux, and (4) the effect on the flux in the rotor structure caused by the shifting of the resultant flux waveform from its no-load position with respect to the field poles. As a rule, none of these effects is important in cylindrical-rotor synchronous devices, though some of them may be appreciable in salient-rotor synchronous devices. For example, omission of the first two effects usually is not serious, because the leakage fluxes usually are small and their paths coincide with the main flux for only a small part of the main flux path. For a considerable portion of their path lengths, the leakage fluxes are in air. Therefore, they are affected little by saturation and the stator-winding leakage reactance X_{ℓ} is assumed to be constant.

Since the OCC- or NLC-curve can be thought of as $|\mathbf{E}_{\mathcal{R}}| = f(\mathcal{F}_{\mathcal{R}})$, for each operating point on the OCC- or NLC-curve, such as point L, there is a corresponding point on the ZPFC-curve, such as point K, Fig. 7.2, for which (1) terminal voltage $|\mathbf{V}_L|$, (2) the field- (rotor-) winding current $I_{\mathcal{F}}$, and (3) the equivalent field- (rotor-) winding mmf $\mathcal{F}_{\mathcal{F}}$ can be written, with the aid of Eqs. (7.13) and (7.14), and Eq. (7.17) with $\mathcal{N}_{\mathbf{R}}$ divided out, as

$$|\mathbf{V}_L| = |\mathbf{E}_{\mathcal{R}}| - |jX_{\ell}\mathbf{I}_{\mathrm{G}}| \tag{7.18}$$

$$|\mathcal{F}_{\mathcal{F}}| = |\mathcal{F}_{\mathcal{R}}| + |-\mathcal{F}_{\mathcal{A}}| \tag{7.19}$$

and

$$I_{\mathcal{F}} = I_{\mathcal{F}(\mathrm{EQ1})} + I_{\mathcal{F}(\mathrm{EQ2})} \tag{7.20}$$

Eq. (7.18) is plotted along the $|\mathbf{V}_{L(\mathrm{OC})}|$-axis of Fig. 7.2. Similarly, Eqs. (7.19) and (7.20) are plotted along the $\mathcal{F}$-axis and its equivalent I_{F}-axis of Fig. 7.2, correspondingly.

It is essential to note the location (in Fig. 7.2) of the quantities contained in Eqs. (7.18) through (7.20) and to compare $|jX_{\ell}\mathbf{I}_{\mathrm{G}}|$ and $I_{\mathcal{F}(\mathrm{EQ1})}$ to lines $\overline{AB}$ and $\overline{BC}$, correspondingly.

These observations enable us to draw the following conclusion. If (1) the open-circuit characteristic, that is, the OCC- or NLC-curve, is identical to $|\mathbf{E}_{\mathcal{R}}| = f(\mathcal{F}_{\mathcal{R}})$ under load, (2) the stator-winding leakage reactance is constant, that is,

$X_\ell = k$, and (3) the stator-winding resistance is zero, that is, $R_S = 0$, then the ZPFC-curve would be a curve of exactly the same shape as the OCC- or NLC-curve shifted down a distance $\overline{AB} = |jX_\ell \mathbf{I}_G|$ and to the right a distance $\overline{BC} = I_{\mathcal{F}(EQ1)}$. The similarity of the shapes of these two curves suggests that one could find experimentally the leakage reactance X_ℓ and the stator-winding mmf $\mathcal{F}_{\mathcal{Q}}$ if one were to find a triangle, such as triangle KLM of Fig. 7.2, which would fit everywhere between the OCC- or NLC- and the ZPFC-curves. Experimentally, the OCC- or NLC-curve can be measured from the origin to well into saturation. However, the ZPFC-curve can only be adequately measured in the neighborhood of the knee of the curve. The ZPFC-curve can then be completed by extrapolation down to the point P on the I_F-axis, Fig. 7.2. This is accomplished by marking off on the $|\mathbf{I}_{G(SC)}|$-axis the rated stator-winding current $|\mathbf{I}_{G(RTD)}| = k_1$ (i.e., the stator-winding current at which the zero power-factor test is taken) and reflecting off the SCC-curve to the I_F-axis to yield the point $I_{\mathcal{F}0}$. Note that the value of $I_{\mathcal{F}0}$ can be obtained by following the thin line identified by the two clear arrowheads originating at $|\mathbf{I}_{G(SC)}| \equiv |\mathbf{I}_{G(RTD)}| = k_1$ of the $|\mathbf{I}_{G(SC)}|$-axis.

The geometric construction of the triangle KLM, known as the **Potier triangle**, in honor of its inventor, is as follows: (1) select a point K above the knee of the ZPFC-curve, say at rated terminal voltage, that is, $|\mathbf{V}_L| \rightarrow |\mathbf{V}_{L(RTD)}|$, Fig. 7.2, (2) mark off the horizontal line $\overline{KO'} \equiv R_1$, where R_1 is the distance from P to O, (3) through point O' draw line $\overline{O'L}$ parallel to the AGC-curve at an angle ψ intercepting the OCC- or NLC-curve at point L, (4) draw the vertical line $\overline{LM}$, and (5) draw the line $\overline{LK}$.

It is concluded earlier in this section that the OCC- or NLC-curve and the ZPFC-curve are similar in shape and that

$$\overline{AB} = |jX_\ell \mathbf{I}_G| \tag{7.21}$$

and

$$\overline{BC} = I_{\mathcal{F}(EQ1)} \tag{7.22}$$

Since, however, $\overline{AB} = \overline{LM}$ and $\overline{BC} = \overline{MK}$, it follows that

$$|jX_\ell I_G| = \overline{LM} \tag{7.23}$$

and

$$I_{\mathcal{F}(EQ1)} = \overline{MK} \tag{7.24}$$

Therefore, the leakage reactance X_ℓ and the equivalent field amperes for the stator windings $I_{\mathcal{F}(EQ1)}$ can be calculated correspondingly from the sides $\overline{LM}$ and $\overline{MK}$ of the Potier triangle LMK, Fig. 7.2.

It turns out, however, that the saturated curves relating the airgap voltage $\mathbf{E}_{\mathcal{R}}$ to the resultant mmf $\mathcal{F}_{\mathcal{R}}$, that is, $|\mathbf{E}_{\mathcal{R}}| = f(\mathcal{F}_{\mathcal{R}})$, at zero power-factor are different for the cylindrical-rotor and the salient-rotor synchronous devices for high values of field- (rotor-) winding currents I_F. In fact, under load the saturation curve of the salient-rotor device is always lower than, or to the right of, that of the cylindrical-rotor device. In other words, the OCC- or NLC-curve for the cylindrical-rotor device does not change with load. However, for the salient-rotor device, the OCC- or NLC-curve *shifts* to the right, or down, as load is increased. Such an OCC- or NLC-curve is designated by a **dashed-line** curve in Fig. 7.2. Note that the lines $\overline{LM}$ and $\overline{\Lambda M}$, which are measures of the stator leakage reactances for (1) the **no-load** (cylindrical- and/or salient-rotor devices) and (2) the **under-load** (salient-rotor devices) conditions, are related by Eq. (7.25). That is,

$$\overline{LM} > \overline{\Lambda M} \tag{7.25}$$

Therefore, for the cylindrical-rotor synchronous devices for which the saturation characteristic **does not shift** from the OCC- or NLC-curve, the Potier triangle provides an **accurate measure** of the leakage reactance by measuring the line $\overline{LM}$. However, for the salient-rotor synchronous device (particularly those that have long slim poles and hence rather large flux leakage) for which the saturation characteristic **does shift** from the OCC- or NLC-curve as load changes, the Potier triangle provides an **inaccurate measure** of the leakage reactance. This is shown by Eq. (7.25), which states that the lines $\overline{LM}$ and $\overline{\Lambda M}$ must satisfy the conditions $\overline{LM} > \overline{\Lambda M}$.

Since the power output capability of electromechanical energy devices is directly dependent upon the degree to which the material is worked magnetically and electrically, it turns out that **for a high output such devices are operated in the saturation region of their characteristics**. Therefore, it can be assumed that the stator leakage reactance can be measured fairly accurately if the Potier triangle is constructed at high levels of saturation. Once the Potier triangle is constructed, the **leakage reactance** is calculated by solving Eq. (7.18) for X_ℓ and extracting the relevant information from Fig. 7.2. Thus

$$X_\ell = \frac{|\mathbf{E}_{\mathfrak{R}}| - |\mathbf{V}_L|}{|\mathbf{I}_G|} \equiv \frac{\overline{LM}}{|\mathbf{I}_G|}, \qquad |\mathbf{I}_G| \equiv |\mathbf{I}_{G(RTD)}| = k_1 \tag{7.26}$$

Many authors refer to the leakage reactance of Eq. (7.26) as the **Potier reactance** and designate it as X_p.

The leakage reactance of Eq. (7.26) is used in conjunction with the **saturation factor** k to calculate the *saturated synchronous reactance*, designated as X_σ, of a given synchronous device regardless of whether its rotor is smooth or salient. This reactance X_σ is used to model accurately, when saturation effects of the iron are taken into account, the synchronous device when its operating point lies along the knee of the OCC- or NLC-curve, as it usually does. A good indication of the degree of saturation of the magnetic circuit is given by the airgap voltage $\mathbf{E}_{\mathfrak{R}}$ or its associated resultant flux $\Phi_{\mathfrak{R}}$. Thus if a synchronous generator is assumed to be delivering rated power at rated voltage and a specified power factor, the actual airgap voltage $\mathbf{E}_{\mathfrak{R}}$ is represented by point L on the saturation curve and is produced by the field- (rotor-) winding current $I_F \equiv I_{\mathfrak{F}(EQ2)}$. It is worthwhile to note here that if there were no saturation of the iron, the voltage $|\mathbf{E}_{\mathfrak{R}}|$ would have been induced by a smaller field- (rotor-) winding current $I_F \equiv I_{\mathfrak{F}(AG)}$, Fig. 7.2. Note that the values of $I_{\mathfrak{F}(EQ2)}$ and $I_{\mathfrak{F}(AG)}$ can be obtained by following the dashed lines identified by the **four fully shaded arrowheads** resulting from the dashed line originating at $|\mathbf{V}_{L(OC)}| = |\mathbf{E}_{\mathfrak{R}}|$ and reflected off the AGC-curve and the OCC- or NLC-curve.

In essence, then, it can be said that the field- (rotor-) winding current required to overcome the reluctance of the airgap for the resultant flux $\Phi_{\mathfrak{R}}$ is $I_{\mathfrak{F}(AG)}$ and that the current required to overcome both the reluctance of the airgap and the iron is $I_{\mathfrak{F}(EQ2)}$. That is, the degree of saturation prevailing under these conditions is represented by the ratio of the two currents $I_{\mathfrak{F}(EQ2)}$ and $I_{\mathfrak{F}(AG)}$ or, similarly, by the ratio of the two voltages $|\mathbf{E}^{U}_{G(AG)}|$ and $|\mathbf{E}_{\mathfrak{R}}|$. These two voltages are determined by reflecting off the OCC- or NLC-curve and the AGC-curve the dashed line originating at $I_F = I_{\mathfrak{F}(EQ2)}$ and characterized by the **half-shaded arrowheads**. Accordingly, the saturation factor is defined as

$$k \equiv \frac{I_{\mathscr{F}(EQ2)}}{I_{\mathscr{F}(AG)}} = \frac{|\mathbf{E}^{U}_{G(AG)}|}{|\mathbf{E}_{\mathscr{R}}|} \tag{7.27}$$

Utilization of (1) the leakage or Potier reactance X_ℓ, Eq. (7.26), and (2) the saturation factor k, Eq. (7.27), enables us to calculate the **saturated synchronous reactance** X_σ of the cylindrical- and salient-rotor synchronous device. Thus utilization of the leakage or Potier reactance enables us to write the equation for the **unsaturated synchronous reactance**, Eq. (7.6), with the aid of Eq. (5.48), as

$$X_{\sigma U} = X_\ell + X_\phi \tag{7.28}$$

Since a large part of the path of the leakage fluxes is in air, saturation has **less** effect on the leakage or Potier reactance than it has on the magnetic reactance because the path for the flux that is associated with X_ϕ is mostly through iron and the only air path is the relatively short length of the airgap. In addition, X_ϕ is several times **larger** than X_ℓ. Therefore, the saturated synchronous reactance X_σ can be written as

$$X_\sigma = X_\ell + \frac{X_\phi}{k} \tag{7.29}$$

Solution of Eq. (7.28) for X_ϕ and substitution for X_ϕ into Eq. (7.29) yields

$$X_\sigma = X_\ell + \frac{X_{\sigma U} - X_\ell}{k}, \sigma = \text{S, d} \tag{7.30}$$

Note that the *saturated synchronous reactance*, Eq. (7.30), is given in terms of the measured values of (1) the **unsaturated synchronous reactance** $X_{\sigma U}$, Eq. (7.6), (2) the **leakage** or **Potier reactance** X_ℓ, Eq. (7.26), and the *saturation factor* k, Eq. (7.27). If the magnetic state of the iron of a given synchronous device remains fixed, as defined by the saturation factor k, the synchronous reactance of the device would have the value specified by Eq. (7.30), and its open-circuit characteristic would be represented by the straight line $\overline{OLO}$, designated as the LOCC-curve, Fig. 7.2, drawn through the origin and the point L on the saturation curve, that is, the OCC- or NLC-curve, at the airgap voltage $|\mathbf{E}_{\mathscr{R}}|$. In other words, the synchronous device has been **linearized** at the magnetic state corresponding to the resultant flux $\Phi_{\mathscr{R}}$. Stating the same thing differently, the synchronous device is **equivalent** to an unsaturated device whose magnetization curve is the straight line $\overline{OLQ}$, that is, the LOCC-curve, Fig. 7.2.

According to this approximation, the *linearized saturated synchronous reactance* $X_\sigma{}^L$ at rated loading voltage $|\mathbf{V}_L{}^L| \equiv |\mathbf{V}^L_{L(RTD)}|$ is defined as

$$X_\sigma{}^L \equiv \frac{|\mathbf{V}^L_{L(RTD)}|}{|\mathbf{I}^L_{G(SC)}|} \doteq X_\sigma \tag{7.31}$$

Correspondingly, the generated voltage $\mathbf{E}_\xi{}^L$, L≡Linearized, for the linearized synchronous generator, now can be written as

$$\mathbf{E}_\xi{}^L = \mathbf{V}_L{}^L + (R_S + jX_\sigma{}^L)\mathbf{I}_G \tag{7.32}$$

The reflection of the magnitude of the generated voltage $|\mathbf{E}_\xi{}^L|$ off the LOCC-curve enables us to determine (1) the field- (rotor-) winding current $I_{\mathscr{F}}{}^L$ (or its equivalent) mmf $\mathscr{F}_{\mathscr{F}}{}^L$ and (2) the generated voltage $|\mathbf{E}_\xi|$ read on the $|\mathbf{V}_{L(OC)}|$-axis, designated by the reflection of the $I_F \equiv I_{\mathscr{F}}{}^L$ or $\mathscr{F}_{\mathscr{F}}{}^L$ line off the OCC- or NLC-curve, Fig. 7.2.

Example 7.1

The open-circuit and short-circuit characteristics of a Y-connected, three-phase, 20-pole, 60-Hz, 72-kVA, 415.69-V (line-to-line), synchronous generator are exhibited in Fig. 7.1.1. For rated voltage and power factor 0.8 LGG, find: (1) the linearized (approximate) saturated synchronous reactance $X_\sigma{}^{\mathrm{L}}$, (2) the linearized generated voltage $|\mathbf{E}_\xi{}^{\mathrm{L}}|$, (3) the field- (rotor-) winding current $I_{\mathscr{F}}^{\mathrm{L}}$, (4) the generated voltage $|E_\xi|$, and (5) the voltage regulation VR.

Solution

The approximate synchronous reactance $X_\sigma{}^{\mathrm{L}}$ is calculated with the aid of Eq. (7.31). The rated loading voltage $|\mathbf{V}_L{}^{\mathrm{L}}|$ is calculated as

$$|\mathbf{V}_L{}^{\mathrm{L}}| = \frac{|\mathbf{V}_{\mathrm{LL}}|}{\sqrt{3}} = \frac{415.69}{\sqrt{3}} = \underline{240.00 \text{ V}} \tag{1}$$

This voltage defines the point L on the OCC- or NLC-curve through which the LOCC-curve is drawn. The linearized short-circuit current $|\mathbf{I}^{\mathrm{L}}_{\mathrm{G(SC)}}| = 315$ A is obtained by reflecting off the SCC-curve. Thus $X_\sigma{}^{\mathrm{L}}$ is calculated as

$$X_\sigma{}^{\mathrm{L}} = \frac{|\mathbf{V}^{\mathrm{L}}_{\mathrm{L(RTD)}}|}{|\mathbf{I}^{\mathrm{L}}_{\mathrm{G(SC)}}|} = \frac{240}{315} = \underline{\underline{0.76\ \Omega}} \tag{2}$$

The linearized generated voltage $\mathbf{E}_\xi{}^{\mathrm{L}}$ is calculated with the aid of Eq. (7.32) and the input data. That is,

$$\mathbf{E}_\xi{}^{\mathrm{L}} \doteq \mathbf{V}_L{}^{\mathrm{L}} + jX_\sigma{}^{\mathrm{L}}\mathbf{I}_{\mathrm{G}} \tag{3}$$

The generated current $|\mathbf{I}_{\mathrm{G}}|$ is calculated from the input data as

$$\mathbf{I}_{\mathrm{G}} = \frac{\mathrm{kVA}}{3} \times \frac{1}{|\mathbf{V}^{\mathrm{L}}_{L(\mathrm{RTD})}|} = \frac{72{,}000.00}{3 \times 240.00} = \underline{100.00 \text{ A}} \tag{4}$$

Appropriate substitution from Eqs. (1) and (2), the input data, and Eq. (4) into Eq. (3) yields

$$\begin{aligned}\mathbf{E}_\xi{}^{\mathrm{L}} &= \mathbf{V}_L{}^{\mathrm{L}} + jX_\sigma\mathbf{I}_{\mathrm{G}} = 240e^{j\cos^{-1}(0.8)} + j0.76 \times 100.00 \\ &= 240e^{j36.87^\circ} + j76 = 192.00 + j144.00 + j76.00 \\ &= 192.00 + j220.00 = \underline{\underline{292.00e^{j48.89^\circ}}} \text{ V}\end{aligned} \tag{5}$$

Reflection of the linearized voltage $|\mathbf{E}_\xi{}^{\mathrm{L}}|$ off the LOCC-curve, Fig. 7.1.1, yields (1) the field- (rotor-) winding current $I_{\mathscr{F}}^{\mathrm{L}}$ on the I_{F}-axis and (2) the generated voltage $|\mathbf{E}_\xi|$ on the $|V_{L(\mathrm{OC})}|$-axis by a double reflection off the OCC-curve. That is,

$$I_{\mathscr{F}}^{\mathrm{L}} = \underline{\underline{4.38 \text{ A}}} \tag{6}$$

and

$$|\mathbf{E}_\xi| = \underline{\underline{248.00 \text{ V}}} \tag{7}$$

Finally, the voltage regulation is calculated with the aid of Eqs. (7) and (1) as

$$\mathrm{VR} = \frac{|\mathbf{E}_\xi| - |\mathbf{V}_L{}^{\mathrm{L}}|}{|\mathbf{V}_L{}^{\mathrm{L}}|} \times 100\% = \frac{248.00 - 240.00}{240.00} \times 100\% = \underline{\underline{3.33\ \%}} \tag{8}$$

Inspection of Fig. 7.2 and Eq. (7.27) and Eq. (7.30) reveal that the saturation factor k and the saturated synchronous reactance X_σ are functions of the airgap voltage $|\mathbf{E}_{\mathscr{R}}|$. These functional relationships are exhibited in Fig. 7.7. If the

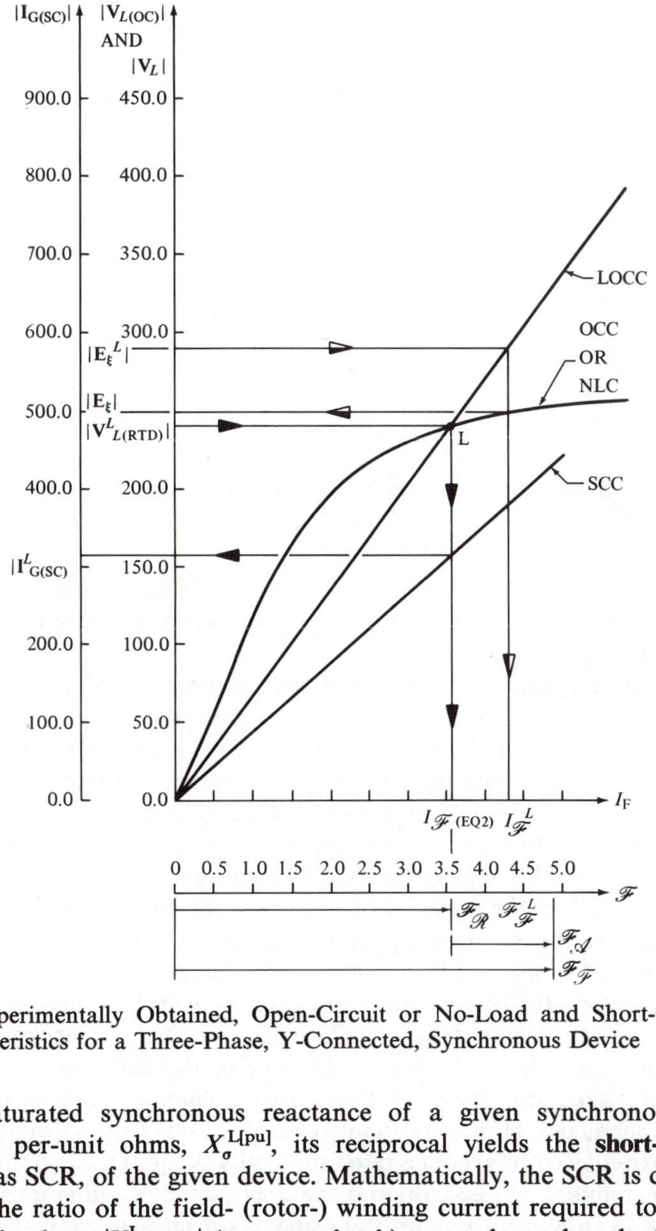

Fig. 7.1.1 Experimentally Obtained, Open-Circuit or No-Load and Short-Circuit Characteristics for a Three-Phase, Y-Connected, Synchronous Device

linearized saturated synchronous reactance of a given synchronous device is expressed in per-unit ohms, $X_\sigma^{L[pu]}$, its reciprocal yields the **short-circuit ratio**, abbreviated as SCR, of the given device. Mathematically, the SCR is defined, from Fig. 7.2, as the ratio of the field- (rotor-) winding current required to produce the rated no-load voltage $|\mathbf{V}^L_{L(RTD)}|$ (on **open-circuit**) at rated speed to the field- (rotor-) winding current required to produce the rated short-circuit, stator-winding current $|\mathbf{I}_{G(SC)}|$ (on **short circuit**). That is,

$$\text{SCR} \equiv \frac{I_{\mathcal{F}}\ \{|V^L_{L(OC)}| \equiv \text{RATED}\}}{I_{\mathcal{F}}\ \{|I_{G(SC)}| \equiv \text{RATED}\}} \equiv \frac{I_{\mathcal{F}(EQ2)}}{I_{\mathcal{F}0}} \equiv \frac{1}{X_\sigma{}^L}\Bigg|_{X_\sigma{}^L \equiv X_\sigma^{[pu]}} \doteq \frac{1}{X_\sigma^{[pu]}} \tag{7.33}$$

where, from Fig. 7.2, $I_{\mathcal{F}(EQ2)}$ designates the field- (rotor-) winding current that would produce rated no-load terminal voltage $|\mathbf{E}^L_{\xi(RTD)}|$ on the LOCC-curve (i.e., the point of intersection between the linearized open-circuit characteristic, i.e., the

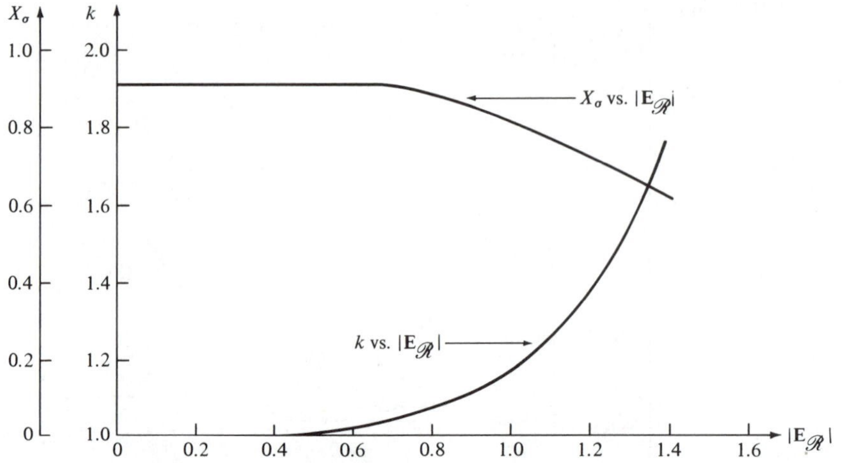

Fig. 7.7 Curves of the Saturation Factor k and the Saturated Synchronous Reactance X_σ as a Function of the Airgap Voltage $|\mathbf{E}_{\mathfrak{R}}|$ of a Synchronous Device

LOCC-curve, and the open-circuit characteristic, i.e., the OCC-curve), and $I_{\mathfrak{F}}$ designates the field- (rotor-) winding current that would produce rated short-circuit stator current $|\mathbf{I}_{G(RTD)}|$.

The short-circuit ratio is a measure of the physical size of a given synchronous device rated at a given kilovoltampere power factor and speed. To gain some idea as to the influence of the physical size of a given synchronous device on its SCR, consider the device in which the length ℓ of the airgap is doubled while the stator winding and the rest of the dimensions of the stator iron remain unchanged. If the reluctance $\mathfrak{R} = \ell/\mu A$ of the iron were negligible, the no-load field- (rotor-) winding current $I_{\mathfrak{F}(EQ2)}$ should be practically doubled to produce the same voltage as before. Doubling the length of the airgap reduces the unsaturated value of the synchronous reactance $X_{\sigma U}$ of the stator winding to one-half its original value. [Note that $\lambda = Li = N\Phi = N(Ni/\mathfrak{R})$ yields $L = N^2/\mathfrak{R} = N^2\mu A/\ell$ and $L\omega_E = X$.] Therefore, only about one-half the original resultant flux $\Phi_{\mathfrak{R}}$ would be necessary to produce rated short-circuit current $|\mathbf{I}_{G(RTD)}|$. Since this one-half value of the flux now traverses twice the original length of the airgap, the value of the field- (rotor-) winding mmf $\mathfrak{F}_{\mathfrak{F}}$ required for rated, short-circuit, stator-winding current $|\mathbf{I}_{G(RTD)}|$ is practically unchanged. However, about twice the field- (rotor-) winding current $I_{\mathfrak{F}}$ or mmf $\mathfrak{F}_{\mathfrak{F}}$ is required to produce rated no-load voltage $|\mathbf{E}^L_{G(RTD)}|$ since the mutual inductance is reduced to one-half, and the field (rotor) winding must be increased in size if the heating is to remain the same. As a consequence, the synchronous device must be made larger to accommodate the larger winding.

Example 7.2

The open-circuit, short-circuit, and zero power-factor characteristics of a Y-connected, three-phase, six-pole, 60-Hz, 294.45-V (line-to-line), 85.68-kVA, synchronous generator are exhibited in Fig. 7.2.1. The numerical divisions of the three-scales, (1) $0 \leqslant |\mathbf{I}_{G(SC)}| \leqslant 900$ A, (2) $0 \leqslant |\mathbf{V}_{L(OC)}| \leqslant 450$ V, and (3) $0 \leqslant |I_F| \leqslant 5$ A, are essential for completing this example. For a rated phase terminal voltage $|\mathbf{V}_{L(RTD)}| = 170.00$ V and a minimum field- (rotor-) winding current of $I_{\mathfrak{F}0} = 1.90$ A required to initiate the ZPFC-curve, calculate: (1) the unsaturated synchronous reactance

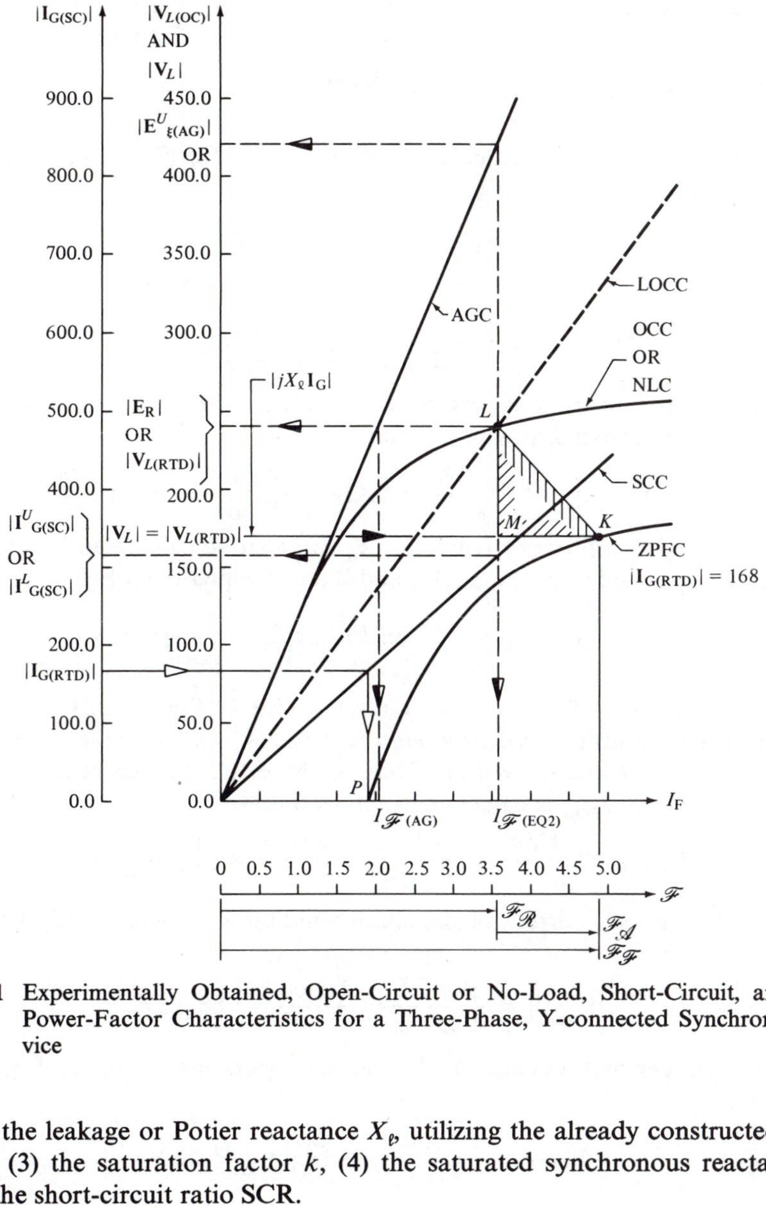

Fig. 7.2.1 Experimentally Obtained, Open-Circuit or No-Load, Short-Circuit, and Zero Power-Factor Characteristics for a Three-Phase, Y-connected Synchronous Device

$X_{\sigma U}$, (2) the leakage or Potier reactance X_ℓ, utilizing the already constructed Potier triangle, (3) the saturation factor k, (4) the saturated synchronous reactance X_σ, and (5) the short-circuit ratio SCR.

Solution

The utilization of Fig. 7.2.1 is essential in doing this example. Therefore, continuous reference to it, and in particular to the Potier triangle and the operating point L, is an integral part of the solution of this example. Note that it is essential to know how to construct the Potier triangle.

The unsaturated synchronous reactance $X_{\sigma U}$ is calculated with the aid of Eq. (7.6) and data obtained from Fig. 7.2.1 as

$$X_{\sigma U} = \frac{|\mathbf{E}^U_{\xi\,(AG)}|}{|\mathbf{I}^U_{G(SC)}|} = \frac{422}{315} = \underline{\underline{1.34\ \Omega}} \qquad (1)$$

The leakage or Potier reactance X_ℓ is calculated with the aid of Eq. (7.26) and data, that is, $|\mathbf{E}_{\mathcal{R}}|$, $|\mathbf{V}_L|$, and $|\mathbf{I}_{G(RTD)}|$, obtained from Fig. 7.2.1 as

$$X_\ell = \frac{|\mathbf{E}_{\mathcal{R}}| - |\mathbf{V}_L|}{|\mathbf{I}_G|} = \frac{240.00 - 170.00}{168.00} = \underline{\underline{0.42\ \Omega}} \tag{2}$$

The saturation factor k is calculated with the aid of Eq. (7.27) and data obtained from Fig. 7.2.1 as

$$k = \frac{I_{\mathcal{F}(EQ2)}}{I_{\mathcal{F}(AG)}} = \frac{3.60}{2.05} = \underline{\underline{1.76}} \tag{3}$$

or

$$k = \frac{|\mathbf{E}^{U}_{\xi\,(AG)}|}{|\mathbf{E}_{\mathcal{R}}|} = \frac{422}{240} = \underline{\underline{1.76}} \tag{4}$$

The saturated synchronous reactance X_ℓ is calculated with the aid of Eq. (7.30) and data obtained from Eqs. (1), (2), and (3) or (4) as

$$X_\sigma = X_\ell + \frac{X_{\sigma U} - X_\ell}{k} = 0.42 + \frac{1.34 - 0.42}{1.76} = \underline{\underline{0.94\ \Omega}} \tag{5}$$

The saturated synchronous reactance X_σ can also be calculated by utilizing the approximation provided by Eq. (7.31) and data obtained from Fig. 7.2.1 as

$$X_\sigma \doteq X_\sigma{}^{L} = \frac{|\mathbf{V}^{L}_{L(RTD)}|}{|\mathbf{I}^{L}_{G(SC)}|} = \frac{240}{315} = \underline{\underline{0.76\ \Omega}} \tag{6}$$

To calculate the SCR that corresponds to point L, Fig. 7.2.1, it is essential to calculate the unsaturated synchronous reactance X_σ and leakage or Potier reactance X_ℓ in per-**unit** quantities first. Thus the base full-load current of the stator-winding current $|\mathbf{I}_{G(RTD)}|$ is calculated first, as follows:

$$|\mathbf{I}_{G(RTD)}| \equiv \frac{\text{kVA}}{3} \times \frac{1}{|\mathbf{V}_{L(RTD)}|} = \frac{85{,}680}{3} \times \frac{1}{170} = \underline{168.00\ \text{A}} \tag{7}$$

Then the short-circuit current of the stator winding in **per-unit** $I^{U[pu]}_{G(SC)}$ is calculated as

$$|\mathbf{I}^{U\,[pu]}_{G(SC)}| = \frac{|\mathbf{I}^{U}_{G(SC)}|}{|\mathbf{I}_{G(RTD)}|} = \frac{315.00}{168.00} = \underline{\underline{1.88}}\ \text{pu A} \tag{8}$$

Similarly, the **per-unit** voltage $|\mathbf{E}^{U[pu]}_{\xi(AG)}|$ at the operating point L, Fig. 7.2.1, is calculated as

$$\mathbf{E}^{U}_{\xi\,(AG)} = \frac{|\mathbf{E}^{U}_{\xi\,(AG)}|}{|\mathbf{V}_{L(RTD)}|} = \frac{422}{170} = \underline{2.48}\ \text{pu V} \tag{9}$$

The unsaturated synchronous reactance in **per-unit** ohms $X^{[pu]}_{\sigma U}$ is calculated as

$$X^{[pu]}_{\sigma U} = \frac{|\mathbf{E}^{U\,[pu]}_{\xi\,(AG)}|}{|\mathbf{I}^{U\,[pu]}_{G\,(SC)}|} = \frac{2.48}{1.88} = \underline{1.32}\ \text{pu}\ \Omega \tag{10}$$

Similarly, the leakage or Potier reactance can be calculated in **per-unit** ohms as follows:

$$X^{[pu]}_\ell = \frac{|\mathbf{E}^{[pu]}_{\mathcal{R}}| - |\mathbf{V}^{[pu]}_L|}{|\mathbf{I}^{[pu]}_{G\,(RTD)}|} = \frac{(|\mathbf{E}_{\mathcal{R}}| - |\mathbf{V}_L|)/\mathbf{V}_{L(RTD)}|}{|\mathbf{I}_G|/|\mathbf{I}_{G(RTD)}|}$$

$$= \frac{(240 - 170)/170}{168/168} = \underline{0.41}\ \text{pu}\ \Omega \tag{11}$$

Now the saturated synchronous reactance in **per-unit** ohms $X_\sigma^{[pu]}$ is calculated with the aid of Eq. (7.30) and data obtained from Eqs. (10), (11), and (3) or (4) as

$$X_\sigma^{[pu]} = X_\ell^{[pu]} + \frac{X_{\sigma U}^{[pu]} - X_\ell^{[pu]}}{k} = 0.41 + \frac{1.32 - 0.41}{1.76} = \underline{0.93} \text{ pu } \Omega \tag{12}$$

As stated earlier in the paragraph preceding Eq. (6), the saturated linearized synchronous reactance $X_\sigma{}^L$ is calculated with the aid of Eq. (7.31) and data obtained from Fig. 7.2.1 and/or Eq. (7). In **per-unit** ohms, this reactance is calculated as

$$X_\sigma^{L[pu]} = \frac{|\mathbf{V}_{L(RTD)}^{L\,[pu]}|}{|\mathbf{I}_{G(SC)}^{L[pu]}|} = \frac{|\mathbf{V}_{L(RTD)}^{L}|/|\mathbf{V}_{L(RTD)}^{L}|}{|\mathbf{I}_{G(SC)}^{L}|/|\mathbf{I}_{G(RTD)}|} = \frac{240/240}{315/168} = 0.53 \text{ pu } \Omega \tag{13}$$

Finally, the short-circuit ratio SCR of the synchronous generator is calculated with the aid of Eq. (7.33) and data obtained from Eq. (13) or Fig. 7.2.1 as

$$\text{SCR} = \frac{1}{X_\sigma^{L[pu]}} = \frac{1}{0.53} = \underline{\underline{1.88}} \tag{14}$$

or

$$\text{SCR} = \frac{I_{\mathcal{F}(EQ2)}}{I_{\mathcal{F}0}} = \frac{3.6}{1.9} = \underline{\underline{1.88}} \tag{15}$$

or approximated, utilizing data from Eq. (12), as

$$\text{SCR} = \frac{1}{X_\sigma^{[pu]}} = \frac{1}{0.93} = \underline{1.08} \tag{16}$$

7.3 MEASUREMENT OF THE STATOR COIL RESISTANCE R_S

Usually the resistance of the stator winding R_S for a synchronous device is ignored. However, when accurate modeling of synchronous devices is desired, measurement of the mechanical power P_M required to drive the synchronous generator during the short-circuit test provides information regarding the losses caused by the stator-winding current $\mathbf{I}_G$ and therefore the stator-winding resistance R_S. The stator-winding resistance has two components, namely $R_{S(AC)}$ and $R_{S(DC)}$. The power required to drive the synchronous generator during the short-circuit test should be equal to the sum of: (1) the **frictional losses** P_{FL}, (2) the **windage losses** P_{WL}, and (3) losses caused by the stator winding current, designated as P_{SCLL}^T, collectively known as the **total short-circuit load losses**. Mathematically, these powers are related as

$$P_M = P_{FL} + P_{WL} + P_{SCLL}^T \tag{7.34}$$

The frictional and windage losses may be determined from simple experiments which will not be discussed here. Therefore, the **total short-circuit load losses** are calculated from Eq. (7.34) as

$$P_{SCLL}^T = P_M - (P_{FL} + P_{WL}) \tag{7.35}$$

The **total short-circuit load losses** P_{SCLL}^T are comprised of (1) the **power losses due to the ac resistance** of the stator windings, $3R_{S(AC)}|\mathbf{I}_{G(SC)}|^2$, (2) the **core losses** due to (a) **the stator flux leakage**, $P_{CL}(\Phi_{S\ell})$ and (b) **the resultant flux**, $P_{CL}(\Phi_{\mathcal{R}})$. These powers are related mathematically as

$$P_{\text{SCLL}}^{\text{T}} = 3R_{\text{S(AC)}}|\mathbf{I}_{\text{G(SC)}}|^2 + P_{\text{CL}}(\Phi_{\text{S}\ell}) + P_{\text{CL}}(\Phi_{\Re}) \tag{7.36}$$

Similarly, the **power losses due to the ac resistance of the stator windings** are comprised of the power losses due to (1) **skin effect** phenomena in the stator windings, P_{SE}, (2) **eddy currents** in the stator windings, P_{EC}, and (3) **dc resistance** of the stator windings, $3R_{\text{S(DC)}}|\mathbf{I}_{\text{G(SC)}}|^2$. These powers are related mathematically as

$$3R_{\text{S(AC)}}|\mathbf{I}_{\text{G(SC)}}|^2 = P_{\text{SE}} + P_{\text{EC}} + 3R_{\text{S(DC)}}|\mathbf{I}_{\text{G(SC)}}|^2 \tag{7.37}$$

Substitution of Eq. (7.37) into Eq. (7.36) yields

$$P_{\text{SCLL}}^{\text{T}} = P_{\text{SE}} + P_{\text{EC}} + 3R_{\text{S(DC)}}|\mathbf{I}_{\text{G(SC)}}|^2 + P_{\text{CL}}(\Phi_{\text{S}\ell}) + P_{\text{CL}}(\Phi_{\Re}) \tag{7.38}$$

Subtraction of the dc resistance losses from the total short-circuit load losses yields the **stray load losses** P_{SLL}. P_{SLL} is considered to have the same value under normal load conditions as it does under short-circuit conditions. Mathematically, P_{SLL} is written as

$$\begin{aligned} P_{\text{SLL}} &\equiv P_{\text{SCLL}}^{\text{T}} - 3R_{\text{S(DC)}}|\mathbf{I}_{\text{G(SC)}}|^2 \\ &= P_{\text{SE}} + P_{\text{EC}} + P_{\text{CL}}(\Phi_{\text{S}\ell}) + P_{\text{CL}}(\Phi_{\Re}) \end{aligned} \tag{7.39}$$

The dc resistance losses can be computed if the dc resistance of the stator winding is measured using an ohmmeter or wheatstone bridge and corrected for the temperature of the winding during the test. If the conductors are made from a given metal, dc resistance correction is achieved utilizing Eq. (7.40).* That is,

$$\frac{R_{T(\text{DC})}}{R_{\tau(\text{DC})}} = \frac{T_0 + T}{T_0 + \tau} \tag{7.40}$$

In Eq. (7.40), (1) $R_{\tau(\text{DC})}$ designates the dc resistance of the winding *to be determined* at temperature τ, (2) $R_{T(\text{DC})}$ designates the dc resistance of the winding at the *known* temperature T (room temperature, 20°C), and (3) $T_0 = 234.50$°C, 225.00°C for copper and aluminum, correspondingly. Finally, the per-phase ac resistance of the stator winding, $R_{\text{S(AC)}}$, is calculated with the aid of Eq. (7.36) **based on the assumptions** that (1) $P_{\text{CL}}(\Phi_{\text{S}\ell}) \doteq 0$, (2) $P_{\text{CL}}(\Phi_{\Re}) \doteq 0$, and (3) the total short-circuit load losses are a function only of the rated stator winding current $\mathbf{I}_{\text{G(SC)}}$. That is,

$$R_{\text{S(AC)}} = \left.\frac{P_{\text{SCLL}}^{\text{T}}/3}{|\mathbf{I}_{\text{G(SC)}}|^2}\right|_{|\mathbf{I}_{\text{G(SC)}}| \equiv |\mathbf{I}_{\text{G(RTD)}}|} \tag{7.41}$$

If the total short-circuit load losses $P_{\text{SCLL}}^{\text{T}}$ are expressed in watts per phase and the current $|\mathbf{I}_{\text{G(SC)}}|$ is expressed in amperes, then $R_{\text{S(AC)}}$ is expressed in ohms. However, if the total short-circuit load losses $P_{\text{SCLL}}^{\text{T}}$ and the stator-winding current $|\mathbf{I}_{\text{G(SC)}}|$ are expressed in per-unit quantities, then $R_{\text{S(AC)}}$ is expressed in per-unit ohms, designated as $R_{\text{S(AC)}}^{[\text{pu}]} = (P_{\text{SCLL}}^{\text{T[pu]}}/3)/|\mathbf{I}_{\text{G(SC)}}^{[\text{pu}]}|^2$. Usually it is sufficiently accurate to find the value of $R_{\text{S(AC)}}$ at rated current and then assume that it remains constant throughout its utilization.

Inspection of Eq. (7.37) suggests that the effects of $R_{\text{S(DC)}}$ are included in the effects of $R_{\text{S(AC)}}$. Therefore, it can be concluded that the effective resistance R_{S} of

*IEEE Test Procedure for Synchronous Machines, no. 115, 1965 (Institute of Electrical and Electronics Engineers, 345 East 47th Street, New York, N.Y. 10017).

the stator winding is given by Eq. (7.41), and can be written as

$$R_S \equiv R_{S(AC)} = \left.\frac{P_{SCLL}^{T}/3}{|\mathbf{I}_{G(SC)}|^2}\right|_{|\mathbf{I}_{G(SC)}|=|\mathbf{I}_{G(RTD)}|} \tag{7.42}$$

Example 7.3

For the synchronous generator of Example 7.2 at rated stator-winding current $|\mathbf{I}_{G(SC)}| = 168$ A and at the temperature of 20°C, the total short-circuit losses P_{SCLL}^{T} are equal to 3.6 kW while $P_{CL}(\Phi_{S\ell})$ and $P_{CL}(\Phi_{\Re})$ are negligible. The dc resistance of the stator winding at this temperature is $R_{S(DC)} = 0.0315$ Ω/phase.

Calculate: (1) the stator-winding resistance R_S in ohms/phase, (2) the stator-winding resistance R_S in per-unit ohms/phase, (3) the total dc losses P_{DCT}, and (4) the total losses $P_{SE} + P_{EC}$ due to the skin effect and eddy currents. Repeat parts (3) and (4) if it is known that the stator-winding composition is copper, its operating temperature is 50°C, and the total short-circuit load losses P_{SCLL}^{T} do not change.

Solution

The stator-winding resistance R_S in ohms/phase is calculated with the aid of Eq. (7.42) and input data as

$$R_S = \left.\frac{P_{SCLL}^{T}/3}{|\mathbf{I}_{G(SC)}|^2}\right|_{\mathbf{I}_{G(SC)}\equiv\mathbf{I}_{G(RTD)}} = \frac{3600/3}{168^2} = \underline{\underline{0.0425\ \Omega}} \tag{1}$$

In per-unit ohms the resistance is calculated with the aid of Eq. (7.42), given input data and input data from Example 7.2, as

$$\begin{aligned} R_S^{[pu]} &= \left.\frac{P_{SCLL}^{T[pu]}/3}{|\mathbf{I}_{G(SC)}^{[pu]}|^2}\right|_{\mathbf{I}_{G(SC)}=\mathbf{I}_{G(RTD)}} \\ &= \frac{(P_{SCLL}^{T}/3)/[\text{kVA}/3]}{|\mathbf{I}_{G(SC)}|^2/|\mathbf{I}_{G(RTD)}|^2} = \frac{3.6\times10^3/85.68\times10^3}{1^2} \\ &= \underline{\underline{0.0420}}\ \text{pu}\ \Omega \end{aligned} \tag{2}$$

The total dc losses P_{DCT} are calculated with the aid of input data as

$$\begin{aligned} P_{DCT} &= 3R_{S(DC)}|\mathbf{I}_{G(SC)}|^2\big|_{|\mathbf{I}_{G(SC)}|=|\mathbf{I}_{G(RTD)}|} = 3\times0.0315\times168^2 \\ &= \underline{\underline{2667.17}}\ \text{W} \end{aligned} \tag{3}$$

The total losses $P_{SE} + P_{EC}$ due to skin effect and eddy currents, respectively, are calculated with the aid of Eq. (7.37), input data, and Eq. (3) as

$$P_{SE} + P_{EC} = P_{SCLL}^{T} - P_{DCT} = 3600.00 - 2667.17 = \underline{\underline{932.83}}\ \text{W} \tag{4}$$

The dc resistance at 50°C is calculated with the aid of Eq. (7.40), input data, and the knowledge that $T_0 = 234.50$°C as

$$R_{\tau(DC)} = \left(\frac{T_0+\tau}{T_0+T}\right)R_{T(DC)} = \left(\frac{234.50+50}{234.50+20}\right)\times0.0315 = \underline{0.0352\ \Omega} \tag{5}$$

Finally, the total losses $P_{SE} + P_{EC}$ due to skin effect and eddy currents, respectively, are calculated with the aid of Eq. (7.37) and Eq. (5) as

$$\begin{aligned} P_{SE} + P_{EC} &= P_{SCLL}^{T} - 3R_{\tau(DC)}|\mathbf{I}_{G(SC)}|^2 \\ &= 3600.00 - 3\times0.0352\times168^2 = \underline{\underline{619.55}}\ \text{W} \end{aligned} \tag{6}$$

7.4 MEASUREMENT OF THE STATOR COIL *d*-AXIS REACTANCES

Although the indirectly measured saturated synchronous reactance X_σ, $\sigma = \text{S}$, Eq. (7.30), and the resistance R_S, Eq. (7.42), of the stator winding are extremely accurate for modeling the **smooth-rotor** synchronous device, the information is inaccurate as well as inadequate if one is interested in modeling the salient-rotor synchronous device. That is, in addition to the measurement of the resistance R_S, Eq. (7.42), a more accurate measurement of the synchronous reactance X_σ, $\sigma = \text{d}$, is needed, as well as the measurement of the parameters X_d', X_d'', X_q, X_q', and X_q'', or, at least, the measurement of the parameters X_d' and X_d''.

The reactances of the salient-rotor synchronous device are calculated utilizing the experimentally measured variables of Eqs. (6.316) through (6.318), where $|\mathbf{E}_\text{G}|$ is the measured RMS value of the **no load, line-to-neutral**, generated voltage and $|\mathbf{I}_\text{F}''|$, $|\mathbf{I}_\text{F}'|$, and $|\mathbf{I}_\text{F}|$ are the RMS values of the **subtransient, transient**, and **steady state currents** respectively, as extracted from the recorded oscillogram of Fig. 6.46(c). The points of intersection that the current envelopes of Fig. 6.46(c) make with the i_V-axis at $t = 0$ can be determined more accurately by plotting (on semilogarithmic paper) the excesses of the two current envelopes at given times $\omega_\text{E} t = \omega_\text{E} t_0, \omega_\text{E} t_1, \omega_\text{E} t_2, \ldots$. These excesses are defined as $\Delta i'' \equiv i_\text{F}'' - i_\text{F}'$ and $\Delta i' \equiv i_\text{F}' - i_\text{F}$ and are plotted on the logarithmic scale while angle, that is, $\omega_\text{E} t$, is plotted on the linear scale, Fig. 7.8. When the work is done carefully, both curves, that is, $\Delta i''$

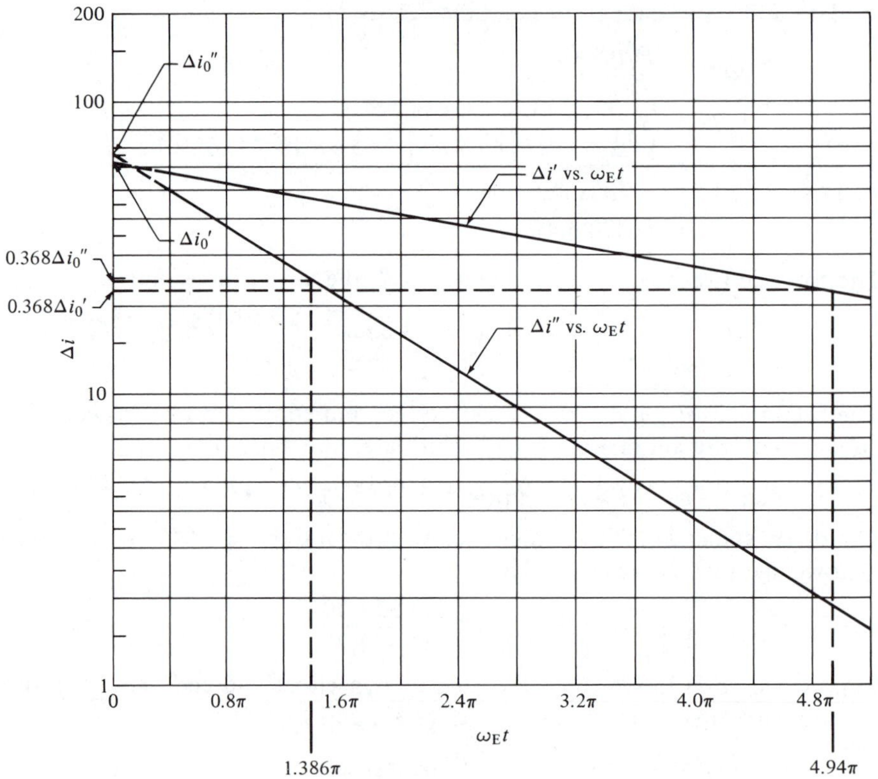

Fig. 7.8 The Excesses, $\Delta i'' \equiv i_\text{F}'' - i_\text{F}'$ and $\Delta i' \equiv i_\text{F}' - i_\text{F}$, vs. Angle for the Current Envelopes of Fig. 6.46(c) Are Plotted on Semilogarithmic Scales: Exponential Decay

versus $\omega_E t$ and $\Delta i'$ versus $\omega_E t$, approximate the straight lines very closely. This illustrates the essentially exponential nature of the decay of the currents $i_F''(t)$ and $i_F(t)$ in Fig. 6.46(c). The straight-line portions in Fig. 7.8 when extended to the Δi-axis yield the Δi-axis intersection currents, designated as $\Delta i_0''$ and $\Delta i_0'$. If $\Delta i_0'$ [or $(\Delta i_0' + \Delta i_0'')$] is added to the maximum instantaneous value of the sustained current $i_{AC}(t)$, the result is the maximum value of the transient current $\sqrt{2}\,|\mathbf{I}_F'|$ (or the subtransient current $\sqrt{2}\,|\mathbf{I}_F''|$). Thus, with the aid of Figs. 7.8 and 6.46(c), the subtransient current $|\mathbf{I}_F''|$, the transient current $|\mathbf{I}_F'|$, and the steady state current $|\mathbf{I}_F|$ of Eqs. (6.316) through (6.318) are calculated as follows:

$$|\mathbf{I}_F''| = |\mathbf{I}_{AC}| + \frac{\Delta i_0' + \Delta i_0''}{\sqrt{2}} \tag{7.43}$$

$$|\mathbf{I}_F'| = |\mathbf{I}_{AC}| + \frac{\Delta i_0'}{\sqrt{2}} \tag{7.44}$$

and

$$|\mathbf{I}_F| = |\mathbf{I}_{AC}| \tag{7.45}$$

Example 7.4

For a faulted 75-kVA, eight-pole, three-phase, salient-rotor, Y-connected, synchronous generator, an oscillogram of the fault current of phase *A* resembling that of Fig. 7.4.1 was recorded in the laboratory. Calculate: (1) the subtransient reactance X_d'', (2) the transient reactance X_d', and (3) the synchronous reactance X_d of the given synchronous generator.

Solution

The generated voltage of the salient-rotor synchronous generator is calculated with the aid of the input data and the oscillogram of Fig. 7.4.1 as

$$|\mathbf{E}_G| = \frac{\text{kVA}/3}{i_{ACMAX}/\sqrt{2}} = \frac{75\times 10^3/3}{250/\sqrt{2}} = \frac{25\times 10^3}{176.78} = \underline{141.42\ \text{V}} \tag{1}$$

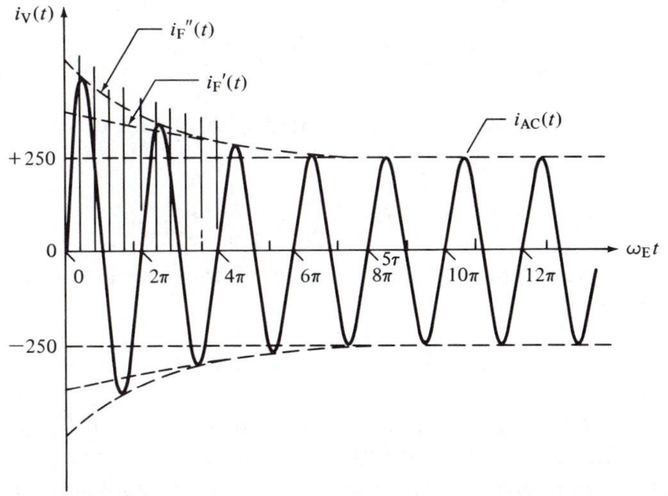

Fig. 7.4.1 Oscillogram of Fault Current of Phase *A* for Example 7.4; Note That the Frequency Is Not Drawn to Scale

Table 7.1 TABULATION OF $\omega_E t$, $\Delta i'$, AND $\Delta i''$ FOR EXAMPLE 7.4 AS EXTRACTED FROM FIG. 7.4.1.

$\omega_E t$	$\Delta i'$	$\Delta i''$
0	—	—
0.4π	56.250	50.000
0.8π	51.875	37.500
1.2π	48.125	28.125
1.6π	44.750	21.250
2.0π	40.625	15.625
2.4π	37.500	12.187
2.8π	35.000	9.062
3.2π	32.500	6.600
3.6π	30.000	5.000
4.0π	27.600	3.750
4.4π	25.500	2.800
4.8π	23.437	2.100
5.2π	21.500	1.580

To calculate the reactances X_d'', X_d', and X_d, the currents $|\mathbf{I}_F''|$, $|\mathbf{I}_F'|$, and $|\mathbf{I}_F|$ of Eqs. (7.43), (7.44), and (7.45) must first be calculated utilizing the oscillogram of Fig. 7.4.1. The data for $\Delta i''$ versus $\omega_E t$ and $\Delta i'$ versus $\omega_E t$ are tabulated in Table 7.1 and are plotted explicitly in Fig. 7.8.

Extrapolation of the curve $\Delta i''$ versus $\omega_E t$ yields the Δi-axis intersection, given as

$$\Delta i_0'' = \underline{66.00 \text{ A}} \tag{2}$$

Similarly, extrapolation of the curve $\Delta i'$ versus $\omega_E t$ yields the Δi-axis intersection, given as

$$\Delta i_0' = \underline{62.00 \text{ A}} \tag{3}$$

Now the subtransient current $|\mathbf{I}_F''|$ is calculated with the aid of Eq. (7.43) and Eq. (2) as

$$|\mathbf{I}_F''| = |\mathbf{I}_{AC}| + \frac{\Delta i_0'' + \Delta i_0'}{\sqrt{2}} = \frac{250}{\sqrt{2}} + \frac{66+62}{\sqrt{2}} = \underline{267.29 \text{ A}} \tag{4}$$

Similarly, the transient current $|\mathbf{I}_F'|$ is calculated with the aid of Eq. (7.44) and Eq. (3) as

$$|\mathbf{I}_F'| = |\mathbf{I}_{AC}| + \frac{\Delta i_0'}{\sqrt{2}} = \frac{250}{\sqrt{2}} + \frac{62}{\sqrt{2}} = \underline{220.62 \text{ A}} \tag{5}$$

Finally, the steady state current $|\mathbf{I}_F|$ is calculated with the aid of Eq. (7.45) and input data as

$$|\mathbf{I}_F| = |\mathbf{I}_{AC}| = \frac{250}{\sqrt{2}} = \underline{176.78 \text{ A}} \tag{6}$$

Once the generated voltage $|\mathbf{E}_G|$ and the subtransient current $|\mathbf{I}_F''|$, the transient current $|\mathbf{I}_F'|$, and the steady state current $|\mathbf{I}_F|$ are known, the subtransient reactance X_d'', the transient reactance X_d', and the synchronous reactance X_d are calculated with the aid of Eqs. (6.316) through (6.318), Eq. (1) and Eqs. (4) through (6), respectively. Thus

$$X_d'' = \frac{|\mathbf{E}_G|}{|\mathbf{I}_F''|} = \frac{141.42}{267.29} = \underline{\underline{0.53\ \Omega}} \tag{7}$$

$$X_d' = \frac{|\mathbf{E}_G|}{|\mathbf{I}_F'|} = \frac{141.42}{220.62} = \underline{\underline{0.64\ \Omega}} \tag{8}$$

and

$$X_d = \frac{|\mathbf{E}_G|}{|\mathbf{I}_F|} = \frac{141.42}{176.78} = \underline{\underline{0.80\ \Omega}} \tag{9}$$

If the envelopes of $i_F''(t)$ and $i_F'(t)$, Fig. 6.46(c), do not decay exponentially, then the excess $\Delta i_V \equiv i_F'' - i_F$ plotted on a logarithmic scale, with $\omega_E t$ plotted on a linear scale, does not yield a straight line as suggested in Fig. 7.8. Instead, Δi_V versus $\omega_E t$ yields a curve as in Fig. 7.9. Note that the rapid decrement of the first few cycles in an actual oscillogram is neglected, as suggested by the dashed portion of the curve. The intersection with the Δi-axis, however, can be determined very accurately by the following two-step procedure. First, the straight-line portion of the Δi_V versus $\omega_E t$ curve is extended to the Δi-axis. This Δi-axis intersection is designated as $\Delta i_0'$ and the dashed straight line designates the $\Delta i'$ versus $\omega_E t$ of Fig. 7.8 and thus the exponentially decaying envelope $i_F'(t)$ (ignoring steady state) of Fig. 6.46. Second, the excess $\Delta i'' \equiv \Delta i_V - \Delta i'$ is plotted as a function of $\omega_E t$, Fig. 7.9. When the line $\Delta i''$ versus $\omega_E t$ is extended to the Δi-axis, the resulting intersection is designated as $\Delta i_0''$. When $\Delta i_0''$ is added to $\Delta i_0'$, the sum yields the Δi-axis intercept

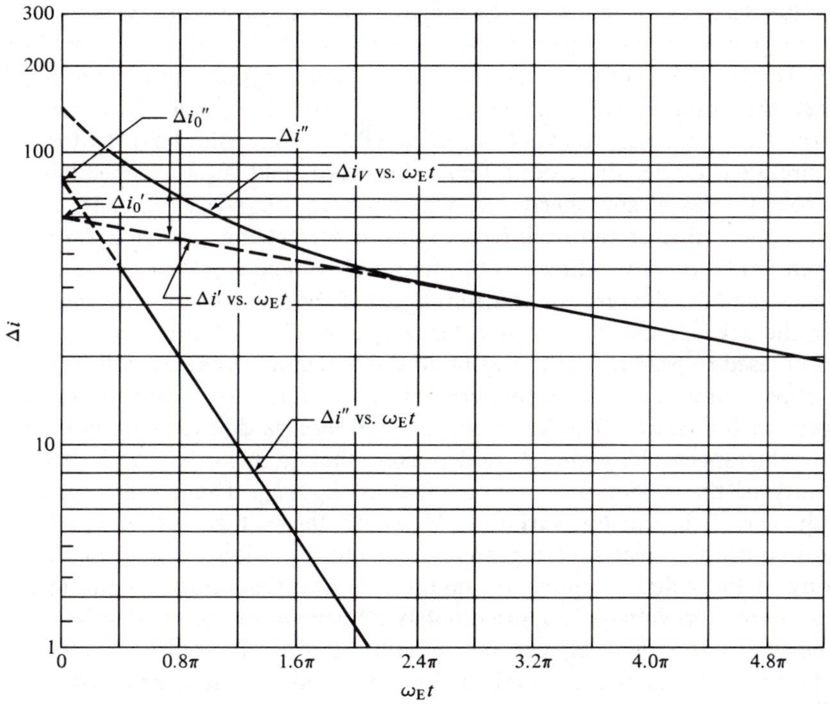

Fig. 7.9 The Excesses, $\Delta i'' \equiv \Delta i_V - \Delta i'$ and $\Delta i_V \equiv i_F'' - i_F$, vs. Angle for the Current Envelopes of Fig. 6.46(c) Are Plotted on Semilogarithmic Scales: Nonexponential Decay

of the Δi_V versus $\omega_E t$ curve. Note that once the Δi-axis intersections, that is, $\Delta i_0'$ and $\Delta i_0''$, are determined, the theory associated with Eqs. (7.43) through (7.45) is valid.

Usually the resistance R_S of the stator winding is **ignored** when modeling the salient-rotor synchronous device. If, however, greater accuracy is desired, the resistance R_S, as measured utilizing Eq. (7.42), is included in the model of the device at hand.

It should be remembered that the current of Fig. 6.46(c) is the current that exists in the synchronous generator operating at **no-load** before the fault occurs. In addition, it should be remembered that in a faulted system the generated current $\mathbf{I}_G$ is comprised only of the component that **lags** the generated voltage at no-load by 90°, Fig. 6.44, that is, $|\mathbf{I}_G| \to I_d$.

7.5 MEASUREMENT OF THE STATOR COIL *q*-AXIS REACTANCES

When the q-axis reactances of the synchronous device are required, special techniques are employed to ensure that only the q-axis magnetic and electric circuits are utilized. There are two common methods used for the measurement of X_q. They are (1) **the slip test** and (2) **the maximum-lagging-current test.**

Slip Test

For this test the device is run as a **lightly loaded synchronous generator** with (1) balanced polyphase voltages of the type $e_{Sj}(t) = 0.2\sqrt{2}\,|\mathbf{E}_S|\cos(\omega_E t \pm q120°)$, $j = A, B, C$ and $q = 0, 1, 2$, of the correct phase sequence applied to the stator windings and (2) the rotor driven mechanically at a speed ω_R *slightly lower* than the synchronous speed ω_S, that is, $\omega_R < \omega_E/(P/2)$ with the field (rotor) winding open-circuited. Under these conditions, the stator field $|\mathbf{B}_{ST}|$ glides slowly past the field (rotor) poles at slip speed $s\omega_S$, where $s = (\omega_R - \omega_S)/\omega_S$. Fig. 7.10 exhibits the general appearance of the oscillograms taken for (1) the voltage $e_F(t)$ induced in the open field (rotor) winding, (2) the terminal voltage $e_{SA}(t)$ of the stator winding marked A, and (3) the response current $i_{SA}(t)$ of the same stator winding.

For the sake of clarity, a much larger value of slip frequency is shown than would be used in practice, as the value of the reactance obtained will be too low for large slips. Similarly, the modulation of the voltage and current waveforms is exaggerated for clarity. The large current variation is due to the inequality of the device reluctances along the d- and q-axes, that is, $\mathcal{R}_d < \mathcal{R}_q$, and therefore, the inequality of the device inductances, that is, $L_d > L_q$. The voltage variation is a consequence of the current variation. However, the voltage variation, which is due to the line drop, is less severe than the current variation because of the inherent property of the voltage supply to remain constant. **One must not fail to note here that the excitation voltage is approximately 20 percent of the rated voltage $|\mathbf{E}_S|$.**

When the stator field $|\mathbf{B}_{ST}|$ is in line with the axes of the field (rotor) poles, the impedance of the device is equal to the d-axis value. One-quarter of a slip cycle later, Fig. 7.10, the stator field $|\mathbf{B}_{ST}|$ is in line with the q-axis, and the impedance is equal to the q-axis value. If the stator-winding resistance R_S is neglected, the d-axis

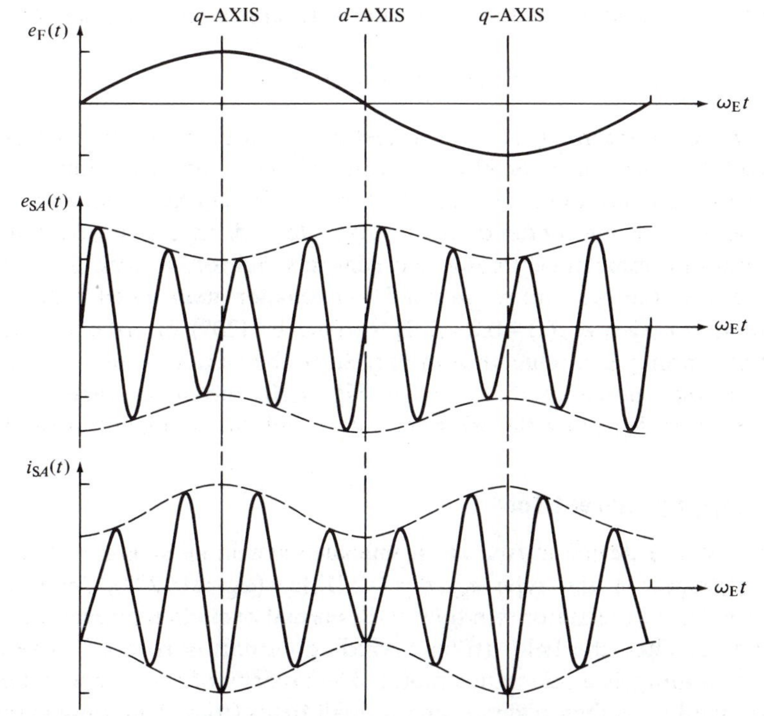

Fig. 7.10 Experimentally Obtainable Oscillograms of (1) The Open Field (Rotor) Winding Voltage $e_F(t)$, (2) the Stator-Winding Impressed Voltage $e_{SA}(t)$, and (3) the Response Current $i_{SA}(t)$ in the Stator Winding of Phase A. Note That the Frequency of Excitation Is Not Drawn to Scale

unsaturated synchronous reactance X_{dU} equals the **maximum ratio** of the per-phase terminal voltage $e_{SA}(t)$ of the phase marked A to the response current $i_{SA}(t)$ of the same phase. This occurs when the instantaneous voltage $e_F(t)$ induced in the field (rotor) winding is in its **zero value**. That is,

$$X_{dU} \equiv \left. \frac{e_{SAMAX}}{i_{SAMIN}} \right|_{e_F(t) \equiv 0} \tag{7.46}$$

Similarly, the q-axis unsaturated synchronous reactance X_{qU} is the **minimum ratio** of the per-phase terminal voltage $e_{SA}(t)$ to the response current $i_{SA}(t)$ of the same phase. This occurs when the instantaneous voltage $e_F(t)$ induced in the field (rotor) winding is in its **maximum value**. That is,

$$X_{qU} \equiv \left. \frac{e_{SAMIN}}{i_{SAMAX}} \right|_{e_F(t) \equiv \text{MAXIMUM}} \tag{7.47}$$

Although voltage or current meters may be used to obtain the corresponding values for the variables of Eqs. (7.46) and (7.47), their **pointers are subject to overswing**, with resulting error in their readings. To avoid such errors **oscillograms** of these variables should be taken.

Best results can be obtained by first finding the ratio X_{qU}/X_{dU} from the slip test, that is, Eqs. (7.47) and (7.46), respectively, and then use the value of X_σ, $\sigma = \text{d}$, obtained from the **open-circuit, short-circuit**, and **zero power-factor** tests, Eq. (7.30), to calculate the unsaturated value of the q-axis synchronous reactance X_q. That is,

$$X_q = \frac{X_{qU}}{X_{dU}} \times X_\sigma, \sigma = d \qquad (7.48)$$

The principal shortcoming of the slip test is that large errors may be produced by the effects of currents induced in the rotor circuits, such as damper windings, unless the slip s is made to be very small. However, it may be difficult to meet this condition because of the tendency of the rotor to lock into step and run as if it were the rotor of a reluctance* motor that runs in synchronism with the stator field $|\mathbf{B}_{ST}|$. Therefore, the test must be made at **smaller than rated** values of the stator-winding voltages $e_{SA}(t) = 0.2\sqrt{2}\,|\mathbf{E}_S|\cos(\omega_E t \pm q120°)$ in order to reduce the likelihood of running as a reluctance motor. Since the values of the reactances X_{dU} and X_{qU} depend on the oscillograms of Fig. 7.10, the accuracy of the results depends to a great degree on the accuracy with which the oscillograms can be read.

Maximum-Lagging-Current Test

For this test the device is run as an **unloaded synchronous motor** with normal polyphase voltages of the type $e_{SA}(t) = \sqrt{2}\,|\mathbf{E}_S|\cos(\omega_E t \pm q120°)$, $j = A, B, C$ and $q = 0, 1, 2$, applied to the stator windings with **normal excitation** applied to the field (rotor) winding. Then the field- (rotor-) winding current is reduced to zero. Now the device is running as a reluctance motor. The polarity of the dc excitation in the field (rotor) winding is then reversed and a small field- (rotor-) winding current I_F^- is caused to appear, having the opposite direction when compared to the normal current I_F^+ that existed originally. The presence of this negative field- (rotor-) winding current I_F^- causes an increase in the response currents of the stator windings $i_{SA}(t)$, $i_{SB}(t)$, and $i_{SC}(t)$. By increasing the field- (rotor-) winding current I_F in small increments, the **maximum stable stator-winding current** $|\mathbf{I}_{SA}| \equiv |\mathbf{I}_{SAMAX}|$ is found. If the field- (rotor-) winding current I_F were to increase beyond the value that causes the maximum stable stator-winding current $|\mathbf{I}_{SAMAX}|$, the device would fall out of step momentarily. However, it would pull back into synchronism after *skipping* one pole. It is shown in reference [3] that the q-axis synchronous reactance is given as

$$X_q \equiv \left. \frac{|\mathbf{E}_{SA}|}{|\mathbf{I}_{SA}|} \right|_{|\mathbf{I}_{SA}| \equiv |\mathbf{I}_{SAMAX}|} \qquad (7.49)$$

In Eq. (7.49), $|\mathbf{E}_{SA}|$ is the RMS value of the impressed voltage in phase A and $|\mathbf{I}_{SA}| \equiv |\mathbf{I}_{SAMAX}|$ is the RMS value of the response current in phase A of the stator winding when the device is on the point of, but has not yet started, skipping poles.

The maximum-lagging-current test is **preferred** to the slip test because it can be made at **rated impressed voltages** $e_{SA}(t) = \sqrt{2}\,|\mathbf{E}_S|\cos(\omega_E t \pm q120°)$, $j = A, B, C$ and $q = 0, 1, 2$, and yields the **saturated value** of the q-axis synchronous reactance X_q. Remember that the saturated value of the d-axis synchronous reactance is obtained utilizing Eq. (7.30) and the theory associated with it.

Before leaving this section it should be mentioned that the q-axis subtransient and transient reactances X_q'' and X_q', respectively, analogous to the d-axis subtransient and transient reactances X_d'' and X_d', respectively, can also be obtained experimentally by performing (1) the **modified slip test** and (2) the **stationary rotor test with short-circuited field (rotor) winding.**

*For the development of the reluctance motor, please see Section 5.10.

Modified Slip Test

With the rotor magnetized in the q-axis, that is, at the instant of maximum field- (rotor-) winding voltage $e_F(t)$, positive or negative, Fig. 7.10, the excitation voltages $e_{SA}(t) = \sqrt{2}\,|\mathbf{E}_S|\cos(\omega_E t \pm q120°)$, $j = A, B, C$ and $q = 0, 1, 2$, applied to the stator windings are suddenly disconnected and the **decay of the terminal voltages** of the device is recorded oscillographically. Utilization of the process used to obtain Fig. 7.8 enables us to extrapolate these terminal voltages to the Δe-axis, ignoring the first few cycles of rapid decay. If $|\mathbf{E}_{SA0}|$ designates the extrapolated response voltage of phase A and $|\mathbf{E}_{SA}|$ designates the impressed voltage of phase A just before the stator windings are opened, then according to reference [3], the q-axis transient reactance X_q' is given by Eq. (7.50) as

$$X_q' = \frac{|\mathbf{E}_{SA}| - |\mathbf{E}_{SA0}|}{|\mathbf{E}_{SA}|} \times X_q \tag{7.50}$$

where X_q is the q-axis synchronous reactance obtained by utilizing the **slip test**. It should be obvious, therefore, that this method of evaluating X_q' suffers the drawbacks of the slip test.

Stationary Rotor Test with Short-Circuited Field (Rotor) Winding

A single-phase voltage, that is, $e_{SA}(t) = \sqrt{2}\,|\mathbf{E}_S|\cos\omega_E t$, is applied **across two of the stator terminals**, that is, **across two of the phases of a Y-connected synchronous device connected in series**. When the rotor is slowly rotated until the induced field- (rotor-) winding current $i_F(t)$ is a minimum, i_{FMIN}, the axis of the pulsating stator winding mmf $\mathcal{F}_{\mathcal{Q}}$ is aligned with the q-axis. The per-phase subtransient synchronous reactance of the q-axis, X_q'', is then obtained by taking one-half the ratio of the RMS value of the impressed voltage $|\mathbf{E}_{SA}|$ to the RMS value of the response current $|\mathbf{I}_{SA}|$ in the two windings connected in series, when $i_F(t) \equiv i_{FMIN}$. That is,

$$X_q'' = 0.5 \times \left.\frac{|\mathbf{E}_{SA}|}{|\mathbf{I}_{SA}|}\right|_{i_F(t) \equiv i_{FMIN}} \tag{7.51}$$

Note that the same process can be used to obtain the subtransient synchronous reactance of the d-axis, X_d''. However, in this case the rotor position is adjusted so that the induced field- (rotor-) winding current $i_F(t)$ is a maximum, i_{FMAX}. Under these conditions the per-phase subtransient reactance of the d-axis, X_d'', is given by Eq. (7.52). That is,

$$X_d'' = 0.5 \times \left.\frac{|\mathbf{E}_{SA}|}{|\mathbf{I}_{SA}|}\right|_{i_F(t) \equiv i_{FMAX}} \tag{7.52}$$

7.6 MEASUREMENT OF PERTINENT TIME CONSTANTS

Before closing this chapter it should be pointed out that there is a *time constant* associated with each subtransient and transient reactance of the d- and q-axes. These time constants are defined as* (1) $\tau_d'' \equiv$ **d-axis, short-circuit, subtransient time constant**, (2) $\tau_d' \equiv$ **d-axis, short-circuit, transient time constant**, (3) $\tau_q'' \equiv$ **q-axis,**

*In the literature the letter "T" is used with appropriate subscripts to designate these time constants rather than the letter τ.

Table 7.2 TYPICAL PARAMETER VALUES FOR SYNCHRONOUS DEVICES

	Generators			Motors	
	Smooth-Rotor				Synchronous
Parameter	Solid	Laminated	Salient-Rotor	Salient-Rotor	Condenser
X_d''	0.10	0.10	0.23	0.35	0.25
X_d'	0.20	0.20	0.35	0.50	0.60
X_d	1.10	1.10	1.00	1.10	1.60
τ_d''	0.035	0.035	0.035	0.035	0.035
τ_d'	1.00	1.00	1.80	1.40	2.00
τ_d	0.15	0.15	0.15	1.50	0.15

short-circuit, subtransient time constant, and (4) $\tau_q' \equiv$**q-axis, short-circuit, transient time constant**. These time constants are *measures of the decay* of the corresponding envelopes, and each is equal to the time required for the corresponding envelope to decay to $1/e$, or 0.368, of its initial value, Fig. 7.8. **These time constants are utilized in transient analysis of synchronous devices, a subject beyond the scope of this text.**

In addition to the above four time constants, there is a fifth time constant defined as $\tau_a \equiv$**armature time constant.** This time constant is equal to the time required for the dc component of the fault current $i_V(t)$, Fig. 6.46(a), to decay to $1/e$, or 0.368, of its initial value.

Typical values of the most important parameters for different types of synchronous devices are exhibited in Table 7.2. The reactances are given in per-unit ohms, with **the rating of each device used as base**. Reasonable variation on either side of these values may be expected for any particular synchronous device.

Example 7.5

Utilize the data and other relevant information of Example 7.4 and calculate: (1) the d-axis, short-circuit, subtransient time constant τ_d'' (or T_d'') and (2) the d-axis, short-circuit, transient time constant τ_d' (or T_d'). How would you calculate the armature time constant τ_a (or T_a)?

Solution

Inspection of Fig. 7.8 yields the $\omega_E t$ values when $\Delta i = 0.368\ \Delta i_0'$ and $\Delta i = 0.368\ \Delta i_0''$: (1) $\omega_E t \equiv \omega_E \tau_d'' = 1.386\pi$ and (2) $\omega_E t \equiv \omega_E \tau_d' = 4.94\pi$, respectively. The d-axis subtransient time constant τ_d'' (or T_d'') is obtained from the first reading as follows:*

$$\tau_d'' \text{ (or } T_d'') = \frac{1.386\pi}{\omega_E} = \frac{1.386\pi}{2\pi \times 60} = \underline{\underline{0.01155 \text{ s}}} \tag{1}$$

Similarly, the d-axis transient time constant τ_d' (or T_d') is obtained from the second reading as follows:

$$\tau_d' \text{ (or } T_d') = \frac{4.94\pi}{\omega_E} = \frac{4.94\pi}{2\pi \times 60} = \underline{\underline{0.04177 \text{ s}}} \tag{2}$$

*Note that this value does not correspond to that of Table 7.2, the reason being that the hypothetical waveform of Fig. 6.46(c) is used only to demonstrate the procedure.

To calculate the armature time constant τ_a (or T_a), we must have the waveform of the stator current of a given phase offset by the dc component. This curve would be similar to that of Fig. 6.46(a). If we were to define $\Delta i_{DC} = i_V - i_{AC}$ and plot Δi_{DC} versus $\omega_E t$ on the scales of Fig. 7.8, following the procedures of Example 7.4 for the plots of $\Delta i''$ versus $\omega_E t$ and $\Delta i'$ versus $\omega_E t$, we would be able to calculate the Δi-axis intercept, designated as Δi_0^{DC}. Reflection of 0.368 Δi_0^{DC} off the Δi_{DC} versus $\omega_E t$ curve would yield $\omega_E t \equiv \omega_E \tau_a = k_a$.

The armature time constant τ_a (or T_a) would be obtained as follows:

$$\tau_a \text{ (or } T_a) = \frac{k_a}{\omega_E} = \underline{\underline{\frac{k_a}{2\pi \times f_E} \text{ s}}} \tag{3}$$

7.7 SUMMARY

In this chapter a concerted effort is made to provide the student with the capability to measure (1) the circuit parameters $\mathbf{E}_G$, R_S, and X_S for the model of the smooth-rotor synchronous device and (2) the circuit parameters $\mathbf{E}_{\mathscr{G}}$, X_d, X_d', X_d'', X_q, X_q', and X_q'' for the model of the salient-rotor synchronous device. In addition, the same effort is made to familiarize the student with the technique of measuring the time constants associated with the dynamic modeling of synchronous devices; that is, (1) the d-axis, short-circuit, subtransient time constant τ_d'', (2) the d-axis, short-circuit, transient time constant τ_d', and (3) the armature time constant τ_a.

In closing, again, it is noted that the theory set forth here is substantiated with the results of five numerical examples worked out in detail.

PROBLEMS

7.1 The open-circuit and short-circuit characteristics of a Y-connected, three-phase, 24-pole, 60-Hz, 75-kVA, 450-V (line-to-line), synchronous generator are exhibited in Fig. 7.1.1. For rated voltage and power factor 0.85 LGG, calculate: (1) the linearized (approximate) saturated synchronous reactance X_σ^L, (2) the linearized generated voltage $|\mathbf{E}_\xi^L|$, (3) the field- (rotor-) winding current $I_{\mathscr{F}}^L$, (4) the generated voltage $|\mathbf{E}_\xi|$, and (5) the voltage regulation VR.

7.2 The open-circuit, short-circuit, and zero power-factor characteristics of a Y-connected, three-phase, eight-pole, 60-Hz, 300-V (line-to-line), 100-kVA, synchronous generator are exhibited in Fig. 7.2.1. The numerical divisions of the three-scales, (1) $0 \leqslant |\mathbf{I}_{G(SC)}| \leqslant 900$ A, (2) $0 \leqslant |\mathbf{V}_{L(OC)}| \leqslant 450$ V, and (3) $0 \leqslant |I_F| \leqslant 5$ A, are essential for completing this problem. For a rated phase terminal voltage $|\mathbf{V}_{L(RTD)}| = 200$ V and a minimum field- (rotor-) winding current of $I_{\mathscr{F}0} = 2$ A required to initiate the ZPFC-curve, calculate: (1) the unsaturated synchronous reactance $X_{\sigma U}$, (2) the leakage or Potier reactance X_ℓ, utilizing the already constructed Potier triangle, (3) the saturation factor k, (4) the saturated synchronous reactance X_σ, and (5) the short-circuit ratio SCR.

7.3 For the synchronous generator of Problem 7.2 at rated stator-winding current $|\mathbf{I}_{G(SC)}| = 180$ A and at the temperature of 30°C, the total short-circuit losses P_{SCLL}^T are equal to 4.2 kW while $P_{CL}(\Phi_{S\ell})$ and $P_{CL}(\Phi_{\mathscr{R}})$ are negligible. The dc resistance of the stator winding at this temperature is $R_{S(DC)} = 0.04$ Ω/phase. Calculate: (1) the stator winding resistance R_S in ohms/phase, (2) the stator winding resistance R_S in per-unit ohms/phase, (3) the total dc losses P_{DCT}, and (4) the total losses $P_{SE} + P_{EC}$ due to the skin effect and eddy currents. Repeat parts (3) and (4) if it is known that the stator-winding composition is copper, its operating temperature is 60°C, and the total short-circuit load losses P^T_{SCLL} do not change.

7.4 For a faulted 80-kVA, 10-pole, 50-Hz, three-phase, salient-rotor, Y-connected, synchronous generator, an oscillogram of the fault current of phase *A*, resembling that of Fig. 7.4.1, was recorded in the laboratory. Calculate: (1) the subtransient reactance X_d'', (2) the transient reactance X_d', and (3) the synchronous reactance X_d of the given synchronous generator.

7.5 Utilize the data and other relevant information of Problem 7.4 and calculate: (1) the *d*-axis, short-circuit, subtransient time constant τ_d'' (or T_d'') and (2) the *d*-axis, short-circuit, transient time constant τ_d' (or T_d'). How would you calculate the armature time constant τ_a (or T_a)?

REFERENCES

1. Del Toro, V., *Electromechanical Devices for Energy Conversion and Control Systems*. Prentice-Hall, Englewood Cliffs, N.J., 1968.
2. Elgerd, O. I., *Electric Energy Systems Theory: An Introduction*. McGraw-Hill, New York, 1971.
3. Fitzgerald, A. E., and Kingsley, C., Jr., *Electric Machinery*, 2d ed. McGraw-Hill, New York, 1961.
4. Gourishankar, V., and Kelly, D. H., *Electromechanical Energy Conversion*, 2d ed. Intext, New York, 1973.
5. Kosow, I. L., *Electric Machinery and Control*. Prentice-Hall, Englewood Cliffs, N.J., 1964.
6. Lansdorf, A. S., *Theory of Alternating Current Machinery*. McGraw-Hill, New York, 1955.
7. Majmudar, H., *Electromechanical Energy Converters*. Allyn & Bacon, Boston, 1965.
8. Matsch, L. W., *Electromagnetic and Electromechanical Machines*, Intext, New York, 1972.
9. McFarland, T. C., *Alternating Current Machines*. Van Nostrand, New York, 1948.
10. Neuenswander, J. W., *Modern Power Systems*. Intext, New York, 1971.
11. O'Kelly, D., and Simons, S., *Introduction to Generalized Electrical Machine Theory*. McGraw-Hill, London, 1968.
12. Stevenson, W. D., Jr., *Elements of Power Systems Analysis*, 3d ed. McGraw-Hill, New York, 1975.
13. Thompson, H. A., *Alternating Current and Transient Circuit Analysis*. McGraw-Hill, New York, 1955.
14. Weedy, B. M., *Electric Power Systems*. Wiley, London, 1972.

A FREQUENCY-DOMAIN ANALYSIS

A.1 INTRODUCTION

The objectives of this appendix are (1) to provide the fundamental definitions of complex numbers and complex quantities, (2) to set forth the principles governing the mathematical operations of complex numbers, and (3) to set forth the concepts of impedance, admittance, complex power, and average and effective values of the essential quantities used in the field of power systems engineering.

A.2 COMPLEX NUMBERS

There are three fundamental reference frames commonly encountered in power engineering work: (1) the one-dimensional, (2) the two-dimensional, and (3) the three-dimensional reference frames.

1. The one-dimensional reference frame is defined by the axis of the real numbers, symbolized by the x-axis (Re-axis). Pictorially it is represented by a horizontal line extending from minus infinity to plus infinity, on which the appropriate units are marked off on either side of the origin, that is, the center of the x-axis (Re-axis) indicated as the 0-point, Fig. A.1. Any real number is represented by a point on this axis. Thus the real number 5, for example, in Fig. A.1 represents the distance from the point in question to the origin of the one-dimensional reference frame.

2. The complex (two-dimensional) reference frame is defined by two axes, that is, the axis of the real numbers, represented by the Re-axis, and the axis of the imaginary numbers, represented by the Im-axis, intersecting each other at 90°. The

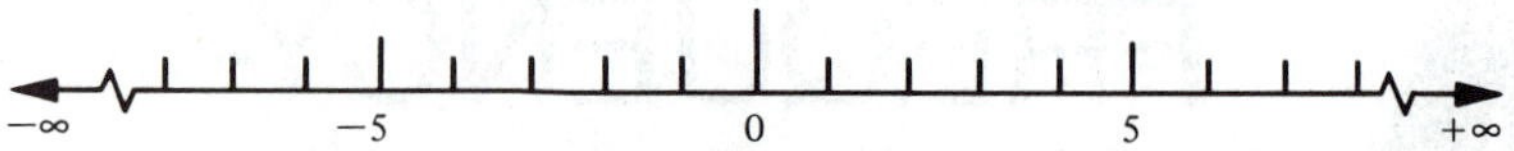

Fig. A.1 One-Dimensional Reference Frame

position of the Im-axis is represented by the operator $j \equiv (0,1)$. This operator locates the Im-axis 90° ahead of the Re-axis. The intersection of two lines defines a plane. Thus the intersection of the Re- and Im-axes defines the **complex plane**, Fig. A.2.

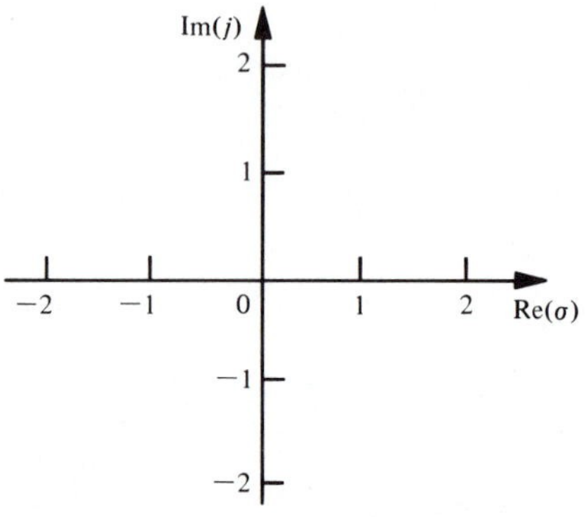

Fig. A.2 Two-Dimensional Reference Frame

3. The three-dimensional reference frame is defined by three axes, that is, the σ-, $j\omega$-, and $|\mathbf{Z}|$-, or $|\mathbf{Y}|$-axes, intersecting each other at 90° and constituting a right-handed system— $\sigma \times j\omega = |\mathbf{Z}|$ or $|\mathbf{Y}|$, Fig. A.3, that is, the cross product of the vectors σ and $j\omega$. This frame enables plotting of $|\mathbf{Z}|$ or $|\mathbf{Y}|$ versus ω, $\sigma = k_i$ or $|\mathbf{Z}|$ or $|\mathbf{Y}|$ versus σ, $\omega = k_j$ $i = j = 1, 2, \ldots, n$.

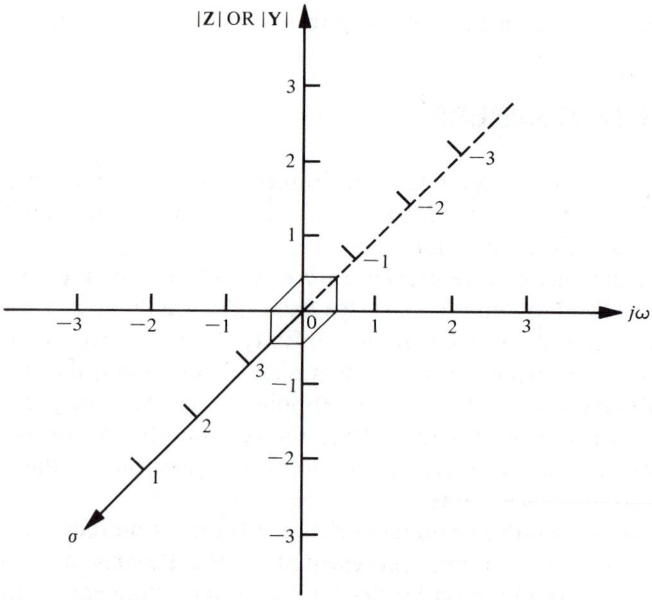

Fig. A.3 Three-Dimensional Reference Frame

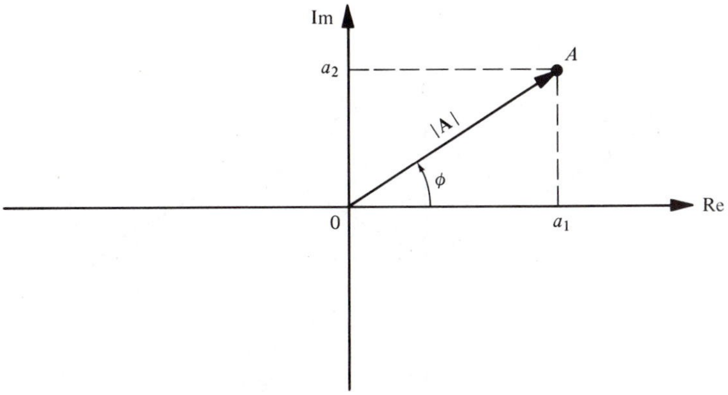

Fig. A.4 Complex Number Representation

Any complex number **A** is represented by a point A on the complex plane, Fig. A.4. The point A can be located if the following information is available: (1) *the length of line segment* $OA \equiv |\mathbf{A}|$ and (2) the *angle* of departure of the line segment, that is, *the angle* ϕ *that* OA *makes with the Re-axis measured positive in the counterclockwise direction*. The point A could also be located if the Re- and Im-axes projections of the point A, that is, the line segments $0a_1$ and $0a_2$, are available. With either of these two sets of information available, the complex number **A** is defined as follows:

$$\mathbf{A} \equiv |\mathbf{A}|e^{j\phi} \text{ (exponential form)} \tag{A.1}$$

$$\mathbf{A} \equiv a_1 + ja_2 \text{ (rectangular form)} \tag{A.2}$$

By expanding the exponential in a MacLaurin series

$$\begin{aligned}\mathbf{A} &= |\mathbf{A}|e^{j\phi} \\ &= |\mathbf{A}|\left[\frac{(j\phi)^0}{0!} + \frac{(j\phi)^1}{1!} + \frac{(j\phi)^2}{2!} + \frac{(j\phi)^3}{3!} + \frac{(j\phi)^4}{4!} + \frac{(j\phi)^5}{5!} + \cdots\right] \\ &= |\mathbf{A}|\left[1 + j\phi + j^2\left(\frac{\phi^2}{2!}\right) + j^3\left(\frac{\phi^3}{3!}\right) + j^4\left(\frac{\phi^4}{4!}\right) + j^5\left(\frac{\phi^5}{5!}\right) + \cdots\right]\end{aligned} \tag{A.3}$$

The j-operator represents a 90° rotation. Thus

$$j \cdot j = (-1,0) \tag{A.4}$$

Eq. (A.4) suggests a 180° rotation of a linear segment of magnitude 1, that is, $\mathbf{A} = 1e^{j\phi}$. Thus, keeping in mind that each j represents a 90° positive rotation of a linear segment enables us to write Eq. (A.3) as

$$\begin{aligned}\mathbf{A} &= |\mathbf{A}|e^{j\phi} \\ &= |\mathbf{A}|\left[\left(1 - \frac{\phi^2}{2!} + \frac{\phi^4}{4!} + \cdots\right) + j\left(\phi - \frac{\phi^3}{3!} + \frac{\phi^5}{5!} + \cdots\right)\right]\end{aligned} \tag{A.5}$$

These can be recognized as the series expansion of sine and cosine, which enables us to write Eq. (A.5) as

$$\mathbf{A} = |\mathbf{A}|e^{j\phi} = |\mathbf{A}|\cos\phi + j|\mathbf{A}|\sin\phi \tag{A.6}$$

Comparison of Eqs. (A.1), (A.2), and (A.6) yields a relationship between the two sets of variables $|\mathbf{A}|$ and ϕ, and a_1 and a_2. Thus if one set of these variables is known, the other can be determined very easily. Using trigonometric properties

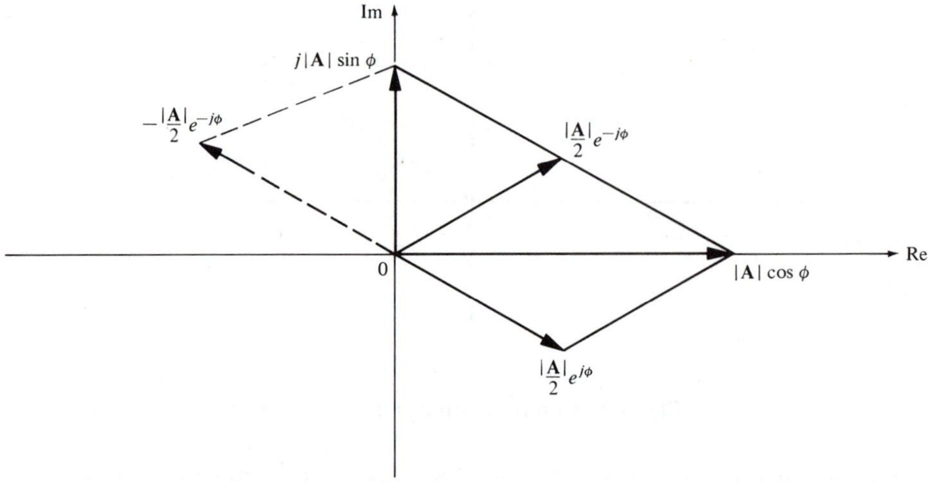

Fig. A.5 Complex Plane Representation of Sinusoidal Quantities as Functions of the Exponential Quantities $e^{j\phi}$ and $e^{-j\phi}$

and the Pythagorean theorem yields

$$a_1 = |\mathbf{A}|\cos\phi \tag{A.7}$$

$$a_2 = |\mathbf{A}|\sin\phi \tag{A.8}$$

or

$$|\mathbf{A}| = \left[a_1^2 + a_2^2\right]^{1/2} \tag{A.9}$$

$$\phi = \tan^{-1}\left(\frac{a_2}{a_1}\right)^{\dagger} \tag{A.10}$$

Substituting $-\phi$ in Eq. (A.6) yields

$$\mathbf{A} = |\mathbf{A}|e^{-j\phi} = |\mathbf{A}|\cos\phi - j|\mathbf{A}|\sin\phi \tag{A.11}$$

Addition of Eqs. (A.6) and (A.11) yields

$$2|\mathbf{A}|\cos\phi = |\mathbf{A}|e^{j\phi} + |\mathbf{A}|e^{-j\phi}$$

or

$$|\mathbf{A}|\cos\phi = \frac{|\mathbf{A}|e^{j\phi}}{2} + \frac{|\mathbf{A}|e^{-j\phi}}{2} \tag{A.12}$$

Subtraction of Eqs. (A.6) and (A.11) yields

$$j2|\mathbf{A}|\sin\phi = |\mathbf{A}|e^{j\phi} - |\mathbf{A}|e^{-j\phi}$$

or

$$j|\mathbf{A}|\sin\phi = \frac{|\mathbf{A}|e^{j\phi}}{2} + \left(\frac{-|\mathbf{A}|e^{-j\phi}}{2}\right) \tag{A.13}$$

Thus Eqs. (A.12) and (A.13) enable expression of the sinusoidal quantities $|\mathbf{A}|\cos\phi$ and $j|\mathbf{A}|\sin\phi$ in terms of the complex numbers (exponential quantities) $|\mathbf{A}|e^{\pm j\phi}/2$. These relationships are exhibited graphically in Fig. A.5. These two equations enable us to write any sinusoidal forcing function (of time with any

† Actually, a four quadrant inverse, $\tan^{-1}(a_2, a_1)$ is required.

phase angle) as the sum of two exponential functions (of time) with imaginary exponents and with complex coefficients.

Example A.1

Given $\mathbf{B}_{11}=4+j3$, find the exponential form of the complex number $\mathbf{B}_{11}$. (See Fig. A.1.1.)

$$|\mathbf{B}_{11}| = [4^2+3^2]^{1/2} = [16+9]^{1/2} = [25]^{1/2} = \underline{5}$$

$$\phi = \tan^{-1}(\tfrac{3}{4}) = \tan^{-1}(0.75) = \underline{36.8°}$$

$$\underline{\underline{\mathbf{B}_{11}=5e^{j36.8°}}}$$

Example A.2

Given $\mathbf{B}_{33}=-4-j3$, find the exponential form of the complex number $\mathbf{B}_{33}$.

$$|\mathbf{B}_{33}| = [(-4)^2+(-3)^2]^{1/2} = [16+9]^{1/2} = [25]^{1/2} = \underline{5}$$

$$\phi = \tan^{-1}\left[\frac{-3}{-4}\right] = 180° + \tan^{-1}(\tfrac{3}{4}) = 180° + 36.9° = \underline{216.9°}$$

Note: In the process of the evaluation of the angle of the complex number $\mathbf{B}_{33}$, attention should be paid to the fact that point $\mathbf{B}_{33}$ is located in the third quadrant of the complex plane. It should be emphasized here that a point B can be located in any of the quadrants. Under those circumstances, the magnitude would remain

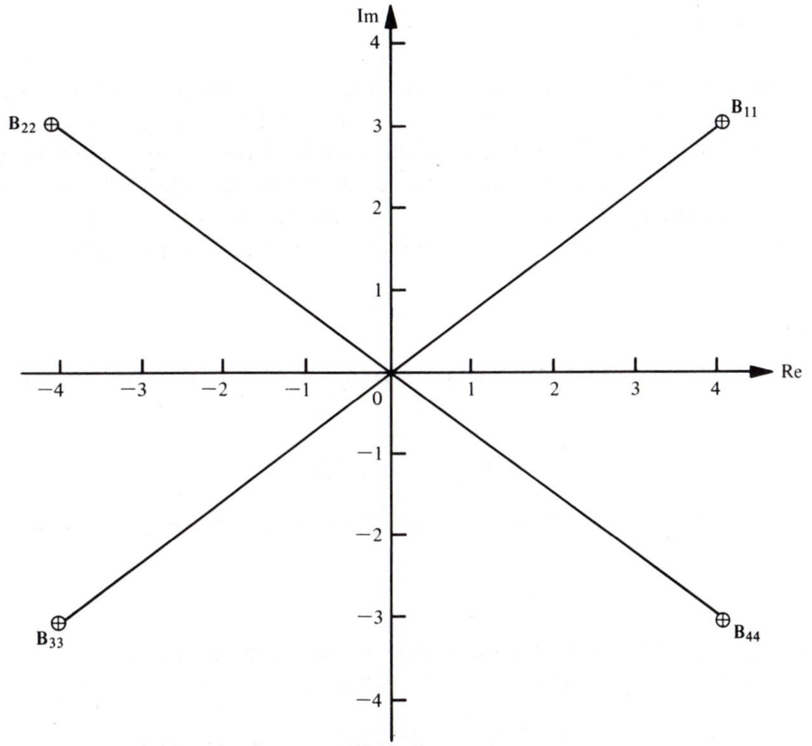

Fig. A.1.1 Magnitude and Angle of the Complex Numbers $\mathbf{B}_{11}$, $\mathbf{B}_{22}$, $\mathbf{B}_{33}$ and $\mathbf{B}_{44}$ Resulting from the Motion of Point B But Occupying the Same Relative Position at the First, Second, Third and Fourth Quadrants of the Complex Plane

the same but the angle would change appropriately. Thus

$$\text{3d quadrant:} \qquad \mathbf{B}_{33} = -4 - j3 = \underline{\underline{5e^{j216.9°}}}$$

$$\text{2d quadrant:} \qquad \mathbf{B}_{22} = -4 + j3 = \underline{\underline{5e^{j143.1°}}}$$

$$\text{4th quadrant:} \qquad \mathbf{B}_{44} = 4 - j3 = \underline{\underline{5e^{j323.1°}}}$$

Example A.3

Given $\mathbf{A} = 10e^{j45°}$, find the rectangular form of the complex number **A**.

$$a_1 = 10\cos 45° = 10(0.707) = \underline{7.07}$$

$$a_2 = 10\sin 45° = 10(0.707) = \underline{7.07}$$

$$\underline{\underline{\mathbf{A} = 7.07 + j7.07}}$$

Example A.4

Given $\mathbf{B} = 100e^{j156°}$, find the rectangular form of the complex number **B**.

$$a_1 = 100\cos 156° = 100\cos(180° - 24°) = -100\cos 24° = \underline{-91.4}$$

$$a_2 = 100\sin 156° = 100\sin(180° - 24°) = 100\sin 24° = \underline{40.7}$$

$$\underline{\underline{\mathbf{A} = -91.4 + j40.7}}$$

The transformation from the rectangular form to the exponential form, and vice versa, of complex numbers can be readily simplified by the use of the natural trigonometric function tables and a pictorial sketch of the complex number one is working with. The reason is that these tables would yield angles $0° \leqslant \phi \leqslant 90°$ and always measured off the Re-axis. This fact necessitates the location of the complex number at hand and to the appropriate quadrant before any calculations are pursued.

Example A.5

Given $\mathbf{C} = 100e^{j225°}$, find the rectangular form of **C** using the trigonometric function tables and Fig. A.5.1.

$$\mathbf{C} = -70.7 - j70.7$$

Notice that *unless* the quadrant is established first, the calculations will yield erroneous results.

Example A.6

Given $i(t) = |\mathbf{I}|\cos(\omega_E t + \alpha)$, find the exponential form of $i(t)$.
Utilization of Eq. (A.12) and Fig. A.5 enables us to write

$$i(t) = |\mathbf{I}|\cos(\omega_E t + \alpha) = \underline{\underline{\frac{|\mathbf{I}|e^{j(\omega_E t + \alpha)}}{2} + \frac{|\mathbf{I}|e^{-j(\omega_E t + \alpha)}}{2}}}$$

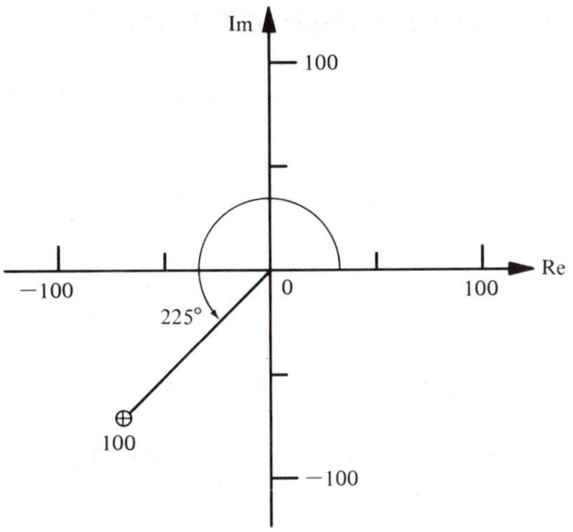

Fig. A.5.1 Quadrant Location of the Complex Number $\mathbf{C} = 100e^{j225°}$ with Its Relative Position, with Respect to the Re-axis, Explicitly Exhibited

A.3 SCIENTIFIC CALCULATOR USE FOR EXPONENTIAL-RECTANGULAR AND RECTANGULAR-EXPONENTIAL TRANSFORMATION

Examples A.1 through A.6 demonstrate the mathematical process of transforming the exponential form of complex numbers to the rectangular form and vice versa. These transformations, however, are **automatic**, as far as **signs** and **angle magnitudes** are concerned, when one uses any of the available scientific calculators:

1. Exponential-rectangular conversion: The objective here is to convert a given number from its exponential form, $|\mathbf{A}|e^{j\phi}$, to its rectangular form, $a_1 + ja_2$. The transformation is based on Fig. A.4 and Eq. (A.6), and Eqs. (A.7) and (A.8); it entails simply entering the available information $|\mathbf{A}|$ and ϕ into the calculator, performing the suggested mathematical operation, and reading the resulting information correctly.
2. Rectangular-exponential conversion: Here the objective is to convert a given number from its rectangular form, $a_1 + ja_2$, to its exponential form, $|\mathbf{A}|e^{j\phi}$. The transformation is based on Fig. A.4 and Eq. (A.6), and Eqs. (A.9) and (A.10); it entails simply entering the available information a_1 and a_2 into the calculator (the appropriate signs should be included in these numbers), performing the suggested mathematical operations, and reading the resulting information correctly.

Note: Depending on the type of scientific calculator available, use of the appropriate **code numbers and key combinations** enables the operator to obtain the forms of the number he or she is seeking automatically. The operator should therefore become familiar with the most efficient conversion capability of the scientific calculator.

A.4 TIME VARIATION OF COMPLEX NUMBERS

Multiplying $|\mathbf{A}|e^{j\phi}$ by $e^{j\omega_E t}$ gives a time-varying complex number.

$$\begin{aligned}\mathbf{A}(t) &= |\mathbf{A}|e^{j\phi}e^{j\omega_E t} \equiv |\mathbf{A}|e^{j(\phi+\omega_E t)}\\ &\equiv |\mathbf{A}|\cos(\omega_E t+\phi)+j|\mathbf{A}|\sin(\omega_E t+\phi)\end{aligned} \tag{A.14}$$

where

$$|\mathbf{A}|\cos(\omega_E t+\phi) = \mathrm{Re}\{|\mathbf{A}|e^{j(\omega_E t+\phi)}\} \tag{A.15}$$

and

$$|\mathbf{A}|\sin(\omega_E t+\phi) = \mathrm{Im}\{|\mathbf{A}|e^{j(\omega_E t+\phi)}\} \tag{A.16}$$

Premultiplying and dividing $|\mathbf{A}|$ by $\sqrt{2}$ and opening the parenthesis of the exponentials in Eq. (A.15) and Eq. (A.16) yields

$$\mathbf{A}(t) = \mathrm{Re}\left\{\sqrt{2}\left(\frac{|\mathbf{A}|}{\sqrt{2}}\right)e^{j\phi}e^{j\omega_E t}\right\} + \mathrm{Im}\left\{\sqrt{2}\left(\frac{|\mathbf{A}|}{\sqrt{2}}\right)e^{j\phi}e^{j\omega_E t}\right\} \tag{A.17}$$

or

$$\mathbf{A}(t) = \mathrm{Re}\{\sqrt{2}\,\mathbf{A}e^{j\omega_E t}\} + \mathrm{Im}\{\sqrt{2}\,\mathbf{A}e^{j\omega_E t}\} \tag{A.18}$$

where

$$\mathbf{A} = \left(\frac{|\mathbf{A}|}{\sqrt{2}}\right)e^{j\phi} \tag{A.19}$$

is defined as the **sinor** or the **phasor** of the time-varying complex number. The quantity $|\mathbf{A}|/\sqrt{2}$ is the RMS value of the magnitude of the time-varying complex quantity. This quantity is of importance because it is measured by instruments designed to measure RMS volts or currents. The angle ϕ is the angle of this quantity or the angle of the time-varying quantity at $t=0$.

These ideas enable writing sinusoidal or cosinusoidal waveforms encountered in the field of electrical engineering in terms of real or imaginary projections. Thus when we are confronted with $v_1(t) = V_m\cos(\omega_E t+\phi)$ or $v_2 = V_m\sin(\omega_E t+\phi)$, we can always write

$$\begin{aligned}v_1(t) &= \mathrm{Re}\{V_m e^{j(\omega_E t+\phi)}\} = \mathrm{Re}\left\{\sqrt{2}\left(\frac{V_m}{\sqrt{2}}\right)e^{j\phi}e^{j\omega_E t}\right\}\\ &= \mathrm{Re}\{\sqrt{2}\,\mathbf{V}_1 e^{j\omega_E t}\}\end{aligned} \tag{A.20}$$

and

$$\begin{aligned}v_2(t) &= \mathrm{Im}\{V_m e^{j(\omega_E t+\phi)}\} = \mathrm{Im}\left\{\sqrt{2}\left(\frac{V_m}{\sqrt{2}}\right)e^{j\phi}e^{j\omega_E t}\right\}\\ &= \mathrm{Im}\{\sqrt{2}\,\mathbf{V}_2 e^{j\omega_E t}\}\end{aligned} \tag{A.21}$$

or writing $v_2(t) = V_m\cos(\omega_E t+\phi-90°) \equiv v_3(t)$ yields

$$\begin{aligned}v_3(t) &= \mathrm{Re}\{V_m e^{j(\omega_E t+\phi-90°)}\} = \mathrm{Re}\left\{\sqrt{2}\left(\frac{V_m}{\sqrt{2}}\right)e^{j(\phi-90°)}e^{j\omega_E t}\right\}\\ &= \mathrm{Re}\{\sqrt{2}\,\mathbf{V}_3 e^{j\omega_E t}\}\end{aligned} \tag{A.22}$$

where

$$\mathbf{V}_1 = \left(\frac{V_m}{\sqrt{2}}\right)e^{j\phi} \equiv |\mathbf{V}_1|e^{j\phi} \tag{A.23}$$

$$\mathbf{V}_2 = \left(\frac{V_m}{\sqrt{2}}\right)e^{j\phi} \equiv |\mathbf{V}_2|e^{j\phi} \tag{A.24}$$

$$\mathbf{V}_3 = \left(\frac{V_m}{\sqrt{2}}\right)e^{j(\phi-90°)} \equiv |\mathbf{V}_3|e^{j(\phi-90°)} \tag{A.25}$$

and $|\mathbf{V}_1|=|\mathbf{V}_2|=|\mathbf{V}_3|\equiv|\mathbf{V}|$. The difference among $\mathbf{V}_1$, $\mathbf{V}_2$, and $\mathbf{V}_3$ is not in the magnitude but in the phase angle. That is, $\mathbf{V}_1$ and $\mathbf{V}_3$ are projections on the axis of the real numbers, while $\mathbf{V}_2$ is a projection on the axis of the imaginary numbers. Thus **when transforming back to the time domain, it is essential that the appropriate axis of projection be reintroduced.**

Example A.7

Given $v_1(t)=141\sin(1000t+53.6°)$, find the phasor referred to (1) the imaginary and (2) the real axis.

1.
$$v_1(t)=\text{Im}\left\{\sqrt{2}\left(\frac{141}{\sqrt{2}}\right)e^{j53.6°}e^{j1000t}\right\}$$

$$\underline{\underline{\mathbf{V}_{1\text{Im}}=100e^{j53.6°}}}$$

2.
$$\begin{aligned} v_1(t) &= 141\cos(1000t+53.6°-90°) \\ &= 141\cos(1000t-36.4°) \\ &= \text{Re}\{\sqrt{2}\,100e^{-j36.4°}e^{j1000t}\} \end{aligned}$$

$$\underline{\underline{\mathbf{V}_{1\text{Re}}=100e^{-j36.4°}}}$$

A.5 OPERATIONS ON COMPLEX NUMBERS

For $\mathbf{A}=a_1+ja_2=|\mathbf{A}|e^{j\phi}$, where

$$\begin{aligned} |\mathbf{A}| &= \left[a_1^{\,2}+a_2^{\,2}\right]^{1/2} \\ \phi &= \tan^{-1}\left(\frac{a_2}{a_1}\right) \\ a_1 &= |\mathbf{A}|\cos\phi \\ a_2 &= |\mathbf{A}|\sin\phi \end{aligned}$$

the fundamental mathematical operations are defined as given in the following paragraphs.

Addition. If $\mathbf{B}=b_1+jb_2$ and $\mathbf{C}=c_1+jc_2$, then

$$\mathbf{A}\equiv\mathbf{B}+\mathbf{C}\equiv(b_1+c_1)+j(b_2+c_2)\equiv|\mathbf{A}|e^{j\phi_\mathbf{A}}$$

Example A.8

$$\begin{aligned} \mathbf{B} &= 5+j6 \\ \mathbf{C} &= 10e^{j45°} = 7.07+j7.07 \end{aligned}$$

$$\underline{\underline{\mathbf{A}\equiv\mathbf{B}+\mathbf{C}=12.07+j13.07}}$$

Subtraction. If $\mathbf{B}=b_1+jb_2$ and $\mathbf{C}=c_1+jc_2$, then

$$\mathbf{D} \equiv \mathbf{B}-\mathbf{C} \equiv (b_1-c_1)+j(b_2-c_2) \equiv |\mathbf{D}|e^{j\phi_\mathbf{D}}$$

Example A.9

$$\mathbf{B} = 5+j6$$
$$\mathbf{C} = 10e^{j45°} = 7.07+j7.07$$
$$\mathbf{D} = \mathbf{B}-\mathbf{C} = -2.07-j1.07$$

Thus **for addition and subtraction, the complex numbers should be expressed in the** *rectangular form* and the rule can be stated as follows: *Addition or subtraction of complex numbers is achieved by adding or subtracting corresponding components*, that is, $\mathbf{B}\pm\mathbf{C}=(b_1\pm c_1)+j(b_2\pm c_2)$.

Multiplication. If $\mathbf{B}=b_1+jb_2$ and $\mathbf{C}=c_1+jc_2$, then

$$\mathbf{P} \equiv \mathbf{B}\cdot\mathbf{C} \equiv |\mathbf{B}|e^{j\phi_\mathbf{B}}|\mathbf{C}|e^{j\phi_\mathbf{C}} = |\mathbf{B}||\mathbf{C}|e^{j(\phi_\mathbf{B}+\phi_\mathbf{C})} \equiv |\mathbf{P}|e^{j\phi_\mathbf{P}} \equiv p_1+jp_2$$

Example A.10

$$\mathbf{A} = 70.7+j70.7 = 100e^{j45°}$$
$$\mathbf{B} = 7.07+j7.07 = 10e^{+j45°}$$
$$\mathbf{P} = \mathbf{A}\cdot\mathbf{B} = 100e^{j45°}\cdot 10e^{j45°} = \underline{\underline{1000e^{j90°}}}$$

Note: Multiplication of complex numbers expressed in rectangular form is sometimes preferable, especially when one of the numbers is a pure real or pure imaginary number or when one of the numbers is predominantly real or imaginary.

Division. If $\mathbf{B}=b_1+jb_2$ and $\mathbf{C}=c_1+jc_2$, then

$$\mathbf{Q} \equiv \frac{\mathbf{B}}{\mathbf{C}} \equiv \frac{|\mathbf{B}|e^{j\phi_\mathbf{B}}}{|\mathbf{C}|e^{j\phi_\mathbf{C}}} \equiv \left(\frac{|\mathbf{B}|}{|\mathbf{C}|}\right)e^{j(\phi_\mathbf{B}-\phi_\mathbf{C})} \equiv |\mathbf{Q}|e^{j\phi_\mathbf{Q}} \equiv q_1+jq_2$$

Example A.11

$$\mathbf{B} = 70.7+j70.7 = 100e^{j45°}$$
$$\mathbf{C} = 7.07+j7.07 = 10e^{j45°}$$
$$\mathbf{Q} = \frac{\mathbf{B}}{\mathbf{C}} = \frac{100e^{j45°}}{10e^{j45°}} = \underline{\underline{10e^{j0°}}}$$

Exponentiation. If $\mathbf{B}=b_1+jb_2=|\mathbf{B}|e^{j\phi_\mathbf{B}}$, then

$$\mathbf{B}^n = (|\mathbf{B}|e^{j\phi_\mathbf{B}})^n = |\mathbf{B}|^n e^{jn\phi_\mathbf{B}}$$

or

$$\mathbf{B}^{1/k} = (|\mathbf{B}|e^{j\phi_\mathbf{B}})^{1/k} = |\mathbf{B}|^{1/k}e^{j\phi_\mathbf{B}/k}$$

Example A.12

$$\mathbf{B}=70.7+j70.7=100e^{j45°}$$

1. $$\mathbf{B}^2 = (100e^{j45°})^2 = \underline{\underline{10000e^{j90°}}}$$

2. $$\sqrt{\mathbf{B}} = (100e^{j45°})^{1/2} = \underline{\underline{10e^{j22.5°}}}$$

Conjugation. The conjugate of the complex number $\mathbf{C}=|\mathbf{C}|e^{j\phi_C}$ is defined as $\mathbf{C}^*=|\mathbf{C}|e^{-j\phi_C}$.

Note: In rectangular form the conjugate of a complex number is obtained by replacing j by $-j$; that is, if $\mathbf{C}=2+j8$, $\mathbf{C}^*=2-j8$.

Example A.13

If $\mathbf{C}=70.7-j70.7=100e^{-j45°}$, then

$$\mathbf{C}^* = 100e^{+j45°} = \underline{\underline{70.7+j70.7}}$$

As a consequence of the conjugation, two special types of operations result.

1. $$\mathbf{A}\cdot\mathbf{A}^*=|\mathbf{A}|e^{j\phi_A}|\mathbf{A}|e^{-j\phi_{A^*}}=|\mathbf{A}|^2e^{j0°}$$

2. $$\frac{\mathbf{A}}{\mathbf{A}^*}=\frac{|\mathbf{A}|e^{j\phi_A}}{|\mathbf{A}|e^{-j\phi_{A^*}}}=\left(\frac{|\mathbf{A}|}{|\mathbf{A}|}\right)e^{j(\phi_A+\phi_{A^*})}=1e^{j2\phi_A}$$

Note that for the operations of multiplication, division, and exponentiation, the complex numbers should be expressed in the exponential form.

A.6 SINUSOIDAL WAVEFORMS

The exponential $|\mathbf{A}|e^{\pm j(\omega_E t+\phi)}$ is usually interpreted in terms of the physical model (with no actual physical significance) of a rotating vector, the direction of rotation being determined by the sign of the exponent. *A positive sign implies counterclockwise (or positive) rotation*, while *a negative sign implies clockwise (or negative) rotation.* For positive rotation the real part of $|\mathbf{A}|e^{j(\omega_E t+\phi)}$ (or the projection on the real axis) varies as the cosine of $\omega_E t$, while the imaginary part (or projection on the imaginary axis) varies as the sine of $\omega_E t$. These ideas are illustrated in Fig. A.6 for

Plate A.1 Panel Mounted Instruments for the Measurement of: (a) the Electric Frequency f_E(Hz) and (b) the Mechanical Frequency [Expressed in Revolutions Per Minute] n_R(RPM). For the Synchronous Device, for Example, These Two Variables are Related Through the Equation $n_R=(120f_E)/P$. (Courtesy of General Electric Co., Schenectady, N.Y.)

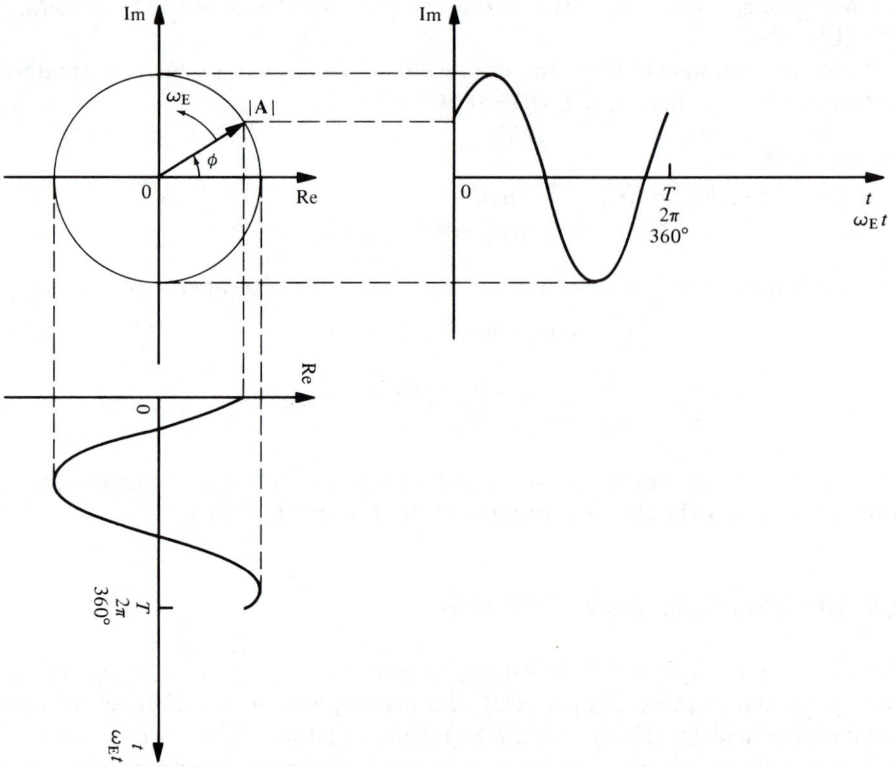

Fig. A.6 A Rotating Vector **A** with Projections on the Re- and Im-axes as Functions of t and $\omega_E t$

the following conditions: at $t=0$ the phase angle is ϕ and the vector $|\mathbf{A}|$ is rotating in the positive direction at an angular velocity ω_E radians/second. Note that when the vector $|\mathbf{A}|$ is along the Re- and Im-axes, it provides the maximum and minimum amplitudes of the sine or cosine waveforms. In general, these maxima and minima do not occur at $t=0$, $T/4$, T and $t=T/2, 3T/4$ as they would if $\phi=0°$.

Fig. A.6 leads to the following definitions pertaining to the sinusoidal waveforms:

1. Cycle: a complete set of all positive and negative values of a Re( ) versus t or Im( ) versus t plot of any variable (or a span of 2π radians or 360 electrical degrees).
2. Frequency: the number of cycles repeated per second. The unit of measurement of the frequency symbolized by f_E is cycles per second, abbreviated as Hz.
3. Period: time required for one cycle of a given waveform. The period is related to the frequency of the waveform by the following equation:

$$T = \frac{1}{f_E} \text{ seconds}$$

4. Angular velocity: the ratio of the radian displacement that corresponds to one cycle, that is, 2π radians over the period. The angular velocity is related to the frequency of the waveform by the following equation:

$$\omega_E = \frac{\Delta\theta}{\Delta T} = \frac{2\pi}{T} = 2\pi f_E \text{ radians/second}$$

Thus utilizing the relations above, $v(t) = V_m \cos(\omega_E t \pm \phi)$ can be written as

$$v(t) = \sqrt{2}\left(\frac{V_m}{\sqrt{2}}\right)\cos\left[\left(\frac{2\pi}{T}\right)t \pm \phi\right] \equiv \sqrt{2}\,|\mathbf{V}|\cos\left[\left(\frac{2\pi}{T}\right)t \pm \phi\right]$$

$$= \sqrt{2}\left(\frac{V_m}{\sqrt{2}}\right)\cos(2\pi f_E t \pm \phi) \equiv \sqrt{2}\,|\mathbf{V}|\cos(2\pi f_E t \pm \phi)$$

A.7 NETWORK CONSIDERATIONS: IMPEDANCE, ADMITTANCE, COMPLEX POWER

Consider the network shown in Fig. A.7. The response current of this network is of similar nature to the forcing function but its magnitude and phase angle are unknown; that is, the general form of the current will be $i(t) = I_m \cos(\omega_E t + \beta)$. It should be noted that the forcing function is explicitly defined; that is, $e(t) = E_m \cos(\omega_E t + \alpha)$ has E_m, ω_E, and α defined in a given problem. However, here the forcing function will be kept in its general form $e(t) = E_m \cos(\omega_E t + \alpha)$ because the generalized theory is being developed.

Utilizing Kirchhoff's voltage law and working on the Re-axis, keeping in mind that Re is an operator and not a variable, leads to (following the rules of integration and differentiation of exponential functions; that is, j is treated as a constant)

1. $$e(t) = \text{Re}\left[\sqrt{2}\left(\frac{E_m}{\sqrt{2}}\right)e^{j\alpha}e^{j\omega_E t}\right] \tag{A.26}$$

2. $$v_R(t) = R\left\{\text{Re}\left[\sqrt{2}\left(\frac{I_m}{\sqrt{2}}\right)e^{j\beta}e^{j\omega_E t}\right]\right\} \tag{A.27}$$

3. $$v_L(t) = L\frac{d}{dt}\text{Re}\left[\sqrt{2}\left(\frac{I_m}{\sqrt{2}}\right)e^{j\beta}e^{j\omega_E t}\right]$$

$$= j\omega_E L\left\{\text{Re}\left[\sqrt{2}\left(\frac{I_m}{\sqrt{2}}\right)e^{j\beta}e^{j\omega_E t}\right]\right\} \tag{A.28}$$

4. $$v_C(t) = \frac{1}{C}\int_0^t \text{Re}\left[\sqrt{2}\left(\frac{I_m}{\sqrt{2}}\right)e^{j\beta}e^{j\omega_E \tau}\right]d\tau + v_C(0)$$

$$= \left[\frac{1}{j\omega_E C}\left\{\text{Re}\left[\sqrt{2}\left(\frac{I_m}{\sqrt{2}}\right)e^{j\beta}e^{j\omega_E t}\right]\right\} - v_C(0)\right] + v_C(0) \tag{A.29}$$

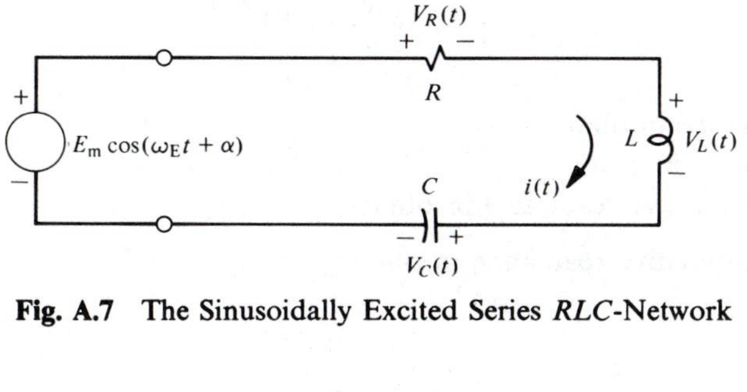

Fig. A.7 The Sinusoidally Excited Series *RLC*-Network

Thus Kirchhoff's law with variables referred to the Re-axis is written as

$$\mathrm{Re}\left[\frac{E_{\mathrm{m}}}{\sqrt{2}}\,e^{j\alpha}e^{j\omega_{\mathrm{E}}t}\right] = R\left\{\mathrm{Re}\left[\sqrt{2}\left(\frac{I_{\mathrm{m}}}{\sqrt{2}}\right)e^{j\beta}e^{j\omega_{\mathrm{E}}t}\right]\right\}$$
$$+j\omega_{\mathrm{E}}L\left\{\mathrm{Re}\left[\sqrt{2}\left(\frac{I_{\mathrm{m}}}{\sqrt{2}}\right)e^{j\beta}e^{j\omega_{\mathrm{E}}t}\right]\right\}$$
$$+\frac{1}{j\omega_{\mathrm{E}}C}\left\{\mathrm{Re}\left[\sqrt{2}\left(\frac{I_{\mathrm{m}}}{\sqrt{2}}\right)e^{j\beta}e^{j\omega_{\mathrm{E}}t}\right]\right\} \tag{A.30}$$

Note that Re, $\sqrt{2}$, and $e^{j\omega_{\mathrm{E}}t}$ appear in every term of Eq. (A.30). Therefore, *cancellation* of all three quantities will not change the equation. However, as these terms are canceled (removed), it is expedient *to store* them in such a way as to be able to retrieve them at will (so, remember where you placed Re, $\sqrt{2}$, and $e^{j\omega_{\mathrm{E}}t}$). Thus Eq. (A.30) is written as

$$\left(\frac{E_{\mathrm{m}}}{\sqrt{2}}\right)e^{j\alpha} = R\left(\frac{I_{\mathrm{m}}}{\sqrt{2}}\right)e^{j\beta} + j\omega_{\mathrm{E}}L\left(\frac{I_{\mathrm{m}}}{\sqrt{2}}\right)e^{j\beta} + \frac{1}{j\omega_{\mathrm{E}}C}\left(\frac{I_{\mathrm{m}}}{\sqrt{2}}\right)e^{j\beta} \tag{A.31}$$

since

$$\left(\frac{E_{\mathrm{m}}}{\sqrt{2}}\right)e^{j\alpha} = |\mathbf{E}|e^{j\alpha} \equiv \mathbf{E}$$

and

$$\left(\frac{I_{\mathrm{m}}}{\sqrt{2}}\right)e^{j\beta} = |\mathbf{I}|e^{j\beta} \equiv \mathbf{I}$$

Eq. (A.31) can be written as

$$\mathbf{E} = R\mathbf{I} + j\omega_{\mathrm{E}}L\mathbf{I} + \frac{1}{j\omega_{\mathrm{E}}C}\mathbf{I} \tag{A.32}$$

Multiplying numerator and denominator of the last term of Eq. (A.32) by j and recalling that $j^2 = -1$ leads to

$$\mathbf{E} = \left[R + j\omega_{\mathrm{E}}L - j\left(\frac{1}{\omega_{\mathrm{E}}C}\right)\right]\mathbf{I} = \left[R + j\left(\omega_{\mathrm{E}}L - \frac{1}{\omega_{\mathrm{E}}C}\right)\right]\mathbf{I} \tag{A.33}$$

Dividing Eq. (A.33) by $\mathbf{I}$ and examining the dimensional units of the resulting equation, that is, volts/ampere, yields

$$\frac{\mathbf{E}}{\mathbf{I}} = R + j\left(\omega_{\mathrm{E}}L - \frac{1}{\omega_{\mathrm{E}}C}\right) \equiv \mathbf{Z} = \left(\frac{|\mathbf{E}|}{|\mathbf{I}|}\right)e^{j(\alpha-\beta)} = |\mathbf{Z}|e^{j\theta_{\mathbf{Z}}} \tag{A.34}$$

where $\mathbf{Z}$ is a complex ohmic quantity defined as *impedance*. Thus

$$\mathbf{Z} = R + j\left(\omega_{\mathrm{E}}L - \frac{1}{\omega_{\mathrm{E}}C}\right) \tag{A.35}$$

where

$\mathbf{Z}$ = impedance in ohms
R = resistance in ohms
$\omega_{\mathrm{E}}L \equiv X_L$ = inductive reactance in ohms
$\dfrac{1}{\omega_{\mathrm{E}}C} \equiv X_C$ = capacitive reactance in ohms

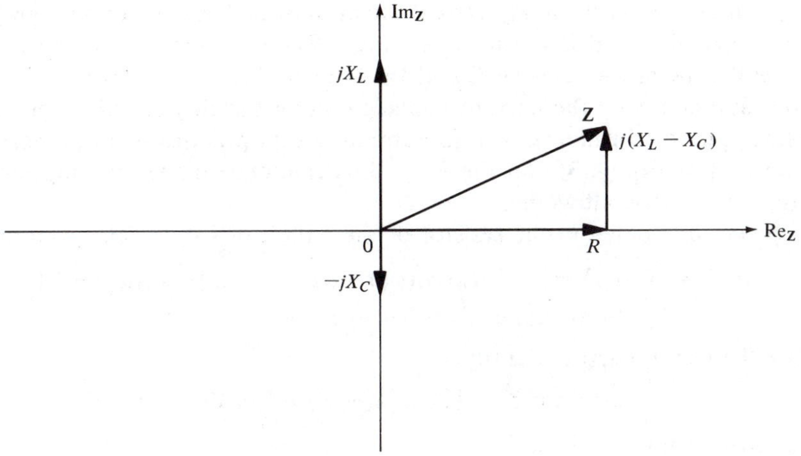

Fig. A.8 Plot of the Generalized Impedance **Z** on the Complex Plane

The complex quantity **Z** can be plotted on the complex plane as shown in Fig. A.8. Multiplying **Z** by the current, that is, $\mathbf{Z}=\mathbf{E}/\mathbf{I}$, yields

$$\mathbf{E} = \mathbf{ZI} = (R+jX)\mathbf{I} \equiv [R+j(X_L-X_C)]\mathbf{I} = \mathbf{V}_R + j(\mathbf{V}_L - \mathbf{V}_C) \equiv \mathbf{V}_R + j\mathbf{V}_X \tag{A.36}$$

Since

$$\mathbf{I} = \left(\frac{I_m}{\sqrt{2}}\right)e^{j\beta} \equiv |\mathbf{I}|e^{j\beta}$$

it is thought that the angle of the current is measured from a complex reference frame which is located at $-\beta$ with respect to the reference frame of Fig. A.8. This is exhibited in Fig. A.9. However, since in the series *RLC*-network, $i(t)$ is common to all elements, **I** is used as the reference. Thus **I** is set along the axis of the real, that is, along Re.

Note that Eq. (A.34) yields the angle of the impedance as $\theta_{\mathbf{Z}} = \alpha - \beta$. Thus lining up the current referred to the current-voltage complex plane and the real axis of the impedance, and therefore lining up with R, is automatic. These ideas suggest that we can use the current of the series *RLC*-network as a reference, that is, let

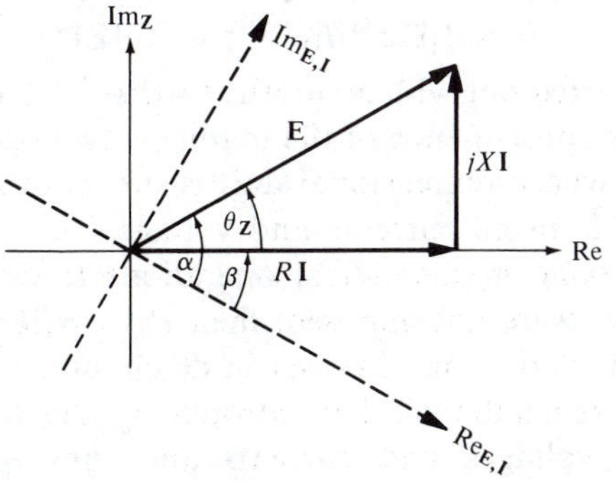

Fig. A.9 Relative Positions of the Impedance and the Voltage–Current Reference Frames and Plots of the Excitation Voltage **E**, Impedance **Z**, Response Current **I** and Complex Power **W** for the Series *RLC*-Network of Fig. A.7

$\mathbf{I}=|\mathbf{I}|e^{j0°}$ and carry on the analysis as if the current had always a phase angle equal to zero. When the load is defined, the angle $\theta_{\mathbf{Z}}=\alpha-\beta=\tan^{-1}(X/R)$ is known; consequently, the angle between $\mathbf{E}$ and $\mathbf{I}$ is known. Also, when defining the forcing function, automatically the current is placed along the $\mathrm{Re}_{\mathbf{Z}}$-axis $\theta_{\mathbf{Z}}$ degrees *behind* the voltage phasor. In fact, this is so automatic that practically all authors exhibit the quantities of Eq. (A.36) on the $\mathrm{Re}_{\mathbf{Z}}-\mathrm{Im}_{\mathbf{Z}}$-reference plane ignoring the $\mathrm{Re}_{\mathbf{E,I}}$–$\mathrm{Im}_{\mathbf{E,I}}$-reference plane altogether.

The power dissipated in the resistor of the circuit of Fig. A.7 is given as

$$\begin{aligned} p(t) &= e(t)i(t) = \left[\sqrt{2}\,|\mathbf{E}|\cos(\omega_{\mathrm{E}}t+\alpha)\right]\times\left[\sqrt{2}\,|\mathbf{I}|\cos(\omega_{\mathrm{E}}t+\beta)\right] \\ &= 2|\mathbf{E}||\mathbf{I}|\cos(\omega_{\mathrm{E}}t+\alpha)\cos(\omega_{\mathrm{E}}t+\beta) \end{aligned} \tag{A.37}$$

Utilizing the trigonometric identity

$$\cos X\cos Y \equiv \tfrac{1}{2}\left[\cos(X+Y)+\cos(X-Y)\right]$$

enables writing Eq. (A.37) as

$$p(t) = |\mathbf{E}||\mathbf{I}|\cos(\alpha-\beta) + |\mathbf{E}||\mathbf{I}|\cos\left[2\omega_{\mathrm{E}}t+(\alpha+\beta)\right] \tag{A.38}$$

Since this power is a time-varying quantity, it will be sensible to calculate the *average power over a period* T, that is, P_{AVG}. Thus

$$P_{\mathrm{AVG}} = \frac{1}{T}\int_0^T p(t)\,dt \tag{A.39}$$

$$= \frac{1}{T}\int_0^T \{|\mathbf{E}||\mathbf{I}|\cos(\alpha-\beta)+|\mathbf{E}||\mathbf{I}|\cos\left[2\omega_{\mathrm{E}}t+(\alpha+\beta)\right]\}\,dt \tag{A.40}$$

or

$$P_{\mathrm{AVG}} = \frac{1}{T}\int_0^T |\mathbf{E}||\mathbf{I}|\cos(\alpha-\beta)\,dt + \frac{1}{T}\int_0^T |\mathbf{E}||\mathbf{I}|\cos\left[2\omega_{\mathrm{E}}t+(\alpha+\beta)\right]dt \tag{A.41}$$

The second term of Eq. (A.41) is 0 (why?). Thus

$$P_{\mathrm{AVG}} = \frac{1}{T}\int_0^T |\mathbf{E}||\mathbf{I}|\cos(\alpha-\beta)\,dt = |\mathbf{E}||\mathbf{I}|\cos(\alpha-\beta) \tag{A.42}$$

But

$$\cos(\alpha-\beta) \equiv \mathrm{Re}\left[e^{j(\alpha-\beta)}\right] = \mathrm{Re}\left[e^{j\alpha}e^{-j\beta}\right] \tag{A.43}$$

Thus Eq. (A.42) can be written as

$$\begin{aligned} P_{\mathrm{AVG}} &= |\mathbf{E}||\mathbf{I}|\cos(\alpha-\beta) = \mathrm{Re}(|\mathbf{E}||\mathbf{I}|e^{j\alpha}e^{-j\beta}) \\ &= \mathrm{Re}(|\mathbf{E}|e^{j\alpha}|\mathbf{I}|e^{-j\beta}) = \mathrm{Re}(\mathbf{EI}^*) \end{aligned} \tag{A.44}$$

Remember that we started out with a current $\mathbf{I}=|\mathbf{I}|e^{j\beta}$. Thus $|\mathbf{I}|e^{-j\beta}\equiv\mathbf{I}^*$, defined as the *conjugate* of $\mathbf{I}$, is a consequence of the mathematical operations imposed for the calculation of P_{AVG}, whose dimensional unit is the watt. Also, it should be noted here that the term $\sqrt{2}$ in all currents and voltages was *imposed* sometime in the past because of the same mathematical operations leading to the calculation of P_{AVG}. If the term $\sqrt{2}$ were not imposed, then $P_{\mathrm{AVG}}=[|\mathbf{E}||\mathbf{I}|\cos(\alpha-\beta)]/2$. This is not consistent with the definition of power in dc circuit analysis, that is, $P=EI$. Of course, it should be remembered that introducing the term $\sqrt{2}$ leads to RMS values of sinusoidal voltages and currents and thus *measurable* voltages and currents with RMS meters.

Eq. (A.44) suggests that $\mathbf{EI}^*$ is an oriented line quantity, that is, a phasor, and therefore it has real as well as imaginary components. However, the quantity $\mathbf{EI}^*$

does not represent a sinusoidal quantity in the same manner as eq. (A.18), so it is **not** a phasor. The quantity **EI*** is defined as **complex power**. The magnitude of the complex power is defined as *apparent power*, and its dimensional unit is the voltampere. The symbol used to designate complex power is **S**. Thus

$$\mathbf{S} = \mathbf{E}\mathbf{I}^* \tag{A.45}$$

since

$$\mathbf{E} = \mathbf{Z}\mathbf{I} \tag{A.46}$$

$$\mathbf{S} = \mathbf{E}\mathbf{I}^* = \mathbf{Z}\mathbf{I}\mathbf{I}^* = \mathbf{Z}|\mathbf{I}|^2 = R|\mathbf{I}|^2 + jX_L|\mathbf{I}|^2 - jX_C|\mathbf{I}|^2 = R|\mathbf{I}|^2 + jX|\mathbf{I}|^2 \tag{A.47}$$

Dividing Eq. (A.46) by **Z** yields

$$\mathbf{E} = \left(\frac{1}{\mathbf{Z}}\right)\mathbf{E} \equiv \mathbf{Y}\mathbf{E} \tag{A.48}$$

where

$$\mathbf{Y} \equiv G + jB \tag{A.49}$$

The following definitions are pertinent:

Y = Admittance in mhos
G = conductance in mhos
B = susceptance in mhos

Taking the conjugate of the current **I** yields

$$\mathbf{I}^* = \mathbf{E}^*\mathbf{Y}^* \tag{A.50}$$

Thus

$$\mathbf{S} = \mathbf{E}\mathbf{I}^* = \mathbf{E}\mathbf{E}^*\mathbf{Y}^* = \mathbf{Y}^*|\mathbf{E}|^2 = G|\mathbf{E}|^2 - jB|\mathbf{E}|^2 = |\mathbf{S}|e^{j\theta_{\mathbf{Y}^*}} = |\mathbf{S}|\cos\theta_{\mathbf{Y}^*} + j|\mathbf{S}|\sin\theta_{\mathbf{Y}^*} \tag{A.51}$$

Also, since $\mathbf{E} = |\mathbf{E}|e^{j\alpha}$ and $\mathbf{I}^* = |\mathbf{I}|e^{-j\beta}$,

$$\begin{aligned}\mathbf{S} = \mathbf{E}\mathbf{I}^* &= |\mathbf{E}||\mathbf{I}|e^{j(\alpha-\beta)} \equiv |\mathbf{E}||\mathbf{I}|e^{j\theta_{\mathbf{Z}}}\\ &= |\mathbf{E}||\mathbf{I}|\cos\theta_{\mathbf{Z}} + j|\mathbf{E}||\mathbf{I}|\sin\theta_{\mathbf{Z}}\\ &= |\mathbf{S}|\cos\theta_{\mathbf{Z}} + j|\mathbf{S}|\sin\theta_{\mathbf{Z}} \equiv P + jQ\end{aligned} \tag{A.52}$$

Summarizing Eqs. (A.45), (A.47), (A.51), and (A.52) yields

$$\begin{aligned}\mathbf{S} = \mathbf{E}\mathbf{I}^* &\equiv P + jQ = |\mathbf{S}|\cos\theta_{\mathbf{Z}} + j|\mathbf{S}|\sin\theta_{\mathbf{Z}} = |\mathbf{S}|\cos\theta_{\mathbf{Y}^*} + j|\mathbf{S}|\sin\theta_{\mathbf{Y}^*}\\ &= \mathbf{Z}|\mathbf{I}|^2 = R|\mathbf{I}|^2 + jX|\mathbf{I}|^2\\ &= \mathbf{Y}^*|\mathbf{E}|^2 = G|\mathbf{E}|^2 - jB|\mathbf{E}|^2\end{aligned} \tag{A.53}$$

where

S = apparent power in voltamperes
P = real power or the energy dissipated in the resistor of the network over one period, that is, $P \equiv P_{\text{AVG}}$, in watts
Q = reactive power in voltamperes reactive, abbreviated as var(s). This power is not dissipated but *enters and exits* a reactive element in intervals of $0.5T$. (See Example A.14, part 12.)

These ideas can be reinforced by noting that for a current $i(t) = I_{\text{m}}\sin(\omega_{\text{E}}t + \psi), \psi = \beta + 90°$, the instantaneous powers to R, L, and C are

$$P_R(t) = Ri(t)^2 = R|\mathbf{I}|^2[1 - \cos 2(\omega_{\text{E}}t + \psi)] \tag{A.54}$$

$$P_L(t) = i(t)L\frac{d}{dt}i(t) = \omega_{\text{E}}L|\mathbf{I}|^2\sin 2(\omega_{\text{E}}t + \psi) \tag{A.55}$$

Plate A.2 Panel Mounted Power Measuring Instruments: (a) Wattmeter and (b) Varmeter. (Courtesy of General Electric Co., Schenectady, N.Y.)

and

$$P_C(t) = i(t)\frac{1}{C}\int i(t)\,dt = \left(\frac{-1}{\omega_E C}\right)|\mathbf{I}|^2 \sin 2(\omega_E + \psi) \tag{A.56}$$

It should be noted that the maximum amplitudes of $P_R(t)$, $P_L(t)$, and $P_C(t)$ are $R|\mathbf{I}|^2$, $X_L|\mathbf{I}|^2$, and $-X_C|\mathbf{I}|^2$, respectively. Note how these amplitudes compare to the power components, that is, $\mathbf{S} = R|\mathbf{I}|^2 + j[\omega_E L - (1/\omega_E C)]|\mathbf{I}|^2$, exhibited in Fig. A.9. (What can you say about parallel RLC-networks?)

Note that $\cos\theta_{\mathbf{Z}}$ or $\cos\theta_{\mathbf{Y}*}$ indicates how much of the apparent power is dissipated in the resistor, that is, $P_{\text{AVG}} = |\mathbf{S}|\cos\theta_{\mathbf{Z}} = |\mathbf{S}|\cos\theta_{\mathbf{Y}*}$. Thus, **$\cos\theta_{\mathbf{Z}}$ or $\cos\theta_{\mathbf{Y}*}$ is defined as power factor.**

$$\text{power factor (PF)} \equiv \cos\theta_{\mathbf{Z}} \equiv \cos\theta_{\mathbf{Y}*} \tag{A.57}$$

Plate A.3 Panel Mounted Power Factor Measuring Instrument—Note That the Instrument Is Calibrated to Distinguish Between Leading, Lagging, and Unity Power Factors—Also, See Pl. B.5. (Courtesy of General Electric Co., Schenectady, N.Y.)

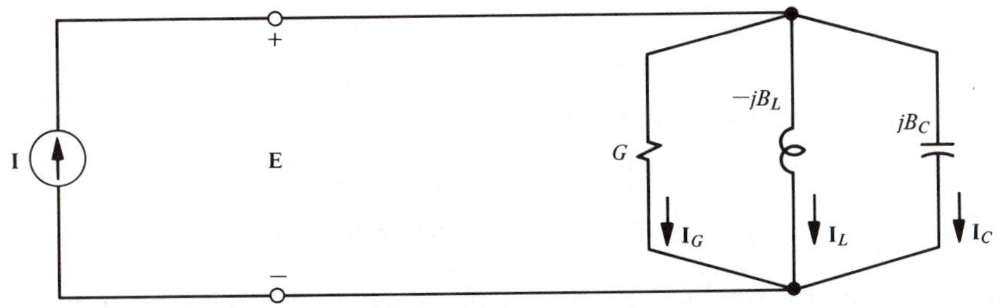

Fig. A.10 The Sinusoidally Excited Parallel *RLC*-Network

The power factor can be leading or lagging and can be adjusted to a desired value. For power-factor adjustment, please see examples A.14 and A.15, part 12, and example A.16, part 4. *The terms leading or lagging are results of the relative position of the phasors* **E** *and* **I**. **If I, in an *RLC*-network, lags E, the power factor is defined as lagging. This suggests that the network is predominantly inductive, Fig. A.8. If, however, I leads E, the power factor is defined as leading. This suggests that the network is predominantly capacitive.** This can be deduced from Fig. A.9 if $|X_C| \gg |X_L|$.

The admittance plane is more useful for elements that are connected in such a way as to have the same voltage across themselves. Such a network, driven by a sinusoidal current generator whose phasor is **I**, is given in Fig. A.10.

The use of Kirchhoff's current law and the plotting of the common quantity $\mathbf{E} = |\mathbf{E}|e^{j0°}$ along the Re-axis, Fig. A.11. yields

$$\begin{aligned}
\mathbf{I} = \mathbf{I}_G + \mathbf{I}_L + \mathbf{I}_C &= \frac{\mathbf{E}}{\mathbf{Z}_R} + \frac{\mathbf{E}}{\mathbf{Z}_L} + \frac{\mathbf{E}}{\mathbf{Z}_C} = \frac{\mathbf{E}}{R} + \frac{\mathbf{E}}{j\omega_{\mathrm{E}} L} + \frac{\mathbf{E}}{1/j\omega_{\mathrm{E}} C} \\
&= G\mathbf{E} + j\omega_E C\mathbf{E} - j\left(\frac{1}{\omega_{\mathrm{E}} L}\right)\mathbf{E} \equiv \mathbf{I}_G + j\mathbf{I}_C - j\mathbf{I}_L \\
&= \left[G + j\left(\omega_{\mathrm{E}} C - \frac{1}{\omega_{\mathrm{E}} L}\right)\right]\mathbf{E} = [G + j(B_C - B_L)]\mathbf{E} \\
&= G\mathbf{E} + jB\mathbf{E} \equiv I_{\mathrm{G}} + jI_{\mathrm{B}}
\end{aligned} \tag{A.58}$$

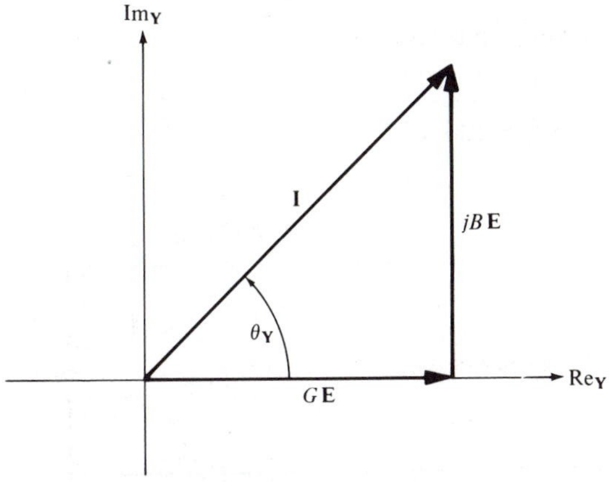

Fig. A.11 The **Y**-Plane with Plots of the Generalized Admittance **Y**, Excitation Current **I**, Response Voltage **E**, and Complex Power **W** for the Parallel *RLC*-Network of Fig. A.10

and

$$\mathbf{S} = \mathbf{Y}^*|\mathbf{E}|^2 = G|\mathbf{E}|^2 - jB|\mathbf{E}|^2 \tag{A.59}$$

Again, the following definitions are pertinent:

$$\omega_E C \equiv B_C = \text{capacitive susceptance in mhos}$$

$$\frac{1}{\omega_E L} \equiv B_L = \text{inductive susceptance in mhos}$$

The voltage and current can be plotted on the admittance plane as in Fig. A.11.

Example A.14

Given the configuration shown in Fig. A.14.1, calculate the following quantities: (1) the impedance of the system **Z**, (2) the admittance of the system **Y**, (3) the response current **I**, (4) the voltage drops across each of the passive elements, that is, $\mathbf{V}_R$, $\mathbf{V}_L$, and $\mathbf{V}_C$, (5) the complex power delivered by the source, **S**, (6) the power dissipated by the system, P, (7) the power factor $\cos\theta_\mathbf{Z}$, (8) the reactive power necessary for normal system operation, Q, (9) the reactive factor, that is, $\sin\theta_\mathbf{Z}$, (10) the time-domain current $i(t)$, (11) the time-domain voltages, that is, $v_R(t)$, $v_L(t)$, and $v_C(t)$, and (12) the nature and numerical value of the circuit element that can be used to correct for unity power factor.

Solution

1. $\mathbf{Z} = 50 + j\left(0.1 \times 10^3 - \dfrac{1}{2\times 10^{-5} \times 10^3}\right) = 50 + j(100-50)$
   $= 50 + j50 = \underline{\underline{70.7e^{j45^\circ}}}\ \Omega$
2. $\mathbf{Y} = \dfrac{1}{\mathbf{Z}} = \dfrac{1}{(70.7e^{j45^\circ})} = \dfrac{10^{-2}e^{-j45^\circ}}{0.707}$
   $= \underline{\underline{1.414 \times 10^{-2}e^{-j45^\circ}}}$ mho
3. $\mathbf{I} = \mathbf{E}\mathbf{Y} = (100e^{j45^\circ})(1.414 \times 10^{-2}e^{-j45^\circ}) = \underline{\underline{1.414}}$ A
4. a. $\mathbf{V}_R = R|\mathbf{I}|e^{j0^\circ} = 50 \times 1.414 = \underline{\underline{70.7}}$ V
   b. $\mathbf{V}_L = jX_L|\mathbf{I}|e^{j0^\circ} = j100 \times 1.414 = \underline{\underline{j141.4}}$ V
   c. $\mathbf{V}_C = -jX_G|\mathbf{I}| = e^{j0^\circ} - j50 \times 1.414 = \underline{\underline{-j70.7}}$ V
5. $\mathbf{S} = \mathbf{E}\mathbf{I}^* \equiv \mathbf{Z}|\mathbf{I}|^2 = (100e^{j45^\circ})(1.414e^{j^\circ})$
   $= (50 + j50)|1.414|^2 = \underline{\underline{100 + j100}}$ VA
6. $P = \text{Re}(\mathbf{E}\mathbf{I}^*) = \text{Re}(\mathbf{Z}|\mathbf{I}|^2) = \underline{\underline{100}}$ W

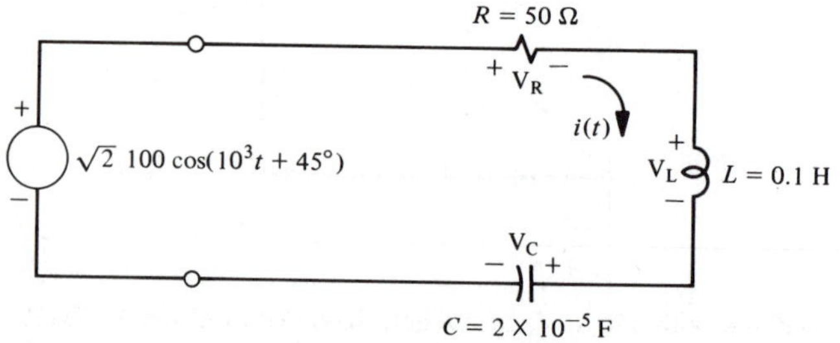

Fig. A.14.1 A Series *RLC*-Network Excited with a Sinusoidal Voltage Forcing Function

7. $PF = \cos 45° = \underline{\underline{0.707}}$
8. $Q = \text{Im}(\mathbf{EI}^*) = \text{Im}(\mathbf{Z}|\mathbf{I}|^2) = \underline{\underline{100}}$ var
9. $RF = \sin 45° = \underline{\underline{0.707}}$
10. $i(t) = \underline{\underline{\sqrt{2}\,(1.414)\cos(10^3t + 0°)}}$ A
11. a. $v_R(t) = \underline{\underline{\sqrt{2}\,(70.7)\cos(10^3t + 0°)}}$ V

    b. $v_L(t) = \underline{\underline{\sqrt{2}\,(141.4)\cos(10^3t + 90°)}}$ V

    c. $v_C(t) = \underline{\underline{\sqrt{2}\,(70.7)\cos(10^3t - 90°)}}$ V
12. *Unity power factor** implies that the impedance angle of the system should be zero, that is, $\theta_{\mathbf{Z}} = 0°$. However, *it is not always economically feasible or technically desirable to correct all the way to unity power factor*. Thus, *an optimum* $\theta_{\mathbf{Z}} \neq 0°$, *that is,* $PF \neq 1.0$, *is determined by comparison of the cost of equipment necessary to change the power factor with the decrease in either the electric bill or the investment in distribution and generation equipment resulting from the correction. A capacitor of* $c = 2 \times 10^{-5}$ *F in series with the other elements will provide* 100% *power factor correction.*

The usual industrial load on a power system operates at a lagging power factor, and in many cases this power factor is low enough so that improvement at the load is economically justifiable. Such power-factor improvement or correction is accomplished by connecting a bank of capacitors in parallel with the load. The size of the bank is chosen so that the power factor of the parallel combination reaches the desired value. The bank of capacitors also absorbs the energy of the system during switching off of the system. If these capacitors were not present, then an extra capacitor would be used to prevent sparking during switching off.

Mathematically, unity power factor for the given system is achieved as follows: (1) find the admittance of the system at its terminals, (2) find the numerical value and nature of the element that should be connected in parallel with the admittance of the system in order to reduce or increase the imaginary part of the system admittance to the desired value. Thus if $\mathbf{Y} = G + jB$ is the admittance of the system, $jB_?$ should be connected in parallel with $\mathbf{Y}$ so that

$$\mathbf{Y}_T = G + j(B + B_?) \equiv G + j0 \tag{A.60}$$

Thus, depending on the nature of B,

$$B_? = \pm|B| \tag{A.61}$$

where

1. $B_? =$ "$-$" suggests that an *inductive element* should be connected across the system.
2. $B_? =$ "$+$" suggests that a *capacitive element* should be connected across the system.

Note that for inductive elements $B_L = 1/\omega_E L$ and for capacitive elements $B_C = \omega_E C$.

From part 2 of Example A.14,

$$\mathbf{Y} = 1.414 \times 10^{-2} e^{-j45°} = 10^{-2} - j10^{-2}$$

*Please see Plates A.3 and B.6.

For unity power factor,

$$\mathbf{Y}_T = 10^{-2} + j\left[-10^{-2} + B_?\right] = 10^{-2} + j0$$

Thus

$$B_? - 10^{-2} = 0$$

or

$$B_? = +10^{-2}$$

Therefore, $B_?$ is capacitive in nature and its numerical value can be found as follows:

$$B_? = \omega_E C = 10^{-2}$$

or

$$C = \frac{10^{-2}}{\omega_E} = \frac{10^{-2}}{10^3} = \underline{\underline{10^{-5}\text{ F}}}$$

To correct for a power factor **other than unity**, proceed in the following manner. Consider Fig. A.12 and assume that the desired power-factor angle is $\theta_{\mathbf{Y}}$ and that $\theta_{\mathbf{Y}} < \phi_{\mathbf{Y}}$. Thus

$$\tan\theta_{\mathbf{Y}} = \frac{B_D}{G} \quad \text{or} \quad B_D = G\tan\theta_{\mathbf{Y}} \tag{A.62}$$

The added amount of admittance and its nature should be such that the total system admittance after the correction is

$$\mathbf{Y}_T = G + jB_{\mathrm{D}} \tag{A.63}$$

Thus the added admittance should be

$$jB_{\text{ADDED}} \equiv -j(B - B_{\mathrm{D}}) \tag{A.64}$$

Once the nature and magnitude of the added admittance is established, the numerical value of the added element is calculated following the same procedure that is followed for unity power-factor correction.

It must be noted that the power-factor correction does not obviate the need for reactive power by the load itself. Despite its advantages from the energy-transfer point of view, lagging reactive power is essential to the operation of many types of

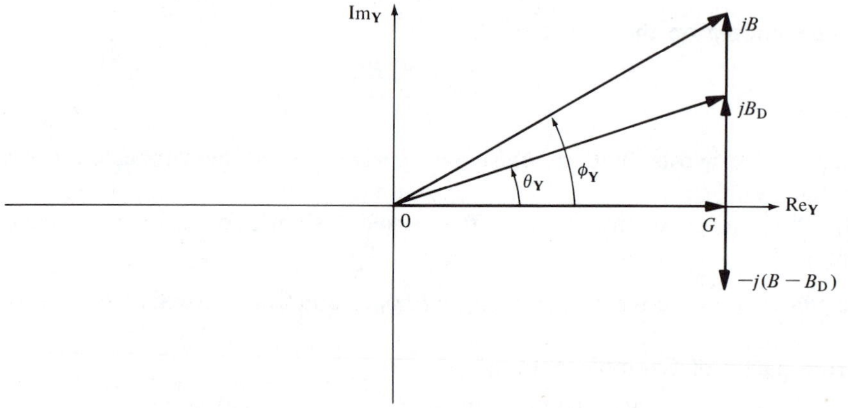

Fig. A.12 Vectorial Diagram of the Generalized Y-Plane Power-Factor Correction Technique

ac equipment, such as motors. *When the power factor is corrected, most of this needed lagging reactive power is supplied locally from the parallel capacitors rather than remotely through the distribution systems from the generators.* The energy oscillation of which reactive power is the quantitative measure [see Eqs. (A.54)–(A.56)] then takes place largely between the load and the *adjacent* capacitors (corrective elements), instead of the load and the *remote* generators. Therefore, the oscillations do not add to the burden of the distribution system and generators or to the copper losses in them. *This feature has a definite financial value.*

Example A.15

Given the configuration shown in Fig. A.15.1. Calculate the following quantities: (1) the admittance of the system **Y**, (2) the impedance of the system **Z**, (3) the response voltage **E**, (4) the currents through each of the passive elements, that is, $\mathbf{I}_G$, $\mathbf{I}_C$, and $\mathbf{I}_L$, (5) the complex power delivered by the source, (6) the power dissipated by the system (7) the power factor $\cos\theta_{\mathbf{Y}*}$, (8) the reactive power necessary for normal system operation, Q, (9) the reactive factor, that is, $\sin\theta_{\mathbf{Y}*}$, (10) the time-domain response voltage $e(t)$, (11) the time-domain currents, that is, $i_G(t)$, $i_C(t)$, and $i_L(t)$, and (12) the nature and numerical value of the circuit element that can be used to correct for unity power factor.

Solution

1. $\mathbf{Y}=\dfrac{1}{5\times 10^{-2}}+j\left(4\times 10^{-2}\times 10^{3}-\dfrac{1}{5\times 10^{-5}\times 10^{3}}\right)$
   $=20+j(40-20)=20+j20=\underline{\underline{28.3e^{j45°}}}$ mho
2. $\mathbf{Z}=\dfrac{1}{\mathbf{Y}}=\dfrac{1}{28.3e^{j45°}}=\underline{\underline{0.0354e^{-j45°}\ \Omega}}$
3. $\mathbf{E}=\dfrac{\mathbf{I}}{\mathbf{Y}}=\mathbf{IZ}=\dfrac{284e^{j45°}}{28.4e^{j45°}}=\underline{\underline{10e^{j0°}}}$ V
4. a. $\mathbf{I}_G=G|\mathbf{E}|e^{j0°}=20\times 10=\underline{\underline{200}}$ A
   b. $\mathbf{I}_C=jB_C|\mathbf{E}|e^{j0°}=j40\times 10=\underline{\underline{j400}}$ A
   c. $\mathbf{I}_L=-jB_L|\mathbf{E}|e^{j0°}=-j20\times 10=\underline{\underline{-j200}}$ A
5. $\mathbf{S}=\mathbf{Y}^*|\mathbf{E}|^2=(20-j20)(10)^2=\underline{\underline{2000-j2000}}$ VA
6. $P=\text{Re}(\mathbf{S})=\underline{\underline{2000}}$ W
7. $\text{PF}=\cos(-45°)=\underline{\underline{0.707}}$

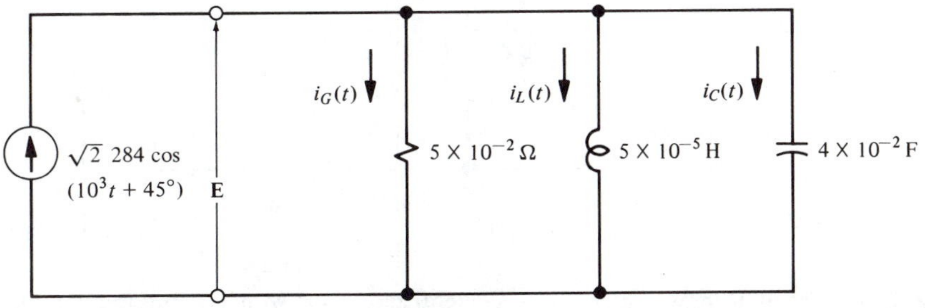

Fig. A.15.1 A Parallel *RLC*-Network Excited with a Sinusoidal Current Forcing Function

8. $Q = \text{Im}(\mathbf{S}) = \underline{\underline{-2000}}$ var
9. $\text{RF} = \sin(-45°) = \underline{\underline{-0.707}}$
10. $e(t) = \underline{\underline{\sqrt{2}\,(10)\cos 10^3 t \text{ V}}}$
11. a. $i_G(t) = \underline{\underline{\sqrt{2}\,(200)\cos 10^3 t}}$ A

    b. $i_C(t) = \underline{\underline{\sqrt{2}\,(400)\cos(10^3 t + 90°)}}$ A

    c. $i_L(t) = \underline{\underline{\sqrt{2}\,(200)\cos(10^3 t - 90°)}}$ A
12. $\mathbf{Y} = 20 + j20 = 20 + j(20 + B_?) \equiv 20 + j0$

Thus

$$20 + B_? = 0$$

or

$$B_? = -20 \text{ (inductive element)}$$

and

$$B_? = \frac{1}{\omega_E L} = 20$$

$$L = \frac{1}{20\omega_E} = \frac{1}{20 \times 10^3} = \frac{10^{-5}}{0.2}$$

$$= \underline{\underline{5 \times 10^{-5} \text{ H}}}$$

Example A.16

Given the configuration shown in Fig. A.16.1. For the system of Fig. A.16.1, where $\omega_E = 1000$ rad/s, $R = 10\ \Omega$, $L = 0.02$ H, and $C = 100\ \mu$F, calculate the following quantities: (1) the power generated by the source, $\mathbf{S}_{GO}$, (2) the power demanded by the plant, $\mathbf{S}_{DO}$, (3) the power factor of the plant, PF_P, (4) the nature and the numerical value of the corrective element one would connect between points AB in parallel with the plant, that is, **just outside the plant**, to correct the power factor of the plant—as far as the source is concerned—to 0.865 LGG, (5) the power generated by the source after the power-factor correction, $\mathbf{S}_{GF}$, and (6)

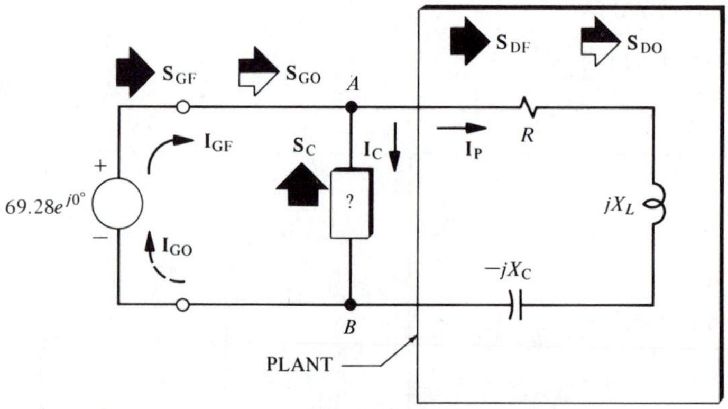

Fig. A.16.1 Circuit Representation of a Generalized Plant with (1) Forcing Function, (2) Ideal Transmission Line, and (3) Parallel Power-Factor Correction Exhibited

the power demanded by the plant after the power-factor correction, $\mathbf{S}_{DF}$. Comment on what was accomplished by the power-factor correction and its advantages, if any.

Solution

1. To calculate the power generated by the source, $\mathbf{S}_G$, we should calculate the response current $\mathbf{I}_G$ first. This necessitates the calculation of the impedance of the plant, $\mathbf{Z}_P$. Therefore, utilization of Eq. (A.35) yields

$$\mathbf{Z}_P = R + j\left(\omega_E L - \frac{1}{\omega_E C}\right) = 10 + j\left(10^3 \times 2 \times 10^{-2} - \frac{1}{10^3 \times 10^{-4}}\right)$$
$$= 10 + j(20 - 10) = 10 + j10 = \underline{14.14e^{j45°}\ \Omega} \tag{1}$$

The response current before power-factor correction, designated as $\mathbf{I}_{G0}$, is calculated with the aid of Eq. (A.48), input data, and Eq. (1) as

$$\mathbf{I}_{G0} = \frac{\mathbf{E}_G}{\mathbf{Z}_P} = \frac{69.28e^{j0°}}{14.14e^{j45°}} = \underline{4.90e^{-j45°}\ \text{A}} \tag{2}$$

Now the power generated by the source, before power-factor correction, designated as $\mathbf{S}_{G0}$, is calculated with the aid of Eq. (A.45), input data, and Eq. (2) as

$$\mathbf{S}_{G0} = \mathbf{E}_G \mathbf{I}_{G0}{}^* = (69.28e^{j0°}) \times (4.90e^{j45°}) = 339.47e^{j45°}$$
$$= \underline{\underline{240.04 + j240.04}}\ \text{VA} \tag{3}$$

2. The power demanded by the plant before power-factor correction, designated as $\mathbf{S}_{D0}$, is calculated with the aid of Eq. (A.47) and Eqs. (1) and (2) as

$$\mathbf{S}_{D0} = \mathbf{Z}_P |\mathbf{I}_{G0}|^2 = (10 + j10) \times (4.90)^2 = \underline{\underline{240.04 + j240.04}}\ \text{VA} \tag{4}$$

3. The power factor of the plant PF_P is calculated with the aid of Eq. (A.57) as

$$\text{PF}_P = \cos 45° = \underline{\underline{0.71\ \text{LGG}}} \tag{5}$$

4. To calculate the nature and the numerical value of the corrective element which, when connected between points AB, Fig. A.16.1, **yields a net power factor of 0.365 LGG for the parallel combination of the corrective element and the plant when viewed from the source terminals**, we must first calculate **and plot** the admittance $\mathbf{Y}_P$ of the plant. Thus with the aid of Eq. (1), Eq. (A.48) yields the admittance of the plant $\mathbf{Y}_P$,

$$\mathbf{Y}_P = \frac{1}{\mathbf{Z}_P} = \frac{1}{14.14e^{j45°}} = 0.07e^{-j45°} = \underline{\underline{0.05 - j0.05}}\ \text{mho} \tag{6}$$

Eq. (6) is plotted in the admittance plane in Fig. A.16.2 as follows. For a the net power factor of 0.865 LGG, the net power-factor angle θ_Y, Fig. A.16.2, is calculated as

$$\theta_Y = \cos^{-1}(0.865\ \text{LGG}) = 30.12° \Rightarrow -30.12° \tag{7}$$

With the aid of Fig. A.16.2, Eq. (7), and Eq. (A.62), we can calculate the desired susceptance B_D, which corresponds to the power factor of 0.855 LGG, as

$$B_D = 0.05 \tan(-30.12°) = \underline{-2.90 \times 10^{-2}\ \text{mho}} \tag{8}$$

Now, in order to have a total system admittance, Fig. A.16.2 (i.e., the admittance $\mathbf{Y}_T$ viewed from the source's terminals after the power-factor correction), given by

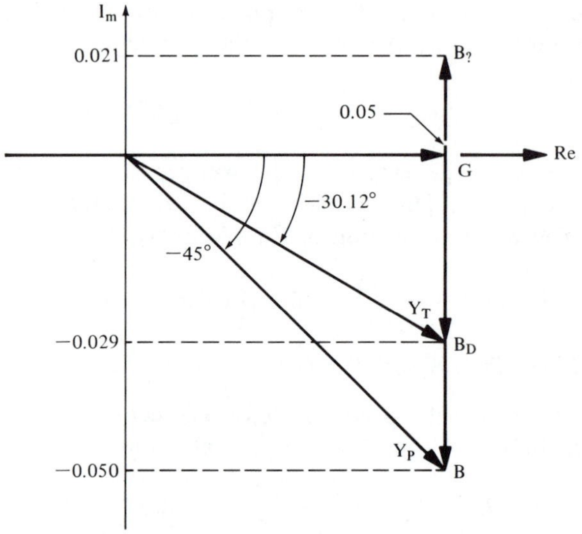

Fig. A.16.2 Y-Plane Plot for the System of Fig. A.16.1

Eq. (A.63) as

$$\mathbf{Y}_T = G + jB_D = 5.00 \times 10^{-2} + j(-2.90 \times 10^{-2}) \tag{9}$$

a susceptance designated as $B_?$ must be connected between points AB in parallel with the plant, Fig. A.16.1. When this susceptance is introduced to Eq. (6), it yields the total system admittance $\mathbf{Y}_T$, designated as

$$\mathbf{Y}_T = 5.00 \times 10^{-2} + j(-5 \times 10^{-2} + B_?) \tag{10}$$

It is clear now that **Eqs. (9) and (10) are identical if and only if the corresponding imaginary parts are identical.** That is,

$$-2.90 \times 10^{-2} \equiv -5 \times 10^{-2} + B_? \tag{11}$$

Solution of Eq. (11) for the susceptance $B_?$ yields

$$B_? = -2.9 \times 10^{-2} + 5 \times 10^{-2} = \underline{\underline{2.1 \times 10^{-2}}} \text{ mho} \tag{12}$$

The theory set forth in the neighborhood of Figs. A.10 and A.12, and the positive sign of Eq. (12), suggest that the susceptance $B_?$ **is capacitive in nature**. Therefore, Eq. (12) can be written, in accordance with the theory following Eq. (A.59), as

$$B_? = \omega_E C \doteq 2.1 \times 10^{-2} \tag{13}$$

Utilization of the input data value for ω_E enables us to solve Eq. (13) for C as

$$C = \frac{2.1 \times 10^{-2}}{10^3} = \underline{\underline{21.0 \times 10^{-6}}} \text{ F} \tag{14}$$

5. After the power-factor correction, the total admittance of the system $\mathbf{Y}_T$, as far as the source is concerned, is given by Eq. (9). It can also be calculated from Eq. (10) with the aid of Eq. (14) and the input data for ω_E. However, the expression for $\mathbf{Y}_T$ is repeated here for clarity and is written as

$$\mathbf{Y}_T = 5.00 \times 10^{-2} - j2.90 \times 10^{-2} = \underline{5.78 \times 10^{-2} e^{-j30.12°}} \text{ mho} \tag{15}$$

Eq. (15) in conjunction with Eqs. (A.48) or (A.58) enables us to calculate the generated current after power-factor correction, designated as $\mathbf{I}_{GF}$, as follows:

$$\mathbf{I}_{GF} = \mathbf{Y}_T\mathbf{E}_G = (5.78\times 10^{-2}e^{-j30.12^\circ})\times(69.28e^{j0^\circ}) = \underline{4.00e^{-j30.11^\circ}} \text{ A} \tag{16}$$

Now Eq. (16) enables us to calculate the generated power, after the power-factor correction has taken place, designated as $\mathbf{S}_{GF}$, as follows:

$$\begin{aligned}\mathbf{S}_{GF} &= \mathbf{E}_G\mathbf{I}_{GF}{}^* = (69.28e^{j0^\circ})\times(4.00e^{+j30.12^\circ}) = 277.12e^{j30.12^\circ}\\ &= \underline{\underline{240.04+j139.21}} \text{ VA}\end{aligned} \tag{17}$$

6. It should be evident that the power demanded by the plant, $\mathbf{S}_D$, has not changed because nothing has changed inside the plant. For the sake of reinforcing understanding, however, we may calculate the power demanded by the plant as follows. The current through the plant is calculated with the aid of input data and Eq. (A.48) and Eq. (1). Thus

$$\mathbf{I}_P = \frac{\mathbf{E}_G}{\mathbf{Z}_P} = \frac{69.28e^{j0^\circ}}{14.14e^{j45^\circ}} = 4.90e^{-j45^\circ} = \underline{3.46-j3.46} \text{ A} \tag{18}$$

For checking purposes the current through the corrective capacitor Y_C, designated as $\mathbf{I}_C$, is calculated with the aid of data and Eqs. (A.48) or (A.58) as follows:

$$\mathbf{I}_C = \mathbf{Y}_C\mathbf{E}_G = (2.1\times 10^{-2}e^{j90^\circ})\times(69.28e^{j0^\circ}) = \underline{1.46e^{j90^\circ}} \text{ A} \tag{19}$$

According to Kirchhoff's current law, the summation of the currents given by Eqs. (18) and (19) should yield the current given by Eq. (16), Fig. A.16.1. That is,

$$\mathbf{I}_{GF} = \mathbf{I}_P + \mathbf{I}_C = 3.46 - j3.46 + j1.46 = 3.46 + j2.00 = \underline{4.00e^{j30.12^\circ}} \text{ A} \tag{20}$$

Clearly Eq. (20) satisfies the assertion made with respect to the currents of Fig. A.16.1.

Now the power demanded by the plant is calculated with the aid of Eq. (A.47) as follows:

$$\mathbf{S}_{DF} = \mathbf{E}_G\mathbf{I}_P{}^* = (69.28e^{j0^\circ})\times(4.90e^{j45^\circ}) = \underline{\underline{240.04+j240.04}} \text{ VA} \tag{21}$$

or

$$\mathbf{S}_{DF} = \mathbf{Z}_P|\mathbf{I}_P|^2 = (10+j10)\times(4.90)^2 = \underline{\underline{240.04+j240.04}} \text{ VA} \tag{22}$$

Inspection of the pertinent powers tabulated in Table A.1 reveals that (1) prior to power-factor correction, the power generated by the source is supplied to the plant in its entirety, that is, $\mathbf{S}_{G0}=\mathbf{S}_{D0}$, and (2) after the power-factor corrections, the power demanded by the load is greater than the power generated by the source, that is, $|\mathbf{S}_{GF}|<|\mathbf{S}_{DF}|$. To be exact, the real parts of the corresponding powers are identical, **but** the imaginary part of the generated power is smaller than the imaginary part of the power demanded by the plant. Therefore, one may pose the question: *Where does the additional imaginary power of j(240.04 – 139.21) = j100.83 var required by the plant come from?*

If we were to calculate the power demanded by the corrective capacitor connected between points *AB*, Fig. A.16.1, utilizing Eqs. (A.53) or (A.59), we would obtain

$$\mathbf{S}_C = \mathbf{Y}_C{}'|\mathbf{E}_G|^2 = (0-j2.1\times 10^{-2})\times(69.28)^2 = \underline{\underline{-101.0}} \text{ var} \tag{23}$$

or

$$\begin{aligned}\mathbf{S}_C &= -j\mathbf{Y}_C|\mathbf{I}_C|^2 = -j\left(\frac{1}{\omega_E C}\right)|\mathbf{I}_P|^2 = -j\left(\frac{1}{10^3\times 21.14\times 10^{-6}}\right)\times(1.46)^2\\ &= \underline{\underline{-j100.83}} \text{ var}\end{aligned} \tag{24}$$

Table A.1 TABULATION OF GENERATED POWER AS WELL AS POWER DEMANDED BY THE PLANT AND THE POWER-FACTOR CORRECTING CAPACITOR

	Generation	Load Demand	Capacitor Demand
Corrective Capacitor Out	$S_{GO}=240.+j240.$	$S_{DO}=240.+j240.$	—
Corrective Capacitor In	$S_{GF}=240.+j139.$	$S_{DF}=240.+j240.$	$S_C=0-j101.0$

The negative sign of the power in Eqs. (23) and (24) suggests that the capacitor generates reactive power rather than absorbs it. In addition, the capacitor generates exactly the amount of power still required by the plant, that is, $j100.83$ var.

This observation enables us to draw the conclusion that the capacitor used to correct the power factor to a given value generates a certain amount of reactive power, which is proportional to the difference between the original and the desired susceptance, Eq. (A.64), and *relieves* the source from the task of generating this amount of reactive power. In addition, the proximity of the capacitor to the plant prevents (and therefore eliminates the extra cost) the **overdesigning** of transmission lines in real systems to handle the flow of reactive power between the plant and the source. This power flow between the source and load takes place because the reactive power is not dissipated in the load but is essential for its operation, Eqs. (A.54) through (A.56), (A.47), and (A.35).

A.8 SUMMARY

This appendix presents the mathematical foundations for power systems engineering. Complex quantities, such as impedance, admittance, and complex power, are derived by utilizing an *RLC*-circuit excited by a sinusoidal forcing function, that is, sinusoidal voltage or current generators. Also, the concepts of *sinor* or *phasor*, average power, reactive power, instantaneous power (flowing into the R, L, or C circuit elements and their relationship to the average or reactive power, i.e., the components of the complex power), power factor, and power-factor correction are set forth. All the concepts are reinforced by numerical examples. *The student is urged to master these concepts*, all of which are essential for the analysis of transformers and induction and synchronous energy converters.

PROBLEMS

A.1 Given $\mathbf{B}_{11}=6+j5$, find the exponential form of the complex number $\mathbf{B}_{11}$.

A.2 Given $\mathbf{B}_{33}=-6-j5$, find the exponential form of the complex number $\mathbf{B}_{33}$.

A.3 Given $\mathbf{A}=20e^{j50°}$, find the rectangular form of the complex number $\mathbf{A}$.

A.4 Given $\mathbf{B}=200e^{j135°}$, find the rectangular form of the complex number $\mathbf{B}$.

A.5 Given $\mathbf{C}=200e^{j240°}$, find the rectangular form of $\mathbf{C}$ by using the appropriate instructions provided in the figure of Example A.5.

A.6 Given $i(t)=120\cos(\omega t+\alpha)$, find the exponential form of $i(t)$.

A.7 Given $v_1(t)=282\sin(1000t+60°)$, find its phasor referred to (1) the imaginary and (2) the real axes.

A.8 If $\mathbf{B}=50+j60$ and $\mathbf{C}=100e^{j45°}$, calculate the sum $\mathbf{A}=\mathbf{B}+\mathbf{C}$.

A.9 If $\mathbf{B}=50+j60$ and $\mathbf{C}=100e^{j45°}$, calculate the difference $\mathbf{D}=\mathbf{B}-\mathbf{C}$.

A.10 If $\mathbf{B}=50+j60$ and $\mathbf{C}=70.7+j70.7$, calculate the product $\mathbf{P}=\mathbf{B}\cdot\mathbf{C}$.

A.11 If $\mathbf{B}=50+j60$ and $\mathbf{C}=70.7+j70.7$, calculate the quotient $\mathbf{Q}=\mathbf{B}/\mathbf{C}$.

A.12 If $\mathbf{B}=141.4+j141.4$, calculate $\mathbf{B}^2$ and $\mathbf{B}^{1/2}$.

A.13 If $\mathbf{B}=141.4+j141.4$, calculate $\mathbf{B}^*$.

A.14 For the figure of Example A.14, set $e_S(t)=\sqrt{2}\,(200)\cos(10^3t+45°)\,V$, $R=60$ Ω, $L=0.2$ H, and $C=1\times10^{-5}$ F and calculate the following quantities: (1) the impedance of the system $\mathbf{Z}$, (2) the admittance of the system $\mathbf{Y}$, (3) the response current $\mathbf{I}$, (4) the voltage drops across each of the passive elements, that is, $\mathbf{V}_R$, $\mathbf{V}_L$, and $\mathbf{V}_C$, (5) the complex power generated by the source, $\mathbf{S}_G$, (6) the power dissipated (demanded) by the system, P_D, (7) the power factor, that is, $\cos\theta_Z$, (8) the reactive power necessary for normal system operation, Q_D, (9) the reactive factor, that is, $\sin\theta_Z$, (10) the time-domain current $i(t)$, (11) the time-domain voltages, that is, $v_R(t)$, $v_L(t)$, and $v_C(t)$, and (12) the nature and numerical value of the circuit element that can be connected across the source in order to correct the system for unity power factor.

A.15 For the figure of Example A.15, set $i_S(t)=\sqrt{2}\,(220)\cos(10^3t+45°)$ A, $R=2\times10^{-2}$ Ω, $L=4\times10^{-5}$ H, and $C=6.5\times10^{-2}$ F and calculate the following quantities: (1) the admittance of the system $\mathbf{Y}$, (2) the impedance of the system $\mathbf{Z}$, (3) the response voltage $\mathbf{E}$, (4) the currents through each of the passive elements, that is, $\mathbf{I}_G$, $\mathbf{I}_C$, and $\mathbf{I}_L$, (5) the complex power generated by the source, $\mathbf{S}_G$, (6) the power dissipated (demanded) by the system, P_D, (7) the power factor, that is, $\cos\theta_{Y*}$, (8) the reactive power necessary for normal system operation, Q_D, (9) the reactive factor, that is, $\sin\theta_{Y*}$, (10) the time-domain response voltage $e(t)$, (11) the time-domain currents, that is, $i_G(t)$, $i_C(t)$, and $i_L(t)$, and (12) the nature and numerical value of the circuit element that can be connected across the source in order to correct the system for unity power factor.

A.16 For the system of Fig. A.16.1, where $\omega_E=1000$ rad/s, $R=20$ Ω, $L=0.04$ H, and $C=100$ μF, calculate the following quantities: (1) the power generated by the source, $\mathbf{S}_{G0}$, (2) the power demanded by the plant, $\mathbf{S}_{DO}$, (3) the power factor of the plant, PF_P, (4) the nature and the numerical value of the corrective element one would connect between points AB in parallel with the plant, that is, **just outside the plant**, to correct the power factor of the plant—as far as the source is concerned—to 0.865 LGG, (5) the power generated by the source after the power-factor correction, $\mathbf{S}_{GF}$, and (6) the power demanded by the plant after the power-factor correction, $\mathbf{S}_{DF}$. Comment on what was accomplished by the power factor correction and its advantages if any.

REFERENCES

1. Balabanian, N., *Fundamentals of Circuit Theory*. Allyn & Bacon, Boston, 1961.
2. Fitzgerald, A. E., Higginbotham, S. M., and Grabel, A., *Basic Electrical Engineering*. McGraw-Hill, New York, 1967.
3. Harris, T., *A-C Circuit Analysis*. McGraw-Hill, New York, 1955.
4. Hayt, W., *Engineering Circuit Analysis*. McGraw-Hill, New York, 1962.
5. Van Valkenburg, N., *Network Analysis*, 2d ed. Prentice-Hall, Englewood Cliffs, N.J., 1964.

B INSTRUMENTATION

B.1 INTRODUCTION

In this appendix the principles of operation of the *electrodynamometer movement* are set forth in some detail. Then the four basic instruments, that is, ammeter, voltmeter, wattmeter, and varmeter, are derived from the electrodynamometer movement. In addition, the *moving-iron* ammeters and voltmeters are discussed. Also, the principle of operation of the *D'Arsonval movement* is briefly discussed and the principle of operation of the ohmmeter is set forth.

B.2 INSTRUMENTATION

There are different types of meter movements whose average deflection is proportional to the square of the RMS value of the applied voltage, current, and so on.

The universal meter movement is known as the **electrodynamometer movement**. It has two independent inputs that permit its use in numerous applications such as ammeters, voltmeters, and wattmeters. The torque produced in this movement arises from magnetic forces exerted between two current carrying coils, Fig. B.1. *One coil is made of heavy wire, having a small number of turns and a very low resistance.* Usually marked with *large solid circles*, this coil is *fixed* and is designated as the *C (current) coil. The second coil is composed of a much greater number of turns of fine wire and has a relatively high resistance.* It is *movable* and is designated as the *P (potential) coil. Ordinarily, the current-coil terminals in an actual instrument are physically larger than the terminals of the potential coil.*

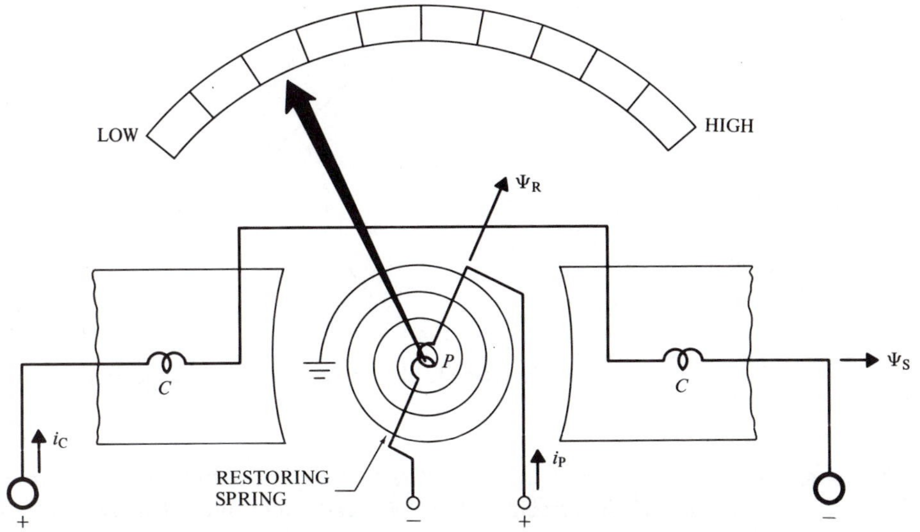

Fig. B.1 The Electrodynamometer Movement

The fixed (C) coil is comprised of two windings that produce a moderately uniform field Ψ_S in the environment of the rotating (P) coil, whose field is Ψ_R. Coil P moves against restoring spiral spring that holds it and the pointer that is mounted rigidly on the moving coil stationary at zero when the two windings are not excited. The pointer indicates the angular displacement, that is, θ_{AVG}, of the rotating coil on a scale that may be calibrated in amperes, volts, watts, or other quantities, depending on how the two coils are energized.

To see how this movement is capable of producing a nonzero average torque when it is used with alternating current, consider the two coils connected in series. The field Ψ_S reverses at the same time that the field Ψ_R reverses. Hence the instantaneous torque remains *unidirectional* and the average deflection is nonzero for nonzero currents.

The key equations that describe the electrodynamometer can be obtained by utilizing the magnetic energy stored in the two coils. The electrostatic field between the two coils is neglected in this analysis. However, it may cause some error if there is an appreciable potential difference between the coils.

For an incremental clockwise displacement $d\theta$ of the coil P, Fig. B.2, the incremental energy is defined as $dW \equiv \mathbf{F}\cdot d\mathbf{S}$. Utilization of the cylindrical coordinate system enables us to write the incremental energy as

$$dW = FR\,d\theta \tag{B.1}$$

or

$$\frac{dW}{d\theta} = FR \tag{B.2}$$

But

$$\tau_\theta \equiv FR \tag{B.3}$$

Thus

$$\tau_\theta = \frac{dW}{d\theta} \tag{B.4}$$

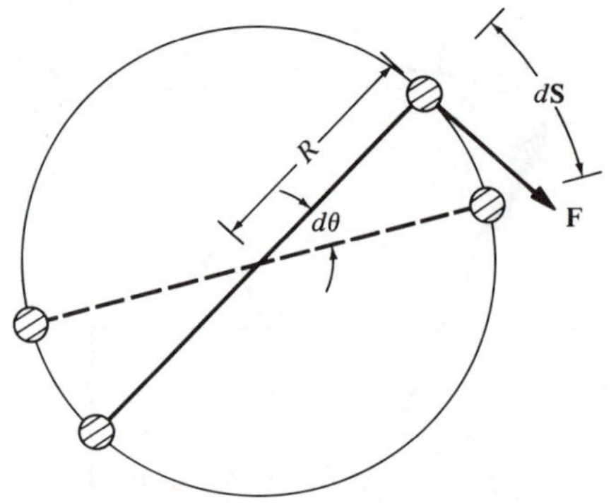

Fig. B.2 Cross-Sectional View of the Rotating Coil Corresponding to the Electrodynamometer Movement

The magnetic energy, however, stored in the two coils is a function of their self inductances L_C and L_P, their currents i_C and i_P, and their mutual inductance M, and is written as

$$W_M = \tfrac{1}{2}L_C i_C^2 + \tfrac{1}{2}L_P i_P^2 + M i_C i_P \tag{B.5}$$

Since i_C, i_P, L_C, and L_P are not functions of the angular displacement between the two coils, that is, θ, but M is a function of θ,

$$\frac{dW}{d\theta} = \frac{dW_M}{d\theta} = \frac{\partial M}{\partial \theta} i_C i_P \tag{B.6}$$

Therefore,

$$\tau_\theta = \frac{\partial M}{\partial \theta} i_C i_P \tag{B.7}$$

Since

$$\tau_{AVG} = \frac{1}{T}\int_0^T \tau_\theta \, dt \tag{B.8}$$

τ_{AVG} becomes

$$\tau_{AVG} = \frac{\partial M}{\partial \theta}\frac{1}{T}\int_0^T i_C i_P \, dt \tag{B.9}$$

The average angular deflection θ_{AVG} can be found from $J\ddot{\theta} + B\dot{\theta} + K_S\theta = \tau$, for the condition $\ddot{\theta} = \dot{\theta} \equiv 0$. Thus

$$\theta_{AVG} = \frac{\tau_{AVG}}{K_S} = \frac{1}{K_S}\frac{\partial M}{\partial \theta}\frac{1}{T}\int_0^T i_C i_P \, dt \tag{B.10}$$

Defining

$$\frac{1}{K} \equiv \frac{1}{K_S}\frac{\partial M}{\partial \theta} \tag{B.11}$$

enables us to write Eq. (B.10) as

$$\theta_{AVG} = \frac{1}{KT}\int_0^T i_C i_P \, dt \tag{B.12}$$

where K is defined as the *instrument constant*. It should be pointed out that the instrument constant is a function of $\partial M/\partial \theta$. Thus in order to achieve proper scale characteristics, the coils must be designed and manufactured carefully.

The movement is confined primarily to low-frequency applications, and its useful frequency range may be extended up to several thousand cycles per second with careful design. The principal high-frequency limitations are imposed by (1) stray coil capacitance, a portion of which varies with angular location of the rotating coil, (2) eddy currents produced in nearby metallic objects, and stray external magnetic fields, because the fixed field, that is, Ψ_S, is not very intense. For these reasons the movement is usually housed inside a laminated magnetic shield. The special housing of the movement and the accurate control of M as a function of θ make instruments utilizing this movement rather expensive. Other disadvantages that should be mentioned in reference with this movement are (1) the extra power required for the operation of the instrument when compared to an instrument with a permanent magnet instead of a fixed coil and (2) the reduction of the sensitivity of an instrument with a fixed coil instead of a permanent magnet. The reduction in sensitivity is due to the nature of the magnetic flux density produced by a fixed coil versus that of a permanent magnet. However, the wide spectrum of applications of the electrodynamometer movement, the best of which is the *wattmeter*, make it extremely popular.

Eq. (B.12) defines the principle of operation of the instruments that measure (1) *current,* (2) *voltage, and* (3) *power.*

The Ammeter

For current measurement the two coils of Fig. B.1 are connected in series and the combination is connected in series with the given load, Fig. B.3.

Since $i_C = i_P \equiv i_L$, Eq. (B.12) becomes

$$\theta_{AVG} = \frac{1}{K}\left(\frac{1}{T}\int_0^T i_L{}^2\,dt\right) \equiv \frac{1}{K}(I_{L[RMS]})^2 \tag{B.13}$$

and

$$I_{L[RMS]} = (K\theta_{AVG})^{1/2} \tag{B.14}$$

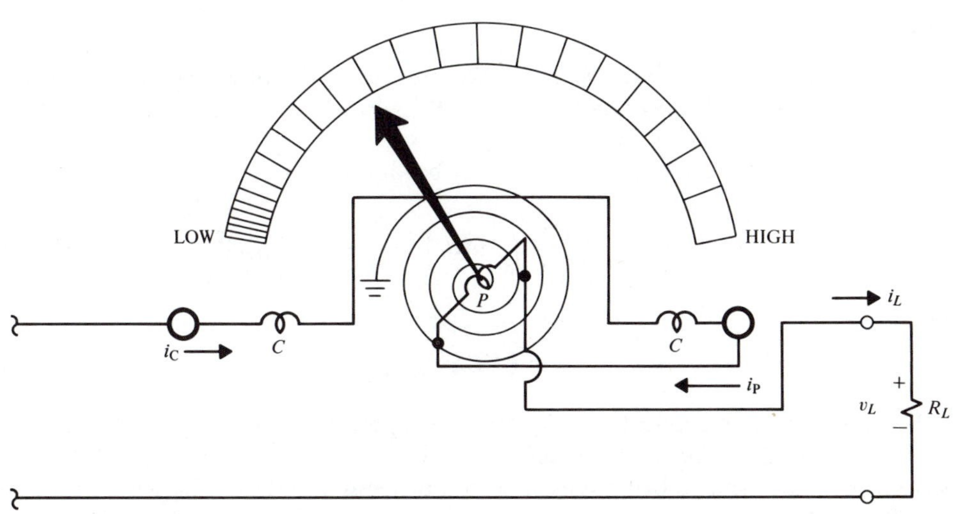

Fig. B.3 Ammeter Connection of the Electrodynamometer Movement [See Pl. B.11(a)]

Plate B.1 Dials of Panel Mounted Ammeters: (a) ac Type and (b) dc Type. (Courtesy of General Electric Co., Schenectady, N.Y.)

Note that the measurement, that is, the average deflection given by Eq. (B.13), *does not depend* on the waveform of the current but **depends** strictly on the RMS value of the alternating current.

Usually the current is delivered to the rotating coil through the restoring spring. For this reason it is used as the shunt coil, and a very high resistor R_M connected in series with the rotating coil serves to limit the current to safe values for the particular operating range of the movement. There is an upper limit of about 0.2 A beyond which the series connection of Fig. B.3 becomes impractical. However, high-current ammeters are available. They are designed to bypass the bulk of the current through suitable *protective shunts* and only a very small portion of the current to be measured goes through the coils of the electrodynamometer movement.

The Voltmeter

For voltage measurements the two coils are connected in series, and both are connected in series with a very high resistance R_M, and the combination is connected in parallel with the given load, Fig. B.4.

Since $i_C = i_P \equiv i$ and $i = v/R_M$, Eq. (B.12) becomes

$$\theta_{AVG} = \frac{1}{KR_M{}^2}\left(\frac{1}{T}\int_0^T v^2\,dt\right) \equiv \frac{1}{KR_M{}^2}(V_{RMS})^2 \tag{B.15}$$

and

$$V_{RMS} = \left(KR_M{}^2\theta_{AVG}\right)^{1/2} \tag{B.16}$$

For the voltmeter applications, $R_M \ggg |X_C + X_P \pm 2X_M|$ and $V_{L[RMS]} = V_{RMS}$.

Note again that the average deflection, Eq. (B.15), *does not depend* on the waveform of the voltage being measured but **depends** strictly on the RMS value of the alternating voltage.

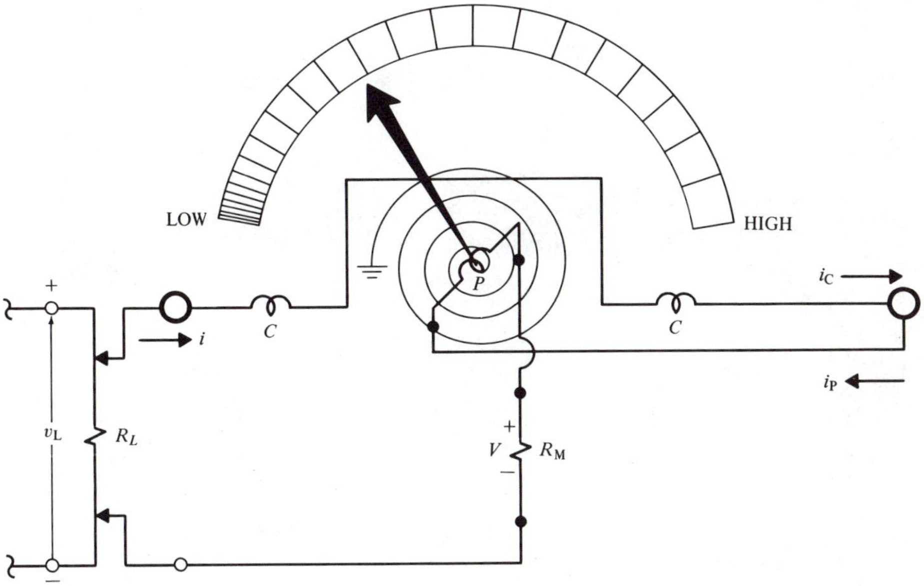

Fig. B.4 Voltmeter Connection of the Electrodynamometer Movement [See Pl. B.11(b)]

The Wattmeter

For power measurements the coils C and P are connected in the following manner:

1. *The fixed coil is connected in series with the given load.*
2. *The moving coil is connected internally in series with a high-value resistor* R_M *and the combination is connected across the given load,* Fig. B.5.

Since $R_M \ggg R_L$, $i_C \simeq i_L$, $v = v_L$ and $i_P = v_L/R_M$, Eq. (B.12) becomes

$$\theta_{AVG} = \frac{1}{KR_M}\left(\frac{1}{T}\int_0^T v_L i_L\,dt\right) = \frac{1}{KR_M}\left(\frac{1}{T}\int_0^T p_L\,dt\right) = \frac{1}{KR_M}(P_{AVG}) \qquad \text{(B.17)}$$

Plate B.2 Dials of Panel Mounted Voltmeters: (a) ac Type and (b) dc Type (Courtesy of General Electric Co., Schenectady, N.Y.)

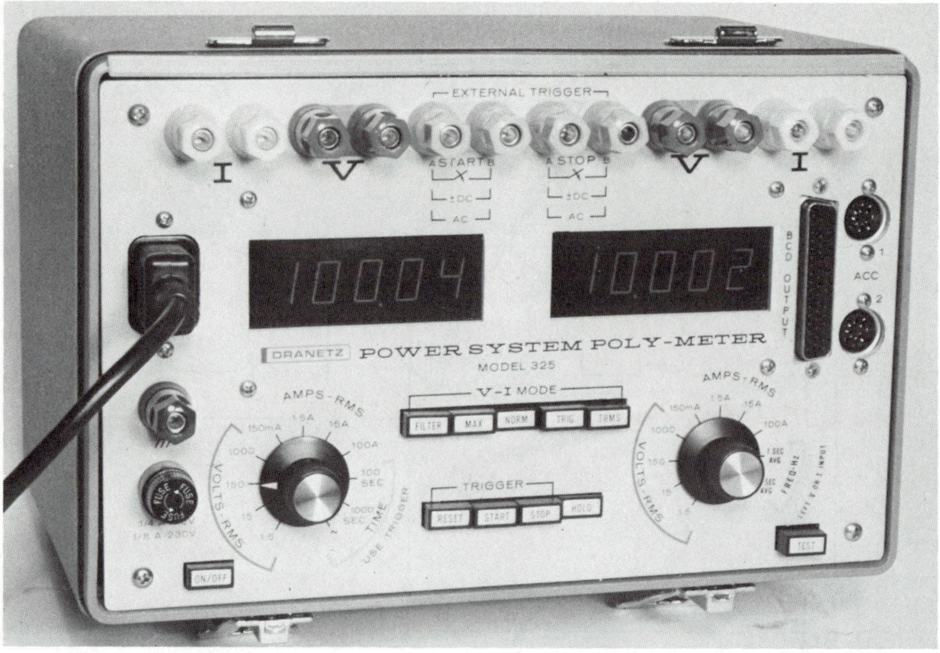

Plate B.3 Model 325 Power System Poly-Meter Provides: (a) Simultaneous Measurement and Display of the Two Independently Selectable Variables, the ac Voltage and the ac Current with Full-Scale Measurement Ranges from 0 to 1000 V RMS and 0 to 100 A RMS; (b) Time-Interval Measurements from 0 to 1000 s (in 0.01 s Increments); and (c) Fast-Frequency-Option Measurement of Frequencies from 20 to 450 Hz with Resolution of 0.001 Hz. (Courtesy of Dranetz Engineering Laboratories, Inc., South Plainfield, N.J.)

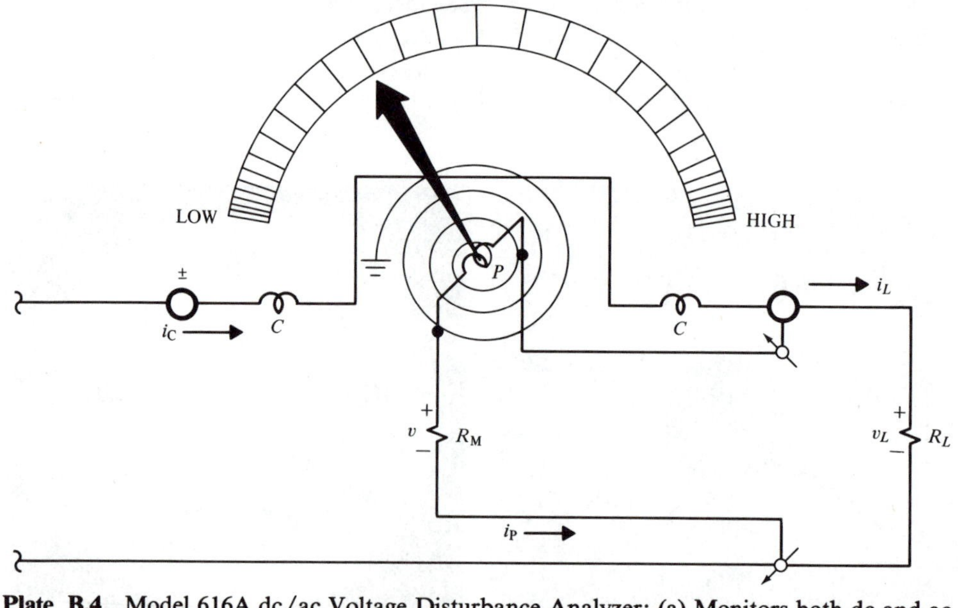

Plate B.4 Model 616A dc/ac Voltage Disturbance Analyzer: (a) Monitors both dc and ac Voltage Levels Simultaneously, (b) Classifies, Analyzes, Stores, and Prints Out Voltage Disturbance Information, and (c) Measures Impulses, Sags and Surges, and Slowly Changing Average Levels. (Courtesy of Dranetz Engineering Laboratories, Inc., South Plainfield, N.J.)

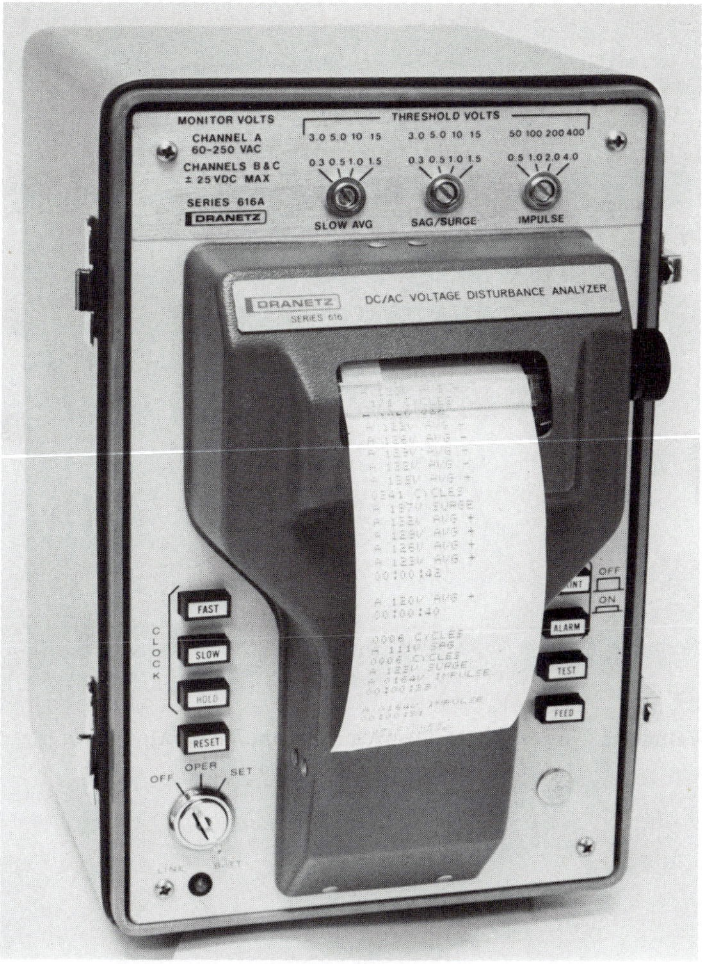

Plate B.5 Model 314 Phase Angle Meter—Measures the Angle Between Two Voltages, Two Currents, or a Voltage and a Current. Can be Used to Check the Polarity, "Rotation," and Phase Displacements, and to Aid in Connecting, Adjusting, and Maintaining the Many Devices in Power Systems That Require Correct Phasing for Proper Operation. (Courtesy of Dranetz Engineering Laboratories, Inc., South Plainfield, N.J.)

and

$$P_{\text{READING}} \equiv P_R = P_{\text{AVG}} = KR_{\text{M}}\theta_{\text{AVG}} = \frac{1}{T}\int_0^T p_L \, dt$$

$$= \frac{R_{\text{M}}}{T}\int_0^T i_{\text{C}} i_{\text{P}} \, dt \tag{B.18}$$

It should be pointed out again that the average deflection of the wattmeter's pointer is *independent* of the waveform of p_L.

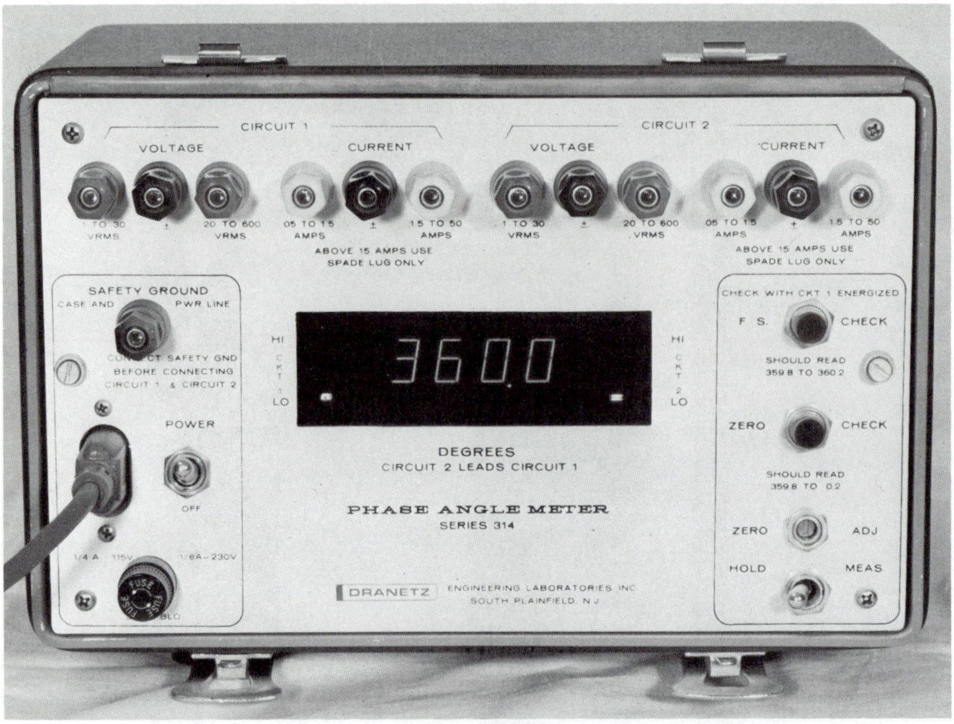

Fig. B.5 Wattmeter Connection of the Electrodynamometer Movement [See Pl. B.11 (c)]

Plate B.6 Dials of Panel Mounted Instruments: (a) ac Wattmeter and (b) Varmeter Resulting from the Insertion of the Proper *RC*-Network in Series with the Potential Coil of Fig. B.5. (Courtesy of Westinghouse Electric Co., Pittsburgh, Pa.)

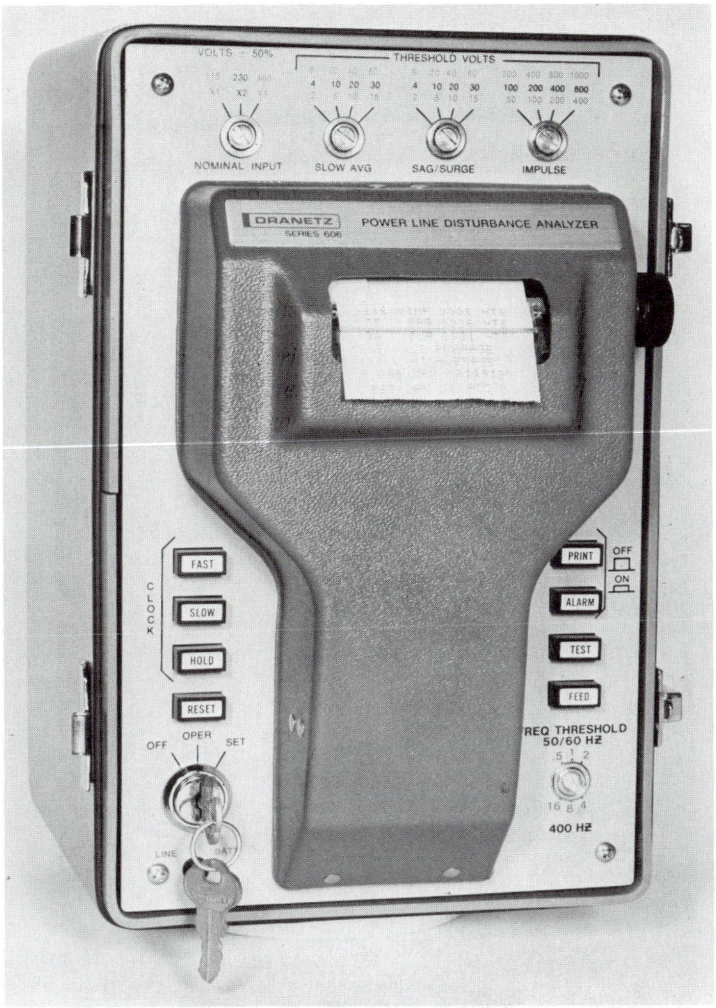

Plate B.7 Model 606 Power Line Disturbance Analyser, for Tracking Down and Identifying Power Line Aberrations; Used to Monitor the Voltage, Frequency, and Waveform of Incoming Commercial Electric Power to: (a) Installed Electric Equipment, (b) Sites where Expansion Is Proposed, (c) Electric Equipment Operating Under Stress Conditions, and (d) Equipment Operating Unattended. (Courtesy of Dranetz Engineering Laboratories, Inc., South Plainfield, N.J.)

The Varmeter

The complex power to a given network is defined in Eq. (A.53) as

$$\mathbf{S} = |\mathbf{V}||\mathbf{I}|\cos\theta_{\mathbf{Z}} + j|\mathbf{V}||\mathbf{I}|\sin\theta_{\mathbf{Z}} \equiv P + jQ \tag{B.19}$$

Since the **wattmeter** responds to

$$P = |\mathbf{V}||\mathbf{I}|\cos\theta_{\mathbf{Z}} \tag{B.20}$$

the **varmeter** should respond to

$$Q = |\mathbf{V}||\mathbf{I}|\sin\theta_{\mathbf{Z}} \tag{B.21}$$

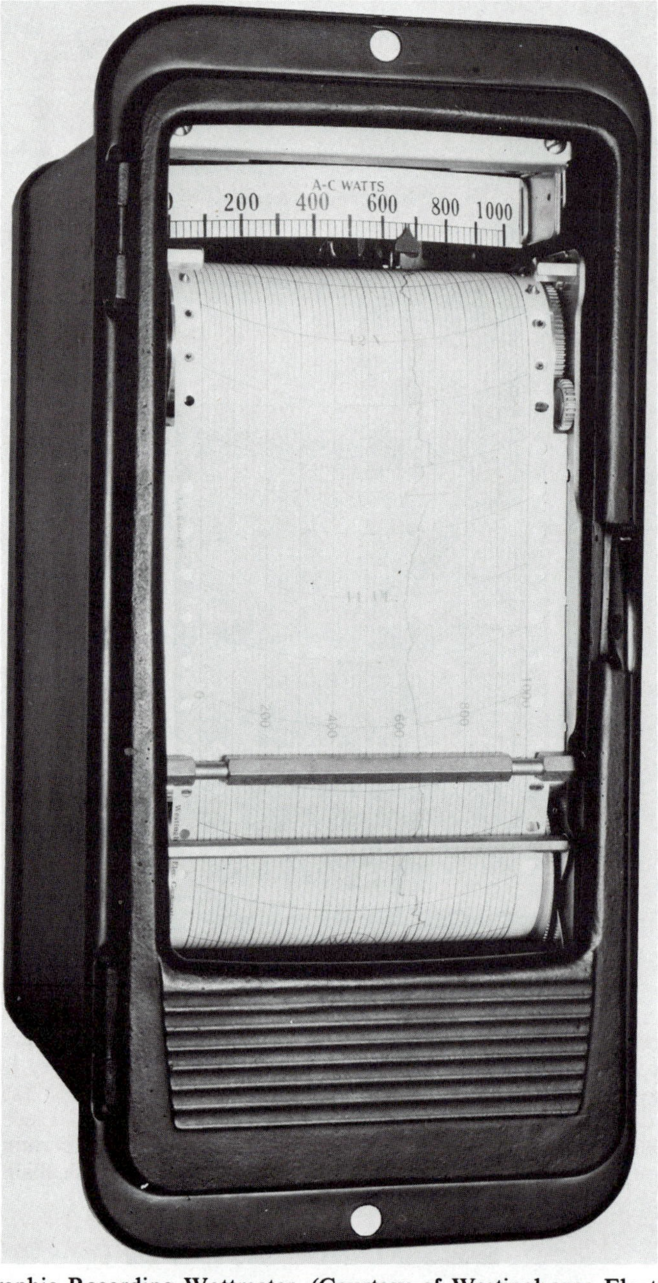

Plate B.8 Graphic Recording Wattmeter. (Courtesy of Westinghouse Electric Co., Pittsburgh, Pa.)

Plate B.9 Single-Phase Watthour Meter—Type I–70–S. (Courtesy of General Electric Co., Schenectady, N.Y.)

However, *since* $\sin\theta_Z = \cos(\theta_Z - 90°)$, *it is possible to make the wattmeter measure vars if the current in the potential coil lags the line voltage by 90°.*

In single-phase networks this required phase shift of the current in the potential coil is accomplished by *inserting a parallel RC-network in series with the potential coil.* This network causes the potential coil current to lead the line voltage by 90°. *By reversing the connections, the required 90° lag is obtained and the wattmeter measures vars.*

For polyphase circuits, the phase-shifted voltages required for reactive power measurement are usually obtained from phase-shifting transformers.

In passing, it should be mentioned that *moving-iron* ammeters and voltmeters, Fig. B.6, are very widely used in ac power applications. The current $i(t)$ in the fixed coil in Fig. B.6 produces a force, **F**, on the plunger that tends to pull it into the coil, and thus change the stored energy $w_L(t) = \frac{1}{2}Li(t)^2$ in the coil, regardless of the polarity of the current $i(t)$. The pointer attached directly to the plunger indicates its angular deflection directly on the scale when equilibrium has been reached between the developed average torque and the spring-restoring torque. The average torque

$$\begin{aligned}\tau_{\text{AVG}} &= \frac{1}{T}\int_0^T \tau_\theta\,dt = \frac{1}{T}\int_0^T \frac{\partial}{\partial\theta} w_L(t)\,dt = \frac{1}{T}\int_0^T \frac{\partial}{\partial\theta}\left[\tfrac{1}{2}Li(t)^2\right]dt \\ &= \frac{1}{2}\frac{\partial L}{\partial\theta}\left[\frac{1}{T}\int_0^T i(t)^2\,dt\right] = \frac{1}{2}\frac{\partial L}{\partial\theta} I_{\text{RMS}}^2 \qquad \text{(B.22)}\end{aligned}$$

depends upon the RMS value of the coil current, Eq. (B.22), for any waveform of the current.

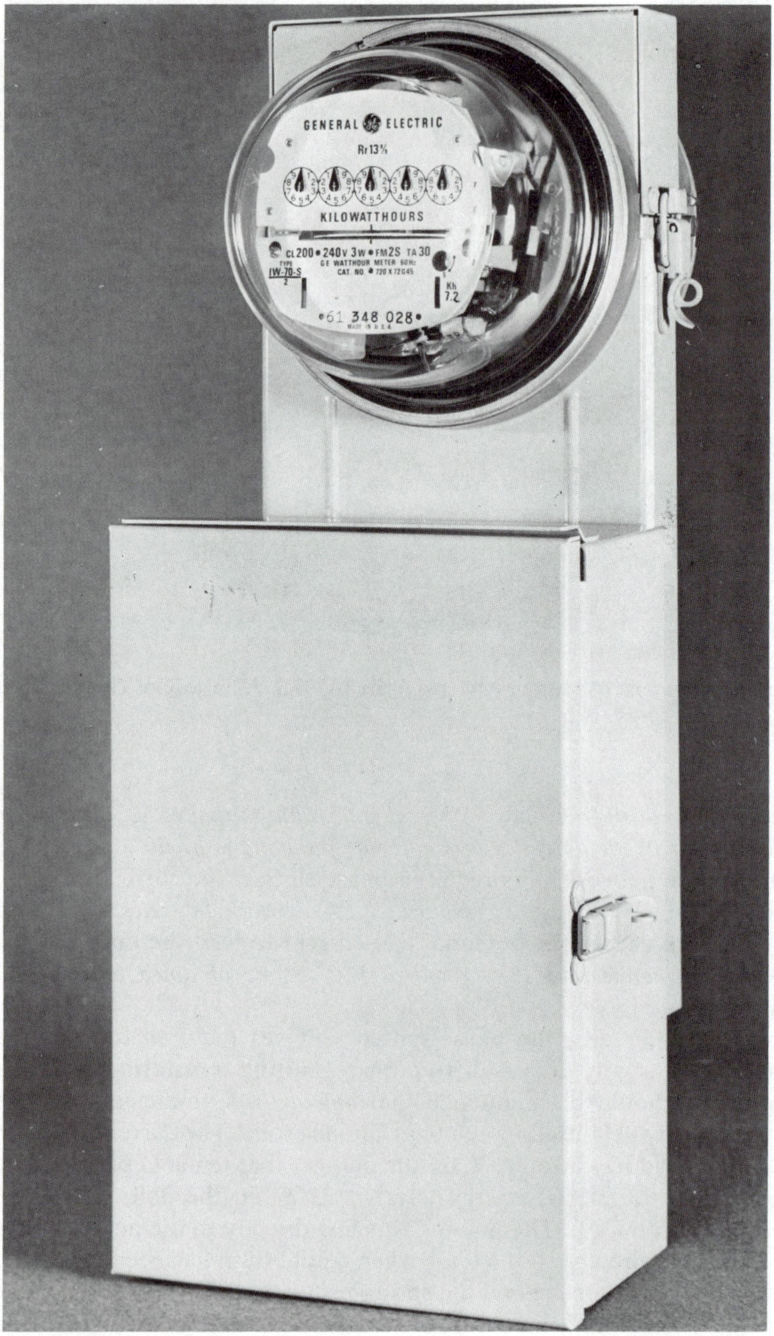

Plate B.10 Magnetic-Tape Demand Recorder, Survey Package Model—Type PDM–75. (Courtesy of General Electric Co., Schenectady, N.Y.)

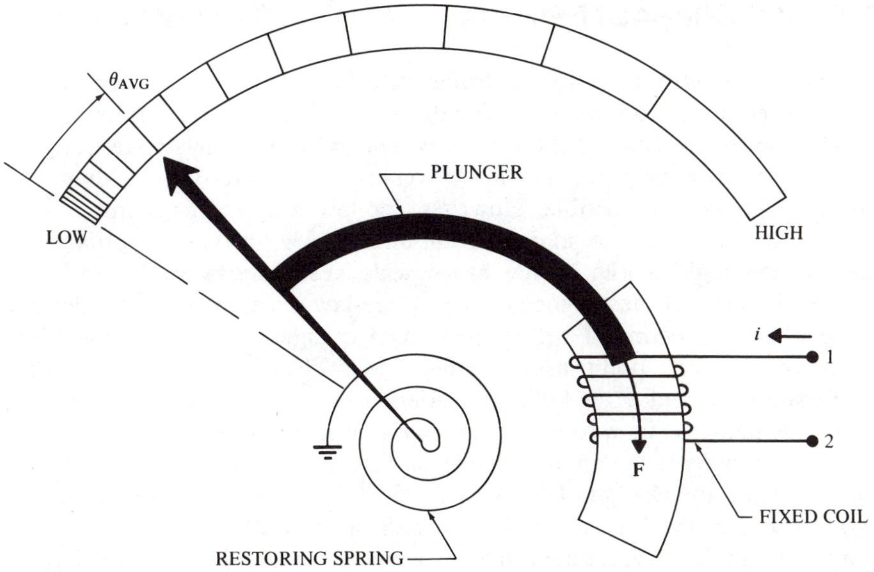

Fig. B.6 Moving Iron Instrument Structure [See Pls. B.11(d) and B.11(e)]

For $\ddot{\theta}=\dot{\theta}=0$ and $K\neq 0$,

$$\tau_{AVG} = \theta_{AVG}K \tag{B.23}$$

Thus

$$\theta_{AVG} = \frac{\tau_{AVG}}{K} = \frac{1}{2K}\frac{\partial L}{\partial \theta} I^2_{RMS} \tag{B.24}$$

This equation suggests that the scale of the instrument, Fig. B.6, almost obeys the square law. However, by proper design of iron vanes, a more *uniform* scale may be achieved especially over the upper portions of the scale by attempting to make $\partial L/\partial\theta$ an inverse function of I_{RMS}. The *compression* of the scale toward the zero end of the scale is unavoidable because $\partial L/\partial\theta$ cannot be made infinite as I_{RMS} approaches zero.

Voltmeters may be constructed out of the ammeter structure, Fig. B.3, by inclusion of a series resistor. However, because the inductance L is large, the increase of the reactance X_L with frequency produces a significant error. This error may be partially overcome, and the range extended up to 2000 or 3000 Hz by shunting a portion of the series resistor with a capacitor C. As the frequency increases, the increased reactance X_L is offset by the decreased reactance X_C. This tends to maintain the total series impedance **Z** nearly constant over a limited frequency range.

Moving-iron instruments are inexpensive, rugged, and capable of accuracies of better than 1 percent in the power frequency range. These advantages overcome the disadvantages, that is, the fact that they are subject to small waveform errors and susceptible to strong external magnetic fields. Thus they are used widely in ac power applications.

B.3 SCALE CHARACTERISTICS AND METER CONNECTIONS

The scale characteristics of the instruments discussed here, ammeter, voltmeter, and wattmeter, are as follows. First, for the ammeter and voltmeter, because of the square-law scale, the coils of the electrodynamometer movement are designed for as uniform a scale as possible. An M versus θ characteristic yielding a fairly uniform upper scale is possible. However, for low-scale readings this is not true because it is impossible to maintain an infinite $\partial M/\partial\theta$ as θ approaches zero. Therefore, **the scale marks in the lower scale are compressed.** Second, for the wattmeter, because of the absence of the square-law scale, the coils are designed to produce a linearly varying M with respect to θ for angular deflections on either side of the M-equal-to-zero point, over a wide range of θ's. M is equal to zero when the angle between Ψ_R and Ψ_S is 90°. This orientation may be made to coincide with the midscale point of the wattmeter. As a consequence the scale of the wattmeter is *uniform* over a wide range about the *midpoint*. However, as Ψ_R approaches Ψ_S, the change of M for any change of θ is very small. This accounts for the *compression* of the scale marks at the *low- and high-scale ends* of the wattmeter.

For an up-scale measurement the instruments should be connected properly. When multiscale instruments are used, the highest scale should be used first. If the deflection is not full but it is relatively low scale (the careful reading of the scale will suggest the available scale to be used), a different scale must be used. As a rule of thumb, try to *utilize full-scale readings for better accuracy.*

Instruments are connected as follows: (1) for a *current measurement* connect the meter in series with the network whose current is being measured, (2) for a *voltage measurement* connect the meter in parallel with the network whose potential is being measured, and (3) for a *power measurement* special attention should be paid because the wattmeter has four available terminals, and *correct connections should be made to these terminals in order to obtain an up-scale reading* on the meter. For clarity it is assumed that the power absorbed by the passive network, Fig. B.7, is to be measured.

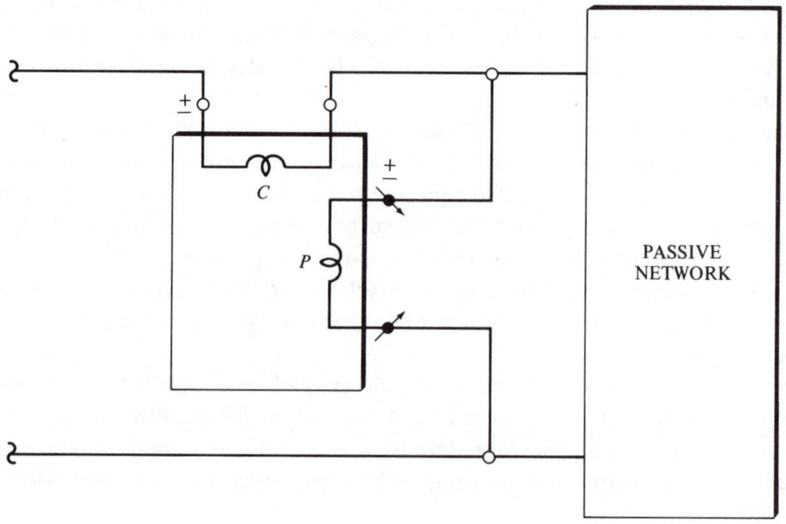

Fig. B.7 Proper Wattmeter Connection for Up–Scale Reading

Note the *connection for up-scale reading*: (1) the *current coil*, marked by heavy circles, is inserted in *series* with one of the conductors connected to the load with the ± *heavy-circle terminal* connected to the **line end**, and (2) the *potential coil*, marked by *arrows*, is inserted between the two conductors preferably on the **load side** of the current coil. The ± arrowed terminal is connected to the side of the unmarked heavy-circle terminal of the current coil.

A reversal of *either* coil, but not both, will cause the meter to try to deflect *down scale*. A reversal of both coils will have no effect on the reading.

It is essential to keep in mind that the connection of the wattmeter is such that *an up-scale reading corresponds to positive, that is, absorbed, power.* If the pointer rests against the down-scale stop, a reversal of the potential coil (reversal of the potential coil is quicker in practice) will yield the up-scale reading of the negative, that is, generated, power.

B.4 THE OHMMETER

There is one more essential measurement in the field of electrical engineering, the measurement of resistance. The measurement is made with an instrument commonly known as the **ohmmeter**. *An ohmmeter is essentially an ammeter calibrated in ohms.* Fig. B.8, shows the circuit of a simple type of an ohmmeter.

The emf E_S represents the emf of a dry cell that is used in the meter. Dry cells possess emf's of a fair order of constancy. The resistance R_S represents the internal resistance of the dry cell. The resistance R_C represents the resistance of the rotating coil. The resistor R is necessary to limit the current to the range of the ammeter element when the *external unknown resistor* R_x approaches zero. A *permanent magnet* provides the fixed magnetic field that is essential for the operation of the instrument. The *permanent magnet, direct-current coil movement* is known as the

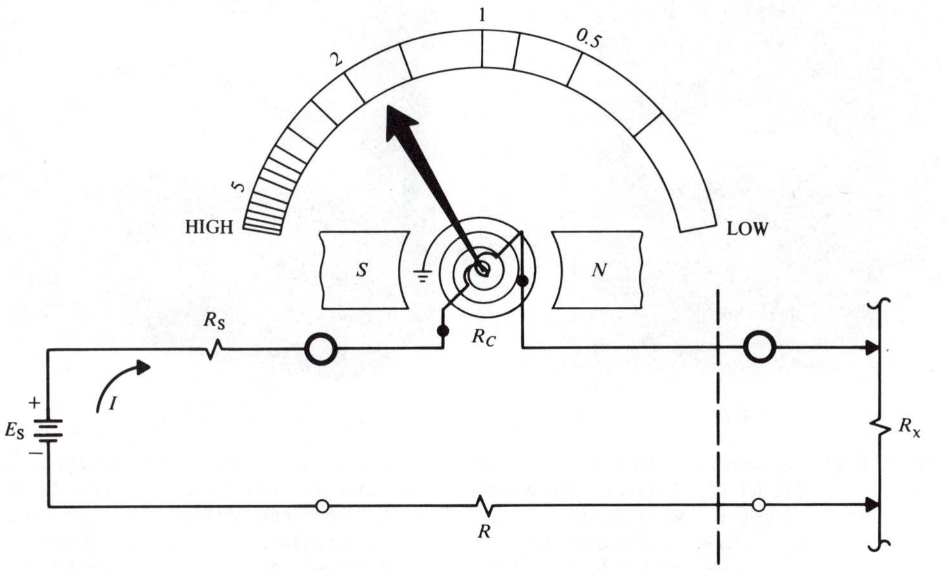

Fig. B.8 The Ohmmeter Connection

(a)

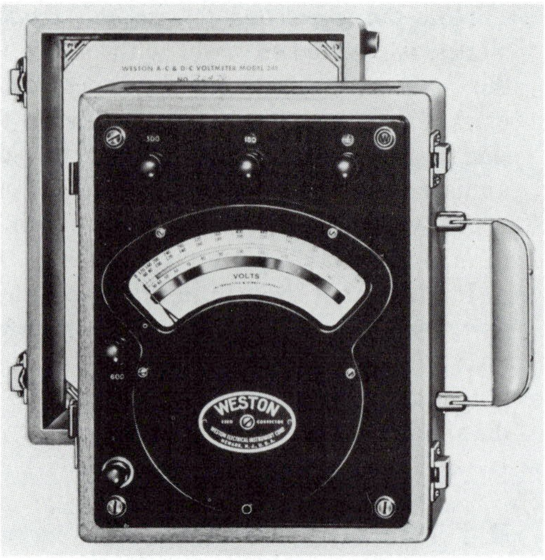

(b)

(c)

(d)

Plate B.11 Laboratory Standard Instruments—ac and dc **Electrodynamometer Type**: (a) Model 370 Ammeter, (b) Model 341 Voltmeter, and (c) Model 310 Wattmeter (When Properly Connected Also Used as Varmeter); **Moving Iron Type** Model 904: (d) ac Voltmeter and (e) ac Ammeter; **Clamp On Type** ac Ammeter–Voltmeter: (f) Model 633 VA–1 and (g) Model 749; and (h) Model 660 Volt Ohm Meter (VOM). (Courtesy of Weston Instruments, Newark, N.J.)

(e)

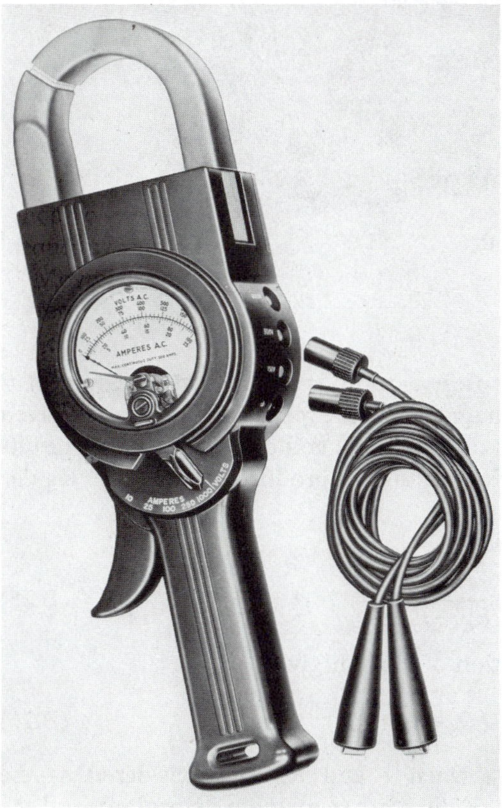

(f)

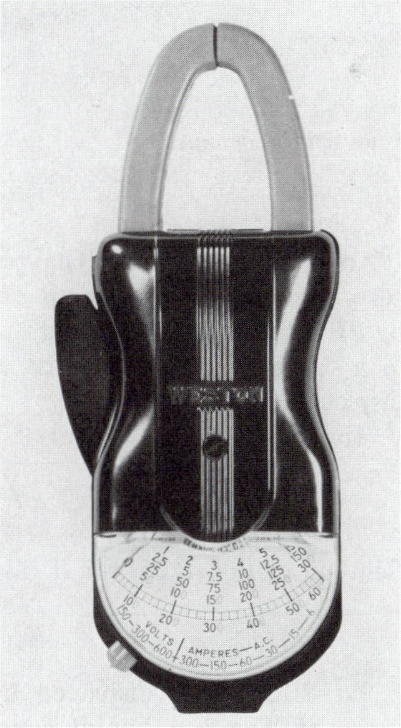

(g)

Plate B.11 (*Continued*)

(h)

Plate B.11 (*Continued*)

D'Arsonval mechanism. The principle of operation is as follows: A current I carrying conductor of length $d\mathbf{L}$ in a magnetic field experiences a force $d\mathbf{F}$ given by $d\mathbf{F}=I\,d\mathbf{L}\times\mathbf{B}$. Integrating this effect on the whole coil of the movement results in a torque that acts against a restoring spring to produce a resulting angular displacement of the pointer.

The current of the circuit of Fig. B.8 is

$$I=\frac{E_S}{R_S+R_C+R+R_x} \tag{B.25}$$

Solution of Eq. (B.25) for R_x as a function of current gives

$$R_x=\frac{E_S}{I}-(R_S+R_C+R) \tag{B.26}$$

Eq. (B.26) is a displaced hyperbola when I and R_x are considered as the variables. The $R_x=0$ point is on the top of the scale (maximum deflection), and the $R_x=\infty$ point is at the bottom (no deflection). A relatively small change in

Plate B.12 600 V Current Transformer—Type JAK–0. (Courtesy of General Electric Co., Schnectady, N.Y.)

Plate B.13 120 V Potential Transformer—Type JVA–0. (Courtesy of General Electric Co., Schenectady, N.Y.)

resistance produces a fairly large change in pointer deflection when the resistance R_x is small, but a relatively large change in resistance is required to produce any appreciable change in pointer deflection when the resistance R_x is large.

B.5 INHERENT ERRORS IN MEASUREMENTS

As a rule of thumb, the more expensive an instrument is the more accurate it is. This is a consequence of the engineering put into the design, development, manufacturing, and testing that precedes production of the instrument. Unfortunately, all these processes are expensive propositions, and if one pays enough, one should get an excellent instrument. However, even if the instruments purchased are high quality, the risk still exists that errors in measurements will occur. E. Frank in Section 5-7 (see Reference [2]) puts it as follows:

> A large class of methods of measurement is that in which the actual deflection of the instrument provides a numerical result. These deflection methods are all highly vulnerable to instrument errors, as contrasted with methods in which a null indication is used as a basis for comparing an unknown quantity with an accurately known standard. Systematic errors owing to shortcomings or defects of the instrument employed in making a measurement are normally to be expected. Therefore, deflection methods are usually avoided when seeking extremely high accuracy. Despite the tremendous improvement achieved over the last few decades, instrument errors are still inevitable, and the experimenter must decide whether these errors are small enough to be tolerated in any given instance.
>
> Many instrument errors are covered under the umbrella of the term "off calibration," which refers to the discrepancy between scale readings and the magnitudes of the quantity that produced the readings. Errors in calibration may be an inherent shortcoming of a new instrument, they may be produced by wear and deterioration of internal elements of the instrument, or they may be induced by abuse in which components of the instrument suffer damage.
>
> For each type of instrument there are an enormous number of items that may produce errors, the details depending upon the particular kind of device. The proficient experimenter will always take precautions to ensure that the instrument he uses is operating normally and that it does not contribute excessive errors. Critical tests of instruments to check their performance and accuracy may be made in many different ways. Faults in instruments may sometimes be detected by simple tests in which the behavior is scrutinized for erraticness, instability, and lack of reproducibility. An easily applied method is to compare the instrument with a similar or better one that is known to be reliable. A recommended method for correcting defects in the case of commercial apparatus is to return it to the manufacturer for calibration and repair.
>
> To illustrate some instrument ills that may plague the experimenter, several items that cause errors in pivoted-coil d'Arsonval instruments, which have been studied in preceding chapters, may be cited. Errors may arise from erratic pivot friction resulting from worn bearings. Even if the pivot and bearings are in new condition, errors may arise from this source in some

instruments if they are used in a physical orientation for which they were not designed. The springs that provide the restoring torque may undergo change with age and use. The shunt of an ammeter or the series resistance of a voltmeter might shift from its correct value because of abuse, such as extreme overload. The magnetic field in the air gap could become permanently altered as a result of having exposed the meter to an intense external magnetic field at some time in its past. Or, other magnetic effects may cause error, such as use of an unmounted instrument designed for steel-panel installations. Corrosion may cause ill effects on delicate metal parts such as coil wire and springs. Finally, a somewhat subtle example of a source of error in a high-quality meter (usually employing a mirror mounted in the plane of the scale to avoid parallax error) may be mentioned. This type of meter usually has slightly irregular scale graduations owing, for example, to small irregularities in the magnetic field. The scale is tailor-made during factory calibration. If the needle on such an instrument becomes bent, it still might be capable of being zeroed with the screw adjustment provided, but then the scale is no longer accurate and recalibration is required.

It is virtually impossible for the experimenter to be acquainted with all the possible difficulties his apparatus may experience. But the competent man should have a general knowledge of the characteristics, capabilities, and limitations of the instrument and should be able to discern when its errors become excessive for the purpose at hand.

Generally, the ammeter and voltmeter provide accurate readings if the resistance of the ammeter coil is zero and the resistance of the voltmeter coil is infinite. However, *the reading of the wattmeter depends on whether the meter is designed to provide compensated readings, that is, accurate readings, or is designed to provide readings to be corrected,* in which case the method of correction should be known.

The method of correction depends on the connection of the meter. For the *orthodox wattmeter connection*, Fig. B.9, the currents and voltages are related as follows:

$$v_L = i_L R_L = i_P R_M \tag{B.27}$$

and

$$i_C = i_L + i_P = i_L\left(1 + \frac{R_L}{R_M}\right) \tag{B.28}$$

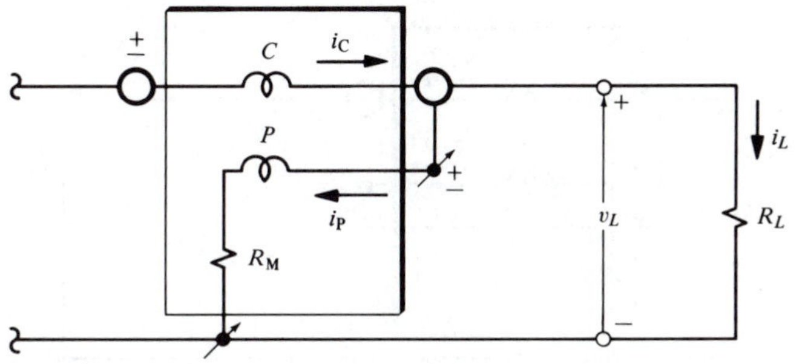

Fig. B.9 The Orthodox, Uncompensated Wattmeter Connection

The wattmeter reading is obtained by direct application of Eq. (B.18). Thus

$$\begin{aligned} P_R &= \frac{R_M}{T}\int_0^T i_L\left(1+\frac{R_L}{R_M}\right)\left(\frac{v_L}{R_M}\right)dt \\ &= \frac{1}{T}\int_0^T i_L v_L\,dt + \frac{R_L{}^2}{R_M T}\int_0^T i_L{}^2\,dt \\ &= \frac{R_L}{T}\int_0^T i_L{}^2\,dt + \frac{R_M}{T}\int_0^T i_P{}^2\,dt \\ &= R_L I^2_{L[RMS]} + R_M I^2_{P[RMS]} \equiv P_L + P_P \end{aligned} \tag{B.29}$$

where P_L is the power dissipated in the load and P_P is the power dissipated in the potential coil. Thus the wattmeter reading P_R *exceeds* the power dissipated in the load by an amount equal to P_P.

For the *unorthodox wattmeter connection*, Fig. B.10, the currents and voltages are related as follows:

$$\begin{aligned} v_L &= i_L R_L \\ i_C &= i_L \end{aligned} \tag{B.30}$$

and

$$i_P R_M = v_L + R_C i_C \tag{B.31}$$

where R_C is the resistance of the current coil or

$$i_P = \frac{v_L + R_C i_C}{R_M} = \frac{i_L}{R_M}(R_L + R_C) \tag{B.32}$$

Again, the wattmeter reading P_R is obtained by direct application of Eq. (B.18). Thus

$$\begin{aligned} P_R &= \frac{R_M}{T}\int_0^T \frac{i_L}{R_M}(R_L+R_C)i_L\,dt \\ &= \frac{1}{T}\int_0^T (R_L+R_C)i_L{}^2\,dt = \frac{R_L}{T}\int_0^T i_L{}^2\,dt + \frac{R_C}{T}\int_0^T i_L{}^2\,dt \\ &= R_L I^2_{L[RMS]} + R_C I^2_{C[RMS]} \equiv P_L + P_C \end{aligned} \tag{B.33}$$

where P_L is the power dissipated in the load and P_C is the power dissipated in the current coil. Thus the wattmeter reading now *exceeds* the power dissipated in the load by an amount equal to P_C.

For a generalized load $\mathbf{Z}_L$ and a sinusoidal forcing function, the wattmeter will read *high* by the amounts P_P and P_C as given by Eqs. (B.29) and (B.33).

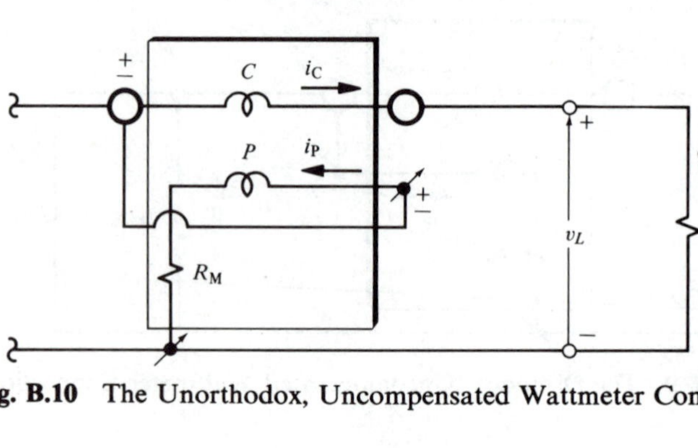

Fig. B.10 The Unorthodox, Uncompensated Wattmeter Connection

To make a choice as to which instrument connection, that is, the connection depicted in Fig. B.9 or Fig. B.10, should be used, one should reason as follows: How can one obtain P_P or P_C? Does P_P or P_C remain constant with variable load conditions? Thus availability and constancy of P_P or P_C are the deciding factors. When the internal impedance of the power source is small as compared to $\mathbf{Z}_L(R_L)$, the connection of Fig. B.9 is used. This connection is preferable because for a constant power source and small R_C and variable $\mathbf{Z}_L(R_L)$, the power dissipated in R_M, that is, P_P, remains constant. This enables easy correction of the different readings that result from varying load conditions by *subtracting* from each reading a constant correction factor. *The correction factor may be calculated by knowing the resistance of the potential coil R_M, which is provided by the manufacturer, and a knowledge of the load voltage.* One approximation of the correction factor may be obtained if the load is temporarily disconnected and the wattmeter is read. The reading is approximately P_P. Under load conditions the current through R_M is different from that of no-load conditions because of the reduction of the voltage across R_M due to the voltage drop across R_C and the internal impedance of the power source. However, if other information is not available, the no-load power reading is an excellent *approximation* of the correction factor.

The connection of Fig. B.10 is *preferred* when no corrections are to be made. The reason is that P_C is small. Also, there are no ways to calculate it since R_C varies due to heating of the current coil, which may carry extremely high currents, and there are no direct ways of measuring them as in the previous case.

There exists commercially a wattmeter that reads *correct* values directly. This wattmeter is comprised of two coils that are connected as shown in Fig. B.11.

Note that the wattmeter has an additional winding, known as a *compensating winding*, CW, connected in series with the *P*-coil but along the axis of the current coil and wound in such a way as to induce a current to the current coil of such magnitude and direction as to *reduce* the reading of the wattmeter by an amount P_P. Therefore, *the wattmeter is capable of reading the actual load power directly.* The effect of the compensating winding on the wattmeter can be checked by disconnecting the load temporarily. For proper compensation, the wattmeter should read zero rather than P_P, as when connected as in Fig. B.9.

In ohmmeters, errors result due to changes in the emf of the dry cell and its internal resistance R_S over long periods of time. The largest error occurs at the top of the current scale, that is, $R_x = 0$. The simplest method to *compensate* for this error, that is, erroneous resistance readings, is to short-circuit the output leads of the ohmmeter—that is, set $R_x = 0$ and, utilizing the zero-adjust potentiometer on

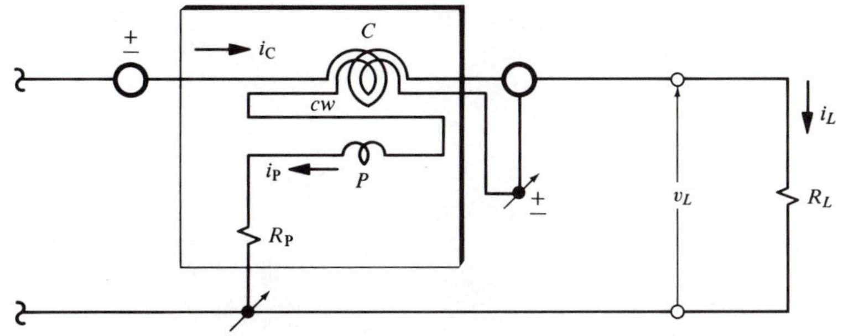

Fig. B.11 The Orthodox, Compensated Wattmeter Connection

the front of the instrument, adjust the resistor R, Fig. B.8, until the meter reads zero ohms. Although this adjustment is not perfect, it is a good first attempt. If the range of adjustment is exceeded and the meter does not read zero when the external leads are shorted and $R_x = 0$, the meter should not be used; it should be serviced.

To maintain measurements within the guaranteed accuracy of the ohmmeter, a limit is set to the range through which R may be adjusted by the zero-resistance adjusting potentiometer.

B.6 SUMMARY

In this appendix the principles of operation of (1) ammeter, (2) voltmeter, (3) wattmeter, (4) varmeter, and (5) ohmmeter are set forth. Equivalent circuits are drawn for each instrument and the distinct differences between each of these instruments are given. Also, the proper way of utilizing these instruments for the appropriate measurements is set forth. A great effort is made to provide adequate information to the reader so that he or she can feel competent in using these instruments. The information here is **not exhaustive** and one should use the literature or electrical measurement texts for supplementary information. However, careful reading of this appendix will prove adequate, especially in understanding the proper connections for wattmeter readings, corrections, and so on.

Possible sources of errors in measurements are discussed in this appendix and methods for correcting these errors are suggested.

REFERENCES

1. Boast, W. B., *Principles of Electric and Magnetic Circuits*. Harper & Row, New York, 1950.
2. Frank, E., *Electrical Measurements*. McGraw-Hill, New York, 1959.
3. Hayt, W. H., Jr., and Kemmerly, J. E., *Engineering Circuit Analysis*. McGraw-Hill, New York, 1962.
4. Kinnard, I. F., *Applied Electrical Measurements*. Wiley, New York, 1956.

C STABILITY THEORY

C.1 INTRODUCTION

The objectives of Appendix C are to (1) set forth the basic definitions of **stability, instability**, and **marginal stability**, and (2) establish procedures for checking whether a given system is **stable, unstable**, or **marginally stable**.

C.2 t-DOMAIN RESPONSE OF THE FIRST-ORDER SYSTEM WITH NO FORCING FUNCTION

The describing equation of Fig. C.1 is obtained through the application of Kirchhoff's current law; that is,

$$\sum_j i\text{'s} = 0$$

This law can be written in terms of circuit variables as

$$i_C(t) + i_R(t) = 0 \tag{C.1}$$

However, the current $i_C(t)$ through the capacitor C is related to the voltage $v_C(t)$ across this capacitor as

$$i_C(t) = C\frac{d}{dt}v_C(t) \tag{C.2}$$

Similarly, the voltage $v_R(t)$ across the resistor R is related to the voltage $v_C(t)$ across the capacitor C as

$$v_R(t) = v_C(t) \tag{C.3}$$

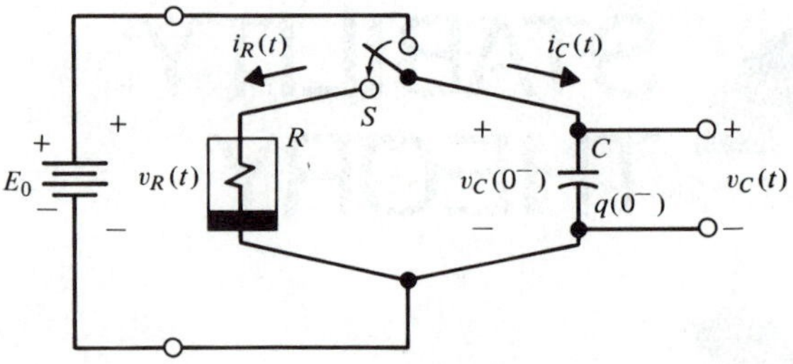

Fig. C.1 A Parallel-connected Network Comprised of a Nonlinear Resistor R and a Capacitor C with an Initial Voltage $v_C(0^-) = E_0$ Across It

Utilization of Ohm's law, that is,

$$i_R(t) = \left(\frac{1}{R}\right)v_R(t) \tag{C.4}$$

and substitution of Eq. (C.3) into Eq. (C.4), and Eqs. (C.2) and (C.4) into Eq. (C.1), yields

$$C\frac{d}{dt}v_C(t) + \left(\frac{1}{R}\right)v_C(t) = 0 \tag{C.5}$$

Note that Eq. (C.5) is a **first-order** differential equation in the dependent variable $v_C(t)$, with **no forcing function.** Such an equation is defined as **homogeneous** and has a solution of the exponential form written as

$$v_C(t) = Ke^{st} \tag{C.6}$$

The coefficient s of t is defined as the **complex frequency** and is designated as $s = \alpha \pm j\omega$.

Substitution of Eq. (C.6), as well as of its first derivative, into Eq. (C.5) and factoring out of Ke^{st} yields

$$\left(Cs + \frac{1}{R}\right)Ke^{st} = 0 \tag{C.7}$$

However, since by definition, $Ke^{st} \neq 0$, Eq. (C.6), the parentheses of Eq. (C.7) must be equal to zero, that is,

$$Cs + \frac{1}{R} = 0 \tag{C.8}$$

This equation is defined as the **characteristic equation,** and the polynomial is the **characteristic polynomial** of Eq. (C.5) and is *designated* as $C(s)$. This polynomial is zero if and only if its root $s = s_1$ is

$$s_1 = -\frac{1}{RC} \tag{C.9}$$

This value of s_1 is known as the **root** or the **characteristic value** of $C(s) = 0$ and is always a function of the system parameters R and C. With this value of s_1, the assumed solution of Eq. (C.5), given as Eq. (C.6), always satisfies the describing equation of the given system [i.e., Eq. (C.5)] no matter what the value of K may be.

Substitution of Eq. (C.9) into Eq. (C.6) yields

$$v_C(t) = Ke^{-(1/RC)t} \tag{C.10}$$

The principle of conservation of charge, Definition D1,* as applied to Eq. (C.10),

*Please see Appendix D, which is comprised of a set of definitions pertinent to this text.

yields

$$v_C(0^+) = Ke^{-(1/RC)\{0\}} = E_0,\ t = 0^+ \tag{C.11}$$

or

$$K = E_0 \tag{C.12}$$

Substitution of Eq. (C.12) into Eq. (C.10) yields

$$v_C(t) = E_0 e^{-(1/RC)t} \tag{C.13}$$

Note that the system's response $v_C(t)$ is written in terms of the initial voltage across the capacitor $E_0 = v_C(0^-)$ and of the circuit parameters R and C.

C.3 s-DOMAIN ANALYSIS OF THE FIRST-ORDER SYSTEM WITH NO FORCING FUNCTION

If we were to use Definition D.3, we would be able to redraw the circuit analog of the parallel-connected RC-system, Fig. C.1, which has no forcing function but has an initial voltage E_0 across the capacitor. This system would be redrawn as a system of the same basic structure but with no initial voltage E_0 across the capacitor. **Instead**, an impulse function $\delta(t)$ current generator of strength CE_0 would be inserted as a forcing function across the given system, Fig. C.2.

Utilization of Kirchhoff's current law and Eqs. (C.2) and (C.4) enables us to write the describing equation of Fig. C.2 in the t-domain as

$$C\frac{d}{dt}v_C(t) + \frac{1}{R}v_C(t) = CE_0\delta(t) \tag{C.14}$$

Now utilization of Definitions D.7 [the initial condition is set equal to zero; i.e., $v_C(0^+)=0$] and D.6, and the Laplace transform pair No. 1 of Table D.1, enables us to write the describing equation of Fig. C.2 in the s-domain, after factoring out $V_C(s)$, as

$$\left(Cs + \frac{1}{R}\right)V_C(s) = CE_0\Delta(s)|_{\Delta(s)\equiv 1} \tag{C.15}$$

Application of Definition D.12 to Eq. (C.15) enables us to write the transfer

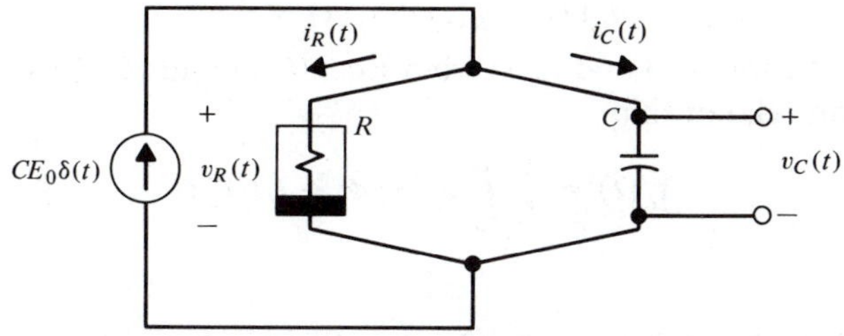

Fig. C.2 System of Fig. C.1 with the Initial Voltage E_0 Across the Capacitor Represented by an Impulse Function Current Generator $\delta(t)$ of Strength CE_0 Connected Across the Given System, That Is, $i_{GC}(t) = CE_0\delta(t)$

function $G_{1ST}(s)$ of the system exhibited in Fig. C.2, for **zero input**,* and thus the system exhibited in Fig. C.1, as

$$G_{1ST}(s) \equiv \frac{V_C(s)}{\Delta(s)} = \frac{CE_0}{Cs+(1/R)} \tag{C.16}$$

Note that the denominator of Eq. (C.16) is a **first-order polynomial in** s. Set equal to zero, this polynomial is defined as the **characteristic equation** of the system at hand, Fig. C.1, and is *designated* as

$$D(s) = Cs + \frac{1}{RC} = 0 \tag{C.17}$$

When $D(s)=0$ is solved for s, in terms of the circuit parameters R and C, it yields the value of $s=s_1$, which satisfies the characteristic polynomial. This value of s, that is,

$$s_1 = -\frac{1}{RC} \tag{C.18}$$

is defined as the **root** or the **characteristic value** of $D(s)=0$ and plays a **vital role** in this system behavior study. This becomes obvious if one were to pursue the solution of Eq. (C.16) for $V_C(s)$. After the factoring out and cancellation of C, and the utilization of the defining relationship $\Delta(s)=1$, Eq. (C.16) yields

$$V_C(s) = \frac{CE_0}{C[s+(1/RC)]} \times 1 = \frac{E_0}{s+(1/RC)} \tag{C.19}$$

The t-domain expression of Eq. (C.19), $v_C(t)=\mathcal{L}^{-1}[V_C(s)]$, has the same value as the *zero-input-response* with initial condition E_0; that is, the initial condition is replaceable by the excitation impulse. Referred to as the **impulse response**, it can be written with the aid of the Laplace transform pair No. 5, Table D.1, as

$$v_C(t) = E_0 e^{-(1/RC)t} \tag{C.20}$$

Again, note that the system's t-domain response $v_C(t)$, Fig. C.1, is written in terms of the initial voltage across the capacitor $E_0 = v_C(0^-)$ (designated as part of the strength of the impulse in Fig. C.2) and of the circuit parameters R and C.

C.4 t-DOMAIN RESPONSE OF THE SECOND-ORDER SYSTEM WITH NO FORCING FUNCTION

The describing equation of Fig. C.3 is again obtained through the utilization of Kirchhoff's current law. Thus

$$i_C(t) + i_R(t) + i_L(t) = 0 \tag{C.21}$$

Proper substitution for $i_C(t)$ and $i_R(t)$ from Eqs. (C.2) and (C.4) as well as for $i_L(t)$ from the defining relationship

$$i_L(t) = \frac{1}{L}\int_0^t v_L(\tau)\,d\tau + i_L(0^+) \tag{C.22}$$

*If the excitation of the given system is an impulse, then, for all $t>0$, there is **no input** or **excitation** to the system.

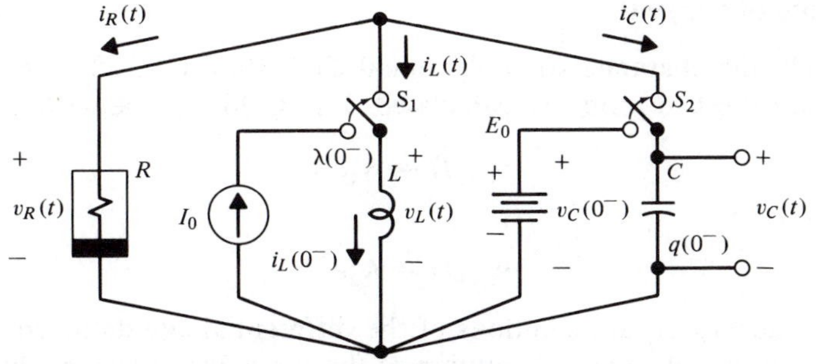

Fig. C.3 A Parallel-connected Network Comprised of a Nonlinear Resistor R, a Capacitor C, with an Initial Voltage $v_C(0^-) = E_0$ Across It, and an Inductor L with an Initial Current $i_L(0^b\text{-}) = I_0$ Through It

where

$$v_L(t) = v_C(t) \equiv v_R(t) \tag{C.23}$$

into Eq. (C.21) yields

$$C\frac{d}{dt}v_C(t) + \left(\frac{1}{R}\right)v_C(t) + \frac{1}{L}\int_0^t v_C(\tau)\,d\tau + i_L(0^+) = 0 \tag{C.24}$$

Differentiation of Eq. (C.24) with respect to time yields

$$C\frac{d^2}{dt^2}v_C(t) + \left(\frac{1}{R}\right)\frac{d}{dt}v_C(t) + \left(\frac{1}{L}\right)v_C(t) = 0 \tag{C.25}$$

Note that Eq. (C.25) is a **second-order differential equation** in the dependent variable $v_C(t)$, with **no forcing function**. Such an equation is also defined as **homogeneous** and has a solution of the exponential form written as

$$v_C(t) = Ke^{st} \tag{C.26}$$

Substitution of Eq. (C.26), as well as its first and second derivatives, into Eq. (C.25) and factoring out Ke^{st} yields

$$\left[s^2 + \left(\frac{1}{RC}\right)s + \frac{1}{LC}\right]Ke^{st} = 0 \tag{C.27}$$

Since by definition $Ke^{st} \neq 0$, Eq. (C.26), the parentheses of Eq. (C.27) must be equal to zero. That is, the **characteristic equation** $C(s) = 0$ of the second-order differential equation, Eq. (C.25), becomes

$$s^2 + \left(\frac{1}{RC}\right)s + \frac{1}{LC} = 0 \tag{C.28}$$

The **roots** of this second-order polynomial in s, known as the **characteristic values** of $C(s) = 0$, can be found as follows:

$$s_{1,2} = -\frac{1}{2RC} \pm j\left[\left(\frac{1}{LC}\right) - \left(\frac{1}{2RC}\right)^2\right]^{1/2} \tag{C.29}$$

The roots s_1 and s_2 of the characteristic polynomial $C(s) = 0$, Eq. (C.28), may be either **distinct** (i.e., $s_1 \neq s_2$) or **not distinct** (i.e., $s_1 = s_2$; in other words, the roots s_1 or s_2 may have a **multiplicity** $k = 2$.) These two cases are now studied in detail.

Distinct Roots of $C(s)=0$

The roots of the characteristic polynomial $C(s)=0$, Eq. (C.28), are distinct, that is, $s_1 \neq s_2$. Once the two roots are calculated, Eq. (C.26) can be written as

$$v_{C1}(t) = K_1 e^{s_1 t} \tag{C.30}$$

and

$$v_{C2}(t) = K_2 e^{s_2 t} \tag{C.31}$$

If both $v_{C1}(t)$ and $v_{C2}(t)$ are solutions of the differential equation, Eq. (C.25), then their sum constitutes the general solution of the same differential equation [i.e., Eq. (C.25)]. Thus

$$v_C(t) = K_1 e^{s_1 t} + K_2 e^{s_2 t} \tag{C.32}$$

To calculate the constants K_1 and K_2, we must obtain two equations that contain the two unknowns K_1 and K_2. Two such equations are obtained as follows. The principle of conservation of charge, Definition D.1, as applied to Eq. (C.32) yields

$$v_C(0^+) = K_1 e^{s_1\{0\}} + K_2 e^{s_2\{0\}} = E_0,\ t = 0^+ \tag{C.33}$$

or

$$K_1 + K_2 = E_0 \tag{C.34}$$

However, since $v_L(t) \equiv v_C(t)$, Fig. C.3, substitution of Eq. (C.32) into Eq. (C.22) yields

$$i_L(t) = \frac{1}{L}\int_0^t (K_1 e^{s_1\tau} + K_2 e^{s_2\tau})\,d\tau + i_L(0^+) \tag{C.35}$$

Integration of Eq. (C.35), remembering that

$$\lim_{t\to 0^+} \frac{1}{L}\int_{-\infty}^t v_L(t)\,dt = i_L(0^+)$$

yields

$$i_L(t) = \frac{1}{L}\left[\left(\frac{K_1}{s_1}\right)e^{s_1 t} + \left(\frac{K_2}{s_2}\right)e^{s_2 t}\right] \tag{C.36}$$

Now application of the principle of conservation of flux linkages, Definition D.2, to Eq. (C.36) yields

$$i_L(0^+) = \frac{1}{L}\left[\left(\frac{K_1}{s_1}\right)e^{s_1\{0\}} + \left(\frac{K_2}{s_2}\right)e^{s_2\{0\}}\right] = I_0,\ t = 0^+ \tag{C.37}$$

or

$$\left(\frac{1}{Ls_1}\right)K_1 + \left(\frac{1}{Ls_2}\right)K_2 = I_0 \tag{C.38}$$

The rewriting of Eqs. (C.34) and (C.38) in matrix form yields

$$\left[\begin{array}{c|c} 1 & 1 \\ \hline \dfrac{1}{s_1 L_1} & \dfrac{1}{s_2 L_2} \end{array}\right]\left[\begin{array}{c} K_1 \\ \hline K_2 \end{array}\right] = \left[\begin{array}{c} E_0 \\ \hline I_0 \end{array}\right] \tag{C.39}$$

Finally, solution of Eq. (C.39) for the constants K_1 and K_2 yields

$$\begin{bmatrix} K_1 \\ \hline K_2 \end{bmatrix} = \frac{Ls_1s_2}{s_1-s_2}\left[\begin{array}{c|c} \dfrac{1}{s_2L} & -1 \\ \hline -\dfrac{1}{s_1L} & 1 \end{array}\right]\begin{bmatrix} E_0 \\ \hline I_0 \end{bmatrix} \tag{C.40}$$

We can simplify Eq. (C.40) if we substitute from Eq. (C.29) and calculate the product s_1s_2 as a function of circuit parameters R, L, and C. Such substitution from Eq. (C.29) yields

$$s_1s_2 = \left(-\frac{1}{2RC}\right)^2 - \left\{\left[\left(\frac{1}{2RC}\right)^2 - \frac{1}{LC}\right]^{1/2}\right\}^2 = \frac{1}{LC} \tag{C.41}$$

Now substitution of Eq. (C.41) into Eq. (C.40) and proper simplification yields

$$\begin{bmatrix} K_1 \\ \hline K_2 \end{bmatrix} = \begin{bmatrix} \dfrac{1}{s_1-s_2}\left(s_1E_0 - \dfrac{I_0}{C}\right) \\ \hline \dfrac{1}{s_1-s_2}\left(-s_2E_0 + \dfrac{I_0}{C}\right) \end{bmatrix} \tag{C.42}$$

Eq. (C.42) enables us to write Eq. (C.32) as

$$v_C(t) = \left\{\frac{[s_1E_0-(I_0/C)]}{s_1-s_2}\right\}e^{s_1t} + \left\{\frac{-[s_2E_0-(I_0/C)]}{s_1-s_2}\right\}e^{s_2t} \tag{C.43}$$

Note that the response $v_C(t)$, Eq. (C.43), of the system, Fig. C.3, is written in terms of the initial voltage across the capacitor E_0, the initial current through the inductor I_0, and the roots or characteristic values of $C(s)=0$, that is, the roots s_1 and s_2 of the characteristic polynomial $C(s)=0$, Eq. (C.28).

Nondistinct Roots of $C(s)=0$

The roots of the characteristic polynomial $C(s)=0$, Eq. (C.28), are not distinct, that is, $s_1=s_2$. Once these two roots are calculated, Eq. (C.32) can be written as

$$v_C(t) = \left\{\sum_{\substack{j=k\\ i=1}}^{\substack{i=k\\ j=1}} K_i t^{k-j}\right\}e^{s_1t} \tag{C.44}$$

For a multiplicity of roots $k=2$, Eq. (C.44) can be written as

$$v_C(t) = (K_1+K_2t)e^{s_1t} = K_1e^{s_1t} + K_2te^{s_1t} \tag{C.45}$$

The constants K_1 and K_2 of Eq. (C.45) are calculated following the same procedure that is followed for the case of distinct roots. That is, application of the principle of conservation of charge, Definition D.1, to Eq. (C.45) yields

$$v_C(0^+) = K_1e^{s_1\{0\}} + \{0\}K_2e^{s_1\{0\}} = E_0,\ t=0^+ \tag{C.46}$$

or

$$E_0 = K_1 + \{0\}K_2 \tag{C.47}$$

Again, if we were to follow the same procedure that is followed in Eqs. (C.35) through (C.37), and integrate by parts, we would obtain the following equation:

$$I_0 = \left(\frac{1}{Ls_1}\right)K_1 + \left(\frac{1}{Ls_2}\right)K_2 \tag{C.48}$$

The rewriting of Eqs. (C.47) and (C.48) in matrix form yields

$$\begin{bmatrix} 1 & 0 \\ \frac{1}{Ls_1} & \frac{1}{Ls_1^2} \end{bmatrix} \begin{bmatrix} K_1 \\ K_2 \end{bmatrix} = \begin{bmatrix} E_0 \\ I_0 \end{bmatrix} \tag{C.49}$$

Finally, solution of Eq. (C.49) for K_1 and K_2 yields*

$$\begin{bmatrix} K_1 \\ K_2 \end{bmatrix} = \begin{bmatrix} E_0 \\ s_1E_0 - \frac{I_0}{C} \end{bmatrix}, \quad \text{where} \quad \frac{1}{C} \equiv Ls_1^2 \tag{C.50}$$

Now Eq. (C.45) can be written, with the aid of Eq. (C.50), as

$$v_C(t) = E_0e^{s_1t} + \left(s_1E_0 - \frac{I_0}{C}\right)te^{s_1t} \tag{C.51}$$

Again, notice the dependence of the response $v_C(t)$, Eq. (C.51), on the initial conditions E_0 and I_0, Fig. C.3, and the circuit parameters R, L, and C, through the characteristic values of $C(s)=0$, that is, the roots s_1 or s_2 of the characteristic polynomial $C(s)=0$, Eq. (C.28).

C.5 s-DOMAIN ANALYSIS OF THE SECOND-ORDER SYSTEM WITH NO FORCING FUNCTION

If we were to use Definitions D.3 and D.4, we would be able to redraw the circuit analog of the parallel-connected *RLC*-system, Fig. C.3, which has no forcing function but which has an initial voltage E_0 across the capacitor and an initial current I_0 through the inductor. This system would be redrawn as a system of the same basic structure but with (1) **no** initial voltage E_0 across the capacitor and (2) **no** initial current I_0 through the inductor. **Instead**, two current generators would be inserted as forcing functions across the given system. One forcing function would be an impulse function $\delta(t)$ of strength CE_0, that is, $i_{GC}(t)=CE_0\delta(t)$; the second would be a step function $u(t)$ of strength $-I_0$, that is, $i_{GL}(t)=-I_0u(t)$, Fig. C.4. Now utilization of Kirchhoff's current law in conjunction with Eq. (C.24) [for which $i_L(0^+)\equiv 0$] enables us to write the describing equation of Fig. C.4 as

$$C\frac{d}{dt}v_C(t) + \left(\frac{1}{R}\right)v_C(t) + \frac{1}{L}\int_0^t v_C(\tau)\,d\tau = CE_0\delta(t) - I_0u(t) \tag{C.52}$$

Utilization of Definitions D.8, D.11, and D.6, with the constraints that $f^{-1}[0^+] = \lambda(0^+) = 0$, and the Laplace transform pairs No. 1 and 2 of Table D.1 enables us

*If we were to set the radical of Eq. (C.29) equal to zero, we would be able to prove that $s_1 = s_2 = -1/2RL, s_1^2 = s_2^2 = 1/LC$, and therefore $1/C = Ls_1^2 = Ls_2^2$.

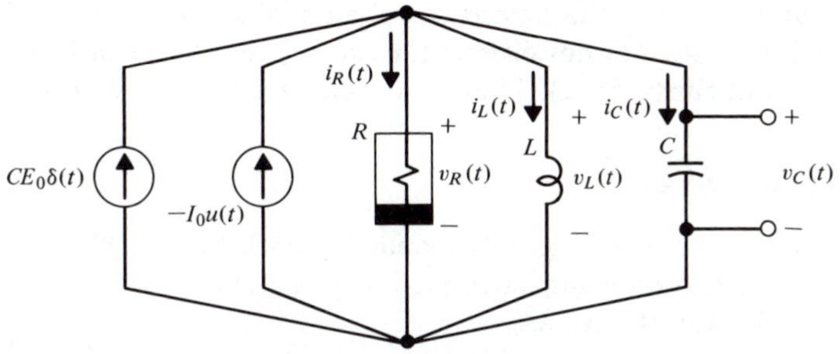

Fig. C.4 System of Fig. C.3 with (1) the Initial Voltage E_0 Across the Capacitor Represented by a Parallel-connected Impulse Function Current Generator $\delta(t)$ of Strength CE_0, That Is, $i_{GC}(t) = CE_0\delta(t)$, and (2) the Initial Current I_0 Through the Inductor Represented by a Step-Function Current Generator $u(t)$ of Strength $-I_0$, That Is, $i_{GL}(t) = -I_0u(t)$

to write Eq. (C.52), after factoring out $V_C(s)$, as

$$V_C(s)\left(Cs + \frac{1}{R} + \frac{1}{sL}\right) = CE_0\Delta(s) - \frac{I_0}{s} \tag{C.53}$$

or

$$V_C(s)\left[s^2 + \left(\frac{1}{RC}\right)s + \frac{1}{LC}\right] = \frac{s}{C}\left(CE_0 - \frac{I_0}{s}\right)\bigg|_{\Delta(s)\equiv 1} \tag{C.54}$$

where

$$\frac{s}{C}\left(CE_0 - \frac{I_0}{s}\right) = E_0 s - \frac{I_0}{C} \equiv E(s) \tag{C.55}$$

Again, utilization of Definition D.12 enables us to write the transfer function $G_{2ND}(s)$ of the system exhibited in Fig. C.4 as

$$G_{2ND}(s) = \frac{V_C(s)}{E(s)} = \frac{1}{s^2 + (1/RC)s + (1/LC)} \tag{C.56}$$

or

$$V_C(s) = \left[\frac{1}{s^2 + (1/RC)s + (1/LC)}\right] \times E(s) = \frac{sE_0 - (I_0/C)}{s^2 + (1/RC)s + (1/LC)} \tag{C.57}$$

Here the denominator of Eq. (C.57) is a second-order polynomial in s. As in the case of the first-order polynomial, Eq. (C.17), when set equal to zero, the denominator of Eq. (C.57) is written as

$$D(s) = s^2 + \left(\frac{1}{RC}\right)s + \frac{1}{LC} = 0 \tag{C.58}$$

It is similarly defined as the **characteristic equation** of the system at hand, Fig. C.4 and thus Fig. C.3. In contrast with Eq. (C.18), however, $D(s)=0$ yields two values, s_1 and s_2, when solved for s. These values are in terms of the circuit parameters R, L, and C, and satisfy the characteristic polynomial $D(s)=0$. Defined as roots or **characteristic values of** $D(s)=0$, s_1 and s_2 play a **vital role** in this system behavior study and are calculated as

$$s_{1,2} = -\frac{1}{2RC} \pm j\left[\frac{1}{LC} - \left(\frac{1}{2RC}\right)^2\right]^{1/2} \tag{C.59}$$

The roots s_1 and s_2 of the characteristic polynomial $D(s)=0$, Eq. (C.58), may be either **distinct** (i.e., $s_1 \neq s_2$) or **not distinct** (i.e., $s_1 = s_2$; in other words, the roots s_1 or s_2 may have a **multiplicity** $k=2$). These two cases are now studied in detail.

Distinct Roots of $D(s)=0$

The roots of the characteristic polynomial $D(s)=0$, Eq. (C.58), are distinct, that is, $s_1 \neq s_2$. Once the two roots are calculated, Eq. (C.57) can be written, with the aid of Definition D.14, Eq. (D.13), as

$$V_C(s) = \frac{E_0 s - (I_0/C)}{(s-s_1)(s-s_2)} = \frac{K_1}{s-s_2} + \frac{K_2}{s-s_2} \tag{C.60}$$

Again, utilization of Definition D.14, Eq. (D.14), enables us to calculate K_1 and K_2 as follows:

$$K_1 = \left.\frac{E_0 s - (I_0/C)}{s-s_2}\right|_{s=s_1} = \frac{s_1 E_0 - (I_0/C)}{s_1 - s_2} \tag{C.61}$$

and

$$K_2 = \left.\frac{E_0 s - (I_0/C)}{s-s_1}\right|_{s=s_2} = \frac{s_2 E_0 - (I_0/C)}{s_2 - s_1} = \frac{-[s_2 E_0 - (I_0/C)]}{s_1 - s_2} \tag{C.62}$$

Utilization of Eqs. (C.61) and (C.62), in conjunction with Laplace transform pair No. 5, Table D.1, enables us to write the t-domain expression of Eq. (C.57), that is, $v_C(t) = \mathcal{L}^{-1}[V_C(s)]$, as

$$v_C(t) = \left\{ \frac{[s_1 E_0 - (I_0/C)]}{s_1 - s_2} \right\} e^{s_1 t} + \left\{ \frac{-[s_2 E_0 - (I_0/C)]}{s_1 - s_2} \right\} e^{s_2 t} \tag{C.63}$$

Notice that the response of the system at hand is written in terms of the initial voltage across the capacitor E_0, the initial current through the inductor I_0, Fig. C.3, and the roots, or characteristic values, s_1 and s_2 of the characteristic polynomial $D(s)=0$, that is, the denominator polynomial of the transfer function $G_{2ND}(s)$, Eq. (C.56).

Nondistinct Roots of $D(s)=0$

The roots of the characteristic polynomial $D(s)=0$, Eq. (C.58), are not distinct, that is, $s_1 = s_2$. Once these two roots are calculated, Eq. (C.57) can be written with the aid of Definition D.15, Eq. (D.15), as

$$V_C(s) = \frac{E_0 s - (I_0/C)}{(s-s_1)^2} = \frac{K_{11}}{s-s_1} + \frac{K_{12}}{(s-s_1)^2} \tag{C.64}$$

Again, utilization of Definition D.15, Eq. (D.16), enables us to calculate K_{11} and K_{12} as follows:

$$\frac{E_0 s - (I_0/C)}{(s-s_1)^2} \times (s-s_2)^2 = \frac{K_{11}}{(s-s_1)} \times (s-s_1)^2 + \frac{K_{12}}{(s-s_1)^2} \times (s-s_1)^2 \tag{C.65}$$

or

$$E_0 s - \frac{I_0}{C} = K_{11}(s-s_1) + K_{12} \tag{C.66}$$

Taking the first derivative of Eq. (C.66) with respect to s yields

$$K_{11} = E_0 \tag{C.67}$$

Evaluation of Eq. (C.66) for $s=s_1$, after the substitution for K_{11} from Eq. (C.67) is effected, yields

$$K_{12} = s_1 E_0 - \frac{I_0}{C} \tag{C.68}$$

Now Eqs. (C.67) and (C.68), in conjunction with Laplace transform pairs No. 5 and 6, Table D.1, enable us to write the t-domain expression of Eq. (C.57), that is, $v_C(t)=\mathfrak{L}^{-1}[V_C(s)]$, as

$$v_C(t) = E_0 e^{s_1 t} + \left(s_1 E_0 - \frac{I_0}{C}\right) t e^{s_1 t} \tag{C.69}$$

Again, notice the dependence of the response $v_C(t)$ on the initial conditions E_0 and I_0, Fig. C.3, as well as the roots, or characteristic values, s_1 and s_2 of the characteristic polynomial $D(s)=0$ of Eq. (C.56).

C.6 STABILITY: CORRESPONDENCE OF THE t-DOMAIN RESPONSE OF FIRST- AND SECOND-ORDER SYSTEMS TO THE ROOT LOCATIONS OF THE CHARACTERISTIC EQUATIONS OR POLYNOMIALS

Inspection of Eqs. (C.7) and (C.16) and (C.27) and (C.56) reveals that the characteristic polynomials $C(s)=0$ and $D(s)=0$ for the corresponding systems (1) look similar and (2) yield identical roots, or characteristic values, Eqs. (C.9) and (C.18) and (C.29) and (C.59). It is obvious, therefore, that $C(s)\equiv D(s)$ and that **both yield the same information about the system they describe**. Consequently, the polynomials $C(s)$ and $D(s)$ can be used interchangeably. Therefore, the term **characteristic equation designated by** $F(s)=0$ will be used hereafter to designate both characteristic polynomials $C(s)$ and $D(s)$, that is, $F(s)\equiv C(s)=D(s)=0$.

Furthermore, inspection of Eqs. (C.9) and (C.18), and (C.29) and (C.59), reveals that the values of the characteristic values, or roots, of the **characteristic equation** $F(s)=0$ depend on **the signs, the relative magnitudes,** and **the combination of the circuit parameters** R, L, **and** C. Note that, although the value of the root of the first-order characteristic equation $F(s)=0$, Eqs. (C.9) and (C.18), depends on the simple product RC, the values of the roots s_1 and s_2 of the second-order characteristic equation $F(s)=0$, Eqs. (C.29) and (C.59), depend on the term $(1/2RC)$ as well as on the terms $(1/2RC)$ and $(1/LC)$ in the radical of Eqs. (C.29) and (C.59). This radical is defined as the **discriminant**.

For the second-order polynomials $F(s)=0$, regardless of which combination of circuit parameters is used, four distinct cases arise. They are defined as (1) overdamped, (2) underdamped, (3) oscillatory, and (4) critically damped. These four cases will be presented in detail through four distinct sets of values for R, L, and C for the system exhibited in Figs. C.3 and/or C.4.

Overdamped Case: $(1/2RC)^2>(1/LC)$

Under these conditions the roots s_1 and s_2 are **real** and **unequal**.

Assume that $R=6\ \Omega$, $L=7$ H, $C=23.80\times10^{-3}$ F, $E_0=50$ V, and $I_0=10$ A. Proper substitution of these values into Eqs. (C.29) and/or (C.59), (C.42) and/or

(C.61), and (C.62) yields

$$s_{1,2} = -\frac{1}{2RC} \pm j\left[\frac{1}{LC} - \left(\frac{1}{2RC}\right)^2\right]^{1/2}$$

$$= -3.5 \pm j\left[6-(3.5)^2\right]^{1/2}\begin{cases} \underline{s_1 = -1} \\ \underline{s_2 = -6} \end{cases} \tag{C.70}$$

$$K_1 = \frac{s_1 E_0 - (I_0/C)}{s_1 - s_2} = \frac{(-1)E_0 - 42I_0}{(-1)-(-6)}$$

$$= -0.2E_0 - 8.4I_0\Big|_{\substack{E_0=50.0\\ I_0=10.0}} = \underline{-94} \tag{C.71}$$

and

$$K_2 = \frac{-[s_2 E_0 - (I_0/C)]}{s_1 - s_2} = \frac{-[(-6)E_0 - 42I_0]}{(-1)-(-6)}$$

$$= 1.2E_0 + 8.4I_0\Big|_{\substack{E_0=50.0\\ I_0=10.0}} = \underline{144.0} \tag{C.72}$$

Substitution of Eqs. (C.70) through (C.72) into Eqs. (C.43) and/or (C.63) yields

$$v_C(t) = \underline{\underline{-94e^{-t} + 144e^{-6t}\text{ V}}} \tag{C.73}$$

Underdamped Case: $(1/2RC)^2 < (1/LC)$

Under these conditions the roots are **conjugate complex**.

Assume that $R = 14.85\ \Omega$, $L = 7$ H, $C = 23.80 \times 10^{-3}$ F, $E_0 = 50$ V, and $I_0 = 10$ A. Proper substitution of these values into Eqs. (C.29) and/or (C.59), and (C.42) and/or (C.61), and (C.62) yields

$$s_{1,2} = -\frac{1}{2RC} \pm j\left[\left(\frac{1}{LC}\right) - \left(\frac{1}{2RC}\right)^2\right]^{1/2}$$

$$= -1.41 \pm j\left[6-2\right]^{1/2} = -1.41 \pm j2\begin{cases} \underline{s_1 = -1.41 + j2} \\ \underline{s_2 = -1.41 - j2} \end{cases} \tag{C.74}$$

$$K_1 = \frac{s_1 E_0 - (I_0/C)}{s_1 - s_2} = \frac{(-1.41 + j2)E_0 - 42I_0}{(-1.42 + j2) - (-1.42 - j2)}$$

$$= \frac{(-1.41E_0 - 42I_0) + j2E_0}{j4}$$

$$= 0.5E_0 + j(0.35E_0 + 10.5I_0)\Big|_{\substack{E_0=50.0\\ I_0=10.0}}$$

$$= 25 + j(18 + 105) = 25 + j123$$

$$= \underline{125.51e^{j78.51^\circ}} \tag{C.75}$$

and

$$K_2 = \frac{-[s_2 E_0 - (I_0/C)]}{s_1 - s_2} = \frac{-[(-1.42 - j2)E_0 - 42I_0]}{(-1.42 + j2) - (-1.42 - j2)}$$

$$= \frac{1.41E_0 + 42I_0 + j2E_0}{j4}$$

$$= 0.5E_0 - j(0.35E_0 + 10.50I_0)\Big|_{\substack{E_0=50.0\\ I_0=10.0}}$$

$$= 25 - j(18 + 105) = 25 - j123$$

$$= \underline{125.51e^{-j78.51^\circ}} \tag{C.76}$$

Substitution of Eqs. (C.74) through (C.76) into Eqs. (C.43) and/or (C.63) yields

$$v_C(t) = 125.51e^{j78.51°}e^{-1.42t}e^{+j2t} + 125.51e^{-j78.51°}e^{-1.42t}e^{-j2t}$$
$$= \underline{125.51e^{-1.42t}\left[e^{j(2t+78.51°)} + e^{-j(2t+78.51°)}\right]} \tag{C.77}$$

Utilization of Eq. (A.12) enables us to write Eq. (C.77) as

$$\underline{\underline{v_C(t) = 251.02e^{-1.42t}\cos(2t+78.51°)\text{ V}}} \tag{C.78}$$

Oscillating Case: $(1/2RC)=0$; $(1/LC)\neq 0$

Under these conditions the roots s_1 and s_2 are **conjugate imaginary**.

Assume $R=\infty\ \Omega$, $L=7$ H, $C=23.80\times10^{-3}$ F, $E_0=50$ V, and $I_0=10$ A. Proper substitution of these values into Eqs. (C.29) and/or (C.59), and (C.42) and/or (C.61), and (C.62) yields

$$s_{1,2} = -\frac{1}{2RC} \pm j\left[\left(\frac{1}{LC}\right) - \left(\frac{1}{2RC}\right)^2\right]^{1/2}$$
$$= 0 \pm j(6)^{1/2} = \pm j2.45 \begin{cases} \underline{s_1 = +j2.45} \\ \underline{s_2 = -j2.45} \end{cases} \tag{C.79}$$

$$K_1 = \frac{s_1E_0 - (I_0/C)}{s_1 - s_2} = \frac{(j2.45)E_0 - 42I_0}{(j2.45)-(j2.45)}$$
$$= \frac{(j2.45)E_0 - 42I_0}{j4.90}$$
$$= 0.5E_0 + j8.57I_0\Big|_{\substack{E_0=50.0 \\ I_0=10.0}}$$
$$= \underline{25 + j85.70} = \underline{89.27e^{j73.74°}} \tag{C.80}$$

and

$$K_2 = \frac{-[s_2E_0 - (I_0/C)]}{s_1 - s_2} = \frac{-[(-j2.45)E_0 - 42I_0]}{(j2.45)-(-j2.45)}$$
$$= \frac{j2.45E_0 + 42I_0}{j4.90}$$
$$= 0.5E_0 - j8.57I_0\Big|_{\substack{E_0=50.0 \\ I_0=10.0}}$$
$$= \underline{25 - j85.70} = \underline{89.27e^{-j73.74°}} \tag{C.81}$$

Substitution of Eqs. (C.79) through (C.81) into Eqs. (C.43) and/or (C.63) yields

$$v_C(t) = 89.27e^{j73.74°}e^{j2.45t} + 89.27e^{-j73.74°}e^{-j2.45t}$$
$$= 89.27\left[e^{j(2.45t+73.74°)} + e^{-j(2.45t+73.74°)}\right] \tag{C.82}$$

Utilization of Eq. (A.12) enables us to write Eq. (C.82) as

$$\underline{\underline{v_C(t) = 178.54\cos(2.45t+73.74°)\text{ V}}} \tag{C.83}$$

Critically Damped Case: $(1/2RC)^2=(1/LC)$

Under these conditions the roots s_1 and s_2 are **real and equal.**

Assume $R=8.58\ \Omega$, $L=7$ H, $C=23.80\times10^{-3}$ F, $E_0=50$ V, and $I_0=10$ A. Proper substitution of these values into Eqs. (C.29) and/or (C.59), and (C.50) and/or

Table C.1 TABULATION OF ALL PERTINENT QUANTITIES ASSOCIATED WITH THE LHP ROOT LOCATION OF $F(s)=0$ FOR FIRST- AND SECOND-ORDER SYSTEMS

Order	Case	s	K	$R=K_1$
First	—	$s_1=-1.5$	100	—
Second	Overdamped	$s_1=-1+j0$ $s_2=-6-j0$	—	-94.00
Second	Underdamped	$s_1=-1.41+j2.00$ $s_2=-1.41-j2.00$	—	$125.51e^{j78.51°}$
Second	Oscillatory	$s_1=0+j2.45$ $s_2=0-j2.45$	—	$89.27e^{j73.74°}$
Second	Critically Damped	$s_1=-2.45+j0$ $s_2=-2.45-j0$	—	—

(C.67), and (C.68) yields

$$s_{1,2} = -\frac{1}{2RC} \pm j\left[\left(\frac{1}{LC}\right)-\left(\frac{1}{2RC}\right)^2\right]^{1/2}$$

$$= -2.45 \pm j(6-6)^{1/2}\begin{cases}\underline{s_1=-2.45}\\ \underline{s_2=-2.45}\end{cases} \qquad \text{(C.84)}$$

$$K_{11} \equiv K_1 = E_0|_{E_0=50} = \underline{50} \qquad \text{(C.85)}$$

and

$$K_{12} \equiv K_2 = s_1E_0 - \frac{I_0}{C} = [(-2.45)E_0 - 42I_0]|_{\substack{E_0=50\\ I_0=10}}$$

$$= -122.5 - 420.0 = \underline{-542.50} \qquad \text{(C.86)}$$

Substitution of Eqs. (C.84) through (C.85) into Eq. (C.51) and/or (C.69) yields

$$\underline{\underline{v_C(t)=50e^{-2.45t}-542.50te^{-2.45t}\text{ V}}} \qquad \text{(C.87)}$$

It is essential to summarize the results of Eqs. (C.9) and (C.18),* (C.13) and/or (C.20),† and Eqs. (C.70) through (C.73), (C.74) through (C.78), (C.79) through (C.83), and (C.84) through (C.87) in Table C.1. These results are plotted on the left-hand side, or unshaded portion of the s-plane, usually designated as the LHP, Table C.2. Note that the **roots, or characteristic values, of the corresponding characteristic equations $F(s)=0$ are plotted on the left-hand side of the s-plane,** usually designated as the LHP, or on the $j\omega$**-axis.** Also, note that the t-domain responses are plotted adjacent to, and to the left of, the corresponding root plots.

*For these two equations, R and C are set as follows: $R=10^3\ \Omega$ and $C=0.665\times10^{-3}$ F.
†For these two equations, E_0 is set as follows: $E_0=100$ V.

Table C.1 (*Continued*)

POSITIVE

K_2	K_{11}	K_{12}	$v(t)$
—	—	—	$v_C(t)=100e^{-1.5t}$
144.00	—	—	$v_C(t)=-94.00e^{-t}+144.00e^{-6t}$
$125.51e^{-j78.51°}$	—	—	$v_C(t)=251.02e^{-1.42t}\cos(2t+78.51°)$
$89.27e^{-j73.74°}$	—	—	$v_C(t)=178.54\cos(2.45t+73.74°)$
—	50	−542.67	$v_C(t)=50.00e^{-2.45t}-542.67te^{-2.45t}$

Examine carefully the shape of these t-domain plots, as well as the relative position of the roots on the s-plane and their relative correspondence to these plots. Both sets of plots are consequences of **positive**-value assignments to the circuit parameters R, L, and C.

If, however, we were to assign **negative** values to the resistance R in the corresponding characteristic equation $F(s)=0$, then the **real parts** of the roots, or the characteristic values, when present, become positive. The plots of such roots are exhibited on the right-hand side, or shaded portion, of the s-plane, usually designated as the RHP, Table C.2. Note that here the $j\omega$-axis of the s-plane is excluded from root location. Under these circumstances, the t-domain plots for each set of roots is exhibited to the right of the corresponding root plots. The pertinent data for these plots is exhibited in Table C.3. One should note, however, that the magnitudes of the voltage components that correspond to roots with positive real parts is smaller than the magnitudes of the voltage components that correspond to roots with **negative** real parts. (Why?)

At this time it is essential for the reader to examine carefully the root location and the corresponding t-domain plots in the **unshaded part of Table C.2** and to compare both to the root location and the t-domain plots in the **shaded part of Table C.2**. Note that **for all cases** but the oscillatory case, **the t-domain plots attain zero magnitude as t approaches infinity for the unshaded part of Table C.2. However, for the shaded part of Table C.2, the t-domain plots attain infinite magnitude as t approaches infinity.**

Now the following question should come to mind: **Does the shape of these t-domain plots have any significance**? To answer this seemingly simple question, one has to examine the source of each voltage $v_C(t)$ at $t=0$. Close examination of Eqs. (C.13), (C.20), (C.43), (C.51), (C.63), and (C.69) reveals that the voltage $v_C(0^+)$ is a consequence of the *energy stored* in the systems of Figs. C.1 and C.3 at $-\infty<t<0^-$. That is, the energy stored in the capacitor of Fig. C.1, $W_{\text{STORED}}=\frac{1}{2}CE_0^2$, and the energy stored in the inductor plus the energy stored in the capacitor of Fig. C.3,

Table C.2 STABILITY VS. INSTABILITY: CORRELATION OF s-PLANE ROOT LOCATION OF $F(s)=0$ AND t-DOMAIN RESPONSE, FIRST- AND SECOND-ORDER SYSTEMS

SYSTEM		Stable System Description		Unstable System Description
		R Positive		*R* Negative
		t-Domain	s-Plane	t-Domain
Order	Case	LHP Response	LHP Root Location ‖ RHP Root Location	RHP Response
First	—	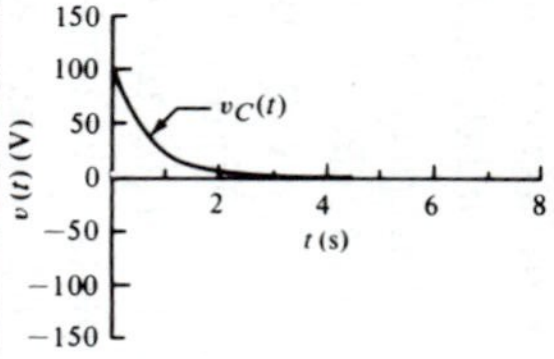	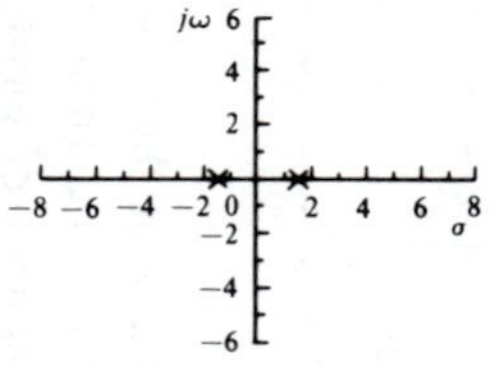	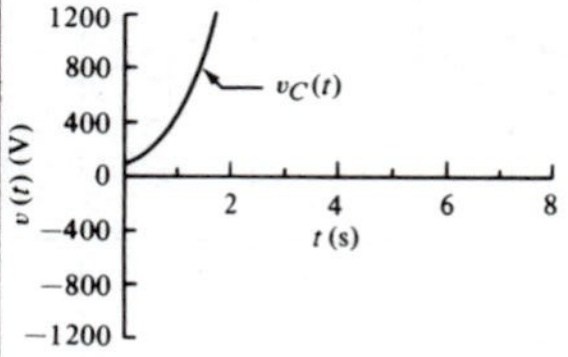
Second	Overdamped	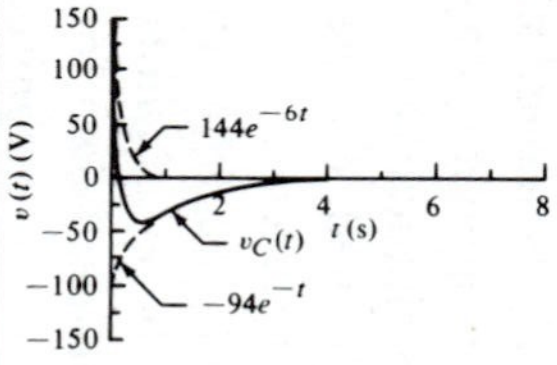	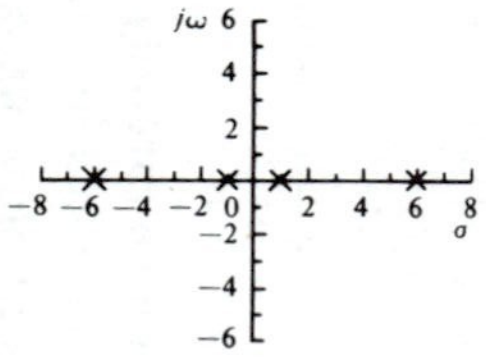	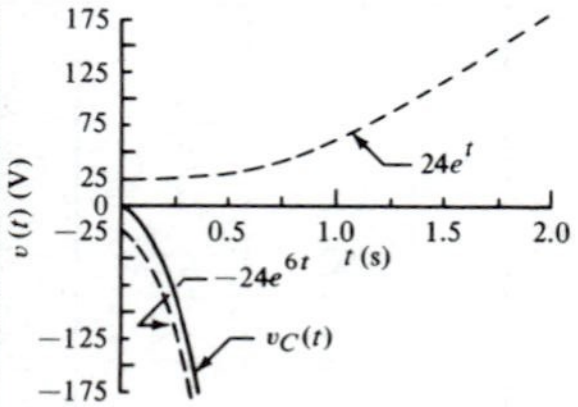

SYSTEM					
	Second	Underdamped	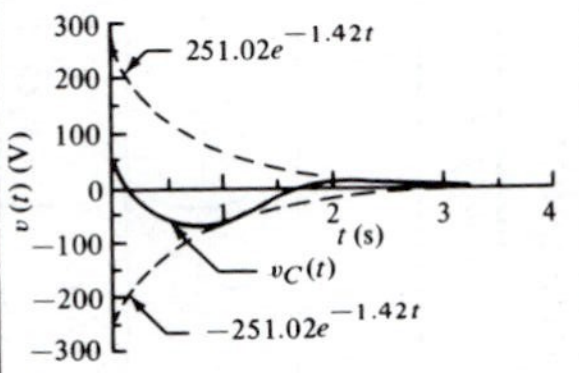	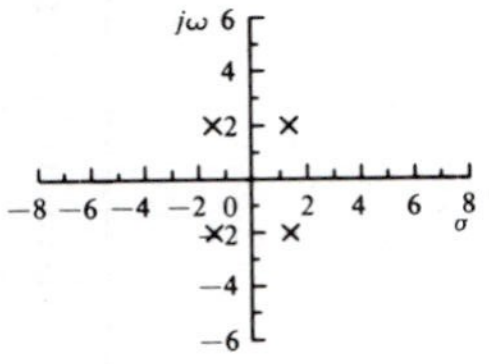	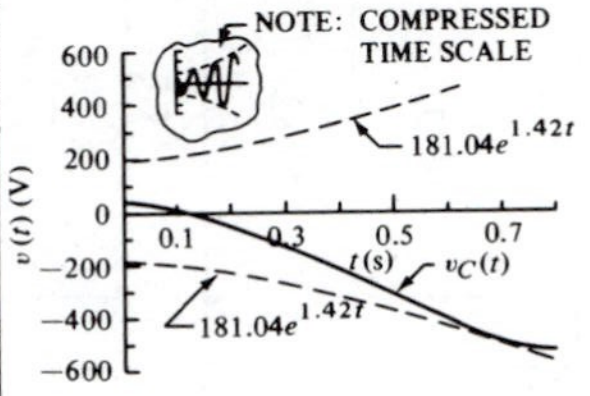
	Second	Oscillatory	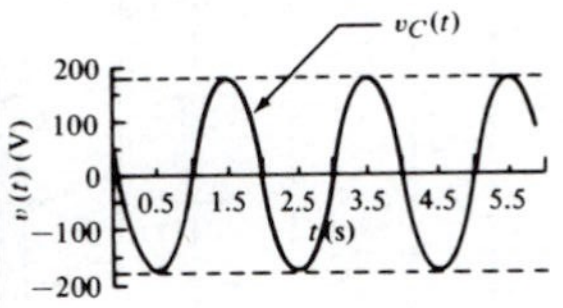	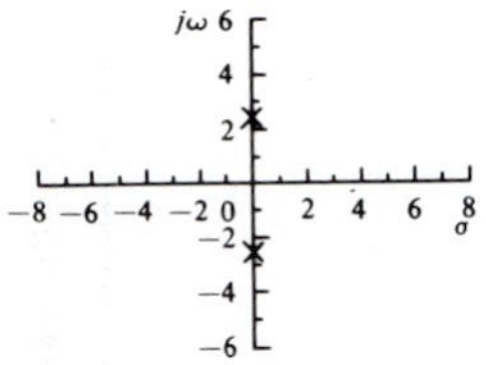	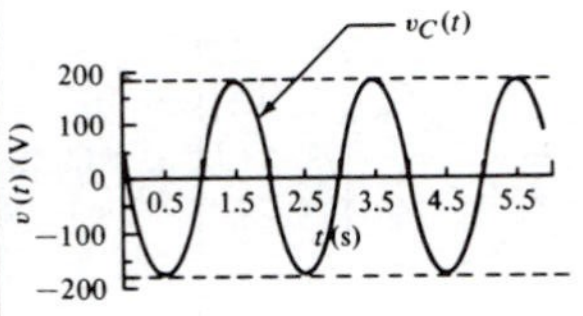
	Second	Critically Damped	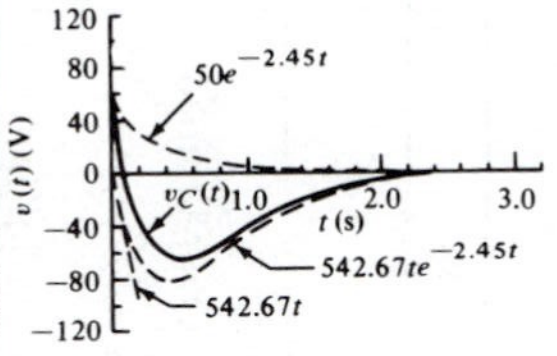	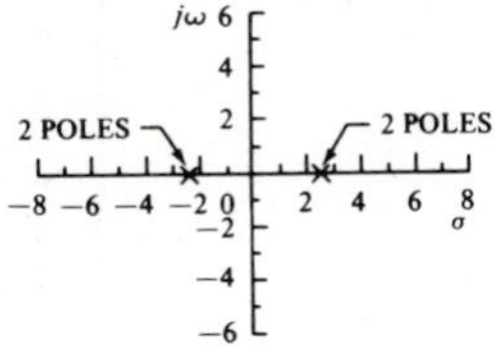	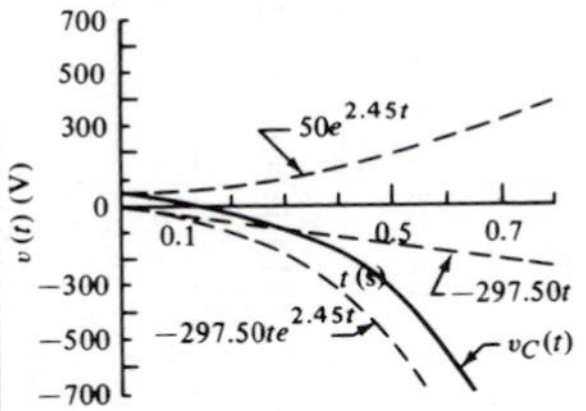

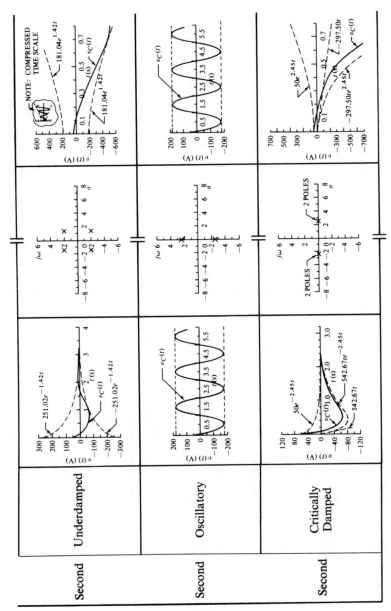

Table C.3 TABULATION OF ALL PERTINENT QUANTITIES ASSOCIATED WITH THE RHP ROOT LOCATION OF $F(s)=0$ FOR FIRST- AND SECOND ORDER SYSTEMS

	Order	Case	s	K	$R=$ K_1
SYSTEM	First	—	$s_1=1.5$	100	—
	Second	Overdamped	$s_1=6+j0$ $s_2=1-j0$	—	-24.00
	Second	Underdamped	$s_1=1.41+j2.00$ $s_2=1.41-j2.00$	—	$90.52e^{j73.97°}$
	Second	—	—	—	—
	Second	Critically Damped	$s_1=2.45+j0$ $s_2=2.45-j0$	—	—

Table C.4 RESPONSE COMPONENT DEPENDENCE ON INITIAL CONDITIONS AND THE ROOTS OF $F(s)=0$

	Order	Case	Magnitude of Response K	K_1	K_2
SYSTEM	First	—	E_0	—	—
	Second	Overdamped	—	$\frac{1}{s_1-s_2}\left(s_1E_0-\frac{I_0}{C}\right)$	$\frac{1}{s_1-s_2}\left(-s_2E_0+\frac{I_0}{C}\right)$
	Second	Underdamped	—	$\frac{1}{s_1-s_2}\left(s_1E_0-\frac{I_0}{C}\right)$	$\frac{1}{s_1-s_2}\left(-s_2E_0+\frac{I_0}{C}\right)$
	Second	Oscillatory	—	$\frac{1}{s_1-s_2}\left(s_1E_0-\frac{I_0}{C}\right)$	$\frac{1}{s_1-s_2}\left(-s_2E_0+\frac{I_0}{C}\right)$
	Second	Critically Damped	—	—	—

Table C.3 (*Continued*)

Negative			
K_2	K_{11}	K_{12}	$v(t)$
—	—	—	$v_C(t)=100e^{1.5t}$
24.00	—	—	$v_C(t)=-24e^{6t}+24e^{t}$
$90.52e^{-j73.97°}$	—	—	$v_C(t)=181.04e^{1.42t}\cos(2t+73.97°)$
—	—	—	
—	50.00	−297.50	$v_C=50e^{2.45}-297.50te^{2.45t}$

Table C.4 (*Continued*)

Components		Poles of $F(s)=0, p_k=\alpha_k\pm j\omega_k$	
K_{11}	K_{12}	$R=$ Positive	$R=$ Negative
—	—	$s_1=-\frac{1}{RC}$	$s_1=\frac{1}{RC}$
—	—	$s_{1,2}=-\frac{1}{2RC}\pm j\sqrt{\frac{1}{LC}-\left(\frac{1}{2RC}\right)^2}$	$s_{1,2}=\frac{1}{2RC}\pm j\sqrt{\frac{1}{LC}-\left(\frac{1}{2RC}\right)^2}$
—	—	$s_{1,2}=-\frac{1}{2RC}\pm j\sqrt{\frac{1}{LC}-\left(\frac{1}{2RC}\right)^2}$	$s_{1,2}=\frac{1}{2RC}\pm j\sqrt{\frac{1}{LC}-\left(\frac{1}{2RC}\right)^2}$
—	—	$s_{1,2}=0\pm j\sqrt{\frac{1}{LC}}$	$s_{1,2}=0\pm\sqrt{\frac{1}{LC}}$
E_0	$s_1E_0-\frac{I_0}{C}$	$s_{1,2}=-\frac{1}{2RC}\pm j0$	$s_{1,2}=\frac{1}{2RC}\pm j0$

$W_{\text{STORED}} = \frac{1}{2}LI_0^2 + \frac{1}{2}CE_0^2$. This dependence of the response voltage $v_C(t)$ on the initially stored energy is expressed by the dependence of the various K's on (1) the initial voltage E_0 across the capacitor and (2) the initial current I_0 through the inductor. Table C.4 exhibits this dependence. Close inspection of Table C.4 suggests that these K's also depend on the roots of the corresponding characteristic equations $F(s)=0$ and therefore on the system parameters R, L, and C.

It should be noted, however, that when $F(s)=0$, the transfer functions $G_{1ST}(s)$, Eq. (C.16), and $G_{2ND}(s)$, Eq. (C.56), or the voltages $V_C(s)$, Eqs. (C.19) and (C.57), become infinity. For this reason the roots of $F(s)=0$ are defined as **poles** of the transfer function at hand and are designated as p_k, $k=1,2,3,\ldots,v$. If the numerator polynomial $N(s)$ of the transfer function at hand, $G_{1ST}(s)$, Eq. (C.16), or $G_{2ND}(s)$, Eq. (C.56), is set equal to zero, that is, $N(s)=0$, the transfer functions $G_{1ST}(s)$ and $G_{2ND}(s)$, or the voltage $V_C(s)$, Eqs. (C.19) and (C.57), become zero. For this reason the roots of $N(s)$ are defined as **zeros** of the transfer function at hand and are designated as z_ℓ, $\ell=1,2,\ldots,u$. As a rule, a given transfer function $G(s)$ has more finite poles than zeros, that is, $v \geqslant u$. **However, when the poles and zeros at zero and infinity are taken into account, the total number of poles is equal to the total number of zeros.** It is clear, therefore, that **the roots of the characteristic equation $F(s)=0$ are the poles of the transfer function** $G(s)$. The poles of the first- and second-order systems studied here are exhibited in the last two columns of Table C.4. Clearly, then, while $F(s)$ is instrumental in calculating the roots of the characteristic equation—that is, the poles of the system's transfer function $G(s)$—$N(s)$ in conjunction with the roots of $F(s)=0$ is instrumental in calculating the magnitudes of the components of the response $v_C(t)$, that is, $K_k, k=1,2,\ldots,v$ [see Table C.4, Eqs. (C.42) with (C.61) and (C.62), and (C.50) with (C.67) and (C.68)].

It should be clear now that the t-domain response plotted in Table C.2 can be written, with the aid of Table D.1, in a compact form as

$$v_C(t) = \sum_{k=1}^{v} K_k e^{-p_k t} \equiv \mathfrak{L}^{-1}\left[\sum_{k=1}^{v} \frac{K_k}{s+p_k}\right] \tag{C.88}$$

where (1) *the poles p_k of $G(s)$ determine the waveform of the time variation of each component of the response* and (2) *the zeros z_ℓ of $G(s)$ determine the magnitude K_k of each component of the response.** These poles and zeros also determine the *phase angle* of the response in the cases of sinusoidal variation. Note that here the response is considered to be the voltage $v_C(t)$ across the capacitors of Figs. C.1 through C.4.

When the limit of the response $v_C(t)$, Eq. (C.88), approaches zero as t approaches infinity, that is,

$$\lim_{t\to\infty} v_C(t) = 0 \tag{C.89}$$

it is said that **the output is bounded for a bounded input**, the bounded input being the initial conditions. If the opposite is true, that is,

$$\lim_{t\to\infty} v_C(t) = \infty \tag{C.90}$$

*Analytically it is not apparent that the zeros of $G(s)$ affect the magnitude of the K's. However, when one draws the vectorial diagram of $K_k = \mathcal{K}(s-s_{pr})[(s-s_{z1})(s-s_{z2})\cdots(s-s_{zu})]/[(s-s_{p1})(s-s_{p2})\cdots(s-s_{pr})\cdots(s-s_{pv})]|_{s=+s_{pr}}$ $\mathcal{K}=$constant, the effect of these zeros becomes clear. For additional information, see M. E. Van Valkenburg, *Network Analysis*, 2d Ed. (Englewood Cliffs, N.J.: Prentice-Hall, 1964), pp. 280–83.

it is said that **the output is unbounded for a bounded input**, the bounded input again being the initial conditions.

In terms of physical systems, then, a system is referred to as **stable** in the sense that excitation due to initial conditions results in a bounded output no matter how large time becomes. On the other hand, a system is referred to as **unstable** in the sense that excitation due to initial conditions results in an unbounded output as time increases. When the output of a system resulting from excitations due to initial conditions oscillates between finite values, as time becomes large, the system is referred to as **marginally stable**. The physical significance of the sustained oscillations can be explained as follows. An initial energy has been assumed present in the system, that is, stored in the inductor and the capacitor, but no means have been provided in the system to dissipate this energy. Therefore the energy is continuously transferred between the capacitor and the inductor unaltered. Actual parallel *RLC*-circuits can be made to have effective values of R so large that any natural undamped sinusoindal response, that is, any sustained oscillation, can be maintained for **years** without applying any additional energy.

Table C.2 clearly shows that (1) when $\mathrm{Re}(p_k)<0$, the poles are located in the LHP and the system associated with these poles is **stable**, (2) when $\mathrm{Re}(p_k)>0$, the poles are located in the RHP and the system associated with these poles is **unstable**, and (3) when $\mathrm{Re}(p_k)=0$ and **nonrepeated**, the poles are located on the $j\omega$-axis and the system associated with these poles is **marginally stable**. Note that **although a single pair of conjugate imaginary poles located on the $j\omega$-axis yields a marginally stable system** (such a system is often referred to as a stable system), **a multiple pair of conjugate imaginary poles located anywhere on the $j\omega$-axis yields an unstable system**. [See material associated with Eq. (C.103).]

C.7 STABILITY STATUS OF *n*th-ORDER SYSTEMS—ROUTH'S STABILITY CRITERION

The evolution of the describing equation of high-order systems becomes obvious if one follows the development of the describing equation of the system* exhibited in Fig. C.5. It is assumed from the outset that the independent variable of the derived equation is the voltage $v_C(t)$ across the capacitor C, Fig. C.5. Utilization of Kirchhoff's voltage law yields the following two equations:

$$L\frac{d}{dt}i_1(t) + Ri_1(t) - Ri_2(t) = e_1(t) \tag{C.91}$$

$$-Ri_1(t) + Ri_2(t) + v_C(t) = e_2(t) \tag{C.92}$$

where

$$i_2(t) = C\frac{d}{dt}v_C(t) \tag{C.93}$$

Substitution of Eq. (C.93) into Eq. (C.92) and solution of the resulting equation for $i_1(t)$ yields

$$i_1(t) = C\frac{d}{dt}v_C(t) + \left(\frac{1}{R}\right)v_C(t) - \left(\frac{1}{R}\right)e_2(t) \tag{C.94}$$

*The word **system** rather than **network** is used henceforth to imply the generality of the derived describing equation used, that is, $v_C(t)\to r(t)$.

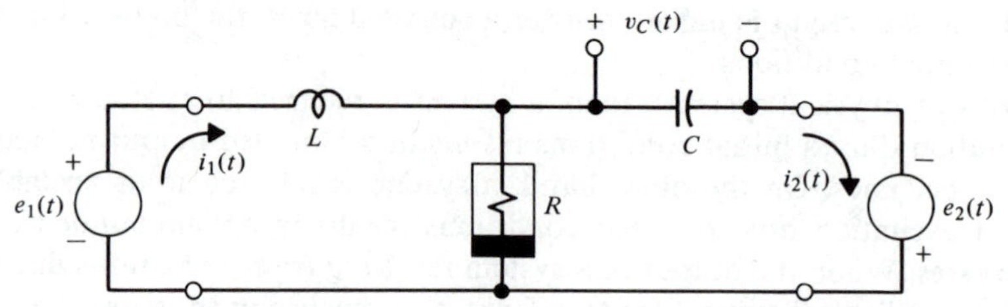

Fig. C.5 A System (Network) Comprised of Two Loops, Two Energy-Storing Elements, and Described by a Second-Order Nonhomogeneous Differential Equation

Finally, substitution of Eqs. (C.94) and (C.93) into Eq. (C.91), regrouping, and cancellation of **appropriate** terms yields

$$\frac{d^2}{dt^2}v_C(t)+\left(\frac{1}{RC}\right)\left(\frac{d}{dt}\right)v_C(t)+\left(\frac{1}{LC}\right)v_C(t)=\left(\frac{1}{LC}\right)e_1(t)+\left(\frac{1}{LC}\right)e_2(t)+\left(\frac{1}{RC}\right)\left(\frac{d}{dt}\right)e_2(t) \tag{C.95}$$

Inspection of Eq. (C.95) and Fig. C.5 suggests that **the order $n=2$ of the system's describing equation is equal to the sum of the network's energy-storing elements.** That is, $n=2$ is equal to the sum of the capacitors and the inductors contained* in the system exhibited in Fig. C.5. This observation suggests that if the sum of the distinct energy-storing elements in a given system is n, then the generalized **nonhomogeneous differential equation** that describes the system can be written as

$$\frac{d^n}{dt^n}r(t)+a_{n-1}\frac{d^{n-1}}{dt^{n-1}}r(t)+\cdots+a_1\frac{d}{dt}r(t)+a_0r(t)=\varepsilon\left[e_1(t),e_2(t),\ldots,\frac{d}{dt}e_1(t),\frac{d}{dt}e_2(t)\ldots\right] \tag{C.96}$$

where:

1. $r(t)$, the dependent variable (the independent variable is t), is defined as the response of the given system.
2. $\varepsilon[\;]$ is defined as the excitation, or the forcing function, of the given system. It may represent a linear combination of voltage sources, or current sources, and/or corresponding derivatives.
3. $a_{n-1}, a_{n-2}, \ldots, a_0$ are defined as the coefficients of the system's describing equation; they are functions of the system parameters and, depending on the system at hand, they could be either **time-independent** or **time-dependent**. However we will only consider time-independent cases below.

*Inductors connected in series are treated as one equivalent inductor, that is, $L_{EQ}=L_1+L_2+\cdots+L_n$, while capacitors connected in parallel are treated as one equivalent capacitor, that is, $C_{EQ}=C_1+C_2+\cdots+C_n$. Similarly, inductors connected in parallel are treated as one equivalent inductor, that is, $L_{EQ}=1/\{(1/L_1)+(1/L_2)+\cdots+(1/L_n)\}$, while capacitors connected in series are treated as one equivalent capacitor, that is, $C_{EQ}=1/\{(1/C_1)+(1/C_2)+\cdots+(1/C_n)\}$.

If we were to assume that (1) the forcing function of the system is zero and (2) its natural response is of the exponential form, that is,

$$v_{nC}(t) = Ke^{st} \tag{C.97}$$

then take the appropriate derivatives of Eq. (C.97), substitute the proper terms into Eq. (C.96), factor out Ke^{st}, and set the resulting parentheses equal to zero, we would obtain the characteristic equation $F(s)=0$ of the given system; that is, a polynomial in s of **degree** n written as

$$F(s) = s^n + a_{n-1}s^{n-1} + a_{n-2}s^{n-2} + \cdots + a_1 s + a_0 = 0 \tag{C.98}$$

Note that the **degree** n of this polynomial is identical to the **order** n of the differential equation that describes the given system. Thus we have the following pertinent theorem:

Theorem C.1

The nth-order characteristic equation of a given system has n roots defined as the poles of the characteristic equation and are designated as $p_1, p_2, \ldots, p_n$. They are related to the characteristic equation of the system as follows:

$$\begin{aligned} F(s) &= s^n + a_{n-1}s^{n-1} + a_{n-2}s^{n-2} + \cdots + a_1 s + a_0 \\ &\equiv (s+p_1)(s+p_2)\cdots(s+p_{n-1})(s+p_n) \end{aligned} \tag{C.99}$$

Each pole of Eq. (C.99) gives rise to a factor of the form $r_k(t) = K_k e^{p_k t}$ in the solution of the homogeneous differential equation [i.e., ε[ ]=0, Eq. (C.96)] and defines, in part, the **dynamics** of the system at hand. The sum of such factors, written as

$$r_n(t) = \sum_{k=1}^{n} K_k e^{p_k t} \tag{C.100}$$

constitutes the solution of the homogeneous differential equation and, therefore, defines the system's dynamic behavior. Therefore, determining the **dynamic behavior, or stability status**, of a high-order system becomes a matter of finding the poles of its characteristic equation $F(s)=0$ [i.e., the roots of the characteristic polynomial $C(s)=0$ of the system's describing equation, Eq. (C.96)].

It is shown earlier in this appendix, Section C.6, that there are **three** possible types of roots for the characteristic equation $F(s)=0$. They are (1) complex roots, (2) real roots, unequal or of multiplicity k, and (3) imaginary roots. **If any of the roots are complex, or imaginary, they must occur in conjugate pairs since this is the only way complex roots can combine to yield real (positive or negative) coefficients of a given system's describing equation, that is,** $a_{n-1}, a_{n-2}, \ldots, a_1, a_0$, Eq. (C.96). Therefore, the following two theorems are pertinent:

Theorem C.2

If two roots of a characteristric equation are complex or imaginary, they constitute a conjugate pair since the coefficients are all real.

Theorem C.3

For an nth-order characteristic equation, (1) if all the roots are complex or imaginary, they occur in conjugate pairs; (2) if n is odd, at least one root is real and the remaining roots are real, or occur in conjugate pairs, or constitute a combination of both; and (3) if n is even, the roots may be real, or may occur in conjugate pairs, or constitute a combination of both.

Since the characteristic equation $F(s)=0$ of a given system encompasses the poles of the system's transfer function $G(s)$—for example, Eqs. (C.16) and (C.56), or Eq. (C.96) for $\varepsilon[\cdot]\rightarrow e(t)\rightarrow E(s)$ and $r(t)\rightarrow R(s)$, and Fig. D.5—it is clear that the poles of $R(s)=G(s)E(s)$ describe the dynamic behavior of the system at hand, with the forcing function $E(s)$, that is, $e(t)$, in effect. Thus the generalized time-domain response of a given system can be written as

$$r(t) = \mathcal{L}^{-1}[G(s)E(s)] = \sum_{k=1}^{v} K_k e^{p_k t} + \sum_{j=1}^{\mu} K_j e^{p_j t} \tag{C.101}$$

where (1) p_k designates the poles that correspond to $G(s)$, that is, the unforced system, (2) p_j designates the poles that correspond to the forcing function $E(s)$, (3) v designates the number of the poles of the transfer function $G(s)$, and (4) μ designates the number of the poles of the forcing function $E(s)$. It is evident therefore from Eq. (C.101) that **the poles of $G(s)E(s)$ determine the waveform of the time variation of the response $r(t)$ and thus the dynamic behavior, or the stability status**, of the system.

Suppose that Eq. (C.101) describes a system for which $k=1$, 2 and $j=0$, $p_1=\alpha$, and $p_2=\beta\pm j\omega$. In accordance with the discussion associated with Eqs. (C.32) and (C.78), Eq. (C.101) can be written as

$$r(t) = K_1 e^{\alpha t} + K_2 e^{\beta t}\cos(\omega t+\phi) \tag{C.102}$$

If $r(t)$ is to be bounded, Eq. (C.89), it is necessary that $\alpha<0$ and $\beta<0$. Therefore, **for a stable system the poles of a transfer function must be located in the left half of the s-plane**, usually designated as LHP. These **poles can never occur in the right half of the s-plane**, usually designated as RHP. The upper bound of such pole location is defined as the $j\omega$-axis, with the additional restriction that **these poles must be simple** at any given point on the $j\omega$-axis. That is, **the existence of the roots $s_{1,2}=\pm j8$ and $s_{3,4}=\pm j8$ is prohibited, if the system they describe is to be stable, while the existence of the roots $s_{1,2}=\pm j8$ and $s_{3,4}=\pm j10$ is permitted.** The reason for this restriction is that such poles of multiplicity $k\geqslant 2$ give rise to time-domain terms that increase as a function of t^{k-1}. Such response is **unbounded**, Eq. (C.90). For example, a double pole on the $j\omega$-axis (i.e., at $s=\pm j\omega$), defined by the Laplace transform pair No. 10, Table D.1, as

$$R(s) = \frac{R_0 s}{(s^2+\omega^2)^2} \tag{C.103}$$

gives rise to the t-domain response

$$r(t) = \mathcal{L}^{-1}[R(s)] = \left(\frac{R_0}{2\omega}\right) t \sin\omega t u(t) \tag{C.104}$$

Note that Eq. (C.104) represents a sinusoid with its amplitude increasing linearly with a slope $R_0/2\omega$, which is unbounded, Fig. C.6.

Similarly, **repeated poles at the origin of the s-plane are also prohibited for a stable system**. The reason for this restriction is that such poles of multiplicity $k\geqslant 2$ also give rise to t-domain terms that increase as a function of $t^{k-1}/(k-1)!$. Such response is also unbounded. For example, a double pole at the origin of the s-plane, defined by the Laplace transform pair No. 4, Table D.1 as

$$R(s) = \frac{R_0}{s^2} \tag{C.105}$$

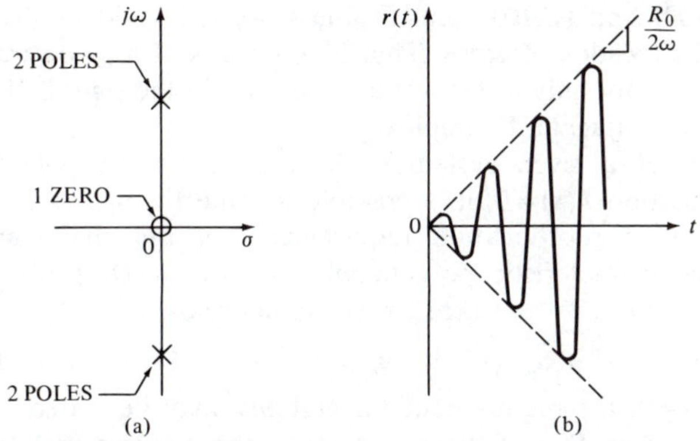

Fig. C.6 The Inverse Laplace Transform of Eq. (C.103) Having the Two Poles and the One Zero Shown in (a) Is Exhibited in (b)

gives rise to the t-domain response

$$r(t) = \mathcal{L}^{-1}[R(s)] = R_0 t u(t) \tag{C.106}$$

Eq. (C.106) represents a ramp of slope R_0, which is unbounded, Fig. C.7.

Multiple poles elsewhere in the left half of the s-plane are permitted since such poles give rise to terms of the form $t^{k-1}e^{-\alpha t}/(k-1)!$. Such terms satisfy the condition

$$\lim_{t\to\infty} \frac{t^{k-1}e^{-\alpha t}}{(k-1)!} = 0 \tag{C.107}$$

which defines a *bounded* response. The following theorem, therefore, is pertinent to the **prohibited** and **permitted** root location in the s-plane.

Theorem C.4

For a given characteristic equation $F(s)=0$, which describes a physically realizable system, multiple-pole location is (1) prohibited on the $j\omega$-axis, or the origin, of the s-plane, as these poles yield t-domain responses that are unbounded; that is, they increase without bound as $t\to\infty$, and (2) permitted elsewhere on the left half of the s-plane, as they yield t-domain responses that vanish as $t\to\infty$.

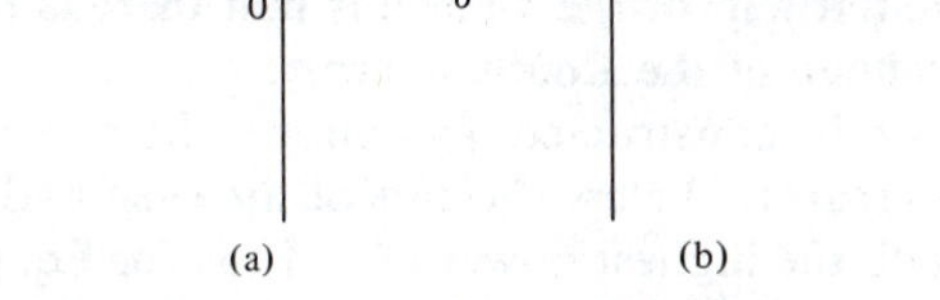

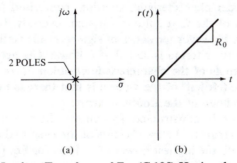

Fig. C.7 The Inverse Laplace Transform of Eq. (C.105) Having the Two Poles Shown in (a) Is Exhibited in (b)

From Eqs. (C.100) and (C.101) (and Table C.4), it is evident that **there are no restrictions on the location of zeros**. That is, the zeros of a given transfer function $G(s)$ can be located not only in the left half but also in the right half of the s-plane, provided they are conjugate if complex.

Since stability of a given system is determined by the pole location of its characteristic equation $F(s)=0$, it is possible to state the necessary conditions for stability in terms of root-location requirements of the characteristic equation $F(s)=0$ [i.e., the characteristic polynomials $C(s)=0$ or $D(s)=0$]. Let this polynomial be Eq. (C.98), rewritten here for convenience as

$$F(s) = s^n + a_{n-1}s^{n-1} + a_{n-2}s^{n-2} + \cdots + a_1 s + a_0 = 0 \tag{C.108}$$

In terms of $F(s)=0$, the requirement for stability may be stated in terms of the simple question: **Does $F(s)=0$ have any roots with positive real (or zero) parts?** Once this question is answered, the stability of the response of the system is determined. The obvious answer to this question, of course, is provided if one were to use the digital computer to determine the roots of $F(s)=0$. Usually, however, scientists, and especially students, are not interested in the numerical value of the roots but rather to a simple answer of "yes" or "no" to the question: Is the given system stable? The problem then becomes as follows: Given the characteristic equation $F(s)=0$, how many of its roots have positive real (or zero) parts? *For a strictly stable system the answer should be none.*

One of the first to investigate this problem was Maxwell in 1868. Many others, such as Routh (England, 1877), Liapunov (Russia, 1892), Hurwitz (Germany, 1895), Lienard and Chipart (France, 1914) and Nyquist (U.S., 1932) followed. Much work was done by each of these men independently from the others, and some confusion has resulted in the scientific world in attributing stability criteria to the proper author. However, the easiest stability criterion, at least for linear systems, is attributed to Routh* and is therefore known as Routh's stability criterion.

The virtue of the criterion is that the stability of a given system can be determined without actually solving the characteristic equation for its roots. Thus if the polynomial $F(s)=0$ of Eq. (C.108) is to have no roots with positive real parts, **it is necessary but not sufficient** that for $F(s)=0$ all the coefficients (1) have the same sign and (2) be present. These two conditions can be checked by inspection. However, they are not sufficient. It is possible that a polynomial with all its coefficients positive and nonzero will have roots in the right half of the s-plane. The **sufficiency** of the criterion is checked by forming the so-called *Routhian array*, from the given nth-order characteristic equation, comprised from $n+1$ **rows**, and checking the elements of the first column for sign reversals. **Lack of sign reversals verifies the sufficiency.** However, **presence of sign reversals verifies the insufficiency.** Thus the Routh stability criterion is stated as follows: **The necessary and sufficient condition that all the roots of the characteristic equation $F(s)=0$, which describes a given system, lie in the left half of the s-plane is that there is no sign reversal in the elements of the first column of the Routhian array.**

The Routhian array is constructed as follows. Start with the characteristic equation $F(s)=0$ and separate the coefficients of the even and odd powers of s into two rows beginning with the highest power of s. Thus for Eq. (C.108), (1) select the

*Stability criteria for nonlinear systems are attributed primarily to Liapunov.

coefficients for the first two rows as

Row 1

$$s^n + a_{n-1}s^{n-1} + a_{n-2}s^{n-2} + a_{n-3}s^{n-3} + a_{n-4}s^{n-4} + \cdots \qquad \text{(C.109)}$$

Row 2

(2) formulate the first two rows of the Routhian array, and (3) complete the array in a specified manner. Thus for $n=5$, the $n+1=6$ rows of the Routhian array are

$$\begin{array}{c|ccc} s^5 & 1 & a_3 & a_1 \\ s^4 & a_4 & a_2 & a_0 \\ s^3 & \alpha_1 & \alpha_2 & \\ s^2 & \beta_1 & \beta_2 & \\ s^1 & \gamma_1 & & \\ s^0 & \delta_1 & & \end{array} \qquad \text{(C.110)}$$

where

$$\alpha_1 = \frac{(a_4)\times(a_3)-(1)\times(a_2)}{a_4} \qquad \text{(C.111)}$$

$$\alpha_2 = \frac{(a_4)\times(a_1)-(1)\times(a_0)}{a_4} \qquad \text{(C.112)}$$

$$\beta_1 = \frac{(\alpha_1)\times(a_2)-(a_4)\times(\alpha_2)}{\alpha_1} \qquad \text{(C.113)}$$

$$\beta_2 = \frac{(\alpha_1)\times(a_0)-(a_4)\times(0)}{\alpha_1} \qquad \text{(C.114)}$$

$$\gamma_1 = \frac{(\beta_1)\times(\alpha_2)-(\alpha_1)\times(\beta_2)}{\beta_1} \qquad \text{(C.115)}$$

and

$$\delta_1 = \frac{(\gamma_1)\times(\beta_2)-(\beta_1)\times(0)}{\gamma_1} \qquad \text{(C.116)}$$

Note that we build the Routhian array by formulating rows of new elements always with the aid of the immediately preceeding two rows. The product of two elements that correspond to separate rows and are joined by a line with **positive slope** has a **positive sign**. However, the product of two elements that correspond to two separate rows but are joined by a line with a **negative slope** has a **negative sign**. The **sum** of such products, taken two at a time, is divided by the first element of the row immediately preceding the row which is built. This process continues until $n+1$ rows are formed. It should be noted, however, that the shape of the Routhian array resembles that of an inverted right triangle.

The Routh stability criterion is applied to the Routhian array by employing **Routh's stability criterion**, stated as follows: **The number of sign changes in the elements that constitute the first column of the Routhian array** (scanned from top to bottom) **is equal to the number of roots the characteristic equation $F(s)=0$ has in the right half of the s-plane.** In other words, a system described by either a differential equation whose characteristic polynomial is $C(s)=0$ or by a transfer function whose characteristic polynomial is $D(s)$—that is, **a system whose characteristic equation is $F(s)=0$—is stable if and only if the elements of the first column of the**

Routhian array are nonzero and have the same sign (all positive, or all negative). **These requirements are both necessary and sufficient** for stability.

The application of Routh's stability criterion is demonstrated in the following two examples.

Example C.1

Assume that the characteristic equation $F(s)=0$ of a given system is given as follows:

$$F(s) = s^4 + 10s^3 + 20s^2 + 30s + 40 \tag{1}$$

What is the stability status of the system?

Solution

The Routhian array is formulated with the aid of Eqs. (C.109) through (C.116) as follows:

$$\begin{array}{c|lll} s^4 & 1 & 20 & 40 \\ s^3 & 10 & 30 & \\ s^2 & 17 & 40 & \\ s^1 & 6.47 & & \\ s^0 & 40 & & \end{array} \tag{2}$$

Notice that all the coefficients of Eq. (1) are present and the first column of the Routhian array, Eq. (2), yields **no change of signs**. Therefore, $F(s)=0$ has no roots in the RHP, and the system described by this characteristic equation is **stable.**

Example C.2

Assume that the characteristic equation $F(s)=0$ of a given system is given as

$$D(s) = 4s^4 + 2s^3 + 6s^2 + 10s + 20 \tag{1}$$

What is the stability status of the system?

Solution

The Routhian array is formulated with the aid of Eqs. (C.109) through (C.116) as follows:

$$\begin{array}{c|lll} s^4 & 4 & 6 & 20 \\ s^3 & 2 & 10 & \\ s^2 & -14 & 20 & \\ s^1 & 12.86 & & \\ s^0 & 20 & & \end{array} \tag{2}$$

Notice that all the coefficients of Eq. (1) are present and the first column of the Routhian array, Eq. (2), exhibits **two changes of sign.** Therefore, $F(s)=0$ has **two roots in the RHP,** and the system described by this characteristic equation is **unstable.**

Although Routh's stability criterion provides information with respect to the stability characteristics of the given system tested, it gives **no information** with respect to the nature, that is, real or complex, and location of the two RHP poles. That is, a characteristic equation $F(s)=0$ [i.e., $C(s)=0$ or $D(s)=0$] with two RHP real or complex roots yields identical results when submitted to a Routh stability criterion test.

Exception to the Rule. Occasionally, when Routh's stability criterion is applied, one of two difficulties may occur in the Routhian array: (1) the **first element** in any one row might be calculated to be zero, while the other elements might be calculated to be nonzero, and (2) all the elements in one row (which could be comprised even of only one element) might be calculated to be zero.

1. If the first element in a given row is zero while the remaining elements are nonzero, the elements of the very next row become infinity, Eqs. (C.111) or (C.112) or (C.113) and (C.114) or (C.115) or (C.116), and Routh's stability test fails. This difficulty is circumvented by either of the two suggested methods: (a) if the zero is substituted by an arbitrarily **small positive number** ε and the test is completed in the usual manner, or (b) if the original polynomial is multiplied by a left-hand singularity, that is, a pole designated as $(s+a)$, and the test is applied to the resultant polynomial $\mathcal{F}(s)=(s+a)F(s)=0$ in the usual manner. Since the singularity $(s+a)$ that is introduced to $\mathcal{F}(s)=0$ belongs to the LHP, Routh's stability criterion remains unaltered.

2. When all the elements in one row are zero, again Routh's stability test fails. This difficulty is circumvented as follows. The elements of the last nonvanishing row are used as coefficients of an auxiliary polynomial in s, designated as $A(s)$. *This polynomial is always of even order and is defined as the* **auxiliary equation** *of the characteristic equation* $F(s)=0$ which describes the system at hand. Once this auxiliary equation is formed, its first derivative with respect to s is taken and the coefficients of the resultant equation are used to replace the corresponding zeros in the zero coefficient row. Once this substitution is complete, the remainder of the Routhian array is completed as described earlier, and Routh's stability criterion is used in the usual manner.

A very useful by-product of the second exception is that the roots of the auxiliary equation $A(s)=0$ define the roots of the characteristic equation $F(s)=0$, of the original system, which are symmetrically displaced about the origin of the s-plane, that is, root pairs that are equal in magnitude but opposite in sign. That is, $F(s)=0$ might have pairs of real roots of equal magnitude but of opposite signs, pairs of imaginary roots of equal magnitude but of opposite signs, or both; or conjugate roots forming a quadrate in the s-plane. Usually, however, such roots constitute a single pair of imaginary roots of equal magnitude but of opposite sign. Such a pair of roots defines **marginal stability**. Therefore, when the characteristic equation $F(s)=0$ is expressed as a function of both s and a *design parameter*, say K, one may use Routh's stability criterion to calculate **both** the design parameter K that causes the system to become marginally stable and the *frequency of oscillation* $\omega[K]$ that corresponds to a given K.

Example C.3

Assume that the characteristic equation $F(s)=0$ of a given system is also a function of the design parameter K (usually defined as **gain constant**) and is given as

$$F(s,K) = 2s^4 + 20s^3 + 48s^2 + 2Ks + 1.6K \tag{1}$$

Calculate the value of K that will cause the system to exhibit *marginal stability*, as well as the frequency of oscillation $\omega[K]$ that corresponds to this K.

Solution

The Routhian array is formulated with the aid of Eqs. (C.109) through (C.116) as follows:

$$
\begin{array}{c|lll}
s^4 & 2 & 48 & 1.6K \\
s^3 & 20 & 2K & \\
s^2 & 48-0.2K & 1.6K & \\
s^1 & \dfrac{64K-0.4K^2}{48-0.2K} & & \\
s^0 & 1.6K & &
\end{array}
\tag{2}
$$

If the system described by the given characteristic equation, Eq. (1), is to exhibit marginal stability, its auxiliary equation $A(s)$ has to be of **even order**. Therefore, the row that should vanish is the row designated as s^1, Eq. (2). When we set the elements of this row equal to zero, we obtain

$$\frac{64K-0.4K^2}{48-0.2K} = 0 \tag{3}$$

Solution of Eq. (3) for K yields

$$64 - 0.4K = 0 \Rightarrow K = 160 \tag{4}$$

Substitution of $K = 160$ to the elements of the row s^2, Eq. (2), yields the auxiliary equation $A(s)$, written as

$$A(s) = (48-0.2\times 160)s^2 + 1.6\times 160 = 16s^2 + 256 \tag{5}$$

Solution of Eq. (5) for $s = \alpha + j\omega$ yields

$$s = 0 \pm j4 \equiv \alpha \pm j\omega \tag{6}$$

Therefore, Eq. (6) gives the system's frequency of oscillation, $\omega[K] = 4$ rad/s. This is the operating frequency of the given system when this system exhibits marginal stability, that is, when $K = 160$.

C.8 SUMMARY

This appendix sets forth the basic theory of stability. Through the use of the basic theory of differential equations and Laplace transforms, it is shown that (a) the characteristic polynomial $C(s) = 0$ derived from the differential equation that describes a given system and (b) the denominator $D(s) = 0$ of the transfer function $G(s)$, which also describes the same system, are identical. Therefore, the t-domain solution of the homogeneous differential equation and the inverse Laplace transform of the transfer function $G(s)$—for zero input—both describing the same system, are also identical. Consequently, **stability, instability**, and **marginal stability** are defined by the correlation of t-domain solutions for the first- and second-order systems to the root location of the corresponding characteristic polynomials $C(s) = 0$ and $D(s) = 0$—defined as the characteristic equation of the system at hand and designated as $F(s) = 0$—and thus the poles of the corresponding transfer functions $G(s)$.

Table C.2 shows that for a **stable** system, and therefore for a **bounded** t-domain response, the roots of the system's characteristic equation $F(s) = 0$, that is, the poles of the system's transfer function $G(s)$, should be located on the **left half** of the

s-plane. Roots located on the **right half** of the s-plane yield **unbounded**, t-domain responses and therefore describe **unstable** systems. **Only single roots of $F(s)=0$ may be located at a given point of the $j\omega$-axis. Single conjugate imaginary** roots located on the $j\omega$-axis of the s-plane yield **sustained oscillations** and therefore describe **marginally stable** systems.

Table C.4 shows that the **zeros** of a given transfer function $G(s)$ **affect only the amplitude of the components of the t-domain response.**Consequently, **the zeros of $G(s)$ can be located anywhere in the s-plane**, even on the right half of the s-plane, as long as they are complex conjugate.

These ideas are extended to nth-order systems. However, in order to avoid the calculation of the n roots for such systems, Routh's stability criterion is introduced. It is demonstrated how this criterion is used to allow one to check the stability status of a given system simply by checking the root location of the characteristic equation, $F(s)=0$, that is, the pole location of the transfer function $G(s)$ of the system at hand. Here it is demonstrated that in addition to the constraints imposed on the root location for the first- and second-order systems, **multiple** conjugate imaginary roots located on the $j\omega$-axis, that is, at $s=\pm j\omega$, and **multiple** roots located in the origin of the s-plane, that is, at $s=0$, yield **unbounded** t-domain responses and therefore describe **unstable systems.**

Before this appendix is closed, it should be made clear that **only the very basic concepts of stability theory are presented here**. However, a concerted effort is made to present these concepts comprehensively. Therefore, the student should invest the time necessary to understand the concepts of **stability, instability**, and **marginal stability** presented. Such concepts are fundamental to the study of power systems engineering.

Those intrigued by the subject of stability theory may venture into the fields of automatic control and system theory for a more complete study of this subject.

REFERENCES

1. Balabanian, N., *Fundamentals of Circuit Theory*. Allyn & Bacon, Boston, 1961.
2. D'Azzo, J. J., and Houpis, C. H., 3d ed., *Feedback Control System Analysis and Synthesis*. McGraw-Hill, New York, 1975.
3. Hayt, W. H., Jr., and Kemmerly, J. E., 3d ed., *Engineering Circuit Analysis*. McGraw-Hill, New York, 1978.
4. Huelsman, L. P., *Basic Circuit Theory and Digital Computations*. Prentice-Hall, Englewood Cliffs, N.J., 1972.
5. Kuo, B. C., 3d ed., *Automatic Control Systems*. Prentice-Hall, Englewood Cliffs, N.J., 1975.
6. Langill, A. W., Jr., *Automatic Control Systems Engineering*. Prentice-Hall, Englewood Cliffs, N.J., 1965.
7. Van Valkenburg, M. E., *Network Analysis*, 2d ed. Prentice-Hall, Englewood Cliffs, N.J., 1964.

D DEFINITIONS AND THEOREMS

The objective of Appendix D is to present certain basic definitions and Laplace transform pairs generally essential in the formulation of the stability theory set forth in Appendix C.

Definition D.1

Principle of Conservation of Charge. The charge $q(t)=Cv_C(t)$ on the capacitor's plates, Fig. D.1, and therefore the voltage $v_C(t)$ across this capacitor, cannot change instantaneously; that is, $q(0^+)\equiv q(0^-)$ and therefore $v_C(0^+)\equiv v_C(0^-)$.

This becomes apparent from the expression

$$q(t) = \lim_{t\to 0}\int_0^t i_C(\tau)d\tau = 0, \quad \text{iff} \quad i_C(t) \neq \infty \tag{D.1}$$

Definition D.2

Principle of Conservation of Flux Linkages. The flux linkages $\lambda(t)=Li_L(t)$ linking the inductor, Fig. D.2, and therefore the current $i_L(t)$ through this inductor, cannot change instantaneously; that is, $\lambda(0^+)\equiv\lambda(0^-)$ and therefore $i_L(0^+)\equiv i_L(0^-)$.

This becomes apparent from the expression

$$\lambda(t) = \lim_{t\to 0}\int_0^t v_L(\tau)d\tau, \quad \text{iff} \quad v_L(t) \neq \infty \tag{D.2}$$

Definition D.3

Any capacitor C with an initial voltage E_0 across its plates, Fig. D.3(a), has an equivalent circuit comprised of the same capacitor C with no initial voltage across it, connected either (1) in series with a *step-function voltage source $u(t)$*, the strength

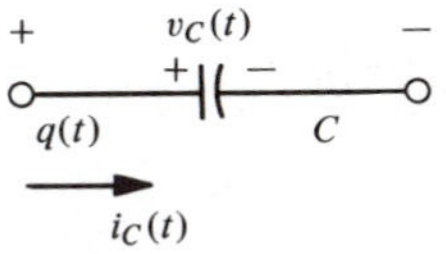

Fig. D.1 A Charged Capacitor with the Pertinent Variables That Constitute Its Describing Equation, $q(t)=Cv_C(t)$

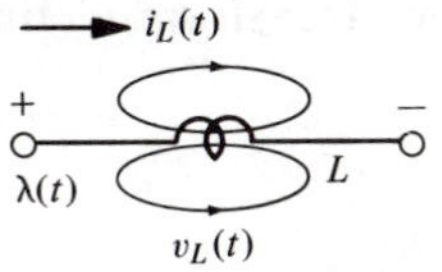

Fig. D.2 An Excited Inductor with the Pertinent Variables That Constitute Its Describing Equation, $\lambda(t)=Li_L(t)$

of which is E_0, Fig. D.3(b), or (2) in parallel with an *impulse-function current source* $\delta(t)$, the strength of which is CE_0, Fig. D.3(c).

Definition D.4

Any inductor L with an initial current I_0 through it, Fig. D.4(a), has an equivalent circuit comprised of the same inductor L with no initial current through it, connected either (1) in series with an *impulse-function voltage source* $\delta(t)$, the strength of which is LI_0, Fig. D.4(b), or (2) in parallel with a *step-function current source u(t)*, the strength of which is I_0, Fig. D.4(c).

Definition D.5

Any function of time having (1) zero value for $t<0$ and (2) a region of convergence, thus satisfying the condition

$$\int_0^{\infty} |f(t)|e^{-\sigma t}\,dt < \infty \tag{D.3}$$

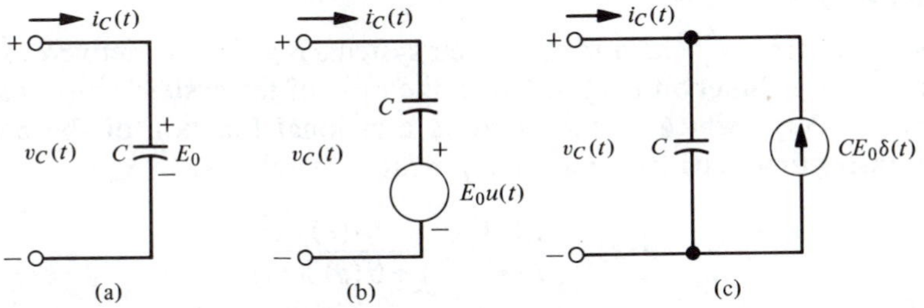

Fig. D.3 (a) An Initially Charged Capacitor, (b) Its Thevenin Equivalent, and (c) Its Norton Equivalent

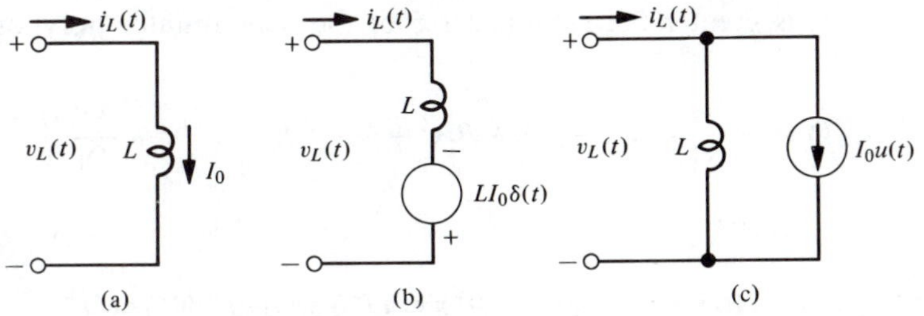

Fig. D.4 (a) An Inductor with an Initial Current Through It, (b) Its Thevenin Equivalent, and (c) Its Norton Equivalent

has a Laplace transform, defined as

$$F(s) \equiv \mathcal{L}[f(t)] = \int_0^\infty f(t)e^{-st}\,dt \tag{D.4}$$

Definition D.6

$$\mathcal{L}[kf(t)] = kF(s) \tag{D.5}$$

Definition D.7

$$\mathcal{L}[k_1f_1(t)+k_2f_2(t)+\cdots+k_nf_n(t)] = k_1F_1(s)+k_2F_2(s)+\cdots+k_nF_n(s) \tag{D.6}$$

Definition D.8

$$\mathcal{L}\left[\frac{d}{dt}f(t)\right] = sF(s) - f(0^+) \tag{D.7}$$

Definition D.9

$$\mathcal{L}\left[\frac{d^2}{dt^2}f(t)\right] = s^2F(s) - sf(0^+) + \frac{d}{dt}f(0^+) \tag{D.8}$$

Definition D.10

$$\mathcal{L}\left[\frac{d^n}{dt^n}f(t)\right] = s^nF(s) - s^{n-1}f(0^+) - s^{n-2}\frac{d}{dt}f(0^+) - \cdots - \frac{d^{n-1}}{dt^{n-1}}f(0^+) \tag{D.9}$$

Definition D.11*

$$\mathcal{L}\left[\int_{-\infty}^{t} f(\tau)\,d\tau\right] = \mathcal{L}\left[\int_0^t f(\tau)\,d\tau + \int_{-\infty}^{0} f(t)\,dt\right] = \frac{F(s)}{s} + \frac{f^{-1}[0^+]}{s} \tag{D.10}$$

Definition D.12

In the s-domain the behavior of a given system, Fig. D.5, is defined in terms of its overall transfer function $G(s)$—that is, the ratio of the system's response $R(s)$ to its excitation $E(s)$—which is expressed as a rational function of the component system transfer functions $\mathcal{G}(s)$ and $\mathcal{H}(s)$, related as follows:

$$\mathcal{G}(s) = \frac{R(s)}{E(s)} = \frac{\mathcal{G}(s)}{1+\mathcal{G}(s)\mathcal{H}(s)} \tag{D.11}$$

Definition D.13

For any rational function $G(s) = N(s)/D(s)$, if the order u of the numerator polynomial $N(s)$ is greater than the order v of the denominator polynomial $D(s)$, then

$$G(s) = \frac{N(s)}{D(s)} = \beta_0 + \beta_1 s + \beta_2 s^2 + \cdots + \beta_{u-v}s^{u-v} + \frac{N_1(s)}{D(s)} \tag{D.12}$$

*$f^{-1}[0^+] \equiv \int_{-\infty}^{0} f(t)\,dt$; $f(t) = i(t) \to f^{-1}[0^+] \equiv q(0^+)$, and $f(t) = v(t) \to f^{-1}[0^+] \equiv \lambda(0^+)$.

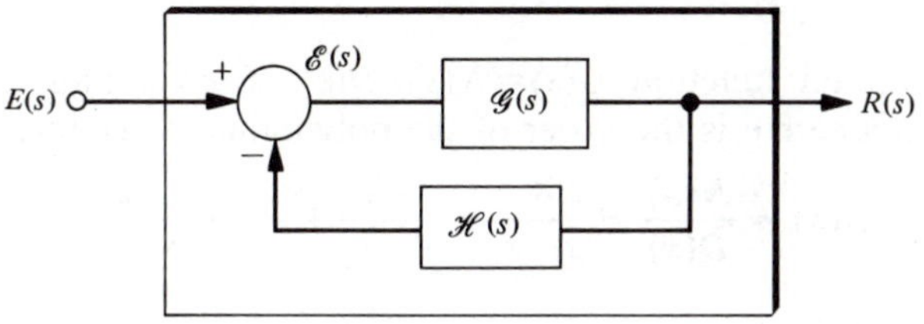

Fig. D.5 *s*-Domain Block Diagram of an Elementary Negative Feedback Control System

Table D.1 PERTINENT LAPLACE TRANSFORM PAIRS

No.	$f(t); f(t)=0, t<0$	$F(s)$
1	$\delta(t)$	1
2	$u(t)$	$\frac{1}{s}$
3	t	$\frac{1}{s^2}$
4	$\frac{t^{k-1}}{(k-1)!}$	$\frac{1}{s^k}$
5	$e^{-\alpha t}$	$\frac{1}{s+\alpha}$
6	$te^{-\alpha t}$	$\frac{1}{(s+\alpha)^2}$
7	$\frac{t^{k-1}e^{-\alpha t}}{(k-1)!}$	$\frac{1}{(s+\alpha)^k}$
8	$\sin \omega t$	$\frac{\omega}{s^2+\omega^2}$
9	$\cos \omega t$	$\frac{\omega}{s^2+\omega^2}$
10	$(\frac{1}{2\omega})t \sin \omega t$	$\frac{s}{(s^2+\omega^2)^2}$
11	$t \cos \omega t$	$\frac{s^2-\omega^2}{(s^2+\omega^2)^2}$
12	$e^{-\alpha t} \sin \omega t$	$\frac{\omega}{(s+\alpha)^2+\omega^2}$
13	$e^{-\alpha t} \cos \omega t$	$\frac{s+\alpha}{(s+\alpha)^2+\omega^2}$

Definition D.14

In a given rational function $G(s)=N(s)/D(s)$, if $v \geqslant u$ and $D(s)=\alpha_0(s-s_1)(s-s_2)\cdots(s-s_v)$, where v is the order of the polynomial $D(s)$, then

$$G(s)=\frac{N(s)}{D(s)}=\frac{K_1}{s-s_1}+\frac{K_2}{s-s_2}+\cdots+\frac{K_v}{s-s_v} \tag{D.13}$$

where

$$K_j=\left[(s-s_j)\times\frac{N(s)}{D(s)}\right]\Bigg|_{s=s_j},\quad j=1,2,\ldots,v \tag{D.14}$$

Definition D.15

In a given rational function $G(s)=N(s)/D(s)$, if $v \geqslant u$ and $D(s)=\alpha_0(s-s_j)^k$, where $k=v$ designates the multiplicity of the repeated root $s=s_j$ ($j=1,2,\ldots$) and n designates the general term, then

$$G(s)=\frac{N(s)}{D(s)}=\frac{K_{j1}}{(s-s_j)}+\frac{K_{j2}}{(s-s_j)^2}+\cdots+\frac{K_{jn}}{(s-s_j)^n}+\frac{K_{jk}}{(s-s_j)^k} \tag{D.15}$$

where

$$K_{jn}=\frac{1}{(k-n)!}\times\frac{d^{k-n}}{ds^{k-n}}\left[(s-s_j)^k\times\frac{N(s)}{D(s)}\right]\Bigg|_{s=s_j},\quad n=1,2,\ldots,k \tag{D.16}$$

REFERENCES

1. Cheng, D. K., *Analysis of Linear Systems*. Addison-Wesley, Reading, Mass., 1959.
2. Hayt, W. H., Jr., and Kemmerly, J. E., 3d ed., *Engineering Circuit Analysis*. McGraw-Hill, New York, 1978.
3. Huelsman, L. P., *Basic Circuit Theory with Digital Computations*. Prentice-Hall, Englewood Cliffs, N.J., 1972.
4. Kuo, B. C., 3d ed., *Automatic Control Systems*. Prentice-Hall, Englewood Cliffs, N.J., 1975.
5. McCollum, P. A., and Brown, B. F., *Laplace Transform Tables and Theorems*. Holt, Rinehart and Winston, New York, 1965.
6. Van Valkenburg, M. E., *Network Analysis*, 2d ed. Prentice-Hall, Englewood Cliffs, N.J., 1964.
7. Scott, R. E., *Elements of Linear Circuits*. Addison-Wesley, Reading, Mass., 1965.

E SOLID STATE POWER ELECTRONICS: THE SILICON CONTROLLED RECTIFIER (SCR) OR THYRISTOR

E.1 INTRODUCTION

The purpose of this appendix is to introduce the reader to the basic theory of thyristor operation and the most basic thyristor circuit, and to provide a comprehensive list of references for the subject of the thyristor, or the Silicon Controlled Rectifier (SCR), and its applications.

E.2 THE SILICON CONTROLLED RECTIFIER (SCR) OR THYRISTOR

The silicon controlled rectifier or thyristor is a *p–n–p–n* structure, that is, a 4-layer, 3-junction device, Fig. E.1(a) with an **input**, an **output**, and a **gate** terminal. Functionally, the thyristor differs from the diode in that forward conduction is inhibited until a firing pulse is applied to the control electrode called the **gate**. Subsequently, its behavior resembles that of the diode until forward conduction ceases, when the thyristor reverts to its inhibited state. The thyristor can then be viewed as a bistable switching device: a short pulse is sufficient to switch the thyristor to its conducting state, which is self-maintained; a lack of this pulse allows this device to regain its nonconducting state.

The behavior of the thyristor is best understood by considering it formed by two complementary transistors, a *p–n–p* transistor and *n–p–n* transistor having a common collector junction, Fig. E.1(b). Note that the collector of each transistor is cross-coupled to the base of the other, and consequently there is positive feedback action which allows each transistor to drive the other into saturation. When the

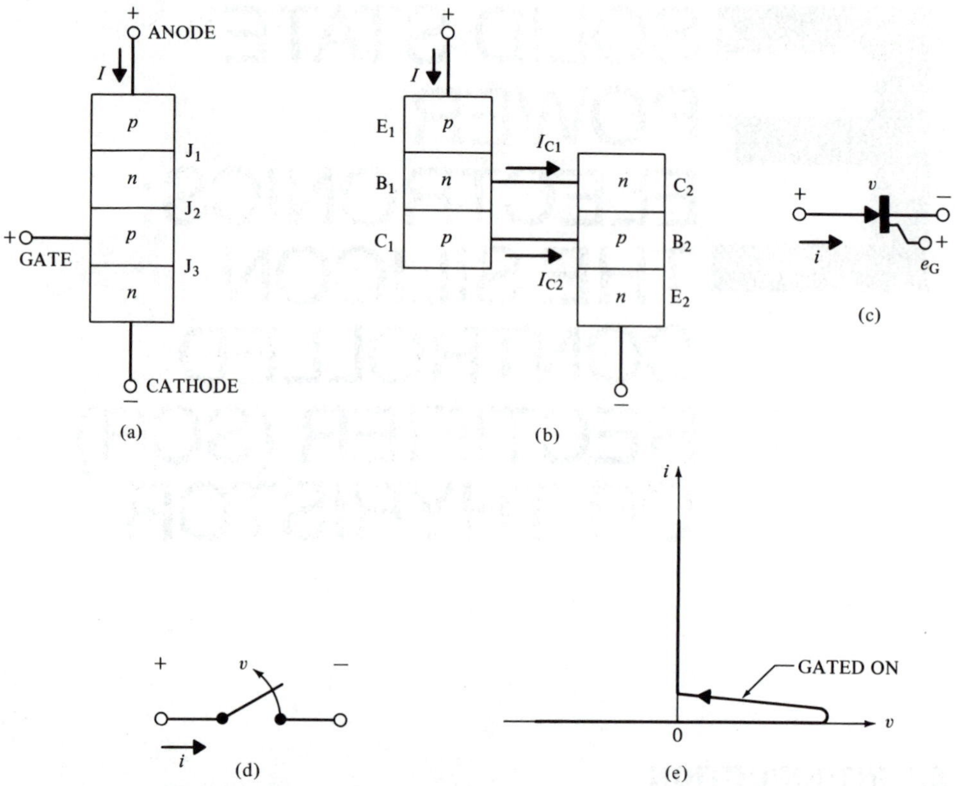

Fig. E.1 Thyristor or Silicon Controlled Rectifier (SCR): (a) Structure, (b) Two Transistor Analog, (c) Symbolic Representation, (d) Circuit Model Switch, and (e) Volt-Ampere Characteristic

thyristor has a positive anode-to-cathode voltage each transistor is biased in the conventional manner and has a base current gain, β, which expresses the ratio of the collector to the emitter currents. Thus, for the p–n–p transistor, β_1 is the function of the hole current injected at the emitter E_1 that reaches the collector C_1. Similarly, for the n–p–n transistor, β_2 is the fraction of the electron current injected at the emitter E_2 that reaches the collector C_2. If I is the current at the anode, the collector currents I_{C1} and I_{C2} are expressed as

$$I_{C1} = \beta_1 I + I_{1CO} \tag{E.1}$$

and

$$I_{C2} = \beta_2 I + I_{2CO} \tag{E.2}$$

where I_{1CO} and I_{2CO} designate the leakage currents of the transistors p–n–p and n–p–n, respectively.

The sum of the two collector currents, Eqs. (E.1) and (E.2), is equal to the current that crosses the common collector junction J_2 of the thyristor, which must be equal to the anode current I. That is,

$$I = I_{C1} + I_{C2} = (\beta_1 + \beta_2)I + I_{CO}|_{I_{CO} \equiv (I_{1CO} + I_{2CO})} \tag{E.3}$$

Solution of Eq. (E.3) for I yields

$$I = \frac{I_{CO}}{1-(\beta_1+\beta_2)} \tag{E.4}$$

If the current gains β_1 and β_2 are small, that is, $(\beta_1+\beta_2)\to 0$

$$I = \left.\frac{I_{CO}}{1+(\beta_1+\beta_2)}\right|_{(\beta_1+\beta_2)\to 0} \doteq I_{CO} \tag{E.5}$$

Hence, the thyristor is in its forward blocking state and remains in this nonconduction or OFF state. When $(\beta_1+\beta_2)\to 1$, regenerative action takes place and

$$I = \left.\frac{I_{CO}}{1-(\beta_1+\beta_2)}\right|_{(\beta_1+\beta_2)\to 1} = \infty \tag{E.6}$$

It is clear that under these circumstances both transistors are saturated and all junctions are forward biased, and the thyristor is in the **conduction** or ON state. It should be obvious that the flow of this high current must be limited by the resistance of the external circuit to which the thyristor is connected.

It is known from transistor theory that the current gain, β, of a transistor is a function of the emitter current. Consequently, if the base-emitter junction of either transistor is forward biased, the sum of the current gains $(\beta_1+\beta_2)$ will increase causing the thyristor to go to its conducting or ON state. Usually, the base of the *n–p–n* transistor is used as the gate terminal into which a pulse is injected in order to bring the thyristor in its conducting or ON state. The thyristor continues to conduct, even with the gate pulse removed as long as the load current exceeds a certain value defined as the **holding** or **sustaining** current. When the loading current is less than this holding or sustaining current, $(\beta_1+\beta_2)<1$, and the thyristor reverts to its nonconducting, or OFF, state. It should be noted that although other mechanisms can be used to trigger a thyristor ON, **in practice gate triggering by a pulse is the usual mechanism employed, and the symbol of the thyristor is that of Fig. E.1(c).**

E.3 BASIC THYRISTOR CIRCUIT PERFORMANCE

The simplest SCR circuit, known as a **single-phase, half-wave thyristor circuit**, Fig. E.2(a), consists of a single thyristor operating into a resistive load, R_L, and a sinusoidal forcing function, e. The circuit for applying the gating pulse, e_G, to the thyristor is not shown. According to the thyristor model of Figs. E.1(c) through E.1(e), however, the thyristor can conduct current only when its voltage is positive, typically during the intervals $\omega_E t=0$ to π; 2π to 3π and so on. In addition, the thyristor must receive a gating pulse, e_G, which we assume is applied at the firing angle α. By definition the firing angle α is measured from the angle that produces the largest load voltage, v_L (Fig. E.2(b)), in this case from $\omega_E t=0$, 2π, 4π, and so on. As shown in Fig. E.2(c) the thyristor conducts from $\omega_E t=\alpha$ to π; $2\pi+\alpha$ to 3π and so on. During the conduction intervals the load voltage, v_L, is equal to the impressed voltage, e. As the firing angle α is increased by the **gating** (or **control**) circuit from 0 to π, the average value of the load voltage V_L^{AVG}, decreases. That is,

$$\begin{aligned} V_L^{\mathrm{AVG}} &= \frac{1}{2\pi}\int_\alpha^\pi \sqrt{2}\,|\mathbf{E}|\sin\omega_E t\,d(\omega_E t) \\ &= \left.\frac{-1}{\sqrt{2}\pi}(|\mathbf{E}|\cos\omega_E t)\right|_\alpha^\pi \\ &= 0.225|\mathbf{E}|(1+\cos\alpha) \end{aligned} \tag{E.7}$$

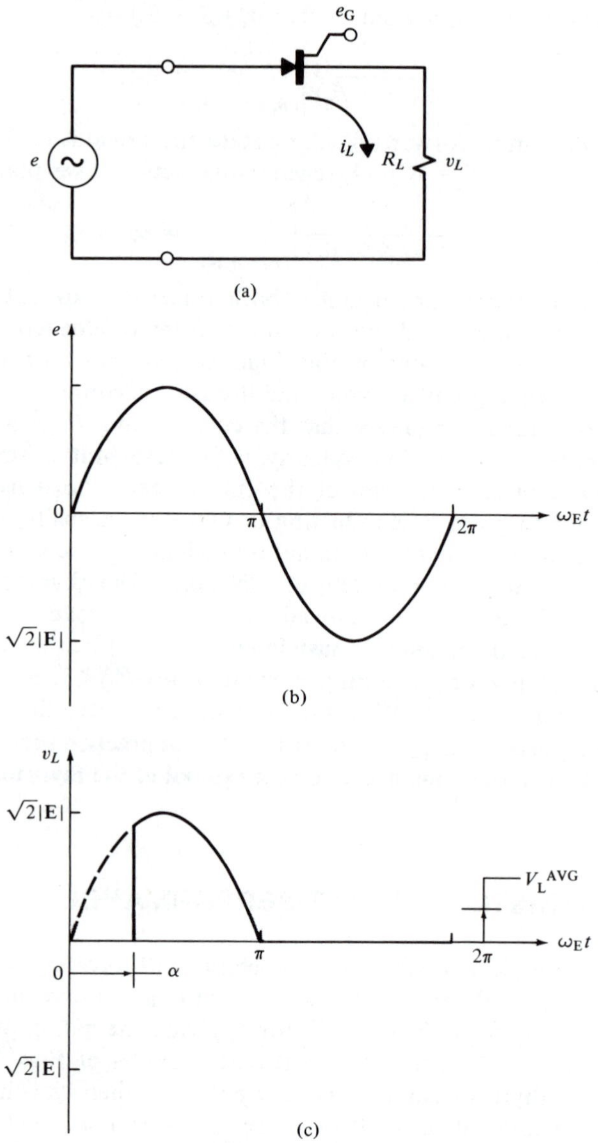

Fig. E.2 Operation of the Single-Phase, Half-Wave Thyristor Circuit: (a) Circuit, (b) Waveform of the Impressed Voltage, and (c) Waveform of the Load Voltage

The relationship between the average loading voltage V_L^{AVG} and the firing angle α is exhibited in Fig. E.3. Note that maximum value of V_L^{AVG} occurs at $\alpha = 0°$; this value corresponds to a half-wave diode operation.

A more efficient system than that exhibited in Fig. E.2 is drawn in Fig. E.4, known as a **three-phase, half-wave thyristor circuit**. It should be obvious that this circuit, in comparison to that of Fig. E.2, provides more voltage pulses per cycle of excitation frequency, and thus, it assures a current i_L over a larger portion of the same cycle.

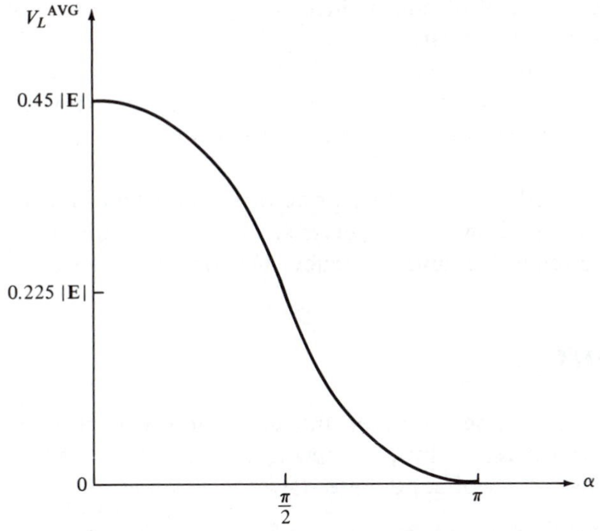

Fig. E.3 Plot of the Average Load Voltage V_L^{AVG} of the Single-Phase, Half-Wave Thyristor Circuit, Fig. E.2(a), as a Function of the Firing Angle α, Eq. (E.7)

If we substitute the resistor of Figs. E.2 or E.4 with the equivalent circuit of the rotor winding of a **separately excited dc motor** and maintain its field current constant, it is obvious (Fig. 2.20) that by controlling the voltage $V_L^{AVG}(\alpha)$ through the control of the firing angle α we can control the **speed** of the dc motor over a wide range.

In contrast, the speed of the **induction** or the **synchronous motor** is a function of ω_E, that is, $\omega_S = \omega_E/(P/2)$. Thus, speed control of these motors is obtained by supplying the stator windings with voltages from adjustable frequency thyristor control circuits. The synchronous motor will follow this frequency directly while the induction motor will follow this frequency within the slip frequency.

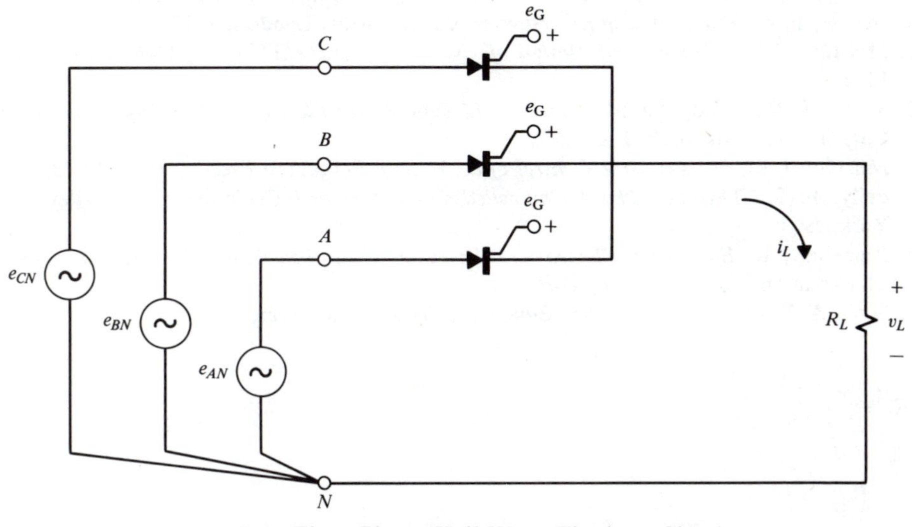

Fig. E.4 Three-Phase, Half-Wave Thyristor Circuit

Although only control of speed through thyristor circuits is mentioned, **torque** and **acceleration** can be adjusted as well.

The control circuits for such tasks consist of relays, contractors, magnetic components, diodes, transistors, and thyristors. The combination of a dc or ac motor with the corresponding control circuit constitutes what is defined as **dc** or **ac drive**, respectively.

Pursuing a detailed analysis of dc or ac drives is beyond the scope of this book. The interested reader, however, is referred to the literature. The list of references provided is comprehensive, and reference [11] constitutes an excellent beginning.

E.4 SUMMARY

The basic theories of operation for the thyristor (SCR) and the basic thyristor circuit are set forth herein. Thus, the **springboard** is established for those that have an interest in the subject of solid state power electronics.

REFERENCES

1. Atkinson, P., *Thyristors and Their Application*. Mills & Boon, London, 1972.
2. Bedford, B. D. and Hoft, R. G., *Principles of Inverter Circuits*. Wiley, New York, 1964.
3. Csáki, F., Ganszky, K., Ipsits, I., and Marti, S., *Power Electronics*. Akadémiai Kiadó, Budapest, 1975.
4. Davis, R. M., *Power Diode and Thyristor Circuits*. Cambridge University Press, London, 1971.
5. Dewan, S. B. and Straughen, A., *Power Semiconductor Circuits*. Wiley, New York, 1975.
6. Ghandi, S. K., *Semiconductor Power Devices*. Wiley, New York, 1977.
7. Harnden, J. D., Jr. and Golden, F. B., *Power Semiconductor Applications, Vol. I: General Considerations*. IEEE Press, New York, 1972.
8. Harnden, J. D., Jr. and Golden, F. B., *Power Semiconductor Applications. Vol. II: Equipment and Systems*. IEEE Press, New York, 1972.
9. Kusko, A., *Solid-State dc Motor Drives*. MIT Press, Cambridge, Mass., 1969.
10. Mazda, F. F., *Thyristor Control*. Newnes-Butterworths, London, 1973.
11. McMurray, W., *Theory and Design of Cycloconverters*. MIT Press, Cambridge, Mass., 1972.
12. Motto, J. W., Editor, *Introduction to Solid State Power Electronics*. Westinghouse Electric Corporation, Pittsburgh, PA, 1977.
13. Murphy, J. M. D., *Thyristor Control of ac Motors*. Pergamon Press, Oxford, 1973.
14. Pelly, B. R., *Thyristor Phase Controlled Converters and Cycloconverters*. Wiley, New York, 1971
15. Ramshaw, R. S., *Power Electronics: Thyristor Controlled Power for Electric Motors*. Chapman and Hall, London, 1975.
16. Scoll, A. W., *Cooling of Electric Equipment*. Wiley, New York, 1974.

INDEX

INDEX

E

F

G

H

I

N

S

U

V

80 81 82 83 9 8 7 6 5 4 3 2 1